BARRON

A TEXTBOOK OF HUMAN ANATOMY

Roger C. Crafts, Ph.D., obtained his doctorate from the Columbia University College of Physicians and Surgeons. He is a Professor of Anatomy at the University of Cincinnati College of Medicine. He taught at the Boston University School of Medicine for nine years before moving to Cincinnati in 1950 to head the Department of Anatomy there, a position he held for twenty-nine years. Among the societies of which Dr. Crafts is a member is the American Association of Anatomists.

SECOND EDITION

A TEXTBOOK OF HUMAN ANATOMY

ROGER C. CRAFTS

UNIVERSITY OF CINCINNATI
COLLEGE OF MEDICINE

A WILEY MEDICAL PUBLICATION
JOHN WILEY & SONS
New York • Chichester • Brisbane • Toronto

Design by Wanda Lubelska

Library of Congress Cataloging in Publication Data:

Crafts, Roger Conant, 1911–
   A textbook of human anatomy.

   (A Wiley medical publication)
   Bibliography: p.
   Includes index.
   1. Anatomy, Human. I. Title.   [DNLM:
1. Anatomy.   QS4.3 C885t]
QM23.2.C7 1979   611   78-11424
ISBN 0-471-04454-7

Printed in the United States of America

10 9 8 7 6 5 4 3

To My Students

PREFACE
TO SECOND EDITION

The enthusiastic response from those who have used this textbook, particularly my own students, has been very encouraging. But, in spite of many changes during the fourteen printings, there comes a time for updating, a time to take advantage of the many suggestions made by my own students as well as those elsewhere, a time to consider the suggestions of my colleagues both in Cincinnati and in other schools.

The format of this second edition has not departed greatly from the successful one used in the first. General descriptions of each region have been followed by more detailed presentations of the structures in the same region; the description of each major area of the body has been terminated by a systemic summary of that area.

Clinical relevance has been increased but no attempt has been made to create a clinical or surgical text. A judicious amount of such reference has been combined with phylogeny and ontogeny to enable the student to understand the anatomy as well as to know its clinical importance.

To help the many students entering medical college with a limited biological background, additional material has been added entitled "Beginning Embryology." The embryology presented in each region springs from this beginning and, I hope, will give the student an understanding of why the body has its definitive arrangement. I want to thank Dr. John DeSesso for advice on this addition and Pamela DeSesso for the illustrations.

Finally, this textbook has been written with the following in mind. There are three ways one can learn structure. First, one can attempt to memorize the material; any knowledge obtained in this fashion will last three to four days at most. Second, one can visualize the anatomy, determining the function of each structure based on this visual picture; knowledge obtained in this manner is quite lasting. But third, and most important, one can understand not only the functional anatomy but also how the structure or area developed; knowledge obtained in this manner is difficult to forget.

R. C. C.

PREFACE
TO FIRST EDITION

The crowded medical curriculum has made it necessary to cover some subjects in less detail than in the past. Acceleration has also increased the importance of a text as a teaching instrument; compactness, functional organization, and clarity are now imperative. This text was developed to provide the student with such a teaching tool.

A textbook on gross anatomy should emphasize visualization and understanding. As an aid to visualization, therefore, this book describes the body regionally. At the beginning of each region the anatomy involved is described as one sees it on the cadaver; these portrayals are given the title of "General Description" and are accompanied by halftone illustrations resembling those in a regional atlas. Frequently, the illustrations are presented as a series depicting deeper and deeper views of the particular region under study. The "General Descriptions" are followed by more detailed presentations of muscles, nerves, blood vessels, and lymphatics found in that region. These sections of the text are accompanied by many diagrammatic line drawings, which should make comprehension of these structures easier.

Systemic summaries have been provided after part of the body has been described in its entirety. Although repetitious, they should help the student review the part as a whole.

Thus, this text combines features usually found in separate sources. It contains descriptions of dissected regions reminiscent of some dissecting guides, illustrations of regions from surface to depth typical of regional atlases, shortened descriptions and line drawings of individual body parts similar to the shorter works on this subject, and summaries that bring in the better features of the books developed around a systemic plan.

The body is described in the following order: (1) back, (2) thoracic and abdominal walls, (3) thoracic cavity, (4) abdominal cavity, pelvic cavity, and perineum, (5) lower limb, (6) head and neck, and (7) upper limb. This order has been found to possess certain advantages. It allows students to start their dissection on a region where their inadequate technique does relatively little harm, and it gives them contact with the spinal cord at an early stage; this is very helpful to their learning of the peripheral nervous system. Study of the thoracic and abdominal walls immediately after the back brings the entire trunk into focus. Exposure to the cavities relatively early in the course allows the student to learn the gross anatomy of viscera before most microscopic anatomy classes, if taught simultaneously, have started on organology. This study leads very nicely, via the blood vessels and nerves, into the lower limb. By this time the student has become more expert, and the study of the head and neck is made easier than if presented earlier. The neck leads naturally into the upper limb.

There is no general agreement on the proper order in which a cadaver should be dissected. Although the order in this text may not be the same as the one used in dissection in a particular course, each part of the body is described in its entirety, and the student should have no difficulty using the text for any order of dissection. Some repetition was required to achieve this flexibility.

References to the literature have been used sparingly, and no attempt has been made to give credit to the hundreds of anatomists from whom the information was drawn. Those given credit are ones substantiating material that is not presented in the usual manner, aid the student in understanding the structure involved, or challenge the usual concept of the structure.

The Parisian International Nomenclature, adopted in 1955, has been used throughout this text, with the traditional terms in parentheses when considered important.

The success of an anatomy text is dependent upon the illustrations. To Mr. Herbert W. Fall goes my profound gratitude for a job well done. He performed his art with great patience and welcomed the author's exacting suggestions. Similar thanks go to Mr. Barney Pisha for his guidance. I also thank Mrs. Robert T. Binhammer for her many hours of typing. And great appreciation goes to Dr. Robert T. Binhammer for devoting many hours to reading the text and checking the pictures. I owe a debt of gratitude to Mr. Ellsworth Cochran, Professor of Medical Illustration, and his staff for much of the labeling done on the illustrations, to Dr. Robert D. Mansfield, for his willingness to check all clinical statements, and to Dr. Harold J. Schneider for the x-ray films.

R. C. C.

CONTENTS

1
GENERAL REVIEW

2
BACK

3
THORACIC AND ABDOMINAL WALLS

4
THORACIC CAVITY

5
ABDOMINOPELVIC CAVITY AND PERINEUM

6
LOWER LIMB

7
HEAD AND NECK

8
UPPER LIMB

1

GENERAL REVIEW

Since many students studying human anatomy have taken preparatory courses in college, textbooks in this discipline are written in a manner that assumes a certain amount of knowledge on the part of the reader. The following sections, under the heading of General Review, should be a part of a student's armamentarium before attempting study of the more detailed aspects of human anatomy. If you have studied embryology and comparative anatomy, the first chapter should serve to refresh your memory; if you have not taken such courses, this chapter should be read quite carefully. This text presumes such knowledge on the part of the student.

TERMINOLOGY

Anatomy is the science of the structure of the animal body and the relations of its various parts. The term anatomy is derived from a Greek word meaning "to cut up"; in the past the word "anatomize" was frequently used instead of the more modern term "dissect." This term "anatomy" was formerly limited to structures seen with the naked eye, but, with the invention of the microscope and other techniques for observing structure, the term has spread to encompass much more than can be seen with the eye alone.

Microscopic anatomy is that which can be seen with the microscope and includes the study of cells (cytology), tissues (histology), and organs (organology). The **electron microscope** has provided much further detail than can be seen with a light microscope, and histochemical techniques have related microscopic structure to chemistry. **Vital microscopic techniques** allow a study of living anatomy, particularly of the vascular system, and **tissue culture** also has provided a knowledge of living cells. The central nervous system is usually treated separately in a special course of study called **neuroanatomy,** and study of development of the embryo is called **developmental anatomy** or **embryology.** Although references will be made to microscopic, develop-

1

mental, and neuroanatomy, this book will concentrate on the gross anatomy of the human body.

Gross anatomy can be related to specific fields of endeavor: **Applied anatomy** links anatomy to medicine in general, while **surgical anatomy** is the same study in reference to the field of surgery. **Comparative anatomy** is concerned with structural relations of one animal to another. **Pathologic anatomy** is a description of the effects of disease on structure. Gross anatomy itself can be studied regionally (**regional anatomy**), where a definite region of the body is investigated no matter what systems are present, or by systems (**systemic anatomy**). **Surface anatomy** is a study of superficial landmarks for the detection of internal structures; the term **topographic anatomy** is used to denote relations of one structure to another. This text, as pointed out in the Preface to First Edition, is organized regionally but with systemic summaries placed after the description of each major part of the body.

HISTORY

Terminology in any science is a problem, and the field of anatomy has not escaped confusion. Before 1895 over 30,000 terms for structures in the human body had appeared, but this number included many duplications. In 1895 the German Anatomical Society met in Basle, Switzerland, and approved an official list of approximately 5000 terms which was called the *Basle Nomina Anatomica* or *B.N.A.* These terms were all based on Latin, for publication use, each country to simplify for teaching purposes.

In 1933 the British met in Birmingham and revised many terms in the list, later known as the *Birmingham Revision (B.R.)*; and in 1935 the Germans met in Jena and changed a few words, this revision being designated the *J.N.A.* The terms adopted as a result of these meetings, however, met with local acceptance only, resulting in considerable confusion.

At the Fourth International Congress of Anatomists, held in Milan in 1936, it was decided that the confusion justified another attempt at clarity. A committee on terminology was formed, but unfortunately World War II made international cooperation difficult. This committee was revived in 1950 at the Fifth International Congress of Anatomists, and in 1955 at the Sixth International Congress in Paris a mild revision of the *Basle Nomina Anatomica* was accepted, being slightly modified at the

Seventh International Congress held in New York City in 1960. One of the main contributions of this revision, known as the *Paris Nomina Anatomica,* is the elimination of all proper names from anatomical terminology. Although it will be difficult for those familiar with these proper names to abandon them, it is hoped that every effort will be made to use the more descriptive terms adopted. All anatomy texts are being changed to this terminology as they are revised.

TERMS OF DIRECTION

Man, in the anatomical sense, is always described as standing with eyes looking forward and with the upper limbs hanging down at the sides of the body with the palms facing forward. This is called the **anatomical position.** It is important to remember that, although the cadaver is on a table, it is always described as if standing in the anatomical position. The difference between this posture and that of the four-footed animal requires a corresponding shift in terms of direction, much used in anatomical description. While the "head" end of a cat or dog is the anterior end and the "tail" end is posterior, changing to the upright stature makes the anterior and posterior sides of the human the "front" and "back" of the body, respectively (Fig. 1-1).

The terms of direction used in describing the human body are as follows:

Superior and inferior. **Superior** is defined as a direction toward the head, while **inferior** is the opposite, toward the feet. **Cranial** is a correct substitute for superior, but the opposite term **caudal** should be avoided in human anatomy. Words such as "above" for superior and "below" for inferior are acceptable if used correctly. They should be avoided, however, for they are often used carelessly for "anterior" and "posterior."

Anterior and posterior. **Anterior** means toward the front of the body, while **posterior** is defined as toward the back. The terms **dorsal** for posterior and **ventral** for anterior are equally correct and are synonymous with those terms.

Medial and lateral. **Medial** is defined as toward the midline or median plane* of the body and **lateral** as away from the median plane.

*Median plane is an imaginary vertical plane passing from front to back and dividing the body into right and left halves.

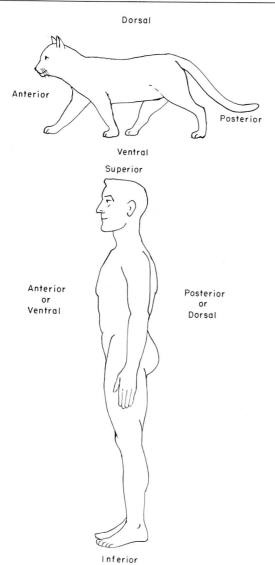

Figure 1-1. *Changes in terms of direction induced by upright stature.*

Internal and external. These are defined as toward the inside or toward the outside of the body.

Proximal and distal. **Proximal** means toward the point of attachment to the body and **distal** away from this point of attachment. These are used primarily with the limbs.

Superficial and deep. These terms, meaning toward and away from the surface, are very useful in describing regional anatomy. They must be used with some care, however, for they can lead to error if not used sen-

sibly. For example, if one is looking at the back, it is obvious that the skin on the back is **superficial** to the vertebral column and, contrariwise, the column is deep to the skin; however, if you are attempting to visualize the back from the anterior side of the body, it is equally obvious that the vertebral column is not **deep** to the skin of the back but superficial. One usually describes a part of the body such as the back from the posterior view, however, so the problem should not arise. It is wise to limit the use of these terms to structures close to the surface.

Afferent and efferent. These terms are used almost exclusively to describe vessels or nerves, an **afferent** vessel being defined as one with blood or lymph flowing toward an organ, an **efferent** vessel as one carrying blood or lymph away from the organ. For example, lymph nodes have afferent lymphatics bringing lymph toward them and efferent lymphatics carrying lymph away. A single vessel may be afferent or efferent, depending on the point of reference. A vessel carrying blood from the heart to the lungs is an efferent vessel of the heart but an afferent vessel of the lung. Afferent nerves conduct impulses toward the central nervous system, and efferent nerves carry impulses away from the central nervous system; these are sensory and motor nerves respectively.

RELATIONS

One of the most important aspects of anatomical studies is the position of one organ in reference to another; this is called the **relations** of the organ or topographic anatomy of the structure. Once again, common sense should prevail to limit relations to those structures that are fairly close to the organ involved.

SECTIONS

There are special terms that are concerned with types of cuts made in the body (Fig. 1-2):

Transverse or cross section is a cut at right angles to the long axis of the body. Such a section separates the body into upper and lower portions.

Longitudinal section is a vertical cut in the body parallel to the median plane. It separates the body into right and left portions; these portions may be unequal in size.

Sagittal section is named after the sagittal suture of the skull. It is exactly the same as a longitudinal section,

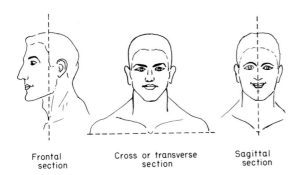

Frontal section   Cross or transverse section   Sagittal section

Figure 1-2.   *Various sections of the body.*

that is, parallel to the median plane. However, common usage has led some to think of a sagittal section as one exactly in the median plane. The terms "midsagittal" and "parasagittal" have been introduced because of this confusion.

Frontal or coronal section was named for the coronal suture in the skull. It is a vertical section at right angles to the median plane. It divides the body into anterior and posterior portions.

___

It should be noted that sections of organs or structures are described in relation to the structure itself rather than to the axes of the body. For example, a transverse section of an artery cuts across the vessel at right angles to its long axis and has nothing to do with the long axis of the body as a whole.

SKIN

The tough, movable skin has other roles besides merely forming a covering for the body:

1. The sensory components in the skin are very important in acquainting one with the environment; nerve endings of various types respond to temperature, pain, and touch. Loss of these sensations is very troublesome; patients with such losses are constantly injuring themselves in many ways.

2. The skin plays an important role in maintaining a nearly constant temperature, which is so necessary to human beings. Perspiration and constant evaporation of this fluid from the surface of the body have a cooling effect. Of course, this is accomplished by means of the sweat glands, which abound in the skin. The skin also has many capillaries, which, when dilated, allow heat to escape from the body; conversely, when vasoconstriction occurs, heat is conserved.

3. The epidermal layer of the skin is important in preventing escape of fluid from the body.

Structurally, the skin is divided into a deep layer, the **dermis,** and a superficial layer, the **epidermis** (Fig. 1-3). The **epidermis** is made up of four layers of cells, which, from superficial to deep, are called stratum corneum, stratum lucidum, stratum granulosum, and stratum germinativum. The **stratum corneum** is made up of dead cells, which are rubbed off gradually and replaced by cells from deeper layers. This layer is thick in places subject to considerable wear and tear, such as the palms of the hands and the soles of the feet. The **stratum lucidum** separates these cells from the **stratum granulosum,** so called because the cytoplasm of the cells in this area contains many deeply staining granules that determine skin coloration. Rapid cell division occurs in the **stratum germinativum,** and it is this layer that produces those more superficially placed.

The dividing line between epidermis and dermis is not a smooth and regular one, for the dermis projects into the epidermis, forming dermal papillae. The **dermis** (Fig. 1-4) is made up of loose connective tissue with many white collagenous and elastic fibers. The direction in which fiber bundles course in the dermis is different for various areas of the skin. These patterns have been determined and are called **tension lines;** it has been shown that incisions following these lines have less tendency to gap than those that cut across them. The dermis is much thicker than the epidermis and contains many blood vessels, nerve endings, sebaceous glands, and arrector pili muscles. The **sebaceous glands** and **arrector pili muscles** are associated with hair follicles, which start as hair papillae deep to the dermis and then completely traverse both layers of the skin. The hairs take a slanting direction as they penetrate the skin, rather than being perpendicular to it. The **sebaceous glands** are the glands secreting the oily material (fat, keratohyalin granules, keratin, and cellular debris) associated with hair, while the arrector pili

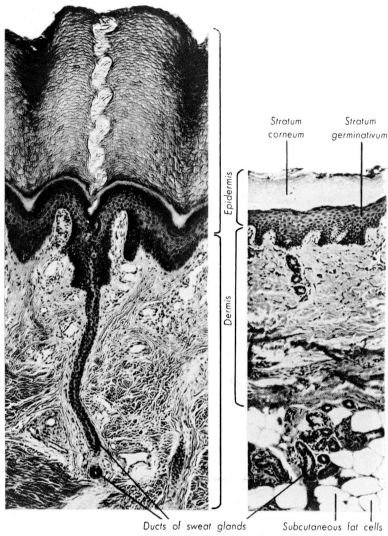

Stratum corneum   *Stratum germinativum*

Epidermis

Dermis

Ducts of sweat glands   Subcutaneous fat cells

Figure 1-3. *Photomicrographs of vertical sections of skin, showing thick epidermis from finger tip at left and thinner epidermis from dorsal surface of finger at right.* × 94. *(From Bailey,* Textbook of Histology, *courtesy of Williams & Wilkins Co.)*

muscles erect the hairs. These muscles are placed on the under side of the hairs and, when contracted, pull the base of the hair down, making the visible portion of the hair move up, thus becoming erect. At the same time the muscles tend to pull on the skin superficial to the muscle, producing the well-known "goose flesh." The **sweat glands** are coiled glands that are also deep to the dermis; their ducts penetrate both layers of the skin to open onto the surface. Both the sweat glands and the arrector mus-

cles are under control of the sympathetic portion of the autonomic nervous system and therefore respond to states of anxiety or fear. The many **nerve endings** are pictured in Figure 1-4. There are free nerve endings for pain reception, discs and corpuscles for touch, end bulbs for cold, and pacinian corpuscles for pressure.

The dermis of the skin blends with the superficial fascia, and it is this fascia that gives the skin the great mobility it possesses in most areas of the body.

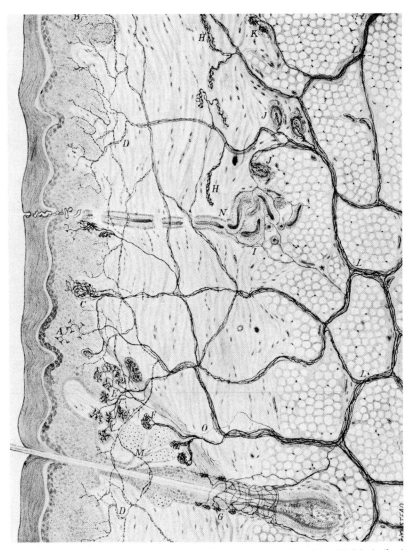

Figure 1-4. *Diagram of nerve endings and other structures in the skin: (A) Merkel's discs for touch; (B) free nerve endings for pain; (C) Meissner's corpuscles for touch; (D) nerve fibers for pain; (E) Krause's end bulbs for cold; (F) nerve endings for warmth; (G) nerve endings and fibers on hair follicle for touch; (H) Ruffini's endings for pressure; (I) sympathetic fibers innervating a sweat gland; (J) Pacinian corpuscles for pressure; (K) Golgi-Mazzoni endings for pressure; (L) nerve trunks; (M) sebaceous gland; (N) sweat gland; (O) sympathetic fibers to arrector pili muscle. (From Woollard, Weddell, and Harpman, J. Anat. 74 [1940], courtesy of Cambridge University Press.)*

SUPERFICIAL AND DEEP FASCIA

Removal of the skin reveals a layer of varying thickness called **superficial fascia** (Fig. 1-5). This connective tissue layer is a mesh of fibers that enclose varying amounts of fat. Although this layer is thicker than the deep fascia, it is not so dense in its consistency. In males this layer, in thin individuals, does not prevent one from seeing the outlines of muscles; in the female it tends to be thicker, to fill in depressions in such a way that muscles are usually not visible, which accounts partially for the difference in the male and female body contour.

The **deep fascia** is a thin but dense layer of connective tissue just deep to the superficial fascia. The deep fascia closely invests muscles and varies in thickness in different parts of the body. It is found not only on the external surface of muscles but on all surfaces. If muscles were removed and fascia left intact, a honeycomb-like arrangement would be revealed, as seen in Figure 1-6. The deep fascia serves the important function of holding the body together. Of equal importance is its role in relation to infectious agents, by serving either as a barrier or as a pathway; although infectious agents will finally penetrate a fascial plane if left unattended, the fascia does serve as a temporary deterrent to their spread. Contrariwise, infectious agents will follow fascial planes and spread rapidly within the confines of a fascial compartment. A knowledge of fascial planes allows intelligent

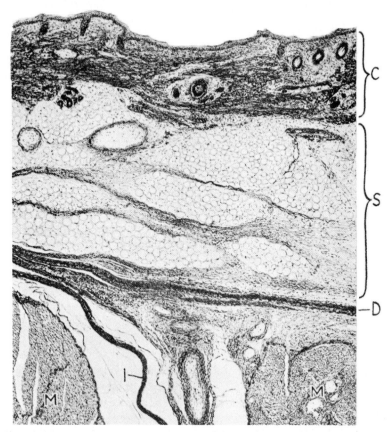

Figure 1-5. *Section through the integument of the leg of a human fetus aged approximately 7 months: (C) skin; (S) superficial fascia; (D) deep fascia; (I) intermuscular septum; (M) muscle. (From Clark,* Tissues of the Body, *courtesy of Oxford University Press.)*

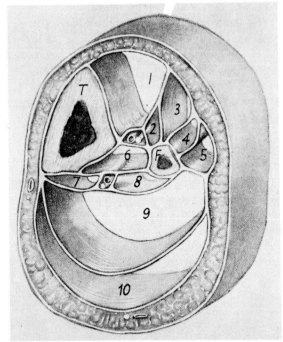

Figure 1-6. *Diagram of a section through the leg at the middle of the calf in which the fascial planes have been preserved after removal of the muscles: (T) tibia; (F) fibula; the numbers represent the various muscles of the leg. (From Clark,* Tissues of the Body, *courtesy of Oxford University Press.)*

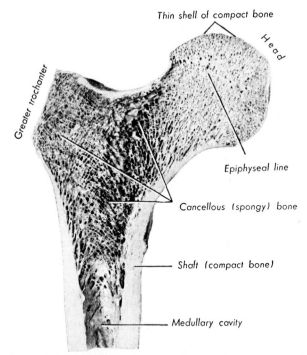

Figure 1-7. *Longitudinal section through the upper end of a femur of an adult male. (From Bailey,* Textbook of Histology, *courtesy of Williams & Wilkins Co.)*

treatment of infections. Surgeons make use of fascia in closing incisions, especially when muscles are cut in transverse section and stitches in the muscle will not hold.

SKELETAL SYSTEM

A common misconception is that bone is a dead substance. This is far from the truth, for bone is very much alive and is a structure that grows, changes in shape, and is subject to disease.

Bone (Fig. 1-7) consists of an organic framework of cells and tissues that would be quite bendable if it were not for the inorganic salts that are laid down in it. It can take the shape of a solid **compact bone** or possess holes around which are spicules of bone; the resemblance to a sponge in the latter instance gives rise to the term **spongy bone** for this type. These spicules of bone have been found to line up in the direction of greatest stress. All bones are covered on the outside by a thin layer of **periosteum,** which is important, for if saved during surgery and replaced, new bone will develop from it.

As mentioned above, bone is quite labile and must be protected from constant wear. Accordingly, the end of a bone taking part in the formation of a joint is covered with cartilage. We shall see later, when joints are studied, that bones are protected even from the constant wear of muscles by fluid-containing sacs called **bursae;** these are particularly numerous around joints where tendons are likely to put pressure on bones (Fig. 1-8). Bones have a blood supply (nutrient arteries), venous drainage, and a sensory nerve supply.

The bones have several **functions:** (1) they serve as a framework for the body and give the body its general shape; (2) they are very important as structures upon

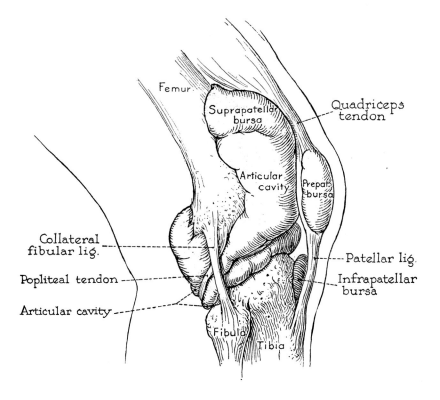

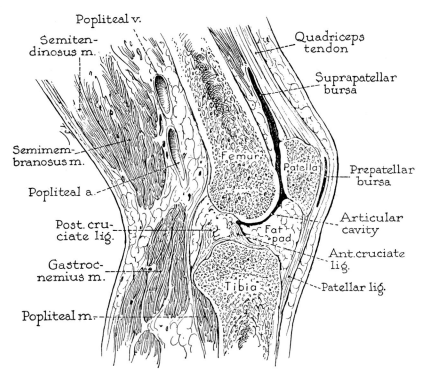

Figure 1-8. *The bursae of the right knee, lateral view, showing the extent they may be distended by fluid. Lower figure shows a sagittal section of the right knee. Do not attempt at this time to learn names of structures labeled.* (From Jones and Shepard, Manual of Surgical Anatomy, *courtesy of W. B. Saunders Co.*)

which muscles can act to induce movements; (3) they serve to protect the easily injured internal organs or viscera; (4) they serve as storehouses for calcium and phosphorus, which are essential to the welfare of the body; and (5) the marrow is an important location for blood cell development.

Bones can be **classified** according to (1) how they develop, (2) their location in the body, or (3) their shape. **Developmentally,** they are grown either from membrane (**membranous bone**), or from preformed cartilage (**cartilaginous bone**). In the former case, bone is formed from the periosteum in layer after layer (Fig. 1-9); this occurs in flat bones such as those of the skull. In cartilaginous bones, ossification centers develop in the preformed cartilage before bone is formed. In a long bone, for example (Fig. 1-10), an ossification center forms in the shaft (the diaphysis) and at each end (the epiphysis). As bone is formed, the ossification centers approach one another; the line of cartilage remaining between the bone formations is called an **epiphyseal** line. These are important, for growth in length occurs at these lines of cartilage; they are visible in x-rays and should not be confused with fractures. When these epiphyseal lines of cartilage no longer exist, the lines are considered closed and no further growth occurs. The endocrine glands are very much involved with the phenomenon of growth, midgets resulting from a lack of growth hormone from the hypophysis and gigantism occurring when too much of

this hormone is released. If this latter condition arises after the above-mentioned epiphyseal lines are closed, the individual exhibits unequal growth, especially of the jaw, hands, and feet—a condition known as acromegaly. A lack of thyroxin in youth also leads to a dwarf condition, but this person is likely to be misshapen in contrast to the well-proportioned individuals suffering a loss of growth hormone. Figure 1-11 presents a listing of the times when ossification starts and when lines close, which may be consulted as a reference.

Regionally, the bones are divided into an **axial skeleton,** consisting of the skull, vertebral column, ribs, sternum, and hyoid bone; and an **appendicular skeleton,** which consists of both limbs. It should be noted that the latter includes the clavicle and the scapula as parts of the upper limb and the coxal bones as part of the lower limb.

Figure 1-12 is a drawing of a skeleton with most of the bones labeled.

Bones, when classified according to **shape,** are long, flat, irregular, or sesamoid. The last group, the sesamoid bones, perhaps should not be in this classification, for they are simply bones that develop inside tendons. The patella is the classic example of such a bone.

On account of pressures and pulls of muscles and other structures, bones develop many irregularities in their surface structure. The following terms are used to denote **elevations:**

> **Process**—a general term for any elevation.
> **Crest**—an elongated elevation.
> **Hamulus**—a hook-shaped process.
> **Spine**—a sharp, pointed process.
> **Tubercle** or **tuberosity**—a rough, rounded elevation.
> **Condyle**—a large rounded eminence for articulation with another bone.
> **Epicondyle**—an elevation on a condyle.

Depressions can take various forms:

> **Alveolus**—a deep and narrow pit.
> **Fovea**—a small depression.
> **Fossa**—a shallow but wider depression.
> **Sulcus**—a groove or furrow.
> **Meatus**—a canal.
> **Incisura**—a notch.
> **Hiatus**—a slitlike gap or cleft.

A cavity in a bone is an **antrum, air cell,** or **sinus,** and a hole through a bone is a **foramen.**

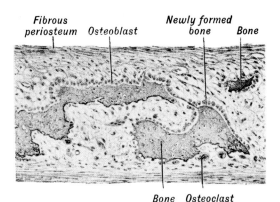

Figure 1-9. Bone formation from membrane. Cross section through the primordium of the parietal bone of a 4-month embryo. × 100. An osteoblast is a bone-forming cell; an osteoclast serves to absorb and remove osseous tissue. (After Schaffer, from Maximow and Bloom, A Textbook of Histology, *courtesy of W. B. Saunders Co.)*

Periosteal bud   Primitive   Bone collar
                      marrow cavity

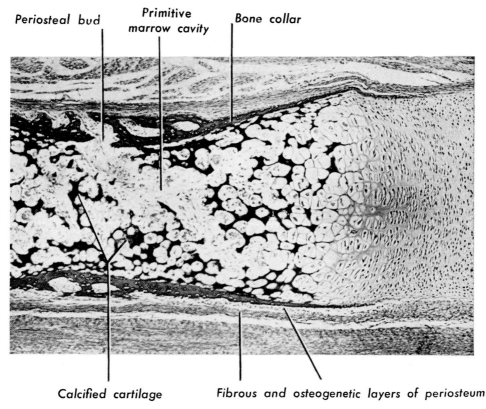

Calcified cartilage   Fibrous and osteogenetic layers of periosteum

Figure 1-10. *Bone formation from cartilage. Metacarpal bone of human fetus, 10 to 12 weeks old. Cartilage to right, new bone formation to left. (From Bailey,* Textbook of Histology, *courtesy of Williams & Wilkins Co.)*

There is considerable **variation** in bone structure among both the races and the sexes, and even between individuals of the same sex and race. Females usually have smaller bones than males. Furthermore, since they are usually less muscular, the projections caused by muscle pull are of lesser degree. The sex of a skeleton can usually be determined, the pelvis, as will be shown later, being the part that is most revealing. Individual variation occurs because of size and age. Naturally, a tall individual will have longer bones than a short person, and a heavy, thickset individual may have heavy bone structure. Furthermore, increased age induces a decrease in resilience.

Fractures occur frequently and are of many varieties. Some of the more common are these: (1) simple or compound, depending upon whether the bone fragments are exposed; (2) complete or incomplete, depending on degree of extension of fracture through the bone; (3) intra- or extracapsular, inside or outside the joint capsule; (4) evulsion, tearing off of a bony prominence; and (5) spontaneous, occurring without apparent trauma as a result of disease.

Bones are susceptible to most of the harmful agents that affect the soft tissues of the body. In addition to fractures, bones respond to nutritional, metabolic, and endocrine disturbances as well as to conditions related to environment and heredity. Furthermore, the cellular components of bone give rise to a variety of tumors.

SCHEMA OF OSSIFICATION OF THE SKELETON
I. TOTAL SKELETON

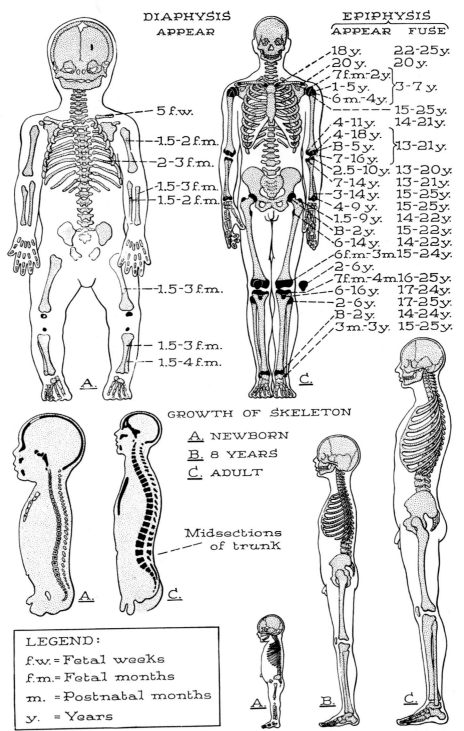

Figure 1-11. *Skeletal development. For reference only. (From Patten's Human Embryology by C. E. Corliss, © 1976 by McGraw-Hill, Inc. Used with permission of McGraw-Hill Book Company.)*

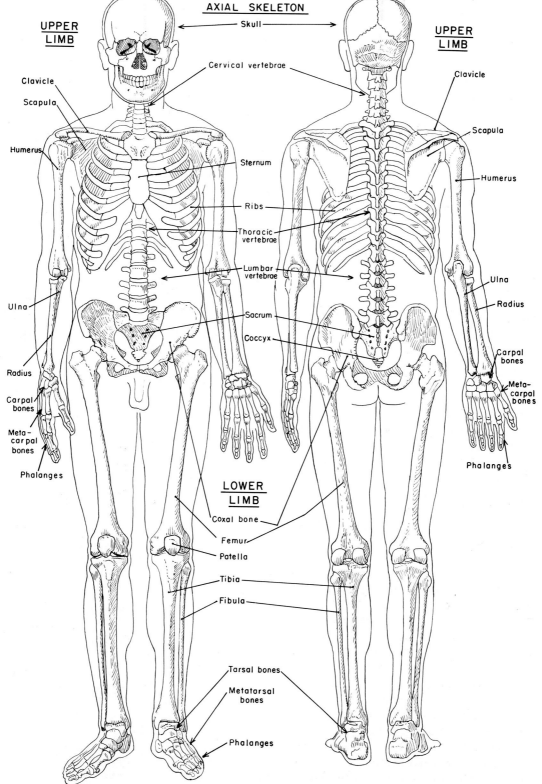

AXIAL SKELETON

13

UPPER LIMB

Skull

Cervical vertebrae

Clavicle

Scapula

Humerus

Sternum

Ribs

Thoracic vertebrae

Lumbar vertebrae

Ulna

Radius

Carpal bones

Meta-carpal bones

Phalanges

Sacrum

Coccyx

UPPER LIMB

Clavicle

Scapula

Humerus

Ulna

Radius

Carpal bones

Meta-carpal bones

Phalanges

LOWER LIMB

Coxal bone

Femur

Patella

Tibia

Fibula

Tarsal bones

Metatarsal bones

Phalanges

Figure 1-12. The skeleton.

JOINTS

Whenever two bones come together, whether movement is allowed or not, a joint is formed. The many joints in the body can be divided into three main groups:

1. Fibrous
2. Cartilaginous
3. Synovial

Typical examples of each are shown in Figure 1-13. Joints are subject not only to structural injuries and dislocations but also to inflammatory conditions of various kinds.

   1.   Fibrous joints are of two types: **syndesmosis** (*Gr.,* a binding together + condition) and **suture** (*L.,* a sewing together). In the former type, two bones are separated by a small amount of fibrous material, a slight amount of movement being allowed. An example of this type of articulation is the tibiofibular joint. **Sutures** are immovable and are found only in the skull. Although sutures contain a thin layer of fibrous tissue, this tissue disappears with increasing age and the sutures become closed. The joining edges of the bones take different forms, some overlapping, some serrated, while others are in the form of pegs and sockets.

   2.   When two bones are separated from each other by cartilage, they are considered to form a **cartilaginous joint. Such joints are also of two types: **synchondrosis** (*Gr.,* together + cartilage + condition) and **symphysis** (*Gr.,* a growing together). The former is a union between two bones separated by cartilage, and the perichondrium of the cartilage is directly continuous with the periosteum of the bone. Many such joints between various parts of bones are present in youth but disappear in adulthood. This type of joint persists between the first rib and sternum, but the length of the cartilage does allow a certain amount of movement. **Symphyses** are found only in the midline of the body, occurring between the bodies of the vertebrae, between body and manubrium of the sternum, and between the two pubic bones (pubic symphysis).

   3.   Synovial joints are the freely movable joints and are lined by a synovial membrane, which secretes synovial fluid. The bones are protected by cartilage, which is not covered by the synovial membrane. The whole joint

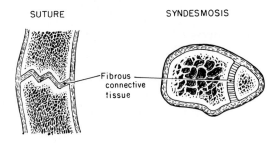

FIBROUS JOINTS

SUTURE   SYNDESMOSIS

Fibrous connective tissue

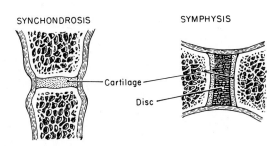

CARTILAGINOUS   JOINTS

SYNCHONDROSIS   SYMPHYSIS

Cartilage

Disc

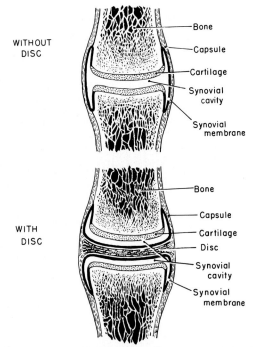

SYNOVIAL JOINTS

WITHOUT DISC

Bone

Capsule

Cartilage

Synovial cavity

Synovial membrane

WITH DISC

Bone

Capsule

Cartilage

Disc

Synovial cavity

Synovial membrane

Figure 1-13.   *The three main types of joints in the body.*

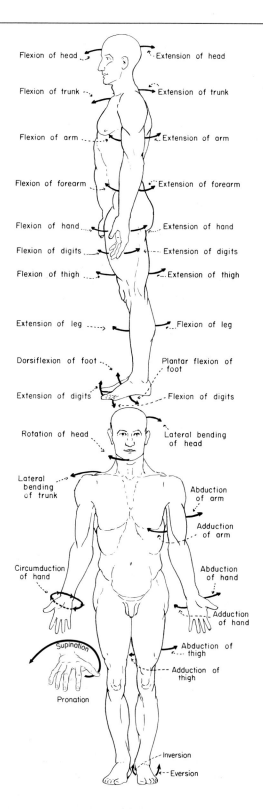

and its synovial sac is contained within a tough connective tissue capsule. These joints may be single or double, the latter being divided into its two parts by an intervening disc, which is directly continuous with the capsule.

There are many types of freely movable joints, differing from one another because of the type of movement allowed. Therefore, it is necessary to describe the types of movement in freely movable joints before describing the joints themselves. These movements are pictured in Figure 1-14.

1. Flexion (L., bending) is defined as a decrease in the angle made by two bones. For example, the upper limb held in the anatomical position makes an angle of 180 degrees at the elbow; if the hand is brought up to the shoulder, the angle is decreased. This is bending at the elbow, flexion at this joint, or flexion of the forearm.

2. Extension (L., stretching out) is the opposite of flexion and is therefore an increase in the angle between two bones. Using the same example, if the limb is brought back to the anatomical position, or the arm is straightened out, the angle at the elbow is increased and the forearm is extended.

3. Abduction (L., away from + draw) is defined as movement away from the midline of the body. If the upper limb is moved laterally from the anatomical position, abduction at the shoulder joint results.

4. Adduction (L., toward + draw) is the opposite of abduction, or movement toward the midline of the body.

5. Circumduction (L., around + draw) is actually a combination of the movements already described. It consists of flexion, abduction, extension, and adduction in sequence. If this movement occurs at the shoulder joint, the arm will be brought forward, then laterally, then posteriorly, and then medially. This will result in the distal end of the limb going around in a circle and, since the arm is attached at the shoulder, the upper limb moves in the form of a cone, with the base distally placed and the tip of the cone proximally located.

6. Rotation is movement of a twisting nature along a single axis and differs from circumduction in that the limb has to remain on a straight line.

Common sense has to be used in interpreting these movements in some areas of the body. For example, the

Figure 1-14. Diagram of the major movements of the body.

wrist can be flexed by moving the hand forward, and extended by returning to the anatomical position. However, the hand can be moved further posteriorly and an angle is actually being decreased in size; although this is further extension, it fits the definition of flexion and is frequently called dorsiflexion. Movements at the shoulder joint are difficult to interpret if the definitions are followed strictly. Abduction moves the arm, forearm, and hand away from the midline until the horizontal plane is reached, but further movement brings the limb back to the midline again, which fits the definition of adduction. In spite of this, the whole movement is considered to be abduction. Flexion of the arm moves it anteriorly; extension, posteriorly; and adduction, medially toward the trunk. Flexion of the vertebral column occurs when the body is bent forward, extension when straightened to the upright position. Other problems in movement will be dealt with where appropriate.

To return to the synovial, freely movable joints, they are of the following types:

1. Plane or gliding. In this situation movement is simply the gliding of one bone on another. Movement between the wrist bones, between the articular facets of the vertebrae, between the tarsal bones, and at the radioulnar joints are good examples.

2. Hinge or ginglymus. This is a joint that allows movement similar to that allowed by a hinge on a door, flexion and extension being the only movements permitted. The best examples are the elbow, ankle, and interphalangeal joints. The knee and temporomandibular joints also are hinge joints, but since some gliding motion is allowed, they are known as modified hinge joints.

3. Pivot or trochoid. This is a joint that allows rotation only. The joint between the first and second cervical vertebrae, the atlas and axis (the atlantoaxial joint), is such an articulation.

4. Ball-and-socket. This is a joint made by a ball of one bone fitting into a socket on another. The shoulder and hip joints are the only joints of this type in the body. This is a universal joint, for it allows all movements.

5. Condyloid. This is similar to the ball-and-socket joint except that the structural configuration does not allow rotation. The metacarpophalangeal joints are good examples; flexion, extension, abduction, adduction, and therefore circumduction, can occur in these joints, but no rotation.

6. Ellipsoidal. This is similar to the condyloid joint except for the shape of the bones; in this type the head of one bone is oval-shaped and fits into a similarly shaped fossa. Example—the wrist joint.

7. Saddle. This is a joint between two surfaces, one of which is concave and the other convex, the latter surface being in the shape of a saddle. The carpometacarpal joint of the thumb is such a joint.

LIGAMENTS

As mentioned before, the freely movable joints are lined with a synovial membrane, which is surrounded by a strong connective tissue capsule. The capsule is further strengthened by the presence of strong fibrous cords extending from one bone to another; such structures are called ligaments. Some of these ligaments developed as part of the joint capsule and are actually thickenings in the capsule; these are known as intrinsic ligaments. Others are extrinsic and have resulted from modification in location and attachments of tendons. In other words, a muscle may have passed the joint in lower animals and inserted on the bone below, but in the human this same muscle may insert at a point proximal to the joint and that part of the tendon which traversed the joint may have changed to a ligament.

BURSAE

As pointed out in the discussion of bones, wherever a muscle or tendon of a muscle might bring pressure on a bone, the muscle is separated from the bone by a fluid-containing sac called a bursa. These bursae are lined by a synovial membrane exactly the same as those found lining joints and contain a similar fluid. Some of these bursae are connected directly with the synovial cavities of the joints. Bursae may become inflamed—a bursitis.

MUSCULAR SYSTEM

There are three types of muscle in the body: (1) smooth or nonstriated, (2) cardiac, and (3) skeletal or striated (Fig. 1-15). All function by contracting.

MUSCLE TISSUE

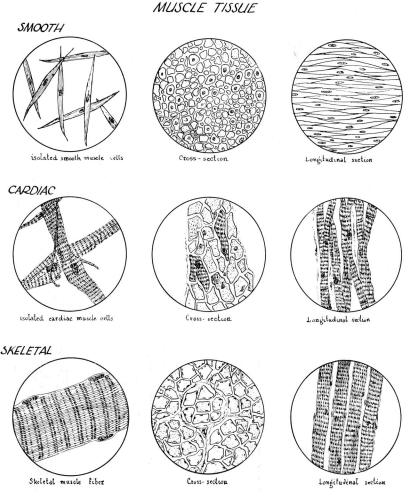

Figure 1-15. *Diagrammatic chart of muscle types. (From Kendall,* Microscopic Anatomy of Vertebrates, *courtesy of Lea and Febiger.)*

Smooth muscle occurs in most viscera and in all blood vessels as far as the arterioles and venules; there is no muscle in capillaries. As we shall see in the chapter on the nervous system, this smooth muscle is involuntary muscle and is innervated by the autonomic nervous system. **Cardiac muscle** is found only in the heart, as the name implies. It has a characteristic histological appearance, and it also is innervated by the autonomic nervous system. The rest of the muscle mass of the body makes up the **skeletal musculature,** which is under voluntary control.

STRUCTURE

The skeletal muscles of the body are of many shapes, depending upon how the muscle fasciculi are distributed around the tendon, that portion of a muscle which attaches it to bone. Muscles of various shapes are shown in Figure 1-16. Tendons are usually round structures, but can take any shape from a small round cord to a wide tendinous sheet. The latter form is an aponeurosis or aponeurotic tendon. Muscles may also attach to bone by fleshy attachments; this simply means that the connective

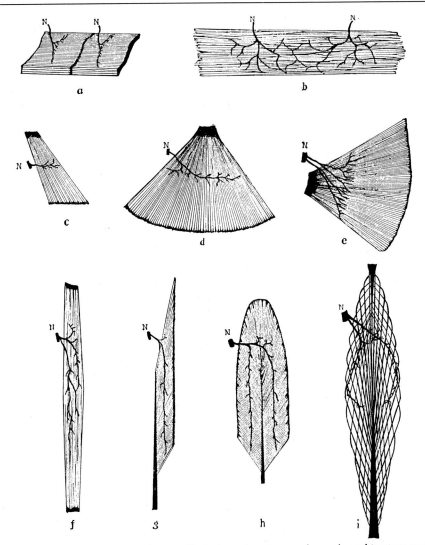

Figure 1-16. *Diagrammatic outlines to illustrate various types of muscle architecture and the relations of the main nerve branches: (a) two segments of a single muscle; (b) portion of a sheet-like muscle with two nerve branches and intramuscular nerve plexus; (c) typical quadrilateral muscle with nerve passing across the muscle about midway between the tendons; (d, e) two triangular muscles with different types of innervation; (f) long ribbon-like muscle with interdigitating fiber bundles; (g) unipenniform muscle; (h) bipenniform muscle; (i) typical fusiform muscle. (From* Patten's Human Embryology *by R. E. Corliss, © 1976 by McGraw Hill, Inc. Used with permission of McGraw-Hill Book Company.)*

tissue surrounding the various fasciculi of the muscle attaches to the periosteum of the bone directly rather than by a tendon.

MUSCLE ACTION

Muscles act (1) by pulling two bony points closer to one another, (2) as a sphincter, or (3) by pulling on the skin. They do not push. Muscles usually have attachments to bones in such a manner as to bridge across a joint, and when the muscle contracts, movement occurs at this joint. These attachments are called **origins** and **insertions,** the origin of a muscle being at the more fixed end and the insertion at that point which is the more movable. It is difficult to determine the origin and insertion of muscles when the two ends move to an equal degree. Many muscles utilize a lever action to accomplish their effect. As can be seen in Figure 1-17, the situation as it exists in human beings is not one that provides great strength, but it does induce great movement for a rather conservative amount of muscle contraction.

The **main action** of single muscles is determined in several ways:

1. Direct observation and manipulation. This can be done right on the cadaver and consists of pulling on muscles and using one's own mechanical sense to determine what would happen when the muscle shortens. Actions of most muscles can be solved in this simple manner.

2. Nerve injuries. All skeletal muscles require an intact nerve supply in order to function. Patients occasionally present certain nerve injuries that allow one to observe the deficiencies resulting therefrom. If the injured nerve innervates a single muscle, the action of that muscle may be determined. Many nerves innervate a group of muscles; in such a case the action of the group may be revealed.

3. Electrical production. Electrodes may be placed in a single muscle and wired to a potentiometer. If the victim of such needling is then asked to perform certain acts and the muscle is used in this act, the electricity evolved from the muscle will be picked up by the potentiometer. This process, called **electromyography,** has been very helpful in solving several problems on muscle action. However, even this method has been found lacking in various respects. It was soon found that individual variations were still apparent; for example, a typist or violin player uses muscles in the forearm differently from a person who has been performing hard labor.

Another vexing problem in determining muscle action is the fact that a single muscle will have different actions, depending on the position of the body or upon the physical condition of the person. For example, muscles about the hip joint will perform differently when a person is sitting and when he or she is standing in the anatomical position. Muscles that are not usually utilized in respiration under normal conditions can become involved if a condition of oxygen want exists; in fact, any muscle attached to the ribs will become a respiratory muscle under these conditions, origins becoming insertions and insertions becoming origins. It is best for the student of anatomy to think of muscle actions with the body in the anatomical position and under normal conditions, and this book will limit itself in almost all cases to a description of muscle action in this manner.

There are very few muscles that act alone; almost all movements are accomplished by group action. Such muscles working together to produce a movement are **synergists;** conversely, muscles that have opposite actions are **antagonists.** Striated muscles are innervated in such a manner as to allow amazing control over our muscle action. This is called coordination and is accomplished by lower brain centers, such as the cerebellum, having an effect on movement initiated by the cerebral cortex. If the forearm, for example, is partially flexed, it can be held in this position. This means that the state of contraction in

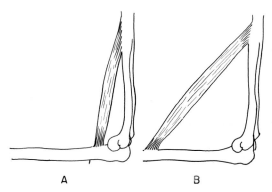

Figure 1-17. (A) Muscle attachment as found in nature: Maximum amount of movement for small amount of muscle contraction, but muscle is not powerful. (B) Powerful muscle, but a minimum amount of movement for a maximum muscle contraction.

the flexors of the forearm is the same as that in the extensors, the antagonists. If the state of contraction is increased in the flexors and decreased to a corresponding degree in the extensors, continued flexion will occur. People who have lost this delicate balance possess tremors.

NOMENCLATURE

Muscles have been named because of their shape, location, action, attachments, direction, their contrasting features, their number of parts, or for combinations of these. The trapezius and the rhomboids are obviously named for their shape, while the levator scapulae and levator palpebrae superioris are named for their action in lifting the scapula and upper eyelid respectively. The pectoralis major and minor muscles are named because of their location in the pectoral region and also for their contrast in size. The sternocleidomastoideus muscle has attachments to the sternum, the clavicle, and the mastoid process of the temporal bone. In accordance with their Latin meanings, the rectus abdominis must be a straight muscle located in the abdominal wall, and the biceps brachii and triceps brachii are muscles located in the brachium (*L.*, arm), and possess two and three parts respectively.

EMBRYOLOGY

All striated or skeletal muscles are derived from **myotomes** of the **mesodermal somites** or from **branchial arches.** The former make up the somatic musculature while the latter are branchiomeric. Figure 1-18B diagrams the myotomes and the muscle primordia associated with the branchial arches.

The human develops these arches in the upper end of the pharynx. These correspond to the gill arches found in fishes. As we will see later, there are **pouches** formed between these arches inside the pharynx, and **clefts** formed between the arches externally. We will also see that each arch has its own cartilage, blood vessel, and nerve as well as the muscle primordium.

A nerve becomes associated with each myotome as well as with each branchial muscle primordium, and even though these muscles migrate some distance from the original site, the nerves elongate and maintain this relationship.

VASCULAR AND NERVE SUPPLY

These will be described in detail in subsequent chapters. Obviously, every muscle must have an arterial supply, a

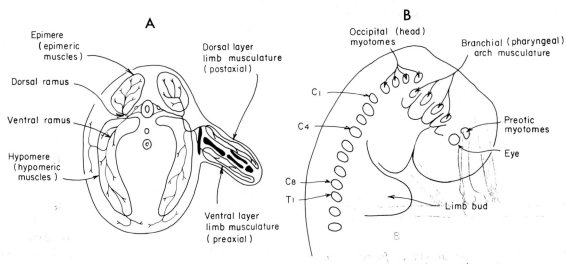

Figure 1-18. *(A) Cross section of an embryo showing the division of trunk musculature into epimeres and hypomeres, and limb musculature into dorsal muscles (postaxial or posterior to axis of limb) and ventral muscles (preaxial or anterior to axis of limb). (B) Myotomes and muscle primordia in the head and body of an embryo in the seventh week.*

venous drainage, and lymphatic drainage; as mentioned previously, no muscle can function without a motor (efferent) nerve supply. Afferent nerve fibers from muscles are called proprioceptive fibers and, combined with afferent impulses from ligaments, are responsible for our being aware of the position of our bodies. Good examples of this are that we can tell when the hand is flexed whether we can see that it is or not, and (a test used clinically by reaching under the bedsheet) that a patient is aware of the fact that his toe is being flexed if his proprioceptive pathways are normal.

The muscular system suffers from infections and from traumatic afflictions such as muscle strain, and naturally will not function properly under conditions that interfere with its blood or nerve supply.

NERVOUS SYSTEM

The nervous system controls the other systems of the body and is responsible for our being aware of our environment. It is extremely important clinically both from a functional and from an anatomical point of view. Its afflictions are reflected in the other systems of the body, and afflictions in the other systems of the body are reflected in the nervous system. These pathological involvements may be traumatic, infectious, or neoplastic. The majority of patients will complain of pain.

The nervous system is divided into a **central nervous system** made up of the brain and the spinal cord, and a **peripheral nervous system** consisting of 12 pairs of cranial nerves emerging from the brain stem and 31 spinal nerves arising from the spinal cord. The cranial nerves are:

I.	Olfactory	**VII.**	Facial
II.	Optic	**VIII.**	Vestibulocochlear
III.	Oculomotor	**IX.**	Glossopharyngeal
IV.	Trochlear	**X.**	Vagus
V.	Trigeminal	**XI.**	Spinal accessory
VI.	Abducens	**XII.**	Hypoglossal

A glance at Figure 7-28 on page 499 at this time might be helpful.

The spinal nerves are divided into **8 cervical, 12 thoracic, 5 lumbar, 5 sacral,** and **1 coccygeal;** they do not correspond exactly to the vertebrae, for there are 7 cervical and 4 coccygeal vertebrae.

The structural unit of the nervous system consists of a nerve cell and its process, a **neuron.** A neuron has a **body** in which the nucleus is placed, and processes that extend out from this body. These processes are **dendrites** and an **axon.** Nerve impulses travel toward the cell body via the dendrites, and away from the cell body in the axon. As can be seen in Figure 1-19, neurons are of many shapes, some possessing very short dendrites and axons, others very long processes, while still others may have one process of great length and another that is extremely short. A collection of neuron cell bodies within the central nervous system is a **nucleus;** a similar collection outside is a **ganglion.** Circuits are formed in the nervous system by the axon of one neuron connecting with the dendrites of another, this connection being called a **synapse.** A collection of these axons or dendrites inside the central nervous system forms a **tract;** a similar group outside is a **nerve.**

Note that the spinal cord (Fig. 1-20) in cross section exhibits a butterfly-shaped central portion, the **gray matter,** consisting of cell bodies and their processes. The gray matter can be divided into areas, that dorsally placed being called the **dorsal horn,** that ventrally placed, the **ventral horn.** The area between these horns is intermediate in position and can be divided into **intermediomedial** and **intermediolateral cell columns.**

The **white matter,** made up of tracts that in turn are made up of dendrites and axons of the cells, surrounds the gray matter. The white matter is divided into **dorsal, lateral,** and **ventral columns.**

Each spinal nerve is attached to the spinal cord by two roots, a **dorsal** or **sensory root** and a **ventral** or **motor root.** The dorsal root exhibits a swelling called a **dorsal root ganglion,** in which are located the cell bodies of the sensory nerves. Note that the two roots join to form a **spinal nerve.** Each spinal nerve divides into a branch, the **dorsal ramus,** coursing to the back, and a branch, the **ventral ramus,** that follows the body wall to the anterior surface of the body.

As can be seen in Figure 1-20B, pathways are provided (**white** and **gray rami communicantes**) so that fibers can course between a spinal nerve and a series of ganglia located just lateral to the bodies of the vertebrae; these ganglia are called **sympathetic ganglia** and extend

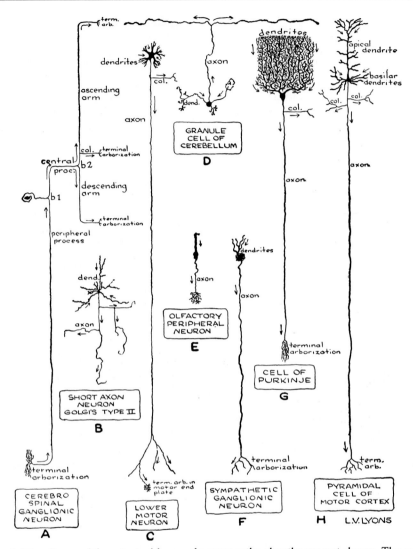

Figure 1-19. *Some of the principal forms of neurons; the sheaths are not shown. The axons, except in B, are shown much shorter in proportion to the size of the body and dendrites than they actually are. The direction of conduction is shown by the arrows; col., collateral branch; proc., process; term. arb., terminal arborization. (From Bailey,* Textbook of Histology, *courtesy of Williams & Wilkins Co.)*

from the neck to the pelvic cavity forming a chain of ganglia—the **sympathetic chain.**

In addition, nerves called **splanchnic nerves** connect these chain ganglia with other ganglia located anterior to the abdominal aorta—**preaortic ganglia.** These preaortic ganglia are located not only anterior to the aorta but also at the bases of its major branches.

All nerves are composed of a great many fibers;

these fibers have been classified into several categories depending on whether they are sensory (afferent) or motor (efferent); whether they are associated with the body wall and limbs (somatic), viscera and visceral structures (visceral), or with muscles derived from the so-called branchial arches (branchial); or whether they are involved with the special senses (special afferent). In more detail:

A. SOMATIC REFLEX

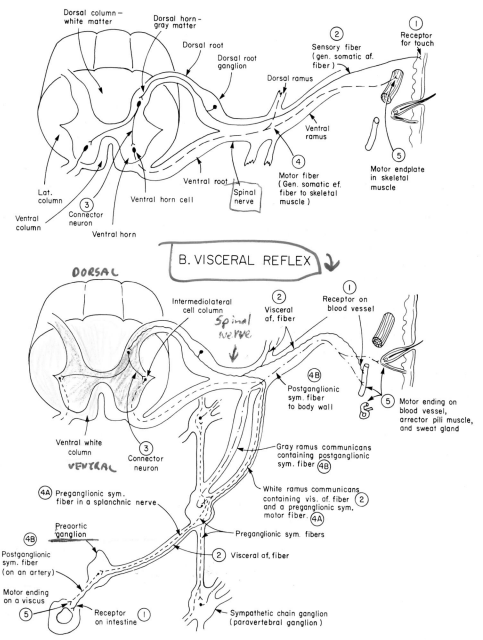

Figure 1-20. *Reflex arcs showing the five components: (1) afferent nerve ending, (2) afferent sensory limb, (3) connecting cell, (4) efferent or motor limb, (5) motor end-plate. (A) a somatic reflex arc, (B) a visceral reflex arc.*

1.   Fibers concerned with the soma or body as a whole.   The sensory fibers of this group are called **general somatic afferent** fibers and are concerned with exteroceptive sensations such as touch, pain, and temperature, and also proprioceptive sensations from ligaments, tendons, and muscles that give one a sense of position of the joints and related structures. The motor fibers are classified as **general somatic efferent** fibers and innervate all skeletal muscles except those derived embryologically from the branchial arches (see immediately below).

2.   Fibers concerned with branchial musculature.   As mentioned in the section on muscles, the human embryo develops branchial or pharyngeal arches that correspond to gill arches found in fishes (Fig. 1-18B). Each arch has a muscle primordium associated with it and a cranial nerve innervating this primordium and all its derivations. The efferent fibers to these branchial muscles are called **branchial efferents,** and proprioceptive impulses from these muscles travel along **branchial afferent** fibers.

3.   Those concerned with viscera, glands, smooth muscle, and cardiac muscle.   The afferent fibers in this group are called **general visceral afferents** and are involved with interoceptive sensations. The motor division is the autonomic nervous system, and these efferent fibers are **general visceral efferents,** which innervate smooth muscle, cardiac muscle, and glands.

4.   Fibers concerned with special senses.   These are **special afferents** for taste, smell, sight, and hearing.*

Fibers concerned with branchial musculature and the special senses are all located in cranial nerves.

The basic functional unit of the nervous system is the **reflex arc.** Figure 1-20 depicts two such reflex arcs—a **somatic** from the body wall and limbs, and a **visceral** from viscera and visceral structures.

Concentrating on the simpler somatic arc first (Fig. 1-20A), if a person touches a hot stove, the heat stimulates a nerve ending in the skin, and impulses travel along the dendrite of a sensory nerve to a dorsal root ganglion. In sensory ganglia of this type the dendrite joins the axon before reaching the cell body. There is, therefore, only one process entering the cell body; these are unipolar cells in contrast to the usual bipolar arrangement (see Fig. 1-18A). *There is no synapse in these ganglia.* The impulses continue into the spinal cord via the axons of these sensory cells. These axons synapse with dendrites of cells located in the dorsal part of the gray matter of the cord which are designated as connector neurons, and the axons of the latter cells synapse with dendrites of cells in the ventral part of the gray matter. The impulses continue through the ventral horn cell and out through its axon, to end in motor end-plates in some muscle. The muscle is thereby stimulated, and the finger is removed from the hot stove by contraction of that muscle. This circuit (1) over the **afferent nerve ending,** (2) over the **afferent sensory limb,** (3) through the **connecting cell,** (4) out the **efferent** or **motor limb,** (5) to a **motor end-plate—** the five components of a reflex arc—is covered in a fraction of a second. These afferent fibers also have connections with circuits leading to the brain, where the stimulus of pain will be recognized. However, one's finger would be removed from the hot stove before realization of a sensation of excess heat occurred.

We will postpone the visceral reflex arc until after the autonomic nervous system is described, immediately below.

AUTONOMIC NERVOUS SYSTEM

The general visceral efferent fibers, by definition, make up the autonomic nervous system; this system innervates glands, cardiac muscle, and smooth muscle. Since all blood vessels except capillaries have smooth muscle, this system has a wide distribution. It is divided into two parts, the **sympathetic** and the **parasympathetic** nervous systems. The **sympathetic** portion arises from cells in the thoracic and upper lumbar regions of the spinal cord (T1 to L3) and is often called the thoracolumbar part of the autonomic nervous system. The **parasympathetic** portion arises from cells in the brain stem and sacral region of the spinal cord. It is found in cranial nerves III, VII, IX, and X (oculomotor, facial, glossopharyngeal, and vagus) and in sacral nerves 2, 3, and 4; it is frequently called the craniosacral part of the autonomic nervous system.

A few **rules** for this system are helpful:

1.   The autonomic nervous system is always a **two-neuron hookup.** In other words, the first neuron starts in the central nervous system, either in the brain

---

*Others divide these into special visceral afferent fibers for taste and smell, and special somatic afferent fibers for sight and hearing, but this seems to be of little value.

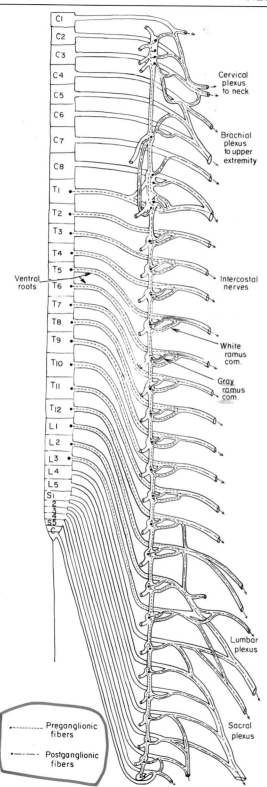

Cervical
plexus
to neck

Brachial
plexus
to upper
extremity

Ventral
roots

Intercostal
nerves

White
ramus
com.

Gray
ramus
com.

Lumbar
plexus

Sacral
plexus

-------- Preganglionic
fibers

-·-·- Postganglionic
fibers

stem or in the spinal cord, and travels to a ganglion; a synapse occurs here with the second neuron, which continues to the organ to be innervated.

2. The first neuron in this arrangement is called a **preganglionic fiber** or neuron, and the second is considered to be **postganglionic.**

3. The ganglia of the **sympathetic** system are located **some distance away** from the organ to be innervated. With the exception of four ganglia in the head, the ganglia of the **parasympathetic** system are located **within the walls** of the organ to be innervated.

4. Postganglionic neurons of the sympathetic system are long and follow arteries or other nerves to get to their destination; postganglionic fibers of the parasympathetic system are very short (except in the head) since the cells are already in the organ to be innervated.

Although the entire body is stimulated simultaneously by the autonomic nervous system, it is helpful for descriptive purposes to divide it into innervation to the body walls and limbs, and to the internal organs or viscera.

FIBERS TO WALLS AND LIMBS. Figure 1-20B can be used to trace pre- and postganglionic sympathetic fibers found in a typical spinal nerve. The cell body of the **preganglionic neuron** is located in the gray matter of the spinal cord halfway between the dorsal and the ventral columns of cells and in a lateral position; this area is the **intermediolateral cell column.** The axon from this cell leaves the cord through a ventral root and then reaches a spinal nerve. It leaves the spinal nerve by a **white ramus communicans** to reach a **sympathetic chain ganglion.** A synapse occurs here, and the axon of the **postganglionic neuron** reaches the spinal nerve again via a **gray ramus communicans** and continues with the spinal nerve to innervate glands and smooth muscle in the territory covered by that particular nerve. The route just described is the one taken to innervate glands and smooth muscle in the entire body wall and the limbs (Fig. 1-21).

Mnemonic: sp. cord + nerves: white = inner outer

Figure 1-21. *The autonomic nervous system to body walls and limbs. Note that the preganglionic fibers arise only in the thoracic and upper lumbar regions, but reach all chain ganglia because of the cervical and pelvic extensions. Postganglionic fibers then reach all nerves. Note also that it is entirely sympathetic.*

We have already seen that the sympathetic system arises in the thoracic and upper lumbar regions only. Therefore, preganglionic sympathetic fibers will be found only in ventral roots that correspond to this level and, consequently, white rami communicantes will be found only in this same area—from T1 to L3. To reach parts of the body superior and inferior to these levels, the fibers arise at the proper spinal cord levels and then either ascend into the neck via the cervical extension of the chain or descend into the pelvis by the pelvic extension. As can be seen in Figure 1-21, the sympathetic nerve supply to the neck area can be traced as follows: Preganglionic fibers start in the intermediolateral cell column of the upper thoracic segments of the spinal cord and traverse a ventral root to reach the spinal nerve; they then travel to the proper sympathetic chain ganglion via a white ramus communicans. Instead of synapsing at this location, the preganglionic fibers ascend in the sympathetic chain and synapse in a cervical ganglion. Postganglionic fibers join a regular nerve in this area via a gray ramus communicans. The same general principle is followed to reach the pelvis and lower limb. *It is easily seen that all the spinal nerves have gray rami communicantes, but only the thoracic and upper lumbar nerves have white rami.* All preganglionic fibers give off acetylcholine at the nerve endings (cholinergic fibers); the postganglionic fibers to sweat glands also give off acetylcholine, but all others produce norepinephrine (adrenergic fibers).

To date no parasympathetic supply to the body wall or limbs has been described. Nevertheless, when sympathetic fibers are stimulated, a vasoconstriction occurs in the skin and a vasodilation in the skeletal muscles. Since both actions are brought about by adrenergic fibers, a concept of alpha and beta receptor sites has been introduced, the alpha sites causing a vasoconstriction and the beta a vasodilation. The anatomy of such sites is difficult to visualize; certainly a vasoconstriction will occur upon contraction of the smooth muscle in arterioles but there is no structural mechanism to cause a vasodilation. In some manner these beta receptor sites must inhibit the vasoconstriction. Sweat glands (cholinergic fibers) and arrector pili muscles (adrenergic fibers) also respond to sympathetic stimulation; thus, sweating and erection of hairs associated with fright.

FIBERS TO VISCERA. Figure 1-22 shows the autonomic nerve supply to viscera. **Preganglionic sympathetic** fibers originate in the same thoracic and upper lumbar segments as just described for the trunk and limbs. Note that postganglionic fibers **to head and neck** structures arise in the cervical chain ganglia. Postganglionic fibers **to the heart** arise in the cervical ganglia and also in the first four or five thoracic sympathetic chain ganglia. These same segments supply the respiratory system as well.

Some preganglionic sympathetic fibers reach the sympathetic chain ganglia, although they do not synapse there but rather continue in splanchnic nerves to the preaortic ganglia located anterior to the aorta and at the bases of its major branches. A synapse occurs here and **postganglionic** fibers then follow the arteries to get to the viscera. These **splanchnics** are called **greater, lesser,** and **least** splanchnics in the thoracic region, **lumbar** splanchnics in the lumbar region, and **sacral** splanchnics in the pelvic area.

The **parasympathetic** supply to the viscera can be seen to be cranial nerves III, VII, and IX to structures in the head, and the **vagus** to thoracic and abdominal viscera as far as the transverse colon; the rest of the colon and the pelvic organs receive their parasympathetic innervation from the **pelvic nerve** from S2, 3, and 4. These are preganglionic fibers; postganglionic fibers are in the walls of the organs innervated (except the head structures).

The two systems are more or less antagonistic to one another as far as viscera are concerned. The parasympathetic system can be thought of as keeping the body eased down to a calm, even pace. Contrariwise, the sympathetic system is used for emergencies and stimulates the body into increased activities which, if continued over long periods, would be deleterious. For example, the sympathetic increases the heart rate, the parasympathetic decreases it; the sympathetic dilates the pupil, the parasympathetic constricts it; the sympathetic stops peristalsis, the parasympathetic increases it.

Figure 1-22. The autonomic nervous system to viscera. Note the double innervation to most structures, and that the sympathetic ganglia are some distance away from the organ to be innervated, while the parasympathetic ganglia (with the exception of those in the head) are close to or in the walls of the organ.

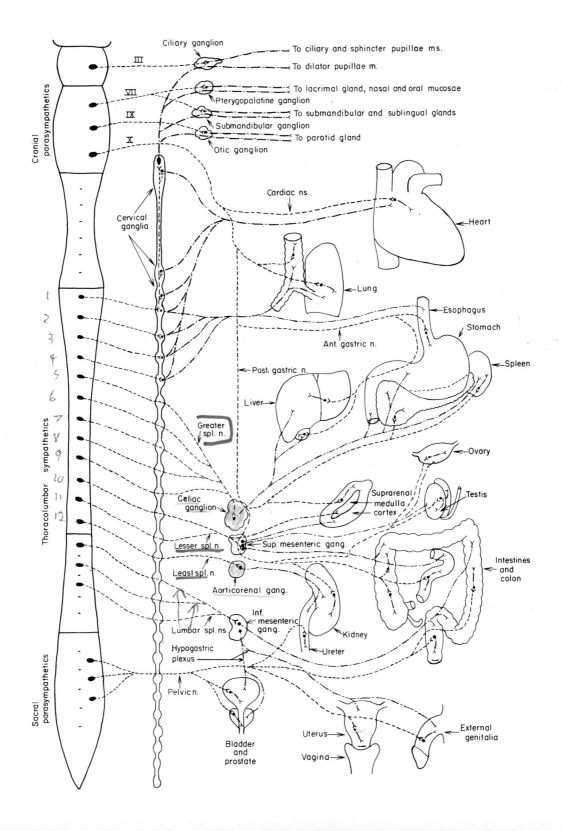

Ciliary ganglion

III — To ciliary and sphincter pupillae ms.

— To dilator pupillae m.

VII — To lacrimal gland, nasal and oral mucosae

Pterygopalatine ganglion

IX — To submandibular and sublingual glands

Submandibular ganglion

X — To parotid gland

Otic ganglion

Cranial parasympathetics

Cardiac ns.

Heart

Cervical ganglia

Lung

Esophagus

Stomach

Ant. gastric n.

Post. gastric n.

Spleen

Liver

Greater spl. n.

Ovary

Suprarenal medulla cortex

Testis

Celiac ganglion

Lesser spl. n.

Sup. mesenteric gang.

Least spl. n.

Aorticorenal gang.

Intestines and colon

Kidney

Lumbar spl. ns

Inf. mesenteric gang.

Ureter

Hypogastric plexus

Pelvic n.

Thoracolumbar sympathetics

Sacral parasympathetics

Bladder and prostate

Uterus

Vagina

External genitalia

Further details of this system will be presented whenever it is involved with any particular region of the body.

Although this autonomic nervous system is extremely important, we must not forget the equally important **visceral afferent fibers** that accompany these motor fibers. A visceral reflex arc (Fig. 1-20B) is more complicated than the somatic. Afferent fibers start as nerve endings in various visceral organs or from blood vessels in the body wall and limbs. These afferent fibers from the body wall simply follow the dorsal or ventral rami to reach a dorsal root and a dorsal root ganglion, and thence the dorsal gray column of the spinal cord. Those from viscera follow the blood vessels associated with the organ, pass through the preaortic ganglia, follow a splanchnic nerve to the chain ganglia, and pass through these ganglia to follow a white ramus communicans to reach the spinal nerve. These fibers then follow the dorsal root and ganglia to reach the dorsal gray column of the spinal cord. Connector neurons connect these afferent fibers with the preganglionic cell bodies located in the intermediolateral cell column. The efferent limb of this reflex arc has already been described as part of the autonomic nervous system.

Although pain from viscera is usually quite vague, diffuse, and difficult to locate, a great deal of such pain is referred to the body wall. This is the result of these visceral afferent fibers entering the spinal cord at the same level as somatic afferent fibers from the body wall and limbs. Pain from the heart being felt on the anterolateral regions of the chest wall and down the medial side of the left arm and hand, and pain of appendicitis being felt around the umbilicus are examples of referred pain. Knowledge of such referred visceral pain is absolutely mandatory in effective diagnostic work; the required information will be presented with each organ as it is described.

CIRCULATORY SYSTEM

Every cell of the body must have nourishment to survive. Utilization of this nourishment by each cell and the pro-

duction of its own particular function, taken together, are called **metabolism.** Oxygen is needed for metabolism to take place, and degradation products have to be eliminated from the body. The circulatory system is essential to these activities. It is especially important clinically, for it is subject to traumatic and inflammatory diseases, and is the system presently responsible for more deaths than any other. Affliction of this widespread system has an effect on the physiology of the entire body.

HEART

The **heart** is a muscular organ that pumps blood by contracting approximately 70 times a minute. The human heart is a four-chambered organ consisting of right atrium and ventricle, and left atrium and ventricle (Fig. 1-23). Unoxygenated blood returns from the body and enters the **right atrium** from the superior and inferior venae cavae, and from the heart musculature itself. This blood then enters the **right ventricle** by passing through the **right atrioventricular orifice,** which contains the **right atrioventricular (tricuspid) valve.** Blood then is forced into the **pulmonary artery,** which carries it to the **lungs,** passing the **pulmonary semilunar valve** on the way. Aeration takes place, and the oxygenated blood returns to the **left atrium** via the **pulmonary veins.** The blood then enters the **left ventricle** through the **left atrioventricular orifice,** which contains the **left atrioventricular (bicuspid, mitral) valve.** The blood is then forced into the **aorta,** passing the **aortic semilunar valve** in doing so. The blood courses to all parts of the body from this aorta and its branches.

Owing to the presence of the **conduction system** of the heart, the two atria contract together, and then the two ventricles. Therefore, blood is being forced into the two ventricles at the same time, and blood is being forced into the pulmonary artery and aorta simultaneously. The customary two main **heart sounds** are caused by the two atrioventricular valves closing when the ventricles contract (systole), thus preventing blood from returning into the atria, and by the two semilunar valves closing when blood in the aorta and pulmonary artery tries to return to the ventricles in diastole, the interval of rest between contraction. The actual cause of these heart sounds is still problematical. It is thought that the valve flaps coming together plus the contraction of the heart muscle induce a minor part of the sound, with the major portion being due

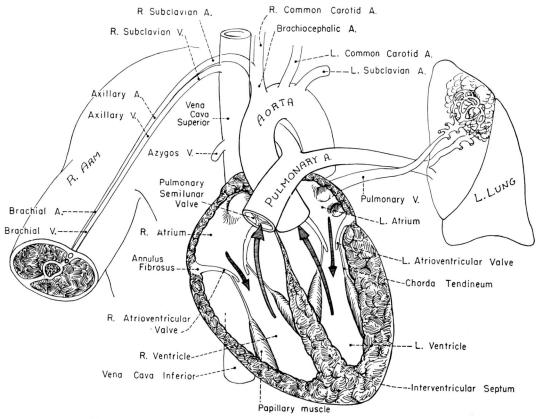

Figure 1-23. *The plan of the circulation; the chambers of the heart, the great vessels and representative pulmonary and systemic circuits. (From* Essentials of Human Anatomy, *third edition, by Russell T. Woodburne. Copyright © 1965 by Oxford University Press, Inc. Reprinted by permission.)*

to the vibrations set up in the blood when the various valves suddenly close.

ARTERIES
(Fig. 1-24)

An artery is defined as any vessel carrying blood away from the heart; this term is independent of the type of blood carried. Arteries have an inner lining of **endothelium,** which, when combined with connective tissue and an elastic layer, is designated as the **tunica intima.** This is surrounded by the **tunica media,** consisting of smooth muscle and elastic tissue; and this in turn is covered with connective tissue called **tunica adventitia.** The aorta branches into arteries, and these arteries, after

innumerable branchings, become small and are called **arterioles.** When the smooth muscle is finally lost from these arterioles, the channel is defined as a **capillary.** Oxygen and nutritional products pass from the capillary to lymph, and thence into the cells.

An important aspect of arteries is the fact that they contain no valves such as are found in many veins and in lymphatics. This allows blood to flow in either direction in arteries, a fact of considerable importance if, for any reason, a main arterial channel is gradually occluded. If such an obstruction occurs, blood can flow through a branch of the artery that emerges proximal to the obstruction and connects (anastomoses) with a second branch that emerges at a point distal to the occlusion. In this manner blood returns to the main arterial channel distal to the

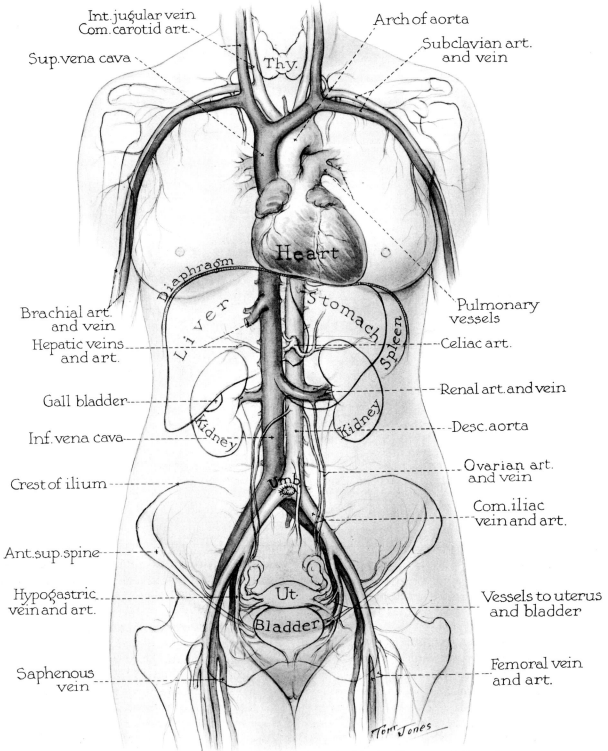

Int. jugular vein
Com. carotid art.

Arch of aorta

Sup. vena cava

Subclavian art.
and vein

Thy.

Brachial art.
and vein

Diaphragm

Liver

Stomach

Spleen

Heart

Pulmonary
vessels

Hepatic veins
and art.

Celiac art.

Gall bladder

Kidney

Kidney

Renal art. and vein

Inf. vena cava

Desc. aorta

Crest of ilium

Umb.

Ovarian art.
and vein

Com. iliac
vein and art.

Ant. sup. spine

Hypogastric
vein and art.

Ut.

Bladder

Vessels to uterus
and bladder

Saphenous
vein

Femoral vein
and art.

Tom Jones

Figure 1-24. *Heart and principal blood vessels in the female. (Courtesy of Camp International, Inc.)*

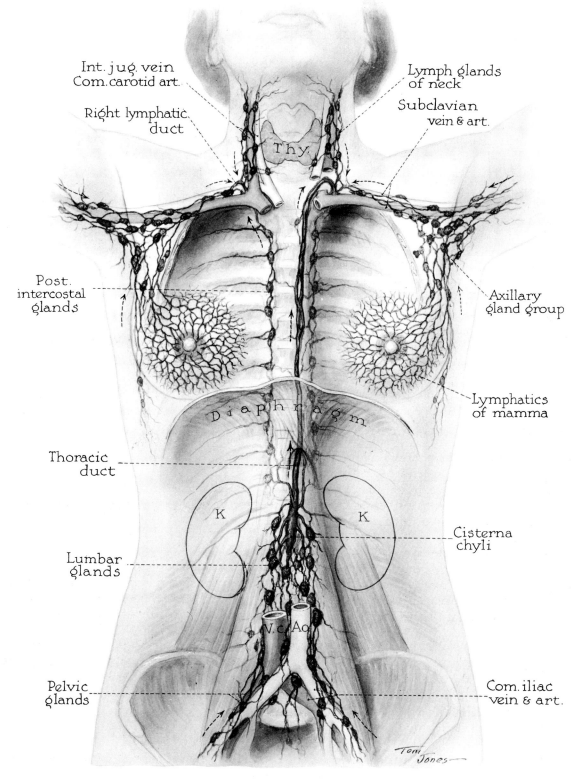

Figure 1-25. *The lymphatics. (Courtesy of Camp International, Inc.)*

obstruction and circulation to the organ or structure is maintained. This is known as a **collateral circulation** or **pathway,** the branches involved enlarging to accommodate the increased amount of blood flowing through them. These collateral pathways are not likely to develop if the occlusion occurs suddenly.

VEINS

Veins are the vessels that carry blood toward the heart. They have the same three layers as arteries, but the walls are thinner. Many of the veins have **valves,** which aid in returning the blood to the heart against gravity by preventing backward flow during diastole; this is particularly true of veins below the level of the heart. Byproducts of metabolism pass from the cells to the lymph, and thence into the capillaries. When smooth muscle returns to the capillaries, they become **venules;** venules become **veins,** which empty into the two **venae cavae,** which return blood to the heart. The liver intervenes between the veins draining many of the abdominal organs and the inferior vena cava; this is the **hepatic portal circulation.**

NERVE SUPPLY

The circulatory system is under control of the autonomic nervous system. The heart rate is increased by the sympathetic division and decreased by the parasympathetic. A vasoconstriction occurs when the smooth muscles in the arterioles are stimulated, and a vasodilation results from interruption of this stimulus. The sympathetic portion of the autonomic nervous system causes a vasoconstriction in skin and viscera, but a vasodilation in muscles. The capillary bed is of much greater volume than the aorta; if complete vasodilation should occur, drastic results would follow, for a person would bleed to death intravascularly.

LYMPHATICS
(Fig. 1-25)

As mentioned above, each cell is bathed in lymph, which intervenes between the circulatory system and the cells themselves. This lymph leaves the blood and finally returns to the blood via innumerable lymphatic vessels. These vessels from the entire lower part of the body and those from the upper left side finally coalesce into the

thoracic duct, which drains into the venous system in the neck where the subclavian joins the internal jugular to form the left brachiocephalic vein. The upper half of the right side of the body drains into the comparable location on the right side.

Lymphatics have many **valves.** Structures called **lymph nodes** intervene between the organs drained and the blood stream, and these act as sieves to stop bacteria, cancer cells, and any other foreign matter from entering the blood. These are naturally very important to us, and *the location of nodes draining any particular structure in the body should be learned.* Infections in these nodes are clues to the original site of infection, and they must be removed if they happen to drain a cancerous area.

OTHER SYSTEMS

All the remaining systems of the body will be described completely, and no particular preliminary information is necessary.

BEGINNING EMBRYOLOGY

As mentioned in the preface, the goal in presenting the material in this section is to provide a springboard from which we may initiate the development of the various organs and systems as an aid to understanding the definitive structure of the human body. We are not attempting to present development for the sake of that discipline; there are many textbooks of embryology available, several of the newer ones having been written for students in the health care disciplines.

MITOSIS AND MEIOSIS

Figure 1-26 presents the well-known sequence that occurs during **cell divisions.** Each of the 46 chromosomes (44 autosomes plus 2 X chromosomes in the female, and

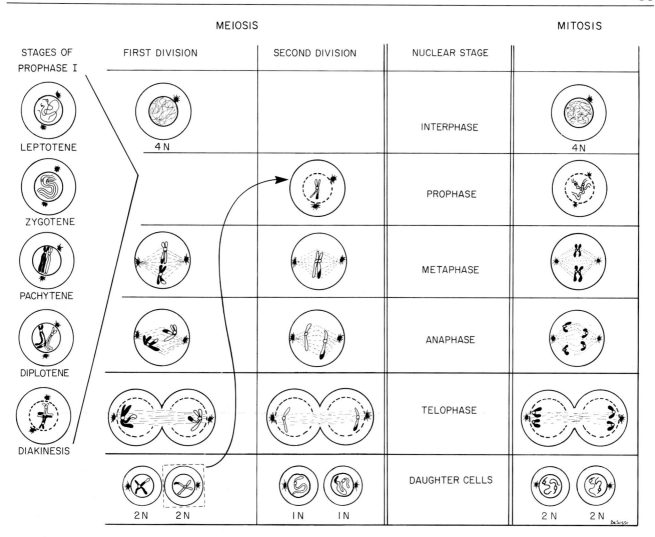

Figure 1-26. *A diagrammatic comparison of the processes of meiosis and mitosis. For simplicity, a diploid number of two chromosomes is illustrated. Note that meiosis utilizes two consecutive cell divisions with no intervening interphase to produce four daughter cells, each with a haploid number of chromosomes. The process of mitosis, on the other hand, involves only one cell division and results in two daughter cells, each possessing the diploid number of chromosomes. The prophase of the first meiotic division is long and complex; it is during this stage that interchange of chromosomal material between homologous chromosomes occurs. To conserve space, the fate of only one daughter cell of the first meiotic division is illustrated. Note that both cells at interphase have developed double the amount of genetic material (4N), but that mitosis leads to two cells, each of which contains the diploid amount of genetic material (2N), while meiosis results in a haploid amount of genetic material (1N). Fertilization restores the diploid amount of genetic material.*

44 autosomes plus 1 X chromosome and 1 Y chromosome in the male) divides, including the important DNA molecule. In this process each daughter cell contains the proper number of chromosomes—46.

Since each individual starts by a sperm fertilizing an ovum, it is obvious that the number of chromosomes must be halved; otherwise each fertilized ovum would have 92 chromosomes. The process whereby the chromosomes are reduced is called **meiosis.** This process also is depicted in Fig. 1-26. Thus each ovum will contain 22 autosomes plus 1 X chromosome, while one half of the sperm will contain 22 autosomes plus 1 X chromosome and the remaining half will contain the same number of autosomes plus 1 Y chromosome. In this way not only is the number of chromosomes halved but a method has been created for assuring an approximately equal number of females and males. If a sperm containing an X chromosome fertilizes the ovum, an XX-female results; if a Y chromosome initiates the individual, an XY-male is created.

Figure 1-27 depicts the 46 chromosomes in human cells. Such a **karyotype** shows that the chromosomes are actually 23 pairs, one of each pair from the female parent and the other from the male parent. Such karyotypes are useful clinically, because it has been discovered that some individuals have an abnormal number of chromosomes. These abnormal conditions result from faulty meiosis whereby the chromosomes do not reduce in a normal fashion. **Trisomy** is the name used to indicate a condition in which three chromosomes exist instead of the usual pair, and **monosomy** refers to one chromosome rather than the pair. These conditions can occur in the autosomes or in the sex chromosomes. **Down's syndrome** results from trisomy of chromosome 21, while trisomy of 17 or 18 results in congenital heart conditions, mental retardation, flexion of the digits and hands, and ears positioned lower than usual. Trisomy of chromosomes 13–15 results in deafness, cleft lip and palate, and eye defects in addition to mental retardation and cardiac anomalies. Most of these children die at a young age.

The conditions whereby the number of sex chromosomes are altered in number are of even greater interest. If the sex chromosomes in the ovum fail to split properly, one ovum may contain XX and another have no (0) sex chromosome. One can easily visualize the several conditions that may arise if each of the above is fertilized by an X-sperm or a Y-sperm. **Klinefelter's syndrome,** occurring only in males and featuring testicular atrophy,

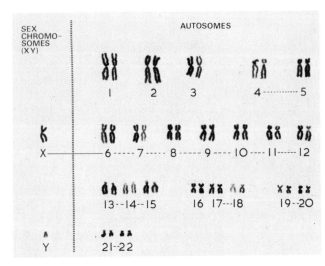

Figure 1-27. *The human male karyotype arranged in homologous pairs. Since the male karyotype is shown, the sex chromosomes are represented by an X and a Y. In the female two identical X chromosomes would be present. (From* Gray's Anatomy, *35th British Edition, courtesy of Churchill Livingstone, Edinburgh, Scotland.)*

gynecomastia, and sterility, results from an XXY condition. **Turner's syndrome,** found in females, is an example of an X0 condition. This condition features a lack of ovaries as well as mental retardation, webbed neck, skeletal defects, and edema of the limbs. **Triple XXX** conditions result in mental retardation and defective menses; these patients exhibit an infantile condition.

Obviously the above examples represent only a small number of possible conditions that may result from faulty meiosis. This, coupled with genetic mutations, merely indicates the importance of chromosomes in the field of medicine.

OVULATION TO IMPLANTATION

OVULATION AND FERTILIZATION

The relationship of the ovaries to the uterine tubes and the uterus itself is shown in Figure 5-65 on page 295.

The ovary goes through a cycle (Fig. 1-28) approxi-

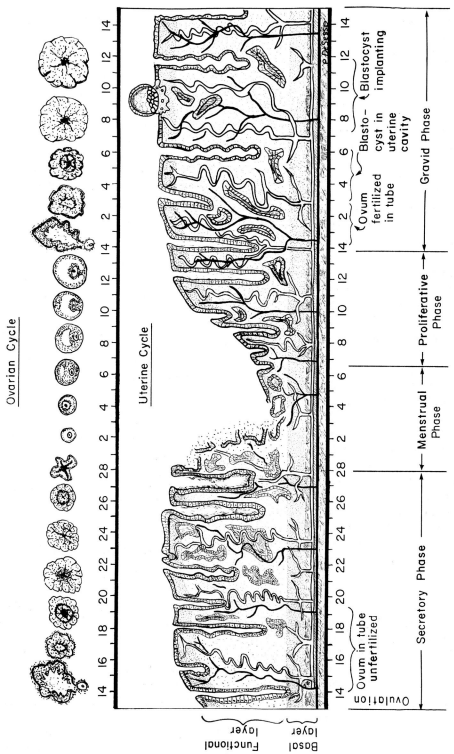

Ovarian Cycle

Uterine Cycle

Blastocyst implanting

Blasto-cyst in uterine cavity

Ovum fertilized in tube

Gravid Phase

Proliferative Phase

Menstrual Phase

Secretory Phase

Ovum in tube unfertilized

Ovulation

Functional layer

Basal layer

Figure 1-28. *Diagram depicting the changes that occur in the ovary and uterus during the latter half of one menstrual cycle and the start of pregnancy during the subsequent cycle. The numbers after ovulation and fertilization represent the days of gestation.*

mately every 28 days in a mature woman. Under the influence of gonadotrophic hormones from the hypophysis (pituitary gland), a group of follicles develops in each ovary, each follicle containing an ovum. In most cycles only one of these follicles develops to the point where ovulation, extrusion of the ovum plus surrounding cells called cumulus oophorus cells, occurs 14 ± 1 days before the onset of the next menstrual period. At the time of ovulation, the ovum is undergoing its second meiotic division.

The follicular cells rapidly change in character and a corpus luteum, so-called because of a yellowish pigment in the cells, develops. The corpus luteum releases progesterone which, in turn, changes the character of the lining of the uterus (endometrium) to nourish a developing ovum, if it is fertilized by a sperm.

The extruded ovum plus the cumulus oophorus is picked up from the surface of the ovary by the fimbriae of the uterine tube. Through ciliary and muscular action, the ovum is moved down the uterine tube toward the uterus, shedding its cumulus oophorus on the way. If fertilization does not occur, the ovum will continue into the uterus and be extruded into the vagina. The corpus luteum will cease its progesterone production, the endometrium of the uterus will break down, and menstruation will occur. If fertilization does occur, however, it will be accomplished while the ovum is still in the uterine tube. This process of **fertilization** is depicted in Figure 1-29. Note that after the sperm has penetrated the zona pellucida (clear area) of the ovum and fusion of their membranes has occurred, two pronuclei are formed, and cell division is initiated. Thus, the two-cell stage **zygote** is formed while still in the uterine tube; this occurs approximately 30 hours after fertilization. If these two cells become separated, identical twins are formed.

CLEAVAGE

The zygote goes through rapid cell divisions, in a process called **cleavage,** until it resembles a mulberry—the **morula** (Fig. 1-29). Even at this stage it is known that those cells in the inside—the **inner cell mass**—will eventually form the embryo itself, and the lining cells will form the so-called **trophoblast** which later develops into the **placenta.** The morula reaches the uterine cavity at about the 12- to 16-cell stage and about three days after fertilization.

FORMATION OF BLASTOCYST AND IMPLANTATION

The blastocyst (Fig. 1-29) is formed by a secretion of fluid into the interior of the morula, this fluid arising from the trophoblast. As the fluid accumulates, the cells of the trophoblast multiply and spread out on the inside surface of the zona pellucida; the zona pellucida is stretched out to a thin membrane and disappears. Thus, we have a structure with a thin wall and an inner cell mass surrounding a fluid-filled cavity, the **blastocoele** (Fig. 1-29).

Approximately one week after fertilization the blastocyst attaches to the uterine wall, the cells of the trophoblast penetrating the spaces between the cells lining the endometrium of the uterus and ultimately destroying these cells (Fig. 1-30). This **implantation** occurs on the anterior or posterior walls of the uterus. By the twelfth day after fertilization, the blastocyst is completely embedded and the point of penetration completely closed by growth of the endometrium. The trophoblast continues to grow rapidly and later combines with mesoderm to form the **chorion,** the extraembryonic membrane that protects the developing embryo. It is the fetal part of the placenta.

Because the ovum is temporarily free in the pelvic cavity, although on the surface of the ovary, transit down the tube sometimes fails; such ova can be fertilized and subsequent development can occur on the surface of the ovary or on the peritoneal lining of the pelvic cavity. More common is a condition in which the fertilized ovum implants in the tube—a **tubal pregnancy**—a precarious situation for the mother.

TWO-LAYERED GERM DISK

In the inner cell mass, the cells become organized into an embryonic disk made up of columnar cells toward the pole of the blastocyst and smaller scattered cells toward the blastocoele. The columnar cells are **ectoderm** and the smaller cells the **entoderm.**

At this same time a cavity appears above the ectodermal disk—the **amnionic cavity**—which will be the environment of the embryo and fetus until birth (Fig. 1-31).

At 9–10 days after fertilization, the most conspicuous advance is in the size of the amnionic cavity and the formation of the **yolk sac** by proliferation of the entoderm. In addition, a layer of cells—the **cytotro-**

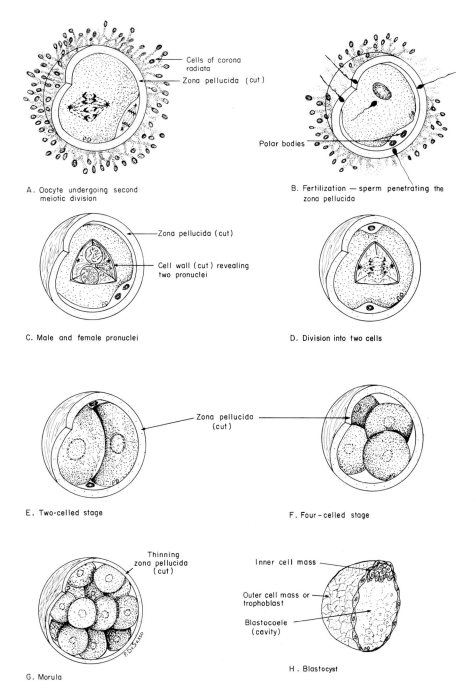

Figure 1-29. *Fertilization, cleavage, and formation of the blastocyst. (A) The oocyte undergoing its second meiotic division; (B) fertilization of the oocyte—note the two polar bodies; (C) the cell wall has been cut to reveal the male and female pronuclei; (D) start of cleavage into two-celled stage (E), four-celled stage (F), and finally into the formation of a morula (G); (H) the blastocyst with the inner cell mass (the embryoblast) and the outer cell mass (the trophoblast). Note that the zona pellucida has disappeared.*

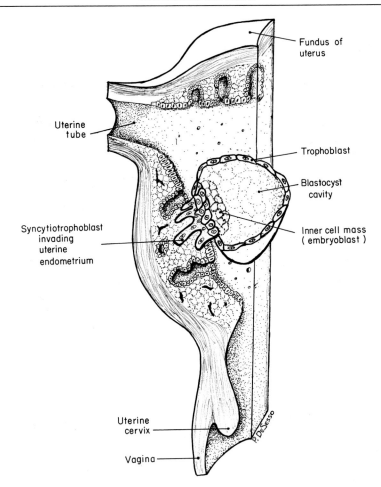

Fundus of uterus

Uterine tube

Trophoblast

Blastocyst cavity

Syncytiotrophoblast invading uterine endometrium

Inner cell mass (embryoblast)

Uterine cervix

Vagina

P.DESESSO

Figure 1-30. *Diagram depicting the process of implantation. This occurs on approximately day 8 (one week after fertilization). Note the syncytiotrophoblast penetrating the endometrium. Although the diagram depicts implantation occurring on the lateral wall of the uterus, almost all occur either on the anterior or on the posterior walls.*

phoblast—lines the inside surface of the large multinucleated trophoblast. **Extraembryonic mesodermal** cells can be seen scattered in the area between the cytotrophoblast and the developing yolk sac; this area is the developing **extraembryonic coelom (exocoelom)**. On day 12 (Fig. 1-32A) the extraembryonic coelom is further along in development and the future somatic and splanchnic mesodermal layers are evident.

On day 13 the yolk sac seems to occupy less space. This has resulted from a pinching off of part of the yolk sac by the developing entoderm to form the definitive yolk sac (Fig. 1-32B), the separated portion finally disintegrating. The mesodermal cells have become organized into a distinct layer close to the cytotrophoblast (the **somatic mesoderm**) and another on the yolk sac (the **splanchnic mesoderm**). Furthermore, there is a layer of these cells surrounding the amnion, and a concentration of similar cells into which a projection from the entoderm—the **allantoic bud**—is growing. This is the site of the future body stalk (Fig. 1-32).

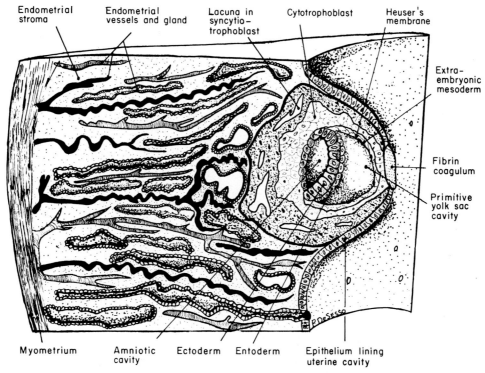

Endometrial stroma — Endometrial vessels and gland — Lacuna in syncytio-trophoblast — Cytotrophoblast — Heuser's membrane — Extra-embryonic mesoderm — Fibrin coagulum — Primitive yolk sac cavity

Myometrium — Amniotic cavity — Ectoderm — Entoderm — Epithelium lining uterine cavity

Figure 1-31. *A 9-day blastocyst. Note the syncytiotrophoblast penetrating the stroma of the uterine endometrium, and the cytotrophoblast on the surface toward the embryo. Inside the cytotroblast is the extraembryonic mesoderm and deep to that is Heuser's membrane lining the primitive yolk sac. The embryoblast has formed an ectoderm and amniotic cavity, and an entoderm.*

All of the above has occurred within the second week of development. We have a two-layered embryonic disk.

FORMATION OF THIRD LAYER

The formation of the third (mesodermal) layer of the embryoblast occurs during the third week. The first sign of this development is the appearance of a streak in the ectoderm on the surface toward the amnion—the **primitive streak** (Fig. 1-33). A small pit known as **Hensen's node** occurs at the cephalic end of this streak. A slight elevation can be seen along the edges of the streak and around the node.

Figure 1-33 also presents the appearance of the germ disc when cut transversely through the primitive streak. It is apparent that new spherical cells are being formed at the edges of the streak and at the node. Ultimately these mesodermal cells, formed by the ectodermal cells on the surface migrating toward the streak and node, fill in the area between the ectoderm and entoderm not only laterally but also cephalad and caudad as well (Fig. 1-34). However, at both ends of the embryo the ectoderm and entoderm have become inseparably fused—the **pharyngeal** and **cloacal membranes,** respectively—and the mesodermal cells merely surround these fused areas.

Those cells migrating directly cephalad from the

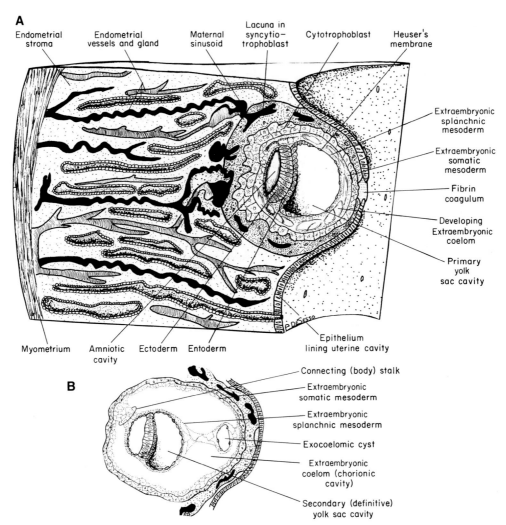

A

Endometrial stroma

Endometrial vessels and gland

Maternal sinusoid

Lacuna in syncytio-trophoblast

Cytotrophoblast

Heuser's membrane

Extraembryonic splanchnic mesoderm

Extraembryonic somatic mesoderm

Fibrin coagulum

Developing Extraembryonic coelom

Primary yolk sac cavity

Epithelium lining uterine cavity

Myometrium

Amniotic cavity

Ectoderm

Entoderm

B

Connecting (body) stalk

Extraembryonic somatic mesoderm

Extraembryonic splanchnic mesoderm

Exocoelomic cyst

Extraembryonic coelom (chorionic cavity)

Secondary (definitive) yolk sac cavity

Figure 1-32. *(A) A 12-day human blastocyst. Note that the trophoblastic lacunae are now in communication with maternal sinusoids in the endometrium. The extraembryonic mesoderm has started to vacuolate to form the extraembryonic coelom. The entodermal cells are beginning to form a layer on the inside surface of Heuser's membrane forming the primary yolk sac. The ectoderm has become more columnar. (B) By day 13 the developing entoderm has pinched off part of the yolk sac forming the definitive yolk sac cavity and the exocoelomic cyst, which later disintegrates. The extraembryonic splanchnic and somatic mesodermal layers are more definitive in their arrangements.*

A

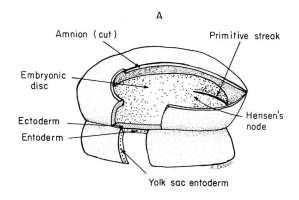

Amnion (cut)

Primitive streak

Embryonic disc

Ectoderm

Entoderm

Hensen's node

Yolk sac entoderm

B

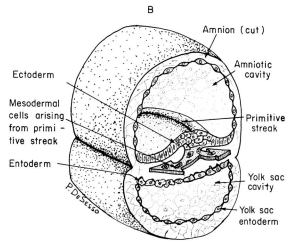

Ectoderm

Mesodermal cells arising from primitive streak

Entoderm

Amnion (cut)

Amniotic cavity

Primitive streak

Yolk sac cavity

Yolk sac entoderm

Figure 1-33. Formation of the third germ layer—the mesoderm. (A) Amnion has been cut to reveal the embryonic disc in which the primitive streak has appeared. Note the two layers—the ectoderm and the entoderm—and the continuity of these layers onto the amnion and yolk sac, respectively. (B) Cross section through the amniotic cavity, embryonic disc, and the yolk sac showing how the mesoderm is formed from ectodermal cells making up the primitive streak.

node form a rod of cells that ultimately fuse with the underlying entoderm. The combined mesoderm/entoderm finally breaks away from the underlying entoderm, which soon fills in the vacated space. This rod of cells is the **notochord,** a characteristic of development in all vertebrates (Fig. 1-35).

The mesoderm in the area of the germ disc is known as the **intraembryonic mesoderm.** However, the migrating mesodermal cells do not stop at the border of the germ disc but continue into the area of the amnion and

the yolk sac, thus meeting the extraembryonic mesodermal tissue lining the extraembryonic coelom.

During this third week the sausagelike **allantois** elongates and grows farther into the connecting stalk (Fig. 1-35); it remains a rudimentary structure in the human, but will be associated with development of the urinary bladder.

Thus, a three-layered embryo now exists. Continued growth finally produces an embryonic disk with a broad end cephalad and a narrower end caudad with the future mouth (pharyngeal membrane) and anogenital openings (cloacal membrane) cephalad and caudad respectively.

DIFFERENTIATION OF GERM LAYERS AND ESTABLISHMENT OF BODY FORM

We have reached the fourth week of development. The period from this week to the eighth is known as the **embryonic period.** During this time the rudiments for all major organ systems are laid down.

Although the following events occur pretty much simultaneously, for descriptive purposes we will study them by germ layer.

ECTODERM

Under the influence of the notochord, the ectoderm immediately above is transformed into the **neural plate.** This plate shows rapid growth particularly at the cephalic end. Due to proliferation of the underlying tissue, the neural plate forms a **neural groove** flanked by elevations on either side of the midline. These elevations fuse in the midline forming a tube, starting in the future neck region and extending in both cephalic and caudal directions. This is the future brain and spinal cord. The cepahalic opening of this neural tube closes on day 25; the caudal on day 27. The cephalic end becomes much larger than the remaining portions of the neural tube, forming large head folds that project cephalad beyond the area of the

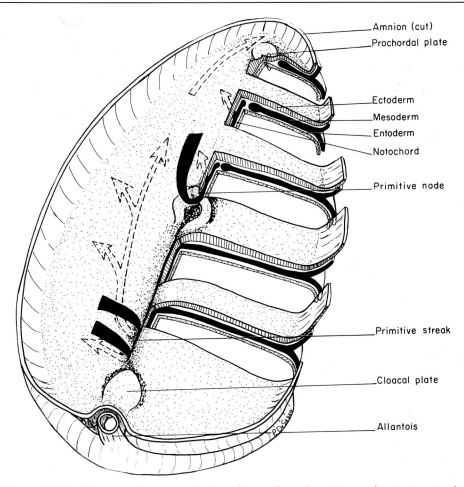

Amnion (cut)
Prochordal plate

Ectoderm
Mesoderm
Entoderm
Notochord

Primitive node

Primitive streak

Cloacal plate

Allantois

Figure 1-34. *Diagrammatic representation of mesoderm formation and migration in the trilaminar germ disc. Dorsal surface cells invaginate through the primitive streak and primitive node to form the mesodermal germ layer; the mesoderm migrates laterally and cephalically filling the space between the ectoderm and the entoderm except in the regions of the prochordal and cloacal plates. The midline axis of the embryo from the primitive node anteriorly to the prochordal plate is occupied by the notochord. (Modified after Tuchmann-Duplessis.)*

original germinal disk. This process is depicted in Figures 1-36 and 1-37. Outgrowths from this neural tube form the various cranial and spinal nerves.

At this time two other ectodermal derivatives are detectable—the **otic** and **lens placodes.**

Thus, we have derived from ectoderm the central nervous system, the peripheral nervous system, and the sensory epithelium of the sense organs. Among the other derivatives of the ectoderm are the epidermal layer of the skin including hair, nails, mammary and sebaceous glands; the hypophysis (pituitary gland); external part of mouth and anal canal; epithelium of the nose and paranasal sinuses; nasolacrimal duct; the conjunctiva, cornea, and lens of the eye as well as internal ear; and the enamel of the teeth (see Fig. 1-41).

MESODERM

We left the mesoderm as an intermediate layer of cells spreading out between the overlying ectoderm and the

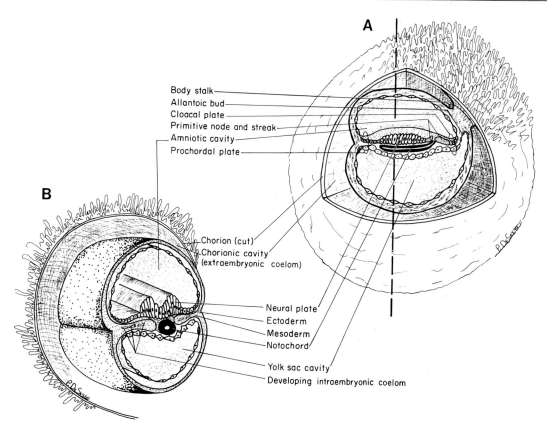

Body stalk
Allantoic bud
Cloacal plate
Primitive node and streak
Amniotic cavity
Prochordal plate

Chorion (cut)
Chorionic cavity
(extraembryonic coelom)

Neural plate
Ectoderm
Mesoderm
Notochord

Yolk sac cavity
Developing intraembryonic coelom

Figure 1-35. *Diagrams illustrating the relations of the notochord in a human embryo of 18 days of gestation. (A) Midsagittal section depicting the relationship of the notochord to the prochordal plate anteriorly and the primitive (Hensen's) node caudally. Note the allantoic bud growing into the body stalk. (B) A transverse section through the same embryo at the level of the dotted line in A. Note the relationship of the notochord to the mesoderm and developing neural plate.*

underlying entoderm (Fig. 1-34). This mesoderm becomes organized into a thickened area close to the midline of the embryo—the **paraxial mesoderm**—and a thin area laterally called the **lateral plate** (Fig. 1-38). Gradually spaces develop in the lateral plate to produce a layer on the yolk sac—the **visceral** or **splanchnic mesodermal layer**—and another layer continuous with that lining the outside surface of the amnion—the **somatic** or **parietal mesoderm.** This new cavity—the **intraembryonic coelomic cavity**—at this stage is directly continuous with the extraembryonic coelomic cavity. Also at this stage, a new thickening of the lateral plate that is close to the paraxial mesoderm occurs, the so-called **interme-**

diate mesoderm. Thus, we have the paraxial, intermediate, splanchnic, and parietal mesoderms.

The paraxial mesoderm soon forms into clumps—the **somites.** They start developing in the cephalic region and gradually progress caudally until there are 42 to 44 in all—4 occipital, 8 cervical, 12 thoracic, 5 lumbar, 5 sacral, and 8 to 10 coccygeal. The first occipital and the last 5 to 7 coccygeal somites disappear in humans. It is not too difficult to connect the occipital somites with future structures in the head and the remaining with the vertebral column. As we will see, the somites are associated with the entire skeletal system and the muscles that move these bones. The somites develop in such a regular fash-

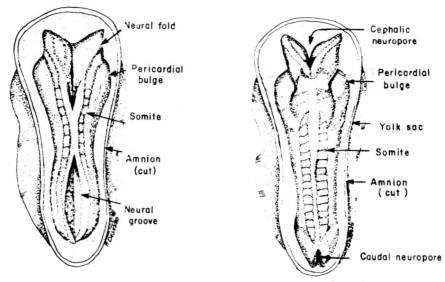

Figure 1-36. *Formation of the neural tube. (A) Dorsal view of an embryo on approximately day 22 of development. Note that the neural tube has formed at the presumptive cervical region, while the cranial and caudal regions still exhibit neural folds. (B) Dorsal view of an embryo on approximately day 23 of development. Note the appearance of additional somites. The neural tube has lengthened and is open only at cephalic and caudal ends. Compare to Figure 1-37. (Redrawn after Langman.)*

ion that the age of the embryo is usually associated with them, e.g., the 2-somite stage, the 4-somite stage, etc.

Approximately at the fourth week the somites differentiate into several portions. The cells located ventrally and medially migrate toward the notochord and are collectively called the **sclerotome.** These cells form connective tissue and its derivatives such as cartilage and bone (vertebrae and skull plus associated ligaments). The remaining dorsal portion of the somite is called the **dermatome.** Cells are formed on the medial side of the dermatome and are collectively called the **myotome.** From these names it is easily deduced that the dermatomal cells will form the dermis, the deeper layer of the skin, and the myotome will develop into skeletal muscles. It is interesting that a nerve grows out of the developing spinal cord to each of these dermatomes and myotomes. The relationship is established at this time, and no matter what happens to these future areas of the dermis or these developing skeletal muscles, this nerve relationship is maintained. Often the nerves are required to elongate considerably in order to do so.

The intermediate mesoderm develops into the definitive excretory portions of the urinary system and the stroma and duct system of the genital systems.

The parietal and splanchnic layers of the mesoderm develop into the layers of the pericardial, pleural, and peritoneal linings of those particular cavities. The parietal, as the name implies, forms the layer lining the walls of these cavities, and the splanchnic, as this name also implies, covers the viscera in these cavities. In addition, the splanchnic layer will produce the smooth muscles found in the walls of the various viscera.

Wherever there is mesoderm there is likely to be blood cells and vessels in state of formation. This is true in the mesoderm of the developing placenta, in the stalk (umbilical cord), and in the wall of the yolk sac. The mesoderm in the embryo is no exception. At approximately the third week mesodermal cells near the midline and close to the entoderm differentiate into blood cells and blood vessels, those centrally placed forming the cells and those more peripherally placed forming the vascular walls. The dorsal aortae (paired at first) result from this

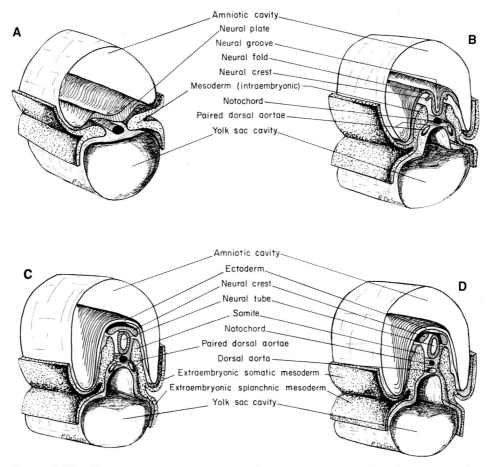

Figure 1-37. *Diagrammatic representations of transverse sections through progressively older embryos to illustrate the development of the neural tube. The neural plate (A) is gradually converted to the neural groove (flanked by neural folds) (B); fusion of the neural folds (C) results in formation of the neural tube. Neural crest cells, initially near the edges of the neural folds (B), form a transitory layer of cells between the surface ectoderm and newly closed neural tube (C). They soon migrate to the location of the adult dorsal root ganglia (D).*

process. Such development also occurs cephalad to the pharyngeal membrane. A pocket of mesodermal cells at this location differentiate into the future heart (Fig. 1-39).

We will see later that in addition to the connective tissue, cartilage, and bone (the skeleton); striated muscles; heart and blood vessels plus the blood cells; the excretory system; the visceral layers of pericardium, pleura, and peritoneum; the mesoderm also provides the reproductive organs, the cortex of the suprarenal glands, and the spleen (see Fig. 1-41).

ENTODERM

We left the entoderm spread out on the inside surface of the germ disk and forming the yolk sac. To understand how this layer forms the gut tube and all its derivatives, we must visualize the cephalic and caudal infoldings as well as that which occurs laterally. This is depicted in Figure 1-39.

By this process the heart, which developed cephalad to the head fold, comes to lie in the area closer to its

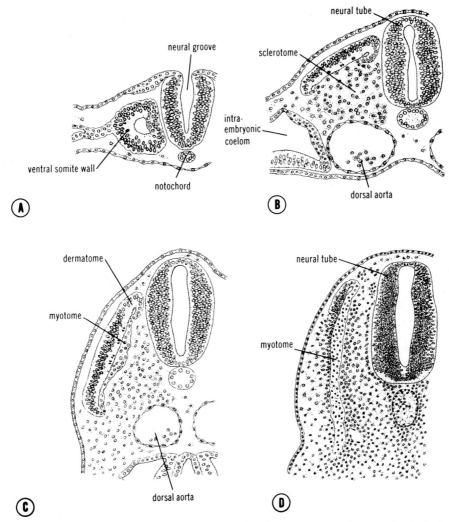

Figure 1-38. *A series of transverse sections of embryos at successive ages to illustrate the development of a somite. (A) Mesodermal cells adjacent to the notochord and neural groove are arranged around a small lumen to form a somite. (B) The sclerotome arises from cells of the ventromedial wall of the somite which proliferate and migrate toward the notochord. (C) The dorsal wall of the somite (the dermatome) gives rise to the medially placed myotome. (D) The cells of the dermatome migrate toward the ectoderm to form the dermis; note the continued growth of the myotome. (From* Langman's Medical Embryology, *3rd edition, courtesy of Williams and Wilkins Co., Baltimore.)*

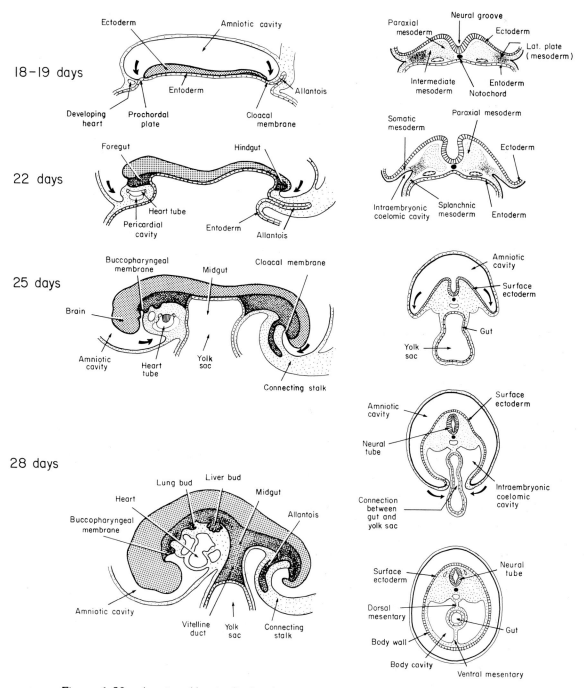

Figure 1-39. *A series of longitudinal and transverse sections of embryos at various stages in development to demonstrate cephalocaudal and lateral infoldings. Note how this determines the definitive positions of various structures. (Redrawn after Langman.)*

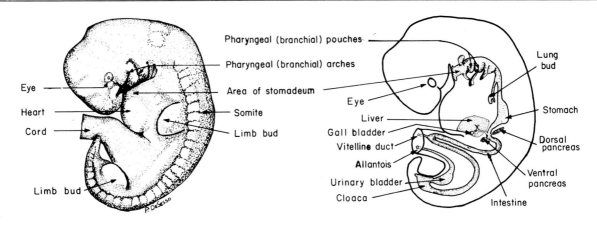

A. External Appearance   *B. Entodermal Derivatives*

Figure 1-40. *(A) External appearance of an embryo at the beginning of the fifth week of development. (B) Schematic drawing of the entodermal tube and its derivatives. (Redrawn after Langman.)*

definitive position; the allantois becomes incorporated into the body; and the opening to the yolk sac becomes conspicuously narrowed.

At this same time the lateral infolding actually makes a tube out of the body. This maneuver finally separates the intraembryonic coelom from the extraembryonic coelom (Fig. 1-39).

The gut tube resulting from this can at this time be divided into **foregut, midgut** (associated with the remaining yolk sac), and **hindgut.** We will find this information useful because these three portions have distinct blood and nerve supplies. We will also see later that many structures develop from this entodermal gut tube. The respiratory system, the liver and pancreas, and parts of the urinary bladder are examples. Of particular importance at this time is the development of the **pharyngeal pouches** (see Fig. 1-40B). These bilateral outpocketings of the foregut, which ultimately will produce the auditory tube, the palatine tonsils, the thymus, and parathyroids, are associated with indentations or **clefts** on the external surface of the embryo. (In fish the membranes thus formed between the pouches on the inside and the clefts on the outside break down to form the gills; this occurs occasionally in the human). Associated with each such arrangement is a cartilaginous bar, striated muscles, blood vessels (the aortic arches), and a cranial nerve.

These so-called **branchial arches** are important in our understanding of the definitive structure of the body.

Figure 1-41 shows the derivations of the three germ layers.

EXTERNAL APPEARANCE

Figure 1-40A shows the external appearance of an embryo in the fifth week of development. Note that the limb buds have appeared. Figure 1-40B shows such an embryo in sagittal section. The size of such an individual is important for comprehension of many future changes. This embryo is only 7 mm. long, a little over a quarter of an inch, from crown to rump. Perhaps this will make what seems like major movements comprehensible; the distances are miniscule.

We will leave the embryo at this stage of development but will be referring to it repeatedly.

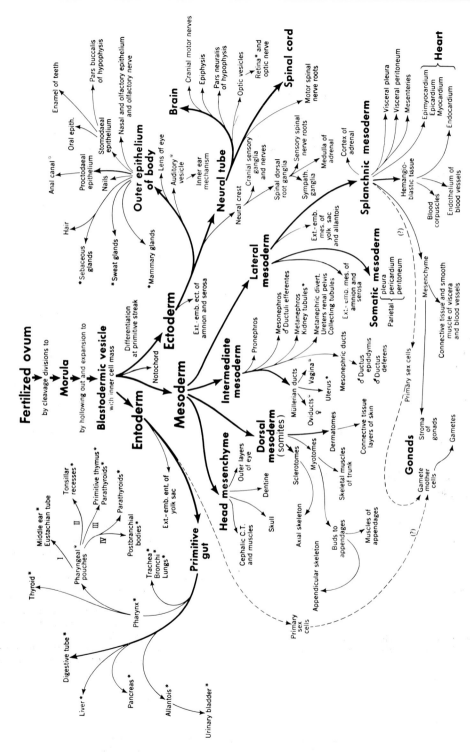

Figure 1-41. Chart showing the derivation of various parts of the body by progressive differentiation and divergent specialization. Note how the origin of all organs can be traced back to the three primary germ layers. (From Patten's Human Embryology by C. E. Corliss, Copyright 1976 by McGraw-Hill, Inc. Used with permission of the McGraw-Hill Book Company.)

VARIATIONS

The first three months are considered to be the formative period of the embryo. The remaining six months, known as the fetal period, are a time of growth.

From the mere sketch of the development of an embryo presented, it should be evident that variations can occur. In fact, variations from one person to another, or even from one side to the other side in the same body, are more easily understood than the so-called normal conditions.

Variation comes in several degrees and it is often difficult to put labels on it. There are "normal variations" that occur very frequently, and "abnormal variations" or "anomalies" that occur less frequently. If a person is born with visible gross structural defects, these defects are called "congenital malformations."

The percentage of cases born with malformations varies with the races, or even with the person or hospital reporting such cases. In a recent worldwide survey of about 20 million births,* the clinic records claimed 1.26% exhibited malformations. Pediatricians, examining these same records, found 4.5% with malformations. When children at 6 and 12 months of age are examined, the percentage may double.

The study of the causes of malformations is called **teratology.** At first it was thought that all such abnormal development was genetic in origin, but when it was discovered that German measles would produce such conditions, it was soon found that many deficiencies and chemicals could be involved. In fact, the present popular use of drugs can be catastrophic.

All of this occurs during the embryonic period and can develop before a woman is aware she is pregnant.

---

*W. P. Kennedy, 1967, Epidemiological aspects of the problem of congenital malformations, In *Birth Defects Original Article Series,* edited by D. Bergsma, p. 1.

GENERAL REFERENCES

APPLETON, A. B., W. J. HAMILTON, and L. SIMON. 1971. *Surface and Radiological Anatomy.* Heffer, Cambridge.

AREY, L. B. 1965. *Developmental Anatomy.* W. B. Saunders Co., Philadelphia.

BALLARD, W. W. 1964. *Comparative Anatomy and Embryology.* The Ronald Press Co., New York.

BASMAJIAN, J. V. 1962. *Muscles Alive—Their Functions Revealed by Electromyography.* Williams & Wilkins Co., Baltimore.

CLARK, W. E. LE GROS. 1971. *The Tissues of the Body.* Clarendon Press, Oxford.

CORLISS, C. E. 1976. *Patten's Human Embryology.* McGraw-Hill, New York.

DUCHENNE, L. B. 1959. *Physiology of Motion,* translated and edited by E. B. KAPLAN. W. B. Saunders Co., Philadelphia. Reports on a collection of thousands of human beings with nerve paralyses.

FRAZER, J. E. 1940. *The Anatomy of the Human Skeleton.* Churchill, London.

GALLAUDET, B. B. 1931. *A Description of the Planes of Fascia of the Human Body.* Columbia University Press, New York.

GARDNER, E. 1952. *The Anatomy of the Joints.* American Academy of Orthopedic Surgery, Instructional Course Lectures, Vol. 9.

GILLILAN, L. A. 1954. *Clinical Aspects of the Autonomic Nervous System.* Little, Brown and Co., Boston.

HAMILTON, W. J., S. D. BOYD, and H. W. MOSSMAN. 1972. *Human Embryology.* Heffer, Cambridge.

KUNTZ, A. 1953. *The Autonomic Nervous System.* Lea and Febiger, Philadelphia.

LANGMAN, J. 1975. *Medical Embryology.* Williams & Wilkins Co., Baltimore.

LOCKHART, R. D. 1949. *Living Anatomy. A Photographic Atlas of Muscles in Action and Surface Contours.* University Press, Glasgow.

McMINN, R. M. H. and R. T. HUTCHINS. 1977. *Color Atlas of Human Anatomy.* Year Book Publishers, Chicago.

MONTAGNA, W. 1974. *The Structure and Function of the Skin.* Academic Press, New York.

PEPPER, O. H. P. 1949. *Medical Etymology.* W. B. Saunders Co., Philadelphia.

QUIRING, D. P. 1949. *Collateral Circulation.* Lea and Febiger, Philadelphia.

SINGER, E. 1935. *Fasciae of the Human Body and Their Relation to the Organs They Envelop.* William Wood, Baltimore.

TRUEX, R. C., and C. E. KELLNER. 1948. *Detailed Atlas of the Head and Neck.* Oxford University Press, New York.

TURNER, C. D. 1966. *General Endocrinology.* W. B. Saunders Co., Philadelphia.

WARWICK, R., and P. L. WILLIAMS. 1973. *Gray's Anatomy.* 35th British Ed., W. B. Saunders Co., Philadelphia.

WHITE, J. C., R. H. SMITHWICK, and F. A. SIMEONE. 1952. *The Autonomic Nervous System.* Macmillan Co., New York.

WRIGHT, W. L. 1928. *Muscle Function.* Hoeber, New York. A study of normal and paralyzed patients.

YOFFEY, J. M., and F. C. COURTICE. 1956. *Lymphatics, Lymph and Lymphoid Tissue.* Harvard University Press, Cambridge.

2

BACK

The back, in spite of being one of the most important areas of the body clinically, has long suffered from neglect, by both doctor and patient. Injury to the back is very frequent in industrial accidents and is one of the most difficult cases to handle medically. Many persons suffer from back ailments due, in part, to the increase in strain and weight-bearing that results from development of the erect posture.

BONY LANDMARKS

Before any region of the body can be studied intelligently, basic information on bony landmarks must be learned. Accordingly, parts of two cranial bones (the occipital and the temporal), the vertebrae, parts of the ribs, lateral end of the clavicle, the scapula, and part of the ilium will now be described (Fig. 2-1).

OCCIPITAL BONE (IN PART)

The outer surface of the occipital bone is the part involved in studying the back (Fig. 2-2). This bone can be divided into a **basilar part** located anterior to the large foramen magnum, a **squamous portion** located posteriorly, and a **lateral (condylar) part** lateral to the foramen. The **squamous portion** presents several markings. In the midline, extending directly posteriorly from the foramen magnum, is a crest called the **external occipital crest,** which ends in the **external occipital protuberance.** Extending laterally from this protuberance are two ridges named the **superior nuchal lines.** Two other ridges—the **inferior nuchal lines**—extend laterally from a point halfway between the foramen magnum and the external occipital protuberance. The most prominent feature of the **lateral portions** are the **occipital condyles,** which articulate with the superior condyles of the atlas—the first cervical vertebra. Just posterior to these condyles are the **condylar canals,** which serve as an important connecting pathway between veins in the deep part of the neck and the venous sinuses inside the

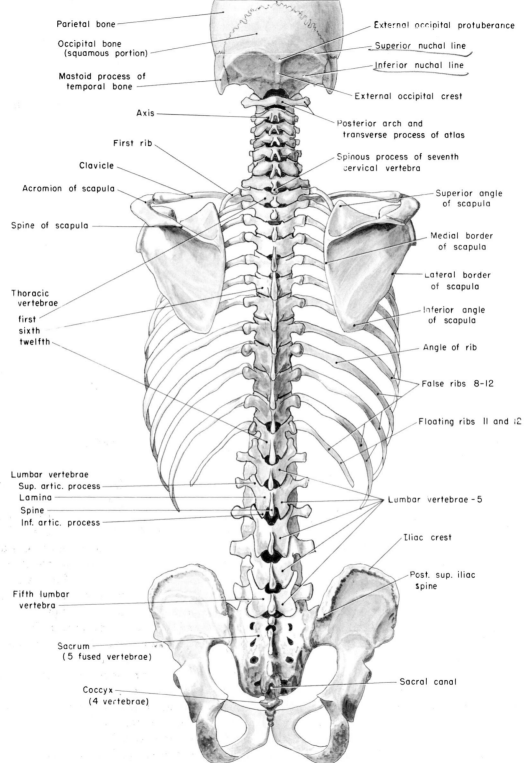

Parietal bone

Occipital bone
(squamous portion)

Mastoid process of
temporal bone

Axis

First rib

Clavicle

Acromion of scapula

Spine of scapula

Thoracic
vertebrae

first
sixth
twelfth

Lumbar vertebrae
Sup. artic. process
Lamina
Spine
Inf. artic. process

Fifth lumbar
vertebra

Sacrum
(5 fused vertebrae)

Coccyx
(4 vertebrae)

External occipital protuberance

Superior nuchal line

Inferior nuchal line

External occipital crest

Posterior arch and
transverse process of atlas

Spinous process of seventh
cervical vertebra

Superior angle
of scapula

Medial border
of scapula

Lateral border
of scapula

Inferior angle
of scapula

Angle of rib

False ribs 8-12

Floating ribs 11 and 12

Lumbar vertebrae - 5

Iliac crest

Post. sup. iliac
spine

Sacral canal

Figure 2-1. *Bones involved in a study of the back.*

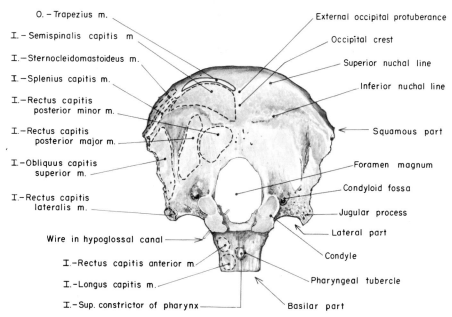

O. – Trapezius m.
I. – Semispinalis capitis m
I. – Sternocleidomastoideus m.
I. – Splenius capitis m.
I. – Rectus capitis posterior minor m.
I. – Rectus capitis posterior major m.
I. – Obliquus capitis superior m.
I. – Rectus capitis lateralis m.
Wire in hypoglossal canal
I. – Rectus capitis anterior m.
I. – Longus capitis m.
I. – Sup. constrictor of pharynx

External occipital protuberance
Occipital crest
Superior nuchal line
Inferior nuchal line
Squamous part
Foramen magnum
Condyloid fossa
Jugular process
Lateral part
Condyle
Pharyngeal tubercle
Basilar part

Figure 2-2.   External surface of the occipital bone with muscle attachments outlined on the left side. Do not attempt at this time to learn that portion of the bone which is anterior to the foramen magnum.

cranium. The **basilar portion** and the inside of the occipital bone will be described in the chapter on the head and neck.

TEMPORAL BONE (IN PART)

The only portion of the temporal bone to be considered now is the **mastoid process,** a large protuberance just posterior to the external ear (Fig. 2-1).

VERTEBRAE

The **vertebral column** (Fig. 2-1) consists of thirty-three vertebrae. There are seven cervical vertebrae in the neck, twelve thoracic in the upper back, five lumbar in the lower part of the back, five sacral fused into one bone (the sacrum), and four coccygeal vertebrae, which may or may not be fused.

Figure 2-3 shows the appearance of a **typical thoracic vertebra.** Such a vertebra consists of a large rounded **body** located anteriorly, from which a **vertebral**

arch extends posteriorly. The arch is made up of two **pedicles** extending posteriorly from the body, and two **laminae** completing the arch. The foramen so formed is the **vertebral foramen** and contains the important spinal cord and its coverings. At the point where the pedicles and laminae join are two projections, one on each side, called **transverse processes;** at the point where the two laminae come together, a single **spinous process** extends more or less posteriorly. Each vertebra articulates with the one above and below; accordingly, there are two **articular processes** containing two **articular facets** superiorly placed and two similar processes and facets inferiorly placed. The cervical, thoracic, and lumbar vertebrae have features that distinguish one from the other, but each set tends to resemble the one lower down as it approaches it. In other words, the lowest cervical vertebra resembles the upper thoracic; the lowest thoracic, the lumbar. Of course, there is no trouble in distinguishing the sacrum or coccyx.

The first and second cervical vertebrae, i.e., the atlas and axis, do not resemble any other vertebrae. The **atlas**

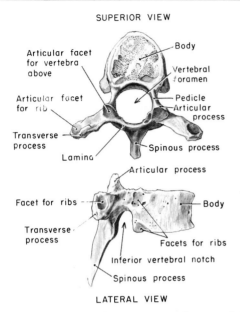

Figure 2-3. *Superior and lateral views of a typical thoracic vertebra.*

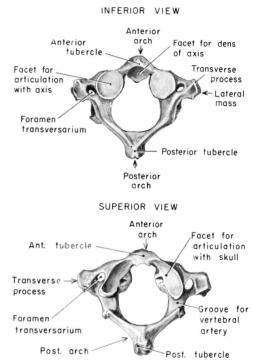

Figure 2-4. *Inferior and superior views of the first cervical vertebra, the atlas.*

(Fig. 2-4) has no body and no spinous process. It consists simply of two **lateral masses** connected by two rather thin **arches.** The anterior arch is smaller and contains a small elevation in the midline, the **anterior tubercle,** and a **facet** on its posterior aspect for articulation with the dens of the axis. The posterior arch has a similar tubercle, the **posterior tubercle,** and grooves for the vertebral artery and the first cervical nerve. The lateral masses have **superior facets** for articulation with the occipital bone and **inferior facets** for articulation with the axis. The transverse processes are large, are not bifid as are those of the regular cervical vertebrae, and possess a large foramen for transport of the vertebral artery and vein. The **axis** (Fig. 2-5) consists of a **body, transverse processes, spinous process, arch,** and a very conspicuous projection extending superiorly from the body—the **dens.** The transverse processes contain the usual foramina; the spinous process is large and bifid, with each projection widely separated. This vertebra articulates with the atlas in three places: the two articular facets with the two inferior articular facets of the atlas, and the dens with the anterior arch of the atlas. Figure 2-6 shows the atlas and axis together.

Cervical vertebrae 3 to 6 (Fig. 2-7) are similar. They possess small bodies and rather delicate arches, spinous processes that are bifid, transverse processes that also are bifid (**anterior and posterior tubercles**) in addition to possessing a foramen, and a triangular vertebral foramen. Articular processes are fairly large, and the facets face superiorly and inferiorly.

The **seventh cervical vertebra** (Fig. 2-7) differs slightly in that it exhibits a longer spine and a larger transverse process. This is the first spine that can be felt in the upper back.

The **thoracic vertebrae** (Fig. 2-8) gradually increase in size as the lumbar region is approached. The **bodies** are heart-shaped, the **laminae** are flat and broad, the **spinous processes** slant inferiorly so that they overlap, the **vertebral foramina** are circular, and the **pedicles** are short. The lower thoracic vertebrae tend to take on the characteristics of the upper lumbar vertebrae. The most conspicuous feature of the thoracic vertebrae that differentiates them from others is the presence of **facets for**

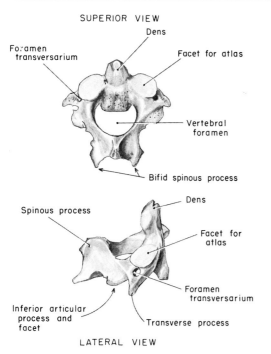

SUPERIOR VIEW

Dens

Foramen transversarium

Facet for atlas

Vertebral foramen

Bifid spinous process

Spinous process

Dens

Facet for atlas

Foramen transversarium

Inferior articular process and facet

Transverse process

LATERAL VIEW

Figure 2-5. *Superior and lateral views of the second cervical vertebra, the axis.*

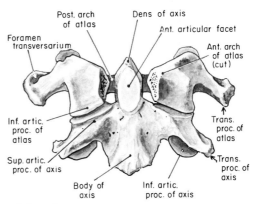

Post. arch of atlas

Dens of axis

Ant. articular facet

Foramen transversarium

Ant. arch of atlas (cut)

Trans. proc. of atlas

Inf. artic. proc. of atlas

Sup. artic. proc. of axis

Body of axis

Inf. artic. proc. of axis

Trans. proc. of axis

Figure 2-6. *Anterior view of atlas and axis. The anterior tubercle of the atlas has been removed to show the relationship of the anterior arch to the dens of the axis.*

articulation with the ribs (Fig. 2-8). Each body articulates with the rib of the same number near the superior aspect of the body, and the bodies of the upper eight or nine thoracic vertebrae also articulate with the rib below by a facet near the inferior end of the bodies. In other words, the first rib articulates with the first thoracic vertebra only; the second rib articulates with the lower part of the first thoracic vertebra and the upper part of the second. This same arrangement holds until the ninth thoracic vertebra is reached, where the inferior facet may be missing, as it is in the remaining thoracic vertebrae. Therefore, all thoracic vertebrae have a superior and an inferior facet except the lowest three or four. The ribs also articulate at their tubercles with the transverse processes (Fig. 2-8) of all the thoracic vertebrae except the last two. The **transverse processes** therefore possess **facets** for this articulation with the rib of the same number. The transverse processes decrease in size and lateral projection as the column is descended until the twelfth thoracic vertebra is reached (Fig. 2-1), the transverse processes on this vertebra being replaced by small tubercles. The articular facets change from the superior-inferior plane of

the cervical vertebrae to a more anteroposterior plane; this is important in understanding movement of the vertebral column.

Lumbar vertebrae (Fig. 2-9) can be distinguished very easily from cervical or thoracic vertebrae by the absence of foramina in the transverse processes and of facets for rib articulation. In addition, they are much larger than the vertebrae above. The **bodies** are quite large, the **pedicles** short and heavy, and the **transverse processes** quite long and slender for a massive vertebra.

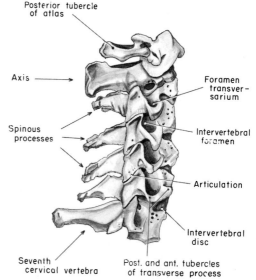

Posterior tubercle of atlas

Axis

Spinous processes

Seventh cervical vertebra

Foramen transversarium

Intervertebral foramen

Articulation

Intervertebral disc

Post. and ant. tubercles of transverse process

Figure 2-7. *Lateral view of cervical vertebrae.*

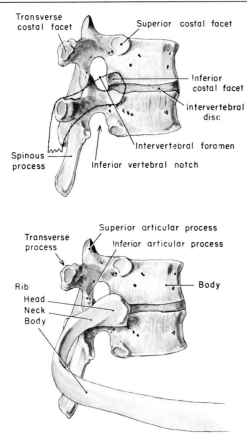

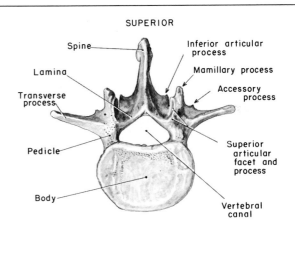

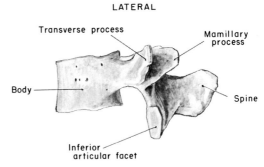

Figure 2-8. *Thoracic vertebrae showing manner of articulation with the ribs.*

Figure 2-9. *Superior and lateral views of a lumbar vertebra.*

The **spinous process** is stocky and blunt, and projects directly posteriorly, a fact that we will see is important in lumbar puncture. The **articular processes** are very large and possess a rounded elevation projecting superiorly called the mamillary process. The **articular facets** face medially and laterally. Small **accessory processes** can be seen between the mamillary and transverse processes.

The **sacral vertebrae** are fused into one bone called the sacrum (Fig. 2-10). The **sacrum** is broad superiorly and tapers down inferiorly, so it can be described as having a **base** superiorly and an **apex** inferiorly. Ventrally it exhibits a **median portion** and two **lateral parts.** It is concave anteriorly and convex posteriorly. On the **anterior surface** the median portion is separated from the lateral by four foramina for passage of ventral rami of the sacral nerves (see Fig. 2-12). The median portion pre-

sents transverse lines that separate the five vertebrae. The convex **posterior surface** has a **midline crest** in the median portion, is separated from the lateral portion by another set of foramina for the passage of dorsal rami of the sacral nerves (Fig. 2-12). The lateral portions possess an articular surface for articulation with the coxal bones (see Figs. 2-1 and 2-10). The **vertebral canal** continues through the sacrum, and this bone has facets for articulation with the facets of the fifth lumbar vertebra and a terminal facet for articulation with the coccyx.

The **coccygeal vertebrae** (Figs. 2-1 and 2-25) are usually fused into one bone also. The first vertebra is of some size but the remaining three are small rounded bones of various sizes and shapes. The vertebral canal does not continue into these vertebrae.

The vertebral column as a whole will be described later.

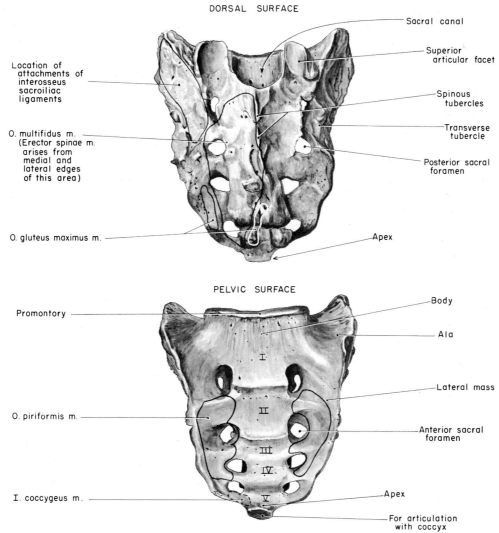

DORSAL SURFACE

Sacral canal

Superior articular facet

Spinous tubercles

Transverse tubercle

Posterior sacral foramen

Location of attachments of interosseus sacroiliac ligaments

O. multifidus m. (Erector spinae m. arises from medial and lateral edges of this area)

O. gluteus maximus m.

Apex

PELVIC SURFACE

Promontory

Body

Ala

Lateral mass

O. piriformis m.

Anterior sacral foramen

I. coccygeus m.

Apex

For articulation with coccyx

Figure 2-10. *Dorsal and pelvic surfaces of sacrum with muscle attachments outlined.*

RIBS (IN PART)

A portion of the ribs should be studied at this time (Figs. 2-1 and 2-8). As mentioned before, each rib (with the exception of the first and the last three or four) articulates with the vertebra of the same number and the one above; therefore, two **facets** can be seen on the **heads** of the second to the seventh or eighth ribs. The **neck** is slightly constricted and presents a **costal tubercle,** which has a rough portion and a facet for articulation with transverse processes. Two or three inches further laterally, the rib exhibits a roughened area called the **costal angle** to which muscles attach. Further details about the ribs will be presented in Chapter 3.

CLAVICLE (IN PART)

The **clavicle,** a bone fractured more frequently than any other, can be felt along its entire length. It articulates with the sternum medially and with the acromion of the scapula laterally. The lateral end of this bone (Fig. 2-1) is the part to be considered at this point; it has **superior and inferior surfaces** and **posterior and anterior borders.** (The clavicle is pictured also in Figure 3-1.)

SCAPULA (IN PART)

The **scapula** is a triangular bone that is attached to the clavicle laterally at the acromion process but otherwise is suspended by muscles. The parts involved with the back are shown in Figure 2-1. The scapula has **medial, lateral, and superior borders** and **superior, inferior, and lateral angles.** The **spine** of the scapula separates the **supraspinatus fossa** from an **infraspinatus fossa** and extends from the medial margin laterally to the enlarged **acromion process.** This ridge or spine gradually increases in height as it proceeds laterally.

ILIUM (IN PART)

A portion of the **ilium** (Fig. 2-1) should be studied at this time. The sacrum articulates with the ilium, which presents projections posteriorly called the **posterior superior iliac spines.** The **iliac crest** starts posteriorly at this superior spine and flares superiorly, laterally, and anteriorly. The remaining parts of the coxal bone will be described in Chapter 6.

SURFACE ANATOMY
(Fig. 2-11)

No matter how obese a person may be, a **median furrow** can be seen in the back. In muscular and lean persons, the **scapula** can be outlined and the **acromion** and **spine** easily palpated. The outline of the **trapezius** muscles can be detected, and the bulge on either side of the midline in the lower part of the back is made by the massive **erector spinae** muscles. Palpation easily reveals the **iliac crest,** the **ribs,** and the **spinous processes** of the vertebrae from the seventh cervical to the sacrum.

It is helpful to realize that the base of the spine of the scapula is at the level of the third thoracic vertebra, while the inferior angle of the scapula is even with the seventh thoracic vertebra; the crest of the ilium is at the level of the spinous process of the fourth lumbar vertebra, while the posterior superior iliac spine is at the level of the second sacral vertebra.

SUPERFICIAL AND
DEEP FASCIA

Removal of the skin of the back reveals the **superficial fascia.** This fascia varies in thickness from individual to individual, depending on the obesity of the person. This layer tends to be thicker in females than in males, and in both sexes will be thicker in the lumbar part of the back than in other regions. Careful removal of this fascial layer reveals the thin but dense **deep fascia,** which rests directly upon the muscles and actually surrounds each muscle.

CUTANEOUS NERVES
AND VESSELS

Coursing in this superficial fascia are the cutaneous nerves (Fig. 2-13) and vessels. They pierce the muscles and deep fascia covering the muscles and then course in a lateral and slightly inferior direction to reach the skin. Cutaneous nerves in the upper part of the back are found piercing the muscles and entering the superficial fascia close to the spines of the vertebrae. In the lower part of the back, the cutaneous nerves are some 3 to 4 inches away from the midline. This cannot be accounted for until we attain a knowledge of a typical segmental spinal nerve.

Figure 2-12 shows such a **segmental nerve.** Each nerve is attached to the spinal cord by two **roots,** one **dorsal** and one **ventral;** the **dorsal root,** containing a ganglion known as the **dorsal root ganglion,** is the **sen-**

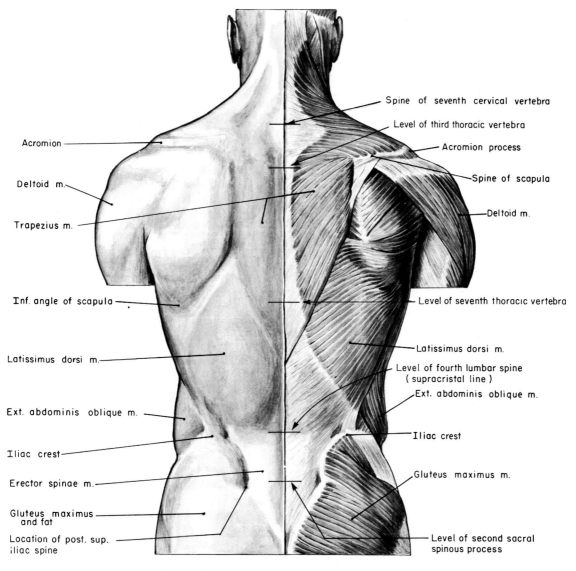

Spine of seventh cervical vertebra

Level of third thoracic vertebra

Acromion process

Spine of scapula

Deltoid m.

Level of seventh thoracic vertebra

Latissimus dorsi m.

Level of fourth lumbar spine (supracristal line)

Ext. abdominis oblique m.

Iliac crest

Gluteus maximus m.

Level of second sacral spinous process

Acromion

Deltoid m.

Trapezius m.

Inf. angle of scapula

Latissimus dorsi m.

Ext. abdominis oblique m.

Iliac crest

Erector spinae m.

Gluteus maximus and fat

Location of post. sup. iliac spine

Figure 2-11. *Surface anatomy of the back—male.*

sory root,* while the **ventral** is the **motor root.** The two roots converge as they emerge from an intervertebral foramen and form a **segmental** or **spinal nerve.** The **gray** and **white rami** are attached to the nerve at this

point,† and then the spinal nerve divides into two branches, one coursing posteriorly into the back and known as the **dorsal ramus** of the spinal nerve, and the other continuing anteriorly as the **ventral ramus** of the

*J. C. Hinsey, 1933, The functional components of the dorsal roots of spinal nerves, *Quart. Rev. Biol.* **8**: 457. This is a comparative approach to the problem of the presence of efferent fibers in the dorsal roots. The author claims that dorsal roots in mammals are purely sensory, there being no fibers in them of spinal cord origin.

†D. Duncan, 1943, The roots of spinal nerves, *Science* **98**: 515. Since all spinal nerves have dorsal and ventral roots, and a gray ramus communicans from a sympathetic ganglion, and thereby contain sensory, motor, and sympathetic fibers, the author feels each nerve should be regarded as having three roots: dorsal, ventral, and gray (sympathetic).

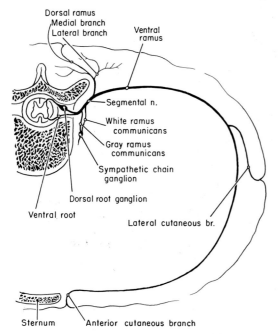

Figure 2-12. A typical segmental (spinal) nerve and its branches.

spinal nerve. This latter portion of the spinal nerve continues around the body, giving off a **lateral cutaneous branch** and, finally, an **anterior cutaneous branch.**

The **dorsal ramus** continues posteriorly and divides into **medial** and **lateral branches.** In the upper half of the body the lateral branch remains in the muscles, while the medial branch supplies muscles and then continues to the skin; in the lower part of the body the medial branch remains in the muscles, while the lateral branch innervates muscles and then continues to become a cutaneous branch. Therefore, the cutaneous nerves found in the upper part of the back are the medial branches, while those in the lower part of the back are the lateral branches.

Figure 2-13 shows the cutaneous nerves in the back. It is interesting to note that some of the cutaneous nerves in the cervical region have a distinctive pathway, and that not all the dorsal rami are represented in the skin area. The **first cervical nerve** has no sensory component ex-

cept upon rare occasions. The medial branch of the dorsal ramus of the **second cervical nerve** is a very large cutaneous nerve, called the **greater occipital nerve.** It ramifies over the scalp on the posterior part of the head. The medial branch of the dorsal ramus of the **third cervical** also reaches the more inferior aspect of the scalp. Cutaneous branches of the dorsal rami of the **fourth** to **sixth cervical** nerves do not exhibit any unusual characteristics. The **seventh** and **eighth cervical** nerves make no contribution to the back region.* Cutaneous branches of the **thoracic nerves** are quite usual in their distribution, those in the upper half of the body being the medial branches of the dorsal rami, and those of the lower part of the body the lateral branches. Cutaneous branches of the dorsal rami of the **first three lumbar** nerves are called the **superior cluneal** (*L.*, buttock) nerves, and enter the superficial fascia just lateral to the erector spinae muscles to be distributed inferiorly over the gluteal region. The dorsal rami of the **fourth** and **fifth lumbar** nerves do not have cutaneous distribution.† The dorsal rami of the **first three sacral** nerves have cutaneous branches, which are called the **middle cluneal** nerves and are distributed over the gluteal region as well as on the posterior surface of the sacral region. The dorsal rami of the **fourth** and **fifth sacral** nerves, and the **single coccygeal** nerve, do not divide into medial and lateral branches. A cutaneous nerve is formed from this complex, however, which is distributed in the region of the coccyx. The **inferior cluneal** nerves will be described with the lower limb.

It should be understood that these cutaneous nerves are not purely sensory nerves; they all contain sympathetic fibers of the autonomic nervous system to sweat glands, smooth muscles in blood vessels, and arrector pili muscles.

Each of these nerves is accompanied by similar branches of **cutaneous arteries and veins;** in fact, the vascular system has a distribution in the back similar to that of a regular segmental nerve (Fig. 2-14). Dorsal branches to the back are obtained from the vertebral artery in the neck, intercostal branches of the aorta in the thoracic region, lumbar branches of the aorta in the lower

---

*It seems that the skin that would have been innervated by these branches has simply disappeared in human beings.

†The preceding footnote is applicable here as well.

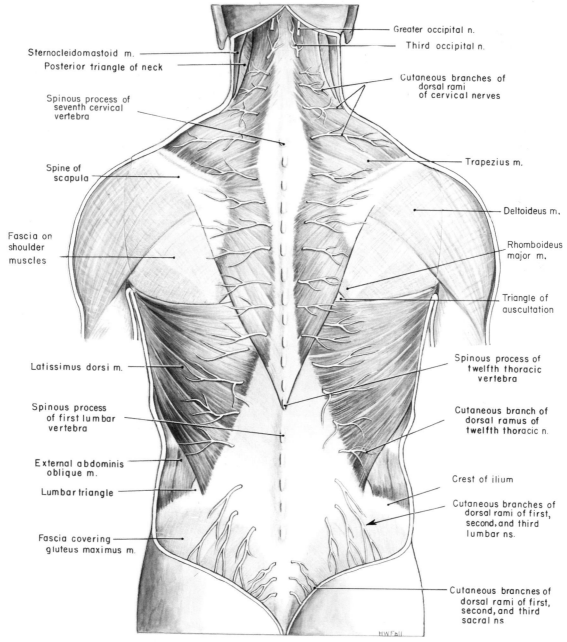

Figure 2-13. *Back musculature I. The cutaneous nerves are shown emerging from the appendicular muscles.*

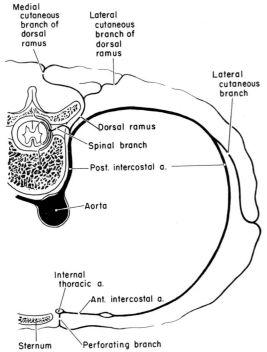

Figure 2-14. A typical segmental artery and its branches (an intercostal artery is pictured).

part of the back, and cutaneous branches of the lateral sacral arteries in the sacrococcygeal region.

Figure 2-30 provides a more realistic picture of the course taken by these cutaneous nerves and arteries.

BACK MUSCLES

Muscles in this text will be described according to their location, origin, insertion, main action, and nerve supply. *Enough information will be presented on origins and insertions to allow an understanding of the main action of each muscle. In this connection the student should realize that the insertion is usually more informative than the origin, unless the muscle happens to bridge over two joints.*

All striated or skeletal muscles are derived from myotomes of the mesodermal somites or from branchial

arches. The former make up the somatic musculature while the latter group are branchiomeric.

Each myotome has its own spinal nerve, which remains with this particular myotome no matter where migration may take it. The myotomes divide longitudinally to form the back or dorsal musculature (epaxial trunk muscles; see Fig. 2-15) and the intercostal and ventrolateral trunk muscles (hypaxial trunk muscles), and the nerves divide in a similar manner. These myotomes may undergo other changes besides this longitudinal splitting. There may be a tangential splitting, a fusion of portions of successive myotomes, a migration of certain portions, and degeneration of parts or entire myotomes.

The limb buds develop in such a way that there is a dorsal limb musculature (postaxial) and a ventral limb musculature (preaxial).

In summary, the nonbranchial skeletal muscles are all derived from the myotomes, and the ultimate arrangement in the human depends upon the many changes that occur in the primary arrangement. It is important to realize that the innervations to the myotomes occur early, and that these nerves follow the myotomes or parts thereof no matter what happens to them. Therefore, the nerve supply is a good clue to the embryology of a particular muscle.

Further details about the above will be given in the systemic summary of the back. For the present it seems easier to divide the muscles of the back into three functional groups: those concerned with the upper limb, those involved mainly with respiration, and those acting mainly on the back and head.

MUSCLES CONCERNED WITH UPPER LIMB

The most superficial muscles in the back are those belonging to the upper limb:

1. Trapezius
2. Latissimus dorsi
3. Levator scapulae
4. Rhomboideus major and minor
5. Serratus anterior

1 and 2. Trapezius and latissimus dorsi muscles. Removal of the superficial and deep layers of fascia in the back reveals two very large muscles, one located in the superior part and the other in the lower part

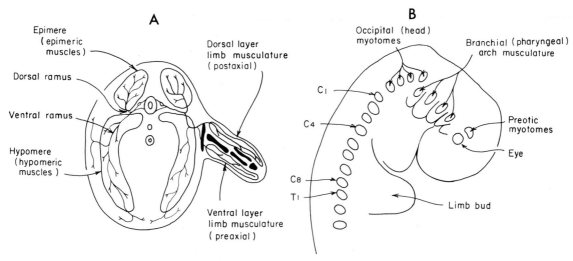

Figure 2-15. *(A) Cross section of an embryo showing the division of trunk musculature into epimeres and hypomeres, and limb musculature into dorsal muscles (postaxial or posterior to axis of limb) and ventral muscles (preaxial or anterior to axis of limb). (B) Myotomes and muscle primordia in the head and body of an embryo in the seventh week.*

of the back (Fig. 2-13). The superiorly placed muscle, the **trapezius,** has muscle bundles converging upon the scapula from the head, neck, and thoracic region; the lower muscle, the **latissimus dorsi,** exhibits fibers coursing in a lateral and superior direction toward the upper end of the humerus, a bone of the upper limb. These two muscles cover nearly the entire back. These muscles are actually concerned with the upper limb and not with the back proper; they are muscle masses that have migrated to the back.

The **trapezius** muscle **originates** from the medial end of the superior nuchal line, the external occipital protuberance, the ligamentum nuchae (see Fig. 2-25 to obtain a good idea of this ligament), and the spines of the seventh cervical and all thoracic vertebrae. The inferior portion of this muscle **inserts** on the base of the spine of the scapula,* the middle portion on the superior surface of the spine of the scapula and the acromion process, and the superior portion on the lateral third of the clavicle. This muscle has several important functions. One of its most important **actions** is to actually hold the upper limb onto the trunk; when the muscle acts as a whole, it tends

to adduct the scapula or bring it toward the midline. If the superior fibers alone contract, the scapula is raised; and if the inferior fibers alone contract, the scapula is lowered. Another action of the trapezius muscle is to aid in rotation of the scapula in such a manner that the inferior angle is moved laterally. It is essential in complete abduction at the shoulder that the glenoid cavity of the scapula be tipped superiorly; the deltoid muscle (which forms the rounded portion of the shoulder—see Fig. 2-11) can abduct the humerus to a horizontal level, but further abduction requires a rotation of the scapula, as shown in Figure 2-16. Patients with nerve injuries to the trapezius muscle have great difficulty in abducting the humerus above the horizontal plane, partly because the scapula is not held firmly against the deeper structures.†

The **latissimus dorsi** muscle is a very broad, flat, triangular muscle in the lower part of the back, which is overlapped by the lower part of the trapezius muscle. It takes **origin** from spinous processes of the lower six thoracic vertebrae, from spinous processes of all the lumbar and upper sacral vertebrae, and from the medial part of the iliac crest. There are additional slips of origin from

---

*A bursa intervenes between the tendon and the base of the spine.

†Note the actions of the serratus anterior muscle on page 66.

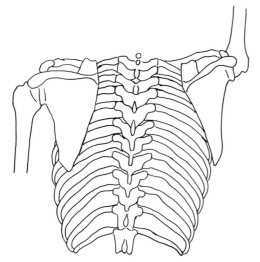

Figure 2-16. *Diagram showing rotation of the scapula in complete abduction at the shoulder joint.*

the lower four ribs and a small slip of attachment to the inferior angle of the scapula. The **insertion** of this muscle is usually not seen in this particular dissection. It does attach to the humerus in such a way that contraction of this muscle pulls the arm posteriorly, or extends it, and rotates it medially; this large muscle is used in such actions as chopping wood or swimming with the crawl stroke.

Two small **triangles** should be noted at this time (Fig. 2-13): (1) The **lumbar triangle** is bounded medially by the latissimus dorsi muscle, laterally by a muscle just lateral to the latissimus dorsi, the external abdominis oblique muscle, and inferiorly by the iliac crest. It is a fairly important triangle, for hernias and infections occasionally are presented here. (2) The **triangle of auscultation** is bounded by the trapezius muscle, the latissimus dorsi muscle, and the medial border of the scapula. The floor is formed by the rhomboideus major muscle, but when the scapula is moved anteriorly, the sixth and seventh ribs can be seen; this is a good location for listening to internal structures.

If the trapezius and latissimus dorsi muscles are dissected and reflected laterally, the **nerve** and blood supply to these muscles is seen (Fig. 2-17). The trapezius muscle has a double **nerve supply,** from the spinal accessory nerve (eleventh cranial) and from the ventral rami of the third and fourth cervical nerves. The actual role of these

nerves has not been determined completely, but it is thought that the cervical nerves may be sensory and that the spinal accessory nerve may be the motor supply to the muscle.* The upper border of the trapezius muscle forms the posterior border of the posterior triangle of the neck. The spinal accessory nerve is quite exposed in that it crosses the posterior triangle of the neck in the superficial layer of the deep fascia. It then reaches the superior border of the trapezius muscle and ramifies on the deep surface throughout its length; the nerve may enter the muscle independently, or it may join with the third and fourth cervical nerves and all three nerves ramify on the muscle as a single nerve.

The **blood supply** in this region is confusing because of frequent variation. An artery called the **transverse cervical** (Fig. 2-18) arises from the thyrocervical trunk, which is a branch of the subclavian artery. This artery finally reaches this region of the back and usually divides into two branches, one of which ramifies on the deep surface of the trapezius muscle, and the other on the deep surface of the next layer of muscles, the rhomboid muscles. The branch coursing on the deep surface of the trapezius is the **superficial branch,** while the other is the **deep branch** of the transverse cervical artery. For approximately half the cases, the superficial branch is the only one that arises from the thyrocervical trunk in the manner just described; when this occurs, the name of **superficial cervical** artery is given to this branch, and the deep branch then arises independently from the third part of the subclavian artery and is known as the **descending scapular artery.**

The **nerve** and **artery** to the **latissimus dorsi** muscle are called the **thoracodorsal nerve** and **artery.** This

---

*K. N. Corbin and F. Harrison, 1939, The sensory innervation of the spinal accessory and tongue musculature in the rhesus monkey, *Brain* 62: 191. Evidence is presented indicating that proprioception from the trapezius, sternocleidomastoid, and the tongue musculature is carried via the cervical nerves, leaving the spinal accessory and hypoglossal nerves to these muscles as the motor supply, at least in the monkey.

W. L. Straus and A. B. Howell, 1936, The spinal accessory nerve and its musculature, *Quart. Rev. Biol.* 11: 387. A survey of the puzzling double nerve supply to the trapezius and sternocleidomastoid muscles. The authors feel that the cervical nerves are merely branchial fibers that have shifted to follow a spinal pathway.

J. McKenzie, 1962, The development of the sternomastoid and trapezius muscles, *Contrib. Embryol.* 37: 121. On the basis of a study of the human embryos of the Carnegie Institute, this author feels that the sternocleidomastoid and trapezius muscles are partially derived from myotomes. This would imply that the nerve supply from the third and fourth cervical nerves is motor as well as sensory.

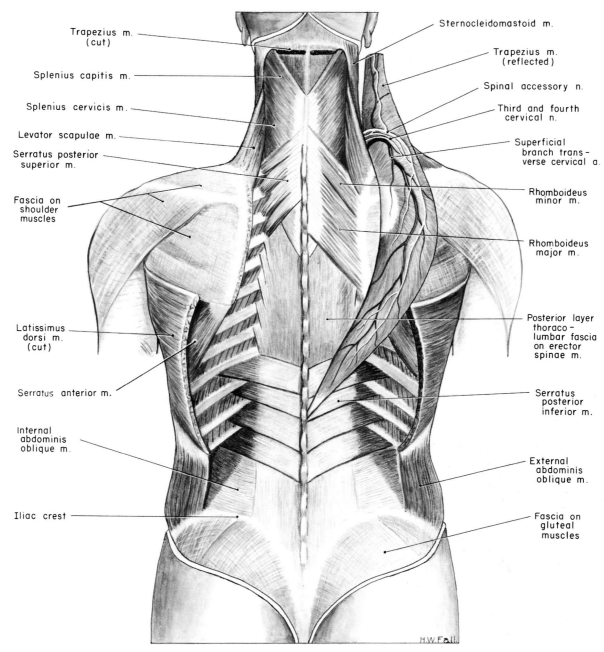

Figure 2-17. *Back musculature II. Latissimus dorsi muscle has been cut, and the trapezius muscle reflected to show its nerve and blood supply on the right side but removed on the left.*

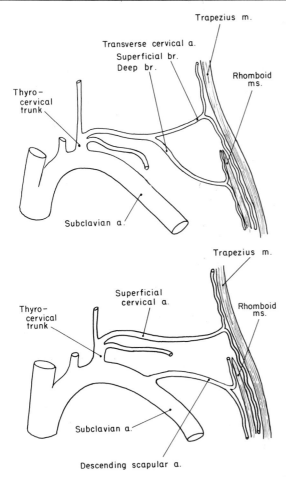

Figure 2-18. *Diagram of the two most frequently found arrangements of the blood supply to the trapezius and rhomboid muscles. Note the difference in names in the two arrangements.*

nerve is a branch of the brachial plexus (ventral rami forming a plexus of nerves that innervate the upper limb); the artery is a branch of the subscapular from the axillary artery. Both will be described in Chapter 8.

You have now observed **three rules about the anatomy of the body:** (1) *Nerves and arteries are usually found on the deep side of the muscle to be innervated.* (2) *Nerves, once they have joined a muscle, remain with this muscle no matter where it migrates.* (3) *Arteries exhibit many variations from the so-called normal; they are quite variable in their arrangement and one side of the*

body need not resemble the other; textbook descriptions are based on the appearance of the majority of cases.

---

3 and 4.   Levator scapulae and rhomboideus major and minor muscles.   Reflection of the trapezius muscles reveals the **levator scapulae** and **rhomboideus major** and **minor** muscles (Fig. 2-17). The most superiorly placed of the three, the levator scapulae, is seen high in the neck, while the rhomboids, so called from their shape, are located between the vertebral column and the scapula. These muscles also are concerned with movement of the upper limb and are not muscles of the back proper.

The **levator scapulae arises** from the transverse processes of the first four cervical vertebrae, and **inserts** on the superior angle of the scapula. Its **main action** is obviously to raise the scapula. Its **nerve supply** is from ventral rami of the third to fifth cervical nerves.

The **rhomboideus major** and **minor muscles** are actually one sheet of muscle but are described separately. The **minor arises** from the inferior end of the ligamentum nuchae and the spine of the seventh cervical vertebra and **inserts** on the medial margin of the scapula just inferior to the point of attachment of the levator scapulae muscle. The **rhomboideus major arises** from the first four thoracic spines and **inserts** on the medial margin of the scapula inferior to the base of the spine. In **action,** these muscles, in addition to actually holding the scapula onto the trunk, adduct the scapula and rotate it in such a manner that the inferior angle is moved medially. The **nerve supply** to the rhomboid muscles is by a specially named nerve, the dorsal scapular nerve, from the ventral ramus of the fifth cervical nerve. This nerve crosses the root of the neck and enters the deep surface of the rhomboid muscles. The deep branch of the transverse cervical artery occupies a similar position.

5.   Serratus anterior.   Although only a portion of this muscle is usually seen in a dissection of the back, its complementary action to the trapezius in rotating the scapula laterally in abduction of the arm warrants its inclusion at this point. It is illustrated in Figure 8-39 on page 705. This muscle is located on the lateral wall of the chest. Its **origin** is anteriorly placed and consists of eight digitations from the upper eight ribs. It hugs the lateral chest wall and **inserts** on the superior angle, the medial margin, and the inferior angle of the scapula on its ventral or anterior surface. The attachment to the inferior angle is particularly massive and is the part seen in this dissection (Fig. 2-19). Its **main action** is to pull the scapula forward in actions such as pushing, and, as just mentioned, to rotate the inferior angle laterally. Its **nerve supply** is by the long thoracic nerve, a branch of the brachial plexus. If this nerve is injured, the scapula is not held snugly to the chest wall, resulting in a condition called winged scapula. Under this condition, total abduction of the arm cannot be accomplished. It is obvious that total abduction of the arm (above the horizontal) cannot occur if either the trapezius or serratus anterior muscles are denervated.

RESPIRATORY MUSCLES

When the appendicular muscles are reflected, two thin muscles are revealed that are concerned with respiration, the serratus posterior superior and inferior (Fig. 2-17); although not visible at this time, another set of muscles, the levatores costarum, are located in the back and are also involved with respiration:

1. Serratus posterior superior
2. Serratus posterior inferior
3. Levatores costarum

1. Serratus posterior superior. This muscle is a thin structure lying deep to the rhomboid muscles. It **arises** from the ligamentum nuchae, the last cervical spine, and the first three thoracic spines; it **inserts** on ribs 2 to 5. It is easily seen that the main **action** of the serratus posterior superior muscle is to lift the ribs, and it is **innervated** by ventral rami of the first three or four thoracic nerves.

2. Serratus posterior inferior. This **arises** from the last two thoracic and the first two lumbar spines; it **inserts** on the lower four ribs. This muscle lowers the ribs or holds them in position against the upward pull of the diaphragm when it contracts. The serratus posterior inferior is **innervated** by ventral rami of the last four thoracic nerves.

3. Levatores costarum muscles. Although these muscles are located in a plane deep to the erector spinae muscles (Fig. 2-21), they will be described here since they are involved with respiration. These small muscles (Fig. 2-22) **arise** from the tips of the transverse processes and course laterally and inferiorly to **insert** either on the rib

below or, by skipping one rib, on the second rib below. They are, therefore, divided into **levatores costarum longi** and **breves** muscles. Although the other respiratory muscles in the back and elsewhere are innervated by ventral rami, recent evidence indicates that these muscles are actually innervated by branches of the dorsal rami.*

MUSCLES OF BACK PROPER

SPLENIUS MUSCLES. Reflection of the trapezius muscle from the neck region reveals a large muscle, covering the posterior surface of the neck, which is called the splenius (*L.*, bandage) muscle. Cutting the serratus posterior superior further reveals this muscle, as can be seen in Figure 2-19. This muscle layer is actually composed of two muscles, **splenius capitis** and **cervicis**. It is very narrow inferiorly and gradually becomes wider as it stretches superiorly to attach to the head and transverse processes of the cervical vertebrae. This muscle complex **arises** from the lower half of the ligamentum nuchae, and from spines of the seventh cervical and first five thoracic vertebrae; the capitis portion **inserts** on the mastoid process of the temporal bone and on the lateral end of the superior nuchal line, while the cervicis part **inserts** on the transverse processes of the first three or four cervical vertebrae. This muscle is **innervated** by the fourth to the eighth cervical nerves, and because muscles of the back proper are innervated by dorsal rami, it is the dorsal rami of these nerves. **Acting** together, these muscles will pull the head posteriorly; if only one side contracts, the head and neck are moved so that the face is turned toward the same side.

THORACOLUMBAR FASCIA AND ERECTOR SPINAE MUSCLE. After the appendicular muscles, the two serratus muscles, and the splenius capitis and cervicis have been reflected, the erector spinae muscle is reached. This cannot be seen, however, because of the presence of the **thoracolumbar fascia** (Fig. 2-19). This fascia, to which the tendons of the serratus posterior inferior and the latissimus dorsi muscles are firmly blended, extends from the iliac crest and sacrum to the thoracic region, being much heavier in the lumbar region. The layer exposed at this stage of the dissection is the **posterior layer** of the

---

*A. B. Morrison, 1954, The levatores costarum and their nerve supply, *J. Anat.* 88: 19.

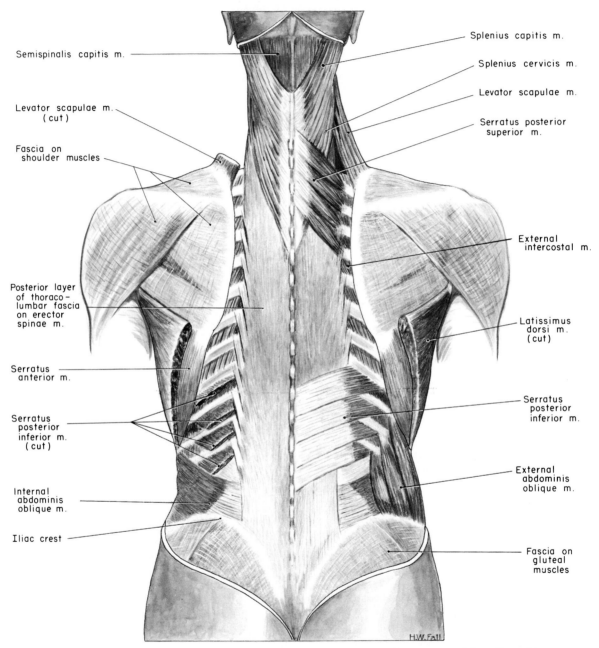

Semispinalis capitis m.

Levator scapulae m. (cut)

Fascia on shoulder muscles

Posterior layer of thoraco-lumbar fascia on erector spinae m.

Serratus anterior m.

Serratus posterior inferior m. (cut)

Internal abdominis oblique m.

Iliac crest

Splenius capitis m.

Splenius cervicis m.

Levator scapulae m.

Serratus posterior superior m.

External intercostal m.

Latissimus dorsi m. (cut)

Serratus posterior inferior m.

External abdominis oblique m.

Fascia on gluteal muscles

H.W.Fall

Figure 2-19. *Back musculature III. Appendicular muscles have been removed from the right side to show respiratory muscles, which have been removed on left side.*

thoracolumbar fascia; it extends from the spines of the vertebrae laterally as far as the angles of the ribs, and covers the erector spinae muscle.

Although not visible at this time, there are other layers of this fascia that are attached to tips of the transverse processes of the lumbar vertebrae (Fig. 2-20). The **middle layer** of **thoracolumbar fascia** is in relationship with the posterior surface of a muscle called the quadratus lumborum (seen in Fig. 2-22), while the anterior layer occurs on the anterior surface of this muscle. Because the quadratus lumborum muscle extends between the iliac crest and the last rib only, the anterior and middle layers of this fascia have a more limited extent than the posterior layer. All layers fuse into one layer lateral to the erector spinae muscle, and here one of the abdominal muscles, the transversus abdominis, takes origin. As found in the body, this thoracolumbar fascia is as much an aponeurotic tendon as it is fascia.

If the posterior layer of the thoracolumbar fascia is reflected, the **erector spinae** muscle is revealed, as can be seen in Figure 2-21. This muscle is large and fills in the groove between the spines of the vertebrae medially and

e[...]
into [...]
posterio[...]
crum, and [...]
this muscle c[...]
should be visuali[...]
individual muscles.

The outer or more [...]
the **iliocostalis** group; it [...]
borum, iliocostalis thoracis, [...]
and these parts are located as [...]
consist of many fasciculi, which hav[...]
the ribs and finally terminate by insertir[...]
verse processes of the fourth, fifth, and sixt[...]
tebrae.

The middle column of muscles is the **long[...] group** and is divided into a large **longissimus thorac[...]** smaller **longissimus cervicis,** and an equally small **lon[...] gissimus capitis.** This group has attachments mainly to accessory and transverse processes of the vertebrae. The

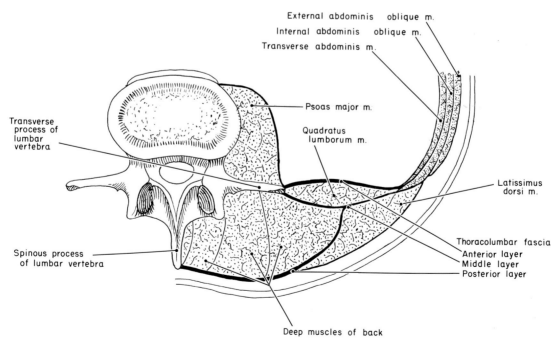

Figure 2-20. Transverse section of the body wall, showing attachments of thoracolumbar fascia (diagrammatic).

the angles of the ribs laterally. It extends from the sacrum and iliac crest all the way to the back of the head. The ...ine parts. After a **common origin** from iliac crest, ...r sacroiliac ligaments, dorsal surface of the sa- ...spines of the sacral and all lumbar vertebrae, ...urses superiorly in three columns and ...ed and learned as columns, not as nine

laterally placed column is called ...consists of **iliocostalis lum-** and **iliocostalis cervicis,** ...e names imply. They ...e attachments to all ...g onto the trans- ...h cervical ver-

ngle expanse of powerful muscles aris- ...m all the transverse processes of the ...e and inserting on all the spines from C2 ...spinalis cervicis has a strong attachment ...ne axis.

...)inalis capitis is a very powerful muscle ...ack of the neck just deep to the splenius ...es from the transverse processes of the ...cic vertebrae and the articular processes of ...cervical vertebrae and **inserts** onto the ...between the superior and inferior nuchal

...these muscles are all arising from transverse ...l inserting on spinous processes of vertebrae ...cated, or onto the head in the case of the ...capitis, it is easy to see that they will have an ...ion on the vertebral column to those more ...located. The transversospinal group pull the ...rd the transverse processes of the same side, ...)re, turn the body to the opposite side. If both ...unison, they will obviously extend the verte- ...n. The semispinalis capitis, being attached to ...has an action different from that of the other ...When these muscles act in unison they pull the ...teriorly; if one muscle contracts, the head is ...)steriorly and turned so that the chin moves ...)e same side.

students ...

cles and simply walk, they will ...
these muscles contract to keep us erect.

These muscles, being of the back proper, are all **innervated** by dorsal rami of most of the spinal nerves.

TRANSVERSOSPINAL MUSCLE GROUP. If the spinalis portion of the erector spinae muscle is removed and the other portions reflected laterally, another muscle layer will be seen in the rather narrow groove between transverse processes and spinous processes of the vertebrae. Because of these attachments they are known as the **transversospinal** group of muscles, consisting of:

1. Semispinalis muscles
2. Multifidus muscle
3. Long and short rotators

1. Semispinalis muscles. These muscles are the most superficial of this group. They start in the lower thoracic region and form a continuous column of muscles all the way to the back of the head. They are divided into **semispinalis thoracis, cervicis,** and **capitis** muscles (Fig. 2-21). The thoracis and cervicis muscles should be

2. Multifidus. If one removes the semispinalis group of muscles, another set of muscles is found, the **multifidus,** which also originate on transverse processes and insert on spinous processes. This group of muscles is made up of many fasciculi that are shorter than those in the semispinalis group (Fig. 2-22). This muscle **arises** from the dorsal surface of the sacrum, mamillary processes of the lumbar vertebrae, transverse processes of the thoracic vertebrae, and articular processes of the lower four cervical vertebrae; it is **inserted** onto the spines of the vertebrae up to that of the second cervical. The multifidus muscle is particularly heavy in the lumbar region. This muscle has an **action** similar to that just stated for the semispinalis muscle, namely, to rotate the body to the opposite side. If both muscles act in unison, they aid in extending the vertebral column.

3. Rotators. If the multifidus muscle is removed (Fig. 2-22), it can be seen that there are still shorter muscles in this groove between transverse and spinous processes of the vertebral column. This last group, called the

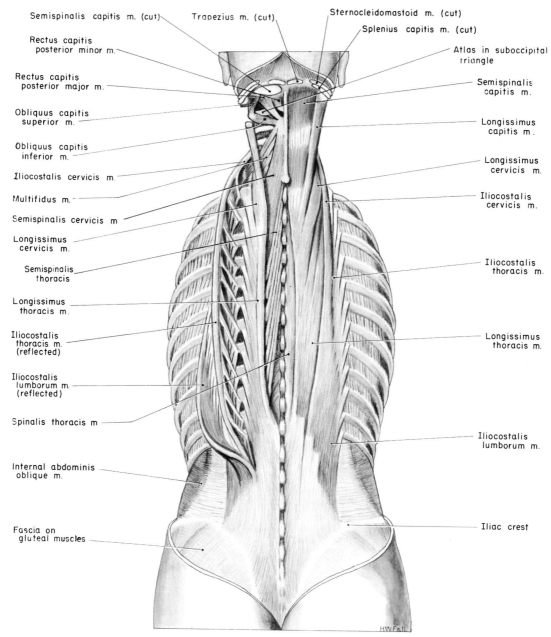

Semispinalis capitis m. (cut)

Rectus capitis
posterior minor m.

Rectus capitis
posterior major m.

Obliquus capitis
superior m.

Obliquus capitis
inferior m.

Iliocostalis cervicis m.

Multifidus m.

Semispinalis cervicis m

Longissimus
cervicis m.

Semispinalis
thoracis

Longissimus
thoracis m.

Iliocostalis
thoracis m.
(reflected)

Iliocostalis
lumborum m.
(reflected)

Spinalis thoracis m

Internal abdominis
oblique m.

Fascia on
gluteal muscles

Trapezius m. (cut)

Sternocleidomastoid m. (cut)

Splenius capitis m. (cut)

Atlas in suboccipital
triangle

Semispinalis
capitis m.

Longissimus
capitis m.

Longissimus
cervicis m.

Iliocostalis
cervicis m.

Iliocostalis
thoracis m.

Longissimus
thoracis m.

Iliocostalis
lumborum m.

Iliac crest

HWFall

Figure 2-21. *Back musculature IV. Musculature after removal of thoracolumbar fascia.*

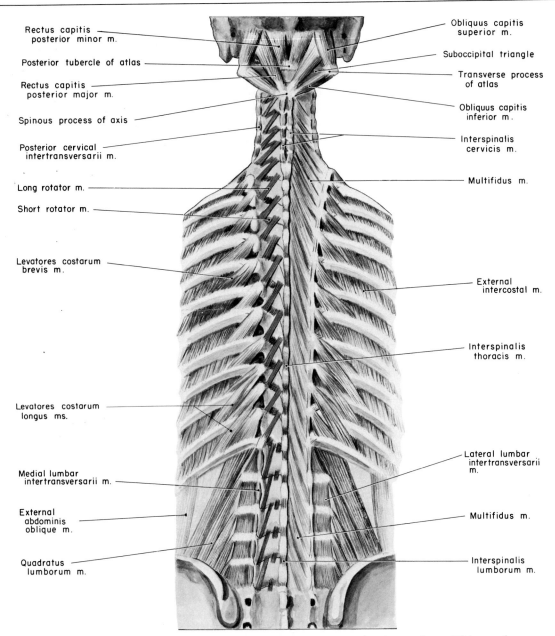

Rectus capitis posterior minor m.

Posterior tubercle of atlas

Rectus capitis posterior major m.

Spinous process of axis

Posterior cervical intertransversarii m.

Long rotator m.

Short rotator m.

Levatores costarum brevis m.

Levatores costarum longus ms.

Medial lumbar intertransversarii m.

External abdominis oblique m.

Quadratus lumborum m.

Obliquus capitis superior m.

Suboccipital triangle

Transverse process of atlas

Obliquus capitis inferior m.

Interspinalis cervicis m.

Multifidus m.

External intercostal m.

Interspinalis thoracis m.

Lateral lumbar intertransversarii m.

Multifidus m.

Interspinalis lumborum m.

Figure 2-22. *Back musculature V. The deepest back muscles are shown: the multifidus on the right side and the long and short rotators on the left.*

rotatores muscles, is found from the sacrum to the cervical region. Each **arises** from a transverse process and either **inserts** on the lamina of the vertebra above (rotatores breves) or skips one vertebra and **inserts** on the lamina of the second vertebra above (rotatores longi). These muscles have the same action just mentioned for the multifidus, namely, to turn the body toward the opposite side.

It is easily seen that the transversospinal musculature by necessity has been made up of muscle fasciculi that have become shorter and shorter as one travels deeper into the back. Figure 2-23 shows that the individual fasciculi of the semispinalis muscles arise on a transverse process and, skipping five or six vertebrae, insert on the sixth or seventh vertebra above; that those of the multifidus muscle arise on a transverse process, skip three vertebrae, and insert on the fourth vertebra above. We have just seen that the rotatores arise from a transverse process and either skip a vertebra or insert on the vertebra immediately above. All these muscles have had identical action and have identical nerve supply, dorsal rami of the spinal nerves.

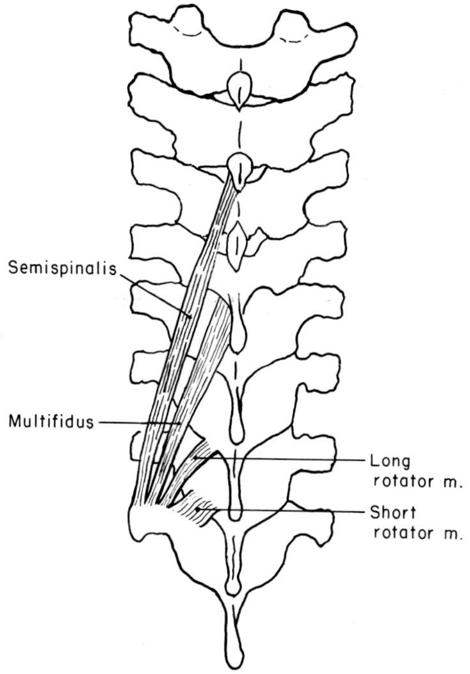

OTHER MUSCLES. For the sake of completeness, it should be mentioned that there are small **interspinalis** muscles located between spinous processes of the vertebrae (Fig. 2-22). These are paired in the cervical region and are attached to the bifid spinous processes. There are also muscles in the cervical and lumbar regions coursing between transverse processes. These muscles, the **intertransversales** muscles, are arranged in pairs in both of these regions. The same muscles in the thoracic region are poorly developed.

With this massive musculature, it is easy to understand why back strains are frequent. They are usually of sudden onset and are very painful.

Figure 2-23. *Diagram showing number of vertebrae spanned by the majority of muscle fasciculi of transversospinal musculature.*

Semispinalis

Multifidus

Long rotator m.

Short rotator m.

SUBOCCIPITAL REGION

The frequency of so-called whiplash has made the anatomy of the suboccipital region quite important clinically.

GENERAL DESCRIPTION

You have already learned that the trapezius muscle is the most superficial muscle on the back of the neck (Fig. 2-13). Just deep to this muscle are found the splenius capitis and cervicis, and deep to these the semispinalis capitis muscle (Figs. 2-19 and 2-21). If these muscles are reflected, the suboccipital region is revealed (Fig. 2-24), so called because it is under or inferior to the occipital bone.

In the region between axis and skull the muscles on each side are arranged in such a manner as to form a triangle, which is known as the **suboccipital triangle.** The obliquus capitis inferior muscle forms the **inferior border** of the triangle, the obliquus capitis superior the **lateral border,** and the rectus capitis posterior major the **medial border.** This triangle **contains** the first cervical nerve, vertebral artery, and a complex of veins. The re-

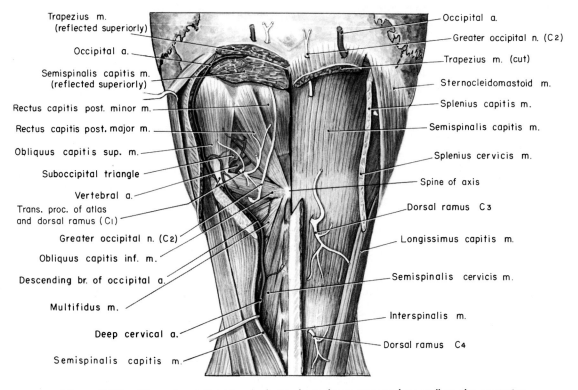

Trapezius m.
(reflected superiorly)

Occipital a.

Semispinalis capitis m.
(reflected superiorly)

Rectus capitis post. minor m.

Rectus capitis post. major m.

Obliquus capitis sup. m.

Suboccipital triangle

Vertebral a.

Trans. proc. of atlas
and dorsal ramus (C1)

Greater occipital n. (C2)

Obliquus capitis inf. m.

Descending br. of occipital a.

Multifidus m.

Deep cervical a.

Semispinalis capitis m.

Occipital a.

Greater occipital n. (C2)

Trapezius m. (cut)

Sternocleidomastoid m.

Splenius capitis m.

Semispinalis capitis m.

Splenius cervicis m.

Spine of axis

Dorsal ramus C3

Longissimus capitis m.

Semispinalis cervicis m.

Interspinalis m.

Dorsal ramus C4

Figure 2-24. *The suboccipital triangle (veins have been removed as well as the posterior atlanto-occipital membrane—see Fig. 7-107). See Figure 2-25 to obtain a sense of depth.*

lationship between the artery and nerve is important because pathology of the artery can put pressure on the nerve which, in this case, is caught between the artery and the arch of the atlas. The **rectus capitis posterior minor** muscle does not take part in formation of the triangle but can be seen just medial to the rectus capitis major. The **greater occipital nerve** is seen crossing the triangle on its way to the scalp, but the **occipital artery** is lateral to the triangle. This triangle is just deep to the semispinalis capitis and just superficial to the ligaments joining the axis and the atlas, and both axis and atlas to the occipital bone.

MUSCLES

The muscles in the suboccipital region, all deep to the semispinalis capitis muscle, are:

1. Obliquus capitis inferior
2. Obliquus capitis superior

3. Rectus capitis posterior major
4. Rectus capitis posterior minor

1. The **obliquus capitis inferior** muscle **arises** from the spine of the axis and **inserts** on the transverse process of the atlas; its **main action** is to turn the head toward the same side.

2. The **obliquus capitis superior** muscle **arises** from the transverse process of the atlas and **inserts** on the occipital bone on the lateral part of the region between the nuchal lines.

3. The **rectus capitis posterior major** muscle **arises** from the spine of the axis and **inserts** on the occipital bone inferior to the inferior nuchal line.

4. The **rectus capitis posterior minor** muscle **arises** from the posterior tubercle of the atlas and inserts on the occipital bone just medial to the insertion of rectus capitis posterior major.

The last three muscles have a **similar action,** namely, to pull the head posteriorly and to turn the head toward the same side. If muscles on both sides work simultaneously, the head will be pulled directly posteriorly. The **nerve supply** is by branches of the first cervical nerve.

NERVES

The **dorsal rami of the first four cervical nerves** are seen in this region (Fig. 2-24).

The **first cervical nerve** is interesting because it supposedly has no sensory component; it is very difficult to locate any dorsal root ganglion on this nerve. How proprioceptive impulses from muscles innervated by this nerve enter the spinal cord is problematical. It emerges from the spinal cord, appears between the atlas and the occipital bone, continues posteriorly between the vertebral artery and the posterior arch of the atlas, and immediately branches to innervate the semispinalis capitis muscle as well as those mentioned above.

The **dorsal ramus** of the **second cervical nerve** has both a sensory and a motor component, as do dorsal rami of the other spinal nerves. It divides into a lateral and a medial branch; the lateral branch, typical of all lateral branches of dorsal rami in the upper half of the body, innervates several of the cervical muscles. The medial branch, the **greater occipital nerve,** appears between obliquus capitis inferior and semispinalis cervicis muscles, and turns superiorly to pass superficial to the muscles forming the triangle; in this position it is deep to the semispinalis capitis muscle, which it pierces close to the skull; it may or may not pierce the trapezius muscle, and then ramifies on the posterior part of the head. This is a sensory nerve to the scalp.

The **dorsal rami** of the **third** and **fourth cervical nerves** can be seen between semispinalis cervicis and longissimus cervicis muscles. The lateral branches innervate all muscles in this area except the trapezius and levator scapulae; medial branches become cutaneous on the back of the neck, C3 reaching the scalp, where it is called the **third occipital nerve.**

ARTERIES

The **arteries** in this region are three in number:

1. Vertebral
2. Occipital
3. Deep cervical

1. The **vertebral artery** has already been mentioned as being located in the triangle; at this position it is taking a medial course from the foramen in the transverse process of the atlas to the foramen magnum (Fig. 2-24). Further details on the course of this artery are presented in Figure 2-25.

2. The **occipital artery** has to reach this posterior location from the anterior side of the neck and, therefore, is not in the same location as the greater occipital nerve. It courses in a groove just deep to the mastoid process of the temporal bone, continues posteriorly between the obliquus capitis superior and the splenius capitis, still remaining close to the skull (Fig. 2-24), becomes superficial to semispinalis capitis and trapezius muscles, or pierces the latter, and ramifies on the posterior surface of the head. It is one of the main sources of blood for the scalp. One of the branches of this artery is the **descending branch,** which descends across the obliquus capitis inferior muscle and anastomoses with the **deep cervical** artery.

3. The **deep cervical artery,** a branch of the costocervical trunk of the subclavian, is located deep to semispinalis capitis, lateral to semispinalis cervicis and medial to the longissimus capitis muscle, where it anastomoses with the descending branch of the occipital. Branches of the vertebral artery also take part in this important anastomosis as well as the superficial branch of the transverse cervical artery.*

VEINS

Veins in this area form a massive plexus of vessels called the **suboccipital plexus.** Branches of this plexus make an important communication with the dural venous sinuses in the cranial cavity via the condyloid canal, the vertebral vein starts in this plexus, and the plexus is the superior limit of a similar plexus of veins that is found throughout the entire length of the vertebral column.

This area reveals two more concepts of human anatomy: (1) *Arteries from quite different sources join*

---

*The ascending branch of the inferior thyroid artery also is involved in this anastomosis. This vessel will be studied with the head and neck (Fig. 7-138, page 662).

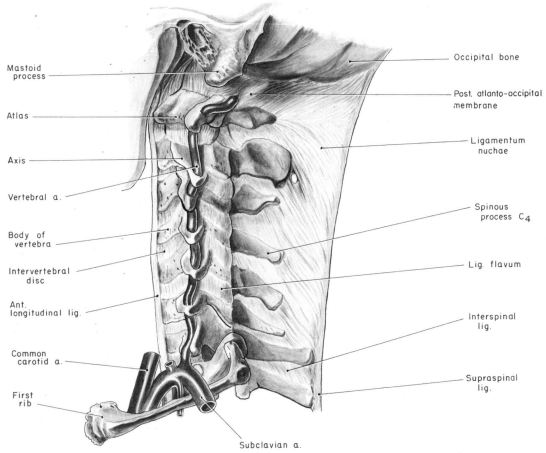

Mastoid
process

Atlas

Axis

Vertebral a.

Body of
vertebra

Intervertebral
disc

Ant.
longitudinal lig.

Common
carotid a.

First
rib

Subclavian a.

Occipital bone

Post. atlanto-occipital
membrane

Ligamentum
nuchae

Spinous
process C₄

Lig. flavum

Interspinal
lig.

Supraspinal
lig.

Figure 2-25.   *The course of the vertebral artery; the ligamentum nuchae.*

one another (anastomose) via relatively large connections. When main channels are occluded, these anastomotic pathways can enlarge and carry the blood around the occlusion coursing in a reverse direction in one of the vessels. This is possible because arteries have no valves. *(2) Veins are quite variable in their connections and arrangement, often forming several channels (venae comitantes) or venous plexuses.* These venous complexes often become important pathways for spread of infection and cancer.

VERTEBRAL COLUMN AS A WHOLE

The vertebral column as a whole is much more important than any single vertebra or type of vertebra. Clinically, the vertebral column and spinal cord are very important parts of the back.

If one looks at the **anterior surfaces** (Fig. 2-26) of the thirty-three vertebrae forming the column, the bodies are seen to increase in size as one proceeds inferiorly. In

Figure 2-26.   *Anterior, posterior, and lateral views of the vertebral column.*

ANTERIOR **POSTERIOR** **LATERAL**

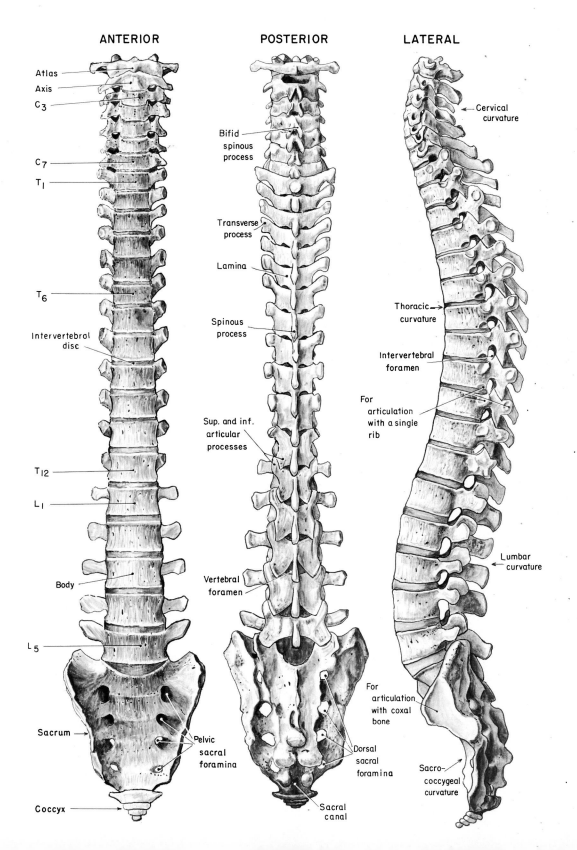

Atlas

Axis

C₃

C₇

T₁

T₆

Intervertebral disc

T₁₂

L₁

Body

L₅

Sacrum

Pelvic sacral foramina

Coccyx

Bifid spinous process

Transverse process

Lamina

Spinous process

Sup. and inf. articular processes

Vertebral foramen

For articulation with coxal bone

Dorsal sacral foramina

Sacral canal

Cervical curvature

Thoracic curvature

Intervertebral foramen

For articulation with a single rib

Lumbar curvature

Sacro-coccygeal curvature

addition, it can be seen that there is considerable space between each vertebra and the next one below or above. This space is filled in by the **articular discs,** which together make up almost one-third of the entire length of the column. These **articular discs** (Fig. 2-27) are white fibrocartilaginous structures, which are very dense in the peripheral part, the **annulus fibrosus,** and soft and pulpy in the center, the **nucleus pulposus.** This nucleus pulposus can herniate through the fibrous portion and put pressure on the nerves. This is a so-called **herniated disc.**

As one looks at the **posterior surface** of the vertebral column, it is seen that the flattened cervical **transverse processes** change rapidly to the rather large thoracic type. These in turn gradually decrease in size until no processes can be seen at all on the last two thoracic vertebrae. The transverse processes of the lumbar vertebrae are long and amazingly slender. Differences in the appearance of the **spinous processes** in the various regions of the column are striking, as is the difference in the size of the space between the laminae. The bifid cervical spinous processes change rapidly into the long, thin, sloping variety found in the thoracic region; these in turn change to the short, blunt type found in the lumbar region. The sloping nature of the thoracic spines just mentioned makes the **interlaminal** space in the thoracic region very small, while this space is rather large in the lumbar region. Advantage is taken of this large space in the procedure of a lumbar puncture to obtain cerebrospinal fluid or to give spinal anesthesia.

The **lateral view** is the most revealing (Fig. 2-26). It is easily seen that there is a **curve** in the cervical region that is concave posteriorly, a thoracic curve concave anteriorly, a lumbar curve concave posteriorly, and a sacrococcygeal curve concave anteriorly. The thoracic and sacrococcygeal curves are primary and are found in the newborn; the cervical curve is secondary to holding the head up and the lumbar secondary to standing in the erect posture. These acquired curvatures are due to changes in the configuration of the articular discs rather than in the vertebrae; the vertebrae, held in apposition in a column, exhibit no such curvatures. At approximately 60 years of age and beyond, the articular discs lose their elasticity, resulting in the decreased height and increased stoop of the elderly. The thoracic curvature can be quite excessive, due to disease of the bodies in that region; this can lead to a hunchbacked condition. Any increase in this curvature is known as kyphosis. An increase in the lumbar curvature is known as a lordosis. A lateral curvature is a scoliosis.

This lateral view also shows the many articular surfaces between the vertebrae, a prime location for inflammatory diseases with a predilection to attack joints. This view also shows the intervertebral foramina between each two vertebrae for the transit of the spinal nerves and vessels.

LIGAMENTS

The ligaments between two vertebrae are:

1. Supraspinal
2. Interspinal
3. Ligamenta flava
4. Posterior longitudinal
5. Anterior longitudinal
6. Intertransverse

After all muscles have been removed from the posterior surface of the vertebral column, one can see that the vertebrae are held together by **ligaments** (Figs. 2-28 and 2-29). The most superficial ligament, the **supraspinal,** connects the tips of the spines; the superficial fibers span several spines but the deeper fibers connect one spine with the next. This ligament is continuous with the ligamentum nuchae in the neck. Deep to these are found the **interspinal** ligaments, connecting the base of one spine to the same region on the adjacent vertebra. The laminae are connected to one another by ligaments with a yellowish cast, the **ligamenta flava.** These ligaments are found in pairs, one for each side of the body, and aid in

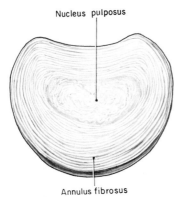

Nucleus pulposus

Annulus fibrosus

Figure 2-27. *An intervertebral disc.*

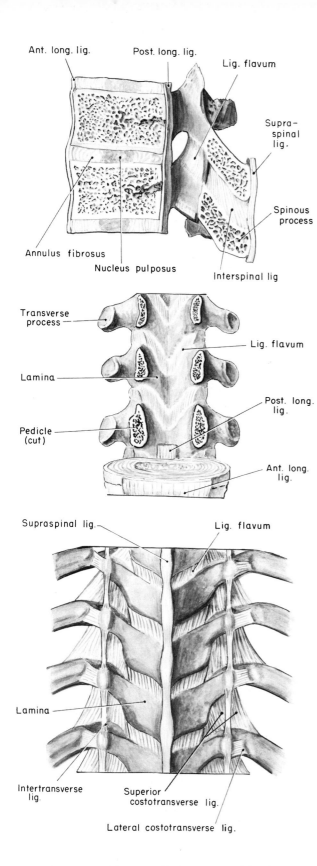

Ant. long. lig.

Post. long. lig.

Lig. flavum

Supra-spinal lig.

Spinous process

Annulus fibrosus

Nucleus pulposus

Interspinal lig

Transverse process

Lig. flavum

Lamina

Post. long. lig.

Pedicle (cut)

Ant. long. lig.

Supraspinal lig.

Lig. flavum

Lamina

Intertransverse lig.

Superior costotransverse lig.

Lateral costotransverse lig.

forming the posterior wall of the vertebral canal. Laterally these ligaments fuse with the articular capsules and actually form one boundary of the intervertebral foramina. The anterior wall of the vertebral canal is formed by the **posterior longitudinal ligament,** which extends the entire length of the vertebral column. There is a similar ligament on the anterior surface of the vertebral bodies, the **anterior longitudinal ligament.** Small **intertransverse ligaments,** between the transverse processes, complete the picture; they are intermingled with the intertransverse muscles in the cervical and lumbar regions, and practically replace the muscles in the thoracic region.

The ligaments connecting the atlas and axis with one another and with the skull demand special description. However, they are rarely seen in dissecting the back, and so will be omitted at this time and included in the description of the head and neck (pages 598–600).

MOVEMENT

As mentioned before, each vertebra is connected with its adjacent members by the discs between the bodies and by articular facets. Bending is allowed by the discs, and a gliding motion occurs between the facets. Although there is only a little movement allowed between any two vertebrae, the accumulated movement throughout the entire length of the column is considerable. Note Plates 1–5 on pages 82, 83, 84, 86, and 87.

Movement is freer in **the cervical region** than elsewhere. The facets are horizontally placed, so that flexion, extension, and rotation are accomplished easily. Although the facets in **the thoracic region** are placed in such a manner that considerable freedom would be allowed, there is relatively little movement at this location on account of the presence of the ribs, the thinness of the articular discs, and the length of the spinous processes. Although the **lumbar vertebrae** are massive and heavy, considerable movement is allowed in this region. The facets are sagitally placed which allows considerable flexion, extension, and lateral bending, but little rotation. Although the ligaments are very elastic and tend to restore the vertebral column to a resting position after movement, most of the ligaments tend to restrict move-

Figure 2-28. *Ligaments of the intervertebral and costovertebral articulations.*

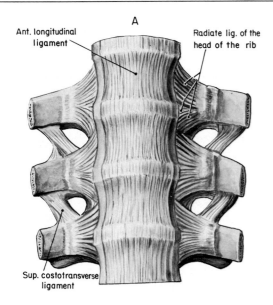

A

Ant. longitudinal ligament

Radiate lig. of the head of the rib

Sup. costotransverse ligament

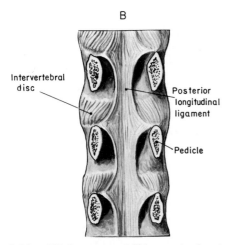

B

Intervertebral disc

Posterior longitudinal ligament

Pedicle

Figure 2-29. *(A) Anterior and (B) posterior longitudinal ligaments of the vertebral column.*

ARTERIES

Spinal arteries arise from the vertebral artery in the cervical region, from posterior branches of the intercostal arteries in the thoracic region, from dorsal branches of the lumbar and the iliolumbar arteries in the lumbar region, and from the lateral sacral arteries in the sacrococcygeal region. (See Fig. 5-64, page 294, to aid in visualizing the last two arteries.) These branches enter the vertebral canal through each intervertebral foramen and immediately divide into three branches (Fig. 2-30). One of these is the **neural branch,** which penetrates the dura mater and divides into anterior and posterior **radicular** (**root**) **branches** which follow each nerve root; they will be seen later. The two **osseous branches** enter the vertebral canal and connect with the artery above and below to form longitudinally running channels; one of these channels is found just lateral to the posterior longitudinal ligament and the other along the ligamenta flava. Transverse connections from one side to the other occur with regularity throughout the length of the vertebral column. Branches from these arteries supply the vertebrae. The exact course of these arteries is important to the surgeon; if a portion of a vertebra is to be removed, the blood supply to the part remaining must be maintained.

VEINS

The veins draining the vertebrae are:

1. Internal vertebral plexus
   a. Anterior
   b. Posterior
2. Basivertebral
3. Intervertebral
4. External vertebral plexus
   a. Anterior
   b. Posterior

The **veins** (Fig. 2-30) form two massive plexuses, one external and one internal. The **internal plexus** is distributed in a fashion similar to that described for the arteries. There is a longitudinally running **anterior internal plexus** just lateral to the posterior longitudinal ligament, and a **posterior internal plexus** coursing on the ligamenta flava. Veins running transversely connect both sides, and the whole complex receives branches from veins draining the vertebrae, called the **basivertebral**

ment. Flexion of the vertebral column is restricted by the posterior longitudinal ligament as well as by others, while the anterior longitudinal ligament and the spines tend to stop overextension.

Horizontal slipping of one vertebra on the next below occurs easily in the cervical region (thus the effectiveness of the hangman) but results in fracture if it occurs below this level.

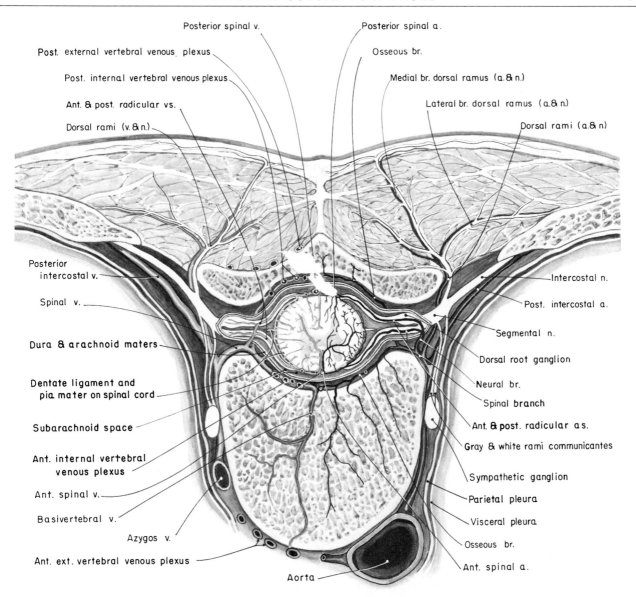

Posterior spinal v.

Posterior spinal a.

Post. external vertebral venous plexus

Osseous br.

Post. internal vertebral venous plexus

Medial br. dorsal ramus (a. & n.)

Ant. & post. radicular vs.

Lateral br. dorsal ramus (a. & n.)

Dorsal rami (v. & n.)

Dorsal rami (a. & n.)

Posterior intercostal v.

Intercostal n.

Spinal v.

Post. intercostal a.

Dura & arachnoid maters

Segmental n.

Dorsal root ganglion

Dentate ligament and pia mater on spinal cord

Neural br.

Spinal branch

Subarachnoid space

Ant. & post. radicular a s.

Gray & white rami communicantes

Ant. internal vertebral venous plexus

Sympathetic ganglion

Ant. spinal v.

Parietal pleura

Basivertebral v.

Visceral pleura

Azygos v.

Osseous br.

Ant. ext. vertebral venous plexus

Ant. spinal a.

Aorta

Figure 2-30. *Transverse section of a vertebra, the spinal cord, and back musculature. The meninges are shown as well as the spinal nerves (the inferior course naturally taken by the nerve roots has been omitted). The spinal arteries and branches are presented on the left side of the body, the veins on the right. Meningeal nerves to dura and adjoining bones and ligaments have been omitted; they follow the vessels.*

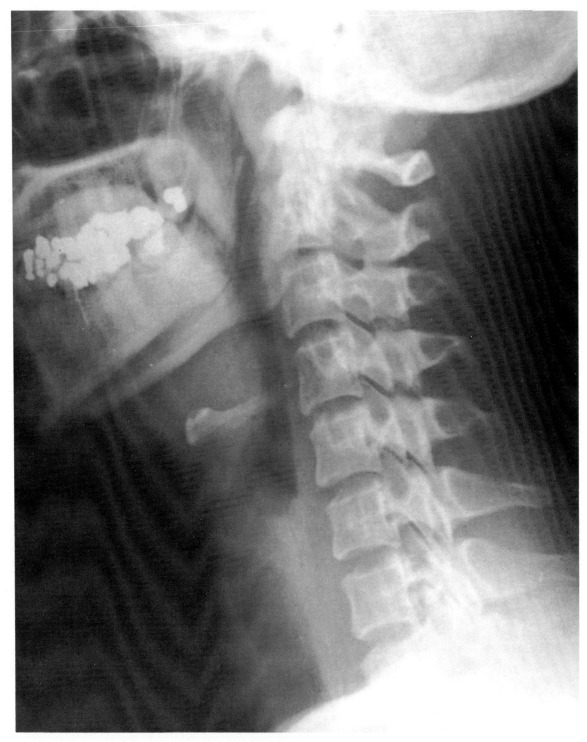

Plate 1. *Lateral view of normal cervical vertebrae. Note direction taken by the articular facets.*

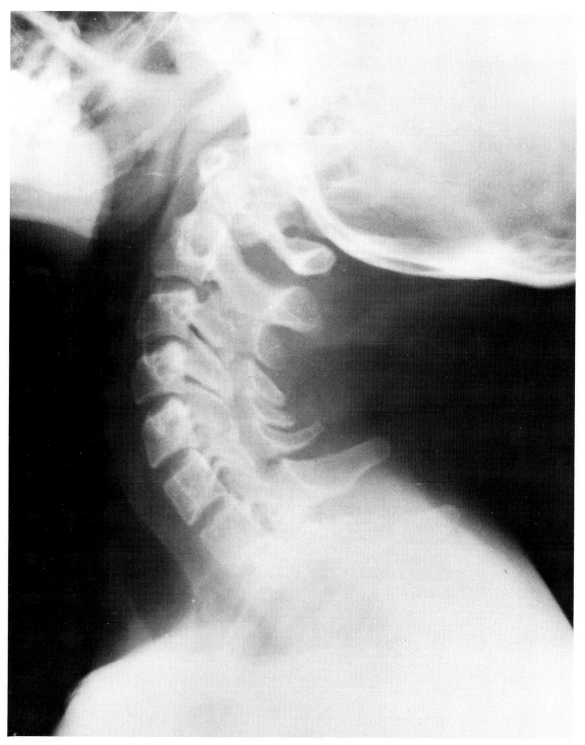

Plate 2. Lateral view of normal cervical vertebrae when head is hyperextended. Note how discs are required to adjust to this movement.

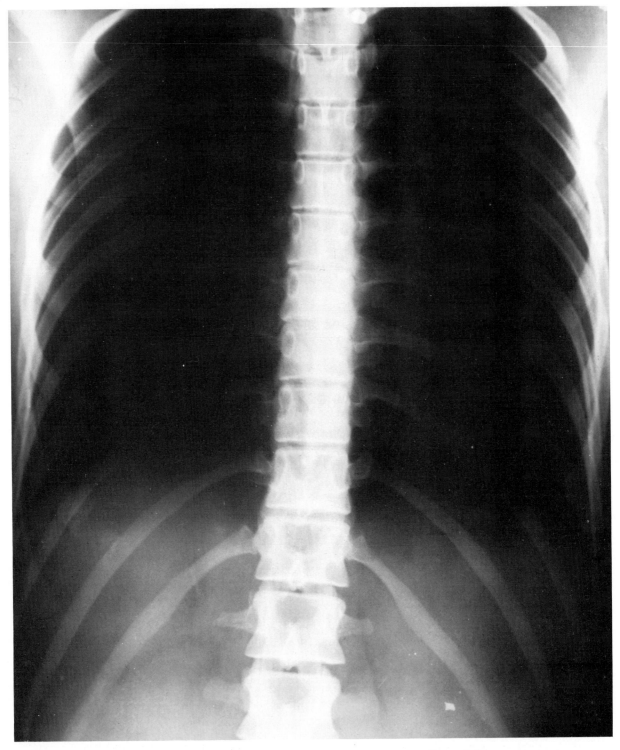

Plate 3. *Normal thoracic vertebral column. Compare to Figure 2-26.*

veins. These plexuses connect with the dural sinuses inside the cranium. This internal network of veins finally drains into **intervertebral spinal veins,** which exit through each intervertebral foramen.

The **external vertebral plexus** (Fig. 2-30) of veins consists of a network of veins on the anterior surface of the bodies of the vertebrae, the **anterior external plexus,** and another winding around the spinous and transverse processes of the vertebrae, the **posterior external plexus.** These external branches also drain into the corresponding intervertebral (spinal) veins. The intervertebral (spinal) veins, in turn, drain into the vertebral veins, intercostal veins, lumbar veins, or lateral sacral veins.

Batson (1940, 1942)* has pointed out the importance of these networks in metastasis of cancer from one region of the body to another. These veins have no valves and form a pathway for cancer cells to spread throughout the body without traversing the lungs.

NERVES

Before each spinal nerve divides into dorsal and ventral rami, it gives off a meningeal nerve or nerves, which return into the vertebral canal usually following blood vessels. They are joined by one or two branches from either a sympathetic ganglion or from the gray ramus communicans. These combined afferent and sympathetic fibers continue to give sensory supply to the dura mater, the periosteum of the vertebrae and the ligaments, and motor supply to the smooth muscles in the blood vessels.

COVERINGS OF SPINAL CORD

The central nervous system, the brain and spinal cord, is covered with three membranes, called **meninges.** The outer layer is a very dense fibrous investment known as the **dura mater** (L., hard + mother); the middle layer, a thin nonvascular connective tissue membrane, is the **arachnoid mater** (Gr., web-like + L., mother); the in-

DURA

When a co
tradural space
covered with fat a
arteries and the pos
plexus of veins just desc
the vertebrae. The nerves t
ligaments and bone course o
mater covering the brain is usuall
of two layers,† the outer of which i
teum on the inside of the cranial bone
magnum, the outer layer is said to be conti
periosteum on the outside of the occipital bo
inner layer continues down over the spinal cord.
inside of the vertebral canal as far inferiorly as the s
sacral vertebra. At this point it tapers to a fine but tot
connective tissue strand called the **filum terminale externum** which continues in the sacral canal and finally attaches to the coccyx (Fig. 2-31). Prolongations of the dura extend over the dorsal and ventral roots of the spinal nerves as separate tubes; these tubal prolongations fuse into one channel as the dorsal root ganglion is reached. Immediately thereafter, the dura fuses with the periosteum around the intervertebral foramina and blends with the connective tissue sheath of each spinal nerve. There are loose strands of connective tissue connecting the dura with the posterior longitudinal ligament; these are more numerous in the cervical and lumbar regions than in the thoracic. Opening of the dural sac reveals the small **subdural space,** that space between the dura and arachnoid that is filled with lymph. This lymph flows into lymphatics around the spinal nerves.

ARACHNOID MATER

The **arachnoid** is a very thin and delicate layer that is directly continuous with that covering the brain; it follows

---

*O. V. Batson, 1940, The function of the vertebral veins and their role in the spread of metastases, *Ann. Surg.* 112: 138.

O. V. Batson, 1942, The role of the vertebral veins in metastatic processes, *Ann. Internal Med.* 16: 38.

---

†L. C. Rogers and E. E. Payne, 1961, The dura mater at the craniovertebral junction, *J. Anat.* 95: 586. Contrary to the usual description, these authors find the dura mater to consist of a single layer in the cranium as well as on the spinal cord. They claim there is no evidence of change of the dura at the foramen magnum.

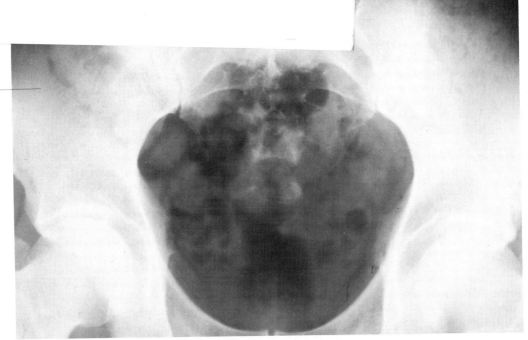

nermost layer, the **pia mater** (L., tender + mother), is a fairly thick vascular connective tissue membrane very ...ntimately attached to the brain or spinal cord.

MATER

...mplete laminectomy is performed, the **ex-**
...is opened. The dura mater can be seen
...d blood vessels. These vessels are the
...rior part of the internal vertebral
...ribed as supplying and draining
... the dura and to surrounding
... these vessels. The dura
... described as consisting
... actually the perios-
... At the foramen
...inuous with the
...e, while the
...t lines the
...econd
...gh

Plate 4.   *Normal lumbar vertebrae. Note large interlaminal spaces. These become even larger if patient flexes the vertebral column. Compare to Figure 2-26.*

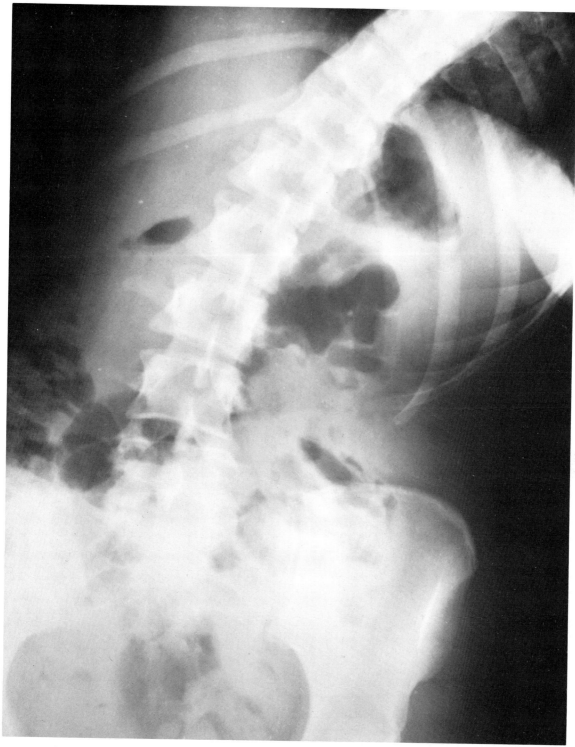

Plate 5. *Normal lumbar vertebral column when trunk is bent laterally. Note how the intervertebral discs are required to adjust.*

the contours of the dura mater throughout the extent of the latter (Fig. 2-31) and sends prolongations over the roots of the spinal nerves as did the dura mater (Fig. 2-30). The space between this layer and the pia mater is the **subarachnoid space** and contains the important **cerebrospinal fluid.** Strands of arachnoid traverse this space to become attached to the pia.* More will be said about this space and fluid after the pia mater and spinal cord have been described.

PIA MATER

The **pia mater** (Figs. 2-31 and 2-32) immediately surrounds the spinal cord and is continuous with the same layer on the brain; it is also continuous over each dorsal and ventral root and blends with the connective tissue sheaths of the spinal nerves. A prolongation of this layer extends laterally from the spinal cord into the subarachnoid space. It does not completely divide the space but has toothlike projections (hence the name **dentate ligament**), which pierce the arachnoid to attach to the dura mater at intervals between the nerve roots. The dentate ligament presents a lunar-shaped free edge between the points of attachment. Because the spinal cord ends at the level between the first and second lumbar vertebrae, the dentate ligament of necessity ends at this point also. The fact that the cord ends at this level also means that this should be the inferior limit of the pia mater, but this layer continues as a thin connective tissue strand, the **filum terminale internum,** to blend with the filum terminale externum, piercing the arachnoid layer to do so.† A thickening of the pia, the **linea splendens,** is

found in the cervical region along the anterior median sulcus, a structure difficult to see from the usual posterior approach to the spinal cord.

SPINAL CORD

The **spinal cord** is directly continuous with the medulla oblongata of the brain stem. It has two **enlargements,** cervical and lumbar, representing the large number of nerve cells required by the upper limbs and lower limbs, respectively. Because the spinal cord does not grow to be as long as the vertebral column, the prolongation of the vertebral column results in an apparent shortening of the cord. It ends at a level between the first and second lumbar vertebrae.‡ In spite of this difference in length between the cord and column, the spinal nerves are not reduced in number. There are thirty-one pairs of nerves, eight of which are cervical, twelve thoracic, five lumbar, five sacral, and one coccygeal. The roots of the nerves do not join the spinal cord as one single root, but in the form of many strands. The area of the cord occupied by strands of a single nerve is a **spinal cord segment.**

In spite of the fact that the cord is not as long as the vertebral column, the roots of the spinal nerves exit from the vertebral canal through intervertebral foramina at the

---

*L. W. Weed, 1917, The development of the cerebrospinal spaces in pig and in man, *Carnegie Contrib. Embryol.* 5: 3. Is there such an entity as a subarachnoid space? Would it be more accurate to combine pia and arachnoid maters into a pia-arachnoid?

†G. L. Streeter, 1919, Factors involved in the formation of the filum terminale, *Am. J. Anat.* 25: 1. This report describes the difference in formation of filum terminale externum from that of filum terminale internum.

---

‡J. H. Needles, 1935, The caudal level of termination of the spinal cord in American whites and American negroes, *Anat. Record* 63: 417. A survey based on 240 cadavers. The termination of the spinal cord "between L1 and L2" is confirmed, 55 per cent ending below the disc, 45 per cent above. More important is the finding that the lowest level was the lower third of the third lumbar vertebra, and that the cord extends further caudally in females than in males.

A. F. Reimann, and B. J. Anson, 1944, Vertebral level of termination of the spinal cord with report of a case of sacral cord, *Anat. Record* 88: 127. Termination "between L1 and L2" still is a safe statement, but this is simply the mean and mode—most fashionable level. Range between upper third of T12 and lower third of L3. The case of the sacral cord, indicated in the title, should be noted.

Figure 2-31. *The distribution of the meninges in the lumbosacral region, showing the dorsal and ventral roots (cauda equina) with their coverings of pia streaming inferiorly to emerge from the proper intervertebral foramen. The enlargement on the left shows a dorsal and a ventral root emerging through the arachnoid and dura via separate openings. The enlargement on the right has been turned to show the distribution of the meninges on the dorsal and ventral roots and how they blend to form the nerve sheath.*

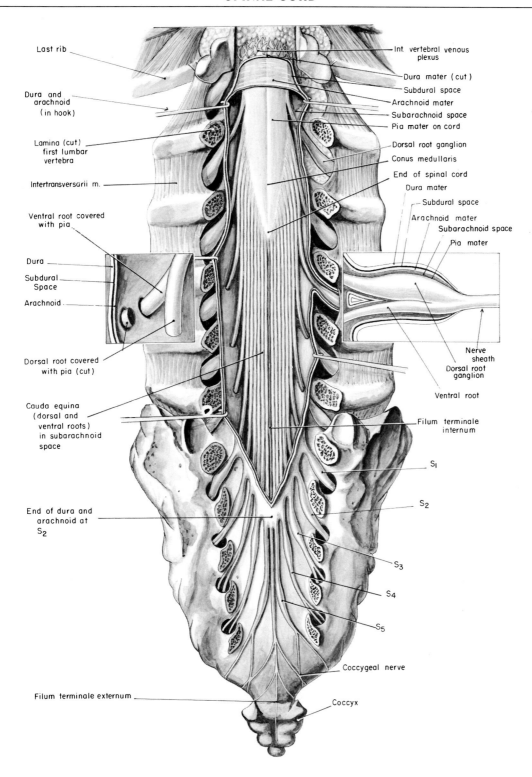

Last rib

Int. vertebral venous plexus

Dura mater (cut)

Subdural space

Dura and arachnoid (in hook)

Arachnoid mater

Subarachnoid space

Pia mater on cord

Lamina (cut) first lumbar vertebra

Dorsal root ganglion

Conus medullaris

End of spinal cord

Intertransversarii m.

Dura mater

Subdural space

Arachnoid mater

Subarachnoid space

Ventral root covered with pia

Pia mater

Dura

Subdural Space

Arachnoid

Nerve sheath

Dorsal root ganglion

Dorsal root covered with pia (cut)

Ventral root

Cauda equina (dorsal and ventral roots) in subarachnoid space

Filum terminale internum

S₁

S₂

End of dura and arachnoid at S₂

S₃

S₄

S₅

Coccygeal nerve

Filum terminale externum

Coccyx

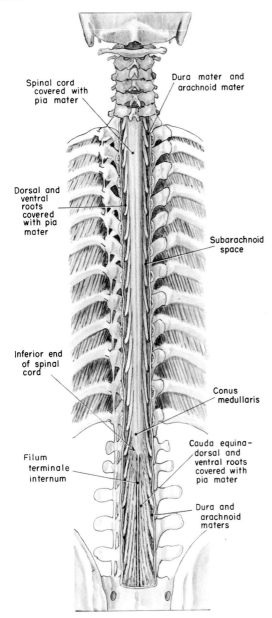

Spinal cord covered with pia mater

Dura mater and arachnoid mater

Dorsal and ventral roots covered with pia mater

Subarachnoid space

Inferior end of spinal cord

Conus medullaris

Filum terminale internum

Cauda equina—dorsal and ventral roots covered with pia mater

Dura and arachnoid maters

Figure 2-32. *The spinal cord in situ after a laminectomy has been performed and the dura and arachnoid maters cut.*

proper vertebral level. This makes it necessary for the roots gradually to become longer and longer as the inferior end of the cord is approached; and since the cord ends at a level between the first and second lumbar vertebrae, this means that the roots of the lower lumbar, sacral and coccygeal nerves are required to descend for some distance to reach their proper intervertebral foramina. This results in a large number of nerve roots, each covered with pia mater, descending through the subarachnoid space in such a manner as to resemble a horse's tail, thus accounting for the name **cauda equina** for these nerve roots. In spite of the fact that the roots gradually become longer as we go inferiorly on the cord, each dorsal root ganglion is in its proper intervertebral foramen. Actually, the cervical ganglia tend to rest at the distal end of their foramina, and the sacral ganglia rest inside the sacral canal.

The first cervical nerve emerges between the skull and the atlas, the second between the atlas and axis, and so on; this means that the cervical nerves emerge above their proper vertebra. As there are eight cervical nerves and only seven cervical vertebrae, a change takes place between the cervical and thoracic vertebral levels. Cervical nerve 8 emerges between the seventh cervical and first thoracic vertebrae; therefore, this nerve and all subsequent nerves emerge just inferior to the proper vertebra. The ganglia, accordingly, are located in these same spaces.

It is often of importance to determine at what vertebral level any one spinal cord segment might be; it is obvious that the segments will not correspond to the proper vertebral levels because the segments do not extend throughout the entire length of the column. The approximate vertebral levels of the spinal cord segments (Fig. 2-33) can be determined by adding one to the cord segments in the lower cervical region, two in the upper thoracic and three in the lower thoracic regions. This would mean that deep to the seventh cervical spine we would find the eighth cervical segment; deep to the sixth thoracic spine, the eighth thoracic segment; and deep to the tenth thoracic spine the first lumbar segment. Since the cord ends at the base of the first lumbar vertebra, all remaining segments would be found deep to the eleventh and twelfth thoracic and first lumbar vertebrae. Clinically, this knowledge is more frequently used in a reverse manner. An area of the skin innervated by a single spinal cord segment is called a **dermatome** (these are diagrammed in

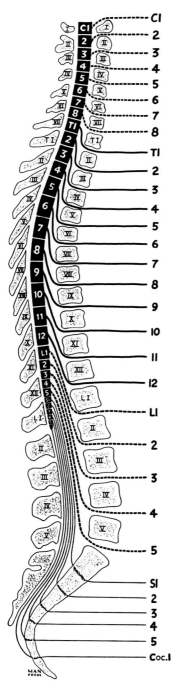

Figure 2-33. *The alignment of spinal segments with vertebrae. (From Haymaker and Woodhall,* Peripheral Nerve Injuries, *courtesy of W. B. Saunders Co.)*

Figure 3-46). If, from cutaneous symptoms and a knowledge of dermatomes, trouble were diagnosed in a certain spinal cord segment, the vertebral level would have to be determined if surgery were necessary. In this case the vertebral level would be found by subtracting instead of adding. For example, if the twelfth thoracic spinal cord segment was involved, this would be found deep to the ninth thoracic vertebra.

Recall that a **dermatome** is different, in most cases, from the distribution of any spinal nerve, for most nerves intermingle with one another to form plexuses before arriving at their destination; therefore, any one nerve may have contributions from several segments of the spinal cord.

SUBARACHNOID SPACE AND LUMBAR PUNCTURE

It is easily seen that the **subarachnoid space,** containing cerebrospinal fluid, is very much enlarged inferiorly because of the absence of the spinal cord. From the second lumbar vertebra to the level of the second sacral vertebra this large space contains the **cauda equina,** in the center of which is the thin **filum terminale internum.** Clinical advantage has been taken of this structural arrangement, in conjunction with the fact that the **interlaminal spaces** in the lumbar region are large. A needle can be inserted just lateral to the supraspinal ligament in the lumbar region and enter the subarachnoid space after piercing the skin, superficial and deep fascial layers, the tendons of the erector spinae and multifidus muscles, ligamentum flavum, the dura, and arachnoid. This is called a **lumbar puncture.** Cerebrospinal fluid can be removed for examination or drugs given for spinal anesthesia. The nerve roots are not pierced by the needle because of the almost impossible task of entering a nerve in such a fluid medium. Lumbar puncture should not be done in regions where the spinal cord is present; perhaps it is fortunate that the overlapping nature of the spinous processes of the thoracic vertebrae makes it very difficult to enter the subarachnoid space at this level.

The cerebrospinal fluid is protective in nature, the

spinal cord being suspended in the middle of this fluid medium. The cord is held in this position by the many nerve roots and the filum terminale internum holding it inferiorly, the attachment to the brain stem holding it superiorly, and the dentate ligaments holding it laterally.

spinal cord (Fig. 2-30) also form longitudinally running channels, three of which are on the anterior surface and three on the posterior surface of the cord. They intermingle profusely and finally drain into the intervertebral veins along with the internal vertebral plexus.

BLOOD VESSELS OF SPINAL CORD

After the vertebral arteries have traversed the foramen magnum, and before they join to form the basilar artery, they give off branches that descend on the spinal cord (Fig. 2-30). The two **anterior spinal arteries** join into one artery anteriorly, but the two **posterior spinal arteries** take independent pathways. The anterior spinal artery is found in the anterior median sulcus of the spinal cord, while the two posterior spinal arteries intertwine around the dorsal rootlets. These arteries would decrease to a very small and inadequate size on the inferior part of the cord if it were not for contributions from branches of the spinal arteries that enter each intervertebral foramen. After sending branches to the vertebrae, these spinal arteries (from vertebral, intercostal, lumbar, iliolumbar, and lateral sacral arteries) give rise to **neural branches** which, in turn, divide into **radicular arteries,** one for each root. These arteries vary in size, and although all nourish the roots, not all reach the anterior spinal artery of the spinal cord. Those that do are larger than the rest and are located at variable levels, being more numerous at the levels of the spinal cord enlargements; one rather large artery occurs in the lower thoracic or upper lumbar region, usually on the left side. These larger radicular arteries* anastomose with the anterior and posterior spinal arteries. Branches from all these arteries to the spinal cord ramify in the pia mater and a prolongation of the pia follows each artery into the cord tissue. The **veins of the**

---

*L. A. Gillilan, 1958, The arterial blood supply to the human spinal cord, *J. Comp. Neurol.* 110: 75. This worker suggests that these large radicular arteries be called ''medullary'' arteries, to be differentiated from those that do not participate in the blood supply to the spinal cord. This work also warns against injury to these large blood vessels of the cord, because the collateral circulation is inadequate.

SYSTEMIC SUMMARY

Having finished the dissection of the back, the student should be able to visualize this region, layer for layer, from the skin to the vertebral column. For example, looking at the back of a person's neck, one should be able to see (from superficial to deep) the skin, superficial and deep fascia, trapezius muscle, splenius muscles, semispinalis capitis muscle, the semispinalis cervicis, and muscles forming the suboccipital triangle, and then the vertebral column. In the upper part of the back the order of structures is skin, fascial layers, trapezius muscle, rhomboid muscles, serratus posterior superior muscle, inferior end of splenius muscles, erector spinae muscles, semispinalis muscles, multifidus muscle, the long and short rotatores muscles, and the vertebral column. The same idea in the lumbar region has already been described when lumbar puncture was discussed. It should not be too difficult to insert arteries and nerves at their proper locations between these layers.

After dissection of an area of the body, it is important for the student to put these regions together—in other words, to review the area from the systemic viewpoint.

BONES AND JOINTS

Now that you are familiar with the anatomy of the back, a rereading of the osseous structures is recommended; the structures should now have more meaning.

Although small portions of the occipital and temporal bones, the scapula and clavicle, and ribs and ilium are involved in muscle attachments, by far the most important osseous structure in the back is the **vertebral column.** These thirty-three bones, fused in sacral and coccygeal regions, give support for the entire upper part of the body, serve as attachments for the ribs, support the skull, and protect the very vital spinal cord. The forerunner of the vertebral column is the notochord, which develops from the deeper cells of the primitive streak and comes to lie between the gut entoderm and the ectodermal neural tube. Cells break away from the medial side of two contiguous mesodermal somites and migrate to surround this notochord. These cells, from two somites, form the centrum of the vertebra, the notochord breaking up into clumps that finally form the nucleus pulposus of the intervertebral discs. Mesenchymal extensions (Fig. 2-34) grow dorsally from the centrum (neural arch primordia) and finally surround the neural tube (future spinal cord). Failure to meet in the midline results in spina bifida. Other extensions grow out ventrally to form the rib primordia. All of these primordia (centrum, neural arch, and ribs) change to cartilage to form a miniature vertebra and ribs. Ossification centers occur in the centrum, in each wing of the neural arch, and in each rib. In addition to spina bifida, other developmental anomalies occur. When this involves a situation where a vertebra is present on one side only, complications inevitably arise.

Although the ribs occur only in the thoracic region, each vertebra has a **costal component,** as indicated in Figure 2-35. It is obvious that the anterior tubercles on the transverse processes of the cervical vertebrae, the transverse processes of the lumbar vertebrae, and the alar portions of the sacrum correspond to ribs in the thoracic region. (Naturally, these parts will have separate ossification centers.) Therefore, mamillary and articular pro-

cesses in the lumbar vertebrae, transverse processes in the thoracic vertebrae, and articular processes in the cervical vertebrae are neural arch derivations.

It is interesting to note how **muscle attachments and muscle innervation reveal this past history** of various parts of the vertebral column. The origin of the multifidus muscle is a good example. It does not arise from all the transverse processes but from the transverse processes in the thoracic region only. The attachments in the lumbar region are on the mamillary processes and in the cervical region to the articular processes. This muscle, therefore, has attachments indicating it is a muscle of the back proper; it is innervated by dorsal rami as might be expected. The intertransversales muscles are of equal interest. In the lumbar region the medial set actually are attached to the mamillary and articular processes, while the lateral sets are situated between the transverse processes. If these bony relationships are true, one would expect that the nerve supply to the medial set of intertransversales muscles in the lumbar region would be by dorsal rami of the spinal nerves, as has been found to be true of the other muscles in the back proper. The nerve supply to the lateral set of intertransversales muscles, because the transverse processes represent ribs, should be typical of that to the intercostal muscles in the thoracic region, namely, ventral rami. This has been found to be true.

The vertebrae are held together by the presence of **discs and ligaments.** The semifluid discs between

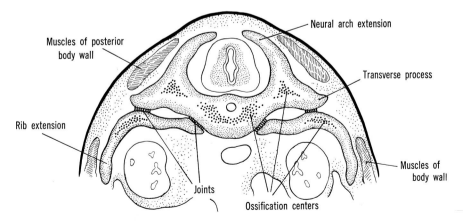

Figure 2-34. *Transverse section of an embryo showing the developing vertebrae and ribs. (From Davies,* Human Developmental Anatomy. *Copyright © 1963, John Wiley & Sons, Inc. After J. E. Frazer,* Manual of Human Embryology.*)*

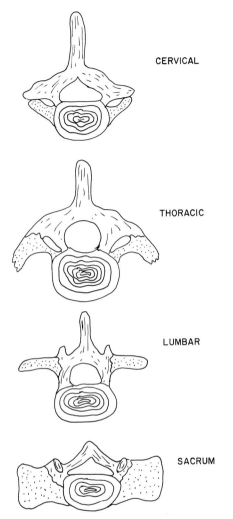

CERVICAL

THORACIC

LUMBAR

SACRUM

Figure 2-35. *Neural arch (line shading) and costal (stippling) components of the vertebrae. (Modified after Patten.)*

there is considerable movement in the column as a whole. The type of motion in any particular region is determined primarily by the direction of the articular facets. Real freedom of movement occurs in only one place in this column, at the atlantoaxial joint, where rotation of the head on the vertebral column occurs.

MUSCLES

No motion would be allowed if it were not for the many muscles in this region. These muscles can be divided into three main groups: (1) those moving the upper limb, (2) those concerned with respiration, and (3) those involved with the vertebral column and skull (epaxial trunk muscles, Fig. 2-36).

The first group consists of the trapezius, rhomboids, levator scapulae, serratus anterior and latissimus dorsi muscles. Except for the latissimus dorsi, which extends and medially rotates at the shoulder joint, they all act on the scapula. This bone is elevated by the upper part of the trapezius and levator scapulae muscles, depressed by the lower fibers of the trapezius plus gravity, adducted by the trapezius and rhomboids, abducted by the serratus anterior, medially rotated by the rhomboids, and laterally rotated by the trapezius and serratus anterior. Other muscles, to be studied at a later time, will be seen to aid in some of these movements.

The muscles concerned with respiration are the two serratus posterior muscles, and the levatores costarum muscles. The ribs are elevated by the serratus posterior superior and the levatores muscles, and depressed by the serratus posterior inferior muscles. The external intercostals can also be seen in the back and play an important role in elevation of the ribs.

The remaining muscles are those of the back proper and are concerned with extension and rotation of the vertebral column and head as well as with maintaining an erect posture. When these muscles on both sides of the body contract in unison, the vertebral column and head are extended directly posteriorly; however, when muscles on one side contract, the body and head are rotated. As far as rotation of the vertebral column is concerned, the erector spinae muscle, being attached mainly to ribs and transverse processes, will rotate the body toward the same side; the transversospinal musculature, originating from transverse processes and inserting on spinous processes, will pull those spinous processes toward the trans-

the bodies of the vertebrae protect us from severe vibrations associated with modern sidewalks and floors. The ligaments—supraspinal, interspinal, flavum, intertransverse, posterior longitudinal, and anterior longitudinal —aid the discs in holding these bones together, and with the neural arches form a canal in which the spinal cord is lodged. Motion between one vertebra and another is limited by the discs already mentioned, by the ligaments, by attachments of the ribs, and by the structural makeup of the vertebrae. However, when the slight movement between the vertebrae is accumulated,

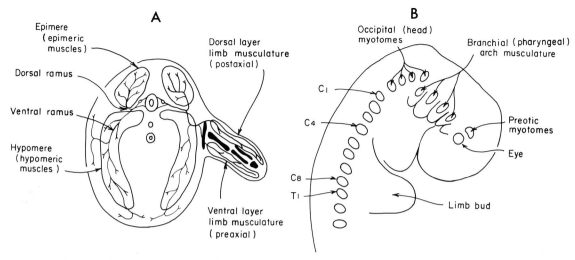

Figure 2-36. (A) Cross section of an embryo showing the division of trunk musculature into epimeres and hypomeres, and limb musculature into dorsal muscles (postaxial or posterior to axis of limb) and ventral muscles (preaxial or anterior to axis of limb). (B) Myotomes and muscle primordia in the head and body of an embryo in the seventh week.

verse processes, thus rotating the body toward the opposite side. (We will see later that the vertebral column is flexed or bent laterally mainly by the abdominal musculature.) If the splenius capitis, semispinalis capitis, longissimus capitis, rectus capitis posterior minor muscles, and the muscles forming the suboccipital triangle all contract on both sides of the body, the head will be extended directly posteriorly. This massive musculature is important in holding the head erect. If the obliquus capitis inferior contracts on one side, the head is turned toward the same side, and we will see later that this muscle aids the large muscle on the side of the neck, the sternocleidomastoid on the opposite side of the body, in performing this act. The remaining muscles, if contracted on one side, also rotate the head to the same side, but will extend and bend it laterally at the same time.

NERVES

The vertebral column contains a vital part of the central nervous system, the **spinal cord.** Shorter than its surrounding vertebral column, it is suspended in cerebrospinal fluid and protected by three meninges: dura, arachnoid, and pia maters. There are thirty-one pairs of **dorsal** and **ventral roots** attached to it, one set for each

segment of the cord. These roots leave the vertebral canal through intervertebral foramina above the vertebra of the same number in the cervical region and below its respective vertebra inferior to the eighth cervical nerve. Because the cord is shorter than the vertebral column and the nerves emerge at their proper vertebral level, the roots are required to stream inferiorly, forming the cauda equina. The dorsal roots, containing sensory or afferent fibers, exhibit ganglia all of which are located in or close to their respective intervertebral foramina. The ventral roots are made up of motor or efferent fibers. These two roots join to form a segmental or spinal nerve. The first branch of a spinal nerve is a **meningeal branch** that returns through the intervertebral foramen to give sensory innervation to the dura and surrounding periosteum and ligaments. One or more small branches from a sympathetic ganglion or a gray ramus communicans join these meningeal nerves to provide motor supply to blood vessels that these nerves follow to reach their destination.

Each spinal nerve almost immediately divides into ventral and dorsal rami. The ventral rami form the cervical and brachial plexuses, intercostal nerves, lumbosacral plexus, and coccygeal nerve and, except for a few branches, will be studied at a later time. The dorsal rami divide into medial and lateral branches, both of which innervate the muscles of the back proper. In the upper part of the body the lateral branches remain in the muscles, but the medial branches continue through the musculature to become cutaneous nerves; in the lower part of the body the reverse arrangement is found. It should be remembered that not all dorsal rami of the spinal nerves have identical distributions. That of C1 is motor only; C2 forms the greater occipital nerve (sensory to the scalp); C3 also reaches the scalp near the external occipital protuberance; C7 and 8, and L4 and 5, are not respresented on the skin at all; L1, 2, and 3 form the superior cluneal nerves, which ramify on the buttock; S1, 2, and 3 form the middle cluneal nerves; and the coccygeal nerve has a cutaneous distribution near the tip of the coccyx.

As just mentioned, dorsal rami are concerned with the musculature of the back proper (epaxial muscles); in contrast, the respiratory muscles, those attached to costal elements of the vertebrae, and the appendicular muscles which happen to be located in the back are innervated by ventral rami (hypaxial muscles). There are two exceptions to this rule: namely, the levatores costarum muscles are definitely concerned with respiration and apparently are innervated by dorsal rami, and the trapezius muscle has a cranial nerve involved with its innervation in addition to ventral rami. The latter exception is explained by the fact that the trapezius muscle is derived from an area caudal to but associated with branchial arches, and such muscles are innervated by the spinal accessory nerve; the former exception would indicate that the levatores costarum muscles may be epaxial in origin.

A typical dorsal ramus of a spinal nerve contains (1) general somatic afferent fibers for pain, temperature, light touch, and proprioceptive impulses; (2) general somatic efferent fibers to skeletal muscles; (3) general visceral afferent fibers from blood vessels; and (4) general visceral efferent fibers (the autonomic nervous system) to smooth muscle in blood vessels, the arrector pili muscles, and sweat glands. Afferent impulses from nerve endings in the skin or elsewhere are carried toward the spinal cord over these dorsal rami to the spinal nerve and thence to a dorsal root ganglion. No synapse occurs here, for this is simply the location of the cell bodies of these afferent neurons. The impulses continue to the spinal cord. In the dorsal gray matter this nerve cell and its processes just described joins with a neuron (connector neuron), which will connect it with a nerve cell in the ventral gray matter, a motor neuron. The impulse continues out a ventral root

to the spinal nerve, to the dorsal ramus again, and thence to a skeletal muscle; if the stimulus has been noxious in character, movement can result by this pathway, called a reflex arc. General visceral afferent and efferent fibers will be described at a later time.

ARTERIES

The arteries to the back arise from the vertebral, intercostals, the lumbar and iliolumbar, and lateral sacral arteries. A **typical segmental artery** has branching similar to that of a segmental nerve; dorsal branches are given off from the above arteries all the way from the skull to the sacrum. The first branch from these dorsal arteries is the spinal artery. This enters an intervertebral foramen and divides into three branches, two of which supply blood to the vertebral column, the third (neural) dividing into radicular branches for each nerve root to supply blood to the spinal cord. Each dorsal branch then continues in company with a segmental nerve to supply blood to the back musculature and skin.

The **vertebral artery,** the first branch of the subclavian artery, enters the foramen on the transverse process of the sixth cervical vertebra, continues through the foramina in the upper five cervical vertebrae, turns medially between the arch of the atlas and the occipital bone, pierces the dura, and goes through the foramen magnum. This artery gives off branches throughout its extent in the neck and gives off the anterior and posterior spinal arteries after traversing the foramen.

Another branch of the subclavian is involved with structures in the back. The **transverse cervical artery** is actually a branch of the thyrocervical trunk of the subclavian; this artery crosses the base of the neck to gain access to the back, divides into a superficial branch, found on the deep side of the trapezius muscle, and a deep branch, which ramifies on the deep side of the rhomboid muscles. This deep branch often arises independently from the third part of the subclavian, in which case it is called the descending scapular artery, a situation that leaves the superficial branch as an equally independent branch called the superficial cervical artery.

VEINS

It has already been seen that veins form a plexus or network rather than exhibiting a single channel, as is usually the case with arteries. The veins in the back follow the arteries in all cases except around the vertebral column. The important **external** and **internal vertebral plexuses,** with their many connections into the cranial, thoracic, and abdominopelvic cavities, should not be forgotten.

LYMPHATICS

Except for lymph nodes in the occipital and postauricular regions, which drain the scalp, there are no major lymph nodes in the back. The superficial structures on the upper part of the trunk drain into lymph nodes in the axilla, while those on the lower part of the body drain into the inguinal nodes. The deeper structures in the back undoubtedly follow the intercostal and lumbar vessels to nodes in the cavities.

CONCEPTS: GENERAL RULES ON STRUCTURE OF THE HUMAN BODY*

Those students who started with and completed a study of the back have been exposed to several concepts or general rules about the human body that should be very helpful in learning the remaining parts.

From the arrangement of the vertebrae, the nerves, and the blood vessels, it is quite obvious that the human being has been derived from, and is, a segmentally arranged animal. We have seen that the skin does indeed serve as an efficient covering of the body; that it has a segmentally arranged sensory nerve supply, a dermatome being an area innervated by one segment of the spinal cord; that the skin contains arrector pili muscles, smooth muscles in blood vessels, and sweat glands that are innervated by the sympathetic portion of the autonomic nervous system, and, therefore, the so-called

*This section is similar to those presented in other chapters of this textbook, but contains special features that relate to the back.

"sensory nerves" to the skin are in reality combined sensory and motor nerves and might better be called "cutaneous nerves"; that the skin is bound down to underlying structures by a layer of superficial fascia containing a varying amount of fat in which these cutaneous nerves course; that these nerves are usually accompanied by an artery and a vein, forming a neurovascular triad.

We have found that many muscles, in their development, migrate to other parts of the body, but carry their previously established nerve supply along with them; that this nerve supply, and the vascular supply as well, usually is found on the deep side of the muscle; that the nerve supply is dependent upon the embryological origin of the muscle; that muscles attach to the osseous elements and by their pull not only cause movement of these osseous structures, and thereby the body, but also cause elevations to develop on these bones; that the main action of a muscle can usually be determined by a simple mechanical sense if the attachments of the muscle are known, the more movable attachment (the insertion) usually being more revealing; that each muscle is surrounded by a layer of deep fascia, and that groups of muscles are often surrounded by a thicker enveloping fascial layer.

We have found also that arteries, in addition to being found on the deep surface of muscles, being segmentally arranged and being accompanied by a vein and a nerve, exhibit frequent variations in pattern on one side of the body or on both sides, and that textbook descriptions are based on that arrangement found in the majority of bodies; that anastomoses of arteries provide important sidetracks around an occluded main channel, a fact made possible by the absence of valves in arteries; that veins are even more variable, often form intricate plexuses, and frequently exhibit multiple (venae comitantes) rather than single channels.

Differentiation between the lymphatic drainage of the skin and the deeper structures of the back has revealed the important fact that lymphatic drainage seems to follow, in a retrograde manner, the vascular pathways.

Study of the spinal cord has revealed that this vital structure not only is protected by bony structures, but is "mothered" by protective meninges and the cerebrospinal fluid; that the second and fourth components of a reflex arc (the afferent and efferent neurons) do indeed reach and leave the spinal cord by dorsal and ventral roots, respectively, and that these roots are attached to each segment of the cord; that the roots join together to form a spinal or segmental nerve, which immediately divides into dorsal and ventral rami for corresponding areas of the body.

The vertebral column has obviously had to adapt itself to an upright stature, doing so by developing secondary cervical and lumbar curvatures. The greater mass of the lumbar vertebrae, compared to those more superiorly located, is typical of constructions whether of the human body or of a building.

Furthermore, the spinal cord being shorter than the vertebral column is a good example of differential growth.

Lastly, I hope you are convinced that the parts of the back you remember best are those you can easily visualize and those you have thought about enough to understand.

3

THORACIC AND ABDOMINAL WALLS

In contrast to the back, medicine has given considerable attention to the thoracic and abdominal walls. It is a vital area clinically; the field of surgery has penetrated these walls for many years; and specialists in internal medicine are constantly relating the walls to structures inside the cavities.

If you have already studied the back, learning of the thoracic and abdominal walls at this time will bring the entire trunk of the body into focus.

BONY STRUCTURES

Once again, some knowledge of the bony structures involved in the area is mandatory if the following descriptions are to be understood.

CLAVICLE

This is an important bone (Fig. 3-1) which can be felt along its entire length. As mentioned previously, it is fractured frequently. It articulates with the sternum medially, and this is the only bony attachment of the upper limb to the trunk; it articulates with the acromion of the scapula laterally. It is responsible for keeping the scapula and shoulder joint in a lateral position away from the important axillary structures. It has a curved form—convex anteriorly at its medial two-thirds and concave anteriorly at its lateral third. The clavicle possesses a **shaft,** and **sternal** and **acromial ends.** The sternal end is rounded and smooth where it articulates with the sternum; the acromial end is rough superiorly but smooth where it articulates with the acromion process of the scapula. The shaft of the clavicle possesses **four surfaces:** superior, inferior, anterior, and posterior. The **superior surface** is smooth throughout the extent of the shaft. The **inferior surface** exhibits a roughened area at the sternal end for attachment of the costoclavicular ligament, a smooth area in the medial third where the subclavius muscle inserts, and a rough area on the lateral third which is the conoid tubercle and trapezoid ridge for attachment of ligaments. The **anterior** and **posterior surfaces** exhibit no particular markings. The lateral third of this bone, being flattened, is said to have superior and inferior surfaces, but **anterior** and **posterior borders** rather than

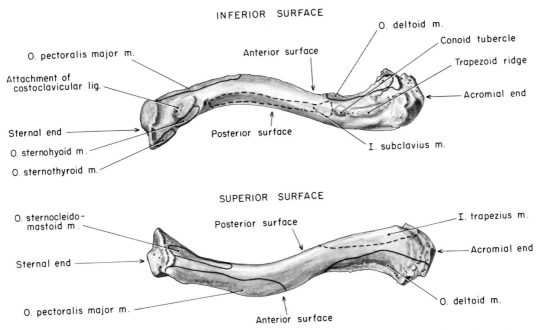

INFERIOR SURFACE

O. pectoralis major m.

Attachment of costoclavicular lig.

Sternal end

O. sternohyoid m.

O. sternothyroid m.

Anterior surface

O. deltoid m.

Conoid tubercle

Trapezoid ridge

Acromial end

Posterior surface

I. subclavius m.

SUPERIOR SURFACE

O. sternocleido-mastoid m.

Sternal end

O. pectoralis major m.

Posterior surface

I. trapezius m.

Acromial end

O. deltoid m.

Anterior surface

Figure 3-1. *Inferior and superior surfaces of the left clavicle, with origins and insertions of muscles outlined.*

surfaces. See Figure 3-1 for the attachment of muscles. Some of these areas are roughened, the degree varying from clavicle to clavicle.

STERNUM

The sternum (Fig. 3-2) is made up of three parts, namely, the **manubrium, body,** and **xiphoid process.**

The **manubrium** is wider than the body and possesses anterior and posterior surfaces, a **suprasternal notch,** and **clavicular notches** for articulation with the clavicles. The first costal cartilages attach to the manubrium just inferior to the clavicular notches. The second costal cartilage attaches partly to the inferior corner of the manubrium and partly to the superior surface of the body of the sternum; there is, therefore, a notch at this point. The point at which manubrium and body join is the **sternal angle;** a line projected directly posteriorly from this point hits on the base of the fourth thoracic vertebra.

The **body** is made up of four parts, often called **sternabrae.** It is two to three times wider than it is thick, and

presents **facets** on its lateral sides for articulation with the second to seventh costal cartilages. The seventh cartilage articulates with both the body and xiphoid process.

The **xiphoid process** is a bone of many shapes and sizes that is the inferior extent of the sternum. The point of attachment of the xiphoid process and the body is the **xiphisternal junction.**

RIBS

There are twelve pairs of ribs (Fig. 3-3) which not only protect vital organs in the thoracic cavity but many in the abdominal cavity as well. The first seven pairs are called **true ribs** because they articulate directly with the sternum; the remaining five pairs are called **false ribs.** Of these false ribs the first three (the eighth, ninth, and tenth) articulate with the ribs above while the last two articulate with nothing anteriorly and are therefore designated as **floating ribs.** The ribs vary greatly in length.

The third to the ninth ribs are typical ribs and will be described first (Fig. 2-8). The **head** of a typical rib has two

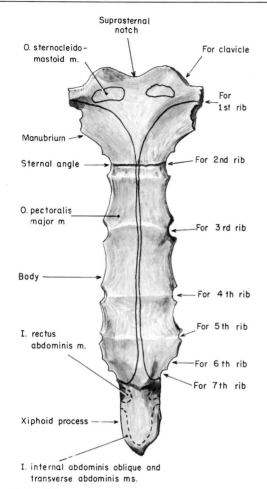

Suprasternal notch

O. sternocleido-mastoid m.

For clavicle

For 1st rib

Manubrium

For 2nd rib

Sternal angle

O. pectoralis major m

For 3rd rib

Body

For 4th rib

For 5th rib

I. rectus abdominis m.

For 6th rib

For 7th rib

Xiphoid process

I. internal abdominis oblique and transverse abdominis ms.

Figure 3-2. *The anterior surface of the sternum, with origins and insertions of muscles outlined.*

facets for attachment to two vertebrae; a crest is found between these two facets. The **neck** is about 1 inch in length. Lateral to the neck a **tubercle** is found, which is rough superiorly for attachments of muscles and ligaments and smooth inferiorly for articulation with the transverse process of a vertebra. The roughened area 2 or 3 inches lateral to the tubercles is the **angle** of the rib; this roughened area is the point of attachment for the iliocostalis muscles. The **shaft** of the rib not only curves to form the lateral wall of the thorax but twists as well. It presents a smooth outer surface which, in the body, is covered by muscles; a smooth inner surface to which is

attached the pleura; a smooth superior border; and a grooved inferior border. Intercostal vessels and nerves course in this groove (note Fig. 3-23). The **anterior end** of the rib presents a small cup-shaped depression for articulation with the cartilaginous portion of the rib.

The **first rib** is very short but wider than the others. On its superior surface are found the scalene tubercle for insertion of the scalenus anterior muscle (a muscle in the neck), a groove anterior to the tubercle for the subclavian vein, and another more pronounced groove posterior for the subclavian artery. It is therefore obvious that the anterior scalene muscle separates the subclavian artery and vein as these vessels leave the thoracic cavity and neck to reach the axilla (Fig. 7-91 on page 576). Its **head** has only one facet, for it articulates with the first thoracic vertebra only, and the **tubercle** is large. There is no costal groove. The **second rib** presents a roughened area about midway from each end for attachment of part of the serratus anterior muscle (Fig. 3-5). The **tenth rib** usually has only one facet on the head, and usually has a facet on the tubercle. The **eleventh rib** is quite short and has only one facet on the head; the other markings are indistinct. The **twelfth rib** is equally short and has one facet and no other distinct markings.

COXAL BONE (IN PART)

Portions of the coxal bone (Fig. 3-4) should be studied at this time. The **iliac crest** starts anteriorly as a bony prominence, the **anterior superior iliac spine,** and continues posteriorly and medially to end as the **posterior superior iliac spine.** The crest between these points is roughened for attachments of muscles presenting **outer and inner lips** and an **intermediate line.** Another part of this bone involved with a description of the abdominal wall is the pubic portion. The **pubic bone** articulates in the midline with its mate on the opposite side of the body, forming the **pubic symphysis.** The portion taking part in this articulation is the **body;** two arches extend laterally from the body as **superior** and **inferior pubic rami.** The superior ramus presents a **pubic tubercle** about 1 inch lateral to the midline and a **pubic crest** between the tubercle and the midline. The **pectineal line** is a ridge extending laterally from the pubic tubercle.

A more detailed description of the coxal bone is given on page 348.

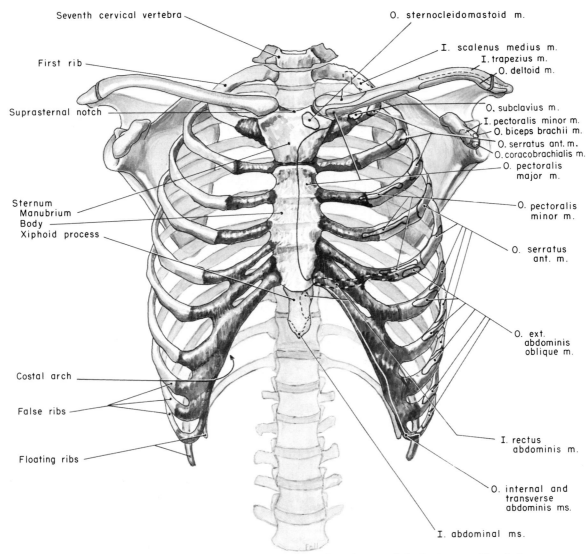

Seventh cervical vertebra

First rib

Suprasternal notch

Sternum
Manubrium
Body
Xiphoid process

Costal arch

False ribs

Floating ribs

O. sternocleidomastoid m.

I. scalenus medius m.
I. trapezius m.
O. deltoid m.

O. subclavius m.
I. pectoralis minor m.
O. biceps brachii m.
O. serratus ant. m.
O. coracobrachialis m.
O. pectoralis
major m.

O. pectoralis
minor m.

O. serratus
ant. m.

O. ext.
abdominis
oblique m.

I. rectus
abdominis m.

O. internal and
transverse
abdominis ms.

I. abdominal ms.

Figure 3-3. *The thoracic cage, showing manner of attachment of ribs to sternum. The darker shaded areas are the cartilaginous portions of the ribs. Origins and insertions of muscles are outlined.*

SURFACE ANATOMY

Many of the structures close to the surface can be seen (Fig. 3-5) by looking at the skin of a lean male body. Starting superiorly, in the midline one sees the **supra-** sternal notch. Continuing in the midline, a groove marking the **sternum** is obvious; this ends as the xiphisternal joint. From this point to the **symphysis pubis** is a median groove marking the **linea alba,** the midline of the abdominal musculature medial to the two **rectus abdominis muscles. The umbilicus** is found two-thirds of the way down on this line.

The **clavicles** can be seen extending laterally from

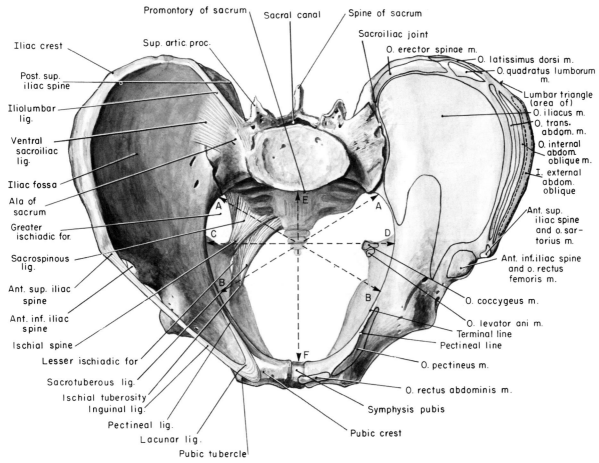

Promontory of sacrum   Sacral canal   Spine of sacrum

Sup. artic. proc.   Sacroiliac joint

Iliac crest   O. erector spinae m.

O. latissimus dorsi m.

Post. sup. iliac spine   O. quadratus lumborum m.

Iliolumbar lig.   Lumbar triangle (area of)

O. iliacus m.

Ventral sacroiliac lig.   O. trans. abdom. m.

O. internal abdom. oblique m.

Iliac fossa   I. external abdom. oblique

Ala of sacrum

Greater ischiadic for.   Ant. sup. iliac spine and o. sartorius m.

Sacrospinous lig.   Ant. inf. iliac spine and o. rectus femoris m.

Ant. sup. iliac spine

Ant. inf. iliac spine   O. coccygeus m.

O. levator ani m.

Ischial spine   Terminal line

Lesser ischiadic for   Pectineal line

Sacrotuberous lig.   O. pectineus m.

Ischial tuberosity   O. rectus abdominis m.

Inguinal lig.

Pectineal lig.   Symphysis pubis

Lacunar lig.   Pubic crest

Pubic tubercle

Figure 3-4. *The male pelvis, showing attachments of muscles on the left and ligaments on the right. At this time concentrate on the pubic symphysis and crest, pectineal line, anterior superior iliac spine, and the iliac crest. Note also the attachments of the inguinal ligament and its continuation as the lacunar and pectineal ligaments.*

the suprasternal notch and can be followed to their articulation with the **acromion processes** of the scapulae. The bulges on either side of the sternum are made by the large **pectoralis major muscles.** The inferior extent of these muscles is usually discernible and they can be followed to the axilla where they form the **anterior wall** or fold of that region. The **nipple** in the male is usually located superficial to the fourth intercostal space and about 4 inches from the midline. Sweeping inferiorly and laterally from the xiphisternal joint is the **costal arch,** the inferior extent of the rib cage. The **interdigitations** between serratus anterior and external abdominal muscles

can be seen in muscular individuals. On either side of the midline in the abdominal region a gracefully curving line, the **linea semilunaris,** marks the lateral edge of the rectus abdominis muscle. The **inguinal groove,** which corresponds in position to the **inguinal ligament,** marks the separation of the abdominal region from the lower limb.

Palpation reveals other structures. Although the first rib is overshadowed by the clavicles, the **second ribs** are easily palpated. They articulate with the sternum at the joint between the manubrium and body of the sternum. The **seventh costal cartilage** is the lowest that attaches directly to the sternum and can also be palpated, as can

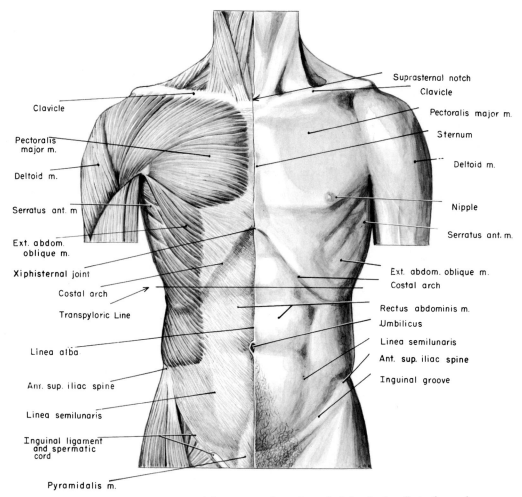

Clavicle

Pectoralis
major m.

Deltoid m.

Serratus ant. m.

Ext. abdom.
oblique m.

Xiphisternal joint

Costal arch

Transpyloric Line

Linea alba

Ant. sup. iliac spine

Linea semilunaris

Inguinal ligament
and spermatic
cord

Pyramidalis m.

Suprasternal notch

Clavicle

Pectoralis major m.

Sternum

Deltoid m.

Nipple

Serratus ant. m.

Ext. abdom. oblique m.

Costal arch

Rectus abdominis m.

Umbilicus

Linea semilunaris

Ant. sup. iliac spine

Inguinal groove

Figure 3-5. *Surface anatomy of the anterior thoracic and abdominal walls in the male.*

the tips of the **eighth, ninth,** and **tenth ribs.** The **eleventh** and **twelfth ribs** can be palpated behind the lowest part of the costal margin. In counting ribs it is wise to start with the second; the first is sometimes difficult to identify, as is the twelfth. Lastly, the **iliac crest** can be felt, barely 4 inches inferior to the last rib.

As an aid in utilizing one's visual picture of the body, it is well to project directly posteriorly to the vertebral column from certain anterior landmarks just mentioned (Fig. 3-6). A line projected posteriorly from the suprasternal notch runs into the upper part of the body of the third thoracic vertebra; from the sternal angle, into the base of the fourth thoracic vertebra; from the xiphisternal joint, into the disc between the ninth and tenth thoracic verte-

brae; from the lowest part of the tenth costal cartilage into the body of the third lumbar vertebra; from the most superior aspect of the iliac crests, into the fourth lumbar vertebra; and from the umbilicus, usually into the fourth lumbar vertebra.

There are certain vertical lines of reference that are also helpful in describing locations of pain, of deeper structures, or of incisions on the skin (Figs. 3-7, 3-8, and 3-9). From the midline of the sternum laterally and, finally, posteriorly, they are: (1) **lateral sternal**—just lateral to the widest part of the sternum, (2) **parasternal**—midway between the lateral sternal and the next line, the (3) **midclavicular**—through the middle of the clavicle, (4) **anterior axillary**—through the anterior

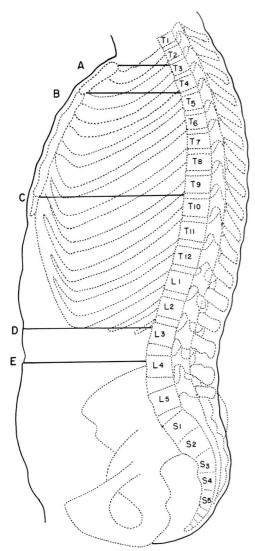

Figure 3-6. *A lateral view of the body, showing the vertebral levels of certain landmarks on the anterior thoracic and abdominal walls: (A) suprasternal notch, (B) sternal angle, (C) xiphisternal joint, (D) subcostal line, and (E) umbilicus.*

zontal lines bisecting these vertical lines (Fig. 3-7). This is done by either of two methods, one of which divides the abdominal wall into four areas, the other into nine. The four areas in the former method are called **quadrants** and are formed by a transverse line through the umbilicus and a vertical line in the midline of the body. These quadrants are known as upper right and left and lower right and left. The nine-area method uses two transverse planes: one between the tubercles on the anterior and lateral side of the iliac crests, the **intertubercular line;** the other, the **subcostal line,** between the inferior limits of the costal arch which is the tenth costal cartilage. The vertical lines are called the **midinguinal lines,** because they bisect the inguinal ligament. These nine regions are, from superior to inferior in the middle part of the abdominal wall, the **epigastric, umbilical,** and **pubic** regions; the laterally placed regions are, from superior to inferior, **hypochondriac, lateral,** and **inguinal** regions. A preferable set of vertical lines are the **semilunar lines;** if these are used, the inguinal canal is in the inguinal region and not divided between two regions.

These areas have proved more useful to clinical work than to descriptive anatomy. The transverse lines mentioned above, and their respective vertebral levels, are more useful, and aid greatly in the study of anatomy. Other lines besides the subcostal and intertubercular are available. The **transpyloric line,** so called because it bisects the pylorus of the stomach, is a line halfway between the suprasternal notch and the symphysis pubis or midway between the xiphisternal joint and the umbilicus. This line is quite useful because it can be visualized easily from the surface and bisects both kidneys, the body of the pancreas, and the left (splenic) flexure of the colon as well as the pylorus. It also bisects the ninth costal cartilage and, if projected to the posterior abdominal wall, hits the disc between the first and second lumbar vertebrae. It will be utilized frequently. The **supracristal line** cuts the body transversely at the superior limits of the iliac crests. It comes quite close to bisecting the umbilicus and is therefore easily visualized. The **interspinous line** bisects the body through the anterior superior iliac spines; this line bisects the sacral promontory and the second sacral spine. All these lines and areas are useful in visualizing through the body; they should be learned thoroughly.

The **female** differs from the male in having a thicker layer of superficial fascia that tends to cover some of the markings mentioned above (Fig. 3-10). Palpation, how-

wall of the axilla, (5) **midaxillary**—through the middle of the axilla, (6) **posterior axillary**—through the posterior wall of the axilla, (7) **scapular**—through the inferior angle, and (8) **paravertebral**—just lateral to the transverse processes of the vertebrae.

The abdominal wall is divided into regions by hori-

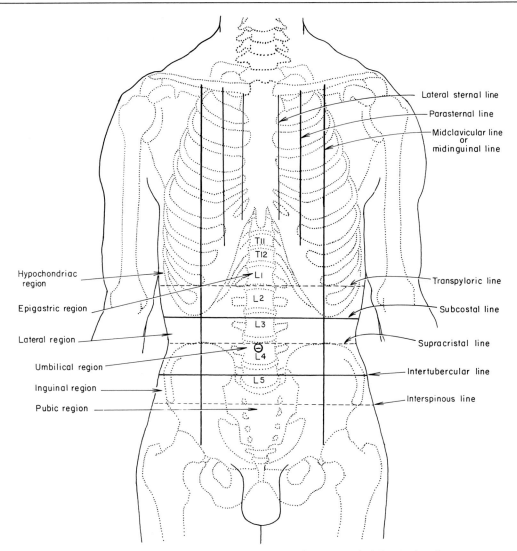

Figure 3-7. Lines of reference on anterior thoracic and abdominal walls.

ever, will reveal them. After puberty, the mammary glands develop to adult size and occupy an area lateral to the sternum and superficial to the pectoralis major muscle; they extend from the second to sixth ribs. This secondary sexual organ is rudimentary in females before puberty and can be used as a landmark for the fourth intercostal space just as in the male. After puberty, the nipple cannot be used in this manner.

The **umbilicus** deserves special comment. Not only does it serve as a guide to the disc between the third and fourth lumbar vertebrae in the young, or the fourth lumbar vertebra in those with less muscle tone and more fat, it is also important clinically. It can contain fecal matter from a persistent vitello-intestinal canal, or urine from a connection with the bladder via a persistent urachus. Both congenital and acquired hernias can occur at the umbilicus, patients talk about pain around it, and veins surrounding it can become dilated (caput medusae) due to obstruction of the portal system of veins, usually associated with disease of the liver.

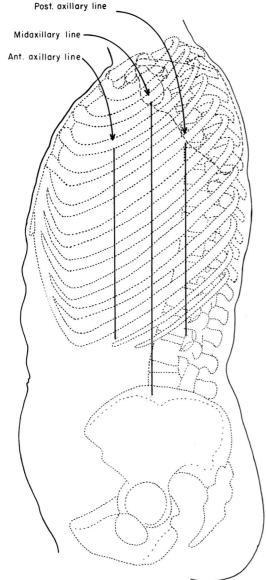

Figure 3-8. Lines of reference on lateral chest wall.

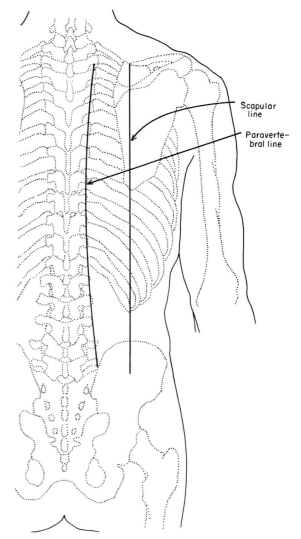

Figure 3-9. Lines of reference on posterior chest wall.

MAMMARY GLANDS

GENERAL DESCRIPTION

The **mammary glands,** or **breasts,** remain rudimentary in the male under normal conditions; hairs may be found at the circumference of the areola that are not present in the female.

After puberty in the female the breast enlarges to its adult size, one frequently being slightly larger than the other. They are contained in the superficial fascia, cover an area extending from the second rib to the sixth intercostal cartilage, and are superficial to the pectoralis major muscle. They tend to overlap this muscle inferiorly to become superficial to the external abdominis oblique and serratus anterior muscles. Breasts enlarge during preg-

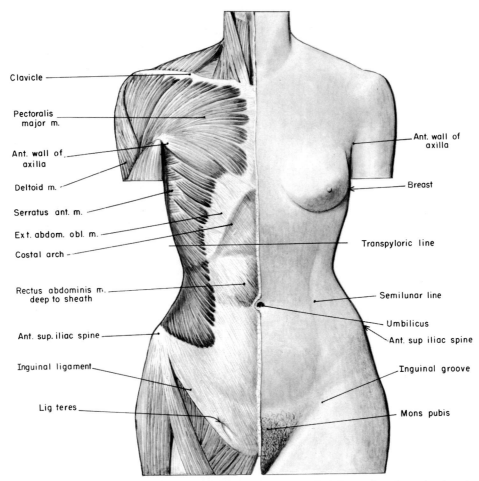

Figure 3-10. *Surface anatomy of the anterior thoracic and abdominal walls in the female. Note that the increased thickness of the superficial fascia and the decreased volume of the muscles tend to decrease prominence of the underlying structures.*

nancy and become pendulous in old age. Figure 3-11 gives an idea of changes in breast contour with age.

The skin of the **nipple** (Fig. 3-12) exhibits wrinkles, and each of the fifteen to twenty compound tubulo-alveolar glands that make up the breast opens onto the nipple by a separate opening. The nipple contains **smooth muscle fibers** that contract on tactile stimulation, inducing greater firmness and prominence. The nipple is surrounded by a pigmented area, the **areola;** this area is pink before pregnancy, turns brown after pregnancy, and remains this color. The skin here contains **tubercles** or **small projections** made by the presence of small areolar glands.

The bulk of the breasts (Figs. 3-12 and 3-13) consists of the aforementioned **tubuloalveolar** glands embedded in fat, the adipose tissue giving the gland its smooth rounded contour. Each gland forms a **lobe** of the breast, and the lobes are separated by connective tissue **septa.** These septa attach to the skin in such a manner as to serve as a clue to mammary tumors; the tumor puts these under stress which, in turn, causes a dimple in the skin. Each of the tubuloalveolar glands opens into a **lactiferous duct** that leads into a more dilated area, the **lactiferous sinus;** constriction then occurs in the lactiferous duct before opening onto the surface of the nipple.

As mentioned previously, cancer of the glandular

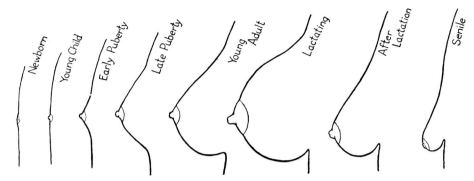

Figure 3-11. *Changes in the breast with age and functional activity.* (*From* Patten's Human Embryology *by C. E. Corliss. Copyright © 1976 by McGraw-Hill, Inc. Used with permission of McGraw-Hill Book Company.*)

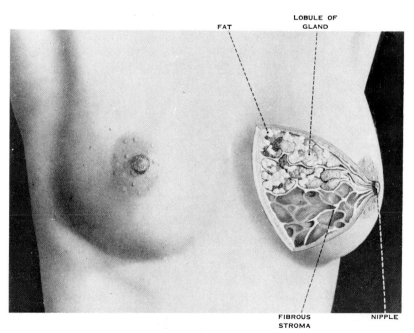

Figure 3-12. *A photograph of the mammary glands. Part of the skin of the left gland has been removed to illustrate structure of the breast.* (*From W. J. Hamilton, ed.,* Textbook of Human Anatomy, *courtesy of Macmillan Co., New York.*)

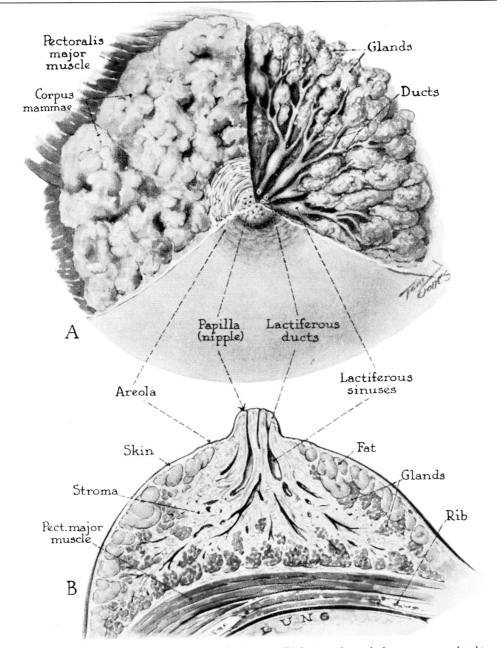

Figure 3-13. *(A) Dissection of the lactating breast. (B) Section through the mammary gland to show its structure and relation to underlying muscles. (Courtesy of Camp International, Inc.)*

tissue causes a dimpling of the skin due to stress being placed on the connective tissue septa. If cancer invades the lactiferous ducts, it can cause an invagination of the nipple.

The breasts are influenced by **hormones.*** Female sex hormones, either directly or through stimulating release of a mammotropic hormone from the hypophysis, cause enlargement of the breasts during pregnancy. After parturition, lactogenic hormone from the hypophysis stimulates secretion from the much-enlarged glandular tissue. The act of suckling has an influence on release of this hormone.

---

*R. P. Reece, 1958, in J. T. Velardo (ed.), *Mammary Gland Development and Function: The Endocrinology of Reproduction,* Oxford University Press, New York.

ARTERIES AND VEINS
(Fig. 3-16)

The blood supply to the breasts is from branches of the **intercostal** arteries and the **perforating** branches of the internal thoracic artery; the third, fourth, and fifth are usually the largest. Branches are also contributed by the **lateral thoracic** artery. The **veins** end in the internal thoracic and the axillary; some branches may reach the external jugular veins.

NERVES
(Fig. 3-16)

The nerve supply to the breasts is from branches of the fourth, fifth, and sixth intercostal nerves and consists of sensory fibers and sympathetic fibers to the smooth muscles in the nipple and blood vessels.

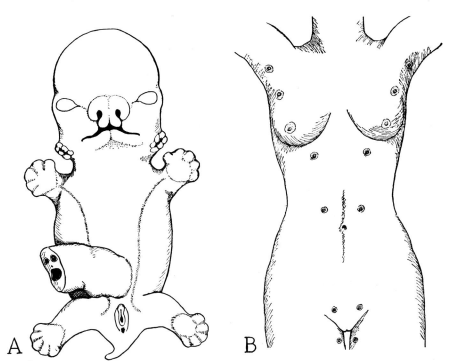

Figure 3-14. *(A) Milk line in fetus. (B) Location of most frequently occurring supernumerary nipples. Note resemblance to milk line in A. (From* Patten's Human Embryology *by C. E. Corliss. Copyright © 1976 by McGraw-Hill, Inc. Used by permission of McGraw-Hill Book Company.)*

LYMPHATICS

As cancer of the breast is a frequent occurrence, the lymph drainage is extremely important. Because of the close association of the breasts with the skin and superficial fascia, the lymphatic drainage can be quite extensive. Although the main drainage is to axillary nodes and to the parasternal nodes along the internal thoracic artery (inside the thoracic cavity), other points of drainage are to pectoral nodes between the pectoralis major and minor muscles, and to subclavicular nodes in the root of the neck deep to the clavicle. If the skin of the breast is involved, the drainage can be much more extensive, reaching the abdominal wall and nodes on the opposite side of the body.

The breasts of the adult female develop from a line of glandular tissue that is found in the fetus (Fig. 3-14). Because of this origin, accessory nipples and glandular tissue may be found along these milk lines, which extend from the clavicular to the inguinal regions. The glands are derived from ectoderm, while the connective tissue stroma is mesodermal in origin.

PECTORAL REGION

We enter now on a study of that portion of the anterior chest wall that is concerned with movement of the upper limb.

GENERAL DESCRIPTION

Upon removal of the skin, the **platysma muscle** can be seen to extend to a point inferior to the clavicle, as indicated in Figure 3-15. In addition, the **supraclavicular nerves,** branches of the cervical plexus, also extend onto the chest. Several anterior **cutaneous branches** of **intercostal nerves** and **perforating branches** of the **internal thoracic artery** can be seen piercing muscle and fascia just lateral to the sternum. The intercostal nerves give off a **lateral branch** that becomes cutaneous also; these branches divide further before emerging from the muscles, and therefore appear as two branches, one coursing

anteriorly and the other posteriorly. They are accompanied by **lateral cutaneous branches** of the **intercostal arteries.** The arterial supply in this lateral region is supplemented by the **lateral thoracic artery,** a branch of the axillary.

Removal of the superficial fascia reveals the large **pectoralis major muscle** (Fig. 3-16), the muscle forming the large bulge on the anterior chest wall. The division between its clavicular and costosternal parts is readily seen; that between the costosternal and abdominal portions is not seen as easily. The **cephalic vein** marks the dividing line between this muscle and its neighbor, the deltoideus.

If the pectoralis major is reflected laterally, as in Figure 3-17, the **pectoralis minor muscle** can be seen through a layer of deep fascia. This fascia extends from the clavicle, where it surrounds the subclavius muscle, to the pectoralis minor muscle, surrounds it, and then continues into the axilla, as diagramed in Figure 3-18. That portion between the clavicle and pectoralis minor muscles is designated as the **clavipectoral fascia;** medially, it is attached to the first costal cartilage and fascia on the first intercostal space, and laterally to the corocoid process. The clavipectoral fascia can be seen to be **pierced** (1) by the **cephalic vein** on its way to drain into the axillary, (2) by the **lateral pectoral nerve** innervating the clavicular part of the pectoralis major muscle, and (3) by the **thoracoacromial artery,** a branch of the axillary. The **medial pectoral nerve** can be seen piercing the pectoralis minor, which it innervates, before entering the substance of the costosternal and abdominal portions of the pectoralis major.

Removal of the clavipectoral fascia (Fig. 3-19) reveals the small subclavius muscle hugging the inferior surface of the clavicle, the axillary artery and vein, and portions of the brachial plexus.

MUSCLES

The muscles of the pectoral region are:

1. Pectoralis major
2. Pectoralis minor
3. Subclavius

1. Pectoralis major (Fig. 3-16). The clavicular portion of the pectoralis major muscle **arises** from the medial half of the clavicle, the costosternal portion from

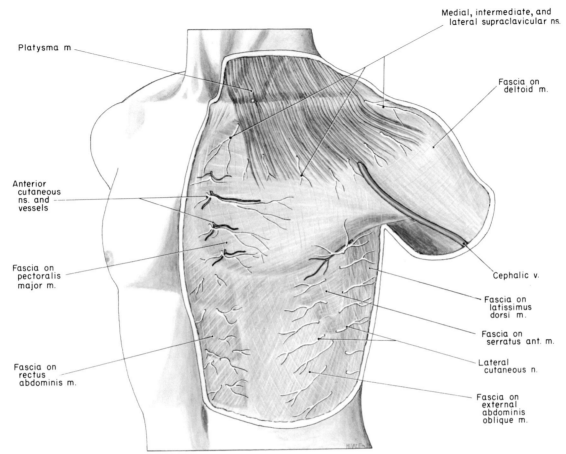

Platysma m

Medial, intermediate, and lateral supraclavicular ns.

Fascia on deltoid m.

Anterior cutaneous ns. and vessels

Fascia on pectoralis major m.

Cephalic v.

Fascia on latissimus dorsi m.

Fascia on serratus ant. m.

Lateral cutaneous n.

Fascia on rectus abdominis m.

Fascia on external abdominis oblique m.

Figure 3-15. *Pectoral region I—after removal of the skin and superficial fascia. (Note that the platysma muscle is actually in the superficial fascia.)*

the anterior surface of the sternum and costal cartilages of the upper six ribs, and the abdominal portion from the aponeurotic tendon of the external abdominis oblique muscle. The **insertion** is on the crest of the greater tubercle of the humerus, the clavicular portion overriding the costosternal part. Its **main action** is to flex, adduct, and medially rotate at the shoulder joint, an act that brings the arm across the chest. Its **nerve supply** is the medial and lateral pectoral nerves from the medial and lateral cords of the brachial plexus.

2. The **pectoralis minor** muscle (Fig. 3-19) lies deep to the pectoralis major and **arises** from the external surface of three ribs—usually the second, third, and fourth—and converges superiorly and laterally to **insert** on the coracoid process of the scapula. Its **main action** is

to lower the shoulder, its **nerve supply** the medial pectoral nerve (see Fig. 8-40 on page 707).

3. **Subclavius** (Fig. 3-19). The subclavius muscle is a small muscle **arising** from the first rib at the point of its junction with its cartilage, coursing laterally, and **inserting** on the inferior surface of the clavicle. Its **main action** is to lower the clavicle, and its **nerve supply** is a special nerve from the brachial plexus called the nerve to the subclavius.

ARTERIES

The arteries of this region are small but numerous; they are:

1. Perforating branches of internal thoracic

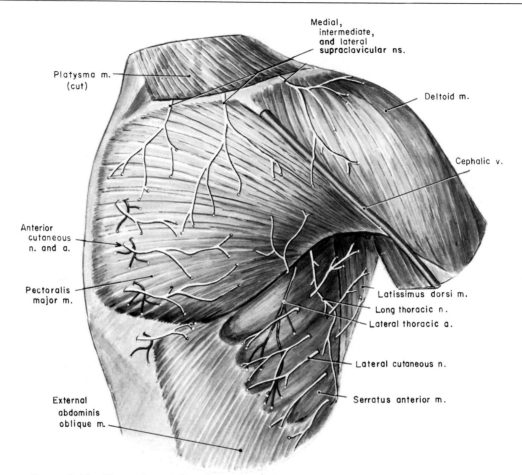

Figure 3-16. *Pectoral region II—after removal of superficial and deep fascial layers. Note how the pectoralis major muscle twists before inserting on the humerus.*

2. Lateral cutaneous of intercostal
3. Lateral thoracic of axillary
4. Thoracoacromial of axillary

1. Perforating (Figs. 3-15 and 3-16). These are branches of the internal thoracic artery. They appear close to the sternum and usually divide into short medial and longer lateral branches. The latter branches of the third, fourth, and fifth perforating arteries are enlarged in the female to supply the breast.

2. Lateral cutaneous (Fig. 3-16). These are found more laterally placed and are branches of the intercostal arteries. The main stem is not visible, but their anterior and posterior terminal branches can be seen emerging between the heads of the serratus anterior muscle. Several of these lateral branches are enlarged in the female to supply the breast.

3. Lateral thoracic (Figs. 3-16 and 3-19). This artery is a branch of the axillary. It courses inferiorly and slightly anteriorly on the surface of the serratus anterior muscle. This artery also has mammary branches in the female.

4. Thoracoacromial artery (Figs. 3-17, 3-18, and 3-19). After piercing the clavipectoral fascia, this artery immediately divides into four branches: (1) **pectoral,** to the pectoralis major and minor; (2) **acromial,** coursing laterally in a plane superficial to the coracoid process to reach the acromion process; (3) **clavicular,** turning medially to supply blood to the subclavius muscle and the

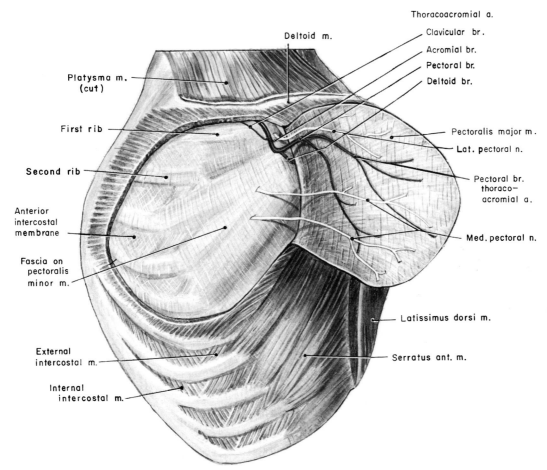

Figure 3-17. *Pectoral region III—after reflection of the pectoralis major muscle. Note the fascia covering the pectoralis minor muscle.*

sternoclavicular joint; and (4) **deltoid,** often arising from one of the other branches, accompanying the cephalic vein, and supplying both pectoralis major and deltoid muscles.

VEINS

All of the above arteries are accompanied by veins. One vein in this region, the **cephalic,** has no accompanying artery. This vein drains the superficial parts of the arm and courses in the groove between the deltoid and pectoralis major muscles. It empties into the axillary vein. It is often found wanting.

NERVES

The nerves found in the pectoral region are:

1. Anterior and lateral cutaneous
2. Long thoracic
3. Lateral pectoral
4. Medial pectoral

1. Cutaneous (Figs. 3-15 and 3-16). The **anterior cutaneous** nerves are distributed in a fashion similar to the arteries. They differ in that the nerves are the terminal branches of the intercostals while the arteries arise from the internal thoracic. The **lateral cutaneous**

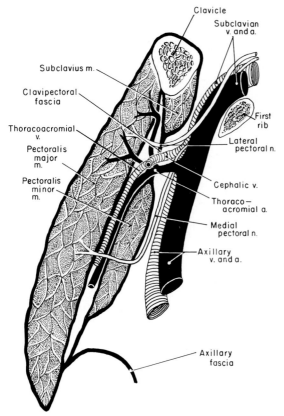

Figure 3-18. *Section through pectoral region showing structures piercing the clavipectoral fascia and their relation to the pectoral muscles. (Modified after Cunningham.)*

branches are branches of the intercostals and divide into anterior and posterior branches just as the arteries do.

    2. The **long thoracic** nerve accompanies the lateral thoracic artery (Figs. 3-16 and 3-19). Its origin from the brachial plexus is not visible; it courses inferiorly and slightly anteriorly on the surface of the serratus anterior muscle, which it innervates. The relation of this nerve to its muscle violates one of the principles in anatomy that nerves to muscles are found on their deep surfaces.

    3. The **lateral pectoral** nerve is named for the fact that it arises from the lateral cord of the brachial plexus and courses to the pectoral region. It appears in this region by piercing the clavipectoral fascia (Figs. 3-17 and 3-18). It immediately enters the clavicular portion of the pectoralis major muscle to innervate it.

    4. **Medial pectoral** (Figs. 3-17 and 3-18). This

nerve is obviously named in the fashion just described for the lateral pectoral nerve; it arises from the medial cord of the brachial plexus and enters the pectoral region but remains deep to the clavipectoral fascia and the pectoralis minor muscle; it then pierces this muscle, innervating it on the way, and enters the deep surface of the sternocostal and abdominal portions of the pectoralis major. There may be more than one branch of this nerve.

    The nerve supply to the subclavius muscle is not seen at this time.

    It should be noted that all the muscles in this region are concerned with the upper limb, the clavicle and scapula being parts of the appendicular skeleton. They are therefore innervated by branches from the brachial plexus because that plexus innervates all muscles developed from the superior limb bud no matter where they may migrate.

THORACIC CAGE

Although a knowledge of the structure of the thoracic cage is mandatory to understand the mechanism of respiration, this part of the body has become even more important because of the comparatively recent extensive surgery in the thoracic cavity.

GENERAL DESCRIPTION

The bony thoracic cage is **composed** of the vertebral column posteriorly, ribs laterally, and sternum anteriorly; it has an opening superiorly called the superior outlet, a much larger opening inferiorly, and open areas between the ribs known as intercostal spaces. The inferior outlet is filled in by the diaphragm and the intercostal spaces by the intercostal muscles and associated structures. The superior outlet, unlike the other openings, has no specific structure closing the thoracic cage off from the neck. Structures in the thoracic cavity and the neck are directly continuous with one another, and it is these structures that fill in this opening.

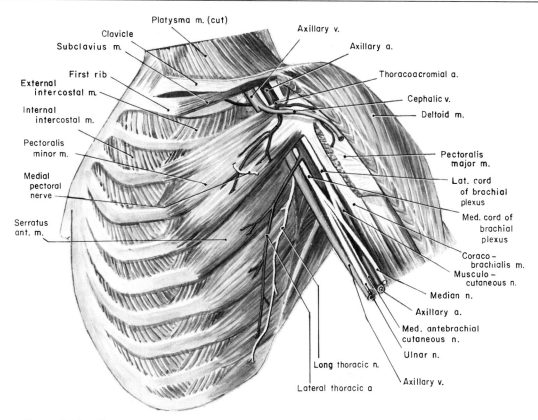

Figure 3-19. *Pectoral region IV—after removal of deep fascia. Note how the pectoralis minor muscle divides the vital structures in the axilla into three parts: proximal, deep, and distal to this muscle. The external intercostal membrane has been removed.*

When muscles on the chest that are concerned with the upper limb are removed, the **intercostal spaces** and their contents are brought to view (Fig. 3-20). The most superficial muscle is the **external intercostal**. This muscle can be identified as the muscle with its fasciculi directed from lateral to medial as it fills in the space from the rib above to the rib below. Anteriorly and medially this muscle is wanting, being replaced by a fibrous sheet called the **external intercostal membrane.**

Deep to this muscle is the internal **intercostal muscle;** this muscle occupies the same space as the external intercostal but the muscle fasciculi course in the opposite direction, that is, inferiorly and laterally. In approximately the midaxillary line this muscle changes to a connective tissue membrane, the **internal intercostal membrane,** which continues around to the vertebral column; it is

similar to that just described as the anterior extension of the external intercostal muscle.

Deep to the internal intercostal muscle is a **set of three muscles** that have often been described as three parts of one muscle. Anteriorly is the **transversus thoracis muscle** (Fig. 3-23), connected by a thin membrane to the **intercostalis intimus muscle** which continues posteriorly as far as the angle of the ribs, where it ends in a free border. The third part of this inner complex of muscles is known as the **subcostal muscle;** this is composed of small slips of muscle near the angles of the ribs that span two or more intercostal spaces. The intercostal nerves and vessels course deep to the internal intercostals and superficial to the inner complex. If the dissection is done at the usual location on the chest wall at about the costochondral articulation, these structures are

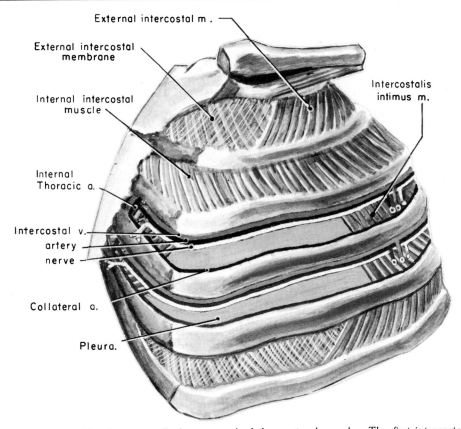

Figure 3-20. *The thoracic wall after removal of the pectoral muscles. The first intercostal space shows the external intercostal muscle and the external intercostal membrane. These have been removed in the second space to show the internal intercostal muscle. In the third space the latter muscle has been removed to reveal the intercostal vein, artery, and nerve (enlarged and more easily seen in this picture than on the body) coursing superficial to the inner complex of transversus thoracis and intercostalis intimus muscles. The third component of this inner complex—subcostal muscles—cannot be seen in this drawing.*

located deep to the internal intercostal muscle and superficial to the membrane connecting the transversus thoracis and intercostalis intimus muscle. As the abovementioned membrane is not easily seen, the nerves and vessels seem to lie directly on the pleura. They are located in a costal groove on the inferior side of each rib, as shown in Figure 3-24. The collateral vessels are located on the superior side of the rib below. The thoracic cavity can be approached, therefore, by cutting in the middle of an intercostal space without cutting nerves or vessels. Ribs are removed, however, for major thoracic surgery.

MUSCLES

The muscles that fill in an intercostal space are:

1. External intercostal
2. Internal intercostal
3. Inner complex
   a. Transversus thoracis
   b. Intercostalis intimus
   c. Subcostal

1. External intercostal. This is the most superficial muscle in the intercostal spaces. Its many fasciculi arise from the rib above and insert on the rib below. It reaches as far as the tubercles of the ribs posteriorly and to the costochondral joint anteriorly; from this point on to the sternum, the muscle has disappeared and a fibrous sheet—**the external intercostal membrane**—is all that remains. Its **main action** is to elevate the ribs in inspiration. **Nerve supply** is the proper intercostal nerve.

2. **Internal intercostal.** This muscle occupies the same area as just mentioned for the external intercostals but at a deeper level. Its fasciculi arise from the rib above and course inferiorly to insert on the rib below. It ends posteriorly in a membrane (**internal intercostal membrane**) in the same manner as found for the external intercostal muscles anteriorly. Its **main action** is problematical, but most feel that the interchondral portion aids in inspiration and the interosseous portion is involved with expiration. The **nerve supply** is the proper intercostal nerve.

3. **Inner complex.** (a) The **transversus thoracis** muscle is actually inside the thoracic cavity. This muscle **arises** from the xiphoid process and the body of the sternum and **inserts** on the second to the sixth costal cartilages. Its **main action** is to lower the ribs in forced expiration. (b) The **intercostalis intimus** muscle has fasciculi that have the same attachments as the intercostal muscles and have the same action. This muscle extends posteriorly as far as the angles of the ribs. (c) The **subcostal** muscles are small strips of muscle found near the angles of the ribs. The fasciculi have the same course as the internal intercostal muscles but skip one rib. Their **main action** is to aid in inspiration. The **nerve supply** to these three muscles is the proper intercostal nerve.

We will see later that this three-layered arrangement of muscles is very similar to the arrangement found in the abdominal wall. More detail on the actions of the intercostal muscles is given on pages 124 and 125.

NERVES

The only nerves involved with the thoracic cage are the intercostals. The **intercostal nerves** (Fig. 3-21) and vessels take similar courses throughout the intercostal spaces. After giving off the **dorsal ramus,** the **ventral ramus,** or **intercostal nerve,** continues laterally between the parietal pleura and internal intercostal membrane; the

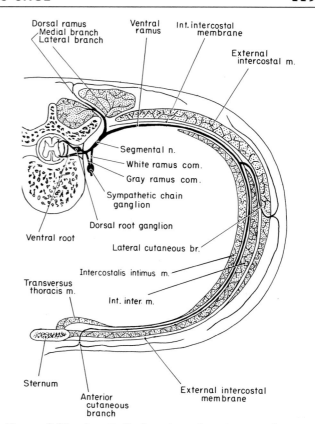

Figure 3-21. Longitudinal section of an intercostal space, showing the course and distribution of an intercostal nerve.

nerve appears to be in contact with the external intercostal muscle but actually the membrane intervenes. At about the angle of each rib, it gives off a **collateral branch** that descends to the superior surface of the rib below. The main nerve continues in the intercostal groove on the under surface of the rib between the intercostalis intimus and the internal intercostal muscle. It gives off its lateral cutaneous branch, continues anteriorly and medially between the internal intercostal muscle and the membrane joining the intercostalis intimus with the transversus thoracis muscle, and then between the transversus thoracis and internal intercostal muscles. It ends by turning anteriorly, piercing the internal intercostal muscle, the external intercostal membrane, and pectoralis major muscle to divide into a short medial and a longer lateral cutaneous branch.

Only the third, fourth, fifth, and sixth nerves can be called typical. The **first thoracic nerve** joins the brachial plexus and sends only a small intercostal branch, which has no lateral cutaneous branch and frequently no anterior cutaneous branch. The **second intercostal nerve** differs in that its lateral cutaneous branch ramifies on the arm and is called the **intercostobrachial nerve.** The **seventh** to **eleventh intercostal nerves** do not remain in their respective intercostal spaces but leave their spaces at a point where the costal cargilages turn superiorly. Passing deep to these cartilages, they enter the abdominal wall musculature. In the intercostal space they were between the internal intercostal muscle and the intercostalis intimus; they continue into the abdominal wall in this same plane—between the internal abdominis oblique and transversus abdominis muscle. The **twelfth nerve** is the **subcostal,** which takes almost its entire course in the abdominal wall and will be described later. It should be recalled that each intercostal nerve connects to the sympathetic chain via white and gray rami communicantes and that visceral afferent and efferent nerves are contained therein as well as somatic afferent and efferent fibers.

ARTERIES

The following arteries are found in this region:

    1. Posterior intercostal
    2. Internal thoracic

    1. A **posterior intercostal artery** (Fig. 3-22) takes a course similar to that of a typical intercostal nerve; it maintains the same relations to the intercostal muscles. It arises from the descending aorta, enters an intercostal space, gives off a collateral branch that goes to the superior surface of the rib below, and continues in the groove on the under surface of the rib in a position superior to the nerve; it gives off a lateral cutaneous branch. Anteriorly, it differs from the nerve.
    2. At approximately the costochondral juncture, each posterior intercostal artery meets with an **anterior intercostal branch** of the **internal thoracic artery,** an artery coursing inferiorly from the subclavian, and found deep to the internal intercostal muscle and superficial to the transversus thoracis muscle; the anterior cutaneous vessel (perforating branch) arises from the latter artery rather than from a continuation of the intercostal. The collateral branch also joins with an anterior intercostal branch from the internal thoracic artery, as shown in Fig.

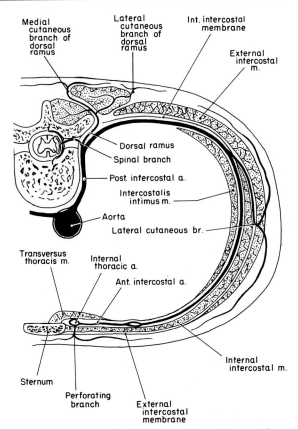

Figure 3-22. *Longitudinal section of an intercostal space, showing the course and distribution of an intercostal artery.*

3-23. The above description is accurate for the third, fourth, fifth, and sometimes the sixth intercostal arteries. **The first two** spaces differ in that the posterior intercostals arise from the costocervical trunk of the subclavian artery (see Fig. 7-93 on page 578). The arteries in the **seventh** to the **ninth spaces** join the **musculophrenic** branch of the internal thoracic rather than the main artery. The **last two intercostals** accompany the intercostal nerves into the abdominal wall.
    More detailed relations of these arteries around the vertebral column will be given later when the thoracic cavity is described.

VEINS

Intercostal veins are located in the same position as the arteries and nerves but in a position superior to both.

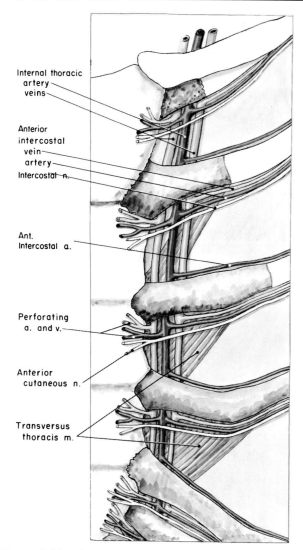

Figure 3-23. *Anterior view of the internal thoracic artery, showing how the anterior perforating and anterior intercostal arteries arise. Note that the anterior perforating nerves have a different arrangement.*

Therefore, from superior to inferior are found the intercostal vein, artery, and nerve.

The first intercostal veins on both sides drain into either the vertebral vein or the brachiocephalic vein (see Fig. 4-9, page 162). In addition, the second and third veins on the left side usually drain into the left brachiocephalic. All remaining intercostal veins drain into the azygous system of veins located in the posterior part

of the thorax (see Fig. 4-44, page 199), which drains into the superior vena cava. There is considerable variation in this pattern.

LYMPHATICS

The lymphatics of the skin in the thoracic region drain into the axillary and internal thoracic nodes, but the intercostal spaces drain posteriorly into nodes along the vertebral column, as well as anteriorly into the internal thoracic nodes.

Figure 3-24 is a series of transverse sections through an intercostal space starting close to the vertebral column, another at the midaxillary line, then at the midclavicular line, and finally close to the sternum. They should serve to summarize the anatomy of the thoracic wall.

ARTICULATIONS OF THORACIC CAGE

The muscles forming the walls of the thoracic cavity are muscles of respiration; the mechanisms involved in respiration should be studied at this time, in spite of the fact that muscles involved in abdominal breathing have not been considered. Respiration involves movement of the ribs and sternum, so a knowledge of the articulations involved is mandatory. These are:

1. Costovertebral
   a. Of head
   b. Costotransverse
2. Sternocostal
3. Interchondral
4. Costochondral
5. Sternal

1. Costovertebral. The heads of the second to the ninth ribs articulate with the bodies of two vertebrae and the intervening disc, and the tubercles articulate with transverse processes.

a. The joints involving the heads of the ribs are of

122

the synovial gliding type. The capsule is strengthened anteriorly by **radiate ligaments** (Fig. 3-25) which consist of three parts: one attaching to each vertebra, and a middle one connecting to a disc. **Intra-articular ligaments** stretch between the crest of the head of each rib to the disc and actually divide the synovial cavity into two parts; this does not occur at locations where the head of the rib articulates with only one vertebra, namely, at the first, tenth, eleventh, and twelfth costovertebral joints.

    b. The joints of the tubercles with the transverse processes are synovial articulations and possess a capsule and ligaments (Fig. 3-26). That ligament stretching from a rib to the transverse process of the vertebra above is a **superior costotransverse ligament;** it has anterior and posterior laminae. The (**inferior**) **costotransverse ligament** (not visible in Fig. 3-26) is located between the neck of the rib and the transverse process of its own vertebra. The **lateral costotransverse ligament** connects a tubercle to the tip of a transverse process.

    2. Sternocostal (Fig. 3-27). These articulations are between the costal cartilages of the first seven ribs and the sternum. The first rib has no joint cavity; the perichondrium of the cartilage is attached directly to the periosteum of the sternum. The remaining joints are similar to those between the heads of the ribs and the vertebral column. The **ligaments** consist of a **capsule** and **radiate ligaments** on the anterior and posterior surfaces of the ribs and sternum, and of **intra-articular ligaments** that split the synovial cavity into two portions. This split always occurs between the second rib and sternum but is not usually present in other articulations.

    3. The **interchondral** articulations occur between the sides of costal cartilages of ribs 6 to 10; these are simple gliding joints possessing a capsule and a few ligamentous fibers.

    4. Costochondral. These joints occur between the end of each bony rib and its costal cartilage. The rounded end of the cartilage fits into a groove at the end of each rib; no movement occurs here for the articulation is firmly fixed, the periosteum of the bony rib being directly continuous with perichondrium of the cartilage.

    5. Sternal. The sternum is not a solid structure

Figure 3-24. *Transverse sections of an intercostal space: close to the vertebral column, in the midaxillary line, midclavicular line, and close to the sternum.*

CLOSE TO VERTEBRAL COLUMN

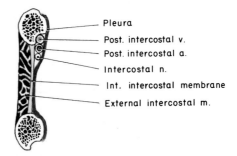

- Pleura
- Post. intercostal v.
- Post. intercostal a.
- Intercostal n.
- Int. intercostal membrane
- External intercostal m.

MIDAXILLARY LINE

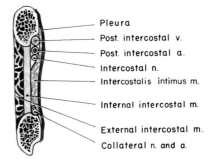

- Pleura
- Post. intercostal v.
- Post. intercostal a.
- Intercostal n.
- Intercostalis intimus m.
- Internal intercostal m.
- External intercostal m.
- Collateral n. and a.

MIDCLAVICULAR LINE

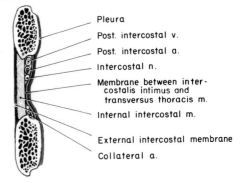

- Pleura
- Post. intercostal v.
- Post. intercostal a.
- Intercostal n.
- Membrane between intercostalis intimus and transversus thoracis m.
- Internal intercostal m.
- External intercostal membrane
- Collateral a.

CLOSE TO STERNUM

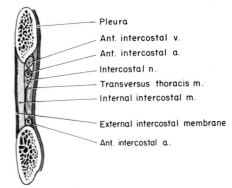

- Pleura
- Ant. intercostal v.
- Ant. intercostal a.
- Intercostal n.
- Transversus thoracis m.
- Internal intercostal m.
- External intercostal membrane
- Ant. intercostal a.

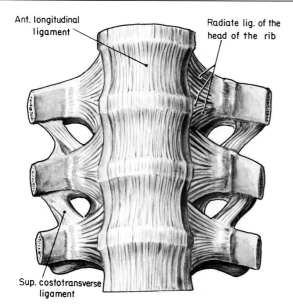

Figure 3-25. *Ligaments attaching the ribs to the vertebral column—anterior view.*

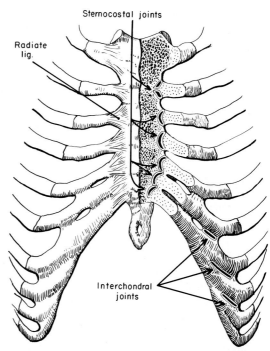

Figure 3-27. *Anterior view of sternocostal and interchondral articulations showing the synovial cavities exposed by a coronal section of the sternum and cartilages on the left side. Intraarticular ligaments are always present in the second joint but usually absent from the others.*

but has joints between its various parts. The fibrocartilaginous joints between any two parts of the body or between the xiphoid process and the body are not permanent, but that between the manubrium and the body does persist.

MECHANISM OF RESPIRATION

The vital process of respiration involves bringing air into close contact with the circulatory system so that oxygen can enter the blood and carbon dioxide can leave it. This close contact is brought about in the lungs; air is brought into this organ—**inspiration**—and carbon dioxide is exhaled—**expiration.** The process by which this occurs is known as the **mechanism of respiration.**

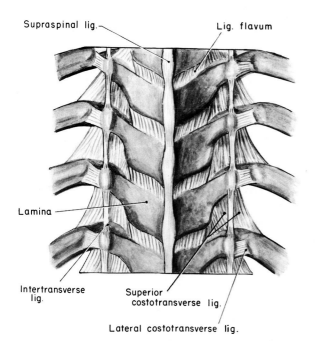

Figure 3-26. *Ligaments of the costovertebral articulations—posterior view.*

The thoracic cavity is lined on the inside with a thin serous membrane, called **pleura,** which is continuous with a similar layer on the lung. The former is **parietal pleura** because it lines the walls, and the latter **visceral** because it is on the viscus—the lung. These two layers of pleura are in contact with one another and maintain this intimacy because of a negative pressure between them and the positive atmospheric pressure in the lung itself. If the thoracic cavity changes in size, the lung will change accordingly because of this contact between the two pleuras. Therefore, inspiration simply involves increasing the size of the thoracic cavity so that air will rush into the lungs; conversely, expiration is caused by decreasing the size of the thoracic cage.

INSPIRATION

The thoracic cavity can be enlarged in three directions: inferiorly, laterally, and anteriorly. The enlargement **inferiorly** is accomplished by the diaphragm. This dome-shaped muscle in contraction pushes down on the abdominal viscera, and an increase in the vertical dimension of the thoracic cavity results. The thoracic cage is increased in size **anteriorly** by movement of the ribs around an axis that passes through the head, neck, and tubercle of each rib (Fig. 3-28). Any movement around this axis thrusts the ribs and sternum anteriorly, owing to the obliquity of the ribs. Increase in **lateral** diameter of the cage is accomplished by a pail-handle type of movement of the third to sixth ribs around an axis that passes through the sternum and posterior end of the shaft of the ribs.

These increases in anteroposterior and lateral diameters are accomplished by the intercostal musculature pulling each rib superiorly toward the rib above, the levatores costarum muscles pulling superiorly on the ribs in such a manner as to pull the anterior ends and sternum superiorly and anteriorly, and the serratus posterior superior muscle elevating the upper ribs. The scalenus muscles (located in the neck) are involved in normal respiration in many subjects;* they pull the first and second ribs superiorly. The serratus posterior inferior, pulling in-

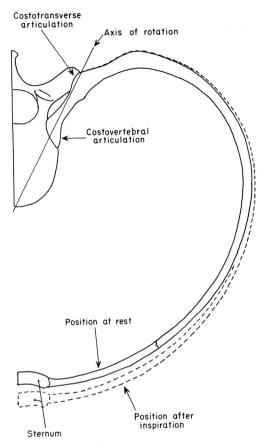

Figure 3-28. *Diagram illustrating increase in anteroposterior dimension of the thoracic cavity during inspiration. Note that the axis of rotation is through the head, neck, and tubercle of the rib.*

feriorly and posteriorly on the lower ribs, counteracts the superior and medial pull of the diaphragm when it contracts.

There has been considerable confusion concerning action of the intercostal muscles. The most recent work of Jones, Beargie, and Pauly (1953), utilizing electromyographic techniques, indicates that all intercostal muscles are used in inspiration but to a much smaller degree than previously supposed.† Others feel that the interosseous

---

*E. J. M. Campbell, 1955, The role of the scaleni and sternomastoid muscles in breathing in normal subjects, *J. Anat.* 89: 378. Electromyographic methods revealed that the scaleni muscles are used in normal respiration, especially in the erect posture. The sternocleidomastoid muscle is not used under normal conditions.

---

†D. S. Jones, R. J. Beargie, and J. E. Pauly, 1953, An electromyographic study of some muscles of costal respiration in man, *Anat. Record* 117: 17. Contrary to the usual concept, this work suggests that the intercostal muscles, in normal respiration, simply maintain the ribs a constant distance apart while the cavity is enlarged superiorly and inferiorly.

portions of the internal intercostals are used in expiration, while still others think that the intercostals in the upper spaces simply maintain the spaces leaving real action to the intercostals in the lower spaces. Even under conditions demanding increased respiration, movement caused by contractions of any one set of intercostal muscles is very slight; however, as was found in movement of the vertebral column, when the small movement of each rib is accumulated, considerable overall movement does occur. (The intercostal muscles also serve to maintain the thoracic wall and are used in restoring the body to a normal position after lateral bending of the trunk.)

In summary, inspiration at rest would seem to be accomplished primarily by an increase in size of the thoracic cage inferiorly by the diaphragm (diaphragmatic or abdominal inspiration); however, increases anteriorly and laterally by the intercostals, levatores costarum, serratus posterior superior muscles, and, in some individuals, by the scalenus muscles (thoracic inspiration) probably do occur.

EXPIRATION

Expiration under conditions of rest is accomplished by the elastic recoil of the thoracic cage and the action of the transversus thoracis muscles. As mentioned above, the interosseous portions of the internal intercostals may play a role in expiration. **Abdominal expiration** occurs by a contraction of the abdominal muscles; this increased intra-abdominal pressure forces the diaphragm superiorly and thereby decreases the size of the thoracic cavity. In prolonged expiration, as in singing a sustained note, the abdominal muscles gradually contract while the diaphragm slowly relaxes, and enough tension is maintained in the intercostal muscles to maintain a normal contour of the thoracic walls.

Quiet breathing tends to be mostly of the abdominal variety, with relatively little movement of the thoracic cage; pregnancy changes this ratio. Under conditions of severe anoxia many other muscles come into play in inspiration. In fact, any muscle having attachments to the ribs—such as the sternocleidomastoids, erector spinae, the scalenes—will act on the thoracic cage whether these attachments be origins or insertions. Under these conditions the whole thoracic cage, including the first two ribs, is raised considerably.

Those continuing at this time with the thoracic cavity should turn to page 153.

ABDOMINAL WALL

SURFACE ANATOMY

This is found on pages 102–106.

SUPERFICIAL FASCIA

The superficial fascia on the anterior abdominal wall can be quite thick, as this is one of the fat depots of the body. On the upper part of the abdomen it is similar to that found elsewhere, but on the lower part it is divided into **superficial fatty** and **deep membranous** layers.* The fatty layer is continuous with this same layer on the upper abdomen and thorax, on the back, in the perineal region, and on the lower limb. The membranous layer, on the other hand, has definite attachments, as can be seen in Figure 3-29. Superiorly, it is simply fused with the fatty layer; inferiorly and laterally, it is attached to the iliac crest, to the deep fascia of the thigh about ½ inch inferior to the inguinal ligament, to the deep fascia on the muscles on the medial side of the thigh, to the margin of the pubic arch, and finally to the posterior edge of the urogenital diaphragm (which will be seen later to be the boundary line between the urogenital and the anal triangles of the perineum). Medially, this fascia is fused with the deep fascia in the midline at the linea alba and to the symphysis pubis, and forms the superficial part of the suspensory ligament of the penis (fundiform ligament), the fibers of which surround the penis to become continuous with the dartos tunic of the scrotum. (We will see later that deep fascia forms the deeper part of the suspensory ligament of the penis.) This membranous fascia is continuous with a similar layer on the penis and scrotum in the male or labia majora in the female.

---

*C. E. Tobin and J. Benjamin, 1949, Anatomic and clinical re-evaluation of Camper's, Scarpa's, and Colles' fasciae, *Surg. Gynecol. Obstet.* 88: 545. This report denies the existence of two distinct fatty and membranous layers of superficial fascia in the lower abdominal wall.

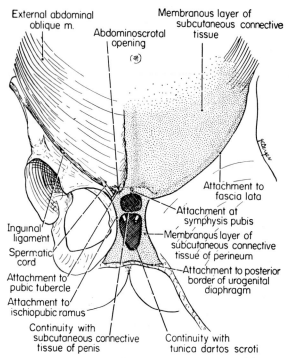

External abdominal
oblique m.

Membranous layer of
subcutaneous connective
tissue

Abdominoscrotal
opening

Attachment to
fascia lata

Attachment at
symphysis pubis

Membranous layer of
subcutaneous connective
tissue of perineum

Attachment to posterior
border of urogenital
diaphragm

Inguinal
ligament

Spermatic
cord

Attachment to
pubic tubercle

Attachment to
ischiopubic ramus

Continuity with
subcutaneous connective
tissue of penis

Continuity with
tunica dartos scroti

Figure 3-29. The attachments and continuations of the membranous layer of superficial fascia of the abdomen and perineum. (From Essentials of Human Anatomy, *third edition, by Russell T. Woodburne. Copyright © 1965 by Oxford University Press, Inc. Reprinted by permission; modified and redrawn after Cunningham.)*

From these attachments, it can be seen that, if the finger is placed deep to this membranous layer of superficial fascia and superficial to the deep fascia on the external abdominis oblique muscle, or, in other words, between these two fascial layers, it can be pushed inferiorly into the scrotum in the male or the labia majora in the female. If, through injury, urine extravasates into this space deep to the membranous fascia in the perineum, it will have free entrance into the scrotum and penis and thence into the anterior abdominal wall via this opening; it will not have access to the anal region or the thigh. Note Figure 5-78 on page 317.

There is no fatty layer of the superficial fascia in the **scrotum.** The membranous layer of superficial fascia in the scrotum has smooth muscle fibers scattered in it that give the scrotum a brownish color and are responsible for the wrinkled appearance. This is the **dartos** (*Gr.,*

skinned) **muscle;** when combined with the fascia it is called the **dartos tunic** of the scrotum. These smooth muscles are under control of the sympathetic part of the autonomic nervous system and respond to temperature. Contracting in cold and relaxing in warm temperatures, they are responsible for maintaining the testes at a more or less constant temperature. This is an important feature, for sperm are sensitive to prolonged changes in temperature. These muscle fibers are also found in the labia majora of the female. The evolution of the membranous layer of the superficial fascia is problematical, but it is interesting to think that it may have been derived from animals, such as cows, which carry heavy udders. This fascia will be discussed again when the perineal region is studied.

The **cutaneous nerves** of the abdominal wall are pictured in Figure 3-30. They are the lateral and anterior cutaneous branches of the **seventh to the eleventh intercostal nerves, subcostal nerves,** and **iliohypogastric** and **ilioinguinal** branches of the lumbar plexus. The lateral branches of the seventh to the eleventh nerves appear in the midaxillary line inferior to those from the higher intercostals. The lateral branch of the subcostal leaves the abdominal wall and enters the gluteal region by crossing the crest of the ilium near its anterior end, and the same branch of the iliohypogastric nerve takes a similar path by crossing the crest at a more posterior position; both of these branches will be studied again with the lower limb. The ilioinguinal nerve has no lateral cutaneous branch. The anterior cutaneous branch of the seventh thoracic nerve enters the fascia just inferior to the xiphoid process and is distributed in this area; the tenth thoracic nerve appears in the region of the umbilicus; the eighth and ninth nerves are distributed in expected positions between these points. The anterior branch of the iliohypogastric nerve is found entering the fascia just superior to the superficial inguinal ring (see Fig. 3-30) of the external abdominis oblique muscle and is distributed in a region just superior to the pubis; anterior branches of the eleventh thoracic nerve and subcostal nerve are found at the proper levels between this area and that of the tenth thoracic nerve around the umbilicus. The ilioinguinal nerve provides the cutaneous innervation for the skin over the pubis, on the anterior wall of the scrotum in the male or on the labia majora in the female.

A knowledge of the above description of the distribution of cutaneous nerves is very important clinically

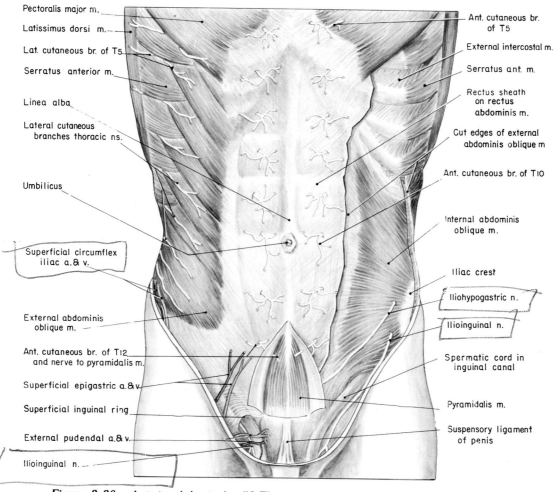

Pectoralis major m.

Latissimus dorsi m.

Lat. cutaneous br. of T5

Serratus anterior m.

Linea alba

Lateral cutaneous
branches thoracic ns.

Umbilicus

Superficial circumflex
iliac a.& v.

External abdominis
oblique m.

Ant. cutaneous br. of T12
and nerve to pyramidalis m.

Superficial epigastric a.& v.

Superficial inguinal ring

External pudendal a.& v.

Ilioinguinal n.

Ant. cutaneous br.
of T5

External intercostal m.

Serratus ant. m.

Rectus sheath
on rectus
abdominis m.

Cut edges of external
abdominis oblique m

Ant. cutaneous br. of T10

Internal abdominis
oblique m.

Iliac crest

Iliohypogastric n.

Ilioinguinal n.

Spermatic cord in
inguinal canal

Pyramidalis m.

Suspensory ligament
of penis

Figure 3-30. *Anterior abdominal wall I. The cutaneous nerves are shown. The superficial and deep fasciae have been removed from the right side of the body to show the external abdominis oblique muscle. The external abdominis oblique muscle has been removed from the left side to show the internal abdominis oblique muscle.*

because pain from the organs inside the cavities is referred to the surface. The patient's source of pain will be from inside the cavity but he or she will report pain on the surface to the physician. Naturally, the physician thinks in reverse after making sure there is no actual reason for pain at the surface area indicated. The actual cause of referred pain is still problematical but certainly is involved with the fact that nerves from several areas of the body enter the spinal cord at the same segmental level. Such referred pain can be from a visceral afferent nerve to a somatic afferent nerve, from a somatic afferent nerve to

another somatic afferent, or from one branch of a nerve to another branch of that same nerve. The majority of your patients will complain of pain.

The **cutaneous blood vessels** will be described after the structure of the abdominal wall has been studied. The only vessels usually found in cadavers in the superficial fascia are the **superficial epigastric** and **superficial circumflex iliac** arteries, branches of the femoral artery, which enter the abdominal wall in the inguinal region (Fig. 3-30).

The lymphatic drainage of the lower part of the abdomen is to the inguinal nodes; drainage of the upper part is to nodes in the axilla.

GENERAL DESCRIPTION

The abdominal wall covers a rather large area between the xiphoid process and costal margin superiorly, the vertebral column laterally and posteriorly, and the iliac crest and pubis inferiorly. The abdominal musculature fills in this entire area.

When all superficial fascia is removed from the abdominal wall, the **external abdominis oblique muscle** can be seen covered with a thin layer of deep fascia.* This muscle (Fig. 3-30, right side of body) fills in the entire lateral and anterior surfaces of the abdomen, its muscular portion occupying the lateral area only; almost the entire anterior portion is made up of a large tendinous aponeurosis that meets the same muscle of the opposite side in the midline, thus forming a whitish area called the **linea alba.** The muscle fasciculi are seen to course in the same direction as the external intercostals, namely, inferiorly and medially. When the above muscle is reflected, the **internal abdominis** oblique muscle is seen (Fig. 3-30, left side of body) with its fibers coursing superiorly and medially. Reflection, in turn, of the internal oblique muscle reveals that the many nerves and vessels in this region course between this and the next muscle, the **transversus abdominis** (Fig. 3-31). The latter muscle, with its fibers running in a manner implied in its name, fills in the same territory as the outer two muscles.

Another important muscle remains to be mentioned. If a longitudinal cut is made through the aponeurotic tendons about two inches from the midline, the **rectus abdominis** muscle is revealed (Fig. 3-31). This muscle stretches from the pubic symphysis to the costal arch on either side of the midline. It has **tendinous intersections** running horizontally across the muscle that divide it into segments. The lateral border of the rectus muscle curves medially as it approaches the pubic crest; the line formed is the **semilunar line.** If the rectus muscle is lifted out of its sheath, the posterior wall of this covering can be seen to vary in thickness in different regions. This is due to the manner of formation of the rectus sheath by the tendons of the external, internal, and transversus abdominis muscles.

These three muscles form a sheath both anterior and posterior to the rectus abdominis muscle—the **rectus sheath** (Fig. 3-32). The external oblique muscle forms the anterior wall over its entire length. The internal oblique muscle splits at the lateral edge of the rectus muscle, one lamina aiding the external oblique in forming the anterior wall and another forming the posterior wall as far as a point halfway between the umbilicus and the pubis. At this point, both layers of the internal oblique are found on the anterior surface of the rectus muscle. The transversus abdominis muscle forms the posterior sheath of the rectus muscle down to the same point where it also courses anterior to the muscle. It is quite apparent that the anterior wall of the rectus sheath is complete, while the posterior wall extends inferiorly only to a point halfway between the umbilicus and the pubis. An arched line occurs where the transition from posterior to anterior sheath occurs; this is known as the **arcuate line.** This line is often indistinct, however. Although there is considerable variation,† the rectus sheath can be summarized as follows. The anterior wall of the rectus sheath is formed throughout its entire length by the external abdominis oblique muscle and the anterior lamina of the internal oblique muscle; these are joined by the posterior lamina of the internal oblique and the transversus abdominis at the level of the arcuate line. The posterior wall of the sheath is incomplete being absent below the arcuate line; superior to the line it is formed by the posterior lamina of the internal oblique muscle and the tendon of the transversus abdominis muscle.

To complete the abdominal wall, the transversus abdominis muscle and the inferior end of the rectus abdominis muscle are lined on their deep surfaces by deep fascia which in this location is called **transversalis fascia.** Deep to this fascia is found a loose layer of connective tissue designated as **subserous fascia,** and then the **peritoneum.**‡ The subserous fascia is named for its posi-

---

*C. B. McVay and B. J. Anson, 1938, Fascial constituents in the abdominal, perineal, and femoral regions, *Anat. Record* 71:401. Claim is made for direct continuation of deep (muscle) fascia on external abdominis oblique, thigh muscles, and muscles in the superficial pouch of the perineum in the female.

†C. B. McVay and B. J. Anson, 1940, Composition of the rectus sheath, *Anat. Record* 77: 213. These authors found the usual stylized description of the rectus sheath in only 2 of 56 bodies studied.

‡We will see later that transversalis fascia is fairly complex because it is not limited to the transversus abdominis muscle. Browne (*Lancet* 1: 460, 1933) makes a plea for a sensible concept of the fascia between peritoneum and the abdominal wall musculature. He would combine transversalis, subserous, and the thin connective tissue which is part of the peritoneum into a single "abdominal connective tissue."

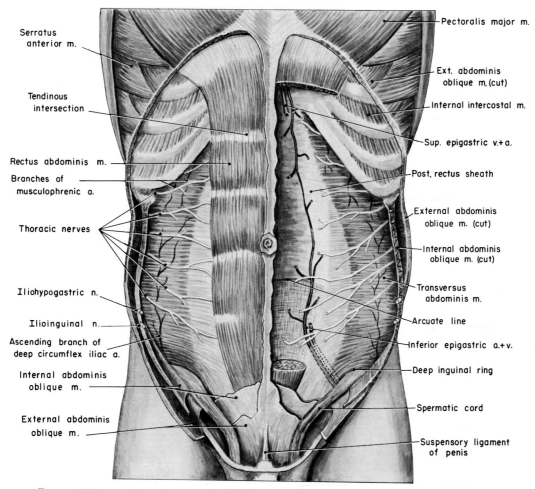

Figure 3-31. *Anterior abdominal wall II. The external and internal abdominis oblique muscles have been removed from the right side of body to show the transversus abdominis muscle. On the left side the rectus abdominis muscle has been removed to show the posterior rectus sheath. Note that the nerves course between the internal and transverse abdominis muscles. The iliohypogastric and ilioinguinal nerves pierce the internal oblique muscle approximately at the point cut.*

tion beneath (external to) the peritoneum when viewed from inside the peritoneal cavity. The abdominal wall is easily understood, whether one looks at it from the outside or the inside. The skin outside can be compared to the peritoneum inside and the subserous fascia inside to the superficial fascia outside. Therefore, starting outside, the order of structures from superficial to deep is skin, superficial fascia, deep fascia, and abdominal musculature; starting inside, the order is peritoneum, subserous fascia, deep fascia (transversalis fascia), and abdominal musculature.

It does not take much imagination to realize the importance of the above to the surgeon.

MUSCLES

The muscles forming the abdominal wall are:

1. External abdominis oblique
2. Internal abdominis oblique
3. Transversus abdominis
4. Rectus abdominis
5. Pyramidalis

1. External abdominis oblique (Fig. 3-30). This muscle **arises** in a toothlike manner from the fifth to the twelfth ribs, interdigitating with the origin of the serratus anterior and latissimus dorsi muscles. The fibers course inferiorly and medially to **insert** by fleshy attachment to the outer lip of the anterior half of the iliac crest, and by aponeurosis to the xiphoid process, linea alba, symphysis pubis, pubic crest, and pectineal line. Its posterior border is free and is not attached to the vertebral column. In the area between the anterior superior iliac spine and the pubic tubercle, the external oblique muscle turns under in such a manner as to form a strong ligament. This is called the **inguinal ligament.** It lies in the groove on the skin surface that marks the boundary line between abdominal wall and thigh and can be felt as a tight elastic band. The medial end of this ligament not only attaches to the pubic tubercle but reflects laterally along the pectineal line as the **lacunar ligament.** A further extension of this ligament along the pectineal line is the **pectineal ligament** (Fig. 3-4 on page 103). A rather inconstant ligament, the **reflected inguinal ligament,** reflects from the inguinal ligament at its medial end and courses superiorly and medially deep to the external oblique muscle to reach the linea alba, where it fuses with fibers from the opposite external oblique muscle. Another conception of this ligament has it coursing in a downward direction from the other side of the body. Since it is called the reflected inguinal ligament, perhaps it is better to think of it as reflecting superiorly from the inguinal ligament. There is an opening in the aponeurotic tendon just superior and lateral to the attachment of the inguinal ligament to the pubic tubercle. This is the **superficial inguinal ring** and the edges of the opening are known as **medial** and **lateral crura.** Intercrural fibers can be seen at the lateral end of the opening bridging between the two crura. This external ring has resulted from the fact that the spermatic cord in the male and the round ligament in the female traverse it to reach the scrotum or labia majora (see Fig. 3-30).

2. The **internal abdominis oblique** (Fig. 3-30) muscle **arises** from the thoracolumbar fascia, the anterior two-thirds of the iliac crest and from the lateral two-thirds of the inguinal ligament. The upper fibers course superiorly and medially to **insert** on ribs 7 to 12 along the entire costal arch, the xiphoid process, and linea alba; the middle fibers course horizontally to **insert** on the lower end of the linea alba; and the lowest fibers curve inferiorly

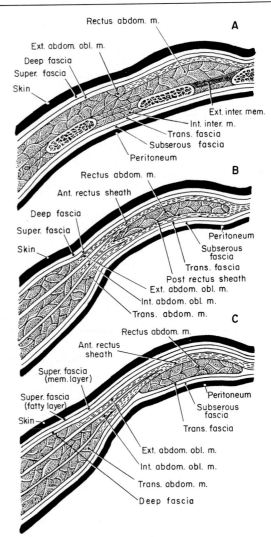

Figure 3-32. *Diagrammatic representation of the rectus sheath: (A) through the rectus near its insertion, (B) superior to the umbilicus, and (C) through the rectus at a point inferior to the arcuate line.*

to **insert** on the pubic crest and pectineal line. Some of those inferior fibers extend along the spermatic cord and surround the testis and epididymis. Known as the **cremasteric muscle,** it is used to elevate the testis under certain conditions. It is much more powerful in other animals. Nerve supply to these fibers is the genitofemoral nerve.

3. Transversus abdominis (Fig. 3-31). This muscle occupies this same territory deep to the above muscles; it **arises** from the costal cartilages on the entire costal arch, from the thoracolumbar fascia, the anterior two-thirds of the inner lip of the iliac crest, and the lateral third of the inguinal ligament. All fibers course transversely except the most inferior, which curve inferiorly. They **insert** on the xiphoid process, linea alba, pubic crest and pectineal line. The attachment to the pectineal line is fused with the internal oblique muscle at this point to form the **falx inguinalis** or **conjoined tendon.**

4. **Rectus abdominis** (Fig. 3-31). This straight muscle **arises** from the pubic crest and **inserts** on the cartilages of the fifth, sixth, and seventh ribs. This is difficult to visualize but the cartilages are close together at this region and the muscle insertion is practically horizontal. Indicative of the segmental origin and structure of the body, this muscle is divided into four or five sections by tendinous **intersections.** One occurs at the level of the umbilicus and another at the inferior end of the xiphoid process; a third occurs at a point equidistant between these two and another can occur between the umbilicus and pubis.

5. **Pyramidalis** (Fig. 3-30). A small, inconstant muscle, the pyramidalis may be found inside the rectus sheath and anterior to the inferior end of the rectus muscle. It **arises** from the pubic crest, extends to a variable height and **inserts** on the linea alba. It **tightens** the linea alba when it contracts; its **nerve supply** is the subcostal nerve.

From the above it is easily seen that, although the external, internal, and transversus abdominis muscles occupy the same territory and have the same attachments, they do differ from each other. The difference in origins and insertions is easily visualized; the difference in distribution in the rectus sheath, in the inguinal region, and along the vertebral column also should be noted.

The role of these abdominal muscles in respiration has already been described; in addition they are used to increase intra-abdominal pressure in forced expiration, defecation, micturition, and parturition. They also play a role in flexing the vertebral column and in bending from side to side.* The **nerve supply** is by the seventh to elev-

---

*W. F. Floyd and P. H. S. Silver, 1950, Electromyographic study of patterns of activity of the anterior abdominal wall muscle in man, *J. Anat.* 84: 132. This study indicates that the oblique and transverse abdominal muscles are used in straining while the rectus abdominis is limited to flexion of the vertebral column.

enth intercostals, the subcostals, and the iliohypogastric, and ilioinguinal nerves.

It is interesting to **compare** the abdominal muscles with those forming the thoracic wall. The external intercostal muscles correspond to the external abdominis oblique, the internal intercostals to the internal abdominis oblique, and the transversus thoracis plus the intercostalis intimus plus the subcostals correspond to the transversus abdominis muscle. The direction of the fasciculi correspond, and the nerves and blood vessels course between similar layers. Occasionally a longitudinally placed muscle on the anterior chest wall is found—the **sternalis muscle**—that corresponds to the rectus abdominis muscle in the abdomen.

NERVES

From the distribution of the cutaneous nerves, it was seen that the following nerves are involved with innervation of the abdominal wall:

1. Seventh to eleventh intercostals
2. Subcostal
3. Iliohypogastric
4. Ilioinguinal

1. **Seventh to eleventh intercostals** (Fig. 3-31). It is obvious that these intercostal nerves do not remain in their proper intercostal spaces. Indeed, the seventh to the eleventh leave their spaces and enter the abdominal wall, coursing deep to the costal cartilages; they continue between the transversus abdominis and internal abdominis oblique muscles until the lateral edge of the rectus sheath is reached. They then pierce the sheath, course posterior to the rectus muscle and then pierce it and the anterior wall of the rectus sheath to become **anterior cutaneous nerves.** They give off branches throughout their course to the abdominal muscles.

2. **Subcostal nerve** (T12). This nerve is difficult to visualize, because it is in relation to structures not as yet dissected or studied. See Fig. 5-41 on page 261. Although it starts in the thoracic cavity posterior to the most superior end of the psoas major muscle, it soon enters the abdominal cavity by passing posterior to the edge of the diaphragm. It courses anterior to the quadratus lumborum muscle and posterior to the kidney to enter the transversus abdominis muscle (Fig. 5-41); it continues

between the transversus abdominis and the internal abdominis oblique muscles to enter the rectus sheath just as described for the other thoracic nerves.

3. The **iliohypogastric nerve** (Figs. 5-41, 3-31, and 3-30) is a branch of the lumbar plexus (L1) and also appears between the psoas major and quadratus lumborum muscles; it pierces the posterior aponeurosis of the transversus abdominis muscle, and continues between this muscle and the internal abdominis oblique. It pierces the latter muscle about 1 inch superior to the anterior superior iliac spine and becomes cutaneous about 1 inch superior to the superficial inguinal ring. It gives off branches to the abdominal musculature along its entire course. Its **lateral cutaneous branch** courses into the gluteal region. The iliohypogastric nerve can be severed in an approach to a diseased appendix. If this occurs, a weakness of the musculature around the inguinal canal occurs which may predispose to the development of an inguinal hernia.

4. The **ilioinguinal nerve** (Figs. 5-41, 3-31, and 3-30) also arises from the lumbar plexus (L1) and follows a course similar to that of the iliohypogastric but is more inferiorly placed. It has no lateral branch, pierces the internal abdominis muscle more anteriorly, and then, passing between the internal and external abdominis muscles where they form the anterior wall of the inguinal canal (a canal through the abdominal muscles for the spermatic cord or ligamentum teres of uterus), enters the inguinal canal and escapes through the external inguinal ring. Besides innervating the muscles between which it courses, it sends cutaneous branches to the skin over the pubis, on the anterior surface of the scrotum or to the labia majora, and sends one cutaneous branch to the thigh. Curiously enough, it does not innervate the cremaster muscle (a muscle found surrounding the spermatic cord and testes) in spite of supplying the internal oblique muscle from which the cremaster is derived. This nerve is always encountered in a surgical repair of an inguinal hernia; it should be preserved.

ARTERIES AND VEINS

The following arteries are diagramed in Figure 3-33:

1. Musculophrenic
2. Lumbars
3. Superficial epigastric
4. Superficial circumflex iliac
5. Deep circumflex iliac
6. Inferior epigastric

1. **Musculophrenic.** The arteries of the abdominal wall might be expected to follow the nerves. This does not occur. Instead of leaving the intercostal spaces and entering the abdominal wall as do the nerves, the seventh, eighth, and ninth intercostal arteries anastomose with one of the terminal branches of the internal thoracic—the **musculophrenic** artery. This artery courses along the costal arch giving off branches to the diaphragm and to the abdominal wall in the region of the costal arch. The tenth and eleventh intercostal arteries do enter the abdominal musculature.

2. The **lumbar** arteries, which are quite short and enter the abdominal musculature for only a short distance, anastomose with the tenth and eleventh intercostals just mentioned and with the following arteries.

3 and 4. The inferior part of the abdominal wall has contributions from two branches of the femoral artery in the thigh, namely, the **superficial epigastric** which crosses the inguinal ligament about at its midpoint, and the **superficial circumflex iliac** which courses laterally just inferior to the inguinal ligament before entering the abdominal region superior to the iliac crest (Fig. 3-30). This artery also anastomoses with the lumbar arteries just described, and with the next artery, the deep circumflex iliac.

5. The **deep circumflex iliac** artery is a branch of the external iliac artery and is similar in its distribution to the superficial circumflex iliac artery, but gives off a large ascending branch that ascends as high as the costal margin. It courses for a short distance in the subserous fascia, pierces the transversalis fascia, and continues just posterior to the inguinal ligament. After reaching the anterior superior iliac spine, it pierces the transversus abdominis muscle and ramifies between it and the internal oblique muscle. Its ascending branch is also distributed between these muscles. The lateral aspect of the abdominal wall, therefore, obtains its blood supply from branches of the musculophrenic, the lumbars, the superficial and deep circumflex iliac, and the superficial epigastric arteries.

6. The largest artery to the abdominal wall is the **inferior epigastric** branch of the external iliac artery (Figs. 3-31 and 3-33). This branch arises just deep to the inguinal canal, pierces the transversalis fascia in its course

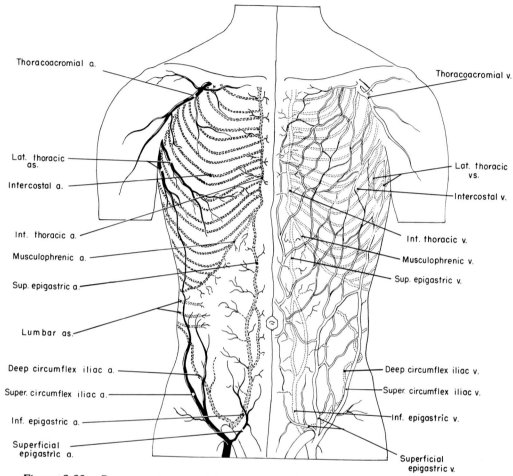

Figure 3-33. *Diagrammatic presentation of the arterial supply (right side of body) and venous drainage (left side of body) of the abdominal wall.*

superiorly and medially, reaches the deep surface of the rectus abdominis muscle, and enters the rectus sheath by passing anterior to the arcuate line. It ramifies on the deep surface of the rectus muscle and anastomoses with branches of the **superior epigastric** artery, the other terminal branch of the internal thoracic artery, which also ramifies on the deep surface of the rectus abdominis muscle and within this muscle after entering the rectus sheath at its superior extent. The blood supply to the anterior and medial aspect of the abdominal wall is, therefore, from the epigastric vessels.

The veins take similar courses as just described for the arteries but are much more profuse in their branching.

LYMPHATICS

Lymphatics of the muscular wall follow the arteries in a retrograde manner to the external iliac, lumbar, and internal thoracic nodes.

CLINICAL ASPECTS

Incisions are frequently made in the abdominal wall, and the anatomy involved is important. Preservation of nerves is mandatory, for muscles are paralyzed when these are cut. The **midline** incision through the linea alba is good in that no nerves are cut and the vascularity is at a minimum. This last point, however, leads to poor healing of this incision. Many surgeons prefer a **midrectus** incision or one between that and a midline incision, a **paramedian** incision. Such an incision, if made superior to the arcuate line, goes through skin, superficial fascia (both layers), and then the fused layers forming the anterior wall of the rectus sheath. Then the rectus muscle is retracted laterally and the posterior wall of the sheath cut, then the transversalis fascia, subserous fascia, and lastly the peritoneum. If such an incision were inferior to the arcuate line, there would be no posterior wall of the rectus sheath. Oblique incisions in a **more lateral location** can take advantage of the muscle arrangement, incisions through the muscles being made in the direction of the fibers in each separate muscle. Such incisions repair easily with almost no complications. The nerves have to be carefully avoided in such incisions, however. Long **transverse** incisions across the abdominal wall are frequent and cut the rectus abdominis transversely. In spite of cutting the muscle in this manner, such incisions heal well if nerves and vascular supply are carefully maintained.

DESCENT OF TESTES

The **testes** do not develop in their definitive positions but migrate to the scrotum during embryonic life (Fig. 3-34).* They start in the upper lumbar region as ridges of **mesodermal tissue** on the posterior wall of the coelom, and cover the medial part of the mesonephros. This, in turn, is covered with a layer of peritoneum containing

---

*L. J. Wells, 1943, Descent of the testes: anatomical and hormonal considerations, *Surgery,* 14: 436. A summary of thoughts on this subject to this date—1943.

germinal cells. The seminiferous tubules develop from this epithelial tissue, and the stroma of the testis from the underlying mesoderm. The abovementioned connections with the mesonephros form the ducts of the testis and epididymis, and the mesonephric duct forms the vas deferens (Fig. 5-89 on page 339).

In addition, a mesodermal band develops retroperitoneally which is attached inferiorly to the skin forming the scrotum and, after penetrating the abdominal muscles, superiorly to the testis and the peritoneum. This is the **gubernaculum testis.** Due to an unequal growth pattern and to an unexplained shortening of the gubernaculum, the peritoneum and testis are pulled down into the scrotum. This peritoneum is the **processus vaginalis.**

This results in a tube of peritoneum—**processus vaginalis**—projecting from the abdominal cavity, through the abdominal musculature (via the inguinal canal), and into the scrotum. The testis continues in its descent, accompanying the processus vaginalis, and comes to lie to one side of its most inferior extent. The testis then embeds itself into the processus vaginalis and thereby obtains two layers of peritoneum, which become the **visceral** (on the testis) and **parietal layers** of the **tunica vaginalis** (Fig. 3-34). The part of the processus vaginalis between the testis and the inside of the abdominal musculature degenerates to a thin connective tissue strand. Because the **vas deferens** remains attached to the testis and epididymis, and the arteries, veins, nerves, and lymphatics were connected with the testis before its descent, these structures remain attached after the descent. They ascend the scrotum, traverse the inguinal canal, and enter the abdominal cavity. These structures make up what is known as the **spermatic cord.**

The processus vaginalis and testis, in their descent through the inguinal canal, pick up fascial layers from each of the abdominal muscles forming this canal. Therefore, the testis and processus are surrounded by three layers of fascia. These are best studied after the inguinal canal is described.

Figure 3-34. *Three dissections showing the formation of the processus vaginalis and the relation of the testis to this process: (A) fetus of about 20 weeks, (B) fetus in seventh month, and (C) fetus in ninth month (left testis has been rotated through 90° to expose epididymis). Normally the processus vaginalis degenerates leaving testis covered by two layers of tunica vaginalis. (From* Patten's Human Embryology *by C. E. Corliss. © 1976 by McGraw-Hill, Inc. Used with permission of McGraw-Hill Book Company.)*

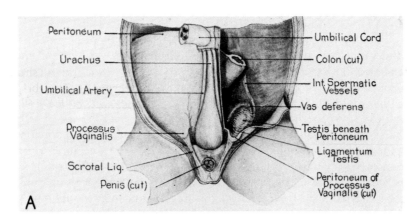

Peritoneum — Umbilical Cord

Urachus — Colon (cut)

Umbilical Artery — Int. Spermatic Vessels

— Vas deferens

Processus Vaginalis — Testis beneath Peritoneum

Ligamentum Testis

Scrotal Liq. — Peritoneum of Processus Vaginalis (cut)

Penis (cut)

A

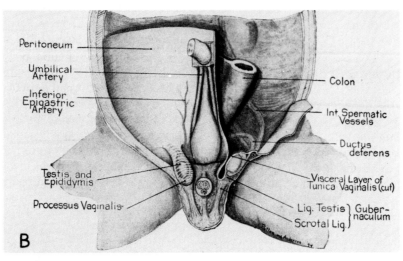

Peritoneum —

Umbilical Artery — Colon

Inferior Epigastric Artery — Int. Spermatic Vessels

— Ductus deferens

Testis and Epididymis — Visceral Layer of Tunica Vaginalis (cut)

Processus Vaginalis — Liq. Testis } Guber-
Scrotal Liq. } naculum

B

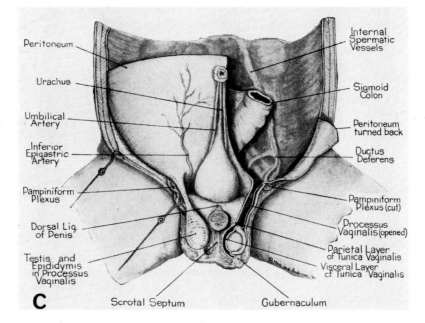

Peritoneum — Internal Spermatic Vessels

Urachus — Sigmoid Colon

Umbilical Artery — Peritoneum turned back

Inferior Epigastric Artery — Ductus Deferens

Pampiniform Plexus — Pampiniform Plexus (cut)

Dorsal Liq. of Penis — Processus Vaginalis (opened)

Testis and Epididymis in Processus Vaginalis — Parietal Layer of Tunica Vaginalis
Visceral Layer of Tunica Vaginalis

Scrotal Septum — Gubernaculum

C

The descent of the testes is usually concluded by the eighth month of intrauterine life, but occasionally a testis will remain inside the abdominal cavity or in the inguinal canal. This is known as **cryptorchidism.** It may be cured by spontaneous descent if there is no obstruction, or by surgery if there is obstruction. It is important to correct the situation before puberty, for sperm are not viable for long periods at body temperature. Small portions of the processus vaginalis may persist and, probably due to injury, may secrete large quantities of serous fluid. This condition is known as a **hydrocele.**

INGUINAL CANAL see p. 138

Growth of the abdominal muscles around the gubernaculum results in a passageway through the inferior part of the abdominal wall called the **inguinal canal.** This imaginary canal is considered to have a **roof** superiorly, a **floor** inferiorly, an **anterior wall,** and a **posterior wall.** The opening already described in the external oblique muscle is the **superficial ring,** and the **deep ring** is about 2 inches lateral and slightly superior, but at the deeper peritoneal level. The canal therefore takes a slanting course through the abdominal muscles from lateral to medial.

An understanding of the boundaries of the inguinal canal is dependent upon knowledge of the abdominal musculature in this region. In way of review, the **external abdominis oblique** muscle formed the **inguinal ligament,** which stretched from the anterior superior iliac spine to the pubic tubercle. A continuation of the medial attachment of this ligament laterally along the pectineal line forms the **lacunar** and **pectineal ligaments,** and a reflection of the ligament superiorly and medially formed the **reflected inguinal ligament.**

Now, if the spermatic cord is placed inside the inguinal canal, reference can be made to it (Fig. 3-35). It can be seen that the inguinal ligament and the lacunar ligament will be inferior to the cord, thereby forming the floor of the inguinal canal. Because the reflected inguinal ligament spreads superiorly and medially, it will be posterior

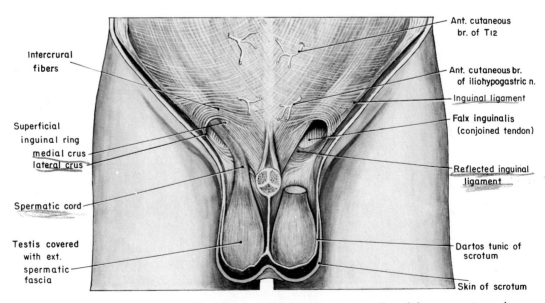

Figure 3-35. *Inguinal canal I—showing walls of the inguinal canal, and the spermatic cord, epididymis, and testis and their coverings. Right side of body—the external abdominis oblique muscle and the external spermatic fascia. Left side—the spermatic cord has been removed to show parts of the posterior wall and floor of the inguinal canal.*

✱ see P·130

to the medial end of the cord and aid in forming the posterior wall of the canal. The aponeurosis of the external oblique muscle covers the spermatic cord throughout its extent in the inguinal canal, and thereby aids in forming the anterior wall of the inguinal canal. The external abdominis oblique muscle, therefore, aids in forming the anterior wall of the entire length of the canal, the entire floor, and the posterior wall of the medial end of the canal.

The internal abdominis oblique muscle (Fig. 3-36) originates from the lateral two-thirds of the inguinal ligament. These fibers course superiorly and cover the spermatic cord, thereby aiding in formation of the anterior wall of the canal. They then continue medially, and these arched fibers aid in forming the roof, because they are superior to the cord in this location. The internal oblique fibers then continue medially and inferiorly to insert on the pectineal line; in this position they are posterior to the reflected inguinal ligament and also posterior to the cord. They therefore aid in forming the posterior wall of the canal. In summary, the internal abdominis oblique muscle aids in forming the anterior wall of the lateral part of

the canal, the roof, and the posterior wall of the medial end of the canal.

The inferior end of the transversus abdominis muscle (Fig. 3-36) arises from the lateral third of the inguinal ligament; the muscle fibers take a path similar to those of the internal oblique muscle, except that these fibers are more lateral in position; this is due to the difference in origin of the two muscles from the inguinal ligament. The transversus is lateral to the internal ring and does not take part in formation of any of the anterior wall of the canal. Its fibers continue medially in the form of an arch, thus joining the internal oblique in forming the roof, and continue medially and inferiorly to insert onto the pectineal line. These inferior fibers are usually fused with those of the internal oblique. This forms the falx inguinalis or conjoined tendon. The transversus abdominis muscle, therefore, forms part of the roof and part of the posterior wall of the medial end of the inguinal canal.

From this description, it is obvious that the conjoined tendon and the reflected inguinal ligament form the posterior wall of the medial third of the canal which leaves the lateral two-thirds of the canal with no posterior wall.

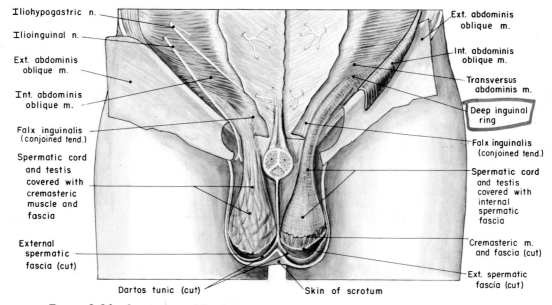

Figure 3-36. *Inguinal canal II—showing walls of the inguinal canal, and the spermatic cord, epididymis, and testis and their coverings. Right side of body—the internal abdominis oblique muscle and the cremasteric muscle and fascia. Left side—the transversus abdominis muscle and internal spermatic fascia.*

This area is filled in by an extension of the transversalis fascia, and the internal ring is bounded by this same fascia. We will see in a moment how an acquired inguinal hernia pushes through this weak posterior wall.

The **boundaries** of the inguinal canal may be summarized as follows (Fig. 3-37)[*]

Anterior wall
 1. External oblique over entire length
 2. Internal oblique on lateral one-third

Posterior wall
 1. Transversalis fascia over entire length
 2. Falx inguinalis (conjoined tendon) of internal oblique and transversus on medial one-third
 3. Reflected inguinal ligament of external oblique on medial one-third

Roof
 1. Arched fibers of internal oblique and transversus muscles

Floor
 1. Inguinal ligament of external oblique
 2. Lacunar ligament of external oblique

SPERMATIC CORD
AND ITS COVERINGS

As mentioned in the description of the descent of the testis, the processus vaginalis, accompanied by the testis, made its way through the abdominal musculature. Since the testis had already obtained its ducts from the mesonephros, the epididymis and vas deferens were pulled down into the scrotum also. In addition, the testis carried its blood supply, venous drainage, nerve supply, and lymphatics along with it. Therefore, the inguinal

---

*Y. Appajee, 1945, On the anatomy of the inguino-hypogastric and inguino-femoral regions, *Indian J. Surg.* 7: 113. This author disagrees with almost all of our concepts on the structure of the inguinal region. He feels there is no inguinal canal as usually described; no changes in rectus sheath in lower abdomen; and the conjoined tendon and other accessory ligaments not mentioned in this text are, in reality, merely thickenings in the transversus abdominis muscle.

 B. J. Anson, E. H. Morgan, and C. B. McVay, 1960, Surgical anatomy of the inguinal region based upon a study of 500 body halves, *Surg. Gynecol. Obstet.* 111: 707. A very good lesson in variation, for the student.

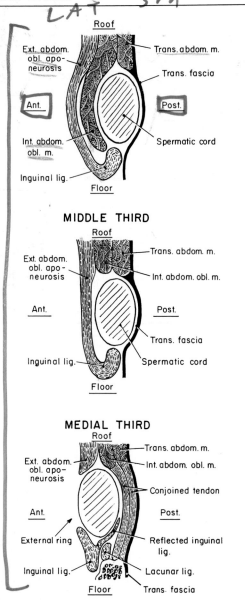

Figure 3-37. *Transverse sections through the lateral, middle, and medial thirds of the inguinal canal showing composition of its walls (diagrammatic).*

canal contains these structures which, together, make up the **spermatic cord.**

 As the processus vaginalis made its way through the abdominal wall, it picked up **fascial layers** and carried them down into the scrotum (Fig. 3-40). Since the testis and attached structures are just deep to the peritoneum in the subserous fascial layer, they also are covered by these same fascial layers (Figs. 3-38 and 3-40). As the processus pushed through what became the internal inguinal ring, it pushed through the posterior wall of the inguinal canal and thereby picked up a layer of transversalis fascia; this layer is called the internal spermatic fascia, and it covers the spermatic cord throughout its entire extent in the inguinal canal. The processus then continued and picked up a layer of fascia and muscle from the **internal oblique muscle;** these are the **cremasteric fascia and muscle,** this fascia joining the spermatic cord approximately in the middle of the inguinal canal (Figs. 3-34 and 3-40). Although some animals can pull the testes into the comparative safety of the abdominal cavity by contracting the cremaster muscle, most men cannot accomplish this much movement. In fact, the cremaster muscle performs at reflex level, its contractions as a result of scratching the

medial surface of the thigh being the well known **cremasteric reflex.** Some claim it will respond to externally applied cold. The outer layer of fascia, the **external spermatic fascia,** was picked up from the **external abdominis oblique aponeurosis** (Figs. 3-39 and 3-40) and therefore starts at the external ring.

The spermatic cord, as mentioned before, consists of the vas deferens and its artery, and of the blood and nerve supply, venous drainage, and lymphatic drainage of the testis and epididymis. These structures are embedded in loose connective tissue that corresponds to the subserous fascia. The **artery to the vas deferens** is a special branch from the arteries to the bladder; this artery reaches the vas deferens while in the pelvic cavity. The **artery to the testis** is the testicular originating from the abdominal aorta. The **veins** form a pampiniform plexus and ultimately drain into the renal vein on the left side

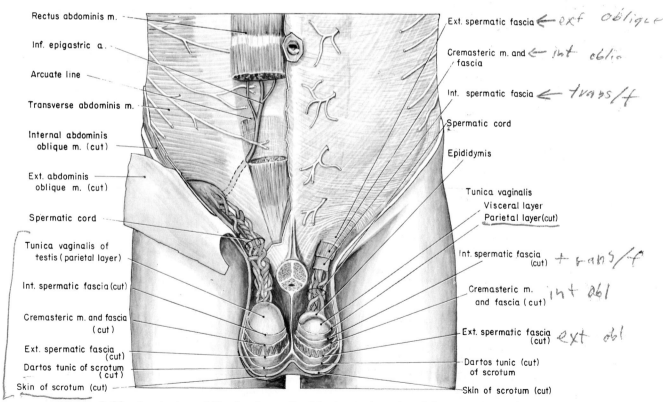

Figure 3-38. *Inguinal canal III—showing walls of the inguinal canal, and the spermatic cord, epididymis, and testis and their coverings. Right side of body—the spermatic cord and its contents devoid of fascial coverings. Left side—a summation of the fascial layers of the spermatic cord, epididymis, and testis.*

LATERAL THIRD

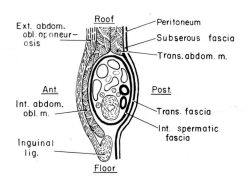

Roof
Ext. abdom. obl. aponeurosis
Peritoneum
Subserous fascia
Trans. abdom. m.
Ant.
Int. abdom. obl. m.
Post.
Trans. fascia
Int. spermatic fascia
Inguinal lig.
Floor

MIDDLE THIRD

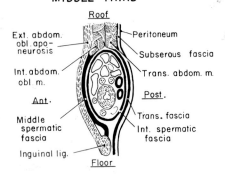

Roof
Ext. abdom. obl. aponeurosis
Peritoneum
Subserous fascia
Int. abdom. obl. m.
Trans. abdom. m.
Ant.
Post.
Middle spermatic fascia
Trans. fascia
Int. spermatic fascia
Inguinal lig.
Floor

MEDIAL THIRD

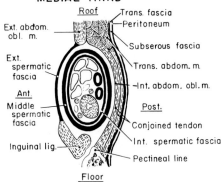

Roof
Ext. abdom. obl. m.
Trans. fascia
Peritoneum
Subserous fascia
Ext. spermatic fascia
Trans. abdom. m.
Int. abdom. obl. m.
Ant.
Middle spermatic fascia
Post.
Conjoined tendon
Inguinal lig.
Int. spermatic fascia
Pectineal line
Floor

IN SCROTUM

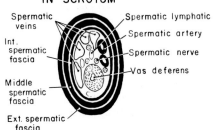

Spermatic veins
Spermatic lymphatic
Int. spermatic fascia
Spermatic artery
Spermatic nerve
Vas deferens
Middle spermatic fascia
Ext. spermatic fascia

and the inferior vena cava on the right side. The **lymphatics** drain into nodes in the abdominal cavity known as the lumbar chain of nodes located next to the aorta on the left side and inferior vena cava on the right. One **nerve** is the **genital branch of the genitofemoral** from the lumbar plexus (L1, 2) which innervates the cremaster muscle and gives sensory supply to the scrotum. Figure 5-40 on page 260 shows many of these structures. The **nerves to the testis** itself are visceral afferent and efferent fibers arising from the tenth and eleventh thoracic segments of the spinal cord; the epididymis receives similar fibers. These fibers join the aortic and renal plexuses before following the testicular artery to the testis and epididymis. The rather excruciating pain associated with a blow to the testes is difficult to explain. It certainly is visceral in nature and quite inclusive. Perhaps the fact that the visceral afferent fibers from this organ return to spinal cord segments T10 and T11, where similar fibers from much of the gut also join the spinal cord (the midgut), accounts for its inclusiveness.

It should be noticed that all these structures are arising from or going to the region where the testis started in its development. This is an important **concept** to obtain for it is true of many other regions of the body. *A structure obtains its nerves, arteries, veins, and lymphatics where it starts in its development; if it migrates to another region, it carries these structures along with it.*

INGUINAL HERNIA

An understanding of the origin of the fascial layers covering the spermatic cord and testis is helpful to an under-

Figure 3-39. Diagrammatic presentation of the walls of the inguinal canal and the layers of the spermatic cord in the lateral, middle, and medial thirds of the canal and in the scrotum. An attempt has been made to indicate at which point each layer is added. Note that the cremasteric (middle spermatic) fascia is added to the anterior surface of cord before it is applied to the posterior surface.

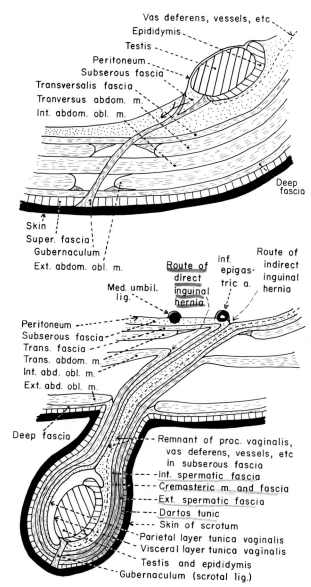

Vas deferens, vessels, etc.
Epididymis
Testis
Peritoneum
Subserous fascia
Transversalis fascia
Tranversus abdom. m.
Int. abdom. obl. m.

Deep
fascia

Skin
Super. fascia
Gubernaculum
Ext. abdom. obl. m.

Route of
direct
inguinal
hernia

Inf.
epigas-
tric a.

Route of
indirect
inguinal
hernia

Med. umbil.
lig.

Peritoneum
Subserous fascia
Trans. fascia
Trans. abdom. m.
Int. abd. obl. m.
Ext. abd. obl. m.

Deep fascia

Remnant of proc. vaginalis,
vas deferens, vessels, etc
in subserous fascia
Int. spermatic fascia
Cremasteric m. and fascia
Ext. spermatic fascia
Dartos tunic
Skin of scrotum
Parietal layer tunica vaginalis
Visceral layer tunica vaginalis
Testis and epididymis
Gubernaculum (scrotal lig.)

Figure 3-40. *Diagrammatic presentation of manner in which spermatic cord and testis obtain their fascial layers: (A) Gubernaculum attached inferiorly to the skin of the scrotum and superiorly to the testis and epididymis. Note that this diagram depicts the gubernaculum as being surrounded by the abdominal muscles in their development. (B) The processus vaginalis and testis have been guided through the inguinal canal by the gubernaculum and have picked up fascial layers from the transversalis fascia (internal spermatic fascia), the internal abdominis oblique muscle (cremasteric muscle and fascia or middle sper-*

standing of the layers of fascia found covering various types of hernias. The inferior epigastric artery, the inguinal ligament, and 'the lateral border of the rectus abdominis muscle form the boundaries of a triangle—the **inguinal triangle**—the artery being lateral, the ligament inferior, and the muscle medial.

Any hernia occurring in this triangle is designated as a **direct inguinal hernia,** so called because of directly traversing the triangle and, thereby, the posterior wall of the inguinal canal. These hernias are of the acquired type, in contrast to the congenital type, and are due to a weakness of the posterior wall of the canal lateral to the falx inguinalis (conjoined tendon). As previously pointed out, the posterior wall at this region is made up of transversalis fascia, covered only by the accompanying subserous fascia and peritoneum. Abdominal contents put pressure on this wall and will push peritoneum, subserous fascia, and transversalis fascia ahead of the herniating mass. It will enter the inguinal canal alongside the spermatic cord distal to the origin of the internal spermatic fascia, because this fascia joins the spermatic cord at the internal ring. However, it will be proximal to the cremasteric fascia, for this joins the posterior surface of the cord from the conjoined tendon which is medial to the hernia. Therefore, this herniating mass, with the layers it has picked up, will start down the inguinal canal between the internal spermatic and cremasteric fascial layers (Fig. 3-40).

Hernias that enter the inguinal canal lateral to the inferior epigastric artery, and therefore outside or lateral to the triangle, are called **indirect inguinal hernias.** The great majority of this type have a congenital origin, due to failure of the processus vaginalis to close off properly. This, of course, provides a tube which the herniating viscus can follow into the scrotum; the layers covering such a herniating viscus would be the same as covered the processus vaginalis originally (Figs. 3-34 and 3-40).

Hernias can occur elsewhere in the abdominal cavity. **Umbilical hernias** result from the failure of tissue growth to close off this region after the umbilical cord is

matic fascia), and the external abdominis oblique muscle (external spermatic fascia). The processus vaginalis has degenerated to a connective tissue cord. Note routes taken by direct and indirect inguinal hernias. The majority of indirect inguinal hernias result from failure of the processus vaginalis to close off.

tied.* Hernias occur at the lumbar triangle—**lumbar hernias**—and, as we shall see later, hernias occur through the diaphragm—**diaphragmatic hernias.** Another type, the femoral hernia, is that found in the femoral canal to be studied later. This hernia often pro-

duces a swelling near the inguinal ligament and is easily confused with an inguinal hernia.

The pathways taken by these hernias are more easily visualized if the inside of the lower anterior abdominal wall is studied (Fig. 3-41). The peritoneum in this region is thrown into folds. The median fold extends superiorly from the superior surface of the bladder and covers the remains of the obliterated urachus of the embryo; this is the **median umbilical fold** or **ligament.** Just lateral to the median fold can be seen two folds that cover the obliterated umbilical arteries, which changed to connective tis-

---

*G. M. Wyburn, 1953, Congenital defects of the anterior abdominal wall, *Brit. J. Surg.* 40: 553. The author divides defects in the umbilical region into four groups: (1) nonformation of the umbilical cord; (2) persistence of the physiological hernia; (3) nonclosure of the umbilical ring; and (4) defects of the parietes in the umbilical region.

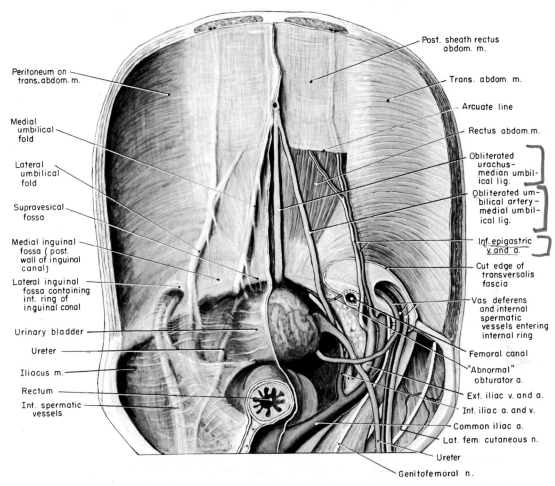

Figure 3-41. *The anterior abdominal wall seen from inside the abdominal cavity. On the right side of the body the peritoneum, subserous fascia, and transversalis fascia (except where it forms the posterior wall of the inguinal canal) have been removed. The fascia and peritoneum are intact on the left. A direct inguinal hernia will occur in the medial inguinal fossa, an indirect in the lateral inguinal fossa.*

sue after the umbilical cord was tied; these are the **medial umbilical folds.** Another set of folds in the peritoneum, the **lateral umbilical folds,** are caused by the inferior epigastric vessels on their way to and from the rectus sheath. Three **fossae** are thereby formed on each side of the midline by these folds: the one between the median and medial folds, being superior to the bladder, is named the **supravesical fossa;** that between the medial and lateral folds, the **medial inguinal fossa;** and that lateral to the lateral fold, the **lateral inguinal fossa.** Because the lateral fold over the inferior epigastric artery is just medial to the internal ring of the inguinal canal, an indirect inguinal hernia is found in the lateral inguinal fossa, while a direct inguinal hernia will be found in the medial inguinal fossa.

TESTIS AND EPIDIDYMIS

The **testis** (Figs. 3-42 and 3-43) produces sperm and is an endocrine organ secreting male sex hormones (andro-

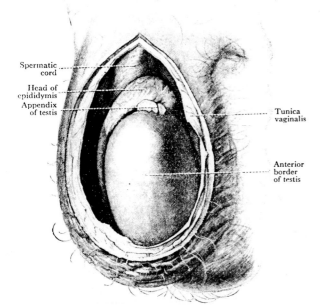

Figure 3-42. *Right testis and epididymis within tunica vaginalis, exposed by removal of the anterior wall of the scrotum. (From Cunningham,* Manual of Practical Anatomy, *courtesy of Oxford University Press.)*

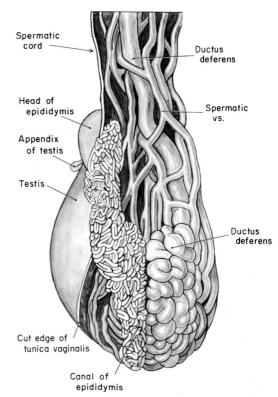

Figure 3-43. *Left testis and epididymis seen from behind, after removal of parietal part of tunica vaginalis. (Modified after Cunningham.)*

gens) into the bloodstream. It is a pale ovoid organ with a greater diameter from superior to inferior than from anterior to posterior; the anteroposterior diameter is greater than that from side to side. It is covered by the **visceral layer** of the **tunica vaginalis** except where the epididymis is attached to it superiorly and posterolaterally. Blood vessels and nerves enter the testis at this posterior **bare area** (Fig. 3-43).

Just deep to the visceral layer of the tunica vaginalis is the capsule of the testis, the **tunica albuginea. Connective tissue septa** extend into the testis from this capsule, dividing the testis into lobes. Spermatozoa, produced by the **seminiferous tubules** in these lobes, enter the head of the **epididymis** by passing through a complicated set of tubles (Fig. 3-44). They first enter a so-called **straight tubule.** These straight tubules form a network of tubules called the **rete testis.** The rete is joined to the

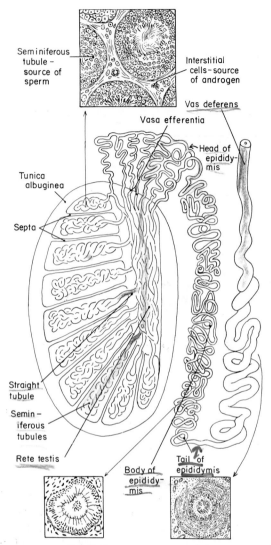

Seminiferous tubule – source of sperm

Interstitial cells – source of androgen

Vas deferens

Vasa efferentia

Head of epididymis

Tunica albuginea

Septa

Straight tubule

Semin- iferous tubules

Rete testis

Body of epididymis

Tail of epididymis

Figure 3-44. *The tubular arrangement of the testis, epididymis, and vas deferens (diagrammatic).*

head of the epididymis by efferent ducts—the **vasa efferentia.**

The **epididymis** consists of this **head,** which is attached to the testis by the ducts just mentioned, a **body,** which is separated from the testis by a space—the **sinus** of the epididymis—and a **tail,** which is attached to the testis by connective tissue. Sperm gradually move inferiorly in the coiled tubule of the epididymis until the tail is reached. Sperm are stored in this organ.

At time of orgasm sperm enter the **vas deferens,** which is attached to the tail of the epididymis, and move along the vas deferens to the **ejaculatory duct,** which enters the **prostatic urethra.** At this point the sperm are mixed with fluid from the **seminal vesicles,** the **prostate,** and **bulbourethral glands** and become motile; this combined fluid is called **semen,** which flows through the remaining urethra during ejaculation.

Because of the method by which the reproductive systems develop from the mesonephric duct (Wolffian duct) in the male and the paramesonephric duct (Mullerian duct) in the female (see Figs. 5-89 and 5-90 on pages 339 and 340), small remnants can be found in close association with the testis and epididymis. These are called the appendix of the testis (derived from the cranial end of the paramesonephric duct) and located on the superior end of the testis (Fig. 3-43); and the appendix of the epididymis, and the paradidymis (persisting tubules of the mesonephros) located on the superior and inferior ends of the epididymis, respectively. When we study the prostatic urethra, we will find a remnant of the caudal end of the paramesonephric duct—the prostatic utricle. In addition, when we study the reproductive organs in the female, we will find similar remnants of the male mesonephric tubules and mesonephric ducts. All of these appendages can become malignant.

Although the vascular and nerve supply to the testes, and the lymphatic drainage, were described under the section on the spermatic cord, these are important enough to repeat. The artery to the testis is the testicular artery, originating from the abdominal aorta. The veins form a panpiniform plexus (Fig. 3-43) and drain into the renal vein on the left side and into the inferior vena cava on the right (Fig. 5-40). The lymphatics drain into nodes located next to the abdominal aorta on the left side and the inferior vena cava on the right. The nerves from and to the testis are visceral afferents and efferents arising from the tenth and eleventh thoracic segments of the spinal cord. All of these facts, including testicular descent, are related to the embryology of the testes and are important clinically, especially the lymphatic drainage. Cancer of the testis or of the appendices of the testis or epididymis is difficult to handle.

SCROTUM

In addition to the parietal layer of the tunica vaginalis, the testis, epididymis, and the spermatic cord are all covered with the three fascial layers already mentioned, namely, the external spermatic, cremasteric, and internal spermatic fascial layers. These are considered part of the testis, not part of the scrotum.

The **scrotum** (Figs. 3-38 and 3-42), a pendulous sac that hangs down from the perineal region posterior to the penis, consists only of **skin** and the **dartos tunic** of smooth muscle and fascia. The skin is wrinkled and possesses a considerable number of hairs. The dartos tunic is a continuation of the membranous layer of superficial fascia in the anterior abdominal wall; we will see later that it is also continuous with a similar layer in the perineum posterior to the scrotum. The dartos differs from these other layers in possessing smooth muscle fibers that contract in response to cold, therefore bringing the testes closer to the body, and relax under conditions of warmth, thus moving the testes away from the body. This smooth muscle has a reddish-brown color. The dartos tunic is divided into two **compartments** that have no connection with one another (Fig. 3-38). Since the fatty layer of superficial fascia in the abdomen is not represented, the scrotum contains no fat in its walls.

Although only the anterior surface of the scrotum is usually dissected at this time, arteries and nerves for the entire scrotum will be mentioned for the sake of completeness. This organ will be studied again with the perineal dissection. The **arteries** to the scrotum are derived from the external and internal pudendal arteries. The former are two in number, are branches of the femoral artery in the thigh and are called **superficial** and **deep external pudendal** arteries; they are distributed on the anterior surface of the scrotum. The **scrotal branches** from the **internal pudendal** are distributed on the posterior surface.

The **sensory nerves** for the anterior surface are the **ilioinguinal** and **genital branch** of the **genitofemoral,** and for the posterior side the **scrotal branches** of the **posterior femoral cutaneous** and of the **perineal,** the latter being a branch of the **pudendal nerve.** The **sympathetic nerves** to the dartos muscle in the scrotum accompany these sensory nerves.

A glance at Figure 6-10 on page 357 and at Figure 5-75 on page 314 might make the above more comprehensible.

The scrotum, having been derived from the skin, has **lymphatic drainage** into the inguinal lymph nodes in the groin. Note that this is not similar to the lymphatic drainage of the testes.

HOMOLOGOUS STRUCTURES IN THE FEMALE

The female has homologous structures to those described for the male. The **ovary** develops from the same genital ridge in the upper lumbar region, and a **gubernaculum** develops in the same manner as in the male but the attachments are to the skin of the labia majora inferiorly and the peritoneum, ovary, and the developing uterus superiorly. In growth of the embryo, the ovary is pulled inferiorly to the brim of the pelvis; from this point on the gubernaculum increases in length; it becomes (1) the **ligamentum teres (uteri),** which can be followed from the labia majora, through the inguinal canal, and into the pelvic cavity where it attaches to the fundus of the uterus, and (2) the **ligament of the ovary** extending from the ovary to the fundus of the uterus just superior to the attachment of the ligamentum teres. Both of these structures are shown in Figures 5-10 and 5-65. Therefore, the **inguinal canal** in the female has the same structure as in the male but has the rather small ligamentum teres instead of the spermatic cord coursing through it. Figure 3-45 shows a series of dissections of the inguinal canal in the female. Since the processus vaginalis does not descend through the inguinal canal in the female, the ligamentum teres will not be invested with the fascial layers found in the spermatic cord of the male. Occasionally, the connection of the gubernaculum to the uterus is wanting; in this case, the ovary descends into a labium majus. Also, in the female, the processus vaginalis can extend into the inguinal canal; when this persists, it is called the **canal of Nuck.**

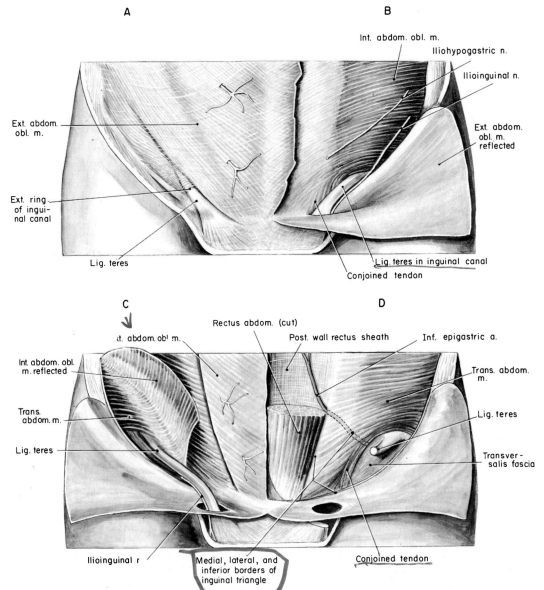

A

Ext. abdom. obl. m.

Ext. ring of inguinal canal

Lig. teres

B

Int. abdom. obl. m.

Iliohypogastric n.

Ilioinguinal n.

Ext. abdom. obl. m. reflected

Lig. teres in inguinal canal

Conjoined tendon

C

.t. abdom. obl. m.

Int. abdom. obl. m. reflected

Trans. abdom. m.

Lig. teres

Ilioinguinal r

Medial, lateral, and inferior borders of inguinal triangle

D

Rectus abdom. (cut)

Post. wall rectus sheath

Inf. epigastric a.

Trans. abdom. m.

Lig. teres

Transversalis fascia

Conjoined tendon

Figure 3-45.   *The inguinal canal in the female. (A) Showing external ring and ligamentum teres emerging through this ring to reach labium majus. (B) External oblique reflected, showing that it formed the anterior wall over the entire length of the canal, and revealing the internal abdominis oblique muscle aiding in forming the anterior wall of the lateral one-third, the roof, and the posterior wall (conjoined tendon) of the medial one-third. (C) The internal oblique reflected to show that the transversus abdominis muscle arises too far laterally to form any of the anterior wall, but does aid in formation of the roof and posterior wall of the medial one-third (conjoined tendon). (D) Ligamentum teres has been cut to show transversalis fascia forming the posterior wall of approximately the lateral two-thirds of the canal.*

The labia majora are homologous to the scrotum and contain some smooth muscle fibers that correspond to the dartos musculature; the ligamentum teres (uteri) and the ligament of the ovary correspond to the gubernaculum of the male; and the ovaries and testes are homologous. Since there is no descent of an organ through the inguinal canal in the female, it is usually less subject to injury; there are relatively few inguinal hernias in the female. More detailed comparisons between male and female structures will be made when the remaining parts are studied.

SYSTEMIC SUMMARY

BONES

In the study of the thoracic and abdominal walls, the **ribs** and **sternum** of the axial skeleton, and the **scapula, clavicle,** and **parts of the coxal bones** of the appendicular skeleton were investigated. Since these bones were adequately described on pages 99–101, it is not necessary to repeat these descriptions. However, because you have become acquainted with the structure of this area, a review of the bones at this time will make them have functional significance. This is particularly true of the muscle attachments that are shown in Figures 3-1, 3-2, 3-3, and 3-4.

The clavicle is the only attachment of the upper limb to the axial skeleton and serves the very important function of holding the shoulder laterally away from the vital axillary structures. The ribs and sternum protect the thoracic viscera, serve for muscle attachments whose contractions cause the vital life-giving movements involved in respiration, and serve an active role in blood formation. In fact, the sternum is one site of puncture when marrow is needed for diagnostic purposes, the ilium being another.

JOINTS

The joints studied in this region were (1) **costovertebral,** (2) **sternocostal,** (3) **interchondral,** (4) **costochondral,** and (5) **sternal.** These were described on pages 121–123,

and repetition is unnecessary. These are all involved with movements that occur during respiration. The costovertebral articulation is double in nature: between the head of the rib and bodies of the vertebrae, and between the tubercles and transverse processes. These articulations are arranged in such a manner that movements in them cause the ribs to be thrust anteriorly and laterally, thereby increasing the size of the thoracic cavity. Because the first ribs and the costomanubrial joints are relatively immobile, and the sternum is required to move anteriorly, the body has to move on the manubrium at this sternal joint. Furthermore, there is movement allowed at the sternocostal joints. Movement at the costochondral and interchondral joints is very limited and relatively unimportant.

It should be noted that movements occur in the above joints with gross movements of the trunk and that the innumerable joints in the vertebral column come into play as well.

FASCIA

It is interesting to note that there are additional features in the **superficial fascia** in both the thoracic and abdominal regions that are not found in other parts of the body, namely, the mammary gland in the thoracic region and the membranous layer of superficial fascia in the abdominal region.

The **mammary gland** consists of approximately twenty-two compound tubuloalveolar glands embedded in fat, each with a separate opening on the nipple, and separated from one another by connective tissue septa. The most important clinical factor concerning the breasts is the fact that they are so prone to cancerous growths; the lymphatic drainage is particularly important. If the cancer is in the glandular part of the organ, the drainage is into the pectoral, internal thoracic, subclavicular, or axillary nodes; if in the skin, it can course down the abdominal wall or even cross to the opposite side.

The abovementioned connective tissue septa are important in cancer detection, because growths put pressure on these septa, which, in turn, cause a dimpling of the skin on the breast.

The mammary glands are under endocrine control, starting to enlarge in the female at puberty. Further enlargement occurs during pregnancy and lactation usually occurs following childbirth.

The superficial fascia in the lower abdomen is divided into **fatty** and **membranous layers.** The fatty layer is the same as such superficial fascia elsewhere in the body, but the membranous layer is peculiar to this region. It is continuous with a similar layer around the penis, on the scrotum, and in the perineum in the male and in the labia majora in the female. It is attached just inferior to the inguinal ligament, to the muscles on the medial surface of the thigh, to the inferior pubic rami, and to a line that stretches across the perineum between the anal and urogenital regions. This fascia is shown in Figure 3-29. Its main importance clinically is the fact that extravasation of urine in a plane deep to this fascia will be limited in its spread to the scrotum, penis, and abdominal wall; it will not go into the thigh or enter the anal region. See Figure 5-78 on page 317 for a demonstration of the above.

Deep fascia will be mentioned in the description of the muscular system.

MUSCLES

As was found in the back, the most superficial muscles on the chest wall are concerned with the upper limb. The **pectoralis major** flexes, adducts, and medially rotates the humerus, and the **pectoralis minor** and **subclavius** muscles both lower the shoulder. The **serratus anterior** muscle, which has a large attachment to the trunk, was described in the chapter on the back. All these muscles are derived embryologically from growth of the limb bud. Because the muscles of this limb bud obtain their nerve supply early in development from branches of the brachial plexus, these same nerves continue to innervate these muscles after they have migrated elsewhere. These same muscles will be considered again when the upper limb is studied.

When the appendicular musculature was removed from the back, muscles concerned with respiration and the back proper remained. A similar situation holds for the chest. The remaining muscles are the **intercostals, subcostals,** and **transversus thoracis** muscles making up the thoracic walls, and the **rectus abdominis, external and internal abdominis oblique, transversus abdominis,** and the **pyramidalis** muscle in the abdominal wall.

Each intercostal space is filled in with muscles, the most superficial of which is the external intercostal, the fasciculi of which course inferiorly and medially. The anterior part of this muscle is a membrane only. The internal intercostal muscle lies deep to the former muscle; its fibers

course inferiorly and laterally and it is reduced to a membrane posteriorly. A third layer of muscle consists of three parts: the transversus thoracis anteriorly, the intercostalis intimus laterally, and the subcostalis muscles posteriorly. These three layers quite effectively fill in the intercostal spaces.

A glance at a skeleton reveals immediately that there is a large expanse to the abdominal wall, particularly anteriorly. This area, outlined by the costal arch, xiphoid process, linea alba, pubic bone, inguinal ligament, iliac crest, and vertebral column, is filled in by these muscles in an interesting manner, the external oblique coursing inferiorly and medially, the internal oblique superiorly and medially, the transversus abdominis transversely, and the rectus abdominis vertically.

The segmentally arranged rectus abdominis muscle is surrounded by the tendons of the other abdominal muscles in such a manner as to create a sheath—the **rectus sheath.** In the upper part of the abdominal wall, the sheath is formed by the external oblique passing anterior to the rectus, the internal oblique splitting and sending a layer both anterior and posterior, and the transversus abdominis muscle remaining posterior to the muscle. Approximately a hand's breadth below the umbilicus, the sheath changes in character, becoming deficient posteriorly. This results from the posterior lamella of the internal abdominis oblique tendon and the entire tendon of the transversus abdominis passing anterior to the rectus muscle joining the anterior lamella of the internal oblique and the tendon of the external oblique muscles.

In the groin, these muscles are arranged in a manner that allows the descent of the testis into the scrotum and that permits the spermatic cord in the male and ligamentum teres in the female to traverse the abdominal wall without undue pressure being put upon them. This arrangement, called the **inguinal canal** (described on pages 136–138), is a location of weakness in the abdominal wall. Hernias occur here with frequency, a direct inguinal hernia passing through the inguinal triangle into the medial inguinal fossa to push into the posterior wall of the canal, while an indirect inguinal hernia remains lateral to the inguinal triangle and enters the internal ring of the canal through the lateral inguinal fossa. Direct hernias are usually acquired, while indirect are almost always congenital in origin.

As might be expected, the abdominal muscles are lined with deep fascia, that on the superficial surface of the external abdominis oblique muscle being intimately fused with the muscle, while that on the inside is, as we will see later, a rather unusually arranged layer designated as transversalis fascia. To complete the wall, the internal oblique is surrounded by fascia and the external oblique fascia is covered with superficial fascia and skin, while the transversalis fascia is lined on its inside surface by subserous fascia and peritoneum. The aforementioned layers of deep fascia were carried down into the scrotum by the descending processus vaginalis and testis as they passed through the inguinal canal. The external spermatic fascia (the outer layer surrounding the spermatic cord) was derived from the fascia on the external abdominis oblique muscle, the cremasteric muscle and fascia (the middle layer) from the internal oblique, and the internal spermatic fascia (the inner layer) from the transversalis fascia.

It is of interest to compare the muscles in the thoracic wall with those in the abdominal wall; this can be done as follows:

Thoracic Wall	Abdominal Wall
External intercostal	External abdominis oblique
Internal intercostal	Internal abdominis oblique
Subcostals, intercostalis intimus, and transversus thoracis	Transversus abdominis
Sternalis	Rectus abdominis

The sternalis is an inconstant muscle found alongside the sternum superficial to the pectoralis major muscle. It occurs in about 4 per cent of individuals and when innervated by the intercostals it probably represents a thoracic rectus muscle.

These muscles are involved in respiration, in movement of the vertebral column, and in vital functions such as micturition, defecation, and parturition.

RESPIRATION.* Inspiration demands an enlargement of the thoracic cavity. Increase in lateral and in anterior directions results, under normal conditions, from contraction of the intercostal musculature, aided by the serratus posterior superior, the levatores costarum muscles, and, in some individuals, by the scaleni muscles. Increase in an inferior direction is accomplished by contraction of the diaphragm, the inward pull on the lower ribs being counteracted by the quadratus lumborum and serratus posterior inferior muscles.

---

*E. J. M. Campbell, 1958, *The Respiratory Muscles and the Mechanics of Breathing,* Lloyd-Luke, London.

Expiration under normal conditions consists of an elastic recoil of all structures to a normal position. However, some feel that the interosseous portions of the internal intercostal muscles, particularly in the lower intercostal spaces, may play an active role. When forced expiration occurs, such as in prolonged blowing, the abdominal wall musculature contracts, forcing the diaphragm superiorly. The external abdominis oblique, the internal abdominis oblique, and transversus abdominis muscles are very much involved in this act but the rectus abdominis is of questionable value. The transversus thoracis muscle also aids in forced expiration.

MOVEMENT OF VERTEBRAL COLUMN. The muscles involved in extending and rotating the vertebral column are considered in the chapter on the back. If you have studied this area, re-reading the summary found on page 94 would be helpful at this time. Flexion of the vertebral column is induced by contraction of all of the muscles forming the abdominal wall, with the rectus abdominis muscle being the most important; pulling the costal margin toward the pubic bone certainly will result in a flexion of the vertebral column. Considerable movement of the vertebral column is permissible in a lateral direction—a lateral bending of the trunk. If we ignore, for the present, the amount of this movement that is occurring in the hip joints, lateral bending of the vertebral column is accomplished by all the muscles attached to the vertebral column as well as those filling in the area between the ribs and iliac crest. Those doing most of the work would be the external and internal abdominis oblique muscles; they are assisted by the rectus abdominis anteriorly, and the quadratus lumborum, the erector spinae, and the transversospinal muscles posteriorly. The psoas muscle (Fig. 5-40 on page 260) may assist if no movement occurs in the hip joint.

MICTURITION, DEFECATION, AND PARTURITION. These three vital processes are aided by an increase in intra-abdominal pressure; this is accomplished by a simultaneous contraction of the abdominal muscles on both sides of the body, while holding the breath.

NERVES

In contrast to the back, the nerves to the thoracic and abdominal walls are all ventral rami of the spinal nerves. The ventral rami of the fifth, sixth, seventh, eighth cervical and first thoracic nerves form the **brachial plexus** and this plexus innervates all muscles concerned with the upper limb. Accordingly, the medial and lateral pectoral nerves to the pectoralis major and minor, the nerve to the subclavius muscle (and we might include the long thoracic nerve to the serratus anterior muscle) are all branches of this plexus.

The nerves involved in the thoracic and abdominal walls proper are the **eleven intercostals,** the **subcostals,** and the **iliohypogastric and ilioinguinal** branches of the lumbar plexus. (If the scrotum is considered, the genitofemoral branch of the lumbar plexus also must be included.) A typical spinal or segmental nerve, after giving off a dorsal ramus to the muscles of the back proper, continues around the body wall to the midaxillary line, gives off a lateral cutaneous branch, continues to the anterior surface, and gives off an anterior cutaneous branch near the midline of the body. It is of interest to note that the anterior cutaneous branch of the seventh thoracic nerve is located just inferior to the xiphoid process, the tenth thoracic nerve appears near the umbilicus, the iliohypogastric innervates the skin just superior to the pubis, and the others are distributed in expected areas between these landmarks. These are important facts to remember, for visceral pain from internal organs is referred to these areas on the body wall.

Figure 3-46 depicts the **dermatomes** of the body. It should be noted that these dermatomes on the thoracic and abdominal walls are quite similar to the nerve distribution, indicating that the nerves maintain their individuality rather than forming plexuses.

A comparison between the course and relations of the nerves in the intercostal spaces and those in the abdominal wall is interesting. If the external abdominis oblique muscle is compared with the external intercostals, the internal abdominis oblique with the internal intercostals, and the transversus abdominis muscle with the combined subcostal, intercostalis intimus and transversus thoracis muscles, as was done under the systemic summary of the muscles, the nerve relations in the abdominal wall are similar to those in an intercostal space. Just as the intercostal nerves course superficial to the inner complex of subcostal, intercostal intimus, and transversus thoracis muscles and deep to the internal intercostal muscle, the nerves in the abdominal wall course between the transversus abdominis muscle and the internal abdominis oblique muscle.

Special features about several of these nerves might be mentioned. The first intercostal is a small nerve, because the main contribution of the first thoracic nerve is to the brachial plexus; there is no lateral cutaneous branch on this nerve. The lateral cutaneous branch of the second intercostal is called the intercostobrachial nerve, because it contributes to the cutaneous nerve supply of the arm. Intercostal nerves 3 to 6 have no peculiarities and are typical intercostal nerves. The seventh to the eleventh intercostals do not remain in their intercostal spaces but contribute to the nerve supply of the abdominal wall. The lateral cutaneous branches of the subcostal and the iliohypogastric are distributed in the gluteal region, while that of the ilioinguinal is wanting.

All of these nerves, from the first thoracic down to the first lumbar, are connected with the sympathetic chain by white and gray rami communicantes. Preganglionic sympathetic fibers arising in the intermediolateral cell column of the spinal cord follow ventral roots to each spinal nerve and then travel to the sympathetic chain via a white ramus communicans; as indicated in the section of the introductory chapter on the nervous system, these preganglionic fibers may (1) go right through this chain ganglion and synapse in a ganglion above or below its particular level, (2) go through this ganglion and travel to a preaortic ganglion via splanchnic nerves, or (3) synapse in this chain ganglion. In the last instance, where a synapse occurred, postganglionic fibers may travel via a splanchnic nerve, travel up or down the chain, or may join the same nerve via a gray ramus communicans. Therefore, any spinal nerve will contain postganglionic sympathetic fibers to innervate blood vessel musculature, sweat glands, and arrector pili muscles.

Each spinal nerve has general somatic afferent fibers for pain and temperature, light touch, deep pain, and proprioception from muscles; general somatic efferent fibers for motor innervation of skeletal muscles; general visceral afferent fibers from blood vessels; and general visceral efferent fibers (autonomic nervous system) to the blood vessels, sweat glands, and arrector muscles just mentioned.

ARTERIES

The arteries to the thoracic and abdominal walls resemble the nerves in that there are intercostal vessels and lumbar vessels that take a similar pathway to that of the intercos-

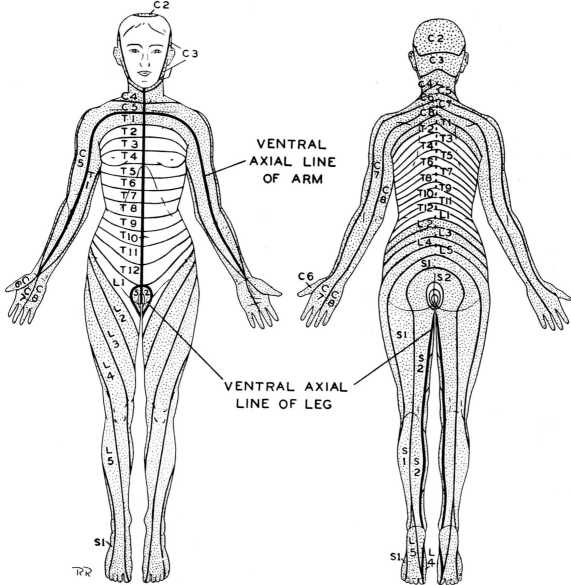

Figure 3-46. *Dermatome chart of the human body. (From J. J. Keegan and F. D. Garrett, Anat. Record 102: 409; courtesy of the authors and the Wistar Institute.) Note that these authors claim that the dorsal rami of C7 and 8, and L4 and 5 reach the skin; this is contrary to the thoughts of others (see page 60).*

tal and lumbar nerves. They differ from the nerve supply in that there is a large midline component derived from extreme superior and inferior ends of the walls in question with which the regular segmental arteries anastomose (Fig. 3-33). This midline arterial channel is made up of the internal thoracic artery—a branch of the subclavian in the neck—and its **musculophrenic** branch, which follows the costal arch, and the **superior epigastric** branch which enters the rectus sheath to anastomose with the **inferior epigastric** branch of the external iliac.

To be more specific concerning the segmental or spinal arteries, that to the first and second intercostal spaces is derived from the **costocervical trunk** of the subclavian artery; this anastomoses with the internal thoracic artery. Of the remaining **intercostal arteries,** all arise from the thoracic aorta but do not have a similar termination anteriorly. The second to the sixth inclusive anastomose with branches of the internal thoracic artery, the seventh to ninth join with the musculophrenic artery, and the tenth and eleventh intercostals leave their intercostal spaces and ramify in the abdominal wall, as do the subcostals. The anterior branches of the five **lumbar arteries** are, in contrast to the posterior branches, comparatively small and do not extend very far around the abdominal wall. This deficit is made up by the **superficial** and **deep circumflex** branches of the femoral and **external iliac** arteries respectively, and by the **superficial epigastric** branch of the femoral.

As might be expected, the arterial supply to the appendicular muscles in the chest is from a branch of an artery of the limb, the **thoracoacromial** branch of the axillary.

VEINS

The veins have a close resemblance to the arteries. A few differences might be mentioned. The veins draining the first intercostal space on the right side join the right brachiocephalic, and that on the left side the left brachiocephalic; and the second and third spaces on the left side also drain into this vessel. The remaining intercostal veins on the right side drain into the azygos vein, while those on the left side drain into the hemiazygos and accessory hemiazygos veins. The latter two veins drain into the axygos, which ultimately drains into the superior vena cava. The lumbar veins drain into a vertically running vein called the ascending lumbar or into the inferior vena cava. These ascending lumbar veins drain into the azygos system of veins just mentioned, and may have connections with the renal veins on one or both sides of the body. A glance at Fig. 5-48 on page 274 will explain the above.

LYMPHATICS

The skin lymphatics drain into two sets of nodes. Those from the lower abdominal region drain into the **inguinal** nodes, those from the upper part of the trunk wall into the **axillary** nodes. Some areas along the sternum drain into the **sternal** nodes along the internal thoracic artery.

The lymphatics of the muscular walls themselves follow the distribution of the arteries. The lower abdominal wall drains into lymph nodes along the **external iliac** artery, the lateral part of the abdominal wall into the **lumbar** nodes, the upper central area of the abdominal and thoracic walls into the **sternal** nodes, and the **intercostal** spaces into nodes along the vertebral column in the thorax.

CONCEPTS: GENERAL RULES ON STRUCTURE OF THE HUMAN BODY

On page 97 some general rules are presented on the structure of the human body, which can be applied as well to this part of the body. Reading that section at this time would be advantageous to those who started with the thoracic and abdominal walls. Furthermore, the abdominal wall provides another fine example of how a detailed knowledge of anatomy is useful in medicine.

4

THORACIC CAVITY

GENERAL FEATURES

The thoracic cavity is that area of the body **bounded** posteriorly by the vertebral column, ribs, and intercostal muscles; laterally by the ribs and intercostal muscles alone; anteriorly by the ribs, intercostal muscles, and sternum; inferiorly by the diaphragm; and superiorly by an imaginary plane just superior to the first rib. This cavity is divided into **three portions** (Fig. 4-1) consisting of the two **pleural cavities** and a centrally located **mediastinum** (*L.,* intermediate). It contains many structures essential to life. Physicians are constantly listening to these structures and visualizing them at the same time. Advances in surgery have made relations in this area of great importance.

The lungs in their development from the entodermal tube push their way into the coelom, which is lined with a serous membrane. They do this as one might push one's fist into an inflated balloon. The fist, under these conditions, is covered intimately with one layer of balloon and this layer is pushed up against the outer layer of the balloon. Thus, the lung is covered very intimately by a layer of pleura—the **visceral pleura**—and this is in contact with a layer of pleura lining the walls of the thoracic cavity—the **parietal** (*L.,* wall) **pleura.** The **pleural cavity** is the potential space between these two layers.

The two **pleural cavities** approach to within a few millimeters of each other just posterior to the body of the sternum (Figs. 4-2A and 4-3). However, the pleural cavities on either side remain distinct from each other, an important fact when one considers that if they were joined, one would not be able to collapse the lung clinically on one side without collapsing it on the other. Just before reaching the xiphoid process the two pleural sacs separate from each other, the one on the left side coursing more laterally than the one on the right side. This leaves a so-called **bare area** of the heart or pericardium that is not covered with pleura (Fig. 4-2A). This is an important fact clinically, because it allows the heart to be approached without collapsing the lungs. Inferiorly, the pleura on either side crosses deep to the seventh rib an-

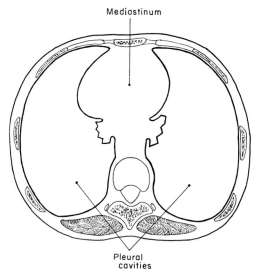

Figure 4-1. *Diagrammatic transverse section of the thoracic cavity showing its divisions into two laterally placed pleural cavities and one centrally located mediastinum.*

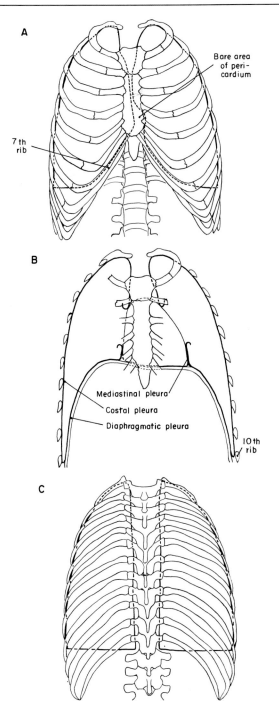

teriorly, the tenth rib in the midaxillary line (Fig. 4-2B), and the twelfth rib posteriorly (Fig. 4-2C). It extends into the neck for a distance of approximately 2 inches superior to the first rib and 1 inch superior to the clavicles.

The parietal pleura on the mediastinal structures is the **mediastinal pleura,** that on the ribs and intercostal spaces the **costal pleura,** and that on the diaphragm the **diaphragmatic pleura.** The lungs and their coverings of the visceral pleura do not completely fill the pleural cavity. There is an area inferiorly, known as the **pleural sinus,** that does not contain lung tissue; this allows the expansion of the lung that necessarily occurs with violent exercise.

The parietal pleura is continuous with the visceral pleura at the hilum of the lung, that area where nerves, vessels, and lymphatics enter and leave the lung. This reflection occurs both on the anterior and posterior surfaces of the structures at the hilum. These anterior and posterior layers have a tendency to extend about 2 inches inferiorly before ending in a sharp edge. This inferior extension is known as the **pulmonary ligament** (see Fig. 4-40 on page 193).

The visceral pleura, an intimate part of the lungs, will be described later with these organs.

The remaining area, that between the two pleural cavities, is the **mediastinum** (Fig. 4-4). Although a defi-

Figure 4-2. *The pleural reflections as seen (A) from the front, (B) in a frontal section of the chest, and (C) from the back. Note that the inferior limit of the pleura crosses the seventh rib anteriorly, the tenth rib laterally, and the twelfth rib posteriorly.*

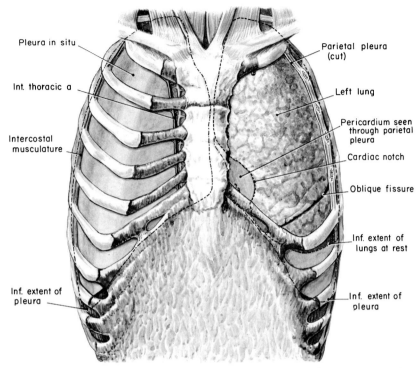

Pleura in situ

Int. thoracic a

Intercostal musculature

Inf. extent of pleura

Parietal pleura (cut)

Left lung

Pericardium seen through parietal pleura

Cardiac notch

Oblique fissure

Inf. extent of lungs at rest

Inf. extent of pleura

Figure 4-3. *The pleura and lungs in situ. All muscles have been removed on the right side of body to show the parietal pleura in position. This parietal pleura has been removed on the left side of the body to show the lung covered with visceral pleura. Dot-dash line gives outline of pleural cavity; dashed line is outline of lung.*

nite structure is described embryologically as a mediastinum, there is no such structure in the adult. The mediastinum is an area in the thoracic cavity bounded by the pleural cavities laterally, the sternum anteriorly, the vertebral column posteriorly, the diaphragm inferiorly, and the thoracic inlet superiorly. The mediastinum is divided into superior and inferior mediastinal areas by a line drawn from the sternal angle (where the manubrium and the body of the sternum join) anteriorly to the base of the fourth thoracic vertebra posteriorly. The inferior mediastinum is further divided into anterior, middle, and posterior mediastinal compartments. These areas are easily visualized if the pericardium and its contents are considered the middle mediastinum, the area anterior to it (between pericardium and sternum) as anterior mediastinum, and everything posterior to it (between pericardium and vertebral column) as contained in the posterior mediastinum.

ANTERIOR THORACIC WALL

When the intercostal musculature is removed in the area lateral to the sternum, the internal thoracic artery and veins (usually more than one) can be seen coursing between the costal cartilages and the transversus thoracis muscles (Figs. 3-23 on page 121, and 4-5). When portions of the sternum and ribs are removed one can look at the inside or deep surface of the anterior thoracic cage (Fig. 4-5). Reflection of the parietal pleura reveals the transversus thoracis muscle in greater extent, with the internal thoracic artery and veins coursing between this muscle and the costal cartilages about ½ inch lateral to the sternum.

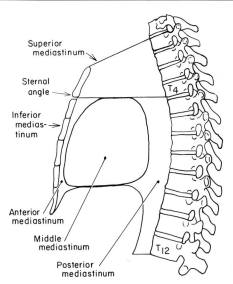

Figure 4-4. *The mediastinal compartments.*

INTERNAL THORACIC VESSELS

The internal thoracic arteries arise from the subclavian arteries, the one on the right starting at a more lateral position than the one on the left. They are both in contact with the pleura, coursing between it posteriorly and the subclavian veins anteriorly (Fig. 4-6). They then proceed inferiorly and come in contact with the first rib; from this point inferiorly they course just deep to the costal cartilages about ½ inch from the sternal border, pass anterior to the transversus thoracis muscle, and end at the level of the sixth intercostal space by dividing into their terminal branches. These arteries are also pictured in Figures 4-45 and 4-46 on pages 202 and 203.

The branches of these arteries are:

1. Pericardiacophrenic
2. Mediastinal
3. Anterior intercostals
4. Perforating
5. Musculophrenic
6. Superior epigastric

1. Pericardiacophrenic (Figs. 4-45 and 4-46 on pages 202 and 203).   This artery arises quite high and follows the course of the phrenic nerve. That on the right side is found on the anterolateral surface of the superior vena cava, then on the lateral surface of the pericardium, and finally on the superior surface of the diaphragm. It

gives branches to the pleura, thymus gland, pericardium, and diaphragm. The left artery takes a similar course but differs from the right in coursing on the aorta rather than the superior vena cava. This description makes sense if you will examine the figures mentioned above.

2.   The **mediastinal** arteries are small branches supplying the connective tissue in the anterior mediastinum and the remnants of the thymus gland.

3.   The **anterior intercostals** (Fig. 4-5) have two branches for each of the first six intercostal spaces, one just inferior to each rib and the other just superior to the rib below. They anastomose with the posterior intercostal arteries, as shown in Figure 4-5.

4.   The **perforating** arteries, one for each intercostal space, perforate the internal intercostal muscle, external intercostal membrane, and pectoralis major muscle to ramify in the superficial fascia and reach the skin. The branches from the second, third, and fourth spaces are large in the female and supply the breast.

5.   The **musculophrenic** artery (Fig. 3-33 on page 133) is one of the two terminal branches of the internal thoracic artery. It arises at approximately the level of the sixth intercostal space, follows the costal arch inferiorly, and gives off two branches to each of the seventh, eighth, and ninth intercostal spaces. It then pierces the diaphragm and continues into the abdominal musculature as a small branch. This artery supplies the pericardium and diaphragm as well as the intercostal and abdominal musculature.

6.   The **superior epigastric** artery (Fig. 3-33 on page 133) is the other terminal branch of the internal thoracic artery and is actually considered its continuation. It passes through the diaphragm to enter the rectus sheath (Fig. 3-31 on page 129). It continues inferiorly just posterior to the rectus abdominis muscle, enters the substance of the muscle, and anastomoses with the inferior epigastric artery. (This anastomosis is important in cases in which the aorta has become partially or wholly occluded.) It gives off small branches to the diaphragm, peritoneum, and skin, and a small branch of this artery follows the ligamentum teres (hepatis) (obliterated umbilical vein) to the liver (see Fig. 5-1 on page 214).

The **internal thoracic vein** is formed from a combination of the musculophrenic and superior epigastric veins and is in the form of a concomitans as far as the third intercostal space; at this point it usually becomes a single vessel and empties into the brachiocephalic vein.

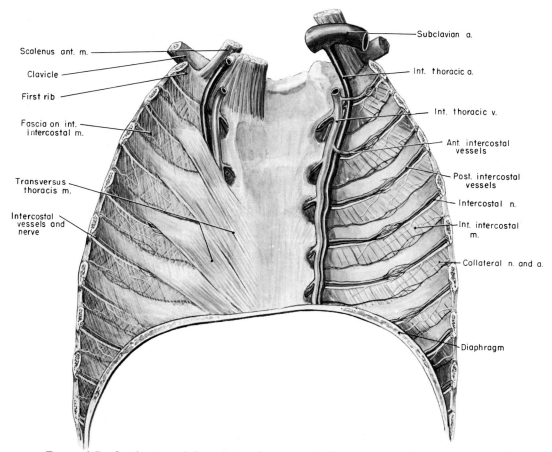

Figure 4-5. *Inside view of the anterior thoracic wall. The transversus thoracis muscle and fascia have been left intact on the left side of body, removed on the right side. The vein may be a concomitans.*

These veins and their tributaries drain the same territory as just described for the arteries; they possess numerous valves.

TRANSVERSUS THORACIS MUSCLE

This muscle is made up of several slips arising from the posterior surface of the inferior end of the body of the sternum and from the xiphoid process (Fig. 4-5). They fan out laterally and superiorly to insert on the second to the sixth costal cartilages. Their main action is to lower the ribs in expiration. Their nerve supply is from the second to sixth intercostal nerves. This muscle plus the intercostalis intimus muscle and the subcostalis muscles is often called the inner complex.

ANTERIOR AND SUPERIOR MEDIASTINAL COMPARTMENTS

GENERAL DESCRIPTION

Removal of the sternum and ribs has opened the mediastinum, and all of the anterior and the anterior part of the superior mediastinum can be seen. The anterior mediastinum contains only fat and connective tissue in addition to the internal thoracic vessels and transversus thoracis muscle already mentioned.

The superior mediastinum is a complex area difficult to visualize. The structures are confined to a small space and relations are confusing unless one at first learns an overall plan. Roughly speaking, the superior mediastinum contains, from anterior to posterior (Fig. 4-7), (1) thymus or its remnants, (2) large veins, (3) arteries—aorta and its branches, (4) the trachea, and (5) the esophagus. Each of these will now be described without detailed relations. After this basic architecture has been mastered, relations can be learned and smaller structures can be added.

THYMUS GLAND

This gland consists of two elongated lobes closely bound by fibrous tissue (Fig. 4-6). In the fetus, it is a large organ occupying considerable space in the thoracic cavity and extending superiorly into the neck. It continues to grow until puberty, after which it decreases in size and finally degenerates into two elongated fatty bodies. This organ contains a large number of lymphocytes and is considered to be the source of the so-called t-lymphocytes so important to autoimmune phenomena. It has been associated with the endocrine system, although its function as a ductless gland is quite obscure at the present time. Recent evidence indicates it may be the source of **lymphopoietin,** a substance that stimulates the development of lymphocytes in the lymph nodes. Its **arterial supply** is by branches of the internal thoracic arteries; **venous drainage** is into the brachiocephalic veins. It is **innervated** by the vagus (parasympathetic) and the sympathetic systems, although, other than their role in innervating blood vessels, the function of these nerves is problematical. Its **lymphatic drainage** is to nodes lo-

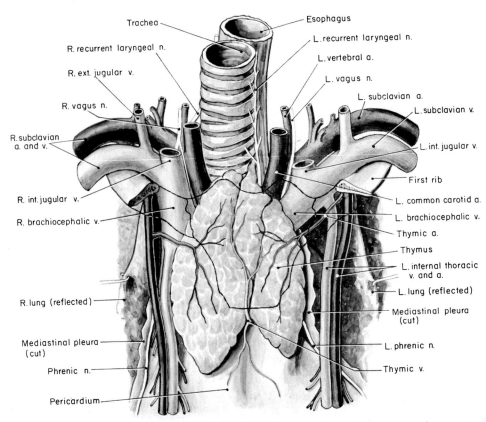

Figure 4-6. *The thymus gland as seen in the child before puberty. This structure degenerates after puberty to two elongated fatty bodies. (Modified from Grant.)*

cated along the internal thoracic artery, at the bifurcation of the trachea, and in the root of the neck.

BRACHIOCEPHALIC VEINS

These veins (Fig. 4-7), located just posterior to the thymus, are formed in the neck by union of the **subclavian** and **internal jugular** veins. Both **brachiocephalic** veins join to form the **superior vena cava,** which empties into the **right atrium** of the heart.

The **right brachiocephalic vein** starts just posterior to the medial end of the right clavicle; it descends almost vertically into the thoracic cavity to meet the left brachiocephalic vein at a point directly posterior to the right first costal cartilage. More of this vein is located in the neck than in the thoracic cavity. Its **tributaries** are, in the neck, the subclavian and internal jugular, the vertebral and first posterior intercostal, and, in the thoracic cavity, the internal thoracic and inferior thyroid veins. Thymic veins may also drain into this vessel.

The **left brachiocephalic** vein also starts at a point just posterior to the medial end of the clavicle; it courses to the right and slightly inferiorly, to join the right brachiocephalic. Its **tributaries** are the same as found on the right side except for the superior intercostal vein which drains the second and third (sometimes more) intercostal spaces on the left side.

SUPERIOR VENA CAVA

This large vein (Fig. 4-7), approximately 2 inches in length, starts by the union of the right and left brachiocephalics at the level of the first right costal cartilage. It courses vertically and ends by piercing the pericar-

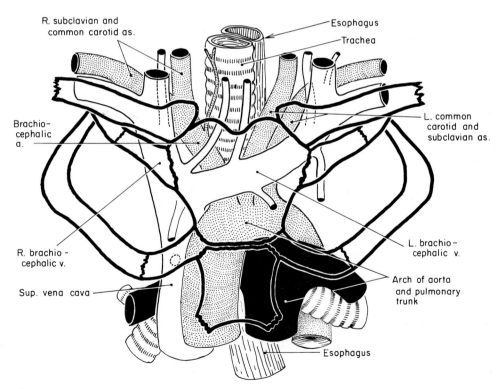

Figure 4-7. *Diagrammatic presentation of the large structures in the mediastinum superior to the heart. Note the arrangement is veins, arteries, trachea, and esophagus from anterior to posterior.*

dium and entering the right atrium of the heart at the level of the third right costal cartilage. **Tributaries** are the azygos vein, which will be described later, and occasional mediastinal veins.

AORTA AND PULMONARY ARTERIES

Posterior to the veins just described is a complex of arteries consisting of the **aorta** and its branches, and the **pulmonary trunk and arteries.** Although the first part of the aorta and the pulmonary trunk and arteries are not actually in the superior mediastinum, they and the rest of the aorta are intimately intertwined with one another and can be visualized as a unit.

AORTA.   **The thoracic portion** of the **aorta** is divided into (1) the **ascending aorta,** (2) the **aortic arch,** and (3) the **descending aorta.** As just mentioned the ascending aorta is actually located in the middle mediastinum. The aortic arch is entirely in the superior mediastinum. The descending aorta is located in the posterior mediastinum and will be described with that region.

The **ascending aorta** (Fig. 4-7) starts as a continuation from the left ventricle, extends superiorly for approximately 2 inches, and joins the aortic arch. This ascending portion of the aorta, in company with the pulmonary trunk, is surrounded by pericardium. At its origin are three dilations—the **aortic sinuses**—opposite the cusps of the aortic semilunar valve. The **branches** of the ascending aorta are the important right and left coronary arteries, which give blood supply to the heart itself and will be described with that organ.

The **aortic arch** (Fig. 4-7) starts as a continuation of the ascending aorta and ascends slightly but at the same time turns posteriorly and then inferiorly after reaching the vertebral column. It actually makes two curves, one concave inferiorly, the other with the concavity toward the right and posterior.

The **branches** of the aortic arch are:

1. Brachiocephalic
2. Left common carotid
3. Left subclavian

1.   The **brachiocephalic** is a large artery that arises from the first part of the aortic arch, courses superiorly and toward the right, and divides into the right common carotid and right subclavian arteries at a level just poste-

rior to the sternoclavicular joint on the right side. It usually has no additional branches, but when a thyroid ima artery is present, it may arise from this artery.

2.   The **left common carotid** arises from the aortic arch just to the left of the brachiocephalic. It continues superiorly through the superior mediastinum and into the neck. There are no branches of this artery in the thorax.

3.   The **left subclavian artery** is the third branch of the aortic arch. It also courses superiorly and to the left to enter the root of the neck, thus leaving the superior mediastinum. This artery has no branches in the superior mediastinum.

The above branches are important in distinguishing between the aorta and the pulmonary trunk on an x-ray; the pulmonary trunk has no such branches.

PULMONARY ARTERIES.   The **pulmonary trunk** (Fig. 4-7) is a continuation from the right ventricle of the heart, and carries unoxygenated blood to the lungs; it is covered with pericardium in company with the ascending aorta, and takes a posterior course as it goes superiorly; it divides into the right and left pulmonary arteries. It is anterior to the ascending portion of the aorta, because it crosses in front of this vessel. The **right pulmonary artery** courses to the right, posterior to the ascending aorta and the superior vena cava, to enter the right lung. The **left pulmonary artery,** in its course to the left lung, passes inferior to the arch of the aorta and anterior to the descending portion of the aorta. The left pulmonary artery is shorter than the right.

LIGAMENTUM ARTERIOSUM

This is the remains of a large vessel—the **ductus arteriosus**—that connected the root of the left pulmonary artery to the aorta in the fetus. This duct allowed the major part of the pulmonary blood, which was the oxygenated blood, to enter the aorta and thus be distributed to the whole body without the necessity of traversing the nonfunctioning lungs. Normally, it closes after birth so that unoxygenated blood (since the umbilical cord has been tied off) can be sent to the lungs for aeration. Occasionally this duct remains patent; it can be tied off surgically.

TRACHEA AND ESOPHAGUS

The **trachea** (Fig. 4-7) is the organ just posterior to the arterial complex just described, and this passageway, serving for conduction of air from the larynx to the bronchi, is just anterior to the esophagus (Fig. 4-7). Both these structures will be described in more detail later in this chapter.

Summarizing, although the most anterior portion of the superior mediastinum appears as fat and connective tissue only, this area in youth contains the thymus gland. Just posterior to this gland or its remains is found the left brachiocephalic vein, which can be seen joining the right brachiocephalic to form the superior vena cava. Posterior to this large vein is the arch of the aorta with its three branches, (1) the brachiocephalic, (2) left common carotid, and (3) left subclavian. The pulmonary trunk leaves the heart at a position to the left of the ascending part of the aorta and soon divides, the right passing just posterior to the aorta and the superior vena cava to reach the lung, and the left passing anterior to the descending aorta. The trachea is posterior to these vessels and anterior to the esophagus, which, in turn, is just anterior to the vertebral column. These relations are diagramed in Figure 4-7. Study of the transverse sections of the superior mediastinum presented in Fig. 4-8 should be helpful at this time.

All nerves and lymphatic ducts were omitted from the above descriptions. If you have this basic architecture in mind, winding the nerves around and between these structures will be simple.

PHRENIC NERVES

These nerves arise in the cervical region of the spinal cord, mainly from the fourth cervical segment, but with frequent additions from the third and fifth. They descend into the thoracic cavity, reach the diaphragm and provide motor and proprioceptive innervation to the diaphragm

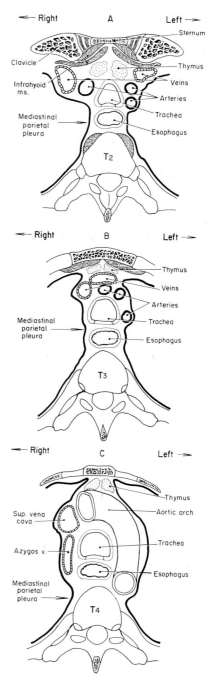

Figure 4-8. *Transverse sections (diagrammatic) through the superior mediastinum at levels of T2, T3, and T4, showing relations of major structures. Nerves and small vessels have been omitted. You are looking superiorly. (Redrawn from Jamieson.)*

and sensory innervation to the pericardium and pleura. Branches of these phrenic nerves perforate the diaphragm, to ramify on the abdominal surface; because of this arrangement they have been given credit for innervating abdominal organs that are close to the diaphragm. The long route from the cervical region to the diaphragm arose from the fact that the diaphragm in its initial development was opposite the cervical vertebrae. With subsequent growth of the body the diaphragm exhibited a relative descent and the phrenic nerves were forced to elongate.

The **right phrenic nerve** (Figs. 4-9 and 4-45 on page 202) enters the thoracic cavity by following the right brachiocephalic vein. It follows the right side of this vein, the right side of the superior vena cava, the right side of

the pericardium, and the right side of the thoracic portion of the inferior vena cava, finally to reach the diaphragm. It is accompanied by the pericardiacophrenic vessels and is covered by pleura throughout its extent in the thoracic cavity.

The **left phrenic nerve** (Figs. 4-9, 4-10, and 4-46 on page 203) enters the thoracic cavity between the left brachiocephalic vein anteriorly and the pleura posteriorly. It immediately comes into contact with the left side of the arch of the aorta where it is superficial to the left vagus nerve, and then follows the pericardium to the diaphragm in a manner similar to that described for the right phrenic nerve.

The main difference between the two phrenic nerves in the thoracic cavity is that the right is in relation to the

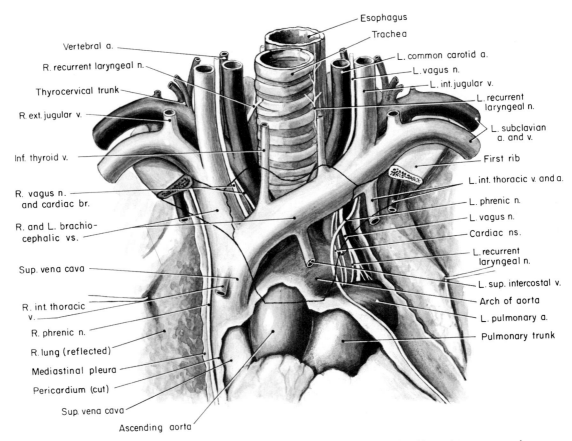

Figure 4-9. *Superior mediastinum I. The remnants of the thymus gland have been removed and the pleura and pericardium cut away. Note how the pericardium covers the main vessels entering or leaving the heart.*

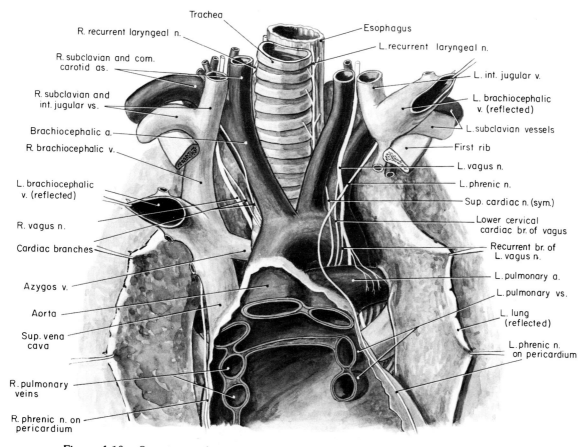

Figure 4-10. Superior mediastinum II. The left brachiocephalic vein has been cut and re-flected to reveal the arch of the aorta and its branches. The heart has been removed from its pericardium, and the parietal pleura has been removed.

lateral surface of the superior vena cava, whereas the left is on the lateral surface of the aorta.

The phrenic nerves are important clinically. The diaphragm can be temporarily paralyzed in cases of tuberculosis by crushing the nerves. The impulses in these nerves are responsible for violent contraction of the diaphragm (hiccoughs). The fact that these nerves are associated with the cervical region of the cord, which also innervates the shoulder region, accounts for the fact that abscesses involving the diaphragm, or structures just inferior to the diaphragm, are often felt in the shoulder—an example of referred pain. This is due to the fact the nerves from the shoulder—the supraclaviculars—enter the spinal cord at the same level as the phrenic. Some feel that these nerves are also responsible, through their in-

nervation of the pericardium, for a considerable amount of apparent pain from the heart.

VAGUS NERVES

The very important vagus nerves enter the thoracic cavity through the superior inlet, course through the superior mediastinum, and continue inferiorly posterior to the roots of the lungs. They continue inferiorly on the surface of the esophagus and follow this organ into the abdominal cavity.

The **right vagus** (Figs. 4-9 and 4-45 on page 202) enters the superior mediastinum by coursing between the right brachiocephalic vein and the brachiocephalic artery to reach the right side of the trachea. It continues inferiorly and posteriorly on the trachea to course posterior to the root of the right lung. In its course it has the pleura, the right lung, and the azygos vein on its lateral side. In the superior mediastinum it gives off **branches** to (1) the esophagus, (2) the deep cardiac plexus, and (3) the anterior pulmonary plexus and thence to right lung.

The **left vagus** (Figs. 4-9, 4-10, 4-11, and 4-46 on page 203) enters the thoracic cavity posterior to the left brachiocephalic vein and anterior to the left subclavian artery. It continues to the arch of the aorta, remains on its left side, and proceeds to the posterior surface of the root of the left lung. In its course it is in contact with pleura on its lateral side, is crossed by the phrenic nerve, is medial to the highest intercostal vein, and is posterolateral to the nerves descending to form the superficial cardiac plexus. In its course through the superior mediastinum, the left vagus nerve gives off **branches** (1) to the esophagus, (2) to the trachea, (3) to the deep cardiac plexus, (4) to the lungs via the pulmonary plexus, and (5) to the larynx—the recurrent laryngeal nerve.

RECURRENT LARYNGEAL NERVE

This nerve (Fig. 4-11) arises from the left vagus while in the superior mediastinum, apparently courses inferior to the arch of the aorta, but in reality around the ligamentum arteriosum, gains the groove between the trachea anteriorly and the esophagus posteriorly, and proceeds superiorly to innervate intrinsic muscles of the larynx in addition to the trachea and esophagus. It gives off cardiac branches to the cardiac plexuses. The close relation between this nerve and the aorta accounts for

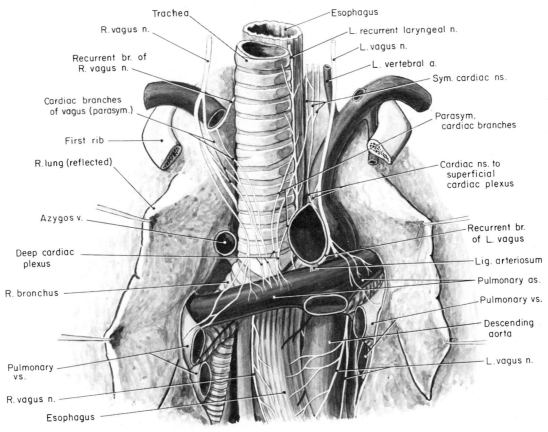

Figure 4-11. Superior mediastinum III. The arch of the aorta has been removed as has the pericardium and the parietal pleura.

coughs resulting from diseased conditions of the aortic arch. (The **right recurrent nerve** courses around the subclavian artery and therefore is not located in the thoracic cavity.)

CARDIAC PLEXUSES

There are two plexuses of nerves in the superior mediastinum involved with innervation of the heart. The **superficial cardiac plexus** is located in the concavity of the arch of the aorta, and the larger **deep cardiac plexus** is located on the anterior surface of the bifurcation of the trachea (Fig. 4-11). Because the heart, during development, is located opposite the cervical vertebrae, the nerves to these plexuses tend to arise from the cervical as well as the upper thoracic region. These plexuses have both sympathetic nerves from the sympathetic chain ganglia and parasymphatetic nerves from the vagus nerves contributing to their formation.

The **sympathetic nerves** arise from each of the cervical sympathetic ganglia (the superior, middle, and inferior sympathetic cardiac nerves respectively) and from the first four or five thoracic sympathetic ganglia (Fig. 4-12). All of these cardiac branches course along the anterior surface of the vertebral column and join the esophagus and then the trachea to get to the deep cardiac plexus, except for the superior cardiac branch from the left superior cervical ganglion; this branch courses on the left side of the aortic arch to join the superficial cardiac plexus. These are all postganglionic fibers. The preganglionic cell bodies are located in the intermediolateral cell column of the first four or five thoracic segments of the spinal cord. Axons from these cells (preganglionic fibers) leave the spinal cord via the ventral roots of these segments, and traverse white rami communicantes to reach the sympathetic ganglia associated with these nerves. Some of the fibers synapse with postganglionic cell bodies and fibers at these ganglia but other fibers pass through the thoracic ganglia without synapsing and travel superiorly in each chain to reach the cervical sympathetic ganglia, where they synapse with similar cells and fibers. The postganglionic fibers pass through the cardiac plexuses and continue to the pulmonary artery and the aorta, which are just anterior to the plexus, to reach the coronary vessels and follow these throughout the heart; they also innervate the sinoatrial and atrioventricular nodes of the conduction system of the heart, which will

be described later in this chapter.* The sympathetic system increases heart rate and causes a vasodilation of the coronary vessels. (Note that this is contrary to the general rule that the sympathetic system causes a vasoconstriction in viscera; however, common sense would indicate that a heart that is working harder would need more blood.) A majority of the sensory fibers from the heart are believed to course in these sympathetic nerves rather than in the vagus nerves.

The **parasympathetic nerve supply** to the heart is from branches of the vagus nerves. On each side of the body the vagus nerves send branches to the cardiac plexuses from the cervical region, from the recurrent laryngeal nerves, and from the vagus nerve as it traverses the superior mediastinum. The **right vagus** gives off (1) a **superior cardiac branch,** which originates rather high in the neck, (2) **inferior cardiac branch,** which comes from the vagus in the lower part of the neck, (3) **cardiac branches from the recurrent laryngeal nerve,** which course posterior to the right subclavian artery, and (4) **thoracic cardiac branches,** which arise from the vagus as it is traversing the superior mediastinum. All these branches contain preganglionic fibers from cell bodies located in the dorsal nucleus of the vagus in the brain stem; they follow the trachea to join the deep cardiac plexus, which they traverse to synapse with cells located in ganglia within the heart itself.

The **left vagus** has a similar set of branches except for the thoracic cardiac branches; these are replaced by numerous branches from the more inferiorly located left recurrent laryngeal nerve. All these fibers from the left side join the deep cardiac plexus except the inferior cervical cardiac branch, which joins the left superior sympathetic cardiac branch to form the superficial cardiac plexus (Fig. 4-12). These preganglionic fibers also traverse the cardiac plexuses and synapse in ganglia located in the heart itself.

Preganglionic parasympathetic fibers form pulmonary plexuses on the pulmonary arteries and then join the aorta to reach the coronary arteries in a manner similar to that just described for the sympathetic system. The

---

*W. A. Statler and R. A. McMahon, 1947, The innervation and structure of the conductive system of the human heart, *J. Comp. Neurol.* 87: 57. Nerves are limited to sinoatrial and atrioventricular nodes, and blood vessels; those to nodes follow conduction system, but terminate before reaching cardiac muscle.

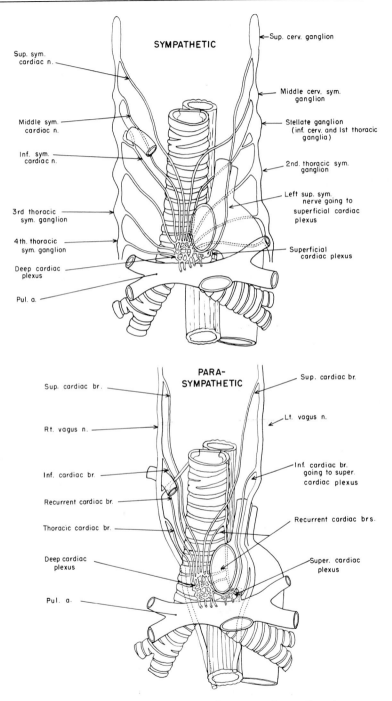

SYMPATHETIC

Sup. sym. cardiac n.

Middle sym. cardiac n.

Inf. sym. cardiac n.

3rd thoracic sym. ganglion

4th. thoracic sym. ganglion

Deep cardiac plexus

Pul. a.

Sup. cerv. ganglion

Middle cerv. sym. ganglion

Stellate ganglion (inf. cerv. and lst thoracic ganglia)

2nd. thoracic sym. ganglion

Left sup. sym. nerve going to superficial cardiac plexus

Superficial cardiac plexus

PARA-SYMPATHETIC

Sup. cardiac br.

Rt. vagus n.

Inf. cardiac br.

Recurrent cardiac br.

Thoracic cardiac br.

Deep cardiac plexus

Pul. a.

Sup. cardiac br.

Lt. vagus n.

Inf. cardiac br. going to super. cardiac plexus

Recurrent cardiac brs.

Super. cardiac plexus

Figure 4-12. *Diagram of the contributions to the cardiac plexuses.*

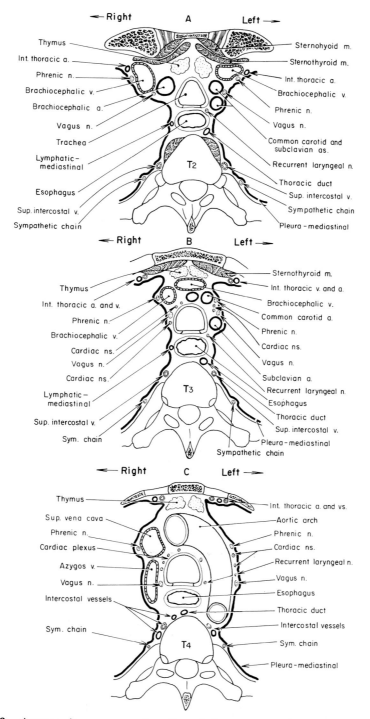

Figure 4-13. *A series of transverse sections through the superior mediastinum similar to those presented in Fig. 4-8 but with the various nerves and vessels in their proper positions. Some of the structures presented that are in the posterior part of the superior mediastinum will be described in the section on the posterior mediastinum. You are looking superiorly. (Redrawn after Jamieson.)*

parasympathetic system decreases heart rate, but the action on the coronary vessels is still problematical.

Although most afferent fibers from the heart are thought to be located in the sympathetic nerves, and thence return to the upper thoracic region of the spinal cord, the vagus nerves are sensory to small collections of tissue in the superior mediastinum known as aortic bodies;* these bodies respond to oxygen tension in the arterial blood and control respiratory rate in this manner.

In summary, the heart's vascular system and conduction system are innervated by both sympathetic and parasympathetic nervous systems; there are sensory fibers in addition. The **sympathetic system** arises from the first four or five thoracic segments; these preganglionic fibers synapse in the three cervical ganglia and the upper four or five thoracic. Postganglionic fibers course to the cardiac plexuses and proceed to the heart; they increase heart rate and cause a vasodilation of the coronary vessels. The **parasympathetic system** starts in the dorsal motor nucleus of the vagus nerves located in the brain stem. These preganglionic fibers proceed to the cardiac plexuses and thence to the heart itself where they synapse with postganglionic cell bodies and fibers; they decrease the heart rate. Most pain from the heart is carried in **visceral afferent fibers** contained in the sympathetic nerves. Pain from the heart is referred to the medial side of the left forearm and hand. This area is covered by nerves from the eighth cervical and first thoracic spinal cord segments, the segments that receive afferents from the heart. The vagus carries visceral afferent fibers from the aortic bodies for reflex control of respiration and from associated arteries for reflex control of heart output.

Figure 4-13 is the same series of transverse sections

---

*The aortic bodies are small structures found in the superior mediastinum, closely associated with structures derived from the fourth branchial arches. The two on the right side are located on the lateral side of the brachiocephalic artery and near the bifurcation of the pulmonary artery, the latter being related to the ascending aorta and left coronary artery; these are innervated by the right vagus nerve. The two on the left side are found in the concavity of the aortic arch near the vagus nerve and on the superior surface of the ligamentum arteriosum; these are innervated by the left vagus nerve. They respond to the oxygen tension of the blood flowing through them. A decrease in oxygen tension reflexively increases rate of respiration and heart rate.

Wherever these bodies are related to the arteries, the vagus nerves also enter the arterial walls. These nerve endings respond to arterial pressure; when stimulated by an increase in pressure, heart output is decreased. These actions are similar to those of the carotid body and sinus (see pages 593 and 594).

of the superior mediastinum as pictured in Figure 4-8, with the various nerves and vessels added. Some of the structures presented in this figure that are in the posterior part of the superior mediastinum will be described in the section on the posterior mediastinum.

MIDDLE MEDIASTINUM

PERICARDIUM

The heart and the beginning of the great vessels are surrounded by a fibroserous sac called the **pericardium** (Fig. 4-14). The **outer fibrous layer** is thick and strong while the **inner serous layer** is quite thin. This serous layer is continuous with a similar layer of serous membrane on the surface of the heart (epicardium), and the space between these two serous layers is the pericardial cavity (Figs. 4-15 and 4-16). These layers are continuous with one another along a line running from the inferior vena cava, right pulmonary veins, superior vena cava, and left pulmonary veins at one end and the aorta and pulmonary trunk at the other end. When the heart is removed, the serous layer has to be cut along these lines of reflection (Fig. 4-17). Because of these reflections, two sinuses are formed. One of these, the **transverse sinus,** lies between the pulmonary trunk and aorta superoanteriorly and the atria and pulmonary veins inferoposteriorly. The right end of this sinus opens between the aorta and the superior vena cava, while the left end is seen between the pulmonary trunk and left pulmonary veins. Although this sinus is small under normal conditions, it is large enough to insert a finger. The **oblique sinus** is more difficult to visualize. However, if the reflections over the veins mentioned earlier are noted, it is seen that they form a C shape or a hooked arrangement (Fig. 4-17). The oblique sinus lies in the concavity of the hook and is in direct communication with the rest of the pericardial cavity. These two serous layers—**visceral** on the heart and **parietal** on the fibrous portion of the pericardium—lubricate each other, so that no binding of the heart occurs. Inflammation of these membranes (pericarditis) can result in adhesions between the layers. This usually results in hypertrophy of the heart, for each

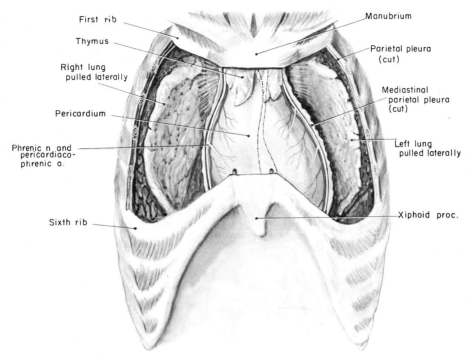

First rib
Thymus
Right lung
pulled laterally
Pericardium
Phrenic n. and
pericardiaco-
phrenic a.
Sixth rib
Manubrium
Parietal pleura
(cut)
Mediastinal
parietal pleura
(cut)
Left lung
pulled laterally
Xiphoid proc.

Figure 4-14. Pericardium in situ *after removal of portions of sternum and ribs. The parietal pleura has been cut and removed from the surface of the pericardium. Dotted lines indicate the location of the medial border of the pleura. The lungs are reflected laterally.*

contraction of the heart under these conditions lifts the pericardium and all structures to which it is attached.

The pericardium has definite **attachments,** and is contained within the middle mediastinum. It is firmly attached to the diaphragm and to the sheath of the inferior vena cava inferiorly, and to the contents of the posterior mediastinum (esophagus, aorta, and accompanying structures) and the four pulmonary veins posteriorly. Its reflections from the great vessels give it an equally firm attachment superiorly. Laterally, it is covered with pleura of the right and left pleural cavities. Anteriorly, the attachments are less definite but connective tissue strands—the **superior** and **inferior sternopericardial ligaments**—do attach the pericardium to the upper and lower parts of the body of the sternum.

The **blood supply** of the pericardium consists of branches of arteries that are close to it, namely, pericardiacophrenic, descending aorta, superior phrenic, and internal thoracic. The **veins** follow the arteries; but those that correspond to the branches of the aorta return to the azygos system of veins (a system of veins, located in the posterior part of the thoracic cavity into which the intercostal veins drain; see Fig. 4-44 on page 199).

The **nerves** to the pericardium are the phrenics and intercostals for general sensation, and the autonomic nervous system (via the esophageal plexus) to the blood vessels.

The **lymph** drains into nodes located in the anterior and posterior mediastinal compartments.

The pericardium is more easily understood if the development is reviewed (Figs. 1-39 on page 47, and 4-18).

HEART (*IN SITU*)

When the pericardium is opened, it can be seen that the heart is twisted in such a manner that the **base,** that part to which the great vessels attach, is facing superiorly, posteriorly, and toward the right shoulder, and that the **apex** (opposite the great vessels) points inferiorly, an-

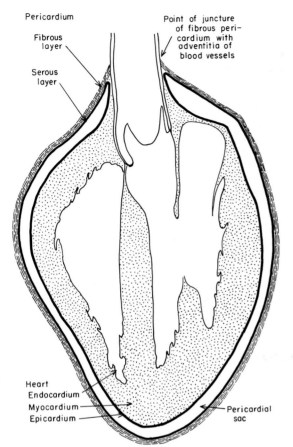

Figure 4-15. *Diagrammatic presentation of the pericardium, showing its fibrous and serous layers and the continuity of the serous layer with a similar layer on the heart (epicardium). Pericardial sac is exaggerated.*

teriorly, and to the left. The heart presents **three surfaces:** (1) the sternocostal, (2) diaphragmatic, and (3) left, and **three borders** which in actuality are the borders of the sternocostal surface; these three borders are (1) lower, (2) right, and (3) left. The heart is a four-chambered organ, and there is evidence of this on the surface. The **coronary sulci** are grooves on the surface between the atria and the ventricles; the **anterior and posterior interventricular sulci** are similar grooves between the right and left ventricles. The **interatrial groove** is not pronounced.

By far the greater portion of the **heart** as seen *in situ*

(Fig. 4-16) is the right ventricle. The right atrium is visible and a small portion of the left ventricle makes up the apex of the heart. The only part of the left atrium visible from the anterior side is the tip of the auricular appendage. The superior vena cava can be seen entering the right atrium, and although the inferior vena cava is not evident with the heart *in situ,* a slight movement of the heart will reveal this vessel. When one realizes that the superior vena cava, the right atrium, and the inferior vena cava make a straight line from superior to inferior, it is easy to see how the heart is twisted inside the pericardium. The pulmonary trunk can be seen leaving the superior end of the right ventricle, and the aorta is visible in a position between the superior vena cava on the right and the pulmonary trunk on the left. The heart itself may be covered with a layer of fat, which varies in thickness from individual to individual. In certain pathological conditions this fat will be found to penetrate into the heart musculature. If there is little fat, the right coronary artery can be seen coursing in the right coronary sulcus and its marginal branch can usually be seen on the lower border; anterior cardiac veins are located in this same region. The inferior end of the anterior interventricular branch of the left coronary artery can usually be seen in the interventricular sulcus accompanied by the great cardiac vein.

The **relation of the heart to the ribs and sternum** is extremely important clinically (Figs. 4-16 and 4-19). We have already seen the relation of the **great vessels** to the superior mediastinum, which automatically makes them posterior to the manubrium of the sternum. This makes the heart an organ that rests much higher in the chest than most people think. The **right border** is formed by the base of the superior vena cava and the right atrium, and extends from the second costal cartilage to the sixth, approximately ½ inch to the right of the sternum. The **left border** represents the union of the sternocostal and left surfaces of the heart; this border is not sharp and is often called the obtuse margin of the heart. It is formed by the auricle of the left atrium for a slight portion at its superior end, and by the left ventricle. It extends from the point where the second left costal cartilage joins the sternum to a point in the fifth left intercostal space approximately 3 to 3½ inches from the midline. In the male, this corresponds to a point approximately 1 inch medial and inferior to the nipple. To follow the left border of the heart, a line joining these two points should form a curve with its concavity

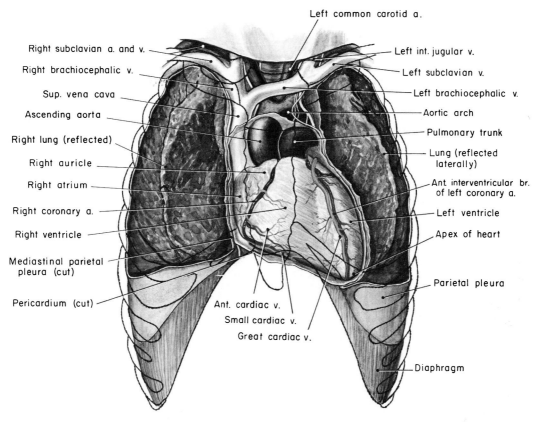

Right subclavian a. and v.
Right brachiocephalic v.
Sup. vena cava
Ascending aorta
Right lung (reflected)
Right auricle
Right atrium
Right coronary a.
Right ventricle
Mediastinal parietal pleura (cut)
Pericardium (cut)

Left common carotid a.
Left int. jugular v.
Left subclavian v.
Left brachiocephalic v.
Aortic arch
Pulmonary trunk
Lung (reflected laterally)
Ant. interventricular br. of left coronary a.
Left ventricle
Apex of heart
Parietal pleura
Diaphragm

Ant. cardiac v.
Small cardiac v.
Great cardiac v.

Figure 4-16. *Heart and lungs (reflected) after cutting the parietal pericardium and parietal pleura. The thin visceral layer of pericardium (epicardium) and the visceral pleura are intimate parts of these organs and are still intact. Note the relation of the heart and great vessels to the sternum.*

toward the midline. The **lower border** is formed where the sternocostal surface joins the diaphragmatic; this is a rather acute margin and is formed by the right atrium, right ventricle, and a small portion of the left ventricle. It extends from the inferior end of the right border to the inferior end of the left, in other words, between the sixth costal cartilage on the right side about ½ inch from the sternum, across the xiphisternal articulation, to the fifth intercostal space on the left side approximately 3 to 3½ inches from the midline. It should be understood that the position of the heart varies with the position, sex, and general build of the patient; it is lower when standing than when lying down. In fact, some claim the lower border of

the heart is 4.5 cm. below the xiphisternal joint, a claim difficult to believe when one realizes the diaphragm arises from this point. The relation of the valves to the chest wall will be described later (page 182).

HEART (REMOVED)

When the heart is removed from the pericardium, the inferior vena cava is brought into view and the four large pulmonary veins bringing blood from the lungs into the left atrium can be seen. The majority of the base of the heart is formed by this left atrium.

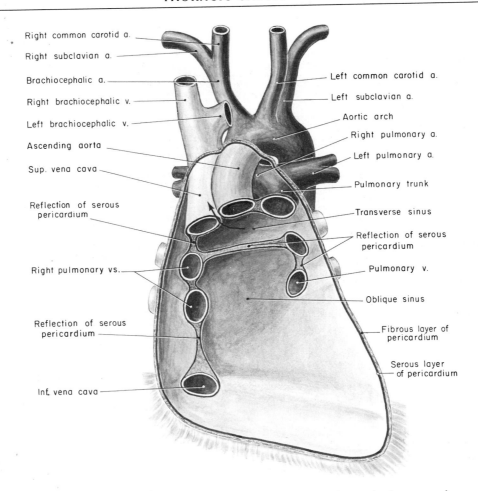

Right common carotid a.
Right subclavian a.
Brachiocephalic a.
Right brachiocephalic v.
Left brachiocephalic v.
Ascending aorta
Sup. vena cava
Reflection of serous pericardium
Right pulmonary vs.
Reflection of serous pericardium
Inf. vena cava

Left common carotid a.
Left subclavian a.
Aortic arch
Right pulmonary a.
Left pulmonary a.
Pulmonary trunk
Transverse sinus
Reflection of serous pericardium
Pulmonary v.
Oblique sinus
Fibrous layer of pericardium
Serous layer of pericardium

Figure 4-17. *Pericardium after removal of the heart. The arrow indicates the free access from the transverse sinus to the potential space between the aorta and superior vena cava. These vessels are not attached to each other.*

CORONARY CIRCULATION

Clinically speaking, there is nothing more important concerning the cardiovascular system than the circulation to the heart itself; known as the **coronary** (*L.,* like a crown) **circulation,** it is essential to life.* A thrombosis (*Gr.,* clot + condition) of a large component of this circulation usually results in death; small areas can be obliterated and recovery usually ensues.

The **left coronary artery** (Figs. 4-20, 4-21, and 4-22) starts in the ascending aorta in the left sinus of the aortic semilunar valve. It enters the atrioventricular sulcus on the left side and divides into its circumflex and anterior interventricular branches. The **anterior interventricular** branch is usually quite large and follows the interventricular septum or sulcus. It courses around the inferior margin of the heart and anastomoses with the posterior interventricular branch of the right coronary artery. This artery supplies blood to both right and left ventricles. The **circumflex branch** continues around the heart in the atrioventricular sulcus and terminates on the posterior surface of the heart by anastomosing with the right coro-

*L. Gross, 1921, *The Blood Supply to the Heart,* Hoeber, Inc., New York.

T. N. James, 1961, *Anatomy of the Coronary Arteries,* Hoeber, Inc., New York.

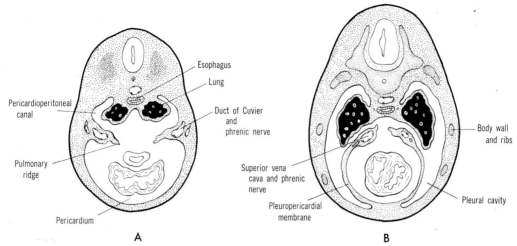

Figure 4-18. *Two transverse sections through the thoracic region of an embryo showing the development of the pleural and pericardial cavities. The relation of the lung buds to the pericardioperitoneal canals is shown in A, and further development of the pleural cavities in B. The extension of the pleural sac into the thoracic wall splits off a layer of the thoracic wall (pleuropericardial membrane) which ultimately forms the fibrous pericardium. Because this extension occurred in a plane lateral to the ducts of Cuvier (common cardinal veins—see Fig. 4-34 on page 185) and the phrenic nerves, these nerves and the derivatives of the ducts (superior vena cava on the right, and the ligament of the left vena cava) are contained in the fibrous pericardium. (From Davies,* Human Developmental Anatomy. *Copyright © 1963, John Wiley & Sons, Inc.)*

nary artery. It may give off large branches to the left ventricle and several smaller branches to the left atrium.

The **right coronary artery** (Figs. 4-20 and 4-22) arises in the right sinus of the aortic valve and enters the atrioventricular sulcus between the right atrium and right ventricle. It courses to the right, giving off a **marginal branch** near the lower border of the heart, continues around to the posterior part of the atrioventricular sulcus, and terminates as the **posterior interventricular branch.** It anastomoses with both the circumflex and anterior interventricular branches of the left coronary artery. The right coronary artery gives blood to the structures on the right side of the heart, but contributes to the left ventricle via its posterior interventricular branch.

These anastomoses of the coronary arteries are by very fine branches only; they will not save the patient with either a gradual or a sudden occlusion of a main vessel. They do increase in number with advancing age, but functionally the smaller branches of the coronary circulation are end arteries.

The **main veins** of the heart are called the great, small, and middle cardiac veins. The **great cardiac** starts (Fig. 4-20) in the anterior interventricular sulcus and ascends to the atrioventricular sulcus between the left atrium and the left ventricle; it then follows the course taken by the circumflex branch of the left coronary artery and ends on the posterior side of the heart in a dilation called the **coronary sinus.** This in turn empties into the right atrium of the heart. The coronary sinus also receives the **middle cardiac vein** which is located in the posterior interventricular sulcus (Fig. 4-21) and corresponds with the posterior interventricular branch of the right coronary artery. The **small cardiac vein** (Figs. 4-20 and 4-22) starts at the lower border of the heart in company with the marginal branch of the right coronary artery; it then courses to the atrioventricular sulcus on the right side, travels to the posterior side of the heart, and empties into the coronary sinus near its termination. These vessels drain the right and left ventricles and the left atrium. Many veins draining the right atrium and ventricle are short and

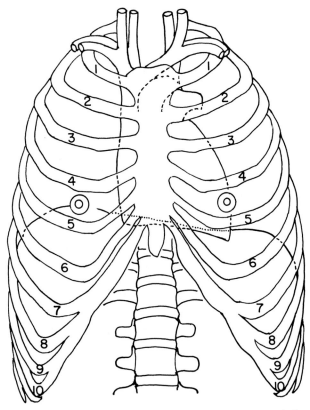

Figure 4-19. *Relation of the heart to the sternum and ribs.*

empty into the right atrium directly (**anterior cardiac veins**).

One other vein should be mentioned. The **oblique vein** is a twig that lies posterior to the left atrium and empties into the coronary sinus. This vein is a terminal part of the left common cardinal vein of the embryo; it frequently communicates with the highest intercostal vein on the left side (see Fig. 4-22).

The **myocardial capillary circulation** has been the subject of considerable interest and much confusion. It has been explored by injecting various dyes into the coronary arteries under various degrees of pressure. Because of these differences in pressure, different results have been obtained. Nevertheless, from this work it has been discovered that a considerable amount of blood entering the coronary arteries does not return to the coronary sinus but enters the cavities of the heart directly, and, from injection of particulate matter into the chambers, that these particles have arrived in the coronary

sinus. Because these particles were too large to go through the capillary network, they must have entered the coronary veins from the chambers directly. From this evidence it has been shown that, in addition to the capillary network, there are the following types of vessels in the myocardium of the heart (Fig. 4-22): (A) minute veins called venae minimi, or Thebesian veins, which start in the wall of the heart itself and empty into the various chambers, (B) channels that connect coronary arteries and veins without going through the capillary network, and (C) direct connections between the chambers and the coronary veins. Some claim* there are direct connections (D) from the coronary arteries through the heart wall to the chambers, but this is denied by Grant and Viko.† These additional channels are extremely important, because they may serve to bring blood to small areas where the regular arterial channel has been occluded; this function probably depends upon a gradual occlusion of the coronary vessels, however.

CHAMBERS OF HEART

RIGHT ATRIUM (Figs. 4-20 and 4-23). The walls of the right atrium are not particularly thick when compared to those of the left ventricle. The inside of the walls presents a smooth appearance at the point of entrance of the superior and inferior vena cavae, but a roughened appearance elsewhere. The roughened area, with its **musculi pectinati,** extends into the auricular appendage. The smooth area represents that part of the heart derived embryologically from the **sinus venosus**—that part of the caudal end of the original heart tube into which the umbilical and vitelline (yolk sac) veins drained. The ridge formed on the anterior surface of the atrium extending between the superior and inferior vena cavae and dividing the smooth part of the wall from the rough is the **crista terminalis;** the site of this crista is marked by a sulcus on the outside surface of the atrium, the **sulcus**

---

*A. A. Luisada, 1961, *Development and Structure of the Cardiovascular System,* Blakiston, New York.

†R. T. Grant and L. E. Viko, 1929, Observations on the anatomy of the Thebesian vessels of the heart, *Heart* 15: 103. These workers report that Thebesian vessels connect with (1) other similar channels, (2) the capillary network, and (3) directly with the coronary veins. They feel that they do not connect with the coronary arteries other than via the capillary bed.

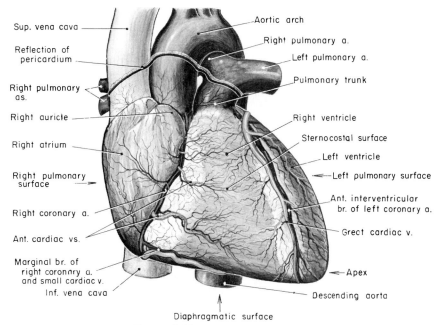

Sup. vena cava

Reflection of pericardium

Right pulmonary as.

Right auricle

Right atrium

Right pulmonary surface

Right coronary a.

Ant. cardiac vs.

Marginal br. of right coronary a. and small cardiac v.

Inf. vena cava

Aortic arch

Right pulmonary a.

Left pulmonary a.

Pulmonary trunk

Right ventricle

Sternocostal surface

Left ventricle

Left pulmonary surface

Ant. interventricular br. of left coronary a.

Great cardiac v.

Apex

Descending aorta

Diaphragmatic surface

Figure 4-20.   Heart—anterior view.

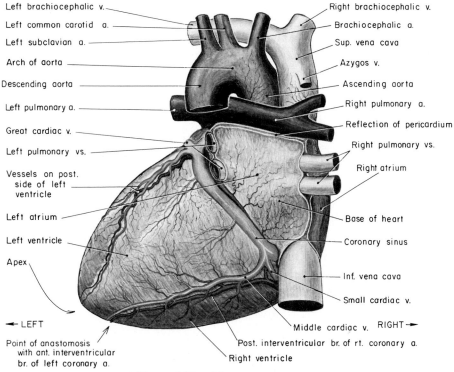

Left brachiocephalic v.

Left common carotid a.

Left subclavian a.

Arch of aorta

Descending aorta

Left pulmonary a.

Great cardiac v.

Left pulmonary vs.

Vessels on post. side of left ventricle

Left atrium

Left ventricle

Apex

← LEFT

Point of anastomosis with ant. interventricular br. of left coronary a.

Right brachiocephalic v.

Brachiocephalic a.

Sup. vena cava

Azygos v.

Ascending aorta

Right pulmonary a.

Reflection of pericardium

Right pulmonary vs.

Right atrium

Base of heart

Coronary sinus

Inf. vena cava

Small cardiac v.

RIGHT →

Middle cardiac v.

Post. interventricular br. of rt. coronary a.

Right ventricle

Figure 4-21.   Heart—posterior view.

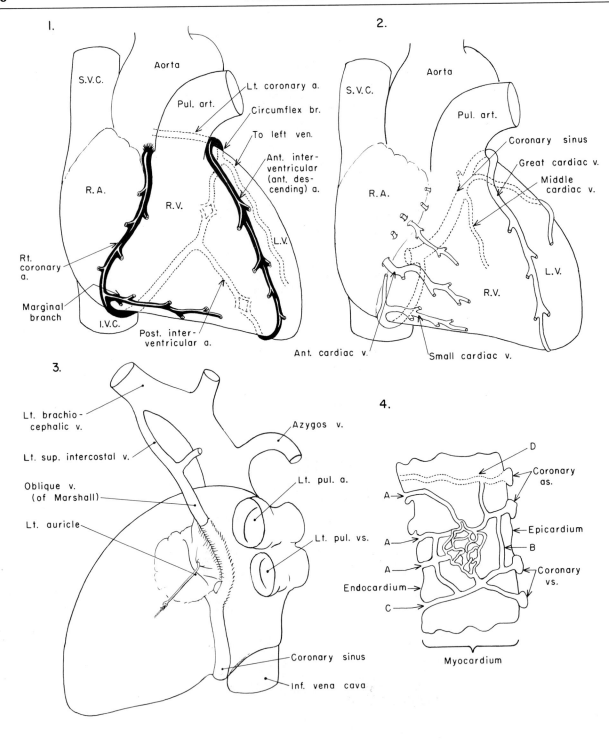

1.

S.V.C.

Aorta

Lt. coronary a.

Pul. art.

Circumflex br.

To left ven.

Ant. inter-
ventricular
(ant. des-
cending) a.

R.A.

R.V.

L.V.

Rt.
coronary
a.

Marginal
branch

I.V.C.

Post. inter-
ventricular a.

2.

S.V.C.

Aorta

Pul. art.

Coronary sinus

Great cardiac v.

Middle
cardiac v.

R.A.

L.V.

R.V.

Ant. cardiac v.

Small cardiac v.

3.

Lt. brachio-
cephalic v.

Azygos v.

Lt. sup. intercostal v.

Lt. pul. a.

Oblique v.
(of Marshall)

Lt. pul. vs.

Lt. auricle

Coronary sinus

Inf. vena cava

4.

D

Coronary
as.

A

A

Epicardium

B

A

Coronary
vs.

Endocardium

C

Myocardium

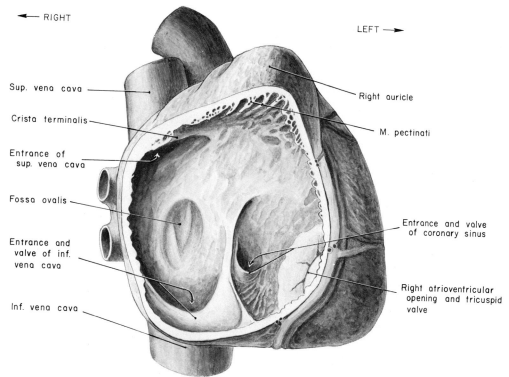

← RIGHT

LEFT →

Sup. vena cava

Crista terminalis

Entrance of
sup. vena cava

Fossa ovalis

Entrance and
valve of inf.
vena cava

Inf. vena cava

Right auricle

M. pectinati

Entrance and valve
of coronary sinus

Right atrioventricular
opening and tricuspid
valve

Figure 4-23. The right atrium.

terminalis. The entrance of the superior vena cava is obvious, as well as that of the inferior vena cava. There is a ridge just to the left of the entrance of the inferior vena cava that varies in size; this is the **valve of the inferior vena cava.** Just to the left of this valve remnant is an opening into the atrium from the **coronary sinus;** this opening also is protected by a small valve. To the left of this opening is the large **atrioventricular opening,** the entrance from the right atrium to the right ventricle. Approximately equally distanced between the superior and

inferior venae cavae is a **fossa** that represents the location of the embryonic **foramen ovale,** the communication between the right and the left atria of the embryo.

RIGHT VENTRICLE (Figs. 4-24 and 4-28). The walls of the right ventricle are considerably thicker than those of the right atrium. On opening this ventricle, the **atrioventricular opening** is seen on the right side of the chamber and the **interventricular septum** to the left. Superiorly, the ventricle opens into the **pulmonary**

Figure 4-22. *Diagrammatic presentation of (1) coronary arteries, (2) coronary veins, (3) the oblique vein (remnant of the left superior vena cava which functioned until the left brachiocephalic vein made connection with the right superior vena cava) (redrawn from Grant, Atlas of Human Anatomy), and (4) the circulation within the myocardium exhibiting (A) Thebesian veins making direct connections between the chambers of the heart and the capillary plexus, (B) direct connections between branches of coronary arteries and veins, (C) direct connections between chambers and coronary veins, and (D) disputed direct connections between chambers and coronary arteries. The anastomoses indicated in (1) are by very fine branches only that will not save a patient with either a gradual or sudden occlusion of a main artery.*

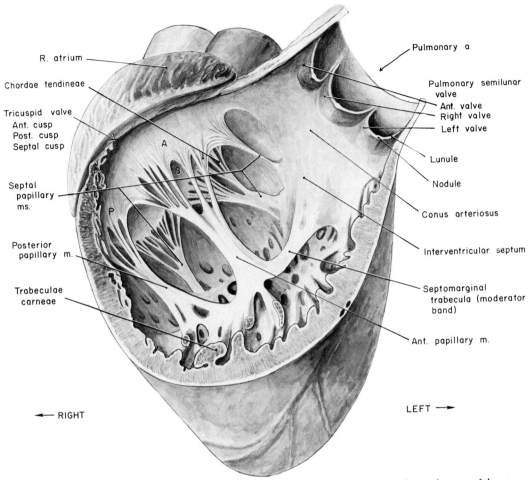

Figure 4-24. *The right ventricle. A, P, and S are the anterior, posterior, and septal cusps of the tricuspid valve.*

trunk. The atrioventricular opening is closed off during contraction of the ventricles (systole) by the **tricuspid valve.** This valve consists of **three cusps** (anterior, posterior, and septal) that are attached by tendinous cords—**chordae tendineae**—to muscular projections from the walls of the ventricle, the **papillary muscles.** Each papillary muscle sends chordae tendineae to halves of contiguous valves (Fig. 4-25). There is an **anterior papillary muscle** that arises from the sternocostal wall of the right ventricle and a **posterior (inferior) papillary muscle** that is attached to the diaphragmatic surface of this ventricle. The expected third papillary muscle (**septal**) is usually wanting and is replaced by small elevations to which the

chordae tendineae are attached. In approximately 40 per cent of cases an elevation extends from the interventricular septum to the base of the anterior papillary muscle. This is the **septomarginal trabecula** or **moderator band** and contains fibers or portions of the conduction system of the heart. When the right ventricle contracts, the cusps fill in the right atrioventricular opening. The cusps themselves are prevented from being driven into the right atrium by the contraction and steadying influence of the papillary muscles and the chordae tendineae. If it were not for these attachments, these cusps would be quite ineffective in preventing blood from flowing back into the right atrium. The **pulmonary semilunar** valve

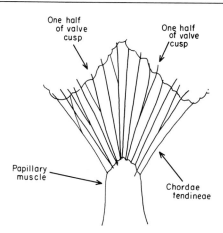

Figure 4-25. The arrangement of the chordae tendineae. Note that each papillary muscle is attached to contiguous halves of two valves by means of these cords.

consists of three valves called anterior, right, and left, respectively; there are **nodules** in the middle of each valve and the lunar shape of the valves on either side of the nodule accounts for the term "semilunar." When the ventricle contracts, blood forces these valves against the wall of the pulmonary trunk. When the ventricle is relaxed (diastole), the blood is prevented from flowing back into the right ventricle by these valves coming together in the midline and forming a barrier. The pulmonary trunk is, at first, anterior to the aorta and soon divides into the right and left pulmonary arteries to take unoxygenated blood to the lungs.

LEFT ATRIUM (Figs. 4-21 and 4-26). The oxygenated blood returns from the lungs to the heart via four **pulmonary veins.** These veins empty into the left atrium of the heart.

The left atrium has the thinnest walls of any of the chambers of the heart; its surface on the inside tends to be smooth except in the auricular appendage. The **fossa ovalis** is evident and its inferior margin is often called the **valvula.** The rather large opening from this chamber to the left ventricle is the **left atrioventricular opening.** Another opening leads into the auricle.

LEFT VENTRICLE (Figs. 4-20, 4-21, 4-27, and 4-28). The walls of the left ventricle are the thickest of any of the chambers. This correlates with the greater work re-

quired of this ventricle when compared with the other chambers; after all, it requires greater effort to push blood through the whole body than it does to send it to the lungs only, as is done by the right ventricle. This, in turn, requires more work than simply forcing blood from atria into the ventricles.

When the left ventricle is opened, the **left atrioventricular opening** is seen on the left side while the **interventricular septum** is to the right. The left atrioventricular opening is protected by two large cusps (anterior and posterior) of the **bicuspid** or **mitral valve.** These cusps are constructed in a manner similar to that found on the tricuspid valve with **papillary muscles** and **chordae tendineae.** Superiorly, the **aortic semilunar valve** can be seen, and it is of similar structure to that described for the semilunar valve in the pulmonary trunk except that the valves tend to be thicker and heavier. The openings of the right and left **coronary arteries** are located in the right and left sinuses of the ascending aorta respectively.

The bicuspid valve prevents blood from returning into the left atrium upon contraction of the left ventricle, and the semilunar aortic valve prevents blood from returning into the left ventricle in diastole.

The walls of both right and left ventricles have an extremely rough appearance; these are endothelial-covered strands of muscle in the myocardium, which can take the form of ridges, bridges, or papillary muscles. These elevations are called **trabeculae carneae.**

The **sequence of events in contraction of the heart** is as follows. Unoxygenated blood enters the right atrium and, simultaneously, oxygenated blood from the lungs enters the left atrium. In diastole, blood flows through the right and left atrioventricular openings from the right atrium to the right ventricle and from the left atrium to the left ventricle. Although the atria do contract, this blood will flow from the atria into the ventricles without contraction of the atria since the ventricles, in diastole, are both empty. After the ventricles are filled with blood, they contract (systole); blood is forced into the pulmonary trunk from the right ventricle and into the aorta from the left ventricle because the tricuspid and bicuspid valves close off the atrioventricular openings. The contraction of the ventricular musculature seems to "wring" the ventricles in a twistlike movement. This results from the spiral-

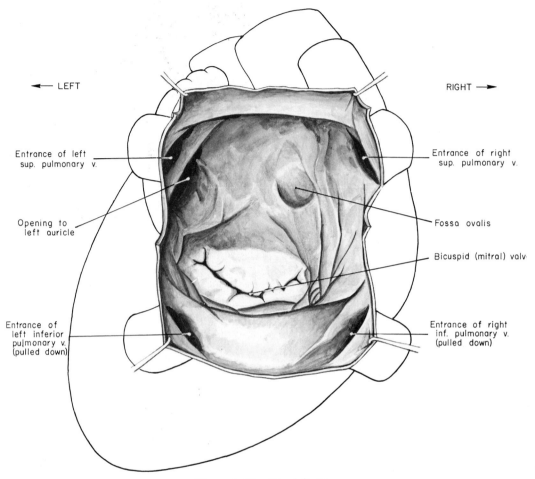

← LEFT

RIGHT →

Entrance of left
sup. pulmonary v.

Entrance of right
sup. pulmonary v.

Fossa ovalis

Opening to
left auricle

Bicuspid (mitral) valv·

Entrance of
left inferior
pulmonary v.
(pulled down)

Entrance of right
inf. pulmonary v.
(pulled down)

Figure 4-26.　*The left atrium.*

like nature of the heart musculature (Fig. 4-28). Both sets of semilunar valves are pushed against the walls of the respective arteries to allow the blood to flow by; the closing of these valves prevents blood from returning to the ventricles. The unoxygenated blood in the pulmonary artery is propelled to the lungs to be oxygenated, while the oxygenated blood in the aorta is delivered to the whole body. During systole, unoxygenated blood from the whole body has again been filling the right atrium and oxygenated blood from the lungs has been filling the left atrium, and the cycle starts anew.

The **heart sounds** are due primarily to fluid vibrations resulting from closure of the valves. Closure of the tricuspid and bicuspid (atrioventricular) valves is respon-

sible for the first heart sound; the closing of the pulmonary and aortic (semilunar) valves is responsible for the second.

The sequence of the atrioventricular beat is maintained by the conduction system of the heart.

FIBROUS FRAMEWORK
(Fig. 4-29)

In addition to the connective tissue found in the walls of the heart associated with the endocardium, myocardium, and epicardium, there are heavy concentrations of fibrous tissue at the bases of the pulmonary trunk and aorta, and at the two atrioventricular openings. These fi-

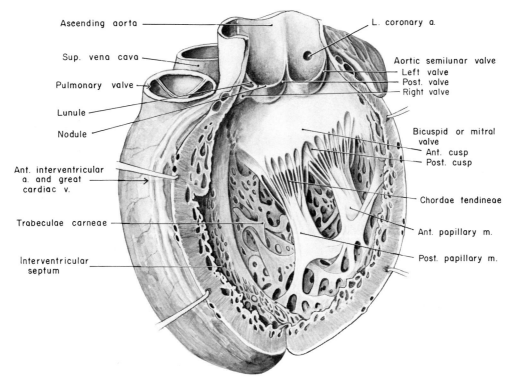

Ascending aorta

Sup. vena cava

Pulmonary valve

Lunule

Nodule

Ant. interventricular a. and great cardiac v.

Trabeculae carneae

Interventricular septum

L. coronary a.

Aortic semilunar valve
Left valve
Post. valve
Right valve

Bicuspid or mitral valve
Ant. cusp
Post. cusp

Chordae tendineae

Ant. papillary m.

Post. papillary m.

Figure 4-27. The left ventricle.

brous rings (Fig. 4-29) serve as a framework to which the muscles of the heart wall attach and as a firm base for attachment of the cusps of the various valves.

The fibrous rings successfully separate the muscles of the atrial walls from those forming the walls of the ventricles. The conduction system bridges these fibrous septa.

CONDUCTION SYSTEM OF HEART (Figs. 4-30, 4-31)

Both sympathetic and parasympathetic fibers enter the superficial and deep cardiac plexuses located in the superior mediastinum, and from these plexuses fibers travel to the coronary vessels and to the heart muscle itself via the pulmonary trunk and the aorta. Many of these nerve fibers end in a neuromuscular tissue called the **sinoatrial node*** (Fig. 4-30); this structure is about ½ inch long and

quite thin, and lies in the atrial wall between the upper part of the crista terminalis and the floor of the sulcus terminalis, that groove on the surface of the heart that corresponds to the crista terminalis on the inside of the heart. Impulses set up in this node spread through both atria via modified conduction tissue. In addition, the **atrioventricular node** is also stimulated; this node is embedded among the myocardial fibers in the septum between the two atria just superior to the opening of the coronary sinus (Fig. 4-31). The **atrioventricular bundle**† starts in this node and goes to the interventricular septum. Here it divides into **right and left branches** that run inferiorly in the two sides of the ventricular septum (Fig. 4-31); they break up and ramify in the walls of the ventricles. A large bundle of this tissue frequently extends from the septum to the base of the anterior papillary muscle in the right ventricle forming the **septomarginal**

---

*T. N. James, 1961, Anatomy of the human sinus node, *Anat. Record* 141: 109. Both gross and microscopic anatomy of the sinus node presented based on 300 human hearts.

†E. W. Walls, 1945, Dissection of the atrioventricular nodes and bundles in the human heart, *J. Anat.* 79: 45. A report on the rather difficult feat of dissecting the conduction system in the human heart.

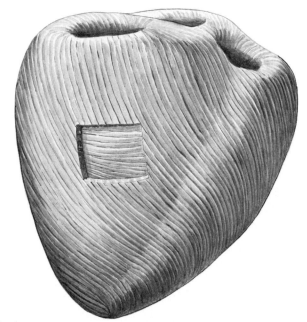

Figure 4-28. *The spiral-like nature of the myocardium of the ventricles.*

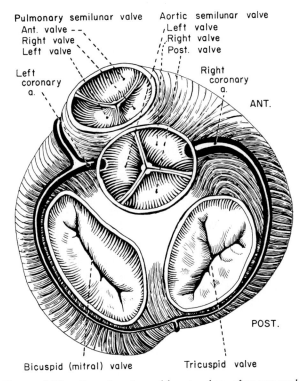

Figure 4-29. *Superior view of heart valves after removal of the atria.*

trabecula or **moderator band.** Since the muscles of the atria and the ventricles are separated by a connective tissue band, impulses must go from the atria to the ventricles via this conduction system. If anything interferes with this system, **arrhythmias** may result.

The sympathetic system increases the heart rate while the parasympathetic system retards it.

TOPOGRAPHIC ANATOMY

The ability to visualize the outline of the heart and position of the valves in relation to the surface of the body is extremely important. This knowledge is used every time a person listens to the heart with a stethoscope.

The outline of the heart was described on page 170. The valves of the heart can also be related to the anterior chest wall (Fig. 4-32).

If a line is drawn from the point of attachment of the *left* second costal cartilage to the sternum inferiorly to the point of attachment of the sixth costal cartilage of the *right* side to the sternum, this approximately represents the atrioventricular groove. The valves, when projected to the surface, lie just to the left of this line. The pulmo-

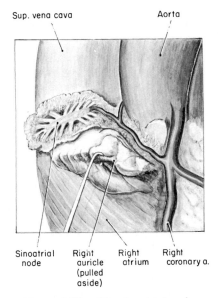

Figure 4-30. *The sinoatrial node.*

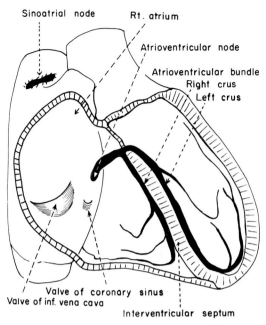

Figure 4-31. *The atrioventricular node and bundle.*

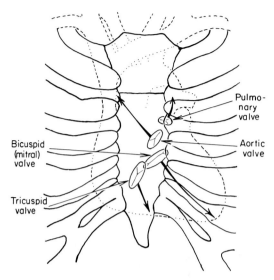

Figure 4-32. *The relation of the heart valves to the sternum and ribs. Arrows indicate where each valve is heard most clearly.*

nary valve is at the point of junction between the third left costal cartilage of the sternum and the aortic valve is just inferior to this point, opposite the third intercostal space. The bicuspid or mitral valve is close to the point of attachment of the fourth left costal cartilage to the sternum, and the tricuspid valve is projected to the midline of the body of the sternum opposite the fourth intercostal space. All valves are a bit lower than described when the patient stands.* One might expect that the best place to listen to these valves would be immediately on the above points. However, it has been found that the sound is conducted along other structures to the surface. The sound of the bicuspid valve is conducted along the left ventricle and is heard best in the fifth intercostal space at the location of the apex of the heart, for it is this apex that is closest to the surface. In the same manner, the tricuspid valve is heard best where the right ventricle is closest to the surface, which is at the lower left quarter of the body of the sternum. The sound of the aortic valve is conducted along

the aorta and is heard best at the point of attachment of the right second costal cartilage with the sternum. The pulmonary valve is heard best at the second intercostal space on the left side, which is almost directly anterior to this valve.

FETAL CIRCULATION

Since the lungs in the fetus are not functioning to aerate the blood, this role being accomplished by the placenta, there is no need of a large volume of blood going to these organs. Contrariwise, there is a need for blood to get to the placenta; this is accomplished by the umbilical cord.

The **circulation of blood in the fetus** is as follows (Fig. 4-33). Blood is returned from the placenta via the **umbilical veins** and is completely oxygenated. This blood follows along the umbilical vein, enters the **liver,** and then is shunted via the **ductus venosus** to the **inferior vena cava,** where it is mixed with unoxygenated blood returning from the lower part of the body. This mixture of blood then enters the **right atrium,** but instead of going into the right ventricle via the right atrioventricular opening, it proceeds through the **foramen ovale** into the **left atrium** thereby bypassing the lungs. This blood then enters the **left ventricle** and thence into

---

*E. Lachman, 1946, The dynamic concept of thoracic topography. A critical review of present day teaching of visceral anatomy, *Am. J. Roentgenol.* 56: 419. Levels of viscera given in many texts are based on recumbent cadavers while x-rays of the upright living person exhibit visceral levels considerably lower.

from the internal iliac arteries. This blood then courses through the umbilical cord to the placenta and becomes aerated once again.

It is easily seen from this description that there is no really pure oxygenated blood in the fetus except in the umbilical vein.

At birth or soon afterwards important changes take place:* the foramen ovale closes, becoming the fossa ovalis; the ductus arteriosus closes, becoming the ligamentum arteriosum; the ductus venosus closes, becoming the ligamentum venosum; and, with the tying of the umbilical cord, the umbilical vessels have no functional role, the umbilical arteries becoming the medial umbilical ligaments found in folds of the same name (review Fig. 3-41 on page 142), and the umbilical vein becoming the ligamentum teres (hepatis) contained in the free edge of the falciform ligament (see Fig. 5-2 on page 215). After these changes occur, the circulation has to proceed as it does in the adult and travel to the lungs for purification via the pulmonary arteries.

Complications would arise if the foregoing changes did not occur. For example, if the ductus arteriosus or the foramen ovale should remain patent,† the blood from the right and left sides of the heart could easily mix. Fortunately, new techniques in surgery have been found to alleviate these conditions.

EMBRYOLOGY OF THE HEART

Figures 4-34 through 4-37 present a brief summary of the development of the endocardial heart tube, with its cranial arterial end and caudal venous complex, into the definitive heart. With this complicated twisting of the original tube, the division into four chambers, and the complicated spiral nature of the separation of the arterial end into the aorta and pulmonary trunk, coupled with the

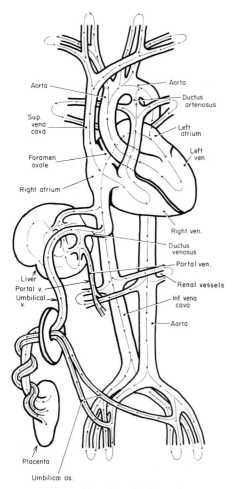

Figure 4-33. *The fetal circulation.*

the **aorta** to be distributed to the whole fetus. At the same time unoxygenated blood is returned to the **right atrium** from the upper part of the body via the **superior vena cava,** and this blood does continue through the **right atrioventricular opening** into the **right ventricle.** It then leaves the right ventricle and enters the **pulmonary trunk.** This blood will bypass the lungs by going through the large **ductus arteriosus,** which takes the blood directly into the **aorta** and thence to the body. Therefore, blood from the upper part of the body, the lower part of the body, and the placenta has been mixed, has bypassed the lungs, and is now in the aorta, whence it is distributed to the whole body. The blood then returns to the placenta via the **umbilical arteries,** which branch off

*A. E. Barclay, Sir Joseph Barcroft, D. H. Barron, K. J. Franklin, and M. M. L. Pritchard, 1941, Studies on the fetal circulation and of certain changes that take place after birth, *Am. J. Anat.* 69: 383. An interesting paper showing how direct x-ray cinematography can be used to follow blood flow. These workers report that closure of the foramen ovale is associated with respiration while closure of the ductus arteriosus coincides with ligation of the umbilical cord.

  D. H. Barron, 1944, Changes in the fetal circulation at birth, *Physiol. Rev.* 24: 277. A review of this problem up to 1944.

†B. M. Patten, 1931, The closure of the foramen ovale, *Am. J. Anat.* 48: 19. This paper points out that probe patencies occur in approximately 25 per cent of hearts and should be considered as a normal variant.

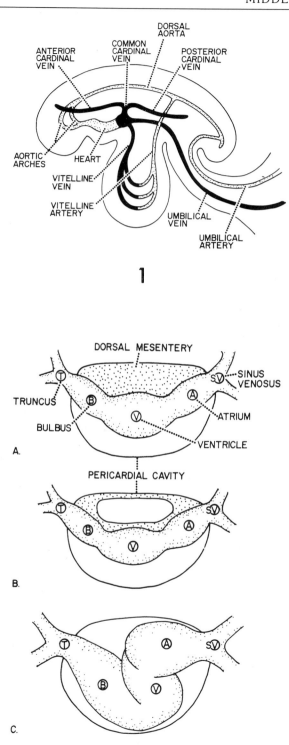

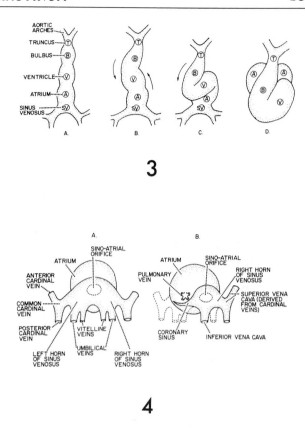

Figure 4-34. *Development of heart I. (1) Embryo at the end of the fourth week, illustrating the arrangement of the major intraembryonic and extraembryonic blood vessels. All major arteries and veins occur in pairs but only vessels on one side are shown. (2) Side views of the heart tube suspended in the pericardial cavity: (A) heart tube suspended by dorsal mesentery, (B) breakdown of the dorsal mesentery leaving heart suspended at cephalic (arterial) and caudal (venous) ends, and (C) bulbus and ventricle bulge ventrally due to rapid growth of the tube. (3) Ventral view of heart tube showing its various stages of development. (4) Dorsal view of heart tube; note the relationship of the sinus venosus to the dorsal wall of the atrium. In A the sinus venosus opens into the middle of the single atrium. Note also the paired umbilical, vitelline, and common cardinal veins. In B the opening between the sinus venosus and atrium has shifted to the right side leading to atrophy of the left side and the left vessel as well. Note the pulmonary veins developing as an outgrowth of the left side of the atrium. (From Crowley, L. V.: An Introduction to Clinical Embryology. Copyright © 1974 by Year Book Medical Publishers, Inc., Chicago. Used by permission.)*

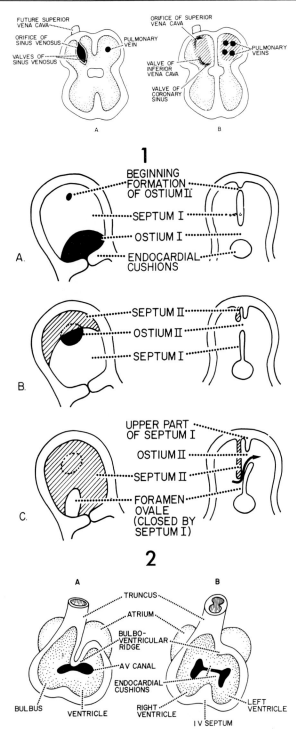

changes that must occur at birth, it is certainly easier to understand the abnormal than it is to understand why so many possess so-called normal hearts.

CLINICAL ASPECTS

In addition to developmental diseases,* the heart is susceptible to several other maladies. In fact, trouble in the cardiovascular system is responsible for more deaths than in any other system. Pathology in the heart can be divided into conditions affecting the pericardium, the myocardium, the endocardium including the valves, the coronary arteries, or the conduction system. Furthermore, many diseases of other systems have an effect on this organ.

All parts of the heart can become infected resulting in pericarditis, myocarditis, or endocarditis, and the myocardium can be invaded by fat. Angina is a term used for pain from the heart and is usually due to an anoxemia of the myocardium. Rheumatic fever is responsible for unusually destructive lesions of the heart, particularly of the valves, and the aortic semilunar valve is prone to diseases that affect the vascular system as a whole. The

---

*H. B. Taussig, 1947, *Congenital Malformations of the Heart*, The Commonwealth Fund, New York.

Figure 4-35. Development of heart II—showing internal changes. (1) Inside surface of dorsal aspect of heart: (A) the openings of the sinus venosus and the pulmonary vein (compare with 1 in Fig. 4-34) and (B) parts of dorsal walls derived from sinus venosus (crosshatched). Note also the gradual development of the four chambers. (2) Stages in the partitioning of the atrium. Illustrations on the left are lateral views of the developing interatrial septum; those on the right are frontal sections. Note that septum I descends toward the endocardial cushion creating ostium I which is gradually obliterated. Ostium II develops in this same septum. Septum II then descends, covering but not obliterating ostium II; its curved inferior edge forms the edge of the foramen ovale. Blood flows through the foramen ovale and then through ostium II. When blood pressure in the two atria becomes equal, the two septa fuse leaving a fossa ovalis. (3) Ventral views of the heart showing the partitioning of the ventricles. In A a single atrioventricular canal is present; in B the single canal is made double by the developing interventricular septum. (From Crowley, L. V.: An Introduction to Clinical Embryology. Copyright © 1974 by Year Book Medical Publishers, Inc., Chicago. Used by permission.)

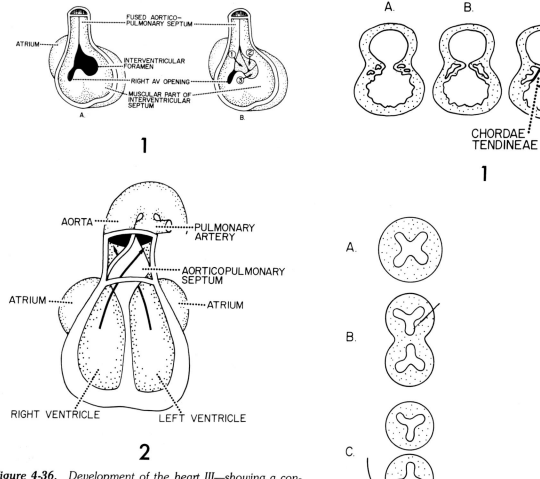

1

2

CHORDAE TENDINEAE   PAPILLARY MUSCLE

1

2

Figure 4-36. Development of the heart III—showing a continuation of the partitioning of the ventricle and also the division of the truncus into pulmonary and aortic parts. (1) Completion of the interventricular partition. Note in B that the final completion results from a fusion of the inferior margins of the aorticopulmonary septum (labeled 1 and 2) and tissue from the dorsal endocardial cushion (labeled 3), the latter forming the membranous portion of the interventricular septum. (2) The development of the spiral septum between the pulmonary trunk and the aorta. This results in the blood from the right ventricle entering the pulmonary trunk and blood from left ventricle entering the aorta. *(From Crowley, L. V.: An Introduction to Clinical Embryology. Copyright © 1974 by Year Book Medical Publishers, Inc., Chicago. Used by permission.)*

Figure 4-37. Development of the heart IV. (1) Formation of the atrioventricular valves. (2) Formation of the semilunar valves. Arrow indicates rotation of the apex of the heart to the left, thereby changing the relationship of the two valves to each other (see Fig. 4-29). *(From Crowley, L. V.: An Introduction to Clinical Embryology. Copyright © 1974 by Year Book Medical Publishers, Inc., Chicago. Used by permission.)*

coronary arteries can become sclerotic; thrombosis of these vessels is common. Cardiac arrhythmias occur as a result of interference with the conduction system. The heart can hypertrophy from several causes.

RESPIRATORY SYSTEM

When the thoracic cage is increased in size by muscle action, the parietal pleura follows the change in position. The lungs are covered with a layer of visceral pleura, and the space between the parietal and visceral layers is the pleural cavity. Since there is a negative pressure in this cavity, the lung, with its covering of visceral pleura, also expands in size. This results in air entering the lungs through the nose or mouth, pharynx, larynx, trachea, and bronchi. Of the above structures, the trachea, bronchi, and lungs are found in the thoracic cavity.

TRACHEA
(Figs. 4-9, 4-10, 4-11, 4-38)

This hollow structure, through which air passes, is kept open because of the many incomplete cartilage "rings" forming part of its walls; these U-shaped "rings" of hyaline cartilage are present anteriorly and laterally but are incomplete posteriorly. Originating in the neck, the trachea enters the superior mediastinum through the superior inlet of the thoracic cavity. At the inferior extent of the superior mediastinum it divides into right and left bronchi, which course to their respective lungs. A ridge is formed in the midline on the inside of the trachea at this point of division into right and left bronchi. This is the **carina,** easily seen when using a bronchoscope. The trachea varies in length in adults from 3½ to 5 inches; it is approximately ¾ inch in diameter. The trachea divides at a lower vertebral level in the standing patient.

Relations. Anterior—most of the structures dissected to this point are found anterior to the trachea. This includes the thymus gland, brachiocephalic veins and their tributaries, superior vena cava, aortic arch and its brachiocephalic branch, and branches of the vagus nerve coursing inferiorly to join the deep cardiac plexus found on the anterior surface of the trachea near its bifurcation into the two primary bronchi. **To the right**—the azygos vein and pleura. **To the left**—the arch of the aorta, left subclavian artery, and left common carotid artery. **Posterior**—(some of the following structures have not been studied as yet but are mentioned at this time for the sake of completeness; a glance at Figs. 4-42, 4-43, 4-45, and 4-46 would be helpful) the esophagus is not directly posterior to the trachea but more on the left side than on the right; to the left and slightly posterior, between it and the esophagus, is found the left recurrent laryngeal nerve; posterior and even more to the left is found the thoracic duct; posterior and to the right are found several intercostal arteries.

The trachea receives its motor **nerve supply** from the autonomic nervous system, the parasympathetic portion being derived from branches of the vagus nerve and the sympathetic portion from the first four thoracic segments of the spinal cord. The parasympathetic fibers cause a vasodilation, secretion of mucous glands, and a contraction of the smooth muscles in the trachea; the sympathetic system has an opposite effect. The majority of afferent fibers are found in the vagus nerves.

The **arteries** to the trachea are branches of the inferior thyroid artery, a branch of the thyrocervical trunk of the subclavian. The **veins** terminate in the plexus of veins around the thyroid, and in the inferior thyroid vein.

The **lymph nodes** draining the trachea are found anterior and lateral to the trachea near its bifurcation.

BRONCHI
(Fig. 4-11 on page 164)

The **primary bronchi** extend inferiorly and laterally from the trachea to enter the hilum of the lungs. The right bronchus is larger and more vertical than the left and is shorter than the left bronchus. The left bronchus, approximately twice the length of the right, takes a more horizontal course to reach the left lung because the trachea is slightly more to the right than to the left, and the heart is more on the left side, which forces the hilum of the left lung to be farther away from the midline than is that of the right. This difference in direction of the bronchi results in more foreign bodies entering the right lung than the left.

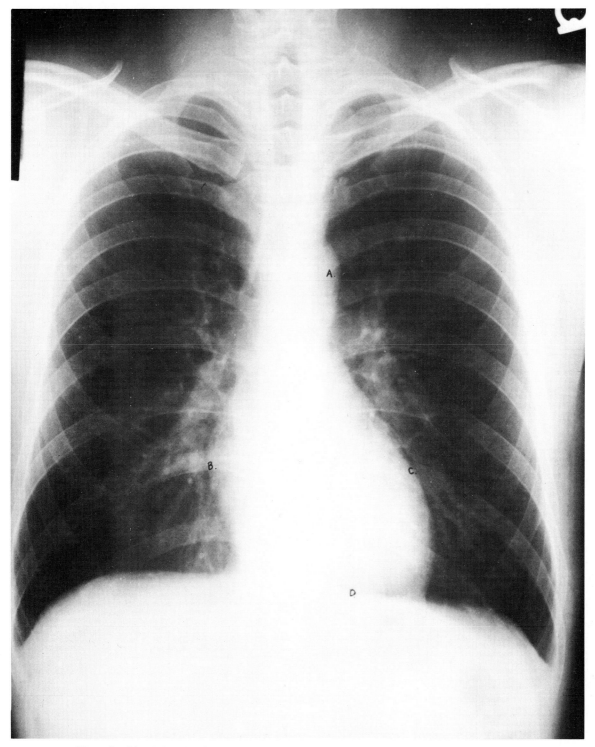

Plate 6. *Normal male thoracic cavity: (A) arch of aorta, (B) right border of heart, (C) left border, and (D) inferior or diaphragmatic border. Compare with Figure 4-19.*

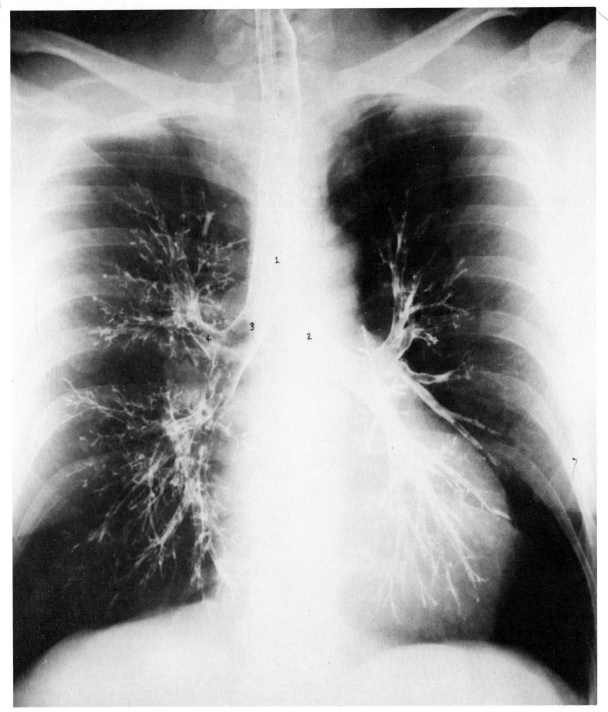

Plate 7. *The trachea and bronchi, outlined by opaque material: (1) trachea, (2) left bronchus, (3) right bronchus, and (4) eparterial bronchus. Note how the right bronchus takes a more vertical course and is shorter than the left bronchus. Note also the outline of the heart, which is larger than normal (hypertrophied).*

Relations.   Before studying the following relations, the architecture of the superior mediastinum should be recalled.

The bifurcation of the trachea occurs approximately at the level of the fifth thoracic vertebra. The **right bronchus** is posterior to the right pulmonary artery which, in turn, is posterior to the superior vena cava. This bronchus is anterior to the right vagus nerve and the right bronchial artery. The **left bronchus** is posterior to the left pulmonary artery, which winds around to gain the superior side of the bronchus. This bronchus is anterior to the left vagus nerve and to the esophagus. After crossing the esophagus, it crosses anterior to the descending aorta. This bronchus is also accompanied by a bronchial artery.

The right bronchus gives off an **eparterial** (above the artery) **bronchus** to the upper lobe and subsequently divides into two bronchi for the middle and inferior lobes. The left bronchus divides into two bronchi for the superior and inferior lobes of that lung. A pulmonary plexus of nerves is found on the posterior side of each bronchus with many branches coursing onto the anterior surface.

The **arteries** to the bronchi are two in number on the left side, both arising from the descending aorta, and one on the right side branching from either the first right aortic intercostal or from the more superiorly placed of the left bronchial arteries. The veins from the bronchi end in the axygos vein on the right side and in the hemiazygos vein on the left side (Fig. 4-43).

The **lymphatics** end in profuse glandular tissue on the anterior surface of each bronchus.

The **nerves** are from the vagus and sympathetic through the pulmonary plexuses. They have the same effect as described for the trachea.

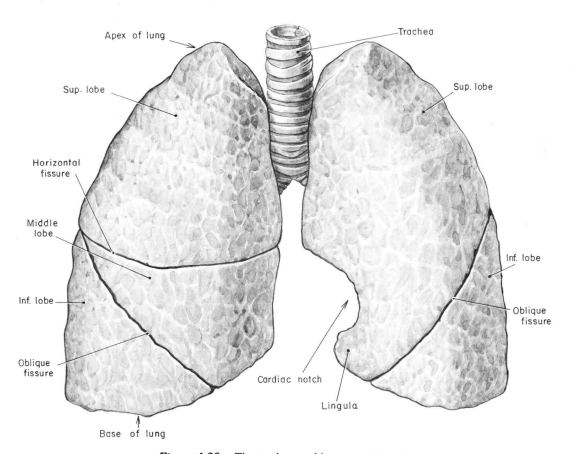

Figure 4-38. *The trachea and lungs—anterior view.*

LUNGS
(Figs. 4-38, 4-39, 4-40)

The lungs are very large organs that occupy the greater part of the space in the thoracic cage. The **right lung** is composed of **three lobes,** a superior, an inferior, and a middle. They are separated from each other by **oblique** and **horizontal fissures.** The left lung, smaller in volume because more of the heart is on the left side, has **two lobes,** a superior and an inferior, which are separated by an **oblique fissure.** Both lungs have a **base** inferiorly and an **apex** which extends for a considerable distance above the first rib. Both lungs have **costal, diaphragmatic,** and **mediastinal surfaces** in contact with the ribs, diaphragm, and mediastinal structures respectively. Structures entering and leaving the lungs do so in one region, the **hilum.** Contrary to the usual impression, a greater volume of the lung is located posteriorly than anteriorly.

The lungs are covered with a serous membrane, the **visceral pleura,** which is directly continuous with the parietal pleura at the hilum of the lung. This visceral pleura is intimately attached to the lung and extends down into the fissures separating the lobes.

RIGHT LUNG.   The lungs can be outlined on the chest wall with considerable accuracy (Fig. 4-39), a point of importance in diagnostic procedures and surgery of the thoracic region.

The **right lung** extends to a point close to the midline medially, follows a curved line inferiorly passing the sixth rib anteriorly, the eighth rib in the midaxillary line, and the tenth rib posteriorly. The apex is rounded and projects into the neck at least 1 inch superior to the clavicle. The base is concave, due to the upward projection of the diaphragm, and the edges of the base are quite sharp where the lung extends into the triangular space between the body wall and the diaphragm. The oblique fissure of

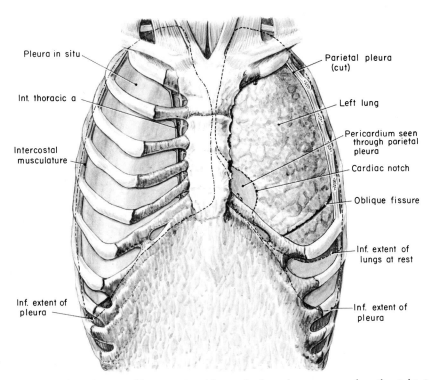

Figure 4-39.   *The pleura and lungs in situ. All muscles have been removed on the right side of the body to show the parietal pleura in position. This parietal pleura has been removed on the left side of the body to show the lung covered with visceral pleura. Dot-dash line gives outline of the pleural cavity; dashed line is outline of the lung.*

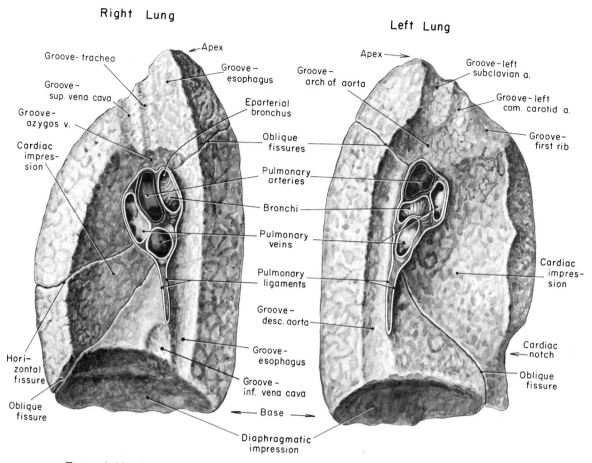

Right Lung

Groove- trachea
Groove-
sup. vena cava
Groove-
azygos v.
Cardiac
impres-
sion
Apex
Groove-
esophagus
Eparterial
bronchus
Oblique
fissures
Pulmonary
arteries
Bronchi
Pulmonary
veins
Pulmonary
ligaments
Groove-
desc. aorta
Groove-
esophagus
Groove-
inf. vena cava
Base
Hori-
zontal
fissure
Oblique
fissure

Left Lung

Apex
Groove-
arch of aorta
Groove-left
subclavian a.
Groove-left
com. carotid a.
Groove-
first rib
Cardiac
impres-
sion
Cardiac
notch
Oblique
fissure

Diaphragmatic
impression

Figure 4-40.   Medial surfaces of the right and left lungs. (Nerves and bronchial arteries have been omitted.) Compare with Figure 4-11 on page 164, and with Figures 4-45 and 4-46.

the right lung follows the sixth rib. This fissure starts approximately 2½ inches inferior to the summit of the lung opposite the second thoracic spine and continues inferiorly and anteriorly to end opposite the sixth or seventh costal cartilage about two inches from the midline. The **horizontal fissure** of the right lung follows approximately the course taken by the fourth rib. The superior and middle lobes are examined clinically from the anterior side while the inferior lobe should be studied from the posterior side of the body.

Although the lungs collapse if the pleural cavity is entered, if they are hardened *in situ* they show **impressions** of the structures with which they come in contact. These impressions are of no significance as far as function

of the lungs is concerned, but they are a good indication of their relations. The right lung (Figs. 4-38 and 4-40), on its inferior surface, is in contact with the diaphragm and the large concavity results from the superior projection of that muscle. On the posterior and lateral surfaces the lung shows impressions for the ribs and intercostal spaces. On the medial surface there is a large impression for the heart and there is a groove where the lung contacts the inferior vena cava. The superior vena cava also makes an impression on the right lung superior to the hilum and a groove for the azygos vein can be seen arching just superior to the hilum. There is a small depression superiorly for the esophagus and another area that comes in contact with the trachea; these are near the apex of the lung superior to the groove for the arch of the azygos vein. Posterior to the hilum of the lung is an elongated groove marking the point of contact with the continuation of the esophagus.

The **hilum of the right lung** (Fig. 4-40) contains the bronchus, pulmonary arteries, pulmonary veins, various branches from the pulmonary plexus, bronchial arteries and veins, and the lymphatics. The hilum of the right lung differs from the left in that there are three lobes in the right lung, which results in an extra bronchus. That to the upper lobe arises in a position superior to the pulmonary artery. This is the eparterial bronchus. The relation of the structures in the hilum of the lung can best be visualized if the relations of the bronchi are kept in mind. The bronchi are posterior to the pulmonary arteries and the pulmonary veins take an anterior and inferior position. The bronchial arteries course on the posterior surfaces of the bronchi.

LEFT LUNG.   The left lung (Figs. 4-38 and 4-40) is very similar to the right except that it is smaller in volume because of the heart and has only two lobes, superior and inferior. Its anterior border is notched (the **cardiac notch**) because of the heart and this notch results in a tongue-like projection, which is called the **lingula**. These differences make the medial surface outline of the left lung quite irregular compared with that of the right lung. Otherwise, the outlines of the two lungs are similar.

The **impressions** made by various structures on the costal and diaphragmatic surfaces of the left lung are the same as those found on the right; the impressions on the medial or mediastinal surface are quite different. There is a very large cardiac impression which is anterior and inferior to the hilum. A large groove starts just superior to the hilum, continues posterior to it and the whole length of the lung for the arch and descending aorta respectively. Grooves are made by the left subclavian artery. This lung does not come in contact with the trachea or the esophagus.

The **hilum of the left lung** contains the same structures as are found on the right, except for the eparterial bronchus. The pulmonary artery tends to wind around the left bronchus to obtain a superior position to it; therefore, the pulmonary artery is the most superior structure in the hilum of the left lung, and the bronchus lies just inferior to it. The pulmonary veins are found anterior and inferior to both structures just as mentioned for the veins on the right side.

BRONCHOPULMONARY SEGMENTATION.*   It has been found that each lobe of the lung can be divided into segments. This is of some importance since a segment of a lobe can be removed surgically rather than the more drastic removal of an entire lobe. The bronchi, as named, and the various segments of the right and left lung are pictured in Figure 4-41. Each bronchus is accompanied by a branch of the pulmonary artery, but the veins drain more than one segment.

The eparterial bronchus to the **right upper lobe** is divided into three: **apical, posterior,** and **anterior bronchi.** These go to three segments of the superior lobe which bear the same names. The **middle lobe** of the right lung is divided into **medial** and **lateral segments** and the bronchi have the same names. The **inferior lobe** has five segments; one, which is quite superiorly located, is naturally named the **superior segment,** and the base is divided into **anterior, medial, lateral, and posterior basal segments.**

Unfortunately, the left lung differs in that there are only two main lobes. The main bronchus to the superior lobe is divided into two bronchi which correspond to those bronchi to the superior and middle lobes on the right side. The **upper part** of the **superior** lobe is divided into **apical, anterior,** and **posterior segments** which is exactly the same as found in the upper lobe on the right side. The bronchi to these segments, however, differ in that the bronchi to the apical and posterior segments are combined into a single **apical-posterior bronchus.** The **lower part of the superior lobe** on the left side is divided into **superior** and **inferior segments,** rather than the medial and lateral found in the middle lobe of the right lung. The **inferior lobe** of the left lung has the same arrangement as found on the right, namely, a **superior,** and **medial, anterior, posterior,** and **lateral basal segments,** but differs in that the bronchus to the anterior and medial basal segments is combined. This bronchopulmonary segmentation is of great importance to the radiologist and to the thoracic surgeon.

LYMPHATIC DRAINAGE.   Lymphatics draining the lungs empty into pulmonary lymph nodules which are located at points of bifurcation of the bronchi inside the

---

*C. L. Jackson and J. F. Huber, 1943, Correlated applied anatomy of the bronchial tree and lungs with a system of nomenclature, *Dis. Chest* 9: 319.

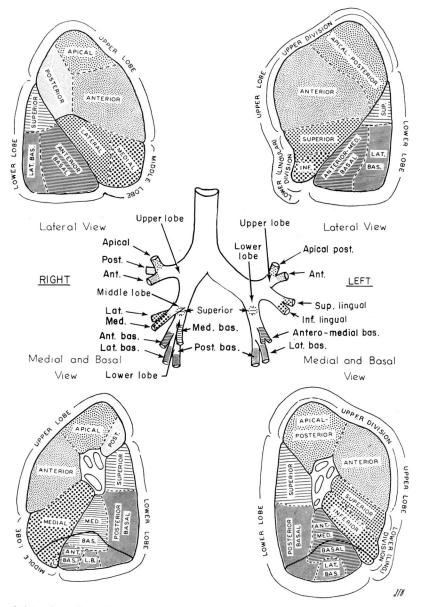

Figure 4-41. *Bronchopulmonary segmentation. (From J. F. Huber, 1949, Practical correlative anatomy of the bronchial tree and lungs, J. Nat. Med. Assoc. 41:49, courtesy of National Medical Association.)*

lung itself. From these nodules the lymphatics continue to nodes located in the hilum of the lung and thence into tracheobronchial nodes which lie on all sides of the trachea near its bifurcation; these nodes are extensive. They drain into right and left bronchomediastinal lymph trunks which empty into the point of junction between the internal jugular and subclavian veins on each respective side.

NERVES.   The nerve supply to the lung is by branches from the pulmonary plexuses. These, as mentioned previously, are composed of branches of the vagus nerve and of sympathetic branches from the first four thoracic segments. These nerves not only innervate the smooth muscle found in the vascular system in the lungs but also the smooth muscle found in the bronchi and bronchioles. The sympathetic system causes a vasoconstriction, a dilation of the bronchioles, and decreases mucus secretion, while the parasympathetic system has an opposite action, i.e., a vasodilation, constriction of the bronchioles, and an increase in gland secretion. The above—a vasoconstriction accompanying a dilation of the bronchioles under sympathetic stimulation and the reverse with parasympathetic stimulation—may seem contradictory if one thinks of more air and more blood as a logical combination in an emergency, and the opposite situation under normal conditions. However, it has been claimed that the more rapid blood flow that accompanies a vasoconstriction enhances the exchange of oxygen and carbon dioxide with the dilated bronchioles.

Afferents accompany the vagus nerve; they should not be confused with sensory nerves from the parietal pleura which are quite different.

The mechanism whereby air can be brought into the lung and carbon dioxide exhaled is described on page 123.

PULMONARY CIRCULATION
(Figs. 4-11, 4-45, 4-46)

As mentioned in the description of the heart, the pulmonary trunk extends superiorly from the right ventricle. This pulmonary trunk is anterior to the beginning of the aorta but soon twists around to a position that is to the left and slightly posterior to the arch of the aorta. It divides into **right** and **left pulmonary arteries;** this point of bifurcation is just anterior to the left bronchus which is, in turn, anterior to the descending aorta. The left pulmonary artery follows the left bronchus but soon takes a position superior to it as it enters the lung. The right pulmonary artery is longer than the left and courses toward the right lung at a point just inferior to the bifurcation of the trachea. In this position it is anterior to many lymph glands and to the esophagus. As it courses to the right it divides into branches that follow respective bronchi into the lung. The right pulmonary artery is posterior to the ascending aorta and also posterior to the superior vena cava.

As the pulmonary arteries enter the lung tissue, they divide in a fashion similar to that described for the bronchi. When the bronchi finally reach their termination in the small bronchioles and, lastly, the air sacs, the pulmonary arteries end in capillaries that form a very rich vascular network between the walls of contiguous alveoli. These pulmonary arteries carry unoxygenated blood to the lungs to be aerated. The branches of the pulmonary artery, for the most part, remain on the posterior surface of the bronchi.

The **pulmonary veins** are formed from the massive capillary network described above. They usually ramify in a manner similar to that of the bronchi but more on their anterior surfaces. They gradually unite to form **two main pulmonary veins on each side.** These empty into the left atrium as four distinct openings but may be combined on either side. They transmit oxygenated blood to the heart where it will be pumped to the whole body.

It should be noted that the lungs have a **double circulation** from (1) pulmonary arteries and (2) from bronchial arteries. The blood in the pulmonary circulation does not give blood supply to the lung tissue itself; this is done by the bronchial arteries. Although some of the blood taken to the lungs by the bronchial arteries is drained by the pulmonary veins, bronchial veins do exist. They drain into the azygos vein on the right side and into the accessory hemiazygos on the left.

CLINICAL ASPECTS

The respiratory system as a whole is quite susceptible to disease. Possibly the constant direct exposure to the outside world, with its increased air pollution, is partially responsible. We all know how prone this system is to viral, bacterial, and fungal infections. Rhinitis, sinusitis, pharyngitis, laryngitis, tracheitis, bronchitis, and pneumonia are common occurrences. Other infections such as diphtheria and tuberculosis also occur but are not as common as heretofore. The respiratory system is quite susceptible to cancer and much has been written in an attempt to correlate air pollution, including smoking, with the high incidence of lung cancer. Emphysema is a condition in which, for one reason or another, the lung cannot get rid of its air and becomes overdilated. Foreign bodies get into the lungs very frequently, the majority entering the right lung because of the straighter course taken by the right bronchus. Respiratory diseases are common problems in industrial medicine. Other disease conditions of the lungs are congestion, pulmonary edema, embolism, and atelectasis (collapse of lung tissue). The respiratory system is quite susceptible to allergic conditions, asthma being a common occurrence. The pleura is also subject to infection—pleurisy—and empyema is a term denoting a purulent infection of these membranes. Pneumothorax is the name given to the condition in which the lung has collapsed due to entrance of air into the pleural cavity, thereby eliminating the negative pressure necessary to keep the lung expanded.

POSTERIOR MEDIASTINUM AND POSTERIOR PORTION OF SUPERIOR MEDIASTINUM

GENERAL DESCRIPTION

If the pericardium is removed, the posterior mediastinal compartment and its contents are revealed (Fig. 4-42 and 4-43). This area is usually limited to that anterior to the vertebral column; structures found on the sides of the vertebral column are, therefore, not included in it. The

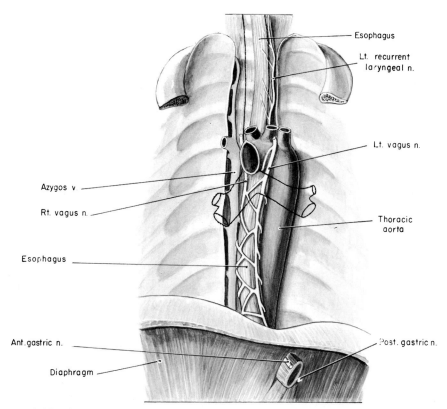

Esophagus

Lt. recurrent laryngeal n.

Lt. vagus n.

Azygos v.

Rt. vagus n.

Thoracic aorta

Esophagus

Ant. gastric n.

Post. gastric n.

Diaphragm

Figure 4-42. *Posterior mediastinum as seen upon removal of the pericardium. The trachea and bronchi have been outlined.*

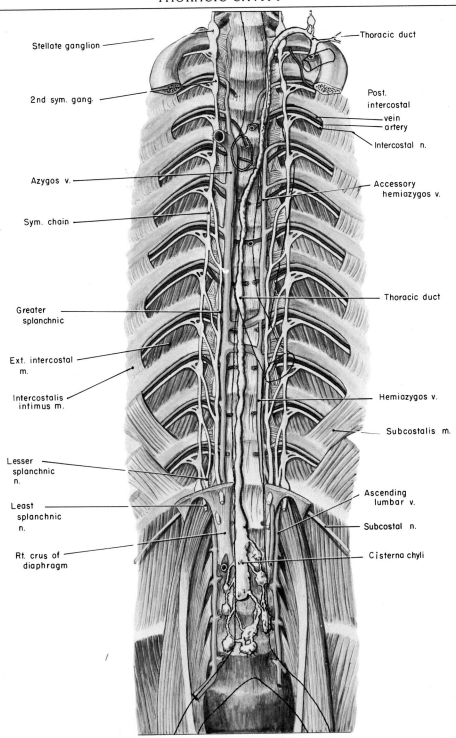

Stellate ganglion

2nd sym. gang.

Azygos v.

Sym. chain

Greater
splanchnic

Ext. intercostal
m.

Intercostalis
intimus m.

Lesser
splanchnic
n.

Least
splanchnic
n.

Rt. crus of
diaphragm

Thoracic duct

Post.
intercostal
vein
artery
Intercostal n.

Accessory
hemiazygos v.

Thoracic duct

Hemiazygos v.

Subcostalis m.

Ascending
lumbar v.

Subcostal n.

Cisterna chyli

Figure 4-43. *Contents of posterior mediastinum posterior to the esophagus and aorta. The esophagus and aorta have been outlined.*

boundaries are the pericardium anteriorly, the vertebral column posteriorly, the mediastinal pleura on each side, the superior mediastinum superiorly, and the diaphragm inferiorly.

With a few exceptions, the structures in the posterior mediastinum course longitudinally.

The largest structure in the posterior mediastinum is the **descending aorta.** This is found in the left side of the compartment and is in contact with the parietal pleura on that side. Just to the right of the aorta is the **esophagus.** Inferiorly, the latter organ tends to cross the aorta on its anterior side. The left and right **vagus nerves** are found on the anterior and posterior sides of the esophagus but not usually as distinct nerves, for they frequently form a network that ramifies on all surfaces of the tube. On the right side, posterior to the esophagus and in contact with the mediastinal pleura and the vertebral column, is the **azygos vein,** into which drain the intercostal veins of the right side of the thoracic wall. A similar set of vertically running veins is found on the left side. These are the **hemiazygos** and **accessory hemiazygos** veins, which drain the intercostal veins of the left side of the chest wall and finally empty into the azygos vein (Fig. 4-44). Between the azygos vein and the descending aorta, and posterior to the esophagus, lies the important **thoracic duct.** The aorta gives off **several intercostal arteries** to each side of the chest wall. Since the aorta is on the left side and in contact with the mediastinal pleura, the left intercostal arteries disappear from view. The right intercostals cross the vertebral column to reach the proper intercostal space. As might be expected, they occupy a position posterior to all structures mentioned.

The **sympathetic chains** are located along the sides of the bodies of the vertebrae and are not considered to be located in the mediastinum. The **greater splanchnic branches** gradually approach the anterior surface of the vertebral column, however, and are considered to be inside the inferior end of the posterior mediastinum.

If the large veins, the large arteries, and trachea in the superior mediastinum are reflected, these structures found in the posterior mediastinum are seen to continue into the posterior part of the superior mediastinum.

ESOPHAGUS

The **esophagus** (Figs. 4-42 and 4-45) is a muscular tube lined with stratified squamous epithelium that serves to

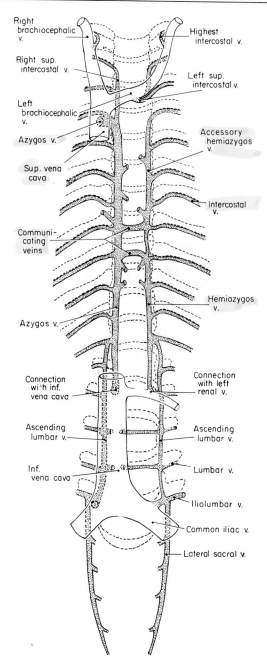

Figure 4-44. *The azygos system of veins. Note how the ascending lumbar veins of the abdominal cavity form a connection between the common iliac veins and this system.*

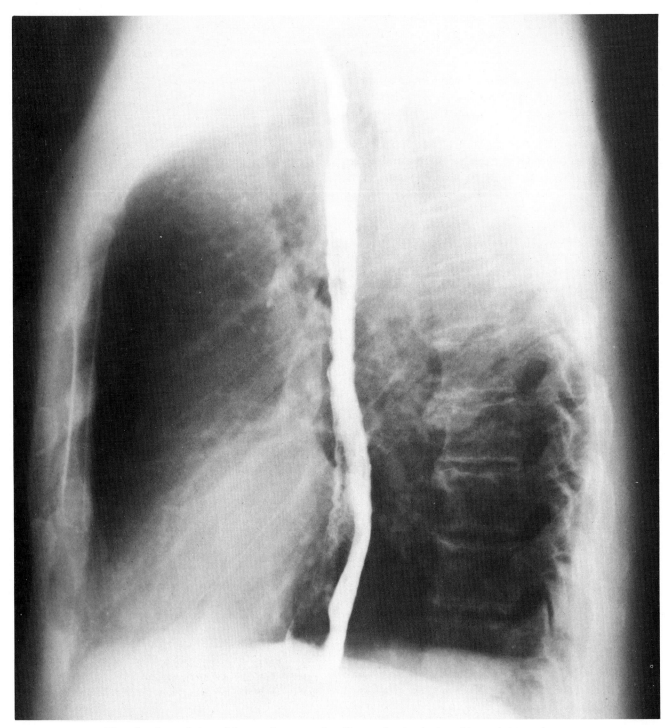

Plate 8. *Lateral view of the esophagus, outlined by swallowing barium. Note relation to heart and vertebral column.*

conduct food from the pharynx to the stomach after it has been chewed and mixed with saliva in the mouth and swallowed.

The esophagus is a long tube about ½ to ¾ inch in diameter that extends from the pharynx superiorly to the stomach inferiorly. It is located in the neck and in the thoracic and abdominal cavities. It is not a straight tube but has several small **curvatures.** In the superior mediastinum it tends to curve slightly to the left and then gradually to the right as it passes the arch of the aorta. Near its inferior end it takes a turn to the left to traverse the esophageal hiatus (in the diaphragm) and join the stomach located in the abdomen.

Relations. The esophagus is **anterior to** the vertebral column in its entire extent and is practically in contact with it except when passing anterior to the descending aorta near the diaphragm. It is also anterior to the thoracic duct, the azygos vein, and several intercostal vessels. It is **posterior to** the larynx in the neck, to the trachea in the neck and superior mediastinum, to the bifurcation of the trachea at the borderline between the superior and inferior mediastinal compartments, to many lymph nodes, and to the heart and pericardium. **On its right,** the esophagus is in contact with the right lung throughout most of its extent except where the azygos vein intervenes. **On the left** side, it is in relation to the subclavian and common carotid arteries, with the arch of the aorta, and the descending aorta. The latter structure separates the esophagus from the left lung. As mentioned previously, the aorta comes to occupy a position posterior to the esophagus near the diaphragm.

The esophagus is richly supplied with a network of **nerves,** the left vagus nerve being on its anterior surface in the posterior mediastinum while the right vagus nerve is on its posterior surface; this relation is due to rotation of the stomach to the right. The recurrent laryngeal nerve on the left side is located in the groove between it and the trachea. Branches from the sympathetic trunks, the greater splanchnic nerves, recurrent laryngeal, and vagus nerves themselves innervate the esophagus. The skeletal muscle found in the upper third of the esophagus is derived embryologically from the same muscle mass as the pharyngeal and laryngeal muscles. We will see later that all such muscles are innervated by the vagus (or bulbar portion of spinal accessory) nerve. Peristaltic waves of the esophagus are due to contraction of the smooth muscles in the esophagus; since these waves cease if the vagus

nerves are cut, these smooth muscles are under parasympathetic control. Although a physiologic cardiac sphincter, located between the inferior end of the esophagus and the stomach, has been described, the anatomic nature of this structure has not been resolved. Suggestions have been an elevation of the mucous membrane at the inferior end of the esophagus, a thickening of the circular layer of muscle, strands of muscle from the lesser curvature of the stomach looping around the esophagus, and either ligamentous or muscular attachments to the diaphragm. That such a sphincter exists has been revealed by the hesitation seen radiologically in barium entering the stomach from the esophagus. We all know that these affairs can be reversed when nauseated; tactile stimulation of the pharynx initiates a vomiting or gag reflex. Afferent fibers from the esophagus are found in the vagus and in the sympathetic nerves, the former being the impulses that reach higher centers for sensation.

The **arterial supply,** from superior to inferior, consists of branches of the left inferior thyroid artery, the bronchial arteries, several branches directly from the aorta, and, most inferiorly, from the left gastric artery from the celiac axis in the abdominal cavity. The phrenic arteries also contribute to the blood supply of the esophagus. The **venous drainage** is to vessels that correspond to the arteries. One important point should be realized; the lower end of the esophagus is a point of communication between the systemic venous circulation and the portal venous circulation (see Fig. 5-23 on page 238). Increased pressure in the portal veins due to liver disease can cause these connecting veins to dilate resulting in **esophageal varices.**

The **lymphatics** from the esophagus end in glands close to it throughout its entire extent.

DESCENDING AORTA

Another large structure in the posterior mediastinum is the descending portion of the thoracic aorta (Figs. 4-42 and 4-46). This is a direct continuation of the aortic arch and it courses through the posterior mediastinum and diaphragm to become the abdominal aorta.

The branches of the descending aorta are:

Visceral
  1. Bronchial

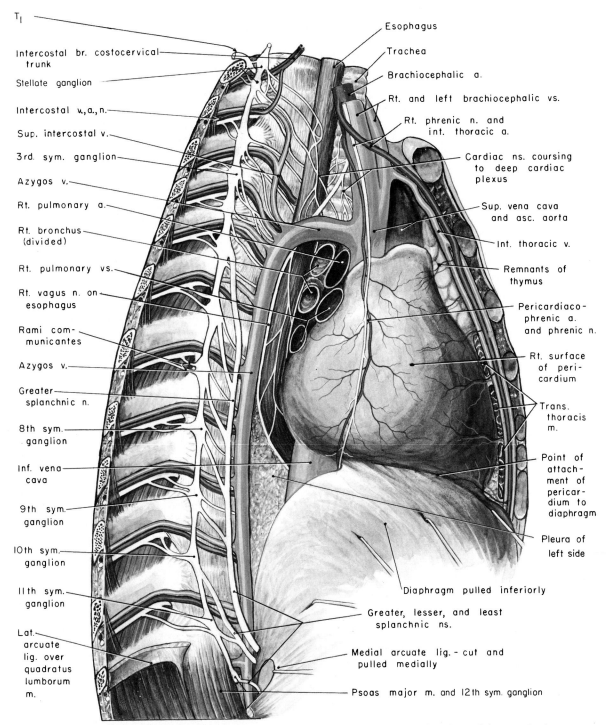

T_I

Intercostal br. costocervical trunk

Stellate ganglion

Intercostal v.,a.,n.

Sup. intercostal v.

3rd. sym. ganglion

Azygos v.

Rt. pulmonary a.

Rt. bronchus (divided)

Rt. pulmonary vs.

Rt. vagus n. on esophagus

Rami communicantes

Azygos v.

Greater splanchnic n.

8th sym. ganglion

Inf. vena cava

9th sym. ganglion

10th sym. ganglion

11th sym. ganglion

Lat. arcuate lig. over quadratus lumborum m.

Esophagus

Trachea

Brachiocephalic a.

Rt. and left brachiocephalic vs.

Rt. phrenic n. and int. thoracic a.

Cardiac ns. coursing to deep cardiac plexus

Sup. vena cava and asc. aorta

Int. thoracic v.

Remnants of thymus

Pericardiaco-phrenic a. and phrenic n.

Rt. surface of pericardium

Trans. thoracis m.

Point of attachment of pericardium to diaphragm

Pleura of left side

Diaphragm pulled inferiorly

Greater, lesser, and least splanchnic ns.

Medial arcuate lig. - cut and pulled medially

Psoas major m. and 12th sym. ganglion

Figure 4-45. *The mediastinal structures plus structures on the lateral surface of the vertebral column as seen from the right side. Note that the parietal pleura has been removed.*

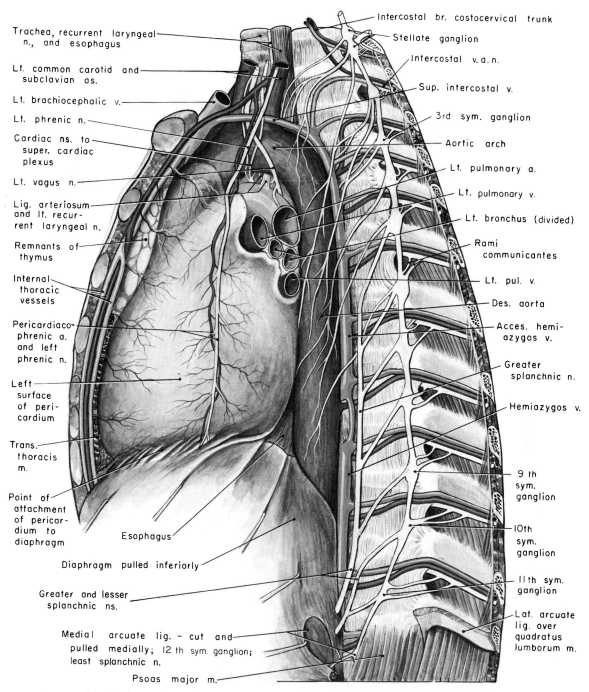

Trachea, recurrent laryngeal n., and esophagus

Lt. common carotid and subclavian as.

Lt. brachiocephalic v.

Lt. phrenic n.

Cardiac ns. to super. cardiac plexus

Lt. vagus n.

Lig. arteriosum and lt. recurrent laryngeal n.

Remnants of thymus

Internal thoracic vessels

Pericardiaco- phrenic a. and left phrenic n.

Left surface of peri- cardium

Trans. thoracis m.

Point of attachment of pericar- dium to diaphragm

Esophagus

Diaphragm pulled inferiorly

Greater and lesser splanchnic ns.

Medial arcuate lig. – cut and pulled medially; 12 th sym. ganglion; least splanchnic n.

Psoas major m.

Intercostal br. costocervical trunk

Stellate ganglion

Intercostal v. a. n.

Sup. intercostal v.

3rd sym. ganglion

Aortic arch

Lt. pulmonary a.

Lt. pulmonary v.

Lt. bronchus (divided)

Rami communicantes

Lt. pul. v.

Des. aorta

Acces. hemi- azygos v.

Greater splanchnic n.

Hemiazygos v.

9 th sym. ganglion

10th sym. ganglion

11th sym. ganglion

Lat. arcuate lig. over quadratus lumborum m.

Figure 4-46. *The mediastinal structures plus structures on the lateral surface of the vertebral column as seen from the left side. Note that the parietal pleura has been removed.*

2. Esophageal
3. Pericardial
4. Mediastinal

Parietal

5. Posterior intercostal
6. Subcostal
7. Superior phrenic

1.   The **bronchial** arteries are quite variable in their origin. There are usually two bronchial branches from the aorta and both course to the left lung. One arises in a more superior position than the other—thus, upper and lower branches. They arise from the anterior surface of the aorta, proceed inferiorly and laterally to gain the posterior surface of the left bronchus. They follow the bronchial tree throughout the lung. They also supply small branches to the pericardium and the esophagus. It will be remembered that the right bronchial artery does not usually arise from the aorta but from one of the intercostal arteries on the right side, usually the third (the first intercostal branch of the aorta—see below). These bronchial arteries give nourishment to the lung parenchyma.

2.   The **esophageal** branches are small but significant. There are usually four or five such branches that arise from the anterior surface of the aorta and course inferiorly to reach the esophagus. These branches form an anastomotic chain connecting with the esophageal branches of the inferior thyroid artery superiorly, and similar branches of the superior phrenic and the left gastric inferiorly.

3.   The **pericardial** arteries consist of small branches that supply the posterior surface of the pericardium.

4.   The **mediastinal** arteries are small branches to lymph nodes and fat.

5.   The **posterior intercostal** arteries (Figs. 4-45 and 4-46) consist of nine pairs of posterior intercostal vessels. They branch from the posterior surface of the aorta and proceed to the third to the eleventh intercostal spaces, the first two spaces being served by the costocervical trunk of the subclavian artery. The branches on the right side are longer than those on the left, due to the position of the aorta on the left side. The first several branches take a superior slant to reach the upper intercostal spaces. These arteries are posterior or deep to other structures in the area such as the azygos system of veins, the sympathetic chains and their branches, and the thoracic duct.

6.   Small **subcostal** vessels correspond to the intercostals but course just inferior to the twelfth rib.

7.   The **superior phrenic** branches are relatively small vessels that branch from the most inferior end of the thoracic aorta and supply the posterior portion of the diaphragm only. They do anastomose with branches of the musculophrenic arteries on the anterior surface and the pericardiacophrenics on the lateral sides.

Relations.   The descending aorta is anterior to the vertebral column and the hemiazygos system of veins. It is posterior to the left bronchus and lymph glands in its superior extent, to the pericardium and the esophagus in its most inferior portion. On its right side will be found the esophagus with its plexus of nerves, the thoracic duct, and the azygos vein, while the lung is the large structure found on its left side.

THORACIC DUCT

The **thoracic duct** (Fig. 4-43) starts in the abdominal cavity just anterior to the first or second lumbar vertebra in a dilated structure called the **cisterna chyli.** This duct continues superiorly in a position anterior to the vertebral column and intercostal vessels of the right side, and between the azygos vein on the right and the descending aorta on the left. It is posterior to the esophagus until it reaches the inferior aspect of the superior mediastinum. It then curves slightly to the left and occupies a position just to the left of the esophagus; it is still anterior to the vertebral column although not in the absolute midline. It continues into the neck and drains into the point of junction of the internal jugular and subclavian veins on the left side (see Fig. 7-84 on page 564). This duct drains lymph from the entire lower half of the body and from the left side of the upper half of the body.

A smaller lymphatic channel is usually found on the right side of the esophagus in the superior mediastinum. This drains into the junction of the internal jugular and subclavian veins on the right side.

AZYGOS SYSTEM OF VEINS

This system of veins (Fig. 4-44) exhibits many variations. Usually the **azygos** vein, which is on the right side of the body, commences from the posterior surface of the inferior vena cava in the abdominal cavity. It proceeds superiorly, passing through the aortic hiatus in the diaphragm, and ascends in the posterior mediastinum to the superior mediastinum, curves anteriorly at a point superior to the root of the right lung, and empties into the superior vena cava. Its **tributaries** are occasionally the first and second lumbar veins on the right side (which usually drain into the ascending lumbar vein as depicted in Fig. 4-44), the subcostal vein, the phrenic vein, the intercostal veins of the lower eight spaces, the right superior intercostal from the second and third spaces, the right bronchial vein, the esophageal, pericardial, and mediastinal veins, and, perhaps most important, the hemiazygos and accessory hemiazygos veins from the left side of the body. The azygos vein is anterior to the vertebral column in its entire extent and anterior to the right intercostal arteries. It is posterior to the esophagus and to the distal end of the thoracic aorta. The thoracic duct and descending aorta are on its left side while the right lung is on its right side.

The **hemiazygos vein** usually starts in communication with the left renal vein. It continues superiorly on the left side of the vertebral column through the aortic hiatus of the diaphragm along the bodies of the thoracic vertebrae until the level of the eighth or ninth vertebra is reached. At this point it turns sharply to the right, posterior to the aorta and thoracic duct, and empties into the azygos vein. Its **tributaries** are occasionally the first two lumbar veins of the left side (see above for right side), the left subcostal vein, and the lower three or four left intercostal veins.

The **accessory hemiazygos** vein is one that occupies a similar position on the left side of the thoracic vertebrae but in the superior part of the thoracic cavity. It drains into either the hemiazygos or directly into the azygos vein by crossing the vertebral column at approximately the eighth thoracic vertebral level. The **tributaries** of this vein will vary with the length and distribution of the highest intercostal vein on the left side. If the latter vein drains two intercostal spaces the remaining spaces from three to seven will be drained by the accessory hemiazygos; if the highest intercostal drains four intercostal spaces, the number of spaces drained by the accessory hemiazygos will be decreased accordingly.

This whole azygos system is subject to a great deal of variation. It is important to note that it connects with vessels in the abdominal cavity. If, by chance, the inferior vena cava became occluded, these vessels would provide a possible channel for blood to get around this occlusion and provide drainage for the lower half of the body.

SYMPATHETIC CHAINS

On either side of the vertebral column, close to the heads of the ribs and covered throughout their extent in the thoracic cavity by pleura, are the **sympathetic chains** (Figs. 4-45 and 4-46). These chains of ganglia start in the upper thoracic region and continue inferiorly to the point where they pierce the diaphragm to enter the abdominal cavity. There are usually twelve ganglia on each side of the thoracic cavity although the number may vary because of fusion of contiguous ganglia. This chain of ganglia extends superiorly into the neck, inferiorly into the abdominal cavity and into the pelvis. Although the ganglia are located on the heads of the ribs in the upper thoracic region, approximately at the fifth rib the chains tend to course slightly anteriorly and come to lie on the sides of the bodies of the vertebrae. They are all connected to each intercostal nerve by white and gray rami communicantes.

Branches from the first four thoracic ganglia enter into formation of the cardiac plexuses and send fibers to the lungs, trachea, bronchi, and esophagus as well as the heart. A large nerve arises from the fifth to the ninth thoracic segments—the **greater splanchnic nerve**—and takes an anterior course on the bodies of the vertebrae anterior to the intercostal arteries and close to the lateral side of the hemiazygos vein on the left side (Figs. 4-45 and 4-46). After giving branches to the descending aorta and esophagus, it pierces the crus of the diaphragm and ends in the celiac ganglion, a preaortic ganglion close to the celiac artery, a branch of the abdominal aorta (see Fig. 5-41 on page 261). Although difficult to see because of the bulging diaphragm, a nerve arises from the tenth and eleventh thoracic ganglia—the **lesser splanchnic nerve**—that proceeds inferiorly, pierces the crus of the diaphragm, and ends in the superior mesenteric ganglion at the base of the superior mesenteric artery. The nerve arising from the twelfth thoracic ganglion is the **least splanchnic nerve** and ends in the aorticorenal ganglion at the base of the renal artery.

You will recall that preganglionic fibers from the thoracic and upper lumbar regions of the cord start at the intermediolateral cell columns, traverse ventral roots, and join the sympathetic ganglia via the white rami communicantes. These preganglionic fibers may synapse in these ganglia and proceed back to the nerves once again via the gray rami communicantes, or they may not synapse in these ganglia and proceed superiorly or inferiorly in the sympathetic chain to synapse in contiguous ganglia. Other fibers synapse in the sympathetic chain ganglia and postganglionic fibers branch directly from these ganglia to reach their destination. The splanchnic nerves consist mostly of preganglionic fibers. These are fibers that have traversed the sympathetic chain ganglia without synapse; they continue to the celiac, superior mesenteric, and aorticorenal ganglia (Fig. 5-41 on page 261), where the synapse with postganglionic fibers occurs. The postganglionic fibers then reach their destination by following arteries, as this system is likely to do.

The student does not need a great deal of imagination to realize the importance of these sympathetic chains. They have control over such vital functions as heart rate, peristalsis, vasoconstriction, etc.

INTERCOSTAL SPACE

At this time, the student should review the anatomy of an intercostal space. The posterior end of an intercostal space viewed from inside the thoracic cavity (Figs. 4-45 and 4-46) shows the nerves and vessels just external to the pleura. These structures actually lie between the pleura and the internal intercostal membrane. They disappear from view by coursing outside of the subcostalis muscles present and the intercostalis intimus muscle. The sympathetic chain is anterior to the intercostal artery and vein, and the white and gray rami communicantes join the intercostal nerve which lies inferior to the vein and the artery. The relation of the three structures in the intercostal spaces is vein, artery, nerve, from superior to inferior.

SUMMARY OF SUPERIOR MEDIASTINUM

Now that structures in the posterior mediastinum have been followed into the superior mediastinum (see page 157), the latter compartment has finally been studied in its entirety and should be reviewed at this time.

The superior mediastinum, bordered superiorly by the thoracic inlet, inferiorly by a line drawn from the sternal angle to the base of the fourth thoracic vertebra, anteriorly by manubrium of the sternum, posteriorly by the vertebral column, and laterally by the pleural cavities is a very complex area. If the student will realize that the large structures from anterior to posterior are (1) the brachiocephalic veins, (2) the complex of the pulmonary arteries (not actually in the superior mediastinum) and the arch of the aorta and its branches, (3) the trachea, and (4) the esophagus, the smaller structures can be put into the superior mediastinum in proper relationship to these. For example, the right phrenic nerve courses along the right brachiocephalic vein, superior vena cava, right atrium, and pericardium to reach the diaphragm. The left phrenic nerve, with its accompanying vessels, courses along the left brachiocephalic vein, the left subclavian artery, arch of the aorta, and pericardium to reach the diaphragm. The right vagus is located on the trachea between that structure and the esophagus. It is just posterior to the superior vena cava and the right brachiocephalic vein. It is medial to the azygos vein and then courses posterior to the root of the lung. The left vagus courses on the left common carotid artery, curls around the arch of the aorta, and continues onto the anterior surface of the descending aorta and posterior to the root of the left lung. Its recurrent laryngeal branch takes a turn around the ligamentum arteriosum and the arch of the aorta to ascend in the groove between the esophagus and the trachea on the left side. Cardiac branches from both vagus nerves will course across the trachea to reach the deep cardiac plexus which lies on the anterior surface of the trachea near its bifurcation. Two cardiac nerves will course inferior to the arch of the aorta to reach the superficial cardiac plexus which lies in the concavity of the arch. The only remaining structures are the thoracic duct, which remains on the anterior surface of the vertebral column, posterior and slightly to the left of the esophagus; the azygos system of veins, which is close to the vertebral column; and many lymphatics and lymph glands.

If the student will now look at Figures 4-44, 4-43, 4-42, 4-11, 4-10, 4-9, and 4-6, in that order, the superior mediastinum can be reconstructed.

AORTIC ARCHES

The asymmetry of structures in the thoracic cavity, such as the difference between the course of the recurrent laryngeal nerves on both sides of the body, is due to the development of the aortic arches. Figure 4-47 shows in graphic form what happens to each of the six paired aortic arches in the developing fetus. Actually those of importance to an understanding of this region are arches 4 and 6. It will be noted that while the sixth arch on the left side persists in the form of the ductus and ligamentum arteriosum, on the right side this arch disappears. The inferior laryngeal nerves were inferior to the fifth and sixth arches and, because both drop out on the right side, the nerve recurs around the remains of the fourth arch, which is the subclavian artery on the right side. Since the ductus arteriosus (sixth arch) ends up as a ligamentum arteriosum on the left side, the recurrent laryngeal nerve persists in going around this particular structure.

SYSTEMIC REVIEW

In dissecting a cadaver, one is liable to forget that structures were once living, dynamic parts of the body. In reality, there are very few parts of the body that are motionless for any length of time. This is true of bone as well as of other tissues.

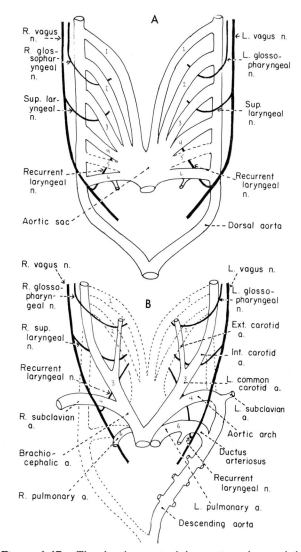

Figure 4-47. *The development of the aortic arches and the relation of the vagus and glossopharyngeal nerves to these arches. Structures dotted have disappeared. Note in A that the fifth arch never develops in humans. B shows definitive arrangement. Note that the recurrent laryngeal nerve on the right side courses inferior to the subclavian artery (the fourth arch) while that on the left side remains inferior to the ductus arteriosus (the sixth arch), the future ligamentum arteriosum.*

The contents of the thoracic cavity are moving constantly. The lungs are expanding and decreasing in size as a result of changes in size of the thoracic cage, and are allowed to do this because of the lubricating effects of the pleura. The heart is another organ that takes advantage of a serous membrane. This structure contracts constantly, wringing the blood from its chambers approximately seventy times a minute year after year.

Other structures in the mediastinum exhibit a small amount of motion. The esophagus has contractions that start at one end and continue to the opposite end, the aorta has to expand when blood is forced into it from the left ventricle, blood is flowing in the great vessels, and even the thoracic duct has some motion. To these we can add the repeated intermittent contraction anywhere there is smooth muscle, such as in arterioles, venules, and bronchioles.

BONES AND JOINTS

The bones involved in the thoracic cavity are the **ribs** and **sternum.**

The twelve **ribs** on either side join with the sternum to form the bony framework of the thorax. This bony framework also overhangs the abdominal contents. The ribs articulate with two vertebrae posteriorly except the first, the tenth, eleventh, and twelfth; these four articulate with one vertebra only. The tubercles of the ribs articulate with transverse processes of the vertebrae. Seven of the ribs articulate directly with the sternum and are therefore called true ribs, while five do not do so and are called false. The eleventh and twelfth are free at their anterior end and are called floating ribs.

The **sternum** possesses a manubrium superiorly, a body, and a xiphoid process inferiorly. The clavicles articulate with the manubrium of the sternum and this is the only bony attachment of the upper limb to the trunk.

The joints involved in this region are the **costovertebral** and **costosternal articulations.** These joints are synovial joints, except for the articulation of the first rib with the sternum. Although there is not a large amount of movement of any one rib, their accumulated movement allows for a considerable increase in size of the thoracic cavity. The ribs not only move superiorly in an anterior direction, but also can be raised in a lateral direction due to the method of articulation with the vertebral column.

MUSCLES

The muscles involved in the thoracic cavity are the intercostals, the transversus thoracis, and possibly the diaphragm, although the diaphragm will be studied in more detail from the abdominal approach. The intercostal spaces contain (1) the **external intercostal** muscles, whose fibers extend downward and medially, and which end anteriorly as the external intercostal membrane, (2) the **internal intercostal** muscle, whose fibers course inferiorly and laterally, and which ends posteriorly as an intercostal membrane, and (3) an **inner complex** consisting of the subcostal muscles—small muscles that often bridge more than one space in the posterior part of the thorax, the intercostalis intimus muscle, and transversus thoracis muscle that arises from the inside surface of the sternum and inserts on the ribs just deep to the anterior chest wall. These muscles can be compared with the three-layered abdominal wall as follows: (1) external intercostal, and external abdominis oblique, (2) the internal intercostal, and internal abdominis oblique, (3) subcostal + intercostalis intimus + transversus thoracis muscles, and transversus abdominis. If the division is made in this manner, the arteries and nerves course in the intercostal spaces in the same planes as in the abdominal wall.

These muscles are involved in respiration. The transversus thoracis portion of the inner complex aids in forced expiration, and some feel that the interosseous portions of the internal intercostals are also involved; the remaining muscles are involved in inspiration, the diaphragm by increasing the superior-inferior diameter and the intercostals by raising the ribs and thereby increasing the side to side and anteroposterior diameters.

NERVES

The nervous system in the thorax can be divided into that going to the walls of the thoracic cavity and that involved with the visceral contents.

That portion innervating the walls consists of the intercostal nerves and the phrenic nerve to the diaphragm. It should be noted that these nerves are all concerned with respiration. The **phrenic** nerves start in the neck from the fourth cervical segment (and often from the third and fifth as well) of the spinal cord and proceed into the thoracic cavity; they follow mediastinal structures just deep to the pleura to reach the diaphragm which they innervate. This long route has been caused by the fact that the diaphragm, when it first started to develop, was opposite the cervical vertebrae and as the subsequent growth of the body occurred it demanded an increase in length of the phrenic nerves. The right phrenic nerve follows the right brachiocephalic vein, the superior vena cava, and the pericardium to get to its destination, while the left follows the subclavian artery, crosses the aorta, and courses along the pericardium to get to the left side of the diaphragm. Each **intercostal** nerve emerges from its proper intervertebral foramen and enters an intercostal space. After giving off a dorsal ramus, the ventral ramus (the intercostal nerve) courses between the parietal pleura and the internal intercostal membrane, or between a subcostal muscle, if it happens to be present, and the internal intercostal membrane. Approximately at the scapular line, the nerve courses between the intercostalis intimus and internal intercostal muscles in which plane it continues until, close to the midline anteriorly, it reaches the transversus thoracis muscle. It then proceeds between the latter muscle and the internal intercostal, pierces the internal intercostal muscle and the external intercostal membrane to become a cutaneous nerve. It gives off a lateral cutaneous nerve at approximately the midaxillary line, a nerve that divides into anterior and posterior branches. These nerves innervate the intercostal muscu-lature, give the cutaneous nerve supply to the skin asso-ciated with the intercostal spaces, and also innervate the parietal pleura in the area of this space. All of the inter-costal nerves do not remain in the intercostal spaces for those in spaces 7–12 continue into the abdominal wall. The fact that these nerves innervate the parietal pleura is an important clinical point. This means that any pain from the parietal pleura will be somatic in nature—clear, sharp, and definite.

The thoracic walls also receive innervation from the **sympathetic nerves.** The sympathetic chains are found along the sides of the vertebral column in the thoracic and abdominal cavities, and extensions from these chains extend into the cervical region and into the pelvic or sa-cral region. All of the thoracic nerves and the upper three lumbar nerves have connections with these chain ganglia called white and gray rami communicantes. Preganglionic sympathetic fibers exit through their proper ventral roots, enter spinal nerves, and then white rami communicantes to join the sympathetic chain ganglia. A synapse occurs in those ganglia and postganglionic fibers return to the spi-nal nerve via gray rami communicantes and follow these nerves to reach smooth muscles in blood vessels, arrector pili muscles, and sweat glands in the area supplied by that particular nerve.

The rest of the nervous system in the thoracic cavity is involved with general visceral afferent and efferent fi-bers from and **to the viscera** in the thoracic cavity. The sympathetic portion takes the same pathway just de-scribed to reach the sympathetic chain ganglia. Here some of the fibers synapse immediately while others travel superiorly or inferiorly in these chains to other gan-glia before synapsing. Postganglionic fibers from the cer-vical ganglia and the upper four thoracic ganglia are in-volved in innervation of the respiratory system, heart, and esophagus. These fibers innervate the upper part of the esophagus directly as they do the trachea; however, they form plexuses before innervating the heart and the lungs, which are called the cardiac and pulmonary plexuses, respectively. These sympathetic fibers increase the heart rate, cause a vasodilation in the coronary circulation, de-crease contractions in the esophagus, and induce a vaso-constriction in the respiratory system as well as decreasing secretion of the mucous glands and dilating the bron-chioles in this system.

The remaining sympathetic ganglia of the thoracic cavity give rise to the **splanchnic nerves**—the greater splanchnic from the fifth to the ninth, the lesser splanch-nic from the tenth and eleventh, and the least splanchnic from the twelfth. These nerves are made up primar-ily of preganglionic sympathetic fibers as synapses have not occurred in the chain ganglia for these particular nerves. Fibers arise from these nerves to innervate the esophagus and other structures, but the majority of fibers travel to the celiac, superior mesenteric, and aorticorenal ganglia, where a synapse occurs. These ganglia are in the abdominal cavity close to arteries of the same name. Postganglionic fibers then continue on the surface of these arteries and their branches to reach the organs to be innervated, which will be studied later.

The **parasympathetic system** in this area is carried in the right and left vagus nerves and is involved with viscera only; no parasympathetic fibers have been de-scribed to the body wall. The right vagus nerve enters the thoracic cavity from the neck region on the right bra-chiocephalic vein and then reaches the area between the superior vena cava and the trachea. It then proceeds to the posterior surface of the hilum of the right lung and then onto the esophagus where it breaks up into a plexus. If the right vagus remains distinct it is found on the posterior surface of the esophagus. The left vagus takes a similar route except that it is associated with the left common carotid, the arch of the aorta, the posterior surface of the hilum of the left lung, and the anterior surface of the esophagus. The vagus nerves give off a recurrent laryngeal nerve that recurs around the subclavian artery on the right side and around the ligamentum arteriosum and the arch of the aorta on the left. In addition, the vagus nerves give off branches in the neck and in the thorax that contribute to the cardiac plexuses, and each of the recurrent laryngeal nerves also contribute fibers to these plexuses. Naturally, the vagus nerve will send many fibers into the lung tissue, and because it forms a network on the esophagus it is obvious that this organ is also innervated by these nerves. The vagus decreases the heart rate, increases peristaltic waves in the esophagus, and causes a vasodilation, secretion of mucous glands, and a constriction of the bronchioles in the respiratory system.

One should not forget that there are **visceral afferent fibers** from the viscera. It is believed that the majority of these fibers course with fibers of the sympathetic portion of the autonomic nervous system and therefore enter the spinal cord in the upper thoracic region. Undoubtedly there are some visceral afferent fibers associated with the vagus nerves as well, one such group carrying sensory impulses from the aortic sinus, thereby reflexly controlling heart output. Pain from the heart is referred to the ulnar side of the left forearm and hand, an important clinical point.

Perhaps we should repeat the important concept that the parietal pleura and pericardium are innervated in a somatic sensory manner by intercostal nerves or the phrenic nerve (pain will be sharp and definite) while the visceral layers on the heart and lungs—epicardium and visceral pleura, respectively—will be innervated by visceral afferent fibers (pain will be dull and vague and may be referred to the body wall where it will resemble somatic pain). We will see later that this same concept will apply to the abdominal cavity as well.

ARTERIES AND VEINS

The vascular system in the thoracic cavity consists of the heart surrounded by its pericardium, the great vessels leading to and emerging from the heart, the brachiocephalic veins and superior vena cava and their tributaries including the azygos system of veins, the thoracic aorta and its branches, and the internal thoracic vessels. The heart and its covering have been described in detail on pages 168 to 188 and there is no particular advantage in repeating this description in this section.

The **brachiocephalic veins** are formed by the joining of the internal jugular and subclavian veins on either side of the body. The right brachiocephalic vein descends in the superior mediastinum rather directly, while the left vein crosses from the left to right side in a horizontal fashion, being in a plane that is deep to the remains of the thymus gland and superficial to the aorta and its branches. The tributaries of these veins vary from side to side. In addition to the two vessels that form this vein, the right brachiocephalic vein drains the vertebral, the first posterior intercostal, the internal thoracic, the inferior thyroid, and thymic veins. The left vein has a superior intercostal vein that drains the second and third intercostal spaces in addition to the same tributaries as mentioned for the right brachiocephalic. These two veins combine to form the **superior vena cava,** which empties into the right atrium of the heart. The superior vena cava also drains the **azygos vein,** which drains the intercostal spaces and posterior abdominal wall on the right side; the hemiazygos and accessory hemiazygos veins drain the same territory on the left side, and then empty into the azygos vein. Since the azygos system of veins connects with the renal veins on the left and the inferior vena cava on the right, it forms a possible collateral circulation around an occlusion of the inferior vena cava.

Blood leaves the right ventricle of the heart via the **pulmonary trunk** and **arteries,** which conduct blood to the lungs. The pulmonary trunk is anteriorly placed and is partially covered by the pericardium. It soon divides and is in close relationship to the aorta. The right artery courses to the right lung passing posterior to the ascending aorta and inferior to the aortic arch. The left crosses anterior to the descending aorta to reach the left lung. The **ligamentum arteriosum,** remains of the ductus arteriosus, stretches between the left pulmonary artery and the arch of the aorta.

Blood is returned from the lungs to the heart by way of **four pulmonary veins.** They open directly into the left atrium which is the most posterior part of the base of the heart.

The **aorta** is the large vessel that conducts blood from the left ventricle to all parts of the body. It consists of a short ascending portion which gives origin to the two important coronary arteries; the arch which gives rise to the brachiocephalic, left common carotid, and left subclavian branches; and the descending portion which has esophageal, bronchial, pericardial, and mediastinal (visceral) branches; and intercostal, subcostal, and phrenic (parietal) branches.

The remaining vessels are the two **internal thoracic arteries** which arise from the subclavian arteries in the neck, and descend through the superior and anterior mediastinal compartments in a plane just deep to the sternum and ribs. This vessel provides branches to the pericardium and diaphragm (pericardiacophrenic), the mediastinum (mediastinal), intercostal spaces (anterior intercostals), anterior chest wall (perforating), and terminates by dividing into musculophrenic and superior epigastric branches.

ESOPHAGUS

The digestive system in the thorax consists of the **esophagus.** It enters into the superior inlet of the thorax in a plane anterior to the vertebral column and posterior to the trachea. It is slightly on the left side of the midline but then curves to the right side as it passes the arch of the aorta. It continues inferiorly in a position anterior to the vertebral column, posterior to the pericardium, and to the right of the descending aorta. It then crosses anterior to the distal end of the aorta to go through the esophageal opening of the diaphragm to attach to the stomach. Its blood supply is from branches of the inferior thyroid artery in the neck; the bronchial, esophageal, and superior phrenic branches of the descending aorta; and the left gastric branch of the celiac in the abdominal cavity. The venous drainage of this organ is into the azygos system of veins in the thoracic cavity and into the inferior thyroid veins in the neck. These veins make an important connection with the left gastric veins in the abdominal cavity. Its nerve supply is by the autonomic nervous system via the sympathetic ganglia from the first to the ninth thoracic segments of the spinal cord (sympathetic) and by branches of the vagus nerves (parasympathetic).

TRACHEA, BRONCHI, AND LUNGS

The respiratory system in the thoracic cavity consists of the **trachea, two bronchi,** and **two lungs.** These have

been described completely on pages 188 to 197, and there is no particular advantage in repeating this description in this section.

It might be helpful to realize once again that an ability to visualize the relation of the primary bronchi to the pulmonary arteries will automatically provide the proper relations of the structures in the hila of the lungs to each other, because it is relatively easy to realize the pulmonary veins will be anterior and inferior in position. The primary bronchi would have to be posterior to the pulmonary arteries. This leaves one fact to memorize—that

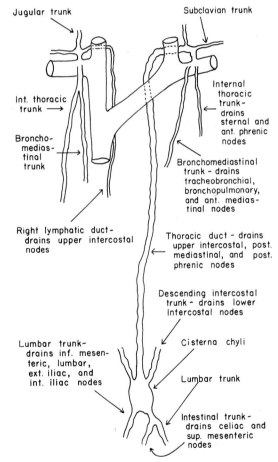

Figure 4-48. *Diagrammatic summary of the lymphatic drainage of the thoracic and abdominopelvic cavities. Lymphatic drainage of the thoracic walls follows arteries retrograde except diaphragm; viscera drain to three large groups of nodes: (1) anterior mediastinal, (2) posterior mediastinal (posterior to pericardium), and (3) tracheobronchial (near bifurcation of trachea).*

the left pulmonary artery gains a superior position to the left bronchus just before entering the lung.

The structure of the respiratory system is correlated well with its function of bringing air into contact with the circulatory system. The mucous membrane lining this system prevents the organ from drying as a result of the constant passing of air, and the ciliated epithelium prevents mucus from going down too far into the lungs. The cartilage ''rings'' prevent these air passages from collapsing. The bronchi divide repeatedly to form a respiratory tree that finally terminates in minute air cells which are in intimate contact with capillaries of the pulmonary vessels.

Air is brought into the lungs as follows. The thoracic cavity is enlarged by action of the respiratory muscles, the thoracic walls (including the parietal pleura) moving anteriorly and laterally while the diaphragm moves inferiorly. Because there is a negative pressure in the pleural cavity, the visceral layer of pleura and its lung follows the parietal pleura, thereby enlarging the lung and drawing air into the respiratory passages. Expiration consists of the reverse of this process.

LYMPHATICS

The **lymph nodes** in the thoracic cavity are very abundant. They occur anterior to the pericardium, in the superior mediastinum especially near the sides of the trachea and around its bifurcation, in lung roots (hilar), and in the interior of the lungs themselves. They are also located posterior to the pericardium in the posterior mediastinum particularly along the aorta and at the vertebral end of the ribs and intercostal spaces. The nodes along the internal thoracic artery and those along the vertebrae at the vertebral end of the intercostal spaces obviously drain the thoracic cage. The other organs drain into nodes in the location of that particular organ. The drainage of the lung deserves special mention in that there is lymphoid tissue inside the lung tissue, hilar nodes in the hilum of the lung, and nodes around the roots. Special mention should also be made of the thoracic duct which courses through the posterior mediastinum and superior mediastinum to end in a confluens at the point of junction of the internal jugular and the subclavian veins on the left side. This **thoracic duct** drains lymph from the entire lower half of the body and from the left side of the upper half of the body. The thoracic duct is extremely important and must be kept in mind in all surgical procedures in this vicinity. The upper half of the right side of the body drains into the same location on the right side. Figure 4-48 presents a summary of the lymphatic channels in the thoracic cavity.

Those students who studied the thoracic cavity immediately after the thoracic wall might review the breast (page 107) and the pectoral region (page 112).

5

ABDOMINOPELVIC CAVITY AND PERINEUM

Before starting this section, so obviously relevant to medicine, the surface anatomy presented in Figures 3-5, 3-7, and 3-10 on pages 104, 106, and 108 should be reviewed. Reference will be made repeatedly to the four quadrants of the abdominopelvic cavity, making the umbilicus an important landmark for general topography. The transpyloric line is easily visualized, bisecting the body at a plane halfway between the xiphisternal joint and the umbilicus, once again emphasizing the importance of the umbilicus, at least in those patients who are not too obese. We will also find that structures in the abdominal cavity can be related to vertebral levels.

ABDOMINOPELVIC CAVITY
IN SITU

When one opens the abdominal cavity (Fig. 5-1), the most prominent structure seen is a fat-laden apron-like

greater omentum which hangs down from the stomach and transverse colon to reach almost to the pelvic cavity. This structure, actually made up of layers of peritoneum surrounding blood vessels and fat, covers the **small intestine** (Fig. 5-2) except for small areas on the sides. It is said to have a function in limiting spread of infection in the abdominal cavity. More superior, in the right upper quadrant, the **liver** can be seen projecting down just inferior to the costal margin, and some of the stomach is visible in the upper left quadrant. The **transverse colon** (Fig. 5-2) is the large structure stretching transversely from one side to the other; this structure divides the abdominal cavity into superior and inferior portions.

The student should become very familiar with the general topography of the abdominal cavity. There is no better way to do this than to feel the organs before any dissection is done. If the liver is lifted and the stomach pulled inferiorly, it can easily be seen that the **esophagus** does indeed pierce the diaphragm and join the upper end of the stomach. The **stomach** itself, with its **fundus, body, and pyloric portions,** is quite apparent. The **pyloric sphincter** leads into the first part of the duodenum. This

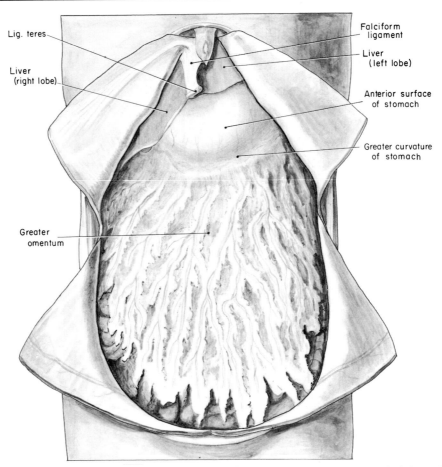

Figure 5-1. *The abdominal cavity I. The greater omentum is shown suspended from the greater curvature of the stomach and transverse colon (not seen).*

is the only part of the duodenum that is visible at this time, since the remaining portions are retroperitoneal (in back of the peritoneum). Some manipulation of the small intestine will reveal the start of the **jejunum,** and this leads gradually into the third part of the small intestine, the **ileum.** The ileum can be followed to the lower right quadrant of the abdominal cavity, where it joins the large dilated **cecum,** the first part of the **large intestine** (Fig. 5-3). The **appendix** is suspended from the inferior aspect of the cecum. The colon proceeds superiorly on the right side of the abdominal cavity as the **ascending colon,** reaches the liver and makes a turn called the **right colic flexure,** courses transversely across the abdominal cavity (**transverse colon**) until it reaches the spleen, makes a

turn at this point called the **left colic flexure,** and then descends on the left side of the abdominal cavity as the **descending colon.** The **sigmoid colon** connects the descending colon with the **rectum,** which connects directly with the **anal canal.**

The tip of the **gall bladder** can usually be seen at the inferior edge of the liver, and if the liver is raised slightly, the remaining portion of the gall bladder is revealed (Fig. 5-3).

In the upper left quadrant, posterior to the stomach, is the **spleen.** This organ is usually firmly attached to the diaphragm.

If the small intestine is moved to one side or the other, it can be seen that it is suspended by a **mesentery**

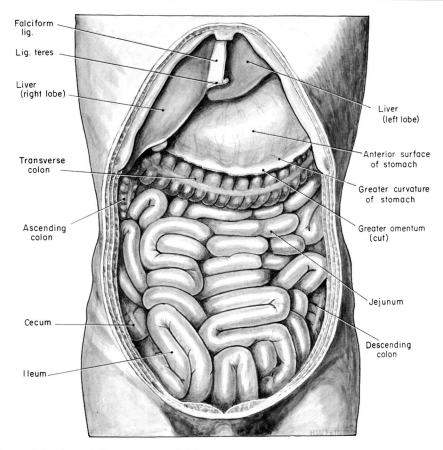

Falciform lig.
Lig. teres
Liver (right lobe)
Transverse colon
Ascending colon
Cecum
Ileum

Liver (left lobe)
Anterior surface of stomach
Greater curvature of stomach
Greater omentum (cut)
Jejunum
Descending colon

Figure 5-2. *The abdominal cavity II. The greater omentum has been removed to show the intestines and transverse colon.*

which attaches to the posterior abdominal wall crossing obliquely downward and to the right. In addition, one can see the outlines of the kidneys. Palpation reveals the **abdominal aorta, inferior vena cava, vertebral column,** and occasionally the **ureters.**

If the small intestine is pulled superiorly out of the pelvic cavity, the various pelvic organs can be seen covered with peritoneum (see Figs. 5-10 and 5-11). The triangular organ just posterior to the pubic bone is the **urinary bladder** which, if full, would extend further into the pelvic cavity. The **rectum** is readily seen in the pelvic cavity in both sexes. If the body is a female, the **uterus** projects into the pelvic cavity between the bladder and the rectum. The **ovaries** also can be found suspended from the superior-posterior surface of the broad liga-

ments, the layers of peritoneum extending laterally from each side of the uterus. In the male, the **prostate** and **seminal vesicles** are not readily seen at this time.

It should be remembered that the viscera in cadavers are not only discolored but very frequently distorted. The stomach may be completely empty and in a contracted state, or it may be dilated with air. In the living state the structures in the abdominal cavity are very mobile, the transverse colon frequently hanging down as far as the pelvic cavity. In addition, all the organs tend to have a lower position when a person stands upright than when a person lies down. In patients it is frequently found that the viscera are lower than the vertebral levels assigned to them by information obtained from cadavers.

Fortunately some organs are relatively stable in posi-

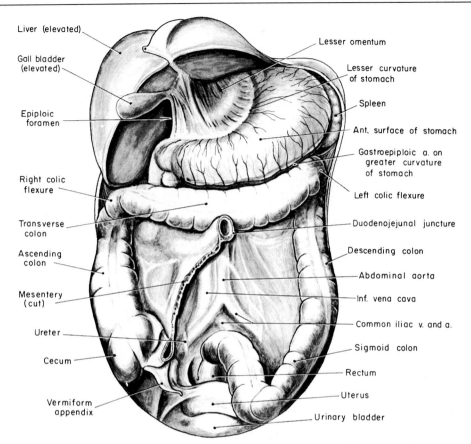

Liver (elevated)

Gall bladder (elevated)

Epiploic foramen

Right colic flexure

Transverse colon

Ascending colon

Mesentery (cut)

Ureter

Cecum

Vermiform appendix

Lesser omentum

Lesser curvature of stomach

Spleen

Ant. surface of stomach

Gastroepiploic a. on greater curvature of stomach

Left colic flexure

Duodenojejunal juncture

Descending colon

Abdominal aorta

Inf. vena cava

Common iliac v. and a.

Sigmoid colon

Rectum

Uterus

Urinary bladder

Figure 5-3. *The abdominal cavity III. The greater omentum and small intestines (jejunum and ileum only) have been removed. Note how this picture reveals the posterior relations of the jejunum and ileum.*

tion, the esophageal entrance to the stomach and the pyloric portion of this organ being good examples, the latter being at the level of the transpyloric line and, obviously, being responsible for its name. The liver and spleen can become enlarged but are relatively well attached to the diaphragm. If normal development of the gut has occurred, the cecum is fairly stable, but the appendix can be in many positions in relation to it. The positions of the right and left colic flexures, under normal conditions, are fairly predictable, the left being bisected by the transpyloric line. The pancreas does not wander and, although the kidneys may descend, they usually are found at proper levels if development has been normal, all three organs being bisected by the transpyloric line. The pelvic organs are relatively stable but the increase in

size and position of the uterus during pregnancy is astonishing. All of the above is dependent upon normal development.

PERITONEUM

The **peritoneum** is a large continuous sheet of serous membrane that lines the walls of the abdominopelvic cavity and is reflected onto the viscera, giving them partial or complete coverings. These reflections are called **omenta, mesenteries,** or **ligaments** and many contain the blood vessels, lymphatics, and nerves passing to and from the abdominal wall to the viscera. The peritoneum secretes a serous fluid that allows the viscera to move, as in peristalsis of the intestines, without adhering to surrounding structures.

Strictly speaking, there is nothing inside the peritoneal cavity except this serous fluid. However, two terms have come into common usage: **intraperitoneal** (inside the peritoneum) and **retroperitoneal** (in back of the peritoneum) (Fig. 5-4). If an organ is completely covered with peritoneum, with the exception of a small area where its mesentery is attached, this organ is considered to be an intraperitoneal structure, although strictly speaking it is not inside the peritoneal cavity. If the organ is partially covered with peritoneum, for example, on its anterior surface only, and the rest of the structure is not covered with peritoneum, this structure is called a retroperitoneal organ. The jejunum and ileum are good examples of intraperitoneal organs; the kidneys are retroperitoneal (Fig. 5-6).

It is imperative that the peritoneum and its reflections be understood in their entirety if one is to comprehend the abdominal and pelvic cavities. One of the best ways to accomplish this is to follow the peritoneum in a superior and inferior direction in the midsagittal plane and in two transverse planes. Usually one selects a transverse plane through the so-called epiploic foramen (of Winslow) and one lower down in the abdominal cavity through the ascending and descending colons. A description will now be made of such trips through the abdominal and pelvic cavities, and then special attention will be given to several regions. Constant reference should be made to Figures 5-5, 5-6, and 5-7.

Starting at the **umbilicus** one can follow the peritoneum superiorly on the **anterior abdominal wall** and reach the **diaphragm;** a sickle-shaped projection of peritoneum (**falciform ligament**) will be seen stretching between the umbilicus and the liver, covering the obliterated umbilical vein (ligamentum teres); from the diaphragm the peritoneum reflects onto the **superior surface of the liver,** forming the **left triangular ligament** (see Fig. 5-8 to understand this); the peritoneum continues onto the **anterior surface of the liver,** around its sharp anterior border, to gain its **inferior surface;** it reflects inferiorly from this inferior surface of the liver to the **lesser curvature of the stomach;** this layer of peritoneum between the liver and the stomach aids in

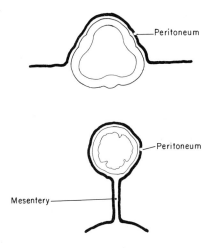

RETROPERITONEAL ORGAN

INTRAPERITONEAL ORGAN

Figure 5-4. *Diagram showing the difference between an intraperitoneal and a retroperitoneal organ. Note that both are actually outside the peritoneal sac.*

forming the **lesser omentum.** The peritoneum can then be followed on the **anterior surface of the stomach** to its **greater curvature,** where the greater omentum is attached; the **greater omentum** can be followed inferiorly on its anterior surface, around its free inferior edge, and then superiorly on its posterior surface; this leads to the inferior edge of the **transverse colon.** The peritoneum continues on the posterior surface of the transverse colon to its own mesentery, the **mesocolon;** the mesocolon leads to the **posterior abdominal wall** at a point where the pancreas is located retroperitoneally. The peritoneum continues inferiorly on the posterior abdominal wall until the **mesentery** of the small intestine is reached; at this point the peritoneum courses to the **small intestine,** continues around this organ, and then back to the posterior abdominal wall once again. It continues down the **posterior abdominal wall,** into the pelvic cavity and onto the anterior surface of the **rectum;** if a male, it then reflects onto the **urinary bladder** forming the **rectovesical pouch.** If a female, the peritoneum reflects onto the **uterus** forming the **rectouterine pouch;** it continues over the uterus and then reflects onto the **urinary bladder** thus forming the **vesicouterine pouch.** The peritoneum then reflects, in both sexes, from the urinary bladder to the

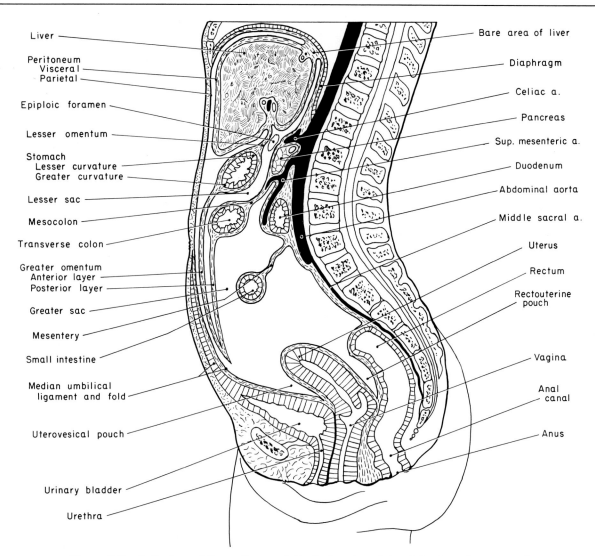

Figure 5-5. *Reflections of the peritoneum I. Sagittal section of the abdominopelvic cavity of the female. In reality the layers of the greater omentum are fused and the anterior layers are attached to the transverse colon, thus forming the gastrocolic ligament. Surgically the lesser sac can be approached through this ligament.*

anterior abdominal wall and follows the **obliterated urachus** to the **umbilicus.**

This trip through the peritoneal cavity has been in the so-called **greater sac** of the abdominal cavity. Because of the so-called rotation of the stomach and small intestines in their development, an area of the peritoneal cavity is cut off called the **lesser sac** or **omental bursa.** If

an incision is made in the lesser omentum (between the liver and the lesser curvature of the stomach) or in the greater omentum close to the greater curvature of the stomach, entrance is gained into this lesser sac. Continuing in the midsagittal plane (Fig. 5-5), the peritoneum can be followed in the lesser sac as follows. Starting on the deep surface of the **lesser omentum,** the peritoneum can

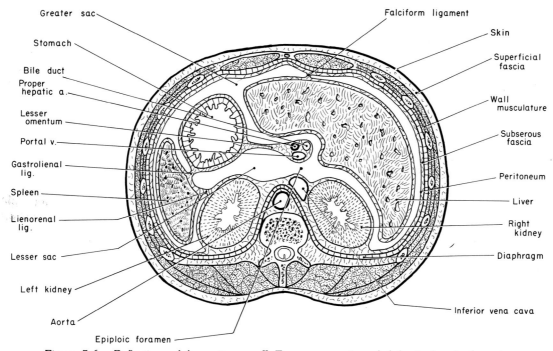

Figure 5-6. *Reflections of the peritoneum II. Transverse section of abdominal cavity through the epiploic foramen. You are looking inferiorly.*

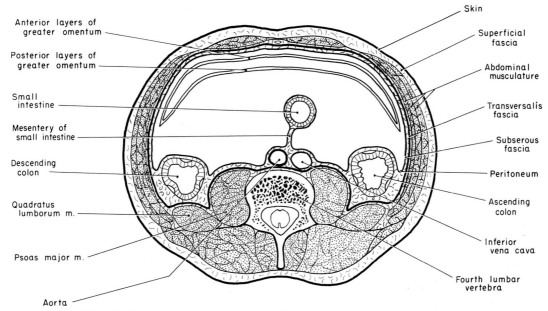

Figure 5-7. *Reflection of the peritoneum III. Transverse section of abdominal cavity at the fourth lumbar vertebra. You are looking inferiorly. In reality there is no such space between the anterior and posterior layers of the greater omentum; they are almost always fused.*

be followed inferiorly to the **lesser curvature of the stomach,** onto the **posterior surface of the stomach** to the **greater curvature;** it continues inferiorly and then superiorly, aiding in the formation of the anterior and posterior layers of the **greater omentum.** Via this greater omentum, it reaches the anterior side of the **transverse colon,** continues on the anterior side of the **transverse mesocolon** to reach the posterior abdominal wall. It then continues superiorly on the **posterior abdominal wall** to reach the **diaphragm,** reflects from the diaphragm to the **posterior surface of the liver (left triangular ligament),** around its posterior edge to the **inferior surface of the liver,** and thence onto the **lesser omentum,** where we started.

You will recall that the ventral mesentery in the embryo was incomplete, due to the presence of the stalk of the yolk sac (see Fig. 1-40 on page 48), and therefore presented a free edge. This free edge still exists in the adult body. During development the duct system of the liver grew into this ventral mesentery, and it and the stomach and duodenum rotated to the right, the free edge of the ventral mesentery persisting as the free edge of the lesser omentum. This double layer of peritoneum (mesentery) does not adhere to the posterior abdominal wall, and, therefore, an area still remains patent between the liver and the duodenum and between the lesser omentum and the posterior abdominal wall. A finger can be placed in this opening, which is called the **epiploic foramen** (of Winslow).

In following the peritoneum transversely through the epiploic foramen (Fig. 5-6), it is convenient to begin on the **anterior abdominal wall** once again. Starting at the **falciform ligament,** the peritoneum can be followed to the right around the abdominal wall to the region of the right kidney, onto the **anterior surface of this kidney,** and then onto the **inferior vena cava.** Continuing to the left through the **epiploic foramen** leads to the anterior surface of the **aorta** and then the surface of the **left kidney.** The peritoneum reflects from this kidney to the **hilum of the spleen,** aiding in formation of the **lienorenal ligament.** From the hilum of the spleen the peritoneum continues to the **stomach** forming the **gastrolienal ligament;** it continues on the **posterior surface of the stomach** to reach the posterior surface of the **lesser omentum,** from which it can be followed around three structures in the free edge of the lesser omentum, namely, the **proper hepatic artery,** the **portal vein,** and

the **bile duct.** It reaches the anterior surface of the lesser omentum in this manner, continues to the **anterior surface of the stomach,** to the **gastrolienal ligament,** around the **spleen** to the **lienorenal ligament,** to the **left kidney,** and thence to the **posterior abdominal wall.** It continues around the left side of the **abdominal wall** to reach the left side of the **falciform ligament.** It then surrounds the liver to reach the right side of the falciform ligament where we started.

Following the peritoneum transversely at a lower point is relatively easy (Fig. 5-7). Starting at the midline of the **anterior abdominal wall,** the peritoneum can be followed to the right to the **posterior abdominal wall** near the **ascending colon.** It reflects over the ascending colon, making this a retroperitoneal organ, continues on the anterior surface of the **psoas major muscle,** the **inferior vena cava,** and the **abdominal aorta** to reach the **mesentery of the small intestine.** It reaches the **small intestine,** surrounds it, and proceeds to the **posterior abdominal wall** and onto the anterior surface of the **psoas major muscle on the left side.** It covers the **descending colon,** and reaches the abdominal wall once again, then follows around the sides of the **abdominal wall** to the midline.

There are several regions in the abdominopelvic cavity deserving special attention. At this time the student should study the reflections of the peritoneum from the posterior abdominal wall to various organs, as shown in Figure 5-8.

From this picture it is easily seen that the liver is attached directly to the diaphragm and to the large vessels of the posterior abdominal wall. This area, which is free of peritoneum, is the **bare area** of the liver. If one follows the falciform ligament superiorly and runs the hand between the liver and the diaphragm to the right side of the ligament, it will contact a reflection of the peritoneum from the diaphragm onto the liver, which can be followed to the right for a considerable distance; this is the **anterior coronary ligament.** If followed completely to the right, a free edge will be found where the peritoneum can then be followed medially again as the **posterior coronary ligament.** The triangle produced by these two layers to the right is the **right triangular ligament.** If the falciform ligament had been followed to the left, it would have followed two layers of peritoneum much shorter than those just described. These two layers form the **left triangular ligament.** These ligaments are

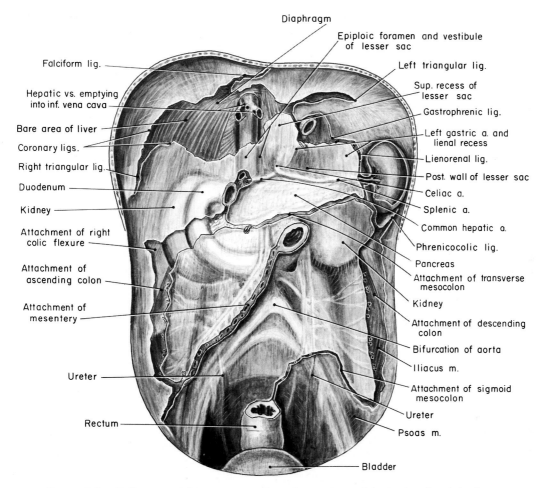

Diaphragm

Epiploic foramen and vestibule of lesser sac

Falciform lig.

Left triangular lig.

Hepatic vs. emptying into inf. vena cava

Sup. recess of lesser sac

Bare area of liver

Gastrophrenic lig.

Coronary ligs.

Left gastric a. and lienal recess

Right triangular lig.

Lienorenal lig.

Duodenum

Post. wall of lesser sac

Kidney

Celiac a.

Splenic a.

Attachment of right colic flexure

Common hepatic a.

Phrenicocolic lig.

Attachment of ascending colon

Pancreas

Attachment of transverse mesocolon

Attachment of mesentery

Kidney

Attachment of descending colon

Bifurcation of aorta

Iliacus m.

Ureter

Attachment of sigmoid mesocolon

Ureter

Rectum

Psoas m.

Bladder

Figure 5-8. *Reflections of the peritoneum from the posterior abdominal wall and diaphragm onto various organs. Careful study of this picture should enhance one's comprehension of the various folds in the peritoneum.*

pictured clearly in Figure 5-8. They aid in holding the liver to the diaphragm.

If one now enters the lesser sac through the epiploic foramen (note Fig. 5-8), one can turn superiorly between the liver and the esophagus as the latter organ comes through the diaphragm. This area, which is posterior to the lesser omentum, is called the **omental recess** or superior recess of the lesser sac. One cannot escape from this recess to the left because of the attachments of the lesser omentum to the diaphragm. By turning the hand inferiorly, remaining posterior to the stomach but still in the lesser sac, one will meet resistance by running into the root of the transverse mesocolon extending transversely across the body at the level of the pancreas (between L1 and L2). If one tries to escape from this cavity to the left, one runs into the reflections of the lienorenal ligament extending from the kidney to the spleen. To the right is the first part of the duodenum. In summary, the **lesser sac** exhibits the **epiploic foramen,** the **vestibule** just inside the entrance, the **superior omental recess** superiorly, the **lienal recess** toward the spleen, and the **inferior omental recess** inferiorly toward the transverse mesocolon. The arteries coursing just deep to the peritoneum on the posterior abdominal wall form ridges.

The lesser omentum (Fig. 5-9) is actually composed of peritoneum stretching inferiorly from the liver to the lesser curvature of the stomach, and, to the right, the same membrane stretches between the liver and the duodenum. This lesser omentum is also called by the dual name of **hepatoduodenal** and **hepatogastric ligaments.** The greater omentum, stretching between the greater curvature of the stomach and the transverse colon, is divided into three parts: that between the stomach and the colon is called the **gastrocolic ligament,** that between the stomach and the spleen the **gastrolienal ligament,** and that between the stomach and the diaphragm the **gastrophrenic ligament.** The area of peritoneum stretching from the kidney to the spleen is the **lienorenal ligament.** The **transverse mesocolon** (Fig. 5-8) stretches across the abdominal cavity from the region of the left colic flexure to that of the right colic flexure. The **phrenicocolic ligament** (Fig. 5-8) is that part of the peritoneum in the area between the diaphragm and the left colic flexure; this area provides a bed, so to speak, for the spleen and is frequently designated as the **sustentaculum lienis.**

The **mesentery** of the small intestine, as mentioned before, crosses obliquely downward and to the right. Figure 5-8 also reveals that the ascending and descending colons are retroperitoneal organs; when removed, two longitudinal areas of the posterior abdominal wall are devoid of peritoneum. The line of attachment of the sigmoid mesentery clearly indicates that this organ connects with the descending colon and with the rectum.

In the female pelvic cavity (Fig. 5-10), the reflection of peritoneum from the rectum to the uterus forms the rectouterine pouch, the most inferior limit of the peritoneal sac; that from bladder to uterus, the **vesicouterine pouch.** The **broad ligaments** extend laterally from the uterus and attach to the lateral pelvic walls, the upper free edges of these broad ligaments containing the **uterine tubes.** The **ovary** is suspended from the posterior-

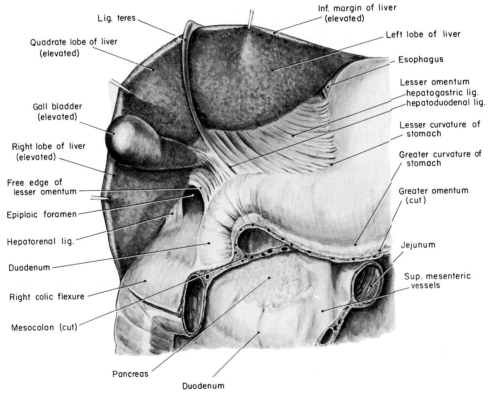

Figure 5-9. The lesser omentum and entrance into lesser sac (epiploic foramen). Note attachments of the greater omentum and the retroperitoneal organs in this area.

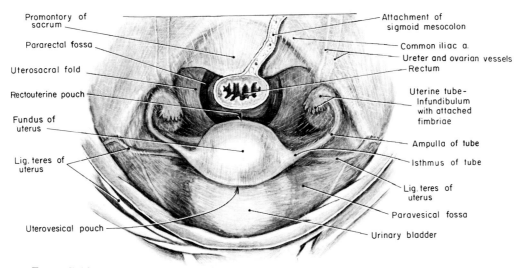

Figure 5-10. *The female pelvic cavity in situ. These structures would normally be covered by loops of ileum.*

superior surface of this ligament by its own mesentery called the **mesovarium.** Of the numerous other ligaments of the uterus, the **uterosacrals** produce ridges in the peritoneum as they stretch between the cervix of the uterus and the sacrum. In the living state there are several **fossae** in the pelvic cavity, especially if the urinary bladder contains some fluid. The fossae on either side of the rectum are the **pararectal fossae,** those on either side of the bladder the **paravesical.** In the male (Fig. 5-11) the pouch between the rectum and the bladder is the **rectovesical pouch,** and this is surrounded by two folds of peritoneum, extending from the sacrum to the base of the bladder, called the **sacrogenital folds.**

Folds of peritoneum occur on the inside of the an-

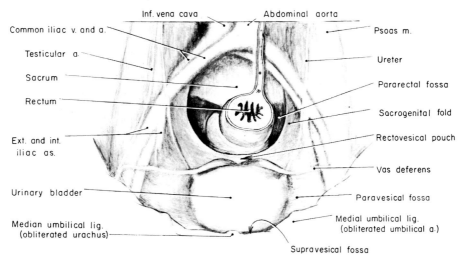

Figure 5-11. *The male pelvic cavity in situ. These structures would normally be covered by loops of ileum.*

terior abdominal wall just superior to the pubis and inguinal canal (Fig. 5-12). (The terminology here can be confusing if the student does not note the difference between median and medial in the following description.) The **median umbilical fold** is in the midline and is made by the obliterated urachus. The two **medial umbilical folds** are just lateral to the bladder and cover the obliterated umbilical arteries. Two **lateral umbilical folds** cover the inferior epigastric arteries. The depression in the peritoneum between the median and medial umbilical folds is the **supravesical fossa,** that between the medial and lateral umbilical folds the **medial inguinal fossa,** and

that lateral to the lateral umbilical fold the **lateral inguinal fossa.** It is immediately apparent, since the inferior epigastric artery forms the lateral umbilical fold, that the medial fossa is medial to the inferior epigastric artery and corresponds to the inguinal triangle (of Hesselbach), where direct inguinal hernias occur, and that the lateral fossa is lateral to this artery and contains the internal inguinal ring, where indirect inguinal hernias occur.

Much deeper recesses can occur in the peritoneum at various sites. One of these locations is at the point at which the retroperitoneal duodenum changes to the intraperitoneal jejunum (Fig. 5-13). These recesses are

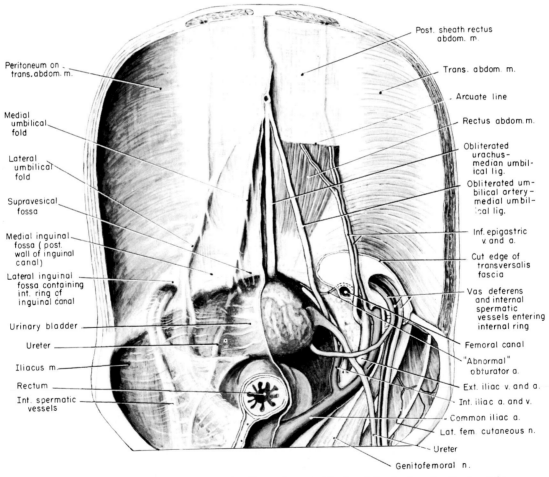

Figure 5-12. *The anterior abdominal wall seen from inside the abdominal cavity. On the right side of the body the peritoneum, subserous fascia, and transversalis fascia (except where it forms the posterior wall of the inguinal canal) have been removed. The fascia and peritoneum are intact on the left side.*

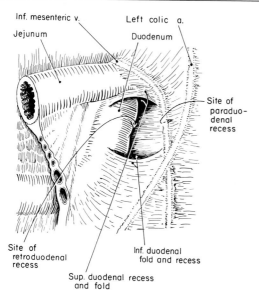

Figure 5-13. Fossae in the peritoneum around the duo-denojejunal juncture. (The duodenojejunal fossa is not shown; see text.)

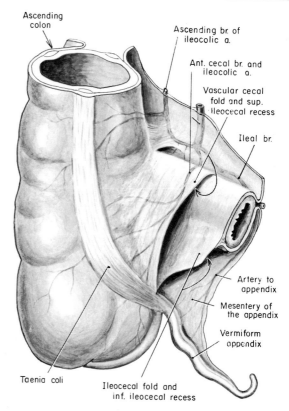

Figure 5-14. The peritoneum around the ileocecal juncture.

sometimes a source of trouble in that hernias can occur at these sites. (1) The **inferior duodenal recess** is on the left side of the ascending portion of the duodenum and its cavity faces superiorly. It is about at the level of the third lumbar vertebra and this particular recess occurs in approximately 75 per cent of bodies. (2) The **superior duodenal recess** is located at a slightly higher level, about opposite the second lumbar vertebra, to the left of the ascending portion of the duodenum and its opening faces inferiorly. This occurs in approximately 50 per cent of bodies. (3) The **duodenojejunal recess** exists in about 20 per cent of bodies. It is bounded superiorly by the pancreas, on the right side by the aorta, and on the left by the kidney. Two other types of recesses, (4) the **paraduodenal** and (5) the **retroduodenal,** are very rarely seen. The paraduodenal recess is in relationship to the ascending portion of the duodenum and has its opening toward the right, and the free edge of the peritoneum bounding its anterior wall contains the ascending branch of the left colic artery. The retroduodenal recess is posterior to the horizontal and ascending parts of the duodenum and anterior to the aorta; its opening faces inferiorly and to the left.

Another region where recesses may occur is near the cecum and the ileocecal juncture (Fig. 5-14). A recess may occur anterior and slightly superior to the ileum, called the **superior ileocecal recess,** and one may occur inferior to the ileum, between it and the appendix, called the **inferior ileocecal recess.** It will be seen that the appendix has a mesentery; the artery to the appendix is contained therein. Another recess, called the **cecal recess,** can occur posterior to the cecum, with the opening facing inferiorly.

Not infrequently an opening may occur in the **sigmoid mesocolon,** and intestines have been known to herniate through this opening, a condition that may be serious if constriction of the arteries to the herniated intestines should occur.

The peritoneum is subject to **infections** (peritonitis) which are always serious in spite of wonderful results with antibiotics. The anatomy of the peritoneum is important in understanding spread of such infections. Another glance at Figure 5-8 will show that there are areas of peritoneum walled off, so to speak, such as that lining the lesser sac or that between the mesentery of the small intestine, the transverse mesocolon, and the ascending colon. It is also easy to visualize the areas lateral to the ascending and descending colons serving as direct pathways superiorly to the liver and diaphragm. That area to the left of the mesentery leads directly to the most inferior part of the peritoneal cavity located in the pelvis. In fact, the relation of the rectouterine pouch to the vagina is important clinically. Finally, that area between the liver and the right kidney, often called the **hepatorenal pouch,** is a point where fluid collects in the patient in bed; this area is drained in surgery where leakage is expected.

The reason why the peritoneum takes its complex arrangement will be given later (page 233).

When studying the anterior abdominal wall, we noted that there was a layer of fatty fascia between the peritoneum and deep fascia appropriately called **subserous fascia.** This is true in other parts of the abdominal cavity as well. Wherever there is peritoneum, there is a layer of this fascia, sometimes very thin as found surrounding the small intestine, but sometimes quite thick as found around the kidneys. The aorta and its many branches and the corresponding veins are found in this fascial layer as well as visceral branches of the nervous system and lymphatics. Whenever this fascia becomes condensed, it is usually mentioned, but you should realize that all the organs found in the abdominopelvic cavity are surrounded by this fascia even when it is not mentioned.

INTESTINES

Now that you have a thorough picture of the peritoneum and its reflections, the learning of the abdominal organs is made much easier. If the greater omentum is reflected superiorly, the abdominal cavity is seen to contain many coils of the small intestine in its central portion, while around its periphery is found the large intestine. It has often been said that the large intestine forms the frame of the picture made by the small intestines. The transverse mesocolon divides the abdominal cavity into superior and inferior portions, the stomach, spleen, and liver being the most conspicuous organs in the upper part of the abdominal cavity, while the intestines are the most conspicuous in the lower part.

SMALL INTESTINE

The **duodenum,** except for a very small portion near the stomach, cannot be seen until other structures are removed; it will be described later.

The small intestine seen in the central part of the abdominal cavity (Fig. 5-2) is made up of the **jejunum** and **ileum,** the second and third portions of the small intestine. Because the small intestine is approximately 22 feet in length and the duodenum makes up approximately 1 foot of this span, it is easily seen that the jejunum and ileum combined are approximately 21 feet long. The jejunum is not as long as the ileum, the former being approximately 8 feet long and the ileum 12 feet. Since the jejunum gradually changes to ileum, the actual point where the two parts meet is not easily found. The jejunum starts at the **duodenojejunal junction,** which lies just to the left of the second lumbar vertebra, and fills the superior and left side of the abdominal cavity, while the ileum is found in the pelvic cavity and in the lower right part of the abdominal cavity. The intestines are freely movable because this very long tube is attached to the posterior abdominal wall by a mesentery which is approximately 10 inches long at its base. We have already seen (Fig. 5-8) that the mesentery of the jejunum and ileum starts a bit to the left of the second lumbar vertebra and proceeds inferiorly and to the right to reach the iliac fossa on the right side. This mesentery, therefore, is in the shape of a large fan.

The small intestine is a muscular tube with outer longitudinal and inner circular layers of smooth muscle. The mucous membrane on the inside is thrown into circular and spiral folds. They are numerous in the distal part of the duodenum and proximal part of the jejunum and gradually diminish in size and number as the terminal end of the ileum is reached. The wall of the jejunum is thicker than that of the ileum; and this is one of the ways in which one can distinguish the jejunum from the ileum, for the former has the feeling of being a much more solid structure. We will note later that the arteries also are a clue to the difference between the jejunum and the ileum, there being more arterial arcades in the region of the

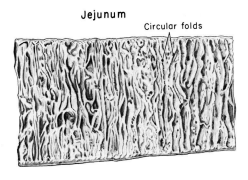

Jejunum

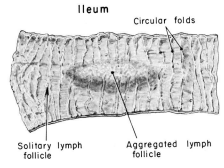

Ileum

Figure 5-15. Appearance of inside surface of jejunum and ileum.

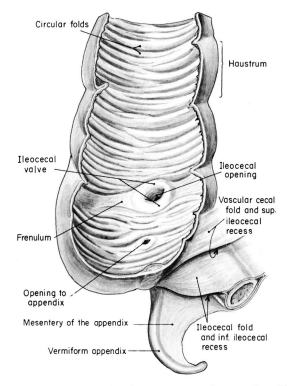

Figure 5-16. Ileocecal valve, cecum, and appendix. (The valve is shown in its open state.)

ileum. Figure 5-15 shows a typical section through the intestine. The ileum presents nodules of lymphoid tissue deep to the mucous membrane at a point opposite the attachment of the mesentery. They may be solitary nodules or large groups called **aggregated nodules.** These nodules are best seen in a young subject, because they tend to atrophy with age.

In summary, the jejunum differs from the ileum in that it occupies a higher area than the ileum, its walls are thicker, it has less fat in the mesentery, it tends to be empty, it contains more circular folds on the inside, there is an absence of aggregated lymph nodules, and there are fewer arterial arcades.

ILEOCOLIC VALVE

The ileum ascends from the pelvic cavity and joins the colon in the lower right quadrant of the abdominal cavity. The opening into the cecum from the ileum is guarded by the **ileocolic valve** (Fig. 5-16). The opening is slitlike with

its long axis in the anterior posterior direction. The lips, one superior and one inferior, project into the cecum and are elevated in such a way as to form a sort of valve. The muscle tone in the ileum allows the valve to act as a sphincter.

LARGE INTESTINE

The **large intestine** is a continuous tube consisting of the **cecum,** the **colon,** the **rectum,** and the **anal canal.** The **cecum** is located in the right iliac fossa. The **colon** ascends on the right side of the abdominal cavity, turns near the liver, crosses the abdominal cavity transversely, makes another turn near the spleen, descends on the left side of the abdominal cavity, and ends as an S-shaped portion called the sigmoid colon. The **rectum** is located in the pelvic cavity, while the **anal canal** is in the perineum, that area inferior to the pelvic cavity.

The cecum and its attached appendix, the transverse colon, and sigmoid colon are completely covered by peritoneum. The remaining portions are usually retroperitoneal. The peritoneum differs from that in the small intestine by the presence of small pouches of fat called **appendices epiploicae.** These are found in all parts of the large intestine except the cecum, rectum, and anal canal. The outer longitudinal muscle coat is incomplete and takes the form of three muscular bands, which course from the base of the appendix along the colon until the rectum is reached, where they tend to fan out into a thin sheet of muscle. These bands (**taeniae coli**) are shorter than the length of the colon and cause sacculations between the muscle bands. These **sacculations** or **haustra** are responsible for the pouched appearance of the large intestine. The circular muscle fibers are in the form of a complete sheet. The mucous coat (Fig. 5-17) is thrown into many crescentic folds, and lymph nodules are scattered in the mucous tissue, being especially numerous in the appendix. Therefore the large intestine differs grossly from the small by the larger size, the presence of taeniae coli and the subsequent sacculations, and the collections of fat (appendices epiploicae).

CECUM (Figs. 5-14 and 5-16).   This blind, pendulous sac is approximately 2½ inches long and 3 inches wide. Being completely covered with peritoneum, it is an intraperitoneal organ, but it does not have a mesentery. Its relations* are as follows. It is anterior to the iliopsoas muscle, the lateral femoral cutaneous and genitofemoral nerves, and the testicular vessels (Fig. 5-40). It is posterior to the many coils of the intestine and the anterior abdominal wall. The ileum is medial to the cecum, and the iliac fossa is lateral.

The **vermiform appendix** (Figs. 5-14 and 5-16) is a worm-like structure, approximately 4 inches in length (although it may vary from 1 to 9 inches), and attached to the posteromedial wall of the cecum about an inch inferior to the entrance of the ileum. This structure may take various shapes and sizes, and it may be located in several different regions. It can hang into the pelvic cavity, it can be directed superiorly behind the cecum (in which case it is called a retrocecal appendix), or it can be directed superiorly and to the left. This structure is completely ensheathed in peritoneum and has a mesentery

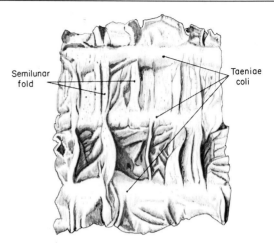

Figure 5-17.   *Inside surface of large intestine.*

called the mesentery of the appendix. This mesentery contains the artery that brings blood to the appendix. If the appendix is hidden, one can find it by following the taeniae coli directly to the base of the appendix. The taeniae coli are spread out on the appendix so that it has a continuous longitudinal muscle layer. Gaps in this muscle layer can lead to spread of infection.

COLON.   This portion of the large intestine consists of the following parts: ascending colon, right colic flexure, transverse colon, left colic flexure, descending colon, and sigmoid colon.

The **ascending colon** (Fig. 5-3) is directly continuous with the cecum and is located on the right side of the abdominal cavity. It ascends until it comes close to the inferior surface of the liver. At this point it takes a turn to the left, making the right colic flexure. The ascending colon is **anterior** to the iliacus, quadratus lumborum, and transversus abdominis muscles; the lateral femoral cutaneous, ilioinguinal, and iliohypogastric nerves; and the right kidney (see Fig. 5-40). It is **posterior** to the anterior abdominal wall and frequently to some of the coils of the small intestine. On its **lateral** side is the abdominal wall; on its **medial** side, the small intestine, psoas major muscle, and a portion of the right kidney.

The **right colic (hepatic) flexure** lies on the right kidney. It is covered with peritoneum except on its posterior side, and the liver is superior and anterior to it (see Fig. 5-18).

---

*Posterior relations of parts of the colon consist of structures not as yet studied, but they will be given at this time for the sake of completeness. A glance at Figure 5-40 on page 260 should be helpful.

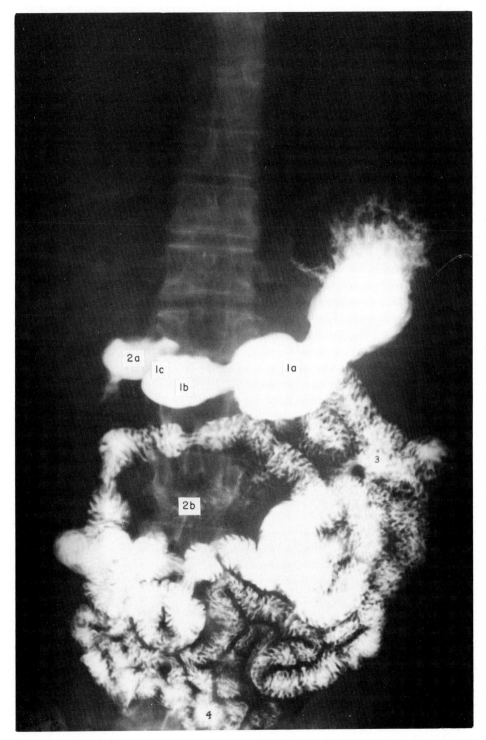

Plate 9. *Stomach and small intestine outlined by barium meal: (1a) body of stomach (note peristaltic waves); (1b) pyloric portion of stomach, (1c) pylorus, (2a) duodenum ("duodenal bulb"), (2b) duodenum (very lightly outlined), (3) jejunum, and (4) ileum.*

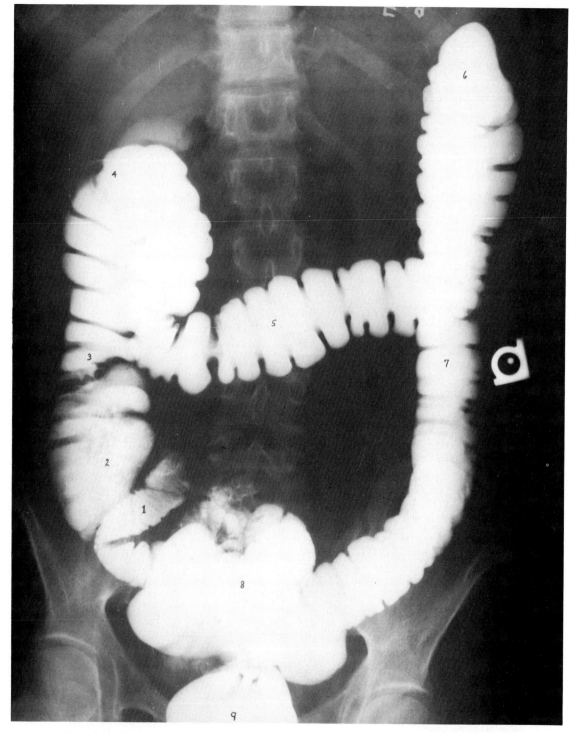

Plate 10. *Outline of large intestine by barium enema: (1) ileum, (2) cecum, (3) ascending colon, (4) right (hepatic) colic flexure, (5) transverse colon, (6) left (splenic) colic flexure, (7) descending colon, (8) sigmoid colon, and (9) rectum. Between them, Plates 9 and 10 give a complete picture of stomach and intestines.*

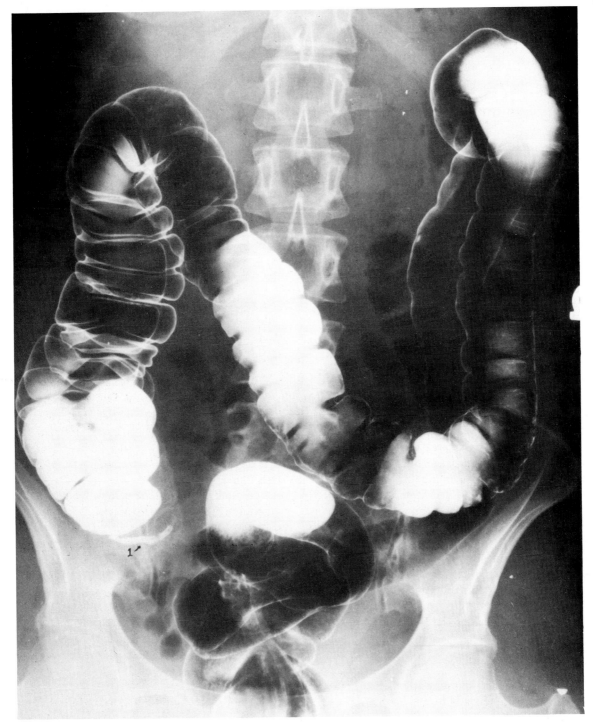

Plate 11. *The large intestine demonstrated by the air contrast method: (1) appendix. Note the low level this transverse colon reaches in this individual.*

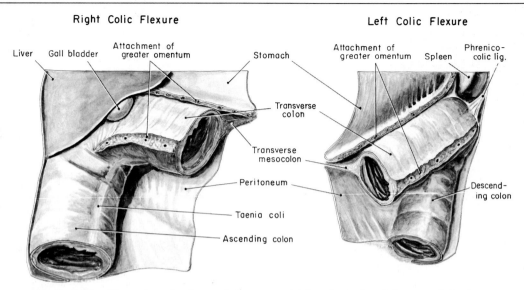

Figure 5-18. *The right colic (hepatic) flexure and left colic (splenic) flexure of the large intestine.*

The **transverse colon** (Fig. 5-3) starts at the right colic flexure just described, crosses the abdominal cavity, and ends on the left kidney just inferior to the spleen. This structure has a mesentery called the **mesocolon** and is therefore an intraperitoneal organ. The **relations** of the transverse colon are extremely variable, since it is quite movable. We have already seen that the anterior surface of the transverse colon is covered by the greater omentum (Fig. 5-1) as it hangs down from the greater curvature of the stomach. Therefore the anterior surface of the mesocolon is the posterior wall of the lesser sac while the posterior layer of the mesocolon is in the greater sac. The stomach and liver are superior to the transverse colon, while the jejunum and ileum are inferior to it.

The **left colic (splenic) flexure** (Fig. 5-18) is higher than the right; the transpyloric line bisects it. It is anterior to the posterior portion of the diaphragm and left kidney, and inferior to the spleen. The peritoneum covers it except posteriorly and, therefore, it is a retroperitoneal organ. The **phrenicocolic ligament** is sometimes present and aids in attaching the colon to the diaphragm.

The **descending colon** (Fig. 5-3) is narrower than the ascending colon. It starts at the flexure just described and continues inferiorly on the left side of the abdominal cavity until it reaches the left iliac fossa. This structure at its commencement is anterior to the diaphragm, and then is anterior to the transversus abdominis, quadratus lumborum, iliacus, and psoas muscles. It is also anterior to the iliohypogastric, ilioinguinal, lateral femoral cutaneous, femoral, and genitofemoral nerves in a fashion similar to that described for the ascending colon (Fig. 5-40). The testicular and external iliac vessels are also posterior to the descending colon. The coils of the small intestine are anterior, while medial to it is the kidney superiorly and the psoas major muscle inferiorly. Laterally the descending colon is in relation to the diaphragm superiorly and the transversus abdominis muscle inferiorly.

The **sigmoid colon** (Fig. 5-3), which has a mesentery and therefore is an intraperitoneal organ, is an S-shaped portion of the large intestine, which starts at the left iliac fossa just lateral to the psoas major muscle and continues into the pelvic cavity. The attachment of its mesentery (Fig. 5-40) shows the rather S-shaped curve taken to reach the rectum.

RECTUM (Figs. 5-10, 5-11, and 5-60). The rectum is approximately 5 inches long. It starts approximately in the middle of the sacrum and follows the curve of the sacrum and coccyx into the pelvic cavity. It is just anterior to the sacrum and coccyx, posterior to the bladder and seminal vesicles in the male or, in the female, to the uterus and vagina. This structure will be studied in greater detail when the pelvic cavity is investigated.

ANAL CANAL (Fig. 5-62).   This structure, connecting the rectum to the anus or anal opening, will be described later in this chapter under the pelvic cavity.

———————————

A comparison of the above with Plates 9, 10, and 11 would be helpful at this time.

———————————

The definitive position of the intestine cannot be understood unless its development is considered. The following should serve as a review of this important process. Figures 5-19, 5-20, and 5-21 should also be helpful.

The intestine, which is attached by a dorsal mesentery only (except for the first part of the duodenum), elongates to such a degree that the abdominal cavity cannot contain it; the intestine, therefore, herniates into the umbilical cord. Further elongation occurs, and when the intestine returns to the abdominal cavity, it goes through a counterclockwise (viewing the embryo from the ventral side) rotation using the superior mesenteric artery as an axis. As a result, the duodenum takes a posterior relation to the mesenteric artery, while the large intestine crosses anterior to it. This leaves the cecum close to the liver; the colon then elongates in such a manner as to bring the cecum into its definitive position in the lower right quadrant. Probably because of the overcrowding of the abdominal cavity, certain parts of the gut are forced against the posterior abdominal wall, and as a result the peritoneum covering the deep surface of the mesentery and the organ, as well as that on the body wall, change to connective tissue; therefore, these particular organs and the vessels and nerves coursing to and from these organs become retroperitoneal in position, this phenomenon occurring to the duodenum, ascending colon, descending colon, and rectum. This leaves the jejunum, ileum, transverse colon, and sigmoid colon with mesenteries that attach to the body wall as shown in Fig. 5-8.

The loop of midgut and hindgut destined to form jejunum, ileum, cecum, and ascending and transverse colons has a temporary connection, through the umbilical orifice, with the remains of the yolk sac. This is known as the vitello-intestinal duct. In about 2 per cent of bodies, remains of this duct—the **diverticulum ilei** (of Meckel)—can be found on the ileum approximately 3 feet from the cecum. It takes the form of a blind sac about 2 inches long, with a lumen approximately the size of that possessed by the ileum. It may have a fibrous connection with the umbilical region, form a cyst, or even open at the umbilicus, inducing a leakage of fecal matter.

Incomplete or abnormal rotation of the gut is not an infrequent occurrence, and can possibly lead to severe complications due to obstruction of the intestine or strangulation of the blood supply to it. If the ascending colon has a mesentery, it tends to cause more complications than when one is present on the descending colon. The former interferes with propeling fecal matter superiorly against gravity. Stagnation may result, the ascending colon may hang into the pelvis, and the weight on the right kidney may induce its descent. Furthermore, this weight may place such a stretch on the superior mesenteric artery that it squeezes the duodenum against the vertebral column resulting in intestinal obstruction.

A useful concept is to think of structures as **"primarily retroperitoneal"** or **"secondarily retroperitoneal,"** the former being organs such as the kidneys that were always retroperitoneal, the latter being organs such as the ascending colon that secondarily become retroperitoneal by the disintegration of peritoneal layers to form fusion fascia.

ARTERIES

The arterial supply to the intestines consists of the (1) superior and (2) inferior mesenteric arteries and their branches. Although there is considerable variation, the pattern found in the majority of cases will be described. The important posterior relations of these vessels are shown in Figure 5-38. They course in the subserous fascia.

1. The **superior mesenteric artery** (Fig. 5-21) supplies the entire embryological midgut. It arises from the anterior surface of the aorta just posterior to the body of the pancreas. It courses inferiorly passing anterior to a small projection of the pancreas (uncinate process), and immediately gives off the inferior pancreaticoduodenal artery (see Fig. 5-36 on page 252). It continues inferiorly, passing anterior to the duodenum, a relationship easily

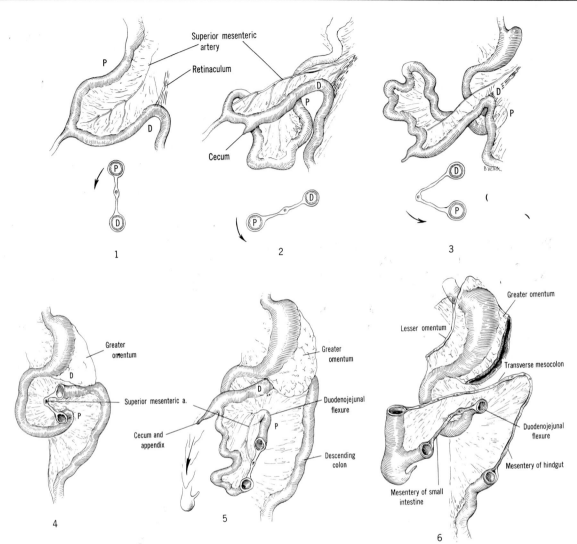

Figure 5-19. *A series of pictures depicting rotation of the gut. P is the proximal (cranial) part of the intestinal loop and D represents the distal (caudal) portion. Note (1) the superior mesenteric artery between these parts of the loop; this artery serves as an axis about which rotation occurs. Rotation has started in a counterclockwise direction (as viewed from the front) in 2 and is continued further in 3 and 4. As a result, the transverse colon lies ventral to the superior mesenteric artery and the duodenum dorsal to this artery, as shown in 5. Further downward growth of the colon (5) causes the cecum and appendix to attain the definitive position in the lower right quadrant of the abdominal cavity (6). Note how the mesenteries have obtained their definitive positions. The retinaculum represents one of two points where the gut is attached to the posterior abdominal wall; it is located at the point where the midgut becomes the hindgut. The other point of attachment (not shown) is the suspensory ligament of the duodenum. (From Davies,* Human Developmental Anatomy. *Copyright © 1963, John Wiley & Sons, Inc. Lower portion after J. E. Frazer.)*

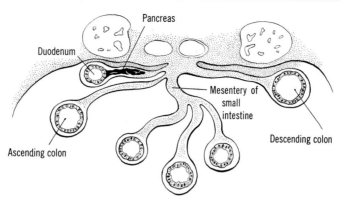

Figure 5-20. *Diagram illustrating how the ascending and descending parts of the colon are forced against the posterior abdominal wall. The peritoneal layers against the posterior abdominal wall disintegrate, leaving these organs and their blood vessels, nerves, and lymphatics in a retroperitoneal position. (From Davies,* Human Developmental Anatomy. *Copyright © 1963, John Wiley & Sons, Inc.)*

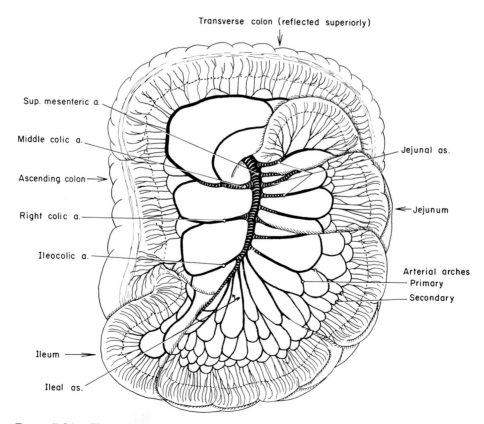

Figure 5-21. *The superior mesenteric artery and its branches. (Inferior pancreaticoduodenal artery is not shown; see Fig. 5-36 on page 252).*

explained by the rotation of the gut; it then continues into the base of the mesentery. Its branches are (not in order of emergence):

    a. Inferior pancreaticoduodenal
    b. 12 to 15 jejunal and ileal branches
    c. Ileocolic
    d. Right colic
    e. Middle colic

    a. The **inferior pancreaticoduodenal** artery (Fig. 5-36) courses to the right and divides into branches that anastomose with the duodenal and pancreatic branches of the superior pancreaticoduodenal artery to be described later.

    b. The **jejunal** and **ileal** branches (Fig. 5-21) are 12 to 15 in number. They immediately enter the mesentery to reach the intestines. These arteries form many loops and arcades before arriving at their destination in the form of straight arteries, and there are more of these arcades in the ileal than in the jejunal region because of the greater length of the mesentery to the ileum. In fact the jejunum has 1 or 2 arterial arcades with straight arteries approximately 1½ inches long, while the ileum has 2 or 3 arcades and straight arteries approximately ½ an inch long. Such facts are an aid to the surgeon in identifying portions of the intestine. The last ileal artery anastomoses with the ileal portion of another branch from the superior mesenteric artery, the **ileocolic.**

    c. The **ileocolic** artery, frequently arising in common with the right colic, courses to the right, across the posterior abdominal wall deep to the peritoneum, and divides into a branch that proceeds to the ascending colon and another that courses along the ileum. This latter artery gives off branches that continue to the cecum and to the appendix, the **artery to the appendix** passing posterior to the ileum and entering the mesentery of the appendix to reach its destination (Fig. 5-14). The branch of the ileocolic to the ascending colon (Fig. 5-21) continues superiorly and anastomoses with a branch of the **right colic artery,** another branch of the superior mesenteric artery.

    d. The **right colic** artery (Fig. 5-21) emerges from the superior mesenteric artery just inferior to the duodenum, courses to the right across the posterior abdominal wall, remaining retroperitoneal in its entire course. After reaching the colon, it divides into ascending and descending branches, which anastomose with the

middle colic and ileocolic arteries respectively. This artery arises frequently in common with the ileocolic.

    e. The **middle colic** artery (Fig. 5-21) arises from the superior mesenteric at the point where it passes deep to the mesocolon. It enters this mesentery to reach the transverse colon. It anastomoses with the right colic and left colic arteries, the latter being a branch of the inferior mesenteric artery.

    2. If the small intestine is swung to the right side of the abdominal cavity, the **inferior mesenteric artery** can be seen through the peritoneum (Fig. 5-8). This arises from the front of the aorta, approximately 1½ inches above its bifurcation into the common iliacs and opposite the third lumbar vertebra. It supplies the embryological hindgut. It courses inferiorly and to the left, deep to the peritoneum, and gives off the following branches:

    a. Left colic
    b. Sigmoidal
    c. Superior rectal

    a. The **left colic** artery divides into several branches, which ascend on the descending colon, and anastomose with branches of the middle colic artery at the left colic flexure (Fig. 5-22).

    b. The inferior mesenteric artery continues inferiorly and gives off several large **sigmoidal** branches, which break up into loops before entering the colon. These vessels anastomose with the left colic artery and the next branch of the inferior mesenteric, the superior rectal.

    c. The **superior rectal** artery reaches the rectum and supplies blood to its upper third only, the remaining two-thirds being supplied by other vessels (see page 305).

The anastomosis of the branches of the ileocolic, middle colic, and the branches of the inferior mesenteric usually forms a continuous arterial pathway along the colon called the **marginal artery** (of Drummond). In spite of this probable anastomosis, it is dangerous to bisect the intestine at any point where its blood supply could be endangered without checking for continuity of these anastomotic vessels. The anastomosis between the middle and left colic arteries is absent in 5% of cases. A so-called critical point (of Sudeck) occurs between the last sigmoidal branch and the superior rectal artery. Surgical transections for cancer of the rectum should be done above the last sigmoidal artery.

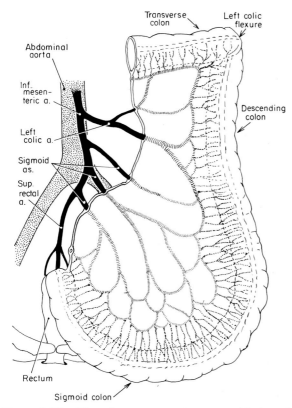

Figure 5-22. *The inferior mesenteric artery and its branches.*

VEINS

The intestines are drained by superior and inferior mesenteric veins, which are important parts of the hepatic portal circulation (Fig. 5-23); they do not drain into the inferior vena cava, but join with the splenic vein to form the portal vein, which transports this blood to the liver. Because these veins drain in this manner, they do not accompany their respective arteries near their point of drainage. This is particularly true of the inferior mesenteric vein, which is longer than the corresponding artery.

The **superior mesenteric vein** courses at the base of the mesentery with the superior mesenteric artery. It drains the jejunum, ileum, appendix, cecum, ascending and transverse colons, and part of the pancreas and duodenum, and has tributaries with the same names as given to the branches of the artery. At a point deep to where the head joins the body of the pancreas, the superior mesenteric vein joins the splenic to form the portal.

The **inferior mesenteric vein** drains the superior third of the rectum, the sigmoid colon, and the descending colon. It courses superiorly on the posterior abdominal wall in a retroperitoneal position and terminates deep to the pancreas by joining the splenic vein. Its anastomosis with the middle and inferior rectal veins (systemic circulation) is important. Liver disease can cause obstruction of the portal veins thus leading to dilations at this point—hemorrhoids (Fig. 5-23).

Figure 5-38 on page 254 shows the many important posterior relations of these vessels. They course in the subserous fascia.

see p. 27

NERVES

Typical of many of the visceral nerves in the body, those to the intestines follow the arteries to reach their destination and are, therefore, in the subserous fascia. These nerves are both sympathetic and parasympathetic. They are diagramed in Fig. 1-22 on page 27. Preganglionic **sympathetic fibers** destined to synapse with postganglionic fibers in the superior mesenteric ganglion, and ultimately to innervate all areas supplied by the superior mesenteric artery (the midgut), start in the intermediolateral cell column in thoracic segments 10 and 11; travel to the sympathetic chain via the ventral roots, segmental nerves, and white rami communicantes; traverse the chain without synapse; and form the lesser splanchnic nerve. These fibers synapse in the superior mesenteric ganglion and are distributed, as just mentioned, by postganglionic fibers traveling on the arteries. The preganglionic fibers destined to reach the inferior mesenteric ganglion, and thence the part of the intestine supplied by the artery of the same name (the hindgut), arise in lumbar segments 1 to 3. They follow the same pathway as just described for the midgut, i.e., ventral roots, segmental nerves, white rami communicantes, to reach the sympathetic chain. Since the inferior mesenteric ganglion is quite small, possibly many of these preganglionic fibers synapse in chain ganglia. These fibers (some preganglionic, some postganglionic) form the lumbar splanchnic nerves, those fibers that did not synapse in the chain ganglia doing so in the inferior mesenteric ganglion. Postganglionic fibers follow the branches of the inferior mesenteric artery to reach their destination.

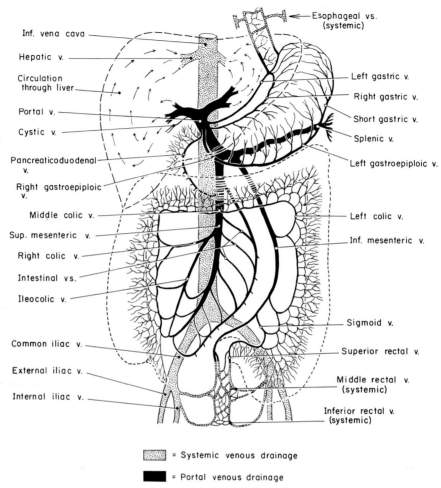

Figure 5-23. *Diagram of the portal circulation. Portal veins in black and systemic veins dotted. Note the important connections between the two systems at the esophagus and on the rectum. See text for other connections between the two systems.*

The **parasympathetic fibers** are preganglionic neurons from the vagus nerve as far along the intestine as supplied by the superior mesenteric artery, the fibers traversing the superior mesenteric ganglia without synapse, and following the branches of the superior mesenteric artery to reach the organ to be innervated. A synapse occurs with postganglionic fibers in the walls of the intestine. The last part of the transverse colon, the descending colon, sigmoid colon, and rectum derive their parasympathetic nerves from the **sacral portion of the parasympathetic system.** These nerves (S2, 3, and 4) reach their destination by (1) following the pelvic nerves

directly to the organ (these branches have occasionally been found to ascend the colon as far as the left colic flexure), or (2) reaching the inferior mesenteric plexus and following the inferior mesenteric artery.

The parasympathetic system stimulates peristalsis and secretion, and causes a vasodilation; the sympathetic system has an opposite effect.

Because patients talk about pain, perhaps of equal if not greater significance are the **visceral afferent fibers** from the intestines. Cutting the intestines induces no sensations, but intense pain can be induced by dilation or violent contraction of this organ, or by involvement of the peritoneum. These afferent fibers travel in the reverse direction as just described for the sympathetic system. Fibers from all areas supplied by the superior mesenteric artery and its branches return to the spinal cord via the lesser splanchnic nerves, traversing the chain, following white rami communicantes, the segmental nerves and dorsal roots (the cell bodies are in the dorsal root ganglia), to reach the spinal cord at segments T10 and 11. All fibers from the hindgut will follow the lumbar splanchnics and return to spinal cord segments L1, 2, and 3 in a similar manner. All of this pain is carried in visceral afferent fibers and is dull and difficult to locate. But most such pain is referred to the skin surface and naturally will be felt wherever nerves T10 or T11, or L1, 2, and 3 are distributed. If pain is originating in the intestines, it would not be surprising if that originating in the jejunum, the ileum, appendix, cecum, ascending colon, and most of the transverse colon might be referred to the abdominal wall around or just inferior to the umbilicus, while that from the descending or sigmoid colons would be felt around the inguinal and pubic regions. If such pain changes to the somatic type, easily recognized and localized, this means that the parietal peritoneum has become involved; if in the area of the appendix, the time for its removal has arrived.

There are visceral afferent fibers associated with the vagus nerve, but these are involved with general physiology of the viscera rather than with pain. Gastric reflexes and sensations of hunger or nausea are examples.

LYMPHATICS

There are innumerable lymph nodes in the mesentery of the jejunum and ileum. These ultimately drain into the large, massive superior mesenteric lymph nodes at the base of the superior mesenteric artery (Fig. 5-46). There are many nodes, as well, along the ileocolic, right colic, and middle colic arteries. Some are close to the colon while others are intermediate in position. All of these nodes ultimately drain into the superior mesenteric nodes just mentioned. Similar nodes are found along branches of the inferior mesenteric artery which ultimately drain into nodes at the base of this artery. From the large superior and inferior mesenteric nodes, the lymph is drained into the cisterna chyli—the origin of the thoracic duct (Fig. 4-43 on page 198).

The lymphatic drainage of the intestines follows the

rule that the lymphatic drainage follows the arteries in a retrograde manner.

FUNCTION

Although not all parts of the intestine have been studied, some attention to function at this time may be helpful. The small intestine serves to digest and absorb food as follows. The partially digested food leaves the stomach and enters the duodenum at intervals. Here it is attacked by intestinal juice from the intestinal glands, bile from the gall bladder and/or liver, and pancreatic juice from the exocrine portion of that gland. The digested food is then propelled along the intestine by peristalsis and absorbed into the venous system and lymphatics through the massive inside surface area of the small intestine, the ileocecal valve preventing too rapid propulsions of the food material from the area of food absorption. The absorbed food enters the portal venous system, which takes this material to the liver, where further digestive procedures occur.

The large intestine exhibits slower peristaltic waves, is more voluminous, and serves to concentrate undigested and unabsorbed food material. It accomplishes this by returning water to the circulation. The lower end of the colon stores the fecal material. When it enters the rectum, it is removed from the body by the act of defecation.

ORGANS SUPPLIED BY CELIAC AXIS

GENERAL DESCRIPTION

You have already seen that the liver occupies the right upper quadrant of the abdominal cavity, while the spleen and the stomach occupy the left upper quadrant (Fig. 5-3). In addition, you observed how the esophagus penetrates the diaphragm and joins the stomach, how the stomach is suspended from the inferior surface of the liver by the lesser omentum, and how the stomach is continuous with the first portion of the duodenum. If the greater omentum is cut away from the greater curvature of the stomach, entrance is gained into the lesser sac. Here, by

way of review (Fig. 5-8), were found the epiploic foramen, the vestibule just inside the opening, the upper
recess of the lesser sac close to the caudate lobe of the
liver, the lienal recess close to the spleen, the reflections
of the lienorenal and the gastrolienal ligaments, and the
reflection of the mesocolon.

Closer observation, through the peritoneum (Fig.
5-8), reveals the **celiac artery** branching from the abdominal aorta and its three main branches: (1) the
splenic coursing to the left to reach the lienorenal ligament, and thence the spleen; (2) the **common hepatic
artery** coursing to the right across the inferior vena cava
to join the duodenum, and then turning superiorly in the
free edge of the lesser omentum as the **proper hepatic
artery** to reach the liver; and (3) the **left gastric** coursing
superiorly on the diaphragm and reaching the esophagus
and thence the lesser curvature of the stomach. The first
two parts of the duodenum also are visible, as is the entire
extent of the pancreas. These structures are all retroperitoneal, as shown in Figure 5-8; they are in the
subserous fascia. These arteries and their branches are
supplying the embryological foregut and its derivatives.

―――――――

The above pathways taken by branches of the celiac
axis exemplify that structures do not course through air or
space. These three vessels start as retroperitoneal structures and become intraperitoneal (between two layers of
peritoneum) by coursing to a retroperitoneal organ or
ligament and utilizing this structure to gain an intraperitoneal position. The common and proper hepatic arteries
utilized the duodenum and the hepatoduodenal ligament
(part of the lesser omentum) to gain entrance to the liver;
the splenic entered the lienorenal ligament to reach the
spleen; the left gastric followed the diaphragm to reach
the esophagus and thence the lesser curvature of the
stomach. All of these vessels originally coursed in the
dorsal and ventral mesenteries; their definitive retroperitoneal position is due to subsequent growth of the
liver and rotation of the stomach, to be described later
(see Fig. 5-27).

―――――――

If the peritoneum forming the posterior wall of the
lesser sac is removed (Fig. 5-24), the **celiac artery** is seen
to emerge from the anterior surface of the aorta immedi-

ately after the latter structure penetrates the diaphragm.
The **inferior phrenic arteries** also arise very close to the
diaphragm and course on its inferior surface. Removal of
this peritoneum also reveals the **pancreas** in its entirety,
with its **head** in the arms of the duodenum, its **body**
crossing the midline, and the **tail** reaching as far as the
spleen. The **duodenum** in all its four parts is seen forming
a C-shaped turn with the concavity to the left (Fig. 5-24).

Removal of the peritoneum forming the lesser
omentum reveals further structures (Fig. 5-25). The **left
gastric artery** can be seen coursing on the lesser curvature of the stomach and anastomosing with the right gastric artery, a branch of the hepatic. In addition, three vital
structures are found in the free edge of the lesser omentum; they are the **portal vein** posteriorly, the **proper hepatic artery** anteriorly and to the left, and the **bile duct**
anteriorly and to the right. These relationships are easily
remembered if one thinks of the direction taken by each
structure to reach the free edge of the lesser omentum.
Further dissection superiorly reveals that the proper hepatic artery divides into **right** and **left hepatic arteries** to
be distributed to the right and left portions of the liver
respectively. The bile ducts make a similar pattern in that
two **hepatic bile ducts** join to form a single duct—the
common hepatic duct; this then joins with the **cystic duct**
from the gall bladder to form the **ductus choledochus** or
bile duct. The portal vein also divides into branches
called the **right** and **left portal veins** before emptying
into the liver. The **gall bladder** can be seen suspended
on the inferior surface of the liver with the fundus of the
bladder visible just inferior to the inferior edge of the liver
itself. The cystic duct is accompanied by arteries and
veins. The relation of these structures to each other exhibits considerable variation, a problem to the surgeon
removing the gall bladder.

STOMACH

The shape of the stomach is difficult to predict, for it can
take many forms (see Plate 9 on page 229), but a glance
at Figure 5-26 will reveal its general structure. It is joined
to the esophagus superiorly and the area of the stomach
around the esophageal entrance is the **cardiac portion.**
The portion to the left of the esophagus, which usually
extends to a more superior position than the entrance of
the esophagus, is the **fundus** of the stomach. The **pyloric
end** is the portion just proximal to the duodenum and is

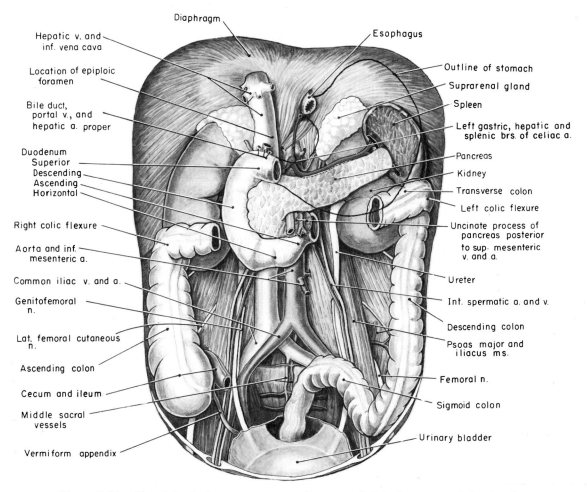

Diaphragm
Hepatic v. and inf. vena cava
Location of epiploic foramen
Bile duct, portal v., and hepatic a. proper
Duodenum
 Superior
 Descending
 Ascending
 Horizontal
Right colic flexure
Aorta and inf. mesenteric a.
Common iliac v. and a.
Genitofemoral n.
Lat. femoral cutaneous n.
Ascending colon
Cecum and ileum
Middle sacral vessels
Vermiform appendix

Esophagus
Outline of stomach
Suprarenal gland
Spleen
Left gastric, hepatic and splenic brs. of celiac a.
Pancreas
Kidney
Transverse colon
Left colic flexure
Uncinate process of pancreas posterior to sup. mesenteric v. and a.
Ureter
Int. spermatic a. and v.
Descending colon
Psoas major and iliacus ms.
Femoral n.
Sigmoid colon
Urinary bladder

Figure 5-24. *The abdominal contents as seen after removal of the liver, stomach (outlined), jejunum, and ileum, and the branches of the superior and inferior mesenteric arteries. The entire peritoneum also has been removed.*

usually delineated from the remaining portion of the stomach, the **body,** by a groove—the **incisura angularis**—that is almost always seen in x-rays as an indentation on the **lesser curvature.** The pyloric portion is divided into the **pyloric antrum** and the **pyloric canal,** which leads to the opening into the duodenum, the **pylorus** (G., gatekeeper). The pylorus is constricted by a **pyloric sphincter.** The long inferior curve of the stomach, opposite the lesser curvature, is the **greater curvature.** The stomacn is an intraperitoneal organ.

The stomach develops as a dilation of the original gut tube and obtains its odd shape as a result of unequal growth. It is positioned in such a manner as to be just proximal to the free edge of the deficient ventral mesentery. By rotation to the right, the right side of the stomach becomes the posterior surface, the left the anterior surface. The ventral mesentery persists to form the lesser omentum, and the dorsal mesentery is pushed to the left (Fig. 5-27) and greatly elongates by a downward growth to form the greater omentum. This rotation also accounts for the posterior and anterior positions of the right and left vagus nerves on the esophagus, for this organ is slightly involved in this twisting. Furthermore, it is easy to visualize that the lesser curvature, which formerly was an-

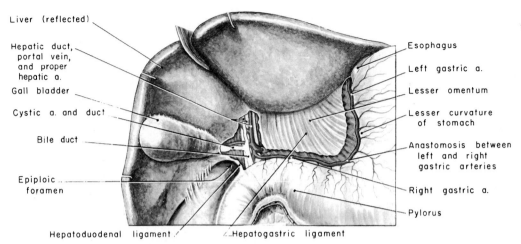

Figure 5-25. *Structures contained in the lesser omentum. The anterior layer of the lesser omentum has been cut, leaving the posterior layer intact.*

teriorly located, will face superiorly and to the right, and the greater curvature, formerly posterior in direction, will face inferiorly and to the left. Because the liver is a comparatively large organ in the embryo and occupies the upper right quadrant of the abdominal cavity, the stomach moves to the left and occupies the left upper quadrant.

The stomach has three layers of **smooth muscle:** an oblique layer in addition to the usual longitudinal and circular layers. It has a mucous membrane, which is elevated into ridges (**rugae**), and the whole surface of the stomach is pitted with **gastric pits** or **glands.** These have three main **types of cells:** chief cells, mucous cells, and parietal cells. The chief cells give rise to digestive en-

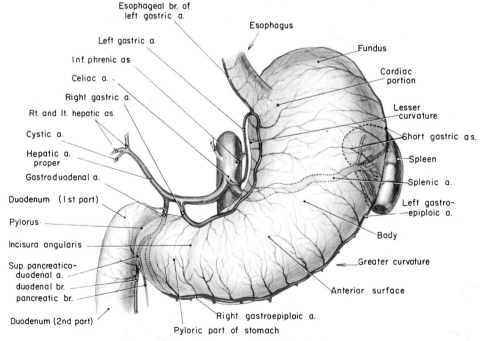

Figure 5-26. *The stomach and its arterial supply.*

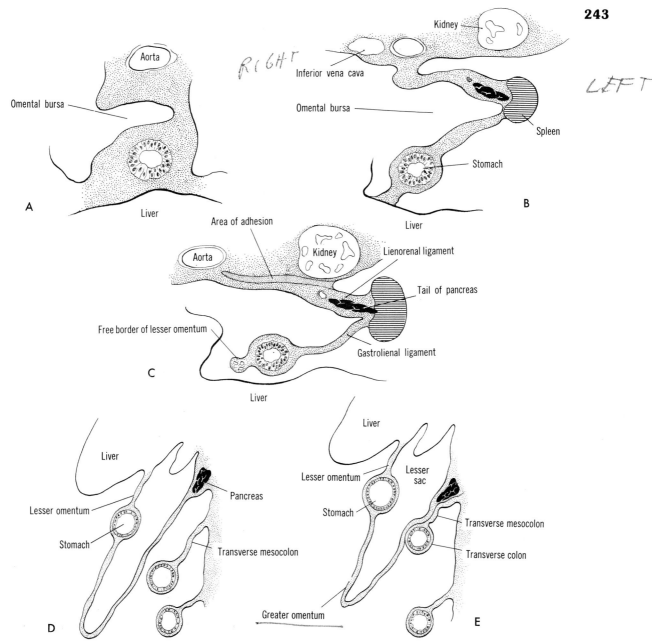

Figure 5-27. *A series of diagrams depicting the "rotation" of the stomach and formation of the lesser sac (omental bursa) and greater omentum. The dorsal mesentery moves to the left with its contained spleen and tail of the pancreas (A and B). That part of the mesentery adhering to the posterior abdominal wall fuses out (C) leaving the mesentery reflecting from the left kidney (lienorenal ligament). The dorsal mesentery also grows inferiorly in a position ventral to the transverse colon (D), thereby elongating the lesser sac. The transverse mesocolon ultimately fuses with a portion of the dorsal mesentery (E), which accounts for the greater omentum extending from the greater curvature of the stomach to the transverse colon. Because of fusion of the layers of the greater omentum, that portion of the lesser sac between its layers is usually obliterated. (From Davies,* Human Developmental Anatomy. *Copyright © 1963, John Wiley & Sons, Inc.)*

zymes, the mucous cells to mucus, and the parietal cells to hydrochloric acid.

The stomach is a very mobile structure* and very frequently, after a heavy meal, will descend much further in the abdominal cavity than one would suspect. In the erect living person, it is lower than usually found in cadavers. The pylorus, on the other hand, is relatively stable and is at the level of the transpyloric line.

Relations. The anterior surface of the stomach is related to the liver to the right, the diaphragm to the left, and the anterior abdominal wall between these two structures (Figs. 5-1 and 5-2). The posterior surface is related to several structures composing what is called the bed of the stomach (Fig. 5-28). Many of these structures have already been mentioned, but, to repeat, the stomach is related posteriorly to the body of the pancreas, a small part of the left kidney superior to the pancreas, the left suprarenal gland, diaphragm, splenic artery, spleen, transverse mesocolon, and the left phrenic artery. Of course, these relations will vary with movement of the stomach.

The stomach is subject to inflammatory conditions (gastritis), its walls may break down (an ulcer), and it is a common site for cancer. This latter disease is difficult to handle because the cancer can become widespread without producing symptoms of its presence. Large portions of the stomach can be removed surgically; the entire organ can be eliminated if necessary. Recall that there is a physiological cardiac sphincter (see page 201) that prevents gastric contents from returning into the esophagus. This is important because gastric acid is deleterious to the esophagus.

Details on blood vessels, nerves, and lymphatics can be found starting on page 253.

---

*R. O. Moody, R. G. Van Nuys, and C. H. Kidder, 1929, The forms and positions of the empty stomach in healthy young adults as shown in roentgenograms. *Anat. Record* 43: 359. There is very little constancy as to the form or proper position of the stomach; its greater curvature can reach as far inferior as the sacrum. The relation of the first part of the duodenum to the pancreas varies accordingly. There would seem to be no such thing as a normal, constant position for the stomach.

SPLEEN

This structure, of variable size and shape, is in the upper left quadrant of the abdominal cavity in contact with the diaphragm (Fig. 5-28). It is an intraperitoneal organ, being covered with peritoneum over its entire extent except for a small area at the hilum. The spleen is suspended from the **gastrolienal ligament** extending from the stomach to the spleen and from the **lienorenal ligament** between the left kidney and the spleen. This organ invariably has attachments to the peritoneum covering the diaphragm. This position of the spleen and its relation to the above-mentioned ligaments is easily understood if the development of this organ is taken into consideration (Fig. 5-27). The spleen develops in the dorsal mesentery and in its growth projects to the left. During rotation of the stomach the dorsal mesentery with its attached spleen is moved to the left. The mesentery is forced against the posterior abdominal wall, the deeper layers change to fusion fascia, and the mesentery ultimately arises from this wall in a more lateral position—actually from the surface of the diaphragm and left kidney. Therefore, the spleen is surrounded by peritoneum and the mesentery (now lienorenal ligament) arises from the left kidney rather than from the midline. This also accounts for the position and course of the splenic artery which was contained in the original dorsal mesentery, which will be described later.

The spleen, a purplish organ, appears wrinkled, and tends to be soft and pulpy. It acts as a filter for the blood and also is one of the organs where **erythrocyte breakdown** takes place. Pigments from this breakdown travel to the liver via the portal circulation, enter the biliary system, and finally reach the lumen of the duodenum. The spleen is hidden from view by the fundic portion of the stomach and is opposite the ninth, tenth, and eleventh ribs, in contact with the diaphragm. Because of this relation it is said to have a **diaphragmatic surface** and a **visceral surface.** The visceral surface is related to four separate structures—stomach, left kidney, left colic flexure, and pancreas—which make impressions upon the spleen as indicated in Figure 5-28. The spleen becomes enlarged under certain conditions; it can be removed if necessary, the most common cause for this surgery being trauma.

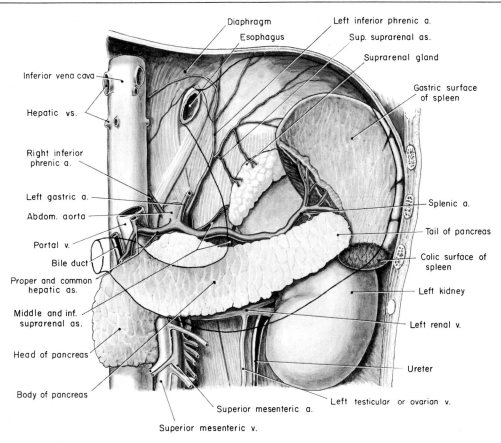

Figure 5-28. The pancreas and spleen. Note how this picture shows the posterior relations (the "bed") of the stomach. The peritoneum has been removed in its entirety.

Details on blood vessels, nerves, and lymphatics start on page 253.

LIVER

The liver is the largest organ in the body and occupies the right upper quadrant of the abdominal cavity (Fig. 5-2); it is in direct contact with the diaphragm. It varies a great deal in shape and size. It is almost entirely protected by ribs, but may project for varying distances inferior to the costal margin.* The liver is an organ essential to life. It has

*R. O. Moody and R. G. Van Nuys, 1928, Some results of a study of roentgenograms of the abdominal viscera, *Am. J. Roentgenol.* 20: 348. Liver and spleen are studied in this report. As usual, organs extend farther inferiorly than is indicated by cadavers.

many functions. Because the portal system of veins drains the intestinal tract and the spleen and carries this blood to the liver, the liver has an opportunity to attack much of the absorbed food material. It is involved with many phases of protein, carbohydrate, and fat metabolism. Since erythrocytes break down in the spleen, the products of this breakdown enter the portal circulation and thus enter the liver. The iron and porphyrin are utilized by the bone marrow in the process of forming more erythrocytes, while the pigments, such as bilirubin, are excreted into the duodenum via the bile ducts. Bile is not entirely a waste product, for it plays an important role in fat digestion. The liver also is involved in erythropoiesis; it is one of the main sources of blood cell formation in the embryo and carries on this vital function until maturity of the bone marrow. The liver maintains this capacity and can substitute for the bone marrow under certain circumstances; this is particularly true in the lower animals.

The liver consists of a very large **right,** a smaller **left, caudate,** and **quadrate lobes** (Fig. 5-30). The left lobe is separated from the caudate lobe by the lesser omentum and from the quadrate lobe by the falciform ligament and ligamentum teres (hepatis). The right lobe is directly continuous with the caudate and quadrate lobes but the gall bladder intervenes between the quadrate and right lobes, while the inferior vena cava is located between the caudate and right lobes. These boundaries are seen best in Figure 5-31.

The liver has two **surfaces,** a diaphragmatic and a visceral. It is obvious that the former is in contact with the diaphragm, while the latter faces inferiorly and is related to other abdominal organs. The diaphragmatic surface is divided into superior, anterior, posterior, and right parts. These are not easily distinguished from one another; combined, they form the dome of the liver (Fig. 5-29).

The **diaphragmatic surface** (Figs. 5-29 and 5-30) exhibits an area of the liver devoid of peritoneum, the **bare area.** This area is surrounded by the **anterior** and **posterior coronary ligaments.** The **inferior vena cava** is located in this bare area and is almost completely surrounded by liver tissue.

The concave **visceral surface** (Fig. 5-31) is more complex. The **right** and **left lobes** are clearly visible, separated by the **caudate lobe** posteriorly and the **quadrate** anteriorly. This surface exhibits **impressions** for other viscera, the **porta** of the liver, where structures enter and leave the organ, and the **gall bladder.**

The **impressions** on the inferior surface of the liver are as shown in Figure 5-31. They are of no functional significance as far as the liver is concerned, but they do give one a very good visual picture of the superior part of the abdominal cavity. The left lobe has a large **gastric area,** a smaller **esophageal area,** and a still smaller **pyloric area;** the right lobe has an area for the **kidney,** for the **duodenum,** and for the **colon.** The quadrate lobe also is in contact with the transverse colon, and this lobe has an area for the pylorus of the stomach.

The point of entrance of the vessels into the liver and the emergence of the bile duct from the liver is the **porta hepatis.** This is a wide cleft about 2 inches long; it is centrally located between the caudate, quadrate, right, and left lobes. It transmits the branches of the portal vein, hepatic arteries, hepatic ducts, nerves, and lymph vessels.*

---

*J. E. Healey and P. C. Schroy, 1953, Anatomy of the biliary ducts within the human liver. *A.M.A. Arch. Surg.* 66: 599. The liver has been found to be made up of distinct segments and the bile ducts, hepatic arteries, and portal veins branch accordingly. The hepatic veins do not follow this pattern.

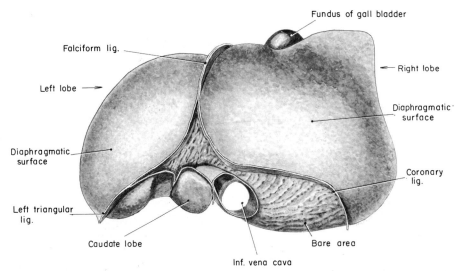

Figure 5-29. Diaphragmatic surface of the liver—superior view.

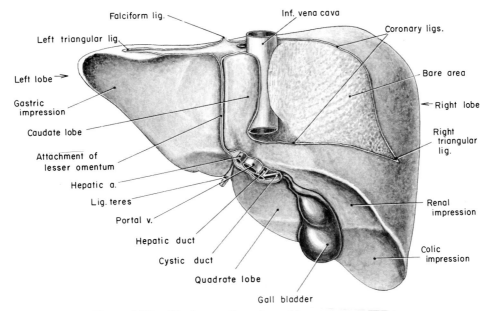

Figure 5-30. Diaphragmatic surface of liver—posterior view.

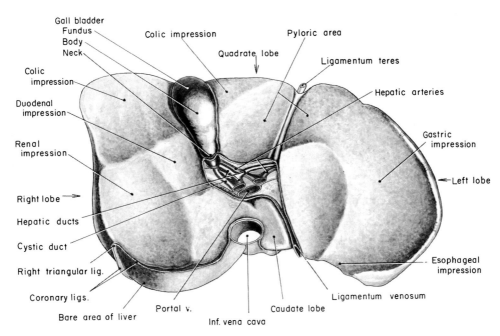

Figure 5-31. Visceral surface of the liver—inferior view.

The **gall bladder** is suspended from the inferior surface of the liver between the right and quadrate lobes. Its duct, the **cystic,** can be seen joining the **hepatic duct** to form the **bile duct.** The gall bladder is described in detail on page 249.

The liver is a heavy organ that almost never leaves its location in the upper right quadrant of the abdominal cavity. Some mechanism must exist for its suspension. We have already discussed the anterior and posterior coronary, and the right and left triangular ligaments. These reflections of the peritoneum from the diaphragm to the liver do play a role. The liver is intimately attached to the diaphragm at its bare area, and its firm attachment to the inferior vena cava probably plays a major role in holding the liver in place. It is also thought that the concave diaphragm and the dome-shaped liver fit together so intimately that a natural suction is produced.

The **ligamentum teres (hepatis),** in the free edge of the falciform ligament, is the obliterated umbilical vein and joins the inferior surface of the liver between the quadrate and left lobes. This obliterated vein before birth brought oxygenated blood from the placenta to the liver. The blood did not enter the liver itself, because of the presence of the ductus venosus, via which it emptied into the inferior vena cava. This ductus venosus closes after birth to form the **ligamentum venosum.** These two ligaments can be seen to join the left portal vein, and the continuity between the ligamentum teres and the ligamentum venosum via this vein can be followed by careful dissection (Fig. 5-32).

The liver originates by a diverticulum from the gut. In brief, this hepatic diverticulum evaginates from the foregut and grows into the ventral mesentery, and through it to the septum transversum (see Fig. 1-40 on page 48). This primordium ultimately forms the bile ducts, gall bladder, and the duct system and parenchyma of the liver itself. The blood sinuses arise by the liver invading the umbilical and vitelline veins. The liver in embryos is a comparatively large organ (due partly to its role in hemopoiesis) and occupies a large portion of the abdominal cavity. This organ later decreases in relative size, the left lobe actually undergoing some degeneration, and ultimately occupies the upper right quadrant only.

The liver is subject to many inflammatory and degenerative diseases. Because of the importance of this organ, these are very troublesome to the patient. If a diseased condition in any way interferes with circulation

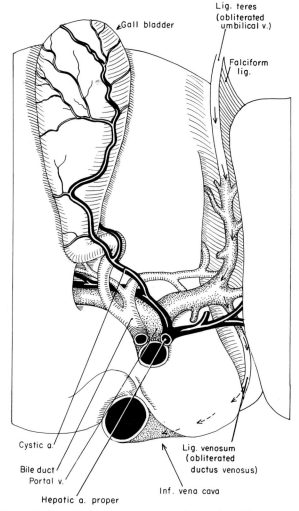

Figure 5-32. *The ligamentum teres (hepatis) and ligamentum venosum on the liver. Arrows indicate the course of the placental blood in the fetus when the ligaments were the umbilical vein and ductus venosus. Note how this route to the inferior vena cava bypassed the liver.*

to the liver, pressure may increase in the portal circulation, which results in dilations of the veins at the points where the portal and systemic venous circulations join; esophageal varices and hemorrhoids may result.

Details on blood vessels, nerves, and lymphatics will be found starting on page 253.

GALL BLADDER

The gall bladder is a pear-shaped organ attached to the inferior surface of the liver (Fig. 5-31). The **fundus** of the gall bladder is in contact with the anterior abdominal wall just inferior to the costal margin. The anterior surface of the **body** of the gall bladder is adherent to the liver by connective tissue. Its posterior surface is covered with peritoneum and is in contact with the very first part of the transverse colon and slightly with the duodenum. The body narrows down into a **neck,** which continues as the **cystic duct.** The cystic duct is about 1 inch long and twisted upon itself. It is not a continuous tube in the ordinary sense, for it possesses valvelike elevations that project into its lumen (Fig. 5-33). It finally joins with the **common hepatic duct** to form the **bile duct** in the free edge of the lesser omentum. The bile duct proceeds inferiorly in a position anterolateral to the portal vein, continues posterior to the first part of the duodenum and the head of the pancreas, and pierces the musculature of the duodenum to empty into its lumen via the **duodenal papilla** (Fig. 5-35). Just before the entrance into the duodenum, the duct is dilated, forming the hepatopancreatic ampulla (of Vater), and the smooth muscle is arranged in such a manner as to form a hepatopancreatic sphincter (of Oddi).

This same point of attachment of the bile duct to the duodenum represents the location of the original hepatic diverticulum just described with the liver. By absorption of gut wall into this diverticulum, the duct of the ventral pancreas becomes an intimate part of it. Its position on the left-posterior side of the duodenum is due to the unequal growth occurring in the duodenum. The gall bladder develops as a secondary outgrowth from this primary diverticulum.

These very vital structures—the bile duct and hepatic ducts—are quite essential to life. The gall bladder, however, is not. The gall bladder is a storage place for bile, which becomes concentrated here on account of water loss. If it is removed because of inflammation and subsequent gallstone formation, bile is sent directly from the liver to the duodenum.

Details on blood vessels, nerves, and lymphatics will be found on page 253.

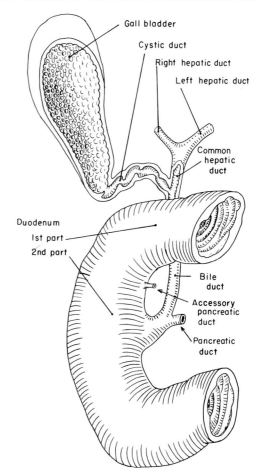

Figure 5-33. *The biliary system. Note the spiral nature of the cystic duct.*

DUODENUM

The **duodenum** (Figs. 5-8, 5-24, 5-33, and 5-35), because of rotation of the gut, became a retroperitoneal organ, thereby losing its mesentery. This arrangement resulted from the duodenum being forced against the posterior abdominal wall, causing that peritoneum posterior to the duodenum (both that on the gut and that on the wall) to disintegrate into connective tissue.

The duodenum is approximately 12 inches in length and turns in the form of a letter C, with the concavity towards the left; the head of the pancreas is in this curve.

The **first part of the duodenum** (pars superior) is about 2 inches long, quite movable, and is attached to the pyloric end of the stomach. It is usually bisected by the transpyloric plane and is opposite the first lumbar vertebra approximately an inch to the right of the midline. It curves slightly posteriorly and to the right and crosses anterior to the portal vein and inferior vena cava. Radiologists tend to use this part of the duodenum as a landmark, calling it the "duodenal bulb." The **second part of the duodenum** (pars descendens) is about 3 inches long. It is anterior to the important renal vessels and the medial margin of the kidney. It ends on the right psoas muscle opposite the third lumbar vertebra. The duodenum then turns to the left, forming the **third part** (pars horizontalis or inferior), which is about 4 inches long. It continues to the left almost transversely and crosses the inferior vena cava and aorta. The **fourth part** (pars ascendens), about 1 inch long, ascends obliquely along the left side of the aorta and head of the pancreas to end on the left psoas opposite the second lumbar vertebra. It then bends abruptly inferiorly to join the jejunum.

Relations. As you have not studied the structures found on the posterior abdominal wall, learning the posterior relations of the duodenum at this time would be a memory feat. However, for the sake of completeness, the posterior relations will be given at this time. Constant reference to Figure 5-38, where the outline of the duodenum is given against the structures found on the posterior abdominal wall, should be of great aid. Because the duodenum varies in its position, its relations also vary; in most cases they are as follows.

First part: **Anterior**—quadrate lobe of liver. **Posterior**—pancreas, portal vein, gastroduodenal artery, bile duct, and, posterior to these structures, the very important inferior vena cava. **Superior**—opening into the lesser sac, which separates the duodenum from the caudate lobe of the liver. The three structures in the lesser omentum (hepatic artery, portal vein, and bile duct) also take a superior position to the duodenum. **Inferior**—head of the pancreas.

Second part: **Anterior**—gall bladder, transverse colon, coils of jejunum. **Posterior**—right renal vessels, right ureter, right kidney, and right psoas major muscle. **Lateral**—right colic flexure and liver. **Medial**—head of the pancreas, bile duct, and pancreatic duct.

Third part: **Anterior**—superior mesenteric vessels, root of mesentery, and coils of jejunum. **Posterior**—right psoas muscle, right ureter, right testicular or ovarian vessels, inferior vena cava, and aorta. **Superior**—head of the pancreas. **Inferior**—coils of jejunum.

Fourth part: **Anterior**—jejunum, and transverse colon. **Posterior**—left testicular or ovarian artery, sympathetic trunk, and left psoas major muscle. **Left side**—jejunum. **Right side**—head of the pancreas.

The duodenum is a muscular tube with an outer longitudinal smooth muscle layer and an inner circular layer. It is lined with mucous membrane the surface area of which is greatly increased by the presence of circular folds that extend into the lumen, and many finger-like projections (villi) on the surface of these folds. Duodenal glands are found in the submucosa, and these give rise to intestinal juice used in digestion. The descending portion of the duodenum exhibits a small elevation or papilla on the medial side approximately 7 to 10 cm. below the pylorus; this major papilla contains the opening of the combined pancreatic and bile ducts (Fig. 5-35). If an accessory pancreatic duct is present, its opening into the duodenum is found approximately 2 cm. superior to that just described, in the middle of a minor papilla. The duodenum, therefore, contains pancreatic juice and bile in addition to its own intestinal juice. All these substances are involved in digestion of food material.

The duodenum, being a retroperitoneal organ (except for its first part), is held in place by the peritoneum. In addition, the fourth part of the duodenum is suspended by a band composed of fibrous tissue and smooth muscle, called the **subtentaculum duodenum (ligament of Treitz)** (Fig. 5-34). This band commences in the connective tissue around the celiac artery and on the right crus of the diaphragm; more specifically, from that part that passes to the left of the esophagus. It descends to the fourth part of the duodenum and serves to suspend this structure. Its smooth muscle component has no known function.*

The duodenum is a common site of ulcer formation. Although most respond to medical treatment, surgery, if necessary, is performed. In these cases the cut end of the duodenum is closed, and the jejunum is connected to the stomach (gastrojejunostomy). This allows the acid food to bypass the remaining duodenum but does not disturb the

---

*I. Jit, 1952, The development and structure of the suspensory muscle of the duodenum, *Anat. Record* 113: 395. This author claims that this muscle is, in reality, two muscles, one attached to diaphragm and another to duodenum; they differ histologically and developmentally.

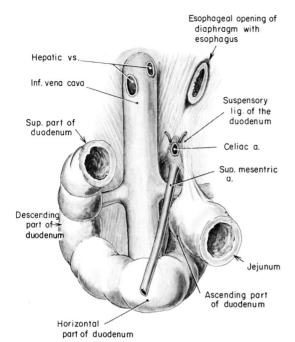

Figure 5-34. *The duodenum and the suspensory ligament of the duodenum. (Peritoneum has been removed.)*

bile and pancreatic ducts. If carcinoma requires the removal of the entire duodenum, the bile and pancreatic ducts as well as the stomach are connected to the jejunum.

Details on blood vessels, nerves, and lymphatics will be found on page 253.

PANCREAS
(Figs. 5-24, 5-28, 5-35, 5-36, 5-37)

The pancreas is divided into a **head,** which extends to the right and lies in the curve made by the duodenum; a **body,** which crosses the midline; and a **tail,** which is in contact with the spleen. It is at the level of the first and second lumbar vertebrae and is bisected by the transpyloric line. A small process of the pancreas is found posterior to the superior mesenteric artery; this is the **uncinate process.**

The pancreas is both an **exocrine** and an **endocrine** organ. Its endocrine portion consists of the pancreatic islets (of Langerhans) that secrete insulin into the blood stream, thus exerting a profound influence on carbohydrate and fat metabolism. The exocrine portion consists of many cells that empty digestive enzymes into the pancreatic duct. This **pancreatic duct** courses the entire length of the pancreas and finally empties into the duodenum after joining the bile duct (Fig. 5-35). There is an **accessory pancreatic duct** as well, which often opens into the duodenum via a separate opening. A glance at Figure 5-37 will show how this accessory duct happened to arise. Since the definitive pancreas is a combination of an embryonic dorsal and ventral pancreas, it is easy to understand why two ducts empty into the duodenum. Actually, the main duct is partly the duct of the dorsal pancreas and partially that contributed by the ventral pancreas. This twisting of the ventral pancreas and its opening into the duodenum on the dorsal side can be explained only on a differential growth basis. By the rotation of the stomach and intestines in their development, this combined pancreas becomes plastered against the posterior abdominal wall; the peritoneum on the deep side disintegrates leaving peritoneum on the anterior side only. Thus, the pancreas becomes a retroperitoneal organ. The body of the pancreas, considerably narrower than the head, is triangular in cross section; thus it has **anterior, posterior,** and **inferior surfaces,** and **superior, inferior,** and **anterior borders.**

Relations. Once again, learning the posterior relations of the pancreas at this time, without a knowledge of the posterior abdominal wall, is difficult. However, constant reference to Figure 5-38 should make up for this deficiency.

Anterior. Starting at the head of the pancreas, the structures in relation to the anterior surface of the pancreas are transverse colon, pyloric-duodenal junction, stomach, and transverse mesocolon. In addition, the gastroduodenal artery and its branches are also anterior to the head of the pancreas.

Posterior. The pancreas is related posteriorly to very vital structures. Starting once again at the head end and going toward the tail, posterior structures are the inferior vena cava, right renal vessels, left renal vein, and a bit of the aorta. The pancreaticoduodenal veins also are posterior to the head. The portal vein, aorta, superior mesenteric artery, the left crus of the diaphragm, the left

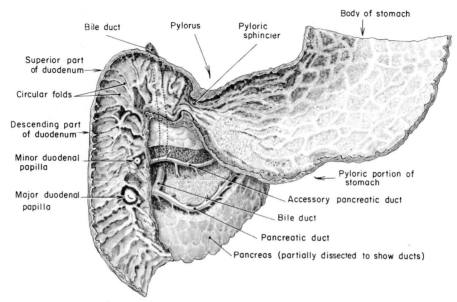

Figure 5-35. Inside surface of the stomach and duodenum, showing the bile duct and pancreatic ducts.

sympathetic trunk and psoas major muscle, left kidney, splenic vein, inferior mesenteric vein, left renal vessels, lower part of the celiac plexus and left suprarenal gland, the left suprarenal vein, and the left inferior suprarenal artery are all posterior to the body. The spleen is posterior to the tail of the pancreas.

Superior. The duodenum is superior to the head (but often is anterior instead) while the hepatic and splenic vessels are superior to the body.

Inferior. The duodenum is inferior to the head and first part of the body of the pancreas, the jejunum inferior to the end of the body and the tail of the pancreas.

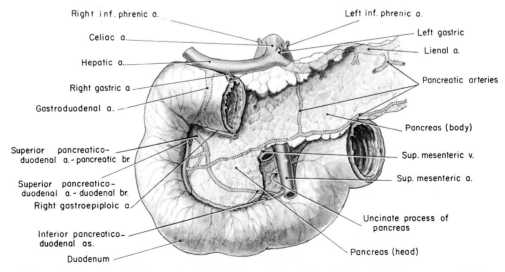

Figure 5-36. The duodenum and its blood supply, including the head of the pancreas. The arteries in this region are quite variable. (Peritoneum has been removed.)

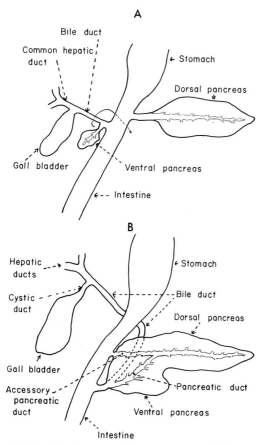

A

Bile duct

Common hepatic duct

← Stomach

Dorsal pancreas

Gall bladder

Ventral pancreas

←-- Intestine

B

Hepatic ducts

← Stomach

Cystic duct

Bile duct

Dorsal pancreas

Gall bladder

Accessory pancreatic duct

Pancreatic duct

Ventral pancreas

Intestine

Figure 5-37. *Development of the pancreas and biliary system. The ventral pancreas and the ductus choledochus (bile duct), because of unequal growth, swing around posterior to the duodenum and join the dorsal pancreas. The duct of the ventral pancreas becomes the main pancreatic duct and the opening of the dorsal pancreas an accessory duct. Several variations of this arrangement may develop.*

Right. The second part of the duodenum is to the right of the head of the pancreas.

Left. The spleen is to the left of the tail of the pancreas.

The pancreas is subject to inflammatory diseases (pancreatitis), endocrine disorders (diabetes mellitus), and cancer of either the islets or the exocrine elements. Removal of the pancreas is a difficult feat because of its many vital relations.

Details of blood vessels, nerves, and lymphatics follow immediately.

ARTERIES

The **celiac axis** arises from the anterior surface of the aorta almost immediately after the latter pierces the diaphragm. It provides blood supply to the embryological foregut and its derivatives. The celiac, at most 1 inch in length, remains retroperitoneal and divides into three main branches—the **common hepatic, left gastric,** and **splenic.** These arteries are shown with the peritoneum in position in Figure 5-8 and with the peritoneum removed in Figures 5-26, 5-28, and 5-36. They are all in the subserous fascia. Although these arteries and their branches are subject to considerable variation,* the majority of cases exhibit the following arrangement:

1. Left gastric
    a. Esophageal branches
2. Common hepatic
    a. Right gastric
    b. Gastroduodenal
        i. Right gastroepiploic
        ii. Superior pancreaticoduodenal
    c. Proper hepatic
        i. Right hepatic and cystic branch
        ii. Left hepatic
3. Splenic
    a. Pancreatic
    b. Short gastrics
    c. Left gastroepiploic.

1. The **left gastric** (Fig. 5-39) courses superiorly on the diaphragm until it reaches the esophagus as the latter organ is piercing the diaphragm. After giving branches to the inferior end of the esophagus, this artery follows along the lesser curvature of the stomach to provide blood to both anterior and posterior surfaces of that organ.

2. The **common hepatic artery** (Fig. 5-39) courses to the right and slightly inferiorly (after giving a branch to the pancreas) to reach the first part of the duodenum. At this point it gives off the **right gastric** and

*N. A. Michels, 1955, *Blood Supply and Anatomy of the Upper Abdominal Organs with a Descriptive Atlas,* Lippincott, Philadelphia. Complete description of the many variations in pattern of arteries in the abdominal cavity.

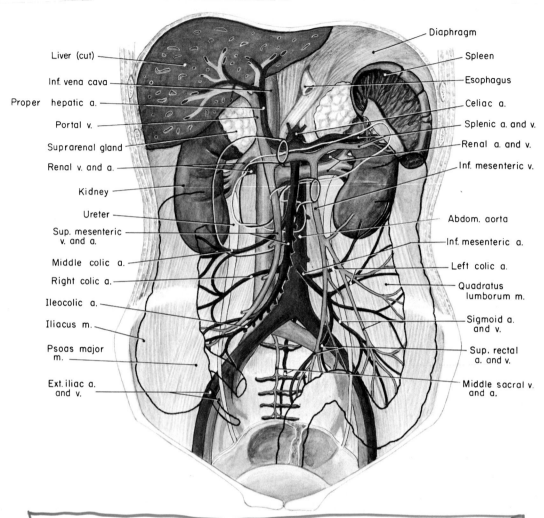

Figure 5-38. *Structures found on the posterior abdominal wall after removal of the gastrointestinal tract and peritoneum. The pancreas, duodenum, and ascending and descending colons have been outlined; the vessels of the colon have been left intact to show their courses across the abdominal wall.*

the **gastroduodenal** arteries. It then swings anteriorly and turns superiorly to enter the free edge of the lesser omentum as the **proper hepatic** artery.

    a. The **right gastric** artery courses on the lesser curvature of the stomach, supplying blood to both anterior and posterior surfaces, and then anastomoses with the left gastric artery.

    b. Gastroduodenal artery (Fig. 5-39). This vessel courses inferiorly in a position posterior to the duodenum

and usually divides into the right gastroepiploic and superior pancreaticoduodenal arteries. (i) The **right gastroepiploic** (Figs. 5-26 and 5-39) artery joins the greater curvature of the stomach and proceeds to the left between the two layers of the greater omentum attached to this greater curvature. It sends many vessels into the greater omentum as well as to both surfaces of the stomach, and finally anastomoses with the left gastroepiploic. (ii) The **superior pancreaticoduodenal** (Fig.

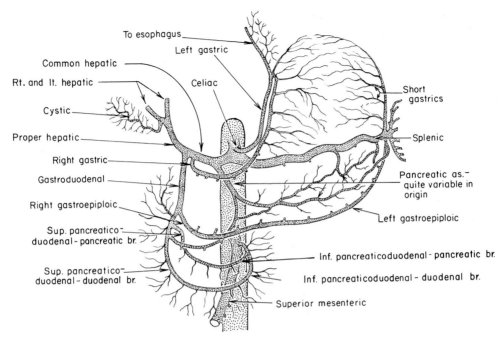

Figure 5-39. *Diagrammatic presentation of the commonest arrangement of the celiac artery and its branches.*

5-39) almost immediately divides into duodenal and pancreatic branches. The **duodenal** branch courses on the anterior surface of the pancreas, between it and the duodenum, supplying blood to both organs. It continues inferiorly along the inside curve of the duodenum and anastomoses with a similar branch of the inferior pancreaticoduodenal artery. The **pancreatic** branch of the superior pancreaticoduodenal artery courses on the posterior surface of the head of the pancreas, giving blood supply to it and also to the duodenum, and then anastomoses with the inferior pancreaticoduodenal artery.

c. **Proper hepatic** (Fig. 5-39). This is actually a continuation of the common hepatic artery. It continues superiorly in the free edge of the lesser omentum in an anteromedial position to the portal vein and bile duct and finally divides into right and left hepatic branches, which enter the liver. (i) The **right hepatic** supplies the right lobe of the liver. This branch quite frequently is replaced by a branch from the superior mesenteric artery which gains this position by coursing posterior to the portal vein. The **cystic** artery, the blood supply to the gall bladder, in approximately 75 per cent of the cases arises from this right hepatic artery and proceeds to the gall bladder, where it divides to course on both superficial and deep surfaces. The cystic artery, however, is quite inconstant in its origin and frequently arises from the right hepatic at a point to the left of the hepatic duct. In this situation it may course either anterior or posterior to the hepatic bile duct. Occasionally there are several branches rather than one, and they may arise from either hepatic artery; sometimes the cystic arises from the proper hepatic artery before the latter has divided into its right and left components, as shown in Figure 5-39. These variations are of great importance to the surgeon, for gall bladder surgery is done frequently. (ii) The **left hepatic** artery gives off a branch to the caudate lobe and frequently one to the quadrate lobe before entering the left lobe.

Before leaving the proper hepatic artery, it should be mentioned that this artery frequently gives off other branches to the duodenum. One, the **retroduodenal artery,** follows the path of the bile duct; this can be a nuisance to the unwary surgeon.

---

3. The third branch of the celiac artery is the **splenic** (Figs. 5-26 and 5-39), which courses to the left until the left kidney is reached, then turns and enters the lienorenal ligament to reach the hilum of the spleen. This vessel gives off (a) many **branches to the pancreas** as it courses to the left, since it is in close relation with the superior border of this organ. Just before entering the spleen it gives off (b) five or six **short gastric vessels** to the fundus of the stomach, and then (c) the **left gastroepiploic artery** to the greater curvature of the stomach. The latter artery anastomoses with the right gastroepiploic from the gastroduodenal artery.

The celiac artery and its branches are retroperitoneal and can be seen to form ridges on the posterior wall of the lesser sac. They continue in a retroperitoneal position until an organ or ligament is reached, also retroperitoneal, upon which they may course to gain a position between two layers of peritoneum and thence to their destination. Originally the celiac artery and its branches were located in the mesenteries and were covered on both sides by peritoneum. The ultimate position of these vessels is explained by the rotation of the stomach and intestines.

VEINS—THE HEPATIC PORTAL SYSTEM

The vessels involved in venous drainage of the organs supplied by the celiac artery and its branches are all parts of the hepatic portal system of veins (Figs. 5-23, 5-32, and 5-38). This system consists of the following veins:*

   A. Superior mesenteric
     1. Ileocolic
       a. Appendicular
     2. Right colic
     3. Middle colic
     4. Jejunal and ileal
     5. Right gastroepiploic
     6. Pancreatic
     7. Pancreaticoduodenals

---

*Although the superior and inferior mesenteric veins were studied in the section on intestines, they have been included again to present the hepatic portal system of veins in its entirety.

   B. Splenic
     1. Short gastrics
     2. Left gastroepiploic
     3. Pancreatic
     4. Inferior mesenteric
       a. Superior rectal
       b. Sigmoidal
       c. Left colic
   C. Portal
     1. Right gastric
     2. Left gastric
     3. Prepyloric
   D. Right branch of portal
     1. Cystic
   E. Left branch of portal
     1. Remains of ductus venosus
     2. Remains of umbilical vein
     3. Paraumbilical

A. The **superior mesenteric** vein drains the appendix, cecum, and ascending colon by its **ileocolic** and **right colic** branches, and most of the transverse colon by way of its **middle colic** branch. Its many **ileal** and **jejunal** branches drain those two parts of the small intestine. In addition the **right gastroepiploic, pancreaticoduodenal,** and **pancreatic** branches drain into this vein just before it joins with the splenic to form the portal. The superior mesenteric vein courses at the base of the mesentery with the superior mesenteric artery. It joins the splenic at a point deep to the pancreas.

B. One of the most important tributaries to the **splenic** vein is the **inferior mesenteric.** This vein, via its **superior rectal, sigmoidal,** and **left colic** branches, drain the appropriate sections of the colon from the area of the left colic flexure to the superior part of the rectum. The superior rectal vein forms the important connection with the systemic veins in the walls of the rectum and anal canal. Several veins from the spleen join to form the **splenic** vein. **Short gastric** veins from the fundic portion of the stomach, the **left gastroepiploic,** and several **pancreatic** veins also drain into this vein. The splenic vein (Figs. 5-23 and 5-38) courses to the right making a groove on the posterior-superior surface of the pancreas.

C. As previously mentioned, the splenic vein joins with the superior mesenteric to form the **portal** (Fig. 5-38). The portal vein has other tributaries besides these two veins that form it. They are (a) the **right gastric,** which courses on the lesser curvature in company with the right gastric artery (it courses to the right and drains into the portal near its origin); (b) the **left gastric,** coursing on the lesser curvature and accompanying the artery of the same name to the location of the celiac artery, where it picks up the hepatic artery and follows it in a position posterior to the lesser sac to reach the portal vein (this vein has the important esophageal branches that connect the portal system with the systemic); and (c) the **prepyloric vein,** which is a small vessel coursing on the anterior surface of the pylorus (it often connects the right gastric with the right gastroepiploic rather than emptying into the portal).

D and E. The **portal** vein, as might be expected, divides into right and left branches before entering the liver. The **cystic** vein from the gall bladder usually drains into the right, but the left has more interesting tributaries (see Fig. 5-32 on page 248). One is the remains of the **umbilical** vein (now ligamentum teres hepatis) and another is the remains of the **ductus venosus**, the ligamentum venosus. Most important are small veins coursing along the ligamentum teres and between the folds of the falciform ligament. These **paraumbilical** veins are so named because they connect with veins that completely surround the umbilicus. Any increase in pressure in the portal circulation may produce dilation of these veins— the **caput medusae.**

In summary, the **hepatic portal system** of veins drains the stomach, the intestines as far as the rectum, the pancreas, the spleen, and the gall bladder and carries this blood to the liver where absorbed nutrients are attacked by the liver. The attachments of this system to the systemic venous system are important clinically. Increased pressure in the portal system for any reason may lead to **esophageal varices,** which may hemorrhage; and **hemorrhoids** in the walls of the rectum and anal canal which may require surgical removal. A **caput medusae** may reveal this pressure. Another connection of the two systems can occur on the posterior abdominal wall due to small connections developing between the retroperitoneal colic vessels and those draining the body wall. Surgical connections between the inferior vena cava and the portal vein (portal-caval shunt) or between the left renal vein and the splenic vein (renolienal shunt) can be performed to relieve this condition.

NERVES
(Fig. 1-22 on page 27)

The **sympathetic** system supplies its fibers to all organs that receive blood supply from the celiac axis via the celiac ganglia. Preganglionic fibers start in the fifth to the ninth segments of the spinal cord, emerge via the ventral roots and proceed to the sympathetic chain ganglia via the white rami communicantes. The majority of fibers do not synapse in these chain ganglia, however, but continue via the greater splanchnic nerves to the celiac ganglia, where they synapse with postganglionic fibers; the latter course to the organs via the respective arteries. The **parasympathetic** system to these organs is derived entirely from the **vagus** nerves. The **left vagus nerve** was last described on the anterior surface of the esophagus; this nerve continues onto the lesser curvature and anterior surface of the stomach as the **anterior gastric nerve.** It enters the free edge of the lesser omentum and gives parasympathetic supply to the liver, gall bladder, and also a few fibers to the duodenum. The **right vagus,** on the posterior surface of the esophagus, not only sends branches to the posterior surface of the stomach— **posterior gastric nerve**—but also a large contribution travels inferiorly to join the celiac ganglia. No synapse takes place at these ganglia, however, for these fibers continue through the ganglia and reach the organs to be innervated via the blood vessels. The synapse in the parasympathetic system occurs in intrinsic ganglia in the walls of the organs to be innervated, and therefore the postganglionic parasympathetic fibers are extremely short.

The function of the nervous system on these viscera is complex and not completely understood. It is complicated by considerable cortical influence over the autonomic nervous system; this is especially true in the case of the stomach. Furthermore, there is a hormonal control over the function of these organs. It must be remembered that the smooth muscle layers in the stomach and duodenum have an inherent ability to contract. The extrinsic fibers alter this ability in one direction or the other. Vagal stimulation (parasympathetic) causes an increase in peristalsis and a vasodilation. In addition, this system will cause an increase in gastric secretion. However, this last function is altered greatly by cortical function. The traditional experiments by Pavlov (1878)* demonstrated that secretion could be conditioned to any stimulus (conditional reflex). Even before this revelation, Beaumont

---

*I. P. Pavlov, 1878, Weitere Beiträge zur Physiologie der Bauchspeicheldrüse, *Pflügers Arch. Ges. Physiol.* 17: 555. The well-known work where the influence of the cerebral cortex over motor activity of the digestive tract was demonstrated—the conditioned reflex.

(1833)* demonstrated the remarkable influence of the emotional condition of the patient on gastric secretion. In fact, ulcer is a condition associated with emotional stress, and the treatment of ulcer demands attention to the whole patient in addition to treatment of the lesion. The parasympathetic system has a similar role on the duodenum. It causes an increase in motility, a vasodilation, and an increase in secretion of the duodenal glands.

The sympathetic system to these organs has an opposite effect, namely, decreased motility, vasoconstriction, and a decrease in secretion.

Sensory fibers to these organs are contained in the greater splanchnic nerves, enter the spinal cord, and proceed to higher centers for the sensation of pain. It should be noted that pain in the gastrointestinal tract is associated with distention or violent contraction. As noted previously, one of the complicating factors of treatment of carcinoma of the stomach is the relatively slow appreciation on the part of the patient of such a condition. Pain from the stomach and duodenum is referred to the epigastric region of the abdominal wall; this might be expected since the fibers of the greater splanchnic nerve innervating the stomach and duodenum arise in segments T7 to T9, and the epigastric region is innervated by intercostal nerves T7 to T9. The vagus nerve also has sensory fibers, but these are primarily associated with reflex control of the organ's function. Nausea is probably carried in the vagus nerve but such sensations can be stimulated by disagreeable odors, motion, etc., which complicate the picture.

Vomiting is a complex function. A vomiting center is located in the medulla oblongata, which is important in this act; when it is destroyed, vomiting cannot occur. This function can be initiated by conditions in the stomach, but also by visual, olfactory, or even psychic stimuli.

Hunger is another phenomenon of a complicated nature. It can be initiated by contractions of the stomach, but there seems to be a hunger center in the medulla oblongata that can be stimulated in other ways. The vagus nerve is the pathway taken to reach this center.

The **pyloric sphincter** is under dual control, the

---

*W. Beaumont, 1833, *Experiments and Observations on Gastric Juice and the Physiology of Digestion,* Harvard University Press, Cambridge, 1929. A reprint of the classical experiments on the hunter named St. Martin, who had a fistula in his stomach which allowed Beaumont to observe directly changes in gastric secretion.

sympathetic closing and the parasympathetic enlarging the opening.

The **biliary system** has an inherent ability to contract and is under hormonal control; the extrinsic nerves control the rate of its activities. The parasympathetic system stimulates contraction of the gall bladder and the duct system, and causes the sphincter (of Oddi) to relax. The sympathetic system has an opposite action. Afferent fibers from the gall bladder and ducts are carried in the greater splanchnic nerves to segments T7–T9, and pain is usually referred to the back or epigastric region (innervated by T7–T9). Gall bladder disease may also irritate the peritoneum on the diaphragm and thus be referred to the right shoulder region via the phrenic nerve.

The **liver** has a dual nerve supply, sympathetic from T7 and T8 via the greater splanchnic nerves and celiac ganglia, and parasympathetic via the vagus. The function of these nerves, other than to blood vessels, is questionable. Many feel that the hepatic cells have no innervation. The liver has no ganglion cells within its substance; the parasympathetic ganglia are located at the hilum.

The **spleen** also has a dual innervation, the sympathetics via the greater splanchnics from spinal cord segments T6 to T8, and parasympathetics via the vagus. Visceral afferents return to segments T6 to T8 and can be referred to the epigastric region. If the diaphragm is involved, referral is to the left shoulder via the phrenic nerve.

The **pancreas** also has a dual nerve supply, but the sympathetic portion (T7 to T9) seems of importance only as regards the blood vessels. The parasympathetic vagal fibers reach the parenchyma of the islet cells as well as the exocrine cells, and stimulation of these fibers induces a release of pancreatic juice and insulin. This nerve control is not essential, however, because of hormonal control. Sensory fibers are carried in the greater splanchnics to spinal cord segments T7 to T9, and are important for registering pain in cases of pancreatitis or carcinoma. The nerves for the head and body are carried in the right splanchnics, while those from the tail are carried on the left side. Pain is referred to the epigastric region.

LYMPHATICS

There is no easy solution to a knowledge of the location of the lymph nodes draining the organs supplied by the celiac axis except to say, once again, that the lymphatics from these organs follow the arterial pattern. Lymph nodes draining the stomach are found in every location where arteries are found, particularly on the lesser and greater curvatures. The lymphatics from the liver end in nodes in the lesser omentum, in the posterior abdominal wall just superior to the pancreas, and even in the thoracic cavity both anterior and posterior to the pericardium. The pancreas has lymphatics ending in nodes that are scattered along all the arteries that send blood to this organ, whereas the spleen has nodes in the gastrosplenic ligament and posterior-superior to the pancreas. They all ultimately drain into the thoracic duct.

POSTERIOR ABDOMINAL WALL
(Figs. 5-40 and 5-41)

GENERAL DESCRIPTION

When the peritoneum, spleen, and all portions of the gastrointestinal tract, including the liver and pancreas, are removed, the retroperitoneal structures on the posterior abdominal wall are revealed (Fig. 5-40). They are covered with a layer of subserous fascia, which varies greatly in thickness from one region to another.

In the midline the **large aorta** is visible. Although the **celiac,** the **superior mesenteric,** and possibly the **inferior mesenteric** arteries remain only as trunks, the **inferior phrenic arteries** to the inferior surface of the diaphragm are visible, the **renal arteries** to the kidneys are quite large, the **testicular or ovarian arteries** can be seen emerging from the anterior surface of the aorta inferior to the renals, and the two terminal branches of the aorta, the **common iliacs,** are equally visible. Branches of the abdominal aorta that have not been mentioned are the lumbar arteries and the middle sacral.

Just to the right of the aorta is the much longer **inferior vena cava.** It is formed by the combination of the **common iliac veins,** and it can be seen to penetrate the diaphragm at a much higher level than did the aorta. The inferior vena cava drains the **right testicular or ovarian vein,** the **renal veins,** and **large veins from the liver**

(hepatic veins) as well as the **right phrenic vein** from the diaphragm, and **right lumbar veins** from the body wall.

Although all the above structures are surrounded by subserous fascia, it is so thin that the structures are quite visible. In contrast, the kidneys and suprarenal glands are surrounded by a capsule of subserous fascia and fat which hides them from view. This is known as the perirenal fat of the kidney and suprarenal. A partition actually separates these two organs from each other; if the kidney should descend, the suprarenal need not necessarily follow.

The **ureters** can be seen descending from the fatty capsules of the kidneys, across the posterior abdominal wall, passing anterior to the iliac vessels to enter into the true pelvis.

The **diaphragm** is a more extensive structure than usually thought and it arches over two muscles—the **psoas major** and **quadratus lumborum**—which form the posterior abdominal wall, the latter filling the area between the iliac crest and the last rib. Lateral to these is the **transversus abdominis muscle** of the abdominal wall. The psoas major continues into the iliac fossa, and then deep to the inguinal ligament, where it disappears from view (it actually inserts on the lesser trochanter of the femur). The lateral part of the iliac fossa is filled in by the iliacus muscle which joins with the psoas major muscle to insert on the lesser trochanter of the femur. Therefore, the posterior body wall is made up superiorly by the diaphragm; centrally by a layer from medial to lateral of the crus of the diaphragm, psoas major, quadratus lumborum, and abdominal musculature; and inferiorly by the psoas major, and iliacus muscles.

The other obvious structures are the branches of the **lumbar plexus,** which can be seen emerging just lateral to the psoas muscle except for one, the genitofemoral, which pierces the muscle and emerges from its anterior surface.

All these structures are shown in Figures 5-40 and 5-41.

SUPRARENAL OR ADRENAL GLANDS

The **suprarenal glands** are a pair of rather small yellowish-brown bodies lying on the medial part of the superior end of each kidney (Fig. 5-40). They are also in contact with the adjoining crus of the diaphragm and are opposite the posterior end of the eleventh intercostal

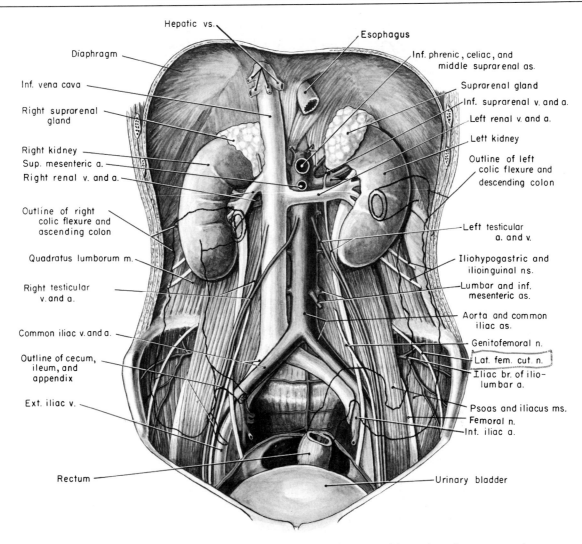

Figure 5-40. *The posterior abdominal wall I. All peritoneum and fascia have been removed. Note outlines of cecum, and ascending, descending, and sigmoid colons.*

space and twelfth rib. The right suprarenal gland is posterior to the liver and partially posterior to the inferior vena cava. It is triangular in shape (Fig. 5-42). The left gland is posterior to the stomach and is semilunar in shape; it actually forms part of the bed of the stomach, being separated from it by the peritoneum and the lesser sac.

The suprarenal glands are divided into two distinct parts—an outer **cortex** and an inner **medulla**. The **adrenal medulla** resembles the sympathetic nervous system in function. It is this structure that secretes epineph-

rine and norepinephrine into the blood stream in times of stress. The **adrenal cortex** is essential to life. It has many functions, the most important of which are a maintenance of electrolyte balance and a control over carbohydrate metabolism. The adrenal cortex is under partial control of the anterior lobe of the hypophysis; after removal of the hypophysis, the part of the adrenal controlling electrolyte balance is not upset, while that controlling carbohydrate metabolism is definitely affected. The hormones from the adrenal cortex affecting electrolyte balance are called

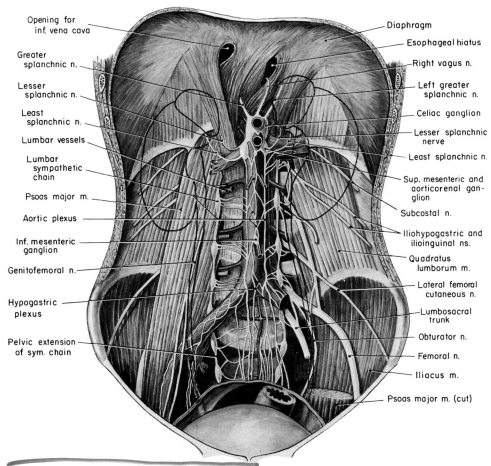

Opening for
inf. vena cava

Greater
splanchnic n.

Lesser
splanchnic n.

Least
splanchnic n.

Lumbar vessels

Lumbar
sympathetic
chain

Psoas major m.

Aortic plexus

Inf. mesenteric
ganglion

Genitofemoral n.

Hypogastric
plexus

Pelvic extension
of sym. chain

Diaphragm

Esophageal hiatus

Right vagus n.

Left greater
splanchnic n.

Celiac ganglion

Lesser splanchnic
nerve

Least splanchnic n.

Sup. mesenteric and
aorticorenal gan-
glion

Subcostal n.

Iliohypogastric and
ilioinguinal ns.

Quadratus
lumborum m.

Lateral femoral
cutaneous n.

Lumbosacral
trunk

Obturator n.

Femoral n.

Iliacus m.

Psoas major m. (cut)

Figure 5-41.   The posterior abdominal wall II. *The peritoneum, all fascia, the inferior vena cava, and the psoas major muscle (left side of body) have been removed. Note how this picture shows the posterior relations of the kidneys, ureters, and psoas major muscle.*

mineralocorticoids; those concerned with carbohydrate metabolism are glucocorticoids.

The suprarenal glands are richly supplied with **blood.** Superiorly there are several branches from the inferior phrenic, a second supply from the aorta, and inferiorly a third from the renal. The single vein drains into the renal vein on the left, and into the inferior vena cava on the right.

The **nerve supply** to the adrenal must be divided into that to the cortex and that to the medulla. The sympathetic nerve supply to the **adrenal cortex** is via the greater and lesser splanchnic nerves (T8–T11). Therefore, the preganglionic fibers start in the eighth to the

eleventh thoracic segments of the spinal cord, enter ventral roots, white rami communicantes to the sympathetic chain, and enter the splanchnic nerves to synapse primarily in the celiac and superior mesenteric ganglia, but some may synapse in the aorticorenal ganglia. Postganglionic fibers follow the above-mentioned arteries to the suprarenal gland. The nerve supply **to the medulla** is different since the adrenal medulla is really modified nervous tissue*; it receives preganglionic fibers

---

*There are certain collections of cells closely related in development to the postganglionic sympathetic neurons. These cells are found in the adrenal medulla, in each sympathetic chain ganglion, associated with

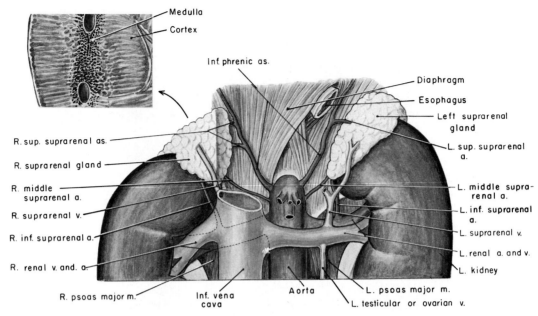

Figure 5-42. The suprarenal (adrenal) glands in natural position, and cut (insert) to show internal structure.

which synapse with cells in the medulla. Thus, the sympathetic innervation of the adrenal medulla is an exception to the general rule that the sympathetic system characteristically synapses in ganglia some distance away from the organ to be innervated. The **parasympathetic system** innervates the adrenal cortex, but does not course to the medulla. Those fibers to the cortex are branches of the vagus nerve. Preganglionic fibers from the brain stem course in the vagus nerve to reach the aorticorenal ganglia. They do not synapse here, but continue to ganglion cells just under the capsule of the gland.

The function of the nerves to the cortex is problematical, for nerves terminating in the parenchyma have not been described; the blood vessels are innervated, and

the sympathetic system causes a vasoconstriction and the parasympathetic a vasodilation.

The **lymphatic drainage** is to nodes along the sides of the glands and those posterior to the pericardium.

As is the case with all endocrine glands, diseases are associated with too little function—hypoadrenocorticalism—or too much secretion—hyperadrenocorticalism. Severe hypoadrenocorticalism is called Addison's disease; hyperadrenocorticalism is often associated with tumors some of which have a virilizing quality. This is very unfortunate in children; it changes secondary sexual characteristics in adult females to those of the male.

KIDNEYS

The **kidneys** are two bean-shaped, rather large organs on either side of the midline in the abdominal cavity (Fig. 5-40). Due to the presence of the liver, the right kidney is lower than that on the left side. The kidneys are smooth organs approximately 4½ inches long, 2½ inches broad, 1½ inches thick. The superior end is nearer the midline than the lower end, and this makes the long axis course slightly laterally from superior to inferior. The anterior

the ovaries or testes, and along the anterior surface of the aorta in association with the sympathetic ganglia and plexuses. This tissue has been called **chromaffin tissue** because of its affinity for chrome salts, or **paraganglia** because of its close association to the sympathetic ganglia. This tissue secretes epinephrine and norepinephrine. Except for the adrenal medulla, degeneration occurs in the first years of life, although tumors of this tissue may develop. Condensations of this tissue anterior to the aorta have been called "aortic bodies," not to be confused with the aortic bodies in the superior mediastinum which serve as afferent centers responding to oxygen tension of the blood.

surface and posterior surfaces are smooth, while the lateral surface is convex and the medial surface is concave. The **hilum** is on the medial side and transmits the renal vessels, nerves, ureter, and lymphatics.

The kidneys are in the upper part of the posterior portion of the abdomen opposite the twelfth thoracic and the first three lumbar vertebrae. The transpyloric line bisects the hilum of the left kidney, but is a little above the hilum on the right kidney. The transpyloric line makes it fairly easy to project the kidneys onto the anterior abdominal wall.

A glance at Figures 5-24, 5-40, and 5-41 will show the important **relations** of the kidneys. On the **right side** the structures found superior to the kidney are the diaphragm, liver, and suprarenal glands; the colon is found inferiorly. Anterior to the right kidney is the liver on the upper part, the colon on the lower part, and a part of the duodenum near the hilum. Posteriorly, the right kidney is in contact with the transversus abdominis, psoas, and quadratus lumborum muscles on the lower part of the organ, and the diaphragm on the upper part. The right kidney is in relation, through the diaphragm, with the eleventh intercostal space and the twelfth rib.

The **left kidney** has similar superior and inferior **relations** to those of the right, but has a large area in contact with the spleen anteriorly and an area near the hilum in relation with the pancreas; there is a small triangular area superior to this that is in contact with the stomach and an area inferior in contact with the jejunum. The anterior relations of the two kidneys are fairly similar if the hepatic and splenic areas are compared and the duodenal area on the right compared with the jejunal and gastric areas on the left. The pancreatic area is the main difference. The posterior relations are also similar, except for the fact that the left kidney is a bit higher and comes into relation with the eleventh rib.

These relations are easily learned, because they are very sensible if the entire region is visualized. Since the surgical approach to the kidney is from its posterior side (thus avoiding the peritoneal cavity), the posterior relations are important to the surgeon.

The **fascial arrangements** around the kidneys demand special attention. Starting on the kidney itself, it has its own **capsule** which, under normal conditions, may be removed easily; under some diseased conditions this capsule adheres to underlying tissue. Just outside the renal capsule is a layer of fat—the **perirenal fat**. This layer is important in maintaining the kidney in its proper position, the kidney being more likely to descend in its absence. Outside this layer of perirenal fat is a special condensation of the subserous fascia called the **renal fascia** (Fig. 5-43). It consists of anterior and posterior layers. The anterior layer blends with the adventitia of the renal vessels and actually can be followed across the midline to the opposite kidney. The posterior layer does not cross the midline but blends with the psoas fascia. Superiorly the two layers of renal fascia fuse with each other and with the diaphragmatic fascia, while laterally a similar fusion between the two layers and the transversalis fascia occurs. In contrast, inferiorly the two layers do not fuse with each other and this accounts for the occasional descent of the kidneys. It should be noted that the suprarenal gland is contained in a special compartment and does not descend with the kidney. All of the above is deep to the peritoneum.

If the kidneys are cut longitudinally through the hilum, it can be seen that there are cortical and medullary areas (Fig. 5-44). The **cortex** contains the **glomeruli,** the tufts of capillaries through which the urine leaves the bloodstream to enter the kidney itself, while the **medulla** shows collections of tubules that form **pyramids;** these pyramids empty into a **minor calyx** via the **renal papillae.** The minor calyces join to form a **major calyx,** which in turn joins with another major calyx to form the **renal pelvis.** The pelvis of the kidney leads directly to the **ureter** itself.

Usually the **renal veins** are anteriorly placed, and the **renal arteries** are posterior, with the ureters posterior to both.* Because the inferior vena cava is on the right side of the abdominal cavity, the renal vein on the right is much shorter than that on the left. As might be expected from the above relations, the left renal vein crosses anterior to the aorta to reach the inferior vena cava, the right renal artery posterior to the inferior vena cava to reach the kidney. The left renal vein drains the left testicular or left ovarian vein, and there is also an important connection between the left renal vein and the azygos system of veins (Fig. 5-48). If the inferior vena cava has to be tied off superior to the point of entrance of the renal veins, these channels just mentioned will frequently serve as a

---

*Accessory renal arteries frequently are found anterior to the renal veins and/or posterior to the ureter. Other branches may arise directly from the aorta.

way of getting blood around the tie. The renal arteries are large structures and the frequency of accessory renal arteries is high because of the segmental supply to the developing kidney. These renal arteries also give a branch to the suprarenal glands.

A rich supply of blood enters the kidney through the glomeruli. Material to be excreted escapes from these capillaries into Bowman's capsule, which surrounds this glomerulus. From here the urine passes into the nephron, and it is in this area that much of the material is returned to the bloodstream. The material being excreted is collected in the collecting tubules and enters the minor calyces, and then the major calyx, to enter into the pelvis of the kidney. This urine is being formed constantly and is forced down the ureter by peristaltic waves. The urine is collected in the urinary bladder and excreted to the outside at convenient intervals via the urethra.

The **sympathetic nervous system** to the kidneys originates as preganglionic fibers in the twelfth thoracic and upper lumbar segments of the cord. Segments as high as T9 may contribute. These fibers then travel to the aorticorenal ganglia, where a synapse occurs, and postganglionic fibers follow the renal arteries to their destination. The kidneys receive **parasympathetic** innervation from the vagal portion of the parasympathetic system, the ganglia being located within the walls of the kidneys themselves. Nerve fibers have been traced directly to all the arterioles of the kidney, including the afferent and efferent arteries of the glomeruli. The sympathetic system causes a vasoconstriction, and the parasympathetic system an opposite effect. The kidney can continue to function without a nerve supply, removal of the sympathetic supply causing a vasodilation and increased blood flow. Sensory fibers from the kidney are mainly from the renal pelvis and accompany the sympathetic nerves; they enter the spinal cord between T12 and L2. Pain is referred to the back near the costovertebral angles. Any sensory fibers contained in the vagus do not reach consciousness; they are for reflex activities.

The **lymphatics** from the kidneys drain into lumbar nodes along the vertebral column and aorta.

The kidneys are subject to several disease conditions

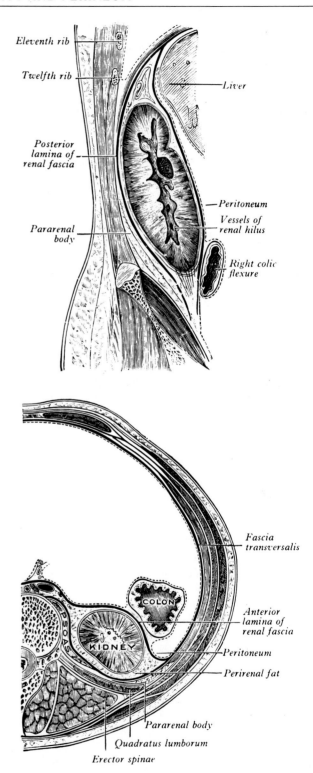

Figure 5-43. *Relations of the renal fascia to the right kidney. (From Gray's* Anatomy, *35th British edition, courtesy of Churchill Livingstone.)*

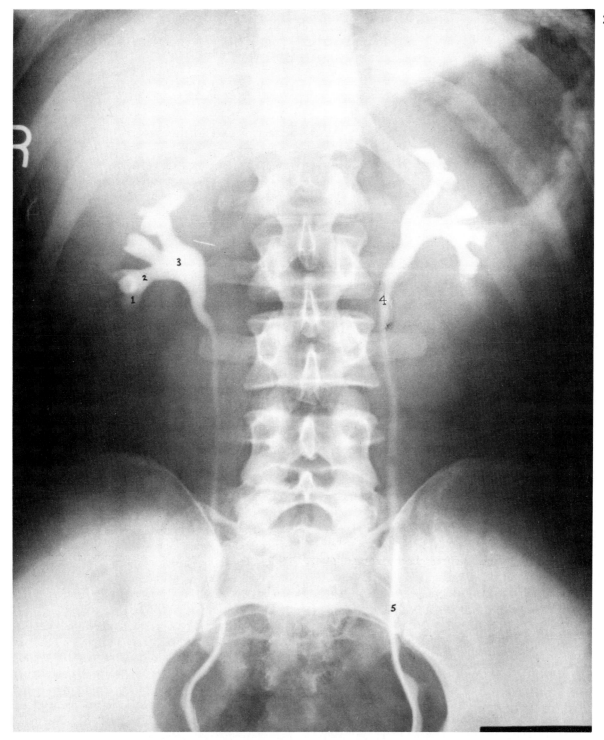

Plate 12.   *Pyelogram of kidney: (1) minor calyx, (2) major calyx, (3) renal pelvis, (4) ureter, and (5) catheter in ureter. Note that left kidney is more superiorly located than right, and observe the course of the ureters.*

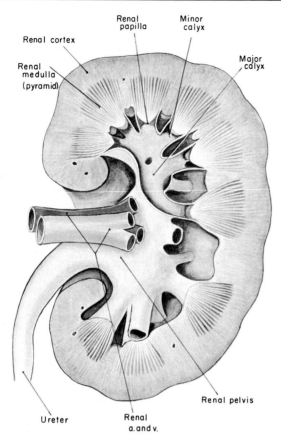

Figure 5-44. *Kidney cut longitudinally to show internal structure.*

but, in addition, may exhibit developmental anomalies. One kidney may be atrophic, and the inferior ends of the kidneys may be fused, forming a horseshoe arrangement. A kidney may even be found in the pelvis, indicative of the organ's original site of development. Accessory kidneys may be present; a double ureter is a frequent finding. Further details on development will be presented after more of the urogenital system has been studied.

URETERS

The **ureters** are two very important tubes extending from the renal pelvises to the urinary bladder (Fig. 5-40). The posterior **relations** of the ureters are fairly similar on both sides. Both ureters start at the renal pelvis, posterior to both the renal veins and arteries. They continue inferiorly, crossing anterior to the psoas major muscles.

They leave these muscles, and cross anterior to the iliac arteries and veins to enter the pelvis. Other relations are not similar on both sides. Anterior to the right ureter is the third part of the duodenum; peritoneum; the right colic, ileocolic, testicular or ovarian vessels; and superior mesenteric vessels in the root of the mesentery. Anterior to the left ureter is the peritoneum, the testicular or ovarian artery, and left colic vessels. The coils of the jejunum and ileum are anterior to both ureters. The very important relations of the ureter in the pelvic cavity will be taken up when this area is studied. These very vital structures must be kept in mind constantly with all abdominal and pelvic surgery.

The ureters receive **blood supply** from the renal artery, from the testicular or ovarian, and from vesical and middle rectal arteries in the pelvis. The **veins** have a similar pattern.

The ureters receive **autonomic innervation** from the sympathetic nervous system, fibers originating in T12 to L2 spinal cord segments. The vagus contributes the parasympathetic fibers to the abdominal portions of the ureters. The ureter can function without a nerve supply; peristalsis will occur. Spasm is induced by stimuli of an unusual nature. Sensory fibers are more important; they return to segments T12 to L2.

The **lymphatic drainage** of the ureters is to nodes nearest them in the abdominal and pelvic cavities.

DIAPHRAGM

The **diaphragm** is an arched musculotendinous partition between the thoracic and abdominal cavities. It domes up far into the thoracic region; it is convex from the thoracic side, concave from the abdominal side. The diaphragm is usually described as if one were looking at it from the abdominal cavity (Fig. 5-45).

The central part is tendinous—the **central tendon**—while the periphery is muscular. The muscular part of the diaphragm inserts on this central tendon. The diaphragm can be divided into **sternal, costal,** and **lumbar portions.** The origin of the sternal part is from the back of the xiphoid process; the costal part from the inner surfaces of the lower six costal cartilages; and the lumbar part from the **lateral** and **medial arcuate ligaments,** which bridge over the quadratus lumborum and psoas major muscles respectively, from the bodies of the upper three lumbar vertebrae by a pair of crura, and also from

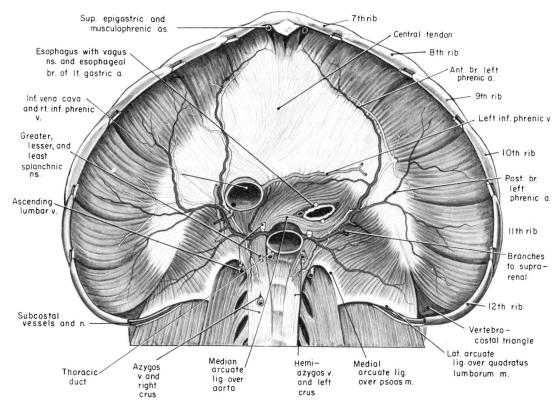

Sup. epigastric and musculophrenic as.

Esophagus with vagus ns. and esophageal br. of lt. gastric a.

Inf. vena cava and rt. inf. phrenic v.

Greater, lesser, and least splanchnic ns.

Ascending lumbar v.

Subcostal vessels and n.

7th rib

Central tendon

8th rib

Ant. br. left phrenic a.

9th rib

Left inf. phrenic v

10th rib

Post. br. left phrenic a.

11th rib

Branches to supra-renal

12th rib

Vertebro-costal triangle

Lat. arcuate lig. over quadratus lumborum m.

Thoracic duct

Azygos v. and right crus

Median arcuate lig over aorta

Hemi-azygos v. and left crus

Medial arcuate lig. over psoas m.

Figure 5-45. *Inferior surface of the diaphragm after removal of peritoneum and fascia. (Note that left inferior phrenic vein often drains into left suprarenal vein.) The branches of the phrenic nerve which penetrate the diaphragm to ramify on its inferior surface have not been labeled.*

the **median arcuate ligament,** which bridges across the aorta.

The **crura** are long tapering tendinous structures attached to the bodies of the lumbar vertebrae. The **right crus** is thicker and a bit longer than the left and is attached to the first, second, and third lumbar vertebrae. If followed superiorly, it is seen to end in fleshy fibers completely surrounding the esophageal opening. The **left crus** is attached to the upper two lumbar vertebrae only; if followed superiorly, it will be seen to end in fleshy fibers passing to the left of the esophageal opening. Frequently the left crus sends muscular fibers to the right of the esophagus, and this slip aids in forming the median arcuate arch over the aorta (Fig. 5-45).

There is usually a gap in the musculature of the diaphragm between the muscle fibers arising from the cartilage of the twelfth rib (costal part) and those from the

lateral arcuate ligament (lumbar part). This is the **vertebrocostal triangle;** it is filled in by pleura on the thoracic side. The diaphragm is **developed** from four distinct sources—a ventral portion, a dorsal portion, and two lateral portions. The ventral portion, between the esophagus and anterior abdominal wall, is derived from the septum transversum; the dorsal portion, that arising from the crura and medial and lateral arcuate ligaments, is derived from cervical somites that grow in between the developing digestive tube and the vertebral column (this is also the origin of the longus capitis and cervicis muscles located on the anterior surface of the cervical vertebrae); the lateral portions are derived from sickle-shaped ridges that appear on the lateral part of the coelom and gradually divide the coelomic cavity by growth in a medial direction. The lateral portions occasionally do not reach the ventral and dorsal portions, and a gap results. This occurs most frequently on the left side and is a source of trouble, for abdominal organs can pass through this opening—a diaphragmatic hernia. The opening can be a complete defect, whereby abdominal organs may herniate into the pleural cavity, or a defect in muscle only, in which case the herniated organ pushes layers of peritoneum and parietal pleura ahead of it. The vertebrocostal triangle is a normal aperture between the dorsal and lateral parts of the developing diaphragm. It should be noted that the diaphragm develops opposite the cervical vertebrae (note the formation of the longus capitis and cervicis muscles as mentioned above), which accounts for the nerve supply to the diaphragm from the cervical region. Subsequent growth of the body caused the apparent descent of the diaphragm.

The **main action** of the diaphragm is in respiration. Contraction of this muscle lowers the dome of the diaphragm and thereby increases the superior-inferior extent of the thoracic cavity. This results in inspiration. It should be noted that this descent of the central tendon demands a descent of the pericardial sac and its contents as well as the abdominal organs attached to the diaphragm, and that this, in turn, depends upon the extensibility of the abdominal musculature. When the dome of the diaphragm has descended as far as possible (abdominal breathing), any further contraction will elevate the ribs and aid in thrusting the sternum anteriorly, thus aiding in thoracic breathing. Furthermore, it should be noted that the diaphragm is involved in all actions that are aided by an increased abdominal pressure; acts such as defecation, micturition, parturition, sneezing, coughing, and even laughing or crying all involve a prior inhalation and contraction of the diaphragm and are accompanied by a closing of the glottis in the larynx. These acts are particularly prominent in weight lifting. Even swallowing is usually preceded by an inhalation and followed by an exhalation. The relation of the muscle fibers forming the right crus of the diaphragm to the esophagus should be noted; they tend to close off the esophagus in inhalation and have an opposite action in exhalation. The relation of these strips of muscle to the so-called physiological cardiac sphincter was discussed on page 201.

There are three important **openings in the diaphragm:** (1) the **aortic,** which transmits the thoracic duct and the azygos vein (the azygos vein may penetrate the right crus) in addition to the aorta; (2) the **esophageal** opening, which encircles the gastric nerves and esophageal branches of the left gastric artery as well as the esophagus; and (3) the **opening for the inferior vena cava,** which contains branches of the right phrenic nerve and a few lymph vessels in addition to the vena cava itself. It should be realized that **other structures pass from the thoracic to the abdominal cavity** and are required to go through or by the diaphragm. From anterior to posterior, they are the **superior epigastric vessels,** passing between parts of the diaphragm arising from the xiphoid process and the seventh cartilage; **musculophrenic vessels,** between slips originating from the seventh and eighth cartilages; the **lower five intercostal nerves accompanied by respective vessels,** between portions arising from the seventh to the twelfth costal cartilages; the **subcostal nerve and vessels,** posterior to the lateral arcuate ligament; the **sympathetic trunk,** posterior to the medial arcuate ligament; the **greater and lesser splanchnic nerves,** piercing the crura; and the least splanchnic, coursing deep to the medial arcuate ligament. The left crus is also pierced by the **hemiazygos vein** and the right crus occasionally by the **azygos vein.**

The **blood supply** to the diaphragm is via the superior phrenic arteries from the thoracic aorta, the inferior phrenic arteries from the abdominal aorta, and branches from the pericardiacophrenic, superior epigastric and musculophrenic branches of the internal thoracic artery. The intercostals also send twigs to the diaphragm. Corresponding veins accompany these arteries and drain into the azygos system of veins, the internal thoracic vein, and the inferior vena cava.

The **nerve supply** to the diaphragm is the phrenic nerve from C3, 4, and 5. The lower intercostals also serve the diaphragm in a sensory manner. It is important to realize that the phrenic nerve, with its sensory fibers as well as motor, does arise in the cervical region, and that nerves from the shoulders enter the cord in this same location. This accounts for the fact that pain from abscesses in the diaphragmatic region is often referred to the tips of the shoulders. Therefore, if the patient complains of pain in the shoulder, the source of trouble may be the diaphragm or its environs. Hiccoughs result from spasmatic contractions of the diaphragm probably caused by irritation to the phrenic nerves.

Lymph nodes draining the diaphragm are on its thoracic side and consist of (1) an anterior group, located near the xiphoid process and the cartilage of the seventh ribs, draining the ventral part of the diaphragm; (2) a middle group, found at a site close to where the phrenic nerves enter this muscle, draining the central and lateral parts of the diaphragm; and (3) a posterior set, located on the posterior sides of the crura, draining the dorsal part of the diaphragm. Note how this follows the arteries to the diaphragm.

AUTONOMIC NERVOUS SYSTEM

The **autonomic nervous system** in the abdominal cavity is continuous with that in the thoracic cavity. The **sympathetic chains** pass posterior to the medial arcuate ligament of the diaphragm to enter the abdominal cavity (Fig. 5-41). They become slightly more anterior on the vertebral bodies as they proceed inferiorly in the groove between the psoas major muscles and the bodies of the vertebrae. The number of ganglia varies, but there are usually four. The chains continue into the pelvic cavity for sympathetic supply to the pelvic body wall and the lower limb in a manner similar to the extension into the cervical region to innervate structures in the head, neck, and upper limb. All chain ganglia are connected to the lumbar and sacral nerves by gray rami communicantes; postganglionic fibers use these rami to reach these nerves and therefore to innervate all smooth muscle and glands in the body wall and lower limb. Contrariwise, since the white rami communicantes serve as a pathway for preganglionic sympathetic fibers to reach the sympathetic chain ganglia, and since there are no preganglionic fibers arising in the lower lumbar or sacral segments of the cord, there are no white rami connected to the lower ganglia in the lumbar sympathetic chain.

The abdominal aorta is surrounded by a plexus of nerves throughout its entire length (Fig. 5-41). Large **celiac ganglia** can be found on either side of the celiac artery in the middle of a dense collection of nerve fibers. These are connected with similar ganglia and plexuses at the base of the superior mesenteric artery (**superior mesenteric ganglia),** and these with others at the base of each renal artery (**aorticorenal ganglia).** It should be realized that these ganglia are very close together; one can cover the origins of the celiac, superior mesenteric, and renal arteries with a half dollar.

This aortic plexus continues inferiorly to the origin of the inferior mesenteric artery, where there is another ganglion (**inferior mesenteric ganglion),** and then leaves the end of the aorta to reach the anterior surface of the sacrum as the **hypogastric plexus.** All these ganglia are known as preaortic ganglia, and are concerned with innervation of the viscera in the abdominal and pelvic cavities.

All that remains is to describe the course taken by preganglionic fibers from the sympathetic chain ganglia (having arrived there from the spinal cord via ventral roots and white rami communicantes) to these preaortic ganglia, and to indicate how postganglionic fibers can reach the organs to be innervated after the synapse has occurred in these ganglia.

Nerves leading from the chain to the preaortic ganglia are called **splanchnic nerves.** There are greater, lesser, least, lumbar, and sacral splanchnics. The **greater splanchnic** nerves arise from the fifth to the ninth thoracic sympathetic ganglia and terminate in the celiac ganglia; the **lesser splanchnics** arise in the tenth and eleventh thoracic ganglia and terminate in the superior mesenteric ganglia; the **least splanchnics** arise from the twelfth thoracic ganglia and terminate in the aorticorenal ganglia; the **lumbar splanchnics** terminate in the inferior mesenteric ganglion and in the hypogastric plexus, where there are scattered ganglion cells; the **sacral splanchnics** arise from the pelvic extension of the sympathetic chain. Most of these sacral splanchnics are postganglionic fibers, having synapsed in the chain ganglia, and innervate blood vessels in the pelvis.

Postganglionic fibers reach the organs to be innervated mainly via the blood vessels, which is easily accomplished since these ganglia just described are all associated with blood vessels. Preganglionic fibers destined to innervate the spleen, for example, would terminate in the celiac ganglia because a branch of the celiac artery reaches the spleen on which postganglionic fibers can course. The hypogastric plexus of nerves continues onto the internal iliac artery and follows its many branches to the pelvic organs.

The various plexuses mentioned above are not exclusively sympathetic in origin. The parasympathetic system, via the vagus nerve and the pelvic nerve, enters these plexuses. However, these nerves do not synapse in the preaortic ganglia but continue to the walls of the organs to be innervated for synapse in parasympathetic ganglia.

LYMPH NODES

The lymph nodes of the abdominopelvic cavity can be divided into visceral and parietal groups. The visceral nodes are located in the various membranes, mesenteries, and ligaments of the abdominal cavity and have been mentioned for each organ described. (Those in the pelvic cavity will be studied later.) Lymphatics lead from these nodes to the parietal (wall) nodes.

The parietal nodes are associated with the arteries. They can be divided into the following groups (Fig. 5-46):

1. External iliac
2. Internal iliac
3. Common iliac
4. Sacral
5. Lumbar
   a. Lateral aortic
   b. Preaortic

The **external iliac, internal iliac,** and **common iliac nodes** are found around the arteries of the same names. The **sacral** are in the anterior concavity of the sacrum in association with the midsacral artery. The **lumbar nodes** are very numerous. The **lateral aortic lumbar nodes** form a chain associated with the inferior vena cava, psoas muscle, and right crus of the diaphragm on the right, and with the aorta, anterior-medial surface of the psoas major muscle, and the left crus on the left. These lateral nodes drain lymphatics from the common iliac nodes, the testes or ovaries, uterine tubes and body of the uterus, the kidneys, suprarenal glands, and the posterior part of the abdominal wall. Efferents from these nodes form trunks,

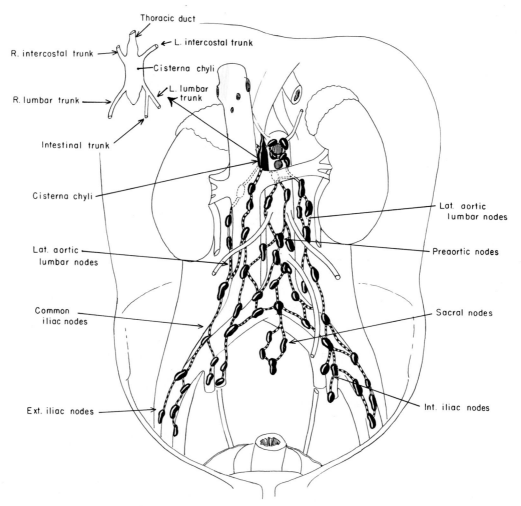

Figure 5-46. Lymph nodes on the posterior abdominal wall. Note that these chains of nodes empty into the cisterna chyli, which is enlarged in the insert. Note also Figure 4-43 on page 198.

which enter the cisterna chyli of the thoracic duct. They also are connected with the other lumbar nodes. The **preaortic lumbar nodes** are on the anterior surface of the aorta and can be divided into celiac, superior mesenteric, and inferior mesenteric groups, located at the origins of these arteries. They drain the organs supplied by these arteries and their branches. Lymphatics from these nodes combine into an intestinal trunk, which drains into the cisterna chyli directly, or into the left lumbar lymph trunk.

The **cisterna chyli** is a dilated structure located on the anterior surface of the upper two lumbar vertebrae between the aorta and the right crus of the diaphragm. The **thoracic duct** passes superiorly from this structure, through the aortic opening of the diaphragm, and ascends through the mediastinum (Fig. 4-43). It ultimately drains into the venous circulation in the neck, where the left internal jugular and left subclavian veins join to form the left brachiocephalic vein.

ABDOMINAL AORTA

The **abdominal aorta*** (Figs. 5-38 and 5-40) is a direct continuation of the thoracic aorta. It starts at the level of the disc between the twelfth thoracic and first lumbar vertebrae, at which point it is situated between the crura of the diaphragm. It passes inferiorly in the midline and ends at approximately the fourth lumbar vertebra by dividing into the two common iliac arteries. The exact vertebral level of this termination is quite variable. At its inferior end it is slightly to the left of the midline. It should be noted that the abdominal aorta is considerably shorter than the inferior vena cava. It lies anterior to the first four lumbar vertebrae and to the left side of the inferior vena cava. Although replacing portions of the aorta with plastic replicas has been quite successful, most abdominal surgery is performed in such a manner as to avoid it. Structures such as the pancreas, third part of the duodenum, left renal vein, and the mesentery are all just anterior to the abdominal aorta.

The **branches of the aorta** can be divided into those

---

*G. J. Baylin, 1939, Collateral circulation following an obstruction of the abdominal aorta, *Anat. Record* 75: 405. A rare obstruction of the aorta just distal to renal arteries. The anastomosis between mesenteric arteries was enlarged to size of aorta, while internal thoracic-epigastric anastomosis was only very slightly enlarged. Vessels nearest the obstruction seem to function to a greater degree than those far away.

supplying the viscera (visceral), those supplying the body wall (parietal), and the terminal branches:

Visceral
1. Celiac
2. Superior mesenteric
3. Middle suprarenal
4. Renal
5. Testicular or ovarian
6. Inferior mesenteric
Parietal
7. Inferior phrenic
8. Lumbar
9. Middle sacral
Terminal
10. Common iliacs

Except for the testicular or ovarian arteries, all of the above visceral arteries (1, 2, 3, 4, and 6) have been described.

5. Although the origin of the **testicular arteries** from the aorta is quite variable, they usually arise as paired arteries from the anterior surface of the aorta inferior to the origin of the renal arteries (Fig. 5-40). Each courses inferiorly and slightly laterally, finally to reach the deep inguinal ring to become a part of the spermatic cord. It traverses the inguinal canal and descends to the upper end of the testis; it is anterior to the vas deferens in position. It divides into branches that supply the epididymis as well as the testis itself. The **relations** of the two arteries differ (Fig. 5-40). The **right testicular artery** is at first anterior to the aorta, crosses anterior to the inferior vena cava and the ureter to reach the anterior surface of the psoas major muscle. It crosses the genitofemoral nerve and the external iliac vessels on its way to the internal ring of the inguinal canal. It is a retroperitoneal structure, and it is crossed anteriorly by the third part of the duodenum, the colic vessels, the root of the mesentery, and by the cecum and appendix near its inferior extent. The **left artery** starts on the anterior surface of the aorta, courses to the left anterior to the sympathetic trunk, ureter, and genitofemoral nerve on the psoas major muscle, and finally crosses the external iliac vessels to enter the internal ring. It is crossed anteriorly by the left colic vessels and the descending colon. The testicular arteries send branches to the ureter and to the cremaster muscle as well as supplying epididymis, vas deferens, and testis. The **ovarian arteries** correspond to the testicular and have the same relations as just described as far as the rim of the pelvis. At

this point each ovarian artery turns medially, anterior to the external iliac artery an inch or so distal to its origin. It proceeds into the pelvic cavity and enters the broad ligament to reach the ovaries. The fold of peritoneum over these vessels is called the **suspensory ligament of the ovary.** The ovarian arteries also give branches to the ureter, uterine tubes, and the round ligament of the uterus, and then continues to supply blood to the uterus itself.

7. The **inferior phrenic arteries** (Fig. 5-45) are the first branches of the abdominal aorta. They ascend on each crus of the diaphragm and, after giving a branch to each suprarenal gland, continue to ramify on the inferior surface of the diaphragm. The right phrenic artery courses posterior to the inferior vena cava, while the left is posterior to the esophagus.

8. The **lumbar arteries** (Fig. 5-47), of which there are four pairs, arise from the posterior surface of the abdominal aorta, course laterally and then posteriorly over the bodies of the corresponding vertebrae. They, in their course, are deep to the sympathetic trunk, psoas major muscle, and the lumbar plexus of nerves, and proceed to the intervals between transverse processes. Each one gives off a **posterior branch,** which is found accompanying the dorsal ramus of each lumbar nerve, and then proceeds laterally, passing posterior to the quadratus lumborum muscle to pierce the transversus and run anteriorly in the abdominal wall between the transversus and internal abdominis oblique muscles. The posterior branches also send a **spinal artery** through each intervertebral foramen to supply the vertebrae as well as the contents of the vertebral canal. The lumbar arteries anastomose with each other and with other arteries supplying the abdominal walls.

9. The **middle sacral artery** (Figs. 5-40 and 5-41) arises from the posterior surface of the aorta also. It usually arises approximately ½ inch superior to the bifurcation of the aorta, continues inferiorly over the front of the lumbar vertebrae, and enters the pelvic cavity on the anterior surface of the sacrum and coccyx. It ends in branches to the coccygeal body.* It is posterior to the hypogastric plexus of nerves, coils of the intestine, and, finally, the rectum. It has branches that would correspond to the fifth lumbar, small branches to the rectum, and others, equally small, that anastomose with lateral sacral branches from the internal iliac arteries.

---

*A small, highly vascular collection of tissue measuring 2–3 mm. in diameter, lying anterior to the tip of the coccyx. It is close to the ganglion impar of the sympathetic chain. Although this tissue resembles that of the carotid body (a structure that responds to oxygen tension in the arterial blood and reflexly controls respiratory rate), its function is unknown.

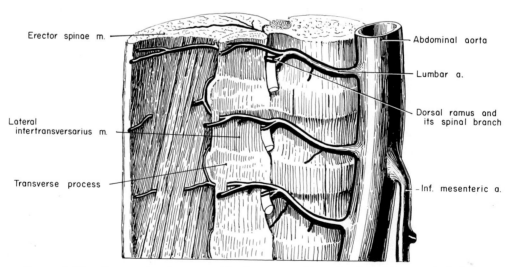

Figure 5-47. Course taken by a typical lumbar artery. (Note that they do not extend all the way around the abdominal wall.)

ILIAC ARTERIES

The abdominal aorta divides into two **common iliac arteries** approximately in front of the fourth lumbar vertebra (Fig. 5-40), a point marked on the abdominal wall by the umbilicus. Each common iliac artery courses inferiorly and laterally for a short distance, and then divides into an external and an internal branch. The right common iliac artery crosses anterior to the beginning of the inferior vena cava and is anterior to its coresponding vein; it is therefore separated by this vein from the psoas major muscle, sympathetic trunk, and bodies of the fourth and fifth lumbar vertebrae. The left common iliac artery is in relation directly with these structures since its vein is medial and slightly inferior to it. Both vessels are posterior to the peritoneum and intestines, and to nerves from the sympathetic trunk to the hypogastric plexus. The branches of the inferior mesenteric artery to the rectum also course anterior to the left common iliac artery. The common iliac arteries have no branches except the terminals already mentioned.

The **external iliac arteries** (Figs. 5-40 and 5-41) arise at the level of the lumbosacral joint as one of the terminal branches of the common iliac. Their origin is also directly anterior to the sacroiliac joint. They course inferiorly and slightly lateral in a position anterior to the psoas major muscle, and end by becoming the femoral artery after coursing posterior to the inguinal ligament. The arteries cross from the medial side to the lateral side of the psoas major muscle in their inferior course. Each artery is anterior to its respective vein at its commencement but gradually becomes lateral to the vein. The peritoneum separates them from the intestines and the left external iliac artery is crossed by the sigmoid colon. Other structures that cross the arteries are the ureters, ovarian vessels, and, more inferiorly, the vas deferens in the male and ligamentum teres (uteri) in the female.

The external iliac arteries have two named branches (Fig. 5-12):

1. Inferior epigastric
2. Deep circumflex iliac

Other unnamed branches supply blood to the surrounding structures. Both of these vessels arise near the end of the artery, just before it becomes the femoral, and are involved with the abdominal wall.

1. The **inferior epigastric** courses superiorly and medially, forming the lateral boundary of the inguinal triangle, and then gains the posterior aspect of the rectus abdominis muscle (Fig. 3-31 on page 129). It continues superiorly to anastomose with the superior epigastric branch of the internal thoracic artery. Its branches are as follows: (a) The **cremasteric** branch supplies blood to the cremaster muscle and the walls of the spermatic cord (this branch is very small in the female and supplies the ligamentum teres). (b) The **pubic** branch courses medially along the inguinal ligament to reach the pubis, where it turns posteriorly to course through the obturator foramen after anastomosing with the obturator branch of the internal iliac. For many cases (probably 40 per cent) the pubic branch is the only obturator artery and is of considerable size. It is important to realize its possible presence when performing surgery in the inguinal region. (c) The **muscular** branches lead to adjoining muscles.

2. The **deep circumflex iliac** artery courses laterally along the inguinal ligament in the fascia formed by the joining of transversalis and iliac fascias. It gives off a branch that anastomoses with the ascending branch of the lateral femoral circumflex artery, a vessel of the lower limb. It then continues along the inner aspect of the crest of the ilium approximately one-half its length and then enters the abdominal musculature; it anastomoses with the lumbar and iliolumbar (a branch of the internal iliac—see Fig. 5-64 on page 294) arteries. A fairly large branch is also given off near the anterior superior iliac spine, which ascends between the internal oblique and transversus abdominis muscles, anastomosing with the lumbar and inferior epigastric arteries. This artery supplies adjoining muscles.

The **internal iliac arteries** will be described after the pelvic structures have been studied.

ILIAC VEINS

The **external iliac vein** (Fig. 5-40) commences as a continuation of the femoral vein as the vessel passes posterior to the inguinal ligament. It is medial to its respective artery. It is at first anterior and then medial to the psoas major muscle. Superficially, it is covered with peritoneum, which separates it from the small intestine. Although each is separated from the structures mentioned that cross the external iliac arteries by the iliac arteries themselves, these structures are in close relation to it. It terminates by joining with the internal iliac vein to form the common iliac vein.

Its **tributaries** are the (1) inferior epigastric, (2) pubic, and (3) deep circumflex veins. These vessels drain into the external iliac near its origin. They have a similar distribution as just described for the arteries of the same name.

The **common iliac veins** (Fig. 5-40) start by the joining of the external and internal iliac veins, course superiorly, and combine to form the inferior vena cava. The right vein takes an almost direct superior course to join the inferior vena cava, while the left has to cross anterior to the vertebral column to reach the right-sided inferior vena cava (Fig. 5-40). These veins are posterior to their respective arteries. The **tributaries** of the common iliac veins are the iliolumbar veins, which course posterior to the psoas major muscle and end in the posterior part of the common iliac vein on each side (Fig. 5-48). In addition, the middle sacral vein forms a comitantes, which unites to form a single vein that ends in the left common iliac vein.

INFERIOR VENA CAVA

The **inferior vena cava** (Fig. 5-40) starts at the level of the fifth lumbar vertebra, slightly to the right side of the midline and to the right side of the aorta, continues superiorly in this position anterior to the vertebral column and the psoas major muscle, and finally pierces the central tendon of the diaphragm and the pericardium to empty into the right atrium of the heart. The inferior vena cava reaches to a much higher vertebral level than the abdominal aorta, being opposite the eighth thoracic vertebra and the sixth costal cartilage at its termination. As with the aorta, most surgery in the abdominal cavity demands being aware of the inferior vena cava at all times. Several important structures have an intimate anterior relation to it. The mesentery crosses it as do the testicular or ovarian vessels. The right suprarenal gland is molded around it. The first and third parts of the duodenum, the bile duct, and the portal vein are also just anterior to the inferior vena cava. Advantage is taken of this relation to the portal vein; it makes portal-caval shunts for portal hypertension relatively easy to accomplish. The inferior vena cava and its covering of peritoneum forms the pos-

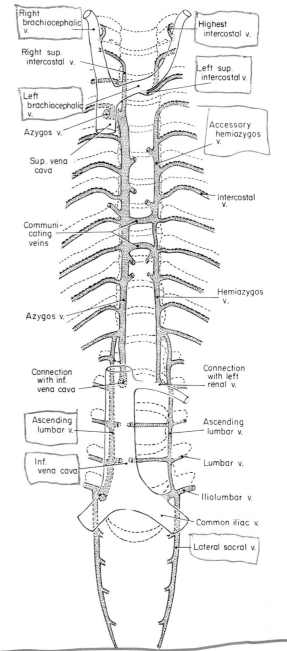

Figure 5-48. *The ascending lumbar veins. Note how these make connections with the azygos system of veins.*

terior boundary of the epiploic foramen. Finally, it should be noted that the inferior vena cava is actually embedded in the liver; therefore, the liver is not only anterior to the vessel, but also on its right and left sides.

The **tributaries** are:

1. Common iliacs
2. Third and fourth lumbar
3. Right testicular or ovarian
4. Renals
5. Right suprarenal
6. Inferior phrenics
7. Hepatic

It should be apparent immediately that the inferior vena cava differs from the aorta in having fewer visceral branches. Since the vessels from the major part of the intestinal tract and associated structures drain via the portal circulation into the liver, there is no need of branches from these viscera entering the inferior vena cava.

1. The **common iliacs** have just been described.

2. **Lumbar veins.** The lumbar veins take a similar course as described for the lumbar arteries. They are superior to their respective arteries in position. Not all the lumbar veins drain into the inferior vena cava. The fifth lumbar vein drains into the iliolumbar vein, which, in turn, empties into the common iliac; the third and fourth lumbar veins end in the inferior vena cava; and the first and second veins usually end in the ascending lumbar vein, but may pass forward between the psoas and the vertebral bodies to end in the abdominal part of the azygos or hemiazygos veins. The **ascending lumbar** vein, just mentioned, is a long anastomosing channel that collects blood from the lateral sacral veins, iliolumbar veins, lumbar veins, and subcostal veins (Fig. 5-48). It begins in the lateral sacral veins and ascends over the pelvic surface and ala of the sacrum. It then gains the anterior surface of the transverse processes of the lumbar vertebrae and ends in the subcostal vein on the side of the twelfth thoracic vertebra. The first two lumbar veins usually drain into this particular vein. If by any chance the inferior vena cava becomes occluded,* these anastomotic

*Keen, J. A. 1941, The collateral venous circulation in a case of thrombosis of the inferior vena cava, and its embryological interpretations, *Brit. J. Surg.* 29: 195. An interesting account of survival by development of alternate blood routes after gradual onset of thrombosis of the entire inferior vena cava.

channels can possibly serve as a means for blood to get around this tie, enlarging greatly to accommodate the increased amount of blood. The anastomotic channels between these veins and the azygos system of veins in the thoracic cavity are quite important. Another point that might be mentioned is the fact that branches of the portal circulation come very close to many of these veins on the posterior abdominal wall. As a result, communications do exist, and this is another location where anastomoses occur between the portal and systemic circulations. These connections can become quite dilated under conditions of portal hypertension.

3. **Right testicular or ovarian vein** (Fig. 5-40). The right testicular vein starts as a pampiniform plexus of veins in the spermatic cord, but at the internal ring forms two vessels, which unite to form a single vein that accompanies its corresponding artery. It soon leaves its artery, because the artery comes from a more medial position from the aorta. In the female the ovarian veins also form a pampiniform plexus from which two veins emerge at the rim of the pelvis. They terminate in the same manner as found for the testicular veins.

4. **Renal veins** (Fig. 5-40). The right renal vein is rather short because the inferior vena cava is on the right side of the abdominal cavity. The left renal vein is considerably longer and courses anterior to the aorta before joining the inferior vena cava. It is inferior to the superior mesenteric artery and the celiac ganglia. This structure is directly posterior to the pancreas. It receives the left suprarenal vein and the left testicular or ovarian veins as well. The right testicular or ovarian vein drains into the inferior vena cava, while the left veins drain into the renal vein.

5. **Right suprarenal vein.** This vein emerges from the hilum of the suprarenal gland and enters the inferior vena cava almost immediately; the left vein drains into the left renal vein rather than the inferior vena cava itself.

6. **Right inferior phrenic vein** (Fig. 5-45). The venae comitantes of the right phrenic artery drain into the inferior vena cava, but those from the left side of the diaphragm may course anterior to the esophagus to drain into the inferior vena cava, or drain into the renal or suprarenal veins.

7. **Hepatic veins.** These structures, usually two in number but occasionally three because of a separate vein from the caudate lobe of the liver, are exceedingly

short because of the close relationship of the liver to the inferior vena cava. These veins are large and drain into the superior end of the inferior vena cava.

These arteries and veins have all been coursing in the subserous fascia.

MUSCULATURE
(Fig. 5-41)

If the visceral structures in the abdominal cavity are removed, the muscles forming the posterior abdominal wall are seen posterior to their fascial layers. This is deep fascia and will be described after the muscles have been considered.

This view (Fig. 5-41) also shows that the distance between the last ribs and the iliac crest is remarkably short, being approximately 3 or 4 inches only. This space is filled in by the **psoas major** and **quadratus lumborum** muscles. The latter muscle blends laterally with the **transversus abdominis** muscle. Superior to this space is the **diaphragm,** and inferior to it the **iliacus** and **psoas** muscles filling in the iliac fossa.

As the diaphragm has been described previously (page 266), the psoas major and minor muscles, the quadratus lumborum, and the iliacus muscles remain to be described at this time.

Psoas major and minor muscles. The psoas major muscle **arises** from the transverse processes and bodies of all the lumbar vertebrae and from the body of the twelfth thoracic vertebra as well. Although the insertion cannot be seen at this time, the muscle does traverse the iliac fossa and enter the lower limb by passing posterior to the inguinal ligament. It joins with the tendon of the iliacus muscle (the two muscles combined are called the iliopsoas) to **insert** on the lesser trochanter of the femur. Its **main action** is to flex the femur at the hip joint; its action in rotation of the hip joint depends upon the position of the limb, medially rotating at first but changing to a lateral rotation upon continued flexion of the thigh. The **nerve supply** to the psoas major muscle is from L2, 3, and 4 directly, but occasionally through branches from the femoral. The **psoas minor** muscle is absent in many bodies, but when present lies on the anterior surface of the psoas major muscle. It **arises** from the twelfth thoracic

and first lumbar vertebrae and **inserts** by a long tendon onto the pectineal line and into the fascia covering the iliopsoas muscle. Its **main action** is to flex the vertebral column and supposedly it plays a role in tightening the iliopsoas fascia. The **nerve supply** to the psoas minor is from the first lumbar segment directly.

Iliacus muscle. This muscle lies in the iliac fossa and **arises** chiefly from the floor of this fossa. It **inserts,** in common with the psoas major, onto the lesser trochanter of the femur. It also serves to flex the femur at the hip joint and will act on rotation as mentioned for the psoas major. Its **nerve supply** is similar to that of the psoas major—L2, 3, and 4 directly, but occasionally through branches from the femoral.

Quadratus lumborum muscle. As mentioned previously, this muscle fills in the space between the last rib and the iliac crest. It **arises** from the iliac crest and **inserts** on the transverse processes of the upper four lumbar vertebrae and on the lower border of the last rib. Its **action** is to fix the last rib in respiration and to serve a minor action in extending the vertebral column and bending it laterally. Its **nerve supply** is from L1, 2, 3, and 4 directly.

TRANSVERSALIS FASCIA

When studying the anterior abdominal wall, we found that the transversus abdominis muscle was lined on its inside surface by a deep fascia, the transversalis fascia. We also found that it was not limited to the transversus abdominis muscle but stretched across the posterior wall of the inguinal canal, even sending down the spermatic cord a prolongation called the internal spermatic fascia.* Furthermore, we found that this fascia spreads across the posterior surface of the rectus abdominis muscle inferior to the umbilicus while the transversus abdominis muscle crosses anterior to the rectus.

Actually, this transversalis fascia forms a lining for the entire abdominopelvic cavity. It is directly continuous with the fascia on the quadratus lumborum and psoas muscles, and then proceeds across the vertebral column

---

*M. A. Hayes, 1950, Abdominopelvic fasciae, *Am. J. Anat.* 87: 119. This report divides abdominal fasciae into (1) migration fasciae, derived from organ migrations producing specializations of the subserous fascia; (2) fusion fasciae, caused by fusion of two primitive mesenteries; and (3) parietal fascia, that which is intrinsic to the structures forming the wall of the developing abdominopelvic cavity (transversalis fascia).

posterior to the aorta and inferior vena cava. It is also continuous with the fascia on the inferior side of the diaphragm. In other words, the diaphragmatic fascia, quadratus lumborum fascia, psoas fascia, and transversus abdominis fascia are all parts of a single layer, the transversalis fascia. The fascia in the iliac fossa, however, seems to be a special aponeurosis on the iliacus muscle and the transversalis fascia is an additional layer. We shall see later that this transversalis fascia is continued into the pelvic cavity. In addition, since blood vessels in the abdominopelvic cavity course in the subserous fascia, which is inside the transversalis fascial sac, they are required to penetrate this transversalis fascia in order to leave this cavity. We shall see that prolongations of this fascia tend to follow the vessels under these circumstances.

We have already learned how fascial layers can serve as a pathway for or a barrier to the spread of infectious agents. It is of clinical interest to note that such infections can spread from the posterior aspect of the thorax to the anterior surface of the thigh by following the fascia on the psoas major muscle.

LUMBAR PLEXUS
(Figs. 5-41 and 5-49)

The **lumbar plexus** is formed by the ventral rami of the upper three lumbar and part of the fourth lumbar nerves; it receives a communication from the subcostal nerve. The fifth lumbar nerve is not included, for the entire ventral ramus of L5 and a portion of L4 join to form a lumbosacral trunk, which aids in formation of the sacral plexus. All these nerves are deep or posterior to the deep fascia in contrast to the arteries that course in the subserous fascia.

The nerves arising from the lumbar plexus are:

1. Iliohypogastric
2. Ilioinguinal
3. Genitofemoral
4. Lateral femoral cutaneous
5. Femoral
6. Obturator

They all arise in association with the psoas major muscle. The genitofemoral nerve emerges through its anterior surface, the obturator at its medial border, and the others

at the lateral border of the psoas muscle. They are all deep to the deep fascia.

1. **Iliohypogastric nerve** (Figs. 5-41 and 3-30 on page 127). This nerve arises from the first lumbar segment of the spinal cord; it then appears at the lateral border of the psoas, courses anterior to the quadratus lumborum muscle, and in this position is posterior to the kidney. It pierces the aponeurosis of the transversus abdominis muscle and proceeds anteriorly and then medially around the abdominal wall between the transversus abdominis and the internal abdominis oblique muscles. It pierces the latter muscle about 1 inch superior to the anterior superior iliac spine and becomes cutaneous about 1 inch superior to the superficial inguinal ring. Its

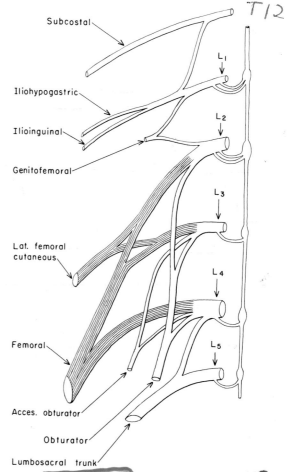

Figure 5-49. Lumbar plexus. *The clear nerves are preaxial, the shaded postaxial (see page 20).*

lateral cutaneous branch is distributed in the gluteal region and will be seen later. It gives off branches to the abdominal musculature along its entire course.

2. **Ilioinguinal nerve** (Figs. 5-41 and 3-30 on page 127). This nerve arises from the first lumbar segment and follows a similar course to that taken by the iliohypogastric, but is more inferiorly placed (Fig. 5-41). It has no lateral branch, it pierces the internal abdominis muscle more anteriorly, and then passing between the internal and external abdominis muscles, where they form the anterior wall of the inguinal canal, enters the inguinal canal, and exits through the superficial inguinal ring. Besides innervating the muscles between which it courses, it sends cutaneous branches to the skin over the pubis, on the anterior surface of the scrotum in the male or a labium majus in the female, and sends one cutaneous branch to the thigh.

3. **Genitofemoral nerve** (Fig. 5-41). This nerve, from the first and second lumbar segments, pierces the psoas major muscle and descends over its anterior surface. It has two branches, namely, the genital branch and the femoral branch. The **genital branch** courses inferiorly, anterior to the inferior end of the external iliac artery. It passes through the deep inguinal ring of the inguinal canal and continues through the canal to innervate the cremaster muscle in the male or the ligamentum teres in the female. It terminates in the skin of the scrotum or a labium majus, and the adjacent thigh. The **femoral branch** descends on the lateral side of the external iliac and femoral arteries and pierces the deep fascia to supply skin on the anterior surface of the thigh.

4. **Lateral femoral cutaneous** (Fig. 5-41). Appearing on the lateral side of the psoas major muscle, this nerve (from L2 and 3) crosses the iliacus muscle and enters the thigh by passing deep to the inguinal ligament.*

5. **Femoral nerve** (Fig. 5-41). The femoral nerve, arising from the second, third, and fourth lumbar spinal cord segments, appears at the lateral edge of the psoas major muscle and follows this lateral edge to its entrance into the thigh. It is lateral to the femoral artery and enters the thigh deep to the inguinal ligament.* It

gives off branches to the psoas major and iliacus muscles while in the false pelvis.

6. **Obturator nerve** (Fig. 5-41). The obturator nerve, from the second, third, and fourth lumbar spinal cord segments, descends in a position anterior to the sacroiliac joint and posterior to the common iliac vessels. It appears on the medial side of the psoas in contrast to the nerves already described. It continues inferiorly on the lateral wall of the pelvis, where it is lateral to the ureter, to the internal iliac vessels, and, in the female, to the ovary and broad ligament. It is posterior to the umbilical artery, but anterior to its accompanying obturator vessels. It leaves the pelvic cavity through the obturator foramen to enter the medial side of the thigh.† Occasionally an **accessory obturator nerve** is found. This nerve arises from the third and fourth lumbar segments and descends along the medial border of the psoas major muscle and enters the thigh on the anterior surface of the superior pubic ramus. This nerve may innervate the pectineus muscle, it may innervate the hip joint, or it may simply substitute for a portion of the obturator nerve.

The **lumbosacral trunk** (Fig. 5-41) is a large nerve formed by the entire ventral ramus of L5 and the descending part of L4. These nerves descend posterior to the psoas major and join the first sacral nerve in the pelvic cavity.

Upon removal of the psoas major muscle, the detailed lumbar plexus can be seen as shown in Figure 5-41.

PELVIC CAVITY

BONY STRUCTURES

Before the pelvic cavity and perineum can be understood, it is mandatory that the student be thoroughly familiar with the bones forming the pelvis. To save words, it is assumed that students will have a pelvis before them

---

*The further distribution of the branches of the lumbar plexus will be studied with the lower limb. The sacral plexus will be described after the pelvic organs have been removed.

†See preceding footnote.

or at least make very good use of the accompanying pictures of the bones forming the pelvis. To comprehend the following description, the student should be careful to hold the pelvis as it occurs in the living person.

The bony pelvis is made up of the two coxal bones, which are actually part of the lower limbs, and the sacrum and coccyx, which are parts of the axial skeleton.

COXAL BONES. The coxal bones are made up of three parts: **ilium, ischium, and pubis** (Fig. 5-50). The ilium is the wing-shaped superior portion of the coxal bone; the ischium the short, blunt posteroinferior portion; and the pubis the anterior inferior portion. These three bones come together as the **acetabulum,** the fossa that articulates with the head of the femur. The acetabulum exhibits a depression, called the **acetabular fossa,** and a notch, the **acetabular notch,** at the base of that fossa. The rest of the acetabulum is called the **lunate surface.**

The **ilium** articulates inferiorly with the pubis and the ischium. This part of the coxal bone flares out into a thin, winglike portion with a concavity on the medial surface and a slight convexity on the lateral surface. The concavity on the medial surface is the **iliac fossa** and the external or lateral surface presents three lines called the **posterior, anterior, and inferior gluteal lines.** The lines separate the areas of origin of the various gluteal muscles. Following these surfaces superiorly, one comes to the **iliac crest.** This crest starts anteriorly with the **anterior superior iliac spine,** proceeds posteriorly and laterally until it reaches the **posterior superior iliac spine;** it has an **inner and outer lip** and an **intermediate line.** In addition to these two superior spines, the ilium possesses an **anterior inferior iliac spine** and a **posterior inferior iliac spine.** The posterior inferior iliac spine is just posterior to the large surface of the ilium that **articulates** with the sacrum.

The superior end of the **ischium** is fused with the ilium and pubis. The ischium is the portion of the coxal bone upon which one sits, and that roughened area located most inferiorly and posteriorly is known as the **ischial tuberosity.** Just superior to the ischial tuberosity is the **lesser ischiadic notch** and, separated from this notch by the **ischiadic spine,** the **greater ischiadic notch.**

The **pubis** articulates with the ilium and the ischium. This part of the coxal bone contains a medial portion—the **body**—and **superior and inferior rami.**

The **inferior ramus** is the portion that joins the ischium, while the **superior ramus** joins the ilium and ischium at the acetabular notch. The superior surface of the body is smooth, while the inferior surface is quite rough because of the attachment of muscles. The articular surface is joined to the pubic bone of the other side by a fibrocartilage as well as ligaments. The **pubic crest** is that portion of the body of the pubis that is most easily felt and ends laterally in an elevation called the **pubic tubercle.** The superior surface of the superior ramus exhibits a sharp edge called the **pectineal line.** This line runs laterally from the pubic tubercle. The two rami of the pubic bone form partial boundaries of a large opening called the **obturator foramen.** The superior edge of this foramen exhibits an **obturator tubercle** and an **obturator sulcus** for passage of vessels and nerves from the pelvis into the lower limb (see obturator membrane, page 282 and Fig. 5-53).

The sacrum and coccyx are described on page 56.

PELVIS AS A WHOLE (Figs. 5-51 and 5-52). The **pelvis** consists of the two coxal bones anteriorly and laterally and the sacrum and coccyx posteriorly. It is divided into a **true and false pelvis.** The **inlet of the true pelvis** is superiorly placed and bounded by the sacral promontory posteriorly, the pubic symphysis anteriorly, and the terminal lines on the sides, the terminal lines being the pectineal lines anteriorly and the arcuate lines posteriorly; the **outlet** is the inferior end and is bounded from anterior to posterior by the pubic arch, inferior ramus of the pubis, ischial tuberosity, lesser ischiadic notch, ischiadic spine, greater ischiadic notch, and the sacrum and coccyx. It will be seen later that ligaments actually form the boundaries of the outlet rather than the two notches mentioned. The **false pelvis** is that part above the inlet to the true pelvis.

JOINTS AND LIGAMENTS OF THE PELVIS

The student needs a knowledge of the ligaments in this area as well as a clear conception of the bony pelvis in order to understand the anatomy in this region.

The sacrum is part of the vertebral column and articulates with the fifth lumbar vertebra at the lumbosacral joint. This has been described under the vertebral column (page 76). The acetabulum of the coxal bone serves as a fossa to contain the head of the femur. This joint will be described with the lower limb. The articulation of the coxal bones with each other in the pubic region (the pubic symphysis), and the articulation between the sacrum and these two coxal bones (the sacroiliac joints) remain. Furthermore, there are very important ligaments that connect the sacrum and coccyx to the ischial portion of the coxal bones. In addition, some attention should be given the articulation of the sacrum and coccyx, and to the obturator membrane.

PUBIC SYMPHYSIS. The pubic symphysis (Fig. 5-53) is a cartilaginous joint that is partially movable. The articular surfaces of each pubic bone are covered with hyaline cartilage and between these two areas of cartilage is interposed a fibrocartilaginous lamina, the **interpubic disc.** This varies in thickness in different areas and is a very dense and firm structure. In lower animal forms a hormone, relaxin, tends to soften this fibrocartilaginous material, allowing the pubic bones to spread apart during parturition. Although this is of a much lesser degree in the human being, it is thought that there is some softening of the fibrocartilage. This interpubic disc is aided by a **superior pubic ligament,** which connects the two pubic bones superiorly, and by the **arcuate pubic ligament,** a thick arch of ligamentous fibers, which connects them inferiorly. The former ligament extends laterally as far as the pubic tubercles. The arcuate ligament forms the superior boundary of the **pubic arch** and continues laterally to be attached to the inferior rami of the pubic bones.

SACROILIAC JOINT. The sacroiliac joints (Fig. 5-54) are important, for they are required to bear great weight. These joints are formed by the articulation between the sacrum and the iliac portion of the coxal bones. This joint is **partly cartilaginous** and **partly synovial** in type. Some movement is allowed. Each articular surface is covered with a thin layer of cartilage, and between these cartilages is found a layer of fibrocartilage. The **ligaments** of this joint are the ventral sacroiliac, dorsal sacroiliac, and

interosseus ligaments. The **ventral sacroiliac ligament** consists of several thin bands that connect the anterior surface of the sacrum to the anterior surface of the ilium. In contrast, the **dorsal sacroiliac ligament** is quite thick and forms the chief attachment of the sacrum and ilium. It consists of bundles that pass between the bones in several directions. The superior part is the **short posterior sacroiliac ligament,** and the fibers take nearly a horizontal course. They pass between the first and second transverse tubercles on the posterior surface of the sacrum to the tuberosity of the ilium. The inferior part of this ligament—the **long posterior sacroiliac ligament**— courses obliquely. It courses from the third transverse tubercle on the posterior surface of the sacrum and attaches to the posterior superior spine of the ilium. (We shall see in a moment that this ligament is intermingled with the sacrotuberous ligament.) The **interosseus sacroiliac ligament** is located deep to the posterior sacroiliac ligament just described. The fiber bundles are very short and connect the tuberosities of the sacrum and the ilium.

Several **other ligaments** serve to attach the coxal bones to the vertebral column. These are the iliolumbar, sacrotuberous, and sacrospinous ligaments. The **iliolumbar ligament** (Fig. 5-54) is attached superiorly to the transverse process of the fifth lumbar vertebra. It radiates inferiorly and laterally and is attached by two main parts to the coxal bone. The inferior portion courses toward the base of the ilium and blends with the ventral sacroiliac ligament. The superior portion is attached to the crest of the ilium anterior to the sacroiliac articulation. The **sacrotuberous ligament** (Fig. 5-54) is a large, heavy ligament that attaches over a wide area to the posterior inferior spine of the ilium, to the fourth and fifth transverse tubercles of the sacrum, and to the inferior part of the lateral margin of the sacrum and the coccyx. It passes obliquely inferiorly and laterally, becoming narrower as it proceeds, to attach to the medial side of the ischial tuberosity. The **sacrospinous ligament** (Fig. 5-54) is much shorter than the sacrotuberous ligament, but is also triangular in form. Its medial attachment is to the lateral margins of the sacrum and coccyx, and its lateral attachment is to the spine of the ischium. It is anterior to the sacrotuberous ligament. These sacrotuberous and sa-

---

Figure 5-50. *Lateral and medial surfaces of the right coxal bone with origins and insertions of muscles indicated.*

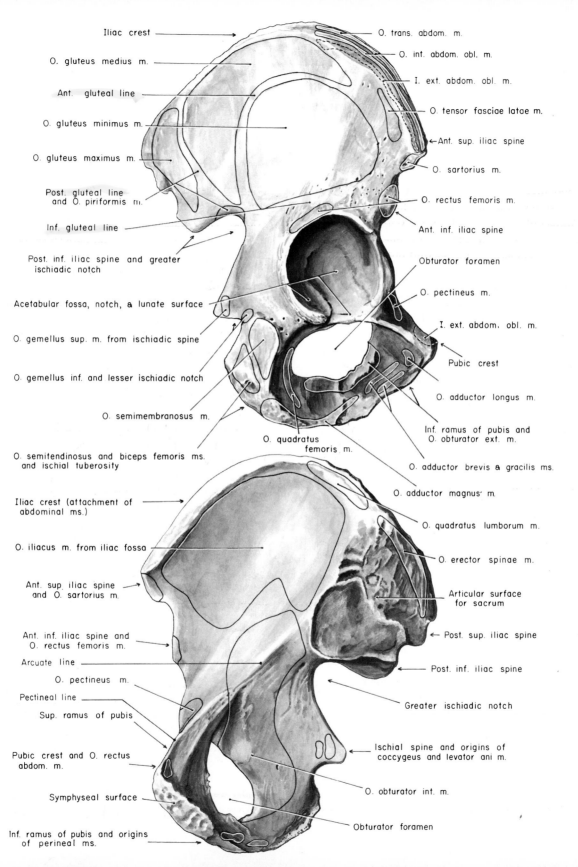

Iliac crest

O. gluteus medius m.

Ant. gluteal line

O. gluteus minimus m.

O. gluteus maximus m.

Post. gluteal line
and O. piriformis m.

Inf. gluteal line

Post. inf. iliac spine and greater
ischiadic notch

Acetabular fossa, notch, a lunate surface

O. gemellus sup. m. from ischiadic spine

O. gemellus inf. and lesser ischiadic notch

O. semimembranosus m.

O. semitendinosus and biceps femoris ms.
and ischial tuberosity

O. trans. abdom. m.

O. int. abdom. obl. m.

I. ext. abdom. obl. m.

O. tensor fasciae latae m.

←Ant. sup. iliac spine

O. sartorius m.

O. rectus femoris m.

Ant. inf. iliac spine

Obturator foramen

O. pectineus m.

I. ext. abdom. obl. m.

Pubic crest

O. adductor longus m.

Inf. ramus of pubis and
O. obturator ext. m.

O. quadratus
femoris m.

O. adductor brevis a gracilis ms.

O. adductor magnus m.

Iliac crest (attachment of
abdominal ms.)

O. iliacus m. from iliac fossa

Ant. sup. iliac spine
and O. sartorius m.

Ant. inf. iliac spine and
O. rectus femoris m.

Arcuate line

O. pectineus m.

Pectineal line

Sup. ramus of pubis

Pubic crest and O. rectus
abdom. m.

Symphyseal surface

Inf. ramus of pubis and origins
of perineal ms.

O. quadratus lumborum m.

O. erector spinae m.

Articular surface
for sacrum

← Post. sup. iliac spine

Post. inf. iliac spine

Greater ischiadic notch

Ischial spine and origins of
coccygeus and levator ani m.

O. obturator int. m.

Obturator foramen

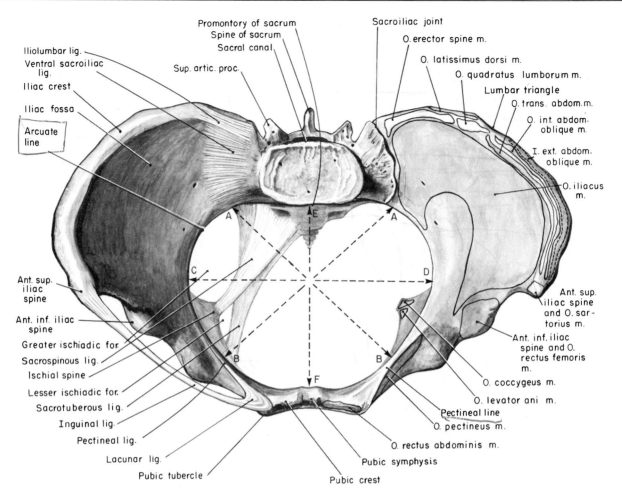

Figure 5-51. *The female bony pelvis, with origins and insertions of muscles indicated on the left and the ligaments shown on the right. A—sacroiliac joint; B—iliopubic eminence; C and D—middle of pelvic brim; E—sacral promontory; F—pubic symphysis.*

crospinous ligaments are in such a position that they make foramina out of the greater and lesser ischiadic notches. They also form part of the boundary of the inferior pelvic outlet.

SACROCOCCYGEAL JUNCTION. This joint is also a cartilaginous joint formed between the inferior articular surface of the sacrum and the base of the coccyx; it is partially movable.* Its disc is much thinner than the disc in the rest of the vertebral column. It is supported by liga-

ments similar to those associated with the rest of the vertebral column. The one that corresponds to the anterior longitudinal ligament is the ventral sacrococcygeal ligament; that which corresponds to the posterior longitudinal and supraspinal ligaments, the deep and superficial portions of the dorsal sacrococcygeal ligament. Lateral sacrococcygeal ligaments connect the transverse processes of the coccyx to the lower lateral angles of the sacrum.

OBTURATOR MEMBRANE. The obturator membrane is a thin but tough aponeurosis that fills in the large bony obturator foramen except for a small area superiorly for

---

*In some cases it is synovial in character, and considerable movement is allowed.

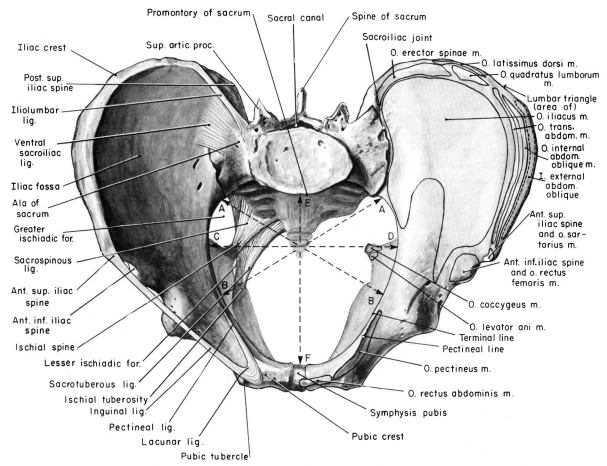

Figure 5-52. *The male bony pelvis, with origins and insertions indicated on the left and ligaments shown on the right. A—sacroiliac joint, B—iliopubic eminence, C and D—middle of pelvic brim, E—sacral promontory, and F—pubic symphysis.*

the passage of the obturator vessels and nerve. This structure, consisting of several interlacing bundles, most of which take a transverse direction, really should be considered part of the coxal bone; it serves for the attachment of the obturator internus and externus muscles, these muscles arising from the inside and outside surfaces, respectively, as well as from the surrounding bone.

SEXUAL DIFFERENCES. Before thinking about the main differences between the female pelvis (Fig. 5-51) and the male pelvis (Fig. 5-52), it is important to realize that there are all degrees of maleness and femaleness. Furthermore, there are racial differences that make measurements in a single sex quite variable. Lines have been

established to measure the diameters of the pelvic inlet; these are shown in Figures 5-51 and 5-52. Anthropologists classify pelves into anthropoid, android, gynecoid, and platypelloid types. Figure 5-51 is a drawing of a gynecoid pelvis while Figure 5-52 is android. Anthropoid has a short transverse (C-D) diameter but a long anteroposterior diameter (E-F), while the platypelloid type is opposite to this with a very long transverse (C-D) diameter and very short anteroposterior (E-F) diameter. Most males are anthropoid or android; most females are android or gynecoid.

In very general terms, the female pelvis, in comparison with the male pelvis, is lighter, the bones are slenderer, and the markings made by the attachments of

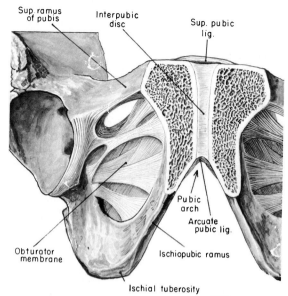

Figure 5-53. *The pubic symphysis, cut to show interpubic disc, and the obturator membrane.*

muscles are less pronounced. The ilium has a greater lateral flair, and the iliac fossa is slightly shallower. The pubic arch is wider in the female, this being one of the easiest ways to differentiate between the male and female pelvis, and the ischial tuberosities are farther apart. The inlet to the true pelvis tends to be oval rather than heart-shaped; the ischial spines do not project into the outlet of the pelvis as they do in the male, and the same is true of the sacrum and coccyx. The promontory of the sacrum is less pronounced in the female. This means that all diameters in the true pelvis are longer in the female than in the male, which allows for childbirth. Looking inferiorly into the true pelvis, one can usually determine instantly whether the pelvis is that of a female or a male.

All of this is of interest to the anatomist or anthropologist, but the obstetrician is interested in comparing the size of the baby's head (fetal cephalometry) with the dimensions of the pelvic outlet (pelvimetry). Therefore, averages are useless; the dimensions of the pelvic outlet on the particular patient giving birth to a baby be-

Figure 5-54. *Anterior and posterior views of the sacroiliac joints. The sacrospinous, sacrotuberous, and iliolumbar ligaments are also shown.*

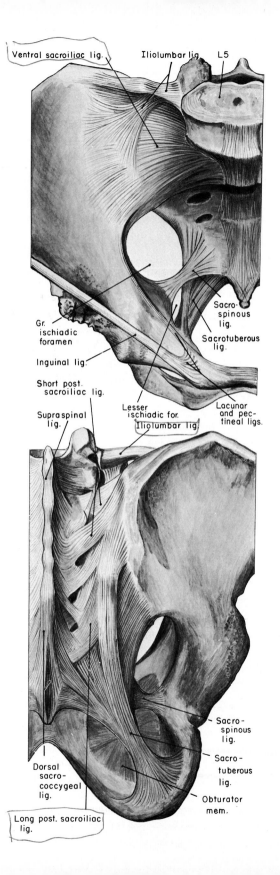

comes important. Additional measurements can be made between the pubic symphysis and the tip of the coccyx or the sacrococcygeal joint, transversely between the inside surfaces of the ischial tuberosities, and obliquely between the junction of the pubic and ischial rami anteriorly to the midpoint in the sacrotuberous ligament posteriorly. These are shown in Figure 5-55. All of these measurements can be determined radiologically or by manual palpation via the vagina. They are longer in most females than in most males.

GENERAL DESCRIPTION

Now that you have studied the bony pelvis, have placed the ligaments on the pelvis, and have filled in the obturator foramen with the obturator membrane, you are ready to begin a study of the contents of this bony pelvis. If the pelvis is held in the correct position, you will note that it is not a bowl-like area that sits directly superiorly-inferiorly, nor is it one that has its direction anteropos-teriorly. As one looks inferiorly through the pelvis, one looks in an inferior-posterior direction. If the pelvic cavity is thought of as a bowl with several holes in the midline, the floor of this bowl will resemble the **floor** of the pelvis (**pelvic diaphragm**) made up of muscles. The area superior to this floor of the bowl or the pelvic diaphragm is the **pelvic cavity proper;** the part inferior and posterior to the bowl is the **perineum,** which will be studied from an inferior approach.

The student should realize that the pelvic cavity is never the empty structure that it resembles during dissection. That area that is not filled with the pelvic organs themselves contains the sigmoid colon, cecum, and the ileum, the amount of sigmoid colon and cecum varying from body to body. If, however, the intestines are removed from the pelvis, the reflections of the peritoneum in the pelvis can be seen.

The **sigmoid colon** continues into the pelvic cavity as the **rectum** and is just anterior to the promontory of the sacrum. The upper two-thirds of the rectum is cov-

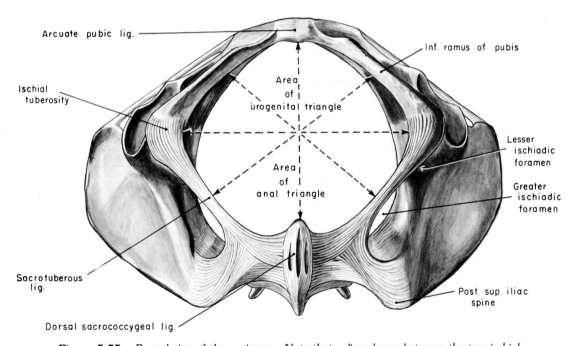

Figure 5-55. *Boundaries of the perineum. Note that a line drawn between the two ischial tuberosities divides this area into anterior urogenital and posterior anal triangles. Dotted lines represent measurements (pelvimetry) which can be compared with the baby's head (fetal cephalometry).*

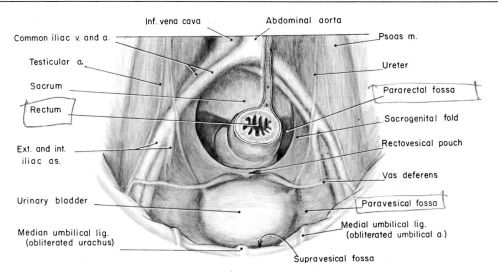

Figure 5-56. *The male pelvic cavity* in situ. *Note various structures showing through the peritoneum (exaggerated).*

ered with peritoneum, making it a retroperitoneal organ. In the male (Fig. 5-56) the peritoneum continues anteriorly to the base of the urinary bladder. It reflects superiorly over the urinary bladder and thence onto the anterior abdominal wall. The fossa formed between the rectum and the urinary bladder is the **rectovesical pouch.** In addition, the peritoneum is distributed over the rectum in such a manner that there is usually a fossa on either side of the rectum—the **pararectal fossae**—and if the bladder is dilated, similar fossae are formed on either side of it—the **paravesical fossae.** Usually a fold of the peritoneum stretches transversely across the pelvic cavity from one lateral side to the other; this is located between the rectum and the urinary bladder and is continuous with folds extending posteriorly to reach the sacrum. These are **sacrogenital plicae or folds.** All of these structures are pictured in Figure 5-56.

In the female (Fig. 5-57) the uterus intervenes between the rectum and the bladder, which divides the pelvic cavity into two pouches, one in which the peritoneum reflects from the rectum to the posterior wall of the vagina and uterus, and another in which the peritoneum is reflected from the anteroinferior (vesical) surface of the uterus to the urinary bladder. These pouches are the **rectouterine** and the **uterovesical pouches,** respectively. The **body** and **fundus** of the **uterus** can be seen; the

cervix, however, extends into the superior end of the vagina and is not visible at this time. The **uterine tubes** extend laterally from the fundus of the uterus as far as the lateral pelvic wall. It is obvious that the peritoneum does not simply reflect over the uterus itself, but also is reflected from the floor of the pelvis superiorly and anteriorly over the uterine tubes, and then back to the floor of the pelvis again (the **broad ligaments**). Therefore, the antiflexed uterus and its accompanying uterine tubes and layers of peritoneum cut the pelvic cavity in the female into an anterior and a posterior portion. The uterus can also be seen to be bent anteriorly, so that the posterior surface is actually a posterosuperior surface and the anterior surface an anteroinferior surface (note Fig. 5-62). This is the normal position for the uterus, and it is approximately at right angles to the direction taken by the vagina.

The peritoneum in the female has a distribution around the rectum similar to that found in the male, and once again there are pararectal fossae formed (Fig. 5-57). In this area two folds of peritoneum can be seen stretching from the sacrum to the base of the uterus. These are folds of peritoneum (uterosacral folds) over the **uterosacral ligaments.** There are **paravesical** fossae on either side of the bladder in the female as well as in the male.

If the uterine tubes are followed laterally until the

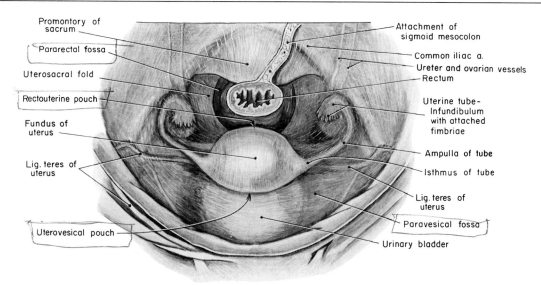

Promontory of
  sacrum
Pararectal fossa
Uterosacral fold
Rectouterine pouch
Fundus of
  uterus
Lig. teres of
  uterus
Uterovesical pouch

Attachment of
  sigmoid mesocolon
Common iliac a.
Ureter and ovarian vessels
Rectum
Uterine tube-
  Infundibulum
  with attached
  fimbriae
Ampulla of tube
Isthmus of tube
Lig. teres of
  uterus
Paravesical fossa
Urinary bladder

Figure 5-57. *The female pelvic cavity* in situ. *Note structures showing through the peritoneum (exaggerated).*

fimbriated end is reached, they will be seen to be in close approximation to an almond-shaped structure hanging onto the posterior-superior aspect of the broad ligaments (Fig. 5-58). These are the **ovaries** and are suspended from the broad ligaments by two layers of peritoneum called the **mesovarium.** In addition to this mesentery, the ovaries can be seen to be attached medially to the uterus by a firm ligamentous structure, the **ligament of the ovary,** and laterally to the lateral pelvic wall by a **suspensory ligament** consisting of a reflection of peritoneum over the vascular system entering and leaving the ovary.

Although the **ureter, vas deferens,** and **ligamentum teres uteri** (round ligament) are not visible through the peritoneum, their locations on the lateral pelvic wall should be noted carefully (Figs. 5-58 and 5-63). The position of the ureters must be taken into consideration at all times during pelvic surgery. The position ordinarily taken by the vas deferens in the male is taken by the round ligament of the uterus in the female.

Now if the student will visualize the pelvic cavity with the peritoneum removed, he or she will see that the midline structures, the structures on the lateral pelvic wall, and those on the anterior surface of the sacrum are all embedded in subserous fascia (Fig. 5-58). Indeed, the subserous fascia in the abdominal cavity is directly con-

tinuous with similar fascia in the pelvis. It continues into the true pelvis, onto the muscles forming the floor of the pelvis, and thence onto the midline organs, where it blends with the capsule of each respective organ.

Removal of the peritoneum in this manner shows that the **sacrogenital folds** in the male actually cover condensations of this subserous fascia, that the **uterosacral ligaments** in the female are similar condensations. Other condensations of this fascia occur at the point where the blood vessels leave the lateral pelvic wall to course along the floor of the pelvis to reach the uterus (the **cardinal ligaments** of the uterus) and anteriorly from the pubis to the prostate in the male (puboprostatic ligaments) or the bladder in the female (pubovesical ligaments) and thence to the uterus (uterovesical ligaments). These ligaments are shown diagrammatically in Figure 5-59.

Removal of the peritoneum forming the broad ligaments also reveals that the **ligamentum teres uteri** and **ovarian ligament** are almost continuous structures; these two structures correspond to the gubernaculum in the male. The broad ligaments can be divided into an area inferior to the mesovarium and an area superior to this structure. The superior area, between the mesovarium and the tube, is the mesosalpinx (mesentery of the tube).

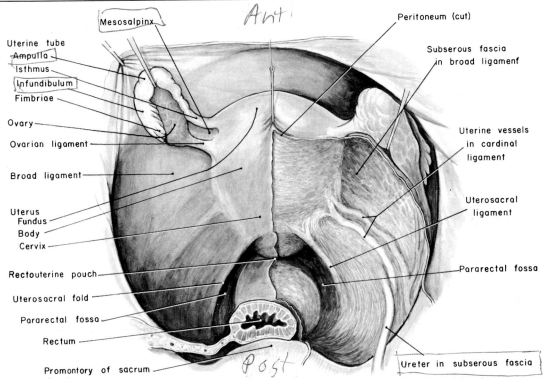

Figure 5-58. *The female pelvic cavity I. The peritoneum is intact on the left and has been removed on the right.*

If the subserous fascia is now dissected, the vital structures coursing in it are revealed.* In the male (Fig. 5-63) the rectum is now denuded of fascia, and near the base of the urinary bladder the **seminal vesicles** can be seen; these have ducts that join the vas deferens to form the **ejaculatory ducts.** These enter the prostate gland and finally empty into the urethra. The prostate gland is located at the base of the bladder. The ureters can be seen following along the lateral pelvic wall and coming into close relation with the vas deferens and seminal vesicles before entering the base of the bladder.

In the **female,** when the rectum is bare of fascia, it is seen that the **rectouterine pouch** in reality is an area between the rectum and the superior end of the vagina (note Fig. 5-62). This is an important fact to realize since the very inferior end of the pelvic cavity is often a place where pus can collect as a result of infections of the peritoneum; it can be drained via the vagina. The **ureters** in the female cross the lateral pelvic wall and course inferior and posterior to the large arteries and veins of the uterus (Fig. 5-60). They also course to the lateral side of the vagina to reach the base of the bladder.

With the exception of the middle sacral artery, which is located on the anterior surface of the sacrum, the **arteries** entering the pelvis follow along the lateral pelvic wall and then the floor of the pelvis to reach the organs to be vascularized (Figs. 5-63 and 5-64). The **veins** take a corresponding pathway in a reverse direction. In fact, the veins form a very profuse network on the floor of the pelvis, which in dissecting will have to be removed. In addition to the vascular system, the nerves forming the visceral afferent and efferent systems are also found in this loose fascia. The hypogastric plexus, on the anterior surface of the sacrum, is made up of contributions from the sympathetic system as well as the parasympathetic.

*C. E. Tobin and J. A. Benjamin, 1945, Anatomical and surgical restudy of Denonvillier's fascia, *Surg. Gynecol. Obstet.* 80: 373. The fascial arrangements on the anterior surface of the rectum, around the seminal vesicles, and on the posterior surfaces of the bladder and prostate are described and the origins discussed.

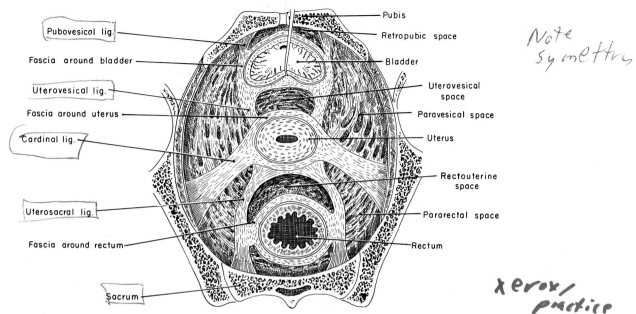

Pubovesical lig.
Fascia around bladder
Uterovesical lig.
Fascia around uterus
Cardinal lig.
Uterosacral lig.
Fascia around rectum
Sacrum

Pubis
Retropubic space
Bladder
Uterovesical space
Paravesical space
Uterus
Rectouterine space
Pararectal space
Rectum

Note symmetry

xerox/practice

innervations

Figure 5-59. *Diagrammatic presentation of the condensations of the subserous fascia which have been called ligaments. Note that this fascia also blends with the capsule of each organ. The male has similar folds and ligaments without an intervening uterus.*

These nerves form a very profuse network of fibers which follow the floor of the pelvis in this loose fascia and reach the organs to be innervated in this manner (Fig. 5-63). The two sympathetic chains can be seen to be prolonged into the pelvic cavity anterior to the sacrum in this same layer (Fig. 5-64).

Now, if the student can visualize the removal of the midline structures in the pelvic cavity, and the removal of vascular and nervous tissue to these organs, he or she will then be able to see the parietal covering of deep fascia (Fig. 5-61). This deep fascia is called parietal pelvic fascia as a general term, but also takes on the name of the particular muscle it covers. If it is on the obturator internus muscle, it is called obturator fascia; if on the levator ani muscle, the supra-anal fascia; if on the coccygeus muscle, the coccygeus fascia, etc. This fascia is continuous with the transversalis fascia in the abdominal cavity.

If the parietal fascia is removed, one can see that the floor and sides of the pelvic cavity are covered by muscles and nerves (Figs. 5-61 and 5-64). Starting posteriorly, the **sacrum** can be seen, and just lateral and anterior to it the many nerves making up the **sacral plexus.** Just deep to these nerves is the **piriformis** muscle, arising from the

sacrum and extending laterally to leave the pelvis via the greater ischiadic foramen. Anterior to this position, the **coccygeus** muscle (S4 and 5) extends from the ischial spine to the lateral surface of the coccyx. Anterior to that the **levator ani** muscle (inferior rectal, S2 and 3, plus twigs from S4 and 5) can be seen arising from the pubis, lateral pelvic wall, and ischiadic spine, and extending inferiorly in a funnel-shaped manner to blend with the organs in the pelvic cavity. This muscle can be divided into **pubovaginal** (levator prostatae), **puborectal, pubococcygeal,** and **iliococcygeal** portions (Fig. 5-61). The pubovaginal portion blends into the vagina in the female or suspends the prostate in the male. Some fibers of the puborectalis muscle not only blend with the rectum but also meet similar fibers from the opposite side in such a fashion as to form a sling for the rectum just posterior to it. The pubococcygeal and iliococcygeal portions attach to the coccyx. The part of the levator ani muscle arising from the lateral pelvic wall is from the fascia on another muscle aiding in forming the walls of the pelvis, the **obturator internus** muscle, and the fascia at this point forms a band called the **arcus tendineus.** The two muscles—the levator ani and the coccygeus muscles on each side—

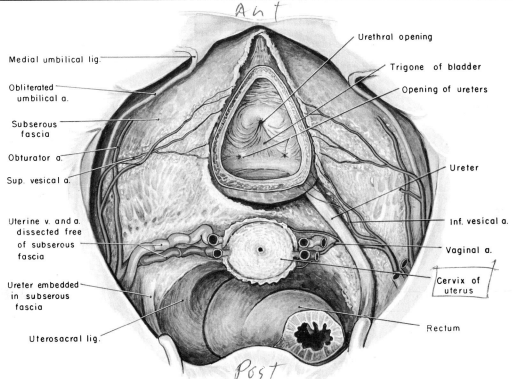

Figure 5-60.   *The female pelvic cavity II. The uterine vessels have been dissected free of the subserous fascia in which they course (the cardinal ligaments) on the left, and have been removed on the right to reveal the course of the ureter. The uterus has been removed to reveal the urinary bladder, which has been cut to show its internal structure.*

form the floor of the pelvis and are designated as the **pelvic diaphragm.**

The **transversalis fascia*** blends with the periosteum at the brim of the true pelvic cavity, but then continues into the bowl on all sides as the parietal **pelvic fascia.**

In summary, if the student can visualize the pelvic floor as being formed of these muscles, with the sacrum located posteriorly and the pubic bones anteriorly; these muscles covered with parietal pelvic fascia which also covers the sacral nerves; the vessels, ureters, vas deferens, and other structures contained in a layer of subse-

rous fascia; the organs, such as the uterus, vagina, prostate, and seminal vesicles, in the midline, covered with a layer of subserous fascia; and then the whole area covered with peritoneum—then he or she is ready to grasp the details of these structures which will now be given.

OVARIES

The **ovaries** in the nullipara are smooth, pink in color, and the shape and size of a large almond. In the elderly person they are small and quite roughened in appearance. The ovaries are **located** between the uterus medially and the lateral pelvic wall laterally, and are suspended from the posterior-superior surface of the broad ligaments by a mesentery—the **mesovarium** (Fig. 5-65) The ovary is attached to the uterus by a derivative of the gubernaculum; this short ligament is the **ligament of the ovary.** The ovary is also suspended laterally from the

*D. Browne, 1933, Some anatomical points in the operation for undescended testicle, *Lancet* 1: 460. An interesting plea for a sensible concept of the fascia between peritoneum and the abdominal wall musculature. The author would combine transversalis, subserous, and the thin connective tissue which is part of the peritoneum into a single "abdominal connective tissue."

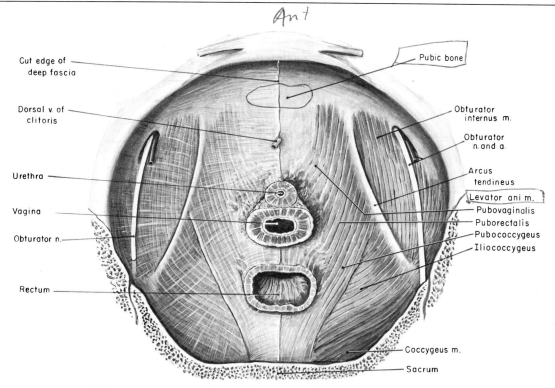

Figure 5-61. *The female pelvic cavity III. The midline pelvic organs have been removed. The anterior part of the pelvic floor is seen, with deep fascia present on the left and removed on the right. Note that removing the subserous fascia has eliminated the blood vessels and the autonomic nerves.*

pelvic wall by the **suspensory ligament of the ovary,** which is made up of peritoneum covering the vascular system entering and leaving this organ. The mesenteries forming the mesovaria are continuous with the germinal epithelial layer on the ovaries themselves.

Since either the ileum or the colon fills in the pelvic cavity, the ovaries are in definite **relation** to either one of these two organs. Each ovary is also close to the lateral pelvic wall and separated only by peritoneum from the umbilical artery, obturator vessels, and nerve on the obturator internus muscle.

The **arteries** to the ovary arise from the aorta just inferior to the origin of the renal arteries (Fig. 5-40). These course retroperitoneally through the abdominal cavity, across the iliac vessels, and approach the ovaries from the lateral side, the peritoneum covering these vessels forming the suspensory ligament of the ovary. The ovarian artery, after supplying the ovary itself, continues

just inferior to the tube, between the layers of the broad ligament, supplying branches to the tube itself, and then anastomoses with the uterine artery, which ramifies on the lateral side of the uterus (Fig. 5-66). The ovarian artery sends branches to the ovarian ligament and to the proximal end of the ligamentum teres as well as to these other structures. The ovarian **veins** form a pampiniform plexus which follows the same course as just described for the arteries except that the left ovarian vein drains into the left renal vein rather than into the inferior vena cava as does the right.

Since the ovary developed in the vicintiy of the kidneys and subsequently migrated, the nerves and blood vessels were carried along. The ovary receives **sympathetic fibers** from the tenth and eleventh thoracic segments via the renal plexus and the ovarian plexus on the ovarian artery. The **parasympathetic fibers** are derived from this same ovarian plexus and are vagal in

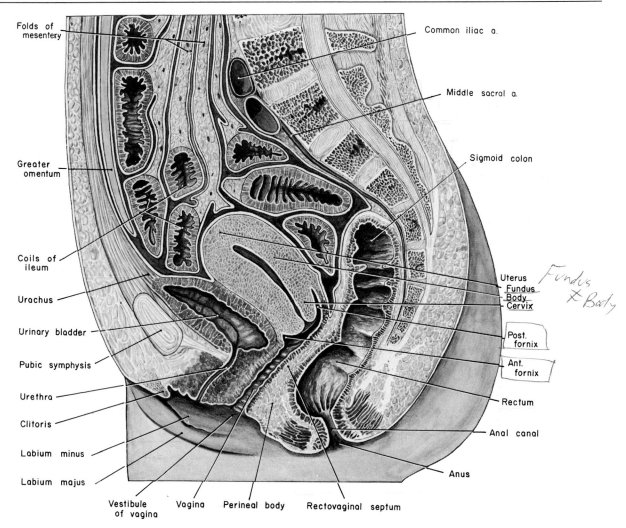

Folds of mesentery
Greater omentum
Coils of ileum
Urachus
Urinary bladder
Pubic symphysis
Urethra
Clitoris
Labium minus
Labium majus
Vestibule of vagina
Vagina
Perineal body
Rectovaginal septum

Common iliac a.
Middle sacral a.
Sigmoid colon
Uterus
Fundus
Body
Cervix
Post. fornix
Ant. fornix
Rectum
Anal canal
Anus

Fundus & Body

Figure 5-62. *Sagittal section of the female pelvic cavity and perineum. Note that the urethra, vagina, and rectum take similar directions, which are opposite to those taken by the uterus and anal canal.*

origin. The **function** of these nerves is problematical; they probably innervate blood vessels only, although Hill,* claims that the nerve supply is necessary for its cyclical function. **Sensory fibers** are more important; they end in the tenth and eleventh thoracic segments. Ovarian pain is referred to the skin innervated by these particular segments of the spinal cord.

The **lymphatics** for the ovary ascend, because of its embryology, to lumbar nodes alongside the inferior vena cava and aorta.

The ovary is responsible for **producing ova** and, in addition, is an **endocrine organ.** Under the influence of the hypophysis, the ovary goes through a **cycle** approximately every 28 days in the nonpregnant adult female. In

*R. T. Hill, 1949, Adrenal cortical physiology of spleen grafted and denervated ovaries in the mouse, *Exptl. Med. Sur.* 7: 86. This work indicates that the nerve supply to the ovary plays a very important role in maintaining its cyclical changes; denervated ovaries apparently secrete adrenal cortical like hormones, for adrenalectomized mice do not live unless such a denervation is performed.

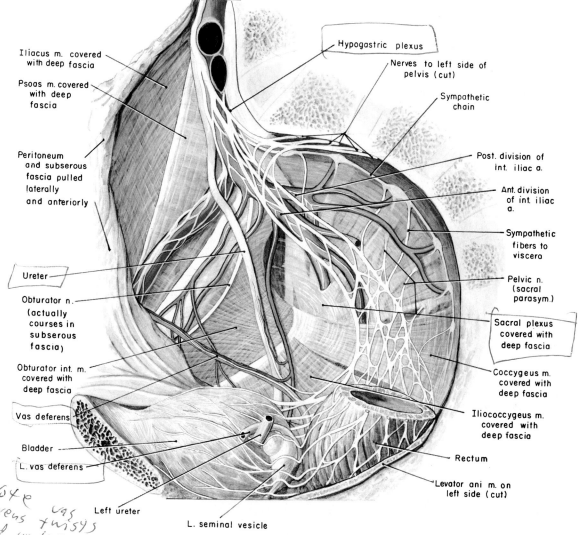

Iliacus m. covered with deep fascia

Psoas m. covered with deep fascia

Peritoneum and subserous fascia pulled laterally and anteriorly

Ureter

Obturator n. (actually courses in subserous fascia)

Obturator int. m. covered with deep fascia

Vas deferens

Bladder

L. vas deferens

Left ureter

L. seminal vesicle

Hypogastric plexus

Nerves to left side of pelvis (cut)

Sympathetic chain

Post. division of int. iliac a.

Ant. division of int. iliac a.

Sympathetic fibers to viscera

Pelvic n. (sacral parasym.)

Sacral plexus covered with deep fascia

Coccygeus m. covered with deep fascia

Iliococcygeus m. covered with deep fascia

Rectum

Levator ani m. on left side (cut)

Note deferens vas around ureta twisys sem/vs "The big difference"

Figure 5-63. *Longitudinal section of the male pelvic cavity after removal of the peritoneum. Note that the autonomic nerves and the blood vessels course in the subserous fascia.*

the first part of the cycle, ovarian **follicles** are stimulated to grow, and the ova are contained in these follicles. Approximately 14 days before the onset of the next menstruation, an ovum bursts through the wall of the ovary, a process called **ovulation.** The site of the follicle is then changed to a **corpus luteum.** When this corpus luteum loses its functional period, it changes to connective tissue and is called a **corpus albicans.** The ovary is stimulated to pass through this cycle by the follicle stimulating and

luteinizing hormones from the anterior lobe of the hypophysis. The follicles produce a hormone—**estrogen**—and the corpus luteum produces **progesterone** as well as a small amount of estrogen. These two hormones are responsible for changes in the endometrial lining of the uterus. This menstrual cycle starts at puberty and ceases at the menopause.

In addition to endocrine imbalances, the ovaries are subject to inflammatory disease. They are also the site of

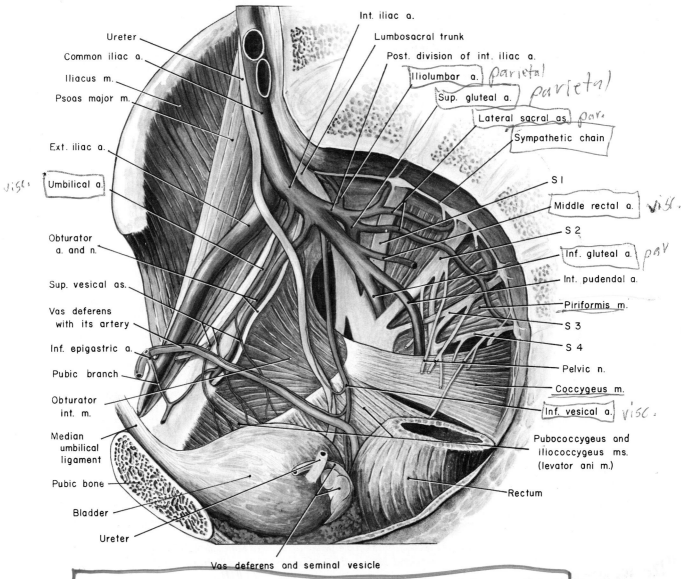

Ureter

Common iliac a.

Iliacus m.

Psoas major m.

Ext. iliac a.

Umbilical a.   *visc.*

Obturator a. and n.

Sup. vesical as.

Vas deferens with its artery

Inf. epigastric a.

Pubic branch

Obturator int. m.

Median umbilical ligament

Pubic bone

Bladder

Ureter

Vas deferens and seminal vesicle

Int. iliac a.

Lumbosacral trunk

Post. division of int. iliac a.

Iliolumbar a.   *parietal*

Sup. gluteal a.   *parietal*

Lateral sacral as.   *par.*

Sympathetic chain

S 1

Middle rectal a.   *visc.*

S 2

Inf. gluteal a.   *par*

Int. pudendal a.

Piriformis m.

S 3

S 4

Pelvic n.

Coccygeus m.

Inf. vesical a.   *visc.*

Pubococcygeus and iliococcygeus ms. (levator ani m.)

Rectum

Figure 5-64. *Longitudinal section of the male pelvic cavity after removal of peritoneum and both subserous and deep fasciae. The vessels, which ordinarily would have been removed, have been left intact.*

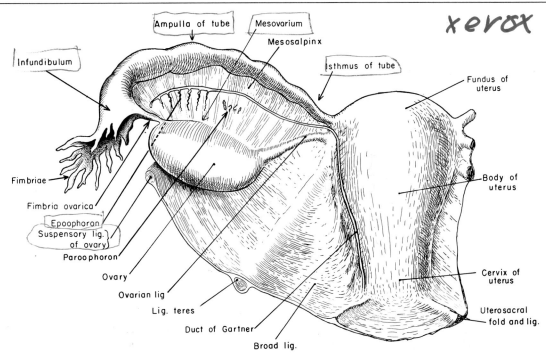

Figure 5-65. *The ovary and broad ligament.*

several types of cancer, some of which are cystic in nature. The greatest number of cysts, however, are due to enlargement of a follicle and are benign.

UTERINE TUBES

The **uterine tube** (Figs. 5-57, 5-58, and 5-65) is approximately 4 inches long and ¼ inch in diameter. It is attached to the fundus of the uterus. The tube extends laterally, enters the free edge of the broad ligament, and curls around the ovary to end in approximation with it on the medial surface of that organ. The medial end is fairly narrow and straight, and is called the **isthmus.** The remainder of the tube, the **ampulla,** is a bit wider and slightly coiled. The abdominal end is the funnel-shaped **infundibulum,** which ends in many finger-like processes called **fimbriae.** The uterine tube forms a direct connection between the peritoneal cavity at one end and the uterine cavity at the other.

The uterine tube is related to the ileum and to the cecum in addition to the organs mentioned above. The latter relationship induces confusion between pathology of the appendix and the uterine tube.

The uterine tube contains **smooth muscle** and is lined with a **ciliated columnar epithelium.** At the time of ovulation, the fimbriated end of the tube picks up the ovum, and by a combined ciliated action and muscle contraction the ovum is pushed down the tube toward the waiting uterus.

It should be noticed that the tubes form two openings in the peritoneal cavity in the female that do not occur in the male. The peritoneal sac in the male is completely enclosed, with no holes in it. This opening to the outside world via the uterine tubes can be a source of danger from infection. In addition, spermatozoa have been known to travel through the uterus to the tubes and to fertilize an ovum on the surface of the ovary or even on the peritoneum. Such fertilized ova may develop, resulting in an ectopic pregnancy. If an ovum is to be impregnated by sperm in the normal fashion, it is usually accomplished as the ovum is traversing the uterine tube. The fertilized ovum then continues into the uterus and embeds itself in the endometrium, and pregnancy ensues. Tubal pregnancies also occur and are a form of ectopic pregnancy that can be quite dangerous to the patient.

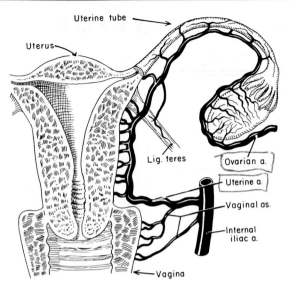

Figure 5-66. *Blood supply to the ovary, uterus, and vagina.*

The **blood supply** to the uterine tube is from the ovarian and uterine vessels (Fig. 5-66).

The **sympathetic nerve** supply to the uterine tube is interesting in that the fimbriated end derives its fibers from the ovarian plexus, while the uterine end receives its sympathetic nerves via the uterine plexus. They arise in spinal cord segments T10 to L1. The **parasympathetic nerves** also seem to be dual in nature. The distal end of the tube is innervated by vagal fibers, and the proximal end (near the uterus) by the pelvic nerve. The exact role of these nerves is problematical. **Sensory fibers** end in segments T10 to L1 and pain may be referred to areas on the skin innervated by these particular segments.

Lymphatics from the uterine tube terminate in two glands alongside the inferior vena cava and the aorta.

The area of the broad ligament between the mesovarium and the tube is the **mesosalpinx.** Several **minor tubules** (Fig. 5-65) can be found in this area that correspond to structures in the male. These tubules are remnants of the mesonephric tubules and correspond to the ducts of the testes and to the lobules of the epididymis. They end in a horizontal tubule called the duct of the **epoophoron,** the latter being a remnant of the mesonephric (Wolffian) duct and corresponding to the canal of the epididymis. It continues as Gartner's duct

(Fig. 5-65). The **paroophoron** consists of a few microscopic tubules, which lie nearer the uterus; these are derived from the mesonephros and correspond to the paradidymis. Abnormal growth can occur in these tubules. The origin of these structures is explained in Figure 5-89 and in the table found on page 344.

The uterine tube is subject to infections (salpingitis), particularly of the venereal variety. The tube is sometimes not patent, which naturally would interfere with fertilization, and tubal pregnancies are quite common.

UTERUS

The **uterus** (Figs. 5-57, 5-58, 5-62, 5-65, and 5-66) is usually described as a pear-shaped organ with a comparatively narrow **cervix** extending into the vagina, a wider **body,** which ends superoanteriorly as the **fundus,** to which are attached the uterine tubes. The cervix is divided into vaginal and supravaginal portions, and the point of juncture between it and the body of the uterus is often called the **isthmus.**

The uterus is a hollow organ, and the lumina of the uterine tubes are directly continuous with the **lumen** inside the uterus. This, in turn, opens into the vagina via the cervix. The lumen inside the uterus has a different shape in different parts. In the cervix it is a fairly small, narrow opening approximately 1 inch in length. Its distal end is called the **ostium** of the uterus. The cavity in the body of the uterus is approximately 1½ inches in length and is triangular in shape.

The walls of the uterus are rather thick and made up of peritoneum (perimetrium), smooth muscle (myometrium), and a glandular lining on the inside (endometrium). Under the influence of estrogen, produced by the follicles of the ovaries, the endometrium of the uterus is stimulated to increase in thickness. The endometrial glands increase considerably in length. After ovulation they become coiled, and glycogen is deposited in the glands as a result of progesterone stimulation from the corpus luteum of the ovary. If the ovum is fertilized, it imbeds itself in the endometrium and menstruation ceases. If, however, fertilization does not occur, the ovum is simply lost into the vagina. In this case the endometrium was built up but did not accomplish its purpose. Therefore this endometrial layer is sloughed off and leaves the body through the vagina (menstruation). The cycle (approximately 28 days in length, but quite variable) then starts anew in order to build up the endometrium of the uterus once again. This will continue until menopause stops the cycle or pregnancy occurs.

The uterus, located between the rectum and the bladder, overhangs the bladder anteriorly (Fig. 5-62). It is normally bent anteriorly in this manner and forms approximately a right angle to the vaginal canal. Its vesical surface is in direct relation to the bladder, and the peritoneum on this surface does not cover the entire uterus. The intestinal surface of the uterus is completely covered with peritoneum and is in direct relation to the ileum. It should be noted that the appendix can hang down into the pelvic cavity very close to the uterus.

Many of the **ligaments of the uterus** have been described previously. There are five paired ligaments. The **broad ligaments** (Fig. 5-65) are extensions of the peritoneum from the margins of the uterus to the lateral pelvic walls; these two layers stretch over the uterine tubes from the floor of the pelvis. In addition to the uterine tube these folds of the broad ligament, as just described, enclose embryological remnants, the ovarian ligament, vessels, lymphatics, and nerves. The **round ligaments** (Fig. 5-57) of the uterus (ligamentum teres uteri) extend laterally from the fundus between the folds of the broad ligament to the lateral pelvic wall, and finally leave the abdominal cavity through the internal ring of the inguinal canal. The two **cardinal ligaments** (Figs. 5-59 and 5-60) extend from the lateral pelvic wall to the cervix of the uterus and are condensations of subserous fascia around the uterine blood vessels (note Power, 1944).* The **uterosacral ligaments** are also condensations of subserous fascia that extend from the sacrum, around the rectum, to the cervix of the uterus. (Similar condensations have been described extending anteriorly from the uterus to the base of the urinary bladder as seen in Figure 5-59; these are **uterovesical ligaments.**) The uterosacral and cardinal ligaments are by far the most important of the ligaments of the uterus. Without these the uterus tends to pass inferiorly through the vagina, a condition called prolapse of the uterus. The round ligaments are of little value, and the reflections of the peritoneum are equally unimportant. The broad liga-

ments probably do serve a purpose in holding the uterus in position. It is important for the uterus to be held in its normal position, for painful menstruation (dysmenorrhea) results from displacement.

The uterus has a rich **blood supply** (Figs. 5-60 and 5-66). The uterine arteries are branches of the internal iliac and enter the uterus near its inferior extent at the base of the broad ligaments. This artery then continues superiorly on the lateral side of the uterus sending branches to it all the way along its course. It anastomoses with the ovarian artery coursing along the uterine tube. The **veins** are a plexus of veins that end in the internal iliac.

The **sympathetic nerves** arise from spinal cord segments T12 and L1 and proceed to the uterus via the hypogastric plexus, and thence on the uterine arteries. The **parasympathetic** supply is via the pelvic nerve from the second, third, and fourth sacral segments of the cord. The role played by these nerves is problematical for the uterus can function when denervated (note Hill and Alpert,† 1961). The **sensory nerve supply** is more important. The nerve endings are found associated with the arteries in the myometrium (very few in endometrium). Those from the peritonealized parts of the uterus follow the sympathetic nerves, via the hypogastric plexus, to spinal cord segments T12 and L1; those from the cervix follow the pelvic nerves to the sacral segments. Uterine pain is referred to the sacroiliac and pubic regions.

The **lymphatics** from the uterus end in numerous glands. They are found on the rectum, anterior to the sacrum, around the iliac arteries, and high up in the abdominal cavity close to the inferior vena cava and the aorta. Lymphatics from this organ are known to course along the ligamentum teres to reach the inguinal nodes.

VAGINA

The **vagina** is the copulatory canal in the female which starts as an opening in the external genitalia posterior to the urethra and anterior to the rectum (Fig. 5-62). The opening to the vagina in the virgin is usually guarded by a **hymen,** two folds of mucous membrane that extend into

---

*R. M. H. Power, 1944, The exact anatomy and development of the ligaments attached to the cervi uteri, *Surg. Gynecol. Obstet.* 79: 390. This report gives the lateral ligaments of the cervix uteri (cardinal) more dignity than "a condensation of subserous fascia" as stated in this text.

†R. T. Hill and M. Alpert, 1961, The corpus luteum and the sacral parasympathetics. *Endocrinology* 69: 1105. A fascinating theory is presented claiming that the uterine endometrium acts as an organ involved in regression of the postovulatory corpus luteum, and that denervation of the uterus prevents this action.

the lumen of the vagina from the lateral walls. The hymen may take the form of a complete membrane, which has to be severed at the first menstruation, a sieve-like structure, or it may be entirely absent. After rupture, the fragments of the hymen persist as small nodules designated as **hymenal caruncles.**

The area between the labia minora just inferior to the hymen is the **vestibule** of the vagina and will be studied in more detail with the description of the external genitalia.

The **vagina proper** extends from the hymen to the cervix of the uterus. It traverses a muscular layer in the perineum—the urogenital diaphragm—and then goes through the pelvic diaphragm by passing the inferior edges of the levator ani muscles. These diaphragms are diagramed in Fig. 5-83 on page 322. Since the cervix extends into the anterior aspect of the superior end of the vagina, the anterior wall is shorter than the posterior. The anterior wall is approximately 3 inches in length, while the posterior wall is 3½ to 4 inches long (Fig. 5-62). The clefts produced by the cervix projecting into the vagina are called **fornices;** there are anterior, posterior, and two lateral fornices, the posterior being by far the deepest. The posterosuperior part of the vagina is covered by peritoneum, but the anterior wall makes no contact with it.

The **relations** of the vagina are important, for many structures in the pelvic cavity can be palpated via this canal. These relations are as follows (Fig. 5-62). **Anterior**—bladder and urethra. **Posterior**—the inferior end of the pelvic cavity, small intestine, rectum, and perineal body (a point in the midline between the vaginal and anal openings—see Fig. 5-62). **Lateral**—broad ligaments, ureters, uterine vessels, pelvic surface of the levator ani muscle, sphincter urethrae muscle, greater vestibular glands, and bulbospongiosus muscles. (The last three structures will be studied with the perineum; they are diagramed in Fig. 5-83 on page 322.) Thus any structure in the inferior part of the pelvic cavity can be palpated via the vagina, particularly if bimanual palpation is used (one hand on the anterior abdominal wall). Structures contained in the broad ligaments can be felt, and when the ovary is increased in size from any pathological condition, it can be felt from the vagina as well. Naturally the cervix of the uterus is also palpable; it is also visible if the vagina is dilated.

The **walls of the vagina** are in contact with each other except where the cervix intervenes. They are made

up of smooth muscle and are lined with stratified squamous epithelium. Ridges (vaginal rugae) are found on the anterior and posterior walls, and several transverse folds extend around the walls of the vagina which connect these ridges. During sexual excitement the vagina elongates in the area of the posterior fornix, produces a clear fluid by transudation, and the distal third undergoes a vasocongestion, all parasympathetic actions. The vestibule of the vagina has two **greater vestibular** or bulbourethral glands* which empty just inferior to the hymeneal caruncles (Fig. 5-80); they serve in lubrication and are also under control of the parasympathetic system.

The **arteries** to the vagina branch from the uterine arteries, from the internal iliac artery directly, and from the middle rectal; the inferior end derives its blood supply from branches of the internal pudendal artery (Figs. 5-76 and 5-80).

The vagina obtains its **sympthathetic nerve** supply from the hypogastric plexus (L1–L3) and its **parasympathetic** from the sacral region of the cord via the pelvic nerve. The inferior end of the vagina is more sensitive than the superior end, receiving its **sensory branches** from the pudendal nerve.

The **lymphatics** drain to glands on the rectum and along the iliac arteries but some of the lymphatics from the inferior end of the vagina may drain into the inguinal glands in the groin.

URETERS

The **ureters,** especially in the female, are extremely important structures and must be located and kept in mind with all pelvic surgery (Figs. 5-58 and 5-60).

The **course** of the ureter is as follows. It enters the pelvic cavity by crossing the external iliac artery near its branching from the common iliac. It follows along the lateral pelvic wall just deep to the peritoneum; in fact, the ureter can be seen as a bulge through the peritoneum. It is anterior to the internal iliac artery and vein. Lateral are the psoas major muscle, the obturator internus muscle, and the levator ani. As the ureter approaches the floor of the pelvis the vessels entering the broad ligament pass

---

*Several small mucous glands, the lesser vestibular, open on the side walls of the vestibule.

Read at cadaver

superior and medial to the ureter, as does the broad ligament itself. The ureter continues inferiorly and after passing deep to these uterine vessels enters the bladder by passing close to the lateral fornix of the vagina. Although the ovary is not in immediate relation, it is very close to the ureter, being anterior and medial to it.

The ureter **in the male** takes a similar course but differs in several ways (Figs. 5-63 and 5-64). There are no uterine vessels to cross anterior to the ureter, and there is no broad ligament to take this same relation to it. The ureter continues in a similar fashion as in the female along the lateral wall, anterior to the iliac vessels, deep to the peritoneum and medial to the levator ani muscle. It continues inferiorly and enters obliquely through the base of the urinary bladder. It is lateral to the vas deferens. The inferior end of the ureters cannot be seen on the posterior surface of the bladder, since they are covered by the vas deferens and the seminal vesicles. The ureters penetrate the bladder wall in an oblique slanting fashion; this is important in preventing reflux of urine into the ureters when the bladder is distended.

The ureters receive small **arteries** from the renal, testicular or ovarian, vesical, and middle rectal arteries.

The ureters receive autonomic innervation from the **sympathetic** nervous system, fibers originating in T12 to L2 spinal cord segments. The vagus contributes the **parasympathetic fibers** to the upper part, the pelvic nerve to the lower part. The ureter can function without a nerve supply; peristalsis will occur. Spasm is induced by stimuli of an unusual nature. **Sensory fibers** return to segments T12 to L2.

The **lymphatics** end in glands closest to it in the pelvis and in the abdominal cavity.

The ureters can become infected—ureteritis. When kidney stones descend from the renal pelvis into the ureter, the pain can be excruciating. Accessory ureters are quite common.

URINARY BLADDER

The **urinary bladder** in cadavers is usually contracted and feels quite firm. It is actually not so in the living state, and its walls can become very thin when the bladder is dilated with urine. The urinary bladder occupies the anterior portion of the pelvic cavity and is just superior and posterior to the pubic bone (Figs. 5-60, 5-62, and 5-67). Note Plate 13 at this time.

When the bladder is empty, it is said to have an **apex,** a **superior surface,** two **inferolateral surfaces,** a **base** or posterior surface, and a **neck.** The apex reaches to a short distance superior to the pubic bone and ends as a fibrous cord, which is a derivative of the urachus (a canal in the fetus connecting the bladder with the allantois). This fibrous cord extends from the apex of the bladder to the umbilicus between the peritoneum and the transversalis fascia; it raises a ridge of peritoneum called the median umbilical ligament. The **superior surface** is the only surface of the bladder covered by peritoneum, although in the male a small part of the base has peritoneal coverings. The superior surface is in relation with the uterus and ileum in the female, and with the ileum and any portions of the colon that happen to be in the pelvic cavity in the male. The base of the bladder faces posteriorly and is separated from the rectum by the vas deferens, seminal vesicles, and ureters in the male, and by the uterus and vagina in the female. The seminal vesicles form a V-shaped structure on the base of the bladder, and the vas deferens enters the cavity of the V. The ureters are also in relation with the vas deferens in this area. The remaining portion of the base is free of these structures and is covered by peritoneum in the male. The **inferolateral** surfaces on each side of the bladder are in relation with the pubic bone, and with the levator ani and obturator internus muscles, but the bladder is actually separated from the pubic bone by the retropubic space, which contains fat. The **neck** of the bladder is the most inferior part next to the urethra. The bladder is a dilatable structure. When filled with urine, the neck remains in position but the superior part rises into the pelvic cavity.

The loose **subserous fascia** of the pelvic cavity is continuous superiorly over the pelvic organs, and this is true for the bladder as well. These condensations of fascia form attachments for the bladder, and those running from the levator ani muscle and the pubic bone to the bladder are called the **pubovesical ligaments** in the female and **puboprostatic ligaments** in the male, the latter name being used because the prostate is located just inferior to the bladder; anything that supports the prostate will in turn fix the urinary bladder.

The **bladder wall** consists of a partial covering of peritoneum, the subserous fascia, a muscular coat made up of intermingling longitudinal and circular fibers of smooth muscle, a submucous layer of connective tissue, and a layer of transitional epithelium.

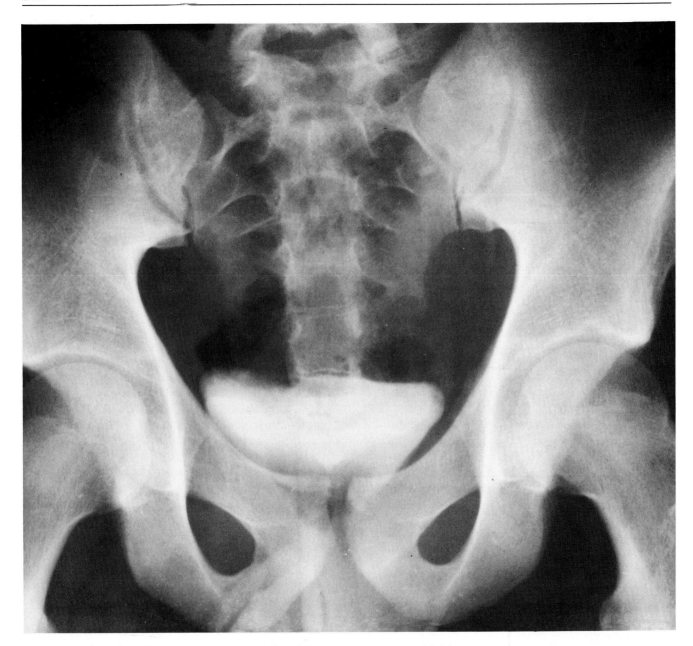

Plate 13.   *Contrast media in urinary bladder.*

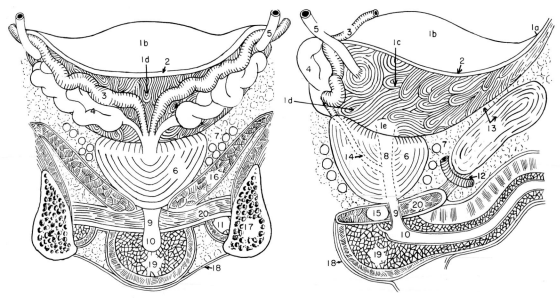

Figure 5-67. *Frontal and sagittal views of the male genitourinary system: (1a) apex of bladder, (1b) superior surface of bladder, (1c) inferolateral surface of bladder, (1d) base of bladder, (1e) neck of bladder, (2) cut edge of peritoneum, (3) vas deferens, (4) seminal vesicles, (5) ureter, (6) prostate, (7) subserous fascia and venous plexus around prostate, (8) prostatic urethra, (9) membranous urethra, (10) spongiose urethra, (11) crus of penis, (12) deep dorsal vein of penis, (13) pubic symphysis and retropubic space, (14) ejaculatory duct, (15) bulbourethral gland, (16) levator ani muscle, (17) inferior pubic ramus, (18) membranous layer of superficial fascia, (19) the bulb, and (20) urogenital diaphragm.*

The interior of the bladder (Figs. 5-60 and 5-68) is wrinkled except at the **trigone,** that area between the openings of the ureters and the opening of the urethra. When the bladder is dilated, all walls become equally smooth in appearance. A ridge is found between the openings of the ureters—the **interureteral** ridge. The trigone of the bladder is an important area in that infections tend to persist at this particular point.

The muscular component of the wall of the bladder—the **detrusor muscle**—consists of three layers of smooth muscle—outer and inner longitudinal layers and a circular layer between them. These layers are not distinct entities for they tend to form an intermingling meshwork. However, the outer longitudinal layer is continuous with a layer of muscle blended with the capsule of the prostate gland in the male or with the vagina in the female. Some fibers, in the male, are said to blend with the outer layer of the rectum, and others blend with the pubovesical and puboprostatic ligaments. The circular and inner longitudinal layers are continued into the urethra. There seems to be no internal sphincter; in fact the muscle arrangement at the neck of the bladder is such as to cause an opening of the urethra rather than a constriction. In addition, a considerable amount of **elastic tissue** is found at the neck of the bladder and this is arranged in such a manner as to aid in keeping the urethra closed.

The **nerves** to the bladder are part of the nerve plexuses found in the pelvic cavity; they follow the arteries to get to the bladder. The **sympathetic** fibers arise in the last thoracic and the first and second lumbar segments of the spinal cord. They innervate the trigone, the ureteral orifices, and the blood vessels. They also carry pain fibers from the bladder. The **parasympathetic** fibers arise in the second, third, and fourth sacral segments. These fibers innervate the detrusor muscle. Afferent fibers from the bladder wall start as stretch receptors. A full bladder stimulates these receptors, the impulses following

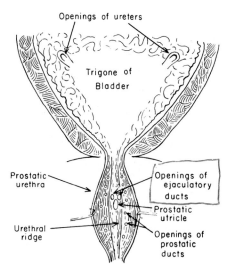

Openings of ureters

Trigone of Bladder

Prostatic urethra

Openings of ejaculatory ducts

Prostatic utricle

Urethral ridge

Openings of prostatic ducts

Figure 5-68. *Diagram of the trigone of the bladder and the prostatic urethra.*

the pelvic nerve back to the sacral region of the spinal cord.

The process of micturition will be described after the urethra has been studied (see page 326).

The **arteries** to the bladder are the superior and inferior vesical branches of the internal iliac artery (Figs. 5-62 and 5-64). There are corresponding **veins** that form a profuse plexus near the base of the bladder; this plexus drains into the internal iliac veins.

The urinary bladder becomes infected easily, particularly in the female. This infection (a cystitis) can ascend the ureters (ureteritis) and reach the renal pelvis (pyelitis). The entire system can be viewed with a cystoscope in both sexes. Fractures of the pelvis often tear the bladder. Many women suffer from a leakage of urine—stress incontinence. Further complications will be mentioned after the prostate and urethra are described.

PELVIC PORTION OF VAS DEFERENS

The student should review the scrotum, the testis, the epididymis, and the first part of the vas deferens at this time. They were described on pages 143 to 145.

After passing through the inguinal canal, the **vas deferens** turns medially and crosses anterior to the iliac vessels to enter the pelvis (Fig. 5-64). On the lateral wall of the pelvis it crosses the umbilical artery, the obturator

nerve and vessels, the vessels to the bladder, and the ureter. It then reaches the sacrogenital fold near the level of the ischial spine, turns medially, loops over the superior end of the seminal vesicle, and reaches the base of the bladder. It continues toward the neck of the bladder medial to the seminal vesicles, joins with the duct of the seminal vesicles to form the ejaculatory duct which, in turn, penetrates the prostate gland to reach the prostatic urethra (Fig. 5-67).

Sympathetic nerve fibers from segments L1 and 2 reach the vas deferens and are responsible for contraction of the smooth muscle fibers during ejaculation. **Parasympathetic** fibers from S2, 3, and 4 also reach the vas deferens and probably serve to inhibit the sympathetic impulses.

The **arteries** to the vas deferens in this region are branches of one of the vesical arteries and from the middle rectal artery. **Veins** drain into the pelvic venous plexus.

SEMINAL VESICLES

Each **seminal vesicle** (Figs. 5-64 and 5-67) consists of a branched, sacculated tube, which is blind at one end and joins with the distal end of the vas deferens to form the ejaculatory duct at the other end. They are about 2 inches long and form a V in their position on the base of the urinary bladder. They are superior to the prostate gland, lateral to the ampulla of the ductus deferens, and lateral to the terminal ends of the ureters. Whereas the anterior surface is attached to the bladder, the posterior surface of the seminal vesicles is in contact with the rectum. Laterally, the seminal vesicles are in relation with the levator ani muscle.

The seminal vesicles contain an alkaline material secreted by the mucosa of this organ. At the time of ejaculation, this fluid joins with spermatozoa being propelled along the vas deferens from the epididymis. The combined fluids enter the prostatic urethra via the ejaculatory ducts.

Sympathetic fibers from L1 and 2 are involved in contraction of the smooth muscle component of the seminal vesicles during ejaculation.

The **arteries** to the seminal vesicles are branches from the vesical and middle rectal branches of the internal iliac artery. Corresponding **veins** drain into the complex pelvic plexus and thence into the internal iliac veins.

Lymphatics end in nodes along the internal iliac arteries on the lateral wall of the pelvic cavity.

EJACULATORY DUCT

The **ejaculatory ducts** (Fig. 5-67) are formed by the union of the vas deferens and the ducts from the seminal vesicles. This occurs immediately superior to the prostate gland near the base of the bladder. These ducts are approximately ¾ inch in length and penetrate the prostate gland to open into the prostatic urethra just to the side of the centrally located prostatic utricle (Fig. 5-68). (The prostatic utricle is homologous to the vagina in the female.)

PROSTATE

The **prostate** gland (Figs. 5-64, 5-67, and 5-68) lies inferior to the neck of the bladder and is in the most inferior aspect of the pelvic cavity. It is a firm structure that has an **apex** inferiorly and a **base** superiorly, and **anterior, posterior,** and **two lateral surfaces.**

The **base** is continuous with the neck of the bladder, being separated from it by a circular groove. The **apex** is in contact with the urogenital diaphragm (a combination of fascia and muscle found in the perineum; see Fig. 5-67) and rests on the pelvic fascia, which forms the superior layer of this diaphragm. The **anterior surface** is separated from the pubic bone by the retropubic space, containing veins and fat, and the puboprostatic ligaments. **Laterally** the prostate is in contact with the levator ani muscles, which aid in its support (the levator prostatae muscles). The **posterior surface** of the prostate is in contact with the most inferior part of the rectum. This relation of the prostate to the rectum is important clinically, for the prostate can be palpated by a finger placed in the rectum through the anal canal. The prostate is surrounded by a dense fascia containing a rich plexus of veins. The surgeon tries to maintain this fascial capsule intact during prostatectomy.

The most important aspect of the prostate is the fact that the urethra penetrates it from its base to its apex (Fig. 5-67). This prostatic urethra (Fig. 5-68) exhibits a midline **crest** posteriorly, in the middle of which is a depression approximately ½ inch deep. This blind sac is the **prostatic utricle** and corresponds to the vagina in the female; the **ejaculatory ducts** open on either side of this

utricle. The glands of the prostate produce an alkaline material which, at the time of ejaculation, empties into the urethra by **several openings** on either side of the urethral crest.

The prostate is divided into **two lateral lobes** and **one middle lobe,** the latter being that portion which lies between the urethra and the ejaculatory ducts; these lobes are not easily separated from one another. The middle lobe, when enlarged, produces a bulging into the trigonal area of the urinary bladder; this may prevent the urethra from closing properly. This allows urine to enter the prostatic urethra making the urge to void almost constant. In addition, when micturition does occur, urine is not completely eliminated. Such residual urine leads to infection. Furthermore, when the entire prostate becomes enlarged, the urethra can become compressed and voiding of urine may be difficult.

Sympathetic fibers from L1 and 2 stimulate the prostatic musculature during ejaculation.

The **arteries** to the prostate are from branches of the inferior vesical and middle rectal branches of the internal iliac, and the plexus of **veins** draining the prostate empty into the internal iliac vein.

Lymphatics end in nodes (1) on the walls of the pelvis, (2) anterior to the bladder, and (3) alongside the iliac arteries.

Cancer of the prostate is a common occurrence. It should be remembered that cells from such growths not only spread by way of lymphatics, but also can reach many areas of the body via the vertebral plexus of veins, the bones being a favorite site of metastasis.

RECTUM AND ANAL CANAL

The **rectum** (Figs. 5-62 and 5-69) is a tube approximately 5 inches in length located just anterior to the sacrum and coccyx and posterior to the bladder, the prostate and seminal vesicles in the male, the vagina and uterus in the female. It serves as a route via which fecal matter can be removed from the body by the process of defecation. It is continuous with the sigmoid colon superiorly and with the anal canal inferiorly. It is a retroperitoneal structure, the peritoneum spreading from its anterior surface to the lateral pelvic walls. The rectum can be divided into

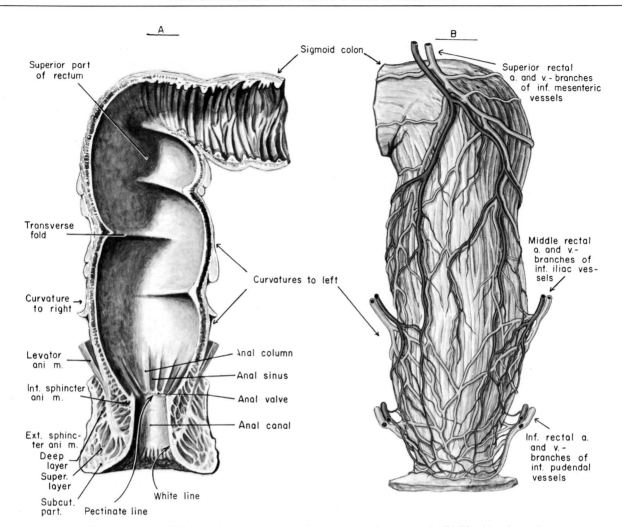

A

Sigmoid colon

Superior part
of rectum

Transverse
fold

Curvature →
to right

Levator
ani m.

Int. sphincter
ani m.

Ext. sphinc-
ter ani m.
Deep
layer
Super.
layer
Subcut.
part. Pectinate line

White line

Curvatures to left

Anal column

Anal sinus

Anal valve

Anal canal

B

Superior rectal
a. and v.- branches
of inf. mesenteric
vessels

Middle rectal
a. and v.-
branches of
int. iliac ves-
sels

Inf. rectal a.
and v.-
branches of
int. pudendal
vessels

Figure 5-69. *(A) A frontal section through the rectum and anal canal. (B) The blood supply and venous drainage of the rectum. Note that branches of both systemic and hepatic portal venous systems are involved.*

thirds; the peritoneum covers the upper and middle thirds of the rectum, but is reflected from the rectum onto the bladder or vagina before it reaches the inferior third.

The **anal canal** is approximately 1½ inches in length and is a connecting link between the rectum superiorly and the external opening—the anus. It forms an angle with the rectum, being directed posteriorly as well as inferiorly. This angle is due to the presence of the puborectalis muscle, that muscle forming the continuous loop posterior to the rectum.

Other muscles associated with the rectum and anal

canal are very important. The rectum has an inner circular layer and an outer longitudinal layer of smooth muscle. The inferior end of the circular fibers is thickened into a sphincter—the **internal anal sphincter** (see Fig. 5-70). The longitudinal layer, in its inferior extent, blends with the circular layer and with a contribution from the levator ani muscle (the puborectalis muscle mentioned above); these longitudinal fibers then continue to attach to the skin around the anus as the **corrugator cutis ani** muscles. These slips of smooth muscle give the anus its puckered appearance. All of the above muscles (except

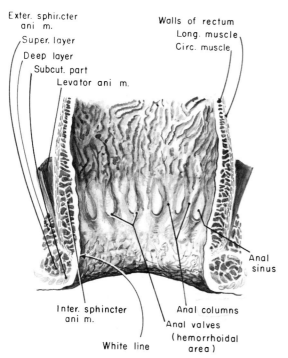

Figure 5-70. *The opened anal canal.*

the puborectalis) are under control of the autonomic nervous system. The **external sphincter** is made up of three parts—subcutaneous, superficial, and deep. They are all skeletal muscle and under voluntary control. The **subcutaneous portion** forms a complete ring around the anus. The **superficial part** arises from the tip of the coccyx and the coccygeal ligament, and continues anteriorly at the sides of the anus to the perineal body, that point between the anus and bulb of the penis in the male, or anus and vagina in the female. The **deep portion** encircles the anal canal; this blends with the puborectalis muscle. The superficial and deep portions encircle the internal sphincter (see Fig. 5-70). The **nerve supply** to the external sphincter is the pudendal nerve from the sacral plexus.

The anatomy of the **inside of the rectum and anal canal** is of importance. The rectum has three **lateral curvatures.** Two of these curvatures bulge to the left and the middle one to the right; these areas of the rectum are separated by **transverse folds** that extend into the lumen. The anal canal exhibits **anal columns,** which are projections into the lumen made by the veins just deep to

the mucous membrane. The depressions between these columns are the **anal sinuses.** Flaps of mucous membrane connect the bases of contiguous columns and bridge across the base of each sinus; these are the **anal valves.** This arrangement results in an irregular line circumventing the anal canal called the **pectinate line.** Fecal matter collects in the sinuses, particularly when they are deeper than usually found, and this can lead to abscesses. Furthermore, anal glands of considerable depth empty in this region and these also can become infected.

At the point in the mucous membrane where the subcutaneous portion of the external sphincter meets the internal sphincter, a wavy so-called **white line** occurs, and an **intersphincteric groove** can be felt. The area between the pectinate line and the white line is known as the **transitional area;** the epithelium is stratified squamous in character. Below the white line the anal canal is lined with true skin, which may be white or brownish in color and contain sweat and sebaceous glands.

The **arteries** to the rectum (Fig. 5-69) and anal canal are the superior, middle, and inferior rectal arteries. The superior rectal, a branch of the inferior mesenteric artery, grossly seems to supply only the upper two-thirds of the rectum. However, these branches penetrate the musculature and course in the submucosal layer, terminating in the anal columns where they anastomose with branches of the inferior rectal artery. The middle rectal, a branch of the internal iliac artery, supplies the lower third of the rectum but is limited to the muscular layer. The inferior rectal, a branch of the internal pudendal in the perineum, supplies the lower end of the anal canal and the immediate area adjacent to the anus. As mentioned above, branches from this artery anastomose with branches of the superior rectal in the anal columns.

The **venous drainage** of the rectum and anal canal are of equal significance owing to the fact that this is a point where the hepatic portal drainage and the systemic venous drainage anastomose. The inferior rectal vein drains the lower part of the anal canal distal to the pectinate line. The middle rectal vein drains the lower one-third of the rectum and upper part of the anal canal but is primarily involved with the muscular walls. The superior rectal vein starts just deep to the mucosa in the anal columns, courses superiorly in the submucosa for about 7.5 centimeters, and then appears grossly as a branch of the inferior mesenteric vein. It is obvious that the point where the two venous systems join is in the anal columns.

This accounts for the internal hemorrhoids occurring at the pectinate line when an increased pressure develops in the hepatic portal system.

The **lymphatic drainage** of the rectum and upper part of the anal canal is into nodes along the lateral wall of the pelvis, along the iliac arteries, and alongside the aorta. The lymphatic drainage of the lower part of the anal canal, however, drains in a manner typical of structures in the area of the external genitalia—to the inguinal nodes in the groin.

The **sympathetic** nerves to the rectum and anal canal are derived from the upper lumbar segments, while the **parasympathetic** system is from the second, third, and fourth sacral segments via the pelvic nerves. **Visceral afferents** from the rectum and anal canal down to the pectinate line return to both lumbar and sacral cord segments. **Somatic afferents** provide sensation for the lower end of the anal canal. Sympathetic fibers cause a decrease in peristalsis and a maintenance of tone in the internal sphincter muscles; the **parasympathetic** nerves increase peristalsis and relax the internal sphincter. The external sphincter of the anal canal, as mentioned previously, is supplied by branches of the pudendal nerve (a branch of the sacral plexus to be studied later) and is under voluntary control. This same pudendal nerve carries the somatic afferent fibers mentioned above. When fecal matter enters the rectum a sense of fullness is felt. During **defecation** impulses to the internal sphincter are inhibited, the parasympathetic system stimulates peristalsis, the external sphincter is voluntarily released, and the intra-abdominal pressure increased by holding the breath and contracting the diaphragm and the abdominal muscles.

The **relations** of the rectum and anal canal are of great importance for surrounding structures can be palpated via this tube. Anterior and posterior relations are shown in Figure 5-64.

The relations of the rectum are as follows. **Posterior**—median and lateral sacral vessels; sympathetic trunks; third, fourth and fifth sacral and coccygeal nerves; piriformis, coccygeus, and levator ani muscles; the lower part of sacrum; coccyx; and coccygeal body. **Anterior in the male**—the upper two-thirds is covered with peritoneum and is in relation with the pelvic colon or ileum; the lower third is related to the bladder, seminal vesicles, vasa deferentia, and prostate. **Anterior in the female**—the upper two-thirds is also covered with peritoneum and is in relation with the pelvic colon or ileum, the vagina, and uterus; the lower third is related to the middle part of the vagina. **Lateral**—pelvic colon or ileum, and the coccygeus and levator ani muscles.

The relations of the anal canal are as follows: **Posterior**—coccygeal body and coccyx. **Lateral**—external sphincter ani muscle and contents of ischiorectal fossa. **Anterior**—perineal body in both sexes, which separates the anus from the vagina in the female and from the root of the penis in the male.

———

Developmentally the anal canal has a double origin: (1) an ingrowth from the skin—the anal pit or proctodeum, and (2) a down growth of the hindgut towards the surface. The anal membrane separates the two areas until, under normal conditions, it breaks down. If it does not, a condition known as imperforate anus occurs. The anal membrane occurs at or just inferior to the pectinate line. It is important to realize that at this line (1) diseases change from those typical of mucous membranes to those characteristically involving skin; (2) the arterial supply changes from branches of the inferior mesenteric to branches of the internal pudendal; (3) the venous drainage changes from tributaries of the hepatic portal system to tributaries of the systemic venous system (the area of hemorrhoids); (4) the afferent nerve supply changes from visceral in nature to somatic (pathology superior to the line is not particularly painful—so-called silent—while that outside is quite painful); and (5) the lymphatics superior to it drain superiorly into pelvic nodes while those inferior drain into the inguinal nodes located in the femoral triangle.

INTERNAL ILIAC VESSELS

The internal iliac artery (Fig. 5-71) courses inferiorly into the pelvic cavity in a position posterior to the ureter and anterior to the corresponding internal iliac vein. The internal iliac artery usually divides into anterior and posterior divisions.

Although the branching of this artery is extremely variable, the following branches usually occur:

Branches of posterior division (PD)
1. Iliolumbar
2. Lateral sacral
3. Superior gluteal

Branches of anterior division (AD)

4. Umbilical and superior vesical
5. Inferior vesical in male; uterine and vaginal in female
6. Middle rectal
7. Obturator *Parietal*
8. Internal pudendal *Parietal*
9. Inferior gluteal

(Note: the letters and numbers in the following description correspond to those in the listing above and in Fig. 5-71.)

The **posterior division (PD)**, lying medial to the sacral plexus, terminates in the large **superior gluteal artery**

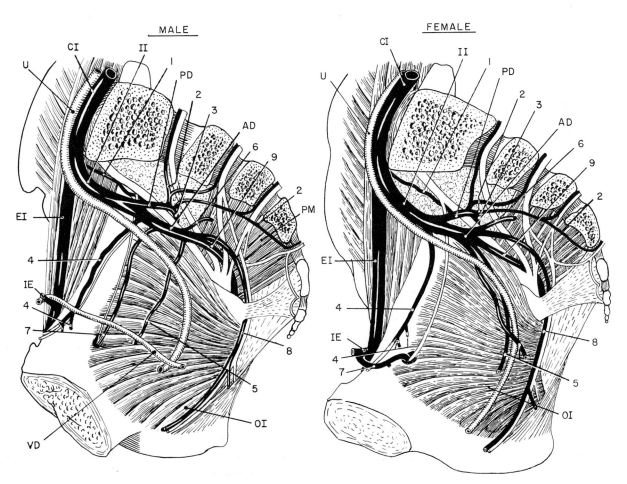

Figure 5-71. *The arteries of the pelvic cavity in the male and female. The pelvic floor (coccygeus and levator ani muscles) has been removed. Note relation of vessels to piriformis muscle (P) as they exit from the pelvic cavity via the greater ischiadic foramen. AD—anterior division, CI—common iliac, EI—external iliac, IE—inferior epigastric, II—internal iliac, OI—obturator internus muscle, P—piriformis muscle, PD—posterior division; U—ureter, VD—vas deferens, I—iliolumbar, 2—lateral sacral, 3—superior gluteal, 4—umbilical and superior vesical, 5—inferior vesical in male or uterine and vaginal in female, 6—middle rectal, 7—obturator—in one arising from the anterior division of the internal iliac and in the other from the inferior epigastric branch of the external iliac—the so-called abnormal obturator artery, 8—internal pudendal, and 9—inferior gluteal. (Modified and redrawn after Jamieson.)*

(3), which exits from the pelvic cavity into the gluteal region through the greater ischiadic foramen in a position superior to the piriformis muscle. The **iliolumbar artery** (1) courses superiorly and laterally between the obturator nerve and the lumbosacral trunk. It is posterior to the common iliac artery. It divides into an **iliac branch,** which ramifies in the iliac fossa, and a **lumbar branch,** which ascends superiorly in a position posterior to the psoas muscle. This latter branch sends a spinal branch into the vertebral canal through the lumbosacral intervertebral foramen. Two or more **lateral sacral arteries** (2) course inferiorly and medially, and enter the four anterior sacral foramina supplying structures in the sacral canal; they then emerge through the posterior sacral foramina and supply muscles and skin on the back of the sacrum. The anterior division (AD) of the internal iliac artery continues inferiorly and divides into the branches listed above. The pattern of the branching cannot be relied upon and is not particularly important. The **umbilical artery** and the **superior vesical** (4) are usually combined into one stem. It continues along the side wall of the pelvis posterior to the obturator nerve and lateral to the ureter and vas deferens in the male or the ligamentum teres in the female. After giving off the superior vesical artery, it becomes the medial umbilical ligament. The **superior vesical arteries** (4) run medially from the umbilical arteries to supply the superior aspect of the bladder. The **inferior vesical artery** (5) courses anteriorly and inferiorly on the surface of the levator ani muscle. It also is crossed by the ureter and vas deferens. It continues to the posterior surface of the bladder and supplies the seminal vesicle, prostate, and ureter in addition to the bladder. It also gives off the **artery to the vas deferens,** which accompanies the vas all the way to the testis and anastomoses with the testicular artery. The **middle rectal artery** (6) courses medially on the lateral pelvic wall and supplies the rectum, prostate, seminal vesicles, and ampulla of the vas deferens. The **obturator artery** (7) courses inferiorly and anteriorly along the side wall of the pelvis in company with the obturator nerve. It leaves the pelvis through the obturator foramen. In approximately 40 per cent of bodies, the artery may arise from the inferior epigastric branch of the external iliac (abnormal obturator artery). In this situation it courses medially in a position posterior to the external iliac vein, and then courses over the brim of the pelvis to enter the obturator foramen. It gives off branches to muscles on the lateral pelvic wall, to the bladder, and a **pubic branch** ramifies

on the pelvic surface of the pubis. The terminal branches of the obturator artery will be described with the lower limb. The **internal pudendal artery** (8) courses inferiorly and slightly posteriorly, and exits from the pelvic cavity into the gluteal region through the greater ischiadic foramen. It is very close to the ischial spine and enters the perineum by coursing through the lesser ischiadic foramen. The **inferior gluteal artery** (9) exits from the pelvic cavity into the gluteal region via the greater ischiadic foramen in a position inferior to the piriformis muscle.

The female differs in the arterial branching because of the addition of the uterus and vagina. The **uterine artery** (5) is of considerable size and runs anteriorly on the levator ani muscle to the base of the broad ligament. It then courses medially between the two layers of the broad ligament, superior to the ureter, and on the lateral fornix of the vagina. It supplies branches to the vagina, and then continues on the lateral sides of the uterus, between the layers of the broad ligament, to the uterine tube; it gives off branches to these organs and anastomoses with the ovarian artery. The inferior vesical artery in the female usually arises from this uterine artery.

The corresponding **veins** form a very profuse plexus before ending in the same branches as just mentioned for the iliac artery, with the exception of the iliolumbar and umbilical arteries which have no corresponding veins. It should be mentioned once again that the portal circulation anastomoses with the systemic veins in the anal columns. It should be noted also that the deep dorsal vein of the penis ends in the prostatic plexus of veins; the dorsal vein of the clitoris has a similar course. The main internal iliac vein lies superior and posterior to its artery.

SACRAL PLEXUS

Although the details of the sacral plexus are best studied with the lower limb, the general location of the plexus as a whole should be noted at this time. The plexus (Figs. 5-64 and 5-71) is posterior and lateral to the parietal pelvic fascia, the internal iliac artery and its branches, and the ureter. It is in contact with the sacrum, the sacroiliac joint, and the piriformis muscle. The lower part of the plexus is in contact medially with the rectum. The gluteal vessels pass between the roots of the plexus, the superior between the lumbosacral trunk and the first sacral nerve and the inferior between the first and second sacral nerves.

We will see later that branches of this plexus innervate muscles in the pelvic cavity. The pelvic nerves, from the ventral rami of the second, third, and fourth sacral segments of the spinal cord, give the parasympathetic supply to the majority of structures in the pelvic cavity.

PERINEUM

GENERAL DESCRIPTION

The **perineum** is the most inferior end of the trunk; it is the region between the thighs and between the buttocks. It is **bounded** anteriorly by the pubic arch; laterally by the pubic and ischial rami combined, the ischial tuberosity, and the sacrotuberous ligament; and posteriorly by the sacrum and coccyx (Fig. 5-72). Its **superior limit** is the pelvic diaphragm consisting of levator ani and coccygeus muscles. If a line is drawn transversely through the point between the anus and the bulb of the penis in the male or between the anus and the posterior end of the vagina in the female, it will extend laterally to the ischial tuberosities. This line, which bisects the **central perineal tendon (perineal body),** divides the perineum into an anterior **urogenital triangle** and a posterior **anal triangle.**

The **anal opening** is located in the anal triangle and is the same in both sexes. The sphincter muscles of the anus keep the opening closed and there are bundles of smooth muscle in the superficial fascia around the anal opening that radiate from the margins of the anus and are attached to the skin around it. This gives the anus a puckered appearance. The smooth muscles are the **corrugator cutis ani** muscles.

The urogenital triangle contains the external genitalia of the male and female.

In the male (Fig. 5-73) the **scrotum** is suspended from the urogenital triangle, and a median raphe is found extending anteriorly from the anus to the posterior surface of the scrotum; this raphe continues around the scrotum to its anterior surface, reaches the base of the penis, and continues on its ventral surface. Anterior to the

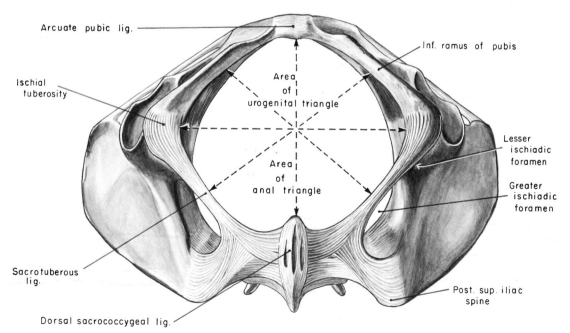

Figure 5-72. *Boundaries of the perineum. Note that a line drawn between the two ischial tuberosities divides this area into anterior urogenital and posterior anal triangles. Dotted lines represent measurements (pelvimetry) which can be compared with the size of the baby's head (fetal cephalimetry).*

scrotum is found the **penis,** which consists of a body and a glans or head. The glans may or may not be covered with a prepuce or foreskin, a layer of loose skin. The urethral orifice can be seen at the end of the penis, and another layer of loose skin—the frenulum—extends back from this orifice a short distance on the ventral side* of the penis. Figure 5-91 on page 341 shows the development of these structures in comparison with the female.

The urogenital triangle in the female contains the **external genitalia** or **vulva** (Fig. 5-74). A pad of fat called the **mons pubis** exists anterior and inferior to the pubic bone. This area after puberty is covered with hair. This pad of fat can be followed posteriorly to two large lips—the **labia majora.** The point where the lips come together anteriorly is the **anterior labial commissure,** and the region where they come together posteriorly the **posterior labial commissure.** These labia majora are also covered with hair. Medial to the labia majora are two fleshy smaller lips called the **labia minora.** These lips are devoid of hair and are usually in contact with one another. The point where the labia minora come together posteriorly is the **frenulum of the labia** while the point where they come together anteriorly is designated as the **frenulum of the clitoris.** The head or glans of the **clitoris** can be seen just at the anterior end of the labia minora and the fleshy fold just anterior to the glans of the clitoris is the **prepuce of the clitoris.** The space between the labia minora is the **vestibule of the vagina.** This has two openings: (1) the opening of the **urethra,** which is located about 1 inch posterior to the head of the clitoris and appears puckered in the living state, and (2) posterior to that the opening to the **vagina.** If a **hymen** is present, it guards the entrance into the vagina from the vestibule and may take the form of a complete membrane, a sievelike structure, or a fleshy shelf. After rupture of the hymen, small fleshy flaps remain, which are the **hymenal caruncles.** Just inferior to the hymenal caruncles on either side are the openings of the **greater vestibular glands.** These glands provide lubrication for the medial sides of the labia minora, being aided in this by small **lesser vestibular glands** opening on the medial sides of the labia minora. The development of these structures is shown in Figure 5-91 on page 341.

---

*Easily understood if one thinks of an animal such as a dog.

ANAL TRIANGLE

The anatomy of the **anal triangle** in both sexes is the same (Figs. 5-75 and 5-76). If the skin is removed, a layer of fatty fascia is found, which is the same as the fatty layer of superficial fascia located in the abdominal wall. It is also continuous with a similar fatty layer in the urogenital triangle. This fat is found to fill in a large wedge-shaped area on either side of the anus and rectum—the **ischiorectal fossa** (Fig. 5-77).

The **boundaries** of the ischiorectal fossa are as follows. **Anterior**—the transverse muscles of the perineum and the fascia that covers them. (If the finger is pushed anteriorly in the ischiorectal fossa, it can be seen that the finger will be superior to the urogenital diaphragm which will be described later; this is the anterior recess of the ischiorectal fossa—9 on Fig. 5-83.) **Posterior**—the sacrotuberous ligament and the overlying gluteus maximus muscle. **Lateral**—the obturator internus muscle with its covering of deep fascia. **Medial**—the levator ani and external sphincter muscles covered with deep fascia—the inferior fascia of the pelvic diaphragm.

The pad of fat in the ischiorectal fossa is said to act as an elastic cushion that gives way when the rectum is full of feces or when the anus is distended during defecation. The fat absorbs the pressure and expands again when the fecal mass has passed.

If the fat is removed from this fossa, the **inferior rectal arteries and nerves** can be seen streaming from the lateral superior aspect of the ischiorectal fossa toward the rectum and anus (Figs. 5-75 and 5-76). These nerves and blood vessels supply innervation and vascularity to the inferior end of the anal canal, the circularly arranged external sphincter ani muscle that surrounds the anal opening, and the skin around the anus. These nerves are important, for incontinence results if they are cut. These nerves and arteries are branches of the **pudendal nerve** and **internal pudendal artery** which course in the layer of fascia on the obturator internus muscle. This canal of fascia is the **pudendal canal.**

In addition to the inferior rectal nerve, a very small branch, the **coccygeal nerve,** can be found close to the posterior part of the external sphincter muscle of the anus close to where this muscle attaches to the coccyx. This is a sensory nerve to the skin over the coccyx.

The **sphincter ani externus muscle** is divided into subcutaneous, superficial, and deep parts (Fig. 5-77); the

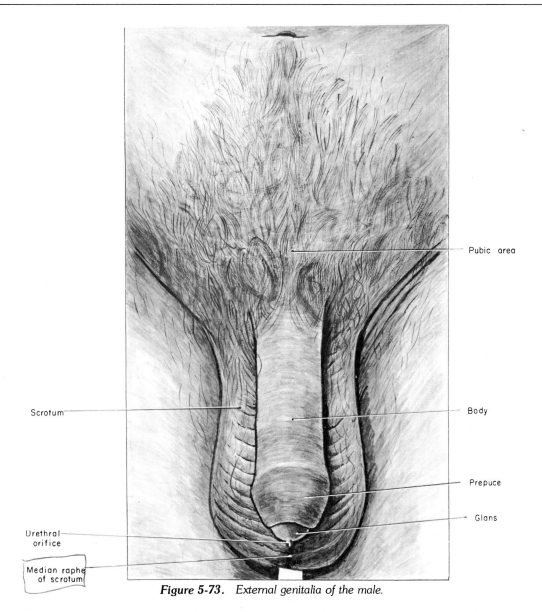

Pubic area

Scrotum

Body

Prepuce

Glans

Urethral orifice

Median raphe of scrotum

Figure 5-73. *External genitalia of the male.*

subcutaneous part completely encircles the anus in a position just deep to the skin. The **superficial portion** has attachments posteriorly to the coccygeal ligament and to the coccyx, and anteriorly to the perineal body, a central fibrous point just anterior to the anus to which several muscles attach. The **deep portion** completely encircles the anal canal; this portion is intimately fused with the puborectal portion of the levator ani muscle. The latter

two parts—superficial and deep—are just lateral to the internal sphincter made up of smooth muscle (see Fig. 5-77). The external sphincter is skeletal muscle under voluntary control; **nerve supply,** as mentioned above, is from inferior rectal branches of the pudendal nerve.

Abscesses occur in the ischiorectal fossa and these can spread to the opposite side since the fat pads are continuous both anterior and posterior to the anus. They

312

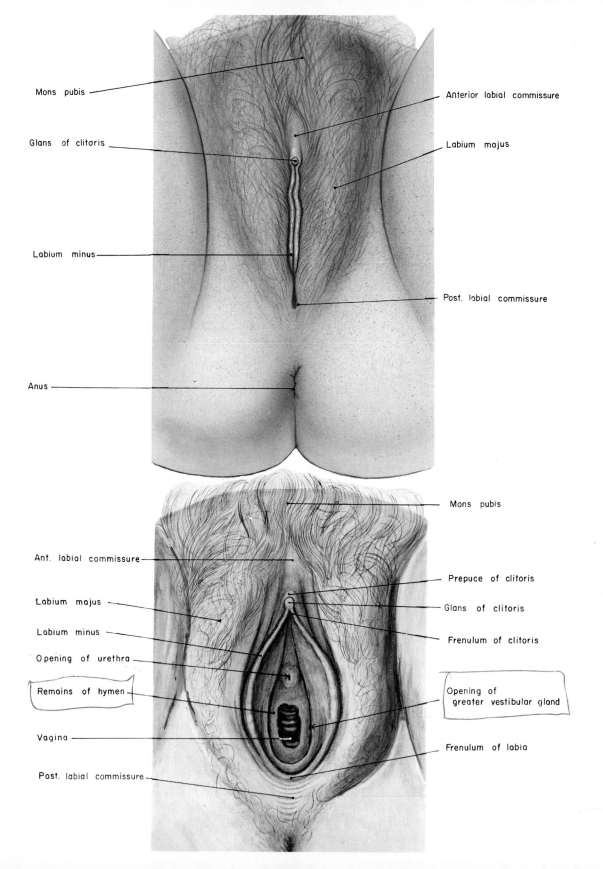

Mons pubis

Glans of clitoris

Labium minus

Anus

Anterior labial commissure

Labium majus

Post. labial commissure

Mons pubis

Ant. labial commissure

Labium majus

Labium minus

Opening of urethra

Remains of hymen

Vagina

Post. labial commissure

Prepuce of clitoris

Glans of clitoris

Frenulum of clitoris

Opening of greater vestibular gland

Frenulum of labia

heal poorly due to the scanty blood supply; they can penetrate the rectum or anal canal between the internal and external sphincter muscles.

UROGENITAL TRIANGLE

This triangle is more complex than the anal area. The architecture of the triangle will be described first, after which the course taken by the nerves and blood vessels will become more comprehensible.

If the skin and fatty layer of superficial fascia are removed from the urogenital triangle of the perineum, a **membranous layer of superficial fascia*** can be seen (Figs. 5-75 and 5-76) to be attached posteriorly to something at the line that divides the urogenital triangle from the anal triangle, and laterally to the pubic rami and the deep fascia on the thigh; this is the continuation of the membranous layer of superficial fascia in the abdominal wall. It is also the same fascial plane that forms the dartos tunic of the scrotum, and this same fascia is continuous onto the penis. The structure to which the fascia is attached posteriorly is the inferior layer of the urogenital diaphragm or the perineal membrane, which is illustrated in Figures 5-79 and 5-80. The attachments of this membranous layer of superficial fascia are shown clearly in Figure 5-78.

If this fascia is removed, the contents of the **superficial space of the perineum** are revealed (Figs. 5-75 and 5-76). In the male the **bulbospongiosus** (or bulbocavernosus) muscle can be seen in the midline, where it covers

the bulb of the penis, the latter being composed of erectile tissue. On the sides, attached to the ischiopubic rami, are the **ischiocavernosus** muscles covering the two crura of the penis, also composed of erectile tissue. Two small muscles run transversely from the perineal body to the inferior pubic rami—the **transversus perinei superficialis** muscles. Each of these muscles is covered with a thin layer of deep fascia. The **action** of the bulbospongiosus muscle is to expel any urine or semen remaining in the urethra after micturition or ejaculation; this is possible since the urethra courses through this bulb. The ischiocavernosus muscles contract after erection of the penis has occurred, and may aid in maintaining this erect condition by compressing the venous outflow of the penis. The action of the transversus perinei superficialis muscles is problematical, but may serve to tense the central point of the perineum so that other muscles attached to it become more effective.

The detailed anatomy of the **female urogenital triangle** is very similar to that of the male if one visualizes the bulb of the penis as split into two halves by the vagina.

If the skin is removed from the labia majora and the fatty layer carefully incised, the labia can be seen to contain an organized process of fat called the **diverticular process** (Fig. 5-76). This is continuous with the fat over the mons pubis and is the structure to which the round ligament of the uterus is attached. If the diverticular process is reflected and the fatty layer of superficial fascia removed in its entirety, the **membranous layer of the superficial fascia** is visible. This is exactly the same layer of fascia as found in the male. In fact, smooth muscle fibers can be found in this layer similar to those found in the dartos tunic of the scrotum.

If this membranous layer of superficial fascia is removed, the contents of the superficial space in the female are revealed (Fig. 5-76). On either side of the labia minora the bulbospongiosus muscles can be seen; these muscles cover the **vestibular bulb,** which is made up of erectile tissue (Fig. 5-80). Each bulb continues anteriorly and actually joins with that of the opposite side near the head of the clitoris. The **greater vestibular glands** are located just posterior to each bulb. (Note that we did not

---

*J. W. Davies, 1934, The pelvic outlet—its practical application, *Surg. Gynecol. Obstet.* 58: 70. A treatise on general fascial arrangements in chest, abdomen, scrotum, pelvic outlet, and around the anus, as well as on the practical importance of these layers in the pelvic outlet; well worth students' time.

C. E. Tobin and J. A. Benjamin, 1944, Anatomical study and clinical consideration of the fasciae limiting urinary extravasation from the penile urethra, *Surg. Gynecol. Obstet.* 79: 195. A good historical summary of the continuity of the membranous layer of superficial fascia in the abdominal wall (Scarpa's fascia), in the scrotum (dartos tunic), and on the penis and in the perineum (Colles' fascia), and an account of the attachments of the deep fascia of the penis (Buck's fascia).

M. W. Wesson, 1953, What are Buck's and Colles' fascia? *J. Urol.* 70: 503. Beautiful demonstrations of the attachments of these fasciae. Figure 5-78 was borrowed from this article.

Figure 5-74. *External genitalia of the female in natural state (above), and with labia spread apart (below).*

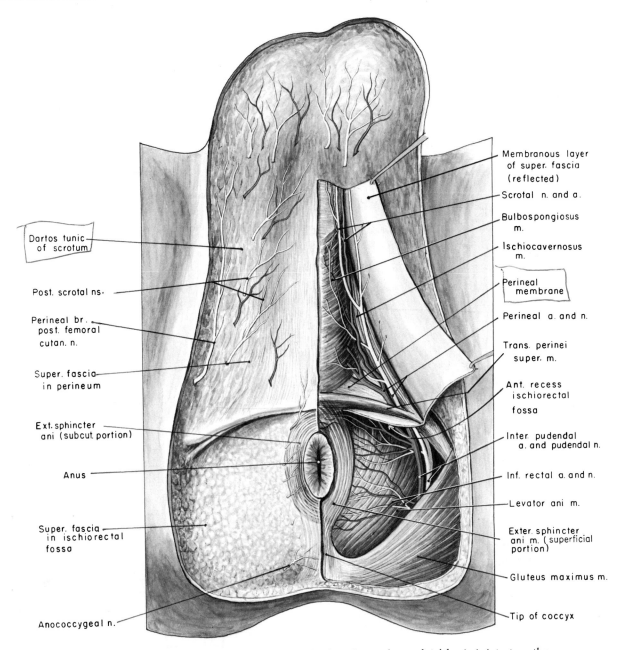

Dartos tunic of scrotum

Post. scrotal ns.

Perineal br. post. femoral cutan. n.

Super. fascia in perineum

Ext. sphincter ani (subcut. portion)

Anus

Super. fascia in ischiorectal fossa

Anococcygeal n.

Membranous layer of super. fascia (reflected)

Scrotal n. and a.

Bulbospongiosus m.

Ischiocavernosus m.

Perineal membrane

Perineal a. and n.

Trans. perinei super. m.

Ant. recess ischiorectal fossa

Inter. pudendal a. and pudendal n.

Inf. rectal a. and n.

Levator ani m.

Exter. sphincter ani m. (superficial portion)

Gluteus maximus m.

Tip of coccyx

Figure 5-75.   *The perineum in the male I. The fatty layer of superficial fascia is intact on the right side of the body, but has been removed on the left. The membranous layer of superficial fascia covering the urogenital triangle has been cut and reflected to show contents of the superficial space. (The deep fascia has been removed from the muscles in this space and from the pudendal canal.)*

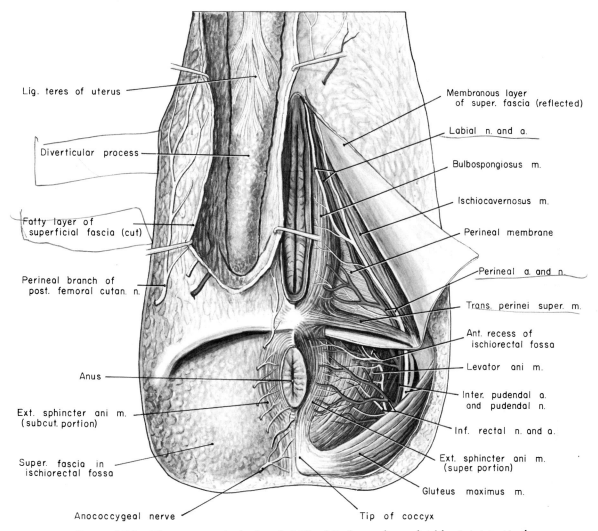

Lig. teres of uterus

Diverticular process

Fatty layer of superficial fascia (cut)

Perineal branch of post. femoral cutan. n.

Anus

Ext. sphincter ani m. (subcut. portion)

Super. fascia in ischiorectal fossa

Anococcygeal nerve

Membranous layer of super. fascia (reflected)

Labial n. and a.

Bulbospongiosus m.

Ischiocavernosus m.

Perineal membrane

Perineal a. and n.

Trans. perinei super. m.

Ant. recess of ischiorectal fossa

Levator ani m.

Inter. pudendal a. and pudendal n.

Inf. rectal n. and a.

Ext. sphincter ani m. (super. portion)

Gluteus maximus m.

Tip of coccyx

Figure 5-76. *The perineum in the female I. The fatty layer of superficial fascia is intact in the anal triangle on the right side of the body, but has been incised in the labium majus to reveal the diverticular process. The fatty layer of superficial fascia has been removed on the left side to reveal the contents of the ischiorectal fossa, and the membranous layer of superficial fascia cut and reflected to show contents of the superficial space. (The deep fascia has been removed from the muscles in this space and from the pudendal canal.)*

find the corresponding bulbourethral glands of the male in the superficial space.) Laterally in the superficial space the **ischiocavernosus muscles** are attached to the ischiopubic rami; these muscles cover the **crura of the clitoris,** which also consist of erectile tissue. The **transversus perinei superficialis muscles** are similar to those

found in the male. The bulbospongiosus muscles form a sort of sphincter around the inferior end of the vagina, the ischiocavernosus muscles may serve to maintain an erect condition of the clitoris upon sexual excitement by compressing the venous return in the crura, and the action of the transversus perinei superficialis muscles is similar to

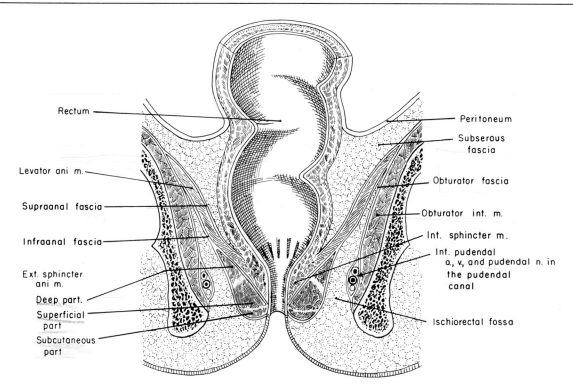

Figure 5-77. *Diagram of a frontal section through the anal triangle of the perineum.*

that of the male—to tense the central point of the perineum to aid action of other muscles.

If the bulb and its bulbospongiosus muscle, and the two crura with their ischiocavernosus muscles, and the transversus perinei superficialis muscles are removed in the male (Fig. 5-79), the superficial space is seen to be bounded superiorly by a heavy layer of fascia, called the **perineal membrane,** which stretches across the urogenital triangle between the two inferior pubic rami. Because the **urethra** in the male is located in the bulb of the penis, such a maneuver will cut through the urethra at right angles; the urethra will be seen to penetrate this perineal membrane.

If the contents of the superficial space in the female are removed, the perineal membrane is seen, just as in the male (Fig. 5-80). It differs from the male in that in addition to the **urethra** we have a larger opening in the perineal membrane for the **vagina.**

If the perineal membrane is now removed, the con-

tents of the deep space are revealed and it can be seen to contain a thin sheet of muscle which, in the male (Fig. 5-81), is actually divided into two parts: an anterior part that surrounds the urethra—**sphincter urethrae muscle**—and a posterior part—**transversus perinei profundus muscle.** The former muscle arises from the inferior rami of the pubis and surrounds the **membranous urethra.** This external sphincter is under voluntary control. Two glands are also found in the deep pouch of the male, one on either side of the urethra. These are the **bulbourethral glands,** which secrete nonviscous material into the urethra at ejaculation to combine with the sperm, seminal vesicle fluid, and prostatic fluid already making up the semen.

If the muscles in the deep space are removed (Fig. 5-81), a layer of deep fascia is found that is continuous with the fascia found on the obturator internus muscle (Fig. 5-83). If this deep fascia, in turn, is removed, the inferior surface of the levator ani muscle is seen.

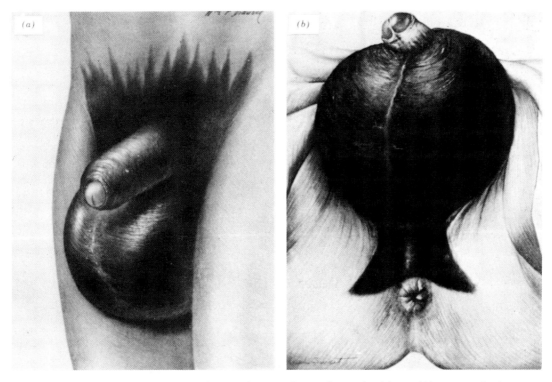

Figure 5-78. *Attachments of the membranous layer of superficial fascia (A) as a result of a hemorrhage from a tear in the corpus spongiosum and (B) as outlined by a torn varicocele. Note that the blood is in the superficial space of the perineum, in the scrotum, penis, and starting superiorly into the abdominal wall. (From Wesson, J. Urol. 70: 503, 1953, courtesy of Williams & Wilkins Co.)*

The deep space in the female (Fig. 5-82) also contains the transversus perinei profundus muscle as well as the sphincter urethrae muscle; in this case, however, the sphincter urethrae muscle surrounds the vagina as well as the urethra. If the muscles of the deep space are removed, the deep fascia superior to these muscles is revealed just as in the male.

It might be reemphasized that the only differences in the male and the female anatomy of the urogenital triangle are the presence of the vagina separating midline structures into two parts, and the location of the bulbourethral glands in the deep space in the male, whereas the corresponding glands in the female (the greater vestibular glands) are in the superficial space.

The layers of the urogenital triangle are, therefore:

1. Skin
2. Fatty layer of superficial fascia
3. Membranous layer of superficial fascia
4. Muscular layer in the superficial space
5. Perineal membrane
6. The muscular layer in the deep space
7. Deep fascia on the superior side of these muscles

These are shown in Figure 5-83. The **superficial space** is the area between layers 3 and 5 in the above list, while the **deep space** is between 5 and 7. The perineal membrane, the muscle layer in the deep space, and the deep fascia superior to this muscle (5, 6, and 7 in the above list) form what is called the **urogenital diaphragm.** Therefore, another name for the perineal membrane is the **inferior layer of the urogenital diaphragm,** and another name for the fascia of layer 7 is the **superior layer of the urogenital diaphragm.** This urogenital diaphragm serves

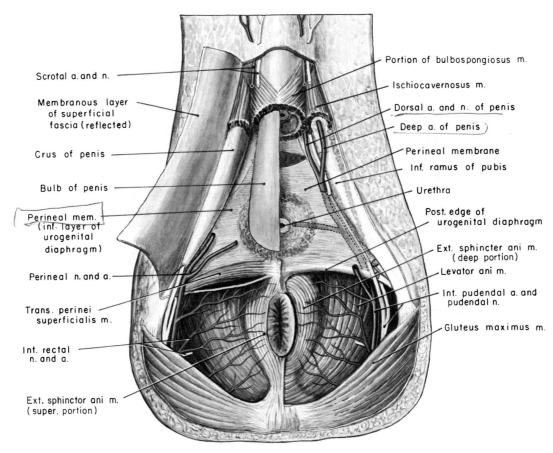

Scrotal a. and n.

Membranous layer
of superficial
fascia (reflected)

Crus of penis

Bulb of penis

Perineal mem.
(inf. layer of
urogenital
diaphragm)

Perineal n. and a.

Trans. perinei
superficialis m.

Int. rectal
n. and a.

Ext. sphinctor ani m.
(super. portion)

Portion of bulbospongiosus m.

Ischiocavernosus m.

Dorsal a. and n. of penis

Deep a. of penis

Perineal membrane

Inf. ramus of pubis

Urethra

Post. edge of
urogenital diaphragm

Ext. sphincter ani m.
(deep portion)

Levator ani m.

Int. pudendal a. and
pudendal n.

Gluteus maximus m.

Figure 5-79. *The perineum in the male II. The bulbospongiosus and ischiocavernosus muscles have been removed from the bulb and crus on the right side, while the crus and bulb and transversus perinei superficialis muscle have been removed on the left.*

to fill the opening in the pelvic diaphragm between the two levator ani muscles, and to give support to the prostate in the male and urinary bladder in both sexes.

Now that you are familiar with the architecture of the urogenital triangle in the male and female, you are ready to follow the nerves and blood vessels to this area. The **blood and nerve supply** to the urogenital triangle in both sexes is by branches of the **internal pudendal artery and pudendal nerve** (Fig. 5-71) aided by the **perineal branch of the posterior femoral cutaneous nerve.**

Although the main role of the **posterior femoral**

cutaneous nerve (S2 and 3) is to provide sensory innervation to the posterior surface of the thigh, its **perineal branch** is possibly more important. Providing cutaneous innervation to a large part of the perineum (Figs. 5-75 and 5-76), it must be considered in anesthesia of this area.

The internal pudendal artery and the pudendal nerve arise from the internal iliac artery and the sacral plexus (S2, 3, and 4), respectively, course inferiorly on the inside surface of the piriformis muscle, leave the pelvic cavity via the greater ischiadic foramen in a position inferior to the piriformis muscle to enter the gluteal region, wind around the ischial spine, and enter the lesser ischiadic foramen to become located in the perineum be-

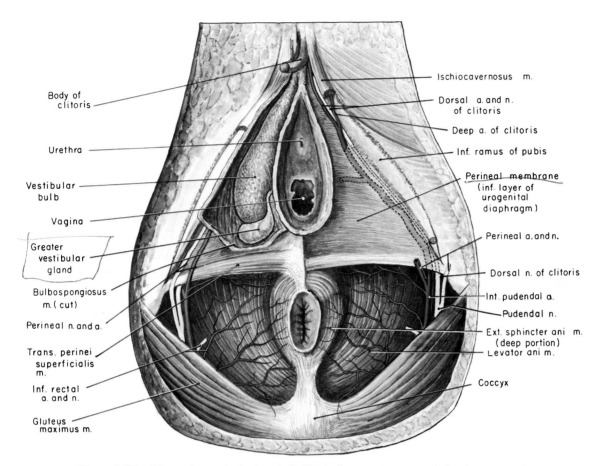

Body of
clitoris

Urethra

Vestibular
bulb

Vagina

Greater
vestibular
gland

Bulbospongiosus
m. (cut)

Perineal n. and a.

Trans. perinei
superficialis
m.

Inf. rectal
a. and n.

Gluteus
maximus m.

Ischiocavernosus m.

Dorsal a. and n.
of clitoris

Deep a. of clitoris

Inf. ramus of pubis

Perineal membrane
(inf. layer of
urogenital
diaphragm)

Perineal a. and n.

Dorsal n. of clitoris

Int. pudendal a.

Pudendal n.

Ext. sphincter ani m.
(deep portion)
Levator ani m.

Coccyx

Figure 5-80. *The perineum in the female II. The bulbospongiosus muscle has been incised on the right side to reveal the vestibular gland and duct, and the bulb. The ischiocavernosus muscle has been removed from the crus of the clitoris. On the left side of the body the entire contents of the superficial space has been removed.*

cause they are now coursing inferior to the pelvic diaphragm. The vessel and nerve continue anteriorly and inferiorly on the obturator internus muscle in a fascial canal—the pudendal canal. You will recall that the first branch of this nerve and artery is the **inferior rectal,** which courses medially through the fat in the ischiorectal fossa to reach the inferior part of the anal canal (Figs. 5-75 and 5-76).

While the internal pudendal artery and pudendal nerve are in the pudendal canal, a **perineal branch** is given off that penetrates the posterior aspect of the membranous layer of the superficial fascia and therefore enters the superficial space (Figs. 5-75 and 5-76). This perineal

nerve and artery immediately divide into **muscular branches** to innervate and give blood supply to the muscles already mentioned and into **scrotal or labial branches,** which carry on to innervate the skin of the posterior surface of the scrotum or labia and give this area a blood supply. In addition, branches of this perineal nerve enter the deep space to innervate the muscles contained therein, entering the space at its posterior edge.

Returning to the main internal pudendal artery and pudendal nerve, the branching unfortunately differs from this point on. The artery continues anteriorly and enters the deep space, where it occupies a lateral position (Figs. 5-81 and 5-82). In addition to **muscular branches** to the

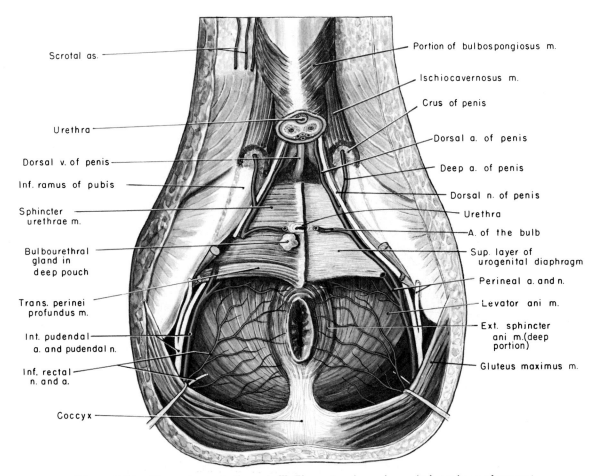

Figure 5-81. *The perineum in the male III. The perineal membrane (inferior layer of urogenital diaphragm) has been removed on the right side of the body to reveal the contents of the deep space. Note that the bulbourethral glands are located in this space in the male. The sphincter urethrae and transversus perinei profundus muscles have been removed on the left to show the pelvic fascia (superior layer of urogenital diaphragm).*

sphincter urethrae and deep transversus perinei muscles contained in this pouch, an **artery to the bulb** emerges, courses medially in the deep space, and then turns inferiorly and penetrates the perineal membrane (inferior layer of urogenital diaphragm), to terminate in the bulb of the penis in the male or the vestibular bulb in the female. The internal pudendal artery continues anteriorly, divides into **deep** and **dorsal** arteries of the penis or clitoris, which then penetrate the perineal membrane to reach these structures. We will see momentarily that these two vessels course on the dorsum of the penis or clitoris (dor-

sal branch) and enter the corpora cavernosa of the penis or clitoris (deep branch) respectively. These vessels are naturally much larger in the male than in the female.

The **pudendal nerve,** in contrast to the artery, terminates before entering the deep space by dividing into the **perineal branch** already mentioned and the **dorsal nerve of the penis or clitoris.** Therefore, the nerve in the deep space is this latter nerve which then penetrates the perineal membrane to course on the dorsum of the penis or clitoris (Figs. 5-81 and 5-82).

The **veins** in this area follow the arteries in a retro-

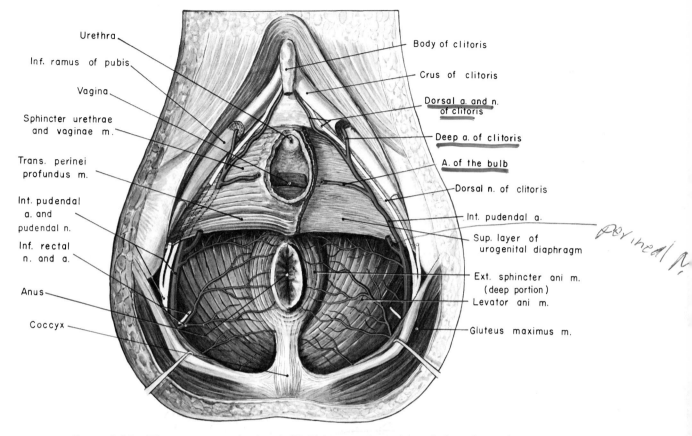

Figure 5-82. *The perineum in the female III. The perineal membrane (inferior layer of urogenital diaphragm) has been removed on the right to reveal the sphincter urethrae and vaginae muscle and the transversus perinei profundus muscle. On the left these muscles have been removed to reveal the pelvic fascia (superior layer of urogenital diaphragm).*

grade direction. We shall see presently that the penis or clitoris has an extra vein—superficial dorsal vein—and that the deep dorsal vein drains directly into the pelvic cavity.

You should now be able to visualize the following structures in the superficial space:

1. Bulb with bulbospongiosus muscle in male, and vestibular bulbs and glands in the female
2. Crura with ischiocavernosus muscles
3. Transversus perinei superficialis muscles
4. Perineal and scrotal or labial nerves
5. Perineal and scrotal or labial arteries and corresponding veins

The following structures can be visualized in the deep space:

1. Transversus perinei profundus muscle
2. Sphincter urethrae muscle
3. Urethra
4. Dorsal nerve of penis or clitoris
5. Internal pudendal artery
6. Arteries to the bulb
7. Bulbourethral glands in the male

As indicated in Figure 5-84, the perineal membrane is a relatively thick structure that joins with the pelvic fascia anteriorly and posteriorly to form an enclosed

MALE   FEMALE

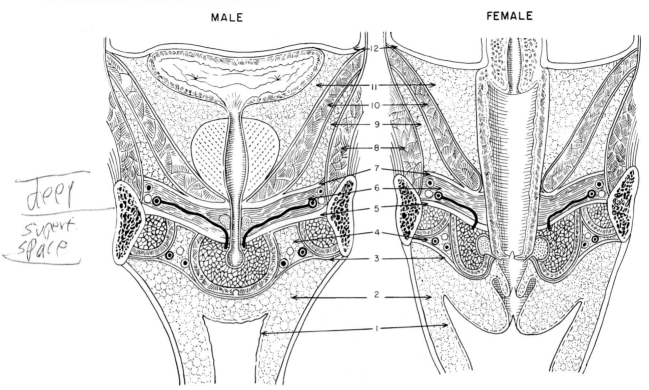

deep
super.
space

Figure 5-83. Frontal sections of the urogenital triangle of the male and female: (1) skin, (2) fatty layer of superficial fascia, (3) membranous layer of superficial fascia, (4) the superficial space containing the bulb(s) and bulbospongiosus muscle(s), the crura and ischiocavernosus muscles, transversus perinei superficialis muscles, greater vestibular glands in the female, perineal nerves and vessels, (5) perineal membrane or inferior layer of urogenital diaphragm, (6) the deep space containing the sphincter urethrae (and vaginae) and transversus perinei profundus muscles, internal pudendal artery and artery of the bulb, and dorsal nerve of the penis or clitoris, (7) pelvic fascia or superior layer of the urogenital diaphragm, (8) obturator internus muscle covered with pelvic fascia, (9) anterior recess of ischiorectal fossa, (10) levator ani muscle covered with pelvic fascia (infra- and supra-anal fascia), (11) subserous fascia, and (12) peritoneum. Note that the superficial space is between 3 and 5, the deep space between 5 and 7, and that 5, 6, and 7 make up the urogenital diaphragm. (Modified after Jamieson.)

space. The anterior edge of this membrane does not reach the pubic symphysis; this opening is utilized by the **dorsal vein of the penis or clitoris** to reach the plexus of veins in the pelvic cavity. Note the direction taken by the urogenital diaphragm; it has superior and inferior surfaces and anterior and posterior borders.

PENIS AND CLITORIS

The **penis** (Figs. 5-73, 5-75, 5-79, and 5-85) consists of a **root** buried in the perineum; a **body,** which is partly in the

perineum, but mainly free; a **glans** or head, the enlarged distal end of the penis, the posterior rim of which is called the corona glandis; a **neck** which is a constricted part proximal to the glans; and the **prepuce,** the loose fold of skin covering the glans and connected to the glans by a loose layer of skin—the frenulum.

The body of the penis consists of a pair of **corpora cavernosa** dorsally and a single **corpus spongiosum** ventrally. These structures are surrounded by fascia and skin. The three structures are composed of erectile tissue enclosed in a dense fibroelastic connective tissue. The

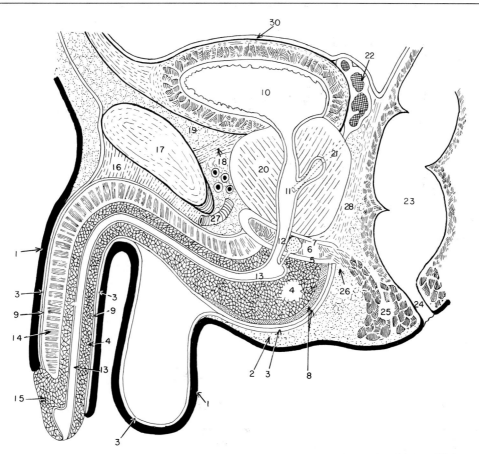

Figure 5-84. *Sagittal section of the male pelvic cavity and external genitalia: (1) skin, (2) fatty layer of superficial fascia, (3) membranous layer of superficial fascia (dartos in scrotum), (4) bulb and corpus spongiosum, (5) perineal membrane (inferior layer of urogenital diaphragm), (6) deep space containing sphincter urethrae and transversus perinei profundus muscles, and the bulbourethral glands, (7) pelvic fascia (superior layer of urogenital diaphragm), (8) bulbospongiosus muscle and its deep fascia, (9) same fascia (deep) on penis (Buck's fascia), (10) urinary bladder, (11) prostatic urethra, (12) membranous urethra, (13) spongiose urethra, (14) septum between corpora cavernosa penis, (15) glans, (16) suspensory ligament of penis, (17) pubic symphysis, (18) pubovesical ligament, (19) retropubic space, (20) prostate, (21) ejaculatory duct, (22) seminal vesicles, (23) rectum, (24) anal canal, (25) anal sphincter musculature, (26) perineal body, (27) deep dorsal vein of penis, (28) rectovesical space, and (30) peritoneum.*

erectile tissue consists of a fine network of fibroelastic tissue, the spaces of which are lined with endothelium; when these spaces are filled with blood, erection ensues. When viewed in cross section, the corpora cavernosa are seen to be divided by a connective tissue septum; this septum is complete in the proximal part of the penis but incomplete in the distal portion. Posteriorly, the corpora

cavernosa diverge and end as the **crura** of the penis, which are located in the superficial space of the perineum and are attached to the inferior pubic rami. Anteriorly the corpora terminate as rounded ends; they play no part in formation of the glans. The posterior part of the corpus spongiosum, which contains the urethra, forms the **bulb** of the penis, which also is located in the

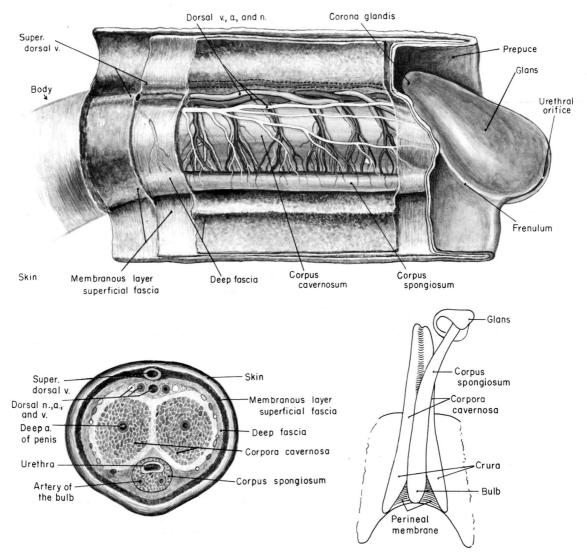

Figure 5-85. *The structure of the penis.*

superficial space of the perineum. Anteriorly, the corpus spongiosum ends as the expanded **glans** penis. The **urethra** enters the deep surface of the bulb after piercing the urogenital diaphragm and goes through the whole length of the corpus spongiosum, and in the glans of the penis lies nearer the ventral than the dorsal surface (see separate description of urethra below).

The skin of the penis is quite thin on the glans and very firmly attached at this location but is quite loose on the rest of the penis. The **superficial fascia** is continuous with the membranous layer of superficial fascia in the abdomen and in the urogenital triangle of the perineum and the scrotum. The **deep fascia** of the penis is continuous with the deep fascia on the muscles of the superficial space in the perineum. This fascia (Buck's fascia) is continuous with the connective tissue septa surrounding the three main parts of the penis. The **suspensory ligament** of the penis is a triangular fibrous cord that is attached to the symphysis pubis and divides into right and left layers that blend with the sheath at the sides of the penis. **Preputial glands** are located between the prepuce and the glans. They secrete broken-down epithelial cells and are modified sebaceous glands.

The **arteries to the penis** are:

1. Internal pudendal
   a. Artery to the bulb
   b. Deep artery to the penis
   c. Dorsal artery to the penis
2. External pudendal

The **artery to the bulb** (Fig. 5-81) arises from the internal pudendal artery, while the latter is in the deep space of the perineum. This vessel courses medially, pierces the perineal membrane, and enters the bulb; it gives blood supply to the corpus spongiosum and the glans of the penis.

The **deep artery** is one of the terminal branches of the internal pudendal (Fig. 5-81). It escapes from the deep space and enters a crus of the penis; this artery gives the blood supply to a corpus cavernosum and courses in the center of this part of the penis. Some branches of this artery empty directly into the cavernous spaces, while others are coiled before supplying blood to the spaces (helicine arteries).

The other terminal branch of the internal pudendal artery, the **dorsal artery of the penis** (Fig. 5-81), runs distally on the dorsum of the penis, deep to the deep fascia, and supplies the fascia and the skin. Its terminal branches end in the prepuce, the frenulum, and the glans.

Additional branches to the skin of the penis come from the superficial external pudendal branch of the femoral artery (to be studied with the lower limb).

The **veins** correspond to the arteries but differ in that the **dorsal vein** is divided into superficial and deep veins. The superficial divides at the proximal end of the penis and drains into the superficial external pudendal veins, but the deep dorsal vein of the penis enters the pelvic cavity and ends in the prostatic plexus of veins. As mentioned previously, this vein enters the pelvic cavity via a small opening posterior to the inferior arcuate ligament and anterior to the urogenital diaphragm.

The **nerves to the penis** can be divided into those to the skin and fascia and those to the substance of the corpora. Those to the skin and fascia are the **dorsal nerve**

of the penis, a branch of the pudendal, the **scrotal branches of the perineal nerve** and **ilioinguinal nerve,** and the **perineal branches of the posterior femoral cutaneous nerve** of the thigh. These nerves, particularly the rich supply to the glans and frenulum, are important in the maintenance of erection. The nerves to the substance of the corpora are **sympathetic fibers** in addition to a branch from the dorsal nerve of the penis to the corpus cavernosum and a branch from the perineal nerve to the corpus spongiosum.

The **lymphatics of the penis** end in inguinal glands in the groin.

The **clitoris** (Fig. 5-86) has a structure very similar to that of the penis except that it is much smaller and does not contain the urethra. It is composed of three portions—the **two corpora cavernosa** and the **corpus spongiosum**—the former two arising from the crura of the clitoris in the perineum, and the corpus spongiosum being an anterior continuation of the bulbs in the superficial space of the perineum. The corpus spongiosum, as in the male, ends in a dilated head or **glans.** The organ is

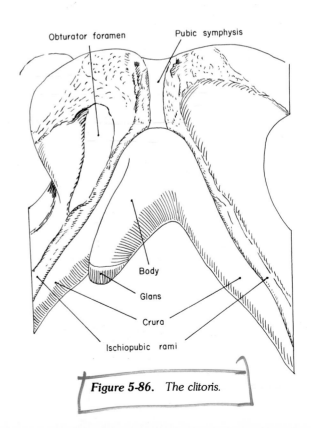

Obturator foramen   Pubic symphysis

Body

Glans

Crura

Ischiopubic rami

Figure 5-86. *The clitoris.*

erectile in nature and its blood and nerve supply are practically identical to that of the penis, the head of the clitoris being particularly sensitive. It has a **suspensory ligament,** which suspends the clitoris from the pubic symphysis. The clitoris takes a sharp bend at the point where the suspensory ligament is attached and heads posteriorly and inferiorly being about 1 inch long. The loose fold of skin over the clitoris is the **prepuce** of the clitoris.

URETHRA—MALE AND FEMALE

The **male urethra** is divided into three parts: (1) **prostatic,** (2) **membranous,** and (3) **spongiose** (Fig. 5-84).

The **prostatic part** is contained within the prostate gland and is a little over 1 inch in length. It starts at the neck of the bladder and traverses the prostate gland from its base to its apex. Posteriorly, it possesses a ridge of tissue called the **urethral crest,** which is a longitudinal elevation in the mucous membrane. The **prostatic ducts** open on either side of the crest, the **prostatic utricle** is found in the center of the crest, and the **ejaculatory ducts** open on either side of the prostatic utricle (Fig. 5-68).

The **membranous part** of the male urethra is rather short, approximately ½ inch in length. It passes through the urogenital diaphragm, and the **bulbourethral glands** are located posterior to it. It naturally starts at the apex of the prostate and pierces the deep fascia, the sphincter urethrae muscle, and the perineal membrane to join the spongiose portion of the urethra.

The **spongiose part** enters the bulb of the penis and traverses the whole length of the corpus spongiosum, opening at the distal end of the penis at the vertical slit-like external orifice of the urethra. A dilation occurs in the bulb—the intrabulbar fossa—and in the glans penis—the fossa navicularis. It should be noted that the urethra takes a sharp bend from an inferior course to an anterior course in the bulb of the penis. Although the bulbourethral glands are located in the deep space, the openings of the ducts of these glands is into the spongiose portion of the urethra. Numerous mucous glands—the **urethral glands**—open into the urethra. In addition, small pits, or **lacunae,** occur, one in the fossa navicularis being rather large—the lacuna magna.

Because the ejaculatory ducts empty into the prostatic urethra, the urethra in the male serves a double purpose; it serves not only as a pathway for urine to leave the bladder during micturition, but also for the passage of semen during ejaculation.

The **female urethra,** in contrast to the male, is rather short being approximately 1½ inches long (Fig. 5-62). It courses inferiorly and slightly anteriorly from the neck of the bladder through the urogenital diaphragm to end in the vestibule of the vagina between the labia minora approximately 1 inch posterior to the glans of the clitoris. It is anterior to the opening of the vagina, and its external orifice has firm raised margins. Very small glands open at the sides of the external urethral orifice. These glands correspond to the prostate in the male. The urethra in the female serves for passage of urine only.

The smooth muscle of the urethra is under control of the **parasympathetic nervous system** (S2, S3, and S4). The external sphincter (sphincter urethrae muscle) is under voluntary control after infancy.

The **lymphatics** of all the external genitalia drain into the inguinal nodes.

The urethra, being open to the outside world, is a prime area for infectious agents, primarily of the venereal variety. Such infections (urethritis) can ascend into the bladder where a cystitis develops. Actually, a cystitis is very common, particularly in women, and the infection tends to travel in both directions from this source, descending into the urethra or ascending into the ureters (ureteritis) to the renal pelvis (pyelitis). The importance of enlargement of the prostate in this regard has already been mentioned (page 303).

MICTURITION

As can be seen in Figure 5-87, while the urinary bladder is filling, the walls of the urethra are close together. This is accomplished by the circularly arranged elastic fibers found surrounding the urethra at the neck of the bladder and in the deep space of the urogenital diaphragm associated with the sphincter urethrae muscle, aided by an elongation of the urethra by a combined downward movement caused by contraction of the sphincter urethrae muscle (voluntary) and an upward movement caused by the pubovesical portion of the levator ani mus-

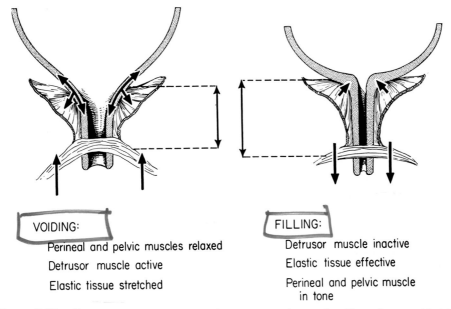

VOIDING:
Perineal and pelvic muscles relaxed
Detrusor muscle active
Elastic tissue stretched

FILLING:
Detrusor muscle inactive
Elastic tissue effective
Perineal and pelvic muscle
in tone

Figure 5-87. *Diagrammatic presentation of events in voiding and in filling of urinary bladder. (After R. T. Woodburne, as published in* Anat. Rec. *141, 11–20, 1961, courtesy of The Wistar Press, Philadelphia.)*

cle (also voluntary). There seems to be no muscular internal sphincter.*

As urine collects in the bladder the walls become thin and stretch receptors are stimulated. The desire to void reaches consciousness. As shown in Figure 5-87, during micturition the urethra is shortened and becomes funnel-shaped. This is accomplished by (1) a relaxation of the pubovesical portion of the levator ani muscle, and (2) by contraction of the detruser muscle (parasympathetic—S2, S3, and S4), this latter action being due to the fact that the inferior portion of the detruser muscle, that portion which is continuous with the urethra, exerts an upward and radial pull on the urethra under these circumstances. After urine has entered the first part of the urethra, the sphincter urethrae muscle is relaxed, allowing a further shortening of the urethra (particularly in the female) and micturition to occur. An increase in abdominal pressure by contraction of the diaphragm and abdominal muscles aids the process. Any urine remaining

in the spongiose portion of the urethra in the male is ejected by contracting the bulbospongiosus muscle.

Relaxation of the detruser muscle and contraction of the sphincter urethrae and pubovesical muscles, aided by the elastic tissue, returns the urethra to its elongated (closed) condition once again.

HUMAN SEXUAL RESPONSE

Sexual response has been divided into four phases: (1) excitement, (2) plateau, (3) orgasm, and (4) resolution.

As the result of either tactile (somatic afferent fibers) or psychic stimuli, the penis or clitoris becomes enlarged and erect. The vestibular bulbs in the female also enlarge. This is known as the excitement phase and is due to a dilatation of the cavernous spaces in the erectile tissue with blood. In addition, the vagina elongates and produces a watery fluid by exudation, and the vestibular glands secrete a fluid to lubricate the vestibule of the vagina. All of this is under control of the parasympathetic nervous system from S2, 3, and 4.

The maintenance of the above state is known as the

*R. T. Woodburne, 1961, The sphincter mechanism of the urinary bladder and urethra, *Anat. Rec.* 141: 11.

plateau phase. Erection and enlargement are maintained by more blood entering than leaving the organs involved. The exact mechanism is not clearly understood, but the enlargement of the corpora cavernosae may put enough pressure on the veins that drain the erectile tissue to accomplish this. The ischiocavernosus muscles have been thought to aid in this process but this has been questioned. Constant stimulation is required to maintain the erect state.

Orgasm in the male occurs in two stages—**emission** and **ejaculation.** During **emission** sperm from the epididymis are propelled through the vas deferens, join fluid from the seminal vesicles, and enter the prostatic urethra via the ejaculatory ducts. Prostatic fluid joins this mixture and fluid from the bulbourethral glands is added in the proximal end of the spongiose part of the urethra. This mixture is called semen. It fills the spongiose portion of the urethra. During **ejaculation** the semen is propelled through the urethra by vigorous contraction of the bulbospongiosus muscle aided by other perineal muscles. Orgasms occur in the female as well but no emission or ejaculation occurs. Both emission and ejaculation are under control of the sympathetic portion of the autonomic nervous system. Because the sympathetic system tends to function as a whole, an orgasm involves many areas of the body. Heart rate, blood pressure, and respiratory rate all increase during orgasm, the time of most intense sexual response.

At the end of the orgasm, which may be multiple in the female, blood leaves the erectile tissue (**resolution phase**) and the organs return to the flaccid state.

In the male, a certain amount of time must elapse before an orgasm can be repeated. With the exception of the somatic afferent fibers involved (touch) and the voluntary supply to the perineal musculature, this entire process is under the control of the autonomic nervous system; in spite of this fact, there is considerable cortical influence over these acts.

The fluid from the prostate, seminal vesicles, and bulbourethral glands serves two purposes: (1) to make the environment of the vagina alkaline in nature and (2) to induce motility on the part of the sperm.*

---

*There is some evidence that sperm are not made motile by this fluid, for they will become motile when removed from the epididymis in the absence of any such fluid. The problem may well be, what keeps the sperm nonmotile in the epididymis?

In a normal individual 1 milliliter of semen contains close to 100 million sperm. Since the normal ejaculate contains from 3 to 4 cc of semen, such a specimen can contain up to 400 million spermatozoa.

REGIONAL REVIEW

You have consumed a considerable number of hours dissecting the body into its various parts. It is now suggested that the reverse be tried, namely, rebuilding the body. This can be done simply by turning the pages throughout this section on the abdominal and pelvic cavities in a reverse direction and studying the pictures.

SYSTEMIC REVIEW

BONES

There is not a great deal of difference between a regional and a systemic approach to the bones in this region. It might be mentioned that two portions of the axial skeleton, the sacrum and the coccyx, join with a single part of each lower limb—the two coxal bones—to form the pelvis. The sacrum consists of five fused vertebrae and the coccyx of four fused vertebrae. They form the posterior aspect of the pelvic cavity and articulate with the two coxal bones. The coxal bones consist of three parts: a superior part, which flares posteriorly and laterally, called the ilium; a posterior part, the ischium; and an anterior part, the pubis. The two pubic bones articulate anteriorly in the form of the pubic symphysis. These three bones articulate with one another and all three take part in the formation of the acetabulum, which articulates with the head of the femur.

The division of pelves into anthropoid, android, gynecoid, and platypelloid is interesting, but the dimensions of the pelvic outlet in these various types is the important item in medicine. Naturally, this is of greater importance in the female than in the male. Three measurements are utilized to compare the pelvic outlet with the size of the fetal head—pubic symphysis to coccyx, junction of pubic and ischial rami to middle of sacrotuberous ligament, and between the inside surfaces of the ischial tuberosities. All of these measurements are greater in most females than in most males.

JOINTS

The joints involved in the abdominal and pelvic cavities are the joints between the vertebrae, the sacroiliac joints, and the pubic symphysis. These have been described on pages 78 and 280 and do not require repeating. The sacroiliac joint is a very firm structure with relatively little movement; since it bears a considerable amount of weight, it is surrounded by heavy and strong ligaments. Many patients complain of sacroiliac trouble but much of this trouble resides in other areas of the vertebral column, particularly the lumbosacral joint. The pubic symphysis contains a disc and has several ligaments that aid in holding the bones together. It is thought that this disc and ligaments can become softened just before parturition so that childbirth will be made easier.

MUSCLES

The muscles involved in this region are:

1. Diaphragm
2. External abdominis oblique
3. Internal abdominis oblique
4. Transversus abdominis
5. Rectus abdominis
6. Quadratus lumborum
7. Psoas major and minor
8. Iliacus
9. Piriformis
10. Obturator internus
11. Pelvic diaphragm
12. Perineal musculature

Besides forming the abdominal wall, the muscles in this list are involved in abdominal breathing; in movement of the vertebral column; in vital processes such as micturition, defecation, and parturition; in movement of the lower limb at the hip joint; in providing a floor for the pelvis and support for the pelvic organs; in maintaining closure of the various openings in the perineum; and in aiding in the process of reproduction.

ABDOMINAL BREATHING. This act was summarized in the section on the thoracic and abdominal walls and should be reviewed at this time (page 149). A more detailed description of the role played by the diaphragm can be found on page 268.

MOVEMENT OF VERTEBRAL COLUMN. This also has been summarized and can be found under the systemic review of the back (page 94), and of the thoracic and abdominal wall (page 149).

MICTURITION, DEFECATION, AND PARTURITION. These three processes are aided by an increase in intra-abdominal pressure; this is accomplished by contraction of the diaphragm and the abdominal musculature on both sides of the body while holding the breath.

MOVEMENT OF LOWER LIMB. The iliacus and psoas major muscles (iliopsoas) are involved in flexing the thigh; their action in rotation seems to depend upon the state of flexion of the hip joint and is not particularly important. The piriformis muscle laterally rotates the hip joint, as does the obturator internus muscle. These muscles will be reviewed again after the lower limb is described.

PELVIC DIAPHRAGM. This diaphragm serves a very important function in providing a floor for the abdominopelvic cavity and in providing support for pelvic organs. Consisting of the coccygeus (a muscle used in moving the tail in lower animals) and the levator ani muscles, this diaphragm forms a sort of sling upon which the midline pelvic organs—the prostate and rectum in the male, and the urethra, vagina, and rectum in the female—rest. The levator ani muscles elevate the anal canal during passage of fecal matter, and aid in sphincter activity at other times.

CLOSURE OF PERINEAL OUTLETS. In addition to the action of the levator ani muscle just mentioned, closure of the anal canal is maintained by an internal sphincter, which is composed of smooth muscle and is an enlarged part of the circular muscle layer in the rectum, and an external sphincter made up of striated voluntary muscle—the external sphincter ani muscle.

There is no necessity to keep the vagina closed, and there are no muscles that accomplish this with any efficiency. However, the inferior edges of the levator ani muscles and the bulbospongiosus muscles do act as a partial sphincter. The urethra is kept closed by elastic tissue at the junction of the bladder and urethra, aided by the pubovesical portion of the levator ani muscles and the sphincter urethrae muscle (voluntary) located in the deep space of the perineum. This is true for both sexes. (The bulbospongiosus muscles in the male serve to eliminate urine remaining in the urethra.)

MUSCLES USED IN REPRODUCTION. The scrotum contains smooth muscle fibers that respond to changes in temperature, contracting to cold stimulation so that the testes are brought closer to the body, and relaxing to a stimulation of warmth. Similar fibers are found in the labia majora of the female but their function is problematical. The cremaster muscle is voluntary in nature and is used by lower animals to pull the testis into the relative protection of the abdominal cavity. This muscle is of little use in man but a cremasteric reflex is useful clinically, and some claim the cremaster muscle will contract if cold is applied to the scrotum. The ischiocavernosus muscles may aid in maintaining erection in the penis or clitoris by preventing outflow of blood from these organs. The entire perineal musculature, as well as the gluteal muscles, are involved in the sexual act, with the bulbospongiosus muscles playing a prominent role. The transversus perinei superficialis and profundus muscles probably serve to tense the central point of the perineum so that other muscles arising from this point can be effective.

NERVOUS SYSTEM

The nervous system in the abdominal and pelvic cavities can be divided into that which is concerned with the abdominal wall and lower limb and that associated with the viscera.

That concerned with the abdominal wall and lower limb consists of the lumbar and sacrococcygeal plexuses. The **lumbar plexus** is composed of ventral rami of nerves from the first four lumbar segments of the spinal cord. (The dorsal rami course to the back musculature and the skin as described previously; they are not part of the plexus.) The nerves given off by this lumbar plexus are the ilioinguinal, iliohypogastric, genitofemoral, lateral femoral cutaneous, femoral, obturator, and a large contribution to the sacral plexus called the lumbosacral trunk.

The **iliohypogastric** nerve proceeds laterally and then anteriorly through the abdominal wall musculature, coursing between the transversus abdominis and the internal abdominis oblique muscle. After giving off a branch to the skin of the gluteal region, it pierces the internal abdominis muscle near the anterior superior iliac spine. It becomes cutaneous just superior to the pubic bone. The **ilioinguinal** nerve takes a similar course but courses through the inguinal canal itself and terminates as a cutaneous nerve on the wall of the scrotum, base of the penis, and the medial side of the thigh in the male and in the corresponding structures, the labia majora and thigh, in the female. The **lateral femoral cutaneous** nerve is a nerve of the lower limb and, after coursing posterior to the inguinal ligament, distributes to the lateral side of the thigh. The **femoral nerve** innervates the psoas major and minor muscles and the iliacus muscle, and then proceeds inferiorly to innervate skin and muscles on the anterior surface of the thigh. The **obturator** nerve courses along the medial side of the psoas major muscle, along the lateral wall of the pelvis, and exits via the obturator foramen; it is also a nerve of the lower limb and innervates skin and muscles on the medial side of the thigh as well as the hip joint. An **accessory obturator** nerve is present occasionally. When present, it arises from the third and fourth lumbar nerves and follows along the medial border of the psoas major muscle. Its course differs from the main obturator nerve in that it crosses anterior to the superior ramus of the pubis instead of going through the obturator foramen. It passes deep to the pectineus muscle, giving it its nerve supply and then sends a branch to the hip joint. The **genitofemoral** nerve is the one nerve that pierces the anterior surface of the psoas major muscle. It divides into a genital branch, which courses inferiorly through the inguinal canal and innervates the cremaster muscle as well as providing sensory branches to the area of the mons pubis, and a femoral branch, which enters the superior aspect of the thigh near the femoral triangle to give cutaneous innervation to the skin in that area.

The **sacral plexus** consists of nerves from the first, second, third, and fourth sacral segments of the spinal cord with a very heavy contribution from the fourth and fifth lumbar nerves and a small contribution from the fifth sacral. The **coccygeal nerves** usually unite to form one coccygeal nerve. This plexus, more properly called the lumbosacral plexus, consists of the ventral rami of these nerves (the dorsal rami course to the back, passing through the posterior foramina in the sacrum). Most of the branches of the sacral plexus are best studied with the lower limb. However, those involved with the perineum must be considered. These are the coccygeal nerve mentioned above, the posterior femoral cutaneous nerve, and the pudendal nerves.

The **coccygeal** nerve provides cutaneous innervation around the coccyx posterior to the anus in both sexes. The **posterior femoral cutaneous** nerve provides a perineal branch that provides cutaneous innervation to the lateral aspect of the perineum in the area of the urogenital triangle.

The **pudendal** nerve exits from the pelvic cavity through the greater ischiadic foramen, winds around the ischiadic spine, and enters the perineum via the lesser ischiadic foramen. It then proceeds anteriorly in a layer of connective tissue (pudendal canal) on the obturator internus muscle along the lateral wall of the ischiorectal fossa. After giving off very important branches to the external sphincter muscle of the anus and the levator ani muscle, it gives off a perineal branch that enters the superficial space of the urogenital triangle, innervates the muscles contained in both spaces, and then pierces the membranous layer of the superficial fascia to become cutaneous on the posterior wall of the scrotum (scrotal branches) in the male or on the labia majora (labial branches) in the female. Going back to the main pudendal nerve, it continues anteriorly, after giving off this perineal branch, as the dorsal nerve of the penis or clitoris. It pierces the fascia forming the posterior boundary of the deep space of the perineum, enters the deep space, and then pierces the perineal membrane to reach the penis or clitoris.

It might be well to summarize the sensory nerve supply to the perineum at this time. All must be taken into consideration in anesthesia of this region. From posterior to anterior we have (1) the coccygeal nerve to the area between the coccyx and anal canal, (2) cutaneous branches of the inferior rectal nerves to the anus and the areas lateral to the anus, (3) the perineal branch of the posterior femoral cutaneous, to the more lateral parts of the urogenital triangle, (4) the sensory branches of the perineal nerves to the scrotum or labia, (5) the ilioinguinal

nerve to the anterior part of the scrotum or labia, (6) the dorsal nerve of the penis or clitoris to those structures, and (7) the genital branch of the genitofemoral nerve to the area of the mons pubis.

The visceral organs of the abdominal and pelvic cavities are innervated by the autonomic nervous system* or the general visceral efferent system. You will recall that the **autonomic nervous system** is made up of two distinct parts, the sympathetic or thoracolumbar division and the parasympathetic or craniosacral division. You will also recall that the **sympathetic system** arises from all the thoracic and first two lumbar segments of the spinal cord. These preganglionic cell bodies are located in the intermediolateral cell column of these segments of the cord and exit via the ventral roots. After reaching the segmental nerve, they exit from that nerve and join a ganglion in the sympathetic chain via white rami communicantes. These sympathetic chains are not limited to the thoracic and upper lumbar regions but continue into the cervical region as a cervical extension and into the lower lumbar and sacral regions as an extension inferiorly to terminate finally in the ganglion impar near the coccyx. Since the preganglionic fibers of the sympathetic system are found only in the thoracic and upper lumbar segments of the spinal cord, there will be white rami communicantes only in this region. However, since the chains do extend superiorly and inferiorly, all of the segmental nerves of the body will be connected with the sympathetic chain via gray rami communicantes, the pathway taken by postganglionic fibers that have synapsed in the chain ganglia to reach the spinal nerve once again. These postganglionic fibers, if they do go back to the regular segmental nerve, will follow these nerves to innervate smooth muscles in blood vessels, arrector pili muscles, and sweat glands in the body wall and the lower limb. Therefore, postganglionic fibers can leave the chain ganglia and join each of the segmental nerves throughout the length of the body.

All the preganglionic fibers do not synapse in these chain ganglia. We have already seen in the thoracic region that nerves continue through these chain ganglia from the fifth to the ninth thoracic segments and form the greater splanchnic nerve, which finally terminates in the

---

*L. A. Gillilan, 1954, *Clinical Aspects of the Autonomic Nervous System,* Little, Brown and Co., Boston.

celiac ganglion at the base of the celiac artery. The nerves from the tenth and eleventh thoracic segments form the lesser splanchnic nerve, which ends in the superior mesenteric ganglion; contribution from the twelfth thoracic segment is the least splanchnic nerve, which ends in the aorticorenal ganglion. Preganglionic sympathetic fibers arising from the intermediolateral cell columns of the first two lumbar segments of the spinal cord emerge via the ventral roots of these two nerves, enter the spinal nerves, and then travel to the sympathetic chain via white rami communicantes. Since the inferior mesenteric ganglion is quite small, it is thought that most of these fibers synapse in the chain ganglia and then continue as lumbar splanchnics. However, some of the fibers do not synapse in the chain ganglia but continue in the lumbar splanchnics and finally synapse with postganglionic fibers in the inferior mesenteric ganglion and in ganglion cells spread along the anterior surface of the aorta and sacrum. Therefore, the sympathetic preganglionic fibers that are destined to innervate the viscera in the abdominal and pelvic cavities are contained in the greater, lesser, and least splanchnic nerves from the thoracic cavity, and in the lumbar splanchnic nerves in the abdominal cavity.

Postganglionic fibers then follow arteries, for the most part, to reach their destination. It is obvious that fibers from the celiac ganglion will follow the branches of the celiac artery to the liver, gall bladder, spleen, stomach, duodenum, and pancreas (the foregut and its derivatives). Those from the superior mesenteric ganglion will follow the superior mesenteric artery to the pancreas, to a portion of the duodenum, jejunum, ileum and all of the colon as far as supplied by that artery (the midgut). The kidney is supplied by postganglionic fibers from the aorticorenal ganglion, as is the superior part of the ureter. Postganglionic fibers from the inferior mesenteric ganglion follow the inferior mesenteric artery and its branches to innervate the parts of the large intestine supplied by this artery (the hindgut). Although some fibers may enter the viscera in the pelvic cavity from the sympathetic chains, by far the great majority of fibers to these organs are derived from the lumbar splanchnic nerves, which form a profuse network of nerves on the anterior surface of the abdominal aorta. These finally end as the hypogastric nerves, which enter the pelvic cavity and join with the profuse plexus of nerves in this region.

As mentioned previously, the **parasympathetic** system is a craniosacral division. The cranial division is found in four of the cranial nerves; the one involved in the abdominal cavity is the vagus. Preganglionic fibers from the brain stem follow the vagus nerve through the neck and thoracic cavity and into the abdominal cavity on the esophagus. These preganglionic parasympathetic fibers are found in the anterior and posterior gastric nerves and follow these nerves to innervate the stomach and to reach the gall bladder, the liver, and a portion of the duodenum. However, a branch from the posterior gastric nerve proceeds inferiorly and posteriorly on the diaphragm itself to reach the celiac plexus. No synapse occurs here, but fibers continue through this plexus and follow the arteries to reach the organs to be innervated. The vagus nerve is considered to innervate all of the viscera in the abdominal cavity as far along as the left colic flexure: in other words, to the area supplied with blood by the celiac and superior mesenteric arteries. In addition, the vagus is also considered to innervate the ovaries and testes since these organs originated high in the abdominal cavity. These fibers follow the ovarian or testicular arteries. As previously mentioned, the preganglionic parasympathetic fibers do not synapse in ganglia away from the organ but continue to the organ via the blood vessels to synapse with ganglia in the walls of the organs themselves; the postganglionic parasympathetic fibers are therefore microscopic in length.

The descending colon, sigmoid colon, rectum, and structures in the pelvic cavity are innervated by the sacral portion of the parasympathetic system. The fibers coming from the second, third, and fourth sacral segments of the spinal cord join to form what is called the pelvic nerve; this nerve sends branches to all of the pelvic viscera (except the ovaries). The parasympathetic fibers to the parts of the intestines just mentioned reach their destination by traveling superiorly in the hypogastric plexus, and then coursing along the branches of the inferior mesenteric artery to all parts of the intestine given blood supply by this artery. Some of these fibers ascend along the posterior abdominal wall, thus bypassing the hypogastric plexus.

Although **visceral afferent fibers** are a bit of an enigma, it is thought that visceral pain does exist and that the visceral afferent fibers are found in the nerves just mentioned. Afferent nerves are quite important, for they frequently carry impulses of pain that are referred to various parts of the body wall, depending upon which segments of the spinal cord receive these sensory fibers. Generally speaking, visceral afferent fibers from the foregut (organs supplied by the celiac artery, i.e., stomach, first two parts of the duodenum, gall bladder, liver, spleen, and pancreas) traverse the celiac ganglia and return to the spinal cord in segments T5 to T9. Pain will be referred to the thoracic or abdominal wall innervated by these segments. Those from the midgut (organs supplied by the superior mesenteric artery, i.e., part of pancreas, last two parts of the duodenum, jejunum, ileum, appendix, cecum, ascending colon, right colic flexure, and most of the transverse colon) traverse the superior mesenteric ganglia and return to spinal cord segments T10 and T11. Pain will be felt anywhere along the path of the nerves from these segments, particularly on the abdominal wall around and just below the umbilicus. Visceral afferent fibers from the hindgut (organs supplied by the inferior mesenteric artery, i.e., a small part of the transverse colon, left colic flexure, descending colon, sigmoid colon, and upper part of the rectum) traverse the inferior mesenteric ganglion and return to the upper lumbar segments of the spinal cord. Pain will be referred to the expected areas of the body wall. Visceral afferents from the kidney and from the upper part of the ureter traverse the aorticorenal ganglia and return to the cord at segment T12. Pain is referred anywhere along the path of the subcostal nerve. Visceral afferent fibers from the pelvic organs may find their way back to L1 and L2, or return to the cord at sacral segments 2, 3, or 4. The patient will usually complain of pain in the lower abdominal wall around the pubis, which would indicate that most pain returns via the lumbar splanchnics. The importance of this information in diagnostic work is obvious.

The specific role of the autonomic nervous system on any particular system will be reviewed with this system.

DIGESTIVE SYSTEM

Little can be added to the descriptions of the digestive system already given in the regional approach. Suffice it to say that the digestive system is well suited to the function it has to perform. The tube—the esophagus—through which the chewed food enters the stomach is attached to the stomach in such a way that food does not easily return into the esophagus. The dilated stomach serves as a very fine temporary storehouse for food that

has not been completely digested. The pyloric sphincter functions in such fashion that the food does not enter into the duodenum until it is further digested, but, when this sphincter does open at intervals, food is allowed to enter the duodenum. In the stomach the food is attacked by hydrochloric acid from the parietal cells and by digestive enzymes from the chief cells. In the duodenum the food is joined with bile from the liver and also by the pancreatic juice, and these two substances continue the process of digestion. The duodenal glands also produce a digestive enzyme that aids in this process. The food is now in a fluid form and is propelled through the many feet of duodenum, jejunum, and ileum by peristaltic movements, movements of these organs being allowed because of the serous fluid on their surfaces produced by the peritoneum. The food is once again attacked by digestive enzymes from the intestinal glands. The fluid-like food is mostly absorbed in the small intestine, but that material that is not absorbable leaves the ileum at intervals through the ileocecal valve. After entering the cecum, undigested food follows along the ascending colon, transverse colon, and descending colon to reach the sigmoid colon. During this time this material is being concentrated by water being returned to the body. At approximately daily intervals the fecal matter enters the rectum. After a certain amount of fecal material has accumulated in the rectum, the reflex is set up that gives one the desire to eliminate this fecal matter and it is removed from the body via the anal opening.

Although the intestine seems to be more massive than needed, this organ is constructed in such a manner as to increase surface area; the rugae and villi in the small intestine, the gastric pits in the stomach, etc., are nature's devices to increase the surface area through which the food can be attacked and can be absorbed into the body.

The pancreas, in addition to producing digestive enzymes that empty into the duodenum via the pancreatic duct, produces insulin via the pancreatic islets. This material is a hormone, is secreted into the blood stream, and is involved in carbohydrate metabolism. The liver, in addition to secreting bile, plays a very vital part in metabolism of carbohydrates and proteins.

The autonomic nervous system plays an important role in the functioning of the digestive system. The sympathetic system stimulates the cardiac, pyloric, and internal anal sphincters to contract but otherwise decreases peristalsis. The parasympathetic system, on the other hand, increases peristalsis throughout the entire gastrointestinal system. In addition, the parasympathetic system stimulates the production of intestinal enzymes and the sympathetic system seems to stimulate the production of mucus. The parasympathetic system also increases the production of bile and causes a contraction of the wall of the gall b aldder.

After feces have moved from the sigmoid colon into the rectum, the distention of the walls sets up afferent impulses that enter the spinal cord in the second, third, and fourth sacral segments and finally reach the brain; these give the desire to remove fecal matter from the rectum, i.e., to defecate. This stimulates the parasympathetic fibers of the pelvic nerves, which increase the tone of the smooth muscle in the bowel and produces a large peristaltic wave, the sphincters of the anus relaxing in order to let the fecal matter exit. The internal sphincter is relaxed by the parasympathetic nervous system, while the external sphincter is voluntarily relaxed. This process is aided by voluntary acts such as a contraction of the diaphragm, closure of the glottis of the larynx, contraction of the abdominal musculature, and a contraction of the levator ani and perineal muscles.

URINARY SYSTEM

There is little advantage to a systemic approach to the urinary system over a regional approach; all that might be accomplished would be to bring continuity to the course and relations of the ureter, information of supreme importance clinically. The student can accomplish this very simply by reading pages 262–266, and then pages 298–302 and 326.

The structure of the urinary system correlates very well with its function. Blood enters the kidneys via the renal arteries, which immediately break up into many branches. These finally divide into small collections of capillaries called glomerular tufts. Fluid moves from these blood vessels into what is called Bowman's capsule, the first part of a kidney unit (the nephron). The fluid flows from this capsule into the proximal convoluted tubule, where certain materials are reabsorbed into the circulation. The resultant fluid then traverses a narrow loop called Henle's loop to reach the distal convoluted tubule where further reabsorption can take place. After this, the urine is collected into collecting tubules which empty into a minor calyx. Several minor calyces join to form a major

calyx, which empties into the pelvis of the kidney. The urine then flows down the ureters to be collected in the urinary bladder and stored there until opportunity is provided for elimination; the bladder, under normal conditions, holds up to 500 cc of urine; under abnormal conditions, much more. It is expelled to the outside world via the relatively short (1½ inches) urethra in the female and the relatively long urethra in the male, the latter being divided into prostatic, membranous, and spongiose portions.

The **autonomic nerve control** of the kidney and the ureter seems to be of little importance. The sympathetic system does cause a vasoconstriction and it therefore controls the amount of urine produced in this manner, but the role played by this system in these particular organs is a minor one. The bladder, on the other hand, is innervated by the autonomic nervous system and the visceral afferent system in a very important manner. Urine enters the bladder from the ureter in a constant and gradual manner, being prevented from refluxing into the ureters by the oblique course of the ureters through the bladder wall. This urine, under normal conditions, does not enter the urethra due to the involuntary sphincteric action of the elastic fibers of the bladder and proximal end of the urethra plus the effects of the voluntary pubovesical portion of the levator ani muscles and the sphincter urethrae muscle in the deep space of the perineum. There is elastic tissue associated with the urethra in the deep perineal space as well as at the neck of the bladder. As the bladder becomes distended the tone in the bladder musculature is increased. Stretch receptors in the bladder wall send impulses to the spinal cord via the pelvic nerve; these reach the brain to give the sensation of a full bladder and a desire to void. At the time of voiding, the pubovesical muscles are relaxed, the detruser muscle contracts thus opening the proximal end of the urethra and forcing urine into the urethra, and the sphincter urethrae muscle is relaxed. The urine is thereby forced out of the bladder, and this reflex is maintained as long as urine is flowing through the urethra. This process can be hastened by a voluntary contraction of the diaphragm and abdominal muscles. Any urine remaining in the urethra in the male is voluntarily ejected by contracting the bulbospongiosus muscle.

The urinary system is likely to become infected, particularly in females, and infection can ascend from the urethra to the kidney itself. An infection in the urethra is

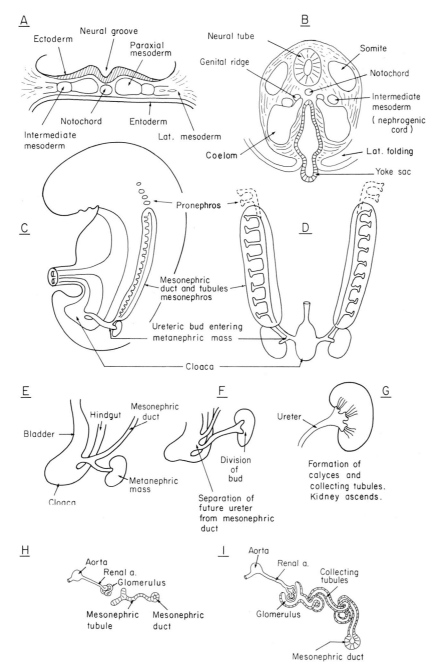

Figure 5-88. *Development of the urinary system. (A) Cross section of the embryo showing the intermediate mesoderm from which the urinary system develops. (B) Position of the intermediate mesoderm (the nephrogenic cord) after lateral folding has neared completion. (C) Longitudinal drawing showing the three stages of development of the urinary system—pronephros (degenerating), mesonephros, and metanephros. (D) Frontal view of same stage depicted in C. (E, F, and G) Development of the ureter, the major and minor calyces, and the collecting ducts of the definitive kidney (note that kidney ascends from pelvic cavity to definitive position in the abdominal cavity). (H and I) Development of the renal arteries and their branches to form glomeruli, which join with the developing duct system.*

called urethritis; in the bladder, cystitis; in the ureter, ureteritis; in the pelvis of the kidney, pyelitis; and in the kidney itself, nephritis.

Figure 5-88 depicts the development of the urinary system.

MALE REPRODUCTIVE SYSTEM

The structure of this system has been covered quite adequately on pages 143–145, 302–303, and 322–326. The following functional considerations may be of aid to the student.

The **testes** descend from a position close to the kidneys in the abdominal cavity, through the inguinal canal, to the scrotum. This position is of advantage because sperm are not viable at body temperature, and the temperature in the scrotum is lower than that found in the body as a whole. The smooth muscle making up the dartos tunic in the walls of the scrotum is responsible for maintaining the proper temperature in the scrotum, contracting when the surrounding temperatures are decreased; this raises the testes closer to the warmer body temperature. The dartos muscle is under control of the sympathetic portion of the autonomic nervous system.

The **testes** play a double role: they produce both spermatozoa and hormones; the latter, called androgens, enter the blood stream and stimulate the accessory sex organs, such as the seminal vesicles, prostate, external genitalia, etc. The production of spermatozoa and androgens, by the seminiferous tubules and the interstitial cells respectively, is under direct control of the anterior lobe of the hypophysis, and the hypophysis, in turn, is controlled by nuclei in the hypothalamus. The hormones given off by the hypophysis that control the testes are the gonadotropic hormones. Spermatozoa are being formed constantly and gradually are pushed into the **epididymis** where they are stored. The epididymis lies alongside the testis in the scrotum. Its head is in direct contact with the testis and is joined to it by the efferent ducts. The body of this organ is separated from the testis by a sinus, and the tail is in direct contact and continuous with the vas deferens. The spermatozoa in the epididymis are nonmotile; the factors involved in keeping them nonmotile during storage are not understood. The vas deferens leads superiorly in the spermatic cord, traverses the inguinal canal, turns medially in a retroperitoneal position, and

enters the pelvic cavity by descending on the lateral pelvic wall. After reaching the region of the prostate it is joined by the duct of the seminal vesicle to form the ejaculatory duct, which empties into the prostatic urethra. The seminal vesicles are located just posterior to the base of the urinary bladder.

Sexual response occurs in four phases—excitement, plateau, orgasm, and resolution. Psychic stimuli or stimulation of the somatic afferent fibers from the external genitalia or the skin areas around the genitalia cause a distention of the cavernous tissue by entry of blood into the penis. This is the excitement phase and results in erection of the penis. Maintenance of erection, the plateau phase, depends on more blood entering the cavernous tissue than drains from it. The distention of the cavernous tissue probably puts enough pressure on the veins to accomplish this. The suggested role of the ischiocavernosus muscles in this process has been questioned. Constant stimulation is necessary to maintain this plateau. Erection is under control of the parasympathetic nervous system.

Orgasm consists of emission and ejaculation. It is under control of the sympathetic nervous system. During emission spermatozoa enter the vas deferens, are joined by fluid from the seminal vesicles; both enter the prostatic urethra where fluid from this gland is added, and the bulbourethral glands complete the mixture, which is called semen. The semen is now located in the spongiose portion of the urethra. Ejaculation is produced by spasmotic contraction of the bulbospongiosus muscle as well as other perineal muscles. This is under reflex voluntary control.

Resolution consists of all organs returning to a resting, flaccid state.

The semen, usually 3 to 5 cc per ejaculate, contains approximately 100 million spermatozoa per milliliter. The function of the various fluids just mentioned is not perfectly clear, although it is said to reduce the acidity of the vagina.

FEMALE REPRODUCTIVE SYSTEM

The structure of the female reproductive system has been adequately covered in the regional approach to the perineum and the contents of the pelvic cavity. The following is a brief functional systemic summary.

The female reproductive system consists of the ovaries, uterine tubes, uterus, vagina, and the external genitalia.

The **ovaries** are two almond-shaped structures, each suspended from the superior-posterior surface of the broad ligaments by its own mesentery (the mesovarium), from the lateral pelvic wall by the suspensory ligament (peritoneum covering the ovarian artery and veins), and from the uterus by a derivative of the gubernaculum, the ligament of the ovary. After puberty, this structure goes through a cycle of activity which is under the control of the hypophysis. As in the male, the hypophysis is controlled by nuclei in the hypothalamus.

Follicle-stimulating hormone from the hypophysis stimulates development and growth of the ovarian follicles and their contained ova. One of these (or more) finally breaks through the ovarian wall in a process called ovulation. These follicles also elaborate a female sex hormone called estrogen. Leaving the ovum at this point for a moment, we find that the follicle is now changed to a **corpus luteum** which is under control of the hypophyseal **luteinizing hormone.** This structure continues to grow and produces a second female sex hormone, progesterone, as well as some estrogen.

The **estrogen** mentioned above stimulates growth of the **endometrial lining** of the uterus during the first half of the cyclic period, while **progesterone** stimulates this endometrium to further growth during the second half of the period. If pregnancy does not ensue, this endometrium is extruded through the vagina in the process of **menstruation.** This combined ovarian-uterine cycle varies in length but approximates 28 days. Since the number of days involved in actual menstruation varies from cycle to cycle and from individual to individual, the first day of bleeding is usually considered as the first day of the menstrual cycle. Since the bleeding period is usually about 4 to 5 days in duration and ovulation is considered to occur approximately 14 days before the onset of the next bleeding period, this leaves something like 9 to 10 days for the follicular phase of the cycle, and approximately 14 days for the progestational phase.

Returning to the **ovum,** which had been extruded onto the surface of the ovary, we see that it enters the **uterine tube** and, by a combined muscular and ciliary action, is carried toward the uterus. If the ovum is to be fertilized by a sperm, it occurs while the ovum is in the tube. The fertilized ovum continues to the uterine cavity and embeds itself in the prepared endometrial lining; in this case menstruation ceases for as long as the pregnancy lasts. If the ovum is not fertilized, it is simply lost in the uterus and vagina. In that case, the prepared endometrium will not be used and so is removed (menstruation), and the cycle starts once again. This reproductive cycle continues until a pregnancy occurs, or until the ovary is no longer able to respond to the hypophyseal hormones and **menopause** occurs.

The **vagina** is a tube that leads from the external genitalia to the cervix of the uterus. It is approximately 3 to 4 inches in length and lies between the rectum posteriorly and the urinary bladder anteriorly. Sperm are deposited in the vagina during coitus; it is also the birth canal. In the virgin, the entrance to the vagina may be guarded by a membrane, the hymen, which can take the form of a complete membrane (which has to be cut surgically at the first menstruation), a sievelike structure, or be entirely absent. Small remains of this structure may persist—hymenal caruncles—and demark the vagina proper from the entrance to the vagina, the vestibule. This **vestibule** is located between the two fleshy folds, the labia minora, which are in turn between the labia majora. The entrance to the vagina is posterior to the opening of the urethra and anterior to the anus.

The **uterus,** consisting of fundus, body, and cervix, is located in the pelvic cavity between the rectum posteriorly and the urinary bladder anteriorly. It is normally flexed anteriorly and is approximately at right angles to the vagina.

Sexual response in the female consists of the same four stages as described for the male. During excitement the clitoris and vestibular bulbs become erect just as does the penis. The vagina elongates and produces a watery fluid by transudation. This serves for lubrication as does the release of fluid from the vestibular glands. Plateau is similar to the male and is dependent on continued stimuli. This is under control of the parasympathetic nervous system. The female has an orgasm just as does the male but with no emission or ejaculation; this is a sympathetic response and includes a contraction of all smooth muscle in this system as well as contraction of perineal and gluteal musculature. The widespread nature of the sympathetic nervous system makes an orgasm an all-inclusive phenomenon. This is true in the male as well. Resolution, as in the male, consists in blood leaving these organs and a return to the resting flaccid state.

Visceral afferent fibers from these organs in both sexes should not be forgotten. Pain is referred to the body wall and can include areas on the skin innervated by spinal cord segments T10 and 11 (testes and ovaries) to L2 or 3.

A table of homologous male and female sexual organs is presented on page 344. A study of Figures 5-89, 5-90, and 5-91 should also be helpful.

SPLEEN AND SUPRARENALS

These structures were adequately covered in the form of a regional approach. See page 259 for the suprarenals and page 244 for the spleen.

ARTERIES

The arteries in the abdominal and pelvic cavities* are branches of the abdominal aorta. They can be divided into those that give blood supply to the walls of the abdomen and those concerned with viscera. They course in the subserous fascia and are, of course, retroperitoneal.

The first branches supplying the walls are the **inferior phrenics;** arising immediately after the aorta pierces the diaphragm, they course on the inferior surface of this structure. The phrenic arteries are not entirely concerned with walls because they do give branches to the suprarenal glands.

There are four sets of **lumbar arteries** that arise from the aorta and these vessels course posteriorly and laterally. They give off a **posterior branch** that penetrates the back musculature in company with a dorsal ramus of the lumbar nerves, supplying this musculature and the skin. Each posterior branch gives off a **spinal branch** that enters an intervertebral foramen and divides into neural and osseous branches. The **neural branch** sends arteries to each dorsal and ventral root (radicular branches), some of which anastomose with the longitudinally coursing anterior and posterior spinal arteries. The **osseous branches** also divide into two branches, one of which courses along the lateral edge of the posterior longitudinal

ligament and anastomoses with similar branches superiorly and inferiorly, and with the artery of the opposite side; and another that courses on the anterior surface (inside surface) of a lamina and ligamentum flavum; this vessel also anastomoses with vessels superiorly and inferiorly, and with the same artery of the other side. The **anterior branch** of a lumbar artery continues through the abdominal musculature to anastomose with other arteries in this area.

The last branch of the aorta concerned with the body wall is the **middle sacral** artery, which takes a course on the anterior surface of the lower lumbar and sacral vertebrae.

The visceral branches of the aorta are the celiac, superior mesenteric, renal, suprarenal, ovarian or testicular, and inferior mesenteric.

The **celiac artery** immediately divides into the left gastric, common hepatic, and splenic branches. The **left gastric** proceeds superiorly, in a retroperitoneal position, onto the diaphragm and finally reaches the esophagus; it turns onto the lesser curvature of the stomach after giving a branch to the inferior end of the esophagus. The **common hepatic** courses to the right and slightly inferiorly in a retroperitoneal position until it reaches the retroperitoneal portion of the duodenum. There it turns superiorly to enter the free edge of the lesser omentum. It gives off a **right gastric** artery, which courses on the lesser curvature of the stomach and anastomoses with the left gastric branch just mentioned. This common hepatic artery also gives rise to a **gastroduodenal** artery, which in turn divides into the **superior pancreaticoduodenal** and the **right gastroepiploic** artery, which courses on the greater curvature of the stomach. The superior pancreaticoduodenal artery almost immediately divides into **duodenal** and **pancreatic** branches. The former courses on the anterior surface of the pancreas between it and the duodenum and feeds both organs. The **pancreatic** branch courses on the deep surface of the pancreas and supplies this organ, and gives some blood to the duodenum as well. Both these vessels anastomose with similar branches from the inferior pancreaticoduodenal artery. (The common hepatic artery may give additional branches directly to the pancreas.) Returning to the main common hepatic artery, it terminates as the **proper hepatic** artery, which courses in the free edge of the lesser omentum. This artery, in turn, terminates by dividing into **right and left hepatic arteries,** which enter the right and

---

*W. H. Hollingshead, 1955, Some variations and anomalies of the vascular system in the abdomen, *Surg. Clin. N. Amer.* 35: 1123. A very readable and brief summary of an important concept.

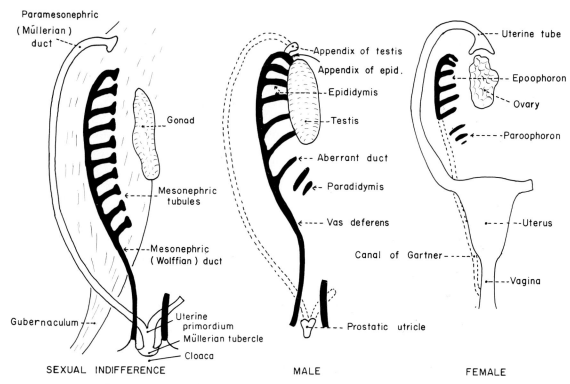

Figure 5-89. *Development of the reproductive systems. The mesonephric duct is the remains of the same duct shown in Figure 5-88. The paramesonephric duct resulted from a longitudinal outfolding of the coelomic wall in the region of the preexisting mesonephric duct. The cephalic end opens into the coelom, an important point in the anatomy of the definitive uterine tube; the caudal end joins tissue called the Müllerian tubercle, which will later develop into the uterine cervix. The gonad develops close to these ducts. The definitive male uses the mesonephric duct (vas deferens) and its branches (epididymis) while the paramesonephric duct degenerates leaving remnants craniad (appendix of the testis and appendix of the epididymis) and caudad (prostatic utricle). The female uses the paramesonephric duct, remnants of the mesonephric duct being the epoophoron, paroophoron, and the canal of Gartner, all found in the broad ligaments of the uterus (see Fig. 5-65).*

left lobes of the liver. The **cystic** artery varies in its origin, but arises most commonly from the right hepatic. Another branch of the common hepatic, rarely mentioned, is the **retroduodenal,** important because it takes a position posterior to the duodenum, very close and parallel to the bile duct. The third branch of the celiac, the **splenic,** courses to the left, retroperitoneal in position, until it reaches the lienorenal ligament; it turns between the folds of this ligament to reach the hilum of the spleen. It gives off several **pancreatic branches, short gastric branches** to the fundic portion of the stomach, and a branch to the

greater curvature of the stomach—the **left gastro-epiploic**—which anastomoses with the right gastro-epiploic from the gastroduodenal branch of the common hepatic.

The **superior mesenteric** artery arises immediately inferior to the celiac and continues inferiorly, posterior to the body of the pancreas, but anterior to the uncinate process of this organ. At this point it gives off the **inferior pancreaticoduodenal** artery, whose branches anastomose with similar branches of the superior pancreaticoduodenal artery, a branch of the common hepatic.

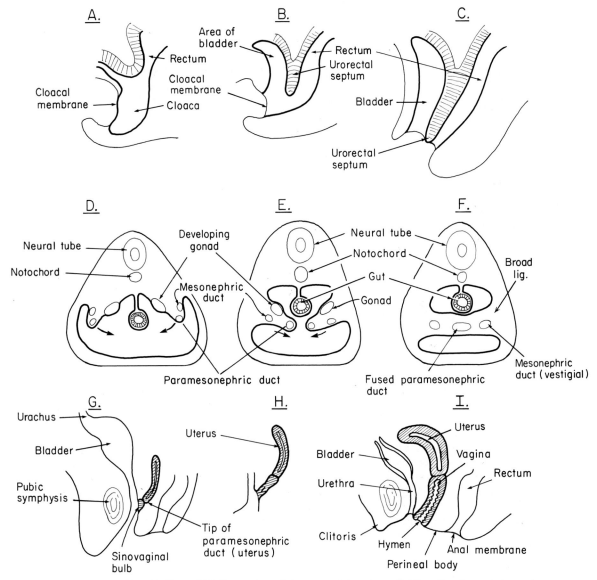

Figure 5-90. *(A, B, and C) Formation of the urorectal septum dividing the cloaca into anal and urogenital areas. (D, E, and F) Formation of the uterus by fusion of the caudal ends of the paramesonephric ducts. (G, H, and I) Growth of the vagina from the sinovaginal bulb and its separation from the urethra by development of a urovaginal septum. (D, E, F, G, H, and I redrawn after Langman.)*

Figure 5-91. *Development of the external genitalia. (From Davies,* Human Developmental Anatomy. *Copyright © 1963, John Wiley & Sons, Inc. After M. H. Spaulding,* The Development of the External Genitalia in the Human Embryo, *courtesy of Carnegie Institution of Washington.)*

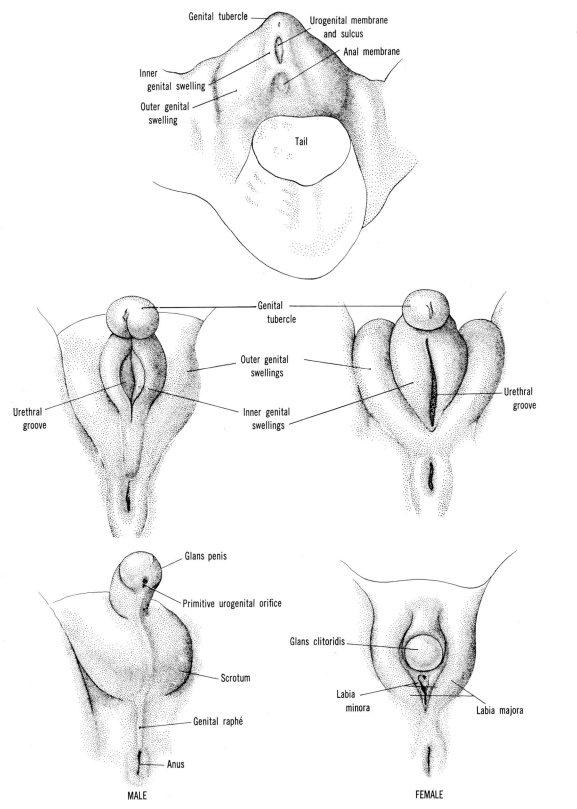

Genital tubercle

Urogenital membrane
and sulcus

Anal membrane

Inner
genital swelling

Outer genital
swelling

Tail

Genital
tubercle

Outer genital
swellings

Inner genital
swellings

Urethral
groove

Urethral
groove

Glans penis

Primitive urogenital orifice

Glans clitoridis

Scrotum

Labia
minora

Labia majora

Genital raphé

Anus

MALE

FEMALE

The superior mesenteric then continues inferiorly, crossing anterior to the duodenum, and immediately gives off many **jejunal** and **ileal** branches. There are considered to be from 20 to 22 such branches, which give a rich blood supply to these two structures. The superior mesenteric artery also gives off an **ileocolic branch,** which courses to the right and supplies blood to the terminal end of the ileum, the appendix, the cecum, and a portion of the ascending colon; a **right colic branch,** which takes care of the remaining portion of the ascending colon; and a **middle colic branch,** which gives blood supply to the transverse colon. (The arrangement of these colic vessels is quite variable.)

The next branches of the aorta are the two **renal** arteries, which emerge from the aorta at a position approximately ½ inch inferior to the location of the superior mesenteric artery. Of these two arteries, the right is the longer, because the aorta tends to be on the left side of the abdominal cavity. These two arteries course directly to the kidney, but give rise to **suprarenal branches** that supply this gland.

We have now mentioned arteries supplying the suprarenal glands from the inferior phrenics and from the renals. These are the superior and inferior suprarenal arteries respectively. In addition to these two sets of arteries, another artery, the **middle suprarenal,** arises from the aorta just superior to the point where the renals are given off.

Inferior to the renals, two small arteries are given off that continue inferiorly, finally to reach the testes in the male or the ovaries in the female; these are the **testicular** or **ovarian** vessels respectively. They arise from this high level because the ovaries and the testes first started to develop near the kidney in the abdominal cavity.

The next visceral branch of the abdominal aorta is the **inferior mesenteric** artery. This artery gives rise to the **left colic** artery, which supplies the descending colon and anastomoses with the middle colic branch of the superior mesenteric artery. The inferior mesenteric also gives rise to several **sigmoid branches** supplying the sigmoid colon. The termination of the inferior mesenteric artery is the **superior rectal** artery, which supplies the rectum. This artery anastomoses with the middle rectal branches of the internal iliac and even more intimately with the inferior rectal branch of the internal pudendal.

The branches of the colic arteries from the superior and inferior mesenteric arteries are arranged in such a manner as to form an anastamotic channel all the way from the cecum to the rectum in a position close to the viscus. This is known as the **marginal artery.** The surgeon can depend on its presence in most areas but should be wary of the connection between the middle and left colic arteries, and particularly suspicious of the anastomosis between the last sigmoidal and superior rectal branches of the inferior mesenteric.

The aorta divides into the two **common iliacs** which in turn divide into the external and internal iliacs. As mentioned for the aorta, the branches of these vessels can, in turn, be divided into branches supplying the body wall or lower limb structures and branches concerned with the viscera. The **external iliac** artery has relatively few branches, a few at its inferior end being the only ones demanding description. These vessels—the inferior epigastric, the deep circumflex iliac, and possibly the obturator artery coming from the inferior epigastric—do give blood supply to the body wall. The **inferior epigastric** artery courses superiorly and slightly medially to reach the lateral edge of the rectus sheath; it penetrates this sheath and continues superiorly on the deep surface of the rectus abdominis muscle, and anastomoses with the superior epigastric branch of the internal thoracic artery. This anastomosis provides an important collateral pathway around an occlusion of the thoracic or abdominal aorta. This inferior epigastric artery forms the lateral boundary of the important inguinal triangle, the site of direct inguinal hernias, and is just medial to the internal inguinal ring, the site of indirect inguinal hernias. If the **obturator** artery arises from the inferior epigastric it is known as an "abnormal obturator" artery; its relation to the femoral canal is extremely important, because surgery in this area may easily cut this rather large artery. The **deep circumflex iliac** artery courses laterally and anastomoses with the lumbar branches of the aorta and also the musculophrenic branch of the internal thoracic artery.

The **internal iliac** artery has many more branches than the external. It is usually considered to divide into anterior and posterior divisions.

The **posterior division** is entirely involved with the body wall or the lower limb. The large termination of the posterior division is the **superior gluteal** artery, which enters the gluteal region through the greater ischiadic foramen. The other branches are the lateral sacral and the iliolumbar arteries. The **lateral sacral** arteries travel to the lateral portions of the sacrum, enter through the anterior foramina, and emerge through the posterior foramina to reach the skin over the sacral region. The **iliolumbar** arteries course laterally, deep to the psoas major muscle, to ramify on the iliacus muscle and around the quadratus lumborum muscle. These vessels also anastomose with the lumbar arteries and the deep circumflex iliac arteries mentioned previously.

The **anterior division** of the internal iliac artery has branches supplying both the body wall or lower limb and the viscera. Those concerned with the body wall are the **inferior gluteal,** which leaves the pelvic cavity inferior to the piriformis muscle and through the greater ischiadic foramen, and the **obturator,** which usually arises from the internal iliac, follows along the lateral pelvic wall, and exits from the pelvic cavity through the obturator foramen to reach the muscles of the lower limb. The **internal pudendal** artery, in both sexes, leaves the pelvic cavity through the greater ischiadic foramen, inferior to the piriformis muscle, courses around the ischial spine, and enters the perineum via the lesser ischiadic foramen. After giving rise to **inferior rectal** branches in the ischiorectal fossa to the external sphincter ani muscle and the anal canal, the internal pudendal artery continues anteriorly in the pudendal canal in the fascia located in the lateral part of the ischiorectal fossa. It gives off a **perineal branch,** which continues into the superficial space, gives blood supply to the muscles therein, and then continues on to the scrotum or the labia majora. The main artery then continues into the deep space, gives off the **artery to the bulb,** and gives blood supply to the muscles in the deep space, and divides into the **deep and dorsal arteries of the penis or clitoris,** which penetrate the perineal membrane to reach those structures.

The **superior vesical, middle rectal, and inferior vesicals** in the male, and the **superior vesical, middle rectal, uterine, and vaginal arteries** in the female are all concerned with viscera. They course medially in the subserous fascia to reach these organs, coursing in a fashion implied by their names. The **uterine** arteries deserve special description. They are surrounded by a thickening of the subserous fascia called the cardinal ligament, travel to the cervix of the uterus, and then ascend on the sides of

this organ between the two layers of the broad ligament. They finally anastomose with the ovarian arteries.

It should be remembered that there are many instances of variation in the origin and course of these arteries.

VEINS

Except for the fact that the **deep dorsal vein of the penis or clitoris** drains directly into the pelvic plexus of veins rather than following the internal pudendal vein, the **veins of the perineum and pelvic cavity** are similar to the arteries. The veins form a much more profuse network and occasionally a pampiniform plexus of veins is present. Many branches from the pelvic cavity join to form the **internal iliac,** which combines with the **external iliac** to form the **common iliac.** The two common iliacs join to produce the **inferior vena cava.**

The **inferior vena cava** courses along the right side of the abdominal aorta in the abdominal cavity. It is considerably longer than the aorta, since it pierces the diaphragm at a much higher level. This is due to the fact that the aorta is just anterior to the vertebral column, while the inferior vena cava pierces the diaphragm closer to its dome. The actual vertebral level of the point of passage of the aorta through the diaphragm is approximately at T12, while the termination of the inferior vena cava is at the level of the eighth thoracic vertebra.

Another difference between the inferior vena cava and the abdominal aorta is the **absence of certain tributaries to the inferior vena cava** that correspond to the celiac, superior mesenteric, and inferior mesenteric branches of the aorta. The absence of these vessels is due to the presence of the portal circulation that drains the spleen and the gastrointestinal tract; the vessels draining these particular organs carry blood to the liver.

Other tributaries corresponding to branches of the aorta are also missing. The **left testicular** or **ovarian vein** does not drain directly into the inferior vena cava, but into the left renal; the **left suprarenal** also drains into the left renal; and the **first and second lumbar veins** drain into the **ascending lumbar veins,** which will be described in a moment. Another difference between the inferior vena cava and the abdominal aorta is the fact that the inferior vena cava receives two or three very large **hepatic** tributaries from the liver.

Therefore, the inferior vena cava receives the **right**

testicular or ovarian vein, renal veins, right suprarenal veins, and the hepatic veins from viscera, and the third and fourth lumbar veins and phrenic veins (left may drain entirely into left suprarenal vein) from body wall structures. It terminates by piercing the dome of the diaphragm and almost immediately entering the right atrium of the heart.

No description of the venous drainage of the abdominal cavity would be complete without mentioning the ascending lumbar veins. These veins are long anastomotic channels that connect the lateral sacral veins, iliolumbar vein, lumbar veins, and the subcostal vein.

Each originates in the lower lateral sacral veins, ascends over the sacrum and the transverse processes of the lumbar vertebrae to end in the subcostal vein on the side of the twelfth thoracic vertebra. These veins usually receive the first two lumbar veins. Because the subcostal veins drain into the azygous system of veins, there is a continuation of venous channels on the posterior aspect of both thoracic and abdominal walls. This can substitute for gradual occlusions of the inferior vena cava.

The portal system of veins consists of those draining blood from the spleen, pancreas, stomach, small intestine, and large intestine including the rectum. The inferior

Homologies of the Sexual Organs
(See Figs. 5-89, 5-90, and 5-91)

Male	Female
Gubernaculum	Ligamentum teres and ligament of ovary
Mesorchium	Mesovarium
Testis	Ovary
Appendix of testis	Fimbriated end of uterine tube
Paradidymis (tubules)	Paroophoron
Paradidymis (collecting duct)	—
Efferent ducts	Epoophoron
Appendix of epididymis	Epoophoron
Duct of epididymis (upper portion)	Epoophoron
Duct of epididymis (lower portion)	—
Ductus deferens (distal part)	Canal of Gartner
Ductus deferens (proximal part)	—
Seminal vesicles	—
Ejaculatory duct	—
Urethra (above prostatic utricle)	Urethra
Colliculus seminalis (urethral ridge)	Hymen
Prostatic utricle	Lower end of vagina
Urethra (distal to prostatic utricle)	Vestibule of vagina
Prostatic glands	Urethral glands
Bulbourethral glands	Greater vestibular glands
Penis	Clitoris
Urethral surface of penis	Labia minora
Corpora cavernosa penis	Corpora cavernosa clitoridis
Corpus spongiosum	Vestibular bulbs
Urethral glands	Lesser vestibular glands
Scrotum	Labia majora
—	Uterine tube other than fimbriae
—	Uterus and upper vagina
—	Broad ligament
Processus vaginalis	Canal of Nuck

mesenteric vein drains into the splenic vein. The splenic vein then joins with the **superior mesenteric** vein to form the **portal vein** itself. After receiving the **right and left gastric, pancreaticoduodenal, and cystic veins** in addition to those mentioned, the portal vein divides into two main divisions and enters the liver. In this manner, food material absorbed from the gastrointestinal tract is carried directly to the liver without going through the entire systemic circulation.

It should be noted that the portal circulation anastomoses in four main places with the systemic venous circulation: (1) at the inferior end of the esophagus, (2) at the anal canal, (3) along the posterior abdominal wall, and (4) along the ligamentum teres of the liver and thence to vessels around the umbilicus. In diseases that cause a portal hypertension, the vessels at these points of juncture may become severely dilated. To understand the position of hemorrhoids, one must remember that the middle rectal vein is mostly limited to the muscle layers of the rectum, and that the primary anastomosis in the submucosa is between the superior rectal (portal system) and the inferior rectal (systemic system) veins. This anastomosis occurs at the anal columns, the area of internal hemorrhoids.

LYMPHATICS

There is no advantage in attacking the lymphatic drainage of each viscus in these cavities systemically. Each has to be learned as an individual organ, but the blood vessels are still a very good clue to the lymphatic channels and the whereabouts of the nodes. There is one feature that is different in this region. All organs possessing mesenteries have small but numerous lymph nodes along the arteries coursing in this mesentery. The lymph then leaves these primary nodes to reach larger secondary nodes at the base of each major branch of the aorta. This same principle holds for those organs that originally had a mesentery, but that later fused to the abdominal wall, such as the ascending and descending colons. Similar small nodes will be located along the colic vessels just as if the mesentery existed.

6

LOWER LIMB

The lower limb consists of the gluteal region, thigh, knee, leg, ankle, and foot. Care should be taken to avoid using the term "leg" to include the entire lower limb.

It is helpful to keep the development of the lower limb in mind while studying this region. Its origin as a limb bud with dorsal and ventral areas and its subsequent medial rotation are exemplified in many of the definitive arrangements. Furthermore, the development of the bipedal posture has created some interesting problems. In contrast to the four-footed animal, where all limbs are used for locomotion, the human being has become upright and uses only the lower limbs in locomotion (except for a natural swinging of the upper limbs), leaving the upper limb for other acts.

The lower limb, as implied above, is used primarily for locomotion; we will attempt to correlate this structure with its action. Furthermore, as one studies the lower limb one should keep in mind that troubles in this area are concerned primarily with vascular deficiencies and difficulties with joints. Varicose veins are prevalent, vascular deficiencies show up in diabetes and arteriosclerosis,

young people seem to have troubles with ankles and knees, while the older people complain about hip troubles.

BONY STRUCTURES

Before you can read intelligently about structures in the lower limb, you must become familiar with the more important bony landmarks. Although associating various structures with the parts of the bones about to be mentioned greatly facilitates learning, some preliminary knowledge is mandatory. You will note that the figures depicting these bones have been labeled with function in mind.

The bones in the lower limb are the coxal bone, femur, patella, tibia, fibula, tarsal bones, metatarsal bones, and the phalanges. Correlated with their role in weight-bearing, many of these bones are large and heavy, especially when compared with corresponding bones in the upper limb.

COXAL BONE

The **coxal bone** (Fig. 6-1) is difficult to describe in words; the student should have a specimen at hand while reading the following. This bone is made up of three parts, **ilium, ischium,** and **pubis.** The three parts are separated in youth but soon fuse into one bone. The large fossa on the lateral surface, about in the middle of the bone, is the **acetabulum;** it is at this point that the three parts of the bone come together. The ilium is that part which extends superiorly from the acetabulum, the ischium inferiorly and posteriorly, and the pubis inferiorly and anteriorly.

The **ilium** possesses a basal portion, called the **body,** and a winglike projection, the **ala.** The body forms part of the acetabulum. The superior edge of the winglike portion is the **iliac crest,** which has **medial** and **lateral lips** and an **intermediate area.** The abdominal muscles attach to the anterior part of this crest, while the quadratus lumborum and erector spinae muscles originate from the posterior end. It starts anteriorly with the **anterior superior iliac spine** and terminates posteriorly in a **posterior superior iliac spine.** Just inferior to the former is another projection, called the **anterior inferior iliac spine,** and still another projection, inferior to the posterior superior iliac spine, is called the **posterior inferior iliac spine.** The lateral surface of the ala is almost entirely concerned with the gluteal muscles and is called the **gluteal surface.** The area of origin of the gluteus maximus is separated from that of the gluteus medius by the **posterior gluteal line,** the area of the gluteus medius from that of the gluteus minimus by the **anterior gluteal line.** An **inferior gluteal line** marks the inferior extent of the gluteus minimus muscle. The internal surface of the ala is called the **sacropelvic surface.** It presents the **iliac fossa** for the iliacus muscle and an articular area for articulation with the sacrum. The roughened area superior to this articular surface is for attachments of sacroiliac ligaments.

The **ischium** forms part of the acetabulum, as do the other two parts of this bone. Posterior to the acetabulum is a pointed projection called the **ischial spine,** the notch superior to it being the **greater ischiadic notch** and the smaller one inferior to it the **lesser ischiadic notch.** The heavy inferior extension of this bone is the **ischial tuberosity.** The **ischial ramus** extends anteriorly from this tuberosity and partially surrounds a large **obturator foramen.** This foramen is almost entirely filled in by an **obturator membrane.**

The **pubis** consists of a **body** and **two rami.** The superior ramus starts in the area of the acetabulum and ends in the body; it exhibits a **pubic tubercle** and a **pectineal line or eminence.** The **inferior ramus** connects the ischial ramus with the body of the pubis. This body has a surface for articulation with the same area of the opposite coxal bone.

The **acetabulum** demands further attention. It exhibits a smooth lunar-shaped surface for articulation with the head of the femur, an **acetabular fossa** for attachment of the ligamentum teres, and an **acetabular notch** for passage of nerves and vessels.

There are many other muscles that have origins or insertions on this bone in addition to those mentioned (see Fig. 6-1). They will be described in proper time.

FEMUR

The **femur** (Figs. 6-2 and 6-3) is a heavy, long bone which can be divided into a **proximal end, body,** and **distal end.**

The **proximal end** presents a rounded **head** that articulates with the acetabulum of the coxal bone. This head has a depression—the **fovea**—for the attachment of the **ligamentum teres,** which aids in holding the head of the femur in the acetabulum. The neck of the femur extends laterally and slightly inferiorly and connects the head with the proximal end of the **body.** Just distal and lateral to the neck can be seen two large projections, the greater and lesser trochanters. The **greater trochanter** projects superiorly and can be seen from the anterior

---

Figure 6-1. *Outside and inside surfaces of the right coxal bone, with origins and insertions of muscles outlined.*

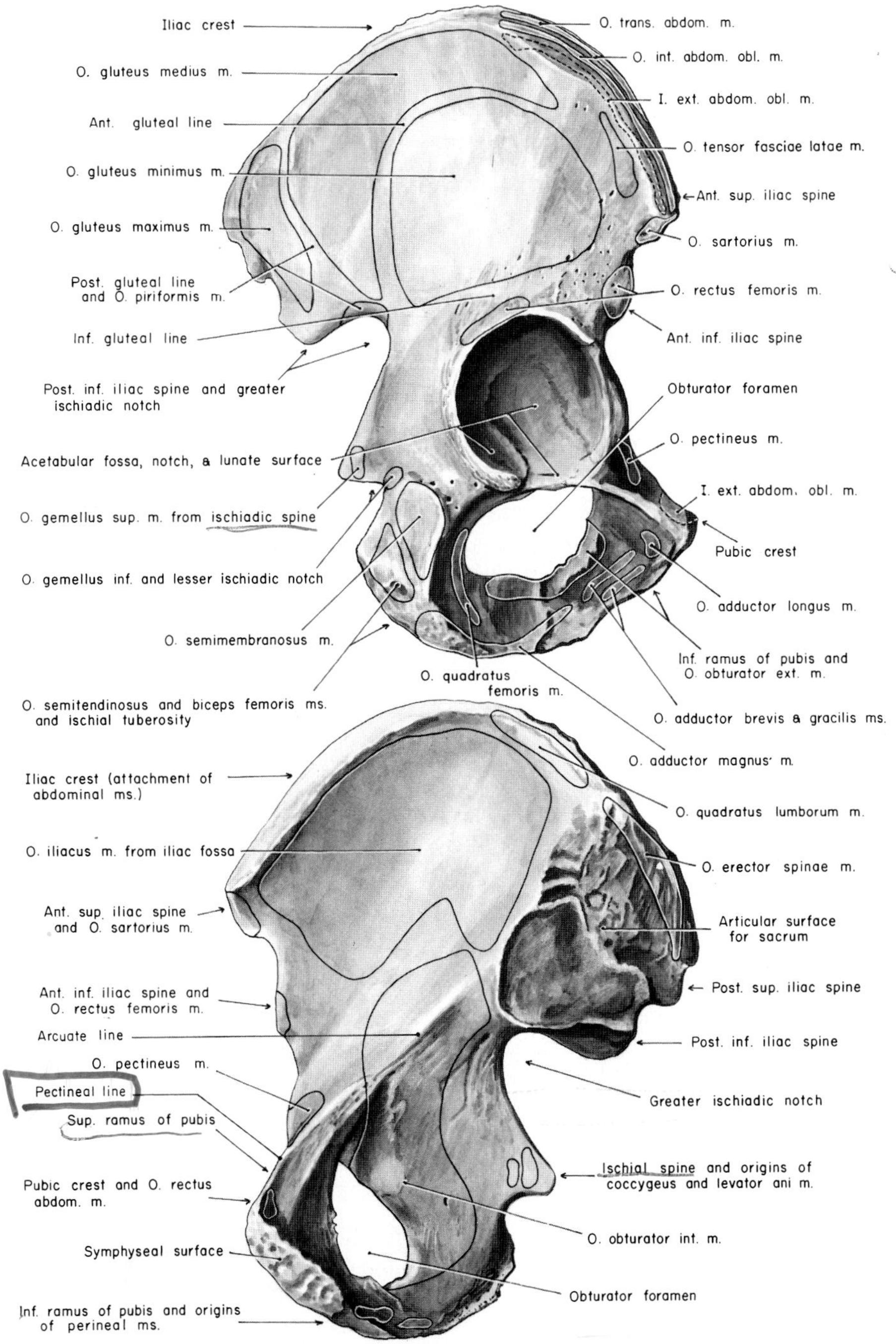

Iliac crest

O. gluteus medius m.

Ant. gluteal line

O. gluteus minimus m.

O. gluteus maximus m.

Post. gluteal line and O. piriformis m.

Inf. gluteal line

Post. inf. iliac spine and greater ischiadic notch

Acetabular fossa, notch, a lunate surface

O. gemellus sup. m. from ischiadic spine

O. gemellus inf. and lesser ischiadic notch

O. semimembranosus m.

O. semitendinosus and biceps femoris ms. and ischial tuberosity

O. trans. abdom. m.

O. int. abdom. obl. m.

I. ext. abdom. obl. m.

O. tensor fasciae latae m.

Ant. sup. iliac spine

O. sartorius m.

O. rectus femoris m.

Ant. inf. iliac spine

Obturator foramen

O. pectineus m.

I. ext. abdom. obl. m.

Pubic crest

O. adductor longus m.

Inf. ramus of pubis and O. obturator ext. m.

O. quadratus femoris m.

O. adductor brevis a gracilis ms.

O. adductor magnus' m.

Iliac crest (attachment of abdominal ms.)

O. iliacus m. from iliac fossa

Ant. sup. iliac spine and O. sartorius m.

Ant. inf. iliac spine and O. rectus femoris m.

Arcuate line

O. pectineus m.

Pectineal line

Sup. ramus of pubis

Pubic crest and O. rectus abdom. m.

Symphyseal surface

Inf. ramus of pubis and origins of perineal ms.

O. quadratus lumborum m.

O. erector spinae m.

Articular surface for sacrum

Post. sup. iliac spine

Post. inf. iliac spine

Greater ischiadic notch

Ischial spine and origins of coccygeus and levator ani m.

O. obturator int. m.

Obturator foramen

Ischiadic spine same side as Post Sup Iliac Spine

Ala.

surface, while the **lesser trochanter** projects posteromedially and, although it can be seen from the anterior surface, is located more posteriorly than anteriorly. On the anterior surface a small roughened line occurs between the trochanters, called the **intertrochanteric line.** On the posterior surface a much larger ridge occurs between these same two trochanters, which is the **intertrochanteric crest.** As can be seen in Figures 6-2 and 6-3, these trochanters serve for attachments of many muscles.

The **shaft** of the **femur** is slightly curved, with the concavity posteriorly. The anterior and lateral surfaces are quite smooth and, except for minute foramina for nutrient vessels, show no particular bony features. On the posterior surface of the shaft, however, two lines can be seen: the **medial** and **lateral lips** of the **linea aspera.** They start superiorly a slight distance apart, become close together in the middle of the femur, but then separate inferiorly. These lines serve for attachment of muscles (Fig. 6-3). They are called the medial and lateral **supracondylar lines** just superior to each of the epicondyles. The superior end of the lateral linea aspera is marked by a more pronounced projection for the attachment of the gluteus maximus muscle; this is the **gluteal tuberosity.**

The **distal end of the femur** exhibits two epicondyles, two condyles, an intercondylar fossa, and popliteal and patellar surfaces. The most prominent structures are the two large, smooth **condyles** of the femur which, if followed onto the anterior surface, can be seen to end as a triangular smooth area that articulates with the patella. The two condyles articulate with the medial and lateral condyles of the tibia. They are separated in the midline by a deep **intercondylar notch.** Just superior to this notch and between the supracondylar lines is the **popliteal surface** of the femur. On either side of the condyles and just superior to each are bony prominences called the **medial and lateral epicondyles,** structures serving for the attachment of muscles. The medial epicondyle is more prominent than the lateral, due to the rather large **adductor tubercle** that serves for the attachment of a large muscle of the thigh, the adductor magnus.

PATELLA

The **patella** (Fig. 6-4) is a sesamoid bone located in the patella tendon. Its **anterior surface** is smooth except for small grooves for the entrance of blood vessels. It is approximately 2 inches in diameter, and its **posterior sur-**

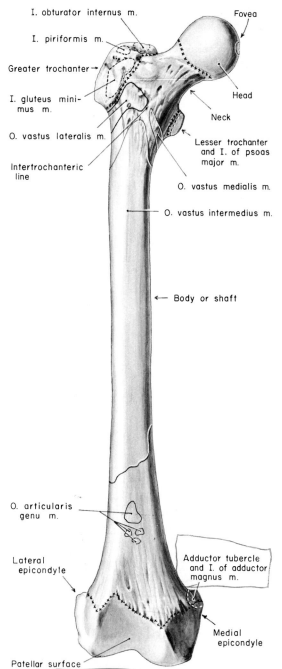

I. obturator internus m.

I. piriformis m.

Greater trochanter →

I. gluteus minimus m.

O. vastus lateralis m.

Intertrochanteric line

Fovea

Head

Neck

Lesser trochanter and I. of psoas major m.

O. vastus medialis m.

O. vastus intermedius m.

Body or shaft

O. articularis genu m.

Lateral epicondyle

Adductor tubercle and I. of adductor magnus m.

Medial epicondyle

Patellar surface

Figure 6-2. *Anterior surface of right femur with origins and insertions of muscles outlined. (Crosses indicate location of epiphyseal lines.)*

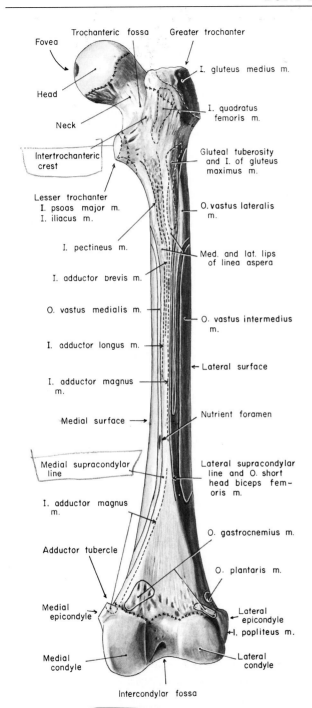

Fovea
Head
Neck
Trochanteric fossa
Greater trochanter
I. gluteus medius m.
I. quadratus femoris m.
Intertrochanteric crest
Gluteal tuberosity and I. of gluteus maximus m.
Lesser trochanter
I. psoas major m.
I. iliacus m.
O. vastus lateralis m.
I. pectineus m.
Med. and lat. lips of linea aspera
I. adductor brevis m.
O. vastus medialis m.
O. vastus intermedius m.
I. adductor longus m.
I. adductor magnus m.
← Lateral surface
Medial surface →
Nutrient foramen
Medial supracondylar line
Lateral supracondylar line and O. short head biceps femoris m.
I. adductor magnus m.
O. gastrocnemius m.
Adductor tubercle
O. plantaris m.
Medial epicondyle
Lateral epicondyle
I. popliteus m.
Medial condyle
Lateral condyle
Intercondylar fossa

Figure 6-3. Posterior surface *of right femur with origins and insertions of muscles outlined. (Crosses indicate location of epiphyseal lines.) The insertion of the obturator externus muscle into the trochanteric fossa is hidden from view by the overhanging greater trochanter.*

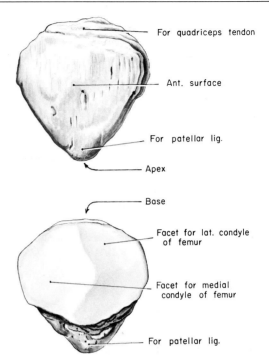

For quadriceps tendon
Ant. surface
For patellar lig.
Apex
Base
Facet for lat. condyle of femur
Facet for medial condyle of femur
For patellar lig.

Figure 6-4. Anterior and posterior surfaces of the right patella.

face is its **articular surface;** it articulates with the anterior surface of the distal end of the femur.

TIBIA

The **tibia** (Figs. 6-5, 6-6, and 6-7) can be divided into a **proximal end, body,** and a **distal end.**

The most prominent structures on its **proximal end** are the two rather flat **medial and lateral condyles** which serve for articulation with the condyles of the femur. Inferior to the articular surfaces the condyles show quite extensive bony projections. The area between the articular surfaces of the condyles is called the **intercondylar eminence;** this eminence projects superiorly into the intercondylar notch of the femur. The lateral condyle has an **articular facet** for articulation with the head of the fibula. On the anterior surface just inferior to the two condyles is a large projection, the **tibial tuberosity;** it is to this tuberosity that the patellar tendon attaches.

The body of the tibia exhibits **anterior, medial, and interosseus margins** between which are the **medial, lateral, and posterior surfaces** (Figure 6-7) makes this

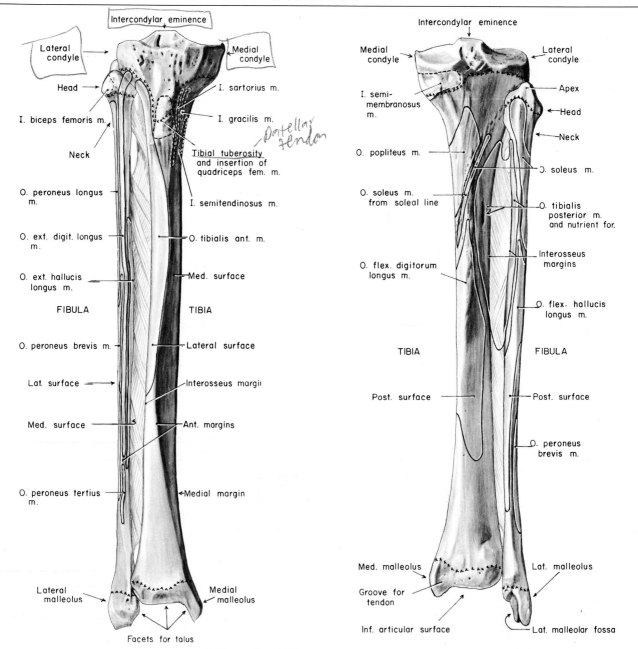

Intercondylar eminence

Lateral condyle

Medial condyle

Head

I. sartorius m.

I. biceps femoris m.

I. gracilis m.

Patellar Tendon

Neck

Tibial tuberosity and insertion of quadriceps fem. m.

O. peroneus longus m.

I. semitendinosus m.

O. ext. digit. longus m.

O. tibialis ant. m.

O. ext. hallucis longus m.

Med. surface

FIBULA

TIBIA

O. peroneus brevis m.

Lateral surface

Lat. surface

Interosseus margin

Med. surface

Ant. margins

O. peroneus tertius m.

Medial margin

Lateral malleolus

Medial malleolus

Facets for talus

Intercondylar eminence

Medial condyle

Lateral condyle

I. semi-membranosus m.

Apex

Head

Neck

O. popliteus m.

O. soleus m.

O. soleus m. from soleal line

O. tibialis posterior m. and nutrient for.

Interosseus margins

O. flex. digitorum longus m.

O. flex. hallucis longus m.

TIBIA

FIBULA

Post. surface

Post. surface

O. peroneus brevis m.

Med. malleolus

Lat. malleolus

Groove for tendon

Inf. articular surface

Lat. malleolar fossa

Figure 6-5. Anterior surface of the right tibia and fibula with origins and insertions of muscles outlined. The interosseus membrane has been included. (Crosses indicate location of epiphyseal lines.)

Figure 6-6. Posterior surface of right tibia and fibula with origins and insertions of muscles outlined. The interosseus membrane has been included. (Crosses indicate location of epiphyseal lines.)

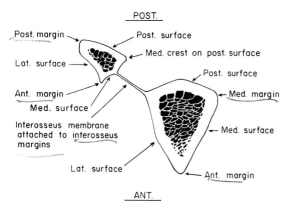

POST.

Post. margin — Post. surface

Lat. surface — — Med. crest on post. surface

Ant. margin — Post. surface

Med. surface — Med. margin

Interosseus membrane attached to interosseus margins — Med. surface

Lat. surface — Ant. margin

ANT.

Figure 6-7. *Transverse section of the right tibia and fibula and interosseus membrane.*

comprehensible). The interosseus or lateral margin is rather sharp and it is to this that the **interosseus membrane** attaches. This membrane is a very firm structure that holds the tibia and fibula together and serves for the attachment of muscles. Other than for nutrient foramina, the rest of the body of the tibia is not remarkable.

The **distal end** exhibits the **medial malleolus,** the **fibular notch,** and **inferior and malleolar facets.** The **medial malleolus** is the structure on the medial side of the bone, which projects more inferiorly than the lateral side. This is the bony projection that can be felt as the so-called ankle bone. The lateral side shows a facet for articulation with the distal end of the fibula. The **articular surface** of the **malleolus** and the **inferior articular surface,** both of which articulate with the talus, are on the inferior surface of the distal end. The **fibular notch** is a groove in the bone on the lateral side into which the fibula inserts.

FIBULA

The **fibula** (Figs. 6-5, 6-6, and 6-7) also can be divided into a **proximal end, body,** and **distal end.**

The proximal end exhibits a **head, neck,** and **articular facet.** The **head** is an enlargement that is attached to the body by a thinner **neck.** The **facet** is the point where the fibula articulates with the tibia.

The **body** exhibits **anterior, posterior,** and **interosseus margins** which divide the shaft of the bone into

medial, lateral, and **posterior surfaces** (Fig. 6-7). Nutrient foramina can be seen along the shaft of the bone.

The **distal end** actually forms the **lateral malleolus** or lateral ankle bone. It exhibits an **articular facet** for articulation with the talus and a **lateral malleolar fossa.**

The fibula is not a weight-bearing bone and serves mainly for the attachment of muscles. It is connected to the tibia by the interosseus membrane as well as by articulations at the proximal and distal ends. The distal end of the fibula actually extends further inferiorly than does the tibia. The distal end of the tibia and fibula wrap themselves around the talus to form the ankle joint.

TARSAL BONES

The **tarsal bones** (Figs. 6-8 and 6-9) are seven in number and consist of the following: a posterior row made up of the **talus** (*L.,* ankle) and **calcaneus** (*L.,* heel); a middle bone, **the navicular;** and an anterior row, the **three cuneiforms** (*L.,* wedge-shaped) and the **cuboid.** It is possibly better to think of these groups of bones from posterior to anterior. On the medial side of the foot, the talus articulates with the navicular; the navicular, in turn, articulates with the three cuneiforms, all three of which articulate with a separate metatarsal bone. On the lateral side of the foot, the calcaneus articulates with the cuboid, which in turn articulates with the last two metatarsal bones. The natural shape of these bones forms two longitudinal arches as well as a transverse arch.

It is not necessary to learn the exact contours and surfaces of each of the individual tarsal bones. However, they should be studied in their natural position in the articulated foot and particular note should be taken of the shape of the calcaneus. If the inferior surface of the calcaneus is observed, you will see that there is a large groove or shallow area in this bone which is extremely important. This area is the vital region where structures from the posterior surface of the leg can enter the plantar surface or sole of the foot. It should be noted also that the tibia and the fibula articulate with the talus; the talus, in turn, articulates with the calcaneus. A medial projection of the calcaneus—the **sustentaculum talus**—is so called because it suspends the talus (see Fig. 6-9). The calcaneus and talus are shaped in such a manner that they project anteriorly and articulate with the navicular and cuboid bones, respectively, at approximately the same plane, a plane called the **transverse tarsal joint.**

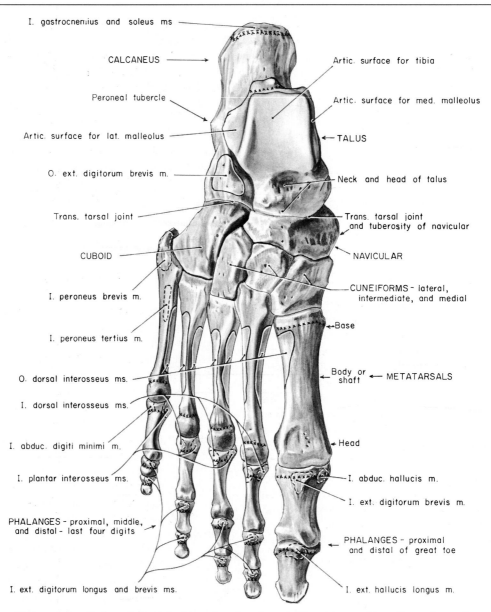

I. gastrocnemius and soleus ms.

CALCANEUS

Peroneal tubercle

Artic. surface for lat. malleolus

O. ext. digitorum brevis m.

Trans. tarsal joint

CUBOID

I. peroneus brevis m.

I. peroneus tertius m.

O. dorsal interosseus ms.

I. dorsal interosseus ms.

I. abduc. digiti minimi m.

I. plantar interosseus ms.

PHALANGES - proximal, middle, and distal - last four digits

I. ext. digitorum longus and brevis ms.

Artic. surface for tibia

Artic. surface for med. malleolus

TALUS

Neck and head of talus

Trans. tarsal joint and tuberosity of navicular

NAVICULAR

CUNEIFORMS - lateral, intermediate, and medial

Base

Body or shaft — METATARSALS

Head

I. abduc. hallucis m.

I. ext. digitorum brevis m.

PHALANGES - proximal and distal of great toe

I. ext. hallucis longus m.

Figure 6-8. *Bones of the right foot—dorsal surface—with origins and insertions of muscles outlined. (Crosses indicate location of epiphyseal lines.) Note how the epiphyseal lines indicate that the bone missing in the great toe is the metatarsal and not a phalanx.*

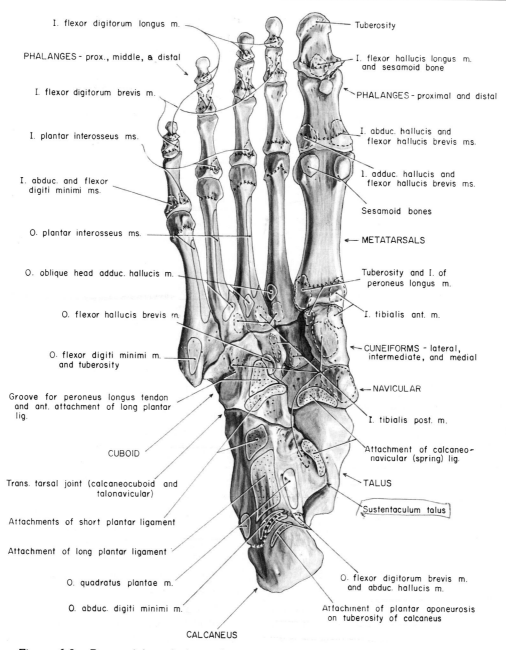

I. flexor digitorum longus m.

PHALANGES - prox., middle, & distal

I. flexor digitorum brevis m.

I. plantar interosseus ms.

I. abduc. and flexor digiti minimi ms.

O. plantar interosseus ms.

O. oblique head adduc. hallucis m.

O. flexor hallucis brevis m.

O. flexor digiti minimi m. and tuberosity

Groove for peroneus longus tendon and ant. attachment of long plantar lig.

CUBOID

Trans. tarsal joint (calcaneocuboid and talonavicular)

Attachments of short plantar ligament

Attachment of long plantar ligament

O. quadratus plantae m.

O. abduc. digiti minimi m.

CALCANEUS

Tuberosity

I. flexor hallucis longus m. and sesamoid bone

PHALANGES - proximal and distal

I. abduc. hallucis and flexor hallucis brevis ms.

I. adduc. hallucis and flexor hallucis brevis ms.

Sesamoid bones

← METATARSALS

Tuberosity and I. of peroneus longus m.

I. tibialis ant. m.

← CUNEIFORMS - lateral, intermediate, and medial

← NAVICULAR

I. tibialis post. m.

Attachment of calcaneonavicular (spring) lig.

← TALUS

Sustentaculum talus

O. flexor digitorum brevis m. and abduc. hallucis m.

Attachment of plantar aponeurosis on tuberosity of calcaneus

Figure 6-9. *Bones of the right foot—plantar surface—with origins and insertions of muscles outlined. (Crosses indicate location of epiphyseal lines.)*

METATARSALS

The **metatarsal bones** (Figs. 6-8 and 6-9) are five in number and possess a **proximal end or base, a shaft,** and a **distal end or head.** The metatarsal bone for the great toe is larger than the others, and its base differs in shape from that of the other metatarsal bones.

PHALANGES

The **phalanges** (Figs. 6-8 and 6-9) can be divided into three sets of bones—a proximal row, middle row, and distal row—except in the great toe, where there are only two phalanges, a proximal and a distal. Each phalanx consists of a **base, body,** and **head.**

Additional facts on the bones of the lower limb will be given as they are needed.

THIGH

SURFACE ANATOMY

The **surface anatomy** of the thigh is not particularly revealing. However, in a muscular, lean individual, projections caused by various muscles and tendons can be seen. These will be described after these muscles have been studied. They are pictured in Figures 6-15, 6-16, and 6-31.

CUTANEOUS NERVES AND VEINS

The **cutaneous nerves** of the thigh are abundant (Figs. 6-10 and 6-11). Those on the anterior surface can be grouped as either lateral to the sartorius muscle or medial to it.

The medial group consists of small nerves that enter the thigh in its superior aspect. Just distal to the inguinal ligament and close to the midline of the thigh are the **femoral branches of the genitofemoral nerve.** More medially located and entering the thigh by traversing the inguinal canal are the cutaneous branches of the **ilioin-**

guinal nerve. Inferior to these branches and on the medial side of the thigh are cutaneous branches of the **obturator nerve** and branches of the **medial cutaneous nerve of the thigh.**

Those nerves lateral to the sartorius muscle pierce the deep fascia along the line formed by that muscle. From superior to inferior, they are: (1) the **lateral cutaneous branch of the subcostal,** which appears very close to the anterior superior iliac spine; (2) the **lateral femoral cutaneous nerve** with its posterior and anterior branches; (3) the **intermediate cutaneous nerves** of the thigh, branches of the femoral nerve that take an intermediate position on the anterior surface of the thigh; and (4) most distal, the **medial cutaneous nerves** of the thigh, also branches of the femoral. The intermediate and medial branches of the femoral provide cutaneous innervation to the anterior surface of the knee as well.

On the posterior surface of the thigh, the main cutaneous nerve is the **posterior femoral cutaneous,** which is deep to the deep fascia and does not completely penetrate it until near the popliteal area posterior to the knee (Fig. 6-11). However, branches of this nerve do penetrate the deep fascia and innervate the skin over the entire posterior surface of the thigh.*

It should be noted that the foregoing is a description of the cutaneous nerve distribution and should not be confused with dermatomes—those areas of the skin innervated by single spinal cord segments. These dermatomes will be presented in the systemic summary of the nerves to the lower limb.

In addition to the cutaneous nerves, **cutaneous veins** can be seen in the superficial fascia (Fig. 6-10). Starting near the medial side and slightly posterior to the knee, a vein courses superiorly, gradually becoming more anterior as it ascends the thigh; this is the **great or long saphenous vein.** It finally empties into the femoral vein by passing through the deep fascia. This vein receives tributaries called the **superficial external pudendals, the superficial circumflex iliac,** and the **superficial epigastric.** These three veins accompany arteries of the same name that also course in the superficial fascia. The superficial circumflex vein drains the lateral and anterior parts of the abdominal wall, the superficial epigastric vein the anterior part of the abdominal wall closer to the midline,

---

*Branches of this nerve also innervate the skin of the buttocks (inferior cluneal nerves). These nerves will be described later.

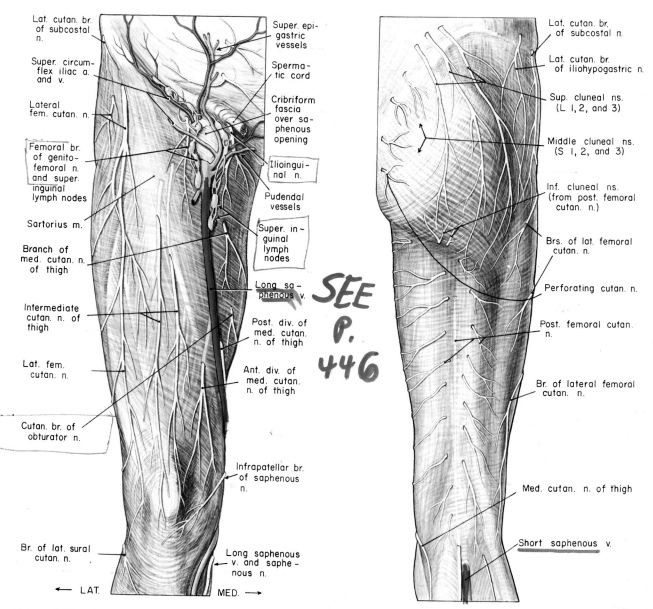

Lat. cutan. br. of subcostal n.

Super. circumflex iliac a. and v.

Lateral fem. cutan. n.

Femoral br. of genitofemoral n. and super. inguinal lymph nodes

Sartorius m.

Branch of med. cutan. n. of thigh

Intermediate cutan. n. of thigh

Lat. fem. cutan. n.

Cutan. br. of obturator n.

Br. of lat. sural cutan. n.

Super. epigastric vessels

Spermatic cord

Cribriform fascia over saphenous opening

Ilioinguinal n.

Pudendal vessels

Super. inguinal lymph nodes

Long saphenous v.

Post. div. of med. cutan. n. of thigh

Ant. div. of med. cutan. n. of thigh

Infrapatellar br. of saphenous n.

Long saphenous v. and saphenous n.

← LAT.   MED. →

SEE P. 446

Lat. cutan. br. of subcostal n.

Lat. cutan. br. of iliohypogastric n.

Sup. cluneal ns. (L 1, 2, and 3)

Middle cluneal ns. (S 1, 2, and 3)

Inf. cluneal ns. (from post. femoral cutan. n.)

Brs. of lat. femoral cutan. n.

Perforating cutan. n.

Post. femoral cutan. n.

Br. of lateral femoral cutan. n.

Med. cutan. n. of thigh

Short saphenous v.

Figure 6-10. *Cutaneous nerves, and superficial veins and lymph nodes on the anterior and medial surfaces of the right thigh. Note relationship of the nerves to the sartorius muscle (shaded in deep to fascia).*

Figure 6-11. *Cutaneous nerves of the gluteal region and on the posterior surface of the thigh—right side.*

and the superficial external pudendal veins drain the anterior surface of the scrotum or labia majora and the dorsal part of the penis or clitoris. The arteries of the same names supply blood to the respective areas.

INGUINAL LYMPH NODES

The superficial **inguinal lymph nodes** (Fig. 6-10) are also located in the superficial fascia. There are two main groups, one vertically placed along the saphenous vein, and one horizontally placed just inferior to the inguinal ligament but in a more lateral position than the former group. These superficial lymph nodes drain (1) the superficial parts of the lower limb, except the parts that are drained by nodes in the popliteal fossa (posterior to knee), (2) the external genitalia and the inferior end of the anal canal, and (3) the abdominal wall inferior to the umbilicus. These lymph nodes drain into the **deep group** of three or four nodes that are located at the superior end of the femoral vein. These deep nodes also receive lymph from the deeper structures of the lower limb.

DEEP FASCIA

The **deep fascia** enveloping the thigh and gluteal region has been given the name of **fascia lata** (*L.*, broad, strong); it has individual characteristics and demands description.

This fascia is weakest on the medial side of the thigh but quite thick and strong on the lateral side. It is **attached superiorly** to the posterior surface of the iliac crest, sacrum, coccyx, sacrotuberous ligament, ischium, inferior pubic ramus and arch, pubic symphysis and crest, and inguinal ligament. **Distally** it is **attached** to the capsule of the knee joint, the condyles of both the femur and tibia, and the head of the fibula. A thickened band of the fascia lata, approximately 1½ inches in width, occurs on the lateral side of the thigh and is known as the **iliotibial band.** As the name implies, it extends from the ilium superiorly to the tibia inferiorly; it has additional distal attachments to the capsule of the knee joint and the patella. Superiorly the tensor fascia lata muscle inserts upon it, as does the gluteus maximus muscle. Actually, the fascia splits and passes on both superficial and deep sides of the gluteus maximus muscle. Extensions of this deep fascia extend centrally to attach to the linea aspera

and supracondylar lines. These are **intermuscular septa.** There is one medially placed—the **medial intermuscular septum**—between the muscles on the anterior surface of the thigh (the extensors) and those medially located (adductors), and one laterally placed—the **lateral intermuscular septum**—between the extensors anteriorly and the flexors located on the posterior side of the thigh. Furthermore, there is a separation between the adductor muscles and the flexors, sometimes called a **posterior intermuscular septum.** These septa obviously divide the musculature of the thigh into three distinct groups, i.e., the extensors of the leg anteriorly, adductors of the thigh medially, and flexors of the leg posteriorly (see Fig. 6-18). The saphenous opening occurs in the fascia lata but deserves the special description immediately following.

SAPHENOUS OPENING

If the superficial fascia is removed from the anterior surface of the thigh, it can be seen that the great saphenous vein courses through a rather large opening in the deep fascia to reach the femoral vein (Figs. 6-10 and 6-12). This is the **saphenous opening,** formed by the fascia lata. Actually, there are two layers of deep fascia in this region, one stretching inferiorly from the inguinal ligament and another investing the muscles on the anterior surface of the thigh. This saphenous opening is formed by a bundle of connective tissue arising from the medial end of the inguinal ligament, coursing laterally and inferiorly and then medially and superiorly (thus forming a circle), but all the time coursing deeper and deeper until the fascia on the pectineus muscle is reached. (Figure 6-14 shows this muscle.) This forms an opening in the outer layer of the fascia lata; it is through this opening that the vein courses. The opening is covered with a layer of deep fascia, the **cribriform** (*L.*, sievelike) **fascia;** the margins of the opening are called the **falciform** (*L.*, sickle-shaped) **margins.** These are quite pronounced laterally but are not as visible on the medial side of the saphenous opening. The sharp edges superiorly and inferiorly located are called the **superior and inferior cornua.**

FEMORAL SHEATH

In reality, the saphenous (*Gr.*, clear, manifest) vein penetrates two layers of deep fascia to reach the femoral vein.

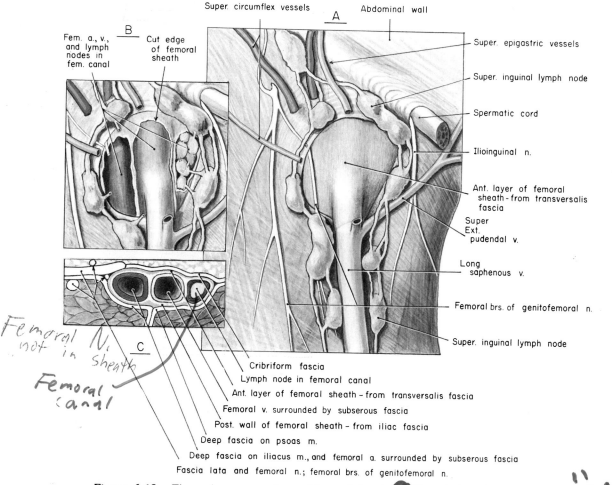

Super. circumflex vessels

A   Abdominal wall

B   Cut edge of femoral sheath

Fem. a., v., and lymph nodes in fem. canal

Super. epigastric vessels

Super. inguinal lymph node

Spermatic cord

Ilioinguinal n.

Ant. layer of femoral sheath - from transversalis fascia

Super Ext. pudendal v.

Long saphenous v.

Femoral brs. of genitofemoral n.

Super. inguinal lymph node

C

Cribriform fascia
Lymph node in femoral canal
Ant. layer of femoral sheath – from transversalis fascia
Femoral v. surrounded by subserous fascia
Post. wall of femoral sheath – from iliac fascia
Deep fascia on psoas m.
Deep fascia on iliacus m., and femoral a. surrounded by subserous fascia
Fascia lata and femoral n.; femoral brs. of genitofemoral n.

Femoral N. not in Sheath (handwritten)

Femoral canal (handwritten)

Figure 6-12. *The saphenous opening and femoral sheath. (A) The cribriform fascia has been removed, revealing the anterior layer of the femoral sheath. (B) The anterior wall has been removed, revealing the three compartments of the femoral sheath containing, from lateral to medial, the femoral artery, femoral vein, and a lymph node in the femoral canal. (C) A transverse section of the femoral sheath and its contents. Note how the vessels are surrounded by subserous fascia, which forms the partitions in the femoral sheath.*

"NAVEL" ↑ femoral canal (handwritten)

The outer layer just mentioned, the cribriform fascia, is superficial to a layer of fascia that covers the femoral artery and vein at this location. This is the anterior wall of the **femoral sheath** (Fig. 6-12). This sheath is approximately 1½ inches long. It has three distinct channels: (1) a lateral containing the femoral artery, (2) an intermediate containing the femoral vein, and (3) a medial, an area usually filled with lymphatics and possibly with a lymph node, the **femoral canal.** The femoral sheath is considered to have

been formed by a pulling down of the layers of fascia from the abdominal wall as the blood vessels left the abdominal cavity to enter the thigh (Fig. 6-13). At the inguinal ligament the transversalis fascia on the transversus abdominis muscle and the iliacus fascia on the iliacus muscle come to a point. As the blood vessels are located in the subserous fascia (between the peritoneum and the deep fascia of the abdominal cavity), they would naturally have to penetrate a layer of deep fascia to get into the

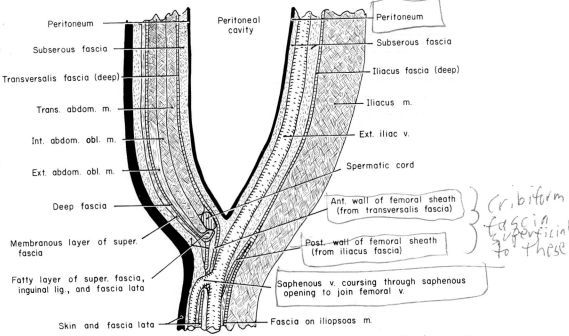

Labels in figure:
Peritoneum
Subserous fascia
Transversalis fascia (deep)
Trans. abdom. m.
Int. abdom. obl. m.
Ext. abdom. obl. m.
Deep fascia
Membranous layer of super. fascia
Fatty layer of super. fascia, inguinal lig., and fascia lata
Skin and fascia lata

Peritoneal cavity

Peritoneum
Subserous fascia
Iliacus fascia (deep)
Iliacus m.
Ext. iliac v.
Spermatic cord
Ant. wall of femoral sheath (from transversalis fascia)
Post. wall of femoral sheath (from iliacus fascia)
Saphenous v. coursing through saphenous opening to join femoral v.
Fascia on iliopsoas m.

Cribiform fascia superficial to these

Figure 6-13. *A diagrammatic longitudinal section through the femoral sheath, showing its formation from transversalis fascia anteriorly and iliac fascia posteriorly. Note that the subserous fascia surrounds the vessels while in the femoral sheath.*

lower limb. Therefore, a prolongation of the transversalis fascia forms the anterior wall of the femoral sheath while a similar extension of the ilacus fascia forms the posterior wall. The vessels, while within the femoral sheath, are surrounded by a prolongation of subserous fascia.* The spinal nerves are not located in the subserous fascia in the abdominal cavity but course outside the deep fascia. Therefore, these nerves do not have to penetrate fascia to get into the lower limb and the femoral nerve is not in the femoral sheath; it is lateral to it.

The **femoral canal** is a source of trouble in that organs in the abdominal cavity can herniate through it. When this occurs, the herniating mass pushes peritoneum and subserous fascia ahead of it, enters the canal, and presents itself in the saphenous opening. Pushing cribriform fascia ahead of it, this herniated mass then tends to turn superiorly in the superficial fascia. The reason for turning superiorly is unknown. Femoral hernia is much more prevalent in females than in males; the wider flair of the hips presumably makes the opening of the femoral canal wider in females than in males. Because the herniated mass turns superiorly, it is easily confused with an inguinal hernia.

FEMORAL TRIANGLE

The **femoral triangle** (Fig. 6-14) is an area on the anterior surface of the thigh just inferior to the inguinal ligament. It is **bounded** superiorly by the inguinal ligament, laterally by the sartorius muscle, and medially by the adductor longus muscle. The medial border of this muscle is usually described as the medial border of the triangle but the triangle is more easily visualized if the lateral border of the adductor longus is used. The **base** of the triangle faces superiorly and the **apex** inferiorly; the apex leads into the **adductor canal.** Unfortunately, this triangle has been described as possessing a floor and a

*C. B. McVay and B. J. Anson, 1940, Aponeurotic and fascial continuities in the abdomen, pelvis, and thigh, *Anat. Rec.* 76: 213. These authors divide the femoral sheath into an inner sheath, derived from subserous fascia, and an outer sheath, derived from transversalis and iliopsoas fasciae. The inner sheath forms the partitions.

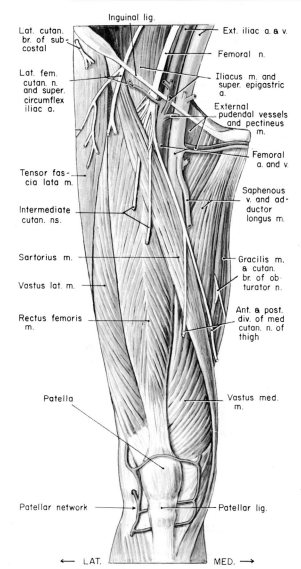

Inguinal lig.

Lat. cutan.
br. of sub-
costal

Lat. fem.
cutan. n.
and super.
circumflex
iliac a.

Tensor fas-
cia lata m.

Intermediate
cutan. ns.

Sartorius m.

Vastus lat. m.

Rectus femoris
m.

Patella

Patellar network

Ext. iliac a. & v.

Femoral n.

Iliacus m. and
super. epigastric
a.

External
pudendal vessels
and pectineus
m.

Femoral
a. and v.

Saphenous
v. and ad-
ductor
longus m.

Gracilis m.
& cutan.
br. of ob-
turator n.

Ant. & post.
div. of med.
cutan. n. of
thigh

Vastus med.
m.

Patellar lig.

← LAT.   MED. →

Figure 6-14. *Anterior and medial surfaces of right thigh I. All deep fascia has been removed, and cutaneous nerves and superficial vessels cut.*

roof although the floor and roof are not inferior and superior as they should be if they are to be given these terms. The floor is, in actuality, the posterior wall of the femoral triangle, while the roof is that which is found superficial to the triangle as we view it from the anterior side. The so-called **roof** consists of the fascia lata and the cribriform fascia mentioned above. The **floor** is made up of the iliopsoas muscle laterally, and the pectineus muscle medially.

The most lateral structure in the femoral triangle is the **lateral femoral cutaneous nerve** as it comes into the area deep to the inguinal ligament and deep to the iliac fascia. The next most medial structure is the large **femoral nerve,** which has a similar position in relation to the iliac fascia. This nerve can be seen to divide immediately into its many branches to innervate the muscles on the anterior surface of the thigh and to give cutaneous nerve supply to the skin on the anterior surface of the thigh and knee. The muscular branches tend to remain deep but the medial cutaneous nerve can be seen to course inferiorly just to the medial side of the long sartorius muscle; it crosses the femoral artery and vein in order to reach this position. The next structure is the large **femoral artery** with its circumflex iliac, superficial epigastric, and superficial and deep external pudendal branches. Medial to this is the **femoral vein** with the saphenous vein draining into it. More superficially located (Fig. 6-10) are the femoral branch of the **genitofemoral nerve** and the **ilioinguinal nerve,** respectively lateral and medial to the saphenous vein. If the femoral vessels are moved to one side, the **profunda artery and vein** can be seen. These deeper vessels are considered to be in the femoral triangle and the artery usually gives rise to the **lateral and medial femoral circumflex** branches.

Therefore, you should be able to visualize in the femoral triangle the femoral artery and vein in the femoral sheath, which contains the femoral canal as well; the branches of these vessels—the superficial circumflex iliac, the superficial epigastric, and external pudendal arteries and veins of the same name; the profunda femoris artery and vein; the femoral nerve and its branches; the lateral femoral cutaneous nerve; the femoral branch of the genitofemoral nerve; the ilioinguinal nerve; lymph nodes and lymph vessels.

ANTERIOR AND MEDIAL SIDES OF THIGH

GENERAL DESCRIPTION

Distal to the area of the femoral triangle, the anterior surface of the thigh is very muscular. The **sartorius** muscle has already been seen to course from the anterior superior iliac spine inferiorly and medially until it disappears on the posterior side of the knee. Lateral to the origin of the sartorius is the **tensor fascia lata** muscle, which attaches to the iliotibial band. The proximal end of the **rectus femoris** muscle appears just deep to the sartorius and tensor fascia lata muscles; this is the rectus femoris portion of the four-headed muscle, the **quadriceps femoris.** Lateral to this is the large **vastus lateralis** portion and medially is the equally large **vastus medialis.** If the rectus femoris is pulled to one side, the **vastus intermedius** can be seen. These four muscles are continuous with the patellar tendon, which surrounds the patella and continues to insert on the tibial tuberosity.

Although the **adductor muscles** are located on the medial side of the thigh, they are nevertheless encountered by any dissection of the anterior surface. The most anterior of the adductor muscles is the large **adductor longus,** and just medial to this is the long, thin **gracilis** muscle. If the adductor longus is cut and reflected, a shorter muscle is seen with fibers coursing in the same general direction; this is the **adductor brevis** muscle (Fig. 6-17). Posterior to the adductor brevis is the very huge, powerful **adductor magnus** muscle. The **obturator nerve,** as it passes through the obturator foramen, divides into anterior and posterior branches. The anterior branch courses anterior to the adductor brevis muscle, between it and the adductor longus, while the posterior branch courses posterior to the adductor brevis, between it and the adductor magnus. The superior end of the adductor magnus is frequently divided off into a separate muscle; when this occurs the name **adductor minimus** is given to this portion.

If the sartorius muscle is pulled to one side, the **adductor canal** is opened (Fig. 6-18). This is a canal that starts at the apex of the femoral triangle and continues inferiorly until an opening is reached in the tendon of the adductor magnus muscle. It is **bounded** anteriorly by the sartorius muscle, laterally by the vastus medialis muscle, and posteriorly by the adductor longus muscle superiorly and the adductor magnus muscle inferiorly. This canal **contains** the femoral artery, femoral vein, the nerve to the vastus medialis muscle, and the saphenous nerve. The saphenous nerve is anterior to the artery and the nerve to the vastus medialis is lateral to it; the femoral vein is posterior to the artery. The relations of the saphenous nerve, femoral artery, and femoral vein to each other make sense in terms of the development of

the lower limb. Just inferior to the inguinal ligament the femoral artery is centrally located with the femoral vein medial and the femoral nerve lateral to it. As the lower limb twists medially in its development, the saphenous branch of the femoral nerve is rotated anteriorly and the femoral vein posteriorly, thus accounting for their definitive positions in the adductor canal.

Some of the structures mentioned above can be seen on the surface of the thigh (Figs. 6-15 and 6-16). On the anterior surface, the elevation produced by the **sartorius** muscle is visible, going from the lateral side superiorly to the medial side of the knee inferiorly. The massive **quadriceps femoris** muscle can be seen just deep to this sartorius muscle; a particularly large prominence is formed medially near the knee by the **vastus medialis.**

On the medial surface of the thigh can be seen an elevation or ridge formed by the **gracilis muscle,** a ridge just posterior to that formed by the sartorius, with which it fuses inferiorly.

A groove between the abdominal wall and the thigh indicates the location of the inguinal ligament, and the tensor fascia lata muscle may produce an elevation, and the femoral triangle a depression.

MUSCLES

Origins and insertions of the many muscles in the lower limb are given in a brief form. In every instance, however, these descriptions will be adequate for an understanding of the main action of the muscle involved. *The insertion is usually more informative than the origin.*

MOVEMENTS OF LOWER LIMB.

Before one can understand the actions of the many muscles in the lower limb, it is mandatory that one be acquainted with the movements allowed at the various joints.

The **hip joint** is a ball-and-socket type that allows all movements: flexion (moving thigh anteriorly), extension (moving thigh posteriorly), abduction (moving thigh away from midline of the body), adduction (moving thigh toward midline of the body), circumduction (a combination of above), and medial (toward midline of body) and lateral (away from midline) rotation (see Fig. 6-19).

These movements occur around three axes: transverse, vertical, and anteroposterior (Fig. 6-20). The transverse axis of flexion and extension, as the name

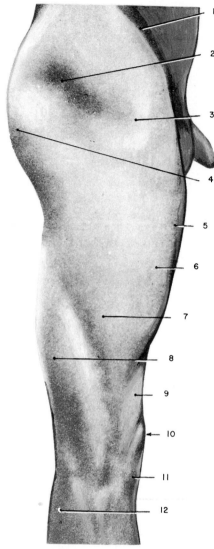

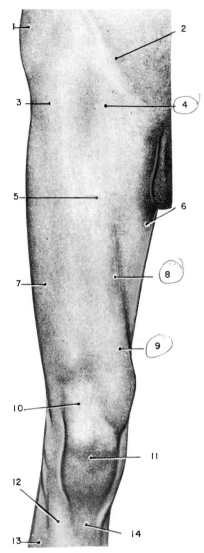

Figure 6-15. Surface anatomy of lateral side of gluteal region and thigh: (1) iliac crest, (2) gluteal depression on gluteus medius muscle, (3) tensor fascia lata muscle, (4) gluteus maximus muscle, (5) rectus femoris muscle, (6) vastus lateralis muscle, (7) iliotibial tract, (8) biceps femoris muscle, (9) quadriceps femoris tendon, (10) patella, (11) patellar ligament, (12) gastrocnemius muscle. (From Appleton, Hamilton, and Tchaperoff, Surface and Radiological Anatomy, courtesy of W. Heffer & Sons, Ltd.)

Figure 6-16. Surface anatomy of the anterior side of thigh: (1) gluteus minimus muscle, (2) inguinal ligament, (3) tensor fascia lata muscle, (4) area of femoral triangle, (5) quadriceps femoris muscle, (6) adductor muscles, (7) vastus lateralis, (8) sartorius muscle, (9) vastus medialis muscle, (10) quadriceps tendon, (11) patella, (12) tibialis anterior muscle, (13) peroneus longus muscle, (14) patellar ligament. (From Appleton, Hamilton, and Tchaperoff, Surface and Radiological Anatomy, courtesy of W. Heffer & Sons, Ltd.)

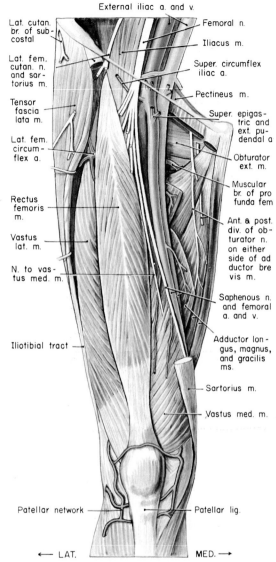

Figure 6-17. Anterior and medial surfaces of right thigh II. The sartorius muscle has been removed to reveal contents of the adductor canal. The pectineus and adductor longus muscles have also been cut to show the obturator externus and adductor brevis muscles.

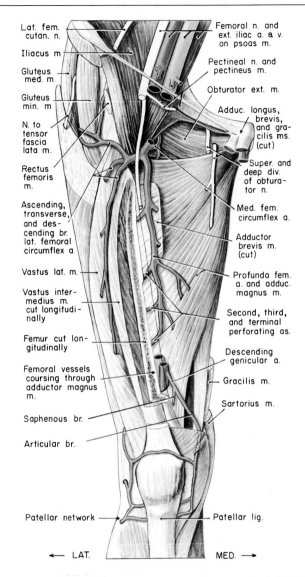

Figure 6-18. Anterior and medial surfaces of thigh III. The course and distribution of the profunda femoris artery are revealed. Note the nerve supply to the tensor fascia lata muscle from the superior gluteal nerve.

implies, extends from medial to lateral through the head of the femur. The vertical axis of medial and lateral rotation continues inferiorly from the head of the femur to just lateral to the medial condyle at its distal end. The anteroposterior axis of abduction and adduction passes through the margins of the acetabulum and the head of the femur. Note that all three axes pass through the head

of the femur.

The **knee joint** (Figs. 6-19 and 6-20) is a modified hinge joint, modified because of a small amount of rotation that is allowed; therefore, although flexion (leg posteriorly) and extension (leg anteriorly) are the main movements at this joint, medial and lateral rotation must be considered. Flexion and extension occur around a trans-

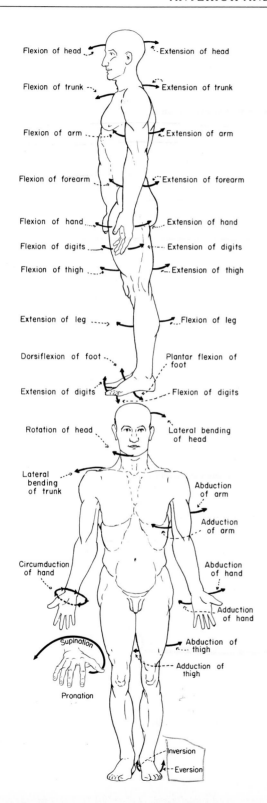

verse axis, but this axis is complicated by the fact that a gliding motion occurs during flexion and extension that involves a rotation on the vertical axis. This vertical axis passes through the middle of the joint, but at a point closer to the medial condyle than to the lateral. Therefore, in normal usage of the knee joint movement occurs about both axes simultaneously.

The **tibiofibular joints** are partially movable joints that play no particularly important roles.

The **ankle joint** (Figs. 6-19 and 6-20) is a hinge joint allowing flexion and extension only. Flexion is movement of the foot superiorly away from the floor while extension is movement of the foot in the opposite direction, toward the floor. A way to avoid confusion concerning movements at this joint is to call flexion "dorsiflexion," because it is a movement toward the dorsum of the foot, and extension "plantar flexion," because it is a movement toward the plantar surface (sole) of the foot. These movements occur on a transverse axis only.

Inversion* of the foot entails raising the plantar surface medial-ward, while eversion* is a similar movement in the opposite direction. These movements take place between the tarsal bones, particularly at the talocalcaneal joint and the transverse tarsal joint (the calcaneocuboid and talonavicular joints), the latter crossing the entire width of the foot. These movements occur on two axes, one through the talus and calcaneus and another through the cuboid and calcaneus. These are pictured in Figure 6-20.

Flexion and extension, and slight abduction and adduction, occur at the **metatarsophalangeal joints,** the former movements occurring around a transverse axis and the latter on a vertical axis from the dorsum of the foot through to the plantar side. There is no rotation in these joints, but circumduction is allowed.

The **interphalangeal joints** are hinge joints that allow only flexion (toes curled) and extension (toes straight) around a transverse axis.

In subsequent sections the main action of the muscles in the lower limb will be stated. These should be related to the axes mentioned above. For example, if a

---

*Inversion is accompanied by adduction and slight extension (plantar flexion) of the foot, while eversion involves the opposite—abduction and slight flexion (dorsiflexion).

Figure 6-19. *Movements allowed in the limbs.*

muscle is obviously located anterior to the hip joint, it will likely be anterior to the transverse axis of this joint and therefore induce a flexion at this joint. Some muscles are difficult to visualize, but if the direction of pull of a muscle is kept in mind, the action can usually be determined.

MUSCLES ON ANTERIOR SIDE. These muscles are:

1. Quadriceps femoris
   a. Rectus femoris
   b. Vastus lateralis
   c. Vastus intermedius
   d. Vastus medialis
2. Articularis genu
3. Sartorius
4. Pectineus
5. Iliopsoas

1. Quadriceps femoris muscle (Figs. 6-17 and 6-18) This muscle is made up of four distinct parts: the rectus femoris, vastus lateralis, vastus intermedius, and vastus medialis. All four parts are located on the anterior surface of the thigh and all **insert** on the tibial tuberosity, which is located on the superior and anterior surface of the tibia (Fig. 6-5). The patella* is a sesamoid bone that develops inside the large and strong patellar tendon by which the quadriceps femoris muscles insert on the tibia. This muscle is a very strong extensor of the leg. The rectus femoris portion also bridges over the hip joint and serves the additional action of flexion of the thigh. Although the insertions of all four parts of this quadriceps muscle are similar, the **origins** are different. The rectus femoris portion is the most superficial head of the quadriceps femoris and arises from the anterior inferior iliac spine and the superior edge of the acetabulum (Fig. 6-1).† It is separated from the capsule of the hip joint by a bursa. The origins of the remaining portions of this muscle are very difficult to give and learn in words; a glance at Figures 6-2 and 6-3 will indicate that the vastus intermedius occupies the anterior surface and the medial and lateral sides of the femur, and that the origin of the vastus

*H. Haxton, 1954, The function of the patella and the effects of its excision, *Surg. Gyn., and Obs.*, 80: 389. This author concludes that the patella does have a function. It seems to increase efficiency of extension by holding the patella tendon anterior to the axis and thereby increasing the pull of the quadriceps femoris muscle.

†Growth of the bone apparently separates the spine from the acetabulum, thus separating the origin into two parts.

lateralis surrounds this area superiorly and laterally, while the origin of the vastus medialis surrounds this area superiorly and medially. All four parts of the quadriceps femoris obtain their **nerve supply** from the femoral nerve, as might be anticipated from the location.

2. Articularis genu. This small muscle is located just anterior to the capsule of the knee joint and deep to the patella tendon. It **arises** by a variable number of small bundles from the inferior part of the anterior surface of the femur (Fig. 6-2). It **inserts** onto the synovial membrane of the knee joint, and its **main action** is to pull the synovial membrane superiorly during extension at the knee joint. The **nerve supply** is by the femoral.

3. Sartorius (*L.,* tailor) **muscle** (Fig. 6-14) This muscle is the most superficial muscle on the anterior surface of the thigh, **arises** from the anterior superior iliac spine (Fig. 6-1), and **inserts** onto the superior end of the medial surface of the tibia (Fig. 6-5). In its insertion it is intimately associated with the tendons of the gracilis and semitendinosus muscles. Because this muscle is posterior to the transverse axis of the knee joint but anterior to the transverse axis of the hip joint, its **main action** is to flex both the leg and the thigh. It also rotates the thigh laterally because of its course from the lateral to the medial side of the thigh. This results in flexion at the knee, and a flexion and lateral rotation at the hip joint, a position often taken by tailors, which accounts for the name of the muscle. **Nerve supply** is the femoral.

4. Pectineus muscle (Fig. 6-14) This is a superficial muscle forming the floor of the medial side of the femoral triangle. It **arises** from the pectineal line on the superior ramus of the pubis (Fig. 6-1) and **inserts** onto the superior half of a line that extends from the lesser trochanter to the linea aspera (Fig. 6-3). Its **main action** is to flex and adduct the thigh since its direction of pull is anterior to the transverse axis and medial to the anteroposterior axis of the hip joint. Its **nerve supply** is from the femoral nerve or sometimes from a branch of the obturator nerve.

5. Iliopsoas muscle (Fig. 6-18) This, in actuality, is two muscles, the **iliacus** and **psoas major.** The **psoas major** is located along the sides of the lumbar vertebrae while the **iliacus** muscle lies in the iliac fossa. Both muscles **insert** by a common tendon onto the lesser trochanter of the femur (Figs. 6-2 and 6-3) and, therefore, have the same action. Because their direction of pull passes anterior to the transverse axis of the hip joint, their

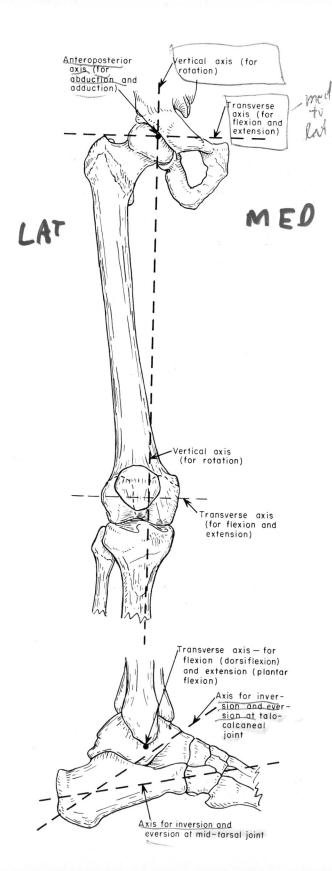

Anteroposterior axis (for abduction and adduction)

Vertical axis (for rotation)

Transverse axis (for flexion and extension)

med to lat

LAT

MED

Vertical axis (for rotation)

Transverse axis (for flexion and extension)

Transverse axis — for flexion (dorsiflexion) and extension (plantar flexion)

Axis for inversion and eversion at talo-calcaneal joint

Axis for inversion and eversion at mid-tarsal joint

main action is to flex the thigh. Another important action is that of bending the trunk anteriorly or laterally, accomplished by contraction of these muscles when both feet are on the ground. The **origin** of the two muscles is entirely different. The psoas major arises from the transverse processes of all the lumbar vertebrae, and from the bodies of these vertebrae, including the twelfth thoracic. The iliacus muscle arises from the floor of the iliac fossa (Fig. 6-1). Frequently a **psoas minor** muscle is present on the anterior surface of the psoas major. When present, this muscle **arises** from the twelfth thoracic and first lumbar vertebrae and the disc between these two vertebrae. It is **inserted** by a long tendon, mainly onto the pectineal line. Its **nerve supply** is from the first lumbar segment directly. The **nerve supply** to the iliopsoas muscle is from the second, third and fourth lumbar segments directly and occasionally through branches of the femoral nerve.

Those who have studied the abdominal cavity are aware of the importance of the fascia covering the psoas muscle in the spread of infection. Actually an infection can reach the femoral triangle from as high as the thoracic cavity by following this fascia.

MUSCLES ON MEDIAL SIDE. These muscles are:

1. Gracilis
2. Adductor longus
3. Adductor brevis
4. Adductor magnus
5. Obturator externus

1. The **gracilis muscle** (Figs. 6-17 and 6-41) is quite superficial in its location and **arises** from the inferior ramus of the pubis close to the symphysis and may actually arise from the symphysis itself (Fig. 6-1). It **inserts** by a long tendon onto the medial surface of the tibia just inferior to the condyle (Fig. 6-5). Its **main action** is to adduct the thigh, and, since the tendon passes posterior to both the transverse and vertical axes of the knee, it will

Figure 6-20. The axes about which motion occurs in the lower limb. If the direction of pull of muscles is related to these axes, their actions become understandable.

in addition flex and medially rotate the leg. The **nerve supply** is the obturator.

2. **Adductor longus muscle** (Fig. 6-14). This muscle is just medial to the pectineus muscle and anterior to the adductor brevis. It **arises** from the body of the pubis (Fig. 6-1) and **inserts** onto the inferior two-thirds of the linea aspera (Fig. 6-3). Its **main action** is to adduct and flex the thigh, the latter action resulting because the direction of pull is anterior to the transverse axis of the hip joint. **Nerve** is the obturator.

3. **Adductor brevis muscle** (Fig. 6-17). This muscle is the key muscle in this location, because the anterior division of the obturator nerve is anterior to it and the posterior division posterior to it. It lies between the adductor longus and the superior end of the adductor magnus. Its **origin** is from the inferior ramus of the pubis (Fig. 6-1) and its **insertion** is onto the lower two-thirds of a line that is inferior to the lesser trochanter; between it and the superior part of the linea aspera (Fig. 6-3). Its **main action** is to adduct and flex the thigh, the latter action resulting because the direction of pull is anterior to the transverse axis of the hip joint. **Nerve supply** is the obturator.

4. **Adductor magnus muscle** (Fig. 6-18). This large muscle **arises** from the body and ramus of the is-chium, and the inferior ramus of the pubis (Fig. 6-1). Its **insertion** is mainly onto the linea aspera, the supracondylar line, and the adductor tubercle (Fig. 6-3). The part of this muscle arising from the pubis is very frequently separated and is called the **adductor minimus.** The **nerve supply** is the obturator and also a branch from the ischiadic nerve. Its **main action** is to adduct the thigh.

5. **Obturator externus muscle** (Figs. 6-17 and 6-18). This muscle hardly lies in the thigh but should be described with the muscles here because technically it is in the thigh. It is located superficial to the obturator membrane. Its **origin** is from the outer surface of this membrane and the adjoining bone (Fig. 6-1). Its **insertion** is onto the trochanteric fossa. As it is posterior to the vertical axis of the hip joint, its **main action** is to rotate the thigh laterally. **Nerve supply** is the obturator.

Many of the muscles just described in the thigh are depicted in cross section in Figure 6-21.

NERVES

The nerves involved with the anterior and medial surfaces of the thigh are (1) the femoral, (2) the lateral cutaneous

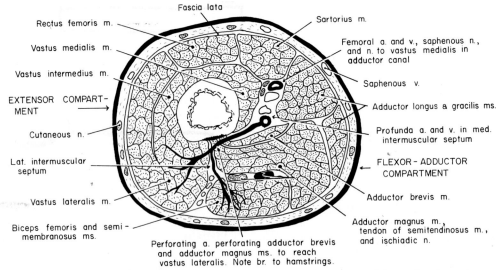

Rectus femoris m.
Vastus medialis m.
Vastus intermedius m.
EXTENSOR COMPARTMENT →
Cutaneous n.
Lat. intermuscular septum
Vastus lateralis m.
Biceps femoris and semi-membranosus ms.

Fascia lata
Sartorius m.
Femoral a. and v., saphenous n., and n. to vastus medialis in adductor canal
Saphenous v.
Adductor longus a gracilis ms.
Profunda a. and v. in med. intermuscular septum
FLEXOR - ADDUCTOR COMPARTMENT
Adductor brevis m.
Adductor magnus m., tendon of semitendinosus m., and ischiadic n.

Perforating a. perforating adductor brevis and adductor magnus ms. to reach vastus lateralis. Note br. to hamstrings.

Figure 6-21. *Transverse section of the right thigh. Note the course of the perforating artery. You are looking superiorly.*

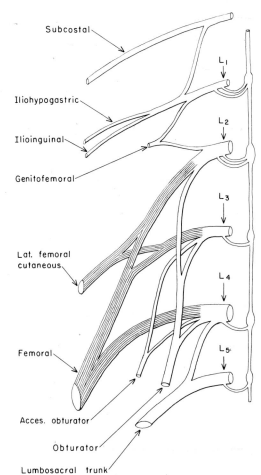

Figure 6-22. *Diagram of the right lumbar plexus. The clear nerves are preaxial, the shaded postaxial. (See page 440 for an explanation of pre- and postaxial).*

Subcostal

Iliohypogastric

Ilioinguinal

Genitofemoral

Lat. femoral cutaneous

Femoral

Acces. obturator

Obturator

Lumbosacral trunk

L_1

L_2

L_3

L_4

L_5

nerve of the thigh, and (3) the obturator. They are branches of the lumbar plexus (Fig. 6-22). Those students who have not studied the abdominal cavity might benefit from a study of Fig. 5-41 on page 261.

FEMORAL.   The **femoral nerve** (Figs. 6-14 and 6-17) arises from the second, third, and fourth lumbar segments of the spinal cord, courses in the substance of the psoas major muscle, and escapes from the lateral surface of the psoas at a point midway between the iliac crest and the inguinal ligament. It gives branches to the psoas and

iliacus muscles and descends between these muscles posterior to the cecum on the right side of the body and the descending colon on the left side. It enters the lower limb posterior to the inguinal ligament and immediately breaks up into its several branches. It is lateral to the femoral artery and vein and is found in the superior part of the femoral triangle. It is not contained within the femoral sheath.

The branches of the femoral nerve are:

1. Medial cutaneous nerve
2. Intermediate cutaneous nerve
3. Saphenous nerve
4. Muscular branches

1.   The **medial cutaneous nerve** (Figs. 6-10 and 6-14) courses inferiorly and slightly medially, crossing anterior to the femoral vessels. It divides into anterior and posterior branches which give cutaneous nerve supply to the anterior and medial surfaces of the thigh; the anterior branch courses across the sartorius muscle, giving off its branches high in the thigh, at an intermediate position, and inferiorly near the knee. The posterior branch will be seen again on the posterior side of the leg.

2.   The **intermediate cutaneous nerve** (Figs. 6-10 and 6-14) divides into medial and lateral branches, and there may be others as well. These branches supply the skin on the distal two-thirds of the anterior surface of the thigh, and continue to join into the patellar plexus.

3.   The **saphenous nerve** (Figs. 6-17 and 6-47) runs inferiorly and enters the adductor canal. It pierces the deep fascia on the medial side of the knee and accompanies the long saphenous vein as far as the medial side of the foot. Just superior to the knee, the saphenous nerve gives off the infrapatellar branch, which joins the patellar plexus giving nerve supply to the skin over the patellar ligament.

4.   **Muscular branches.**   The nerves **to the psoas and iliacus** muscles have already been described. The **pectineal nerve** (Fig. 6-18) arises close to the inguinal ligament, courses medially in a plane that is posterior to the femoral sheath, and reaches the deep surface of the pectineus muscle. This nerve is often found wanting, since this muscle may be innervated by the obturator or the accessory obturator nerve. The branches **to the sartorius** muscle (Fig. 6-14) may be several in number and enter the deep surface of the muscle at different levels;

branches **to the rectus femoris** muscle (Fig. 6-17) are also likely to enter the deep surface of the muscle at different levels. A branch of the latter nerve innervates the hip joint. Branches **to the vastus lateralis** accompany the lateral femoral circumflex artery deep to the rectus femoris muscle. The branch **to the vastus medialis** (Fig. 6-17) courses parallel to the saphenous nerve in the upper end of the adductor canal. It enters substance of the muscle halfway down the thigh. The branches **to the vastus intermedius** enter the superficial surface of the muscle rather than the usual deep entrance. One branch continues to give nerve supply to the articularis genu muscle and the knee joint itself.

In summary, the femoral nerve gives (1) cutaneous innervation to the anterior and medial surfaces of the thigh (except that area covered by the ilioinguinal and genitofemoral nerves) and knee, and to the medial surface of the leg, ankle, and foot; (2) sensory nerves to the hip and knee joints; and (3) motor nerves to the muscles on the anterior surface of the thigh, frequently including the iliopsoas muscle as well as the pectineus. (These nerves demonstrate that the nerve to a muscle is usually on its deep surface.) Injury to the femoral at its origin will cause anesthesia to these cutaneous areas, will greatly decrease flexion of the thigh, and eliminate extension of the leg. The loss of flexion of the leg and lateral rotation of the thigh caused by loss of the sartorius muscle is very minimal; therefore, if the femoral is cut the thigh will be extended and the leg flexed.

LATERAL FEMORAL CUTANEOUS NERVE (Figs. 6-10 and 6-14)

This nerve arises from the second and third lumbar segments, courses in the substance of the psoas muscle, and appears at its lateral border approximately at the level of the iliac crest. This nerve courses anterior to the iliacus muscle and reaches the anterior superior iliac spine by passing posterior to the ascending colon and cecum on the right side and the descending colon on the left. It enters the thigh deep to the inguinal ligament and courses on the anterior surface of the sartorius muscle. It then pierces the deep fascia approximately 3 to 4 inches inferior to the anterior superior iliac spine and gives off posterior and anterior branches. The **posterior branch** innervates the skin over the buttocks and the superior end of the thigh while the **anterior branch** gives nerve supply to the anterolateral surface of the thigh as far inferiorly as the knee.

OBTURATOR (Fig. 6-17).

This nerve arises from the second, third, and fourth lumbar segments. It too courses in the substance of the psoas muscle, and escapes from the psoas muscle, but on its medial surface rather than the lateral. It is at first posterior to the common iliac artery and then courses along the side wall of the pelvis lateral to the internal iliac vessels, ureter, ovary, and broad ligament but anterior to the obturator vessels. It then enters the obturator foramen and leaves the pelvic cavity to enter the thigh where it divides into two branches almost immediately. The **anterior branch,** after sending a branch to the hip joint through the acetabular notch, courses inferiorly deep to the pectineus and adductor longus muscles and anterior to the adductor brevis muscle. It innervates the adductor longus, gracilis, and adductor brevis muscles and occasionally the pectineus. This nerve often sends a branch to the skin along the medial side of the thigh close to the knee joint. The **posterior branch** of the obturator nerve innervates the obturator externus muscle and then pierces it. It courses inferiorly between the adductor brevis anteriorly and adductor magnus posteriorly and gives nerve supply to the magnus and occasionally the brevis. It continues as the articular branch that pierces the adductor magnus muscle and courses along the popliteal artery to the back of the knee joint, finally reaching the synovial membrane of this joint.

An **accessory obturator** nerve is found occasionally. When this nerve is present, it descends along the medial border of the psoas muscle and enters the thigh anterior to the superior ramus of the pubis. It either ends in the pectineus muscle or in the hip joint.

Summarizing, the obturator nerve may give (1) cutaneous innervation to a small area on the medial side of the knee, (2) sensory innervation to the hip and knee joints (with aid from the femoral), and (3) motor innervation to the adductor muscles on the medial side of the thigh, and to the obturator externus muscle. (We will see later that the ischiadic nerve also aids in innervating the adductor magnus muscle.)

When the obturator nerve is injured there is, therefore, a loss of sensation on the medial side of the knee and in the hip and knee joints, and a loss of adduction of the thigh, and a partial loss of flexion and medial rotation of the thigh, since the adductor muscles perform these latter actions to a degree. In walking, the limb tends to swing too far laterally.

ARTERIES

The arteries found on the anterior and medial surfaces of the thigh are the femoral and obturator.

FEMORAL The **femoral artery** (Figs. 6-14 and 6-17) is a continuation of the external iliac; it starts when the external iliac enters the thigh deep to the inguinal ligament. At first, the femoral artery is contained within the femoral sheath and occupies the lateral compartment of this sheath; it is within the femoral triangle. At the distal end of the triangle, it enters the adductor canal and continues in this canal until it penetrates the tendon of the adductor magnus muscle (Fig. 6-18) to enter the popliteal space and become the popliteal artery.

Relations. While in the femoral sheath, the artery is lateral to the femoral vein; it is medial to the femoral nerve although the latter is not contained within the sheath. The psoas muscle is posterior to it, and it is covered anteriorly only by the skin, and the superficial and deep fascial layers, in addition to the femoral sheath. In the femoral triangle, the artery is anterior to the pectineus and adductor longus muscles. At the apex of the triangle, it is posterior to the sartorius muscle. It is at first lateral to the femoral vein but tends to become anterior to it as the artery courses distally. The profunda femoris artery emerges from the posterior side of the artery. In the adductor canal (Fig. 6-21), the artery is anterior to the adductor longus and adductor magnus muscles, medial to the vastus medialis and the nerve to this muscle, and posterior to the sartorius muscle and the saphenous nerve. The femoral vein is posterior to the artery. The positions of these vessels are easily explained on the basis of the medial twisting that occurs during development of the lower limb.

The branches of the femoral artery are:

1. Superficial circumflex iliac
2. Superficial epigastric
3. Superficial external pudendal
4. Deep external pudendal
5. Profunda femoris
6. Muscular
7. Descending genicular

They exhibit considerable variation; the usual arrangement is as follows:

1. Superficial circumflex iliac (Figs. 6-10 and 6-14). This artery arises from the anterior surface of the femoral, pierces the fascia lata, and courses laterally just inferior to the inguinal ligament, reaching as far as the crest of the ilium. It supplies blood to the skin in this area and to the superficial inguinal lymph nodes.

2. Superficial epigastric (Figs. 6-10 and 6-14). This artery arises from the anterior surface of the femoral approximately 1 cm. inferior to the inguinal ligament. It courses through the femoral sheath and the fascia cribrosa and turns superiorly to course anterior to the inguinal ligament. It ascends between the membranous and fatty layers of the superficial fascia and usually reaches as high as the area of the umbilicus. It gives blood supply to the superficial lymph nodes, the superficial fascia, and the skin in the area covered.

3. Superficial external pudendal (Figs. 6-10 and 6-14). This artery arises from the medial side of the femoral artery, pierces the femoral sheath and the cribriform fascia, and courses medially in a position anterior to the spermatic cord in the male or the round ligament in the female. This artery distributes blood to the skin on the inferior part of the abdomen, to the penis, and to the scrotum in the male or a labium majus in the female.

4. Deep external pudendal (Fig. 6-14). This artery arises further along the femoral and courses medially, crossing anterior to the pectineus, adductor longus, and gracilis muscles. It supplies blood to the scrotum and penis in the male and to a labium majus in the female. It usually courses posterior to the spermatic cord or the round ligament.

5. Profunda femoris (Fig. 6-18) This large artery provides almost the entire blood supply to the thigh. It arises from the posterior surface of the femoral while still in the femoral triangle. The artery curves posteriorly and inferiorly, and courses close to the femur in a plane posterior to the adductor longus muscle and anterior to the adductor brevis and magnus muscles. It ends as the fourth perforating artery. The branches of the profunda femoris artery are:

a. Lateral femoral circumflex
b. Medial femoral circumflex
c. Muscular
d. Perforating

a. The **lateral femoral circumflex** artery (Fig. 6-18) courses laterally on the anterior surface of the iliacus muscle; it is deep to the rectus femoris and sartorius muscles. It divides into ascending, descending, and transverse branches. The **ascending branch** courses superiorly deep to the tensor fascia lata muscle. The **descending branch** courses inferiorly deep to the rectus femoris muscle and superficial to the vastus lateralis muscle. This artery continues all the way down the thigh as far as the knee and anastomoses with branches of the popliteal artery. This is an important anastomosis, because it is one of the few anastomotic channels that will take blood around a tie of the femoral artery at a point distal to the emergence of the profunda femoris. The **transverse branch** enters the vastus lateralis muscle. In the substance of this muscle it winds around the femur and anastomoses on the back of the thigh with the medial femoral circumflex, inferior gluteal, and first perforating arteries to take part in what is called the **cruciate anastomosis.** (This anastomosis will be considered when the gluteal region and posterior surface of the thigh are studied.)

b. The **medial femoral circumflex artery** (Fig. 6-18) arises from the medial side of the profunda and immediately courses posteriorly between the psoas and pectineus muscles. It continues to wind around the femur and comes into relation with the obturator externus and adductor brevis muscles. It gives blood supply to these muscles and then gives off its **acetabular branch,** which enters the acetabular notch and courses along the ligament of the head of the femur. The medial femoral circumflex artery then continues and divides into ascending and transverse branches. These branches are best seen when dissecting the gluteal region and posterior surface of the thigh, but will be given here for the sake of completeness. The **ascending branch** (Fig. 6-27) courses along the tendon of the obturator externus muscle and reaches the trochanteric fossa, where it lies anterior to the quadratus femoris muscle, a muscle in the gluteal region. It gives off a branch to the hip joint. The **transverse branch** (Fig. 6-27) courses between the quadratus femoris and adductor magnus muscles to end in muscles on the posterior surface of the thigh and gluteal region. The medial femoral circumflex artery also takes part in the cruciate anastomosis.

c. The **muscular branches** of the profunda femoris supply blood to the muscles in the vicinity.

d. The **perforating arteries** (Figs. 6-18 and 6-21) consist of three or four arteries that pierce the tendon of the adductor magnus muscle and end by entering the substance of the vastus lateralis muscle. Branches of these arteries provide blood to the muscles on the posterior surface of the thigh. The **nutrient artery** of the femur is a branch of the second perforating artery. The first perforating artery anastomoses with the two circumflex vessels and also with the inferior gluteal. The perforating and the gluteal arteries form a vertical line, while the two circumflexes form a horizontal line. This results in a cross, which accounts for the name **cruciate.** (See Fig. 6-33 on page 386.)

6. **Muscular branches.** The muscular branches of the femoral artery give blood supply to the muscles in relation to this artery.

7. **Descending genicular** (Fig. 6-18). This artery arises from the femoral just before the latter passes through the opening in the tendon of the adductor magnus muscle. It immediately divides into a saphenous and musculoarticular branch. (a) The **saphenous branch** accompanies the saphenous nerve to the medial side of the knee. It is distributed to the skin of the superior and medial parts of the leg and anastomoses with branches of the popliteal artery. (b) The **musculoarticular** branch descends in the substance of the vastus medialis muscle anterior to the tendon of the adductor magnus. It reaches the medial side of the knee and also anastomoses with branches of the popliteal. This artery also sends branches around the knee joint, which anastomose with branches from the opposite side; they form a sort of arch superior to the knee joint and supply branches to the knee joint itself.

OBTURATOR. This artery (Fig. 6-23) was previously described as arising from the internal iliac, and coursing anteriorly and inferiorly on the lateral wall of the pelvis to reach the obturator foramen. It passes through this foramen to enter the medial surface of the thigh, where it immediately divides into anterior and posterior branches. These branches completely encircle the foramen deep to the obturator externus muscle. (a) The **anterior branch** of this artery courses along the anterior margin of the foramen and distributes blood to the obturator externus, pectineus, and adductor muscles, including the gracilis. (b) The **posterior branch** courses on the posterior margin of the foramen and divides into two branches, one of which courses anteriorly on the ischial ramus while the other supplies muscles attached to the ischial tuberosity. Branches from this artery enter the hip joint through the acetabular notch. It should be observed that, in a very

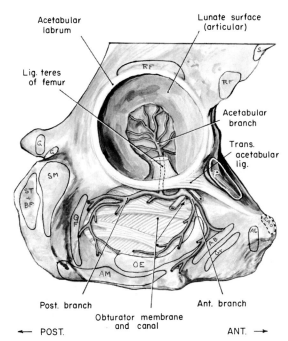

Acetabular labrum

Lunate surface (articular)

Lig. teres of femur

Acetabular branch

Trans. acetabular lig.

Post. branch

Ant. branch

Obturator membrane and canal

← POST.

ANT. →

Figure 6-23. *The distribution of the obturator artery (right side). Origins and insertions of muscles are outlined: (AB) adductor brevis, (AL) adductor longus, (AM) adductor magnus, (BF) biceps femoris, (EAO) external abdominis oblique, (G) gemelli, (Gr) gracilis, (OE) obturator externus, (P) pectineus, (QF) quadratus femoris, (RF) rectus femoris, (S) sartorius, (SM) semimembranosus, (ST) semitendinosus.*

large percentage of cases, the obturator artery does not arise from the internal iliac but from the inferior epigastric branch of the external iliac artery.

Special attention should be given to the blood supply to the head of the femur. Although both the medial femoral circumflex and obturator arteries supply the acetabular fossa and provide small branches that course along the ligamentum teres of the femur, they do not provide the major blood supply to the head of the femur. Indeed, they may not participate at all. The ascending branches of the medial and lateral femoral circumflex arteries both wind around the anatomical neck of the femur, particularly on its posterior surface. Numerous branches from these arteries course along the neck and enter the head of the femur at its base. These arteries

anastomose inside the bone with the medullary artery of the shaft. (The above description applies to the adult. In children the branches coursing along the ligamentum teres are more important.)

You should note that major blood vessels are found on the flexor side of the joints.

VEINS

There are veins that correspond to all of the arteries just described. The additional points that demand some attention are the relations of the femoral vein and the presence of a superficial as well as a deep venous circulation.

The **femoral vein** enters the thigh as a continuation from the popliteal vein by piercing the hiatus in the adductor magnus muscle in company with the femoral artery. It is at first posterior to the artery but gains its medial side as it goes superiorly. It enters the femoral sheath on the medial side of the artery in the middle compartment of that sheath. It terminates by entering the abdominal cavity posterior to the inguinal ligament to become the external iliac vein. Once again, these relationships are easily remembered if one recalls the medial twist occurring during development of the lower limb. The femoral vein has tributaries that correspond to the branches of the femoral artery.

One tributary has no counterpart in the arterial system. This is the **long saphenous vein** (Fig. 6-10), important from the point of view of varicose veins. This vein begins at the medial side of the foot and extends superiorly anterior to the medial malleolus and on the anterior surface of the tibia. It reaches the posterior part of the medial side of the knee and then courses superiorly and anteriorly, pierces the cribriform fascia, and passes through the saphenous opening or hiatus to enter the femoral vein. This vein has many unnamed tributaries and contains numerous valves; when these valves become incompetent, blood accumulates, causing it to dilate (varicose veins). There is also a short saphenous vein in the leg, which will be described later.

Many of the structures found on the anterior and medial sides of the thigh are shown in cross section in

Figure 6-21. Study of this picture at this time should be helpful.

GLUTEAL REGION

SURFACE ANATOMY

The gluteal muscles are covered by a variable amount of fat, which gives the rounded contour to the **buttocks** (Fig. 6-31). The buttocks are **bounded** superiorly by the iliac crest, inferiorly by a fold in the skin called the gluteal fold, and medially by the furrow between the two buttocks. There is no particular boundary laterally except the lateral side of the thigh. A depression—the **gluteal depression**—on the lateral side of the buttocks is just anterior to the gluteus maximus muscle and actually superficial to the gluteus medius muscle (see Fig. 6-25). Just anterior to the gluteal depression is an elevation caused by the tensor fascia lata muscle. The **iliac crest** can be palpated throughout its entire extent from the anterior superior iliac spine to the posterior superior iliac spine. The same can be said for the **sacrum,** the spinous processes of the sacrum, and, in the furrow between the buttocks, the tip of the coccyx. You should become familiar with the structures of the gluteal region in relation to the skin surface because many injections are given in this region.

GENERAL DESCRIPTION

Figure 6-24 presents the bones involved in a study of the gluteal region and outlines the attachments of the many muscles about to be mentioned.

After removing the skin, the **cutaneous nerves** of the gluteal region are easily seen (Figs. 6-11 and 6-25). They are divided into superior, middle, and inferior cluneal nerves. The **superior cluneal nerves** are three in number and are the dorsal rami of the first three lumbar nerves. They course superficial to the iliac crest to reach the gluteal region and cover a large percentage of the area. The **middle cluneal nerves** are the dorsal rami of the sacral nerves, and they enter near the posterior border of the gluteus maximus muscle and cover a small area near the furrow between the buttocks. The **inferior cluneal nerves** are the gluteal branches of the posterior femoral cutaneous nerve, and they appear in the gluteal region by curving around the inferior border of the gluteus maximus muscle.

Removal of the superficial and deep fascial layers reveals the **gluteus maximus** muscle, which is found to be a very thick and heavy muscle (Fig. 6-25). It does not cover the entire gluteal region, for its superior border does not reach all the way to the iliac crest. Between the iliac crest and the gluteus maximus muscle is the **gluteus medius** muscle, and anterior to that is the **tensor fascia lata** muscle. Therefore, as one views the gluteal region the superficial muscles are, from anterior to posterior, the tensor fascia lata, the gluteus medius, and the gluteus maximus (Note Fig. 6-15).

If the gluteus maximus muscle is cut, the smaller muscles in the gluteal region are revealed (Fig. 6-26). The key to this region is the **piriformis muscle,** which enters the gluteal region from the pelvic cavity through the **greater ischiadic foramen.** This pear-shaped muscle is said to be the key to this region because nearly everything entering the gluteal region does so either superior or inferior to it. Superior and lateral to the piriformis is the **gluteus medius** and, if cut (Fig. 6-27), the **gluteus minimus** is revealed; anterior to these two muscles is the **tensor fascia lata** muscle. These three muscles have a similar origin and nerve supply. Inferior to the piriformis muscle are the **superior gemellus** and **inferior gemellus** muscles on either side of the tendon of the **obturator internus** muscle. Next in line, inferiorly, is the **quadratus femoris** muscle, deep to which is the **obturator externus** muscle. Inferior to the quadratus femoris is the **adductor magnus** muscle.

There are many nerves and vessels in this area. The largest nerve is the **ischiadic,** which appears just inferior to the piriformis muscle and courses superficial or posterior to the superior gemellus, obturator internus, inferior gemellus, obturator externus, and quadratus femoris muscles. This is also accompanied by the **posterior femoral cutaneous** nerve, which can be seen close to the ischiadic nerve. **Superior and inferior gluteal** vessels and nerves enter the region superior and inferior to the piriformis muscle and enter the deep surface of the gluteus maximus. Both the superior gluteal nerve and artery send a branch between the gluteus medius and minimus to supply these muscles; in addition, the nerve

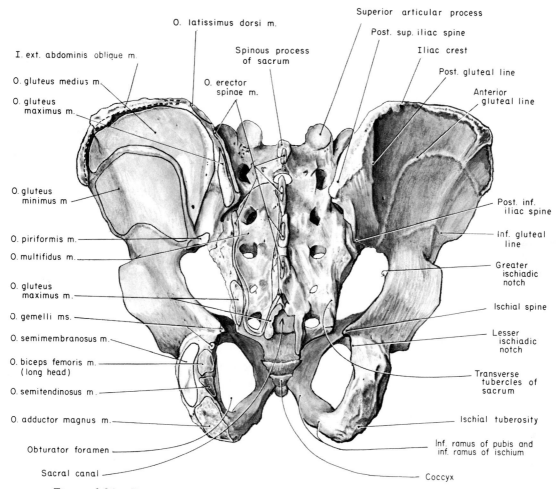

I. ext. abdominis oblique m.

O. gluteus medius m.

O. gluteus maximus m.

O. gluteus minimus m

O. piriformis m.

O. multifidus m.

O. gluteus maximus m.

O. gemelli ms.

O. semimembranosus m.

O. biceps femoris m. (long head)

O. semitendinosus m.

O. adductor magnus m.

Obturator foramen

Sacral canal

O. latissimus dorsi m.

Spinous process of sacrum

O. erector spinae m.

Superior articular process

Post. sup. iliac spine

Iliac crest

Post. gluteal line

Anterior gluteal line

Post. inf. iliac spine

Inf. gluteal line

Greater ischiadic notch

Ischial spine

Lesser ischiadic notch

Transverse tubercles of sacrum

Ischial tuberosity

Inf. ramus of pubis and inf. ramus of ischium

Coccyx

Figure 6-24. *Posterior view of the male pelvis, with origins and insertions of muscles outlined.*

continues to innervate the tensor fascia lata muscle (see Fig. 6-18).

Medially, two other nerves and vessels can be seen entering the gluteal region through the greater ischiadic foramen just inferior to the piriformis muscle, coursing posterior to the sacrospinous ligament, and then leaving the gluteal region via the lesser ischiadic foramen. These nerves and vessels are the **internal pudendals** and are accompanied by the **nerve to the obturator internus** muscle, which also sends a branch to the superior gemellus muscle (Fig. 6-27). The internal pudendal artery and pudendal nerve, by leaving the gluteal region through the lesser ischiadic foramen, enter the perineum.

Other nerves lie deep to the small muscles and will be described presently.

The **medial femoral circumflex artery** (Fig. 6-27), from the profunda femoris, reaches the gluteal region and courses between the quadratus femoris and the adductor magnus muscles. This particular branch is the transverse branch of the medial circumflex. The ascending branch of this artery appears between the quadratus femoris and the obturator externus.

The **veins** of the gluteal region have the same names as the arteries and are quite profuse.

If the superior gemellus, obturator internus tendon, inferior gemellus, and quadratus femoris muscles are bisected or reflected to one side, the small nerve supplying the inferior gemellus, the hip joint, and the quadratus femoris can be followed deep to these muscles (Fig. 6-27). The nerve courses superficial to the obturator externus muscle. The removal of these muscles also reveals the fact that they have been in contact with the capsule of the hip joint.

If a line is drawn from the greater trochanter to the posterior superior iliac spine, it divides the gluteal region into two areas. To avoid injury to vital structures, injections should be made superior and anterior to this line.

MUSCLES

The muscles found in the gluteal region are:

1. Gluteus maximus
2. Gluteus medius
3. Gluteus minimus
4. Piriformis
5. Superior gemellus
6. Inferior gemellus
7. Obturator internus
8. Obturator externus
9. Quadratus femoris
10. Tensor fascia lata

1. Gluteus maximus (Fig. 6-25). This very large muscle is the most superficial muscle in the gluteal region and, except for the gluteus medius and minimus and tensor fascia lata muscles, covers all muscles in the preceding list. It **arises** from the upper part of the external surface of the ilium from an area posterior to the posterior gluteal line, from the posterior surface of the sacrum and coccyx, and from the dorsal surface of the sacrotuberous ligament (Fig. 6-24). This muscle **inserts** onto the iliotibial tract and the gluteal tuberosity of the femur (Fig. 6-3). Its **main action** is to extend and abduct the thigh and rotate it laterally. It also has the effect of tightening the iliotibial tract. This muscle is important in maintaining the erect posture. **Nerve supply** is the inferior gluteal.

2. Gluteus medius (Fig. 6-26). This muscle occupies the superior and anterior portion of the gluteal region. It **arises** from the outer surface of the ilium in an area bounded by the anterior and posterior gluteal lines (Fig. 6-24). It **inserts** onto the greater trochanter of the femur (Fig. 6-3). Its **main action** is to abduct the thigh. Because of the relation of this muscle to the vertical axis of the hip joint, the anterior fibers are said to flex the thigh and to rotate it medially while the posterior fibers are thought to rotate the femur laterally.* The gluteus medius muscle is separated from the trochanter by a bursa. **Nerve supply** is the superior gluteal nerve.

3. The **gluteus minimus** muscle lies deep to the gluteus medius (Fig. 6-27). It **arises** from the outer surface of the ilium between the middle and inferior gluteal lines (Fig. 6-24), and **inserts** on the anterior surface of the greater trochanter (Fig. 6-2), a bursa lying between the bone and muscle. Its **action** is similar to the gluteus medius, namely, in abducting the thigh. The anterior fibers flex and rotate the thigh medially while the posterior fibers rotate it laterally.* **Nerve supply** is the superior gluteal nerve.

4. Piriformis (Figs. 6-26 and 6-27). As mentioned previously, this muscle is the key muscle in the gluteal region, since structures entering the gluteal region do so either superior or inferior to it. Its **origin** is mainly from the anterior surface of the sacrum and the superior margin of the greater ischiadic notch. It **inserts** by a rounded tendon onto the greater trochanter (Fig. 6-2). Its **main action** is to abduct the thigh and rotate it laterally. The **nerve supply** is by the first and second sacral segments directly.

5. Superior gemellus (Fig. 6-26). This small muscle **arises** from the margins of the lesser ischiadic notch (Fig. 6-24) and **inserts** on the tendon of the obturator internus muscle. Its **main action** is to aid the obturator internus muscle in laterally rotating the thigh. **Nerve supply** is a branch from the nerve to the obturator internus muscle.

6. The **inferior gemellus** muscle (Fig. 6-26) lies inferior to the obturator internus tendon and has attachments similar to the superior gemellus. It also aids the obturator internus in laterally rotating the thigh. **Nerve supply** is different from that of the superior gemellus; it is innervated by a special branch from the nerve to the quadratus femoris muscle.

7. Obturator internus (Fig. 6-26). The fleshy part of this muscle lies in the pelvic cavity; its tendon

---

*These two muscles play an important role in walking. By reversing origin and insertion, they prevent too great a tilt of the pelvis when the foot is taken off the ground or floor.

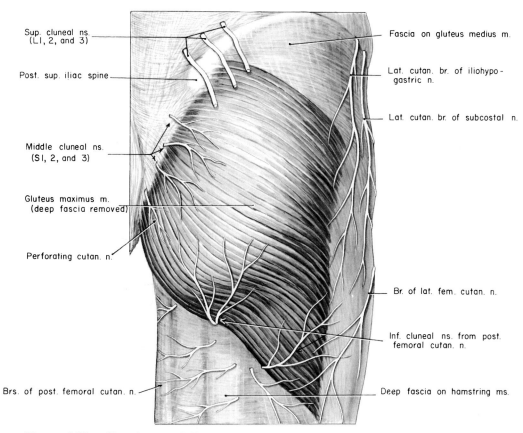

Sup. cluneal ns.
(LI, 2, and 3)

Post. sup. iliac spine

Middle cluneal ns.
(SI, 2, and 3)

Gluteus maximus m.
(deep fascia removed)

Perforating cutan. n.

Brs. of post. femoral cutan. n.

Fascia on gluteus medius m.

Lat. cutan. br. of iliohypogastric n.

Lat. cutan. br. of subcostal n.

Br. of lat. fem. cutan. n.

Inf. cluneal ns. from post.
femoral cutan. n.

Deep fascia on hamstring ms.

Figure 6-25. *Gluteal region I. Deep fascia has been removed from the gluteus maximus muscle (right side).*

enters the gluteal region through the lesser ischiadic foramen by making a right angle turn, and then it continues in the gluteal region as a tendon surrounded by the superior and inferior gemelli. It **arises** from the pelvic surface of the obturator membrane and the margins of the obturator foramen (Fig. 6-1). Its **insertion** is onto the greater trochanter (Fig. 6-2). Its **main action** is to rotate the thigh laterally. **Nerve supply** is the nerve to the obturator internus from the fifth lumbar and first two sacral segments.

8. The **obturator externus** muscle lies partially on the medial side of the thigh and partly in the gluteal region; it is deep to the quadratus femoris muscle (Fig. 6-27). It **arises** from the outer surface of the obturator membrane and the bone forming the margins of the obturator foramen (Fig. 6-1), inserts onto the trochanteric

fossa, and its **main action** is to rotate the thigh laterally. **Nerve supply** is the obturator nerve.

9. **Quadratus femoris** (Fig. 6-26). This muscle **arises** from the lateral margin of the ischial tuberosity (Fig. 6-1) and **inserts** onto the trochanteric crest and a line just inferior to this crest (Fig. 6-3). Its **main action** is to rotate the thigh laterally. **Nerve supply** is the nerve to the quadratus femoris muscle.

10. **Tensor fascia lata** muscle (Fig. 6-17). This muscle can be said to lie on the lateral surface of the thigh or between the gluteal region posteriorly and the front of the thigh anteriorly. It **arises** from the anterior superior iliac spine and the adjacent iliac crest (Fig. 6-1) and inserts into the iliotibial tract, which extends inferiorly to attach to the lateral condyle of the tibia, the capsule of the knee joint, and the patella. Its **main action** is to tense the

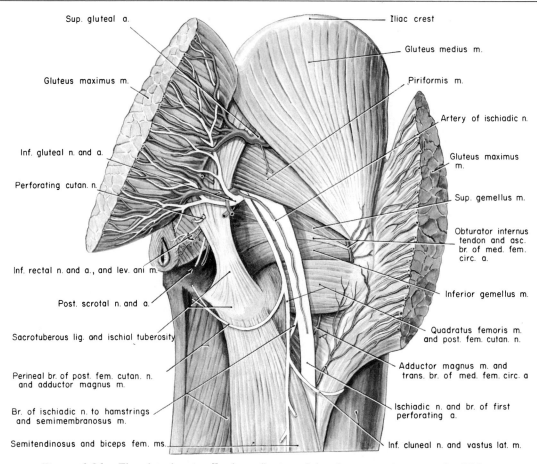

Sup. gluteal a.

Gluteus maximus m.

Inf. gluteal n. and a.

Perforating cutan. n.

Inf. rectal n. and a., and lev. ani m.

Post. scrotal n. and a.

Sacrotuberous lig. and ischial tuberosity

Perineal br. of post. fem. cutan. n. and adductor magnus m.

Br. of ischiadic n. to hamstrings and semimembranosus m.

Semitendinosus and biceps fem. ms.

Iliac crest

Gluteus medius m.

Piriformis m.

Artery of ischiadic n.

Gluteus maximus m.

Sup. gemellus m.

Obturator internus tendon and asc. br. of med. fem. circ. a.

Inferior gemellus m.

Quadratus femoris m. and post. fem. cutan. n.

Adductor magnus m. and trans. br. of med. fem. circ. a

Ischiadic n. and br. of first perforating a.

Inf. cluneal n. and vastus lat. m.

Figure 6-26. *The gluteal region II, after reflection of the gluteus maximus muscle. All fascia has been removed (right side).*

iliotibial tract and, because this tract courses anterior to the transverse axis of the knee joint, it aids in maintaining the knee in the extended position which is important in standing. This muscle also abducts the thigh and rotates it medially, actions made possible since the direction of pull is lateral to the anteroposterior axis and anterior to the vertical axis of the hip joint. If the knee is extended, this muscle aids in flexion at the hip joint. **Nerve supply** is the superior gluteal nerve.

NERVES

Nerves to the gluteal region arise from the sacral plexus; those who have studied it should review the plexus at this time (Fig. 6-28).

The sacral plexus is formed by the lumbosacral trunk and the ventral rami of the upper three and part of the fourth sacral nerves. The roots of the plexus combine and course toward the inferior part of the greater ischiadic foramen. The plexus terminates in two nerves, ischiadic and pudendal, and all the other nerves arise from the anterior or posterior surfaces of the plexus. Those arising from the anterior surface are the pelvic splanchnic nerves, nerves to the quadratus femoris and obturator internus muscles, the nerves to the levator ani and coccygeus muscles, part of the posterior femoral cutaneous, and the perineal branch of the fourth sacral nerve. The nerves arising from the posterior surface of the plexus are branches to the piriformis muscle, the superior and inferior gluteal, the perforating cutaneous, and part of the

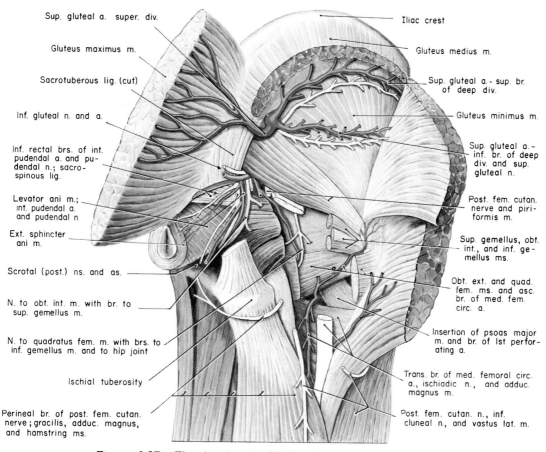

Sup. gluteal a. super. div.

Gluteus maximus m.

Sacrotuberous lig. (cut)

Inf. gluteal n. and a.

Inf. rectal brs. of int. pudendal a. and pudendal n.; sacrospinous lig.

Levator ani m.; int. pudendal a. and pudendal n

Ext. sphincter ani m.

Scrotal (post.) ns. and as.

N. to obt. int. m. with br. to sup. gemellus m.

N. to quadratus fem. m. with brs. to inf. gemellus m. and to hip joint

Ischial tuberosity

Perineal br. of post. fem. cutan. nerve; gracilis, adduc. magnus, and hamstring ms.

Iliac crest

Gluteus medius m.

Sup. gluteal a.- sup. br. of deep div.

Gluteus minimus m.

Sup. gluteal a.- inf. br. of deep div. and sup. gluteal n.

Post. fem. cutan. nerve and piriformis m.

Sup. gemellus, obt. int., and inf. gemellus ms.

Obt. ext. and quad. fem. ms. and asc. br. of med. fem. circ. a.

Insertion of psoas major m. and br. of 1st perforating a.

Trans. br. of med. femoral circ. a., ischiadic n., and adduc. magnus m.

Post. fem. cutan. n., inf. cluneal n., and vastus lat. m.

Figure 6-27. *The gluteal region III. The deepest structures—right side.*

posterior femoral cutaneous. The pelvic splanchnics form the sacral portion of the parasympathetic nervous system; the nerves to the levator ani, coccygeus, and piriformis muscles are short and enter these muscles directly; and the perineal branch of the fourth sacral penetrates the coccygeus muscle, enters the ischiorectal fossa, and reaches the posterior part of the external sphincter ani muscle and the skin overlying it. All the remaining nerves are involved in the gluteal region; they are:

1. Nerve to the quadratus femoris
2. Nerve to the obturator internus
3. Superior gluteal
4. Inferior gluteal
5. Posterior femoral cutaneous and perforating cutaneous
6. Pudendal
7. Ischiadic

1. Nerve to quadratus femoris (Figs. 6-27 and 6-28). Because this nerve arises from the anterior surface of the sacral plexus, one would expect to find the muscle that it innervates most anteriorly placed in the gluteal region. So it is; this nerve arises from the fourth and fifth lumbar and first sacral segments, and leaves the pelvic cavity inferior to the piriformis muscle. It courses inferiorly on the capsule of the hip joint between it and the superior gemellus, obturator internus, inferior gemellus, and quadratus femoris muscles. It gives a branch to the hip joint and to the inferior gemellus muscle and then enters the deep surface of the quadratus femoris.

2. Nerve to obturator internus (Figs. 6-27 and 6-28). This muscle also lies deep in the gluteal region; its nerve arises from the fifth lumbar and first two sacral segments, and emerges from the anterior surface of the

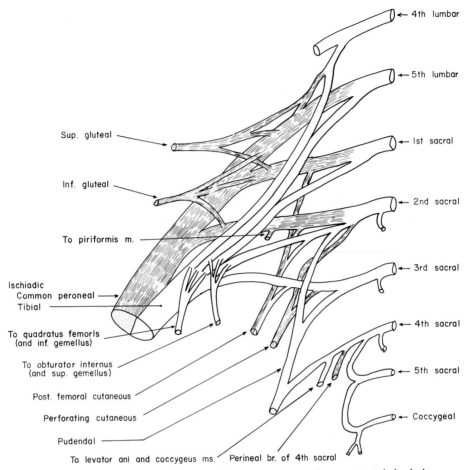

Figure 6-28. The sacral plexus. Preaxial nerves are clear, postaxial shaded.

sacral plexus. It leaves the pelvic cavity through the greater ischiadic foramen inferior to the piriformis muscle, crosses the ischial spine or the sacrospinous ligament, gives a branch to the superior gemellus muscle, and then enters the ischiorectal fossa by passing through the lesser ischiadic foramen. This nerve supplies the obturator internus muscle, entering the fleshy part of the muscle located in the lateral wall of the ischiorectal fossa.

3. Superior gluteal nerve (Figs. 6-27 and 6-28). This arises from the fourth and fifth lumbar and first sacral segments and leaves the pelvic cavity through the greater ischiadic foramen superior to the piriformis muscle. It courses between the gluteus medius and minimus muscles supplying both and then continues to enter the deep surface of the tensor fascia lata muscle.

4. Inferior gluteal nerve (Figs. 6-26 and 6-28). This nerve, from the fifth lumbar and first two sacral segments, leaves the pelvis inferior to the piriformis muscle, traversing the greater ischiadic foramen. It is on the posterior surface of the ischiadic nerve and almost immediately disappears into the substance of the gluteus maximus muscle which it innervates.

5. Posterior femoral cutaneous nerve (Figs. 6-26 and 6-28). This nerve, from the first three sacral segments, leaves the pelvic cavity in company with the inferior gluteal nerve, inferior to the piriformis muscle. It courses deep to the gluteus maximus muscle and is posterior to the ischiadic nerve. This nerve continues along the posterior surface of the thigh deep to the deep fascia; it finally pierces this fascia over the popliteal area and continues to the posterior surface of the leg. The posterior cutaneous nerve of the thigh has gluteal branches, a perineal branch, femoral, and terminal branches. The **gluteal branches**—the inferior cluneal nerves—have already been described as curving around the inferior border of the gluteus maximus muscle to become cutaneous on the inferior aspect of the buttocks.* The **perineal branch** courses medially across the hamstring tendons inferior to the ischial tuberosity. It enters the perineum and ends in the skin of the scrotum or a labium majus. **Femoral and terminal branches** of the posterior cutaneous nerve of the thigh will be described with the posterior surface of the thigh.

6. **Pudendal nerve** (Figs. 6-27 and 6-28). This nerve has already been described on page 318, but it should be reviewed at this time. It arises from the second, third, and fourth sacral segments and leaves the pelvic cavity inferior to the piriformis muscle by traversing the greater ischiadic foramen. It passes posterior to the sacrospinous ligament and then enters the lesser ischiadic foramen to reach the ischiorectal fossa in the perineal region. It courses along the lateral surface of the ischiorectal fossa in the pudendal canal of fascia, and gives off important **inferior rectal branches** to the external sphincter muscle of the anus. The next branch, the **perineal,** comes off from the pudendal while it is in the pudendal canal; it divides into posterior scrotal or labial nerves and muscular branches which innervate all muscles in the urogenital triangle and also the bulb. After emergence of the perineal branch, the pudendal nerve continues as the **dorsal nerve of the penis or clitoris.** It enters the deep pouch and then penetrates the perineal membrane to reach the penis or clitoris.

7. **Ischiadic** (Figs. 6-26 and 6-28). This very large nerve arises from the fourth and fifth lumbar and first three sacral segments. It leaves the pelvic cavity through the greater ischiadic foramen just inferior to the piriformis muscle. It traverses the gluteal region, and in this position it is deep or anterior to the gluteus maximus muscle and superficial or posterior to the superior gemellus, obturator internus, inferior gemellus, obturator externus, and quadratus femoris muscles. It continues into the thigh and will be described in that region.

The **coccygeal plexus** might be described at this time. It consists of the **coccygeal nerve,** with contributions from the fourth and fifth sacral nerves. It lies on the pelvic surface of the coccygeus muscle. It forms a single nerve that innervates the coccygeus muscle and a portion of the levator ani muscle, and then penetrates the coccygeus muscle and the sacrotuberous ligament to be distributed in the skin over the dorsal surface of the coccyx.

ARTERIES AND VEINS

The arteries involved in the gluteal region are:

1. Superior gluteal
   a. Superficial division
   b. Deep division
      i. Superior branch
      ii. Inferior branch
2. Inferior gluteal
3. Internal pudendal
4. Medial femoral circumflex

A study of Fig. 5-71 on page 307 would be helpful before reading the following.

1. **Superior gluteal** (Fig. 6-27). This artery is a continuation of the posterior division of the internal iliac. It leaves the pelvic cavity through the superior part of the greater ischiadic foramen superior to the piriformis muscle. It immediately divides into superficial and deep divisions. (a) The **superficial division** enters the gluteus maximus. (b) The **deep division** divides into **superior and inferior branches,** both of which course between the gluteus medius and minimus muscles. One branch takes a more superior position and reaches as far as the tensor fascia lata muscle where it anastomoses with the ascending branch of the lateral femoral circumflex artery. The inferior branch courses as far as the trochanteric fossa; it anastomoses with the ascending branch of both the medial and lateral femoral circumflex arteries.

2. **The inferior gluteal artery** (Fig. 6-26) is a branch of the anterior division of the internal iliac artery; it traverses the greater ischiadic foramen but is in a position inferior to the piriformis muscle. It courses in the gluteal region between the piriformis muscle superiorly and the superior gemellus muscle inferiorly. It is usually medial to the ischiadic nerve. It anastomoses with the first perforating

---

*One of these cluneal branches, located most medially, often pierces the sacrotuberous ligament before reaching the gluteal skin. In many cases this branch is a separate nerve arising from the posterior surface of the second and third sacral nerves; in this instance it is called a perforating cutaneous nerve (Fig. 6-26).

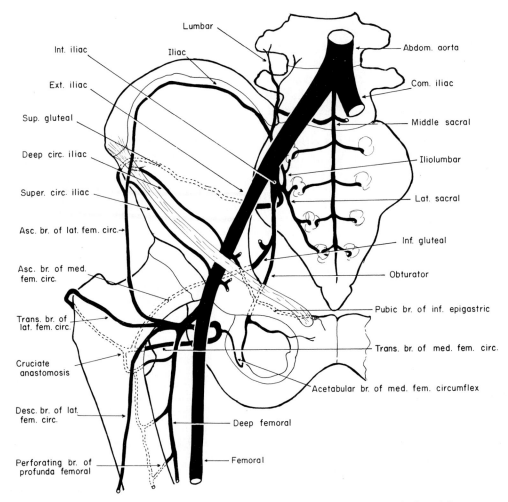

Figure 6-29. *Diagram of arterial anastomoses around the hip joint (right side).*

artery and both medial and lateral femoral circumflex arteries (all branches of the profunda femoris artery) to form the cruciate anastomosis (Fig. 6-33).

The branches of the inferior gluteal artery are: (a) muscular, (b) cutaneous, (c) coccygeal, (d) artery to the ischiadic nerve. (a) The **muscular** provides blood to the muscles of the gluteal region, while (b) **cutaneous** branches reach the skin. (c) The **coccygeal branch** courses medially, pierces the sacrotuberous ligament, and finally terminates on the dorsal aspect of the coccyx. (d) The branch **to the ischiadic nerve** is interesting in that this at one time was the main arterial pathway to the lower limb. Since all main vessels finally locate on the

flexor side of joints, this particular artery was reduced in size and is simply large enough to take care of this particular nerve.

3. Internal pudendal (Fig. 6-27). This artery takes the same pathway as just described for the pudendal nerve. It leaves the pelvis through the greater ischiadic foramen inferior to the piriformis muscle, passes posterior to the sacrospinous ligament, and enters the ischiorectal fossa by traversing the lesser ischiadic foramen. It is then found in the pudendal canal of fascia on the obturator internus muscle. Its branches have been described previously. In review, it gives off inferior rectal branches and scrotal or labial branches while in the ischiorectal fossa, continues into the deep pouch of the urogenital triangle, gives off an artery to the bulb, and finally terminates by dividing into the dorsal and deep arteries to the penis or clitoris which penetrate the perineal membrane to get to their destination.

4. Medial femoral circumflex (Fig. 6-27). The terminal branches of the medial femoral circumflex artery are found in the gluteal region. The **transverse branch** is found between the quadratus femoris and the superior edge of the adductor magnus muscle, and the **ascending branch** ascends deep to the quadratus femoris muscle following the tendon of the obturator externus muscle into the trochanteric fossa.

Veins of a similar name and course accompany these arteries. The important anastomoses of the arteries are shown in Figure 6-29.

LYMPHATICS

The lymphatic drainage for the gluteal region, except for the skin, follows the arteries just described into the pelvic cavity. The first nodes are the internal iliac nodes, which lie along the internal iliac artery and its many branches. Furthermore, the deeper parts of the perineum also drain into the same nodes.

POSTERIOR SURFACE OF THIGH

GENERAL DESCRIPTION

Upon removal of the skin, branches of the **posterior femoral cutaneous nerve** can be seen on either side of the midline in the superficial fascia (Fig. 6-11). The main nerve, however, remains deep to the deep fascia until the popliteal fossa is reached. It appears in the thigh just inferior to the gluteus maximus muscle and continues in the midline of the thigh in its entire course. It gives cutaneous branches to the posterior surface of the leg as well as to the thigh.

Removal of the deep fascia reveals the hamstring muscles: the **biceps femoris** on the lateral side, the **semitendinosus** and **semimembranous** muscles on the

medial side (Fig. 6-30). If these muscles are separated, the large **ischiadic nerve** can be seen; it divides into the **common peroneal** and **tibial** nerves at various levels. **Branches to the hamstring** muscles from the ischiadic arise fairly high in the thigh.

The blood vessels for this area (Fig. 6-31) are derived from two sources: (1) the **transverse branch** of the **medial femoral circumflex** artery descends on the posterior surface of the adductor magnus muscle to join the (2) **four perforating arteries,** branches of the **profunda femoris.** These perforating arteries penetrate the tendon of the adductor magnus muscle, and, after giving branches to the hamstring muscles, end in the vastus lateralis muscle (see Fig. 6-21 on page 368).

SURFACE ANATOMY

The muscles mentioned above can be seen on the surface (Fig. 6-32). The bulge on the medial side is produced by the **semimembranosus** and **semitendinosus** muscles and that laterally by the long head of the **biceps femoris muscle.** The tendons of these muscles are located on the medial and lateral sides of a shallow groove posterior to the knee, the **popliteal fossa.**

MUSCLES

The muscles found on the posterior surface of the thigh are:

1. Biceps femoris
2. Semitendinosus
3. Semimembranosus

Collectively, these muscles are called "hamstrings."

1. Biceps femoris (Figs. 6-30 and 6-31). This double-headed muscle **arises** by its long head from the ischial tuberosity, while the short head arises from the linea aspera (Figs. 6-24 and 6-3). Both heads join to **insert** onto the head of the fibula (Fig. 6-5). Since the direction of pull of the biceps muscle is posterior to the transverse axes of both the knee and hip joints, its **main action** is to flex the leg and extend the thigh. In addition, it will rotate the leg laterally. The **nerve supply** to the biceps is the ischiadic nerve; the long head is innervated by the tibial portion of this nerve while the short head is innervated by the common peroneal portion. This difference in nerve supply is due to the fact that the short

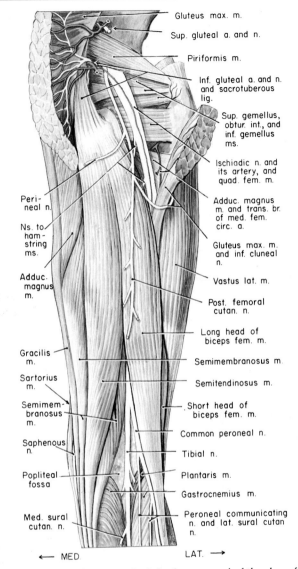

Figure 6-30. *Posterior thigh I, after removal of the deep fascia (right side).*

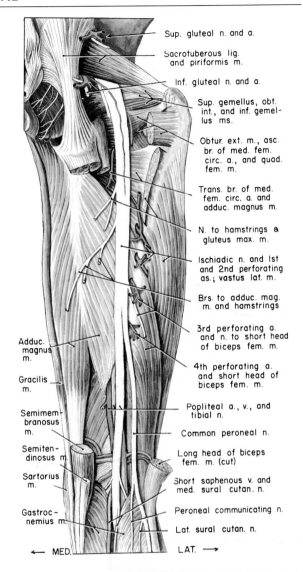

Figure 6-31. *Posterior thigh II, after removal of the hamstring muscles except for the short head of the biceps femoris (right side).*

head has been derived from the same antecedents as the gluteus maximus muscle. In fact, the nerve to the short head may arise from the inferior gluteal nerve.

2. Semitendinosus (Fig. 6-30). This muscle is located on the posterior surface of the semimembranosus and **arises** from the ischial tuberosity close to the point where the long head of the biceps arises (Fig. 6-24). Its long tendon **inserts** onto the medial surface of the tibia

just inferior to the condyle. It is in close association with the tendons of insertion of the gracilis and sartorius muscles being just posterior to them (Fig. 6-5). Its **main action** is to flex and medially rotate the leg and to extend the thigh, actions easily understood if related to the axes of the lower limb (Fig. 6-20 on page 367). **Nerve supply** is the ischiadic.

3. Semimembranosus (Fig. 6-30). This muscle

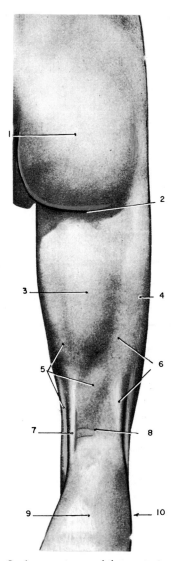

Figure 6-32. *Surface anatomy of the posterior side of thigh:*
(1) buttock, (2) gluteal fold, (3) semitendinosus muscle, (4)
iliotibial tract, (5) semimembranosus muscle, (6) biceps femoris
muscle, (7) semitendinosus muscle, (8) posterior surface of
knee (popliteal area), (9) gastrocnemius muscle, (10) calf.
(From Appleton, Hamilton, and Tchaperoff, Surface and
Radiological Anatomy, *courtesy of W. Heffer & Sons, Ltd.)*

lies deep to the semitendinosus, **arises** from the ischial tuberosity (Fig. 6-24), and **inserts** on a groove on the medial condyle of the tibia (Fig. 6-6). Its **main action** is similar to the semitendinosus in that it flexes and medially rotates the leg, and extends the thigh. The **nerve supply** is the ischiadic.

NERVES

The nerves found on the posterior surface of the thigh are (1) the posterior femoral cutaneous and (2) the ischiadic nerve and its branches.

    1. Posterior femoral cutaneous. This nerve (S1, S2, S3) enters the gluteal region through the greater ischiadic foramen in close association with the large ischiadic nerve. It courses inferior to the piriformis muscle, anterior or deep to the gluteus maximus, and superficial to the obturator internus, the superior and inferior gemelli, and the quadratus femoris muscles. After giving off its **inferior cluneal and perineal branches,** it continues into the thigh where (Fig. 6-11) it lies deep to the deep fascia, penetrating this fascia just proximal to the popliteal fossa. It gives off branches to the skin of the entire posterior surface of the thigh and continues through the popliteal region to reach the skin of the posterior surface of the leg.

    2. Ischiadic nerve. This large nerve (L4, L5; S1, S2, S3) enters the gluteal region through the greater ischiadic foramen and courses inferiorly between the gluteus maximus muscle posteriorly and the obturator internus, with its superior and inferior gemelli, and the quadratus femoris muscles anteriorly. It then escapes from the gluteal region and enters the thigh (Fig. 6-30) where it comes into relation with the hamstring muscles posteriorly and the adductor magnus muscle anteriorly. It is actually in closer relation to the biceps femoris than the semimembranosus or semitendinosus since the biceps crosses the nerve from medial to lateral. Although the division into its terminal branches, the **common peroneal and tibial nerves,** can occur anywhere along the thigh, the usual arrangement is that the ischiadic nerve gives off direct branches to the semitendinosus, semimembranosus, and long head of the biceps; furthermore, the tibial portion gives a branch to the adductor magnus and the peroneal portion innervates the short head of the biceps.

ARTERIES AND VEINS

Although the posterior surface of the thigh contains no artery specifically associated with this region, other arteries do terminate here. These arteries are (1) the medial femoral circumflex, and (2) the perforating branches of the profunda femoris artery.

   1.   Medial femoral circumflex. This artery has been described previously. Its transverse branch (Fig. 6-31) enters the posterior part of the thigh at a point just inferior to the quadratus femoris muscle and just superior to the adductor magnus muscle. It descends on the posterior surface of the latter muscle and gives off branches to the hamstrings.

   2.   The perforating arteries (Figs. 6-18 and 6-31). These vessels enter the posterior region of the thigh by penetrating the tendon of the adductor magnus muscle close to the femur. There are usually four such arteries. They give numerous branches to the hamstring muscles and then terminate in the vastus lateralis muscle.

   As mentioned previously, the first perforating artery joins with the inferior gluteal artery to form a vertical anastomotic channel that anastomoses with a horizontal vascular channel made up of the medial and lateral femoral circumflex vessels. This **cruciate anastomosis** (Fig. 6-33) serves an important role in getting blood around a tie of the femoral artery if it occurs proximal to the point where the profunda femoris artery arises. It is difficult, however, to find all four vessels taking part in this anastomosis.

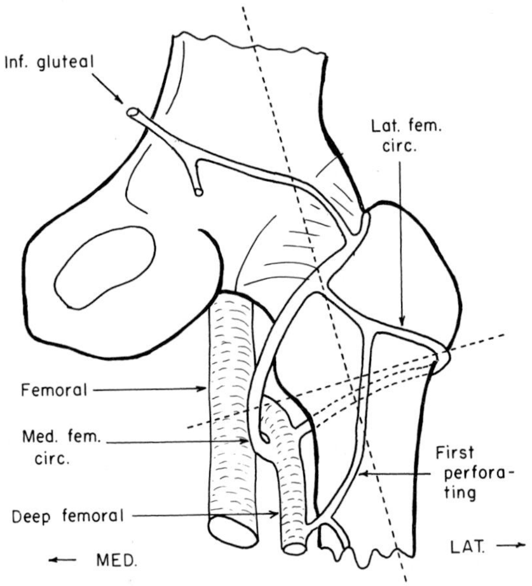

   There are **veins** that correspond to these arteries.

   Many of the structures found on the posterior side of the thigh are shown in Figure 6-21, a cross section of the thigh. Study of this picture at this time should be helpful.

Figure 6-33. *Diagram of the cruciate anastomosis, the upright and transverse portions of the cross being indicated by dotted lines (right side).*

POPLITEAL FOSSA AND POSTERIOR LEG

SURFACE ANATOMY

The hollow area that appears on the posterior surface of the knee is the popliteal space (Fig. 6-32). It is bounded by two tendons superiorly placed and two muscle bellies inferiorly placed, which in a thin, muscular individual are easily seen. Those superiorly located are the tendon of the biceps femoris muscle laterally (Fig. 6-34B) and the semimembranosus and semitendinosus tendons medially (Fig. 6-34A). Inferiorly, the popliteal space is bounded by the two heads of the gastrocnemius muscle. The sartorius and gracilis tendons are more anteriorly placed on the medial side. The elevation caused by the belly of the semimembranosus muscle can be seen in the superior aspect of the popliteal fossa.

   In the leg, the gastrocnemius muscle shows in muscular individuals; and the point where the muscle bellies end and the tendon starts is easily discernible (Fig. 6-34A and B).

GENERAL DESCRIPTION

If the skin is removed from the posterior surface of the knee and leg, one vein and several cutaneous nerves become evident (Fig. 6-35). The vein is the **short saphenous,** which starts on the lateral side of the foot

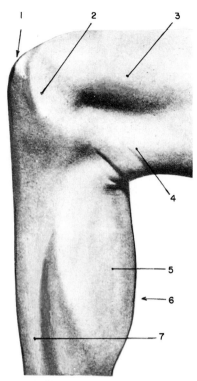

Figure 6-34A. *Surface anatomy of the medial side of the right knee: (1) patella, (2) medial epicondyle, (3) vastus medialis muscle, (4) sartorius, gracilis, and semitendinosus muscles, (5) gastrocnemius muscle, (6) calf, (7) subcutaneous surface of tibia. (From Appleton, Hamilton, and Tchaperoff,* Surface and Radiological Anatomy, *courtesy of W. Heffer & Sons, Ltd.)*

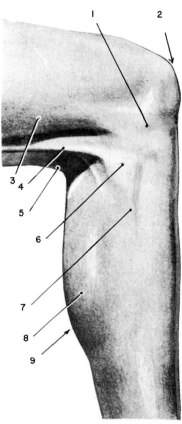

Figure 6-34B. *Surface anatomy of the lateral side of the right knee: (1) knee joint, (2) patella, (3) iliotibial tract, (4) biceps femoris muscle, (5) semitendinosus, (6) head of fibula, (7) peroneus longus muscle, (8) gastrocnemius muscle, (9) calf. (From Appleton, Hamilton, and Tchaperoff,* Surface and Radiological Anatomy, *courtesy of W. Heffer & Sons, Ltd.)*

near the fourth and fifth digits. It courses the length of the foot, turns superiorly at a point posterior to the lateral malleolus, and continues up the posterior side of the leg. It penetrates the deep fascia just inferior to the popliteal region and joins with the popliteal vein in the popliteal fossa. This is an important vein as it is subject to varicosities, as is the long saphenous vein, which is found on the anteromedial surface of the leg.

The cutaneous nerves in this region are the **lateral sural** (*L.,* calf) **cutaneous** and **peroneal communicating branches** of the common peroneal nerve; the **medial sural cutaneous** of the tibial nerve; branches from the **posterior femoral cutaneous nerve,** which reach into the region of the leg; the terminal branches of the **medial cutaneous nerve of the thigh;** and even some of the branches of the **saphenous nerve.** These nerves are ar-

ranged as seen in Figure 6-35. The peroneal communicating and medial sural cutaneous nerve usually join to form the **sural nerve,** which continues distally to the lateral side of the foot in company with the short saphenous vein.

If the deep fascia is removed, the muscles of this area are revealed. The **gastrocnemius** muscle is the most superficial in the leg, and its two heads of origin from the condyles of the femur are very obvious (Fig. 6-36). The **soleus** muscle is of large size but lies deep to the gastrocnemius; if a **plantaris** muscle exists, its slender tendon will be found to be located between gastrocnemius and soleus muscles in the upper part of the leg and along the medial side of the tendo calcaneus in the lower part (Fig. 6-37). The tendon of this plantaris muscle, when torn, becomes very annoying to the patient. The **popliteal space** will now be seen (Fig. 6-38) to be a diamond-shaped area bounded superiorly by the muscle belly of the semimembranosus medially and the biceps femoris laterally, and inferiorly by the two heads of the gastrocnemius muscle, with the plantaris muscle, if present, aiding in the formation of the lateral inferior boundary. Of the important structures in the popliteal fossa, the **nerves** are most superficially located (Fig. 6-38). This would be expected, since the nerves are coming down to this area from the thigh in a posterior location when compared to the vascular system. The division of the ischiadic nerve into the common peroneal and tibial is easily seen, and the branches already mentioned of the common peroneal (the lateral sural cutaneous nerve and the peroneal communicating) can be seen arising on the medial side of this nerve. The branches of the tibial nerve are equally obvious; the medial sural cutaneous nerve arises from the tibial nerve just before the nerve courses deep to the gastrocnemius muscle. It follows the same pathway as the short saphenous vein. The other branches of the tibial nerve in this location are motor branches to the plantaris muscle and to the two heads of the gastrocnemius. The tibial nerve disappears from view as it courses deep to the gastrocnemius muscle, while the common peroneal nerve disappears by piercing the peroneus longus muscle and winding around the fibula to the anterior surface of the leg.

If the nerves are pulled to one side, the main **popliteal vein** is revealed. It is anterior to the nerve but posterior to the popliteal artery. This position would be expected, since we found in the thigh that the vein coursed posterior to the artery as the artery and vein pierced the tendon of the adductor magnus to enter the popliteal fossa. Except for the short saphenous vein, the tributaries of the popliteal vein are similar to the branches of the popliteal artery which will be described in the next paragraph.

The femoral artery, as it courses through the tendon of the adductor magnus muscle, enters the popliteal fossa anterior to the vein and to the tibial nerve. This vessel is called the **popliteal artery** and is in contact with the capsule of the knee joint (Fig. 6-39). The branches of the artery appear on either side and are the **sural,** to the various muscles, and the five **genicular** branches, two of

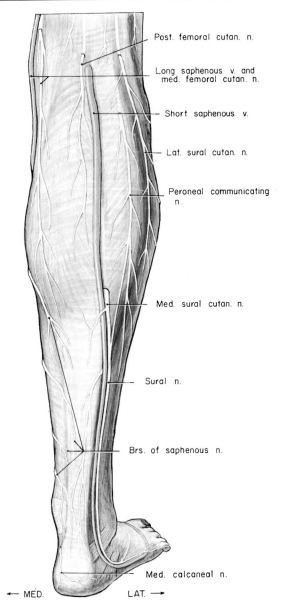

Post. femoral cutan. n.

Long saphenous v. and med. femoral cutan. n.

Short saphenous v.

Lat. sural cutan. n.

Peroneal communicating n.

Med. sural cutan. n.

Sural n.

Brs. of saphenous n.

Med. calcaneal n.

← MED.   LAT. →

Figure 6-35. *Cutaneous nerves and superficial veins of the posterior surface of the right leg.*

which course around the superior and two around the inferior aspect of the knee joint; the **middle genicular** enters the cavity of the knee joint directly. As the popliteal artery is followed inferiorly, the large **anterior tibial** artery is given off and disappears between the fibula and the tibia to gain the anterior surface of the leg.

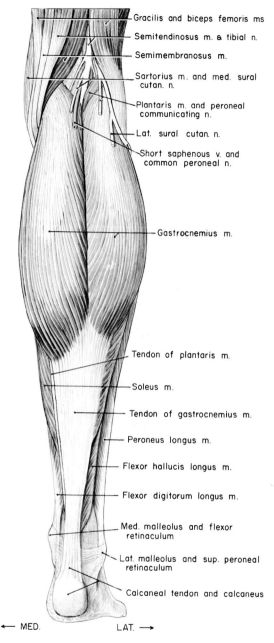

Gracilis and biceps femoris ms

Semitendinosus m. a tibial n.

Semimembranosus m.

Sartorius m. and med. sural cutan. n.

Plantaris m. and peroneal communicating n.

Lat. sural cutan. n.

Short saphenous v. and common peroneal n.

Gastrocnemius m.

Tendon of plantaris m.

Soleus m.

Tendon of gastrocnemius m.

Peroneus longus m.

Flexor hallucis longus m.

Flexor digitorum longus m.

Med. malleolus and flexor retinaculum

Lat. malleolus and sup. peroneal retinaculum

Calcaneal tendon and calcaneus

← MED.   LAT. →

Figure 6-36. Posterior leg I. All deep fascia has been removed as well as the cutaneous nerves and superficial veins (right leg).

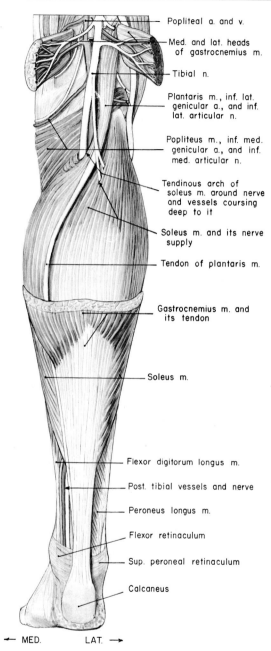

Popliteal a. and v.

Med. and lat. heads of gastrocnemius m.

Tibial n.

Plantaris m., inf. lat. genicular a., and inf. lat. articular n.

Popliteus m., inf. med. genicular a., and inf. med. articular n.

Tendinous arch of soleus m. around nerve and vessels coursing deep to it

Soleus m. and its nerve supply

Tendon of plantaris m.

Gastrocnemius m. and its tendon

Soleus m.

Flexor digitorum longus m.

Post. tibial vessels and nerve

Peroneus longus m.

Flexor retinaculum

Sup. peroneal retinaculum

Calcaneus

← MED.   LAT. →

Figure 6-37. Posterior leg II. The gastrocnemius muscle has been cut to reveal the soleus muscle (right leg).

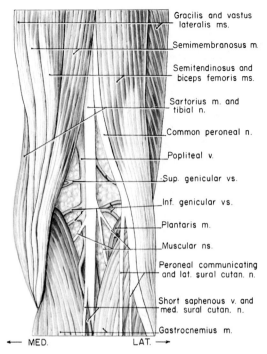

Gracilis and vastus lateralis ms.

Semimembranosus m.

Semitendinosus and biceps femoris ms.

Sartorius m. and tibial n.

Common peroneal n.

Popliteal v.

Sup. genicular vs.

Inf. genicular vs.

Plantaris m.

Muscular ns.

Peroneal communicating and lat. sural cutan. n.

Short saphenous v. and med. sural cutan. n.

Gastrocnemius m.

← MED.   LAT. →

Figure 6-38. Popliteal fossa I, after removal of the muscular fascia (right leg).

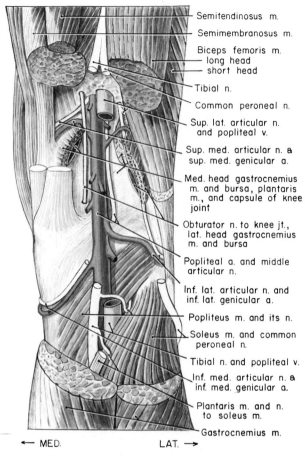

Semitendinosus m.

Semimembranosus m.

Biceps femoris m.
— long head
— short head

Tibial n.

Common peroneal n.

Sup. lat. articular n. and popliteal v.

Sup. med. articular n. a sup. med. genicular a.

Med. head gastrocnemius m. and bursa, plantaris m., and capsule of knee joint

Obturator n. to knee jt., lat. head gastrocnemius m. and bursa

Popliteal a. and middle articular n.

Inf. lat. articular n. and inf. lat. genicular a.

Popliteus m. and its n.

Soleus m. and common peroneal n.

Tibial n. and popliteal v.

Inf. med. articular n. a inf. med. genicular a.

Plantaris m. and n. to soleus m.

Gastrocnemius m.

← MED.   LAT. →

Figure 6-39. Popliteal fossa II, after removal of main nerves and veins. The hamstrings (except for the short head of the biceps) and the gastrocnemius muscles have been cut (right leg).

Another muscle appears deep to the artery; this is the **popliteus muscle,** which takes a position on the inferior aspect of the popliteal fossa and courses superiorly and laterally.

If the gastrocnemius, plantaris, and soleus muscles are removed from the posterior surface of the leg, the tibial nerve and the posterior tibial artery can be seen coursing inferiorly just superficial to the deep muscles on the posterior side of the leg (Fig. 6-40). They would thereby course between the soleus and these deep muscles. The muscles are the **tibialis posterior** in the center flanked by the **flexor hallucis longus** on the lateral side and the **flexor digitorum longus** on the medial side. Branches of the tibial nerve and the posterior tibial artery to these muscles can be seen quite easily. One large artery, the **peroneal,** courses to the lateral side of this area deep to the flexor hallucis longus. It appears again at the region of the ankle. All of the important nerves and muscles that are located on the posterior surface of the leg enter the plantar surface of the foot posterior to the medial malleolus. Usually, the muscles on the lateral side of

the leg are seen at this time as well; these are the **peroneus longus** and **brevis muscles.**

If one looks at the **medial side of the knee,** as shown in Figure 6-41, many of the structures already mentioned will be seen again. Starting anteriorly, the patellar ligament is evident, attaching to the anterior surface of the tibia. Posterior to this, the capsule of the knee joint is seen, and the **inferior and superior medial genicular arteries** coursing on the capsule of the joint. Posterior to that area is the **tibial collateral ligament** of the knee joint and further posterior the three tendons of the sartorius, the gracilis, and the semitendinosus mus-

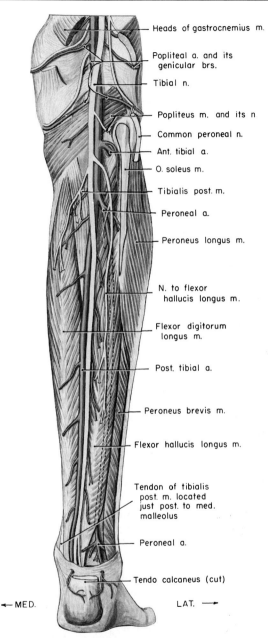

Heads of gastrocnemius m.

Popliteal a. and its
genicular brs.

Tibial n.

Popliteus m. and its n.

Common peroneal n.

Ant. tibial a.

O. soleus m.

Tibialis post. m.

Peroneal a.

Peroneus longus m.

N. to flexor
hallucis longus m.

Flexor digitorum
longus m.

Post. tibial a.

Peroneus brevis m.

Flexor hallucis longus m.

Tendon of tibialis
post. m. located
just post. to med.
malleolus

Peroneal a.

Tendo calcaneus (cut)

← MED.   LAT. →

*Figure 6-40. Posterior leg III. Gastrocnemius and soleus
muscles have been removed to reveal the three deep muscles of
this surface of the leg (right leg).*

cles. On the surface of these tendons are the **long
saphenous vein** and the **saphenous nerve.**

On the **lateral side of the knee joint** (Fig. 6-42), the
patellar ligament and the patella itself can be seen anteriorly. The iliotibial tract, which is just superficial to the
capsule of the knee joint, is just posterior to the patellar
ligament. The **superior and inferior lateral genicular**
arteries course on the capsule of the knee joint. Posterior
to this is the **fibular collateral ligament** and, posterior to
this in turn, the tendons of the biceps femoris muscle. The
common peroneal nerve can also be seen at this location
as it winds around the lateral head of the gastrocnemius
and the fibula to pierce the peroneus longus muscle and
gain the anterior surface of the leg.

MUSCLES

The muscles found on the posterior surface of the knee
and leg are:

1. Gastrocnemius
2. Soleus
3. Plantaris
4. Popliteus
5. Flexor digitorum longus
6. Flexor hallucis longus
7. Tibialis posterior

You will soon note that all muscles in the above list except
item 4 either attach or pass posterior to the transverse axis
of the ankle joint and will therefore plantar flex the foot.

   1. Gastrocnemius. This muscle is the most superficial muscle in the calf of the leg (Fig. 6-36). Its two
heads of origin form the inferior boundary of the popliteal
fossa. The **origin** of the lateral head is from the inferior
end of the supracondylar line and the lateral surface
of the lateral condyle, and the medial head from a
roughened spot just above the medial condyle (Fig.
6-3). Its **insertion** is with the common tendon of the soleus muscle, called the tendo calcaneus, onto the calcaneus (Fig. 6-8). Its **main action** is to plantar flex the
foot and secondarily to flex the leg. **Nerve supply** is the
tibial.

   2. The **soleus** muscle (Fig. 6-37) is almost entirely
covered by the gastrocnemius. It has an **origin** from the
posterior surface of the head and the upper third of the
shaft of the fibula, and also from a line on the upper third
of the tibia that corresponds to the edge of the popliteus

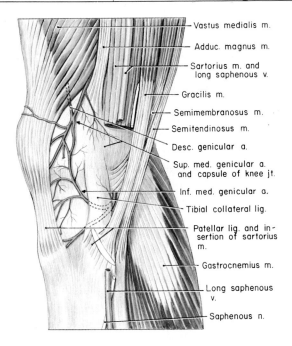

Figure 6-41. *Medial side of the right knee. The sartorius muscle has been cut.*

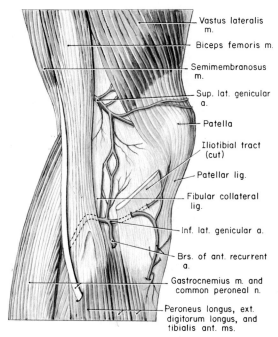

Figure 6-42. *Lateral side of the right knee. The iliotibial tract has been cut.*

muscle (Fig. 6-6). This line is also in relation to the flexor digitorum longus and the tibialis posterior muscles. Its **insertion** is onto the calcaneal bone by the tendo calcaneus (Fig. 6-8). Its **main action** is to plantar flex the ankle joint. **Nerve supply** is the tibial.

3. The **plantaris** muscle is frequently absent but when present lies just superior to the lateral head of the gastrocnemius muscle. Its tendon courses between the gastrocnemius and soleus muscles superiorly and along the medial side of the tendo calcaneus inferiorly (Fig. 6-37). It **arises** from the posterior surface of the lateral condyle (Fig. 6-3) and **inserts** onto the calcaneus. This interesting muscle may represent the proximal end of the flexor digitorum brevis muscle (confined in humans to the sole of the foot) as this muscle arises from the leg in some animals. **Nerve supply** is the tibial.

4. **Popliteus** (Fig. 6-39). This muscle lies in the floor of the popliteal fossa in contact with the capsule of the knee joint. Its **origin** is from the posterior surface of the upper end of the tibia (Fig. 6-6) and it is **inserted** onto the lateral condyle of the femur (Fig. 6-3). The tendon of this muscle is contained within the fibrous capsule of the knee joint. Its **main action** is to cause a rotation at the

knee joint and secondarily to flex the leg. This muscle is said to have an action in unlocking the knee joint when the latter moves from the locked extended position.* This locking mechanism, since the foot is planted on the ground, involves a rotation of the femur on the tibia; the lateral condyle moves anteriorly and therefore this is a medial rotation of the femur. When the popliteus muscle contracts, there is a lateral rotation of the femur, with the lateral condyle being pulled posteriorly. This process unlocks the knee joint in flexion. If the leg is raised, however, the popliteus muscle acts in a reverse direction and causes the leg to be rotated rather than the femur; the rotation is in a medial direction. **Nerve supply** is the tibial.

5. **Flexor digitorum longus** (Fig. 6-40). This muscle **arises** from the posterior surface of the tibia approximately in its middle third (Fig. 6-6). The tendon of this muscle passes posterior to the medial malleolus and enters the plantar surface of the foot. It breaks up into

---

*C. H. Barnett, 1953, Locking at the knee joint, *J. Anat.* 87: 91. The mechanics of this act are discussed. According to this paper, quadriceps is involved in locking, popliteus in unlocking.

four tendons for the lateral four digits and **inserts** onto the base of the distal phalanx of these digits (Fig. 6-9). Its **main action** is to flex the four lateral toes and to plantar flex the foot. **Nerve supply** is the tibial.

6. The **flexor hallucis longus** (Fig. 6-40) muscle **arises** from the middle and distal thirds of the fibula (Fig. 6-6); its tendon also passes posterior to the medial malleolus to enter the plantar surface of the foot, and it **inserts** on the distal phalanx of the great toe (Fig. 6-9). Its **main action** is to flex the great toe and plantar flex the foot. **Nerve supply** is the tibial.

7. **Tibialis posterior** (Fig. 6-40). This is the deepest muscle on the posterior surface of the leg and lies between the flexor digitorum longus and flexor hallucis longus. Its **origin** is from the posterior surface of the upper end of the tibia and the fibula, and the interosseus membrane between these two bony attachments (Fig. 6-6). The tendon of this muscle also passes posterior to the medial malleolus and enters the sole of the foot to have a very wide insertion (Fig. 6-9). Its **main insertion** is onto the navicular and medial cuneiform bones, but it also has attachments to all of the other tarsal bones except the talus and onto the bases of the second, third, and fourth metatarsal bones. We will see later that this wide

insertion aids in maintenance of the arches of the foot. Its **main action** is to invert (turning the plantar surface of the foot medially) and plantar flex the foot. **Nerve supply** is the tibial.

(Figure 6-43) depicts these muscles in cross section and also shows how the interosseus membrane and fascia divide the leg into compartments. Further details are given on page 444.

NERVES

Except for several cutaneous nerves and a branch of the obturator to the knee joint, the nerves in the popliteal fossa and posterior surface of the leg are branches of the **common peroneal** and the **tibial,** the terminal branches of the ischiadic nerve.

COMMON PERONEAL. This nerve arises from the ischiadic approximately in the middle of the thigh although this division may occur all the way up and down the thigh or even at the sacral plexus itself. The common peroneal nerve enters the popliteal space at its apex (Figs. 6-38),

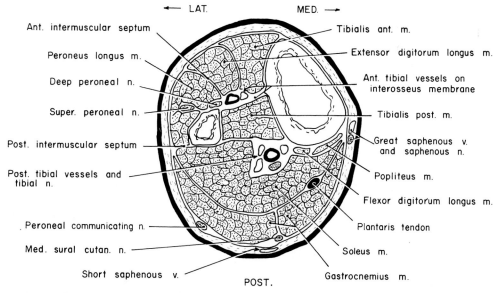

← LAT. MED. →

Ant. intermuscular septum
Peroneus longus m.
Deep peroneal n.
Super. peroneal n.
Post. intermuscular septum
Post. tibial vessels and tibial n.
Peroneal communicating n.
Med. sural cutan. n.
Short saphenous v.

POST.

Tibialis ant. m.
Extensor digitorum longus m.
Ant. tibial vessels on interosseus membrane
Tibialis post. m.
Great saphenous v. and saphenous n.
Popliteus m.
Flexor digitorum longus m.
Plantaris tendon
Soleus m.
Gastrocnemius m.

Figure 6-43. *Diagrammatic transverse section of the leg showing the intermuscular septa (right leg). You are looking superiorly. The section is made through the upper third, superior to the origin of flexor hallucis longus muscle.*

follows the medial side of the biceps femoris muscle, crosses the plantaris, gastrocnemius, and soleus muscles, enters the peroneus longus, and winds around the head of the fibula to arrive on the anterior surface of the leg (Fig. 6-42).

The branches of the common peroneal nerve are:

1. Articular
2. Lateral sural cutaneous
3. Peroneal communicating
4. Superficial peroneal
5. Deep peroneal

1. The **articular branches** (Fig. 6-39) arise quite high in the thigh. Coursing inferiorly, they enter the popliteal fossa and form the **superior and inferior lateral articular** branches, which follow the superior and inferior lateral genicular arteries. A third articular nerve is called the **recurrent articular** and arises from the nerve near its termination into the superficial and deep peroneals. This nerve ascends along with the anterior recurrent artery through the tibialis anterior muscle and continues to the knee joint.

2. The **lateral sural cutaneous** (Figs. 6-35 and 6-38) arises in the popliteal fossa and continues inferiorly in the superficial fascia to supply the skin on the lateral part of the upper half of the leg.

3. The **peroneal communicating branch** (Figs. 6-35 and 6-38) also arises in the popliteal fossa, pierces the deep fascia, continues inferiorly, and usually joins with the medial sural cutaneous branch of the tibial nerve to form the **sural** nerve. This nerve then continues to the foot and ends on the lateral side near the fifth digit.

4 and 5. The two terminal branches—the **superficial and deep branches** of the common peroneal—will be described with the anterior surface of the leg.

The fact that the common peroneal nerve winds around the head of the fibula and is therefore quite superficially located on the lateral surface of the leg makes this nerve prone to injury. It is approximately at bumper height as far as cars are concerned, and frequently this nerve is injured when people are hit by automobiles. The nerve is crushed between the bumper of the car and the bony fibula.

TIBIAL. This nerve is the other terminal branch of the ischiadic and, as was true of the common peroneal nerve, may originate at any level of the thigh (Fig. 6-31). It continues directly inferiorly, enters the apex of the popliteal fossa and gains a posterolateral relationship to the popliteal vein and artery. It continues inferiorly, crosses to the medial side of the artery, crosses the popliteus muscle, and then continues inferiorly superficial to the tibialis posterior muscle and deep to the soleus muscle. It continues inferiorly on the superficial surface of the deep muscles and finally terminates at the level of the flexor retinaculum (a connective tissue membrane serving to hold tendons in place; see Fig. 6-36) by dividing into the **medial and lateral plantar nerves,** which continue to the plantar surface of the foot (note Fig. 6-55).

The branches of the tibial nerve are:

1. Articular
2. Muscular
3. Medial sural cutaneous
4. Medial calcaneal
5. Terminal—medial and lateral plantar

1. The **three articular branches** arise quite high up in the thigh and continue inferiorly, one branch joining each of the superior and inferior medial genicular arteries while the third continues directly inferiorly to enter the posterior surface of the knee joint (Fig. 6-39).

2. Muscular. As might be expected, the tibial nerve is motor to the gastrocnemius, plantaris, soleus, popliteus, tibialis posterior, flexor digitorum longus, and flexor hallucis longus muscles. The branches to the two heads of the gastrocnemius and to the plantaris arise in the middle of the popliteal fossa, while those to the soleus (a second branch arises more distally) and popliteus muscle arise just inferior to the space, deep to the gastrocnemius muscle. The nerve to the popliteus muscle winds around its inferior border to enter the deep surface. Branches to the deep muscles of the leg arise further distally in the leg proper.

3. The **medial sural cutaneous nerve** (Figs. 6-35 and 6-38) pierces the deep fascia, courses in company with the short saphenous vein, and usually joins with the peroneal communicating nerve previously described to form the **sural nerve.**

4. The **medial calcaneal nerve** is a cutaneous branch of the tibial that pierces the flexor retinaculum and innervates the skin on the plantar surface of the heel.

5. The **terminal branches** of the tibial nerve, the medial and lateral plantar nerves, will be described with the plantar surface of the foot.

ARTERIES

The arteries in the popliteal fossa and posterior leg are the popliteal and posterior tibial arteries and their branches.

POPLITEAL.   The **popliteal artery** starts at the superior end of the popliteal fossa where the femoral artery has pierced the tendon of the adductor magnus muscle. It continues inferiorly and divides into the anterior and posterior tibial arteries at the lower border of the popliteus muscle. In the popliteal fossa it is posterior to the femur superiorly, to the capsule of the knee joint in the middle of the fossa, and inferiorly it is posterior to the popliteus and tibialis posterior muscles. The popliteal vein is superficial (posterior) to the artery and the tibial nerve is, in turn, superficial to the vein.

The branches of the popliteal artery are:

1. Muscular
2. Cutaneous
3. Medial superior genicular
4. Lateral superior genicular
5. Middle genicular
6. Medial inferior genicular
7. Lateral inferior genicular

1.   The **muscular** arteries (Fig. 6-39) can be divided into two groups, a superior and an inferior or sural. The **superior** arteries are branches that course mainly to the inferior parts of the adductor magnus and hamstring muscles. These vessels anastomose with the fourth perforating branches of the profunda femoris and serve as a pathway around a tie of the femoral artery anywhere between the profunda superiorly and the descending genicular inferiorly. The **inferior** or **sural** branches supply the muscles inferior to the popliteal space.

2.   Two **cutaneous branches** arise from the popliteal artery, pierce the deep fascia, and supply the skin on the posterior surface of the leg. One of these cutaneous branches usually accompanies the short saphenous vein.

3 and 4.   The **superior genicular arteries** (Figs. 6-39, 6-41, and 6-42) can be described as a pair. They arise on either side of the popliteal vessel and wind around the femur immediately above its condyles. They reach the front of the knee joint in this manner and anastomose with the inferior genicular arteries, thereby forming a plexus on the deep side of the patellar tendon.

5.   The **middle genicular artery** arises just poste-

rior to the knee joint and immediately pierces the capsule to supply the ligaments and the synovial membrane inside the joint.

6 and 7.   The **inferior genicular arteries** (Figs. 6-39, 6-41, and 6-42) are two in number and arise from the popliteal artery at a position deep to the gastrocnemius muscle. The medial inferior genicular artery courses along the superior edge of the popliteus muscle before winding around to the anterior surface. The lateral inferior genicular winds around the knee joint just superior to the head of the fibula.

We have already seen that the descending (supreme) genicular branch of the femoral takes part in the anastomosis around the knee joint, as do the last perforating branch of the profunda and the descending branch of the lateral femoral circumflex. We will see momentarily that other branches also are involved (Fig. 6-44).

The popliteal artery divides into anterior and posterior tibial arteries at the inferior edge of the popliteus muscle. The **anterior tibial** courses for a very short distance until it reaches the tibialis posterior muscle, through which it passes to reach the anterior surface of the leg. Before it crosses the interosseus membrane, it gives off the **posterior recurrent artery** and the **circumflex fibular artery**. The posterior recurrent artery (Fig. 6-44) takes part in the anastomosis around the knee joint and gives a branch to the popliteus muscle. The circumflex fibular artery (Fig. 6-44) courses around the head of the fibula to reach the anterior surface and anastomoses with the lateral and medial inferior genicular arteries. Another artery takes part in the anastomosis around the knee joint which is a branch from the anterior tibial after it has reached the anterior side of the leg. This is the **anterior tibial recurrent** artery and will be described with the arteries on the anterior surface of the leg.

POSTERIOR TIBIAL.   The **posterior tibial artery** (Fig. 6-40) continues inferiorly in a plane deep to the soleus, gastrocnemius, and plantaris muscles and superficial to the tibialis posterior muscle. It courses inferiorly just medial to the tibial nerve and lies along the tendon of the tibialis posterior muscle between the flexor digitorum longus medially and the flexor hallucis longus laterally. It leaves the leg by passing posterior to the medial malleolus to gain the plantar surface of the foot.

The branches of the posterior tibial artery are:

1. Peroneal
2. Nutrient
3. Muscular
4. Posterior medial malleolar
5. Communicating
6. Medial calcaneal
7. Terminal—medial and lateral plantar

1. Peroneal (Fig. 6-40). The **peroneal** artery arises from the posterior tibial just inferior to its commencement and continues inferiorly deep to the flexor hallucis longus muscle. It continues deep to the muscle or in the substance of it to reach the calcaneal region on the lateral side of the foot, where it breaks up into two or three lateral calcaneal branches.

The branches of the peroneal artery are:

a. Muscular
b. Nutrient
c. Perforating
d. Communicating
e. Lateral calcaneal

(a) The **muscular** branches of this artery are distributed to the soleus, tibialis posterior, flexor hallucis longus, and the long and short peroneal muscles. (b) The **nutrient** artery is an artery to the fibula, while the (c) **perforating** artery pierces the interosseus membrane about two inches above the lateral malleolus to reach the anterior aspect of the leg. At this point it anastomoses with the anterior lateral malleolar artery and also the lateral tarsal arteries on the anterior surface of the foot (see Figs. 6-50 and 6-51). This artery is frequently enlarged and sometimes takes the place of the dorsalis pedis artery. (d) The **communicating** branch is simply a branch that joins with a communicating branch of the posterior tibial. (e) The **lateral calcaneal** arteries are the terminal branches of this artery and on the lateral side of the calcaneus communicate with other arteries around the ankle.

2. The **nutrient** artery. The nutrient branch of the posterior tibial is the nutrient vessel to the tibia.

3. The **muscular** branches of the posterior tibial are distributed to the soleus and the deep muscles on the posterior surface of the leg.

4. Posterior medial malleolar. This artery winds around the tibial malleolus and ends in the malleolar network.

5. The **communicating** branch is the branch that

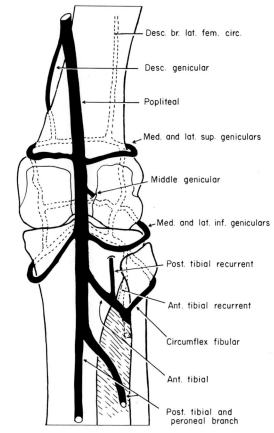

Figure 6-44. *Diagram of arterial anastomoses around the knee joint (right knee) from posterior aspect.*

Labels: Desc. br. lat. fem. circ.; Desc. genicular; Popliteal; Med. and lat. sup. geniculars; Middle genicular; Med. and lat. inf. geniculars; Post. tibial recurrent; Ant. tibial recurrent; Circumflex fibular; Ant. tibial; Post. tibial and peroneal branch

communicates with a similar branch of the peroneal artery.

6. Medial calcaneal. These arteries may be several in number and arise from the posterior tibial just before the artery divides into the medial and lateral plantar arteries, the terminal branches. These arteries take part in the anastomosis with the malleolar arteries and the calcaneal arteries.

7. The **terminal** branches are the medial and lateral plantar arteries, which will be described with the sole of the foot.

VEINS

The **popliteal** and **tibial** veins form a comitans, take the same pathway as the arteries, and have similar tributaries.

The popliteal vein is anterior or deep to the tibial nerve and posterior or superficial to the popliteal artery. The only thing that should be mentioned here is the short saphenous tributary to the popliteal vein; this branch starts near the fifth digit on the lateral side of the foot, continues superiorly on the posterior surface of the leg in the superficial fascia, pierces the deep fascia posterior to the popliteal vein, and empties into this vein. There is, therefore, a superficial and deep set of veins in the leg, as was also found in the thigh.

Figure 6-43 presents in cross section many of the structures found on the posterior surface of the leg.

ANTERIOR AND LATERAL ASPECTS OF LEG, DORSUM OF FOOT

GENERAL DESCRIPTION

Removal of the skin reveals the superficial veins and cutaneous nerves of this region (Fig. 6-45). The **saphenous vein** can be seen starting at the medial side of the foot, coursing anterior to the medial malleolus to reach the anterior surface of the leg, and finally winding around to the posterior surface just inferior to the knee. The saphenous vein is accompanied by the **saphenous nerve,** a branch of the femoral, and this nerve can be followed inferiorly to the medial surface of the foot. The upper part of the anterior surface of the leg, on the lateral side, is innervated by branches of the **lateral cutaneous nerve** of the calf, while the inferior part is supplied by the **superficial peroneal nerve,** which pierces the deep fascia approximately two-thirds of the way down the leg. The ramifications of this nerve give cutaneous supply to the dorsum of the foot and to all sides of the digits except between the big toe and second digit. This particular area, the lateral side of the big toe and the medial side of the second digit, is innervated by a terminal branch of the **deep peroneal nerve.**

Removal of the skin from the anterior surface of the leg and the dorsum of the foot also reveals thickenings in the deep fascia around the ankle joint (Fig. 6-46). These

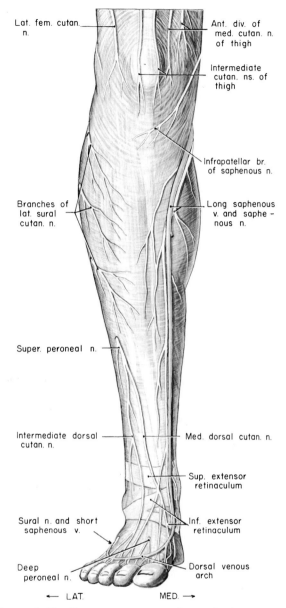

Figure 6-45. *Cutaneous nerves and superficial veins on the anterior leg and dorsum of the foot (right leg).*

thickenings are the **superior and inferior extensor retinacula** and the **superior and inferior peroneal retinacula.** The **superior extensor retinaculum** forms a bandage-like band across the extensor tendons while the **inferior extensor retinaculum** has a Y-shaped arrangement with the stem of the Y laterally placed. These retinacula are extremely important in holding the extensor tendons in position. The **superior and inferior peroneal retinacula** hold the tendons of the peroneus longus and brevis muscles in place (Fig. 6-47). The superior peroneal retinaculum is very close to the lateral malleolus, while the inferior seems to be a continuation of the inferior extensor retinaculum and is located inferior to the lateral malleolus.

Removal of the deep fascia reveals the muscles in this region. Most laterally placed are the two evertors of the foot, the **peroneus longus and brevis.** The muscle belly of the peroneus longus is located in the superior aspect of the leg while that of the peroneus brevis is more inferiorly placed (Fig. 6-48).

Lateral to the "shin bone" but in the midline of the leg are the tibialis anterior muscle and the extensor muscles of the digits. The most medial muscle is the **tibialis anterior** and next to it the **extensor digitorum longus.** The **extensor hallucis longus** is deep to these muscles (Fig. 6-46). The **superficial peroneal nerve** courses between the muscles on the lateral surface of the leg and those on the anterior surface; the nerve actually courses between the peroneus longus and brevis laterally and the extensor digitorum longus medially. The **anterior tibial artery** and the **deep peroneal nerve** are located deep to the muscles on the anterior surface of the leg and are actually in contact with the **interosseus membrane** between the fibula and tibia (Fig. 6-48).

Proceeding inferiorly, if the superior and inferior extensor retinacula are removed, the tendons of the muscles on the anterior compartment of the leg can be seen to separate from each other, the tibialis anterior coursing to the medial side of the sole of the foot, the extensor hallucis longus to the first digit, and the extensor digitorum longus to the last four digits. Between the last two muscles, the extensor hallucis longus and extensor digitorum longus, is the **dorsalis pedis artery** accompanied by the **deep peroneal nerve** (Fig. 6-46). The tendons of the peroneus muscles course posterior to the lateral malleolus to continue into the sole of the foot (Fig. 6-47). The **perforating branch of the peroneal artery** appears be-

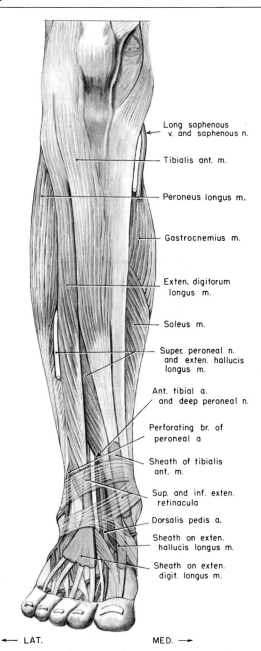

Long saphenous
v. and saphenous n.

Tibialis ant. m.

Peroneus longus m.

Gastrocnemius m.

Exten. digitorum
longus m.

Soleus m.

Super. peroneal n.
and exten. hallucis
longus m.

Ant. tibial a.
and deep peroneal n.

Perforating br. of
peroneal a

Sheath of tibialis
ant. m.

Sup. and inf. exten.
retinacula

Dorsalis pedis a.

Sheath on exten.
hallucis longus m.

Sheath on exten.
digit. longus m.

← LAT.   MED. →

Figure 6-46. *Anterior surface of right leg and dorsum of right foot I, after removal of cutaneous nerves, superficial veins, and deep fascia. Note the retinacula—thickenings in the deep fascia that hold tendons in place.*

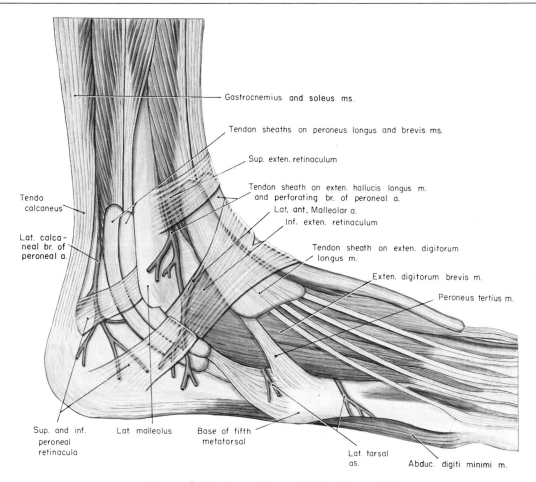

Gastrocnemius and soleus ms.

Tendon sheaths on peroneus longus and brevis ms.

Sup. exten. retinaculum

Tendon sheath on exten. hallucis longus m.
and perforating br. of peroneal a.

Lat. ant. Malleolar a.

Inf. exten. retinaculum

Tendon sheath on exten. digitorum longus m.

Exten. digitorum brevis m.

Peroneus tertius m.

Tendo calcaneus

Lat. calca-neal br. of peroneal a.

Sup. and inf. peroneal retinacula

Lat. malleolus

Base of fifth metatarsal

Lat. tarsal as.

Abduc. digiti minimi m.

Figure 6-47. *Lateral view of right ankle and foot.*

tween the lateral malleolus and the lateral border of the extensor digitorum longus; this artery has numerous branches to the ankle region. Further dissection will reveal the **extensor digitorum brevis muscle** (Fig. 6-48) which lies deep to the extensor digitorum longus and sends tendons to join with those of the latter muscle. In addition, the **dorsal interossei muscles** are visible between the metatarsal bones (Fig. 6-50).

SURFACE ANATOMY

The surface anatomy of the anterior and lateral aspects of the leg and the dorsum of the foot is not particularly revealing (Fig. 6-49). Palpation, however, does show that the anterior border of the tibia is palpable and the medial

surface of the tibia is not covered by muscles for almost the entire extent of the leg. Muscle masses can be felt on either side of the "shin bone," the ones laterally placed being the tibialis anterior, the extensor digitorum longus (the extensor hallucis longus is deep to these muscles), and the peroneus brevis and longus muscles.

If we follow the anterior and lateral muscles inferiorly, their tendons can be felt. In addition to the tendons of the extensor digitorum longus to the last four digits and the peroneus longus and brevis muscles coursing posterior to the lateral malleolus, the tendon of the extensor hallucis longus can also be seen. The medial malleolus is readily visible and if tension is put on the tibialis anterior and posterior muscles, the tendons of these muscles are seen, the tibialis anterior coursing an-

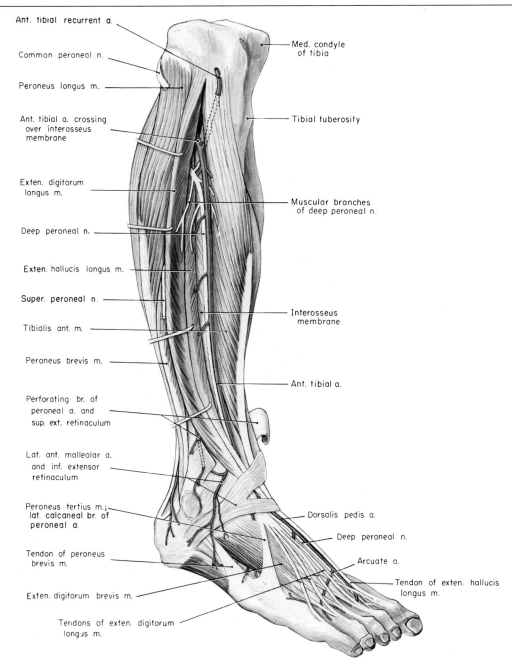

Ant. tibial recurrent a.

Common peroneal n.

Peroneus longus m.

Ant. tibial a. crossing
over interosseus
membrane

Exten. digitorum
longus m.

Deep peroneal n.

Exten. hallucis longus m.

Super. peroneal n.

Tibialis ant. m.

Peroneus brevis m.

Perforating br. of
peroneal a. and
sup. ext. retinaculum

Lat. ant. malleolar a.
and inf. extensor
retinaculum

Peroneus tertius m.;
lat. calcaneal br. of
peroneal a.

Tendon of peroneus
brevis m.

Exten. digitorum brevis m.

Tendons of exten. digitorum
longus m.

Med. condyle
of tibia

Tibial tuberosity

Muscular branches
of deep peroneal n.

Interosseus
membrane

Ant. tibial a.

Dorsalis pedis a.

Deep peroneal n.

Arcuate a.

Tendon of exten. hallucis
longus m.

Figure 6-48. *Anterior surface of right leg and dorsum of right foot II. The muscles have been spread apart to reveal the deeper-lying nerves and arteries.*

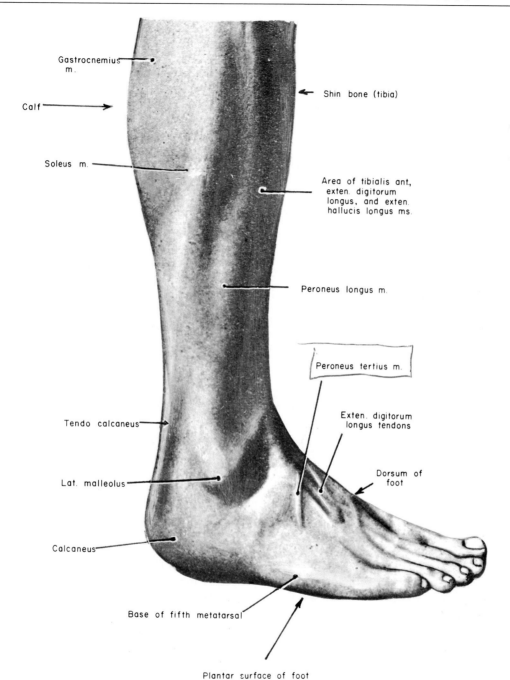

Figure 6-49.   *Surface anatomy of the lateral side of the leg and lateral side of the foot. (From Appleton, Hamilton, and Tchaperoff,* Surface and Radiological Anatomy, *courtesy of W. Heffer & Sons, Ltd.)*

terior to the medial malleolus while that of the posterior is very close to the malleolus and posterior to it.

The base of the fifth metatarsal is palpable on the lateral side of the foot and the bone just posterior to that is the cuboid. The phalanges of the digits are obvious.

MUSCLES

The muscles on the anterior and lateral aspects of the leg and the dorsum of the foot are:

1. Tibialis anterior
2. Extensor digitorum longus
3. Extensor hallucis longus
4. Peroneus tertius
5. Peroneus longus
6. Peroneus brevis
7. Extensor digitorum brevis
8. Dorsal interossei

You will soon note that 1 to 4 in the above list pass anterior to the transverse axis of the ankle joint and therefore dorsiflex the foot while 5 and 6 pass posterior to this axis and plantar flex the foot.

1. Tibialis anterior (Figs. 6-46 and 6-48). This muscle is the largest of those on the front of the leg and is just lateral to the tibia. Its **origin** is from the superior two-thirds of the lateral surface of the tibia (Fig. 6-5) and it **inserts** onto the medial cuneiform bone and the base of the first metatarsal. Since the direction of pull passes anterior to the transverse axis of the ankle joint, and it attaches to the medial side of the foot, its main action is to dorsiflex and invert the foot. **Nerve supply** is by the deep peroneal branch of the common peroneal.

2. Extensor digitorum longus (Figs. 6-46 and 6-48). This muscle lies just lateral to the tibialis anterior. It **arises** from the superior two-thirds of the fibula (Fig. 6-5). It courses inferiorly and breaks up into four tendons that **insert** onto the middle and distal phalanges of the lateral four toes. Because of its location on the dorsal surface of the toes and its passing anterior to the transverse axis of the ankle, its **main action** is to extend the toes and secondarily to dorsiflex the foot. **Nerve supply** is the deep peroneal.

3. Extensor hallucis longus (Fig. 6-48). This muscle is located deep to the two preceding muscles (tibialis anterior and extensor digitorum longus) but can be seen between these muscles if they are separated. It **arises** entirely from the fibula (Fig. 6-5), approximately from its middle third. (Observe that the flexor hallucis longus and the extensor hallucis longus both arise from the fibula. This helps in remembering the rather confusing difference between the muscles on the posterior surface of the leg and those on the anterior surface.) Its tendon has to cross the anterior tibial artery and the deep peroneal nerve to reach the large toe, where it **inserts** onto the base of the distal phalanx. Its **main action** is to extend the great toe and secondarily to dorsiflex the foot, obvious actions since it is located on the dorsal surface of the great toe and passes anterior to the transverse axis of the ankle joint. **Nerve supply** is the deep peroneal.

4. Peroneus tertius (Figs. 6-48 and 6-49). This muscle, though mainly involved in everting the foot, should be described at this time since it receives its **nerve supply** from the deep peroneal nerve along with the muscles just described. It is really a continuation of the extensor digitorum longus; it is frequently absent. Its **origin** is from the distal end of the fibula midway between medial and lateral surfaces (Fig. 6-5). The tendon of this muscle accompanies that of the extensor digitorum longus but finally **inserts** onto the base of the fifth metatarsal bone. Its **main action** is to aid the peroneus longus and brevis in everting the foot but it also plays a secondary role in dorsiflexion of the foot.

5. Peroneus longus (Figs. 6-48 and 6-49). This muscle is located on the lateral side of the leg and the muscle belly is superiorly located while the muscle belly of its partner, the peroneus brevis, is more inferiorly placed. Its **origin** is from the superior two-thirds of the fibula (Fig. 6-5). Its tendon courses inferiorly and winds around the posterior side of the lateral malleolus. It continues in a groove on the plantar surface of the cuboid bone, crosses the foot, and is **inserted** onto the lateral sides of the base of the first metatarsal and the medial cuneiform. (This muscle should be compared with the tibialis anterior, which has insertions on the medial side of the foot just opposite the above attachments.) Because it inserts on the lateral side of the foot and passes posterior to the transverse axis of the ankle, the **main action** of the peroneus longus muscle is to evert the foot and secondarily to plantar flex it. **Nerve supply** is by the superficial peroneal.

6. Peroneus brevis (Figs. 6-47 and 6-48). This muscle is at a deeper plane than the longus and actually anterior to it. It **arises** from the distal end of the fibula

(Fig. 6-5) and its tendon courses posterior to the lateral malleolus, being held in place by the superior and inferior peroneal retinacula. It is **inserted** onto the lateral side of the tuberosity on the base of the fifth metatarsal bone. Its **main action** is to evert the foot and secondarily to plantar flex the foot, actions that are obvious when its attachments and relation to the transverse axis of the ankle are considered. **Nerve supply** is the superficial peroneal.

7. Extensor digitorum brevis (Fig. 6-48). This muscle has its bony origin from the dorsal surface of the calcaneus (Fig. 6-8). It divides into four fleshy bellies and sends tendons across the dorsum of the foot to end on the first four digits—the big toe and the first three small toes. The tendon for the big toe **inserts** onto the dorsum of the base of the proximal phalanx quite independently (it is sometimes called the **extensor hallucis brevis** muscle), while the other three tendons join the tendons of the extensor digitorum longus. The **main action** of this muscle is obviously to extend the first four digits. **Nerve supply** is the deep peroneal.

8. Dorsal interossei (Fig. 6-50). These will be described with the ventral interossei in the section on the plantar surface of the foot.

Whenever a tendon passes over or around a bone, the bone and the tendon are protected by the presence of **tendon sheaths.** These sheaths are double-layered sacs, the inner layer intimately fused with the tendon while the outer layer is separated from the inner by synovial fluid. The sheaths in reality are tubes, the inner layer being continuous with the outer layer at the ends.

On the anterior and lateral sides of the ankle (Figs. 6-46 and 6-47), there is a large sheath for the extensor digitorum longus and peroneus tertius, and a separate sheath for each of the extensor hallucis longus, tibialis anterior, peroneus longus, and peroneus brevis tendons.

These tendon sheaths are of considerable importance; they are subject to infection, and such an infection can spread rapidly throughout the sheath although the sheath acts as a limitation to further spread.

NERVES

The nerves on the anterior surface of the leg and dorsum of the foot are the saphenous, lateral sural cutaneous, superficial peroneal, and deep peroneal nerves.

SAPHENOUS. This nerve, a branch of the femoral, accompanies the long saphenous vein (Fig. 6-45). It first appears in the anterior surface of the leg as it winds around the medial side of the leg just inferior to the knee. It continues inferiorly to finally reach the medial side of the foot. It provides cutaneous innervation to the anteromedial surface of the leg and ankle region.

LATERAL SURAL CUTANEOUS. This is a branch of the common peroneal nerve that arises in the popliteal fossa. It gives cutaneous innervation to the anterolateral surface of the proximal end of the leg (Fig. 6-45).

SUPERFICIAL PERONEAL. This nerve, one of the terminal branches of the common peroneal, pierces the peroneus longus muscle as it winds around the fibula and then descends close to the bone in the groove between the peroneus brevis laterally and the extensor digitorum longus and extensor hallucis longus muscles medially (Fig. 6-46). It pierces the deep fascia in the distal third of the leg and then divides into its terminal branches.

The branches of the superficial peroneal nerve are:

1. Muscular
2. Cutaneous
3. Terminal
   a. Medial dorsal cutaneous
   b. Intermediate dorsal cutaneous

1. The **muscular** branches are given off to the peroneus longus and brevis muscles as the nerve passes these particular muscle bellies.

2. Cutaneous. These are small branches to the lateral side of the leg.

3. Terminal. The superficial peroneal nerve terminates on the distal part of the leg by dividing into its terminal branches—the medial and intermediate dorsal cutaneous nerves.

a. The **medial dorsal cutaneous** nerve (Fig. 6-45) courses inferiorly on the dorsum of the foot and divides into digital branches that innervate the skin on the dorsal surface of the medial side of the great toe and contiguous sides of the second and third toes as far as the base of the first phalanx. (Note that this omits contiguous sides of the first and second toes, an area innervated by the deep peroneal nerve.)

b. The **intermediate dorsal cutaneous** nerve (Fig. 6-45) also proceeds distally on the dorsum of the foot and divides into digital branches. These innervate the dorsal surface of contiguous sides of the third and fourth, and fourth and fifth toes over the proximal and middle phalanges. (Note that this omits the lateral side of the fifth digit, which is innervated by the sural nerve. Occasionally this nerve covers the entire fifth digit and half of the fourth; see page 387 and Fig. 6-35 on page 388.)

DEEP PERONEAL. This nerve, the other terminal branch of the common peroneal, also pierces the peroneus longus muscle but additionally pierces the extensor digitorum longus muscle as it continues inferiorly (Figs. 6-48 and 6-50); it joins the anterior tibial artery remaining on its lateral side. It continues inferiorly, passing deep to the extensor retinacula, continues into the foot alongside the dorsalis pedis artery, and divides into its medial and lateral terminal branches.

The branches of the deep peroneal nerve are:

1. Muscular
2. Articular
3. Terminal
   a. Lateral terminal
   b. Medial terminal

1. The **muscular** branches in the leg innervate the tibialis anterior, extensor digitorum longus, extensor hallucis longus, and peroneus tertius muscles.

2. Articular. This is a small branch to the ankle joint.

3. The deep peroneal nerve ends on the dorsum of the foot by dividing into medial and lateral **terminal** branches (Figs. 6-45 and 6-50).

a. Lateral terminal branch. This nerve courses laterally in a plane deep to the extensor digitorum brevis muscle, innervates this muscle, and continues to give sensory supply to the joints of the foot.

b. The **medial terminal branch** divides and sends branches to the dorsal surface of contiguous sides of the first and second toes over the proximal and middle phalanges. This nerve also sends a branch to the metatarsophalangeal joint of the first toe.

(Note that the cutaneous nerves to the dorsal surface of the digits do not innervate the nails. The ventral nerves innervate this region due to migration of this area from the ventral to the dorsal position. This is important to know when doing surgery in this area.)

In summary, the muscles on the anterior surface of the leg are innervated by the deep peroneal while those on the lateral surface are innervated by the superficial peroneal nerve. The superficial peroneal nerve gives cutaneous supply to the dorsum of the foot and digits except for the area between the great and second toes, which is innervated by the deep peroneal, and the lateral surface of the fifth toe, which is innervated by the sural nerve.

ARTERIES

The arteries on the anterior surface of the leg and dorsum of the foot are the anterior tibial (leg) and dorsalis pedis (foot) arteries and their branches.

ANTERIOR TIBIAL. This artery arises as one of the terminal branches of the popliteal close to the inferior border of the popliteus muscle (Fig. 6-40). The artery continues inferiorly and anteriorly and is surrounded by the two heads of origin of the tibialis posterior muscle. After giving off the posterior recurrent and the fibular circumflex arteries, it passes superior to the interosseus membrane to gain the anterior surface of the leg (Fig. 6-48). It continues inferiorly between the tibialis anterior and extensor digitorum longus muscles and on the interosseus membrane. It then comes in contact with the extensor hallucis longus and continues to the ankle, passing deep to the tendons of the extensor hallucis longus muscle and the extensor retinaculum. The deep peroneal nerve lies on its lateral side, a natural position for this nerve since it enters the anterior surface of the leg from a lateral position. The continuation of this artery is the dorsalis pedis.

The branches of the anterior tibial (after gaining the anterior side of the leg) are:

1. Anterior tibial recurrent
2. Muscular
3. Medial anterior malleolar
4. Lateral anterior malleolar

1. The **anterior tibial recurrent** (Fig. 6-48), which is given off just as the artery comes into the anterior part of the leg, pierces the tibialis anterior muscle and is distributed to the anterior surface of the knee; it takes part in the anastomosis around the knee joint.

2. Muscular. These are short branches to the adjoining muscles.

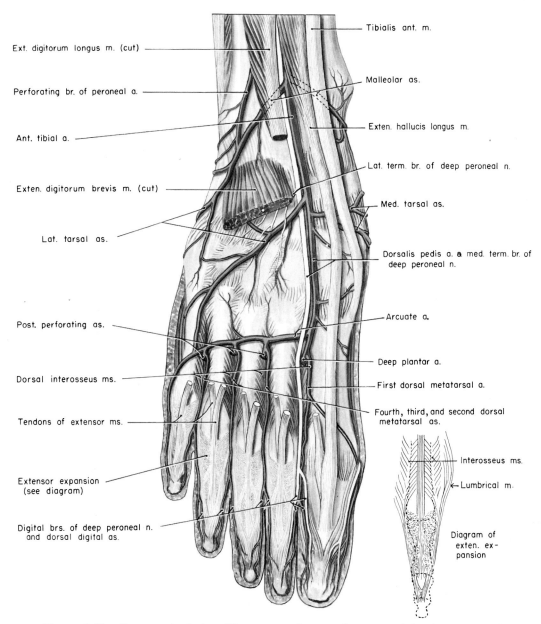

Tibialis ant. m.

Ext. digitorum longus m. (cut)

Malleolar as.

Perforating br. of peroneal a.

Exten. hallucis longus m.

Ant. tibial a.

Lat. term. br. of deep peroneal n.

Exten. digitorum brevis m. (cut)

Med. tarsal as.

Lat. tarsal as.

Dorsalis pedis a. a med. term. br. of deep peroneal n.

Arcuate a.

Post. perforating as.

Deep plantar a.

First dorsal metatarsal a.

Dorsal interosseus ms.

Fourth, third, and second dorsal metatarsal as.

Tendons of extensor ms.

Interosseus ms.

Extensor expansion (see diagram)

Lumbrical m.

Digital brs. of deep peroneal n. and dorsal digital as.

Diagram of exten. expansion

Figure 6-50.   *Dorsum of right foot. The extensor digitorum brevis muscle has been removed. The arcuate artery is often smaller than shown.*

3 and 4. The **medial** and **lateral anterior malleolar** arteries course deep to the tendons of the muscles on the anterior and lateral surfaces of the ankle joint and are distributed to the area of the two malleoli.

DORSALIS PEDIS. This artery (Figs. 6-50 and 6-51) is the continuation of the anterior tibial. It starts on the anterior surface of the ankle midway between the two malleoli. It continues inferiorly and anteriorly on the dorsum of the foot to end in the first intermetatarsal space, where it continues into the sole of the foot and terminates by joining the plantar arch.

The branches of this artery are:

1. Lateral tarsal
2. Medial tarsal
3. Arcuate
   a. Second dorsal metatarsal
   b. Third dorsal metatarsal
   c. Fourth dorsal metatarsal
4. First dorsal metatarsal
5. Deep plantar

1 and 2. The **tarsal** arteries course on the medial and lateral sides of the dorsum of the foot, deep to the extensor digitorum brevis muscles.

3. Arcuate. This artery, after sending a branch deep to the tendon of the extensor hallucis longus muscle to the medial side of the first toe, courses laterally across the bases of the metatarsal bones and gives off three branches, one to each of the second, third, and fourth metatarsal spaces. Each of these arteries divides into digital branches that supply contiguous sides of the dorsal surface of the respective digits. We will see later that each of these dorsal metatarsal arteries is joined to the plantar metatarsals by posterior and anterior perforating arteries. These are diagramed in Fig. 6-62.

4. The **first dorsal metatarsal** artery courses in the first metatarsal space terminating in digital branches for the contiguous sides of the first two toes.

5. Deep plantar. This is simply the continuation of the dorsalis pedis artery as it courses between the first two metatarsal bones to reach the deep plantar arch.

PERFORATING BRANCH OF PERONEAL. This artery appears on the lateral malleolus and its branches anastomose with the lateral malleolar branch of the anterior tibial artery.

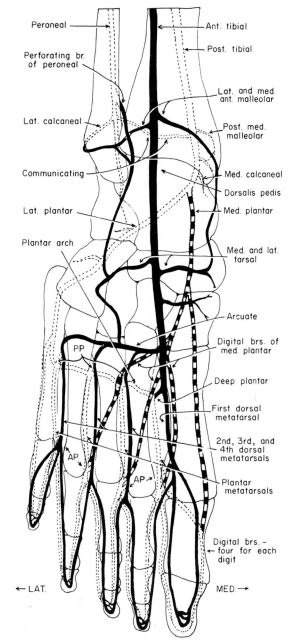

Figure 6-51. *Diagram of arteries in right foot: (AP) anterior perforating, (PP) posterior perforating. Medial plantar artery is often quite small.*

VEINS

There are corresponding veins for each of the arteries mentioned above.

(Figure 6-43) depicts many of the above structures in cross section.

PLANTAR SURFACE OF FOOT

SURFACE ANATOMY

The surface anatomy of the sole of the foot is not particularly revealing. The points that are most likely to rest on the ground when standing are the calcaneal region posteriorly and the pads over the heads of the metatarsals anteriorly. The tips of the toes only would touch the ground, because the toes are curved with a concavity facing inferiorly. The skin over the sole of the foot is extremely thick and serves as a protective pad. Structures coursing from the leg into the sole of the foot do so by passing posterior to the medial malleolus, which is quite obvious (Fig. 6-52).

GENERAL DESCRIPTION

If the skin is removed from the plantar surface of the foot, it will be seen to be covered with a comparatively thick layer of superficial fascia that serves as a cushion against the shock of repeatedly placing the foot on hard surfaces. The cutaneous nerves and vessels, which will be described momentarily, course in this layer to reach the skin.

If this superficial fascia is removed from the plantar surface of the foot, it will be seen that the muscles are covered with a thick layer of deep fascia (Fig. 6-53). This consists of three distinct parts: a medial part to cover muscles of the great toe, another laterally to cover those of the small toe, and a middle portion designated as the **plantar aponeurosis.** All attach posteriorly to the calcaneus and anteriorly to their respective digits, the plantar aponeurosis dividing anteriorly into five distinct strands that cover the tendons going to the respective digits. When the tendons reach the digits, they are contained within fibrous sheaths directly continuous with this plantar aponeurosis. They serve to hold the tendons in place and are thick between the joints but thinner at the joint for purposes of movement. Each tendon is surrounded by a synovial sheath inside this fibrous tube (Fig. 6-54).

The attachments of this deep fascia on the foot divide the foot into compartments, just as found in the hand. However, since the foot is described in terms of layers in this text, details on these compartments will be given after the section on muscles (page 417).

Several cutaneous nerves are visible at this stage in the dissection (Fig. 6-53). Posteriorly, the small **medial calcaneal nerve,** a branch of the tibial nerve, ramifies on the heel. **Several cutaneous branches** appear to the right and left of the plantar aponeurosis and are branches of the medial and lateral plantar nerves respectively. Further distally, the **first digital branch** of the medial plantar nerve pierces the deep fascia just medial to the plantar aponeurosis and is a nerve of considerable size. This is distributed to the medial side of the great toe. In addition, **digital branches** of the medial and lateral plantar nerves can be seen in each of the clefts for the digits, there being one in each cleft. These divide into branches for contiguous sides of the two toes on either side of this particular cleft.

If the deep fascia is entirely removed, as in (Figure 6-54) the first layer of muscles in the sole of the foot is revealed, plus the medial and lateral plantar arteries and nerves. The muscle in the midline is the **flexor digitorum brevis,** which has tendons of insertion on the last four digits. On the lateral side is a rather long and thin muscle, the **abductor digiti minimi,** and on the medial side the equally long **abductor hallucis muscle.** The medial and lateral plantar vessels course between these muscles. The **medial plantar artery** gives off, at approximately one-third of the distance from the heel, a **digital artery** that courses to the medial side of the great toe. The artery then continues and gives off **digital arteries** to the first, second, third, and part of the fourth digits. The **lateral plantar artery** courses in the groove between the flexor digitorum brevis and the abductor digiti minimi and its contributions to the digital arteries are also visible. The **medial plantar nerve,** with its digital branches coursing to the first, second, third, and one-half of the fourth digits, is also visible, as are the digital branches of the **lateral plantar nerve.**

If the muscles forming the superficial layer are removed, the muscles of the second layer are revealed (Fig.

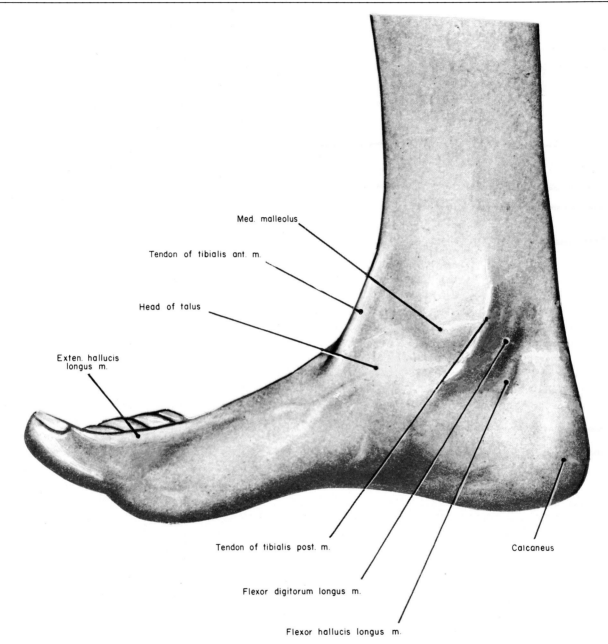

Med. malleolus

Tendon of tibialis ant. m.

Head of talus

Exten. hallucis longus m.

Tendon of tibialis post. m.

Flexor digitorum longus m.

Flexor hallucis longus m.

Calcaneus

Figure 6-52. *Surface anatomy of medial side of right ankle and foot. (From Appleton, Hamilton, and Tchaperoff,* Surface and Radiological Anatomy, *courtesy of W. Heffer & Sons, Ltd.)*

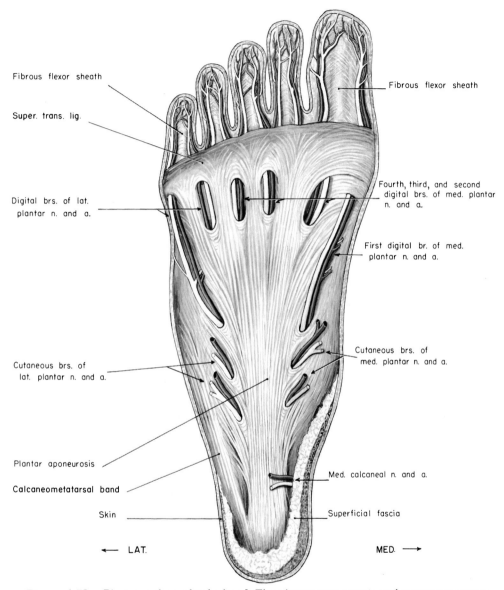

Fibrous flexor sheath

Super. trans. lig.

Digital brs. of lat. plantar n. and a.

Cutaneous brs. of lat. plantar n. and a.

Plantar aponeurosis

Calcaneometatarsal band

Skin

Fibrous flexor sheath

Fourth, third, and second digital brs. of med. plantar n. and a.

First digital br. of med. plantar n. and a.

Cutaneous brs. of med. plantar n. and a.

Med. calcaneal n. and a.

Superficial fascia

← LAT.   MED. →

Figure 6-53. *Plantar surface of right foot I. The plantar aponeurosis and cutaneous nerves and arteries.*

6-56). These are the **flexor hallucis longus tendon,** which was observed previously in the groove between the abductor hallucis and flexor digitorum brevis muscles; the **flexor digitorum longus tendon,** which can be seen entering the sole of the foot from the position posterior to the medial malleolus (Fig. 6-55) and dividing into four tendons for the last four digits; the **quadratus plantae muscle,** which extends along the length of the sole of the foot, coming from the calcaneal region on the lateral side and joining the tendon of the flexor digitorum longus; and also the **lumbrical muscles,** which are located between the respective tendons of the flexor digitorum longus

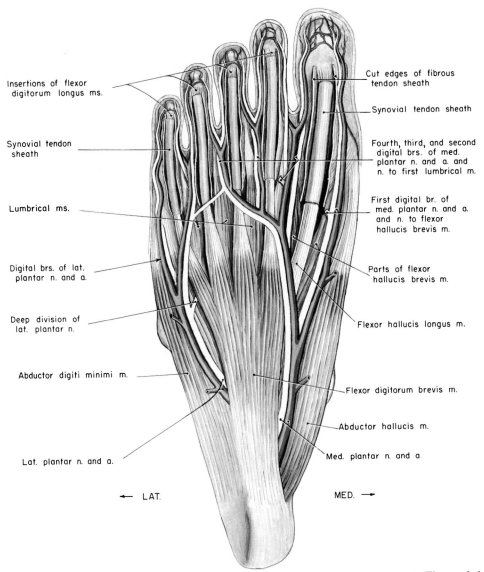

Insertions of flexor
digitorum longus ms.

Synovial tendon
sheath

Lumbrical ms.

Digital brs. of lat.
plantar n. and a.

Deep division of
lat. plantar n.

Abductor digiti minimi m.

Lat. plantar n. and a.

Cut edges of fibrous
tendon sheath

Synovial tendon sheath

Fourth, third, and second
digital brs. of med.
plantar n. and a. and
n. to first lumbrical m.

First digital br. of
med. plantar n. and a.
and n. to flexor
hallucis brevis m.

Parts of flexor
hallucis brevis m.

Flexor hallucis longus m.

Flexor digitorum brevis m.

Abductor hallucis m.

Med. plantar n. and a.

← LAT.   MED. →

Figure 6-54. Plantar surface of right foot II, after removal of plantar aponeurosis. The medial plantar artery is often smaller than shown.

muscles. Removing the first layer of muscles also has revealed the **lateral plantar nerve and artery** in their entire length crossing superficial to the quadratus plantae muscle; the deep branch of this nerve can be seen as well as the deep branch of the lateral plantar artery.

Removal of the second layer of muscles reveals the third layer (Fig. 6-57); these are the **flexor hallucis brevis, flexor digiti minimi brevis,** and **adductor hallucis muscles.** The flexor hallucis brevis is a two-headed muscle, because it has two heads of insertion onto the basal phalanx of the large toe. The flexor digiti minimi brevis is a very small muscle located on the lateral side of the foot, and the adductor hallucis muscle is a two-headed muscle consisting of oblique and transverse heads. The transverse head, consisting of four small muscle strands, is much smaller than the oblique head.

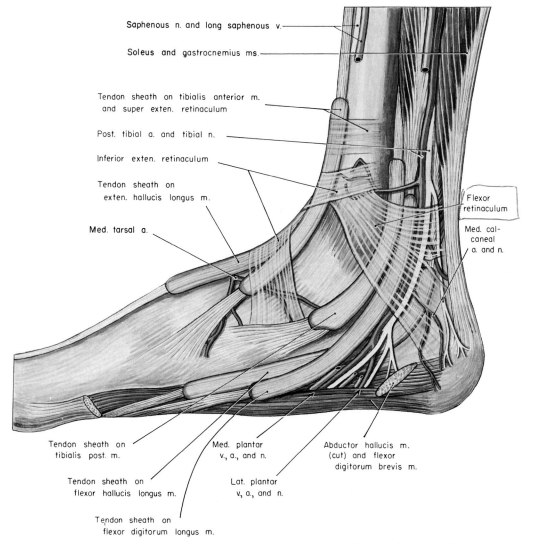

Saphenous n. and long saphenous v.

Soleus and gastrocnemius ms.

Tendon sheath on tibialis anterior m. and super exten. retinaculum

Post. tibial a. and tibial n.

Inferior exten. retinaculum

Tendon sheath on exten. hallucis longus m.

Med. tarsal a.

Flexor retinaculum

Med. calcaneal a. and n.

Tendon sheath on tibialis post. m.

Tendon sheath on flexor hallucis longus m.

Tendon sheath on flexor digitorum longus m.

Med. plantar v., a., and n.

Lat. plantar v, a., and n.

Abductor hallucis m. (cut) and flexor digitorum brevis m.

Figure 6-55. *Medial view of right ankle and foot showing the relations of structures between medial malleolus and calcaneus. Note tendon sheaths and retinacula.*

If the third layer of muscles is removed and a slit is made in the portion of the long plantar ligament that covers the tendon of the peroneus longus muscle, the fourth layer of muscles is revealed (Fig. 6-58) as consisting of the **interossei**, the **tibialis posterior tendon**, and the **peroneus longus tendon.** This dissection also reveals the **plantar arterial arch**, which courses on the inferior surface of the interossei muscles, and the **deep branch** of the lateral plantar nerve, which accompanies this arch

and innervates the interossei muscles as well as others. The rather extensive insertion of the peroneus longus muscle and the tibialis posterior muscle is revealed by this dissection, and it can easily be seen how these muscles form a sling that aids in support of the longitudinal arch of the foot when walking.

All these deeper muscles are located anteriorly on the foot. A large ligament, the long plantar, extends from these muscles (actually the cuboid and cuneiform bones)

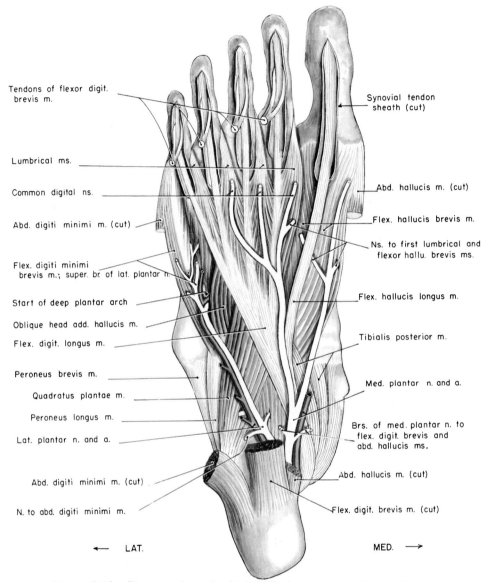

Tendons of flexor digit. brevis m.

Lumbrical ms.

Common digital ns.

Abd. digiti minimi m. (cut)

Flex. digiti minimi brevis m.; super. br. of lat. plantar n.

Start of deep plantar arch

Oblique head add. hallucis m.

Flex. digit. longus m.

Peroneus brevis m.

Quadratus plantae m.

Peroneus longus m.

Lat. plantar n. and a.

Abd. digiti minimi m. (cut)

N. to abd. digiti minimi m.

Synovial tendon sheath (cut)

Abd. hallucis m. (cut)

Flex. hallucis brevis m.

Ns. to first lumbrical and flexor hallu. brevis ms.

Flex. hallucis longus m.

Tibialis posterior m.

Med. plantar n. and a.

Brs. of med. plantar n. to flex. digit. brevis and abd. hallucis ms.

Abd. hallucis m. (cut)

Flex. digit. brevis m. (cut)

← LAT.   MED. →

Figure 6-56. *Plantar surface of right foot III, showing second layer of muscles.*

posteriorly to the calcaneus. Deep to the above are the ligaments connecting the bones of the foot, which will be described later.

MUSCLES

Although there is no necessity to remember the muscles in the sole of the foot by layers, it is convenient for descriptive purposes to divide them in this manner. Fur-thermore, muscles in each layer do have common fea-tures. (The origins of these short muscles in the foot are not as important for an understanding of their action as are the insertions. Furthermore, it is helpful to realize that the metatarsophalangeal joints are of the condyloid vari-ety and allow flexion, extension, abduction, and adduc-tion, and, therefore, circumduction. There is no rotation at these articulations. The interphalangeal joints are hinge joints and allow flexion and extension only.)

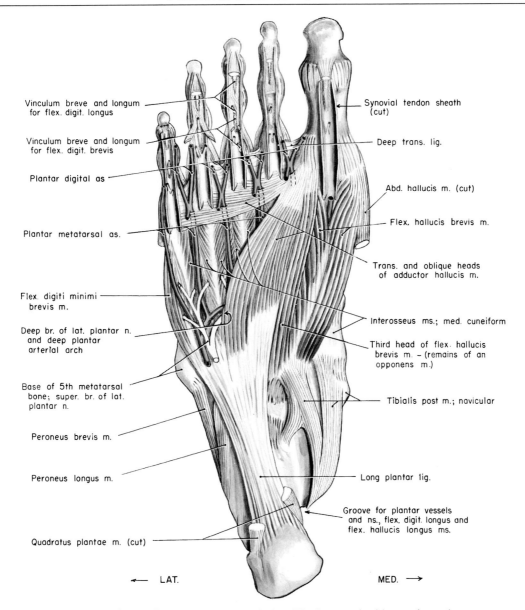

Vinculum breve and longum
for flex. digit. longus

Vinculum breve and longum
for flex. digit. brevis

Plantar digital as

Plantar metatarsal as.

Flex. digiti minimi
brevis m.

Deep br. of lat. plantar n.
and deep plantar
arterial arch

Base of 5th metatarsal
bone; super. br. of lat.
plantar n.

Peroneus brevis m.

Peroneus longus m.

Quadratus plantae m. (cut)

Synovial tendon sheath
(cut)

Deep trans. lig.

Abd. hallucis m. (cut)

Flex. hallucis brevis m.

Trans. and oblique heads
of adductor hallucis m.

Interosseus ms.; med. cuneiform

Third head of flex. hallucis
brevis m. – (remains of an
opponens m.)

Tibialis post m.; navicular

Long plantar lig.

Groove for plantar vessels
and ns., flex. digit. longus and
flex. hallucis longus ms.

←   LAT.   MED.   →

Figure 6-57. *Plantar surface of right foot IV, showing third layer of muscles.*

FIRST LAYER.   The muscles of the first layer are the (1) flexor digitorum brevis, (2) abductor hallucis, and (3) abductor digiti minimi. All three arise from the calcaneus; the flexor inserts on the middle phalanx while both abductors insert on the proximal phalanx.

   1.   Flexor digitorum brevis (Fig. 6-54).   This muscle is located centrally and **arises** from the calcaneus; after dividing into four bellies it **inserts** onto the middle

phalanges of the last four toes. Each tendon splits before attaching to the bone (Fig. 6-56); this allows the tendons of the flexor digitorum longus, which lie deep to those of the flexor digitorum brevis, to escape from this deep location and reach the distal phalanx. Its **main action** is to flex the middle and proximal phalanges of the last four toes. **Nerve supply** is by the medial plantar.

   2.   Abductor hallucis (Fig. 6-54).   This long, thin

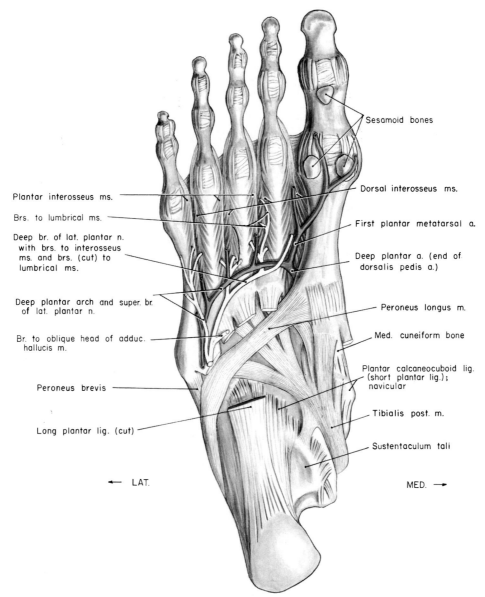

Plantar interosseus ms.

Brs. to lumbrical ms.

Deep br. of lat. plantar n.
with brs. to interosseus
ms. and brs. (cut) to
lumbrical ms.

Deep plantar arch and super. br.
of lat. plantar n.

Br. to oblique head of adduc.
hallucis m.

Peroneus brevis

Long plantar lig. (cut)

← LAT.

Sesamoid bones

Dorsal interosseus ms.

First plantar metatarsal a.

Deep plantar a. (end of
dorsalis pedis a.)

Peroneus longus m.

Med. cuneiform bone

Plantar calcaneocuboid lig.
(short plantar lig.);
navicular

Tibialis post. m.

Sustentaculum tali

MED. →

Figure 6-58. Plantar surface of right foot V. Note "slinglike" arrangements of the tendons.

muscle, located on the medial side of the sole of the foot, **arises** from the calcaneus and **inserts** onto the base of the proximal phalanx of the big toe on its medial side. Its **main action** is to abduct the toe and secondarily flex it. **Nerve supply** is by the medial plantar.

    3. Abductor digiti minimi (Fig. 6-54). This thin muscle **arises** from the calcaneus and is **inserted** onto the

lateral side of the base of the proximal phalanx of the little toe. It **abducts** the little toe and secondarily flexes it. **Nerve supply** is from the lateral plantar.

SECOND LAYER. The second layer of muscles in the sole of the foot consists of tendons of the flexor muscles located on the posterior surface of the leg, plus muscles attached to these tendons. They are the (1) flexor hallucis longus tendon, (2) flexor digitorum longus tendons, (3) quadratus plantae, and (4) lumbrical muscles.

1. Flexor hallucis longus tendon (Fig. 6-56). The muscles of the posterior leg that enter the plantar surface of the foot do so by passing posterior to the medial malleolus, between it and the heel (Fig. 6-55). These muscles, plus the posterior tibial artery and the tibial nerve, are held in place by the **flexor retinaculum.** We have already seen that the flexor hallucis longus arises from the fibula only and, therefore, is laterally placed. As we see it in the groove posterior to the medial malleolus, the tendon of the flexor hallucis longus maintains its natural position and is nearest the heel bone. However, to get from this position, which is most posterior, to the large toe, which is most medial, it has to cross the tendon of the flexor digitorum longus. It does this by passing deep (superior) to it. The tendon is quite superficially located and is actually in the groove between the abductor hallucis and flexor digitorum brevis muscles. This tendon **inserts** on the base of the distal phalanx of the great toe and its **main action** is to flex the toe and secondarily to plantar flex the foot. The tendon is surrounded by a synovial sheath and fastened down to the metatarsophalangeal joint by a fibrous sheath.

2. Flexor digitorum longus tendons (Figs. 6-55 and 6-56). This muscle is most medially placed in the leg but at the ankle is in the middle between the tendon of the tibialis posterior next to the malleolus and the tendon of the flexor hallucis longus next to the heel bone. In order to gain this central position, it has to cross the tendon of the tibialis posterior muscle. It then enters the sole of the foot, crossing superficial to the tendon of the flexor hallucis longus muscle. About in the middle of the foot, it divides into four tendons that continue to each of the last four digits **inserting** on the base of the distal phalanges. The tendons are surrounded by synovial sheaths and held down to the capsule of the metatarsophalangeal joints by fibrous sheaths. The **main action** of this muscle is to flex the toes and, secondarily, to plantar flex the foot.

3. Quadratus plantae (Fig. 6-56). This muscle lies in the central part of the sole of the foot deep to the flexor digitorum brevis. It is a double-headed muscle and has an **origin** from both the medial and lateral sides of the calcaneus, and also from adjacent ligaments. This muscle has a peculiar arrangement in that it **inserts** on the tendon of the flexor digitorum longus muscle. Its **main ac-**

tion is to induce a change in the direction of the pull of the flexor digitorum longus, because this tendon tends to cross the foot in an oblique fashion; the combined action of the quadratus plantae muscle and the flexor digitorum longus creates a pull on the digits that is directly posterior. By working independently, it can flex the distal phalanges of the last four digits. **Nerve supply** is from the lateral plantar.

4. Lumbrical muscles (Fig. 6-56). These are four slender worm-like muscles that **arise** from the four tendons of the flexor digitorum longus muscle and **insert** on the extensor tendons of the toes by passing medial to the respective metatarsophalangeal joints. The **main action** of these muscles is to flex the metatarsophalangeal joint and simultaneously, because of their attachment to the extensor tendons, to extend the two distal phalanges. The first lumbrical is **innervated** by the medial plantar nerve, while the other three are innervated by the lateral plantar nerve.

THIRD LAYER. The third layer of muscles is made up of three relatively short muscles, two for the great toe and one for the fifth digit. They are the (1) flexor hallucis brevis, (2) adductor hallucis, and (3) flexor digiti minimi brevis.

1. Flexor hallucis brevis (Figs. 6-56 and 6-57). This muscle is on the medial side of the foot and **arises** from the cuboid and lateral cuneiform bones, and also from the tendons of the posterior tibialis muscle. It divides into two distinct parts,* the medial part joining the abductor hallucis to **insert** on the medial side of the base of the first phalanx of the great toe. The other portion joins with the adductor hallucis muscle to **insert** on the lateral side of the base of the first phalanx. **Sesamoid bones** are located in each of these tendons. The tendon of the flexor hallucis longus muscle courses between these two attachments. The **main action** of this muscle is to flex the big toe. **Nerve supply** is the medial plantar.

2. Adductor hallucis (Fig. 6-57). This muscle has two heads of **origin,** a large oblique and a small transverse. The oblique head has its bony origin from the bases of the second, third, and fourth metatarsal bones,

---

*A third head occurs occasionally; this is considered to be remains of an opponens muscle. Such an arrangement is shown in Figure 6-57. The human being is the only primate that cannot bring the big toe across the plantar surface of the foot (opposition).

while the transverse head consists of four small strands of muscle that arise from the lateral four plantar ligaments of the metatarsophalangeal joints. The **main action** of these two heads of the adductor hallucis is, as the name implies, to adduct the large toe; secondarily, it will flex the toe as well. **Nerve supply** is lateral plantar.

3. Flexor digiti minimi brevis (Fig. 6-56) This small muscle has its bony **origin** from the base of the fifth metatarsal bone. It is **inserted** by a tendon onto the lateral side of the base of the proximal phalanx of the little toe in company with the abductor digiti minimi muscle. Its **main action** is to flex the small toe and secondarily to abduct it. **Nerve supply** is lateral plantar.

FOURTH LAYER. The fourth layer of muscles in the sole of the foot consists of (1) the interosseus muscles, so called because of their location between the metatarsal bones; (2) the tendon of the tibialis posterior muscle; and (3) the tendon of the peroneus longus muscle. Tendons of both muscles cross from one side of the foot to the other at a point approximately equidistant between the heel and the toes.

1. Interossei (Figs. 6-58, 6-59, and 6-60). These muscles are divided into two distinct sets, the dorsal interossei and the plantar interossei. They are at approximately the same plane, however, and **arise** from the sides of the metatarsal bones and **insert** onto the base of the first phalanx of the last four toes and also on the extensor tendons of these toes. The dorsal interossei arise from adjacent sides of the metatarsals between which they lie, while the plantar interossei arise from the plantar surfaces of the third, fourth, and fifth metatarsals only. There are, therefore, four dorsal and three plantar interossei. Their **main action** is to abduct and adduct the toes at the metatarsophalangeal joints; because of their attachment to the extensor tendons they have an additional action of extending the two distal interphalangeal joints. The dorsal interossei abduct, while the plantar interossei adduct. If the student knows that the axis of the foot is the second digit and that the dorsal interossei are abductors, the insertions need not be memorized. In order to move the second digit away from the axis, it has to be moved both toward the large toe and away from it. Therefore, a dorsal interosseus is needed on both sides of this particular toe. The third toe to be abducted must be moved away from the second toe, and the same is true of the fourth toe. Therefore, the two remaining dorsal in-

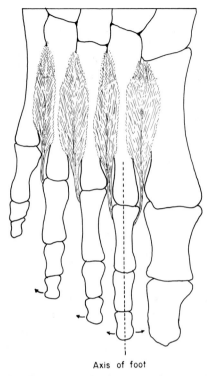

Figure 6-59. Diagram of the bony attachments of the dorsal interossei showing their action in abducting the toes.

terossei will insert on the lateral sides of these toes. This accounts for all four dorsal interossei. There is no need of a dorsal interosseus on the large toe or the little toe because each of these digits has its own abductor. Contrariwise, the plantar or ventral interossei adduct and, since the second digit is on the axis already, there is no way of getting it closer to it. Therefore, we need an interosseus muscle to insert on the medial side of the third toe, the medial side of the fourth toe, and to the medial side of the fifth; this accounts for all three plantar interossei, and all three move the toes toward the axis—the second toe. All interossei are innervated by the lateral plantar nerve.

2. Peroneus longus tendon (Fig. 6-58). The peroneus longus tendon enters the sole of the foot from the lateral side, curving in a groove on the cuboid bone. It then continues medially across the lateral and intermediate cuneiform bones and finally reaches the lateral side of the base of the first metatarsal and the medial cuneiform bone upon which the tendon **inserts**. It therefore goes

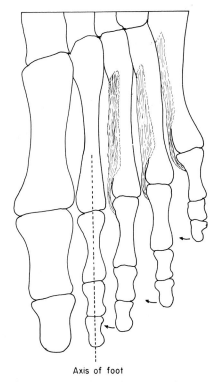

Axis of foot

Figure 6-60. Diagram of the bony attachments of the plantar interossei showing their action in adduction of the toes.

across the entire sole of the foot. The long plantar ligament forms a fibrous sheath for this tendon and holds it in place. As the tendon rounds the cuboid bone to enter the sole of the foot, there is a swelling that contains a **sesamoid bone.** This muscle, in addition to everting the foot, has a role in maintaining the arches of the feet, especially when walking.*

3. **Tibialis posterior tendon** (Fig. 6-58). This

---

*R. L. Jones, 1941. The human foot. An experimental study of its mechanics, and the role of its muscles and ligaments in the support c the arch, *Am. J. Anat.* 68: 1. Long muscles of the leg play a relatively minor role in maintenance of the arches according to this author, not functioning at all when a person is standing still. They do play a role, however, when walking, mainly to maintain proper balance of pressure on each metatarsal bone.

J. V. Basmajian and J. W. Bentzon, 1954, An electromyographic study of certain muscles of the leg and foot in the standing position, *Surg. Gynecol. Obstet.* 98: 662. Contrary to common opinion, the tibialis anterior and peroneus longus muscles in the leg, and the interossei muscles in the foot play no role in maintenance of the arches when a person is standing. The authors caution that they may play such a role when one is walking.

---

muscle occupies an intermediate position on the posterior surface of the leg, but at the ankle joint the tendon tends to become closest to the malleolus or most anterior in position. It enters the sole of the foot, winds around the talus, and then divides into its numerous slips of insertion. The **chief insertion** is onto the navicular and the medial cuneiform bones, but it also has insertion onto the plantar surfaces of the sustentaculum tali, the intermediate and lateral cuneiforms, the cuboid, and the bases of the second, third, and fourth metatarsals. The only bones it does not attach onto are the talus posteriorly and the first and fifth metatarsals anteriorly. In addition to inverting the foot, this muscle, because of its extensive insertion, has an influence in maintaining the arches of the foot when walking.† This tendon, combined with that of the peroneus longus, can easily be seen to form a sling for the inferior aspect of the foot.

COMPARTMENTS, CLEFTS, AND TENDON SHEATHS

Now that you have learned the muscles of the foot, the **fascial compartments** can be readily understood. Frequent reference to Figure 6-61 will be helpful. The arrangement of the deep fascia divides the foot into these compartments, which are important from the point of view of spread of infection. Those who have studied the hand will remember that there is a thenar compartment containing the muscles in that eminence, a hypothenar compartment for the muscles in that area, a midpalmar compartment for the tendons of the forearm muscles, and an adductor compartment for the adductor pollicis muscle. It is not too surprising to find that the foot has a similar arrangement.

The deep fascia forms **medial and lateral intermuscular septa** that divide the foot into a **medial compartment** for the muscles of the great toe (abductor hallucis, flexor hallucis longus and brevis), a **lateral compartment** for the muscles of the fifth digit (abductor digiti minimi, flexor digiti minimi brevis), a **central compartment** (containing the flexor digitorum longus tendons, flexor digitorum brevis, the quadratus plantae, and the lumbrical muscles), and a **deep compartment** (adductor hallucis and interosseus muscles).

---

†See preceding footnotes.

Clefts or spaces also occur in the foot as well as in the hand. They can be placed between muscle compartments or within them. Grodinsky* has described the following clefts in the sole of the foot; a **lateral cleft** between the muscles in the lateral compartment; a **medial cleft** between the muscles in the medial compartment; **three clefts** in the central compartment (one between the plantar aponeurosis and the flexor digitorum brevis, a second between the latter muscle and the quadratus plantae, and a third deep to this muscle); and another **cleft** between the adductor hallucis and interosseus muscles. (Note that a dorsal subaponeurotic cleft occurs deep to the extensor tendons.) Infections can spread throughout a cleft and thence into adjoining spaces, the exact spread having been determined by Grodinsky. These clefts are shown in Figure 6-61.

Infections can also occur in the **tendon sheaths** (Figs. 6-47, 6-54, and 6-55) and spread throughout a sheath with great rapidity. All tendons crossing the ankle joint possess such sheaths. Anteriorly on the dorsum of the foot are three sheaths, one for each muscle, i.e., the extensor hallucis longus, which reaches farther distally than the others, extensor digitorum longus, and tibialis anterior. Laterally there are two, one for the peroneus longus and another for the peroneus brevis. Medially there are three, one each for the flexor digitorum longus, flexor hallucis longus, and tibialis posterior. In addition, each set of flexor tendons to the digits is surrounded by synovial sheaths. That of the big toe reaches proximally as far as the shaft of its proper metatarsal, while those to the other digits reach only as far as the heads of these bones. Those who have studied the hand will note the difference between the structure of flexor tendon sheaths in the hand and those in the foot. Tenosynovitis in the foot is not the same clinical problem as seen in the hand.

NERVES

The tibial nerve, in the area between the medial malleolus and the calcaneus, divides into its two terminal branches, the medial and lateral plantar nerves.

---

*M. Grodinsky, 1929, A study of the fascial spaces of the foot and their bearing on infections, *Surg. Gynecol. Obstet.* 49: 737. Fascial spaces in the foot are described and potential spread of an infection from these spaces determined.

MEDIAL PLANTAR.   The **medial plantar nerve** (Figs. 6-54 and 6-56) corresponds to the median nerve in the hand. It continues into the sole of the foot deep to the abductor hallucis muscle and appears in the groove between this muscle and the flexor digitorum brevis, in company with the medial plantar artery and the tendon of the flexor hallucis longus muscle. It divides into digital branches close to the heel.

The branches of the medial plantar nerve are:

1. Four digital branches
2. Cutaneous branches to sole
3. Muscular

1.   Digital branches. One of these two main branches is long and slender and courses to the medial side of the large toe, while the other contributes three branches that head toward the webs between the first and second, second and third, and third and fourth digits, respectively. These digital nerves break up into branches that course to the contiguous sides of these digits.

2. The **cutaneous branches** to the sole consist of small branches to the medial side of the sole of the foot.

3.   Muscular. Branches arise from the trunk of the medial plantar nerve to innervate the **abductor hallucis and flexor digitorum brevis muscles.** In addition to being cutaneous, the first digital nerve gives rise to a muscular branch to the **flexor hallucis brevis** muscle, while the second digital branch supplies a motor branch to the **first lumbrical.**

The medial plantar nerve, therefore, is the motor nerve supply to the abductor hallucis, flexor digitorum brevis, the flexor hallucis brevis, and the first lumbrical, which means that all of the remaining muscles in the sole of the foot must have their nerve supply from the lateral plantar nerve. It should be noted that the medial plantar nerve gives the cutaneous nerve supply to the first, second, third, and one-half of the fourth digits as well as the medial side of the sole of the foot. Furthermore, these plantar digital nerves innervate the dorsal sides (the nails) of the distal phalanges.

LATERAL PLANTAR.   The **lateral plantar nerve** (Figs. 6-54 and 6-56) corresponds to the ulnar nerve in the hand and is also covered by the abductor hallucis muscle. It courses deep to the flexor digitorum brevis muscle to reach the groove between the latter muscle and the thin

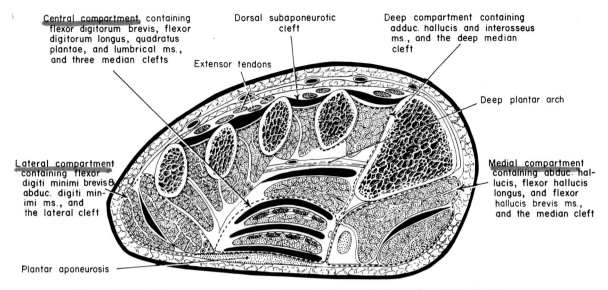

Central compartment containing flexor digitorum brevis, flexor digitorum longus, quadratus plantae, and lumbrical ms., and three median clefts

Dorsal subaponeurotic cleft

Deep compartment containing adduc. hallucis and interosseus ms., and the deep median cleft

Extensor tendons

Deep plantar arch

Lateral compartment containing flexor digiti minimi brevis & abduc. digiti min-imi ms., and the lateral cleft

Medial compartment containing abduc. hal-lucis, flexor hallucis longus, and flexor hallucis brevis ms., and the median cleft

Plantar aponeurosis

Figure 6-61. Transverse section of the right foot (you are looking toward the heel) showing the compartments (surrounded by dotted lines) and clefts. (Redrawn after Grodinsky, 1929, Surg. Gynecol., and Obstet. 49: 737, who calls the median clefts m1, m2, m3, and m4.)

and long abductor digiti minimi muscle. The lateral plantar nerve is accompanied by the lateral plantar artery. It courses superficial or inferior to the quadratus plantae muscle and deep or superior to the flexor digitorum brevis and abductor hallucis muscles. The lateral plantar nerve continues anteriorly and divides into superficial and deep branches.

The branches of the lateral plantar nerve are:

1. From the trunk
   a. Muscular
   b. Cutaneous
2. From superficial branch
   a. Muscular
   b. Cutaneous
3. Deep branch—muscular

1. From the trunk. The branches from the trunk are (a) to the quadratus plantae and to the abductor digiti minimi muscle; it also gives off (b) cutaneous branches to the lateral side of the sole.

2. The **superficial branch** immediately gives off twigs (a) to the flexor digiti minimi brevis muscle and to the third plantar and fourth dorsal interosseus muscles. (b) It then continues to the skin of the entire fifth digit and the lateral half of the fourth digit.

3. The **deep branch** (Figs. 6-57 and 6-58) is entirely muscular. It curves medially at a deep level and courses superficial to the interosseus muscles and the metatarsal bones and deep to the oblique head of the adductor hallucis muscle. It gives off twigs to all of the remaining interossei muscles, to the adductor hallucis and the lateral three lumbricals. This deep branch is accompanied by the deep plantar arterial arch.

The lateral plantar nerve, therefore, gives cutaneous nerve supply to the fifth digit and the lateral half of the fourth, and to the lateral side of the sole of the foot. These nerves also give cutaneous supply to the dorsal surface of the respective terminal phalanges. It is the motor nerve supply to the quadratus plantae, the abductor digiti minimi, the flexor digiti minimi brevis, all of the interosseus muscles, the adductor hallucis and the lateral three lumbricals. In other words, it is the nerve supply to all the muscles not innervated by the medial plantar nerve.

Those who have studied the upper limb will note the resemblance of the above to the nerve supply of the hand.

ARTERIES

The arteries to the sole of the foot are the medial and lateral plantars and their branches.

MEDIAL PLANTAR. The medial plantar artery (Fig. 6-54), often quite small, arises at the distal border of the flexor retinaculum as a terminal branch of the posterior tibial and courses in company with the medial plantar nerve, deep to the abductor hallucis muscle, to gain the groove between it and the flexor digitorum brevis muscle.

The branches of this artery are:

1. Muscular
2. Cutaneous
3. Digital branches that accompany the digital branches of the medial plantar nerve

1. Muscular branches are small branches to adjoining muscles.

2. Cutaneous arteries supply the skin on the medial side of the sole.

3. Digital branches have an arrangement very similar to the digital branches of the nerve. A long slender branch supplies the medial side of the great toe, and other digital branches supply both sides of the second, third, and the medial side of the fourth toe. These digital branches course toward the web between the toes and then split and give branches to contiguous sides.

LATERAL PLANTAR. The lateral plantar artery also arises deep to the flexor retinaculum as a terminal branch of the posterior tibial artery (Figs. 6-54 and 6-58) and courses deep to the abductor hallucis and flexor digitorum brevis muscle to reach the groove between the latter muscle and the abductor digiti minimi muscle. It accompanies the lateral plantar nerve. It continues anteriorly and then swings medially at a deeper level, forming the plantar arch. It finally ends at the base of the first metatarsal, where it joins the terminal end of the dorsalis pedis artery. This plantar arch is deep to the oblique head of the adductor hallucis muscle and superficial to the metatarsal bones and the interosseus muscles.

The branches of the artery are:

1. Muscular
2. Cutaneous
3. Medial calcaneal
4. Terminal—the deep plantar arch

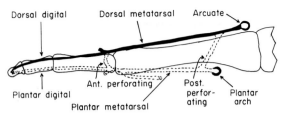

Figure 6-62. *Diagram of the perforating arteries in the foot.*

1. Muscular. These are short branches to muscles on the lateral side of the foot.

2. Cutaneous. Blood supply to the skin on the lateral side of the sole.

3. Medical calcaneal. This artery pierces the abductor hallucis to reach the plantar surface of the heel.

4. Plantar arch. The branches of the plantar arch are confusing. They are the (a) **perforating,** which pass through the lateral three intermetatarsal spaces to join the dorsal metatarsal arteries, and (b) **three plantar metatarsal** arteries, which course to the clefts between the lateral four toes and divide into a pair of **plantar digital** arteries that supply the adjacent sides of the toes. Near their points of division, each sends an **anterior perforating branch** to join the corresponding dorsal metatarsal artery. Furthermore, these arteries are joined by the **digital branches** of the medial plantar artery. This arrangement is diagramed in Figures 6-51 and 6-62. The **first plantar metatarsal artery** arises from the junction between the lateral plantar and the deep plantar arteries (termination of dorsalis pedis artery) and sends a digital branch to the medial side of the great toe. The digital branch for the lateral side of the fifth toe arises from the lateral plantar artery near the base of the fifth metatarsal bone.

It should be noted that the terminal branch of the dorsalis pedis artery, the deep plantar artery, forms one end of the plantar arterial arch while the other is formed by the lateral plantar artery. This means that there is a large anastomosis between the dorsalis pedis artery on the dorsum of the foot and the plantar arteries on the sole of the foot. In addition there are the perforating branches previously mentioned.

VEINS

The arteries in the sole of the foot are accompanied by corresponding veins.

JOINTS

The joints of the lower limb are quite important from a clinical point of view; they should be learned thoroughly. They are:

Sacroiliac
Pubic symphysis
Hip
Knee
Superior and inferior tibiofibular
Ankle
Intertarsal
Tarsal metatarsal
Intermetatarsal
Metatarsophalangeal
Interphalangeal

The sacroiliac joint and the pubic symphysis were previously described on page 280 under the section on the pelvis.

HIP JOINT

The hip joint is formed by the articulation of the **head of the femur** with the **acetabulum of the coxal bone.** This articulation is a synovial joint of the ball-and-socket variety, and all movements are allowed. They occur around three axes: transverse, vertical, and anteroposterior.

The **acetabulum** (Fig. 6-1) is not a smooth fossa; a centrally located pit (**acetabular fossa**) is surrounded by a lunar or C-shaped area, with its opening anteriorly and slightly inferiorly placed. The acetabular fossa is filled with fat. The rim of the acetabulum is deficient both anteriorly and inferiorly. This is the **acetabular notch;** it is bridged by the **transverse acetabular ligament,** which transforms the notch into an acetabular foramen; this foramen serves for the passage of blood vessels and nerves to and from the hip joint (Fig. 6-23).

The depth of the acetabulum is increased by the **labrum acetabulare,** which (in cross section) is a triangular piece of cartilage attached to the edges of the acetabulum and to the transverse ligament. This arrangement aids in keeping the head of the femur in the acetabulum. Another aid of this nature is the **ligamentum teres** of the femur, which is triangular in shape, with its

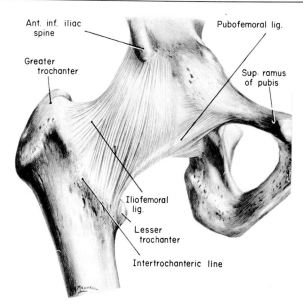

Figure 6-63. *Ligaments of the right hip joint—anterior view. (From Quain,* Elements of Anatomy, *courtesy of Longman Group Ltd.)*

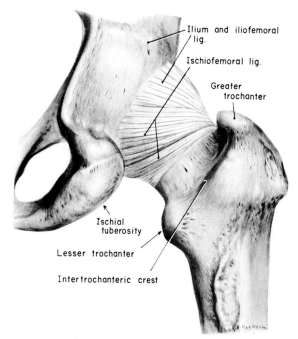

Figure 6-64. *Ligaments of the right hip joint—posterior view. (From Quain,* Elements of Anatomy, *courtesy of Longman Group Ltd.)*

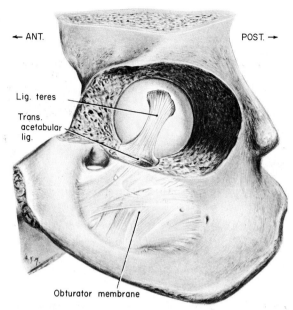

Figure 6-65. *Right hip joint—medial view showing attachments of the ligamentum teres. The acetabulum has been cut away. (From Quain,* Elements of Anatomy, *courtesy of Longman Group Ltd.)*

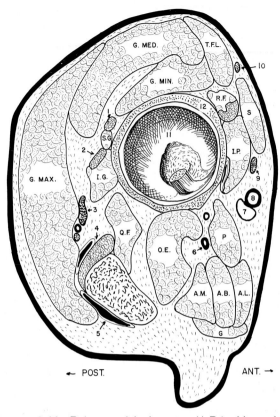

Figure 6-66. *Relations of the hip joint: (A.B.) adductor brevis, (A.L.) adductor longus, (A.M.) adductor magnus, (G) gracilis, (G. MAX.) gluteus maximus, (G. MED.) gluteus medius, (G. MIN.) gluteus minimus, (I.G.) inferior gemellus, (I.P.) iliopsoas, (O.E.) obturator externus, (P) pectineus, (Q.F.) quadratus femoris, (R.F.) rectus femoris, (S) sartorius, (S.G.) superior gemellus, (T.F.L.) tensor fascia lata, (1) piriformis tendon, (2) obturator internus tendon, (3) ischiadic nerve and inferior gluteal vessels, (4) hamstring tendons and bursa, (5) bursa between ischium and gluteus maximus muscle, (6) medial femoral circumflex artery, (7) femoral vein, (8) femoral artery, (9) femoral nerve, (10) lateral femoral cutaneous nerve, (11) acetabulum, (12) capsule of joint. (Redrawn from Jamieson.)*

base attached to the rim of the acetabulum and the transverse ligament, and its apex attached to a shallow pit on the head of the femur (Fig. 6-65).

The hip joint is surrounded by a **capsule,** which has several thickenings in the form of ligaments: (1) iliofemoral, (2) pubocapsular, and (3) ischiocapsular. The **capsular ligament** is quite strong, and particularly so anteriorly. As might be expected, its attachments to the acetabulum are to the bony rim superiorly and posteriorly, and to the labrum and transverse ligament anteriorly and inferiorly. The femoral attachments are to the intertrochanteric line, to the neck close to the trochanters, and to the back of the neck of the femur about ½ inch from the intertrochanteric crest. The fibers making up the capsule are both longitudinal and circular in arrangement. The longitudinal fibers on the anterior side of the capsule form the **iliofemoral ligament** (Fig. 6-63). It is triangular in outline with the base of the triangle attached to the intertrochanteric line and the apex to the iliac portion of the acetabular rim; it is also attached to the inferior surface of the anterior inferior iliac spine. The **pubofemoral and ischiofemoral ligaments** (Figs. 6-63 and 6-64) are

not real ligaments but simply thickenings in the capsule. They are attached to the pubic and ischial portions of the acetabular rim and pass into the capsule of the joint blending with the fibers of same. These ligaments do not actually reach the femur.

The assumption of the upright posture has twisted the capsule and ligamentous thickenings around the hip joint. This arrangement results in considerable freedom

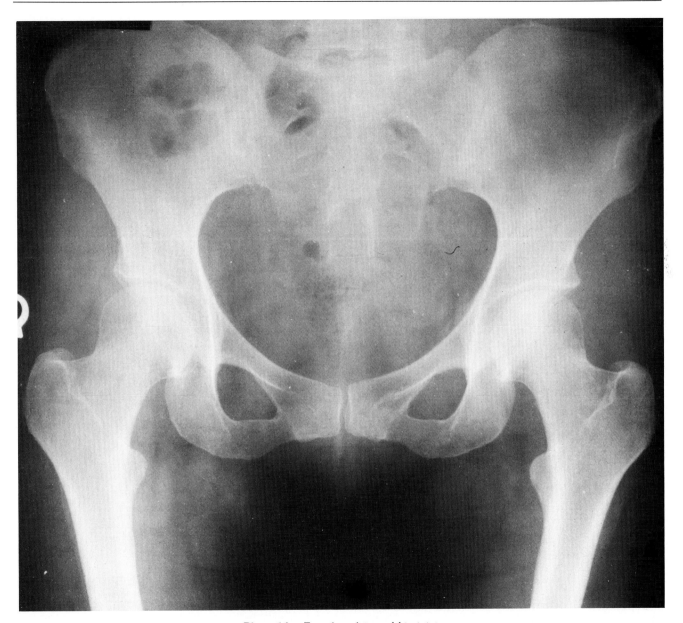

Plate 14. *Female pelvis and hip joints.*

for flexion of this joint but limits extension. If one bends forward and obtains a four-footed stance a greater range of action is allowed.

The head of the femur is covered with cartilage except at the point where the ligament is attached, and the lunate surface of the acetabulum is also covered with cartilage.

The hip joint is lined with a **synovial membrane** that lines the insides of the capsule. It is found on the surfaces of the labrum and is actually on the fat in the acetabular fossa. It is continued over the ligamentum teres, forming a tubular arrangement around this ligament, and is also continued onto the neck of the femur from the points where the capsule is attached to the femur.

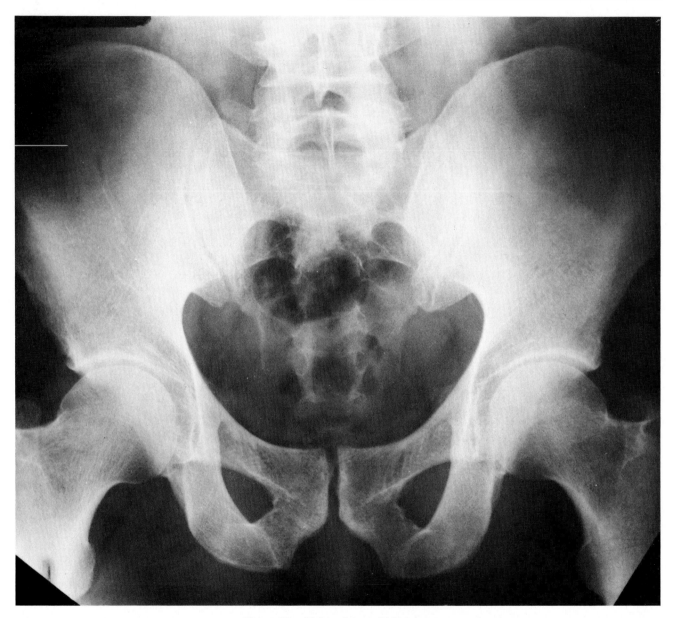

Plate 15.   *Male pelvis and hip joints.*

Bursae.   There are many bursae around the hip joint. One occurs superficially on the greater trochanter. Two are revealed by reflection of the gluteus maximus muscle, a large one between it and the greater trochanter and a small one between the muscle and the ischial tuberosity. There are two between the greater trochanter and the gluteus medius muscle, and one between this bony structure and the gluteus minimus muscle. Others are found between the capsule of the hip joint and various muscles; one for the piriformis, two for the obturator internus, and a very important one deep to the iliopsoas tendon—the iliopectineal bursa. These locations for bursae should not be surprising, for they are likely to occur at any location where a muscle comes into contact with bone or the joint capsule while contracting. Some bursae communicate with the synovial cavity of the joint while others do not.

The hip joint has a **blood supply** made up of branches from the superior and inferior gluteal, the medial and lateral circumflex, and the obturator arteries. As mentioned under the description of the femur, special attention should be given to the blood supply to the head of the femur, a region subject to fracture in many older patients. The main source of arterial supply is not the small branches in the ligamentum teres femoris,* nor from the gluteal branches, but from the branches of the medial and lateral femoral circumflex that ascend through the capsule to reach the bone. The **nerves** are branches from the femoral, the obturator, and also from the nerve to the quadratus femoris muscle.

There are many muscles that have an intimate relation to the capsule of the joint and that aid in maintaining the head of the femur in the acetabulum (Fig. 6-66). Superiorly are the gluteus medius and minimus; anteriorly the rectus femoris, pectineus, and iliopsoas; inferiorly the adductor muscles and obturator externus; and posteriorly the piriformis, superior gemellus, internal obturator, inferior gemellus, and quadratus femoris muscles. Atmospheric pressure also is said to hold the hip joint together.

Compared with the shoulder joint, the hip joint has more stability but less maneuverability.

KNEE JOINT

This joint occurs between the large, rounded condyles of the femur and the much flattened condyles of the tibia. The fibula does not take part in this joint, but it should be remembered that the patella does take part in its formation, articulating with the patellar surface of the femur. It is a synovial joint and mainly of the hinge variety, but a slight amount of rotation does occur; it would be more accurately called a modified hinge joint. It has two axes of movement, a transverse and a vertical.

The knee joint is held together by a capsule that is

*These vessels are important in children.

strengthened by the patellar ligament, and tibial and fibular collateral ligaments; several intracapsular ligaments also aid in maintaining the joint.

Capsule. The capsular ligament completely surrounds the knee joint except for two small gaps through which the synovial cavity communicates with surrounding bursae. The **attachments to the femur** are as follows. Anteriorly, the capsule stretches to either side of the quadriceps tendon and attaches to the anterior surface of the femur, covering a large prolongation of the synovial sac, which extends superiorly. The quadriceps tendon, patella, and patellar ligament substitute for the capsule in the midline anteriorly. Laterally, the capsule attaches to the lines between the epicondyles and the condyles. Posteriorly, the superior edges of the rounded condyles serve as the line of capsular attachment.

Inferiorly, the capsule is attached to the tibia but not the fibula; anteriorly, it reaches down to the tibial tuberosity; laterally, it extends a short distance distal to the articular cartilage; and posteriorly, it dips down to cover the posterior cruciate ligament. There is an opening in the capsule posteriorly for the tendon of the popliteus muscle, the tendon actually penetrating the capsule.

Patellar ligament (Figs. 6-67 and 6-70). The quadriceps tendon, the patella, and the patellar ligament form the anterior wall of the knee joint. Lateral expansions from the quadriceps tendon (retinacula) blend with the capsular ligament.

Tibial collateral ligament (Figs. 6-67, 6-70, and 6-71). This thin band is approximately ½ inch in width and extends from the medial epicondyle of the femur to the side of the tibia. It is separated superficially from the tendons of the sartorius, gracilis, and semitendinosus by a bursa. It is attached to the medial meniscus and damage to the ligament usually results in damage to the meniscus.

Fibular collateral ligament (Figs. 6-68 and 6-70). This ligament is cordlike and extends from the lateral epicondyle of the femur to the head of the fibula. It is very close to the tendon of the popliteus muscle, which is just deep to the ligament, and is separated superficially from the biceps tendon by a bursa. In some lower animals the peroneus longus muscle reaches the femur; this ligament is thought to represent the superior part of this muscle in man. Unlike the tibial collateral ligament, this one is not attached to the lateral meniscus.

Popliteal ligaments (Fig. 6-68). These are thickenings in the posterior part of the capsule and consist of

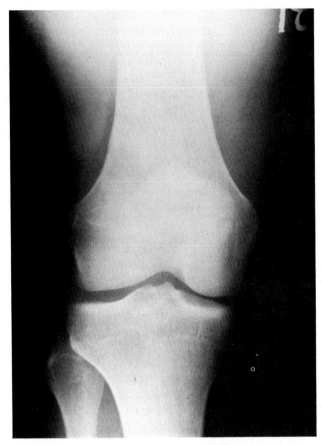

Plate 16. *The normal knee joint.*

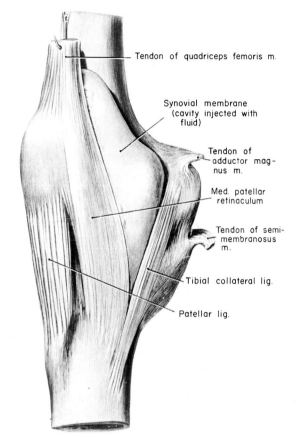

— Tendon of quadriceps femoris m.

Synovial membrane
(cavity injected with
fluid)

Tendon of
adductor mag-
nus m.

Med. patellar
retinaculum

Tendon of semi-
membranosus
m.

— Tibial collateral lig.

— Patellar lig.

Figure 6-67. *Ligaments of right knee joint—anteromedial view. (From Quain,* Elements of Anatomy, *courtesy of Longman Group Ltd.)*

an oblique popliteal and an arcuate popliteal ligament. The **oblique ligament** stretches medially and inferiorly from the lateral condyle to the tendon of the semimembranosus muscle. The **arcuate ligament** also attaches to the lateral condyle, crosses the tendon of the popliteus muscle, and blends with the capsule.

The **intracapsular ligaments** are the semilunar cartilages or menisci, transverse, and two cruciate ligaments.

Menisci (Fig. 6-69) These are two crescent-shaped fibrocartilaginous plates that serve to deepen the shallow fossae on the articular surface of the tibia. They are firmly attached to the intercondylar region of the tibia but slightly movable elsewhere. The lateral meniscus is round in its outline, while the medial appears oval; this point is important when one considers rotation in this joint. If the foot is on the ground, any rotation in the knee

joint is a movement of the femur; if the foot is off the ground, the tibia can move. When the knee joint is completely extended, it is said to become locked. This locking is accomplished by a medial rotation of the femur;[*] unlocking of the joint demands a lateral rotation of the femur, thought to be accomplished by the popliteus muscle. This locking phenomenon allows one to stand for long periods by balancing rather than utilizing great muscular effort. In the locked position, the quadriceps muscles are relaxed, as evidenced by an ability to move the patellae up and down. If a meniscus is cut in cross section,

---

[*]R. W. Haines, 1941, A note on the actions of the cruciate ligaments of the knee joint, *J. Anat.* 75: 373. This author feels that both cruciate ligaments are involved in locking the knee joint.

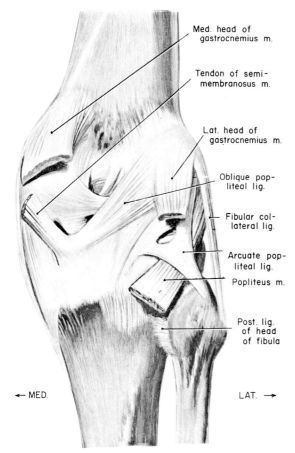

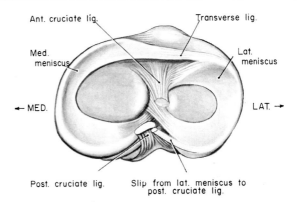

Figure 6-69. *Superior view of the menisci of the right knee joint. (After Gray.)*

Figure 6-68. *Ligaments of right knee joint—posterior view. (From Quain,* Elements of Anatomy, *courtesy of Longman Group Ltd.)*

it is seen to be wedge-shaped, with the wide part peripherally located and the sharp edge inside and projecting into the joint cavity. The lateral meniscus is attached to the intercondylar region just anterior and posterior to the intercondylar eminence. A cord of fibers (Fig. 6-71) stretches from the posterior surface of this meniscus to the posterior cruciate ligament.

The medial meniscus is not as complete as the lateral; its attachments to the intercondylar region are comparatively far apart (Fig. 6-69). Anteriorly it is attached to the lateral meniscus by a band of fibers coursing transversely, the **transverse ligament.**

Cruciate ligaments (Figs. 6-70 and 6-71). These important ligaments actually hold the femur and tibia to-

gether; the name is derived from the fact that they cross one another. The **anterior cruciate ligament** arises from the anterior intercondylar region of the tibia, and courses posteriorly and laterally to attach to the posterior end of the lateral condyle of the femur. This ligament becomes tense in extension of the joint, prevents overextension, and is thought to induce the medial rotation needed to lock the knee joint.* The **posterior cruciate ligament** arises from the posterior part of the intercondylar region of the tibia, courses superiorly in an anterior and medial direction to reach the medial condyle. This ligament tenses in flexion of the joint and prevents overflexion, and may be involved in locking the joint.*

Synovial membrane. Just as in all synovial joints, the synovial membrane does not cover the articular cartilages but does line the inside of the capsule of the knee joint. The knee joint, however, is complicated by the presence of intracapsular structures that give the synovial cavity a peculiar shape. If you will think of the cruciate ligaments as pushing into a sac and thereby obtaining a synovial covering on their anterior and lateral sides, the shape of the synovial sac will be more comprehensible (Fig. 6-72). The membrane attaches superiorly to the condyles of the femur along the edges of the articular cartilages. It spreads out superiorly until the capsule is reached, upon which it reflects and courses inferiorly. Although the superior extent on the sides of the joint is quite limited, anteriorly the sac extends superiorly for

---

*R. W. Haines, 1941, A note on the actions of the cruciate ligaments of the knee joint, *J. Anat.* 75: 373. See remark in preceding footnote.

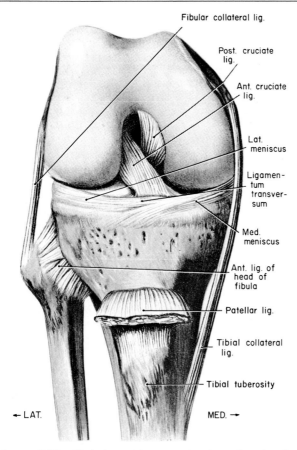

Fibular collateral lig.

Post. cruciate lig.

Ant. cruciate lig.

Lat. meniscus

Ligamentum transversum

Med. meniscus

Ant. lig. of head of fibula

Patellar lig.

Tibial collateral lig.

Tibial tuberosity

← LAT.   MED. →

Figure 6-70. *Right knee joint—anterior view of cruciate ligaments. (From Quain,* Elements of Anatomy, *courtesy of Longman Group Ltd.)*

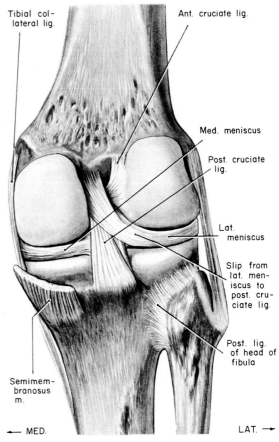

Tibial collateral lig.

Ant. cruciate lig.

Med. meniscus

Post. cruciate lig.

Lat. meniscus

Slip from lat. meniscus to post. cruciate lig.

Post. lig. of head of fibula

Semimembranosus m.

← MED.   LAT. →

Figure 6-71. *Right knee joint—posterior view of cruciate ligaments. (From Quain,* Elements of Anatomy, *courtesy of Longman Group Ltd.)*

quite a distance deep to the articularis genu muscle and quadriceps tendon. Two small membranous flaps (alae) separate this superior projection into the synovial sac proper and the more superiorly located suprapatellar bursa. Laterally the synovial membrane follows the capsule inferiorly to the point of attachment of the capsule to the menisci; the membrane thins out and disappears on these menisci. Posteriorly the cruciate ligaments produce an inward projection of the synovial sac; therefore, the posterior cruciate ligament is not covered by synovial membrane posteriorly but is in contact with the capsule. Anteriorly, the synovial membrane can be followed inferiorly until the patella is reached to which it is attached. The articular surface of the patella is not covered with membrane as is true of all articular surfaces. The synovial

membrane is attached to the inferior edge of the patella and spreads out inferiorly covering the infrapatellar pad of fat. A projection of tissue arises from this fat and attaches to the intercondylar region of the femur in a manner similar to the round ligament of the head of the femur; and, just as with the latter ligament, the synovial membrane covers this, making what is called the **infrapatellar synovial fold.** In the midline, the synovial membrane attaches inferiorly to the anterior part of the intercondylar region of the tibia.

The synovial cavity, as just described, has connections with other projections from the main sac, in addition to the suprapatellar bursa. These are the **bursae** deep to the heads of the gastrocnemius muscle, the bursa deep to

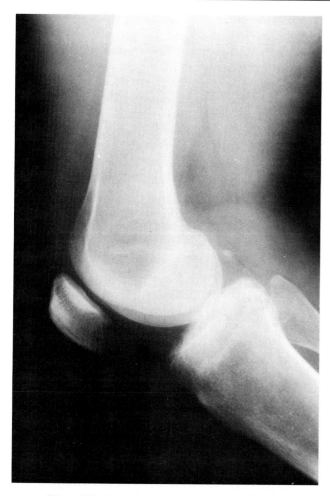

Plate 17. Lateral view of normal knee joint.

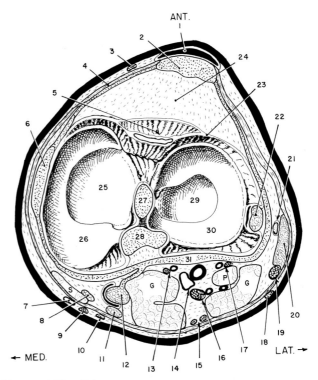

Figure 6-72. *Relations of the right knee joint. You are look-ing inferiorly: (G) gastrocnemius, (P) plantaris, (S) sartorius, (1) prepatellar bursa, (2) patellar ligament, (3) saphenous nerve, (4) deep fascia and expansion of vastus medialis, (5) infrapatellar synovial fold, (6) tibial collateral ligament, (7) long saphenous vein, (8) gracilis tendon, (9) saphenous nerve, (10) medial femoral cutaneous nerve, (11) semitendinosus tendon, (12) semimembranosus tendon and its bursa, (13) articular nerve and genicular artery, (14) popliteal vessels, (15) posterior femoral cutaneous and medial sural cutaneous nerves, (16) tibial nerve, (17) articular nerve and genicular artery, (18) peroneal communicating nerve, (19) common peroneal nerve, (20) biceps femoris tendon, (21) fibular collateral ligament and its bursa, (22) popliteus tendon and its sheath, (23) synovial membrane, (24) infrapatellar pad of fat, (25) medial condyle of tibia, (26) medial meniscus, (27) anterior cruciate ligament cov-ered by synovial membrane, (28) posterior cruciate ligament, (29) lateral condyle of tibia, (30) lateral meniscus, (31) capsule of knee joint. (Redrawn from Jamieson.)*

the semimembranosus muscle, and the bursa that ex-tends along the tendon of the popliteus muscle, which often reaches as far as the head of the fibula.

Other bursae are present around the knee joint but do not communicate with the joint cavity. These bursae are deep to the sartorius, gracilis, and semitendinosus tendons, separating them from the tibial collateral liga-ment; between the tendon of the biceps femoris and fibular collateral ligament; and several associated with the patella itself and its tendon and ligament. Those super-ficially located are between the skin and fascia on the patella, between the fascia and tendon on the patella, between the tendon and the patella itself, between the skin and the patellar ligament, and between the skin and the tibial tuberosity. A deep bursa occurs deep to the patellar ligament.

The **arterial supply** to the knee joint is from numerous branches from the many vessels taking part in the anastomosis around this joint, namely, the five genicular branches of the popliteal, descending branch of the lateral femoral circumflex, the descending branch of the femoral, and recurrent and circumflex fibula branches of the anterior tibial artery.

The **nerve supply** is quite extensive: five articular branches from the ischiadic nerve; the recurrent branch from the common peroneal; a branch from the obturator; and articular branches from the branches of the femoral to the vastus lateralis, vastus intermedius, and vastus medialis muscles.

The knee joint is strengthened by various muscles that surround it (Fig. 6-72) They are the quadriceps femoris anteriorly; the biceps femoris laterally; the sartorius, semitendinosus, and gracilis muscles medially; and the popliteus, gastrocnemius, semimembranosus, and plantaris muscles posteriorly.

TIBIOFIBULAR JOINTS

The **superior tibiofibular joint** (Figs. 6-70 and 6-71) is a synovial joint between the lateral condyle of the tibia and the head of the fibula; a gliding movement occurs. It possesses a capsule that is strengthened by the presence of ligaments anteriorly and posteriorly. The **anterior ligament** consists of two or three bands that stretch superiorly from the head of the fibula to the lateral condyle of the tibia. The **posterior ligament** is a single band that stretches superiorly and medially to reach the tibia; this ligament lies deep to the tendon of the popliteus muscle. This joint is **innervated** by branches from the nerve to the popliteus muscle and from the recurrent genicular.

The **inferior tibiofibular joint** (Figs. 6-73 and 6-75) is a fibrous articulation between the lateral malleolus and the inferior end of the tibia. This joint is quite strong, for the ankle-joint architecture is dependent upon it. It is supported by anterior and posterior inferior tibiofibular ligaments, a transverse tibiofibular ligament, and an interosseus ligament. The **anterior and posterior tibiofibular ligaments** pass from the lateral malleolus medially and superiorly to attach to the tibia. The **in-**terosseus ligament is a very strong band of tissue connecting the tibia to the fibula; this ligament is covered by the tibiofibular ligaments. The **transverse tibiofibular ligament** extends between the most inferior end of the tibia and the malleolar fossa of the fibula; it lies posterior to the ankle joint and plays a role in this articulation; it is almost continuous with the posterior tibiofibular ligament.

Interosseus membrane (Figs. 6-5 and 6-6). This strong membrane is attached to the interosseus margins of the tibia and fibula. It reaches superiorly to the superior tibiofibular joint and inferiorly to the interosseus ligament of the inferior tibiofibular joint. The anterior tibial artery passes superior to the interosseus membrane to gain the anterior side of the leg, and the perforating branch of the peroneal artery passes from posterior to anterior just inferior to the membrane.

ANKLE JOINT

The **ankle joint** (Figs. 6-73, 6-74, and 6-75) is a synovial articulation of the hinge type between the inferior surfaces of the tibia and the malleoli, and the superior surface of the talus. Plantar and dorsiflexion around a transverse axis are the only movements allowed. The joint is surrounded by a capsular ligament; the **capsule** is attached to the margins of the articular surfaces and is reinforced by anterior, posterior, medial, and lateral collateral ligaments. The anterior part of the capsule is quite thin and wide; it extends from the inferior end of the tibia to the anterior edge of the articular surface of the talus. The posterior part of the capsule is equally thin and extends from the inferior end of the tibia and lateral malleolus to the posterior edge of the superior articular surface of the talus. It is also attached to the transverse tibiofibular ligament. The **medial collateral ligament** is thickened and, owing to its shape, is called the deltoid ligament (Fig. 6-74) It is narrow where it is attached to the medial malleolus but fans out to a much wider attachment inferiorly. The anterior fibers attach to the talus and navicular bones and to the spring ligament (calcaneonavicular ligament); the middle fibers course vertically and attach to the sustentaculum tali, part of the calcaneus; and the posterior fibers course inferiorly and posteriorly to the tubercle of the talus. The **lateral collateral ligaments** attach to the lateral malleolus superiorly and to the talus inferiorly and are known as the anterior

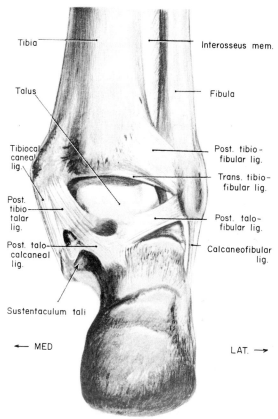

Figure labels: Tibia; Talus; Tibiocalcaneal lig.; Post. tibiotalar lig.; Post. talocalcaneal lig.; Sustentaculum tali; MED; Interosseus mem.; Fibula; Post. tibiofibular lig.; Trans. tibiofibular lig.; Post. talofibular lig.; Calcaneofibular lig.; LAT.

Figure 6-73. *Ligaments around the right ankle joint— posterior view. Posterior part of capsule has been removed. (From Quain,* Elements of Anatomy, *courtesy of Longman Group Ltd.)*

and posterior talofibular ligaments. These ligaments are aided by the **calcaneofibular ligament,** which remains separated from the capsule (Fig. 6-75).

The **synovial membrane** of this joint lines the inside of the capsule in a rather lax arrangement. A fold projects into the joint posteriorly between the lateral malleolus and the articular surface of the tibia. Subcutaneous **bursae** occur on each malleolus.

The **blood supply** is from branches of the anterior tibial, posterior tibial, and peroneal arteries. The joint receives **nerves** from both the tibial nerve and deep branch of the common peroneal.

The many structures in relation to the ankle joint are depicted in (Figure 6-76.)

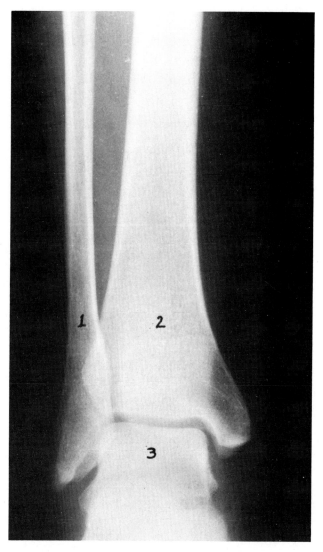

Plate 18. *Anterior view of normal ankle joint: (1) fibula, (2) tibia, (3) talus.*

JOINTS OF FOOT

The joints of the foot are the tarsal, tarsometatarsal, intermetatarsal, metatarsophalangeal, and interphalangeal articulations. As can be seen in (Figures 6-74, 6-75, and 6-77,) these joints are well fortified with ligaments which, fortunately, take the names of the two bones connected, in most cases.

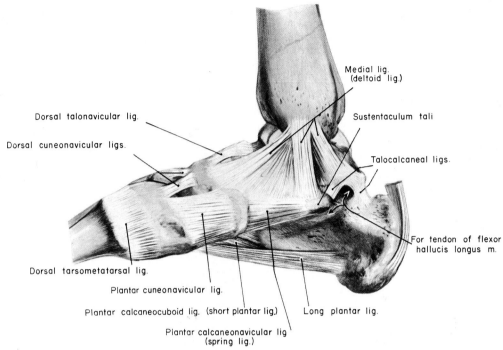

Dorsal talonavicular lig.

Dorsal cuneonavicular ligs.

Medial lig. (deltoid lig.)

Sustentaculum tali

Talocalcaneal ligs.

For tendon of flexor hallucis longus m.

Dorsal tarsometatarsal lig.

Plantar cuneonavicular lig.

Plantar calcaneocuboid lig. (short plantar lig.)

Long plantar lig.

Plantar calcaneonavicular lig (spring lig.)

Figure 6-74. Ligaments around the ankle and foot—medial view. (From Quain, Elements of Anatomy, *courtesy of Longman Group Ltd.)*

One ligament is longer than the others and is located in a more superficial position. This **long plantar ligament** extends from the middle of the calcaneus posteriorly, to the cuboid and the bases of the second, third, and fourth metatarsal bones anteriorly. It aids in making a canal for the tendon of the peroneus longus muscle, and plays an important role in maintaining the arches of the foot.

TARSAL JOINTS. The tarsal joints can be divided into (1) those located posteriorly—subtalar and talocalcaneonavicular joints; (2) the transverse tarsal joint—the combined calcaneocuboid and talonavicular joints; and (3) those anteriorly located—between the navicular, the three cuneiforms, and the cuboid.

  1. The **subtalar joint** (Fig. 6-78) is an articulation between the talus superiorly and the calcaneus inferiorly; the posterior part of the talus articulates with the largest facet on the calcaneus. This is a gliding joint with a capsule and synovial membrane. The capsule is aided by a strong interosseus ligament, which is located in the deep

grooves between the bones, and by the medial and lateral talocalcaneal ligaments. The subtalar joint is not the only articulation of the talus and calcaneus. The **talocalcaneonavicular joint** (Fig. 6-78) is a complex of synovial joints between the talus superiorly and the navicular and calcaneus bones, and the calcaneonavicular ligament (spring ligament) inferiorly. The talus actually rests on the navicular bone anteriorly, on the anterior end of the calcaneus and the calcaneonavicular ligament just posterior to the navicular attachment, and posterior to all on the sustentaculum tali of the calcaneus and the anterior end of the calcaneus itself. This arrangement demonstrates the importance of the calcaneonavicular ligament in maintaining the arches of the feet. Considerable movement of inversion and eversion takes place at this talocalcaneonavicular joint.* Supporting ligaments are the calcaneonavicular ligament just mentioned, the dorsal

_____

*Almost all of the movement of inversion and eversion occurs in the talocalcaneonavicular and transverse tarsal joints.

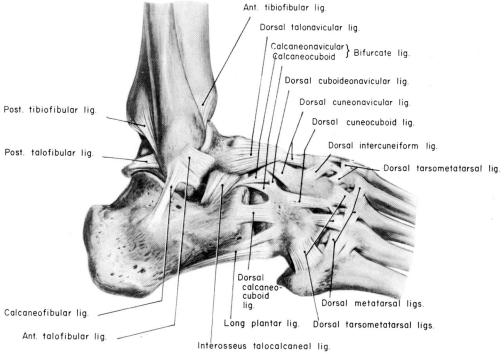

Figure 6-75. *Ligaments around the right ankle and foot—lateral view. (From Quain,* Elements of Anatomy, *courtesy of Longman Group Ltd.)*

talonavicular, and the calcaneonavicular part of the bifurcated ligament which is also located dorsally. The joint is surrounded by a capsule.

2. **Transverse tarsal joint** (Fig. 6-79). This articulation extends across the foot from one side to the other and is the location where the remaining part of inversion and eversion of the foot occurs.* It is a combination of the **calcaneocuboid and talonavicular joints.** They are synovial joints allowing gliding movement. The capsule of the calcaneocuboid joint is strengthened dorsally by the calcaneocuboid part of the bifurcated ligament and by the dorsal calcaneocuboid ligament (Fig. 6-75). Ventrally, the plantar calcaneocuboid (short plantar) ligament is very important. The talonavicular joint is supported dorsally by the dorsal talonavicular ligament, which is quite broad; the plantar calcaneonavicular ligaments (spring ligament) also serve to hold this joint together. Before leaving the transverse tarsal joint, it should

be mentioned that the cuboid and navicular bones are held together with adequate ligamentous support. These bones are not usually in contact with each other.

3. The **anterior intertarsal articulations** (Fig. 6-79) occur between the navicular and three cuneiform bones, between the cuboid and the most lateral cuneiform, and between the cuneiform bones themselves. The **cuneonavicular** joint is supported by dorsal, plantar, and medial ligaments, which take the same name as the joint. The **intercuneiform and cuneocuboid** joints are held together with dorsal, interosseus, and plantar ligaments. Only slight movement occurs in these articulations.

TARSOMETATARSAL JOINTS. Three distinct joints are formed between the tarsal and metatarsal bones. These are synovial joints with a gliding movement allowed. The first joint occurs between the first metatarsal and the medial cuneiform bone; the second between the second and third metatarsals and intermediate and lateral cuneiform bones; and the third between the fourth and

---

*Almost all of the movement of inversion and eversion occurs in the talocalcaneonavicular and transverse tarsal joints.

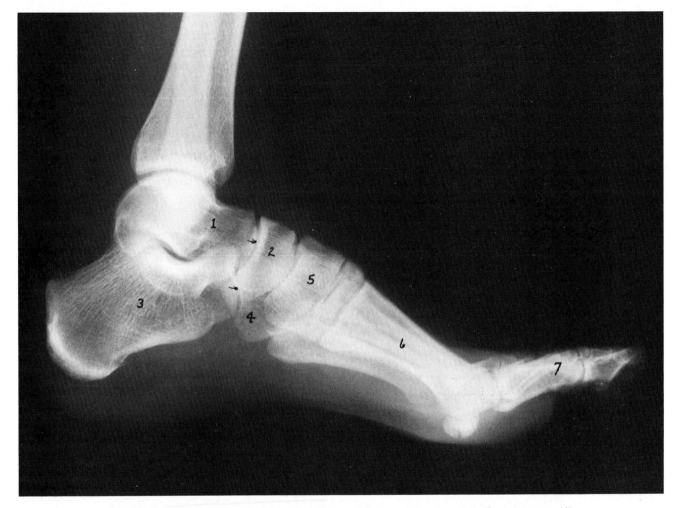

Plate 19. *Lateral view of foot in dorsiflexion: (1) talus, (2) navicular, (3) calcaneus, (4) cuboid, (5) cuneiform bones, (6) metatarsals, (7) proximal phalanges. The arrows indicate the transverse tarsal joint.*

fifth metatarsals and the cuboid. All of these joints are provided with dorsal and plantar ligaments, and some of them with interosseus ligaments. The long plantar ligament also is involved because of its attachment to the second, third, and fourth metatarsal bones. The second digit is the strongest articulation, for it fits into a socket made by the three cuneiform bones; furthermore, it is provided with more ligaments than the other joints. These joints are innervated by the deep peroneal and the medial and lateral plantar branches of the tibial.

INTERMETATARSAL JOINTS. All of the bases of the metatarsal bones are close to each other and the synovial cavity of the tarsometatarsal joints extends a short distance into the intermetatarsal joints. These articulations are strengthened by dorsal and plantar ligaments, and by a complete set of interosseus ligaments. Furthermore, there is a deep transverse ligament of the foot that connects the bases of all the metatarsal bones; this ligament is grooved by the tendons of the long flexor muscle. A slight gliding movement is allowed in these joints.

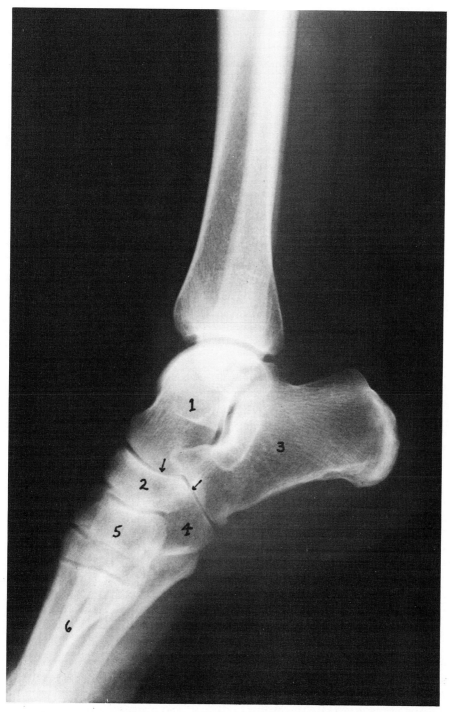

Plate 20. *Lateral view of foot in plantar flexion: (1) talus, (2) navicular, (3) calcaneus, (4) cuboid, (5) cuneiform bones, (6) metatarsals. The arrows indicate the transverse tarsal joint.*

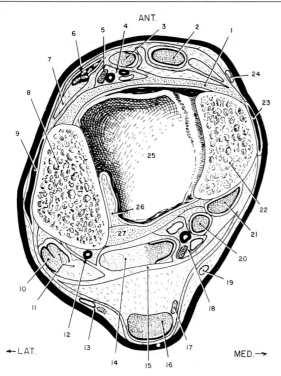

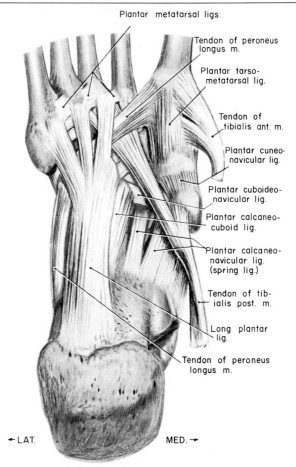

Plantar metatarsal ligs.

Tendon of peroneus longus m.

Plantar tarso-metatarsal lig.

Tendon of tibialis ant. m.

Plantar cuneo-navicular lig.

Plantar cuboideo-navicular lig.

Plantar calcaneo-cuboid lig.

Plantar calcaneo-navicular lig. (spring lig.)

Tendon of tib-ialis post. m.

Long plantar lig.

Tendon of peroneus longus m.

←LAT.   MED.→

Figure 6-76. *Relations of the right ankle joint. You are look-ing superiorly: (1) anterior ligament of capsule, (2) tibialis an-terior tendon and its sheath, (3) extensor hallucis longus and its sheath, (4) anterior tibial vessels, (5) deep peroneal nerve, (6) extensor digitorum longus tendons and sheath, (7) peroneus tertius, (8) lateral malleolus, (9) bursa, (10) peroneus longus tendon and its sheath, (11) peroneus brevis and its sheath, (12) peroneal artery, (13) short saphenous vein and medial sural cutaneous nerve, (14) flexor hallucis longus muscle, tendon, and sheath, (15) deep fascia, (16) tendo calcaneus and its bursa, (17) plantaris tendon, (18) posterior tibial vessels and tibial nerve, (19) branch of short saphenous vein, (20) flexor digitorum longus tendon and sheath, (21) tibialis posterior ten-don and sheath, (22) medial malleolus, (23) bursa, (24) saphenous vein, (25) inferior end of tibia, (26) synovial folds, (27) posterior ligament of capsule. (Redrawn from Jamieson.)*

Figure 6-77. *Ligaments on plantar surface of right foot. One slip of plantar ligament to base of second metatarsal has been removed. (From Quain,* Elements of Anatomy, *courtesy of Longman Group Ltd.)*

are synovial and of the condyloid variety; flexion and extension are the most important actions but abduction and adduction do occur; since these four movements are allowed, circumduction also is possible. There is no rota-tion in these joints.

METATARSOPHALANGEAL JOINTS.   These are ar-ticulations between the rounded heads of the metatarsals and the fossae on the proximal phalanges. The fossae are deepened by the presence of cartilaginous lips on the phalanges. Each joint is provided with a capsule, which is weak dorsally, and two collateral ligaments. These joints

INTERPHALANGEAL JOINTS.   These occur between the phalanges; they have a capsule and collateral liga-ments as just described for the metatarsophalangeal joints. The plantar side is thickened and grooved by the long flexor tendons. Since these are synovial joints of the hinge type, only flexion and extension occur.

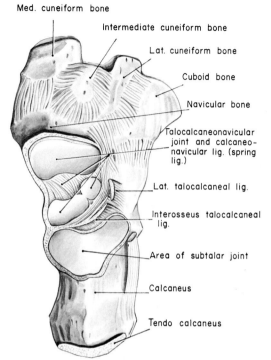

Figure 6-78. *The talocalcaneonavicular joint and subtalar joint—the "bed" for the talus—exposed by removal of the talus (right foot).*

ARCHES OF FOOT

When the normal foot is placed on the ground, all parts do not make contact. The heel posteriorly and the heads of the metatarsal bones anteriorly rest on the ground; these points are connected by an area along the lateral side of the foot that also makes contact. The medial side of the foot between these anterior and posterior points of

Plate 21. *Plantar surface of normal foot: (1) talus, (2) navicular, (3) calcaneus, (4) cuboid, (5a) first cuneiform, (5b) second cuneiform, (5c) third cuneiform, (6) metatarsals, (7) proximal phalanges, (8) middle phalanges, (9) distal phalanges. Note that the great toe has no middle phalange. The arrows mark the transverse tarsal joint.*

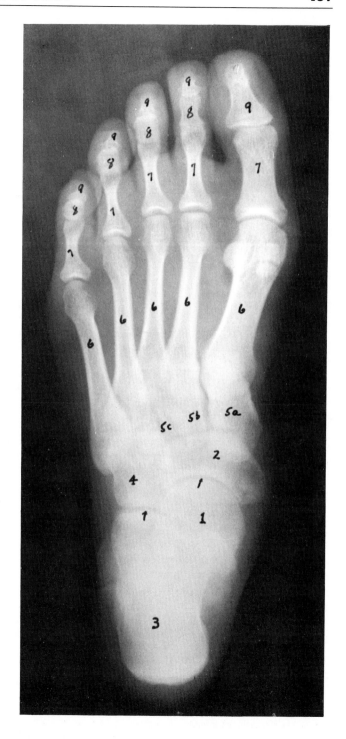

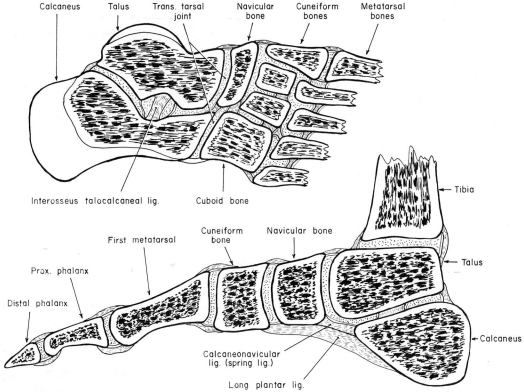

Figure 6-79. *Sections through the tarsal bones showing the synovial cavities (after Quain). The top drawing cuts through the talus and calcaneus longitudinally, but through the remaining bones in an oblique fashion. The lower drawing is a longitudinal section through the great toe. The joint cavities are shown by dots. Note that they coalesce so that there are fewer joint cavities than articulations.*

contact is off the ground. This is due to the formation of the longitudinal and transverse arches.

The **longitudinal arches** are actually two in number, one on each side of the foot.* The skeletal elements forming the medial arch are the calcaneus, talus, navicular, the cuneiforms, and the three medial metatarsal bones; the corresponding arch on the lateral side consists of the calcaneus, cuboid, and lateral two metatarsal bones (see Figs. 6-8 and 6-9).

The **transverse arch** is formed posteriorly by the cuboid and navicular bones, in the intermediate area by the cuneiforms and cuboid bones, and anteriorly by the bases of the metatarsals (Fig. 6-9).

Maintenance of these arches is important for proper foot health. The arches prevent too great a pressure being exerted on vital structures entering and contained within the plantar surface of the foot. There are three methods whereby the arches are preserved: (1) the shape of the bones, (2) ligaments, and (3) muscle action. If the bones of the foot were lined up properly and the above-mentioned contact points made secure, the wedge shape of the bones would form these arches. They are very similar to the rocks used in building archways—narrower at the bottom than at the top. The ligaments are most important; all the ligaments play a role, but special attention should be given to the long plantar, short plantar, and the calcaneonavicular (spring) ligaments. The role played by

---

*J. H. Manter, 1946, Distribution of compression forces in the joints of the human foot, *Anat. Rec.* 96: 313. It has been well established that pressure under the sole is shared by the calcaneus and the heads of all five metatarsals. This paper describes the pressures involved between the tarsal bones, and recommends that a double longitudinal arch be abandoned, since both function as a unit.

muscles is problematical. The peroneus longus and posterior tibialis muscles form a sling under the foot, but these muscles, according to Basmajian and Bentzon,* are not functioning when one is standing still; they may be more important when walking (Jones, 1941).* It is interesting that people who have occupations that demand continued standing have a tendency to develop fallen arches unless their occupation demands walking. Even when walking the long muscles play a relatively small part in arch support; the role played by the short muscles in the foot is also relatively small. It is obvious that standing still puts a great strain on the plantar ligaments.

SYSTEMIC SUMMARY

The limbs first make their appearance in the embryo measuring about 4 mm. in length. The limb buds increase in length in a manner that parallels the long axis of the embryo. Indentations occur that represent the future joint areas, and at the same time the limbs take a more anterior position, their axes being at right angles to that of the trunk. In this situation both the thumb and radial side of the forearm, and the large toe and tibial side of the leg are facing toward the cranium, just as occurs in an animal such as the salamander; the palmar surface of the hand and plantar surface of the foot face medially.

It is quite obvious that something has happened to the limbs in the human, for in the lower limb just studied the great toe and tibia are medially placed while the thumb and radius in the upper limb are laterally placed. This resulted from a 90-degree lateral rotation in the forelimb and a medial rotation in the hindlimb of equal degree. We will see evidences of this rotation in many of the systems about to be described.

BONES AND JOINTS

There is no particular advantage derived from describing the bones of the lower limb at this point, for they were described as a whole at the beginning of this section. You would gain considerably if you would reread those pages,

*See page 417 for these references.

now that you know more about structures attached to and associated with these bones.

The articulations of the lower limb have also been described systemically on pages 421–436, and do not demand repetition. This does not mean they are unimportant; the joints are involved in the complaints of many patients.

Those students who have studied the upper limb should note the differences and similarities between the corresponding parts in the two limbs. There is no doubt about the fact that the coxal bone is attached to the axial skeleton more firmly than attachments found in the upper limb, where the sternoclavicular joint is quite movable. Although the hip and shoulder joints are both freely movable ball-and-socket joints, from observation alone it is easily seen that the hip joint is a more stable articulation than the shoulder joint. The acetabulum is considerably deeper than the glenoid cavity, and the head of the femur is held in the acetabulum by the ligamentum teres, while the head of the humerus has no such attachment. The femur itself is a more massive structure than the humerus. The elbow and knee joints are both hinge joints and, of the two, it would seem that the elbow is a more stable articulation than the knee. The rotation allowed in the knee, in spite of being a locking mechanism, would seem to weaken this joint as far as weight-bearing is concerned. In contrast is the fact that the superior and inferior radioulnar joints in the upper limb allow considerable movement not found in the lower limb. The tibia is a more massive bone than either bone in the forearm, and this bone bears the entire weight of the body, the fibula being a bone for muscle attachment only. Any weight put against the foot, which occurs when one stands, is transferred to the tibia and then directly to the femur. In contrast, weight put against the hand is transferred to the radius, from the radius to the ulna, and from this bone to the humerus. The interosseus membrane is very important in this arrangement for it prevents the upward thrust of the radius and the simultaneous downward thrust of the ulna. So two bones and an interosseus membrane are found in the upper limb, while the same thing is accomplished in the lower limb with a single strong bone. The ankle joint is a hinge joint whereas the wrist is a condyloid articulation that allows considerably more movement than occurs in the ankle. The tarsal bones are large compared with the small carpal bones and these bones in the foot are connected with each other by much stronger ligaments than occur in the wrist. And finally, the long, slender fingers and the opposable thumb are much more effective in performing fine movements than the shorter, less movable toes.

In summary, the lower limb, with its strong articulation with the axial skeleton, its strong hip joint, massive femur, stable and massive tibia, relatively stable ankle joint, massive foot bones, short toes, and strong ligaments, is well suited for locomotion and weight-bearing; while the upper limb, with its more movable attachment to the axial skeleton, freely movable shoulder joint, less massive bone structure, more movable forearm, more movable wrist joint, and long digits and opposable thumb, is well suited to perform the many fine acts that have been partially responsible for the human being maintaining a status well above that of other animals.

Our upper and lower limbs have evolved from being almost identical in the lower animal forms to the specialized limbs just described. There is still room for improvement, however. For example, the knee joint might be more efficient if constucted in a manner that would give it more stability, and the interosseus membrane in the upper limb might be more useful to us if reversed in its direction.

MUSCLES

As the limb buds grow, two muscle masses develop: one dorsal and one ventral (Fig. 6-80). Each of these muscle masses receives nerves from the ventral rami of the segmental nerves associated with the particular somites giving rise to the limb buds. Thinking of the bones as the central axis of the limb, the dorsal muscle mass develops into postaxial muscles, which are the extensors and abductors, and the corresponding branch of the ventral ramus innervating these muscles is called a postaxial nerve; conversely, the ventral muscle mass develops into the preaxial muscles—the flexors and adductors—and these are innervated by preaxial nerves.

While the limbs are in the quadripedal position, the preaxial muscles face in a medial direction in both limbs while the postaxial muscles are on the lateral surface. As indicated previously, the lower limb undergoes a 90-degree rotation medially, which has caused the postaxial muscles to face ventrally (anteriorly) and laterally, the preaxial muscles dorsally (posteriorly) and medially. (This arrangement is the opposite to that found in the upper limb, since that limb rotates laterally.)

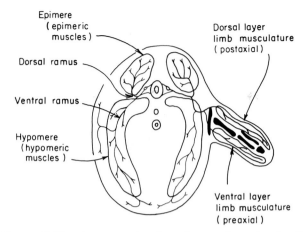

Figure 6-80. *A cross section through an embryo showing the development of a limb bud. Using the bones as the central axis of the limb, note the development of dorsal (postaxial) and ventral (preaxial) muscle masses, and that these muscle masses are innervated by postaxial and preaxial nerves, respectively.*

Therefore, the extensor muscles on the anterior surface of the thigh are postaxial, as is the tensor fascia lata muscle on the lateral surface. In addition, the short head of the biceps femoris muscle is a postaxial muscle that has migrated in such a fashion as to become a flexor rather than an extensor in action. The muscles on the anterior and lateral surfaces of the leg and dorsum of the foot are also postaxial and this extensor action on the digits and eversion (abduction) of the foot are characteristic.

The preaxial muscles in the thigh are those on the medial surface (adductor group) and those on the posterior surface (except short head of biceps). The preaxial muscles continue into the leg being located on the posterior surface. Their action on digits (flexors) and inversion (adduction) of the foot are proper actions for such muscles. Furthermore, all muscles located on the plantar surface of the foot are preaxial in origin.

The alert student will note that movements at the ankle joint were carefully omitted from the foregoing description. The development of the heel is limited to the primates, and the nomenclature given to movements at the ankle joint has been confusing and unfortunate. Movement of the foot superiorly, called flexion, should have been called extension at the ankle joint, and movement inferiorly, called extension, should have been called flexion. This would have simplified understanding of the muscles of the lower limb and made them comparable to the upper limb. The prevalent use of the terms dorsiflexion and plantar flexion for movement at the ankle joint have come into use to avoid this confusion.

Returning to the hip joint, action here has become equally confusing because of the upright posture obtained by the human being. Muscles that would have a particular action in the four-footed stance changed their action when the upright posture was assumed.

Nevertheless, the postaxial muscles in the gluteal region (gluteus maximus, medius, and minimus; piriformis) and the tensor fascia lata are extensors and abductors, which is proper action for postaxial muscles. The preaxial muscles (obturator internus; gemellus superior and inferior; quadratus femoris) are lateral rotators of the hip joint, but unfortunately for our ideas that flexors are preaxial muscles, the psoas and iliacus muscles are postaxial in origin.

It should be noted that the muscles originating from the ilium are dorsal musculature and therefore postaxial, while those arising from the pubis and the ischium are ventral in origin and preaxial.

From the above description it seems obvious that the gluteal nerves, the nerves to the piriformis, iliacus, and psoas major, the femoral nerve, and the common peroneal are all postaxial. The nerve to the obturator internus and superior gemellus, that to the inferior gemellus and quadratus femoris, and the tibial nerve and its branches are all preaxial nerves. This concept will be reexamined in the systemic review of nerves.

MOVEMENTS AT HIP JOINT. Since the hip joint is a ball-and-socket articulation, all movements are allowed. The following is based on the body being in the anatomical position.

Flexion (direction of pull must pass anterior to transverse axis). The most important flexors of the thigh are the iliopsoas, rectus femoris, sartorius, pectineus, and tensor fascia lata muscles. There is a considerable degree of flexion allowed by the hip joint, particularly if the knee is flexed, and contact with the anterior abdominal wall is all that stops it. If the knee is extended, the length of the hamstring muscles determines the amount of flexion allowed.

Extension (direction of pull must pass posterior to transverse axis). The main extensors are the gluteus maximus, semimembranosus, semitendinosus, and the long head of the biceps femoris. After flexion has occurred, or if the body is placed in the quadripedal position, extension can be of considerable magnitude, but this movement is quite limited from the upright anatomical position. Extension is considered complete when the lower limb is in a straight line with the trunk. Further extension is limited by the strong iliofemoral ligament.

Abduction (direction of pull must pass superior to anteroposterior axis). The important abductors of the thigh are gluteus medius, gluteus minimus, and the piriformis. These are assisted by the tensor fascia lata and gluteus maximus. This movement is checked by the pubofemoral ligament and partly by the iliofemoral ligament.

Adduction (direction of pull must pass inferior to anteroposterior axis). This movement is quite strong and is mainly produced by the pectineus, adductor longus, adductor brevis, and the adductor magnus muscles; these are assisted by the gracilis, and the lower part of the gluteus maximus muscle. This movement is limited by contact with the other thigh but can continue beyond this position if flexed; in this case the limiting factor is the iliofemoral ligament.

Circumduction. This is a combination of the four movements described above.

Medial rotation (direction of pull must pass anterior to vertical axis). Possibly due to the fact that the lower limb rotated medially in its development or to the fact of our upright stature, or both, there are no specific medial rotators of the thigh, while there are several muscles that have no other action than that of lateral rotation. All of the following muscles are weak medial rotators (of some we are not certain): the anterior parts of gluteus medius and minimus, tensor fascia lata, ischiocondylar part of the adductor magnus, semimembranosus, and the semitendinosus.

Lateral rotation (direction of pull must be posterior to vertical axis). This movement is accomplished primarily by the gluteus maximus, posterior parts of the gluteus medius and minimus, piriformis, obturator internus, obturator externus, superior and inferior gemelli, quadratus femoris, and the sartorius.

Muscle actions such as the above have often times been determined by position alone. You should not merely accept them but should understand the principle involved by visualizing these muscles on a skeleton. Another thought that may be bothering you is just how valuable all this knowledge is when it is obvious that these motions might be possible when performing gymnastics but rather difficult when both feet are on the ground. We will see in a moment, however, that these motions are important during locomotion, which, after all, is the main function of the lower limb in humans.

MOVEMENTS AT KNEE JOINT.

The knee joint is a modified hinge joint, so called because of the slight rotation allowed.

Flexion (direction of pull must be posterior to transverse axis). The main flexors of this joint are the semitendinosus, semimembranosus, and the biceps femoris; these are assisted by the gastrocnemius and popliteus; some flexor action is also induced by the plantaris, sartorius, and gracilis muscles. A considerable amount of flexion is allowed, being limited by contact with the posterior side of the thigh, the length of the extensors, and particularly by the posterior cruciate ligament.

Extension (direction of pull must be anterior to transverse axis). Extension of the knee joint is produced by contractions of the four parts of the quadriceps femoris muscle: rectus femoris, vastus medialis, vastus intermedius, and vastus lateralis. The articularis genu muscle also assists, but its main action seems to be to lift the synovial membrane during extreme extension so it will not be pinched between the patella and femur. The amount of extension is controlled by the tibial and fibular collateral ligaments and the anterior cruciate ligament.

Rotation (direction of pull is posteriorly and either medial or lateral to vertical axis). When the knee is flexed, rotation of the tibia can occur if the foot is off the ground. Under these conditions, **medial rotation** is caused by the semimembranosus, semitendinosus, gracilis, sartorius, and the popliteus; **lateral rotation** is accomplished by the biceps femoris. If the foot is on the ground, any rotation in the knee joint, of necessity, must be a movement of the femur. In the hyperextended condition the knee becomes locked; this demands a medial rotation, which is accomplished by the medial rotators of the thigh at the hip joint if any actual pull is necessary to accomplish this act; unlocking the joint demands muscle action and the popliteus muscle is usually given credit for accomplishing this act. The sudden collapse that results from pushing a person unannounced in the back of the knee is a simple uncontrolled unlocking of the knee joint.

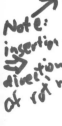
Note: insertion → direction of rotor

MOVEMENT AT ANKLE JOINT.

Movement at the ankle joint is limited to dorsi and plantar flexion, since this joint is of the hinge type.

ALL ANT

Dorsiflexion (direction of pull must pass anterior to transverse axis). This movement is not as strong as plantar flexion, and there are no specific muscles that have this role only. This movement is caused by contraction of the tibialis anterior, extensor hallucis longus, extensor digitorum longus, and the peroneus tertius.

POST LAT

Plantar flexion (direction of pull must pass posterior to transverse axis). This powerful movement is accomplished by the gastrocnemius and soleus muscles, but these are assisted by the tibialis posterior, flexor hallucis longus, flexor digitorum longus, the peroneus longus and brevis, and the plantaris muscles.

MOVEMENT AT TARSAL JOINTS.

The important movements that occur in these joints are inversion and eversion, which occur in the talocalcaneonavicular and transverse tarsal joints for the most part. These movements occur whenever we turn a corner or stand with our feet wide apart.

Inversion (direction of pull must be medial to longitudinal axis of foot). This movement of raising the sole of the foot medially is done primarily by the tibialis anterior and posterior muscles, aided by the flexors of the toes. *M's attach medially on foot*

Eversion (direction of pull must be lateral to longitudinal axis of foot). This movement is done primarily by the peroneus longus and brevis muscles, aided by the peroneus tertius and extensor digitorum longus. *M's attach laterally on foot*

MOVEMENTS OF DIGITS.

The human being has a remarkable set of muscles in the foot that correspond closely to those in the hand. The introduction of shoes has reduced the effectiveness of many of them, and, furthermore, shoes worn by women have distorted the shape of the foot considerably.

Flexion (direction of pull must pass on plantar side of transverse axes). Flexion of the great toe is accomplished by the flexor hallucis longus (terminal phalanx) and flexor hallucis brevis (proximal phalanx); these are aided by the abductor hallucis and the oblique head of the adductor hallucis owing to the common insertion of these muscles with the two heads of the flexor hallucis brevis. Flexion of the remaining digits is accomplished by the flexor digitorum longus (terminal phalanx), flexor dig-

READ

itorum brevis (middle phalanx), lumbricals (proximal phalanx), and the flexor digiti minimi brevis. This last muscle is aided by the abductor digiti minimi, since it inserts in a common tendon with the short flexor of this digit. The quadratus plantae muscle must be added to the list because of its aid to the flexor digitorum longus.

Extension (direction of pull must pass dorsal to transverse axes). This movement is accomplished by the extensor digitorum longus, the extensor digitorum brevis, and the extensor hallucis longus. The lumbrical muscles will extend the middle and terminal phalanges while flexing the proximal member. The interossei also will extend because of their attachment to the extensor expansion of each of the last four digits.

Abduction and adduction. Since the interphalangeal joints are hinge joints, abduction and adduction occur at the metatarsophalangeal joints only. The second digit is considered the axis as far as adduction and abduction are concerned. Knowing this and the fact that the dorsal interossei are abductors, it is obvious that they must be inserted on either side of the second digit to move it away from the axis in either direction, on the lateral side of the third digit, and the lateral side of the fourth; this makes four in all. The large toe has an abductor hallucis so needs no dorsal interosseus, and the same is true for the fifth digit—abductor digiti minimi. The plantar interossei adduct so must be placed on the medial sides of the third, fourth, and fifth digits, making a total of three. The large toe has its own adductor with oblique and transverse heads.

It is of interest to note that all primates except the human have opposable great toes—in other words, can bring the great toe across the sole of the foot.

LOCOMOTION. Locomotion* is a complex act. As shown in Figure 6-81A, taking a step, in this case with the right limb, involves moving it forward (swing phase) while balancing on the left; then balancing on this right limb (stance phase) while swinging the left one forward.

The swing phase can be divided into a period of acceleration, a midswing, and a period of deceleration. During this time it is obvious that we must perform this act in such a manner as to prevent the foot from dragging on the floor.

*A. Steindler, 1935, *Mechanics of Normal and Pathological Locomotion in Man,* Thomas, Springfield.

As the swing forward is made, the hip joint undergoes flexion, a lateral rotation, and an abduction. The flexion is easy to visualize (Fig. 6-81A) but the lateral rotation and abduction demand further explanation. When the swing phase first starts, the hip joint is in a state of medial rotation, and as the swing continues this joint gradually rotates laterally. In actuality this is due to movement of the pelvis to allow for a direct forward movement of the limb (Fig. 6-81B). Naturally when the limb is off the floor, the support for the pelvis is lost. Therefore, it tends to tilt toward the side of the swinging limb; this explains the abduction. Too great a tilt is prevented by contraction of the deeper gluteal muscles on the opposite side, the origins and insertions of these muscles being reversed in this action.

The action at the knee joint during the swing phase is easier to visualize. It starts fairly well extended and during acceleration flexion occurs, which prevents the foot from hitting the floor at midswing. This is followed by an extension during deceleration.

The ankle joint starts the swing phase in a state of plantar flexion since this act was essential in what is called "push off." During acceleration dorsiflexion occurs to prevent the toes from hitting the floor at midswing, and this is followed by a plantar flexion during deceleration.

The stance phase is divided into heel strike, midstance, and push off. For the most part, the actions are just the reverse of those just described for the swing phase.

The hip joint undergoes extension, a medial rotation, and an adduction. These are depicted in Figures 6-81 A, B, and C, respectively. The knee starts this phase almost fully extended, becomes fully extended at midstance and in push off. The ankle joint is slightly plantar flexed at heel strike and the digits are extended. This is followed by a gradual increase in plantar flexion of the ankle so that the foot will not slap down on the floor. Dorsiflexion then occurs at midstance and gradually increases until it is rapidly reversed (plantar flexion) at push off. This push off is accompanied by an extension at the metatarsophalangeal joints and a flexion at the interphalangeal joints. In fact, the great toe is very important in this act. Patients who have lost the last four digits can push off with relative ease, but those who have lost the great toe have considerable difficulty.

The above knowledge is important and should be learned by visualizing the act of walking. Many diagnoses can be made by noting variations from the normal pattern. In the section on nerves we will relate nerve injuries to the above description of walking.

MUSCLE COMPARTMENTS AND CLEFTS. Another way to think of the muscles is in terms of the fascial compartments in which they are located. This becomes particularly important in places where infections are prevalent, such as in the sole of the foot. Infection can be predicted to follow these fascial membranes, and at the same time to be limited temporarily by them.

The first compartment of importance is that containing the iliopsoas muscle. Infection in the abdominal cavity, or even as high as the lower part of the thoracic cavity, within the psoas fascia or the iliacus fascia can appear at the lesser trochanter of the femur.

Although each muscle is surrounded by its own fascia, certain layers are thickened to form intermuscular septa and, therefore, muscular compartments containing groups of muscles. In the thigh there is an extensor compartment anteriorly, an adductor compartment medially, and a flexor compartment posteriorly (Fig. 6-21).

The interosseus membrane divides the leg into anterior (extensor) and posterior (flexor) compartments; in addition there is a lateral (evertor) compartment. The lateral compartment is separated from the anterior by the anterior intermuscular septum and from the posterior compartment by the posterior intermuscular septum. The posterior compartment is divided into a superficial (gastrocnemius, soleus and plantaris) and deep (flexor) compartments (Fig. 6-43).

The sole of the foot is not only divided into compartments but also exhibits clefts or spaces within these compartments. The medial and lateral intermuscular septa divide the foot into a medial compartment for the muscles of the great toe, a lateral for the muscles of the fifth digit, and a central compartment containing the flexor digitorum longus tendons, the flexor digitorum brevis, quadratus plantae, and lumbrical muscles. A deep compartment contains the adductor hallucis and interosseus muscles. The clefts (Fig. 6-61) are located as follows: in the lateral compartment—the lateral cleft; in the medial compartment—the medial cleft; in the central compartment—one between the plantar aponeurosis and the flexor digitorum brevis, a second between the latter muscle and the quadratus plantae, and a third deep to this muscle; and in the space between the adductor hal-

lucis and interosseus muscles—the deep median cleft. In addition, a dorsal subaponeurotic cleft occurs deep to the extensor tendons.

NERVES

The nerve supply to the lower limb is derived from the lumbosacral plexus, which is diagramed in Figures 6-22 and 6-28. These are all ventral rami. Pain from the lower limb is very common.

The **cutaneous nerve supply** to the lower limb is presented most easily by diagram. Although there has been a considerable disagreement on the exact distribution of the cutaneous nerves, Figures 6-82 and 6-83 show two methods of consideration of this item, Figure 6-82 being the distribution of the cutaneous nerves while Figure 6-83 presents the cutaneous innervation according to dermatomes—that region of the skin innervated by a single spinal cord segment. This knowledge is important clinically, for if an area of anesthesia is found to fit one of those presented in Figure 6-82, it is undoubtedly due to a nerve injury; contrariwise, if the anesthesia matches an area indicated in Figure 6-83, it is indicative of spinal cord or nerve root injury. In fact, the exact location of tumors on the spinal cord can often be determined by such information. It should be noted that the pattern of the dermatomes is further evidence of the medial rotation of the lower limb during development.

Under the systemic summary of muscles, it was noted that muscles developed from two main masses, a ventral or preaxial, and a dorsal or postaxial. Accordingly, the ventral rami divide into two branches, one to each muscle mass, which are called preaxial and postaxial nerves respectively. As might be expected, preaxial nerves arise from the anterior or ventral surface of the plexus while the postaxial arise from the posterior or dorsal aspect. The preaxial nerves innervate flexors and adductors (clear in Figures 6-22 and 6-28) while the postaxial nerves innervate extensors and abductors (shaded in Figures 6-22 and 6-28).

The nerve supply to muscles of the lower limb, diagramed in Figures 6-84 through 6-88, is easily remembered. The **inferior gluteal** nerve enters the gluteal region in a position inferior to the piriformis muscle and so will naturally innervate the gluteal muscle most inferiorly placed—the gluteus maximus—while the **superior gluteal** nerve courses superior to the piriformis and inner-

REREAD

The Lower Limb

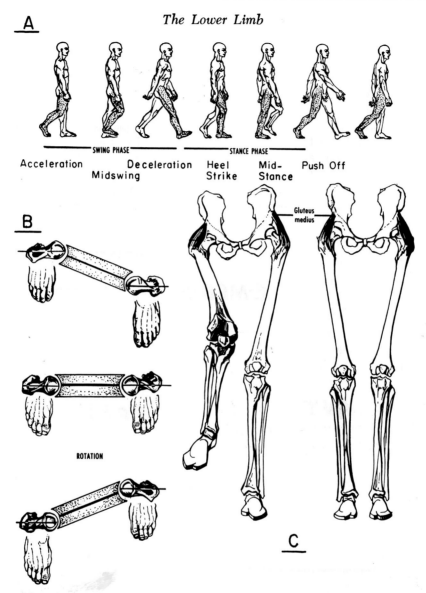

Figure 6-81. *Movements of the lower limb in walking: (A) the swing and stance phases of the right lower limb, (B) rotation at the hip joints, (C) action of the gluteus medius in preventing too great a tilt of the pelvis toward the side of the freely moving limb. (Modified from Gardiner, Gray, and O'Rahilly,* Anatomy, *courtesy of W. B. Saunders Co.)*

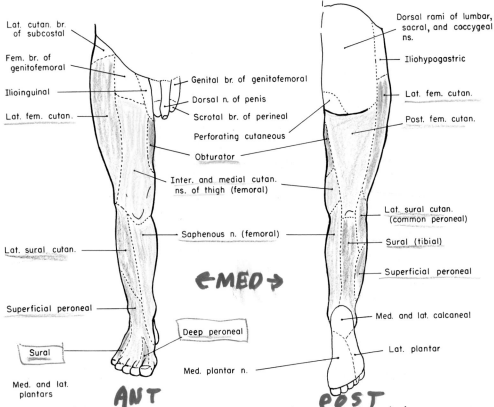

Figure 6-82. Cutaneous nerve distribution to the lower limb.

vates the gluteal muscles most superiorly located—the gluteus medius and minimus; it courses between these two muscles and terminates by innervating the tensor fascia lata muscle. The remaining muscles in the gluteal region are anteriorly placed and are innervated by preaxial nerves arising from the anterior surface of the sacral plexus. These nerves are the **nerve to the obturator internus,** which sends a branch to the superior gemellus, and the **nerve to the quadratus femoris** muscle, which sends a branch to the inferior gemellus. They both enter the gluteal region via the greater ischiadic foramen in a position inferior to the piriformis muscle and then course on the deep surface of the muscles to be innervated.

It is quite obvious that the **femoral nerve** enters the thigh by coursing posterior to the inguinal ligament and innervates the muscles on the anterior surface of the thigh, the **obturator** those on the medial surface by traversing the obturator foramen and dividing into anterior and posterior divisions which course on the anterior and posterior surfaces of the adductor brevis muscle, and the **ischiadic** those on the posterior surface. Two exceptions to this rule are the fact that the pectineus muscle may be innervated by either the femoral or the obturator, and the adductor magnus is innervated by both the obturator and ischiadic.

It is equally obvious that the ischiadic descends on the posterior surface of the thigh deep to the long head of the biceps femoris muscle and divides into tibial and **common peroneal** nerves, and that the latter winds around the fibula to gain the anterior surface of the leg and divides into superficial and deep branches. The most superficial muscles are the evertors and it is logical to have them innervated by the superficial branch; this leaves the remaining muscles on the anterior surface of the leg to be innervated by the deep branch.

Nobody could miss the fact that the muscles on the posterior surface of the leg must be innervated by the **tibial** nerve, since this nerve descends throughout the

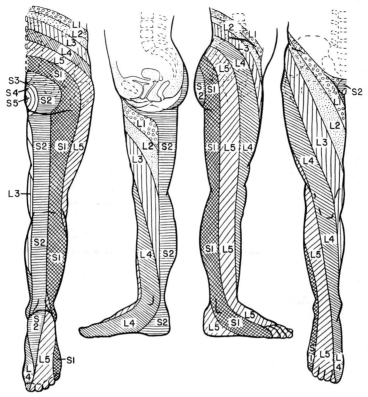

Figure 6-83. *Dermatome chart of the lower limb based on hyposensitivity to painful stimuli after loss of a single nerve root. (From J. J. Keegan and F. D. Garrett, Anat. Record 102: 409, 1948, courtesy of the authors and the Wistar Institute.) Other workers, using different methods, have found distinctly different patterns.*

entire length of the posterior surface of the leg between the superficial and deep muscles. This nerve divides terminally into medial and lateral plantars and here some memorization is required. But since the **medial plantar** nerve innervates fewer muscles (abductor hallucis, flexor hallucis brevis, flexor digitorum brevis, and the first lumbrical), it is easier to learn this list with the knowledge that all the remaining muscles must be innervated by the **lateral plantar** nerve.

The latter nerve does so by coursing deep to the flexor digitorum brevis muscle to reach the superficial muscles on the lateral side of the foot and then sending a deep nerve transversely across the foot to reach the deeper muscles.

It is of interest to note the results of injury to these main nerves of the lower limb. Naturally, whenever the nerve supply to a muscle is lost, that muscle no longer functions and the antagonistic muscles move the affected area away from the injury in almost all cases. For example, if the flexors of the leg are denervated, the extensors will move the leg to the extended position and no flexion can take place.

If the femoral nerve is cut as it enters the thigh, the extensors of the leg will be functionless, so the leg will be flexed. Flexion of the thigh will be weakened due to the loss of the rectus femoris and sartorius, and possibly of the pectineus muscle, but flexion of the thigh will still be possible with the powerful iliopsoas muscle. The effect of such an injury on walking is rather drastic. Although the patient may be able to flex the thigh, the inability to extend the leg is quite debilitating.

If the obturator nerve is cut, the adductor muscles of

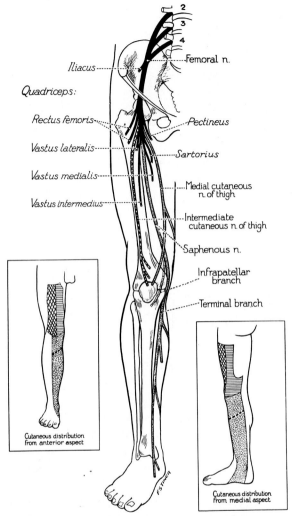

Figure 6-84. *Diagram of the origin and distribution of the femoral nerve. Branches to joints have been omitted. Dotted lines in field of saphenous nerve distribution separate the area covered by the infrapatellar branch from the remaining area. (From Haymaker and Woodhall,* Peripheral Nerve Injuries, *courtesy of W. B. Saunders Co.)*

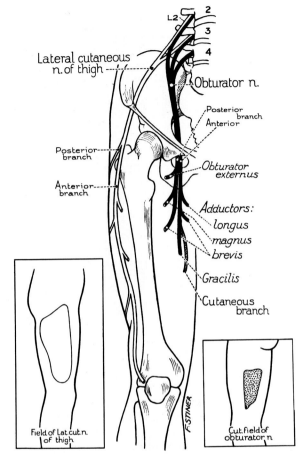

Figure 6-85. *Diagram of the origin and distribution of the obturator and lateral femoral cutaneous nerves. Branches to joints have been omitted. (From Haymaker and Woodhall,* Peripheral Nerve Injuries, *courtesy of W. B. Saunders Co.)*

the thigh will be lost except for a part of the adductor magnus, so the thigh will be abducted. This results in too great a lateral swing in the swing phase of walking.

Cutting the ischiadic results in a very severe muscle loss. The only remaining muscles will be those in the gluteal region, and those on the medial and anterior surfaces of the thigh. Accordingly, the leg would be ex-

tended, there would be a weakening of extension of the thigh, and a total loss of control over the foot.

Injury to the common peroneal nerve is quite frequent because of its location. It winds around the neck of the fibula at about car bumper height and can be crushed between these two hard structures. When this occurs, the extensors and evertors are denervated, so the toes become flexed, and the foot plantar flexed and inverted. If only the superficial branch is involved, the foot will be inverted but other actions would be quite normal. Loss of the deep branch denervates the extensors of the digits and the tibialis anterior, so the foot exhibits a flexion of the toes and a plantar flexion of the foot, and possibly an eversion. Patients with injuries to the common peroneal

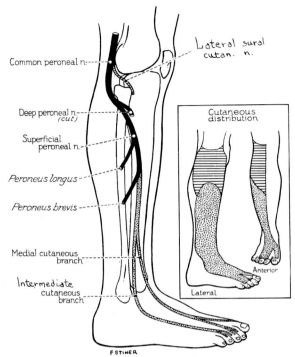

Figure 6-86. *Diagram of the distribution of the superficial peroneal nerve. The peroneal communicating branch has been omitted. (From Haymaker and Woodhall,* Peripheral Nerve Injuries, *courtesy of W. B. Saunders Co.)*

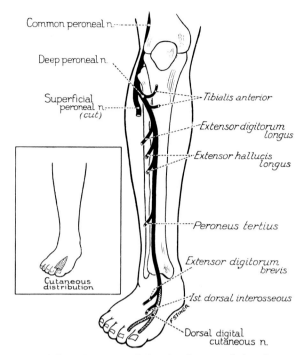

Figure 6-87. *Diagram of the distribution of the deep peroneal nerve. Branch to interosseus muscle is probably afferent. (From Haymaker and Woodhall,* Peripheral Nerve Injuries, *courtesy of W. B. Saunders Co.)*

nerve exhibit an abnormal heel strike. Since the foot is plantar flexed, the toes hit the floor first followed by a banging of the heel on the floor.

Cutting the tibial nerve eliminates almost all plantar flexors of the foot and digits and one of the powerful invertors. The toes would, therefore, be extended, and the foot dorsiflexed and everted. When walking the heel would hit the floor in a normal fashion but would be followed by a slapping on the floor of the anterior part of the foot.

Returning to the hip joint, you will remember that lateral rotation and abduction occurs during the swing phase of walking, and opposite movements during the stance phase. Therefore, injury of the nerve supply to the rotators is bothersome in walking, and injury to the superior gluteal nerve will obviously result in too great a tilt of the pelvis toward the side opposite the injury.

Injury to these nerves also results in a loss of cutaneous innervation to the respective areas involved.

AUTONOMIC NERVOUS SYSTEM. The autonomic nervous system reaches the smooth muscles in the vessels of the lower limb, the sweat glands, and the arrector pili muscles. These structures are innervated entirely by the sympathetic portion of the autonomic nervous system; no parasympathetic fibers have, as yet, been described going to either limb or the trunk. Preganglionic sympathetic fibers start in the intermediolateral cell column of the upper lumbar spinal cord segments, leave the cord via ventral roots, reach the spinal nerves, and then course to the sympathetic chain via white rami communicantes. All of these fibers do not synapse at these ganglia but descend in the chain to ganglia in the lower lumbar and sacral regions where a synapse with postganglionic fibers occurs. These postganglionic fibers reach the nerves of the lumbosacral plexus via gray rami communicantes and are distributed with these nerves to the lower limb. Some of these postganglionic fibers course on the arteries as well.

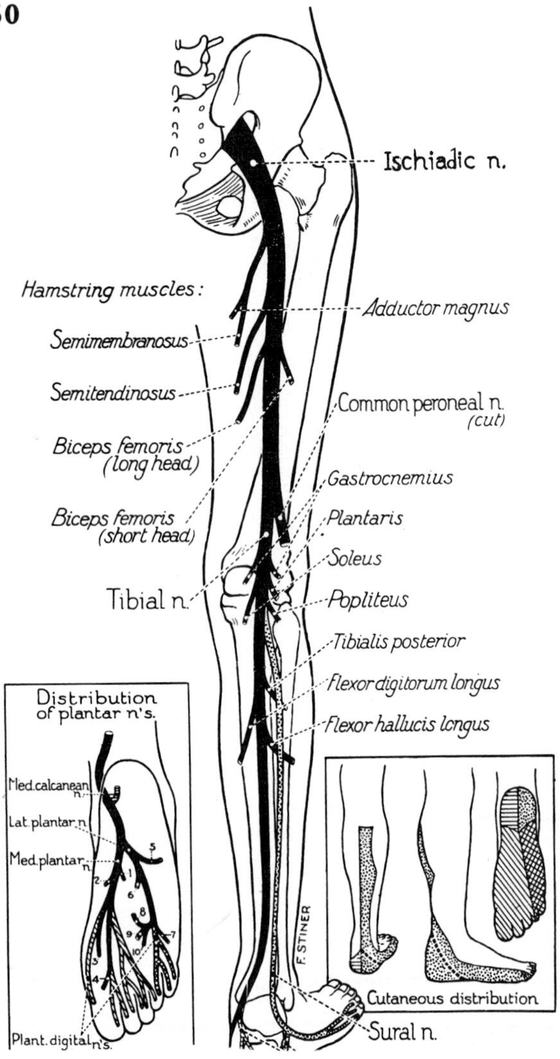

Figure 6-88. *Diagram of the distribution of the ischiadic, tibial, medial, and lateral plantar nerves. Articular nerves have been omitted. The dotted lines represent the transition between sural and tibial nerves. The numbered nerves are (1) to flexor digitorum brevis, (2) to abductor hallucis, (3) to flexor hallucis brevis, (4) to first lumbrical, (5) to abductor digiti minimi, (6) to quadratus plantae, (7) to flexor digiti minimi brevis, (8) to adductor hallucis, (9) to interossei, (10) to second, third, and fourth lumbricals. Note that this diagram shows the sural nerve giving innervation to the skin between the fourth and fifth digits in addition to the lateral side of the fifth digit, a distribution often found. This text has described the sural nerve as being formed by the joining of the medial sural cutaneous nerve and the peroneal communicating branch of the common peroneal. (From Haymaker and Woodhall,* Peripheral Nerve Injuries, *courtesy of W. B. Saunders Co.)*

READ

Stimulation of this system causes a vasodilation in the muscles, a vasoconstriction in the skin, a contraction of the arrector pili muscles, and an increase in activity of the sweat glands. Visceral afferent fibers from the blood vessels are present and follow the nerves and dorsal roots back to the cord.

ARTERIES

There are four main arteries entering the lower limb: the femoral, superior gluteal, inferior gluteal, and the obturator.* They are diagramed in Figure 6-89.

Since the gluteals and obturator vascularize a relatively small area, we will describe these first. The **superior gluteal artery** is the terminal branch of the posterior division of the internal iliac artery. It enters the gluteal region via the greater ischiadic foramen in a position superior to the piriformis muscle. It immediately divides into a **superficial division,** which enters the substance of the gluteus maximus muscle, and a **deep division,** which divides into two branches, both of which course between the gluteus medius and minimus muscles. The **superior** of the two branches courses along the superior edge of the gluteus minimus and continues as far as the deep surface of the tensor fascia lata muscle where it anastomoses with the ascending branch of the lateral femoral circumflex artery. The **inferior** branch courses between the two muscles at a more inferior level and ends in the trochanteric fossa. This artery anastomoses with the ascending branch of both the medial and lateral femoral circumflex arteries.

The **inferior gluteal** artery is the terminal branch of the anterior division of the internal iliac. It courses through the greater ischiadic foramen in a position that is inferior to the piriformis and enters the gluteal region. This artery courses inferiorly deep to the gluteus maximus muscle which it supplies. It has branches to the other surrounding muscles, to the ischiadic nerve, and to the posterior surface of the coccyx. This artery anastomoses with the first perforating branch of the profunda femoris and with the medial and lateral femoral circumflex vessels to form the cruciate anastomosis.

The **obturator** artery may arise from the internal iliac or from the deep epigastric. In either case, it traverses the

---

*H. D. Senior, 1919, On the development of the arteries of the human lower extremity, *Am. J. Anat.* 25: 55. A long but famous paper describing the changes that occur in the arterial channels as the limb bud grows; it makes the large percentage of "normal" arterial patterns difficult to understand.

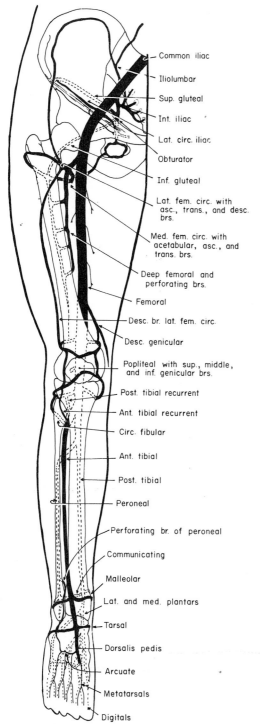

Common iliac

Iliolumbar

Sup. gluteal

Int. iliac

Lat. circ. iliac

Obturator

Inf. gluteal

Lat. fem. circ. with asc., trans., and desc. brs.

Med. fem. circ. with acetabular, asc., and trans. brs.

Deep femoral and perforating brs.

Femoral

Desc. br. lat. fem. circ.

Desc. genicular

Popliteal with sup., middle, and inf. genicular brs.

Post. tibial recurrent

Ant. tibial recurrent

Circ. fibular

Ant. tibial

Post. tibial

Peroneal

Perforating br. of peroneal

Communicating

Malleolar

Lat. and med. plantars

Tarsal

Dorsalis pedis

Arcuate

Metatarsals

Digitals

Figure 6-89. *Summary of the main arteries to the lower limb. More detail can be seen in Figures 6-29, 6-44, 6-51, and 6-62.*

obturator foramen and divides into two branches which course in a circular fashion on the obturator membrane. In the thigh, this artery has **muscular branches** and sends an **acetabular branch** into the acetabular fossa via the acetabular notch. The obturator anastomoses with the acetabular branch of the medial femoral circumflex.

The **femoral** artery is the direct continuation of the external iliac, becoming femoral after passing deep to the inguinal ligament. Typical of major arteries, it is located on the flexor surface of the hip joint—the anterior side. It is contained in the most lateral compartment of the femoral sheath. Almost immediately, three or four small branches emerge from the femoral: (1) the **superficial epigastric** which courses superiorly and enters the abdominal wall remaining in the superficial fascia, (2) the **superficial circumflex iliac** which courses laterally just inferior to the inguinal ligament to reach as far as the anterior superior iliac spine, and (3) the **external pudendal** arteries, one coursing superficial and one deep to the spermatic cord or ligamentum teres (uteri) to reach the scrotum or a labium majus.

The **profunda femoral** artery arises from the femoral near the middle of the femoral triangle. This artery, in turn, gives off the medial and lateral femoral circumflex branches although these branches very often come off from the femoral directly. The **medial femoral circumflex** disappears by coursing between the iliopsoas and pectineus muscles, giving blood supply particularly to the latter; it gives a branch to the **acetabulum,** continues posteriorly to divide into **ascending** and **transverse branches,** the former ending in the trochanteric fossa anastomosing with the superior gluteal, the latter coursing between the quadratus femoris and adductor magnus muscles to end in muscles in this area.

The **lateral femoral circumflex** courses laterally and divides into three branches, an **ascending** which courses superiorly deep to the tensor fascia lata muscle and anastomoses with the superior gluteal artery, a **transverse** which sinks into the vastus lateralis, and a **descending** which descends on the vastus lateralis muscle all the way to the area of the knee joint and anastomoses with branches of the popliteal artery.

The profunda femoral artery continues inferiorly close to the femur and just deep to the adductor longus muscle. It gives off four **perforating** arteries which perforate the tendon of the adductor magnus muscle, give off branches to the hamstring muscles, and then continue, to end in the substance of the vastus lateralis muscle. The first perforating takes part in the cruciate anastomosis, the second is the nutrient vessel of the femur, and the fourth anastomoses with branches of the popliteal artery.

The transverse branch of both femoral circumflex arteries anastomose with branches of the inferior gluteal and first perforating arteries to form a cross—the **cruciate anastomosis.**

To return to the femoral, this large vessel continues inferiorly toward the apex of the femoral triangle and enters the adductor canal. After traversing this canal it pierces the tendon of the adductor magnus muscle to become the popliteal artery, but gives off its descending genicular branch before doing so. This latter branch divides into a branch that accompanies the saphenous nerve to the side of the knee and another that reaches the knee by passing through the vastus medialis muscle.

Once again the main artery, the **popliteal,** is on the flexor side of the joint—in this case the knee. It is deep to its accompanying vein which is, in turn, deep to the nerves. This artery gives off **muscular** branches and five **genicular** branches that form an anastomotic network around the knee joint. The **middle genicular** pierces the capsule to enter the knee joint, the two **superior geniculars** wind around the femur, and the two **inferior geniculars** take a similar course inferiorly. We have already seen that this series of vessels is joined by the descending branch of the lateral femoral circumflex, the descending genicular, and possibly by the last perforating, and we will presently see that it is also joined by three branches of the anterior tibial artery.

The popliteal artery divides into anterior and posterior tibial arteries at the distal edge of the popliteus muscle.

The **anterior tibial** artery almost immediately gives off its **posterior tibial recurrent** artery and then its **circumflex fibular** artery, which winds around the fibula. Both these vessels join in the anastomosis of vessels around the knee joint. The anterior tibial artery then pierces the interosseus membrane to gain the anterior surface of the leg, gives off an **anterior tibial recurrent** branch, which also joins the other vessels around the knee, continues inferiorly in the anterior compartment, and terminates near the ankle as the dorsalis pedis artery. In addition to **muscular** branches, there are two **malleolar** branches, one medial and one lateral, that ramify over the two malleoli.

The **dorsalis pedis** artery continues inferiorly on the dorsum of the foot and terminates by diving into the first

intermetatarsal space to aid in formation of the deep plantar arch. It has **tarsal** branches that ramify on the dorsum of the foot, an **arcuate** branch that courses laterally and sends branches to the clefts between the four small toes, and a **first dorsal metatarsal** artery that courses to the first cleft to reach contiguous sides of the first and second toes. The end of the dorsalis pedis artery, that part that joins the plantar arch, is called the **deep plantar** artery.

To return to the posterior tibial artery, about an inch or so after its origin it gives off its **peroneal branch,** which continues inferiorly deep to the flexor hallucis longus muscle, gives off **muscular** branches and a **nutrient** branch to the fibula, continues inferiorly passing posterior to the inferior tibiofibular articulation, gives off a **perforating branch** that perforates the interosseus membrane and ramifies over the lateral malleolus (anastomosing with the other vessels there), and then terminates by dividing into the **lateral calcaneal** branches.

After giving off the peroneal branch, the **posterior tibial** continues inferiorly between the superficial and deep muscles and in company with the tibial nerve until it reaches the ankle, where it divides into medial and lateral plantar arteries. Its branches are the usual **muscular,** the **nutrient** to the tibia, a **communicating** branch to the peroneal, a **malleolar** branch to the medial malleolus, and a **calcaneal** branch to the medial surface of the heel.

The **medial plantar** artery courses deep to the abductor hallucis muscle and comes to lie between this muscle and the flexor digitorum brevis where, after giving off **muscular** branches, it breaks up into **digital** branches, that course to the clefts between digits 1–4. These branches anastomose with the metatarsal arteries from the deep plantar arch. They also connect with the dorsal digital arteries from the arcuate branch of the dorsalis pedis artery.

The **lateral plantar** artery also courses deep to the abductor hallucis muscle and also the flexor digitorum brevis muscle to gain the groove between this latter muscle and the abductor digiti minimi. It gives off a **medial calcaneal, muscular,** and **digital** for the last toe, and then forms the **deep plantar arch.** This arch courses on the metatarsal bones and interosseus muscles and is accompanied by the deep branch of the lateral plantar nerve. The **perforating** arteries are given off from this arch, as well as the metatarsal arteries, which course toward the clefts of the digits and join with the digitals from the medial and lateral plantar arteries. They also send another set of **perforating** arteries to the dorsal vascular system. Therefore, the plantar and dorsal vessels are connected by two sets of perforating arteries as well as by the continuity of the plantar arch with the dorsalis pedis artery. Each digit is given blood supply by four arteries—two dorsally and two ventrally placed.

A good way to review the arterial supply to the lower limb is by determining how the blood could possibly get around a tie of the main vascular channel at various levels. (Figures 6-29, 6-44, and 6-51) will be helpful in this endeavor.

An initial plan to list the anastomotic channels around ties in these main channels was abandoned for fear the student would attempt to memorize such lists. One should make such lists and then there might be some possibility of remembering these important anastomoses. The process of thinking involved is to recall branches of the main arterial channel that arise proximal to the tie and to determine whether these branches join with any branches of the main channel distal to the tie. Since there are no valves in arteries, blood can flow in the reverse direction and return to the main channel at a point distal to the tie.

For example, suppose a tie were made in the external iliac artery. One would immediately think of the internal iliac and its branches. The superior and inferior gluteal branches come to mind immediately, since they participate in an important anastomosis with the medial and lateral femoral circumflex, which can conduct blood back to the femoral artery distal to the tie. The obturator artery may be involved in two ways: (1) because of its pubic branch anastomosing with the deep epigastric branch of the external iliac at a point distal to the tie, and (2) because the obturator anastomoses with the medial femoral circumflex around the acetabular notch. The iliolumbar branch of the internal iliac and the lumbar arteries of the abdominal aorta anastomose with the deep and superficial circumflex iliac branches of the external iliac and femoral arteries respectively. Another possibility is the joining of the scrotal or labial branches of the perineal branch of the internal pudendal with the superficial and deep external pudendal branches of the femoral.

One should always think of vessels that arise distal to the tie to make sure the answer is complete. For instance, the inferior epigastric comes to mind; since this anastomoses with the superior epigastric branch of the internal thoracic artery, which, in turn, arises from the subclavian, this bypasses our tie very nicely.

Similar thinking should be done after a tie at the point where the external iliac becomes the femoral, the femoral proximal to the emergence of the profunda femoral, the femoral in the adductor canal (the descending branch of the lateral femoral circumflex and the perforating arteries joining with the anastomotic branches around the knee are the only possibilities in this instance), the popliteal, the posterior tibial, and the anterior tibial.

VEINS

There are veins that correspond to the arteries just described. They are often in the form of a comitans rather than a single channel. The main femoral vein lies medial to the artery in the femoral sheath. There are numerous valves in the veins of the lower limb.

One important feature distinguishes the veins from the arteries: there are two sets of veins, one deep (as just mentioned) and another superficially located. Two main superficial channels have been named, one that starts on the medial side of the foot and ascends on the medial side of the leg and thigh to end in the main femoral vein by traversing the saphenous opening. This is the **great or long saphenous vein,** which contains many valves; when these valves become incompetent varicose veins result. The other main channel is the **short saphenous,** which starts on the lateral side of the foot and ends in the popliteal vein in the popliteal fossa. This vein is also subject to varicosities.

LYMPHATICS

The lymphatics of the lower limb end in two groups of nodes, the popliteal and the inguinal.

The popliteal nodes lie along the veins in the popliteal space and drain that portion of the lower limb covered by the short saphenous vein, namely, the lateral surface of the foot and leg. Efferents from these nodes course to the deep inguinal nodes.

The **inguinal nodes** are divided into **superficial** and **deep** sets, the former lying just inferior to the inguinal ligament and alongside the great saphenous vein. These nodes drain the superficial portions of the lower limb except for that area just mentioned as draining into the popliteal nodes, and the external genitalia (except the testes) in both males and females. These superficial nodes, in turn, drain into the deep set that lie along the femoral vein, one usually being found in the femoral canal. These deep nodes drain the efferent lymphatics from the popliteal nodes and the deeper part of the entire lower limb. Efferents from these nodes empty into the external iliac nodes.

It should be noted that the lymphatics in the limb seem to follow the veins rather than the arteries, as found in the cavities.

CONCEPTS: GENERAL RULES ON STRUCTURE OF THE HUMAN BODY*

Those students who started with a study of the lower limb have been exposed to several concepts or general rules about the structure of the human body that should be very helpful in learning the remaining parts.

From the arrangement of the vertebrae, the nerves, and the blood vessels, it is quite obvious that the human being has been derived from and is a segmentally arranged organism. We have seen that the skin does indeed serve as an efficient covering of the body; that it has a segmentally arranged sensory nerve supply, a dermatome being an area innervated by one segment of the spinal cord; that the skin contains arrector pili muscles, smooth muscles in blood vessels and sweat glands that are innervated by the sympathetic portion of the autonomic nervous system, and therefore the so-called "sensory nerves" to the skin are in reality combined sensory and motor nerves and might better be called "cutaneous nerves"; that the skin is bound down to underlying structures by a layer of superficial fascia containing a varying amount of fat in which these cutaneous nerves and superficial veins course.

We have found that many muscles, in their development, migrate to other parts of the body but carry their previously established nerve supply along with

them; that this nerve supply and the vascular supply as well is likely to be found on the deep side of the muscle; that the nerve supply is dependent upon the embryological origin of the muscle; that muscles attach to the osseous elements and by their pull not only cause movement of these osseous structures, and thereby of the body, but also cause elevations to develop on these bones; that the main action of a muscle can usually be determined by a simple mechanical sense if the attachments of the muscle are known, the more movable attachment (the insertion) usually being more revealing, but that there is still considerable discussion about muscle actions, primarily because until recently accurate means of measurement were not available; that each muscle is surrounded by a layer of deep fascia and that groups of muscles are often surrounded by a thicker enveloping fascial layer.

We have also found that arteries, in addition to being accompanied by a vein and a nerve, exhibit frequent variations in pattern on one side of the body or on both sides; that these variations are particularly prevalent in the limbs, since multiple channels develop in a limb bud, most of which disappear with further growth of the limb; that the definitive main arterial channel is located on the flexor side of joints; that anastomoses of arteries, particularly around joints, provide important sidetracks around an occluded main channel, a fact made possible by the absence of valves in arteries; that veins are even more variable, often form intricate plexuses, are divided into a superficial set and a deep set of venous channels in the limbs, and frequently exhibit multiple (a concomitans) rather than single channels.

Differentiation between the lymphatic drainage of various parts of the lower limb has revealed the important fact that lymphatic drainage in the limb seems to follow the venous pathways.

Study of the lower limb has shown us that phylogeny often provides more understanding than ontogeny; that the primary purpose of the lower limb—locomotion—is indeed accomplished in spite of the several problems induced by the change to the upright stature; that the structure of the lower limb, with its massive bones, large and powerful muscles, and comparatively strong articulations, correlates well with its function.

Lastly, I hope you are convinced that the parts of the limb you remember best are those you can visualize easily and those you have thought about enough to understand.

---

*This section is similar to those presented in other chapters of this textbook, but contains special features that relate to the lower limb.

7

HEAD AND NECK

The head and neck are interesting areas of the body, but rather complex. This complexity is due primarily to the fact that there is a great deal of anatomy in a relatively small and compact area. In addition, there is no doubt about the fact that the skull is the most complicated part of the skeleton, that the brain and special senses add to the complexity, and that the substitution of the cranial nerves in all their irregularity for the rather uniform segmental nerves has added to the confusion. This is particularly true since the nerves are coursing in one direction while the arteries are coursing in the opposite direction. The study of the head and neck is a study unto itself; it defies some of the concepts we have already learned.

Although many areas of the head and neck seem to be the property of the specialists—ophthalmologists, otolaryngologists, neurosurgeons, etc.—all physicians must understand head and neck structure to make the original diagnosis. In fact, it is said that over 65 per cent of patients reporting to the family practitioner suffer from upper respiratory infections.

BONES OF THE SKULL*

In former years, students were expected to be able to describe any bone of the skull, and some of them are very complex indeed. Our emphasis will be on the skull as a whole rather than on the individual bones. Nevertheless, a certain amount has to be known about the individual bones to comprehend the anatomy of the head and neck. Accordingly, the individual bones will first be considered briefly, omitting details, and then the skull as a whole will be described.

Do not attempt to memorize this description; familiarize yourself thoroughly enough with the parts mentioned so that they will not be completely strange when mentioned again throughout the description of the

---

*G. R. DeBeer, 1937, *The Development of the Vertebrate Skull*, Clarendon Press, Oxford.

squamous relative to basilar part.

head and neck. Attaching muscles to various processes, and coursing nerves and vessels through the foramina, gives them a functional significance that aids learning. Since this first study is cursory, you should also remember that your preliminary study must be augmented during study of each region of the head and neck.

FRONTAL BONE

The **frontal bone** (Figs. 7-1, 7-2A, 7-3, 7-5, and 7-7) occupies the anterior end of the skull, forms parts of the orbits, and contains paranasal sinuses (Fig. 7-5). It articulates posteriorly with the parietal bones, and inferiorly with the sphenoid, zygomatic, lacrimal, maxillary, ethmoid, and nasal bones. It presents **supraorbital margins** (Fig. 7-1) which form the superior margins of the orbits; each contains a **supraorbital notch or foramen** in which the supraorbital nerve and artery course. The **frontal eminence** is the most pronounced projection of the convex anterior surface. The **zygomatic process** (Fig. 7-2A) is that part which articulates with the zygomatic bone. Two ridges extend superiorly from the zygomatic process; these are the **temporal lines** (Fig. 7-2A), corresponding to the anterior boundary of the temporalis muscle, one of the muscles of mastication. That part of the frontal bone forming the roof of the orbit is the **orbital plate** (Fig. 7-7), and the **nasal part** is that portion just superior to the bridge of the nose. The frontal bone originally consisted of two bones, one on either side of the midline, and a median suture occasionally persists. The grooves on the inside (Fig. 7-7) are formed by the meningeal (*Gr., meninges,* membranes) blood vessels.

PARIETAL BONE

The **parietal bones** (Figs. 7-2A, 7-3, 7-5, and 7-8), as might be expected from the name, are two large bones forming the lateral and superior walls of the cranium. Each parietal bone is concave internally and convex externally, and articulates with its mate medially, the frontal and sphenoid bones anteriorly, the occipital bone posteriorly, and the temporal bone inferiorly. The **temporal lines** (Figs. 7-2 and 7-3) mark the superior limit of the temporalis muscle. The grooves on the inside (Figs. 7-5, 7-7, and 7-8) are formed by the meningeal vessels.

OCCIPITAL BONE

The **occipital bone** (Figs. 7-3, 7-4, 7-5, and 7-7) forms the base of the skull, particularly that part on the posterior basal portion. It contains the opening, the **foramen magnum** (Fig. 7-4), through which the spinal cord joins the brain stem. The bone is divided into parts in relation to this foramen: the **squamous portion** is posterior, the **condylar part** lateral, and the **basilar portion** anterior to the foramen. The basilar part articulates with the sphenoid bone anteriorly and the petrous portions of the temporal bones laterally, while the squamous portion articulates with the mastoid process of the temporal bone laterally and with the parietal bone anteriorly and superiorly. The condylar portion exhibits two **facets** (Fig. 7-4) for articulation with the superior facets of the atlas.

The squamous portion, on the outside surface, shows a midline projection—the **external occipital protuberance** (Figs. 7-4 and 7-5)—and two ridges extending laterally from this protuberance, the **superior nuchal lines** (Fig. 7-4). In addition, ridges can be seen on this portion of the bone that serve as boundaries for areas of muscle attachment. Two lesser lines, the **inferior nuchal lines** (Fig. 7-4), course laterally from the midline about ¾ to 1 inch inferior to the superior nuchal lines.

The inside surface (Figs. 7-5 and 7-7) exhibits two large **fossae** for the cerebellar hemispheres, and an internal occipital crest between these fossae. This surface also has **sulci** where dural venous sinuses are located. This bone aids in formation of the jugular foramen on each side (**jugular processes**), and contains, on each side, the **hypoglossal canal** for the hypoglossal nerve and the **condylar canal** for passage of a vein.

TEMPORAL BONE

The **temporal bone** (Figs. 7-2A, 7-4, and 7-7), located on the lateral side of the head, consists of four portions: a **squamous part,** which extends superiorly and articulates with the parietal and sphenoid bones; a **petrous portion,** which projects into the cranial cavity along the lateral part of its floor; the **mastoid portion,** which is posterior to the external auditory meatus and is a fairly large projection; and the **tympanic part,** which actually contains the external auditory meatus. The squamous portion also contains the **zygomatic process,** which articulates with the

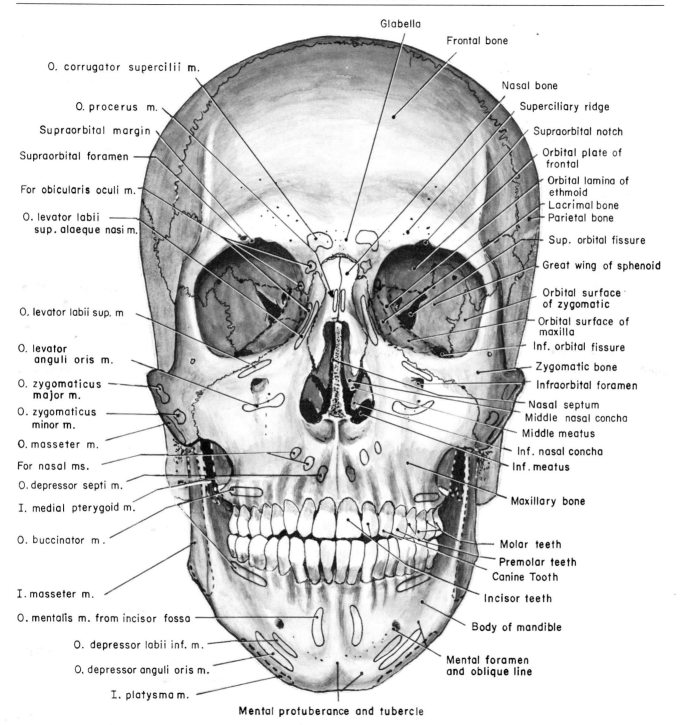

Figure 7-1. Anterior view of skull with origins and insertions of muscles outlined.

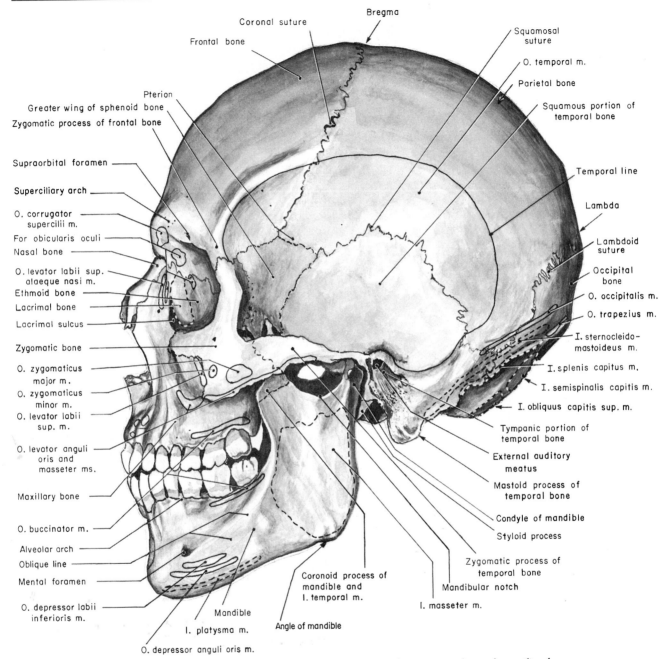

Figure 7-2A. *Lateral view of skull with origins and insertions of muscles outlined.*

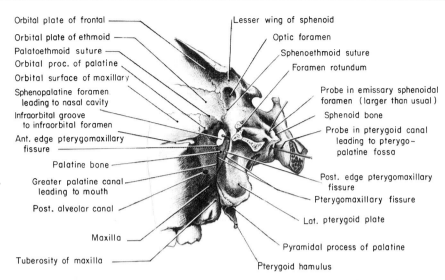

Figure 7-2B. *Lateral view of left pterygomaxillary fissure looking into the pterygopalatine fossa, shown by removal of zygomatic bone, temporal bone, and the greater wing of the sphenoid bone. (From Toldt's* Atlas of Human Anatomy, *courtesy of Macmillan Co., New York.)*

zygomatic bone. The petrous portion is considered to have anterior, posterior, and inferior surfaces, an apex anteromedially, and a base (fused with mastoid process). The mastoid process exhibits a foramen through which courses an emissary vein (connecting the skin and scalp outside with a dural venous sinus on the inside) and a small meningeal artery; inside the mastoid process are **air cells,** which may become infected. The tympanic part, in addition to containing the external auditory meatus, possesses a spicule of bone hanging inferiorly and anteriorly called the **styloid process;** this serves for muscle attachments.

SPHENOID BONE

The **sphenoid bone** (*Gr.,* resembling a wedge—Figs. 7-4, 7-5, 7-6, and 7-7) is a complex butterfly-shaped bone that aids in forming the floor of the cranial cavity and articulates anteriorly with the ethmoid and frontal bones, laterally and posteriorly with the temporal bones, and posteriorly in the midline with the basilar portion of the occipital bone. This bone is said to consist of a **body** and **three pairs of processes—lesser wings, greater wings,** and **pterygoid processes.** The **lesser wings** (Fig. 7-7) are

the most anterior structures and are at a slightly superior plane to the much larger **greater wings,** which make up the floor of the middle fossa of the cranial cavity. The **pterygoid processes** (Fig. 7-5) hang inferiorly from the base of the skull and cannot be seen from inside the cranial cavity. The centrally located body exhibits the **sella turcica** of the sphenoid bone, the location of the hypophysis. The lesser wings are anterior to this sella turcica and contain the **optic foramina** and the **anterior clinoid processes.** The **posterior clinoid processes** are just posterior to the sella turcica. On each side of the sella turcica is an opening called the **foramen lacerum** which in a living person is actually filled in with cartilage.

The greater wing of the sphenoid bone exhibits, from anterior to posterior in a slightly curved line, the **superior orbital fissure, foramen rotundum, foramen ovale,** and **foramen spinosum** (see Fig. 7-7).

The **pterygoid process,** as mentioned previously, is suspended from the base of the sphenoid bone and exhibits **medial and lateral laminae** and a **pterygoid fossa** between these laminae. The **pterygoid hamulus** (Figs. 7-4 and 7-5) projects inferiorly from the medial lamina. It should be noted that the sphenoid bone contains one of the **paranasal air sinuses.**

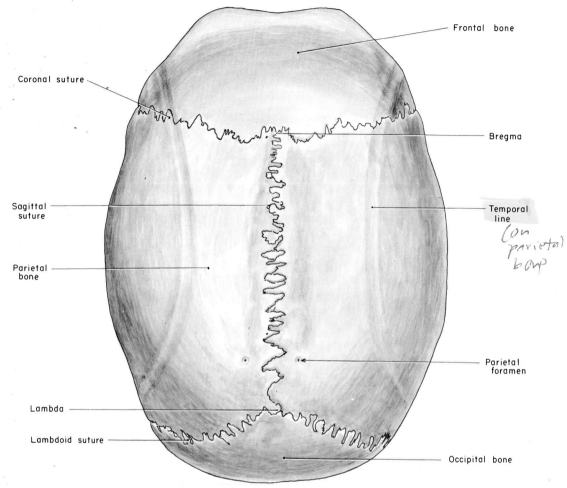

Figure 7-3. *Superior view of skull.*

ETHMOID BONE

The **ethmoid bone** (*L.,* sieve) (Figs. 7-5 and 7-6), which defies description, lies anterior to the sphenoid bone and forms parts of the lateral wall and septum of the nose or nasal cavity. It possesses a **perpendicular plate,** which aids in formation of the nasal septum, a **cribriform plate** (Fig. 7-7), which can be seen in the anterior fossa of the cranial cavity, and a **crista galli** (Fig. 7-7), which is a small crest in the middle of the cribriform plate. It also helps in formation of the orbit (Figs. 7-1 and 7-2), forms the **superior and middle conchae** of the lateral wall of the nose (Fig. 7-6), and contains many **paranasal air**

sinuses. Its **uncinate process** (Fig. 7-6) is a projection inferiorly into the lateral wall of the nasal cavity in the area of the middle meatus (the area under the middle concha).

INFERIOR NASAL CONCHA

The **inferior nasal concha** (*L.,* shell) is a small bone forming the inferior portion of the lateral wall of the nose (Fig. 7-6). It articulates anteriorly with the maxillary, posteriorly with the palatine, and superiorly with the lacrimal and ethmoid bones.

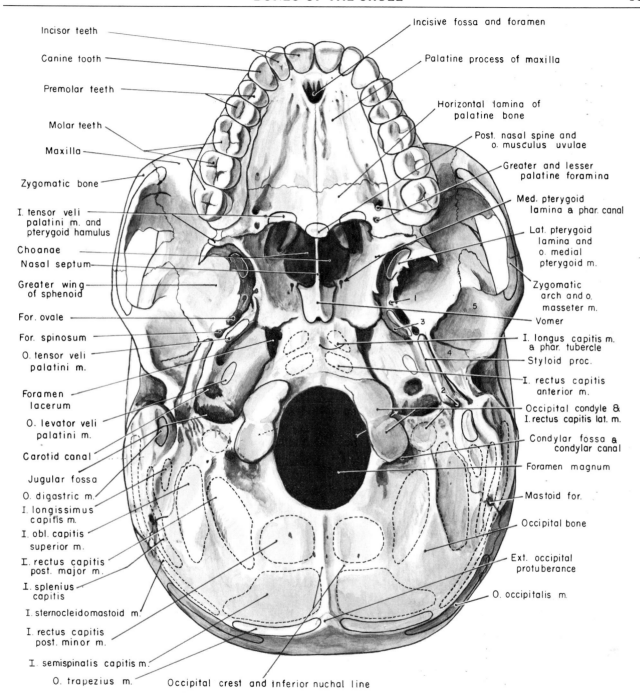

Incisor teeth

Canine tooth

Premolar teeth

Molar teeth

Maxilla

Zygomatic bone

I. tensor veli palatini m. and pterygoid hamulus

Choanae

Nasal septum

Greater wing of sphenoid

For. ovale

For. spinosum

O. tensor veli palatini m.

Foramen lacerum

O. levator veli palatini m.

Carotid canal

Jugular fossa

O. digastric m.

I. longissimus capitis m.

I. obl. capitis superior m.

I. rectus capitis post. major m.

I. splenius capitis

I. sternocleidomastoid m.

I. rectus capitis post. minor m.

I. semispinalis capitis m.

O. trapezius m.

Incisive fossa and foramen

Palatine process of maxilla

Horizontal lamina of palatine bone

Post. nasal spine and o. musculus uvulae

Greater and lesser palatine foramina

Med. pterygoid lamina & phar. canal

Lat. pterygoid lamina and o. medial pterygoid m.

Zygomatic arch and o. masseter m.

Vomer

I. longus capitis m. & phar. tubercle

Styloid proc.

I. rectus capitis anterior m.

Occipital condyle & I. rectus capitis lat. m.

Condylar fossa & condylar canal

Foramen magnum

Mastoid for.

Occipital bone

Ext. occipital protuberance

O. occipitalis m.

Occipital crest and inferior nuchal line

Figure 7-4. *Inferior view of skull with origins and insertions of muscles outlined: (1) emissary foramen (Vesalius), (2) stylomastoid foramen, (3) sulcus for auditory tube, (4) mandibular fossa, (5) mandibular tubercle. Arrow is in hypoglossal canal.*

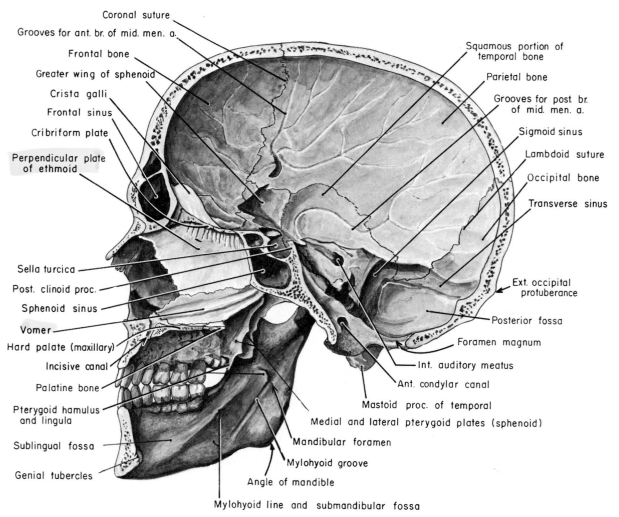

Figure 7-5. Midsagittal section of the skull.

LACRIMAL BONE

The **lacrimal bone** (*L.,* a tear) is a thin, small bone found on the medial side of the orbit (Fig. 7-1); it contains the first part of the **nasolacrimal canal** (Fig. 7-49).

VOMER

The **vomer** (*L.,* ploughshare) is a small bone located in the posterior part of the nasal cavity. It forms the posterior portion of the nasal septum and consists of winglike **alae** in addition to the **septal portion** (Fig. 7-5).

NASAL BONE

The **nasal bone** forms the bony part of the nose (Fig. 7-1). It articulates superiorly with the frontal and posteriorly with the maxillary bones. It also articulates, via the septum, with the ethmoid bone.

MAXILLARY BONE

The **maxillary bone** (*L.,* jawbone) is located in the anterior part of the skull; it forms a large part of the face, particularly that part between the orbits and the upper

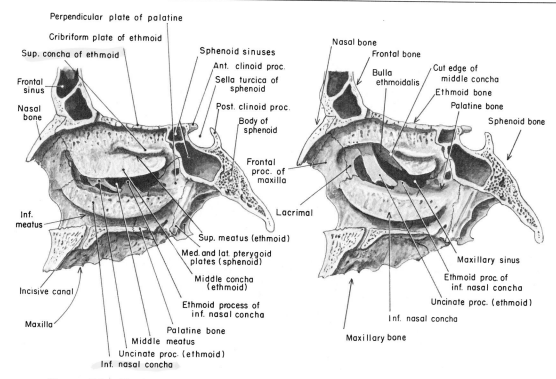

Figure 7-6. *The lateral wall of the nasal cavity. On the right, a portion of the middle concha has been removed.*

teeth. It forms the upper jaw and a great portion of the roof of the mouth (Figs. 7-1, 7-2A, and 7-4). It possesses a **body,** a **frontal process** that articulates with the frontal, lacrimal and nasal bones, a **zygomatic process** that articulates with the zygomatic bone, an **alveolar process** for the attachment of the teeth, and a **palatine process** that articulates posteriorly with the palatine bone in the roof of the mouth. This bone possesses a very large **paranasal sinus.**

PALATINE BONE

The **palatine bone** is difficult to describe, but it does possess a **horizontal portion** (Fig. 7-4), which forms the posterior part of the hard palate, and a **perpendicular portion** (Fig. 7-6), which contains the **palatine canals.** This perpendicular portion is anterior to the pterygoid processes of the sphenoid bone (Fig. 7-2B), and posterior to the ethmoid bone and the inferior nasal concha. It

therefore aids in forming the posterior part of the lateral wall of the nasal cavity.

ZYGOMATIC BONE

The **zygomatic** (*Gr.,* bar) **bone** is the cheek bone and possesses two processes (Figs. 7-1 and 7-2A). The **frontal process** articulates with the frontal bone, and the **temporal process** with the temporal bone. The zygomatic bone also articulates with the maxillary and sphenoid bones.

HYOID BONE

Although the **hyoid** (*Gr.,* U-shaped) **bone** is actually not a bone of the skull, it should be investigated at this time. This bone is not articulated with the skull but simply hangs in the neck region, suspended by muscles and

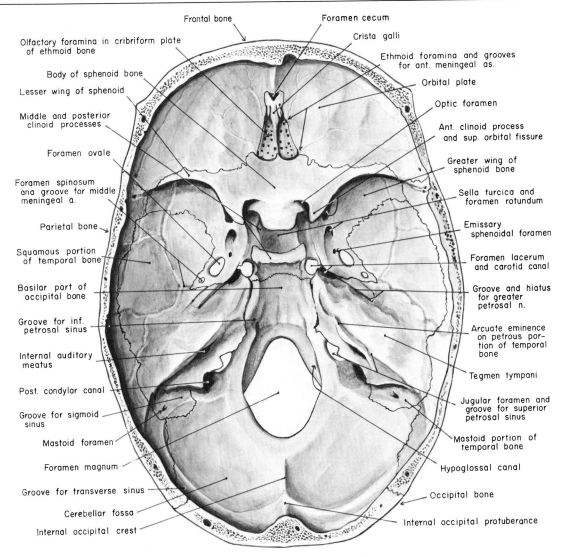

Frontal bone

Foramen cecum

Olfactory foramina in cribriform plate of ethmoid bone

Crista galli

Ethmoid foramina and grooves for ant. meningeal as.

Body of sphenoid bone

Orbital plate

Lesser wing of sphenoid

Optic foramen

Middle and posterior clinoid processes

Ant. clinoid process and sup. orbital fissure

Foramen ovale

Greater wing of sphenoid bone

Foramen spinosum and groove for middle meningeal a.

Sella turcica and foramen rotundum

Parietal bone

Emissary sphenoidal foramen

Squamous portion of temporal bone

Foramen lacerum and carotid canal

Basilar part of occipital bone

Groove and hiatus for greater petrosal n.

Groove for inf. petrosal sinus

Arcuate eminence on petrous portion of temporal bone

Internal auditory meatus

Tegmen tympani

Post. condylar canal

Jugular foramen and groove for superior petrosal sinus

Groove for sigmoid sinus

Mastoid portion of temporal bone

Mastoid foramen

Hypoglossal canal

Foramen magnum

Occipital bone

Groove for transverse sinus

Cerebellar fossa

Internal occipital protuberance

Internal occipital crest

Figure 7-7. *The floor of the cranial cavity.*

ligaments. It possesses a **body,** a **greater horn,** and a **lesser horn** (Fig. 7-102 on page 591).

MANDIBLE

The **mandible** (*L.,* to chew) is the lower jaw; it possesses a **body** anteriorly, **rami** posteriorly, and **coronoid** and **condyloid processes** projecting superiorly from the rami. It contains the lower teeth.

The **body** (Fig. 7-1) possesses a **mental protuberance** in the midline anteriorly, and two **mental tubercles** on either side of the protuberance. Extending posteriorly from these tubercles on each side is a ridge of bone that leads directly to the anterior aspect of each ramus; this is known as the **oblique line,** and serves for attachment of the muscles that depress the lower lip. The **mental foramina,** for nerves and vessels of the same name, are located just above this line. The **incisor fossae** are two rather shallow depressions, between the mental protuberance and the roots of the incisor teeth, which serve as sites of origin for the mentalis muscles.

The inside surface of the body (Fig. 7-5) possesses two small projections on either side of the symphysis menti—the **genial tubercles**—which serve for the attachment of muscles. Just lateral to these tubercles are **sublingual fossae** for glands of the same name; the **mylohyoid lines** are ridges that extend posteriorly from these fossae. The shallow areas inferior to these lines house the submandibular glands.

Each **ramus** has the **mandibular foramen** (Fig. 7-5) on the inside surface and a spicule of bone, the **lingula,** just anterior and inferior to this foramen. The **mylohyoid groove** extends inferiorly from the lingula. The **condyloid processes** possess a head and neck, and are the processes by which the mandible articulates with the temporal bones. Those processes that project superiorly, but in a more anterior position than the condyloid processes, are the **coronoid processes** (Fig. 7-2A), which serve for the attachment of the temporalis muscles.

A **mandibular canal** courses the length of the body through which the inferior alveolar nerves and vessels course.

SKULL AS A WHOLE

ANTERIOR ASPECT

The anterior aspect (Fig. 7-1) of the skull exhibits, from superior to inferior, the forehead, orbits, nasal cavity, facial region (consisting of cheek bones and upper jaw), and lower jaw.

The **forehead** is made up of the frontal bone, which forms part of the superior aspect of the **orbits** and possesses **temporal lines** (just superior to the zygomatic processes), **superciliary arches and eminences,** and **supraorbital notches or foramina.** The center of the frontonasal suture is called the **nasion,** and that area just superior to the nasion, between the superciliary arches, is the **glabella.**

The **orbits** can be said to have a **base** anteriorly, an

apex posteriorly, and four walls—superior, inferior, lateral, and medial. The **superior wall or roof** is made up of the orbital plate of the frontal bone anteriorly and the lesser wing of the sphenoid posteriorly. The **lateral wall** is the greater wing of the sphenoid posteriorly and the zygomatic bone anteriorly. The **inferior wall or floor** is made up of the maxilla anteriorly and medially, the zygomatic anteriorly and laterally, and the orbital process of the palatine bone at the posterior angle. The **medial wall,** from anterior to posterior, is made up of the frontal process of the maxilla, the lacrimal bone, the orbital plate of the ethmoid, and the side of the body of the sphenoid. A fossa is found superiorly and laterally for the lacrimal gland and another smaller fossa superiorly and medially for the trochlea (L., pulley) of the superior oblique muscle. At the apex of the orbit the **optic foramen** can be seen as well as the **superior orbital** fissure. On the inferior aspect of the orbit the **inferior orbital fissure** makes a sizable opening. The **nasolacrimal canal** is located anteriorly and medially, and two small openings, the **zygomatico-orbital foramina,** are found inferiorly and laterally. The infraorbital groove and canal course along the floor of the orbit, and the anterior termination of this canal is the **infraorbital foramen,** located just inferior to the inferior margin of the orbit.

The **nasal region** exhibits the **nasal bones** forming the bridge of the nose, and the **septum** and **inferior nasal concha** are visible from an anterior view.

The **facial region** consists of the **upper jaw** medially (the maxilla), and the **zygomatic or cheek bones** laterally. The **infraorbital foramen** is located just inferior to the orbit; the **zygomaticofacial foramen** is a small opening in the zygomatic bone. The upper teeth are attached to the maxillary bone.

The **lower jaw (mandible)** exhibits a **mental protuberance** in the midline, with **mental tubercles** on either side of it; the **incisor fossae** are on either side of the midline just superior to the protuberance; the **mental foramina** occur approximately an inch from the midline.

LATERAL ASPECT

Looking at the lateral side of the skull (Fig. 7-2A), one sees that it is made up primarily of the frontal, parietal, occipital, temporal, and sphenoid bones forming the cranium; the zygomatic, maxillary, and nasal bones forming the face; and the mandible forming the lower jaw. The **temporal line,** forming the boundary of the origin of the temporalis muscle, starts on the frontal bone near the zygomatic process, curves superiorly across the middle of the parietal bone to reach the suture line between the parietal and occipital bones, and then continues just anterior to the mastoid process. The **external auditory meatus** is very obvious and the **mastoid process** of the temporal bone is just posterior to this opening. The **squamous portion** of the temporal bone extends superiorly from this region, and the **zygomatic process** of the temporal bone forms an archway around an opening for the temporalis muscle. The masseter muscle (muscle of mastication) attaches to this zygomatic process of the temporal bone and the zygomatic bone itself and inserts on the ramus of the mandible. That region deep to the zygomatic process of the temporal bone is known as the **temporal fossa,** and that portion deep to the ramus of the mandible (inferior to the temporal bone) is the **infratemporal fossa.** On the deep surface of the zygomatic bone can be seen a small opening, the **zygomaticotemporal foramen,** which transmits a sensory nerve and vessels to the scalp.

If the mandible is moved to one side, the point of articulation between the maxillary bone and the pterygoid plate of the sphenoid bone can be seen (Fig. 7-2B). The **pterygoid lamina** is visible, as are small foramina in the maxillary bone, the **posterior superior alveolar foramina.** The **pterygomaxillary fissure** can be seen between the pterygoid process posteriorly and the maxillary bone anteriorly (Fig. 7-2B); this leads into the **pterygopalatine fossa,** located between the pterygoid process posteriorly and the palatine bone anteriorly. Although difficult to see, the following foramina lead from the pterygopalatine fossa to other regions:

1. **Sphenopalatine foramen,** which leads medially into the nasal cavity
2. **Pharyngeal canal,** which courses posteriorly and medially into the pharynx itself
3. **Pterygoid canal,** which is directed posteriorly and ends in the superior aspect of the foramen lacerum
4. **Foramen rotundum,** which contains the maxillary division of the trigeminal nerve
5. **Greater palatine canal** and **lesser palatine canals,** which lead inferiorly to reach the roof of the mouth
6. **Inferior orbital fissure,** through which structures reach the floor of the orbit
7. **Pterygomaxillary fissure,** which connects laterally with the infratemporal fossa

These openings are diagrammed in Figure 7-123 on page 621.

Features about the mandible, not clearly seen from an anterior approach, are revealed from a lateral view. Between the free inferior edge of the mandible and the mental foramen is a ridge of bone that can be folllowed posteriorly to the base of the most anterior part of the ramus. This is the **oblique line,** on which the depressors of the lower lip originate. The **angle** of the mandible marks the boundary line between the body and ramus, the latter extending superiorly. Two projections from the ramus are quite obvious, an anterior one—the **coronoid process** onto which the temporalis muscle inserts—and a posterior one—the **condyloid process**—which serves to articulate the mandible with the base of the skull. This process has a head and a constricted neck.

POSTERIOR ASPECT

Looking at the posterior aspect of the skull (Figs. 7-2A, 7-3, and 7-4), one can see the large **occipital bone** taking up a large area in the center, with the **parietal bones** extending superiorly and laterally from it. In addition, the **mastoid processes** of the temporal bone can be seen laterally, and the inside surface of the mandible is also visible.

That point where the sagittal suture and the suture between the parietal bones and the occipital bones, known as the **lambdoid suture,** come together is called the **lambda.** The **external occipital protuberance** is visible in the midline, as are the **superior and inferior nuchal lines,** which extend laterally from the midline **occipital crest.** The roughened areas on the posterior aspect of the occipital bone are areas for muscle attachments.

The **mastoid processes,** which contain air cells, are also used for the attachment of large muscles. Four small foramina can frequently be seen from this view: **mastoid foramina** in the mastoid bones close to the suture between them and the occipital bones, and **parietal foramina** located near the midline close to the sagittal suture. These, as we will see later, are foramina through which pass veins from the outside of the skull to the inside; these are called emissary veins and are important in spread of infections from outside to inside the skull.

The inside of the jaw contains the foramen through which the inferior alveolar nerve and artery course; the **lingula,** a spur just anterior and medial to this opening; the **mylohyoid lines;** and the **sublingual fossae.** Naturally, the teeth can be seen and, near the midline, the **genial tubercles,** which serve for the attachment of muscles.

BASE

The base of the skull (Fig. 7-4) can be divided into anteromedial, anterolateral, posteromedial, and posterolateral areas.

The **anteromedial area** consists of (1) the **hard palate,** made up by the maxillary bones anteriorly and palatine bones posteriorly, and containing the **upper teeth,** the **incisive foramen or foramina** near the incisor teeth, and the **greater and lesser palatine foramina** just medial or slightly posterior to the third molar teeth (all of these foramina serve as a pathway for nerves and vessels supplying the hard and soft palates); (2) the **choanae,** which are the posterior openings of the nasal cavity into the nasal pharynx; (3) and the **pterygoid processes** of the sphenoid bone with their medial laminae (the **hamulus** is attached to the inferior end of each of these laminae), **lateral laminae,** and **fossae** between. In addition, a small portion of the palatine bone, which surrounds the lesser palatine foramen and forms an extension from the hard palate (pyramidal process), is wedged in between the maxillary bone and the pterygoid process of the sphenoid bone. The **pharyngeal canals** are just lateral to the base of the **vomer.** Another small opening, the **emissary sphenoidal foramen,** is located at the base of the lateral lamina just medial to the foramen ovale. This, as the name implies, contains an emissary vein.

The **anterolateral area** is actually in the infratemporal fossa and presents the **foramen ovale** for the mandibular division of the trigeminal nerve, and the **foramen spinosum** for the middle meningeal artery and meningeal nerve.

The **posteromedian area** consists of (1) the **basilar portion** of the occipital bone located anterior to the foramen magnum (this portion serves for the attachment of muscles, the centrally located **pharyngeal tubercle** being the point of attachment of the superior constrictor mus-

cles of the pharynx); (2) the **two occipital condyles** for articulation with the superior facets of the atlas; (3) the **foramen magnum,** through which course the spinal cord and its vessels and meningeal coverings, the spinal accessory nerves, vertebral arteries, and meningeal lymphatics; (4) the **hypoglossal canal** for the twelfth cranial nerve, located just lateral to the occipital condyle; (5) the **condylar canal,** located just posterior to the condyle, a canal serving as a pathway for an emissary vein and veins draining into the suboccipital plexus of veins; (6) the **occipital crest,** a ridge extending from the posterior edge of the foramen magnum to the external occipital protuberance; (7) **superior and inferior nuchal lines,** extending laterally from this crest; and (8) many impressions for attachment of the muscles located in the back of the neck.

The **posterolateral area** shows (1) the **foramen lacerum** between the basilar portion of the occipital bone medially, the petrous portion of the temporal bone laterally, and the sphenoid bone anteriorly—this foramen, in the living state, is filled in with cartilage but still transmits small vessels, nerves, and lymphatics; (2) the groove for the **auditory tube;** (3) the **articular tubercle** and **fossa** for the mandible; (4) the **tympanic plate** of the temporal bone; (5) the **styloid process;** (6) the **stylomastoid foramen** just posterior to the styloid process; (7) the **jugular fossa** and **foramen** for the internal jugular vein and several cranial nerves; (8) the **mastoid canaliculus** (located deep in the jugular fossa), which transmits a branch of the vagus nerve to the external auditory meatus; and (9) the **carotid canal** for the internal carotid artery.

MEDIAN SAGITTAL SECTION

A median sagittal section of the skull (Figs. 7-5 and 7-6), is quite revealing, especially around the nasal cavity. The sphenoid bone with its **sphenoid sinus** and the **superior and middle conchae** of the ethmoid bone are quite obvious. Anterior to this we find the frontal process of the **maxillary bone.** A small part of the **lacrimal bone** is also visible between the maxillary anteriorly and the ethmoid posteriorly. The **nasal bone** forming the bridge of the nose is anterior to the maxillary. The **inferior nasal concha** is seen to be surrounded, anteriorly, by the maxillary bone, which not only forms this part of the lateral wall of the nose but also the anterior part of the hard palate; and, posteriorly, by the **perpendicular plate of the palatine** bone, which also forms the posterior part of the hard palate and lateral wall of the nasal cavity. Posterior to the palatine bone are the **pterygoid laminae** of the sphenoid bone. Figure 7-5 shows the **nasal septum.** It consists of the **perpendicular plate of the ethmoid** superiorly and the **vomer** inferiorly. More details of the nasal region will be taken up in a study of the nasal cavity.

A median sagittal section of the skull also provides a view of the inside surface of the mandible. Starting anteriorly, there are two small projections near the symphysis menti. These are the **genial tubercles,** which serve for muscle attachment. The sublingual fossa for the sublingual gland is just posterior to these tubercles, and the mylohyoid line extends posteriorly from this fossa. The shallow area just inferior to this line is the location of the submandibular gland. The inside surface of the ramus possesses the mandibular foramen for the inferior alveolar nerve and vessels, a projection of bone—the **lingula** —just anterior to this opening, and the **mylohyoid** groove extending inferiorly and anteriorly, this groove being the location of the mylohyoid nerve and vessels.

CRANIAL FOSSAE

The floor of the cranial cavity (Fig. 7-7) is divided into anterior, middle, and posterior cranial fossae.

The **anterior cranial fossa** is formed by the frontal bone anterolaterally, the ethmoid bone in the midline, and posteriorly by the lesser wing of the sphenoid bone. The **crista galli** and the **cribriform plate** of the ethmoid bone can be seen anteriorly and in the midline, and the **foramen cecum** lies just anterior to the crista. The openings in the cribriform plate are for the olfactory nerves, the foramen cecum is the point of origin of the superior sagittal venous sinus, and the crista serves as the anterior attachment of a fold of dura mater called the falx cerebri. Close scrutiny will reveal the **ethmoidal foramina,** through which the anterior and posterior ethmoidal nerves and vessels course, and **grooves for the anterior meningeal vessels.** The large roof of the orbit is located in this fossa and is known as the **orbital plate.**

The **middle cranial fossa** is formed by the greater wing of the sphenoid and the squamous portion of the temporal bone laterally, and the petrous portion of the temporal bone posteriorly. The **sella turcica** of the sphenoid bone (for the hypophysis) is located medially just posterior to the **anterior clinoid processes** and an-

terior to the **posterior clinoid processes.** The **optic foramina,** for the optic nerves and ophthalmic arteries, and the **foramen lacerum** can be seen anterior and posterior to the sella turcica respectively. The middle cranial fossa exhibits a **superior orbital fissure** anteriorly, and the **foramen rotundum, emissary sphenoidal foramen, foramen ovale,** and **foramen spinosum** from anterior to posterior. Although difficult to see, the posterior end of the **pterygoid canal** opens into the foramen lacerum superior to the cartilage in this foramen. The middle cranial fossa also exhibits grooves for the middle meningeal arteries, the **arcuate eminence,** and **tegmen tympani** of the petrous portion of the temporal bone, the groove and hiatus for the greater petrosal nerve (an important branch of the facial), and the **carotid canal** just lateral to the foramen lacerum.

The **posterior cranial fossa** is made up of the occipital bone posteriorly and the posterior surface of the petrous part of the temporal bones anteriorly. It presents the basilar portion of the occipital bone anterior to the foramen magnum, the **foramen magnum** itself, two **fossae** for the large lobes of the cerebellum, the **internal occipital protuberance,** and the **internal occipital crest.** The **grooves for the transverse and sigmoid venous sinuses** of the dura mater are visible, leading directly to the **jugular foramen,** a large opening for the bulb of the internal jugular vein. The **mastoid foramen** opens into the groove made by the sigmoid sinus. The **condylar canal** is just posterior to the jugular foramen, and the **hypoglossal canal** leads laterally away from a point just superior to the foramen magnum. The **internal auditory meatus** is an opening on the medial side of the petrous portion of the temporal bone, and transmits the seventh and eighth cranial nerves as well as the internal auditory artery.

INSIDE SURFACE OF CALVARIA

If one looks at the concave inferior surface of the calvaria (Fig. 7-8), one sees a long **groove in the midline for the superior sagittal (venous) sinus of the dura,** many **grooves for the meningeal vessels,** and **small pits near the midline for arachnoid granulations (granular fovea),** which are involved with circulation of the cerebrospinal fluid. The bones involved are the frontal, parietal, and occipital. The **lambdoid suture,** between the occipital and parietal bones, meets the **midsagittal**

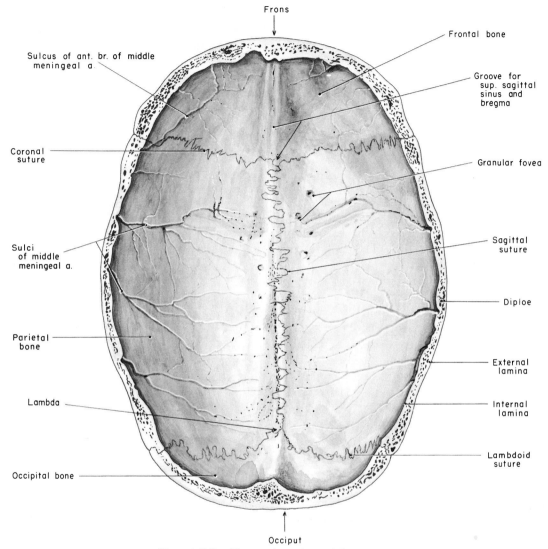

Figure 7-8. *The inside surface of the calvaria.*

suture at the **lambda;** the **coronal suture,** between the frontal and parietal bones, meets the midsagittal suture at the **bregma;** the **coronal sutures** meet the squamous sutures (around the squamous portion of the temporal bone) at the **pterion.**

As mentioned at the beginning of this chapter, the student should review and supplement the foregoing during the study of each region of the head and neck.

SUPERFICIAL LANDMARKS

The superficial landmarks around the face and scalp (Fig. 7-9) are not as revealing of the deeper structures as one might at first suspect. Of course, the **forehead** reveals the shape of the frontal bone, the **eyebrows** the location of the **superciliary arches;** the **eyes** are obvious, as are the **nose** and **mouth.**

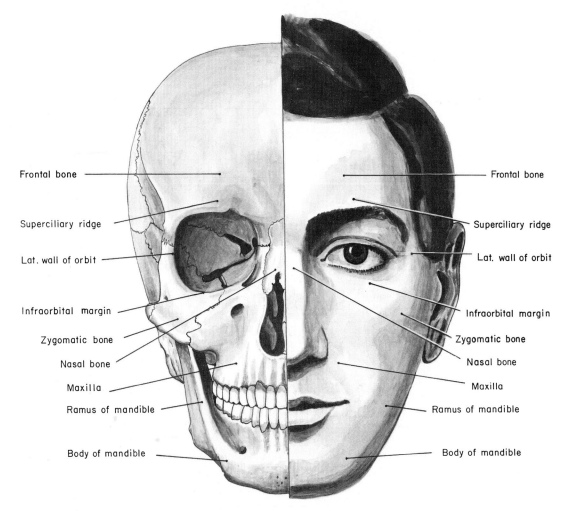

Figure 7-9. *Surface anatomy of the face, with the bony framework revealed.*

Palpation, however, will reveal much more. The edges of the **orbit** can be felt, as can the bony part of the nose. The point where the bony part of the nose and the cartilaginous portion join can easily be determined. The **masseter** muscle can be outlined as well as the **zygomatic arch,** and the whole edge of the **mandible** is revealed. The **superficial temporal artery** is palpable; in fact, the pulse can be taken at this point.

The neck is more revealing (Fig. 7-10). By inspection in the midline, the **chin, hyoid bone, thyroid cartilage, cricoid cartilage, suprasternal fossa,** and **superior edge of the manubrium** of the sternum are all revealed.

The two **sternocleidomastoid muscles** produce bulges that extend from the sternum and medial end of the clavicle to the **mastoid process** of the temporal bone, which also can be palpated. The supraclavicular fossa is located just lateral to the sternocleidomastoid muscle and superior to the clavicle. The inferior limits of the neck are clearly outlined by the fact that the entire length of the **clavicle** and the **acromion process** of the scapula can be felt. Posteriorly, the **external occipital protuberance** is easily located, and the **superior nuchal lines** extend laterally from this protuberance. In children, the **semispinalis capitis muscles** stand out rather prominently and

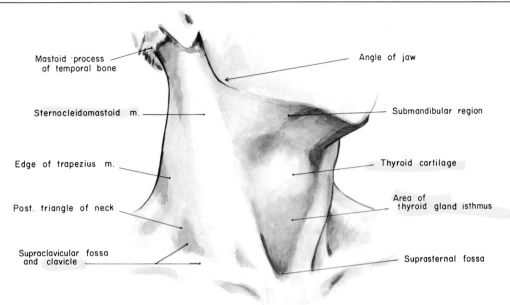

Figure 7-10. *Surface anatomy of the neck.*

the **median groove** between them is therefore equally prominent. This tends to disappear as an individual ages.

There is nothing particularly revealing on the scalp.

We note, as an aid in reference, that the hyoid bone is opposite the third cervical vertebra; the thyroid cartilage, the fourth; and the cricoid cartilage, the sixth.

SCALP AND TEMPLE

GENERAL FEATURES

Although the **scalp,** to the layman, is that part of the head that is usually covered with hair, the term scalp to the anatomist is a bit more inclusive. It starts anteriorly with the supraorbital ridges and continues posteriorly to the external occipital protuberance and the superior nuchal lines. Laterally it extends as far as the temporal lines but is continuous with a thin layer of fascia over the rather heavy fascial covering of the temporalis muscle.

The scalp consists of five layers (Fig. 7-11):*

_____
*Note initial letters in list following.

1. Skin
2. Connective tissue—dense strands and fat
3. Aponeurosis—a tendon stretching between the frontalis and occipitalis muscles (fronto-occipital aponeurosis or galea aponeurotica)
4. Loose connective tissue
5. Pericranium

The skin is not particularly different from that elsewhere, except for the hair which is found over a large proportion of the scalp in most cases. Its firm attachments to an aponeurosis resemble the arrangement found in the palm of the hand and the sole of the foot. The **dense connective tissue** layer is important, for it is in this layer that the nerves, arteries, and veins course. When the scalp is cut, this dense connective tissue tends to hold the wound together but, at the same time, to hold the vessels open, resulting in rather profuse bleeding from scalp wounds. Infections are not common in this layer but those that do occur are quite painful since there is very little give to the surrounding tissue. The **fronto-occipital aponeurosis** consists of a **frontalis muscle** anteriorly and an **occipitalis muscle** posteriorly with an aponeurotic tendon between the two. It arises posteriorly from the superior nuchal lines, the frontalis muscle blends with the skin over the forehead, and it extends laterally to reach the zygomatic bones. It is these muscles that are responsible for movement of the scalp. The **loose connective tissue** separates this fronto-occipital aponeurosis from the periosteum covering the bone, which in this area is called the **pericranium.** This loose connective tissue layer is often the site of infections, and its loose texture allows for rapid and extensive spread of such maladies. Hematomas can occur deep to the pericranium; they stop at suture lines.

The area of the temple demands special attention. This particular area is the point of entrance to the cranial cavity in many surgical procedures. The layers at this point consist of the skin, the same dense strands and fat found over the more superior aspects of the scalp, a layer of thinned-out aponeurosis, the thick and heavy temporal fascia (actually homologous to bone in some animals that have a temporal cave), the temporal muscle, and the pericranium. Actually, since the blood supply to the pericranium in this area is derived from the vascular supply to the temporalis muscle, the muscle and pericranium, or indeed the entire thickness of the bone, are removed as a single layer.

NERVES

The sensory nerve supply to the scalp and temple consists of:

A. Nerves anterior to the ear
1. Supratrochlear
2. Supraorbital
3. Zygomaticotemporal
4. Auriculotemporal
B. Branches posterior to the ear
5. Great auricular
6. Lesser occipital
7. Greater occipital
8. Third occipital

Those in group A are branches of the trigeminal nerve; in group B, of the cervical nerves. At this time a glance at Figures 7-28 and 7-32 on pages 499 and 504 will give a good idea of the origin of the trigeminal nerve and how it penetrates the skull; Figure 7-85 on page 565 will give an idea of the cervical plexus, made up of the ventral rami of the first four cervical nerves.

1. The **supratrochlear nerve** (Figs. 7-12 and 7-13) (so called because it courses superior to the trochlea or pulley of the superior oblique muscle in the orbit)

leaves the orbit, pierces the frontalis muscle, and gives sensory supply to the front of the forehead near the midline.

2. The **supraorbital nerve** (Figs. 7-12 and 7-13), which courses through the superior part of the orbit, is very close to the bony margin of the orbit; it courses either through a supraorbital notch or a foramen, pierces the frontalis muscle, and innervates the skin of the scalp as far back as the vertex. Both of these branches (supraorbital and supratrochlear) are terminal branches of the frontal nerve, which is a branch of the ophthalmic division of the trigeminal nerve.

3. The **zygomaticotemporal nerve** (Figs. 7-12 and 7-13) emerges from the skull through a small foramen in the zygomatic bone and supplies the skin over a small area on the temple. This is a branch of the maxillary division of the trigeminal nerve.

4. The **auriculotemporal nerve** (Figs. 7-12 and 7-13), as the name implies, courses close to the ear and supplies that area of the scalp that is superficial to the temporalis muscle, the temple. It reaches almost as far as the vertex of the head. This particular nerve is a branch of the mandibular division of the trigeminal nerve and actually sends branches to the external ear, the external auditory meatus, and the superficial surface of the tympanic membrane as well as to the skin over the temple.

5. The **great auricular nerve** (Fig. 7-12) is a branch of the cervical plexus arising from the ventral rami of the second and third cervical nerves. It ascends on the sternocleidomastoid muscle and reaches a small area of the scalp just posterior to the ear, as well as the posterior surface of the external ear itself.

6. The **lesser occipital nerve** (Figs. 7-12 and 7-13), also a branch of the cervical plexus from the ventral ramus of C_2, ascends along the posterior border of the sternocleidomastoid muscle to reach the superior part of the external ear and the skin superior and posterior to this area.

7. The **greater occipital nerve** (Figs. 7-12 and 7-13) is the medial division of the dorsal ramus of C_2. As pointed out in Chapter 2, it appears in a position just inferior to the inferior capitis oblique muscle, courses superiorly in a plane superficial to the muscles forming the suboccipital triangle and deep to the semispinalis capitis muscle, and pierces the latter muscle and the trapezius muscle to reach the scalp. It pierces the deep fascia and then ramifies over a large area of the posterior

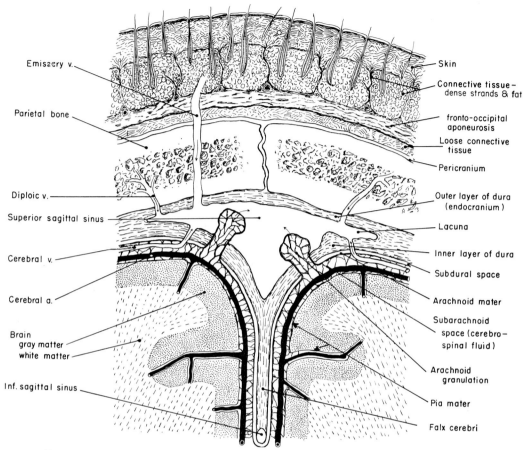

Figure 7-11. A frontal section of the scalp, skull, meninges, and brain (diagrammatic).

part of the scalp, reaching as far as the vertex of the head to meet the branches of the supraorbital and auriculotemporal nerves just described.

8. The **third occipital nerve** (Fig. 7-12) is also the median division of a dorsal ramus, but of C_3 rather than C_2. It pierces the semispinalis capitis and trapezius muscles also and then ramifies on the back of the neck in the skin over the external occipital protuberance. It does not cover a large area of the scalp.

It should be noted that the great auricular and lesser occipital nerves are ventral rami and branches of the cervical plexus, while the greater occipital and third occipital nerves are branches of dorsal rami and, therefore, not branches of the plexus.

The motor nerve to the occipitofrontalis muscle is the seventh cranial nerve (facial) and consists of a small branch posterior to the ear (**posterior auricular nerve** —Fig. 7-15) to the occipitalis muscle, and the **temporal** branches of the facial nerve to the frontalis muscle. This nerve supply is explained by the fact that these muscle bellies are derived from the second branchial arch; the facial nerve innervates all such muscles (see Fig. 7-136 on page 636).

ARTERIES

Before describing the arteries to the scalp, those who have dissected the thoracic cavity will recall that the aortic arch has three branches—(1) brachiocephalic, which divides into the right subclavian and right common carotid, (2) left common carotid, and the (3) left subclavian. The common carotids ascend in the neck and divide into

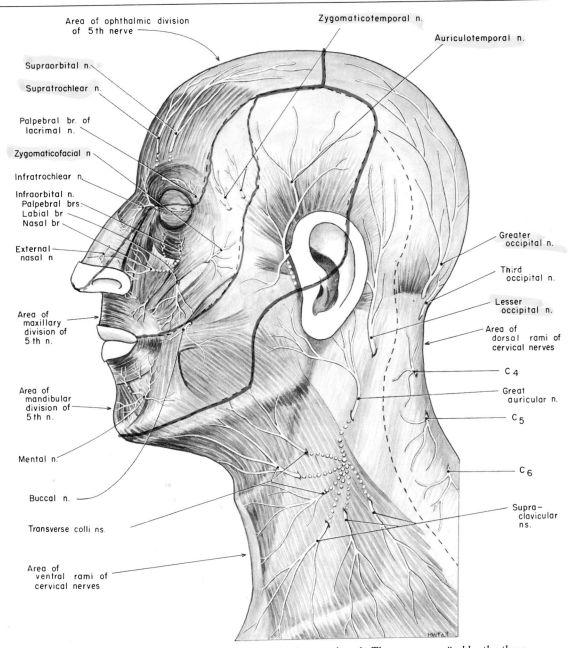

Figure 7-12. *Cutaneous nerves of the scalp, face, and neck. The areas supplied by the three divisions of the trigeminal nerve, and by the ventral and dorsal rami of the cervical nerves are outlined.*

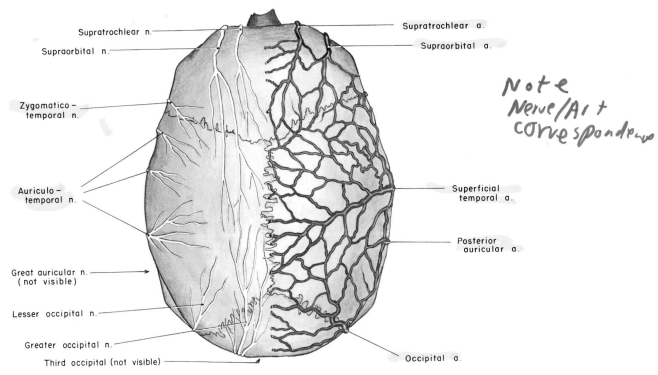

Figure 7-13. *The arteries and cutaneous nerves of the scalp. Note that there are many arteries anastomosing across the midline.*

internal and external branches. The internal carotid has no branches in the neck but penetrates the base of the skull and gives off branches inside the skull, one of which is the ophthalmic which enters the orbit. The external carotid has numerous branches, which supply head and neck structures. The vertebral arteries also participate; they traverse the foramen magnum, give off branches to the spinal cord, and then join to form the basilar on the brain stem.

The arteries to the scalp and temple are:

1. Supratrochlear
2. Supraorbital
3. Superficial temporal
4. Posterior auricular
5. Occipital

1 and 2. The **supratrochlear** and **supraorbital** arteries (Figs. 7-13 and 7-14) are branches of the ophthalmic artery inside the orbit, the ophthalmic, as just mentioned, being a branch of the internal carotid. They take a pathway and distribution similar to the nerves of the same name.

3. The **superficial temporal artery** (Figs. 7-13 and 7-14) is one of the terminal branches of the external carotid artery. After giving branches to the parotid gland, to the external ear, and to the face (transverse facial), it continues superiorly in a position anterior to the external ear. It gives off a branch that pierces the deep fascia over the temporalis muscle and then divides into anterior and posterior branches. These two branches ramify over the entire parietal and temporal regions of the side of the head. The superficial temporal artery is one of the arteries in the body with which the pulse rate can be taken.

4. The **posterior auricular artery** (Fig. 7-14) is a branch of the external carotid; after ramifying among deeper structures, it follows the posterior auricular nerve to the area of the scalp posterior to the ear.

5. The **occipital artery** (Fig. 7-14), also a branch of the external carotid, ramifies over the posterior part of the scalp, in an area similar to that supplied by the greater

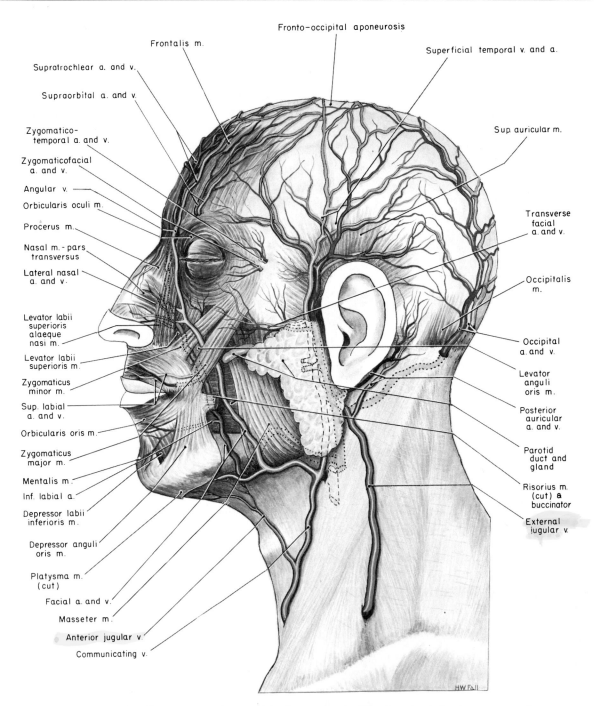

Fronto-occipital aponeurosis

Frontalis m.

Superficial temporal v. and a.

Supratrochlear a. and v.

Supraorbital a. and v.

Sup. auricular m.

Zygomatico-temporal a. and v.

Zygomaticofacial a. and v.

Angular v.

Transverse facial a. and v.

Orbicularis oculi m.

Procerus m.

Nasal m.-pars transversus

Lateral nasal a. and v.

Occipitalis m.

Levator labii superioris alaeque nasi m.

Occipital a. and v.

Levator labii superioris m.

Levator anguli oris m.

Zygomaticus minor m.

Posterior auricular a. and v.

Sup. labial a. and v.

Orbicularis oris m.

Parotid duct and gland

Zygomaticus major m.

Risorius m. (cut) & buccinator

Mentalis m.

Inf. labial a.

External jugular v.

Depressor labii inferioris m.

Depressor anguli oris m.

Platysma m. (cut)

Facial a. and v.

Masseter m.

Anterior jugular v.

Communicating v.

HW Fall

Figure 7-14. *The arteries and veins of the face and scalp. The deep fascia has been removed from the parotid gland and the masseter muscle.*

occipital nerve. Those students who have dissected the suboccipital region will remember that this artery does not take the same course as the greater occipital nerve. It appears in this region deep to the splenius capitis muscle and superficial to the obliquus capitis superior muscle in direct contact with the skull. It then escapes from the medial edge of the splenius capitis muscle to become deep to the trapezius and superficial to the semispinalis capitis. It then pierces the trapezius to ramify on a large area on the posterior part of the scalp. This artery has an important **descending branch,** which anastomoses with branches of the subclavian artery. It also sends two branches **to the meninges,** one through the mastoid foramen and another through the parietal foramen.

VEINS

The veins of the scalp and temple form a pattern that corresponds to that of the arteries just described (Fig. 7-14). The **supratrochlear** and **supraorbital** veins drain the anterior part of the scalp and, after communicating with the ophthalmic veins, join to form the origin of the facial vein. The **superficial temporal** vein drains the entire side of the scalp and joins with the maxillary vein to form the retromandibular vein. The **posterior auricular** vein drains the areas of the scalp posterior to the ear, and descends posterior to the external ear and joins the posterior division of the retromandibular vein to form the external jugular vein. The **occipital vein** drains the occipital and posterior parietal regions and, after piercing the trapezius muscle, joins in formation of the complex plexus of veins in the suboccipital triangle. Therefore, the occipital veins do not take the same course as the corresponding arteries.

At this point, **emissary veins** (Figs. 7-11) should again be mentioned. These are direct connections from outside the cranial cavity to the dural venous sinuses on the inside of the cranial cavity. As just described, the occipital artery sends branches through the mastoid and parietal foramina; these two foramina also serve as pathways for emissary veins. These are important in that infections can spread directly from the outside of the skull to the important dural venous sinuses on the inside.

LYMPH DRAINAGE

The lymphatics (Fig. 7-18) from the occipital area end in nodes in the occipital region close to the attachment of the trapezius muscle, those from the posterior parts of the parietal and temporal regions end in postauricular or mastoid nodes, which lie near the mastoid process of the temporal bone, and the anterior part of the scalp drains into parotid or preauricular nodes on the surface of the parotid gland. There is a possibility of lymph drainage from the forehead to the submandibular region following the facial artery. *Notice how these lymphatics follow in a retrograde fashion the vascular pathways of the scalp.*

FACE

GENERAL DESCRIPTION

Since the scalp starts at the supraorbital ridges, the face is the area inferior to these ridges; it continues inferiorly to the edge of the lower jaw and posteriorly until the ear is reached. Although the term ''face'' to the embryologist means all the facial structures including the bones, to the gross anatomist it is limited to an area approximately ½ inch in depth.

The central part of the face is occupied by the **eyes,** the **nose,** and **mouth,** the latter two structures having openings into the nasal and oral cavities respectively. You can easily determine on yourself that all movement of the face arises in the anterior part. Any movement of the lateral portion of the face is the result of movement around the mouth, eyes, and nose, or of movement of the lower jaw. The muscles moving these centrally located structures are called the **muscles of facial expression** (Figs. 7-12, 7-14 and 7-15). If the skin and superficial fascia are removed from the posterior part of the face, a pad of fat—the **buccal pad**—can be seen to lie between the angle of the mouth and a large muscle, the **masseter muscle,** one of the muscles of mastication, and a muscle that forms a large part of the posterior aspect of the face (Fig. 7-15). Posterior to this muscle is found the **parotid gland,** which fills in that area between the masseter and the ear.

Crossing the masseter muscle from posterior to anterior, at about the level of the lobe of the ear, is the **parotid duct** on its way from the parotid gland to empty into the mouth near the upper second molar tooth. Spreading out in all directions from the parotid gland are the five terminal branches of the **facial nerve** (Fig. 7-15). They cross the masseter muscle and finally reach the muscles of facial expression all the way from the frontalis muscle superiorly to the platysma muscle inferiorly. Several arteries can also be seen (Fig. 7-14). The **superficial temporal** has already been mentioned. Its **transverse facial** branch crosses the masseter muscle usually just superior to the parotid duct. The **facial artery** (Fig. 7-14) is a coiled vessel that curves around the edge of the mandible and courses superiorly just lateral to the corner of the mouth and the nose. As we will see momentarily, it gives off several branches to structures in the face. Just posterior to the artery and usually at a little more superficial plane is the **facial vein**. This vein does not take as coiled a pathway as the artery since it does not course deep to the mandible, and is therefore not stretched when the mouth is opened.

In summary, the lateral part of the face presents the parotid gland just anterior to the ear with the many branches of the facial nerve streaming superiorly, anteriorly, and inferiorly from the area of this gland across the masseter muscle to reach the many facial muscles that move the eyes, nose, mouth, and scalp in the anterior part of the face. The parotid duct also crosses the masseter muscle to reach the mouth. Branches of the facial artery bring blood to the anterior part of the face, while branches of the superficial temporal supply the posterior part. The veins take corresponding courses.

MUSCLES

The muscles of facial expression (derived from the second branchial arch) are attached to the skin, and some of them are located in superficial fascia only. Others do have bony attachments, but the insertion is finally into the skin and superficial fascia. These muscles can be divided into those around the mouth, eyelids, nose, and ear, and in the scalp. They are pictured in Figures 7-14 to 7-16, and the bony origins of many of them are pictured in Figures 7-1 and 7-2A. Constant reference to these latter figures while reading the following will be helpful.

The muscles around the mouth are:

1. Platysma
2. Orbicularis oris
3. Risorius
4. Depressor anguli oris
5. Depressor labii inferioris
6. Mentalis
7. Transversus menti
8. Zygomaticus major
9. Zygomaticus minor
10. Levator labii superioris
11. Levator labii superioris alaeque nasi
12. Levator anguli oris
13. Buccinator

The **platysma** muscle is a thin sheet of muscle tissue in the superficial fascia of the neck. It arises in the fascia on the upper part of the pectoralis major muscle, sweeps superiorly in a plane superficial to the clavicle, and finally reaches the area of the mandible. The more medial fibers insert on the edge of this bone (Fig. 7-1) but the more lateral fibers blend with the muscles around the mouth. This is the same muscle that extends over the entire trunk of animals such as the horse and that allows such animals to move areas of the skin when annoyed by flies, etc. The **orbicularis oris** muscle completely surrounds the mouth; it has no bony attachments. It forms a complete circle and its fibers blend with many of the muscles about to be mentioned. The **risorius** muscles, the so-called laughter muscles, extend posteriorly from the corners of the mouth; these small muscles are quite superficially located and naturally pull the corners of the mouth laterally. The **depressor anguli oris** muscles are located as one might expect from the name. They originate from the mandible very close to the attachment of the platysma muscle (Figs. 7-1 and 7-2A) and insert on the corner of the mouth. The **depressor labii inferioris** muscles lie deep to the depressor anguli oris muscles, arising from the mandible just superior to the depressor anguli oris and inserting onto the lower lip. The **mentalis** is not named for its action but for its location and is a slip of muscle just anterior to the mental foramen. It arises from the mandible at a point just inferior to the incisor teeth (Figs. 7-1 and 7-2A) and passes inferiorly to be inserted into the skin of the chin. It is responsible for raising the skin of the chin as one protrudes the lower lip. The **transversus menti** is found in approximately half of bodies and is a small muscle just under the chin; this muscle forms a sort of sling for it is frequently continuous with the depressor anguli oris muscles.

The **zygomaticus major** muscle, named for its attachment, is fairly sizable, arises from the zygomatic bone (Figs. 7-1 and 7-2A) and inserts into the orbicularis oris muscle near the angle of the mouth. The **zygomaticus**

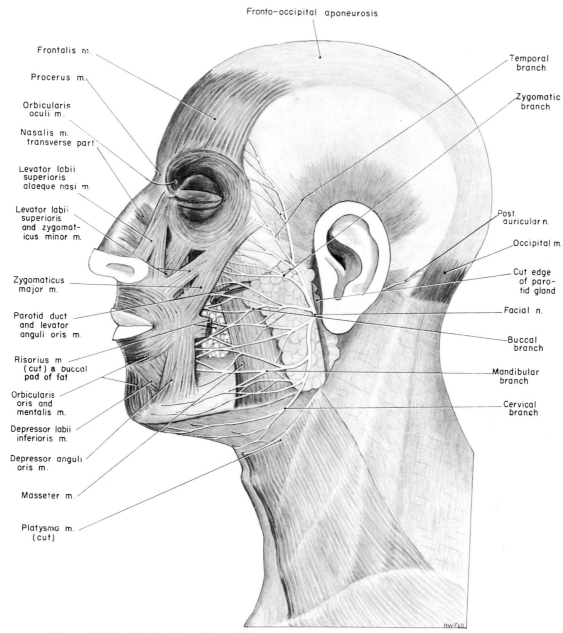

Fronto-occipital aponeurosis

Frontalis m.

Procerus m.

Orbicularis oculi m.

Nasalis m. transverse part

Levator labii superioris alaeque nasi m.

Levator labii superioris and zygomaticus minor m.

Zygomaticus major m.

Parotid duct and levator anguli oris m.

Risorius m. (cut) a buccal pad of fat

Orbicularis oris and mentalis m.

Depressor labii inferioris m.

Depressor anguli oris m.

Masseter m.

Platysma m. (cut)

Temporal branch

Zygomatic branch

Post. auricular n.

Occipital m.

Cut edge of parotid gland

Facial n.

Buccal branch

Mandibular branch

Cervical branch

Figure 7-15. *The facial muscles and their innervation by the facial nerve. The superficial part of the parotid gland has been removed.*

minor is an occasional slip of muscle that seems to be continuous with the orbicularis oculi muscle; it also inserts on the angle of the mouth. These two muscles are responsible for lifting the angles of the mouth in smiling. The **levator labii superioris** arises from the maxilla just superior to the infraorbital foramen (Figs. 7-1 and 7-2A) and is inserted into the upper lip. A muscle just medial to this, the **levator labii superioris alaeque nasi,** arises from the frontal process of the maxilla and is inserted on the ala of the nose as well as on the lip. The **levator anguli oris** arises from the maxilla inferior to the infraorbital foramen (Figs. 7-1 and 7-2A) and inserts on the corner of the mouth; it is located on a deeper plane than the zygomaticus and the levator labii superioris muscles. The **buccinator** muscle arises from the pterygomandibular ligament (a ligamentous band stretched between the pterygoid hamulus superiorly and the mandible, near the posterior end of the mylohyoid line, inferiorly) and from the lateral surfaces of the maxilla and mandible (Figs. 7-1 and 7-2A). It is inserted into the muscles around the mouth, the upper fibers crossing to the inferior lip, the lower fibers to the superior lip. The buccinator muscle serves to press the cheeks against the teeth and prevents food from accumulating in the area between the cheeks and the teeth during mastication.

The muscles around the nose are (Fig. 7-16):

1. Procerus
2. Nasalis
   a. Pars transversus
   b. Pars alaris
3. Depressor septi
4. Dilator naris

The **procerus** muscle is actually an inferior extension from the occipitofrontalis muscle over the bridge of the nose; it is responsible for wrinkling the skin of this structure. The **transverse portion of the nasalis** muscle arises from the maxilla just superior to the fossa above the incisor teeth (Fig. 7-1) and ends in joining its partner from the other side on the dorsum of the nose; it compresses the sides of the nose. The **alar portion** of this muscle is a small slip placed on the lateral sides of the nostril; its action is to augment the transverse portion in decreasing size of the nostril. The **depressor septi** muscle extends from the incisor fossa of the maxilla to the ala and septum of the nose; this naturally would pull the nose inferiorly. You should recall that the **levator labii superioris**

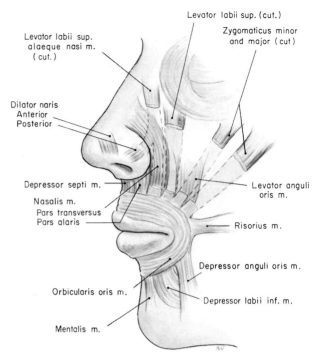

Figure 7-16. Muscles of facial expression around nose and mouth.

alaeque nasi also has an attachment to the ala of the nose; this muscle dilates the nostril and is aided in this act by small slips of muscle placed on the lateral margin of the nostril, the **dilator naris** muscle.

The muscles around the eyelids are:

1. Orbicularis oculi
2. Corrugator supercilii

The **orbicularis oculi** muscle consists of three distinct portions: palpebral, orbital, and lacrimal. The **palpebral portion** of the muscle arises from the bifurcation of the medial palpebral ligaments in the medial corner of the eye and forms a series of concentric curves that are inserted on the lateral palpebral raphe in the lateral corner of the eye. The **orbital portion** is thicker than the palpebral portion, and its fibers form a complete ellipse without any interruption at the lateral palpebral commissure.* The **lacrimal part** (see Fig. 7-49E on page 521) is

_____

*Some fibers extend superiorly into the eyebrow; the fibers are often called the depressor supercilii muscle.

a small portion arising from the orbital surface of the lacrimal bone; it passes posterior to the lacrimal sac and inserts on the medial end of each tarsal plate. This portion serves to put pressure on the lacrimal sac when the eye is closed, and to release this pressure when the eye is opened; this results in a suctionlike action of the lacrimal sac that draws tears into the nasolacrimal apparatus. Rapid blinking will often prevent tears from overflowing onto the cheeks. The palpebral portion tends to close the eye gently as in blinking, while the orbital portion closes the eye firmly as in winking or in protecting oneself from foreign elements. When the latter portion of the muscle is used, its extensions to other neighboring facial muscles cause a wrinkling of the skin around the eyes.

The **corrugator supercilii** muscles are located in a position deep to the eyebrows. They arise from the medial end of the superciliary arch (Fig. 7-1) and are inserted into the skin of the eyebrows in a more lateral position. When they contract they move the skin toward the midline, forming a wrinkling typical of a scowl. These muscles are usually contracted in eye strain. The muscles in the scalp have already been described; they consist of the fronto-occipital aponeurosis.

Three other muscles that should be mentioned are those around the ear. They are the (1) auricularis anterior, (2) auricularis superior, and (3) auricularis posterior. They extend anteriorly, superiorly, and posteriorly from the external ear, as the names imply. These muscles are interesting but of no importance (unless one wishes to entertain socially).

These facial muscles are all innervated by the facial nerve, the seventh cranial nerve. They are extremely important to us. When a person has the nerve supply to these muscles severed, he or she is unable to close the eye, thus causing a drying of the conjunctiva; such patients are unable to control the corner of the mouth and, therefore, unable to prevent drooling. The whole side of the face develops a distinct droop that is quite unattractive. The course and distribution of this facial nerve will be described thoroughly.

PAROTID GLAND

The parotid gland (Fig. 7-14), the largest of the salivary glands, fills in the area between the mandible and the ear. It extends superiorly as far as the zygoma and inferiorly beyond the angle of the mandible. It is in close relationship posteriorly to the sternocleidomastoid muscle. The duct of this gland extends anteriorly, crossing the masseter muscle; it then turns medially at the medial border of this muscle, and pierces the buccinator muscle and the mucous membrane to open into the mouth opposite the upper second molar tooth. It provides a serous fluid which, combined with the product of the submandibular and sublingual glands, is known as saliva. Further relations of the parotid gland will be given when deeper structures have been studied (page 585).

FACIAL NERVE

The **facial nerve** (Fig. 7-15), the great motor nerve to the face, exits from the stylomastoid foramen and, after giving off a combined branch to the posterior belly of the digastric and the stylohyoid muscle (muscles to be studied later), dives into the center of the parotid gland. The nerve actually follows a fascial plane that divides the gland into superficial and deep portions.* Although the facial nerve may exhibit many patterns, these almost invariably show two main stems, thus forming a Y-shaped structure. From these two stems five branches are distributed across the face to finally reach the muscles of facial expression. These branches are, from superior to inferior, the temporal, zygomatic, buccal, mandibular, and cervical. The names are characteristic of the areas through which these branches course to get to their destination: the **temporal** courses in the temporal area to reach the scalp and part of the eyelid muscles; the **zygomatic** crosses the bone of the same name to reach the eyelids and nose; the **buccal** crosses the masseter muscle and the buccal pad of fat to reach the muscles around the mouth; the **mandibular** follows the mandible to reach the chin; and the **cervical** courses in the neck just inferior to the mandible to innervate the platysma muscle. This nerve is of the utmost importance, for if cut, a facial paralysis results. Four tests usually used to check the facial nerve are (1) raising the eyebrows, (2) closing eyes tightly, (3) showing teeth or pursing lips as in whistling, and (4) pulling corners of the mouth laterally and inferiorly to raise the skin on the neck.

---

*H. Bailey, 1948, The surgical anatomy of the parotid gland, *Brit. Med. J.* 2: 245. An interesting account of the bilobed nature of the parotid gland, and of the branches of the facial nerve coursing between these lobes and surrounding the isthmus that connects them.

SENSORY NERVES

Although the posterior part of the face, just anterior to the ear, obtains its cutaneous nerve supply from branches of the cervical plexus, the **trigeminal nerve** is considered to be the great sensory nerve of the face and scalp. The branches of the trigeminal nerve to the scalp (supratrochlear, supraorbital, zygomaticotemporal, and auriculotemporal) have already been described. Cutaneous nerves to the face are pictured in Figure 7-12; they are:

1. Infratrochlear
2. Lacrimal
3. External nasal
4. Zygomaticofacial
5. Infraorbital
6. Buccal
7. Mental

The **infratrochlear nerve** is a branch of the ophthalmic division of the trigeminal nerve; it pierces the orbital septum (a sheet of fascia attached to the rim of the orbit and to the tarsal plates of the eyelids) superior to the medial angle of the eye and ramifies in the skin close to the medial angle of the eye and the bridge of the nose. It obtains its name from the fact that its course is inferior to the trochlear, or pulley, of the superior oblique muscle. In a similar position, piercing the lateral part of the eyelid, is a terminal branch of the **lacrimal nerve.** The lacrimal nerve is also a branch of the ophthalmic division of the trigeminal. The **external nasal nerve** is a terminal branch of the ophthalmic division of the trigeminal and gives cutaneous nerve supply to the anterior part of the nose.

The **zygomaticofacial** nerve, a branch of the maxillary division of the trigeminal, appears through a foramen in the zygomatic bone and covers a small area of the skin superficial to that bone. The larger **infraorbital** nerve, also a branch of the maxillary division of the trigeminal, exits through the infraorbital foramen and divides into three main branches: nasal, labial, and palpebral. These innervate the structures implied by the names.

The remaining nerves, the **buccal** and **mental,** are branches of the mandibular division of the trigeminal. The **buccal** nerve courses on the surface of the buccinator muscle and gives cutaneous nerve supply to the corner of the mouth and cheek, and to the mucous membrane on the inside of the mouth. The **mental** nerve emerges through the mental foramen of the mandible and gives cutaneous nerve supply to the chin.

ARTERIES AND VEINS

Although a small artery enters the face through every foramen, the main arteries to this region are:

1. Facial
2. Transverse facial

1. The **facial artery** (Fig. 7-14) crosses the angle of the mandible just anterior to the masseter muscle. In the face this artery is superficial to the mandible, the buccinator muscle, and the levator anguli oris muscle and courses in the substance of the levator labii superioris alaeque nasi muscle. It is deep to the platysma, risorius, zygomaticus major, and levator labii superioris muscles. The facial vein is posterior to the artery, as is the masseter muscle. The facial artery has the following branches in the face: (a) inferior labial, (b) superior labial, (c) lateral nasal, and (d) muscular. The **inferior labial** courses medially into the lower lip and is distributed between the superficial muscles and the mucous membrane of the lip. The **superior labial** courses medially into the upper lip between the orbicularis oris and the mucous membrane and terminates as a septal branch to the inferior end of the nose. This small branch is often ruptured, resulting in nosebleed, particularly in children. The **lateral nasal** branch courses on the side of the nose; the **muscular** are simply branches to muscles and may be one or several in number. The facial artery is quite coiled in its course due to the fact that it loops around the edge of the mandible. If it were not coiled, it would be severely stretched each time the mouth opened.

2. The remaining main artery to the face is the **transverse facial** branch of the superficial temporal artery (Fig. 7-14). This artery courses to the cheek, across the surface of the masseter muscle. It usually lies between the zygoma and the parotid duct.

The **facial vein** (Fig. 7-14) starts at the medial angle of the eye by the joining of the supratrochlear and the supraorbital veins. It courses inferiorly and slightly posteriorly, superficial to the muscles raising the angle of the mouth and usually separated from its artery by these muscles. It gains a position posterior to the artery, and finally crosses the submandibular gland deep to the platysma muscle and drains into the internal jugular vein. Its tributaries are similar to the branches of the facial artery. Since the vein does not loop around the mandible, it is not put on a stretch when the jaw is opened and is, therefore, not coiled.

LYMPHATICS

The lymphatic drainage of the face is described in the systemic summary (page 486).

NECK— SUPERFICIAL STRUCTURES

Superficial structures in the neck will be described at this time so that superficial aspects of the entire head and neck can be considered as a whole.

The **platysma** muscle covers a large area of the anterior and lateral sides of the neck (Fig. 7-12); when this muscle is removed the deep fascia is exposed (Fig. 7-14). This **deep fascia** starts posteriorly at the ligamentum nuchae, surrounds the trapezius muscle, bridges across an area between the trapezius and the sternocleidomastoid muscle, surrounds the latter muscle, and then continues medially to meet similar fascia from the opposite side. The area between the anterior edge of the trapezius and the posterior edge of the sternocleidomastoid muscle is known as the **posterior triangle of the neck.** That area anterior to the sternocleidomastoid muscle, between it and the midline of the body, is known as the **anterior triangle.** Further details on the triangles of the neck will be taken up when deeper structures are considered.

The most superficial structure seen after the platysma is removed is the **external jugular vein** (Fig. 7-14) and its tributaries. In addition, three sets of **cutaneous nerves,** all branches of the cervical plexus, are also seen piercing this deep fascia (Fig. 7-12). Those coursing superiorly are the **lesser occipital** and **great auricular;** that coursing transversely, the **transverse colli;** and those coursing inferiorly, the **medial, intermediate, and lateral supraclaviculars.**

These cutaneous nerves emerge, through the fascia covering the posterior triangle, just posterior to the sternocleidomastoid muscle. The **lesser occipital** nerve pierces the deep fascia and courses superiorly on the surface of the sternocleidomastoid muscle to reach the scalp in the area posterior to the ear. The **great auricular** nerve pierces the deep fascia inferior to the point where the lesser occipital nerve pierces the fascia, and courses superiorly on the surface of the sternocleidomastoid muscle but in a position anterior to the lesser occipital nerve. It gives branches to the skin posterior to the ear, to the ear itself, and to the lower part of the side of the face superficial to the parotid gland and angle of the mandible. It also supplies the area of the neck on the superior half of the sternocleidomastoid muscle. The **transverse colli** nerve also pierces the deep fascia in the posterior triangle just posterior to the sternocleidomastoid muscle; it proceeds anteriorly, pierces the platysma muscle, and becomes cutaneous on the skin over the anterior triangle of the neck. The **medial, intermediate, and lateral supraclavicular nerves** course inferiorly, pierce the deep fascia about 2 inches superior to the clavicle, and then pierce the platysma muscle to ramify on the skin as far as about 1 inch inferior to the clavicle. Those who have studied the thoracic wall will recall that the area inferior to this point is supplied by the second intercostal nerve; C5–T1 supply cutaneous areas on the upper limb and are not represented on the trunk.

Although the veins in the face and neck are quite variable, the most common arrangement is as follows. The **superficial temporal** vein is joined by the **maxillary vein** (a vein that drains the deep structures in the cheek) to form a vein in back of the mandible, the **retromandibular vein.** This vein then combines with a vein that drains the area of the scalp posterior to the ear (**posterior auricular vein**) to form the **external jugular vein.** The external jugular then courses inferiorly superficial to the deep fascia in the neck, pierces the deep fascia approximately an inch superior to the clavicle, and empties into the subclavian vein. It may be joined by a vein entering from the posterior aspect of the neck, called the **posterior external jugular vein.** Anteriorly, the **facial vein** empties into the **internal jugular vein** and, if an **anterior jugular vein** is present, it drains into the inferior end of the external jugular vein. There are often **communicating branches** between the retromandibular and the facial and between the facial and the anterior jugular veins. These veins are diagramed in Figure 7-17 and also can be seen in Figure 7-14.

The internal jugular vein will be described when deeper structures in the neck are studied.

SYSTEMIC SUMMARY OF SCALP, FACE, AND SUPERFICIAL NECK

Subsequent description will be concerned with deep structures in the head and neck. Hence it is convenient and beneficial to summarize at this time the structures found superficially on the head and neck.

MUSCLES

The superficial muscles in the scalp, face, and neck are shown in Figures 7-12, 7-14, 7-15, and 7-16.

Except for the unimportant muscles associated with the external ear, the muscles of facial expression are located near the midline of the body. The **fronto-occipital** muscle is obviously used to move the scalp. Movement around the eye is accomplished mainly by the **orbicularis oculi** muscle, which is divided into three parts: the palpebral portion in the eyelid itself, which is used in closing the eye as in blinking; the orbital portion, which surrounds the eye outside the eyelids and is used in closing the eye tightly as in squinting; and a lacrimal portion which passes posterior to and attaches to the lacrimal sac and, therefore, induces a sucking action when the eyes are rapidly closed so that tears enter the nasolacrimal apparatus. The most medial end of the orbicularis oculi muscle superior to the eye is attached to the skin of the eyebrow and pulls the eyebrow inferiorly when this muscle is contracted; this is often called the **depressor supercilii.** The **corrugator supercilii** muscle, which pulls the eyebrow toward the midline as in frowning, is closely associated with the superior aspect of the orbicularis oculi muscle. The **procerus** muscle, although considered to be associated with the nose, also will pull the eyebrows medially and inferiorly.

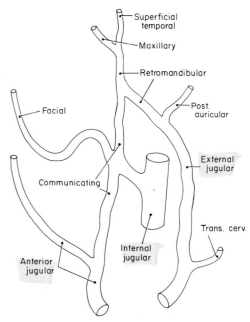

Figure 7-17. *Diagram of the most common arrangement of the veins draining the head and neck (left side of body). The external jugular drains into the subclavian vein, the anterior jugular into the subclavian or the external jugular.*

The muscles concerned with the nose are the **procerus** just mentioned, the transverse and alar parts of the **nasalis** which compress the nares, the **levator labii superioris alaeque nasi** and the **dilator naris** muscles which dilate the nostrils, and the **depressor septi** muscles which pull inferiorly on the septum.

The mouth is completely encircled by the **orbicular oris** muscle; this muscle is responsible for the fine movements associated with acts such as speech. It acts as a sphincter because of its arrangement. The other muscles associated with the mouth blend with this muscle. The **risorius** muscles pull the lateral aspects of the mouth posteriorly, as in a grin. The corners of the mouth are raised by the **zygomaticus major** and **minor** and the **levator anguli oris** muscles; the corner of the mouth is pulled inferiorly by the **depressor anguli oris** muscle and may be assisted by the **platysma.** The upper lip is raised by the actions of the **levator labii superioris** and **levator labii superioris alaeque nasi** muscles; the lower lip is depressed by the **depressor labii inferioris** as well as the **platysma** muscle. This leaves the buccinator and mentalis muscles. The **buccinator** muscle keeps the cheeks in contact with the teeth during the process of mastication, but in addition will aid the risorius muscle and the zygomaticus major in pulling the angles of the mouth posteriorly. The **mentalis** muscle, because of its attachments, will raise the skin on the chin, thereby causing the lower lip to protrude anteriorly.

The **platysma** muscle will raise the skin on the neck as well as pull inferiorly on the lower lip.

MOTOR NERVES

You will recall that the seventh cranial nerve, the **facial** nerve, innervates the muscles that move the second or hyoid gill arch in the dogfish. In the human being, these muscles have migrated throughout the head and neck but, in spite of this, they still receive the facial nerve as their innervation. The muscles derived from the second branchial arch include these muscles of facial expression just described. They are therefore all innervated by the motor nerve of the face, the facial.

The **facial** nerve, after emerging from the stylomastoid foramen, gives a **posterior auricular** branch to the occipitalis portion of the fronto-occipitalis muscle and to muscles associated with the external ear. It then proceeds anteriorly, giving off a branch **to the posterior belly of the digastric** muscle and **to the stylohyoid muscle** (which probably will not have been dissected at this time). The facial nerve continues anteriorly through the substance of the parotid gland. In the gland it divides into its five terminal branches (Fig. 7-15) which stream superiorly across the temporalis muscle, anteriorly across the zygomatic bone and masseter muscle, and inferiorly across the mandible to reach the muscles of facial expression. Although the arrangement of these nerves differs from body to body, usually five terminal branches can be distinguished that take on the names associated with their location. From superior to inferior, they are the **temporal, zygomatic, buccal, mandibular,** and **cervical.**

SENSORY NERVES *Cutaneous*

The cutaneous nerves of the head and neck are derived from five separate sources: (1) the ophthalmic, (2) maxillary, and (3) mandibular divisions of the trigeminal nerve, (4) dorsal rami of cervical nerves C2 through C6, and (5) ventral rami of cervical nerves C2, C3, and C4. The areas

supplied by each of these sources are outlined in Figure 7-12.

The cutaneous nerves derived from the ophthalmic division of the trigeminal nerve are the **supratrochlear, supraorbital, infratrochlear, external nasal,** and **lacrimal.** These nerves innervate an area that includes the anterior part of the scalp, the tip of the nose, and the upper eyelid.

The maxillary division of the trigeminal nerve contributes the **zygomaticotemporal,** the **zygomaticofacial,** and the **infraorbital** branches to the skin. These innervate the lateral side of the nose, the lower eyelid, the upper lip, and a small area superficial to the zygomatic bone.

The **auriculotemporal, buccal,** and **mental** nerves innervating the skin are branches of the mandibular division of the trigeminal. These innervate an area of the skin that covers the lower lip, chin, part of the lower jaw, the area of the cheek anterior to the ear, and a large area of the skin covering the temporalis muscle.

The **greater occipital** nerve, which is the medial branch of the dorsal ramus of C2, the **third occipital** nerve, which is the medial branch of the dorsal ramus of C3, and the medial branches of the **dorsal rami** of C4–C6, all send branches to the skin on the posterior part of the scalp and the posterior part of the neck. The area innervated is outlined in Figure 7-12.

The skin on the remaining portion of the neck (the lateral and anterior sides), on part of the lower jaw, and on the lower part of the parotid gland is innervated by branches of the ventral rami of C2, C3, and C4, portions of the cervical plexus. They are divided into ascending branches (the **great auricular** and the **lesser occipital**), transverse branches (the **transverse colli**), and descending branches (**medial, intermediate,** and **lateral supraclaviculars**). The ascending and transverse branches are derived from the second and third cervical segments of the spinal cord, while the descending branches are derived from the third and fourth segments.

SUPERFICIAL VEINS

Although there are many connections between the various superficial veins of the head and neck, for the sake of description they can be divided into those that drain into the internal jugular vein, those that drain into the suboccipital plexus of veins, and those that drain into the external jugular vein. Many of these vessels are pictured in Figure 7-14.

The anterior part of the face and scalp is drained by the **facial** vein, which originates by the joining of the **supraorbital** and **supratrochlear** veins draining the anterior surface of the scalp. The facial vein continues inferiorly, receiving tributaries from the nose, eyelids, upper lip, lower lip, etc., and finally empties into the internal jugular vein. Posteriorly, the scalp is drained by the **occipital** vein, which drains into the suboccipital plexus of veins in the suboccipital triangle. The remaining portions of the scalp and face are drained by the **superficial temporal** vein and the **posterior auricular** vein. These finally drain into the **external jugular** vein as follows: the **superficial temporal** joins with a vein draining the deeper part of the cheek, the **maxillary** vein, to form the **retromandibular** vein; the retromandibular vein, in turn, joins with the **posterior auricular** vein, which drains the territory just posterior to the ear, to form the **external jugular** vein; the external jugular vein, after receiving the **anterior jugular** and **posterior jugular** veins (if present), empties into the **subclavian** vein.

As mentioned earlier, there are many communications between these veins. Two fairly constant ones should be mentioned, one between the retromandibular vein and the facial vein, and another between the facial vein and the anterior jugular vein. Of much greater importance than the communications just mentioned are those between the superficial veins and the deeper veins in the head and neck. Such veins are those that pass directly from the superficial veins on the scalp through the bone and connect with the dural venous sinuses on the inside of the skull—the **emissary** veins. Still more important than these are other connections with these sinuses, such as those via the ophthalmic veins in the orbit, which connect the veins around the eyes, nose, and upper lip with the cavernous sinus (one of the dural venous sinuses), and the pterygoid plexus of veins deep in the cheek, which also connects the superficial veins of the face with the cavernous sinus. Although not strictly emissary veins, these serve the same purpose of connecting superficial veins with the sinuses in the dura mater. This is of extreme importance, because infections on the surface can spread directly to the venous sinuses. The areas around the nose and upper lip are prone to infections because of the many glands located there and, owing to these connections, this area is a dangerous place in which

to have these infections. These deeper connections will be described in more detail in subsequent chapters.

ARTERIES

The arteries supplying the superficial aspects of the head and neck (Fig. 7-14) can be divided into those derived from the **internal carotid** and those from the **external carotid.**

The **supratrochlear, supraorbital, palpebral, dorsal nasal,** and **external nasal** branches are the only ones derived from the internal carotid. These supply blood to the anterior aspects of the scalp as far posteriorly as the vertex, to the eyelids, and the nose.

The branches of the **external carotid** supplying blood to the cutaneous areas of the head and neck are the **superficial temporal,** which gives blood to the lateral aspects of the scalp as well as the posterior part of the face; the **posterior auricular,** giving blood supply to the area of the scalp posterior to the ear; the **occipital** giving blood supply to the posterior part of the scalp; and the **facial,** supplying blood to the entire anterior region of the face with aid from a small **buccal** branch. These main branches are aided by small arteries that accompany all of the sensory nerves already studied that reach the skin through foramina. There are arteries accompanying the mental, infraorbital, zygomaticotemporal and zygomaticofacial nerves.

LYMPHATICS

There are groups of lymph nodes located in such a fashion as to form a ring completely around the head. Starting anteriorly, the **submental** nodes are located inferior and posterior to the chin; the **submandibular** nodes, inferior to the body of the mandible (some of these nodes extend superiorly into the buccal region of the cheek); the **superficial cervical** nodes, near the angle of the mandible in the superior part of the neck; the **preauricular** or **parotid** nodes, just anterior to the ear; the **postauricular** or **mastoid** nodes, posterior to the ear; and the **occipital** nodes, on the posterior part of the head.

The superficial parts of the head drain into these nodes and the pathways taken by the lymphatic channels are very similar to the vascular patterns just described, once again demonstrating that the lymphatic channels follow the pathway taken by the vascular system (Fig.

7-18). The lymphatics from the area of the scalp deriving its blood supply from the occipital artery drains into the occipital nodes, the area posterior to the ear drains into the postauricular or mastoid nodes, the areas supplied by the superficial temporal artery drain into the preauricular or parotid nodes, and the area supplied by the facial artery drains into the superficial cervical, submandibular, or submental nodes from posterior to anterior.

All these nodes serve as the primary barrier but have subsequent connections with secondary nodes in the neck called the **upper deep cervical** nodes. These deep nodes, in turn, have efferent channels that connect with

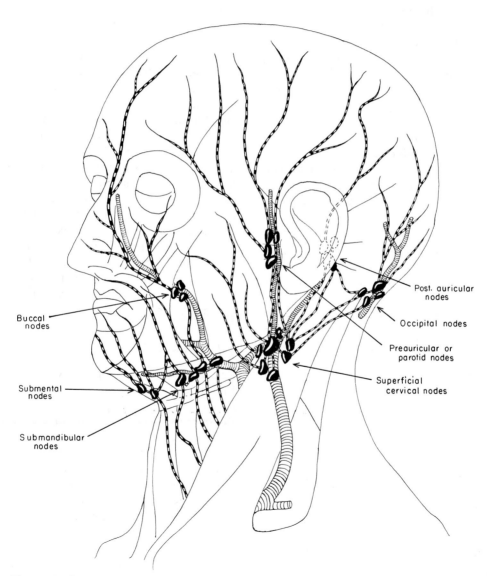

Figure 7-18. *Location of lymph nodes draining the superficial parts of the head and neck. Note how these lymphatics follow the vascular pathways. It should be noted that the drainage of the lower lip can cross the midline. These nodes have connections with a deeper set of nodes.*

the **lower deep cervical** nodes, and efferent pathways from these nodes drain into the venous system at the point where the internal jugular and subclavian veins join.

CONCEPTS: GENERAL RULES ON STRUCTURE OF THE HUMAN BODY*

Those of you who started this course with the back or the upper limb were presented with certain concepts on the structure of the human body that could be utilized in learning the remaining parts. These concepts can also be visualized at this point in the study of the head and neck, and are being repeated for those who started in this region of the body.

We have found these tenets to have general validity: that the skin does play an important role, not only in protecting underlying structures but in providing contact with the outside world; that it contains sweat glands, arrector pili muscles, and small blood vessels, all of which are innervated by the autonomic nervous system; that the sensory nerves, therefore, are combinations of sensory fibers and autonomic (motor) fibers and might better be called "cutaneous" nerves; that the skin is attached to underlying structures by a layer of superficial fascia of varying thickness through which the cutaneous nerves and vessels must course to reach the skin; and that skeletal muscles can be contained in superficial fascia.

We have found that muscles induce movement by contracting, and that this action can usually be determined by inspection, the insertion of the muscle being more informative than the origin; that muscles are surrounded by fascia; that they obtain a nerve supply early and keep this same innervation no matter where they migrate; that many of the muscles are derived from the branchial arches and are innervated by nerves that innervate these arch muscles in the lower animals; indeed, that

---

*This section is similar to those presented in other chapters in this textbook, but contains special features that relate to the head and the neck.

the nerve supply to a muscle is dependent upon its embryology.

We have found also that study of the head and neck is concerned with cranial nerves as well as those of cervical origin; that the great sensory nerve to the face is the trigeminal, while the motor nerve is the facial; that all three divisions of the trigeminal nerve provide cutaneous innervation and that these divisions are aided by dorsal and ventral rami of the cervical nerves; and that nerves are usually accompanied by an artery and a vein, forming a triad.

We have seen that the arteries are not always constant, not even from one side of the body to the other, and that veins are even more variable; that when one vessel is extra large a neighboring vessel may be correspondingly small; that incisions in the skin can be made in such a manner as to preserve both nerve and blood supply; and that arteries anastomose with one another in such a manner as to form important alternate pathways to a structure should one vessel be occluded, this being possible because arteries have no valves.

Lastly, we have clearly demonstrated that the lymphatics tend to follow the vascular pattern in a retrograde fashion, and that collections of nodes are likely to be found around these vessels.

CRANIAL CAVITY

GENERAL APPEARANCE AFTER REMOVAL OF CALVARIA

If the **dura mater** is pried away from the bone during removal of the calvaria, the brain is seen to be covered by this dense layer of connective tissue. The layer seen at this time was firmly attached to the bone, serving as the **endocranium (periosteum)**. This is shown on the left side of Figure 7-19.

The **meningeal arteries** and veins can be seen coming in from the lateral sides and coursing toward the midline. The arteries are branches of the middle meningeal artery, the branches of the anterior and posterior meningeal vessels being hidden from view. If the dura is cut longitudinally in the midline, the **superior sagittal**

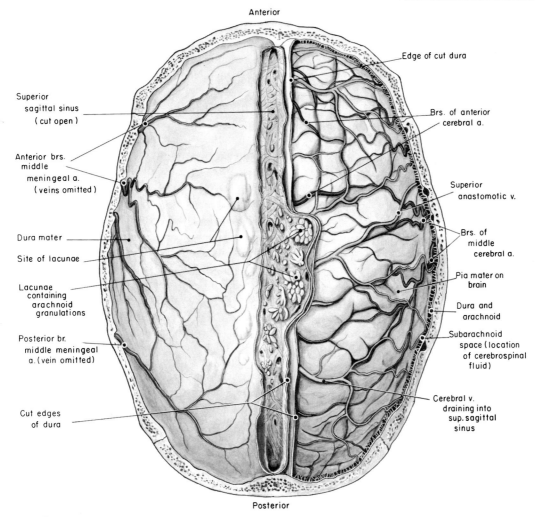

Anterior

Edge of cut dura

Superior
sagittal sinus
(cut open)

Brs. of anterior
cerebral a.

Anterior brs.
middle
meningeal a.
(veins omitted)

Superior
anastomotic v.

Brs. of
middle
cerebral a.

Dura mater

Site of lacunae

Pia mater on
brain

Dura and
arachnoid

Lacunae
containing
arachnoid
granulations

Subarachnoid
space (location
of cerebrospinal
fluid)

Posterior br.
middle meningeal
a. (vein omitted)

Cerebral v.
draining into
sup. sagittal
sinus

Cut edges
of dura

Posterior

Figure 7-19. *The cranial contents seen upon removal of the calvaria. The dura mater is intact on the left; the dura and arachnoid maters have been removed on the right. The superior sagittal sinus has been opened. (The meningeal vessels are actually within the dura mater.)*

sinus is opened. This contains venous blood and has direct connections with lakes or **lacunae** on either side of the sinus (Fig. 7-19). The openings of the **meningeal veins** and **veins from the superior surface** of the brain are visible in the superior sagittal sinus as well as the openings of the lacunae. In addition, small clusters of rounded elevations occur in both the lacunae and the superior sagittal sinus; these are the **arachnoid granulations,** which are actually clusters of arachnoid villi that

project into the sinuses and lacunae by penetrating the dura, and through which cerebrospinal fluid returns to the venous system (note Fig. 7-11). This superior sagittal sinus is one of many such sinuses that contain venous blood; they are located between the two layers of dura mater.

If the dura mater is cut and reflected, a thin **subdural space** is opened; this contains a lymph-like fluid. The thin glistening layer seen as a result of this reflection is the

arachnoid mater. If this layer is reflected, one sees the brain itself, with its intimate meningeal covering, the **pia mater.**

The space between the arachnoid and pia, the **subarachnoid space,** is extremely important, for it is in this area that the cerebrospinal fluid is found. This space* is considerably wider than the subdural and is made even more extensive by the fact that the pia mater extends into the sulci of the brain, while the arachnoid bridges over them. Numerous connective tissue strands extend from the arachnoid to the pia, making the subarachnoid space a meshwork of spaces (Fig. 7-11). Further details on the arachnoid mater will be given during a discussion of the circulation of the cerebrospinal fluid.

The two hemispheres of the brain are separated from one another by a midline inferior projection of the dura mater called the **falx** (*L.,* sickle) **cerebri** (Fig. 7-20). This sickle-shaped projection is attached anteriorly to the crista galli of the ethmoid bone and posteriorly to the superior limit of the **tentorium cerebelli,** the layers of dura forming a tent or covering over the cerebellum. The **inferior sagittal sinus** is located in the free margin of the falx cerebri.

When the brain is elevated the **anterior and middle cranial fossae** are in evidence, but the **posterior cranial fossa** cannot be seen because of the presence of the cerebellum and its dural covering, the tentorium cerebelli (Fig. 7-21).

This tentorium has a free margin anteriorly and medially that surrounds the brain stem. If the tentorium cerebelli is removed, the cerebellum is seen lying in the posterior cranial fossa. Space-occupying lesions of the brain are classified as either supra- or infratentorial depending on their relation to the tentorium.

When the entire brain is removed, all three sets of cranial fossae are quite visible. They are lined with dura mater, and the cranial nerves can be seen penetrating this dura (Fig. 7-32). The middle meningeal arteries are visible in the middle cranial fossa. There is a surgical approach to each fossa—through the frontal bone, through the temporal bone, and through the occipital bone to the anterior, middle, and posterior fossae, respectively.

*L. W. Weed, 1917, The development of the cerebrospinal spaces in pig and in man, *Carnegie Contrib. Embryol.* 5: 3. This work questions the presence of an actual subarachnoid space. From its development it would appear to be an intra-arachnoid space.

DURA MATER

This firm connective tissue covering of the brain lines the entire cranial cavity (Figs. 7-19 and 7-20). It is supposedly made up of two layers,† the outer one serving as the periosteum on the inside of the cranial bones. This outer layer is continuous with the periosteum on the outside of the skull at the foramen magnum and at all of the smaller foramina as well; it also connects to the pericranium by connective tissue strands at the sutures. The so-called inner layer of dura is continuous with that covering the spinal cord and also forms sheaths for the various cranial nerves. The two layers of dura divide from each other at various locations to form venous channels.

There are four **projections** into the cranial cavity from the dura mater, called the falx cerebri, falx cerebelli, the tentorium cerebelli, and the diaphragma sellae. The **falx cerebri** has already been mentioned as being attached to the crista galli anteriorly and the superior part of the tentorium cerebelli posteriorly. The **tentorium cerebelli** (Fig. 7-21) attaches to the arcuate eminence of the temporal bone and projects anteriorly from that to the anterior clinoid processes. Going posteriorly, it is attached to the point on the skull marked by the transverse grooves. The **falx cerebelli** is a small fold that projects anteriorly and superiorly from the internal occipital crest into the notch found in the cerebellum. The **diaphragma sellae** (Fig. 7-34) is a shelflike projection that forms a roof for the sella turcica (*L.,* Turkish saddle), in which the hypophysis (*Gr.,* under + growth) lies. The opening in the diaphragma sellae is for the **infundibulum** (*L.,* funnel), which is the attachment of the hypophysis to the base of the brain.

SINUSES. The dura mater exhibits many spaces that are filled with venous blood; they serve to drain blood from the brain and skull. The named sinuses are:

1. Superior sagittal
2. Inferior sagittal
3. Straight
4. Transverse
5. Sigmoid
6. Occipital

†L. C. Rogers and E. E. Payne, 1961, The dura mater at the craniovertebral junction, *J. Anat.* 95: 586. These workers claim the dura is a single-layered structure here as well as around spinal cord.

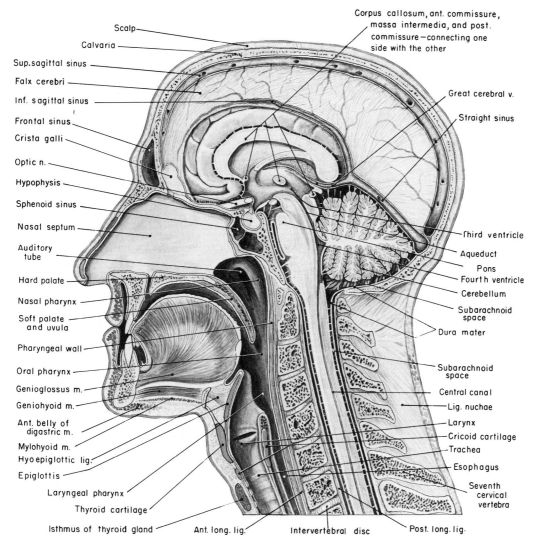

Scalp

Calvaria

Sup. sagittal sinus

Falx cerebri

Inf. sagittal sinus

Frontal sinus

Crista galli

Optic n.

Hypophysis

Sphenoid sinus

Nasal septum

Auditory tube

Hard palate

Nasal pharynx

Soft palate and uvula

Pharyngeal wall

Oral pharynx

Genioglossus m.

Geniohyoid m.

Ant. belly of digastric m.

Mylohyoid m.

Hyoepiglottic lig.

Epiglottis

Laryngeal pharynx

Thyroid cartilage

Isthmus of thyroid gland

Corpus callosum, ant. commissure, massa intermedia, and post. commissure—connecting one side with the other

Great cerebral v.

Straight sinus

Third ventricle

Aqueduct

Pons

Fourth ventricle

Cerebellum

Subarachnoid space

Dura mater

Subarachnoid space

Central canal

Lig. nuchae

Larynx

Cricoid cartilage

Trachea

Esophagus

Seventh cervical vertebra

Ant. long. lig.

Intervertebral disc

Post. long. lig.

Figure 7-20. *A sagittal section of the head and neck. At this time note the falx cerebri; the superior sagittal, inferior sagittal, and straight dural sinuses; and the great cerebral vein.*

7. Cavernous
8. Sphenoparietal
9. Superior petrosal
10. Inferior petrosal
11. Intercavernous
12. Basilar

1. The **superior sagittal sinus** (Figs. 7-19, 7-20, and 7-22) begins at the foramen cecum and sometimes connects with the veins of the nose via this foramen. It

arches posteriorly to end at the confluens of sinuses at the internal occipital protuberance. It usually turns to the right to become the right transverse sinus. The superior sagittal sinus receives veins from the superior part of the brain and some of the dural veins, and communicates with the lacunae on either side of this sinus. It has an important communication with the occipital veins by means of emissary veins passing through the parietal foramina.

2. The **inferior sagittal sinus** (Figs. 7-20 and

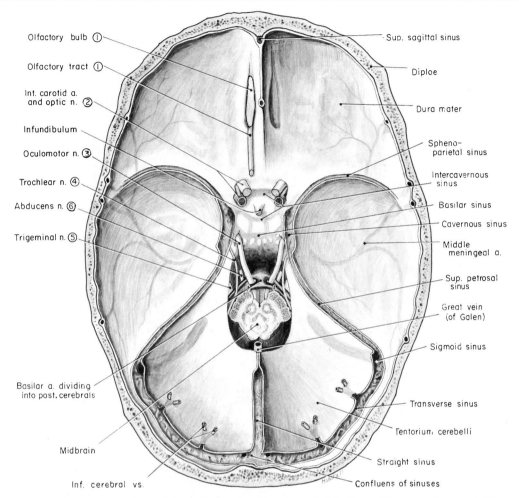

Olfactory bulb ①
Olfactory tract ①
Int. carotid a. and optic n. ②
Infundibulum
Oculomotor n. ③
Trochlear n. ④
Abducens n. ⑥
Trigeminal n. ⑤
Basilar a. dividing into post. cerebrals
Midbrain
Inf. cerebral vs.

Sup. sagittal sinus
Diploe
Dura mater
Spheno-parietal sinus
Intercavernous sinus
Basilar sinus
Cavernous sinus
Middle meningeal a.
Sup. petrosal sinus
Great vein (of Galen)
Sigmoid sinus
Transverse sinus
Tentorium cerebelli
Straight sinus
Confluens of sinuses

Figure 7-21. *The floor of the cranial cavity after removal of the brain. The tentorium cerebelli remains intact and hides the cerebellum from view. Several of the dural sinuses have been opened. See Figure 7-32 for other sinuses not visible in this illustration.*

7-24) lies in the posterior part of the free edge of the falx cerebri. It ends at the anterior edge of the tentorium to empty into the anterior end of the straight sinus. It drains the veins of the falx and a few veins on the cerebral hemispheres.

3. The **straight sinus** (Figs. 7-20, 7-21, and 7-24) is formed by the inferior sagittal sinus just mentioned plus the important great cerebral vein, which drains the deep parts of the brain. The fact that this vein attaches to the comparatively firm dural (straight) sinus makes it liable to injury at childbirth, a fatal occurrence. The straight sinus courses posteriorly and slightly inferiorly at the line where

the falx cerebri joins the tentorium. At the confluens of sinuses the straight sinus usually turns to the left to form the left transverse sinus. This straight sinus drains veins from the occipital lobes of the brain, the cerebellum, and the falx itself.

4. The **transverse sinuses** (Figs. 7-21, 7-23, and 7-24) extend transversely, as might be expected from their names, from the confluens of sinuses at the internal occipital protuberance. As mentioned previously, the left one is usually a continuation of the straight sinus, while the right connects with the superior sagittal sinus. These sinuses are contained in the base of the tentorium cere-

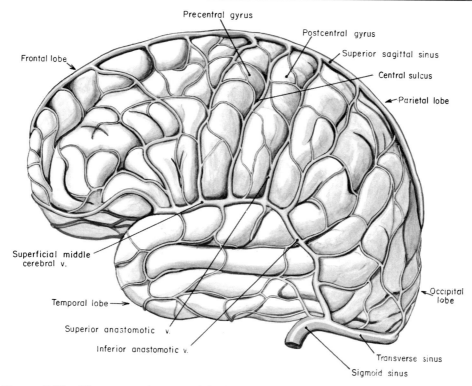

Precentral gyrus

Postcentral gyrus

Superior sagittal sinus

Frontal lobe

Central sulcus

Parietal lobe

Superficial middle cerebral v.

Temporal lobe

Occipital lobe

Superior anastomotic v.

Inferior anastomotic v.

Transverse sinus

Sigmoid sinus

Figure 7-22. The venous drainage of the lateral surface of the brain. The superior sagittal, sigmoid, and transverse dural sinuses have been included in the picture. (Modified after Jamieson.)

belli. When they reach a point where they leave the occipital bone to pass onto the posterior ends of the parietal bones, they are known as the sigmoid sinuses. The occipital diploic vein, pictured in Fig. 7-26, drains into the transverse sinuses.

5. The **sigmoid sinuses** (Figs. 7-21, 7-22, and 7-23) leave the tentorium cerebelli and take a sharp curve anteromedially to end in the internal jugular veins. They are located at the base of the arcuate eminence of the temporal bone. The sigmoid sinuses receive superior petrosal sinuses, veins from the cerebellum and the inferior part of the cerebrum, and the posterior temporal diploic vein (see Fig. 7-26), and they communicate with the superficial veins by emissary veins in the mastoid and posterior condylar canals. The relation of the sigmoid sinus to the mastoid air cells is of particular importance; they are located lateral to the sigmoid sinus and are separated from it by a very thin layer of bone.

6. The **occipital sinuses** (Fig. 7-32) start in several small venous channels one of which is connected with the terminal end of the sigmoid sinus just before it becomes the internal jugular vein. These occipital sinuses continue as small channels that surround the foramen magnum to gain the base of the falx cerebelli where the two channels join to form a single sinus. This sinus empties into the confluens.

7. The **cavernous sinuses** (Fig. 7-21) lie on either side of the sella turcica of the sphenoid bone. They drain the ophthalmic veins and communicate with the pterygoid plexus of veins deep in the cheek by emissary veins that pass through the carotid canal, the foramen ovale, and the sphenoidal emissary foramina. These connections with the ophthalmic veins are extremely important for the veins draining the face have connections with these ophthalmic veins. Therefore, infections on the face, which are frequent, have an indirect connection to the

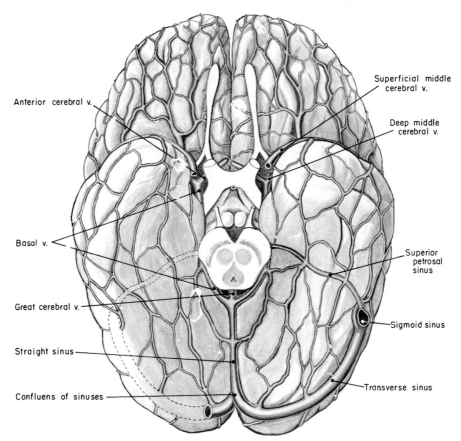

Figure 7-23. *The venous drainage of the base of the brain. The straight, sigmoid, and trans-verse dural sinuses have been included. (Modified after Jamieson.)*

cavernous sinuses. A generalized septicemia can develop. We shall see later that the internal carotid artery and the abducens nerve course through this sinus, and that the oculomotor, trochlear, and parts of the trigeminal nerve are located in the walls of these cavernous sinuses. These sinuses communicate with many of the following sinuses.

8. The **sphenoparietal sinuses** (Fig. 7-21) are located on the inferior surface of the lesser wings of the sphenoid bone. They receive the anterior temporal diploic veins and communicate with meningeal veins. They terminate in the anterior ends of the cavernous sinuses.

9. The **superior petrosal sinuses** (Fig. 7-21) begin at the posterior ends of the cavernous sinuses and, following the attachment of the tentorium cerebelli to the superior edge of the petrous portion of the temporal

bone, end in the sigmoid sinuses. These sinuses receive the inferior cerebral and superior cerebellar veins.

10. The **inferior petrosal sinuses** (Fig. 7-32) also start at the posterior end of the cavernous sinuses. They run posteriorly and inferiorly in the groove between the base of the petrous bone and the basal portion of the occipital bone. They enter the jugular foramina to end in the jugular veins. They drain inferior cerebellar veins and veins from the internal ear.

11. The **intercavernous sinuses** (Fig. 7-21) are a plexus of venous channels that connect the two cavernous sinuses to one another. They lie in the floor of the sella turcica and also occur in the free edge of the diaphragma sellae.

12. The **basilar sinuses** (Fig. 7-21) are located on

the clivus of the skull just posterior to the sella turcica of the sphenoid bone. These basilar sinuses join the inferior petrosal sinuses to each other and communicate with the cavernous sinuses.

All of the dural sinuses ultimately drain into the internal jugular veins; these veins, therefore, drain the very important contents of the cranial cavity. The connections of these sinuses with veins outside the cranial cavity are extremely important clinically. Infection can spread from the outside to the inside via these veins because they have no valves. Figures 7-22, 7-23, and 7-24 show the relations of the brain to these dural sinuses and the many veins of the brain that drain into them.

BLOOD SUPPLY.   The arteries to the dura are the **anterior meningeal** in the anterior cranial fossa, the **middle** and **accessory meningeals** in the middle cranial fossa, and the **posterior meningeals** in the posterior cranial fossa. The anterior and posterior meningeal arteries are quite small, as is the accessory meningeal. By far the greater part of the dura is supplied by the middle meningeal artery.

The **anterior meningeal** is a branch of the ethmoi-dal artery. It supplies the dura on the floor of the anterior cranial fossa.

The **middle meningeal** (Fig. 7-19) is a branch of the maxillary artery (in the infratemporal fossa) and enters the middle cranial fossa via the foramen spinosum. This artery courses anteriorly, laterally, and superiorly in the outer layer of the dura and ends on the greater wing of the sphenoid bone by dividing into anterior and posterior branches. The **anterior branch** courses almost vertically to reach the vertex of the skull. It passes the point where the frontal, parietal, temporal, and the sphenoid (the greater wing) bones come together, the pterion. The course of this artery is along the anterior part of the parietal bone just posterior to the coronal suture. The **posterior** branch courses posteriorly and superiorly to supply the posterior part of the dura mater. It crosses the squamous portion of the temporal bone and the posterior part of the parietal bone. Figures 7-5 and 7-8 show the grooves made by these vessels in the bones of the skull, and Figure 7-35 relates these vessels to the skull and brain.

The **accessory meningeal** artery is a small branch of the maxillary artery, or of the middle meningeal itself, that

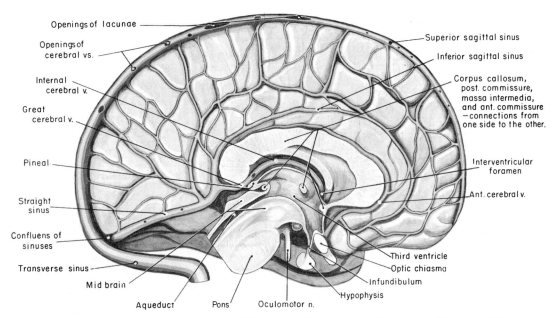

Figure 7-24. *The venous drainage of the medial side and internal structure of the brain. The superior sagittal, inferior sagittal, straight, and transverse sinuses have been included. (Modified after Jamieson.)*

passes through the foramen ovale to supply the dura mater around the cavernous sinus. This branch also supplies blood to the semilunar ganglion, the sensory ganglion of the trigeminal nerve.

The **posterior meningeal** arteries are branches of the ascending pharyngeal, occipital, and vertebral arteries. The branches of the ascending pharyngeal artery enter the posterior fossa of the cranial cavity via the jugular foramen and the hypoglossal canal. The meningeal branch of the vertebral artery enters the dura mater as that artery penetrates the dura to enter the cranial cavity. The branch of the occipital artery giving blood supply to the meninges enters the cranial cavity via jugular and mastoid foramina. All the posterior meningeal arteries are limited to the posterior cranial fossa.

The **meningeal veins** correspond to the arteries just described. Those in the anterior and posterior cranial fossae usually drain into the nearest venous dural sinuses. The middle meningeal vein drains into the pterygoid plexus of veins in the infratemporal fossa.

Injury to the meningeal vessels is frequently associated with skull fractures. Hemorrhages from these vessels usually occur outside (epidural) the dura mater.

NERVES. The sensory nerve supply to the dura is by branches of all three divisions of the trigeminal nerve, and by the vagus and hypoglossal nerves (actually from the first and second cervical nerves). The meningeal contribution from the **ophthalmic division** is twofold. A meningeal branch from the ophthalmic nerve turns posteriorly to reach the tentorium cerebelli and thence the posterior part of the falx cerebri. Meningeal branches of the ethmoidal nerves (Fig. 7-44) supply the anterior cranial fossa and the anterior part of the falx cerebri. Meningeal branches from the **maxillary division** supply the middle cranial fossa. The meningeal branch of the **mandibular division** actually arises from the undivided nerve and enters the infratemporal fossa. It returns to the cranial cavity via the foramen spinosum to be distributed to that part of the dura supplied by the middle meningeal artery. This leaves only the dura in the posterior cranial fossa. This area receives sensory innervation from the **first and second cervical nerves** by way of the vagus and hypoglossal nerves. Pain from the dura can be referred to the face. Figure 7-25 summarizes the sensory nerve supply to the dura.

The blood vessels of the dura are supplied by the

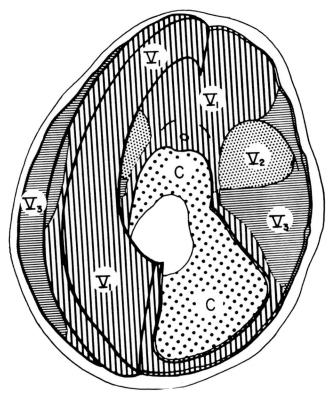

Figure 7-25. *Sensory nerve supply to the dura mater by the three divisions of the trigeminal nerve and the first three cervical nerves. Pain from these areas can be referred to appropriate areas on the face or scalp. (From work of Donald L. Kimmel, Chicago Medical School Quarterly, volume 22, page 24, 1961; courtesy of the Dean of the Chicago Medical School.)*

automatic nervous system in the same manner that all vessels in the head and neck derive their nerve supply. This may be important clinically, for there is a possibility that some headaches are caused by dilation of these vessels.

DIPLOIC VEINS

The skull bones are made up of two tables of compact bone with spongy bone (diploe) between the tables. Veins coursing through this spongy area are called **diploic veins** (Fig. 7-26). Those that have acquired names are (1) the **frontal** diploic veins, which drain into the supraorbital veins, (2) the **anterior temporal** veins, which terminate in the sphenoparietal sinuses, (3) the **posterior**

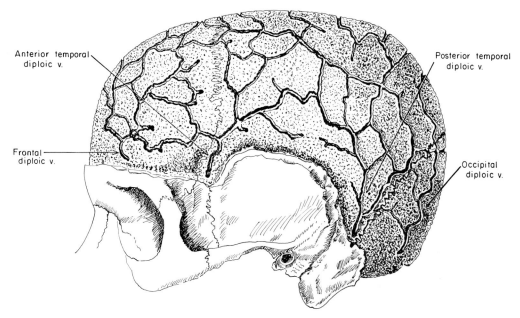

Figure 7-26. *The diploic veins. The outer table of the skull bones has been removed.*

temporal veins, which open into the sigmoid sinus, and (4) the **occipital diploic** veins, which drain into the transverse sinuses. The latter two may open into superficial veins. These diploic veins anastomose freely with each other and also communicate with the meningeal veins internally and the superficial veins externally.

Since the meningeal and superficial arteries (to the scalp) supply blood to the cranial bones, there are no named diploic arteries.

BRAIN

Although the central nervous system is usually studied in a separate course, you should have some acquaintance with the brain at this time. The brain can be divided into the (1) cerebral hemispheres, (2) cerebellum, (3) midbrain, (4) pons, and (5) medulla oblongata.

The **cerebral hemispheres** are made up of elevations (**gyri**) and valleys (**sulci**) between these gyri. The hemispheres are divided into **lobes,** which correspond roughly to the names given the skull bones. As can be seen in Figures 7-27 and 7-29, a vertical sulcus divides the frontal lobe from the parietal; this is called the **central sulcus.** The gyrus just anterior to the central sulcus is called the **precentral gyrus** and is the primary motor area

of the brain. (If a diagram of the human figure is inverted and placed in this gyrus, the body areas will correspond roughly with their motor areas in the gyrus.) The rest of the frontal lobe is used in modifying motor acts. The gyrus just posterior to the central sulcus, the **postcentral gyrus,** is the primary area for receipt of sensations from the body; the rest of the parietal lobe interprets these sensations on the basis of past experience. The posterior end of the brain is the **occipital lobe,** mainly concerned with vision, while the **temporal lobe** is primarily involved with hearing. The frontal lobe is located in the anterior cranial fossa and the temporal lobe in the middle cranial fossa.

The **cerebellum** (Figs. 7-28 and 7-30) is that part of the brain that lies in the posterior cranial fossa. This organ is very important to us in bringing coordination to our motor activities.

Looking at the inferior surface of the brain (Fig. 7-28), that part just superior to the spinal cord is the **medulla oblongata.** Just anterior to the medulla is the large rounded **pons,** which serves to bring cerebellar control into our motor activities. Anterior to the pons is the **midbrain,** in which the most conspicuous structures are the two **cerebral peduncles** that serve as a pathway to and from higher centers. Just anterior to the cerebral

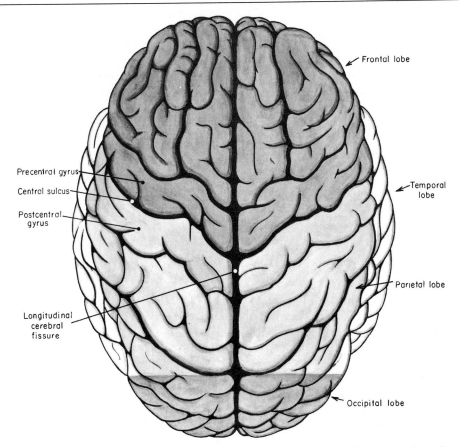

Figure 7-27. *The superior surface of the brain. The various lobes of the brain have been shaded differently.*

peduncles are the two **mamillary bodies** (concerned with smell); the **anterior and posterior perforated substances** are just anterior and posterior to these structures. These perforations are made by small arteries entering this area of the brain. Anterior to these areas is the **infundibulum,** which attaches the **hypophysis** to the brain, and anterior to this is the **optic chiasma.**

The medulla oblongata, pons, and midbrain collectively make up the **brain stem.**

ARTERIES AND VEINS.   The brain receives its arterial supply from two main sources, the **vertebral** and the **internal carotid arteries** (Fig. 7-28).

The two **vertebral arteries** course through the foramen magnum and, after penetrating the dura mater, come to lie on the lateral sides of the medulla oblongata.

These arteries gradually come closer together and join to form a single **basilar artery** which lies on the pons. The basilar artery terminates by dividing into the posterior cerebral arteries.

The vertebral arteries have the following branches: (1) posterior spinal, (2) meningeal, (3) posterior inferior cerebellar, (4) anterior spinal, and (5) small branches to the medulla oblongata (Fig. 7-28). The **meningeal** vessel is given off before the vertebral penetrates the dura mater. The **posterior spinal** artery courses the length of the spinal cord closely associated with the dorsal roots. There is a similar artery from the other side that makes a second posterior spinal artery. The **anterior spinal** arteries combine to form a single artery that courses the length of the spinal cord in the anterior median sulcus. The **posterior inferior cerebellar** artery gives blood sup-

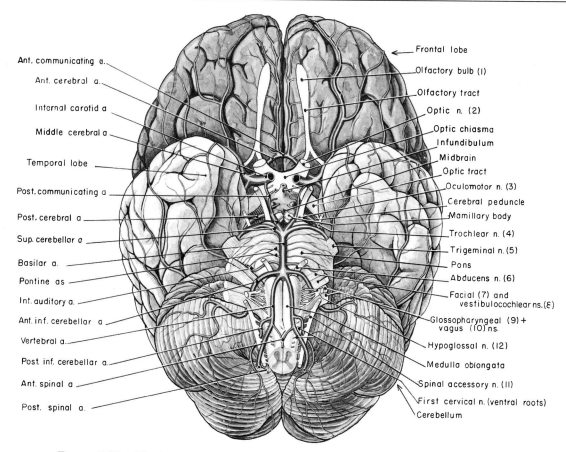

Figure 7-28. *The base of the brain, showing distribution of the cerebral arteries and the emergence of the cranial nerves.*

ply to the posterior and inferior surfaces of the cerebellum.

The **basilar artery** has the following branches: (1) anterior inferior cerebellar, (2) small branches to the pons, and (3) superior cerebellar (Fig. 7-28). The **anterior inferior cerebellar** courses to the inferior surface of the cerebellum, while the **superior cerebellar** artery gives blood supply to the superior surface of that organ. The very important **auditory** artery usually arises from the anterior inferior cerebellar and accompanies the eighth nerve into the internal auditory meatus.

The **posterior cerebral arteries** (Fig. 7-28), the terminal branches of the basilar, wind around the cerebral peduncles and give blood supply to that part of the base of the brain that lies on the tentorium cerebelli. The close

relation of this artery and the superior cerebellar to the third and fourth cranial nerves should be noted; diseases involving these arteries often affect the nerves (note also Fig. 7-30).

The **internal carotid artery** goes through the base of the skull in the carotid canal. It then enters the cavernous sinus and, after turning superiorly, divides into its terminal branches. The branches of the internal carotid artery in the cranium are: (1) small **arteries to the hypophysis** and to the walls of the cavernous sinus and the nerves contained therein, including the semilunar ganglion; (2) the **ophthalmic,** which enters the orbit and will be described later (page 518); (3) the **posterior communicating,** which joins the posterior cerebral; (4) the **choroid,** which is the vessel supplying the choroid plexuses in the ventricles from which the cerebrospinal fluid is secreted; and the terminals, (5) the **middle cerebral** and (6) **anterior cerebral** arteries.

The **middle cerebral** artery passes into the lateral fissure between the parietal and temporal lobes (Figs. 7-28 and 7-29). It sends many branches to the lateral sides of the cerebral hemispheres. In addition to these superficial branches, many central branches pass into the brain substance itself.

The **anterior cerebral artery** courses medially, superior to the optic chiasm, and enters the space between the two cerebral hemispheres (Figs. 7-28 and 7-30). It courses superiorly and then posteriorly, giving blood supply to the medial sides of both hemispheres. This artery also has branches that penetrate the brain substances.

Cerebral arterial circle. The two **anterior cerebral** arteries are connected to each other by a short **anterior communicating** artery. The internal carotid and the posterior cerebral arteries are also connected by the **posterior communicating** arteries. Because of these connections, an arterial circle is formed around the infundibulum that is known as the cerebral arterial circle (of Willis). This circle is pictured in Figure 7-28. It is of some importance in providing a collateral circulatory pathway to the brain when injuries occur to any of the blood vessels in this region. There are numerous arteries arising from this circle that penetrate the brain substance and are extremely important; not only are they small and easily plugged, but they are also end arteries (no collateral circulation) and supply vital areas.

The **superficial veins** on the surface of the cerebral hemispheres drain into the superior sagittal sinus, transverse sinus, sigmoid sinus, cavernous sinus, and the other sinuses as well (Figs. 7-22, 7-23, and 7-24). The veins from the deeper part of the brain join into one, the **great cerebral vein** (of Galen), which empties into the straight sinus (Fig. 7-24).

You should realize that the above is a very brief description of the arteries and veins of the brain. This description will undoubtedly be augmented in a study of the central nervous system.

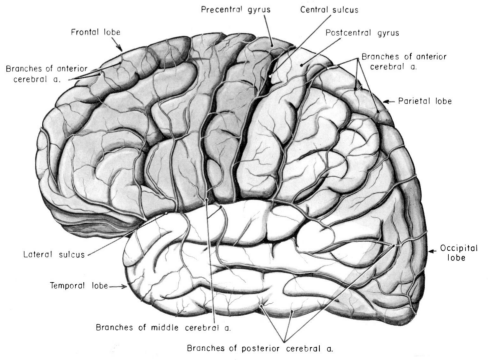

Figure 7-29. *The lateral surface of the brain showing the distribution of the cerebral arteries.*

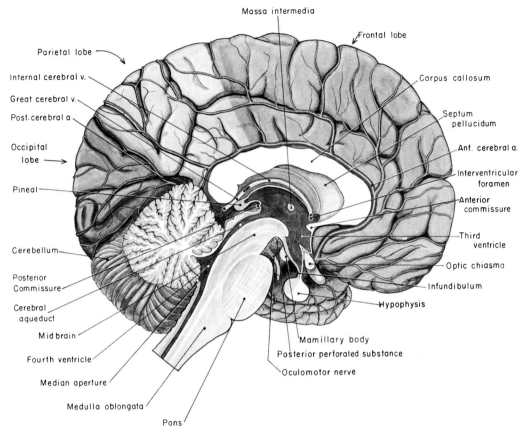

Massa intermedia

Frontal lobe

Parietal lobe

Corpus callosum

Internal cerebral v.

Great cerebral v.

Septum pellucidum

Post. cerebral a.

Ant. cerebral a.

Occipital lobe

Interventricular foramen

Pineal

Anterior commissure

Third ventricle

Cerebellum

Optic chiasma

Posterior Commissure

Infundibulum

Cerebral aqueduct

Hypophysis

Mid brain

Mamillary body

Fourth ventricle

Posterior perforated substance

Median aperture

Oculomotor nerve

Medulla oblongata

Pons

Figure 7-30. *A sagittal section of the brain showing the distribution of the cerebral arteries.*

CEREBROSPINAL FLUID

As mentioned previously, the **subarachnoid space** is filled with **cerebrospinal fluid** which covers the surfaces of the brain and spinal cord. It is also located in the spaces inside the brain and spinal cord—the ventricles and central canal. Figure 7-31 shows the conventional schema of the sites of production, circulation, and exit of the cerebrospinal fluid into the blood stream. According to this theory, the cerebrospinal fluid is given off from a tuft of arteries, the **choroid plexus;** it enters the third and lateral ventricles of the brain from these blood vessels. It then travels via a small aqueduct (**cerebral aqueduct**) into the fourth ventricle (Fig. 7-30). From there it not only enters the central canal of the spinal cord but also leaves the inside of the central nervous system via a central **median aperture** (of Magendie) and two **lateral apertures** (of Luschka) to enter the **subarachnoid space.** The cerebro-spinal fluid circulates inferiorly around the spinal cord and then superiorly to surround the brain stem, cerebellum, and cerebral cortex. The cerebrospinal fluid is emptied into the venous system via **arachnoid villi,*** which are projections of the arachnoid into the lacunae alongside the superior sagittal sinus and into the superior sagittal sinus itself.†

---

*G. B. Hassins, 1930, Villi (Pacchionian bodies) of the spinal arachnoid, *Arch. Neurol. Psychiat.* 23: 65. This author describes arachnoid villi in the spinal cord but denies they play any great role in circulation of the cerebrospinal fluid, either in the area of the brain or spinal cord. He thinks the fluid leaves the "open" subarachnoid space by following perineural root or nerve spaces.

†F. Howorth and G. R. A. Cooper, 1955, The fate of certain foreign colloids and crystalloids after subarachnoid injection, *Acta Anat.* 25: 112. These authors have found, in cats, that more cerebrospinal fluid drains into the veins contained in the subarachnoid space than through the arachnoid villi.

The cerebrospinal fluid serves as a protective layer against vibration. Unfortunately, the small aqueduct or the equally narrow apertures mentioned above can become closed; in this condition, the cerebrospinal fluid continues to be secreted, producing too great a pressure on the parts of the brain related to the various ventricles. In children, this condition (hydrocephalus) results in enlarged heads or, if the bones are formed, in a severe thinning of the cerebral cortex.

The cerebrospinal fluid can become infected, as can the meninges (meningitis).

CRANIAL NERVES

The origins of the twelve cranial nerves can be seen on the base of the brain (Fig. 7-28).

1. The **olfactory tracts and bulbs** are located on the base of the frontal lobes. They are approximately ½ inch apart and on either side of the median sulcus.

2. The **optic chiasma** is easily seen; the optic nerves arise from this chiasma.

3. The **oculomotor nerves** arise between the midbrain and the pons and course between the posterior cerebral and superior cerebellar arteries, an important relation. As the name implies, this nerve innervates muscles that move the eyeball.

4. The **trochlear nerves** (named for the fact that they innervate a muscle that uses a pulley or trochlea in its action) actually arise on the dorsal side of the brain stem (an interesting point), but wind around it to appear between the pons and the temporal lobes. These nerves are very small.

5. The **trigeminal nerves** are large nerves that can be seen arising from the middle of the pons. These nerves are named for the three divisions into which they divide.

6. The **abducens nerves** first appear in the groove between the pons and the medulla oblongata. They are closely related to the anterior inferior cerebellar arteries, also an important clinical point. They innervate a muscle that abducts the eyeball.

7 and 8. The seventh and eighth cranial nerves, the **facial** and **vestibulocochlear nerves,** arise in the groove between the pons and the medulla oblongata but in a more lateral position than the abducens.

9, 10, and 11. The **glossopharyngeal, vagus,** and **spinal accessory nerves** arise from the medulla at a point that is posterior to the inferior olive (a large structure that produces a rounded projection on the surface of the medulla). The glossopharyngeal nerves are involved with the tongue and pharynx, the vagus "wanders" over a large territory, and the spinal accessory nerves, because they actually arise from the spinal cord, ascend the medulla to join the other two nerves.

12. The **hypoglossal nerve** (motor to the tongue) arises by many rootlets from the medulla in a groove between the inferior olive posteriorly and the bulge of the pyramid (a bundle of fibers descending to the cord from the cerebral cortex) anteriorly. The first cervical nerve can also be seen to arise from this same location but a bit more inferiorly placed.

These same cranial nerves can be seen on the floor of the cranial cavity as they pierce the dura mater. It should be remembered that the point where these nerves pierce the dura mater is not always the same as their exit point through the skull bones themselves. Figure 7-32 shows the cranial nerves piercing the dura on the left and traversing their respective bony foramina on the right. Anteriorly the cribriform plate of the ethmoid bone is in evidence, and it is at this location that the bulb of the olfactory nerve is located. The **olfactory nerves** stream inferiorly into the nasal cavity through the many openings in this sievelike plate. The **optic nerves** enter directly into the optic foramina.

The remaining cranial nerves are located in the posterior cranial fossa. The third cranial nerve, the **oculomotor,** penetrates the dura mater just at the anterior end of the attachment of the tentorium cerebelli; it is approximately ½ inch posterior to the optic foramen. The small **trochlear nerve** penetrates the free edge of the dura, which forms the anterior end of the tentorium cerebelli. The **trigeminal nerve** enters the dura mater on the posterior medial side of the arcuate eminence at a point inferior to the attachment of the tentorium, while the **abducens nerve** appears about ¾ inch posterior to the posterior clinoid processes and about ¼ inch medial to the trigeminal nerve. The **facial** and **vestibulocochlear nerves** enter immediately into the internal auditory meatus, which is on the medial side of the petrous portion of the temporal bone. The **glossopharyngeal, vagus,** and **spinal accessory nerves** penetrate the dura mater immediately over the jugular foramen; these nerves accompany the internal jugular vein through this foramen.

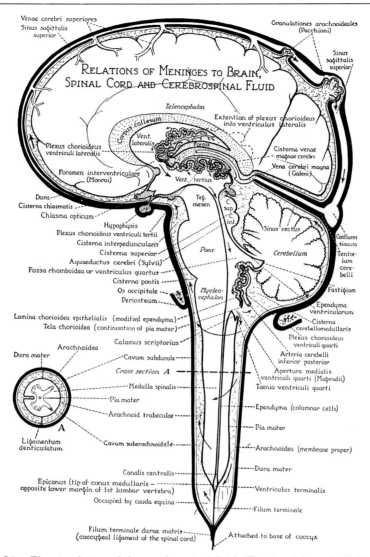

RELATIONS OF MENINGES TO BRAIN, SPINAL CORD AND CEREBROSPINAL FLUID

Figure 7-31. *The circulation of the cerebrospinal fluid. (Reprinted from A. T. Rasmussen, The Principal Nervous Pathways, courtesy of Macmillan Co.)*

The **hypoglossal nerve** enters the dura immediately adjacent to the hypoglossal canal, through which it exits from the skull.

CAVERNOUS SINUS

The **trigeminal nerve,** immediately after penetrating the dura, forms a large ganglion—the **trigeminal gan-** glion—which is located in the **trigeminal cave,** a space between two layers of dura mater. Two of its three divisions—the ophthalmic and maxillary—then enter the lateral wall of the cavernous sinus as depicted in Fig. 7-33. The **oculomotor** and **trochlear** nerves also course in the lateral wall of this sinus but the **abducens** nerve actually traverses the sinus close to the internal carotid artery. These nerves are separated from the blood in the

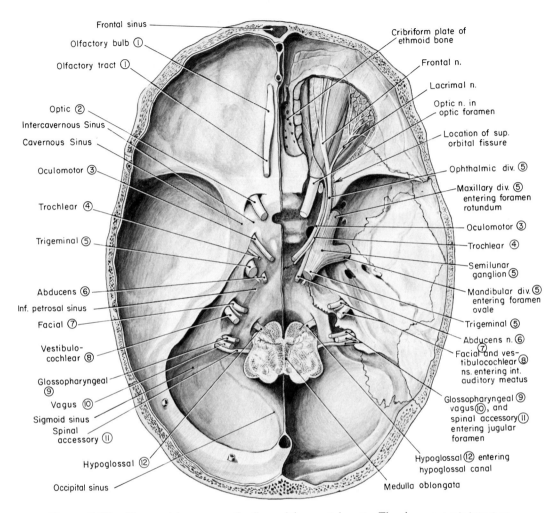

Frontal sinus
Olfactory bulb ①
Olfactory tract ①
Optic ②
Intercavernous Sinus
Cavernous Sinus
Oculomotor ③
Trochlear ④
Trigeminal ⑤
Abducens ⑥
Inf. petrosal sinus
Facial ⑦
Vestibulo-cochlear ⑧
Glossopharyngeal ⑨
Vagus ⑩
Sigmoid sinus
Spinal accessory ⑪
Hypoglossal ⑫
Occipital sinus

Cribriform plate of ethmoid bone
Frontal n.
Lacrimal n.
Optic n. in optic foramen
Location of sup. orbital fissure
Ophthalmic div. ⑤
Maxillary div. ⑤ entering foramen rotundum
Oculomotor ③
Trochlear ④
Semilunar ganglion ⑤
Mandibular div. ⑤ entering foramen ovale
Trigeminal ⑤
Abducens n. ⑥
Facial ⑦ and ves-tibulocochlear ⑧ ns. entering int. auditory meatus
Glossopharyngeal ⑨ vagus ⑩, and spinal accessory ⑪ entering jugular foramen
Hypoglossal ⑫ entering hypoglossal canal
Medulla oblongata

Figure 7-32. *The cranial nerves on the floor of the cranial cavity. The dura mater is intact on the left but has been removed from the right.*

sinus by a thin layer of connective tissue. An intimate knowledge of the relations of these nerves to the dura mater is vital for the neurosurgeon.

After leaving the area of the cavernous sinus, the third and fourth cranial nerves, the ophthalmic division of the fifth, and the sixth cranial nerves enter the orbit through the **superior orbital fissure** and will be studied in the section on the orbit. The maxillary division of the trigeminal nerve leaves the wall of the cavernous sinus and traverses the **foramen rotundum.** The mandibular division leaves the cavity via **foramen ovale.** The

trigeminal ganglion is the site of the cell bodies of the many sensory fibers of the trigeminal nerve. This ganglion is exactly the same as a dorsal root ganglion of a spinal nerve; synapses do not occur in this ganglion. The relatively small **motor root of the trigeminal nerve** is located on the medial side of the mandibular division and on the medial side of the ganglion. This motor root is responsible for the innervation of all muscles that are innervated by the trigeminal nerve.

Another very vital structure located in the cavernous sinus is the **internal carotid artery.** After the artery

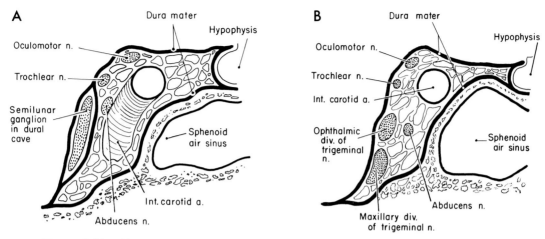

Figure 7-33. *Frontal sections through the cavernous sinus and its contents. (A) Through the semilunar ganglion, (B) more anteriorly after divisions have emerged from the ganglion. Note intimate relation between abducens nerve and the internal carotid artery.*

penetrates the skull via the internal carotid canal, it takes a turn and courses anteriorly through the sinus (at this point it is superior to the foramen lacerum) and then turns superiorly to leave the sinus and divide into its terminal branches.

In addition to showing the intimate relations of these structures to each other, Fig. 7-33 depicts their close relation to the sphenoid paranasal sinus. It should also be noted that these vital structures are lateral relations of the hypophysis and can be injured in tumors of this gland.

HYPOPHYSIS (PITUITARY GLAND)

The **hypophysis** (Figs. 7-20, 7-24, and 7-30) is a small gland about the size of a large pea which is located in the sella turcica of the sphenoid bone. It is connected to the brain by the **infundibulum,** which pierces a small opening in the diaphragma sellae.* It is divided into anterior and posterior lobes. The **posterior lobe** is derived embryologically from the brain and is that portion attached to the brain via the infundibulum. The **anterior lobe** is derived embryologically from an ectodermal outpouching

*G. B. Wislocki, 1937, The meningeal relations of the hypophysis cerebri II. An embryological study of the meninges and blood vessels of the human hypophysis, *Am. J. Anat.* 61: 95. This author could find no pia or arachnoid on the hypophysis; the stalk is surrounded by the subarachnoid space but this does not extend into the sella turcica.

from the roof of the future mouth (Rathke's pouch) and has no direct connection with the inferior surface of the brain. It makes up a large part of the total hypophysis and is flat from superior to inferior and concave posteriorly. The **posterior lobe** fits into this concavity. The **anterior lobe** consists of the **pars distalis, pars infundibularis, and pars intermedia** (Fig. 7-34). The pars distalis is the largest part that is located anterior to the residual lumen, the pars intermedia is that portion located between the residual lumen and the posterior lobe, and the pars infundibularis is a superior extension of the intermedia along the stalk or infundibulum. Confusion has arisen because the presence of the residual lumen between the pars intermedia and pars distalis has lead many to consider the pars intermedia as part of the posterior lobe.

Because tumors of the hypophysis occur in which the gland enlarges to a considerable size, its relations are very important. Superiorly it has an important relation to the optic chiasma and optic tracts, and tumors of this gland frequently produce visual symptoms. Laterally, the gland is related to the internal carotid artery and to other structures in the cavernous sinus, and pressure can be put upon these structures when tumors of this gland exist. The above relations are particularly important for the bones prevent extension of the tumor anteriorly, posteriorly, and inferiorly. The sphenoid air sinus is just inferior and anterior to the pituitary and the bony layer between these two structures is frequently very thin.

These are known as "neighborhood symptoms" produced by pituitary tumors. We will see later that many "constitutional symptoms" occur as well, their characteristics depending on which cell type is producing the tumor.

There are no known **nerves** to the anterior lobe except those following the blood vessels. However, the posterior lobe has a direct connection, as previously mentioned, to the inferior surface of the brain, and this part of the gland is innervated by fibers from neuron cell bodies located in nuclei of the hypothalamus. These neurons contain granules of neurosecretion in their cell bodies and axons and it is thought that the hormones produced by the posterior lobe are actually produced in these neurons and stored in the posterior lobe.

The **blood supply** to the hypophysis is from many small arteries that arise from various vessels forming the cerebral arterial circle. They are divided into a superior and an inferior group, which course to the superior and inferior parts of the stalk and the gland proper. However, this is not the only blood supply to the pars distalis. The arteries entering the infundibulum break up into the regular capillary network. Veins from this network course along the stalk (the hypophyseal-portal system) and enter capillaries in the pars distalis. The pars distalis by this means has an interesting supply of both arterial and venous blood. The hypophyseal-portal system is important in an understanding of hypophyseal function, for the hypothalamus can have a stimulatory or inhibitory influence on the pars distalis via this system. The regular veins from the hypophysis as a whole drain into the cavernous sinuses on either side of the gland.*

Function. The **posterior lobe** secretes two hormones—**oxytocin** and **antidiuretic hormone** (vasopressin). **Oxytocin** has an effect on smooth muscle and causes a contraction of uterine musculature; it is used in the field of obstetrics to cause contractions of the uterus when needed. **Antidiuretic hormone** is a substance that exerts an influence on the kidney tubules to aid in readsorption of water. Absence of antidiuretic hormone causes a disease known as diabetes insipidus in which there is a greatly increased urine output.

---

*J. D. Green, 1951, The comparative anatomy of the hypophysis, with special reference to its blood supply and innervation, *Am. J. Anat.* 88: 225. A fine study of seventy-six species. The hypophyseal-portal system is described as occurring in all vertebrates from anura to the primates. No nerve endings were found in the pars distalis.

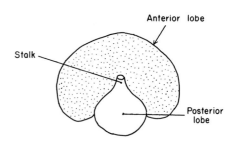

SUPERIOR VIEW

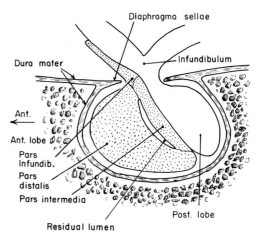

SAGITTAL SECTION

Figure 7-34. *Diagram of the hypophysis.*

The **pars distalis** of the anterior lobe has a tremendous influence on the body, in part directly and in part through its effects on other endocrine glands. It affects growth and carbohydrate metabolism through its **somatotropic hormone;** when this hormone is absent, dwarfism results, and when there is too much of this hormone, gigantism occurs. It has a control over general metabolism by secreting a **thyrotropic hormone,** which regulates the thyroid gland. It affects carbohydrate metabolism through its influence on the adrenal cortex, since it secretes **adrenocorticotropic hormone.** It has two **gonadotropic hormones,** one called follicle-stimulating hormone and the other luteinizing hormone, which have an influence on the ovaries and testes. In addition, the pars distalis of the anterior lobe stimulates secretion by the mammary glands by means of its **lactogenic hormones.** Other hormones, such as the parathyrotropic hormone and hormones to influence fat metabolism, have been postulated but are not supported by sufficient evidence.

That the pars distalis has a decided influence on the other endocrine glands is well established, but the problem of what stimulates the pars distalis is still present. However, it is thought that the products of the other endocrine glands influence it directly and also indirectly by their influence on the hypothalamus. A great deal of ex-

perimental work is currently being conducted to clarify these endocrine interrelations.

RELATIONS OF MENINGEAL ARTERIES AND BRAIN TO SKULL BONES (Fig. 7-35)

Because the central nervous system is usually studied in a separate course, there is a tendency to study the brain as an entity and the skull as another entity and not relate the

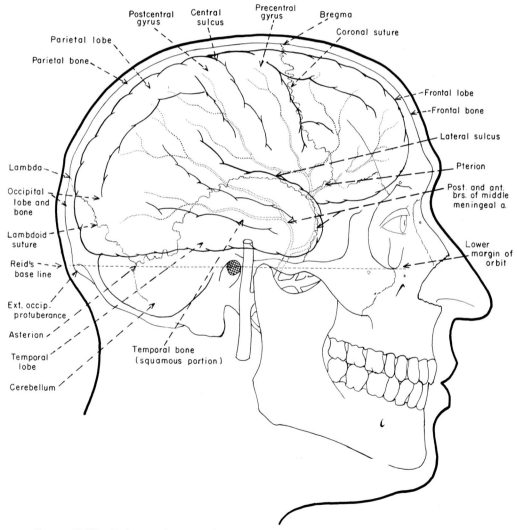

Figure 7-35. *Relations between brain, meningeal arteries, skull, and surface structures. (Modified after Jones and Shepard.)*

two. It is useful to be able to visualize through the superficial structures on the head to the bony landmarks, and from there into the meninges and brain itself.

In order to do this, it is helpful to have certain landmarks in mind. The **orbit, zygomatic bone, zygomatic arch,** and **external auditory meatus** are easily palpated, and a line drawn between the inferior edge of the orbit and the external auditory meatus is known as **Reid's line.** The **mastoid process** and **external occipital protuberance** are also easily located. The point on the superior sagittal suture at which the frontal bone and parietal bone meet is called the **bregma** (Gr., the forepart of the head) and the same point posteriorly, at which the parietal and occipital bones meet, is the **lambda** (the eleventh letter of the Greek alphabet). The suture extending inferiorly and laterally between the parietal and frontal bones is the **coronal suture** (the so-called coronal plane is named for this suture) and that between the parietal and occipital bones is the **lambdoid suture.** The base of the coronal suture runs into an area where the squamous portion of the temporal bone, the parietal bone, the frontal bone, and the greater wing of the sphenoid bones all meet; this is called the **pterion** (Gr., wing). The point where the lambdoid suture meets the parietal mastoid and occipital mastoid sutures is called the **asterion** (Gr., starry).

The **middle meningeal artery** can be projected to the surface approximately in the middle of the zygomatic arch or at a point approximately midway between the lateral edge of the orbit and the external auditory meatus. The division into its anterior and posterior branches occurs approximately an inch superior to Reid's line. The anterior branch continues superiorly and passes the pterion in its course to the region of the vertex of the head, while the posterior branch takes a horizontal course to reach the posterior parietal region of the skull.

The **brain itself** can be outlined by taking a line slightly above Reid's base line as the base of the temporal lobe. The anterior end of the temporal lobe corresponds approximately to the middle meningeal artery, the beginning of the Sylvian fissure or lateral fissure starts at the pterion, the anterior end of the frontal lobe is naturally just superior to the superciliary arches, and the posterior end of the occipital lobe is just superior to the external occipital protuberance. The cerebellum would naturally be inferior to this protuberance. The lobes of the brain (frontal, parietal, occipital, and temporal) do not parallel in size their corresponding bones. The simplest way to

visualize it is to think of the parietal lobe as being much smaller than the parietal bone. Therefore the frontal lobe extends into the region of the parietal bone, the occipital lobe extends into the region of the parietal bone, and the temporal lobe also extends into the parietal bone area. The central sulcus, which divides the parietal lobe from the frontal lobe, is approximately at the vertex of the head.

The base of the brain, where the temporal lobe and the occipital lobe connect, passes directly through the asterion. The transverse sinus courses in this same location and can be traced posteriorly to the external occipital protuberance, which is immediately outside the confluens of the sinuses.

ORBIT*

GENERAL DESCRIPTION

The floor of the anterior cranial fossa is the roof of an interesting part of the body, the orbit. If this bone is removed, the **periorbita** (Fig. 7-36A) lining the inside of the orbit is seen to surround all structures in this area; this is continuous posteriorly with the dura mater at the optic foramen and anteriorly with the periosteum at the edge of the orbit. A layer of the dura also continues over the optic nerve and attaches to the outer covering of the eyeball, the sclera.

The most superior structure in the orbit and one that can clearly be seen through the periorbita is the **frontal branch** of the ophthalmic division of the trigeminal nerve. If the periorbita is now removed, the frontal nerve and the rest of the contents of the orbit are revealed (Fig. 7-36B). If more bone is chipped away posteriorly, the ophthalmic division of the trigeminal nerve can be seen breaking up into branches that course through the superior orbital fissure in order to gain entrance into the orbit. Anteriorly, the frontal nerve divides into the **supratrochlear** and **supraorbital nerves,** which we have already found giving

---

*E. Wolff, 1954, *The Anatomy of the Eye and Orbit,* H. K. Lewis and Co. Ltd., London.

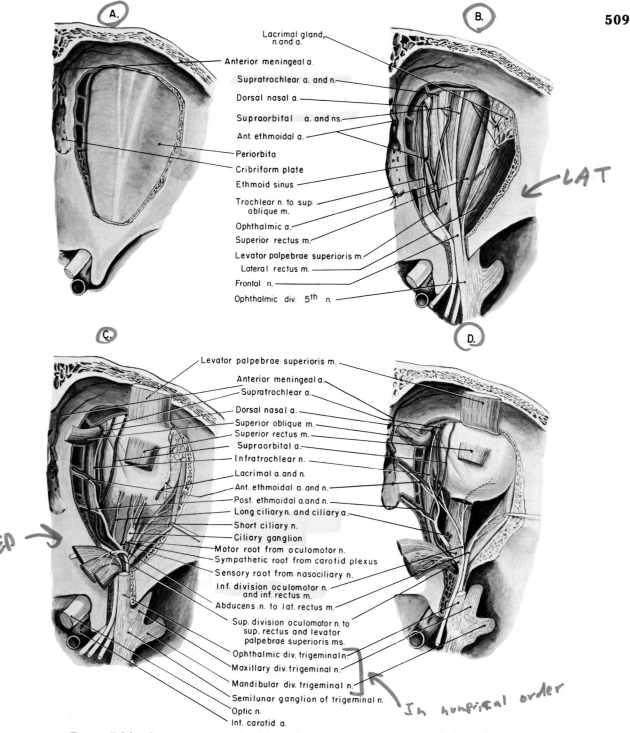

Lacrimal gland, n. and a.

Anterior meningeal a.

Supratrochlear a. and n.

Dorsal nasal a.

Supraorbital a. and ns.

Ant. ethmoidal a.

Periorbita

Cribriform plate

Ethmoid sinus

Trochlear n. to sup. oblique m.

Ophthalmic a.

Superior rectus m.

Levator palpebrae superioris m.

Lateral rectus m.

Frontal n.

Ophthalmic div. 5th n.

Levator palpebrae superioris m.

Anterior meningeal a.

Supratrochlear a.

Dorsal nasal a.

Superior oblique m.

Superior rectus m.

Supraorbital a.

Infratrochlear n.

Lacrimal a. and n.

Ant. ethmoidal a. and n.

Post. ethmoidal a. and n.

Long ciliary n. and ciliary a.

Short ciliary n.

Ciliary ganglion

Motor root from oculomotor n.

Sympathetic root from carotid plexus

Sensory root from nasociliary n.

Inf. division oculomotor n. and inf. rectus m.

Abducens n. to lat. rectus m.

Sup. division oculomotor n. to sup. rectus and levator palpebrae superioris ms.

Ophthalmic div. trigeminal n.

Maxillary div. trigeminal n.

Mandibular div. trigeminal n.

Semilunar ganglion of trigeminal n.

Optic n.

Int. carotid a.

Figure 7-36. *Superior views of the right orbital contents: (A) after removal of the orbital bony roof; (B) the orbital contents in situ after removal of the periorbita; (C) after reflection of the levator palpebrae superioris, superior rectus, and superior oblique muscles; (D) after cutting the optic nerve. (Ophthalmic veins are omitted for the sake of clarity.)*

cutaneous nerve supply to the anterior part of the scalp. Another branch of the ophthalmic division of the trigeminal nerve can be seen to course laterally toward the superior, lateral, and anterior aspect of the orbit; this is the **lacrimal** nerve, which innervates the **lacrimal gland** and then continues through the **orbital septum** (this is shown in Fig. 7-49) to give cutaneous nerve supply to the lateral part of the superior eyelid. This nerve is accompanied by the lacrimal branch of the ophthalmic artery. The most superior muscle, located in the midline of the orbit, is the **levator palpebrae superioris muscle; the** muscle just inferior to this is the **superior rectus.** As one turns to the medial side of the orbit, the **superior oblique muscle** and the small **trochlear nerve** innervating this muscle can be seen. This nerve can be followed posteriorly through the superior orbital fissure in a position medial to the ophthalmic division of the trigeminal. The **ophthalmic artery** appears between the levator palpebrae superioris and superior oblique muscles. This artery gives off **posterior and anterior ethmoidal branches** medially and then the **supraorbital branch,** continues, and finally divides into the **dorsal nasal** and **supratrochlear branches.** The ophthalmic artery, in this location, is accompanied by the **superior division of the ophthalmic vein.**

If the levator palpebrae superioris muscle is cut and the two ends reflected (Fig. 7-36C), the proximal end will reveal the branch of the oculomotor nerve innervating it; this is a branch of the **superior division of this nerve.** If a similar cut is made in the superior rectus and the two ends reflected, a branch from this same nerve can be seen to innervate this muscle as well.

After these muscles have been reflected, the most superficial structure seen is the **nasociliary branch** of the ophthalmic division of the trigeminal nerve, the only remaining branch of this division to be described. The nasociliary nerve is at first on the lateral side of the optic nerve, then crosses the nerve in a position superior to it, and then courses anteriorly along with the ophthalmic artery. One of the first branches of the nasociliary nerve is very small and goes to a small collection of cells about 1 inch posterior to the eyeball and on the lateral side of the optic nerve, the **ciliary ganglion.** This ganglion is a parasympathetic ganglion, and the branch of the nasociliary nerve to it is probably sensory in nature (the sensory root) but may contain sympathetic fibers. The

next branches from the nasociliary are the **long ciliary nerves** which follow along the optic nerve, pierce the capsule around the eyeball, and enter the eyeball to give sensory nerve supply to it; these nerves are also thought to contain sympathetic fibers to the blood vessels of the eyeball. The next two branches of the nasociliary nerve are the **posterior and anterior ethmoidal branches,** which course between the superior oblique and medial rectus muscles. These nerves will be seen to accompany the posterior and anterior ethmoidal branches of the ophthalmic artery. The nasociliary nerve then continues anteriorly to become the **infratrochlear nerve.**

Further branches of the ophthalmic artery can also be seen at this time. There are many branches following the optic nerve to penetrate the fascial sheath around the eyeball. The **central retinal artery** enters the optic nerve; this artery finally ramifies in the retina of the eyeball and is the one viewed with an ophthalmoscope.

The **lateral rectus muscle** is easily seen at this time; its nerve supply, the **abducens nerve,** enters its deep surface.

The portion of the oculomotor nerve remaining, after the branches to the superior rectus and levator palpebrae superioris have been given off, is called the **inferior division of the oculomotor nerve.** This appears just lateral and slightly inferior to the optic nerve (Fig. 7-36D). This nerve gives off a branch to the ciliary ganglion which is the **motor root** to this ganglion, then divides into two branches, one of which courses inferior to the optic nerve and medially to innervate the **medial rectus muscle,** and the other of which continues forward, giving nerve supply to the **inferior rectus** and **inferior oblique muscles.** This portion of the oculomotor nerve is accompanied by the **inferior division of the ophthalmic vein.**

Closer inspection of the **ciliary ganglion** (Figs. 7-36C and 7-37) will often reveal that it has **three roots:** one, already mentioned, from the nasociliary nerve, probably containing sensory fibers; the motor root just described from the oculomotor nerve; and a **sympathetic root** that is obtained from nerves surrounding the ophthalmic artery. In addition, nerves can be seen streaming toward the eyeball from the anterior side of the ciliary ganglion. These nerves are called the **short ciliary nerves** (Figs. 7-36C and 7-37) in contrast to the long ciliary branches of the nasociliary nerve.

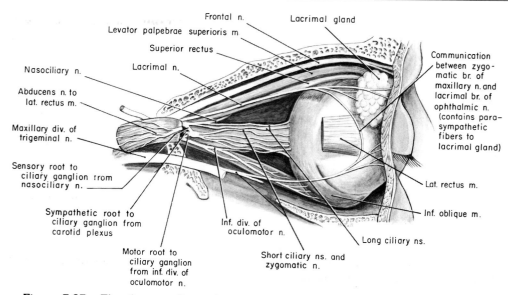

Figure 7-37. *The ciliary ganglion and its connections from the lateral aspect. (The short ciliary nerves have been reduced in number for the sake of clarity.)*

MUSCLES

The voluntary muscles contained in the orbit are:

1. Levator palpebrae superioris
2. Superior rectus
3. Inferior rectus
4. Medial rectus
5. Lateral rectus
6. Superior oblique
7. Inferior oblique

These muscles are all derived from preotic myotomes —see Figure 1-18 on page 20.

The **levator palpebrae superioris** is the most superior muscle in the orbit and lies just inferior to the roof. Its origin is from the roof of the orbit, at a point just anterior to the optic foramen. Its insertion is in the form of a broad aponeurotic sheet that penetrates the orbital septum (see Fig. 7-49) to attach to the anterior surface of the superior tarsal plate, some fibers ending in the skin of the upper eyelid by penetrating the orbicularis oculi muscle and others attaching to the conjunctiva. Laterally this aponeurosis attaches to the bony orbit, which serves as a check on the degree of motion produced by the levator muscle. A layer of smooth muscle fibers is attached to the deep surface of the aponeurotic tendon (behind orbital septum) and to the superior edge of the tarsal plate. The levator palpebrae superioris muscle, with help from the smooth muscle fibers, raises the eyelid or opens the eye. The conjunctiva is not folded under these circumstances, because of the attachment of the levator to the fornix of the conjunctiva.

The nerve supply to the levator palpebrae superioris muscle is the oculomotor nerve; the smooth muscle portion is innervated by the sympathetic nervous system. Injury to either nerve will cause a ptosis (drooping) of the eyelid.

The **four rectus muscles** arise from two tendinous bands that take origin from a tubercle on the lateral side of the superior orbital fissure (Fig. 7-38). These bands, an upper and a lower, course medially to surround the optic nerve and attach to the bone both superior and inferior to the optic foramen. These tendinous bands cross the superior orbital fissure. All four muscles insert on the anterior half of the eyeball. The actions of these muscles become obvious if one realizes that the orbit slants from medial to lateral while the eyeball heads directly anteriorly. As can be seen in Figure 7-39, the direction of pull of a muscle such as the superior rectus on the eyeball is not directly superiorly but superiorly and medially. (The action of these muscles is determined by the direction in which the pupil of the eye will turn.) For the same reason, the inferior rectus muscle does not simply pull the eyeball inferiorly, but pulls it inferiorly and medially. The medial rectus pulls the eyeball medially and the lateral rectus pulls it laterally.

In contrast to the four rectus muscles, the **oblique muscles** both insert on the posterior half of the eyeball (Fig. 7-39). The **superior** oblique muscle arises from the roof of the orbit, medial to the origin of the levator palpebrae superioris muscle. The tendon of this muscle passes through a fibrous ring or pulley that is attached to the side of the orbit at a point that is anteriorly, superiorly, and medially located. The tendon makes a turn at this pulley and continues posteriorly and laterally, to be attached to the posterior part of the eyeball. Because of this ingenious arrangement, the posterior part of the eyeball is pulled superiorly and medially when the muscle contracts. This means, since the eyeball is a spheroid, that the pupil will necessarily move in the opposite direction—inferiorly and laterally. Therefore, the main action of the superior oblique muscle is to move the eye inferiorly and laterally.

The **inferior oblique muscle** arises from the orbital surface of the maxillary bone, just lateral to the nasolacrimal groove. It proceeds laterally under the eyeball and finally inserts on its posterolateral side. This muscle, when it contracts, will naturally pull the posterior half of the eyeball inferiorly and medially. The pupil, therefore, will move superiorly and laterally.

These muscles never work as individuals; there is always teamwork. This group action is presented in Figure 7-40. The muscle primarily involved in turning the eye medially is the medial rectus; this is assisted by the superior and inferior rectus muscles since both of these move the eyeball medially as well as superiorly and inferiorly. The muscle primarily involved in moving the eye laterally is the lateral rectus muscle; this is assisted by the superior and inferior oblique muscles since both of these move the eyeball laterally as well as superiorly and inferiorly. The muscles involved in moving the eye directly superiorly are the superior rectus plus the inferior oblique muscle. The inferior oblique pulls the eyeball superiorly and laterally, while the superior rectus pulls it superiorly and medially; the combined action of the two muscles is to pull the eyeball directly superiorly. The eyeball is

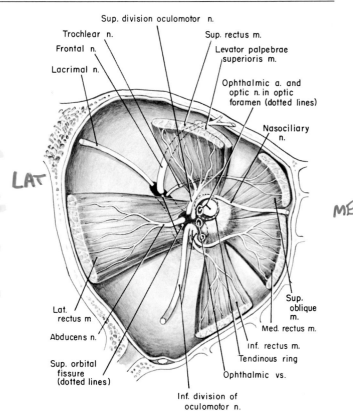

Figure 7-38. *Origins of the orbital muscles, and the relations of the nerves, arteries, and veins contained in the optic foramen and superior orbital fissure to these muscles—right orbit.*

moved inferiorly by the combined action of the inferior rectus pulling the eyeball inferiorly and medially and the superior oblique muscle pulling the eyeball inferiorly and laterally; the two muscles acting synergistically will move the eye directly inferiorly. In certain movements of the eye a real rotation occurs, this motion being determined by a movement of the superior part of the eyeball either medially or laterally. This occurs primarily when one focuses on an object and then tilts the head one way or the other; by a rotation of the eyeball, the image remains erect. A rotation can be made medially by the combined actions of the superior rectus and the superior oblique muscles since both pull the superior part of the eyeball medially. Lateral rotation is caused by the inferior rectus and inferior oblique muscles since both pull the inferior part of the eyeball medially, thus turning the superior part laterally.

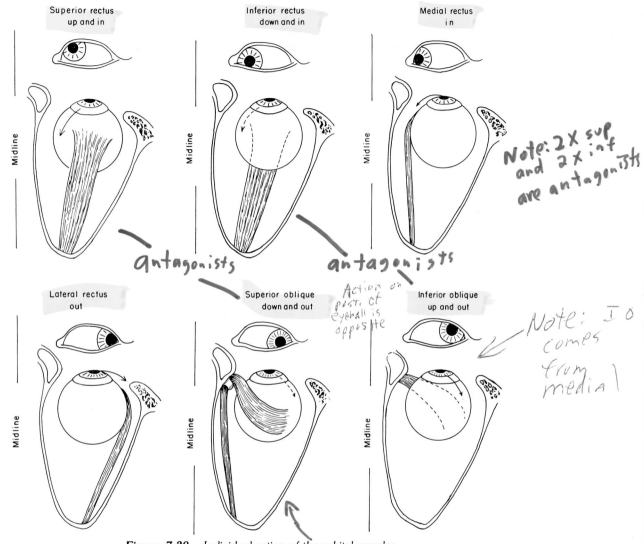

Figure 7-39. Individual action of the orbital muscles.

The superior oblique muscle is innervated by the trochlear nerve, the lateral rectus by the abducens; all the remaining muscles, including the levator palpebrae superioris, are innervated by the oculomotor nerve. Figure 7-40 also has a summarizing figure indicating the muscles responsible for raising the eye superiorly (elevators), inferiorly (depressors), medially (adductors), and laterally (abductors). This diagram may also be used to determine displacement of the eye as a result of nerve injuries.

NERVES

The second cranial nerve or optic, the third or oculomotor, the fourth or trochlear, the ophthalmic division of the fifth or trigeminal, and the sixth or abducens nerves are all involved with structures in the orbit.

Optic nerve.   The optic nerve is a direct continuation of the retina in the eyeball. It is made up of axons from the many cells in the retina; these axons continue through the **optic nerve, optic chiasma,** and **optic tract**

GROUP ACTION

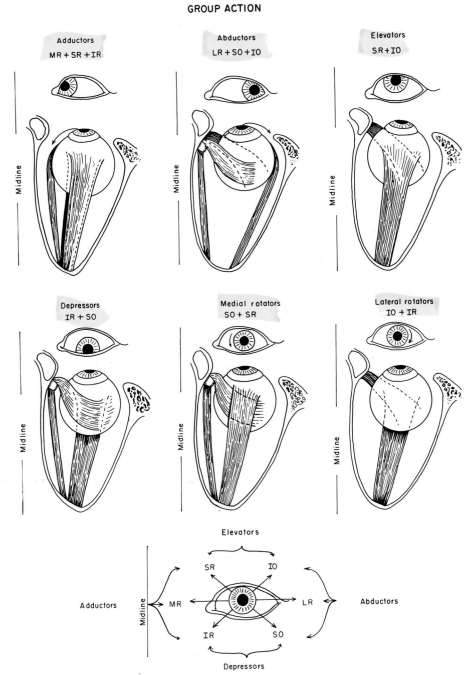

Adductors
MR + SR + IR

Abductors
LR + SO + IO

Elevators
SR + IO

Depressors
IR + SO

Medial rotators
SO + SR

Lateral rotators
IO + IR

Figure 7-40. Group action of the orbital muscles.

to the central nervous system. The optic chiasma is so arranged that the lateral fibers of an optic nerve travel to the optic tract on the same side while the medial fibers cross to the opposite optic tract. Therefore, the axons of the neurons in the lateral half of the retina in the right eye will be in the same tract with the axons from neurons in the medial half of the left eye. Conversely, the left optic tract will contain fibers from neurons in the lateral half of the left eye and the medial half of the right eye. However, the visual fields are opposite to the arrangement just described. Things on the right are seen by the lateral side of the left eye, and the medial side of the right; conversely, objects to the left are received on the lateral side of the right eye and the medial side of the left. Therefore, if the optic nerve were severed, a person would not be able to see anything from the eye with the severed nerve; however, if the right optic tract were injured, the right halves of the retina would be lost and this person would not see anything to his left. This concept is easily understood by a study of Figure 7-41.

The optic nerves are also used in reflex pathways for response to light. These will be described at a later time (page 528). The optic nerve contains special afferent fibers for vision.

Oculomotor nerve. This nerve is divided into two divisions with the following branches:

Superior division
1. To levator palpebrae superioris
2. To superior rectus

Inferior division
1. Motor root to ciliary ganglion
2. To medial rectus
3. To inferior rectus
4. To inferior oblique

The **oculomotor nerve** leaves the anterior end of the cavernous sinus and then divides into the abovementioned **superior** and an **inferior division,** both of which enter the orbit through the superior orbital fissure. The superior division crosses the lateral side of the optic nerve and gives branches to the superior rectus and to the levator palpebrae superioris muscles. The inferior branch courses anteriorly near the floor of the orbit to end in the inferior oblique muscle (Fig. 7-36). It gives a branch on the way to the ciliary ganglion, which is its motor root, and to the medial rectus and the inferior rectus muscles.

The motor root to the ciliary ganglion contains **pre-**

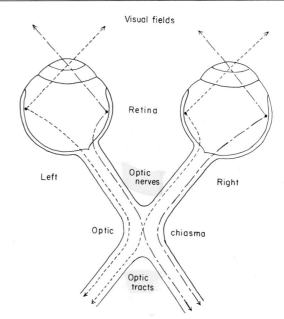

Figure 7-41. *The visual fields.*

ganglionic parasympathetic fibers. The cell bodies of these fibers are located in the brain stem in the visceral part of the oculomotor nucleus. These preganglionic fibers reach the ciliary ganglion via the oculomotor nerve. A **synapse** with postganglionic cell bodies occurs in the ciliary ganglion,* and **postganglionic neurons** travel to the eyeball via the short ciliary nerves to innervate the ciliary muscle (concerned with lens thickness) and the sphincter pupillae muscle.

The oculomotor nerve, therefore, contains **general somatic efferent** fibers for innervation of the levator palpebrae superioris, superior rectus, inferior rectus, medial rectus, and inferior oblique muscles; **general somatic afferent** fibers for proprioception from these particular muscles, although there is a possibility that these proprioceptive fibers finally terminate in the trigeminal nerve via the connections between the oculomotor nerve and the trigeminal nerve in the cavernous sinus; **general visceral efferent fibers** (parasympathetic) to the ciliary muscle and sphincter pupillae muscle; and **general visceral**

---

*G. A. Wolf, Jr., 1941, The ratio of preganglionic neurons to postganglionic neurons in the visceral nervous system, *J. Comp. Neurol.* 75: 235. Evidence, in cats, that preganglionic parasympathetic fibers to ciliary ganglion synapse with more than one postganglionic fiber.

afferent fibers from these same muscles. This nerve is summarized in Figure 7-42.

Trochlear nerve. This nerve enters the orbit via the superior orbital fissure, crosses superior to the oculomotor nerve and to the levator palpebrae superioris muscle, and continues medially to reach the superior oblique muscle (Fig. 7-36). It enters the superior surface of this muscle rather than its inner deep surface.

This nerve contains **general somatic efferent** fibers for innervation of the superior oblique muscle and **general somatic afferent** fibers for proprioception from this muscle. Once again, there is a possibility that proprioceptive fibers in this nerve finally enter the trigeminal nerve through the connections between the trochlear and trigeminal in the cavernous sinus. This nerve is summarized in Figure 7-43.

Ophthalmic division of trigeminal nerve. This nerve has the following branches.

1. Meningeal
2. Lacrimal
   a. To lacrimal gland
   b. To superior eyelid
3. Frontal
   a. Supraorbital
      i. To eyelid
      ii. To scalp
   b. Supratrochlear
      i. To eyelid
      ii. To nose
      iii. To scalp
4. Nasociliary
   a. Sensory root to ciliary ganglion
   b. Long ciliary
   c. Posterior ethmoidal
   d. Infratrochlear
      i. To nose
      ii. To eyelid
      iii. To lacrimal sac
      iv. To lacrimal caruncle
   e. Anterior ethmoidal
      i. Meningeal
      ii. To nasal cavity
      iii. External nasal

The **ophthalmic division of the trigeminal nerve** emerges from the anterior aspect of the cavernous sinus and, after giving off a **meningeal** branch that curves superiorly and then posteriorly to reach the tentorium

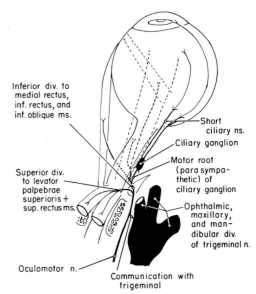

Figure 7-42. Distribution of the oculomotor nerve (diagrammatic).

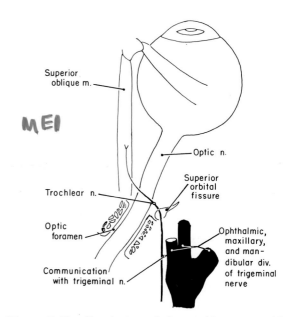

Figure 7-43. Distribution of the trochlear nerve (diagrammatic).

cerebelli to which it gives sensory innervation (Fig. 7-25), divides into **lacrimal, frontal,** and **nasociliary** branches.

The **lacrimal nerve** continues anteriorly through the superior aspect of the superior orbital fissure and tends to course laterally just medial and superior to the lateral rectus muscle (Fig. 7-36). It picks up a filament from the zygomatic nerve (Fig. 7-37) that contains secretory fibers for the lacrimal gland, and continues anteriorly to give sensory innervation to the lacrimal gland and to the conjunctiva and skin of the lateral parts of the superior eyelid.

The **frontal nerve** enters the orbit through the superior orbital fissure, and is the most superior structure in the orbit; it courses anteriorly in a position superior to the muscles (Fig. 7-36). It divides into the supraorbital and supratrochlear nerves. The **supraorbital nerve** passes through the supraorbital notch or foramen and supplies the skin and conjunctiva of the upper eyelid as well as the scalp as far superiorly as the vertex. It also sends branches to the mucous membrane of the frontal sinus. The **supratrochlear nerve** runs anteriorly and slightly medially in a position superior to the pulley of the superior oblique muscle. It leaves the orbit and, after giving branches to the medial side of the upper eyelid and conjunctiva and the root of the nose, it continues on to the scalp to supply the medial part of the forehead.

The **nasociliary nerve** enters the orbit through the superior orbital fissure between the two divisions of the oculomotor nerve (Fig. 7-36C). It courses medially and anteriorly between the superior rectus muscle and the optic nerve. The nerve continues anteriorly to end as the **anterior ethmoidal nerve.** This nerve exits from the orbit through the anterior ethmoidal foramen and enters the cranial cavity at the side of the cribriform plate of the ethmoid bone. It proceeds anteriorly over the cribriform plate, gives off a **meningeal branch,** and leaves the cranium to enter the nasal cavity. Here it gives off branches to the septum and lateral wall of the nose and then continues anteriorly and inferiorly to emerge between the nasal bone and the nasal cartilages and to terminate as the **external nasal branch** supplying the skin on the lower half of the nose.

The nasociliary nerve gives off (1) a branch to the **ciliary ganglion,** which is its sensory root, (2) the **long ciliary nerves,** which course along the optic nerve to enter the posterior part of the eyeball (sensory to eyeball but may contain sympathetic fibers), (3) the **posterior ethmoidal nerve,** which exits from the orbit through the posterior ethmoidal foramen to give sensory supply to the ethmoid and sphenoid sinuses, and (4) the **infratrochlear nerve,** which is actually one of the two terminal branches of the nasociliary and which continues an-

teriorly, leaves the orbit at a point superior to the medial angle of the eye, and supplies the upper half of the nose, the eyelids, the conjunctiva, the lacrimal sac, and the lacrimal caruncle.

The ophthalmic division of the trigeminal nerve is, therefore, a sensory nerve and contains **general somatic afferent** fibers. Although sympathetic fibers do join various branches of this nerve, they are not considered to be a component of them. This nerve is summarized in Figure 7-44.

Abducens nerve. The abducens nerve emerges from the brain stem at the inferior end of the pons, between the pons and the superior end of the medulla oblongata, enters the cavernous sinus where it is in intimate contact with the internal carotid artery, leaves the anterior side of the cavernous sinus, and enters the orbit through the superior orbital fissure between the heads of the lateral rectus muscle. It courses anteriorly along the deep or medial surface of the lateral rectus and innervates this muscle (Fig. 7-36).

The abducens nerve contains **general somatic effer-**

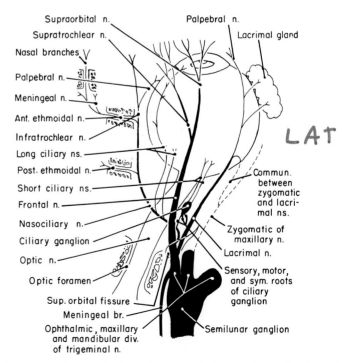

Figure 7-44. *Distribution of the ophthalmic division of the trigeminal nerve (diagrammatic).*

ent fibers to innervate the lateral rectus muscle and **general somatic afferent** fibers for proprioception from this muscle; the latter fibers may join the trigeminal nerve via connections with this nerve in the cavernous sinus. This nerve is summarized in Figure 7-45.

SYMPATHETIC INNERVATION OF EYEBALL. **Preganglionic sympathetic** fibers for structures in the head arise from the upper segments of the thoracic cord. These cells are located in the intermediolateral cell column and exit from the cord via the ventral roots. They then travel to the segmental nerve, and thence to the sympathetic chain via a white ramus communicans, and travel superiorly in the chain without synapsing in the thoracic ganglia. After reaching the superior cervical ganglion, a synapse occurs with postganglionic fibers. These **postganglionic fibers** then travel superiorly on the internal carotid artery; they follow this artery, enter the skull, and follow the ophthalmic branch of the internal carotid. In the orbit, these fibers join the ciliary ganglion as the sympathetic root (when no sympathetic root is found, the sympathetic fibers will be found in the sensory root), pass directly through the ganglion, and enter the eyeball via the short ciliary nerves. This pathway is presented in Fig-

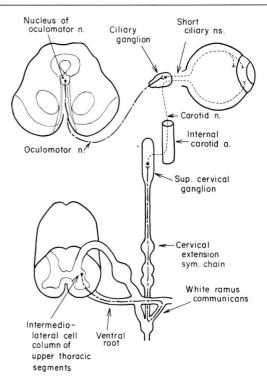

Figure 7-46. *Pathways of the autonomic nerves to the eyeball. Upper part of diagram is parasympathetic; the lower part, sympathetic (diagrammatic).*

ure 7-46. These sympathetic fibers innervate the blood vessels and the dilator pupillae muscle (some authorities feel that sympathetic fibers in the long ciliary branches of the nasociliary nerve aid in innervation of the muscle). Several of these sympathetic fibers have been seen to join the other nerves in the cavernous sinus, which probably accounts for the pathway taken by sympathetic nerves to innervate the smooth muscle component of the levator palpebrae superioris muscle.

ARTERIES

The **ophthalmic artery** (Fig. 7-36) branches from the internal carotid just after the latter artery leaves the cavernous sinus. The ophthalmic is at first lateral to the anterior clinoid process and then enters the optic foramen to lie lateral and slightly inferior to the optic nerve. It enters the orbit and is at first on the lateral side of the optic nerve. It then crosses superior (usually) to the optic

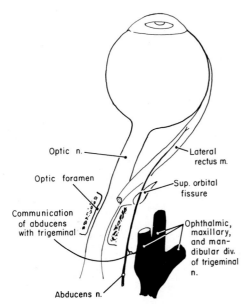

Figure 7-45. *Distribution of the abducens nerve (diagrammatic).*

nerve and proceeds anteriorly on the medial side of the orbit, finally to end by dividing into the dorsal nasal and supratrochlear arteries. It is summarized in Figure 7-47.

The branches of the ophthalmic artery are:

1. Central retinal
2. Meningeal
3. Long and short posterior ciliary
4. Lacrimal
5. Supraorbital
6. Posterior ethmoidal
7. Anterior ethmoidal
8. Muscular
9. Medial palpebrals
10. Dorsal nasal
11. Supratrochlear

1. The **central retinal artery** usually enters the optic nerve at a point distal to the optic foramen. This artery proceeds in the optic nerve and ends in the retina of the eye and is the vessel one sees with the ophthalmoscope (note Fig. 7-53); this is the only vessel in the body that is visible in this manner. If this vessel is occluded, blindness results.

2. The small **meningeal artery** enters the middle

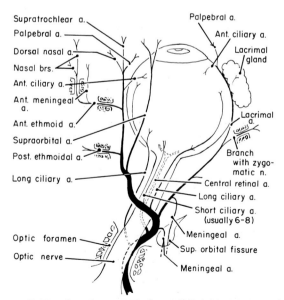

Figure 7-47. Distribution of the ophthalmic artery and its branches. Muscular arteries are not shown. The ciliary arteries are more numerous than indicated (diagrammatic).

cranial fossa by turning posteriorly and traversing the superior orbital fissure.

3. The **long and short posterior ciliary arteries** surround the optic nerve and enter the posterior part of the eyeball and pierce the sclera. The short ciliary vessels are distributed in the choroid layer while the long ciliary vessels proceed anteriorly to give blood supply to the ciliary body (see Fig. 7-55).

4. The **lacrimal artery** branches from the ophthalmic while it is lateral to the optic nerve. It proceeds laterally and anteriorly on the superior surface of the lateral rectus muscle and finally terminates by dividing into glandular and palpebral branches. The **glandular** branches give blood supply to the lacrimal gland and the **palpebral** branches to the lateral sides of the eyelids. This artery gives off the **anterior ciliary** branch to the eyeball and also branches to the muscles.

5. The **supraorbital artery** arises from the ophthalmic at various locations but usually just after it has crossed the optic nerve. It proceeds anteriorly to pass through the supraorbital notch or foramen and gives blood supply to the scalp as far superiorly as the vertex of the head.

6. The **posterior ethmoidal artery** passes through the posterior ethmoidal foramen to supply the mucous membrane of ethmoid and frontal sinuses and the nasal cavity.

7. The **anterior ethmoidal artery** passes through the anterior ethmoidal foramen, enters the cranial cavity, gives off its anterior meningeal branch, and then enters the nasal cavity where it gives off nasal branches to the septum as well as the lateral wall, and finally terminates as the **external nasal vessel** in the company of the external nasal nerve.

8. **Muscular branches** come off at intervals to supply the orbital muscles; these vessels give rise to anterior ciliary branches to the eyeball.

9. The **medial palpebrals** are small branches that give blood supply to the medial sides of both upper and lower eyelids.

10. The **dorsal nasal,** one of the two terminal branches of the ophthalmic artery, gives blood supply to the bridge of the nose.

11. The **supratrochlear** courses superior to the trochlea of the superior oblique muscle and leaves the orbit to course superiorly on the forehead in company with the supratrochlear nerve.

VEINS

The **ophthalmic veins** (Fig. 7-48) have branches that correspond to the arteries just described. They are divided into **superior and inferior divisions,** which combine to empty into the cavernous sinus after traversing the superior orbital fissure. The veins draining the posterior part of the eyeball are called **venae vorticosae.** An important thing about the ophthalmic veins is that they communicate with the pterygoid plexus deep in the cheek and with veins on the scalp and face. Therefore, they serve as a direct connection from the outside surface to the dural sinuses, an important source of trouble if infection should be carried into the sinuses.

LYMPHATICS

Although there are undoubtedly lymphatics in the orbit and lymph nodes draining this territory,* they have not as yet been described.

FASCIA

The **dura mater** (periosteum) is continued into the orbit through the optic foramen and lines the bones of the orbit. This periosteum is directly continuous with the periosteum on the outside of the skull bones at the orbital margins. An extension of this periosteum is attached to both eyelids; this very firm structure is the **orbital septum** (Fig. 7-49) and separates the outside world from the contents of the orbit. The dura is also continuous into the orbit on the optic nerve itself and blends with the outer layer of the eyeball, the sclera. (It should be noted that the arachnoid and pia maters, and thereby the subarachnoid space, also continue onto the optic nerve, all terminating at the point the nerve joins the eyeball.)

The eyeball itself is surrounded by a fascial layer, the **fascia bulbi,** which is loosely attached to the sclera, thus allowing slight movement of the eyeball within it. This fascia covers the eyeball except on the cornea; it is penetrated by the muscles, nerves, and vessels attached to or coursing to or from the eyeball.

---

*M. Foldi, F. Kukan, G. Szeghy, A. Gellert, M. Kozma, M. Poberai, O. T. Zolton, and L. Varga, 1963, Anatomical, histological and experimental data on fluid circulation of the eye, *Acta Anat.* 53: 333. Tying off lymphatics in the neck (dog) induced an edema of the orbital muscles, retrobulbar connective tissue, and lacrimal gland. Furthermore, dye injected into the vitreous humor of the eyeball was found subsequently in the retrobulbar connective tissue and ocular muscles, from which structures it left via lymphatic capillaries.

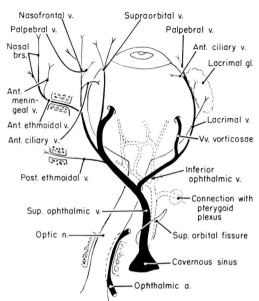

Figure 7-48. *The ophthalmic veins and their tributaries (diagrammatic).*

Every muscle is covered with its own tube of deep fascia. Anteriorly, this fascia is attached to the fascia bulbi. Extensions from this fascia on the lateral rectus and medial rectus muscles attach to the periosteum. These serve as **check ligaments,** preventing too great a movement of the eyeball in a medial or lateral direction. Because the inferior oblique muscle and the inferior rectus muscles are arranged in such a manner that they are inferior to the eyeball, the fascial coverings on these muscles are said to form a sling, the **suspensory ligament of the eyeball;** the eyeball is thought to rest upon this suspensory ligament. The eye is prevented from being raised too far superiorly by a connection between the superior rectus muscle and the levator palpebrae superioris, and too far inferiorly by a connection between the fascia of the inferior rectus and the suspensory ligament.

EYELIDS AND LACRIMAL APPARATUS

As one looks at another person's eye (Fig. 7-49A), one sees a centrally placed dark **pupil** surrounded by the col-

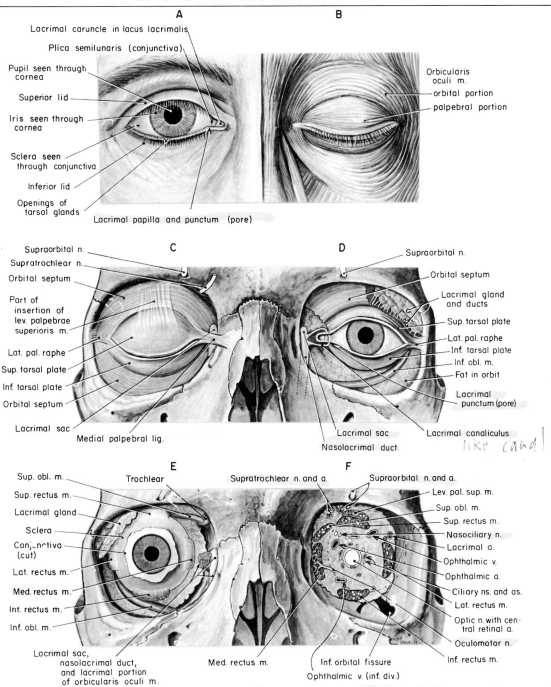

A

Lacrimal caruncle in lacus lacrimalis

Plica semilunaris (conjunctiva)

Pupil seen through cornea

Superior lid

Iris seen through cornea

Sclera seen through conjunctiva

Inferior lid

Openings of tarsal glands

Lacrimal papilla and punctum (pore)

B

Orbicularis oculi m.
— orbital portion
— palpebral portion

C

Supraorbital n.
Supratrochlear n.
Orbital septum

Part of insertion of lev. palpebrae superioris m.

Lat. pal. raphe

Sup. tarsal plate

Inf. tarsal plate

Orbital septum

Lacrimal sac

Medial palpebral lig.

D

Supraorbital n.

Orbital septum

Lacrimal gland and ducts

Sup. tarsal plate

Lat. pal. raphe

Inf. tarsal plate

Inf. obl. m.

Fat in orbit

Lacrimal punctum (pore)

Lacrimal sac
Nasolacrimal duct

Lacrimal canaliculus

E

Sup. obl. m.
Sup. rectus m.
Lacrimal gland
Sclera
Conjunctiva (cut)
Lat. rectus m.
Med. rectus m.
Inf. rectus m.
Inf. obl. m.

Trochlear

Lacrimal sac, nasolacrimal duct, and lacrimal portion of orbicularis oculi m.

Med. rectus m.

F

Supratrochlear n. and a.

Supraorbital n. and a.

Lev. pal. sup. m.
Sup. obl. m.
Sup. rectus m.
Nasociliary n.
Lacrimal a.
Ophthalmic v.
Ophthalmic a.
Ciliary ns. and as.
Lat. rectus m.
Optic n. with central retinal a.
Oculomotor n.
Inf. rectus m.

Inf. orbital fissure
Ophthalmic v. (inf. div.)

Figure 7-49. *Dissection of the eyelids and lacrimal apparatus. (A) The normal appearance of the "eye." (B) Skin has been removed from the eyelids to show orbicularis oculi muscle. (C) Orbicularis oculi muscle has been removed, revealing the orbital septum and tarsal plates of the eyelids. (D) Partial removal of the septum to show lacrimal apparatus. (E) Complete removal of orbital septum and partial removal of orbital fat to show insertions of orbital muscles. (F) Transverse section of orbital contents.*

ored iris; this is, in turn, surrounded by the whites of the eyes, the **sclera.** Anterior to the pupil and iris is the **cornea,** which is directly continuous at its periphery with the sclera. The cornea and sclera are covered with a thin layer of cells, the **conjunctiva,** which is also reflected from the eyeball onto the inside surfaces of the eyelids (Fig. 7-50). The **eyelids** are in contact with each other when the eye is closed and away from each other when the eye is open. The opening made by the eyelids reveals almost the entire iris inferiorly but usually cuts across the iris superiorly.

EYELIDS

The **eyelids** are two folds that, when closed, cover and protect the anterior surface of the eyeball.

The skeleton of the eyelids (Fig. 7-49C) consists of two cartilaginous **tarsal plates,** one for each eyelid, the superior being the larger. These plates are attached medially and laterally to the edges of the orbit by the medial palpebral ligament and the lateral palpebral raphe. They are attached superiorly and inferiorly to an important layer of fascia, the **orbital septum.** As mentioned previously, this is a layer of fascia that is continuous with the periosteum at the edges of the orbit. It is attached to the lateral palpebral raphe as well as to the superior and inferior tarsal plates and forms a barrier between the outside and the inside contents of the orbit. If a sharp instrument is pushed against the eyelid at points superior or inferior to the tarsal plates, entrance is not gained into the orbit because of this orbital septum. If, however, the eyelids are lifted and the probe pushed through the conjunctiva, where is reflects from the eyeball onto the eyelids, entrance to the orbit is easily gained and the insertions of the four rectus muscles onto the sclera of the eyeball are quite visible.

The tarsal plates contain glands that open on the edges of the eyelids. These **tarsal glands** (Fig. 7-50) secrete a viscous material that decreases the rate of evaporation of the lacrimal fluid from the surface of the conjunctiva.

Superficial to these plates are the skin and **orbicularis oculi** muscle (Fig. 7-49A and B). The skin possesses practically no hair and contains very little fat. This lack of fat accounts for the very visible hemorrhage into the skin that results from injury—the common "black eye." The orbicularis oculi muscle is a sphincter lying not only in the eyelids but around the eyelids as well. It is attached medially to the medial palpebral ligament and to the bone adjoining the ligament. There is no attachment laterally except for the palpebral portion, which is attached to the lateral palpebral raphe. It consists of three parts: **palpebral, orbital,** and **lacrimal portions.** The **palpebral portion** lies in the eyelids themselves and is responsible for blinking of the eye. The **orbital portion** is responsible for closing the eye tightly and draws the skin surrounding the eye toward the eye as in squinting. The **lacrimal portion** arises from the lacrimal sac and the bone next to it (Fig. 7-49E). We will see in a moment that this relation is important in the control of tears.

Deep to the tarsal plates and orbital septum is the layer of conjunctiva found lining the inside of the eyelids. This folds back from the lids onto the anterior surface of the eyeball.

The eyelids meet one another medially and laterally at the medial and lateral angles of the eye. The space between the lids when the eye is open is the **palpebral fissure.** The free margin of the eyelids (Fig. 7-49A) possesses eyelashes as far medially as a small elevation, the **lacrimal papilla,** and the opening into the lacrimal papilla, the **punctum.** Medial to this lacrimal papilla the lids are devoid of eyelashes and surround an area called the **lacus lacrimalis.** In the middle of the lacus is found a rounded, fleshy mass of tissue called the **lacrimal caruncle,** the function of which is problematical. This material is made up of skin and glands and is responsible for the white material that often collects in the medial angles of the eyes; just lateral to the lacrimal caruncle is the **plica semilunaris,** a semilunar fold of the conjunctiva. Besides the tarsal glands, the eyelid possesses **sebaceous glands** (Fig. 7-50) for the eyelashes, and also **sweat glands.** Infection of the tarsal glands usually presents itself on the inside of the lid and is a chalazion; infection of the sebaceous glands is external and is a stye.

The **levator palpebrae superioris muscle** is responsible for opening the eyes. This muscle was described on page 511 but will be repeated here for completeness. It arises from the bone superior to the optic foramen and inserts by a broad aponeurosis onto the anterior surface of the superior tarsal plates, some fibers penetrating the orbicularis muscle to attach to the skin of the upper eyelid, and other fibers attaching to the conjunctiva. This tendon has to penetrate the orbital septum. The aponeurosis attaches to the bony orbit laterally which

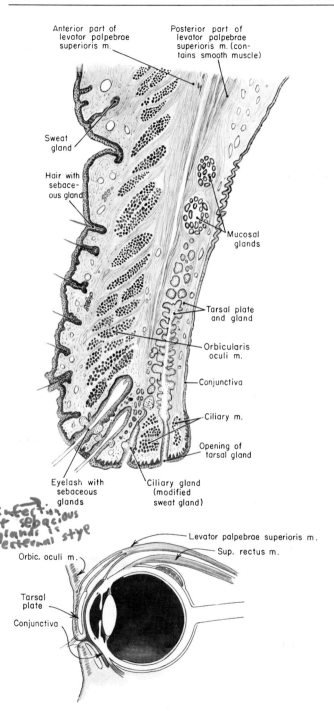

Anterior part of levator palpebrae superioris m.

Posterior part of levator palpebrae superioris m. (contains smooth muscle)

Sweat gland

Hair with sebaceous gland

Mucosal glands

Tarsal plate and gland

Orbicularis oculi m.

Conjunctiva

Ciliary m.

Opening of tarsal gland

Eyelash with sebaceous glands

Ciliary gland (modified sweat gland)

Infection of sebaceous glands is external stye

Orbic. oculi m.

Levator palpebrae superioris m.

Sup. rectus m.

Tarsal plate

Conjunctiva

Figure 7-50. *Transverse section of an eyelid. (Modified after Gray.) Insertion of levator palpebrae superioris muscle (diagrammatic).*

serves as a check to the amount of movement that the levator can induce. A layer of smooth muscle fibers is attached to the deep surface of this aponeurotic tendon (behind orbital septum) and to the superior edge of the tarsal plate. These smooth muscle fibers are innervated by the sympathetic nervous system by way of the superior cervical ganglion. Interruption of the sympathetic fibers anywhere along its path will induce a ptosis (drooping) of the eyelid accompanied by a constriction of the pupil and anhydrosis (lack of sweating); this is known as Horner's syndrome.

The **motor nerves** to the eyelids, besides the sympathetic fibers just mentioned, are the zygomatic branch of the facial to the orbicularis oculi muscle, and the oculomotor nerve to the levator palpebrae superioris muscle. The **sensory nerves** are palpebral branches of the supraorbital, supratrochlear, infratrochlear, and lacrimal branches of the ophthalmic division of the trigeminal nerve, and the palpebral branch of the infraorbital branch of the maxillary division of the same nerve.

The **arteries** of the eyelids are the palpebral branches of the lacrimal, and palpebral branches of the ophthalmic. The veins are similar to the arteries.

Lymphatics from the upper eyelid drain into the superficial parotid and superficial cervical nodes while the lower lid tends to drain into the submandibular nodes.

LACRIMAL APPARATUS

The **lacrimal apparatus** consists of the lacrimal gland and its ducts, conjunctiva, lacrimal puncta and canaliculi, and the lacrimal sac and nasolacrimal duct.

The **lacrimal gland** (Fig. 7-49D) lies in the lacrimal fossa on the superior lateral aspect of the anterior part of the orbit. It is surrounded by fascia and separated from the eyeball by the levator palpebrae superioris and the lateral rectus muscles, the former muscle indenting the gland to such an extent as to divide it into orbital and palpebral portions. It has several ducts that open into the superior fornix of the conjunctiva. Its **blood supply** is by lacrimal branches of the ophthalmic and the **nerve supply** is by the lacrimal branch of the trigeminal, which is sensory but also contains secretory filaments from the zygomatic branch of the maxillary division. These filaments are not part of the trigeminal nerve but ultimately come from the facial nerve. They are responsible for secretion of lacrimal fluid.

When the eyes are closed, the conjunctiva forms a slitlike cavity across the anterior surface of the eyeball and on the insides of the eyelids. Fluid is constantly produced by the lacrimal gland and is spread evenly over the conjunctiva by blinking of the eyelids.

This fluid is collected by two openings, one in each eyelid, called the **puncta lacrimalis.** From these puncta **canaliculi** course medially around the lacus lacrimalis to empty into the **lacrimal sac.** From the lacrimal sac this fluid flows inferiorly in the **nasolacrimal duct** to empty into the inferior meatus of the nasal cavity. Since no paranasal sinuses empty into this meatus any pus or infection around this opening in the inferior meatus can immediately be delegated to trouble in the nasolacrimal duct.

The lacrimal portion of the orbicularis oculi muscle aids in collecting lacrimal fluid. It actually puts pressure on the lacrimal sac when the eyelids are closed and thus reduces its size. With relaxation of the muscle, the sac is enlarged and acts as a suction bulb. This is the reason why an overabundance of tears does not overflow onto the cheek if one rapidly opens and closes the eyes.

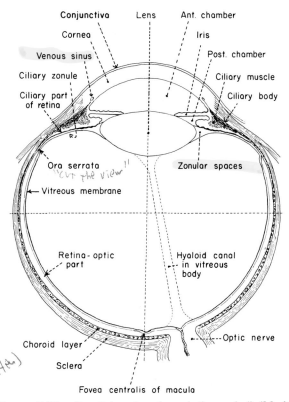

Figure 7-51. *Saggital section through the eyeball. (Modified after Cunningham.)*

EYEBALL*

GENERAL DESCRIPTION

The eyeball (Figs. 7-36, 7-37, and 7-51) is a spheroid located in the anterior part of the orbit. The human eye is about an inch in diameter; it consists of anterior and posterior portions and has three layers. The outer layer of the posterior portion is a tough connective tissue layer, the **sclera.** This is continuous anteriorly with the **cornea,** which is transparent and covers that portion of the eye that is seen from the anterior side as the pupil and iris. The middle or vascular layer is the **choroid layer** and is continuous anteriorly with the **ciliary body** and then the colored portion of the eye, the **iris** (Fig. 7-52).

The internal layer is the **retina** which, in the human, contains a pigmented layer. This is continuous anteriorly

as a nonvisual layer that lines the inside surface of the ciliary body and iris. The point at which the visual or optic part and the ciliary or nonvisual part join is the **ora serrata.**

The posterior four-fifths of the eyeball is occupied by a transparent jelly-like material, the **vitreous body.**

The **lens** is a transparent, biconvex, and circular structure that lies on the anterior surface of the vitreous body and posterior to the iris and pupil. The lens consists of many concentric layers surrounded by an elastic capsule, and is held in position by **suspensory ligaments.**

The portion of the eye between the transparent cornea anteriorly and the iris posteriorly is the **anterior chamber** of the eye and contains **aqueous humor.** This chamber is connected directly with the **posterior chamber** of the eye via the **pupillary opening.** The posterior chamber of the eye is, therefore, between the iris anteriorly and the lens and its suspensory ligament posteriorly.

---

*S. Duke-Elder, 1932, *System of Ophthalmology,* Vol. I. *The Eye in Evolution,* Kimpton, London.

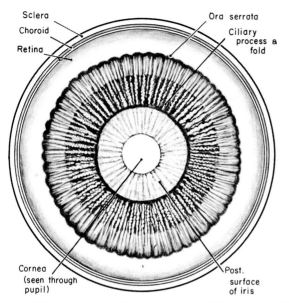

Figure 7-52. *Anterior half of eyeball viewed from within. Vitreous body and lens have been removed.*

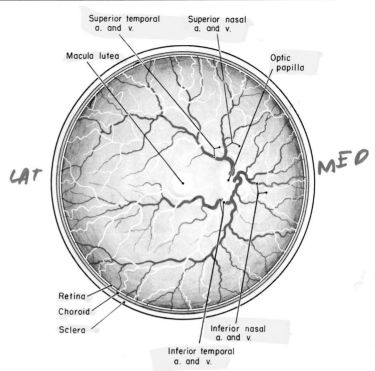

Figure 7-53. *Posterior half of eyeball viewed from within. The vitreous body has been removed.*

SCLERA (Figs. 7-51 and 7-54). This white, opaque, and dense covering of the posterior and lateral parts of the eyeball is continuous anteriorly with the translucent cornea. It is loosely attached to the choroid layer, and is pierced by the optic nerve, ciliary nerves, and numerous blood vessels. The optic nerve pierces the sclera slightly medial and inferior to the posterior pole of the eyeball. This is due to the obliquity of the orbit and the direct anteroposterior position of the eyeball. The short ciliary arteries and nerves enter the sclera around the optic nerve and the long ciliary nerves and arteries pierce it a bit farther away from the optic nerve. Just posterior to the equator of the eyeball, four or five large veins (venae vorticosae) exit through the sclera. The anterior ciliary arteries pierce the sclera anteriorly near the cornea.

CORNEA (Figs. 7-51 and 7-54). The cornea is the transparent anterior covering of the eyeball. Its superficial epithelial layer is directly continuous with the conjunctiva. The point at which the sclera and cornea join is the **corneoscleral junction,** which possesses a small canal, the **sinus venosus sclerae,** which encircles the eye. We will see later (page 526) that this serves as an important connection between the anterior chamber of the eye and the venous system.

CHOROID (Figs. 7-51 and 7-55). This is the vascular layer of the eyeball, for it is made up primarily of blood vessels. These vessels occur in two layers, a superficial, made up of veins that drain into the vena vorticosae already mentioned, and a deeper layer made up of arteries. The choroid is loosely attached to the sclera but quite firmly attached to the retina. In fact, the pigmented layer of the retina often remains attached to the choroid when the retina is removed.

CILIARY BODY (Fig. 7-54). This complex structure is directly continuous anteriorly with the choroid, and the iris is an anteromedial extension from this ciliary body. The body is made up of (1) a posterior part—the **ciliary ring**—which is made up primarily of nerves and vessels, (2) an anterior part—the **ciliary muscle**—which is composed of circular fibers and radiating fibers of smooth muscle, the latter arising from the corneoscleral junction and radiating posteriorly to insert on the ciliary ring, and (3) the third part of the ciliary body, the **ciliary pro-**

cesses. These processes are accumulations of tissue that extend from the choroid posteriorly to the iris anteriorly. They are nodular in appearance, and form the lateral boundary of the posterior chamber of the eye between the iris anteriorly and the anterior surface of the lens posteriorly. They also serve as important attachments for the suspensory ligaments of the lens. The epithelium of the ciliary processes secretes the aqueous humor into the posterior chamber. This flows through the pupil to the anterior chamber and then via aqueous veins to the sinus venosus sclerae; from this sinus the humor flows into the anterior ciliary veins (see insert—Fig. 7-54).

IRIS (Figs. 7-52 and 7-54).   The iris is the colored part of the eye; it projects medially and anteriorly into the aqueous humor and is attached radially to the ciliary body and to the cornea by short pectinate ligaments. The iris is doughnut-shaped; the space between its free edges is the pupil. This heavily pigmented iris possesses two sets of smooth muscles: a radially arranged set that are responsible for enlargement of the pupil (dilator pupillae muscle), and a circularly arranged set that decrease pupil size when contracted (sphincter pupillae muscle).

RETINA (Figs. 7-51 and 7-53).   The all-important retina is made up of an outer pigmented layer and of an inner nerve layer, which is in contact with the vitreous body. The retina is directly continuous with the optic nerve, and anteriorly it changes to a nonvisual character at the **ora serrata** to continue on the inside surfaces of the ciliary body and iris. When viewed with an ophthalmoscope, the point of entrance of the optic nerve shows as an elevation—**optic disc**—which naturally is located medially and slightly inferiorly. In the exact center of the posterior part of the retina is a yellowish spot called the **macula lutea,** which has in its center a dimple, the **fovea centralis.** This is the area of most acute vision.

VITREOUS BODY (Fig. 7-51).   This large transparent mass exhibits a fossa anteriorly—the **hyaloid fossa**—in which the lens rests. The **hyaloid canal** is a channel, leading from the optic nerve to the posterior part of the lens, which, in embryonic life, contained the artery to the lens. This channel persists during adult life but contains no artery and does not interfere with vision.

The vitreous body is surrounded by a transparent

vitreous membrane. This membrane thickens in the area posterior to the ciliary body and is called the **ciliary zonule.** This thickened area possesses furrows produced by the ciliary processes. The ciliary zonule, if followed medially, separates into layers, the posterior layer forming the lining of the indentation in the vitreous body for the lens. The other layers fuse with the capsule of the lens, forming its suspensory ligament. The spaces between the layers just described are the **zonular spaces.** Two of the preceding points are important in understanding accommodation: (1) the vitreous membrane, where it becomes the ciliary zonule, is quite intimately fastened to the ciliary bodies; and (2) the vitreous membrane forms the suspensory ligament of the lens.

ARTERIES AND VEINS (Fig. 7-55)

The eyeball is supplied by three sets of ciliary arteries—(1) short posterior ciliary, (2) long posterior ciliary, and (3) anterior ciliary arteries—and by a very important artery to the retina. The **short posterior ciliary** arteries are numerous branches of the ophthalmic which surround the optic nerve, pierce the sclera close to this nerve, and proceed in the choroid layer. The **long posterior ciliary arteries** are two in number; they also pierce the sclera, proceed anteriorly in the choroid layer, anastomose with the anterior ciliary arteries to form two rings of arteries, one at the base of the iris and another just lateral to the sphincter pupillae muscle. The **anterior ciliary** vessels, just mentioned as anastomosing with the long ciliary arteries, are branches of vessels supplying the recti muscles.

The large veins—**venae vorticosae**—pierce the sclera just posterior to the equator; they arise from a complex of veins in the choroid layer and empty into the ophthalmic veins.

The artery to the retina—**central retinal**—usually enters the optic nerve at some point anterior to the optic foramen. It enters the retina at the optic disc and immediately branches into a superior and inferior vessel. These, in turn, divide into superior nasal and temporal, and inferior nasal and temporal branches. These arteries are end arteries, not anastomosing with any other vessels; they course anteriorly only as far as the ora serrata. The **retinal veins** correspond to the arteries and converge on

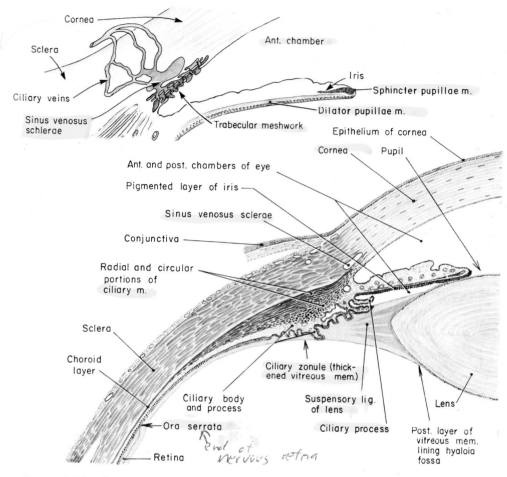

Figure 7-54. *Section through the anterior half of the eyeball. Insert enlarges the sinus venosus schlerae and surrounding structures.*

the optic disc to enter the optic nerve. The typical view seen with an ophthalmoscope is presented in Fig. 7-53.

NERVES

The nerves to the eyeball are the short and long ciliary and the optic nerve. The **optic nerve** is made up of axons of the many cells in the retina and is, of course, responsible for carrying impulses from the retina to the brain. The **short ciliary nerves,** resembling the arteries, are made up of several nerves coursing along the optic nerve; they are branches of the **ciliary ganglion** and course an-

teriorly in the choroid layer until they reach the ciliary body and iris. These short ciliary nerves consist of these fibers: (1) parasympathetic postganglionic fibers from the oculomotor nerve that synapsed in the ciliary ganglion; these nerves innervate the ciliary muscles and the sphincter pupillae muscle; (2) sensory fibers passing through the ganglion to reach the nasociliary branch of the trigeminal nerve; (3) sympathetic fibers that pass through the ganglion without synapsing, and finally innervate the blood vessels of the eyeball and the dilator pupillae muscle of the iris. The **long ciliary nerves** do not traverse the ganglion but branch from the nasociliary nerve and course to the eyeball directly and pierce the sclera to gain the choroid layer, which they follow to the iris. They are mainly sensory, but sympathetic fibers do occur in these nerves and probably aid in innervation of the dilator pupillae muscle.

ALTERING SIZE OF PUPIL

The iris, besides containing columnar pigmented cells which are responsible for its color, also contains smooth muscles, blood vessels, and nerves. The **smooth muscle** fibers are arranged in two parts (Fig. 7-54). The one close to the free edge of the iris is arranged in a circular manner and acts as a sphincter—the **sphincter pupillae muscle.** The **dilator pupillae muscle** consists of many fibers that radiate from the sphincter to the periphery of the iris; when these fibers contract, the pupil is enlarged. The sphincter pupillae muscle is innervated by the parasympathetic portion of the autonomic nervous system carried in the oculomotor nerve. The dilator pupillae muscle is innervated by the sympathetic portion of the autonomic nervous system which was derived by pre- and postganglionic fibers starting in the upper thoracic segments of the spinal cord, as already described.

The **light reflex** is set up somewhat as follows. Light shines into the eye, is picked up by the retina, and the impulses are carried via the optic nerve to brain centers. There is a direct connection between these centers and the nucleus of the oculomotor nerve. The parasympathetic preganglionic cells are stimulated and impulses pass along the oculomotor nerve to the ciliary ganglion. Here the synapse occurs with postganglionic fibers that travel to the sphincter pupillae muscle to constrict the pupil as a response to the bright light.

ACCOMMODATION

Accommodation for near vision consists of three distinct actions: (1) a convergence of the eyes, (2) an increase in the thickness of the lens, and (3) a constriction of the pupil. The increase in thickness of the lens is accomplished by the circularly arranged and radially arranged portions of the ciliary muscle. As mentioned previously, the ciliary processes stick into the furrows of the ciliary zonule which is made up of the vitreous membrane and which is directly continuous with the suspensory ligament of the lens. Therefore, any movement of the ciliary body

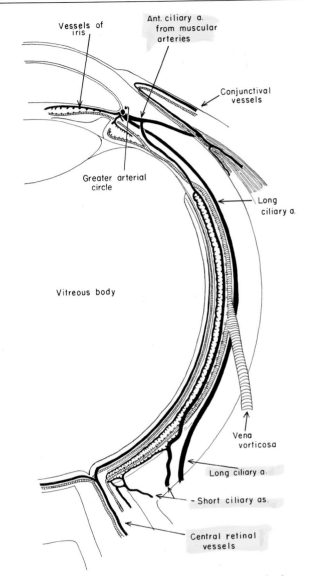

Figure 7-55. Diagram showing the three sets of ciliary arteries to the eyeball. (Modified after Cunningham.)

will produce a movement in the suspensory ligament. The circularly arranged muscle fibers of the ciliary body act as a sphincter and move the ciliary body and suspensory ligaments toward the central axis of the eyeball. The same movement is accomplished when the radially arranged portion of the ciliary muscle is contracted. This muscle has its origin from the comparatively firm corneoscleral junction and inserts on the ciliary zonule.

Therefore, when this muscle contracts, the ciliary zonule is moved toward the central axis of the eye and the suspensory ligament is moved as well. This movement decreases the tension on the suspensory ligament, and the naturally elastic lens tends to thicken. When the eye is at rest, the lens is flattened by the natural tension on the suspensory ligament. Since the eye at rest is adjusted for distant vision, this accommodation or thickening of the lens is done for near vision. The work is done by the ciliary muscle, both portions of which are innervated by the parasympathetic nervous system. The sympathetic portion of the autonomic nervous system apparently plays no role in this action. The third part of the act of accommodation, the closing of the pupil, is of course accomplished by the sphincter pupillae muscle which is also innervated by the parasympathetic system. Convergence of the eyes is effected by a voluntary act of contracting the three muscles used to move the eyes medially—the medial rectus, superior rectus, and inferior rectus. This convergence of both eyes toward an object close to us is possibly the voluntary act that initiates the reflex resulting in accommodation.

The eye is subject to many difficulties. The disease known as glaucoma results when the aqueous humor cannot exit into the venous system. The cornea may become opaque, as can the lens; in the latter case, the disease is called a cataract. The lens can also lose its elasticity, so that a person finally has to wear glasses to aid in accommodation for near vision. The cornea and sclera sometimes are injured from severe blows; the choroid layer can become inflamed and the retina can separate from the choroid layer. A shortening of the eyeball results in farsightedness and an elongation of the eyeball in nearsightedness; astigmatism can result from a faulty curvature of the cornea or the lens, or from a different shape to the eyeball on one side in contrast to the other. In spite of these and other difficulties, the eye is a remarkable organ.

You have now finished study of the optic, oculomotor, trochlear, ophthalmic division of the trigeminal, and the abducens nerves. A systemic review would be helpful at this time. This can be found on pages 647 and 648, and on page 650.

EAR

→ FACIAL NERVE

The **ear** is divided into three parts: the **external ear, middle ear,** and **internal ear** (Fig. 7-56). The internal ear is concerned not only with hearing but with balance as well. As the latter is very intimately associated with the internal ear, the phenomenon of balance will be discussed at the same time as hearing is considered. Whereas the external ear is of little clinical significance except for the occasional plugging of the external auditory meatus with ear wax, and the internal ear is a structure that cannot be helped a great deal clinically, the middle ear is of extreme importance. Not only can an infection spread directly from the nasopharynx via the auditory tube to the middle ear and thence into the mastoid air cells, but this particular part of the ear is involved with conduction of sounds; conduction deafness, a type of deafness that can be cured, results when anything interferes with transference of vibrations through the middle ear.

EXTERNAL EAR

The **external ear** consists of the **auricle** and the **external auditory meatus.**

The **auricles** (Fig. 7-57) are two rather peculiarly shaped pieces of cartilage covered with skin. Although many of the elevations and depressions are given names, it would not seem necessary for the student to know all of them. The outside rim is the **helix** (*L.*, a coil), which ends inferiorly in a fleshy **lobule.** The shelllike hollow of the center of the ear is the **concha; the tragus** (*L.*, goat, because of the hairs that can grow thereon) is the elevation anterior to the external auditory meatus. The **antitragus** and **antihelix** are shown in Fig. 7-57. There are several intrinsic muscles between various parts of the auricle that are of no importance in the human being. Equally unimportant are three extrinsic muscles—the anterior, superior, and posterior auricular muscles. The **motor nerves** to these muscles are branches of the facial; **sensory nerves** to the auricle consist of branches of the auriculotemporal from the trigeminal, and the great auricular and lesser occipital nerves of the cervical plexus. The **blood supply,** as might be expected, is by branches

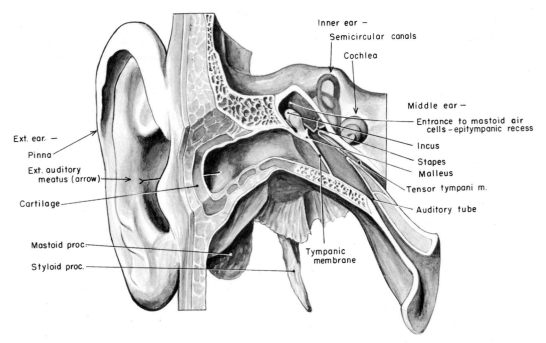

Figure 7-56. *The external, middle, and internal portions of the right ear.*

of the superficial temporal and posterior auricular arteries; the **veins** take the same names and the same distribution. The **lymphatics** end in glands posterior to the ear (posterior auricular or mastoid), and anterior to the ear (the preauricular or parotid nodes).

The **external auditory meatus** (Fig. 7-56) starts at the deepest part of the concha and continues medially until the tympanic membrane is reached. This canal is lined with skin, and the epidermis of the skin continues onto the tympanic membrane. There are many **hairs** present, and the skin has many modified sweat glands that give rise to cerumen or ear wax (**ceruminous glands**). The lateral third of this canal is quite movable and lined with cartilage that is continuous with that of the auricle. The medial two-thirds is a bony canal in the temporal bone. Because of the obliquity of the tympanic membrane the walls are not of the same length; however, the entire canal is approximately one and one quarter inches long. The external auditory meatus is not a straight canal but takes a rather oblique course. It can be straightened if the ear is pulled superiorly and posteriorly. The diameter is not the same in its entire length since the pressure of the lower jaw on the parotid tends to keep the lateral third closed. The **sensory nerve supply** to the meatus is by branches of the auriculotemporal of the trigeminal and the auricular branch of the vagus (and possibly small branches of the facial). This branch of the vagus is of importance in that irritations to the external auditory meatus can often be referred to other areas innervated by the vagus to the point that one can feel nauseated or develop a cough as a result of such irritations. The **arteries** to the external auditory meatus are branches of the superficial temporal and the deep auricular branch of the maxillary (described on page 551). The **veins** are similar to the arteries. The **lymphatics** end in the superficial nodes both posterior and anterior to the auricle.

The **tympanic membrane** (Figs. 7-56 and 7-58) separates the external auditory meatus from the middle ear. It is placed obliquely in such a manner that its lateral surface faces inferiorly and anteriorly as well as laterally. The tympanic membrane is a connective tissue membrane lined on the outside with skin and on the inside by mucous membrane. It is thickened around its periphery where it attaches to the bone. Since the tympanic membrane is always looked at from the lateral side, this surface

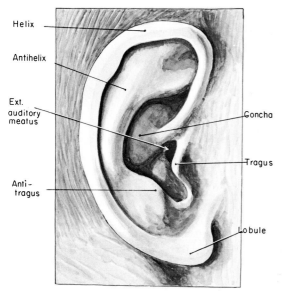

Figure 7-57. The right auricle.

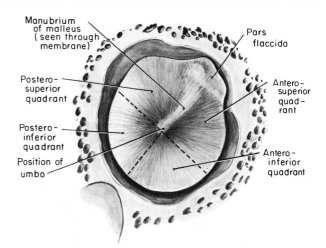

Figure 7-58. The external surface of the right tympanic membrane. Its division into quadrants is shown.

will be described. The tympanic membrane can be divided into **quadrants** with an anterosuperior and an anteroinferior quadrant, a posterosuperior and a posteroinferior quadrant. The two superior quadrants are divided by the elevation produced by the manubrium of the first ear bone, the malleus. Just superior to this elevation produced by the manubrium of the malleus is an elevation produced by the lateral process of this bone. This appears as a white spot when observing the membrane with an otoscope. Extending laterally from this elevation are two folds, the **anterior and posterior malleolar folds.** Superior to these folds is the **pars flaccida** of the tympanic membrane; it is this part that is quite delicate and contains no connective tissue, in contrast to the rest of the tympanic membrane. The most concave area in the tympanic membrane is the point of attachment of the manubrium of the malleus to the central portion; it is just opposite to an elevation on the medial surface of the membrane that is given the term **umbo** (L., knob). A bright reflected cone of light is seen in the anteroinferior quadrant. The **nerve supply** to the tympanic membrane is derived from the auriculotemporal branch of the trigeminal and the auricular branch of the vagus to the outside, and from the tympanic branch of the glossopharyngeal to

the inside. The **arteries*** are derived from the deep auricular branch of the maxillary to the outside and the stylomastoid branch of the posterior auricular and tympanic branch of the maxillary to the inside surface. The **veins** on the superficial surface drain into the external jugular, while those on the deep surface drain into the transverse sinus of the dura and to a plexus of veins around the auditory tube.

MIDDLE EAR

The middle ear is a space in the temporal bone lined with mucous membrane which is directly continuous with mucous membrane lining the auditory tube anteriorly and the mastoid air cells posteriorly. This cavity contains the middle ear bones—the malleus, incus, and stapes—which are responsible for conducting vibrations set up in the tympanic membrane to the internal ear. These middle ear bones (Fig. 7-59) are attached to each other and to the surrounding bones by ligaments and are arranged in such a manner that vibrations set up on the tympanic membrane are relayed to the membrane covering the oval window. Although students should not burden themselves with the detailed structure and attachments of these middle ear bones, the attachment of two muscles is

---

*These nerves and arteries will be described later in this chapter.

532

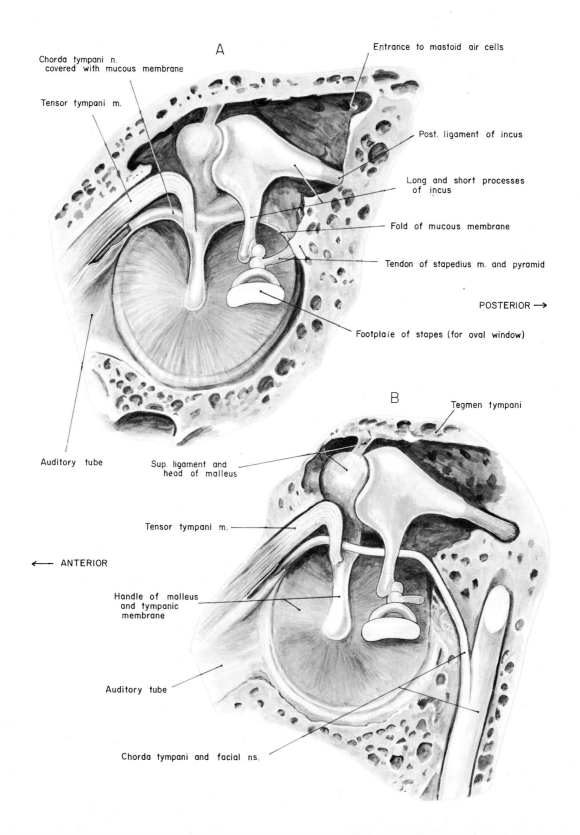

A

Chorda tympani n.
covered with mucous membrane

Tensor tympani m.

Entrance to mastoid air cells

Post. ligament of incus

Long and short processes
of incus

Fold of mucous membrane

Tendon of stapedius m. and pyramid

POSTERIOR →

Footplate of stapes (for oval window)

Auditory tube

B

Tegmen tympani

Sup. ligament and
head of malleus

Tensor tympani m.

← ANTERIOR

Handle of malleus
and tympanic
membrane

Auditory tube

Chorda tympani and facial ns.

of importance. The **tensor tympani muscle** attaches to the manubrium of the malleus, and the **stapedius muscle** inserts on the neck of the stapes. These muscles are brought into play reflexly by sounds that are of great volume. By contracting, they are able to decrease the magnitude of the vibrations received at the tympanic membrane and delivered to the oval window. The tensor tympani muscle, having been derived from the first branchial arch, is innervated by the trigeminal nerve, while the stapedius, derived from the second branchial arch, is innervated by the facial nerve. The **chorda tympani nerve,** a branch of the facial, enters the middle ear posteriorly and exits anteriorly. In its course through the middle ear it is lateral to the incus and medial to the manubrium of the malleus. This nerve ultimately joins the lingual branch of the trigeminal nerve in the infratemporal fossa and it is involved with taste for the anterior two-thirds of the tongue and carries secretory fibers to the submandibular and sublingual glands.

The student should remember that all of these structures mentioned and pictured in the accompanying diagrams are covered with a mucous membrane.

The **middle ear** possesses a **roof,** a **floor,** and **anterior, posterior, medial, and lateral walls.** You should make constant reference to the illustrations while reading the following comments about these walls.

The **roof** requires no lengthy description: it is a thin layer of bone, called the **tegmen tympani,** which separates the middle ear from the middle cranial fossa. It is on the anterior surface of the petrous portion of the temporal bone.

The **floor** consists of a thin layer of bone that separates the middle ear from the **jugular fossa** posteriorly, and from the **carotid canal,** containing the internal carotid artery, anteriorly. These very vital structures are important relations of the middle ear.

The important structures seen on the **posterior wall** (Fig. 7-60) are, from superior to inferior, the entrance to the mastoid air cells called the **aditus antrum,** the **pyramid** through which the **stapedius** muscle enters the cavity to insert on the stapes, and the foramen for the entrance of the **chorda tympani** nerve.

The **anterior wall** (Fig. 7-60) from superior to inferior consists of a canal that contains the **tensor tympani muscle,** the entrance of the **auditory tube,** and the foramen for the exit of the **chorda tympani** nerve.

The **lateral wall** (Fig. 7-59) is made up almost entirely by the **tympanic membrane.**

The **medial wall** (Fig. 7-60) is more complex. Many of the structures on this wall are related to the internal ear which is just medial to this wall. Most anteriorly placed, just at the entrance of the auditory tube, is a prominence—the **promontory**—produced by the first (basal) coil of the cochlea. The mucous membrane covering the promontory contains a plexus of nerves designated the **tympanic plexus.** Posterior to this promontory the **stapes** can be seen in position, covering the **fenestra ovalis** (L., oval window) or **fenestra vestibuli.** The latter name indicates that it is related to the scala vestibuli of the internal ear. The stapedius muscle can be seen attaching to the neck of the stapes. Inferior to the fenestra ovalis is a round opening, covered over by a membrane, called the **fenestra rotunda** (round window) or **fenestra cochleae.** The latter name indicates the relation of the round window with the cochlea of the internal ear. Posterior and superior to these structures is a ridge which is the bony covering of the **facial nerve.** Posterior to this prominence is an elevation produced by the **lateral semicircular canal;** this prominence is located in the **epitympanic recess.** Posterior to this epitympanic recess is the entrance into the **mastoid air cells.**

A glance at Figure 7-64A at this time will make sense out of this rather complex medial wall.

The **auditory tube** starts in the lateral wall of the nasopharynx and ends in the middle ear (Fig. 7-56). It is approximately 1½ inches in length. The part toward the middle ear is bony, while the remaining portion is cartilaginous and wider than the bony portion. In fact, the tube becomes wider as one approaches the nasopharynx (Fig. 7-56). The opening of the bony portion of the auditory tube can be seen on the base of any skull in the petrous portion of the temporal bone just posterior to the foramen spinosum and foramen ovale. The cartilaginous portion would have been located at the same point in a groove between the petrous portion of the temporal bone and the sphenoid bone. It is located just posterior and

Figure 7-59. Structures found on the lateral wall of the middle ear (A) with the mucous membrane intact, (B) with mucous membrane removed—right ear.

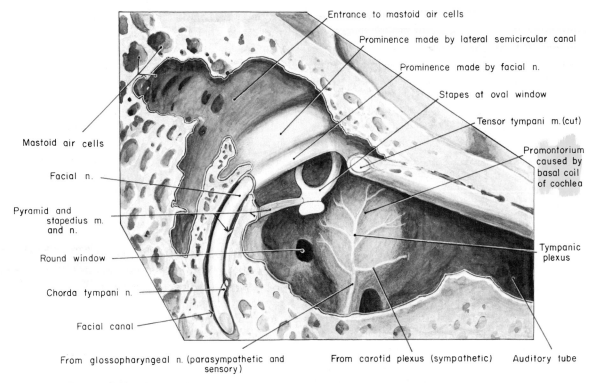

Entrance to mastoid air cells

Prominence made by lateral semicircular canal

Prominence made by facial n.

Stapes at oval window

Tensor tympani m. (cut)

Promontorium caused by basal coil of cochlea

Tympanic plexus

Mastoid air cells

Facial n.

Pyramid and stapedius m. and n.

Round window

Chorda tympani n.

Facial canal

From glossopharyngeal n. (parasympathetic and sensory)

From carotid plexus (sympathetic)   Auditory tube

Figure 7-60. *The medial wall of the middle ear—right side. Note that structures forming the anterior and posterior walls are also visible in this illustration. Compare with Figure 7-64A.*

medial to the foramen spinosum and foramen ovale. As we will see later, a muscle of the palate, the tensor veli palatini, takes origin from the cartilaginous portion of the tube and, since this muscle is used in the process of swallowing, it aids in opening the auditory tube during that process. The opening of the auditory tube in the nasopharynx is located posterior to the nasal cavity and in a direct line with the inferior concha (see Fig. 7-120 on page 618). The mucous membrane lining the tube is innervated by branches from the tympanic plexus and by the pharyngeal branch of the pterygopalatine ganglion, to be described later (page 622).

The auditory tube serves to equalize the pressure on each side of the tympanic membrane. The mucous membrane, in the osseous portion, is covered with ciliated epithelium, while in the cartilaginous portion it contains many mucous glands. Near the pharyngeal opening a considerable amount of lymphoid tissue is present which can be annoying when swollen.

Mention should be made of the **mastoid air cells**

(Fig. 7-61) which, in spite of antibiotics, are of clinical importance. The entrance to the mastoid air cells is via the antrum and this is entered by the aditus antrum which leads posteriorly from the epitympanic recess. Therefore, infection can travel, from the middle ear, by way of the epitympanic recess, aditus antrum, and antrum, into the mastoid air cells. The latter structures vary considerably in number, size, and shape and usually develop to their largest form at puberty. They are lined with the same mucous membrane found lining the auditory tube and middle ear. One very important relation of the mastoid air cells is to the sigmoid sinus. There is a very thin layer of bone separating these cells from the sigmoid sinus and this should be recognized when performing surgery in this area.

INTERNAL EAR

The **internal ear** is a small but complex structure located in the petrous portion of the temporal bone between the

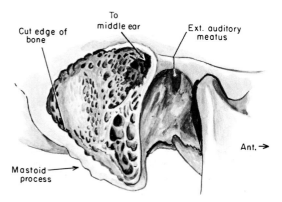

Figure 7-61. *The mastoid air cells—right side.*

middle ear laterally and the internal auditory meatus medially (Fig. 7-62). It consists of two parts—a **vestibular portion** consisting of the semicircular canals, and an **auditory portion,** the cochlea. The internal ear is connected to the brain by the **vestibulocochlear** nerve. If injury occurs to this nerve, nerve deafness results. The internal ear is made up of canals in the bone itself which are filled with **perilymph;** inside the bony canals is a series of membranous canals filled with **endolymph.** The osseous portion is known as the **osseous labyrinth** while the membranous portion is the **membranous labyrinth.** Note in Figure 7-62 that the facial nerve has an intimate relation to the internal ear as well as to the middle ear.

To aid in comprehension of the following description, study of Figure 7-63 would be helpful. This is a diagrammatic depiction of the **osseous labyrinth** with the **membranous labyrinth** contained therein. (The lateral semicircular canal has been omitted for the sake of clarity.) As mentioned, the osseous labyrinth is a series of spaces and canals in the petrous portion of the temporal bone containing **perilymph** (white in Fig. 7-63). The membranous labyrinth, containing **endolymph** (gray), although contained inside the osseous labyrinth, does not occupy the entire space nor have the identical structure. The **osseous labyrinth** consists of a **vestibule,** three **semicircular canals,** a **cochlea,** and two small canals— the **perilymphatic duct** and **vestibular canal.** The

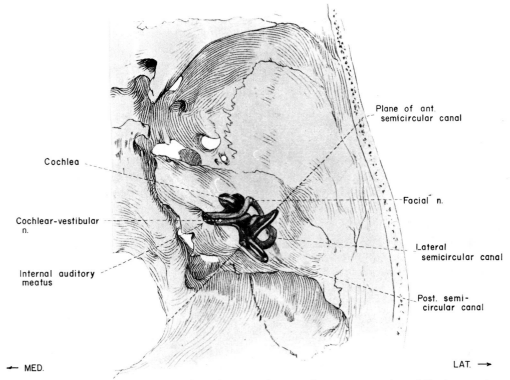

Figure 7-62. *The position of the right internal ear in the petrous portion of the temporal bone. (From Spalteholz,* Hand Atlas of Anatomy, *courtesy of J. B. Lippincott Co.)*

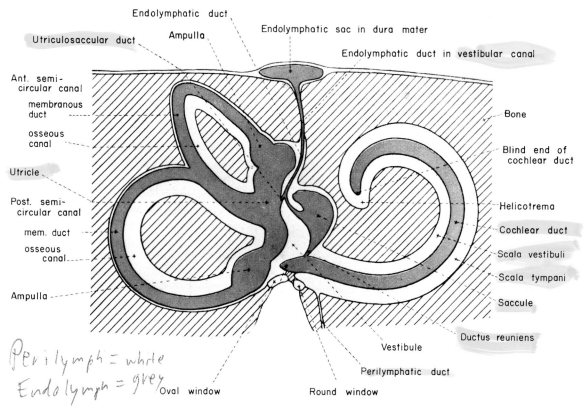

Endolymphatic duct

Utriculosaccular duct

Ampulla

Endolymphatic sac in dura mater

Endolymphatic duct in vestibular canal

Ant. semi-
circular canal

membranous
duct

osseous
canal

Utricle

Post. semi-
circular canal

mem. duct

osseous
canal

Ampulla

Bone

Blind end of
cochlear duct

Helicotrema

Cochlear duct

Scala vestibuli

Scala tympani

Saccule

Ductus reuniens

Vestibule

Perilymphatic duct

Oval window

Round window

Perilymph = white
Endolymph = grey

Figure 7-63. *The membranous labyrinth inside the osseous labyrinth (diagrammatic). The lateral canal and duct were omitted for the sake of clarity. (From Spalteholz,* Hand Atlas of Human Anatomy, *courtesy of J. B. Lippincott Co.)*

membranous labyrinth has a **saccule, utricle,** and **three ampullae of the semicircular canals** located in the vestibule, as well as various ducts connecting these structures. It also has three **semicircular ducts,** a **cochlear duct,** and a **ductus endolymphaticus** leading via the vestibular canal to a sac located subdurally—the **endolymphatic sac.** This supposedly serves as an exit for the endolymph. Note also that the **oval** and **round windows** are associated with the perilymph; that the bony cochlea is divided into two portions—the **scala vestibulae** and the **scala tympani** (note that the round window is at the base of this scala tympani), which are connected with each other at the apex of the cochlea by what is called the **helicotrema,** but at no other point; and that the perilymph also has an exit via the **perilymphatic duct.** A more detailed description follows.

As just mentioned, the **osseous labyrinth** (Fig. 7-64)

consists of a centrally located **vestibule, three semicircular canals,** and a coiled **cochlea.** The **vestibule** is a small, oval cavity whose greatest diameter is 6 mm.; it communicates anteriorly with the cochlea and posteriorly with the semicircular canals. Posteroinferiorly, a narrow canal called the vestibular canal or **aqueduct** (for the ductus endolymphaticus) runs from it to open on the posterior surface of the petrous part of the temporal bone. The **oval window** is in its lateral wall. The **semicircular canals** each form two-thirds of a circle and have a dilation at one end called the **ampulla** (Fig. 7-64). The **anterior semicircular canal** takes an anterolateral plane and is at right angles to the crest of the petrous portion of the temporal bone (Fig. 7-62); the ampulla of this semicircular canal is near the superior part of the vestibule. The **posterior semicircular canal** is in a plane at right angles to the anterior semicircular canal and follows,

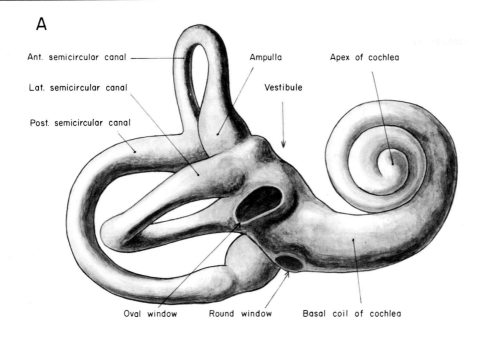

A

Ant. semicircular canal

Lat. semicircular canal

Post. semicircular canal

Ampulla

Vestibule

Apex of cochlea

Oval window   Round window   Basal coil of cochlea

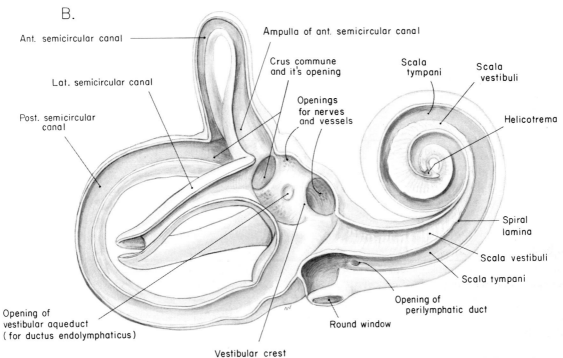

B.

Ant. semicircular canal

Lat. semicircular canal

Post. semicircular canal

Opening of vestibular aqueduct (for ductus endolymphaticus)

Ampulla of ant. semicircular canal

Crus commune and it's opening

Openings for nerves and vessels

Scala tympani

Scala vestibuli

Helicotrema

Spiral lamina

Scala vestibuli

Scala tympani

Opening of perilymphatic duct

Round window

Vestibular crest

Figure 7-64. *(A) The osseous labyrinth—right side. Compare with Figure 7-60. (B) The interior of the right osseous labyrinth.*

in an anteromedial plane, the crest of the petrous portion of the temporal bone; its ampulla enters the postero-inferior part of the vestibule. The **lateral semicircular canal** courses horizontally and has its ampulla just inferior to that of the anterior semicircular canal. It should be noted that the anterior semicircular canal of one side of the body corresponds in plane of direction to the posterior on the opposite side. Several clusters of small holes can be seen in the vestibule; these serve to transmit nerves to various parts of the membranous labyrinth located in the vestibule. The **cochlea** consists of a bony canal which is coiled two and a half to two and three-quarters times around a horizontal central pillar, the **modiolus** (see Fig. 7-66). The basal coil is widest and is responsible for producing the promontory on the medial wall of the middle ear. The cochlea gradually tapers to an **apex.** It is turned on its side, so to speak, and points anterolaterally (Figs. 7-56 and 7-62). The cochlear portion of the vestibulocochlear nerve enters its base (Fig. 7-66). Projecting from the central modiolus throughout this bony canal is a spiral bony shelf, the **spiral lamina,** which starts at the vestibule and continues to the very apex (Fig. 7-64B). A **basilar membrane** stretches from this spiral lamina to the periphery of the cochlear tube, dividing it into two parts, the **scala vestibuli and scala tympani** (Fig. 7-66). Although the scala vestibuli and scala tympani are separated from each other basally, they are in communication at the apex of the cochlea via an opening designated as the **helicotrema** (Figs. 7-63 and 7-64B). The **round window** is located at the base of the scala tympani and there is a small opening, the **perilymphatic duct** (Fig. 7-64B), which opens in this same region and leads through the petrous portion of the temporal bone to open at the anterior margin of the jugular fossa. The perilymph of the osseous labyrinth is thought to connect directly with the subarachnoid space through this canal.

From the foregoing description, it can be seen that the vestibule connects directly with each semicircular canal and with the scala vestibuli of the cochlea. It has no direct connection with the basal part of the scala tympani but does so indirectly via the scala vestibuli and the helicotrema; this is an important point in understanding the process of hearing. As mentioned previously, the osseous labyrinth is filled with perilymph.

The **modiolus** (Fig. 7-66) is not a solid structure but contains a large number of longitudinally running canals for the cochlear nerves and vessels. These canals course for varying distances longitudinally and then turn to form a spiral canal just opposite the spiral lamina. The ganglion cells for the cochlear nerve are located in this spiral canal and form the **spiral ganglion.** As we shall see later, the distal ends of these nerves are attached to the **spiral organ** (of Corti).

The **membranous labyrinth** (Fig. 7-63) is composed of the **utricle** and **saccule,** located in the vestibule, **three semicircular ducts** in the semicircular canals, and the **cochlear duct** in the cochlea. It should be noted that the membranous labyrinth is much smaller in diameter than the osseous labyrinth; it contains **endolymph.** While the cochlear duct and semicircular ducts have the same general shape exhibited by their respective bony channels, the portion of the membranous labyrinth in the vestibule is entirely different than the vestibule itself. The **utricle** is a small sac lying in the posterior superior aspect of the vestibule; the **ampullae** of the semicircular ducts empty into it. The **saccule** is smaller than the utricle and lies in the anterior inferior part of the vestibule. It is connected to the cochlear duct by a small canal, the **ductus reuniens.** Although the saccule and utricle have no direct connections with each other, they are connected by a duct, the **ductus endolymphaticus,** which takes origin from both of these structures. This endolymphatic duct traverses the vestibular aqueduct or canal previously mentioned and ends as a dilation called the **endolymphatic sac,** which is located between the dura and the bone on the medial side of the petrous portion of the temporal bone. The functional aspects of this ending in the epidural space are not understood. The semicircular ducts are attached to the periosteum lining the semicircular canals by connective tissue strands (Fig. 7-65).

The **utricle, saccule,** and **ampullae** of the semicircular ducts demand further attention. The utricle and saccule possess an elevation of cells at certain points—the **macula**—which is lined with hair cells plus supporting cells. The hairs project into a gelatinous material that contains particles—the **otolythic membrane.** These hair cells are arranged in such a manner that any movement of the hairs sends a message to the brain that is relayed to the nerves to the orbital muscles and to all the other skeletal muscles via a tract called the vestibulospinal tract. The utricle seems to respond to centrifugal and vertical acceleration, and the saccule to linear acceleration. Some think the saccule, being more intimately associated with

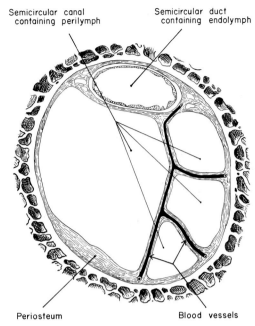

Semicircular canal
containing perilymph

Semicircular duct
containing endolymph

Periosteum

Blood vessels

Figure 7-65. Transverse section of a semicircular canal. (Modified after Gray.)

the cochlear duct, is involved with hearing as well, particularly low tones. Each ampulla of the semicircular ducts contains a crest, on the surface of which are hair cells and supporting cells. These are arranged in such a manner that any movement in the endolymph in the semicircular ducts bends the hairs on the hair cells and a message is sent to the brain via the vestibular portion of the vestibulocochlear nerve. Just as with afferent nerves from the utricle and saccule, those from the ampullae also connect with the skeletal muscles of the body. Thus, we keep our balance.

As mentioned previously, the **cochlea** is divided into two portions by the bony spiral lamina and the attached basilar membrane. In addition, there is a thin **vestibular membrane** that also extends from the osseous spiral lamina to the outside wall of the bony canal. This, with the basilar membrane, cuts off a triangular area from the scala vestibuli—the **cochlear duct;** it contains endolymph. The **spiral organ** is located on the basilar membrane (Fig. 7-66) and projects into the endolymph in the cochlear duct. The cochlear duct follows the cochlea to its apex where it ends blindly.

The **spiral organ** is situated on the basilar membrane and extends from the base to the apex of the cochlea. It projects into the cochlear duct and is therefore bathed with endolymph. When seen in transverse section (Fig. 7-66), it is observed to consist of a row of **inner hair cells** and a row of **outer hair cells** which have hairlike projections into the endolymph. These cells are accompanied by various types of supporting cells. The nerves are a direct continuation of these hair cells. In addition, a membrane—the **tectorial membrane**—is seen to project into the endolymph from the osseous spiral lamina and this is suspended over the hair cells. It is thought that when one of the hair cells touches this tectorial membrane an impulse is carried over the cochlear nerve to the brain where it is recognized as sound.* It is interesting to note that the base of the cochlea is involved with high sounds, while the low sounds involve the area near the helicotrema; this is due to the fact that the basilar membrane is longer near the apex of the coil and shorter toward the base.

Figure 7-67 is an attempt to explain the mechanism whereby vibrations started on the tympanic membrane finally reach the brain to be interpreted as a sound. The cochlea in this diagram has been stretched out or uncoiled. Air waves enter the external auditory meatus and cause a vibration to occur in the tympanic membrane. Vibrations in the tympanic membrane are carried through the malleus, incus, and stapes to the oval window in the vestibule of the osseous labyrinth. If the sounds are too loud, reflex stimulation of the tensor tympani and stapedius muscles will decrease the magnitude of the vibrations at this oval window. The vibrations of the membrane over the oval window caused by the stapes induce vibrations in the perilymph of the scala vestibuli and scala tympani, the membrane of the round window bulging out as the membrane of the oval window bulges in. These vibrations pass through the vestibular membrane into the cochlear duct. One theory of hearing has a specific part of the basilar membrane responding to any given frequency of vibration in the perilymph and endolymph. The stimulated hair cells send impulses to the brain where they are interpreted as a certain sound.

---

*Some feel that the tectorial membrane is attached to the hair cells; in this case a bending of the hairs is supposed to initiate the impulse along the cochlear nerve.

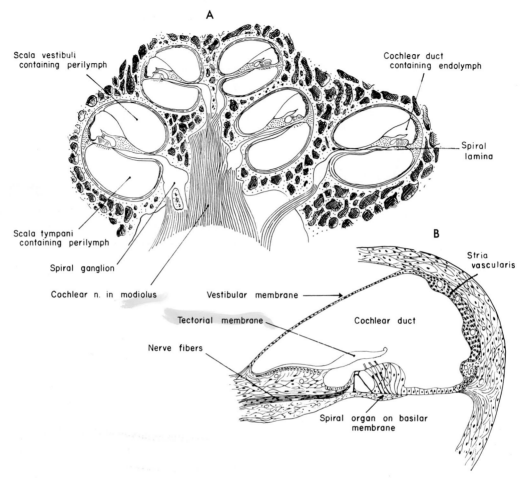

A

Scala vestibuli containing perilymph

Cochlear duct containing endolymph

Spiral lamina

Scala tympani containing perilymph

Spiral ganglion

Cochlear n. in modiolus

B

Stria vascularis

Vestibular membrane

Tectorial membrane

Cochlear duct

Nerve fibers

Spiral organ on basilar membrane

Figure 7-66. *Transverse section of (A) the cochlea and (B) the cochlear duct.*

FACIAL NERVE

No description of the ear would be complete without a consideration of the pathway taken by the facial nerve and its branches, since this nerve is very closely associated with the ear. We have already seen that the vestibulocochlear nerve enters the internal auditory meatus (Fig. 7-62); the cochlear portion turns slightly medially to enter the base of the cochlea and the vestibular portion of the nerve enters the ampullae of the semicircular canals. The facial nerve accompanies these nerves in the internal auditory meatus continuing anteriorly and slightly laterally until the ganglion of the seventh nerve, the **geniculate ganglion,** is reached. Here the nerve makes a sharp bend posteriorly and laterally at a level superior to the middle ear. It then turns inferiorly and the canal for the facial nerve at this point makes a prominence on the medial wall of the middle ear between the middle ear proper and the epitympanic recess (Fig. 7-60). The facial nerve continues inferiorly and finally emerges from the stylomastoid foramen.

The branches of the facial nerve are:

1. Greater petrosal
2. Communication with lesser petrosal
3. To stapedius muscle
4. Chorda tympani
5. Posterior auricular
6. To stylohyoid muscle and posterior belly of digastric
7. Terminal

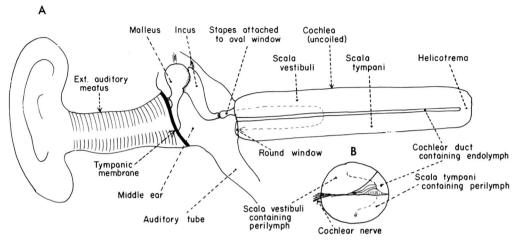

Figure 7-67. *Schematic diagram of the process of hearing. (A) The cochlea is stretched out straight. (B) Transverse section of the cochlea.*

a. Temporal
b. Zygomatic
c. Buccal
d. Mandibular
e. Cervical

1. Greater petrosal (Figs. 7-68 and 7-69). This first branch of the facial nerve, which arises from the geniculate ganglion, turns anteriorly and medially and enters the middle cranial fossa by emerging from the **facial hiatus** on the anterior surface of the petrous portion of the temporal bone. This nerve crosses the foramen lacerum, at which point it is joined by a **deep petrosal branch** from the plexus of sympathetic nerves on the internal carotid artery. After the deep petrosal fibers have joined with the greater petrosal nerve, the nerve becomes the **nerve of the pterygoid canal,** traverses this canal, and enters the pterygopalatine fossa (see Figs. 7-2B and C) to end in the **pterygopalatine ganglion** (Fig. 7-68). The greater petrosal nerve contains important parasympathetic preganglionic fibers and probably sensory fibers for taste around the palate. Preganglionic cell bodies are located in the nucleus of the facial nerve and axons proceed with the facial nerve to the geniculate ganglion. This being a sensory ganglion, no synapse occurs and the fibers simply continue through this to form the greater petrosal nerve, thence to the nerve of the pterygoid canal, and to the pterygopalatine ganglion. Here a synapse oc-

curs between pre- and postganglionic fibers and postganglionic fibers are distributed to the mucous membranes and vessels of the oral cavity, nasal cavity, pharynx, and other regions, including the lacrimal gland. The exact pathways taken by these postganglionic fibers will be described later (page 622).

2. The second branch of the facial nerve is a **communication to the lesser petrosal nerve.** The lesser petrosal nerve is mainly a contribution from the tympanic plexus which is located on the promontory of the medial wall of the middle ear (Fig. 7-68). This tympanic plexus received contributions from the glossopharyngeal nerve and also from the sympathetic nerves and is therefore a combined parasympathetic and sympathetic plexus; it is thought also to contain many afferent fibers. The lesser petrosal nerve enters the infratemporal fossa and synapses with cells in the **otic ganglion.** Further description of this pathway to the parotid gland will be given later (page 550).

3. To stapedius muscle. This small branch of the facial enters the muscle directly.

4. The **chorda tympani** nerve arises from the facial nerve while it is in the facial canal posterior to the middle ear. This branch courses through the middle ear and enters the infratemporal fossa, where it joins the lingual branch of the mandibular division of the trigeminal nerve. It follows this to the submandibular ganglion

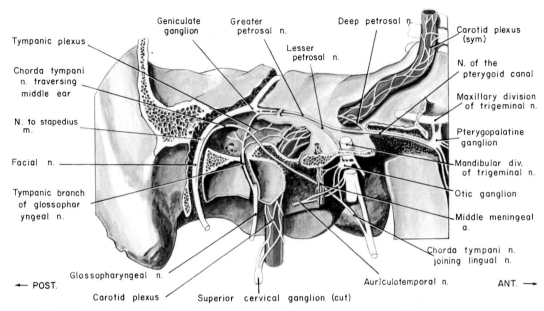

Figure 7-68. *Origin and destination of the petrosal nerves—right side. (Modified after Warren). This is a longitudinal section through the middle ear and adjoining structures.*

where synapses occur, and postganglionic fibers travel from there to the submandibular and sublingual glands. This nerve also contains afferent fibers of taste from the anterior two-thirds of the tongue. The cell bodies of these fibers are located in the geniculate ganglion.

5, 6, and 7. The facial nerve then emerges from the skull via the stylomastoid foramen, gives off its **posterior auricular branch,** its branch **to the posterior belly** of the digastric and the **stylohyoid muscle,** and then divides into the **terminal branches** already studied: the temporal, zygomatic, buccal, mandibular, and cervical.

Therefore, the important facial nerve not only innervates all muscles derived from the second branchial arch, no matter where they are located, but also contains secretory parasympathetic fibers for the sublingual, submandibular, and lacrimal glands; secretory fibers for the mucous membrane in the mouth, nasal cavity, and pharynx; and also has taste fibers from the palate and anterior two-thirds of the tongue. The facial nerve, therefore, contains **branchial efferent fibers** for muscles derived from the second branchial arch; **branchial afferent fibers** for proprioception from these muscles (although these proprioceptive fibers may course with the trigeminal nerve); **general visceral efferent** fibers to the submandibular, sublingual and lacrimal glands, and to the mucous membrane of the oral cavity, nasal cavity and

pharynx; **general visceral afferent** fibers from these same regions; and **special afferent** fibers for taste from the palate and anterior two-thirds of the tongue.

Although there is a communication between the facial and lesser petrosal nerve that might indicate that the facial nerve also gives secretory fibers to the parotid gland, the lesser petrosal nerve seems to obtain its major contribution from the glossopharyngeal nerve via the tympanic plexus. Therefore, the parotid gland is usually considered to be innervated by the glossopharyngeal nerve. The communication with the facial may very well be composed of sensory (taste) fibers.

Many of these branches will be described again and in more detail.

INFRATEMPORAL FOSSA

GENERAL DESCRIPTION

You will be aided greatly in your study of the next few sections of this book if you realize there is a long, centrally

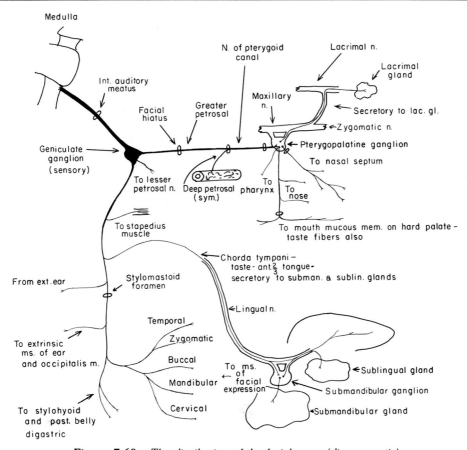

Figure 7-69. *The distribution of the facial nerve (diagrammatic).*

located visceral tube, extending from the base of the skull down through the head and neck and into the thoracic cavity, which lies just anterior to the vertebral column. This visceral tube superiorly is the pharynx; inferiorly it consists of the esophagus posteriorly and the larynx and trachea anteriorly. The nasal cavity leads into the superior portion of this visceral tube, the nasopharynx, and the oral cavity leads into the middle of the tube, the oral pharynx. The inferior end of the pharynx is just posterior to the entrance to the larynx—the laryngeal pharynx. The many structures that we will study in the infratemporal fossa and inferiorly in the neck are located just lateral to this visceral tube and anterior to the vertebral column.

The **infratemporal fossa** is so called because it is located inferior to the temporal bone. It is **bounded** laterally by the ramus of the mandible and medially by the visceral tube just mentioned—in this case the pharynx—and superiorly by the base of the skull. Inferiorly, it is

continuous with the submandibular region. With very few exceptions, all of the structures in this area are involved with the jaw or with the tongue.

Deep to the parotid gland and its duct, the transverse facial artery, and the branches of the facial nerve is a powerful muscle, the **masseter,** which stretches from the zygomatic arch superiorly to the angle of the mandible inferiorly (Fig. 7-70).

If the zygomatic arch and masseter muscle are reflected inferiorly, the **ramus of the mandible** is seen with its **coronoid process** anteriorly and **condyloid process** posteriorly (Fig. 7-71). The space between the two processes is the **mandibular notch** and the masseteric artery and nerve course in this notch. The large, massive **temporal muscle—temporalis**—can be seen inserting on the coronoid process of the mandible and this muscle has a strong insertion on the deep surface of the ramus of the mandible as well. The ramus of the mandible and this

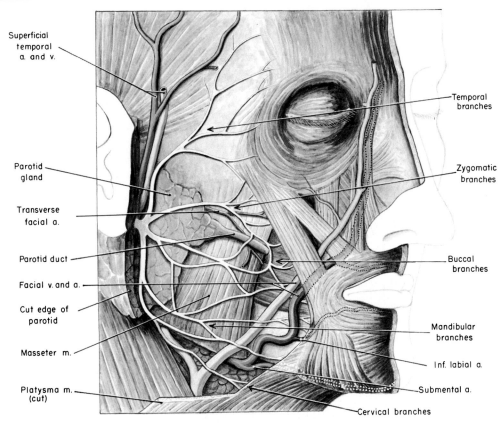

Superficial
temporal
a. and v.

Parotid
gland

Transverse
facial a.

Parotid duct

Facial v. and a.

Cut edge of
parotid

Masseter m.

Platysma m.
(cut)

Temporal
branches

Zygomatic
branches

Buccal
branches

Mandibular
branches

Inf. labial a.

Submental a.

Cervical branches

Figure 7-70. *The masseter muscle covered by the facial nerve, parotid gland and duct, and the transverse facial artery. All fascia has been removed.*

muscle make a continuous layer, which, if removed, allows entrance into the infratemporal fossa.

In about half of all cases, the most superficial structure found in the infratemporal fossa is the **maxillary artery** and its many branches (Fig. 7-72). The artery courses across the **lateral pterygoid muscle.** As might be expected, there are branches for all structures in the vicinity. The first two are small branches to the ear—**deep auricular** and **anterior tympanic.** The next branch is usually the **middle meningeal,** which courses deep to the lateral pterygoid muscle to reach the foramen spinosum. The lower jaw gets the next branch—the **inferior alveolar;** and the remaining branches are associated with the muscles of mastication—**masseter, lateral pterygoid, medial pterygoid,** and two **deep temporals;** and the cheek—the **buccal.** These are all branches of the first two parts of the maxillary artery. The branches of the third part will be described in a subsequent dissection.

If the artery courses deep to the lateral pterygoid muscle, the arterial arrangement is somewhat different (Fig. 7-73). Almost invariably, when this situation occurs, there is a Y-shaped vessel coursing superficial to the lateral pterygoid muscle. One branch of this vessel is the **inferior alveolar** and the **posterior deep temporal** is the other branch of the Y (which usually gives off one of the branches to the ear and also the masseteric branch). The maxillary artery then continues deep to the lateral pterygoid muscle; since it will course very close to the various foramina through which the meningeal vessels pass, the meningeal vessels are quite short.

Several nerves can be seen before the lateral pterygoid muscle is removed. The **buccal nerve** emerges between the two heads of this muscle and courses inferiorly to give sensory fibers to the inside of the cheek as well as the skin surface around the corner of the mouth. The two or three **deep temporal nerves** can also be seen

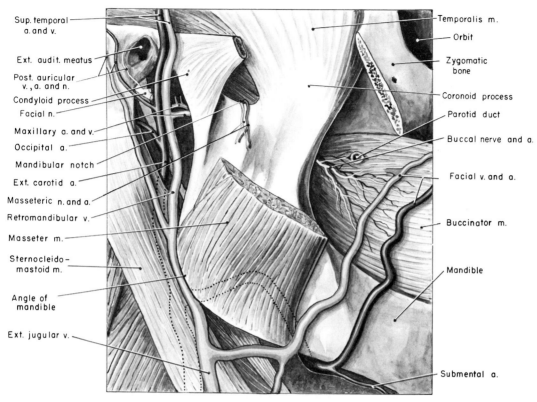

Figure 7-71. *Infratemporal fossa I. The zygomatic arch and masseter muscle have been removed, revealing the ramus of the mandible and the temporalis muscle attached to the coronoid process.*

at this time coursing very close to the bone and entering the deep side of the temporalis muscle. In addition, if the ramus of the mandible is chipped away sufficiently, the **inferior alveolar** and **lingual nerves** can be seen emerging from a position deep to the lateral pterygoid muscle and coursing on the superficial surface of the medial pterygoid muscle.

Now, if the lateral pterygoid muscle is removed (Fig. 7-73), the **medial pterygoid muscle** is seen as are the many branches of the **mandibular division of the trigeminal nerve.** If the artery remained deep to the lateral pterygoid, this large artery can be seen stretching across the infratemporal fossa to reach the pterygomaxillary fissure.

The **mandibular division** of the trigeminal nerve (Fig. 7-73) leaves the skull through the foramen ovale and divides into **anterior and posterior divisions.** Before dividing, small nerves are given off: the **meningeal**

branch, which immediately returns to the cranial cavity through the foramen spinosum to give sensory nerve supply to a very large area of the dura mater, the **medial pterygoid nerve** innervating that particular muscle, and, either directly or from the former branches, branches **to the tensor tympani and tensor veli palatini muscles.** The **anterior division** has the following branches and distribution: the **masseteric** which courses slightly laterally; the **temporal branches** which course superiorly; the nerve to the **lateral pterygoid** which goes directly laterally to that muscle; and the **buccal** branch which courses inferiorly. The **posterior division** is by far the larger component of the nerve and includes the large **lingual nerve,** the equally large **inferior alveolar nerve,** and the **auriculotemporal nerve.** The auriculotemporal nerve, in coursing posteriorly and laterally, usually splits and surrounds the middle meningeal artery. The **mylohyoid nerve** can be seen branching off the inferior alveolar be-

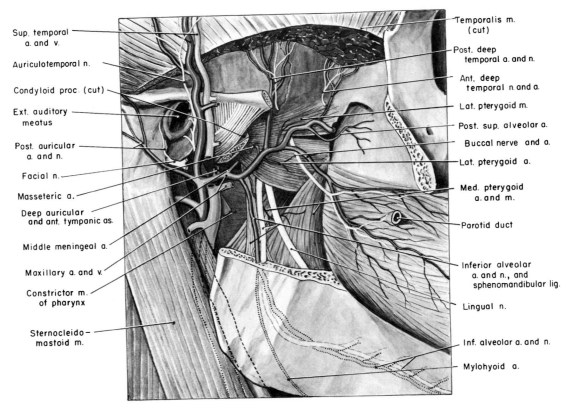

Sup. temporal a. and v.

Auriculotemporal n.

Condyloid proc. (cut)

Ext. auditory meatus

Post. auricular a. and n.

Facial n.

Masseteric a.

Deep auricular and ant. tympanic as.

Middle meningeal a.

Maxillary a. and v.

Constrictor m. of pharynx

Sternocleido-mastoid m.

Temporalis m. (cut)

Post. deep temporal a. and n.

Ant. deep temporal n. and a.

Lat. pterygoid m.

Post. sup. alveolar a.

Buccal nerve and a.

Lat. pterygoid a.

Med. pterygoid a. and m.

Parotid duct

Inferior alveolar a. and n., and sphenomandibular lig.

Lingual n.

Inf. alveolar a. and n.

Mylohyoid a.

Figure 7-72. Infratemporal fossa II. Entrance has been gained into the fossa by removal of the temporalis muscle and the ramus of the mandible.

fore the latter enters the mandibular foramen and the **chorda tympani nerve** can usually be found to enter the posterior side of the lingual nerve. The small **otic ganglion** is suspended on the medial side of the mandibular division as seen in Figure 7-68. This ganglion is difficult to see from the lateral view.

Dissecting layer for layer into the infratemporal fossa, one encounters: (1) skin; (2) the parotid gland and its duct, the transverse facial artery, and branches of the facial nerve; (3) the masseter muscle; (4) ramus of the mandible and temporalis muscle; (5) maxillary artery and its branches; (6) lateral pterygoid muscle; (7) mandibular division of the trigeminal nerve and its many branches; and (8) the medial pterygoid muscle inferiorly and the lateral pterygoid lamina superiorly. Item 4 is the lateral boundary of the infratemporal fossa, while 5, 6, 7, and 8 are contained within the fossa. If one dissects further, the

pharynx will be reached which is the medial wall of the infratemporal fossa.*

It might be well to visualize this area from the reference point of the oral cavity, since anesthetics such as novocaine are injected frequently into the infratemporal fossa, via the mouth, to anesthetize the lower jaw.

MUSCLES OF MASTICATION

The muscles in the infratemporal fossa are mainly the muscles of mastication derived embryologically from the first branchial arch. They are:

---

*F. A. Coller and L. Yglesias, 1935, Infections of the lip and face, *Surg. Gynec. Obst.* 60: 277. This article contains anatomical descriptions of spaces in the face. Important spaces described are (1) mandibular, (2) masticator, (3) superficial and (4) deep temporal, (5) parotid, and (6) lateral pharyngeal.

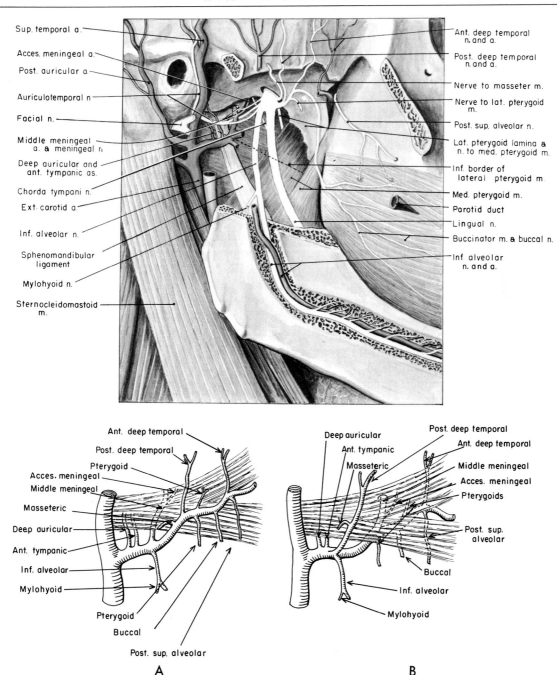

Sup. temporal a.

Acces. meningeal a.

Post. auricular a.

Auriculotemporal n.

Facial n.

Middle meningeal
a. & meningeal n.

Deep auricular and
ant. tympanic as.

Chorda tympani n.

Ext. carotid a.

Inf. alveolar n.

Sphenomandibular
ligament

Mylohyoid n.

Sternocleidomastoid
m.

Ant. deep temporal
n. and a.

Post. deep temporal
n. and a.

Nerve to masseter m.

Nerve to lat. pterygoid
m.

Post. sup. alveolar n.

Lat. pterygoid lamina a
n. to med. pterygoid m.

Inf. border of
lateral pterygoid m.

Med. pterygoid m.

Parotid duct

Lingual n.

Buccinator m. & buccal n.

Inf. alveolar
n. and a.

A

Ant. deep temporal

Post. deep temporal

Pterygoid

Acces. meningeal

Middle meningeal

Masseteric

Deep auricular

Ant. tympanic

Inf. alveolar

Mylohyoid

Pterygoid

Buccal

Post. sup. alveolar

B

Deep auricular

Ant. tympanic

Masseteric

Post. deep temporal

Ant. deep temporal

Middle meningeal

Acces. meningeal

Pterygoids

Post. sup.
alveolar

Buccal

Inf. alveolar

Mylohyoid

Figure 7-73. *Infratemporal fossa III. The lateral pterygoid muscle has been removed. (A) Usual arrangement of branches of maxillary artery when its course is superficial to the lateral pterygoid muscle. (B) Usual arrangement when artery courses deep to the muscle.*

1. Masseter
2. Temporalis
3. Lateral pterygoid
4. Medial pterygoid

1. Masseter (Fig. 7-70). The masseter muscle lies on the ramus of the mandible on its superficial surface. Its **origin** is from the inferior border of the zygomatic arch and its **insertion** is onto the lateral surface of the ramus and the coronoid process of the mandible. Its **main action** is to close the jaw. Its **nerve supply** is from the masseteric branch of the mandibular division of the trigeminal nerve.

2. Temporalis (Figs. 7-71 and 7-72). This muscle is located in the temporal region **arising** from the temporal fossa inferior to the temporal line. It **inserts** on the coronoid process and on the anterior border of the ramus of the mandible on its deep side. Its **main action** is to close the jaw and to retract it after protrusion has been induced by the lateral pterygoid (see below). Its **nerve supply** is from the anterior and posterior deep temporal branches of the mandibular division of the trigeminal nerve. This muscle lies deep to the masseter muscle and the zygomatic arch. Combined with the ramus of the mandible it forms the lateral boundary of the infratemporal fossa.

3. Lateral pterygoid (Fig. 7-72). This muscle is located in the infratemporal fossa deep to the temporalis muscle and superficial to the lateral pterygoid lamina and medial pterygoid muscle. Its **origin** is in two parts since the muscle has two heads; the upper head arises from the greater wing of the sphenoid bone while the lower head arises from the lateral pterygoid lamina. This muscle **inserts** on the anterior surface of the neck of the mandible and also into the capsule of the temporomandibular joint. Its **main action** is to protrude the mandible pulling it toward the opposite side. If these muscles are used alternately the jaw is moved from side to side in a grinding motion such as is used in chewing food. If both muscles contract simultaneously, the mandible is protruded. The muscle also assists in opening the mouth by pulling the capsule and disc of the temporomandibular joint, and the condyloid process anteriorly.* The **nerve supply** to this muscle is the lateral pterygoid branch of the mandibular division of the trigeminal nerve. It might be mentioned that the buccal nerve emerges between the two heads of this muscle.

4. Medial pterygoid (Fig. 7-73). This muscle lies deep in the infratemporal fossa, deep to the lateral pterygoid muscle, and superficial to the muscles involved with the visceral tube already mentioned. This muscle is also a two-headed one, its superficial head **arising** from a tuberosity on the maxilla and the deep head arising from the medial surface of the lateral pterygoid lamina. The superficial portion is quite small while the deep portion is large. This muscle **inserts** on the medial surface of the angle and ramus of the mandible, posterior to the mylohyoid groove. Its **main action** is to raise the mandible and its **nerve supply** is from the medial pterygoid branch of the mandibular division of the trigeminal nerve; this branch arises from the mandibular nerve before it divides into its anterior and posterior divisions.

In chewing, one side predominates over the other. For example, if chewing is being done on the right side, the muscles on that side will contract more forcefully than those on the left side.

NERVES

The **mandibular division of the trigeminal nerve** descends into the infratemporal fossa through the foramen ovale. After giving off four branches, it divides into an anterior and posterior division. The branches of the undivided trunk and of these divisions (Fig. 7-74) are as follows:

Of undivided trunk
   1. Meningeal
   2. Medial pterygoid
   3. To tensor tympani muscle
   4. To tensor veli palatini muscle
Of anterior division
   1. Buccal
   2. Lateral pterygoid
   3. Masseteric
   4. Anterior deep temporal
   5. Posterior deep temporal
Of posterior division
   1. Auriculotemporal

---

*In the resting state, the teeth are not in contact. Raising the jaw so that they are involves the lower joint—the condyloid process and disc. Movement of the jaw below this resting point involves the upper joint—disc and temporal bone—as well.

2. Lingual

3. Inferior alveolar

The **mandibular division** arises from the semilunar ganglion of the fifth nerve, the ganglion being located in its own cave of dura mater, and then continues through the foramen ovale. The **motor division** of the trigeminal nerve, located on the medial side of the main nerve and semilunar ganglion, courses with this division. Therefore, the mandibular nerve is a combined motor and sensory nerve. As already noted, the main nerve is superficial to the medial pterygoid muscle and deep to the lateral pterygoid muscle.

UNDIVIDED TRUNK. The following branches arise from the **undivided trunk:**

1. Meningeal (Fig. 7-73). The meningeal nerve remains very close to the base of the skull, and immediately returns to the cranial cavity via the foramen spinosum in company with the middle meningeal artery. It divides into two branches accompanying the anterior and posterior branches of the middle meningeal artery. It is the sensory nerve to the vast majority of the dura mater, innervating the same territory supplied by the middle meningeal artery.

2. Medial pterygoid (Fig. 7-73). This is the second branch of the undivided trunk of the mandibular division of the trigeminal nerve, is quite short, and immediately enters the medial pterygoid muscle. The otic ganglion encompasses this nerve at its origin. Although the nerve appears to be arising from the ganglion, it has nothing to do with the cells therein. This is a motor nerve to the medial pterygoid muscle and also contains proprioceptive fibers from this muscle.

3. To tensor tympani muscle (Fig. 7-74).* This nerve may arise separately or in common with other nerves. It passes through the otic ganglion and courses directly to the tensor tympani muscle, which is located just medial to the ganglion.

4. To tensor veli palatini muscle (Fig. 7-74).* This nerve may also arise in common with others and also courses through the otic ganglion. It is short since the tensor veli palatini muscle is just medial to the main mandibular division of the trigeminal nerve.

---

*The nerves to the tensor muscles are usually described as branches of the otic ganglion; this is misleading since these nerves have nothing to do with the cells contained in this ganglion.

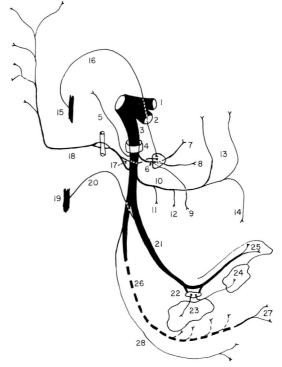

Figure 7-74. *Diagram of the mandibular division of the trigeminal nerve: (1) ophthalmic division, (2) maxillary division, (3) mandibular division, (4) foramen ovale, (5) meningeal branch, (6) branch to otic ganglion which goes through ganglion to 7, 8, and 9, (7) to tensor tympani muscle, (8) to tensor veli palatini muscle, (9) to medial pterygoid muscle, (10) anterior division, (11) to masseter muscle, (12) to lateral pterygoid muscle, (13) to temporalis muscle, (14) buccal, (15) glossopharyngeal nerve, (16) lesser petrosal (preganglionic fibers), (17) postganglionic fibers joining 18, (18) auriculotemporal nerve, (19) facial nerve, (20) chorda tympani, (21) lingual, (22) submandibular ganglion, (23) submandibular gland, (24) sublingual gland, (25) tongue, (26) inferior alveolar, (27) mental, (28) mylohyoid. Connections of otic ganglion to middle meningeal plexus (sympathetic) and with chorda tympani were omitted.*

ANTERIOR DIVISION. The branches arising from the **anterior division** of the mandibular division of the trigeminal nerve are:

1. Buccal (Figs. 7-72 and 7-73). This branch courses laterally and appears in the infratemporal fossa between the two heads of the lateral pterygoid muscle. It sends a branch superiorly to the temporalis muscle and then passes inferiorly and slightly anteriorly superficial to the lower part of the lateral pterygoid muscle, then superficial to the medial pterygoid muscle to ramify on the superficial surface of the buccinator muscle. This nerve (beyond the point of emergence of the branch to the temporalis muscle) is a sensory nerve to the inside of the cheek and to the skin around the corner of the mouth; in its course it is just anterior to the temporalis muscle and just deep to the masseter muscle.

2. The **lateral pterygoid nerve** (Fig. 7-73) is very short and immediately enters the deep surface of that muscle. This is a motor nerve to the muscle, which also carries proprioceptive fibers from it.

3. The **masseteric nerve** (Fig. 7-71) passes laterally in a course superior to the lateral pterygoid muscle, passes through the mandibular notch posterior to the temporalis muscle, and supplies the masseter muscle by entering its deep surface. It also sends a sensory branch to the temporomandibular joint.

4. The **anterior and posterior deep temporal nerves** (Figs. 7-72 and 7-73). These nerves appear at a point superior to the lateral pterygoid muscle, course superiorly into the temporal fossa, and enter the deep surface of the temporalis muscle. These are motor nerves to this muscle, which also carry proprioceptive fibers.

POSTERIOR DIVISION. The branches arising from the **posterior division** of the mandibular division of the trigeminal nerve are:

1. **Auriculotemporal** (Fig. 7-73). This nerve arises by two roots surrounding the middle meningeal artery. They combine, and the nerve passes posteriorly in a plane deep to the lateral pterygoid muscle. It then proceeds between the neck of the mandible and the sphenomandibular ligament (see Fig. 7-75) to ascend posterior to the temporomandibular joint and deep to the parotid gland. The nerve continues superiorly, usually at a deeper level than the superficial temporal artery and slightly posterior to it, and ramifies over the side of the head in a distribution similar to that of the superficial temporal artery. This nerve has branches to (a) the superior part of the auricle, (b) the superior part of the external auditory meatus and the lateral surface of the tympanic membrane, (c) the temporomandibular joint, (d) the parotid gland, (e) the skin over the parotid region, and (f) the skin on the side of the head as previously mentioned. This nerve is mainly sensory to the areas

mentioned, but in addition it carries postganglionic parasympathetic fibers (belonging to the glossopharyngeal nerve) from the otic ganglion to the parotid gland.

2. **Lingual** (Figs. 7-72 and 7-73). This nerve courses inferiorly in a plane superficial to the medial pterygoid muscle and deep to the lateral pterygoid muscle; it then courses between the medial pterygoid muscle and the body of the mandible. It ultimately reaches the tongue and gums, giving general sensory nerve supply for pain, temperature, and touch to the anterior two-thirds of the tongue and the contiguous gums. The lingual nerve is joined by the small **chorda tympani** nerve, which consists of two types of fibers: (1) preganglionic parasympathetic fibers destined to innervate the sublingual and submandibular glands, and (2) special sensory fibers for taste from the anterior two-thirds of the tongue. The chorda tympani nerve is in actuality a branch of the facial nerve that has joined the lingual nerve as a pathway to its destination. The lingual nerve is a sensory nerve containing general somatic afferent fibers; the chorda tympani component of that nerve contains special afferent and general visceral efferent fibers.

3. **Inferior alveolar** (Figs. 7-72 and 7-73). This nerve courses inferiorly, at first in company with the lingual nerve, superficial to the medial pterygoid muscle and deep to the lateral pterygoid muscle. It enters the mandibular foramen after giving off its mylohyoid branch. This nerve is destined to give sensory nerve supply to the teeth and gums of the lower jaw. It terminates as the mental nerve, which emerges from the mental foramen and gives sensory nerve supply to the skin over the chin and the mucous membrane of the lower lip. The **mylohyoid nerve** arises near the mandibular foramen and courses on the medial surface of the mandible between it and the medial pterygoid muscle in the mylohyoid groove. This nerve is destined to give motor supply to the mylohyoid muscle and the anterior belly of the digastric (in submandibular region; see Fig. 7-76); it also contains proprioceptive fibers from these muscles.

The **otic ganglion** (Fig. 7-68) is a flattened, stellate ganglion closely associated with the medial surface of the mandibular division of the trigeminal nerve. It is immediately inferior to the foramen ovale and is just lateral to the cartilaginous portion of the auditory tube. The cells in the otic ganglion are postganglionic parasympathetic neurons. The preganglionic neurons start in the **nucleus salivatorius** of the brain stem and accompany the **glossopharyngeal nerve** through the jugular foramen of the skull. These fibers then enter the middle ear and aid in forming the **tympanic plexus.** They leave the plexus and, with a small branch from the facial, form the **lesser petrosal nerve** which enters the middle fossa of the cranial cavity. This nerve then passes through an unnamed foramen to join the **otic ganglion** in the infratemporal fossa. Here a synapse occurs and postganglionic fibers join the **auriculotemporal** nerve to reach the parotid gland. These are secretory fibers to the gland.

The otic ganglion has communications with other nerves although these connections are not involved with the cells therein. The medial pterygoid nerve and the nerves to the **tensor tympani** and **tensor veli palatini** muscles pass through the ganglion. (It should be noted that the ganglion is immediately lateral to the canal containing the tensor tympani muscle and to the tensor veli palatini muscle; these nerves, therefore, are very short.) The ganglion also has a sympathetic root from the middle meningeal plexus. In addition, a small filament communicates with the chorda tympani nerve in the infratemporal fossa (probably sensory).

In summary, the mandibular division of the trigeminal nerve is motor to the muscles of mastication plus the anterior belly of the digastric, mylohyoid, and to two tensors—tensor veli palatini and tensor tympani. It is sensory to the muscles just mentioned (proprioception), and to the anterior two-thirds of the tongue, the gums and teeth of the lower jaw, temporomandibular joint, and the skin of the lower jaw, cheeks, parotid region, and temporal region; these cutaneous areas are outlined in Figure 7-12. This entire division of the trigeminal nerve is summarized in Figure 7-74.

ARTERIES

The student should not memorize a list of the branches of the first two parts of the maxillary artery, but should think of the structures located in this area. First, the ear is in this vicinity and this recalls two branches, the tympanic and deep auricular. The jaw is an obvious structure and this brings to mind the inferior alveolar. The muscles of mastication, of which there are four, recall five more of the branches since the temporalis muscle usually has two branches. The buccal branch might have to be recalled as an individual, and the middle meningeal and the accessory meningeal might be forgotten.

The branches of the first two parts* of the maxillary artery are:

1. Deep auricular
2. Anterior tympanic
3. Middle meningeal
4. Accessory meningeal
5. Inferior alveolar
6. Posterior deep temporal
7. Pterygoid branches
8. Masseteric branch
9. Buccal
10. Anterior deep temporal

1. Deep auricular (Fig. 7-73). This small artery ascends in the substance of the parotid gland posterior to the temporomandibular joint. It pierces the cartilaginous or bony wall of the external auditory meatus and supplies the lining of the meatus and the lateral surface of the tympanic membrane. This artery also gives a branch to the temporomandibular joint.

2. Anterior tympanic (Fig. 7-73). This artery also courses superiorly and passes posterior to the temporomandibular joint. It enters the tympanic cavity and is distributed on the medial surface of the tympanic membrane. It anastomoses with the stylomastoid branch of the posterior auricular (to be described later), which also supplies the medial side of the tympanic membrane.

3. Middle meningeal (Fig. 7-73). This important artery varies in its origin depending on whether the main maxillary artery courses deep to the lateral pterygoid muscle or superficial to it. If the maxillary artery is superficial, the middle meningeal artery arises before the muscle is reached and therefore is seen arising at a point lateral to the lateral pterygoid muscle. It ascends, in this case, between the sphenomandibular ligament and the lateral pterygoid muscle and enters the foramen spinosum between the two roots of the auriculotemporal nerve. After entering the cranium, it gives rise to several rather unimportant branches and then divides into its **anterior and posterior branches.** The anterior branch crosses the greater wing of the sphenoid bone and reaches a deep groove, or even a canal in the sphenoid angle of the

---

*See page 623 for the third part.

parietal bone close to the pterion (Fig. 7-35). From this point it passes superiorly and very slightly posteriorly to ascend to the vertex of the head. The posterior branch curves posteriorly and takes a horizontal course along the squamous portion of the temporal bone and reaches the posterior part of the parietal bone. It should be recalled that these branches not only supply the dura mater, but are the main blood supply to the inner table of the cranial bones themselves. They frequently hemorrhage as a result of skull fracture; epidural or subdural hemorrhage results.

4. Accessory meningeal (Fig. 7-73).   This may either branch directly from the maxillary artery or be a branch of the middle meningeal. It enters the skull through the foramen ovale and supplies the semilunar ganglion of the trigeminal nerve as well as the dura surrounding it.

5. Inferior alveolar (Figs. 7-72 and 7-73).   This artery descends with the inferior alveolar nerve to the mandibular foramen on the medial side of the ramus of the mandible, which it enters. It courses in the body of the mandible giving many branches to the roots of the lower teeth. The terminal branch is the **mental artery,** which emerges with the nerve of the same name from the mental foramen and supplies the chin. The inferior alveolar artery gives rise to the **mylohyoid artery,** which branches from the artery just before it enters the mandibular foramen. The mylohyoid branch runs in the mylohyoid groove in company with the mylohyoid nerve and ramifies on the inferior surface of the mylohyoid muscle.

6 and 10. Deep temporal (Figs. 7-72 and 7-73).   The temporal branches are usually two in number although there may be more. The anterior and posterior deep temporal arteries ascend between the temporalis muscle and the pericranium. They supply blood to this muscle and give branches to the outer table of the bones in this region.

7. Pterygoid (Fig. 7-73).   The pterygoid arteries vary in number and supply the lateral and medial pterygoid muscles.

8. Masseteric (Figs. 7-71 and 7-73).   This artery passes laterally through the mandibular notch and sinks into the posterior aspect of the masseter muscle. If the maxillary artery courses deep to the lateral pterygoid muscle, the masseteric artery may arise in common with the posterior deep temporal.

9. Buccal (Fig. 7-73).   This artery courses between the heads of the lateral pterygoid muscle and, in company with the buccal nerve, courses inferiorly just anterior to the temporalis muscle, superficial to the lower head of the lateral pterygoid muscle and to the medial pterygoid muscle, and deep to the masseter muscle. It escapes from being deep to the masseter muscle and ramifies on the buccinator.

VEINS

The **veins** in the infratemporal fossa are very extensive and form the pterygoid plexus. This plexus has connections with the cavernous sinus on the inside of the cranium, with vessels on the inside of the orbit, with the pharyngeal plexus of veins, and with the facial vein by way of the deep facial vein. This plexus receives branches that correspond to the branches of the maxillary artery and the plexus finally emerges as the maxillary vein, which combines with the superficial temporal vein to form the retromandibular vein. The connections of this plexus with veins inside the cranial cavity is of considerable importance from the point of view of spread of infection.

TEMPOROMANDIBULAR JOINT (Fig. 7-75)

This joint is a diarthrodial articulation of the hinge or ginglymus variety. It really should be classified as a modified hinge joint since a gliding motion is allowed in addition to the usual movement found in such a joint. The **head** of the condyloid process articulates with the **mandibular fossa** posteriorly and the **articular tubercle** anteriorly. The temporomandibular joint contains a **disc,** which actually separates the head of the condyloid process from the temporal bone. This disc divides the joint into two portions, upper and lower, each of which is lined with a synovial membrane. The **capsular ligament** is by necessity fairly lax, but its lateral portion is strengthened to form the **temporomandibular ligament.** There are two **accessory ligaments** associated with the temporomandibular joint: the **stylomandibular ligament** attaches to the styloid process and to the posterior border

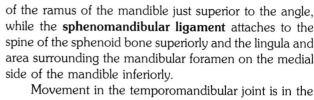

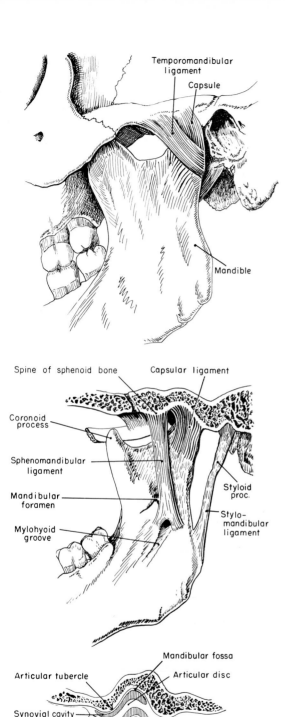

Temporomandibular ligament

Capsule

Mandible

Spine of sphenoid bone

Capsular ligament

Coronoid process

Sphenomandibular ligament

Mandibular foramen

Mylohyoid groove

Styloid proc.

Stylomandibular ligament

Articular tubercle

Mandibular fossa

Articular disc

Synovial cavity

Condyle

Mastoid process

Lat. pterygoid m.

of the ramus of the mandible just superior to the angle, while the **sphenomandibular ligament** attaches to the spine of the sphenoid bone superiorly and the lingula and area surrounding the mandibular foramen on the medial side of the mandible inferiorly.

Movement in the temporomandibular joint is in the manner of a hinge, with gliding in addition. When the lower jaw is at rest, the upper and lower teeth are not in contact. When they are brought into contact, the hingelike motion occurs in the lower part of the joint, between the condyloid process and disc. When the jaw is protruded or lowered beyond the resting point, the upper part of the joint is used as well. The head of the condyloid process and the disc both are pulled anteriorly toward the tubercle of the temporal bone. When retracted, the disc and head slip posteriorly into the fossa. When the mouth is opened widely and very quickly, there is a possibility of the joint becoming dislocated into a position anterior to the articular tubercle. In this case, pressure put on the lower teeth in an inferior and posterior direction usually will cause the jaw to return to its normal position. The person doing this should be careful that he or she is not bitten by the patient.

The muscles responsible for **closing the jaw** are the masseters, medial pterygoids, and temporalis muscles; these are quite powerful. Those responsible for opening the jaw are the lateral pterygoids—which pull capsules, discs, and condyloid processes anteriorly—and other muscles located in the neck between the hyoid bone and the mandible; these will be described later. **Protrusion** of the jaw is produced by the lateral pterygoids, while **retraction** is produced by the posterior part of the temporalis muscles and the deep part of the masseter muscles. Grinding is produced by the lateral pterygoids contracting in alternating fashion so that one side of the jaw is pulled anteriorly while the other side remains posteriorly.

The **vessels** supplying the joint are branches from the superficial temporal and masseteric arteries. The **nerves** are branches from the auriculotemporal and masseteric nerves.

Figure 7-75. The temporomandibular joint. The dotted lines indicate the forward movement of the condyloid process during opening of the mouth; top, lateral view; middle, medial view; bottom, sagittal section through joint.

SUBMANDIBULAR REGION

GENERAL DESCRIPTION

The **submandibular region,** as the name implies, is under the mandible. It is the area between the hyoid bone and the inferior edge of the mandible. If the skin and **platysma muscle** are removed (Fig. 7-76), the first muscle seen is the **digastric.** This is divided into two bellies, an anterior and a posterior, and the tendon between these two bellies loops through a pulley that attaches the muscle to the hyoid bone. The anterior belly is attached to the point of the chin and stretches inferiorly and slightly laterally to the hyoid bone, while the posterior belly stretches posteriorly and superiorly to attach to the

mastoid process of the temporal bone. These two bellies, with the inferior edge of the mandible, form the **digastric triangle.** The most superficial structure located in this triangle is the **submandibular gland,** which not only occupies the triangle but overflows superficial to both bellies of the digastric muscle. **The stylohyoid muscle** can be seen attached to the anterior and superior surface of the posterior belly of the digastric. This muscle arises from the styloid process and inserts on the hyoid bone at the point where the body and great horn of the hyoid bone join. The muscle is perforated by the tendon of the digastric muscle.

If the anterior belly of the digastric is reflected inferiorly and the superficial portion of the submandibular gland removed, the **mylohyoid** muscle is revealed (Fig. 7-77). This muscle fills in the floor of the mouth from the mylohyoid line of the mandible down to the hyoid bone.

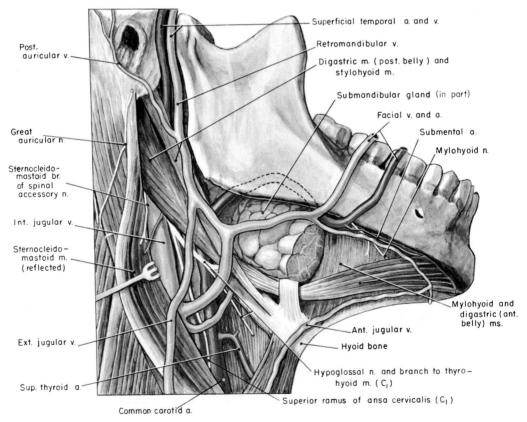

Figure 7-76. Submandibular region I. Skin, platysma muscle, lymph nodes, facial nerve, and part of the submandibular gland have been removed.

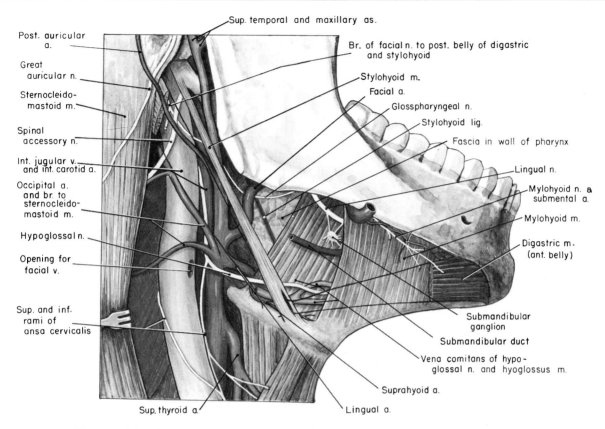

Post. auricular a.

Great auricular n.

Sternocleido-mastoid m.

Spinal accessory n.

Int. jugular v. and int. carotid a.

Occipital a. and br. to sternocleido-mastoid m.

Hypoglossal n.

Opening for facial v.

Sup. and inf. rami of ansa cervicalis

Sup. thyroid a.

Sup. temporal and maxillary as.

Br. of facial n. to post. belly of digastric and stylohyoid

Stylohyoid m.

Facial a.

Glosspharyngeal n.

Stylohyoid lig.

Fascia in wall of pharynx

Lingual n.

Mylohyoid n. a submental a.

Mylohyoid m.

Digastric m. (ant. belly)

Submandibular ganglion

Submandibular duct

Vena comitans of hypo-glossal n. and hyoglossus m.

Suprahyoid a.

Lingual a.

Figure 7-77. *Submandibular region II. The digastric muscle and submandibular gland have been removed.*

Besides lymph glands, the **mylohyoid nerve** and **blood vessels** can be seen on the superficial surface of this muscle, a position contrary to the usual rule.

If the mylohyoid muscle is removed, small muscles on either side of the midline are revealed, the **geniohyoid muscles** (Fig. 7-78). Posterior and lateral to the genio-hyoid muscle is one that is attached to the base of the tongue and to the hyoid bone. This, the **hyoglossus muscle,** has important relations. Besides the deep portion of the **submandibular gland** and the **submandibular duct,** the **lingual nerve, submandibular ganglion, hypoglossal nerve, suprahyoid artery,** and the **vein that accompanies the hypoglossal nerve** are all seen on the superficial surface of the hyoglossus muscle. From superior to inferior, these structures are the lingual nerve, submandibular ganglion, submandibular duct, hypoglos-

sal nerve and accompanying vein, and the suprahyoid artery (a duct, a ganglion, two nerves, and two vessels).

If the geniohyoid muscle is removed, the **genioglossus muscle** arising from the genial tubercle of the mandible and inserting into the tongue can be seen. Just lateral to the genioglossus muscle, between it and the mandible, is the **sublingual gland.**

If the hyoglossus muscle is removed, the **lingual artery** can be seen coursing deep to it. The **dorsal lingual** branches arise at a position deep to the hyoglossus muscle; they are distributed to the dorsum of the tongue. The main artery gives off two terminal branches, the **sublingual artery** to the sublingual gland and the **deep artery** to the tongue, which ramifies on the genioglossus muscle (Fig. 7-80).

Therefore, if one should approach the region just

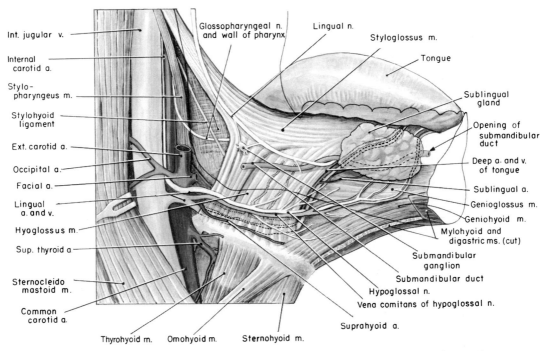

Figure 7-78. *Submandibular region III. The relations of the hyoglossus muscle are shown. (Note the outline of the mandible.)*

inferior to the chin and superior to the hyoid bone near the midline, one would encounter structures in this order: (1) skin, (2) superficial fascia with its contained platysma muscle, (3) deep fascia, (4) anterior belly of the digastric, (5) mylohyoid muscle, (6) geniohyoid muscle, and (7) genioglossus muscle. More laterally one would encounter (1) skin, (2) superficial fascia with the platysma muscle, (3) deep fascia, (4) submandibular gland and the posterior belly of the digastric muscle, (5) the hyoglossus muscle with all its important relations, and (6) the genioglossus muscle. Many of these structures produce markings on the inside surface of the mandible (Fig. 7-5).

GLANDS

SUBMANDIBULAR. The **submandibular gland** (Figs. 7-76 and 7-79) is a large one lying partly deep to the mandible and the medial pterygoid muscle and partly in a superficial position in the digastric triangle. It tends to overlap the digastric muscle. It is enclosed in two layers of fascia, one of which stretches from the hyoid bone to the edge of the mandible and the other from the hyoid bone to the mylohyoid line on the deep surface of the mandible. The portion of the submandibular gland in contact with the mandible is devoid of fascia. The posterior end of the gland is separated from the parotid by the stylomandibular ligament.

The **relations** of the submandibular gland to surrounding structures may be discerned by a study of Figure 7-79B.

The **deep part of the submandibular gland** is a flat and elongated portion that accompanies the submandibular duct. This portion lies deep to the mylohyoid muscle and superficial to the hyoglossus and then the genioglossus muscles. The submandibular duct is on its medial surface as well as the lingual nerve.

The **submandibular duct** courses anteriorly in company with the deep portion of the gland and is on the surface of the hyoglossus and genioglossus muscles; it lies deep to both the superficial and deep parts of the submandibular gland, to the sublingual gland, and to the

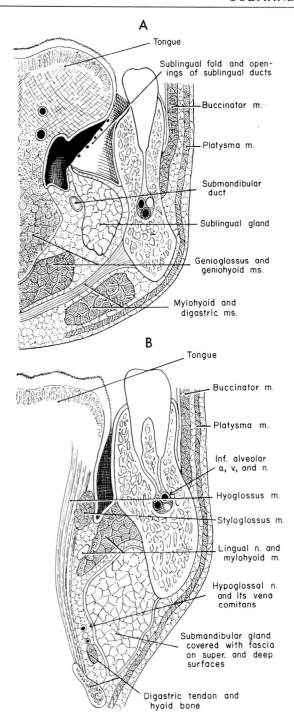

A

- Tongue
- Sublingual fold and openings of sublingual ducts
- Buccinator m.
- Platysma m.
- Submandibular duct
- Sublingual gland
- Genioglossus and geniohyoid ms.
- Mylohyoid and digastric ms.

B

- Tongue
- Buccinator m.
- Platysma m.
- Inf. alveolar a, v, and n.
- Hyoglossus m.
- Styloglossus m.
- Lingual n. and mylohyoid m.
- Hypoglossal n. and its vena comitans
- Submandibular gland covered with fascia on super. and deep surfaces
- Digastric tendon and hyoid bone

Figure 7-79. *Frontal sections through (A) sublingual and (B) submandibular glands showing relations. Note the fascial layers around the submandibular gland. (Modified after Jamieson.)*

mylohyoid muscle. The lingual nerve crosses superficial to the duct and then curves medially passing inferior to it in so doing. The duct opens into the mouth under the anterior end of the tongue at its frenulum.

The **arteries** to the submandibular gland are branches of the submental branch of the facial artery. The secretory **nerve supply** comes from both parasympathetic and sympathetic sources. The parasympathetic supply is from the facial nerve via the chorda tympani nerve (thin watery saliva); the sympathetic fibers reach the gland from the superior cervical ganglion via the arteries (thick mucous saliva).

Lymphatic drainage is to the nodes located on the mylohyoid muscle, the submandibular nodes. A portion of the gland may also drain into the superficial cervical nodes.

SUBLINGUAL.   The **sublingual gland** (Figs. 7-78 and 7-79) is an almond-shaped salivary gland located just inferior to the sublingual fold in the mouth (see Fig. 7-128 on page 627). It has numerous ducts opening on this sublingual fold. It is inferior to the mucous membrane of the mouth, superior to the mylohyoid muscle, medial to the sublingual fossa on the inside of the mandible, and lateral to the lingual nerve, submandibular duct, and the genioglossus muscle (Fig. 7-79A).

Its **nerve supply** is similar to that just described for the submandibular gland, the secretory parasympathetic fibers being derived from the facial nerve via the chorda tympani nerve and the submandibular ganglion, and its sympathetic fibers from the superior cervical ganglion via the lingual and submental arteries. Its sensory nerve supply is the lingual branch of the trigeminal.

The **arteries** to the sublingual gland are those just mentioned, the submental branch of the facial and sublingual branch of the lingual. It has corresponding veins.

Its **lymphatic drainage** is to the submandibular nodes.

MUSCLES

The muscles found in the submandibular region are:

1. Digastric
2. Mylohyoid
3. Geniohyoid
4. Genioglossus

5. Hyoglossus
6. Stylohyoid

1. Digastric (Fig. 7-76). The digastric muscle consists of anterior and posterior bellies with an intermediate tendon. This tendon is considered to be the **insertion** of both parts, and it attaches to the hyoid bone; the anterior belly **arises** from the inferior surface of the mandible close to the midline, and the posterior belly from the mastoid process of the temporal bone. The **main action** of the anterior belly is to raise the hyoid bone or lower the mandible, while that of the posterior belly is to raise the hyoid bone only. This act is done primarily in swallowing. The two parts of this muscle are derived from separate branchial arches and therefore receive different **nerve supplies;** since the anterior belly is derived from the first branchial arch, it is innervated by the trigeminal nerve, the mylohyoid branch of the inferior alveolar branch of the mandibular division being its actual nerve supply. The posterior belly, being derived from the second branchial arch, receives a branch from the facial nerve that arises from the facial just after it emerges from the stylomastoid foramen. It should be noted that the tendon of the digastric muscle perforates the stylohyoid muscle.

2. Mylohyoid (Fig. 7-77). The mylohyoid muscles form a sling across the floor of the mouth. Each is deep to the anterior belly of the digastric and superficial to the geniohyoid muscle. They **arise** from the mylohyoid line of the mandible and **insert** onto the body of the hyoid bone and into a midline raphe that extends from the hyoid bone to the end of the mandible. The **main action** of these muscles is to raise the hyoid bone and tongue in the process of swallowing. The **nerve supply** is the trigeminal nerve through the mylohyoid branch of the inferior alveolar.

3. Geniohyoid (Fig. 7-78). This muscle is just deep to the mylohyoid and superficial to the genioglossus muscle. Its **origin** is from the genial tubercle of the mandible, its **insertion** is onto the body of the hyoid bone, and its **main action** is to raise the hyoid bone in swallowing. Its **nerve supply** is from a branch of the hypoglossal nerve, although these fibers are actually derived from the first cervical segment of the spinal cord; these fibers have simply joined the hypoglossal nerve in order to reach their destination.

4. Genioglossus (Figs. 7-78 and 7-80). This fan-shaped muscle has an **origin** from the genial tubercle

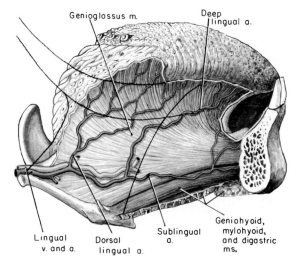

Figure 7-80. The genioglossus muscle with the styloglossus and hyoglossus outlined. Note that the lingual artery courses on this muscle in a plane deep to the hyoglossus.

and an **insertion** onto the entire length of the tongue from its posterior end to its tip. Its **main action** is to protrude the tongue and to pull it inferiorly. Its nerve **supply** is the hypoglossal nerve, the nerve that innervates all muscles of the tongue. The clinical test for intactness of the hypoglossal nerve is protrusion of the tongue. Contrary to the usual finding in nerve injuries, the tongue, when protruded, turns toward the side of the injured nerve. This is explained by the fact that the denervated side of the tongue drags behind, causing the tip of the tongue to turn.

5. Hyoglossus (Fig. 7-78). This muscle lies inferior to the tongue between its base and the greater horn of the hyoid bone. Its **origin** is from the body and greater horn of the hyoid bone and its **insertion** is into the posterior half of the tongue. The **main action** of the hyoglossus muscle is to pull the tongue inferiorly, or if the tongue is raised, to raise the hyoid bone. The **nerve supply** is the hypoglossal nerve. Important structures course on its superficial surface.

6. Stylohyoid (Fig. 7-77). This muscle is closely applied to the posterior belly of the digastric muscle. It **arises** near the base of the styloid process and **inserts** on the hyoid bone near the junction of the body and the great horn. It is perforated by the intermediate tendon of the digastric muscle. Its **main action** is to draw the hyoid bone superiorly and posteriorly during the act of swallowing. The **nerve supply** is by a branch of the facial nerve.

––––––––––

If the hyoid bone is held in place by the infrahyoid muscles, many of these suprahyoid muscles will act in reverse and aid gravity in opening the mouth.

NERVES

The nerves in the submandibular region are:

1. Mylohyoid branch of the inferior alveolar
2. Lingual branch of the mandibular
3. Hypoglossal
4. Digastric branch of facial
5. Stylohyoid branch of facial
6. Parasympathetic to submandibular and sublingual glands
7. Sympathetic to glands

1. The **mylohyoid branch** of the inferior alveolar nerve (Figs. 7-76 and 7-77) arises close to the mandibular foramen and courses along the mylohyoid groove on the mandible to arrive on the inferior (superficial) surface of the mylohyoid muscle. It immediately sends branches to this muscle and to the anterior belly of the digastric. This nerve contains motor fibers to these muscles and proprioceptive afferent fibers.

2. The **lingual branch** of the mandibular division of the trigeminal nerve (Figs. 7-73 and 7-78) is a large nerve that is a branch of the posterior division of the mandibular nerve. It courses inferiorly in the infratemporal fossa initially between the lateral and medial pterygoid muscles, and then between the mandible and the medial pterygoid muscle. It continues under cover of the mucous membrane of the mouth and in relation to the pharynx and the styloglossus muscle. It then courses between the mandible and the hyoglossus muscle, continues anteriorly deep to the mylohyoid muscle, and then terminates between the sublingual gland and the genioglossus muscle. It crosses the submandibular duct in its course. It supplies the anterior two-thirds of the tongue with general sensation and also gives branches to the gums of the lower jaw. The lingual nerve also contains

fibers of the chorda tympani nerve, which will be described momentarily.

The mandibular division of the trigeminal nerve is summarized in Figure 7-74.

3. The **hypoglossal** nerve appears in this region on the superficial surface of the hyoglossus muscle (Figs. 7-77 and 7-78). It hooks around the occipital artery near the inferior border of the digastric muscle and turns anteriorly, crossing the internal and external carotid arteries to gain the deep surface of the digastric and mylohyoid muscles. It then proceeds toward the tongue between the mylohyoid and hyoglossus muscles. The hypoglossal nerve is the motor nerve to all the extrinsic and intrinsic muscles of the tongue. The branches of this nerve to the geniohyoid and other muscles in the neck, which will be studied later, are actually branches from the first cervical segment of the spinal cord. These branches have merely borrowed the hypoglossal as a pathway to their destination and should not be considered as part of the hypoglossal nerve. The geniohyoid muscle, therefore, obtains its nerve supply from C1 via the hypoglossal nerve.

4 and 5. The **digastric and stylohyoid branches** of the facial nerve are short nerves that enter the posterior aspects of the posterior belly of the digastric and the stylohyoid muscle.

6. The **parasympathetic** supply to submandibular and sublingual glands is as follows. Preganglionic cell bodies are located in the salivatory nucleus in the brain stem. These preganglionic fibers continue in the facial nerve, branch off with the chorda tympani nerve, and proceed through the middle ear to the infratemporal fossa to join the lingual nerve. These fibers then branch from the lingual nerve and synapse with postganglionic parasympathetic fibers in the submandibular ganglion, which is located on the superficial surface of the hyoglossus muscle. Postganglionic fibers proceed from this ganglion to the submandibular and sublingual glands. The chorda tympani nerve also contains taste fibers for the anterior two-thirds of the tongue. These fibers follow the chorda tympani proximally in the direction reverse to that just described and the cell bodies of these sensory fibers are located in the geniculate ganglion.

7. The **sympathetic** supply to these glands arrives via the arteries. Preganglionic sympathetic fibers start in the intermediolateral cell column of the upper thoracic region of the cord and proceed to the superior cervical ganglion. After synapsing here, postganglionic fibers proceed on the external carotid artery and follow the branches of the facial and lingual arteries to reach this area.

ARTERIES

The arteries found in the submandibular region are:

1. Submental branch of facial
2. Mylohyoid branch of inferior alveolar
3. Suprahyoid branch of lingual
4. Lingual

1. The **submental branch** of the facial artery (Fig. 7-76) emerges from this artery just as it courses inferior to the edge of the mandible. This small artery proceeds anteriorly in a plane superficial to the mylohyoid muscle and deep to the anterior belly of the digastric.

2. The **mylohyoid branch** of the inferior alveolar artery (Fig. 7-77) proceeds to approximately the same place superficial to the mylohyoid muscle. This small artery branches from the inferior alveolar just as the latter enters the mandibular foramen. The mylohyoid artery follows the mylohyoid nerve along the mylohyoid groove and terminates on the mylohyoid muscle.

3. The **suprahyoid branch** of the lingual (Fig. 7-78) arises just before the latter artery courses deep to the hyoglossus muscle. It proceeds along the superficial surface of the hyoglossus muscle just superior to the hyoid bone. Its existence is variable.

4. The **lingual artery** (Figs. 7-78 and 7-80) courses deep to the hyoglossus muscle, between it and the musculature of the pharynx and genioglossus muscles. While deep to the hyoglossus muscle the lingual gives off the **dorsal lingual** arteries, which ascend between the hyoglossus and genioglossus muscles to supply the dorsum of the tongue and the palatine tonsil. The lingual artery continues anteriorly, giving off the **sublingual artery** to the sublingual gland, and then proceeding to the tip of the tongue as the **deep lingual artery.**

VEINS

The **veins** of this area are similar to the arteries except that there is an additional set of veins accompanying the hypoglossal nerve.

LYMPH DRAINAGE

The lymph drainage in this area is to the submandibular nodes. It should be noted that lymph drainage from the tongue may cross the midline.

POSTERIOR TRIANGLE OF NECK

GENERAL DESCRIPTION

The platysma muscle partially covers the posterior triangle as shown in Figures 7-12 and 7-15. If this muscle is reflected, an area in the neck is seen bounded anteriorly by the sternocleidomastoid muscle, posteriorly by the trapezius muscle, and inferiorly by the clavicle; this area is the **posterior triangle of the neck** (Fig. 7-81). The **external jugular vein** is one of the most superficial structures in this triangle, coursing from superior to inferior across it. The triangle is filled in with a double-layered deep fascia, which is a portion of the fascia enveloping the neck. This fascia starts posteriorly from the ligamentum nuchae, surrounds the trapezius muscle, fuses together across the triangle, and separates again when the posterior border of the sternocleidomastoid muscle is reached. These fascial layers then surround this muscle, come together anteriorly, and meet the fascia from the opposite side in the midline. This is known as the enveloping layer of deep fascia.

Branches of the cervical plexus destined to become cutaneous nerves penetrate this fascia at locations shown in Figure 7-81 in order to reach their destination. All these branches are portions of the cervical plexus, being made up of ventral rami, and appear in the triangle just posterior to the sternocleidomastoid muscle. The **great auricular nerve** turns superiorly and crosses the sternocleidomastoid muscle to ramify around the parotid and external ear. The **lesser occipital** follows the posterior border of the sternocleidomastoid muscle, to ramify on the scalp posterior to the ear, sending some branches to the auricle itself. The **transverse colli nerve** courses anteriorly around the sternocleidomastoid muscle to ramify

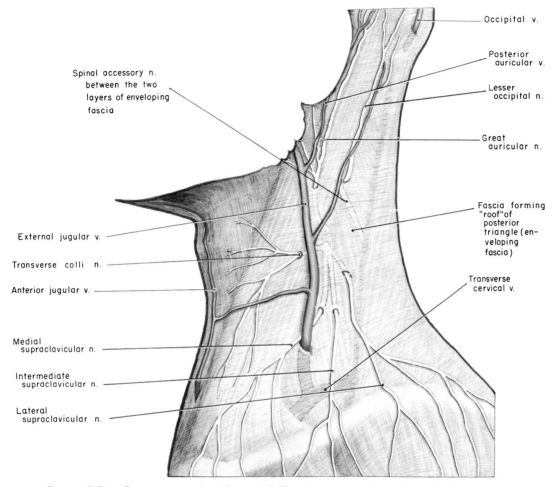

Occipital v.

Posterior
auricular v.

Lesser
occipital n.

Great
auricular n.

Spinal accessory n.
between the two
layers of enveloping
fascia

Fascia forming
"roof"of
posterior
triangle (en-
veloping
fascia)

External jugular v.

Transverse colli n.

Anterior jugular v.

Transverse
cervical v.

Medial
supraclavicular n.

Intermediate
supraclavicular n.

Lateral
supraclavicular n.

Figure 7-81. Posterior triangle of the neck I. The platysma muscle and superficial fascia have been removed.

on the anterior surface of the neck as far as the midline. The **medial, intermediate, and lateral supraclavicular nerves** are cutaneous to the skin on the lower part of the neck, on the shoulder, and to an area extending a short distance inferior to the clavicle.

The **spinal accessory nerve** innervates the sterno-cleidomastoid muscle and the trapezius with aid from branches of C3 and C4 of the cervical plexus. This important nerve pierces the sternocleidomastoid muscle and crosses the posterior triangle of the neck in an inferior and posterior direction to reach the deep surface of the trapezius. It is between the two layers of fascia just de-

scribed and is therefore in a very superficial and vulnerable position in the triangle.

If the enveloping layer of fascia in this posterior triangle is removed, the entire course of the cutaneous nerves just described is revealed (Fig. 7-82). However, the muscles forming the so-called floor of this triangle are not revealed because of a covering of deep fascia. If this fascia is removed (Fig. 7-83), the muscles seen forming the floor of the posterior triangle from superior to inferior are the **splenius capitis, levator scapulae, scalenus posterior** (small portion only), **scalenus medius,** and **scalenus anterior.**

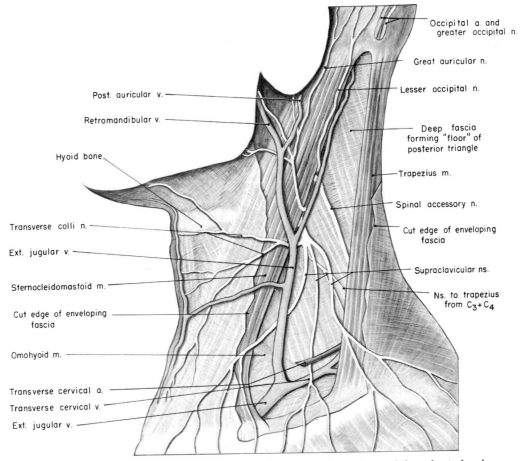

Figure 7-82. *Posterior triangle of the neck II. The enveloping layer of deep fascia has been removed without removing the spinal accessory nerve.*

The inferior belly of the **omohyoid muscle** courses from lateral to medial across the inferior end of the triangle, disappearing deep to the sternocleidomastoid muscle. An artery, the **transverse cervical,** crosses the triangle transversely at a level just superior to the omohyoid muscle, and another, the **suprascapular,** crosses the triangle at a level close to the clavicle.

The **phrenic nerve** is found on the anterior surface of the scalenus anterior muscle. A portion of the **brachial plexus** is seen between the scalenus anterior and scalenus medius muscles, and the **dorsal scapular nerve** to the rhomboids can be seen emerging from the middle of the muscle belly of the scalenus medius muscle.

Therefore, from superficial to deep the layers covering the posterior triangle of the neck are: (1) skin, (2) the superficial fascia with its contained platysma muscle and external jugular vein, (3) the enveloping deep fascia penetrated by the cutaneous branches of the cervical plexus and containing the spinal accessory nerve, (4) deep fascia, and (5) the muscles forming the floor of the triangle. The student should be able to visualize the phrenic and dorsal scapular nerves, brachial plexus, transverse cervical and suprascapular vessels, and omohyoid muscle in their proper positions in this triangle.

MUSCLES

The muscles found in the posterior triangle of the neck are:

1. Splenius capitis
2. Levator scapulae

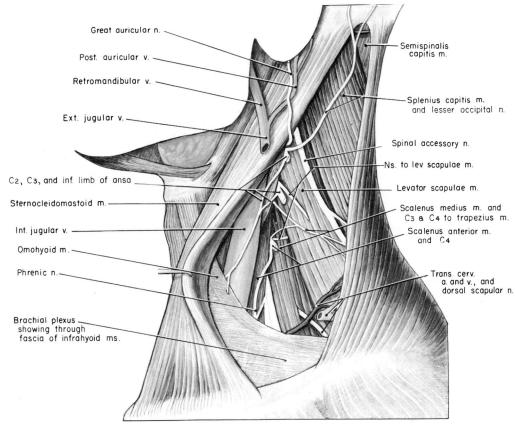

Figure 7-83. *Posterior triangle of the neck III. All deep fascia has been removed.*

3. Omohyoid
4. Scalenus anterior
5. Scalenus medius
6. Scalenus posterior

1 and 2. The **splenius capitis** is described on page 67, and the **levator scapulae** is described on page 66.

3. Omohyoid muscle (Fig. 7-83). The inferior belly of the omohyoid muscle is seen crossing the inferior end of the posterior triangle in an anterior and superior direction. It **arises** from the superior margin of the scapula. It is enclosed in its own layer of fascia, which is deep to the enveloping layer of cervical fascia and superficial to the prevertebral fascia, the fascia that extends across the neck just anterior to the vertebrae (see Fig. 7-86). Its tendon is held to the clavicle and first rib by this special fascia. This muscle leaves the triangle by disappearing deep to the sternocleidomastoid muscle. It will be

described in more detail when the other infrahyoid muscles are studied (page 571).

4. Scalenus anterior* (Figs. 7-83 and 7-84). This muscle is deep to the sternocleidomastoid muscle. It **arises** from the third to the sixth cervical transverse processes and **inserts** on the scalene tubercle and upper surface of the first rib. The insertion of this muscle separates the subclavian artery, which courses deep to the muscle, from the subclavian vein, which courses superficial to it. The phrenic nerve courses on its anterior surface, a location easily approached surgically. Its **main action** is to elevate the first rib in inspiration, but it can work in a reverse direction bending the neck to the same side. Its **nerve supply** is by branches of ventral rami from C4 to C7.

5. Scalenus medius* (Figs. 7-83 and 7-84). This

---

*The scalene muscles are pictured very clearly in Figure 7-94.

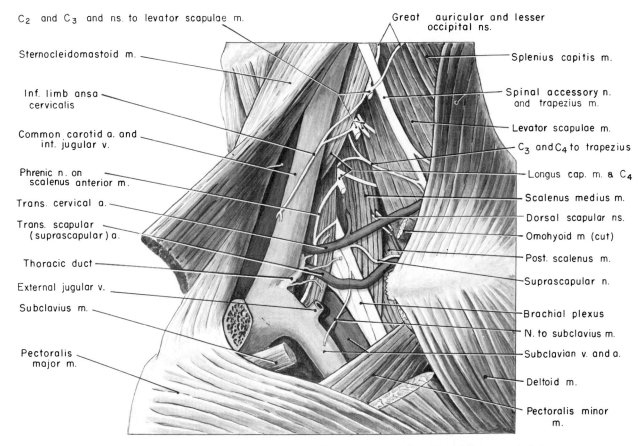

C₂ and C₃ and ns. to levator scapulae m.

Great auricular and lesser occipital ns.

Sternocleidomastoid m.

Splenius capitis m.

Inf. limb ansa cervicalis

Spinal accessory n. and trapezius m.

Common carotid a. and int. jugular v.

Levator scapulae m.

C₃ and C₄ to trapezius

Phrenic n. on scalenus anterior m.

Longus cap. m. & C₄

Trans. cervical a.

Scalenus medius m.

Trans. scapular (suprascapular) a.

Dorsal scapular ns.

Omohyoid m (cut)

Thoracic duct

Post. scalenus m.

External jugular v.

Suprascapular n.

Subclavius m.

Brachial plexus

N. to subclavius m.

Pectoralis major m.

Subclavian v. and a.

Deltoid m.

Pectoralis minor m.

Figure 7-84. *Posterior triangle of the neck IV. Removal of the clavicle has revealed the subclavian vessels and the brachial plexus entering the axilla. These are the vital structures protected by this bone.*

muscle, forming part of the floor of the posterior triangle, **arises** from all seven cervical transverse processes. It inserts onto the superior surface of the first rib at a point between the groove made by the subclavian artery and the tubercle of the rib. Its **main action** is to elevate the first rib in inspiration and, if working in the reverse direction, to bend the neck to the same side. The **nerve supply** is from the ventral rami of C3 to C7.

6. Scalenus posterior (Fig. 7-94). The scalenus posterior muscle is found also in the floor of the posterior triangle, but slightly posterior to the scalenus medius and often blended with it. It **arises** from the fourth, fifth, and sixth cervical transverse processes and **inserts** onto the superior border of the second rib. Its **main action** is to elevate the second rib in inspiration and, if working in

reverse, to bend the neck to the same side. Its **nerve supply** is from the ventral rami of C5, C6, and C7.

The relations of the scalene muscles to such vital structures as the phrenic nerve, brachial plexus, and the subclavian artery and vein make them particularly important. Note Figure 7-91.

NERVES

The nerves found in the posterior triangle of the neck are:

1. Spinal accessory
2. Dorsal scapular
3. Branches of cervical plexus
4. Roots and trunks of brachial plexus

1. Spinal accessory nerve (Figs. 7-82 and 7-83). The spinal accessory nerve, or the eleventh cranial nerve, pierces the sternocleidomastoid muscle fairly high up in the neck, traverses the posterior triangle enclosed between the two layers of enveloping fascia, and disappears deep to the trapezius muscle. It ramifies throughout the entire length of this muscle. The location of this nerve between the two layers of enveloping fascia is explained by the fact that in lower animals the sternocleidomastoid and trapezius muscles were continuous across the area called the posterior triangle in humans. When the muscle tissue disappeared, the nerve remained between the fascial layers. This superficial position of the nerve makes it liable to injury. The spinal accessory nerve innervates the sternocleidomastoid and trapezius muscles, but not alone—it is aided by branches of ventral rami of the third and fourth cervical nerves.

The double nerve supply to these muscles is of interest in that it indicates their possible dual origin. The fact that they are being innervated in a manner typical of somatic musculature (from the cervical nerves) and branchial musculature (spinal accessory nerve) would seem to indicate that this dual origin is indeed correct. However, if one cuts the spinal accessory nerve, the patient is unable to abduct the shoulder joint to a higher plane than horizontal. Therefore, it is thought that the cervical nerves, C3 and C4, may be sensory in character while the true motor nerve to the sternocleidomastoid and trapezius muscles is actually the spinal accessory nerve.[*]

2. Dorsal scapular (Fig. 7-83). This nerve arises from the ventral ramus of C5. It then pierces the scalenus medius muscle, descends deep to the levator scapulae muscle, and ends on the deep surface of the rhomboids. This nerve innervates all three of these muscles. The dorsal scapular nerve is a branch of the brachial plexus and arises from the root of the fifth cervical nerve.

[*]W. L. Straus and A. B. Howell, 1936, The spinal accessory nerve and its musculature, *Quart. Rev. Biol.* 11: 387. A survey of the puzzling double nerve supply to the trapezius and sternocleidomastoid muscles. The authors feel that the cervical nerves are merely branchial nerves that have shifted to follow a spinal pathway.

K. N. Corbin and F. Harrison, 1939, The sensory innervation of the spinal accessory and tongue musculature in the rhesus monkey, *Brain* 62: 191. Evidence is presented indicating that proprioception from the trapezius, the sternocleidomastoid, and the tongue musculature is carried via the cervical nerves, leaving the spinal accessory and hypoglossal nerves to these muscles as the motor supply, at least in the monkey.

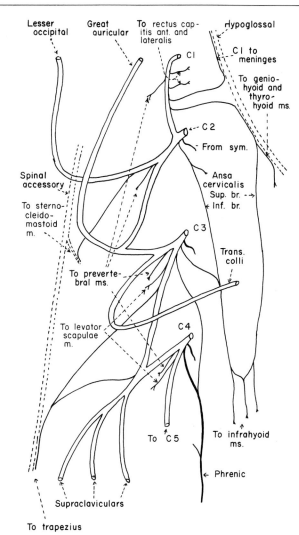

Figure 7-85. *Diagram of the cervical plexus—ventral rami of C1 to C4. The cutaneous nerves are clear; the communicating and muscular nerves are dark.*

3. Branches of cervical plexus (Fig. 7-85). The branches of the cervical plexus found in the posterior triangle are the **lesser occipital, great auricular, transverse colli,** and **supraclavicular nerves,** all of which are cutaneous. The nerves coursing superiorly are from C2 and C3, those coursing inferiorly from C3 and C4. The **phrenic nerve** arises chiefly from the fourth cervical segment although frequently contributions are made from C3 and C5. It courses inferiorly on the anterior surface of the scalenus anterior muscle deep to its fascial covering.

It, therefore, courses between the scalenus anterior posteriorly and the sternocleidomastoid anteriorly. The transverse cervical and suprascapular arteries are superficial to the nerve, as is the omohyoid muscle. The phrenic nerves, after leaving the scalenus anterior, descend into the thoracic cavity anterior to the pleura, which extends into the neck region. They continue through the thoracic cavity to innervate the diaphragm. Sensory components in this nerve also are quite important. Not only do they contain sensory fibers from the pericardium, but pain originating in the diaphragmatic area can be referred to the shoulder because of the close association of the phrenic nerves with the supraclaviculars. **Accessory phrenic nerves** occur occasionally. These usually join the regular phrenic at the inlet to the thoracic cavity.

4. Brachial plexus (Fig. 7-84). The only parts of the brachial plexus concerned in the study of the posterior triangle are the roots and trunks, and at this time the student should simply note that the plexus is found between the scalenus anterior and medius muscles. The entire plexus is diagramed in Fig. 8-40 on page 707.

ARTERIES AND VEINS

The main arteries found in the posterior triangle are both branches of the thyrocervical trunk, which arises from the subclavian artery.* These branches are:

1. Transverse cervical
2. Suprascapular

1. Transverse cervical (Figs. 7-83 and 7-84). The transverse cervical artery courses laterally and slightly posteriorly in a plane anterior to the brachial plexus and anterior to the scalene muscles. It reaches the levator scapulae muscle and divides into its superficial and deep branches. The superficial branch courses superficial to the levator scapulae muscle and ramifies on the deep surface of the trapezius. The deep branch courses inferiorly on the deep surface of the levator scapulae and rhomboid muscles. Occasionally the deep branch arises directly from the third part of the subclavian artery, in which case it is called the descending scapular artery, and the transverse cervical is called the superficial cervical artery.

2. Suprascapular (Fig. 7-84). This artery courses

*See Figure 7-97 on page 582 for the origin of the above arteries.

laterally and slightly posteriorly at a more inferior level than that of the transverse cervical. It is anterior to the phrenic nerve and the scalenus anterior muscle, and also to the subclavian artery as well as the brachial plexus. It then courses to the superior margin of the scapula and crosses superficial to the suprascapular ligament and enters the supraspinous fossa. After ramifying in this fossa, it descends through the great scapular notch to end in the infraspinous fossa.

The **veins** take similar names and pathways and empty into the internal jugular vein.

LYMPH NODES

The inferior belly of the omohyoid muscle divides the deep cervical nodes into lower deep cervicals inferior to this muscle and upper deep cervicals superior to the muscle. As we will see later, these nodes are located along the internal jugular vein and therefore just lateral to the centrally located visceral tube. They not only drain many of the posteriorly located parts of this tube, but secondarily drain the superficially located nodes as well.

CERVICAL FASCIA

The deep fascia in the neck can be divided into the enveloping layer, pretracheal fascia, prevertebral fascia, the carotid sheath, and several additional unnamed layers (Figs. 7-86 and 7-87).

The **enveloping fascia** encloses all the neck structures except the platysma muscle, which is superficial to it. If one starts posteriorly at the ligamentum nuchae, and follows this fascia around the neck, one sees that it at first encloses the trapezius muscle; at the anterior edge of the trapezius muscle the two fascial layers fuse and bridge across the posterior triangle; when they reach the posterior border of the sternocleidomastoid muscle, the two layers separate to enclose this muscle and then proceed to the midline anteriorly to meet the same fascia from the other side of the body.

Inferiorly, this fascia is distributed as follows. Posteriorly, near the midline, the fascia proceeds inferiorly on

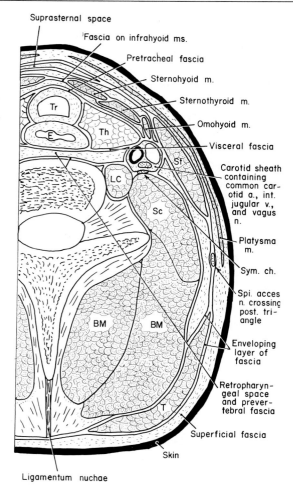

Figure 7-86. *Diagram of a transverse section of the neck, showing the distribution of the cervical fascia (spaces exaggerated): (Tr) trachea, (E) esophagus, (Th) thyroid gland, (St) sternocleidomastoid muscle, (LC) longus capitis muscle, (Sc) scalene muscles, (BM) back muscles, (T) trapezius muscle.*

the back musculature; more laterally, the fascia tends to follow the insertions of the trapezius muscle, being attached to the spine and the acromion process of the scapula and the lateral portion of the clavicle; in the region of the inferior end of the posterior triangle it attaches to the clavicle itself, and since the sternocleidomastoid muscle attaches to the clavicle and the sternum, it attaches to these bones near the midline. Actually, just superior to the manubrium of the sternum the two fascial planes separate and produce the **suprasternal space.**

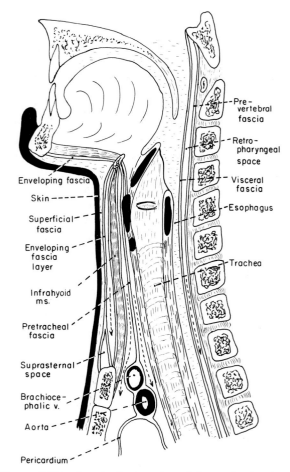

Figure 7-87. *Longitudinal section of the neck, showing the distribution of the cervical fascia (diagrammatic). Dotted arrows indicate direction of spread of infection (spaces exaggerated).*

Superiorly, this fascia is attached, from posterior to anterior, to the external occipital protuberance, the superior nuchal line, the mastoid process of the temporal bone, the angle of the mandible, and the inferior border of the mandible; this leaves a space between the angle of the mandible and the mastoid process of the temporal bone unaccounted for; here the fascia extends superiorly, surrounds the parotid gland, and then blends with the fascia covering the masseter and temporalis muscles.

The **prevertebral fascia** is a transverse extension of fascia from one enveloping layer to the other. This fascial plane is found anterior to the prevertebral musculature. This obviously divides the neck into an area posterior to the prevertebral fascia and another anterior to it. The muscles found posterior to the fascia are all surrounded by extensions from this fascia and the enveloping fascia. The prevertebral fascia is attached to the carotid sheath. Superiorly, the prevertebral fascia is attached to the base of the skull, while inferiorly, it continues into the thoracic cavity, still remaining anterior to the vertebral column. It should be noted also that a tunnel of fascia extends into the upper limb along the brachial plexus and the blood vessels which is really a continuation of this prevertebral fascia.

The **carotid sheath** is a tube of fascia extending superiorly and inferiorly, surrounding the internal jugular vein, the common carotid and internal carotid arteries, and the vagus nerve. It extends from the root of the neck all the way to the base of the skull. The external carotid artery and its branches, of course, have to penetrate this sheath to get to their destinations.

The **pretracheal fascia** is a layer of fascia that surrounds the visceral tube on its anterior surface. This starts posteriorly by blending with the carotid sheath and then spreads anterior to the larynx and trachea. A continuation of this fascia surrounds the posterior surface of the visceral tube and is found on the constrictor muscles of the pharynx superiorly and the esophagus inferiorly. The pretracheal fascia attaches to the hyoid bone superiorly, and an extension carries on from the hyoid bone, around the submandibular gland, to the inferior edge of the mandible; inferiorly, this fascia continues into the thoracic cavity and blends with the fascia on the great vessels. That fascia found on the posterior side of the pharynx and esophagus attaches to the base of the skull superiorly and continues into the thoracic cavity, remaining posterior to the esophagus at all times. The portion of this fascia which is found on the posterior walls of the visceral tube and which extends anteriorly and blends with the fascia on the buccinator muscle is called the **buccopharyngeal fascia.** In addition, one usually finds one or more connections between that found on the visceral tube and the prevertebral fascia.

One remaining fascial layer or layers need to be described—**those surrounding the infrahyoid muscles.** These layers of fascia are seen to be separate from both the enveloping layer and pretracheal fascia. They actually lie between these two layers of fascia, stretching from side to side, surrounding the infrahyoid muscles. They attach to the hyoid bone superiorly and to the sternum inferiorly, following the infrahyoid muscles to their points of attachment. (Figure 7-88 shows these infrahyoid muscles and the fascial layers just described.)

From the above description it is easily seen that these fascial planes are important in the spread of infection (Fig. 7-87). If an infection occurs between the enveloping layer of fascia and that surrounding the infrahyoid muscles, the infection will usually stop at the superior edge of the sternum and clavicle. If, however, the infection should occur between this fascia surrounding the infrahyoid muscles and the pretracheal fascia, it can spread into the thoracic cavity to a position anterior to the pericardium. If the infection is deep to the pretracheal fascia, it can follow the trachea and esophagus to the thoracic cavity and terminate in a position posterior to the pericardium. The space between the prevertebral fascia and the fascia on the posterior portion of the visceral tube is behind the pharynx and is therefore called the **retropharyngeal space.*** If infection occurs in this area, it can spread inferiorly into the thoracic cavity in a plane just anterior to the vertebral column and posterior to the esophagus.†

APPROACH TO THYROID GLAND

GENERAL DESCRIPTION

The sternocleidomastoid muscle, the inferior border of the mandible, and the midline form the boundaries of the **anterior triangle of the neck.** This area actually can be divided into four triangles as follows: a muscular triangle bounded by the superior belly of the omohyoid, the lower end of the sternocleidomastoid, and the midline; a

---

*Most authors describe a thin layer of fascia—**alar fascia**—that separates the retropharyngeal space into two parts, one just posterior to the pharynx and another just anterior to the vertebrae.

†F. A. Coller and L. Yglesias, 1937, The relation of the spread of infection to fascial planes in the neck and thorax, *Surgery* 1: 23. This account of the cervical fascia gives very good evidence for its importance in the spread of infection into the thoracic cavity.

M. Grodinsky and E. A. Holyoke, 1938, The fasciae and fascial spaces of the head, neck and adjacent regions, *Am. J. Anat.* 63: 367. A thorough and detailed description of the cervical fascia and spaces.

carotid triangle bounded by the posterior belly of the digastric, the superior belly of the omohyoid, and the superior end of the sternocleidomastoid; a digastric triangle, already described, bounded by the two bellies of the digastric muscle and the inferior border of the mandible; and a submental triangle bounded by the anterior belly of the digastric, the hyoid bone, and the midline. Actually, as far as the anterior triangle is concerned, most people think of the region inferior to the mandible, which contains both digastric and submental triangles, as the **submandibular region;** the muscular triangle as the **approach to the thyroid gland;** and the **carotid triangle** as an entity in itself.

As one approaches the thyroid gland, one can see that the skin has transverse creases, a fact utilized by the surgeon in making incisions. After the skin and platysma muscle have been removed, the enveloping layer of deep fascia is found just deep to two or three large veins (Fig. 7-88). These veins are the **anterior jugular** and the **communicating branch** between tne anterior jugular inferiorly and the facial vein superiorly. Both the anterior jugular and the communicating branch pierce the deep fascia and empty into the jugular arch, which empties, in turn, into the external jugular vein.

If the veins and the enveloping layer of deep fascia are removed, the **infrahyoid muscles** are seen, enclosed

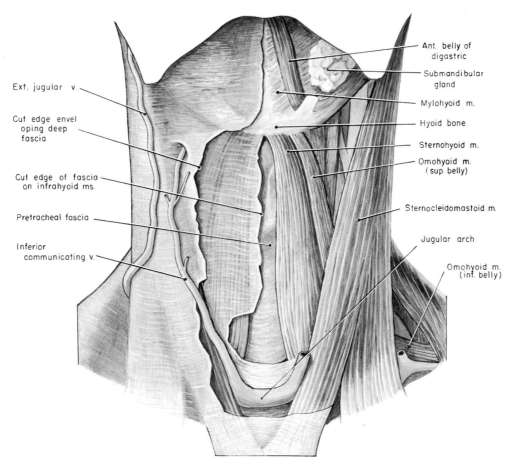

Figure 7-88. *Anterior triangle of the neck I. On the right side of the body, the enveloping layer of fascia has been cut to reveal the deep fascia on the infrahyoid muscles. On the left side, all muscular fascia has been removed, leaving the pretracheal (visceral) fascia visible in the midline.*

in their fascial layers, streaming inferiorly from the hyoid bone, to disappear deep to the two sternocleidomastoid muscles (Fig. 7-88). The muscle closest to the midline is the **sternohyoid,** and lateral to that is the **superior belly of the omohyoid** muscle; these muscles are contained in a fascial layer of their own. If they are cut and reflected, the **sternothyroid** muscle will appear, stretching inferiorly from the thyroid cartilage; this also is covered with a fascial layer of its own. If this muscle is reflected, it brings one down upon the **pretracheal fascia** covering the larynx and trachea (Fig. 7-88). If this is cut and reflected, the **inferior thyroid vein** can be seen in the midline, coursing inferiorly from the thyroid region to empty into the brachiocephalic vein. Deep and just lateral to this vein are the **cricoid cartilage** and the **cricothyroid muscles,** the **first tracheal ring,** the lobes of the **thyroid gland,** and the **trachea.** Superior to the vein, the large **thyroid cartilage** and the membrane from the hyoid bone to the

thyroid cartilage—the **thyrohyoid membrane**—also can be visualized.

If all three infrahyoid muscles are removed, the remaining portion of the thyroid gland is revealed (Fig. 7-89). On its anterior surface are the **superior thyroid arteries** and the **superior thyroid vein.** Contrary to the arterial pattern, there is a **middle thyroid vein,** which drains into the internal jugular. If the cricothyroid muscle is followed superiorly, it can be seen to be continuous with the **thyrohyoid muscle,** coursing between the thyroid cartilage and the hyoid bone.

Therefore, in the midline, from superior to inferior, are the hyoid bone, thyrohyoid membrane, thyroid cartilage, cricothyroid membrane, cricoid cartilage, cricotracheal ligament, and cartilages of the trachea. The thyrohyoid muscle is located on either side of the thyroid cartilage, and the cricothyroid muscles stretch between the thyroid and cricoid cartilages. Lateral and inferior to

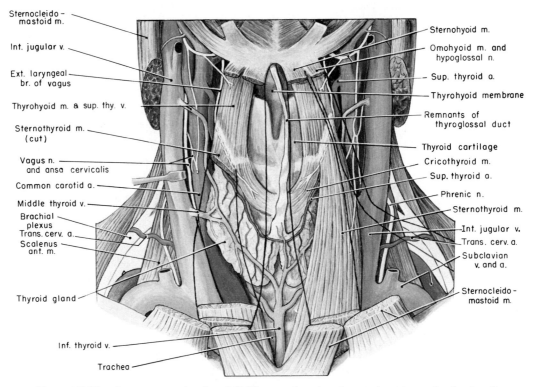

Figure 7-89. Anterior triangle of neck II. The omohyoid and sternohyoid muscles (outlined) have been removed from the left side of the body to reveal the thyrohyoid and sternothyroid muscles. Removal of the sternothyroid muscle (outlined on right side of body) brings the thyroid gland into view. All fascia has been removed.

the cricoid cartilage and anterior to the trachea is the U-shaped thyroid gland, with the superior thyroid artery and vein coursing on its anterior surface and the inferior thyroid vein coursing inferiorly in the midline (Fig. 7-89).

At this time you should note that the carotid sheath and its contents extend from the base of the skull, throughout the neck, to the thoracic inlet surrounded by its fascia, located just anterior to the vertebral column and lateral to the centrally located visceral tube. Many of the nerves and vessels about to be described will be related to this carotid sheath and its contents—the common carotid artery medially, the internal jugular vein laterally, and the vagus nerve posterior to the two vessels.

The anterior wall of the carotid sheath contains the **superior branch of the ansa** (*L.,* handle) **cervicalis** as it descends from the hypoglossal nerve (Fig. 7-89). The **inferior branch** of this loop may lie between the artery and vein, and therefore disappear into the fascia, or may go around the lateral side of the vein. The branches from the ansa to the omohyoid, sternothyroid, and sternohyoid muscles course inferiorly to enter the deep surface of the inferior end of these particular muscles. If the omohyoid and the sternocleidomastoid muscles are reflected, the carotid sheath and its contents can be followed inferiorly.

MUSCLES

The muscles associated with the approach to the thyroid gland are:

1. Sternocleidomastoid
2. Sternohyoid
3. Omohyoid
4. Sternothyroid
5. Thyrohyoid
6. Cricothyroid

These muscles are all named for their attachments.

1. Sternocleidomastoid (Fig. 7-88). This muscle forms the ridge on the neck extending from the manubrium and medial end of the clavicle to the mastoid process of the temporal bone. The sternal head **arises** from the manubrium, while the clavicular head arises from the medial third of the clavicle. These two heads join and **insert** on the mastoid process of the temporal bone. The main action of this muscle is to pull on the head in such a manner that the chin turns to the opposite side. The

nerve supply is from the spinal accessory nerve and the second cervical nerve of the cervical plexus. Paralysis of this muscle causes the head to turn toward the side of injury, which is contrary to results obtained from denervating the majority of muscles.

2. Sternohyoid * (Fig. 7-88). This muscle is located on the anterior surface of the neck and **arises** from the posterior surface of the manubrium and the sternal end of the clavicle. It **inserts** on the body of the hyoid bone. Its **main action** is to depress the hyoid bone, and its nerve supply is from the first, second, and third cervical segments via the ansa cervicalis.

3. Omohyoid * (Figs. 7-88 and 7-83). This muscle is divided into superior and inferior bellies with a central tendon. The inferior belly arises from the superior margin of the scapula and inserts into the intermediate tendon, which lies deep to the sternocleidomastoid muscle. This tendon is bound down by its own fascia to the posterior surface of the clavicle and the first rib. The superior belly ascends almost vertically to be inserted onto the body of the hyoid bone. Its **action** is to depress the hyoid bone. Some have speculated that the intimate relation of the omohyoid fascia to the internal jugular vein aids in preventing collapse of this vessel under conditions of continued deep breathing. Its **nerve supply** is C1, C2, and C3 via the ansa cervicalis. This muscle is probably more useful to the lower animals.

4. Sternothyroid * (Fig. 7-89). Lying deep to the sternohyoid muscle in the lower part of the anterior triangle of the neck, this muscle **arises** from the posterior surface of the manubrium, and **inserts** onto the side of the thyroid cartilage at the location of the oblique line. (It is easy to remember which muscle—the sternohyoid or sternothyroid—lies deep, for the sternohyoid muscle could not reach the hyoid bone if it were deep to a muscle attached to the more inferiorly placed thyroid cartilage.) The **main action** of the sternothyroid is to pull the larynx inferiorly, and its **nerve supply** is C1, C2, and C3 via the ansa cervicalis.

5. Thyrohyoid (Fig. 7-88). This is a continuation of the sternothyroid muscle and is located superior to it. It

---

*It should be noted that the sternohyoid, omohyoid, and sternothyroid are called infrahyoid muscles and serve in opening the mouth, since it is important to maintain the hyoid bone in a steady state in forcefully performing this act. The infrahyoid muscles hold the hyoid bone in place, while the suprahyoid muscles actually pull the jaw inferiorly.

arises from the oblique line on the thyroid cartilage and **inserts** onto the body and greater horn of the hyoid bone. Its **main action** is to elevate the larynx in swallowing. Its **nerve supply** is from C1, the branch coursing with the hypoglossal nerve to reach the muscle.

6. Cricothyroid muscle (Fig. 7-89). This muscle is actually an intrinsic muscle of the larynx and will be described with that organ.

ARTERIES

The arteries involved in an approach to the thyroid gland are the (1) left and (2) right common carotids and (3) the superior and (4) inferior thyroids.

1. The **left common carotid** (Fig. 7-89) arises directly from the arch of the aorta and, after passing through the superior mediastinum in the thoracic cavity, enters the neck region. It continues superiorly in a plane anterior to the prevertebral muscles (which are illustrated in Fig. 7-94) and just lateral to the visceral tube in the neck until a level corresponding to the superior edge of the thyroid cartilage and the lower part of the third cervical vertebra is reached. At this point it divides into internal and external carotid arteries. In the neck the common carotid is enclosed in the carotid sheath along with the internal jugular vein (lateral) and the vagus nerve (posterior).

In addition to the structures mentioned above, the left common carotid is related anteriorly to the middle thyroid vein, the thyroid gland itself, the infrahyoid muscles, the anterior jugular vein, and the sternocleidomastoid muscle, all structures coming into play in performing a thyroidectomy. We will see later that this artery is also related posteriorly to such vital structures as the sympathetic trunk, the thoracic duct, the vertebral artery, and the inferior thyroid artery.

2. The **right common carotid** (Fig. 7-89) differs from the left in arising from the brachiocephalic artery and entering the neck in a more horizontal fashion from the midline of the body. It is separated slightly from the inferior end of the internal jugular vein, and there is no relation to the thoracic duct.

3. The **superior thyroid** artery (Fig. 7-89) is the same on both sides of the body. It arises from the anterior surface of the external carotid close to its origin. In fact, in appearance it very often seems as though this

artery were branching from the superior end of the common carotid. It courses inferiorly and slightly medially to end on the anterior surface of the thyroid gland, anastomosing with the same artery from the opposite side on the isthmus of the gland. It has the following branches: (a) infrahyoid, (b) superior laryngeal, (c) sternocleidomastoid, (d) cricothyroid, (e) glandular, and (f) muscular. The **infrahyoid** branch courses along the inferior border of the hyoid bone deep to the thyrohyoid muscle. The **superior laryngeal** branch courses anteriorly deep to the thyrohyoid muscle, and then pierces the thyrohyoid membrane in company with the internal branch of the superior laryngeal branch of the vagus nerve; this artery supplies the mucous membrane, muscles and ligaments of the larynx. The **sternocleidomastoid** branch turns laterally and crosses the common carotid artery to enter the muscle. The **cricothyroid** branch courses anteriorly on the surface of the cricothyroid muscle. The **glandular** branches supply the thyroid gland as indicated, and the **muscular** branches are several to adjoining muscles. Several of these branches are demonstrated on Figure 7-102.

4. Inferior thyroid (Fig. 7-92). This artery courses superiorly in a position just anterior to the scalenus anterior muscle, and then courses medially, passing posterior to the carotid sheath and its contents, but anterior to the vertebral artery. It reaches the thyroid gland and terminates by coursing on its posterior surface. Its branches will be described later (page 584). It anastomoses with the superior thyroid artery and with both arteries from the opposite side of the body.

VEINS

The superficial veins of the head and neck have been described (page 485).

The veins involved in the thyroid area are (1) the superior, (2) middle, and (3) inferior thyroid veins, and (4) the large internal jugular vein. These are shown in Figure 7-89.

1. Superior thyroid vein. This vein drains the anterior surface of the thyroid gland, courses superiorly in a plane deep to the omohyoid muscle, an drains into the internal jugular vein.

2. The **middle thyroid** vein has no counterpart in an artery. It also drains the anterior surface of the gland, takes a short course laterally in a plane deep to the sternocleidomastoid muscle, and drains into the internal jugular vein.

3. The **inferior thyroid** vein drains both surfaces of the thyroid gland, courses inferiorly in the midline, and drains into the left brachiocephalic vein.

4. Internal jugular. This large vein starts at the base of the skull and courses inferiorly in the carotid sheath just lateral to the internal and then common carotid arteries. It is just anterior to the vertebral column and lateral to the centrally located visceral tube. In the region of the thyroid gland, it is crossed by the sternocleidomastoid, omohyoid, sternohyoid, and sternothyroid muscles, and often by the ansa cervicalis and its branches to the infrahyoid muscles.

NERVES

The nerves in the region of the thyroid gland are:

1. Ansa cervicalis
2. External laryngeal branch of the vagus
3. The recurrent laryngeal branch of the vagus

1. The **ansa cervicalis** (Fig. 7-89) is a loop of fibers that arise from C1, C2, and C3. The fibers from C1 join the hypoglossal nerve and then branch off to supply the geniohyoid and thyrohyoid muscles; in addition, these same fibers form the superior root of the ansa cervicalis, which passes inferiorly on the anterior wall of the carotid sheath. This meets with another descending contribution from C2 and C3—the inferior root of the ansa cervicalis—forming a loop, the ansa. Branches from this loop then give nerve supply to the infrahyoid muscles—the omohyoid, sternohyoid, and sternothyroid.

2. The **external laryngeal** nerve courses inferiorly in a position just lateral to the thyrohyoid muscle, turns medially in a plane deep to the sternothyroid muscle, and ends in the substance of the cricothyroid muscle, which it innervates. The intimate relation of this nerve to the superior thyroid artery should be noted; it is often caught in a ligature of the artery.

3. The **recurrent laryngeal** nerves are extremely important since they innervate most of the intrinsic muscles of the larynx. They are located just posterior to the thyroid gland and are in danger during thyroidectomies. They are best seen in the next dissection (root of neck).

THYROID AND PARATHYROID GLANDS

THYROID.   The thyroid gland is a U-shaped structure located in the neck (Fig. 7-89) on either side of the larynx and trachea. It consists of two large lobes connected by an isthmus that crosses the second, third, and fourth rings of the trachea. It is covered by pretracheal fascia, which holds it onto the larynx and trachea. Each lobe is approximately 2 inches in length, with an apex extending superiorly. It varies in shape and size, usually being larger in females.

Because the thyroid gland develops as an inferior projection from the base of the tongue, remnants of the gland tend to project superiorly from the isthmus along this descending pathway. This **pyramidal lobe** is often connected directly with a **thyroglossal duct,** which extends superiorly to the hyoid bone, usually on the left side of the midline. From the hyoid bone the duct leads to the foramen cecum of the tongue. Remnants of thyroid tissue can be found anywhere along this pathway.

The **relations** of the thyroid gland are as follows: **Anterior**—skin, platysma muscle, enveloping fascia; sternocleidomastoid muscle; sternohyoid, sternothyroid, and omohyoid muscles and their fascial layers; and pretracheal fascia. **Posteromedial**—larynx and trachea; esophagus; recurrent laryngeal nerve; and parathyroid glands, which are essential to life. **Posterolateral**—carotid sheath and its contents; sympathetic chain.

Blood is supplied by the superior and inferior thyroid arteries, the former coursing on its anterior surface and anastomosing with the same artery of the other side, while the latter courses on its deep surface. The **venous drainage** is done by three vessels—the superior, middle, and inferior thyroid veins. The first two drain into the internal jugular vein while the inferior vein drains into the left brachiocephalic. These veins are pictured in Figure 7-89.

The thyroid gland is **innervated** by branches from the sympathetic chain and also by branches of the vagus. Their function, other than innervating blood vessels, is problematical.

Lymph nodes draining the thyroid gland are located on either side of the trachea—paratracheal nodes—and secondarily the deep cervical nodes located inferior to the omohyoid muscle and along the internal jugular vein.

The thyroid is an endocrine gland. Its main **function** is the regulation of basal metabolic rate, which involves all parts of the body. In addition, it is associated with the phenomenon of growth. As is customary with endocrine organs, it can secrete too little or too much hormone. Too little in childhood leads to cretinism; in adulthood, to myxedema. Thyrotoxicosis is the condition in which too much hormone is released. This gland can increase in size through a lack of iodine—a goiter. The thyroid can be surgically removed, destroyed with radioactive iodine, or inhibited by drugs.

PARATHYROIDS. The parathyroid glands (Figs. 7-92 and 7-112) are pairs of small yellowish structures located on the deep surface of the lateral lobes of the thyroid. Their exact location varies, as does their number (see Fig.

7-90). They are difficult to see in cadavers, and may be anywhere from the pharynx to the superior mediastinum. These glands are derived from the third and fourth branchial pouches, those from the fourth ending up at a more superior level than those from the third. The usual location is on the posterior surface of the thyroid, embedded in its capsule, the lower pair near the inferior ends of the lobes, while the upper pair is approximately at the level of the cricoid cartilage.

The **arterial supply** is by branches of the inferior thyroid arteries, often a guide to their location, and their **nerve supply** is similar to that of the thyroid and just as questionable in function.

The parathyroids are endocrine glands that **function** to regulate calcium metabolism. They are essential to life.

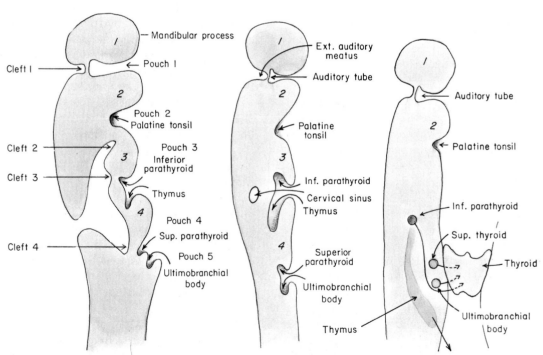

Figure 7-90. *Diagrammatic representation of the derivatives from the pharyngeal clefts and pouches. Note that clefts 2, 3, and 4 are buried and become the cervical sinus. Cysts can form from this sinus and may even open to the outside surface; if such cysts become infected they should be removed. Note also that the parathyroid gland in pouch 3 comes to lie inferior to that from pouch 4. The intimate relation of the inferior parathyroid to the thymus explains why the thymus gland can occasionally pull the inferior parathyroid gland into the thoracic cavity. The ultimobranchial body probably becomes incorporated into the thyroid gland. (Redrawn after Starch and Langman.)*

Before finishing the area of the thyroid gland, some attention should be given to the performance of a **tracheotomy.** The technique may vary considerably depending on the circumstances—whether in a surgical operating room or far removed from such an area with no time to reach it. A tracheotomy may be done superior or inferior to the isthmus of the thyroid gland. In either case one must penetrate, via a vertical incision, the skin, superficial fascia, enveloping fascia, fascia connecting the infrahyoid muscles, the pretracheal fascia, and then the trachea. If one is penetrating the trachea above the isthmus, between it and the cricoid cartilage, the isthmus should be pulled inferiorly. If one is penetrating the trachea inferior to the isthmus—the inferior thyroid vein(s) may be bothersome—the isthmus should be elevated. The neck should be extended in either approach. Any hollow tube can be used in an emergency, a tracheotomy tube being used under hospital conditions. The isthmus may be cut surgically, if necessary, but the superior thyroid arteries must be ligated.

ROOT OF NECK

GENERAL FEATURES

The region called the root of the neck, which connects the contents of the thoracic cavity with those in the neck, is best understood if you have a concept of (1) the thoracic inlet made up of the manubrium of the sternum anteriorly, the ribs laterally, and the vertebral column posteriorly; (2) the insertion of the scalenus anterior muscle upon the first rib; (3) the extension of the dome of the lung well into the neck region; (4) the midline position of the esophagus just anterior to the vertebral column; and (5) the trachea just anterior to the esophagus (Fig. 7-91). The structures leading to and from the thoracic cavity and neck take a vertical course in this area, while structures leading to and from the thoracic cavity and upper limb take a more horizontal and lateral position crossing the first rib.

The **scalenus anterior** muscle is extremely important, for many structures have an intimate relation to it (Figs. 7-91 and 7-94).

Assuming that the clavicle has been disarticulated at the sternoclavicular joint and reflected laterally, the **sternohyoid** and **sternothyroid** muscles are seen first just deep to the inferior end of the sternocleidomastoid muscle (Fig. 7-88), which was partially removed when the clavicle was reflected. If these muscles, in turn, are removed, the contents of the carotid sheath can be seen coursing superiorly into the neck (Fig. 7-89). These structures—the **internal jugular vein** located laterally, the **common carotid artery** medially, and the **vagus nerve** posterior and between the two vessels—are located just lateral to the **trachea** and **esophagus,** the midline structures. The **subclavian vein** joins the internal jugular to form the **brachiocephalic vein,** and this subclavian vein can be seen to be coursing superficial to the scalenus anterior muscle as it passes the first rib on its way to the root of the neck.

The **subclavian artery,** on the other hand, courses posterior or deep to the scalenus anterior muscle to reach the upper limb (Figs. 7-91, 7-92, and 7-94). The **vertebral artery** is given off first and takes an almost vertical course to enter into the foramen on the transverse process of the sixth cervical vertebra. The **thyrocervical trunk** with its **transverse cervical, suprascapular,** and **inferior thyroid** vessels can be seen emerging just before the anterior scalenus muscle is reached. These vessels course superficial to the scalenus anterior muscle. The **costocervical trunk** (Fig. 7-93) arises from the posterior side of the vessel and immediately turns superiorly and posteriorly over the dome of the pleura and divides into its **deep cervical** and **intercostal** branches. The **internal thoracic** artery can be seen branching deep to the scalenus anterior muscle and coursing inferiorly in a position deep to the subclavian vein to enter the thoracic cavity.

The **thoracic duct** on the left side ascends in the groove between the esophagus and prevertebral muscles to a point approximately 1 inch superior to its termination in the notch made by the joining of the internal jugular and subclavian veins. It then courses laterally in a plane posterior to the carotid sheath and its contents, and then inferiorly to drain as indicated. The same area on the right side of the neck contains several lymphatic vessels that drain the upper right half of the body.

The **phrenic nerve** is also in evidence in this dissection, coursing on the anterior surface of the scalenus anterior muscle and escaping from these muscles near their

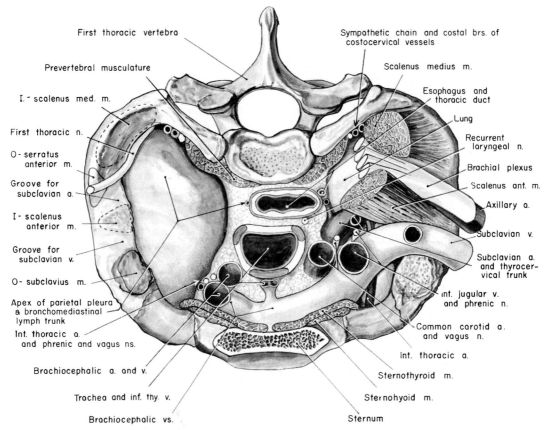

First thoracic vertebra

Prevertebral musculature

I.- scalenus med. m.

First thoracic n.

O- serratus
anterior m.

Groove for
subclavian a.

I- scalenus
anterior m.

Groove for
subclavian v.

O- subclavius m.

Apex of parietal pleura
a bronchomediastinal
lymph trunk

Int. thoracic a.
and phrenic and vagus ns.

Brachiocephalic a. and v.

Trachea and inf. thy. v.

Brachiocephalic vs.

Sympathetic chain and costal brs. of
costocervical vessels

Scalenus medius m.

Esophagus and
thoracic duct

Lung

Recurrent
laryngeal n.

Brachial plexus

Scalenus ant. m.

Axillary a.

Subclavian v.

Subclavian a.
and thyrocer-
vical trunk

Int. jugular v.
and phrenic n.

Common carotid a.
and vagus n.

Int. thoracic a.

Sternothyroid m.

Sternohyoid m.

Sternum

Figure 7-91. *Structures in the thoracic inlet.*

inferior ends to enter the thoracic cavity between the sub-clavian artery and subclavian vein (Fig. 7-91). One of the most posteriorly located structures in this area is the **sympathetic chain** and its middle and inferior cervical ganglia. These ganglia are related (Fig. 7-94) to the vertebral artery, the inferior one being located posteromedial to it. Very frequently the inferior cervical ganglion is combined with the first thoracic ganglion to form a rather large ganglion called the **stellate ganglion**. These ganglia are posterior to the dome of the pleura. The **recurrent laryngeal nerves** are found in slightly different places on the two sides, since its origin on the left side is much more inferiorly located than on the right. The recurrent nerve on the left is located in the groove between the trachea and the esophagus, while that on the right takes a more horizontal course to reach this same groove.

One of the most important aspects of the anatomy of

the root of the neck is the fact that the dome of the **pleura** and the **lung** extend superiorly for some distance into the neck region. It is considerably higher than the superior edge of the manubrium and the first rib, which marks the inlet of the thorax. Posteriorly, however, the lung does not reach a point superior to the first rib. This phenomenon is due to the inferior slant of the first rib as it curves around from posterior to anterior.

STERNOCLAVICULAR JOINT

This joint (Fig. 7-95) is a double arthrodial joint consisting of the articulation between the sternal end of the clavicle and the articular disc, and between this same disc and the superior and lateral part of the manubrium and cartilage of the first rib. This is the only bony attachment of the upper limb to the axial skeleton.

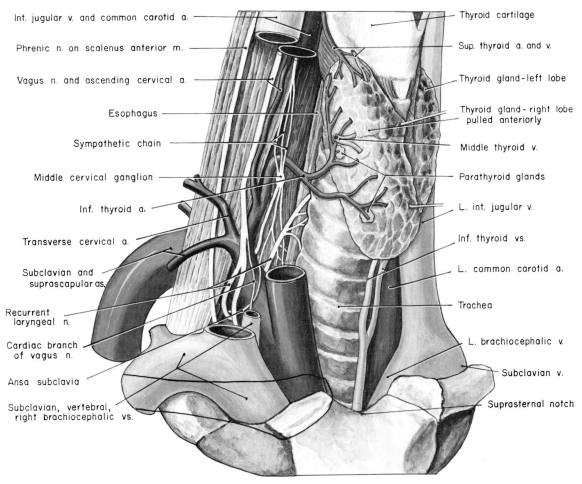

Int. jugular v. and common carotid a.

Phrenic n. on scalenus anterior m.

Vagus n. and ascending cervical a.

Esophagus

Sympathetic chain

Middle cervical ganglion

Inf. thyroid a.

Transverse cervical a.

Subclavian and
suprascapular as.

Recurrent
laryngeal n.

Cardiac branch
of vagus n.

Ansa subclavia

Subclavian, vertebral,
right brachiocephalic vs.

Thyroid cartilage

Sup. thyroid a. and v.

Thyroid gland-left lobe

Thyroid gland - right lobe
pulled anteriorly

Middle thyroid v.

Parathyroid glands

L. int. jugular v.

Inf. thyroid vs.

L. common carotid a.

Trachea

L. brachiocephalic v.

Subclavian v.

Suprasternal notch

Figure 7-92. Root of the neck on the right side of the body. The clavicle is outlined. The right lobe of the thyroid is pulled anteromedially to show its posterior surface. Note the intimate relationship of the inferior thyroid artery to the recurrent nerve and sympathetic chain.

The ligaments involved in this articulation are:

1. Articular capsule
2. Anterior sternoclavicular ligament
3. Posterior sternoclavicular ligament
4. Interclavicular ligament
5. Costoclavicular ligament
6. Articular disc

The **articular capsule,** as in other joints, is thick in some regions and thin in others; the thicker regions are actually the locations of the ligaments. The thick parts of

the capsule are located anteriorly and posteriorly, while the thin parts are located superiorly and inferiorly.

The **anterior sternoclavicular ligament** spreads from the clavicle medially and slightly inferiorly to attach to the manubrium of the sternum. This is accompanied posteriorly by a similar ligament, the **posterior sterno-clavicular ligament.** These ligaments are rather firm and thick structures.

The **interclavicular ligament** extends from the medial end of one clavicle to the medial end of the other. It bridges across the notch of the manubrium and therefore

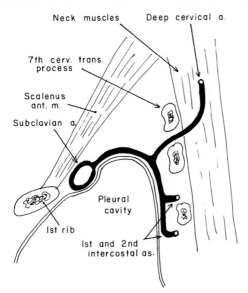

Neck muscles   Deep cervical a.

7th cerv. trans. process

Scalenus ant. m.

Subclavian a.

Pleural cavity

1st rib

1st and 2nd intercostal as.

Figure 7-93. *Origin, course, and distribution of the costocervical trunk and its branches. (Modified after Jamieson.)*

is continuous from one clavicle to the other, just superior to the manubrium. This ligament has connections with both articular discs, and the **articular disc** stretches from these ligaments inferiorly to be attached to the cartilage of the first rib.

The **costoclavicular ligament** is a firm, short, and flat ligament that extends from the costal cartilage of the first rib superiorly and slightly laterally to the inferior surface of the medial end of the clavicle.

It can be seen from glancing at Figure 7-95 that the ligaments and articular disc are arranged in such a manner as to prevent the sternal end of the clavicle from escaping from the rather shallow fossa on the manubrium. The very fact that the articular disc is attached to the cartilage of the first rib prevents the sternal end of the clavicle from leaving the fossa and terminating in the suprasternal notch. In addition, the anterior and posterior sternoclavicular ligaments prevent medial displacement of the clavicle. The angle of the costoclavicular ligament does not aid in this although it would be very difficult, since this ligament is quite short, for the sternal end of the clavicle to travel any great distance.

As mentioned previously, the sternoclavicular joint represents the only bony attachment of the upper limb to the axial skeleton. Great movement is allowed. Since the shoulder can be elevated, depressed, or moved anteriorly

or posteriorly, this joint allows circumduction. It is thought that a certain amount of rotation also takes place in this joint, which would make it almost a ball-and-socket type of articulation.

The **nerve supply** is from the medial supraclavicular nerve.

ACROMIOCLAVICULAR JOINT

This is a freely movable joint and is the articulation between the lateral end of the clavicle and the acromion process of the scapula (Fig. 7-96).

The ligaments associated with this joint are:

1. Articular capsule
2. Superior acromioclavicular
3. Inferior acromioclavicular
4. Articular disc
5. Coracoclavicular (accessory)
   a. Conoid
   b. Trapezoid

The **articular capsule** completely surrounds the joint and is strengthened superiorly and inferiorly by the **superior and inferior acromioclavicular ligaments.** The **articular disc** in this joint is attached to the superior part of the capsular ligament and projects into the cavity. This disc is usually incomplete and frequently is found wanting.

Although the **coracoclavicular ligaments** are not actually part of the acromioclavicular articulation, they are usually described with this joint and serve to maintain the clavicle and the acromion process in juxtaposition. They also are situated in such a manner as to prevent the scapula from moving medially, which is the main function of the clavicle. The very vital structures in the axilla must be protected from pressure, and the clavicle does maintain the scapula in a normal lateral position. These ligaments are attached to the base of the coracoid process and to the underside or inferior surface of the clavicle. The lateral one is **trapezoid** in shape while the medial one is **conoid** in appearance, and these ligaments are often called by these names.

The movements allowed in this joint are of the gliding variety. As the scapula is moved anteriorly and posteriorly, a certain amount of rotation also occurs.

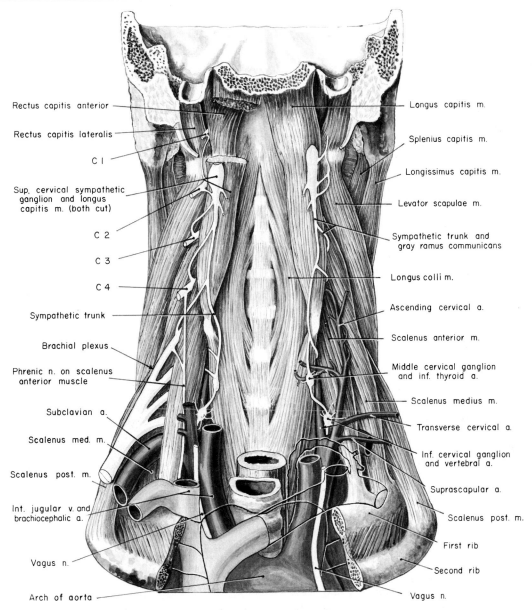

Rectus capitis anterior

Rectus capitis lateralis

C 1

Sup. cervical sympathetic ganglion and longus capitis m. (both cut)

C 2

C 3

C 4

Sympathetic trunk

Brachial plexus

Phrenic n. on scalenus anterior muscle

Subclavian a.

Scalenus med. m.

Scalenus post. m.

Int. jugular v. and brachiocephalic a.

Vagus n.

Arch of aorta

Longus capitis m.

Splenius capitis m.

Longissimus capitis m.

Levator scapulae m.

Sympathetic trunk and gray ramus communicans

Longus colli m.

Ascending cervical a.

Scalenus anterior m.

Middle cervical ganglion and inf. thyroid a.

Scalenus medius m.

Transverse cervical a.

Inf. cervical ganglion and vertebral a.

Suprascapular a.

Scalenus post. m.

First rib

Second rib

Vagus n.

Figure 7-94. *The scalene and prevertebral muscles, the relation of the subclavian vessels to these muscles, the branches of the subclavian artery, and the sympathetic chain. (Inferior cervical ganglion is lateral to vertebral artery in this specimen.)*

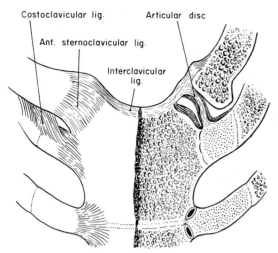

Figure 7-95. *The sternoclavicular joint. Right side intact; left side with a frontal cut.*

It might be mentioned that the clavicle is the bone most frequently fractured. An important role of this bone, as already mentioned, is to protect the very vital structures lying just deep to it. Of equal importance is to realize that a fractured clavicle can do just the opposite—provide real trauma to these same vital structures, thus the pulling of the shoulders posteriorly as a first aid measure in such fractures.

The coracoacromial ligament is usually described with the shoulder joint—page 754.

MUSCLES

The important muscles in the root of the neck are the three scalene muscles—anterior, middle, and posterior—and the inferior end of the sternocleidomastoid. The last muscle has been given considerable attention, and the scalene muscles were previously described. This description demands repetition. They are important to an understanding of the root of the neck; many structures are related to them.

The **scalenus anterior** muscle (Fig. 7-94) is located in the root of the neck deep to the sternocleidomastoid muscle. It **arises** from the third to the sixth cervical transverse processes, and **inserts** on the scalene tubercle on the superior surface of the first rib. Its **nerve supply** is by direct branches from ventral rami of C4 to C7. It is con-

sidered by many to be an important respiratory muscle since its **main action** is to elevate the first rib.

The following structures are related to this muscle. **Anterior**—phrenic nerve; suprascapular, ascending cervical, and transverse cervical arteries; inferior belly of the omohyoid muscle; thoracic duct; subclavian, anterior jugular, and internal jugular veins; and the sternocleidomastoid muscle. **Posterior**—subclavian artery, brachial plexus, and scalenus medius.

The **scalenus medius** muscle (Fig. 7-94) was studied with the posterior triangle of the neck since it is considered to form part of the floor of this triangle. It **arises** from the transverse processes of C1 to C7, and **inserts** on the superior surface of the first rib posterior to the subclavian artery and just anterior to the tubercle of the rib. The nerve supply is the ventral rami of C3 to C7 directly. It **elevates** the first rib.

The **scalenus posterior** muscle (Fig. 7-94) is located posterior to the medius and quite intimately fused to it at times. It **arises** from the fourth to the sixth cervical transverse processes and **inserts** on the second rib. It elevates this rib and is innervated by ventral rami of C5 to C7.

NERVES

Although most of the nerves in the root of the neck either have been described or will be studied with the upper limb, the student should be sure to connect the phrenic nerve, sympathetic chain, and vagus nerve, as seen in the neck, with these same structures in the thoracic cavity. If this latter area has been dissected, a review of Figures 4-9, 4-10, and 4-11 found on pages 470, 471, and 473 should be helpful at this time.

The **phrenic nerve** (Figs. 7-92 and 7-94) arises from C4 with possible contributions from C3 and C5. It immediately gains the anterior surface of the scalenus anterior muscle and proceeds inferiorly in a plane deep to the sternocleidomastoid muscle. It continues into the thorax by leaving the scalenus anterior muscle and coming into contact with the pleura; in this position it is posterior to the brachiocephalic veins. The left phrenic actually comes into contact with the subclavian artery when leaving the anterior surface of the scalenus anterior muscle, while the right does not. The omohyoid muscle and the transverse cervical and suprascapular arteries are anterior relations of the phrenic nerve.

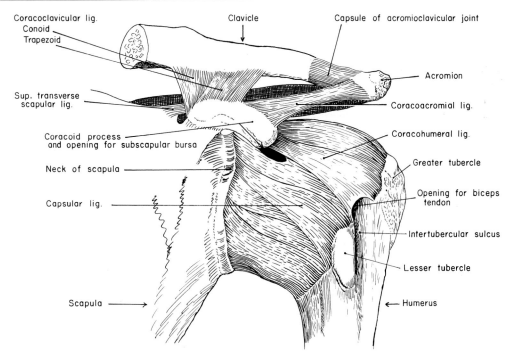

Coracoclavicular lig.
Conoid
Trapezoid
Clavicle
Capsule of acromioclavicular joint
Acromion
Sup. transverse
scapular lig.
Coracoacromial lig.
Coracohumeral lig.
Coracoid process
and opening for subscapular bursa
Greater tubercle
Neck of scapula
Opening for biceps
tendon
Capsular lig.
Intertubercular sulcus
Lesser tubercle
Scapula
Humerus

Figure 7-96. *Acromioclavicular joint and associated ligaments, and the capsule of the shoulder joint.*

Both phrenic nerves proceed through the thoracic cavity (page 208) to reach the diaphragm. They contain both motor and sensory fibers. The motor fibers innervate the diaphragm, while the sensory fibers give branches to the pleura and pericardium as well as to the diaphragm. Some authorities claim that the phrenic nerve gives sensory nerve supply to such organs as the liver and inferior vena cava; this may account for the referred pain from phrenic and/or hepatic abscesses to the shoulder region via the supraclavicular nerves.

The **sympathetic chain** (Fig. 7-94) continues superiorly from the thorax into the neck, remaining just anterior to the prevertebral musculature and posterior to the carotid sheath—actually in the fascia forming its posterior wall. Three ganglia are usually described, but there may be more. The **inferior cervical ganglion** is in close relation with the origin of the vertebral artery, being posteromedial to the artery and anterior to the seventh cervical transverse process and neck of the first rib. When this inferior cervical ganglion combines with the first thoracic

ganglion (which it frequently does), it is called the stellate ganglion. The **middle cervical ganglion,** if low in position, is close to the vertebral artery near the point where it enters the foramen transversarium of the sixth cervical vertebra, but if high, close to or above the inferior thyroid artery.[*] Ligation of the inferior thyroid artery must be done with care. These ganglia are much smaller than the **superior cervical ganglion,** which is described on page 597.

---

[*]M. Axford, 1928, Some observations on the cervical sympathetic in man, *J. Anat., 62*: 301. This work emphasizes that the superior and inferior ganglia are fairly constant, while the middle cervical ganglion is rather variable in location. If low in position, it is close to the vertebral artery; if high, close to the inferior thyroid artery.

R. F. Becker and J. A. Grunt, 1957, The cervical sympathetic ganglia, *Anat. Record* 127: 1. Due to the marked variability of the middle cervical sympathetic ganglion, these authors recommend that four cervical sympathetic ganglia be described: superior, middle, vertebral, and inferior. The middle and vertebral ganglia correspond to the "high" and "low" middle ganglia of others and are located above the inferior thyroid artery and just lateral to the vertebral artery respectively.

The **ansa subclavia** is a loop around the subclavian artery connecting the middle and inferior cervical ganglia.

The **vagus nerves** (Figs. 7-91, 7-92, and 7-94) are important nerves that continue into the thoracic cavity by remaining posterior to the vessels in the carotid sheath. They course posterior to the large veins but anterior to the subclavian arteries. The right continues inferiorly by following the trachea, while the left is required to pass around the arch of the aorta. The further wandering of these nerves is described on pages 209 and 332.

SUBCLAVIAN ARTERIES

The **left subclavian** arises directly from the arch of the aorta (Fig. 4-10). It is posterior to the common carotid artery and just lateral to the trachea. It courses superiorly and then turns laterally to pass between the pleura of the lung posteriorly and the scalenus anterior muscle anteriorly (Fig. 7-94). It then crosses the first rib to become the axillary artery at the rib's outer border. The artery is divided into three parts: the first part from its origin to the scalenus anterior muscle, the second part posterior to the scalenus anterior, and the third part from this muscle to the outer border of the first rib.

The first part of the left subclavian artery has some important structures in an anterior relation to it. The vagus and phrenic nerves, as well as the cervical cardiac branches of the left vagus and left sympathetic chain, are all on the anterior surface of this vessel. In addition, several vessels are also anterior to the subclavian—the left common carotid artery, the vertebral vein, and the left brachiocephalic. The thoracic duct is also anterior to this first part of the artery. The second part of the subclavian is covered by the scalenus anterior muscle. The third part has mainly vessels as intimate anterior relations although the nerve to the subclavius muscle is an anterior relation. The vessels are the suprascapular artery and vein, transverse cervical and external jugular veins, and the subclavian vein.

The **right subclavian artery** begins at a point posterior to the sternoclavicular joint as one of the two terminal branches of the brachiocephalic artery. It takes the same general course as the left, curving laterally in a position anterior to the pleura and lung, and posterior to the scalenus anterior muscle; it is divided into three parts just as found on the left side.

The **relations** of the right subclavian artery are simi-

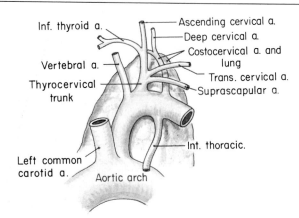

Figure 7-97. *Diagram of the subclavian artery and its branches (left side).*

lar to the left subclavian except that the phrenic nerve does not make contact with it and the thoracic duct is replaced by several lymph vessels. The more transverse course taken by the recurrent laryngeal nerve on this side brings it into closer relation to the subclavian than seen on the left side.

The branches of the subclavian artery are the same on the two sides. They exhibit considerable variation,* the usual arrangement being:

1. Vertebral
2. Thyrocervical trunk
3. Internal thoracic
4. Costocervical trunk

These are diagramed in Figure 7-97.

1. Vertebral (Figs. 7-94, 7-97, and 7-98). This artery arises from the subclavian about ½ inch medial to the scalenus anterior muscle and takes a vertical course to disappear into the transverse process of the sixth cervical vertebra. Near its commencement it is anterior to the pleura and lung and posterior to the common carotid artery. The thoracic duct courses anterior to it, between it and the contents of the carotid sheath, while the inferior cervical ganglion is posteromedial and the middle cervical ganglion just medial to the artery. The vertebral artery

---

*G. H. Daseler and B. J. Anson, 1959, Surgical anatomy of the subclavian artery and its branches, *Surg. Gynecol. Obstet.* 108: 149. The many variations of the branches are presented; based on 400 specimens.

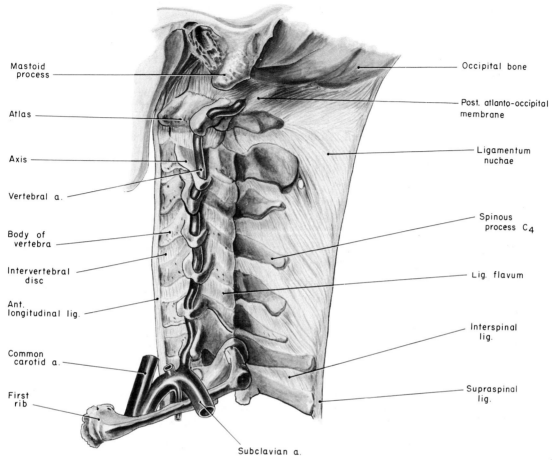

Mastoid process
Atlas
Axis
Vertebral a.
Body of vertebra
Intervertebral disc
Ant. longitudinal lig.
Common carotid a.
First rib
Subclavian a.

Occipital bone
Post. atlanto-occipital membrane
Ligamentum nuchae
Spinous process C₄
Lig. flavum
Interspinal lig.
Supraspinal lig.

Figure 7-98. *Course of the vertebral artery (see also Fig. 7-107).*

continues (Fig. 7-98) through the foramina in the transverse processes of the upper six cervical vertebrae. After passing through the transverse process of the atlas, it turns medially and posteriorly on the posterior arch of the atlas, pierces the dura mater, and enters the foramen magnum. It terminates at the lower border of the pons by combining with the same artery on the other side to form the basilar artery.

The vertebral artery gives off the following branches: (a) **muscular** branches supply surrounding muscles; (b) **spinal** branches enter the intervertebral foramina and follow the roots to aid the spinal arteries in giving blood supply to the spinal cord; (c) **meningeal** branches are given off before the vessel pierces the dura and supply part of the posterior cranial fossa; (d) **posterior spinal** descend on the posterior side of the spinal cord and maintain two longitudinal channels that course in and out of the dorsal nerve roots in a winding fashion; (e) the **anterior spinal** branch unites with the vessel of the other side to course down the anterior surface of the spinal cord in the anterior median fissure; and (f) the **posterior inferior cerebellar** arises near the terminal end of the vertebral and courses to the inferior surface of the cerebellar hemisphere. In addition, there are small branches to the medulla oblongata and branches that anastomose in the posterior part of the neck with the deep cervical and descending branch of the occipital.

2. Thyrocervical trunk (Figs. 7-94 and 7-97). This trunk arises from the subclavian artery just medial to the scalenus anterior muscle. It divides into three branches: (a) inferior thyroid, (b) suprascapular, and (c) transverse cervical.

a. Inferior thyroid (Fig. 7-92). This artery courses superiorly in a position just anterior to the scalenus anterior muscle, and then courses medially, passing posterior to the carotid sheath and its contents, but anterior to the vertebral artery. It reaches the thyroid gland and terminates by coursing on its posterior surface. It gives off the following branches: **ascending cervical,** which courses superiorly in a position anterior to the prevertebral muscles, supplying these muscles, and sending spinal branches through the intervertebral foramina to aid in giving blood supply to the spinal cord; **inferior laryngeal,** a blood vessel that accompanies the recurrent laryngeal nerve to the larynx; **muscular** branches to adjacent muscles; **tracheal** branches; **esophageal** branches; and **terminal** branches to the gland itself. The inferior thyroid artery anastomoses with the superior thyroid artery from the external carotid and with both arteries from the opposite side of the body. Its relationship to the sympathetic chain and the recurrent laryngeal nerve should be recalled.

b. Suprascapular. This artery courses laterally and posteriorly across the root of the neck. It is deep to the sternocleidomastoid muscle, clavicle, and trapezius muscle, but superficial to the scalenus anterior muscle, phrenic nerve, third part of the subclavian artery, and brachial plexus. It finally reaches the superior border of the scapula, where it crosses the superior transverse scapular ligament to enter the supraspinatus fossa. It gives off no branches in the neck.

c. Transverse cervical. This artery will be familiar to those who have studied the back or upper limb. This vessel also courses laterally and anterior to the scalenus anterior muscle; it is deep to the sternocleidomastoid and omohyoid muscles. It crosses the brachial plexus and reaches the anterior border of the levator scapulae muscle. Here it divides into two branches: a **superficial** branch, which crosses the levator scapulae muscle and ramifies on the deep surface of the trapezius muscle, and a **deep** branch, which descends along the medial border of the scapula deep to the levator scapulae muscle and the rhomboids. The deep branch of this artery frequently arises from the third part of the subclavian as an independent branch; in this case it distributes in a manner similar to that just described for the deep branch of the transverse cervical artery and is called the **descending scapular** artery, and the remaining superficial branch is known as the **superficial cervical** artery (see Fig. 2-18 on page 66).

3. Internal thoracic (Figs. 7-98 and 4-6 on page 158). This artery arises just medial to the scalenus anterior muscle but courses in an inferior direction in contrast to the thyrocervical trunk. It lies posterior to the internal jugular or subclavian vein, the origin of the brachiocephalic vein, and clavicle, but anterior to the pleura and lung. This artery, as described on page 156, continues into the thoracic cavity just posterior to the ribs and anterior to the transversus thoracis muscle, and gives off pericardiacophrenic, mediastinal, anterior intercostal, and perforating branches before it terminates by dividing into the musculophrenic and superior epigastric arteries. Those who have studied the abdominal wall will remember that this makes an important anastomosis with the inferior epigastric branch of the external iliac.

4. Costocervical trunk (Figs. 7-93 and 7-97). This artery, on the left side, arises from the first part of the subclavian at the medial border of the scalenus anterior muscle, but on the right side it usually branches from the second part. It passes superiorly and posteriorly over the apex of the pleura, coursing posterior to the scalenus anterior muscle. When it reaches the neck of the first rib it divides into two branches: (a) a **superior intercostal** artery, which descends anterior to the neck of the first rib and divides to form the posterior intercostal arteries of the upper two intercostal spaces, and (b) the **deep cervical** which ascends in the back of the neck between the semispinalis cervicis and semispinalis capitis muscles. This vessel anastomoses with the descending branch of the occipital artery.

SUBCLAVIAN VEINS

The **subclavian vein** starts at the outer edge of the first rib as a continuation of the axillary. It courses medially just posterior to the subclavius muscle and the clavicle, and anterior to the subclavian artery, being separated from it by the scalenus anterior muscle. It ends by joining with the internal jugular vein to form the brachiocephalic. The only tributary to the subclavian vein is the external jugular. It should be noted that the numerous branches of the subclavian artery are not represented as tributaries of the subclavian vein. These veins are present, but drain into other vessels. The vertebral, first posterior intercostal, inferior thyroid, and internal thoracic drain into the brachiocephalic veins rather than into the subclavian, the middle thyroid vein drains into the internal jugular vein, and the transverse cervical and suprascapular veins drain into the inferior end of the external jugular vein.

THORACIC DUCT

If you have not studied the cavities, you may not be aware of the fact that the thoracic duct starts in the abdominal cavity as a dilation called the **cisterna chyli,** which is located on the anterior surface of the first two lumbar vertebrae between the aorta and the right crus of the diaphragm (Fig. 4-43). It courses superiorly through the aortic opening of the diaphragm, ascends in the posterior mediastinum near the midline of the body between the thoracic aorta on the left and the azygos vein on the right, lying posterior to the esophagus. At the fifth thoracic vertebral level the duct inclines to the left and ascends in the superior mediastinum just to the left of the esophagus. The duct ascends into the root of the neck to a point higher than its termination, then curves laterally in a plane posterior to the contents of the carotid sheath, and then courses inferiorly anterior to the branches of the thyrocervical trunk and phrenic nerve to terminate in the crotch between the internal jugular and subclavian veins on the left side of the body.

In the root of the neck (Figs. 7-84 and 7-94) it is at first to the left of the esophagus and anterior to the longus colli muscle—a prevertebral muscle. When it turns laterally, it is at a higher level than the dome of the pleura and reaches the seventh cervical vertebra. It is anterior to the vertebral artery and the sympathetic trunk. Near the termination of the thoracic duct it is joined by the **left jugular lymph trunk,** which drains lymph from the left half of the head and neck, the **left subclavian trunk,** which drains lymph from the left upper limb, and the **left mediastinal trunk,** which drains lymph from the left half of the thorax. These channels may enter the subclavian or brachiocephalic veins independently rather than joining with the termination of the thoracic duct.

The three lymph trunks mentioned above are found on the right side of the body and drain corresponding territories. The right jugular and right subclavian trunks may combine to form what is called the **right lymphatic duct.** The **right jugular lymph trunk** usually ends in the

inferior end of the internal jugular vein, the **right subclavian** into the subclavian vein, while the right **mediastinal lymph trunk** ends in the brachiocephalic vein. The thoracic duct drains lymph from the entire lower half of the body and from the left side of the upper half of the body. The corresponding vessels on the right side drain the right upper half of the body.

The thoracic duct is a very important structure, and its course and relations should be learned thoroughly.

CAROTID TRIANGLE AND PAROTID BED

GENERAL DESCRIPTION

The **carotid triangle** (Fig. 7-99), bounded by the sternocleidomastoid muscle posteriorly, the anterior belly of the omohyoid muscle medially and inferiorly, and the posterior belly of the digastric and stylohyoid muscle superiorly, is well named. The most prominent structure in this triangle is the **common carotid artery** and its division into **internal** and **external carotids,** this division occurring approximately at the level of the superior border of the thyroid cartilage. In addition, several branches of the external carotid are very conspicuous.

Assuming that the very superficial structures, such as the platysma muscle, the branches of the cervical plexus, the cervical branch of the facial nerve, lymph nodes, and any communicating veins have been removed, the abovementioned branches of the external carotid are easily identified. Contained in his triangle are the **superior thyroid** coursing inferiorly, the **lingual** coursing anteriorly, the **facial** coursing superiorly and slightly medially, the **ascending pharyngeal** coursing superiorly, and the **occipital** coursing posteriorly and superiorly (Fig. 7-99). **Veins** corresponding to these branches are also contained in this triangle. They drain into the internal jugular vein, which is hidden by the sternocleidomastoid muscle.

The **carotid body** is found in the crotch made by the origin of the internal and external carotid arteries; the

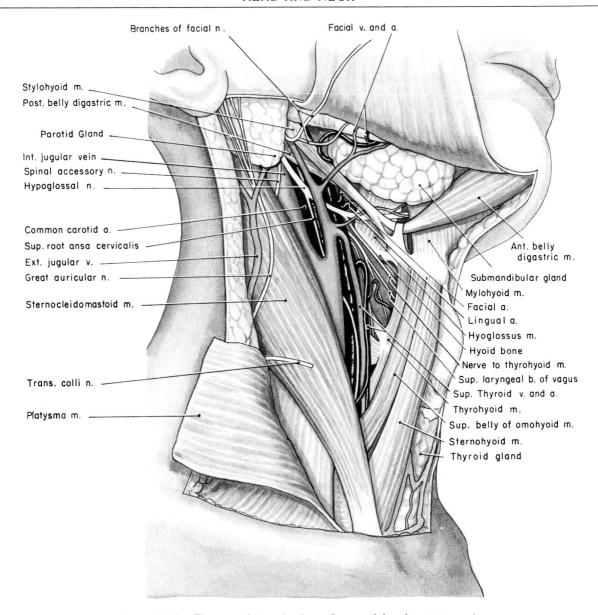

Branches of facial n.

Facial v. and a.

Stylohyoid m.
Post. belly digastric m.

Parotid Gland

Int. jugular vein
Spinal accessory n.
Hypoglossal n.

Common carotid a.
Sup. root ansa cervicalis
Ext. jugular v.
Great auricular n.

Sternocleidomastoid m.

Trans. colli n.

Platysma m.

Ant. belly
digastric m.

Submandibular gland
Mylohyoid m.
Facial a.
Lingual a.
Hyoglossus m.
Hyoid bone
Nerve to thyrohyoid m.
Sup. laryngeal b. of vagus
Sup. Thyroid v. and a.
Thyrohyoid m.
Sup. belly of omohyoid m.
Sternohyoid m.
Thyroid gland

Figure 7-99.   *The carotid triangle after reflexion of the platysma muscle.*

origin of the internal carotid is dilated to form the **carotid sinus.**

The **hypoglossal nerve** crosses both the internal and external carotid arteries in its course to the tongue, actually curving around a sternocleidomastoid branch of the occipital artery to gain this position (Fig. 7-100). It gives

off the superior root to the **ansa cervicalis** in this triangle. In addition, both **internal** and **external laryngeal branches** of the **superior laryngeal branches** of the **vagus** are found in the carotid triangle, the internal being closely associated with the superior laryngeal branch of the superior thyroid artery, and the external

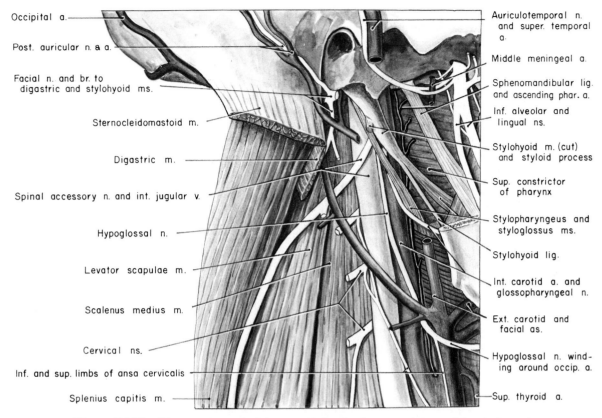

Occipital a.

Post. auricular n. & a.

Facial n. and br. to digastric and stylohyoid ms.

Sternocleidomastoid m.

Digastric m.

Spinal accessory n. and int. jugular v.

Hypoglossal n.

Levator scapulae m.

Scalenus medius m.

Cervical ns.

Inf. and sup. limbs of ansa cervicalis

Splenius capitis m.

Auriculotemporal n. and super. temporal a.

Middle meningeal a.

Sphenomandibular lig. and ascending phar. a.

Inf. alveolar and lingual ns.

Stylohyoid m. (cut) and styloid process

Sup. constrictor of pharynx

Stylopharyngeus and styloglossus ms.

Stylohyoid lig.

Int. carotid a. and glossopharyngeal n.

Ext. carotid and facial as.

Hypoglossal n. winding around occip. a.

Sup. thyroid a.

Figure 7-100. *The superior end of the carotid sheath, showing relations of the ninth, tenth, eleventh, and twelfth cranial nerves to the carotid arteries and the internal jugular vein. The vagus nerve cannot be seen, since it is deep to the large vessels. Note that these structures form the medial relations of the parotid gland. All fascia has been removed.*

with the superior thyroid artery itself; as mentioned previously, ligatures of the superior thyroid artery must not include this nerve.

If the sternocleidomastoid muscle is pulled laterally, the entire **carotid sheath** and its contents are revealed, the **sympathetic trunk** being located in its posterior wall.

If the parotid gland is removed, thus revealing its bed, and the posterior belly of the digastric cut and reflected, the internal jugular vein, vagus nerve, and internal carotid artery can be followed superiorly to the base of the skull (Fig. 7-100). They course deep to the styloid process and the muscles arising from this process—the **stylohyoid, stylopharyngeus,** and **styloglossus.** The **vagus nerve** (Fig. 7-112) lies posterior and between the two large vessels in this position as well as inferiorly. The

spinal accessory nerve crosses the internal jugular vein and then proceeds in a posterior direction to reach the sternocleidomastoid muscle while the **glossopharyngeal nerve** crosses the internal carotid artery and winds around the stylopharyngeus muscle to reach its destination. The **hypoglossal nerve** also courses between the vein and the artery and proceeds inferiorly, as mentioned previously, crossing the anterior surface of the internal carotid artery, hooking around the sternocleidomastoid branch of the occipital artery, and crossing superficial to the external carotid artery to reach the superficial surface of the hyoglossus muscle. All these nerves are obviously penetrating the carotid sheath to take these courses.

The **external carotid artery** and its many branches course superiorly in an anterior relation to the internal carotid artery. If the contents of the carotid sheath are pulled laterally, the **cervical extension of the sympathetic chain** can be seen contained within the fascia forming the posterior wall of the carotid sheath. If this is followed superiorly, the large **superior cervical** ganglion can be found (Fig. 7-112).

It should be noted that this area just described is medial to the parotid gland and the ramus of the mandible, lateral to the pharynx, and anterior to the vertebral column.

Figure 7-100 illustrates the structures forming the medial relations of the parotid gland or its bed. This gland fills in the area between the mandible anteriorly and the sternocleidomastoid and digastric muscles and the mastoid process of the temporal bone posteriorly. It is obvious that the glandular tissue is adjacent to the styloid process and muscles arising from it; in fact, a portion of the gland is found anterior to the process and another portion posterior to it. The superficial temporal artery and vein also are deep to the gland and may actually penetrate it.

MUSCLES

The muscles in this region are:

1. Digastric
2. Stylohyoid
3. Styloglossus
4. Stylopharyngeus

1. Digastric (Fig. 7-76). This muscle is located inferior and slightly deep to the mandible. It is divided into two bellies with an intermediate tendon. The posterior belly **arises** from the inferior surface of the mastoid process of the temporal bone and is **inserted** into the intermediate tendon. The intermediate tendon, in turn, is attached to the body of the hyoid bone by a pulley-like band of fascia. The tendon also perforates the stylohyoid muscle. The anterior belly sweeps superiorly and anteriorly to attach to the mandible. Its **main action** is to draw the hyoid bone superiorly and posteriorly. The posterior belly of the digastric muscle is innervated by the facial nerve, the anterior by the trigeminal.

2. Stylohyoid (Fig. 7-76). The stylohyoid muscle is deep to the angle of the mandible. It **arises** from the styloid process near its base and **inserts** onto the hyoid bone near the junction of the body and the great horn. It surrounds the intermediate tendon of the digastric mus-

cle. It **pulls** the hyoid bone posteriorly and superiorly during the act of swallowing and is innervated by the facial nerve.

3. Styloglossus (Figs. 7-78, 7-100, and 7-102). This muscle is also deep to the mandible and located on the side of the tongue. It **arises** from the styloid process close to the tip and inserts into the side of the tongue, its fibers interlacing with those of the hyoglossus muscle. Its **main action** is to pull the tongue posteriorly and superiorly, and its **nerve supply,** typical of all muscles of the tongue, is by the hypoglossal nerve.

4. Stylopharyngeus (Figs. 7-100, 7-102, and 7-113). This muscle **arises** from the styloid process near its base and **inserts** onto the posterior border of the thyroid cartilage and the side of the pharynx. Its **main action** is to elevate the larynx and pharynx in swallowing and its **nerve supply** is the glossopharyngeal nerve; this is the only muscle innervated by this nerve.

You have undoubtedly noted the marked variability in nerve supply to the muscles just described, and are probably wondering how you can remember them. It all makes sense, however, if one recalls the embryology of these particular muscles. Several are derived from branchial arches. Muscles from the first arch are innervated by the trigeminal; those in the above list from the second (stylohyoid and posterior belly of digastric) by the facial; and that from the third arch (stylopharyngeus) by the glossopharyngeal. These are all branchial efferent fibers. The styloglossus is probably developed from head somites, along with the other muscles of the tongue, and thereby gets its nerve supply from the hypoglossal; these are general somatic efferent fibers (see Fig. 1-18 on page 20).

ARTERIES

If the carotid sheath is followed superiorly (Fig. 7-77), the common carotid is seen to divide into the **external carotid,** with its many branches, and the **internal carotid.**

A dilation can be seen at the end of the common carotid and commencement of the internal carotid arteries known as the **carotid sinus.** Specialized cells in the walls of this sinus respond to arterial blood pressure, reflexly controlling heart output. The **carotid body** is located in the concavity made by the origins of the internal and external carotid arteries. This responds to oxygen tension in the arterial blood and reflexly controls respiratory and heart rates. (See also page 593).

INTERNAL CAROTID ARTERY. This artery (Fig. 7-100) is the same on the two sides of the body. It arises, at a level that corresponds to the superior border of the thyroid cartilage, by bifurcation of the common carotid. It ascends anterior to the prevertebral musculature and just lateral to the pharynx, to reach the base of the skull where it courses through the carotid canal of the temporal bone. It then crosses superior to the foramen lacerum (which is filled in with cartilage), enters the cavernous sinus, turns superiorly at a point medial to the anterior clinoid process, posteriorly over the cavernous sinus, and then superiorly to end by dividing into anterior and middle cerebral arteries.

The branches of the internal carotid are (note that there are none in the neck):

In the carotid canal
  1. Caroticotympanic
  2. Pterygoid
In the cranial cavity*
  1. To walls of cavernous sinus, trigeminal ganglion, and hypophysis
  2. Ophthalmic
  3. Posterior communicating
  4. To choroid plexus
  5. Anterior cerebral
  6. Middle cerebral

1. The small **caroticotympanic** artery enters the posterior inferior wall of the middle ear and ramifies on the walls of this structure.

2. The **pterygoid** branch also is very small; it accompanies the greater petrosal nerve.

Relations. The internal carotid artery is enclosed in the carotid sheath with the internal jugular vein lateral to it and the vagus posterior and between the two vessels. The detailed relations of the internal carotid artery will not be given. If, however, the student will get the general relations of the internal carotid artery, the internal jugular

---

*The branches of the internal carotid in the cranium have already been described on page 499.

vein, and the vagus nerve, as a group, well in mind, the other structures in this area will be related to them. It should be stated once again that these three structures are enclosed in the carotid sheath and are just anterior to the prevertebral musculature and just lateral to the visceral tube, in this case the pharynx. From superior to inferior, these structures are deep to the styloid process, stylohyoid muscle, stylopharyngeus muscle, posterior belly of the digastric, and the sternocleidomastoid muscle.

EXTERNAL CAROTID ARTERY. The external carotid artery (Figs. 7-101 and 7-102) begins as the remaining terminal branch of the common carotid opposite the upper border of the thyroid cartilage. It courses superiorly deep to the sternocleidomastoid muscle, the posterior belly of the digastric, the stylohyoid muscle, and the parotid gland. It is superficial to the superior constrictor of the pharynx (Fig. 7-102), styloid process, and the stylopharyngeus muscle.

The branches of the external carotid artery are:

  1. Superior thyroid
  2. Ascending pharyngeal
  3. Lingual
  4. Facial
  5. Occipital
  6. Posterior auricular
  7. Maxillary
  8. Superficial temporal

The following paragraphs give a formidable list of branches. They are learned easily if you will visualize the area in which each main vessel courses. You will find a branch for most structures passed. Knowledge of the course and surrounding structures will tell you the names of the branches.

1. Superior thyroid. (Figs. 7-89 and 7-102). This vessel was described on page 572.

2. Ascending pharyngeal (Figs. 7-100 and 7-102). This vessel arises from the deep side of the external carotid near its origin. It courses superiorly between the external and internal carotid arteries and along the side of the pharynx as high as the base of the skull. It has the following branches: (a) pharyngeal, (b) prevertebral, (c) meningeal, (d) inferior tympanic, and (e) palatine. The **pharyngeal** are several branches to the pharynx, tonsil, and auditory tube. The **prevertebral** gives branches to the prevertebral muscles and adjoining lymph nodes.

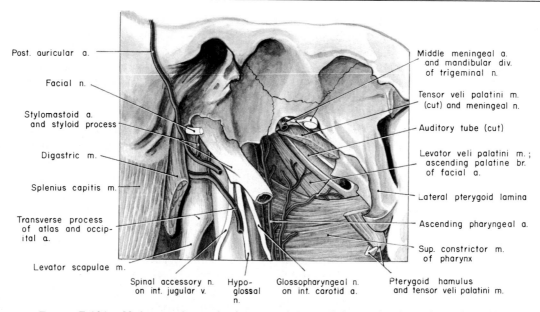

Post. auricular a.

Facial n.

Stylomastoid a.
and styloid process

Digastric m.

Splenius capitis m.

Transverse process
of atlas and occip-
ital a.

Levator scapulae m.

Middle meningeal a.
and mandibular div.
of trigeminal n.

Tensor veli palatini m.
(cut) and meningeal n.

Auditory tube (cut)

Levator veli palatini m.;
ascending palatine br.
of facial a.

Lateral pterygoid lamina

Ascending pharyngeal a.

Sup. constrictor m.
of pharynx

Pterygoid hamulus
and tensor veli palatini m.

Spinal accessory n.   Hypo-   Glossopharyngeal n.
on int. jugular v.   glossal   on int. carotid a.
                        n.

Figure 7-101. *High carotid triangle showing relations of the ninth, eleventh, and twelfth cranial nerves to the internal carotid artery and internal jugular vein. (The vagus is deep to the artery and vein and so cannot be seen.)*

The **meningeal** are small branches that pass through the hypoglossal canal and the jugular foramen to reach the meninges in the posterior cranial fossa. The **inferior tympanic** accompanies the tympanic branch of the glossopharyngeal nerve to the medial wall of the middle ear, and the **palatine** follows the levator palati muscle, to the soft palate. For a small vessel, it certainly has a wide distribution.

3. **Lingual** (Figs. 7-78, 7-80, and 7-102). This artery arises from the external carotid opposite the tip of the greater horn of the hyoid bone. It forms a loop on the surface of the middle constrictor muscle of the pharynx, at which point it is crossed by the hypoglossal nerve. It then courses anteriorly along the superior border of the hyoid bone, still on the middle constrictor, and then on the genioglossus. It courses deep to the hyoglossus muscle. It continues anteriorly and superiorly on the surface of the genioglossus muscle and along the anterior border of the hyoglossus to reach the base of the tongue. It gives off its sublingual branch to the sublingual glands, and then continues as the deep lingual artery to the tip of the tongue. Its branches are (a) the suprahyoid, (b) dorsal lingual, (c) sublingual, and (d) deep lingual. The **suprahyoid** branch courses along the superior border of the

hyoid bone superficial to the hyoglossus muscle as it is given off before the hyoglossus muscle is reached; this vessel is rather inconstant. The **dorsal lingual** may be either one or two vessels that ascend deep to the hyoglossus between this muscle and the genioglossus and supply the dorsal region of the tongue and the tonsil. The **sublingual** artery courses forward between the mylohyoid and genioglossus to reach the sublingual gland, while the **deep lingual** is merely the continuation of the lingual to the tip of the tongue (note Fig. 7-80).

4. **Facial** (Figs. 7-102 and 7-14). The facial artery arises just superior to the lingual, or in common with it. It ascends deep to the posterior belly of the digastric and the stylohyoid muscles, and deep to the angle of the mandible. It makes a groove in the posterior end of the submandibular gland and courses between it and the medial pterygoid muscle. It curves around the inferior border of the mandible and then courses in the face about ½ inch from the corner of the mouth and deep to some of the facial muscles; it ends as the angular artery, which continues to the medial corner of the eye. This vessel is tortuous, the coiling being necessary to adjust to the opening of the jaw and the wide movements of the cheeks. Its branches are (a) ascending palatine, (b) ton-

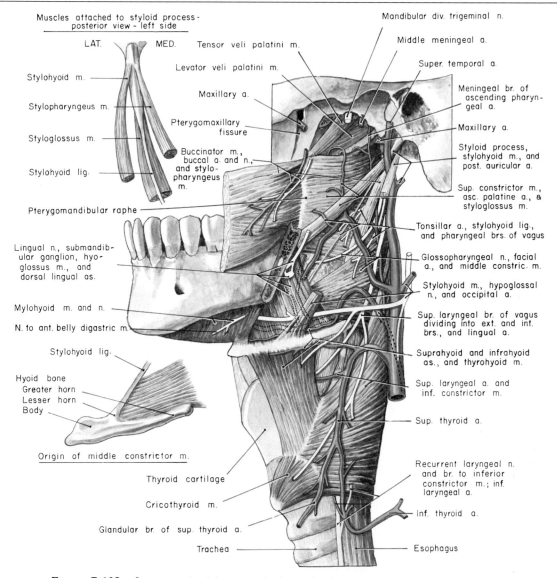

Muscles attached to styloid process - posterior view - left side

LAT.   MED.

Stylohyoid m.

Stylopharyngeus m.

Styloglossus m.

Stylohyoid lig.

Pterygomandibular raphe

Lingual n., submandibular ganglion, hyoglossus m., and dorsal lingual as.

Mylohyoid m. and n.

N. to ant. belly digastric m.

Stylohyoid lig.

Hyoid bone
Greater horn
Lesser horn
Body

Origin of middle constrictor m.

Thyroid cartilage

Cricothyroid m.

Glandular br. of sup. thyroid a.

Trachea

Tensor veli palatini m.

Levator veli palatini m.

Maxillary a.

Pterygomaxillary fissure

Buccinator m., buccal a. and n., and stylopharyngeus m.

Mandibular div. trigeminal n.

Middle meningeal a.

Super. temporal a.

Meningeal br. of ascending pharyngeal a.

Maxillary a.

Styloid process, stylohyoid m., and post. auricular a.

Sup. constrictor m., asc. palatine a., & styloglossus m.

Tonsillar a., stylohyoid lig., and pharyngeal brs. of vagus

Glossopharyngeal n., facial a., and middle constric. m.

Stylohyoid m., hypoglossal n., and occipital a.

Sup. laryngeal br. of vagus dividing into ext. and int. brs., and lingual a.

Suprahyoid and infrahyoid as., and thyrohyoid m.

Sup. laryngeal a. and inf. constrictor m.

Sup. thyroid a.

Recurrent laryngeal n. and br. to inferior constrictor m.; inf. laryngeal a.

Inf. thyroid a.

Esophagus

Figure 7-102. Lateral view of the visceral tube and its blood and nerve supply. Inserts show origins of the styloid muscles, and the origin of the middle constrictor muscle of the pharynx.

sillar, (c) glandular, (d) submental, (e) inferior labial, (f) superior labial, (g) lateral nasal, (h) muscular, and (i) angular. The **ascending palatine** (Fig. 7-102) courses superiorly on the superior constrictor muscle of the pharynx to reach the base of the skull. It divides into two branches one of which follows the levator veli palatini muscle to the soft palate, while the other curves over the superior border of the superior constrictor muscle and

supplies the soft palate, palatine tonsil, and auditory tube. The **tonsillar** branch (Fig. 7-102) courses superiorly and pierces the superior constrictor muscle of the pharynx, to end in the palatine tonsil. **Glandular** branches supply the submandibular gland and the **submental** artery arises where the facial turns around the inferior border of the mandible. This vessel runs anteriorly deep to the mandible as far as the symphysis. The remaining branches are all located in the face and are distributed as indicated by their names; they were described previously (page 482).

5. Occipital artery (Figs. 7-77 and 7-100). This artery arises from the external carotid near the posterior belly of the digastric muscle. It courses superiorly and posteriorly deep to the digastric muscle and grooves the mastoid process. Continuing posteriorly, it is superficial to the obliquus capitis superior and semispinalis capitis muscles, and deep to the splenius capitis muscle. After escaping from the medial border of the splenius capitis, it continues onto the posterior surface of the scalp. The branches of the occipital artery are (a) muscular, (b) meningeal, (c) descending, and (d) terminal branches. The **muscular** branches consist of two to the sterno-cleidomastoid muscle. The hypoglossal nerve hooks around one of these muscular branches. The **meningeal** branches enter the skull through the jugular and mastoid foramina. The **descending** branch arises from the artery on the surface of the obliquus capitis superior muscle. It continues to the lateral border of the semispinalis capitis and divides into superficial and deep branches, which course on corresponding surfaces of this muscle (Fig. 2-24 on page 74). It anastomoses with the deep cervical branch of the costocervical trunk, the ascending cervical branch of the thyrocervical trunk, and the superficial branch of the transverse cervical artery. The **terminal** branches course on the posterior surface of the scalp.

6. Posterior auricular (Figs. 7-77 and 7-100). This vessel arises immediately superior to the posterior belly of the digastric and courses superiorly and posteriorly close to the styloid process. It continues between the mastoid process and the external auditory meatus, and ends in occipital and auricular branches. It has **branches to the parotid gland** and adjacent muscles, and a **stylomastoid** branch, which enters the stylomastoid foramen and follows the facial canal to end finally in twigs to the medial wall of the middle ear, mastoid air cells, stapedius muscle, and the middle ear bones. The **auricular** branch supplies both surfaces of the external ear, and the **occipital** branch supplies the adjacent scalp.

7. Superficial temporal (Fig. 7-14). This artery begins deep to the neck of the mandible as one of the two terminal branches. It ascends anterior to the ear and ends by dividing into its **anterior** and **posterior** branches. The

branches of this artery have already been described on page 486.

8. Maxillary artery (Fig. 7-73). This artery, the other terminal branch of the external carotid, has been described on page 551. The third part of the artery will be discussed when the nasal cavity is described.

VEINS

The **superficial veins** of the head and neck have already been described on page 485.

The **internal jugular vein** starts as a continuation of the sigmoid sinus at the jugular foramen. It continues inferiorly in the carotid sheath just lateral to the internal carotid and common carotid arteries, and anterolateral to the vagus nerve. It remains just anterior to the prevertebral musculature in its entire descent to join the subclavian vein to form the brachiocephalic vein. It is just lateral to the visceral tube in its entire extent, being lateral to the pharynx superiorly, to the larynx in the middle of the neck, and to the trachea and esophagus in the inferior part of the neck. The following structures are superficial to the internal jugular vein—the styloid process with the stylopharyngeus and stylohyoid muscles; parotid gland; posterior belly of the digastric; sternocleidomastoid muscle; omohyoid, sternohyoid, and sternothyroid muscles; and many other nerves and vessels that you will gradually learn as you study these respective structures.

The tributaries of the internal jugular vein are:

1. Inferior petrosal sinus
2. Lingual
3. Pharyngeal
4. Facial
5. Occipital
6. Superior thyroid (already described; Fig. 7-89)
7. Middle thyroid (already described; Fig. 7-89)

1. Inferior petrosal sinus. This leaves the skull through the anterior part of the jugular foramen and joins the jugular bulb of the internal jugular vein.

2. Lingual veins. These start on the dorsum and sides of the tongue and pass posteriorly along the same course as taken by the lingual artery. The vena comitans of the hypoglossal nerve is often a vein of some size, which accompanies the hypoglossal nerve and therefore is superficial to the hyoglossus muscle, while the lingual veins course deep to it. The vena comitans of the hypoglossal nerve may empty into the internal jugular vein directly, or indirectly via the lingual veins.

3. Pharyngeal veins. These are the drainage of a pharyngeal plexus on the outer surface of the pharynx. After receiving communications from the posterior meningeal veins and the vein of the pterygoid canal, this plexus finally ends in the internal jugular vein.

4. Facial vein. This vessel has already been described on page 482, but does continue down across the face from the angle of the eye, to empty finally into the internal jugular vein.

5. Occipital veins. These vessels take the same course as the occipital artery and empty into the internal jugular vein on occasion; usually, however, the occipital vein empties into the suboccipital plexus of veins in the suboccipital triangle.

6. Superior thyroid vein. It starts in the substance and on the anterior surface of the thyroid gland and has tributaries that correspond to the branches of the superior thyroid artery. (See also page 572 and Fig. 7-89.)

7. Middle thyroid vein. This vessel collects blood from the inferior part of the thyroid gland and from part of the larynx and trachea, and finally courses posterolaterally to empty into the internal jugular vein opposite the thyroid gland. (See also page 573 and Fig. 7-89.)

NERVES

The nerves involved in this area are:

1. Facial
2. Glossopharyngeal
3. Vagus
4. Spinal accessory
5. Hypoglossal
6. Sympathetic chain.

1. Facial (Fig. 7-100). The only branches of the facial nerve involved in this area are those to the posterior belly of the digastric and the stylohyoid muscles. The facial nerve, after emerging from the stylomastoid foramen, curves anteriorly around the lateral side of the styloid process and the internal jugular vein, and sinks into the parotid gland. It remains lateral to all structures lying deep to the parotid gland. The branch to the posterior belly of the digastric arises just after the nerve has

emerged from the stylomastoid foramen, and immediately sinks into the muscle. The branch to the stylohyoid muscle often arises in common with the branch to the digastric, but has a slightly longer course, entering the stylohyoid muscle close to its middle.

2. Glossopharyngeal (Figs. 7-100, 7-101, and 7-102). This nerve arises from the superior end of the medulla oblongata, leaves the cranial cavity via the jugular foramen, and, after sending its tympanic branch into the middle ear, descends between the internal carotid artery medially and the internal jugular vein laterally. It is deep to the styloid process and the muscles arising from it. It then curves anteriorly between the internal and external carotid arteries and curves around the stylopharyngeus muscle, coming to lie on its superficial surface. It then proceeds anteriorly to reach the palatine tonsil and the base of the tongue. It has two **ganglia,** an upper and a lower, in which its sensory cells are located.

The branches of the glossopharyngeal nerve are:

   a. Tympanic
   b. To carotid body and sinus
   c. To stylopharyngeus muscle
   d. Pharyngeal
   e. Tonsillar
   f. Lingual

a. Tympanic. You will recall that the tympanic branch of the glossopharyngeal nerve enters the tympanic cavity to ramify on the promontory of the medial wall of the middle ear. It provides sensory fibers to the middle ear and contributes to the formation of the lesser petrosal nerve, which innervates the parotid gland via the otic ganglion (Fig. 7-68).

b. Nerve to carotid body and sinus. The **carotid sinus** is a dilation that occurs at the origin of the internal carotid artery or where the common carotid artery divides into its internal and external branches. This structure has sensory receptors that respond to blood pressure. If the pressure is low, a reflex is initiated here that results in an increase in heart rate and output, thus elevating blood pressure. The **carotid body*** is a small

---

*R. D. Dripps, Jr., and J. H. Comroe, Jr., 1944, The clinical significance of the carotid and aortic bodies. *Am. J. Med. Sci.* 208: 681. An article worth reading. These "chemoreceptors" (carotid and aortic bodies) should not be confused with the "pressoreceptors" in the carotid sinus and aortic arch.

piece of tissue, found in the concavity formed by the origins of the internal and external carotid arteries, which is also equipped with sensory endings; these endings respond to the oxygen tension of the blood. A reflex is started when the oxygen tension of the blood decreases, which results in an increase in heart and respiratory rates. The **carotid nerve** arises from the glossopharyngeal as it passes the internal carotid artery, and descends directly inferiorly to reach the carotid sinus and body. This nerve often is joined by a branch of the vagus nerve and a branch from the superior cervical sympathetic ganglion.

c. Stylopharyngeal. This nerve enters the muscle as the glossopharyngeal winds around its lateral surface. This is the only muscle innervated by the glossopharyngeal nerve; it is derived from the third branchial arch.

d. Pharyngeal. These branches consist of small twigs that penetrate the superior pharyngeal constrictor muscle to reach the mucous membrane of the pharynx; these are sensory nerves. Another pharyngeal branch of the glossopharyngeal nerve seems to join with the pharyngeal branch of the vagus.

e and f. Tonsillar and lingual branches. These are the terminal branches of the glossopharyngeal nerve. They are given off while the nerve is deep to the hyoglossus muscle and in the wall of the pharynx, and send twigs to the epiglottis and soft palate as well as to the tonsil and tongue. These are sensory nerves, the branch to the tongue containing special afferent fibers for taste as well as general somatic afferent fibers for general sensation.

In summary (Fig. 7-103), the glossopharyngeal nerve contains branchial efferent fibers to the stylopharyngeus muscle derived from the third branchial arch; branchial afferent fibers for proprioception from this muscle; general visceral efferent fibers to the parotid gland and general visceral afferent fibers from the parotid gland, carotid body and sinus, tympanic cavity, mastoid air cells, auditory tube, pharynx, and posterior third of the tongue; and special afferent fibers for taste from the posterior one-third of the tongue. The ganglia of the glossopharyngeal nerve are in the jugular foramen.

3. Vagus nerve (Figs. 7-102 and 7-112). This nerve arises from the superior end of the medulla oblongata, very close to the origin of the ninth nerve. It also enters the jugular foramen and courses inferiorly in the carotid sheath posterior to the combined internal jugular

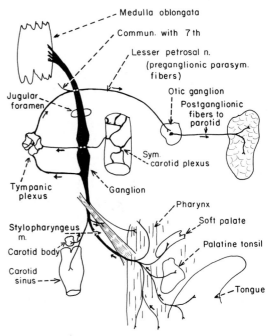

Figure 7-103. *Diagram of the glossopharyngeal nerve and its branches.*

vein and internal and common carotid arteries. It also has two ganglia, an upper and a lower, immediately inferior to the skull. It courses throughout the neck in this same relative position, and then passes anterior to the subclavian artery to enter the thoracic cavity.

The branches of the vagus nerve in the head and neck are:

a. Meningeal
b. Auricular
c. Pharyngeal
d. Superior laryngeal
e. Cardiac
f. Recurrent laryngeal

a. Meningeal (Fig. 7-104). This branch arises from the upper ganglion and supplies the dura mater in part of the posterior cranial fossa. Because this is out of character for the vagus nerve, some feel this branch may actually be derived from the first cervical segment of the spinal cord.

b. Auricular (Fig. 7-104). Arising from the upper ganglion, it passes into a small opening in the lateral wall of the jugular fossa and finally emerges through the tym-

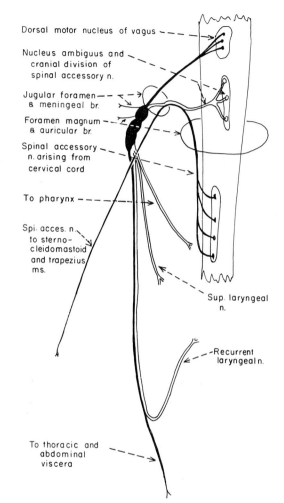

Dorsal motor nucleus of vagus

Nucleus ambiguus and
cranial division of
spinal accessory n.

Jugular foramen
& meningeal br.

Foramen magnum
& auricular br.

Spinal accessory
n. arising from
cervical cord

To pharynx

Spi. acces. n.
to sterno-
cleidomastoid
and trapezius
ms.

Sup. laryngeal
n.

Recurrent
laryngeal n.

To thoracic and
abdominal
viscera

Figure 7-104. *Diagram of vagus and spinal accessory nerves. The meningeal branch is probably derived from the first cervical nerve.*

part in the formation of the pharyngeal plexus. These branches are distributed to the muscles of the pharynx and to the muscles of the soft palate except for the tensor veli palatini, which is innervated by the fifth nerve. In addition, branches from the pharyngeal nerve can be found joining the glossopharyngeal nerve or passing separately to the carotid body and sinus.

d. **Superior laryngeal** (Figs. 7-102 and 7-112). This branch arises from the lower ganglion and runs inferiorly and anteriorly between the internal carotid artery laterally and the superior constrictor muscle of the pharynx medially. It divides into **internal** and **external laryngeal branches** on the surface of the middle constrictor of the pharynx. The external continues inferiorly on the inferior constrictor in a plane deep to the superior thyroid artery, an important relationship, and the thyroid gland. It gives partial nerve supply to the inferior constrictor muscle and then ends in the cricothyroid muscle. The internal branch runs inferiorly and anteriorly and passes through the thyrohyoid membrane. It gives nerve supply to the mucous membrane of the larynx as far inferior as the vocal cords, and to the membrane on the posterior surface of the epiglottis and the valleculae between the epiglottis and the larynx. These fibers serve as the afferent end of the cough reflex. It has been suggested that this nerve innervates part of the transverse arytenoid muscle (see page 614).

e. **Cardiac branches** (Fig. 7-92). There are two cardiac branches of the vagus nerve arising in the neck, one fairly high up and the other low down. On the right side, both these branches follow the vagus nerve inferiorly and usually pass posterior to the subclavian artery to enter the thoracic cavity. Here they course on the trachea to reach the deep cardiac plexus located on the bifurcation of this organ. The superior cardiac branch of the left vagus takes the route just described and ends in the deep cardiac plexus. The lower cardiac branch, however, does not end in the deep cardiac plexus, but travels inferiorly and then crosses anterior to the aortic arch to end in the superficial cardiac plexus. A review of Figure 4-12 on page 166 would be helpful.

f. **Recurrent laryngeal nerves** (Figs. 7-91, 7-92, and 7-102). These nerves arise differently on the two sides of the body. On the right, the recurrent nerve arises just inferior to the subclavian artery, curls around this artery and courses superiorly in the groove between the trachea and the esophagus. It finally reaches the trachea

panomastoid fissure on the side of the skull. It supplies the skin on the back of the external meatus, the skin that lines the lower part of the meatus, and the lower part of the tympanic membrane. This becomes a nerve of some importance in irritations of this region; pain from this area can be referred to the larynx, inducing a cough, or to the gastric area, inducing nausea.

c. **Pharyngeal** (Figs. 7-102 and 7-112). The pharyngeal nerve arises from the lower ganglion and runs inferiorly and anteriorly between the external and internal carotid arteries to end in the side of the pharynx, taking and supplies all the muscles of the larynx except the cricothyroid (and possibly the transverse arytenoid) and the mucous membrane from the vocal cords down. Its relation with the inferior thyroid artery should be recalled. The recurrent laryngeal nerve on the left side does not take the same pathway, but curls around the aortic arch and ligamentum arteriosum to turn superiorly and course in the groove between the trachea and the esophagus on the left side. This difference between the right and left recurrent laryngeal nerves is due to the differences in development of the aortic arches on the two sides of the body. These recurrent laryngeal nerves are of extreme importance and, because they course close to the thyroid gland, must be watched at all times during surgery in this area. If the recurrent laryngeal nerves are cut, all control of the vocal cords is lost and, because the cricothyroid muscle and possibly the transverse arytenoid muscle are the only ones with nerve supply, they tend to bring the vocal cords closer together. The patient, therefore, may suffocate.

In summary (Fig. 7-104), if the cranial portion (see below) of the spinal accessory nerve is included, the vagus nerve in this area has general somatic afferent fibers to the external auditory meatus and tympanic membrane; general visceral efferent fibers to the trachea and heart (those who have studied the thoracic and abdominal cavities know that it has a wide distribution in these areas); general visceral afferent fibers from these same organs, from the mucous membrane lining the larynx, and possibly from the carotid body and sinus; branchial efferent fibers to muscles of the pharynx, soft palate, and larynx derived from the fourth and sixth branchial arches; branchial afferent fibers for proprioception from these muscles; and special afferent fibers for taste around the epiglottis.

4. Spinal accessory (Figs. 7-100 and 7-101). This nerve is composed of two parts: (a) a cranial portion and (b) a spinal portion. The cranial portion is found in the vagus nerve and is actually distributed with this nerve, the fibers innervating the pharynx and larynx as just described under the vagus nerve. The spinal accessory nerve as dissected consists of the spinal portion of the accessory nerve only. This complexity can be explained as follows.

The spinal portion of the accessory nerve arises in the anterior gray column of the spinal cord just posterior to the origin of the ventral roots of the upper five cervical

nerves. They emerge from the side of the spinal cord between the ventral and dorsal roots and unite to form one or two cords, which ascend through the foramen magnum and then turn laterally to enter the jugular foramen. The cranial root arises from the medulla oblongata, and the cell bodies are actually located in the inferior end of the nucleus ambiguus. They emerge from the side of the medulla inferior to the vagus, course laterally to the jugular foramen and join the spinal portion of the accessory nerve. These nerves, however, do not remain intact, but separate. The cranial root joins with the vagus nerve and is distributed with this nerve as mentioned previously; this accounts for the pharyngeal and laryngeal branches of the vagus nerve. The spinal root descends between the internal carotid artery and the internal jugular vein; it becomes more anterior as it courses inferiorly and then crosses anterior to the internal jugular vein and enters the superior end of the sternocleidomastoid muscle. It then reappears at the posterior side of the sternocleidomastoid, and crosses the posterior triangle to reach the trapezius muscle.

In summary (Fig. 7-104), the spinal portion of the spinal accessory nerve contains branchial efferent fibers to the sternocleidomastoid and trapezius muscles, and branchial afferent fibers of proprioception from these muscles. If the cranial portion of the spinal accessory nerve is included in this nerve, this portion of the nerve will contain branchial efferent fibers to muscles of the pharynx and larynx and branchial afferent fibers of proprioception from these same muscles; this nerve also contains general visceral afferent fibers to the mucous membrane on the inside of the larynx.*

5. Hypoglossal nerve (Figs. 7-100, 7-101, and 7-105). The twelfth cranial nerve arises from the superior end of the medulla oblongata from a long series of roots emerging in a vertical row. The bundles unite and pass through the hypoglossal canal. It at first passes laterally in a position posterior to the internal carotid artery, and then forward between the internal jugular vein and the carotid artery, coursing to the lateral side of the vagus nerve in so doing. It continues inferiorly and appears between the two vessels, becoming superficial to them at the angle of the mandible. It then loops around a ster-

---

*Note Straus and Howell, 1936, and Corbin and Harrison, 1939, cited on page 565.

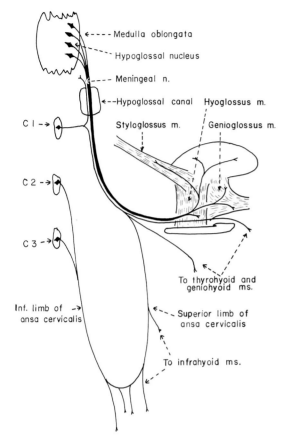

Figure 7-105. *Diagram of the hypoglossal nerve.*

nocleidomastoid branch of the occipital artery, passes anterior to the external carotid artery, and then disappears deep to the digastric and stylohyoid muscles. It continues anteriorly on the hyoglossus muscle, between it and the mylohyoid, and is distributed to both extrinsic and intrinsic muscles of the tongue.

The branches of the hypoglossal nerve are meningeal, a contribution to the ansa cervicalis, a thyrohyoid branch, geniohyoid branch, and the terminal muscular branches. The contribution to the ansa cervicalis, the meningeal, and the branches to the thyrohyoid and geniohyoid muscles are not strictly part of the hypoglossal nerve. They are branches of the first cervical nerve, which have used the hypoglossal nerve as a pathway, and therefore, should not be considered as part of the twelfth nerve. The remaining branches—the terminal branches —are the only true branches of this nerve.

Meningeal (Fig. 7-105). These are minute filaments given off while the nerve is in the hypoglossal canal. They pass proximally to supply part of the dura mater in the posterior cranial fossa. These are contributions from the cervical nerves.

Terminal branches (Fig. 7-78). The two terminal branches are separate fibers that are distributed to the styloglossus, hyoglossus, genioglossus, and intrinsic muscles of the tongue.

Therefore, since the tongue musculature is considered to be derived from somites, the hypoglossal nerve has general somatic efferent fibers for tongue musculature and general somatic afferent fibers for proprioception from these muscles.

6. Sympathetic chain (Fig. 7-94). The sympathetic chain finally terminates superiorly in the relatively large **superior cervical ganglion** found imbedded in the fascia forming the posterior wall of the carotid sheath. It is at the level of the second cervical vertebra. This ganglion is considered to be a fusion of the sympathetic primordia from the upper four cervical segments of the body.

The superior cervical ganglion has the following connections (Fig. 7-106):

   a. Internal carotid nerve
   b. Branches to upper three or four cervical spinal nerves
   c. Pharyngeal branches
   d. Nerves to external carotid artery
   e. Carotid branches
   f. Superior cardiac nerve

   a. The **internal carotid nerve.** This group of postganglionic fibers immediately joins the internal carotid artery and travels superiorly on this artery. It gives branches into the tympanic plexus, a contribution—the deep petrosal—that joins with the greater petrosal to form the nerve of the pterygoid canal, and it has connections to each of the cranial nerves found in the cavernous sinus.

   b. **Branches to the upper three or four cervical spinal nerves,** which are the gray rami communicantes to these nerves. In this way sympathetic fibers are delivered to the cervical nerves and distributed to the territory supplied by these nerves. Because the cervical region of the spinal cord has no sympathetic cells, there are no white rami communicantes associated with the cervical nerves.

   c. **Pharyngeal branches.** There are usually sev-

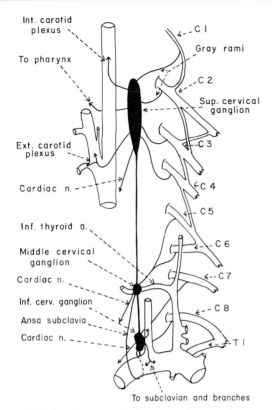

Int. carotid plexus
To pharynx
C I
Gray rami
C 2
Sup. cervical ganglion
Ext. carotid plexus
C 3
Cardiac n.
C 4
C 5
Inf. thyroid a.
C 6
Middle cervical ganglion
Cardiac n.
C 7
Inf. cerv. ganglion
Ansa subclavia
C 8
Cardiac n.
T I
To subclavian and branches

Figure 7-106. *Diagram of the cervical sympathetic ganglia and their connections. (Modified after Jamieson.)*

eral of these nerves, which leave the medial side of the ganglion and course toward the pharynx. They join in the formation of the pharyngeal plexus with the glossopharyngeal and vagus nerves.

d. Nerves to the external carotid artery. There are usually two of these nerves, which are quite large, and the postganglionic fibers on these arteries are distributed to all parts of the body where the branches of the external carotid artery ramify.

e. Carotid branches. These are branches to the carotid body and sinus and probably serve a visceral motor function.

f. Superior cardiac nerve. This arises by two or three filaments from the ganglion and proceeds inferiorly in the fascia forming the posterior wall of the carotid sheath. Inferiorly, the branches on the two sides differ in their course. On the right side, the superior cardiac branch passes either posterior or anterior to the subcla-

vian artery and joins the deep cardiac plexus by coursing on the anterior surface of the trachea. On the left side this branch joins the superficial cardiac plexus by crossing the anterior surface of the arch of the aorta (*see* Fig. 4-12 on page 166).

———

If the skull is disarticulated from the atlas, the skull and the attached pharynx can be reflected anteriorly, thus opening the space between the pharynx and vertebral column—the retropharyngeal space. If this is done, the relationships of the seventh, ninth, tenth, eleventh, and twelfth nerves to the contents of the carotid sheath and to the branches of the external carotid artery can be seen more easily than before (*see* Fig. 7-112). The nerves in the carotid triangle are difficult to visualize, particularly in reference to their relations to each other. However, they do pierce the carotid sheath near the base of the skull, and their relation to the internal carotid artery and internal jugular vein can be remembered very easily and gives one a good insight into this area.

VISCERAL TUBE

The visceral tube consists of the pharynx, larynx, trachea, and esophagus.

Since most dissections of the pharynx involve reflecting it anteriorly, which demands cutting through the atlanto-occipital joint, the ligaments connecting the axis to the atlas, and both atlas and axis to the skull, will be described at this time. Furthermore, this procedure also provides access to the prevertebral muscles. This entire area is becoming more important with the many complaints of pain as the result of collisions in cars—the whiplash condition.

ATLANTOAXIAL AND ATLANTO-OCCIPITAL JOINTS

The ligaments involved are:

Axis to atlas
   1. Anterior longitudinal ligament

2. Ligamenta flava (posterior atlantoaxial)
3. Transverse ligament
4. Ligamentum nuchae

Axis to skull
1. Ligamentum nuchae
2. Longitudinal bands of ligamentum cruciatum
3. Apical and alar ligaments of dens
4. Membrana tectoria

Atlas to skull
1. Anterior and posterior atlanto-occipital membranes

The **supraspinal ligaments** connect directly with the **ligamentum nuchae,** a large and strong ligament, which increases in anterior-posterior dimension as it approaches the skull, to be attached to the external occipital protuberance and crest (Fig. 7-98). It is approximately 2 inches in depth at its superior end and serves for attachment of many muscles as well as to aid in holding the head erect. The **ligamenta flava** connect the axis with the atlas, but a ligament similarly placed between the atlas and the occipital bone receives the name of **posterior atlanto-occipital membrane** (Fig. 7-107).

The **posterior longitudinal ligament** is made up of a thin superficial layer and a thicker deep layer. The superficial layer attaches to the occipital bone; the deeper fibers attach to the axis and then continue to the skull as the **tectorial membrane** (Fig. 7-108).

If this tectorial membrane is removed, special ligaments are seen connecting the superiorly projecting dens of the axis with surrounding structures (Fig. 7-109). These ligaments form a cross—**cruciform ligament**—and consist of a strong **transverse ligament** extending between tubercles on the inside surface of the atlas, an extension superiorly—**superior longitudinal fasciculus**—to the occipital bone, and another inferiorly—**inferior longitudinal fasciculus**—to the axis. The transverse part is posterior to the dens and forms a ring, with the aid of the atlas anteriorly, in which the dens pivots during movement at this joint. There is a synovial joint between the ligament and the dens.

If the superior fasciculus of the cruciform ligament is removed, the superior surface of the dens is visible. The dens has three ligaments: the **apical ligament** extends directly superiorly and attaches to the edge of the foramen magnum, and the two **alar ligaments** extend superolaterally and also attach to the edge of the foramen magnum.

Moving to the anterior side of the vertebral column,

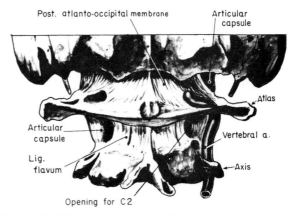

Figure 7-107. *Posterior view of ligaments attaching axis to atlas and atlas to the skull.*

the anterior longitudinal ligament ascends as far as the atlas and is directly continuous with the anterior atlanto-occipital membrane, which attaches to the occipital bone (Fig. 7-110). If this membrane is removed, the tubercle of the anterior arch of the atlas is seen. The dens is posterior to this tubercle, the two bones being separated by a joint cavity.

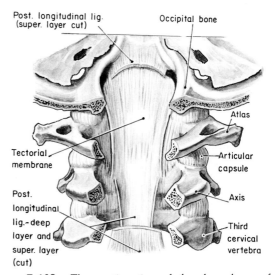

Figure 7-108. *The continuation of the deep layer of the posterior longitudinal ligament—the tectorial membrane. The superficial layer of the posterior longitudinal ligament has been cut.*

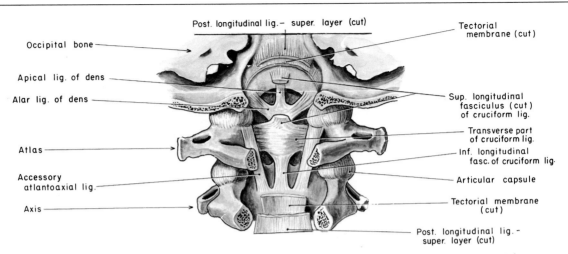

Figure 7-109. *Cruciform ligament and ligaments of the dens as seen after removing tectorial membrane.*

PREVERTEBRAL MUSCLES

Also on this anterior surface of the cervical vertebrae one can see the **prevertebral muscles** (Fig. 7-111). These consist of:

1. Longus colli
2. Longus capitis
3. Rectus capitis anterior
4. Rectus capitis lateralis

1.   Longus colli.   This muscle lies on the anterior surface of the cervical vertebrae, and is **attached** by tendinous slips to the vertebral bodies and transverse processes, and to the anterior tubercle of the atlas. **Nerve supply** is ventral rami of C3 to C8. Its **main action** is to bend and rotate the neck to the same side.

2.   Longus capitis.   This muscle lies on the anterior surface of the upper cervical vertebrae only. Its **origin** is from the anterior tubercles of the transverse processes of cervical vertebrae 3 to 6. (Note that this is exactly the same as the origin of the scalenus anterior muscle, the two muscles forming a continuity.) **Insertion** is onto the basilar portion of the occipital bone close to the pharyngeal tubercle (see Fig. 7-4). **Nerve supply** is ventral rami C1 to C4. **Main action** is to flex head and turn face slightly to the same side.

3.   Rectus capitis anterior.   This small muscle lies deep to the longus capitis muscle. Its **origin** is from the anterior surface of the atlas. **Insertion** is onto the basilar portion of the occipital bone (Fig. 7-4). **Nerve supply** is ventral ramus of C1. **Main action** is to flex the head.

4.   Rectus capitis lateralis.   This muscle is lateral in position to the rectus capitis anterior. Its **origin** is from the transverse process of the atlas. **Insertion** is onto the

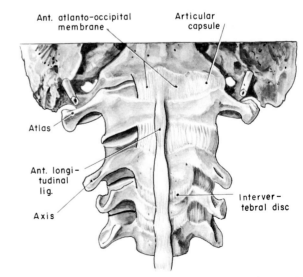

Figure 7-110. *Anterior view of ligaments attaching axis to atlas and atlas to the skull. The articular capsules have been omitted from right side of body.*

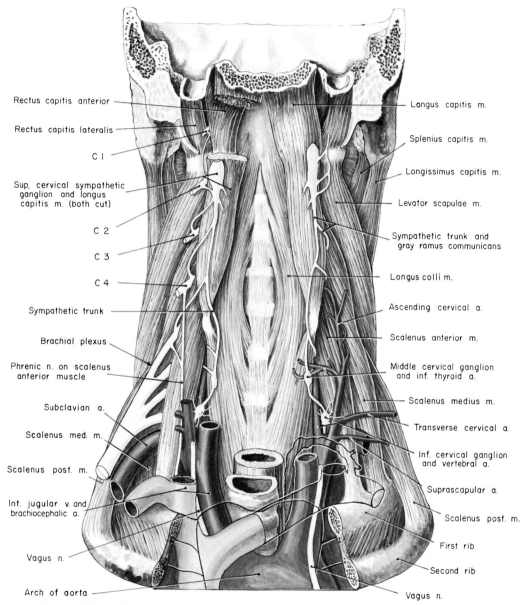

Rectus capitis anterior

Rectus capitis lateralis

C I

Sup. cervical sympathetic ganglion and longus capitis m. (both cut)

C 2

C 3

C 4

Sympathetic trunk

Brachial plexus

Phrenic n. on scalenus anterior muscle

Subclavian a.

Scalenus med. m.

Scalenus post. m.

Int. jugular v. and brachiocephalic a.

Vagus n.

Arch of aorta

Longus capitis m.

Splenius capitis m.

Longissimus capitis m.

Levator scapulae m.

Sympathetic trunk and gray ramus communicans

Longus colli m.

Ascending cervical a.

Scalenus anterior m.

Middle cervical ganglion and inf. thyroid a.

Scalenus medius m.

Transverse cervical a.

Inf. cervical ganglion and vertebral a.

Suprascapular a.

Scalenus post. m.

First rib

Second rib

Vagus n.

Figure 7-111. *The scalene and prevertebral muscles, the relation of the subclavian vessels to these muscles, the branches of the subclavian artery, and the sympathetic chain. (Inferior cervical ganglion is lateral to vertebral artery in this specimen.)*

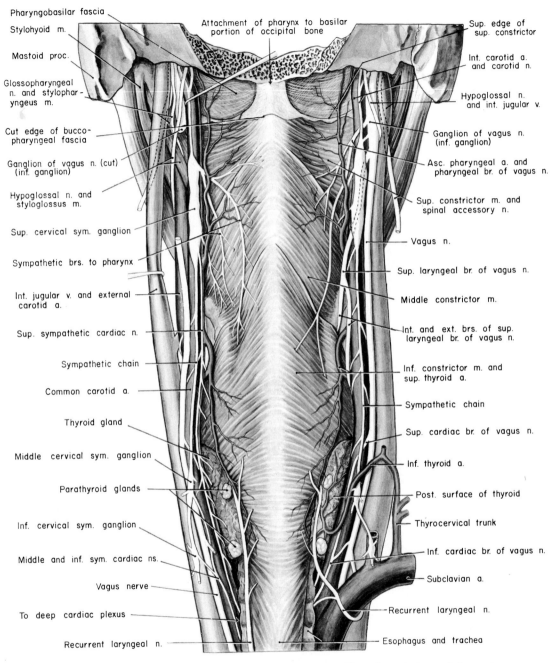

Pharyngobasilar fascia

Stylohyoid m.

Mastoid proc.

Glossopharyngeal n. and stylopharyngeus m.

Cut edge of bucco-pharyngeal fascia

Ganglion of vagus n. (cut) (inf. ganglion)

Hypoglossal n. and styloglossus m.

Sup. cervical sym. ganglion

Sympathetic brs. to pharynx

Int. jugular v. and external carotid a.

Sup. sympathetic cardiac n.

Sympathetic chain

Common carotid a.

Thyroid gland

Middle cervical sym. ganglion

Parathyroid glands

Inf. cervical sym. ganglion

Middle and inf. sym. cardiac ns.

Vagus nerve

To deep cardiac plexus

Recurrent laryngeal n.

Attachment of pharynx to basilar portion of occipital bone

Sup. edge of sup. constrictor

Int. carotid a. and carotid n.

Hypoglossal n. and int. jugular v.

Ganglion of vagus n. (inf. ganglion)

Asc. pharyngeal a. and pharyngeal br. of vagus n.

Sup. constrictor m. and spinal accessory n.

Vagus n.

Sup. laryngeal br. of vagus n.

Middle constrictor m.

Int. and ext. brs. of sup. laryngeal br. of vagus n.

Inf. constrictor m. and sup. thyroid a.

Sympathetic chain

Sup. cardiac br. of vagus n.

Inf. thyroid a.

Post. surface of thyroid

Thyrocervical trunk

Inf. cardiac br. of vagus n.

Subclavian a.

Recurrent laryngeal n.

Esophagus and trachea

Figure 7-112. *Posterior view of pharyngeal constrictors. Note the relations of the various nerves to the internal carotid artery and internal jugular vein, and to the visceral tube. The pharyngeal plexus is composed of a mixture of the nerves shown on both sides of the middle constrictor muscle.*

occipital bone just posterior to the jugular fossa (Fig. 7-4). **Nerve supply** is ventral ramus of C1. **Main action** is to bend head to same side.

PHARYNX

The **pharynx** is a muscular tube extending from the base of the skull superiorly to the esophagus inferiorly (Figs. 7-102, 7-112, and 7-113). It possesses posterior and lateral walls (although the lateral walls are usually quite narrow), but no anterior wall since it is directly continuous anteriorly with the nasal cavity, the oral cavity, and the larynx. The pharynx serves the double role of a passageway for air and for food. The passage for air is always open except during the process of swallowing. It should be noted that the air passage in the pharynx, between the nasal cavity and the larynx, crosses the passage for food between the oral cavity and the esophagus. Therefore,

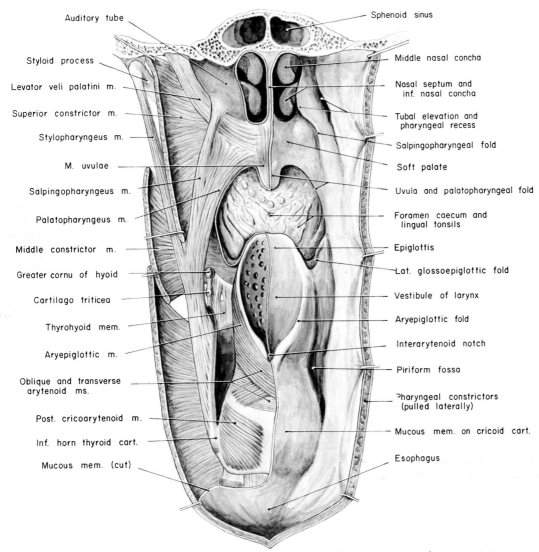

Figure 7-113. *Interior of the pharynx—posterior view. The mucous membrane is intact on the right side, removed from the left.*

one must hold one's breath during swallowing; disobedience leads to an immediate coughing spasm. The pharynx is located in the midline anterior to the vertebral column and the prevertebral musculature. The carotid sheath and its contents (internal jugular vein, common carotid artery, and vagus nerve inferiorly; internal jugular vein, ninth, tenth, eleventh, and twelfth nerves, and internal carotid superiorly) course throughout the extent of the neck, from the base of the skull to the thoracic inlet, just lateral to this visceral tube (Fig. 7-112).

The **walls** of the pharynx consist of five layers: (1) **mucous membrane,** which is composed of columnar epithelium, some areas being ciliated, while others are not; (2) **submucosa,** containing a profuse plexus of veins, many lymph nodules, and mucous glands; (3) a layer of **fascia**—the pharyngobasilar—located between the submucosa and muscular layer, which is quite firm, which fills in the gaps in the walls of the pharynx, and which is firmly attached to the base of the skull; (4) the **muscular layer;** and (5) the **buccopharyngeal fascia,** which lines the outside of the pharynx, forming the anterior boundary of the retropharyngeal space.

If the buccopharyngeal fascia is removed, the three constrictor muscles of the pharynx can be seen—the **superior, middle, and inferior constrictors** (Figs. 7-102 and 7-112). The entire inferior constrictor muscle is visible, but it overlaps the middle constrictor, thus covering a portion of this muscle, and the middle constrictor, in turn, overlaps the superior constrictor. This overlapping is important when one considers the fact that food has to pass on the inside of this tube. The muscles are arranged in such a manner that food will not penetrate between any two muscles.

Nasal pharynx (Fig. 7-113). This part of the pharynx is most superiorly located; it is just inferior to the base of the skull. The posterior wall contains a condensation of lymphoid tissue—**pharyngeal tonsils**—which can be a considerable nuisance when enlarged.

The lateral wall contains the opening of the **auditory tube,** located posterior to the nasal cavity almost in a direct line with the inferior nasal concha (see Fig. 7-120 on page 618). Just posterior to the auditory tube is an elevation that separates the tube from a fairly deep fossa—the **pharyngeal recess.** Extending inferiorly from this elevation, which is formed by the cartilaginous portion of the auditory tube, is a fold of mucous membrane—the **salpingopharyngeal fold**—over a mus-

cle of the same name (Fig. 7-113). This muscle arises from the auditory tube (salpinx = tube; *Gr.,* trumpet) and inserts on the thyroid cartilage. It raises the pharynx during swallowing.

The anterior wall is absent since the pharynx and nasal cavity are directly continuous with each other via the choanae. The choanae are oblong openings of considerable size—approximately 1 inch long and ½ inch wide—through which one can see the inferior part of the nasal cavity (Fig. 7-113).

The **soft palate** can be considered as providing a partial floor for the anterior part of the nasal pharynx. This very flexible structure is attached to the posterior part of the hard palate and laterally to the walls of the pharynx; it extends into the cavity of the pharynx, sloping inferiorly. Its convex superior surface is directly continuous with the floor of the nasal cavity, while its concave inferior surface forms part of the roof of the mouth. A midline structure—the **uvula**—projects further posteriorly than does the rest of the soft palate.

More detail will be given on the soft palate under the description of the mouth (see page 633).

Auditory tube (Figs. 7-113 and 7-114). Before leaving the nasal pharynx, we should give the auditory tube further attention. This important structure connects the nasal pharynx with the middle ear and thence with the mastoid air cells. It is responsible for maintenance of a similar pressure on the inner surface of the tympanic membrane as exists on its outer surface. It is approximately 1½ inches long and consists of a medial cartilaginous portion and a lateral bony part. The cartilaginous portion is about 1 inch in length, leaving ½ inch for the bony canal.

This canal takes a course that is lateral, posterior, and slightly superior in direction. The medial cartilaginous portion is considerably wider than the bony portion, the cartilage taking the shape shown in Fig. 7-120B with the larger portion medially placed. This portion is in immediate relation to the tensor and levator muscles of the palate (Fig. 7-101 and 120B), and these muscles and the salpingopharyngeus muscle are responsible for opening the tube. This cartilaginous portion, as mentioned previously, is responsible for the elevation just posterior to the opening of the tube; it is located in a groove on the base of the skull between the petrous part of the temporal bone and the greater wing of the sphenoid bone.

The lateral bony portion is located between the tym-

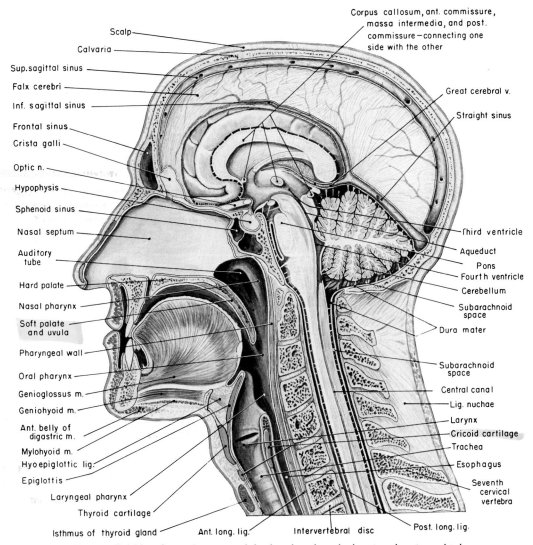

Figure 7-114. *Sagittal section of the head and neck showing the visceral tube.*

panic and petrous portions of the temporal bone. It is much smaller than the cartilaginous portion, although the part of the canal with the smallest lumen—the isthmus—occurs at the point where the two portions join.

The entire auditory tube is lined with the same epithelium as found in the pharynx, and this epithelium continues into the middle ear and onto its contents, and into the mastoid air cells.

Oral pharynx (Figs. 7-113, 7-114, and 7-115). This part of the pharynx is located posterior to the mouth.

Its posterior wall is similar to that found in the nasal portion.

Anteriorly and laterally, however, the appearance is quite different. The most prominent structure seen is the base of the **tongue,** which is located in the mouth. The **epiglottis** is equally obvious, but this, strictly speaking, is in the laryngeal portion of the pharynx. There is an area between these two structures, however, that is part of the oral pharynx.

The midline fold extending from the base of the

tongue to the epiglottis is the **median glossoepiglottic fold;** the two laterally placed folds are the **lateral glossoepiglottic folds,** and the valley-like areas between the two folds are the **valleculae epiglotticae** (note Fig. 7-130).

A fold of tissue extends inferiorly from the sides of the soft palate to blend with the side walls of the pharynx. This fold covers the **palatopharyngeus** muscle and takes this same name. Another pair of folds—**palatoglossal**—are located just anterior to the above folds and contain muscles of the same name—palatoglossus muscles. These two palatoglossal folds form an arch, which bounds the **isthmus faucium** or entrance to the throat. Therefore, anything located posterior to these arches is in the pharynx. You can see these structures by looking into your own mouth.

Palatine tonsils (Fig. 7-115) Two important structures—the **palatine tonsils**—are located in the tonsillar fossa between the palatoglossal and palatopharyngeal arches just described. In fact, tonsillar tissue forms a **ring** around the entrance from the oral cavity into the oral pharynx. Posteriorly tonsillar tissue is found on the pharyngeal wall, and anteriorly similar tissue can be found on the base of the tongue—the lingual tonsils. Between these two, and placed laterally, are the palatine tonsils.

The palatine tonsils project into the pharynx for a variable distance, depending on their size. The medial surface of these structures is covered with mucous membrane, which follows small crypts into the tonsillar tissue. The lateral surface of the tonsil is covered by a connective tissue capsule attached to the superior constrictor muscle of the pharynx but easily separated from it unless repeated infections have occurred. The close relation of the tonsil bed to the facial artery and to the internal carotid artery should be noted.

The **blood supply** to the palatine tonsils is very rich (Fig. 7-115). They receive branches from every artery in the vicinity. Branches are contributed from the **ascending pharyngeal** branch of the external carotid, from the **lesser palatine** branches of the maxillary (to be studied), from the **dorsal lingual** artery, and from the **ascending palatine** and **tonsillar** branches of the facial artery. The **veins** end in the pharyngeal plexus. The **nerves** are sensory in nature and are branches of the glossopharyngeal and the lesser palatine branches of the pterygopalatine ganglion (to be studied).

The **lymphatics** end in nodes near the angle of the mandible—the upper deep cervicals—one particular gland being named the jugulodigastric.

Laryngeal pharynx (Figs. 7-113 and 7-114). This part of the pharynx decreases rapidly in size from superior to inferior. Its posterior wall is directly continuous with that in the other parts of this organ and presents no changes. The anterior wall, on the other hand, is quite different. As indicated by the name laryngeal, the anterior wall of the laryngeal portion of the pharynx is concerned mainly with the larynx. In actuality, the anterior wall is made up of (1) the entrance to the larynx, (2) the mucous membrane on the cricoid cartilage, and (3) the piriform fossae located on each side of the above. The entrance to the larynx is bounded by the tip of the **epiglottis,** and by the two folds, which can be followed posteriorly and inferiorly to two projections apparently attached to the cricoid cartilage. These are the **arytenoid cartilages,** and since the folds mentioned above stretch between these and the epiglottis, they are called **aryepiglottic folds.** An **interarytenoid notch** occurs in the midline between the two cartilages.*

The **piriform recess** is a fairly deep hollow between the arytenoid cartilages and aryepiglottic folds medially and the side wall of the pharynx laterally. Actually, in this area the lateral wall of the pharynx consists of the hyoid bone, thyrohyoid membrane, and the thyroid cartilage.

MUSCLES.   The muscles of the pharynx are:

1. Superior constrictor
2. Middle constrictor
3. Inferior constrictor
4. Stylopharyngeus
5. Salpingopharyngeus
6. Palatopharyngeus

1. Superior constrictor (Figs. 7-102 and 7-112). This muscle **arises** from the pterygoid hamulus, from the pterygomandibular raphe, from the mandible near the posterior end of the mylohyoid line, and from the side of the tongue. The muscular fibers curve posteriorly to be **inserted** into the median raphe which is prolonged superiorly to the base of the skull to be attached onto the **pharyngeal tubercle** on the basilar part of the

---

*More detail will be given on these parts of the larynx when this organ is described.

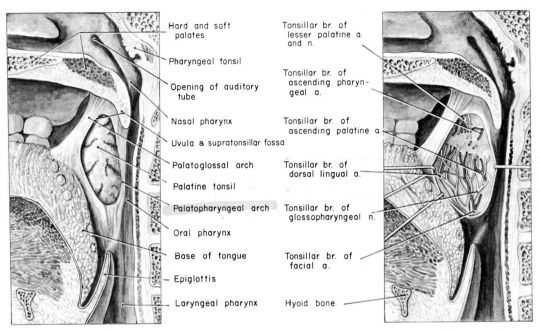

Hard and soft palates

Pharyngeal tonsil

Opening of auditory tube

Nasal pharynx

Uvula & supratonsillar fossa

Palatoglossal arch

Palatine tonsil

Palatopharyngeal arch

Oral pharynx

Base of tongue

Epiglottis

Laryngeal pharynx

Tonsillar br. of lesser palatine a. and n.

Tonsillar br. of ascending pharyngeal a.

Tonsillar br. of ascending palatine a.

Tonsillar br. of dorsal lingual a.

Tonsillar br. of glossopharyngeal n.

Tonsillar br. of facial a.

Hyoid bone

Figure 7-115. *On the left, the palatine tonsil surrounded by the palatoglossal arch anteriorly and palatopharyngeal arch posteriorly. On the right the tonsillar fossa is shown after removal of the palatine tonsil.*

occipital bone (Fig. 7-112). The interval between the superior border of the muscle and the base of the skull is closed by fascia.

2. Middle constrictor muscle (Figs. 7-102 and 7-112). This muscle **arises** from the whole length of the superior border of the greater horn of the hyoid bone, from the lesser horn, and from the inferior end of the stylohyoid ligament. The fibers course posteriorly to be **inserted** into the median raphe.

3. Inferior constrictor muscle (Figs. 7-102 and 7-112). The inferior constrictor is thicker than the other two and **arises** from the sides of the cricoid and thyroid cartilages. From these origins fibers course posteriorly to be **inserted** into the median raphe. This muscle can be divided into two portions—thyropharyngeus and cricopharyngeus—depending upon the origin from the thyroid or the cricoid cartilage. The most inferior fibers, which are horizontal in position, blend with the circular fibers of the esophagus. This is the narrowest part of the pharynx. During swallowing the thyropharyngeus acts in unison with the superior and middle constrictors in propulsion, while the cricopharyngeus, which acts as a

sphincter, must be relaxed. Failure to accomplish this relaxation can lead to acute respiratory infections.

4. Stylopharyngeus (Fig. 7-113). This muscle is long and slender, **arises** from the base of the styloid process, and courses inferiorly along the side of the pharynx between the superior and middle constrictor muscles. It then continues inferiorly external to the mucous membrane and **inserts** onto the posterior border of the thyroid cartilage along with the palatopharyngeus muscle.

5. Salpingopharyngeus (Fig. 7-113). This muscle **arises** from the inferior part of the auditory tube, passes inferiorly, and joins the palatopharyngeus muscle to **insert** on the posterior border of the thyroid cartilage.

6. Palatopharyngeus (Fig. 7-113). This muscle may be considered a muscle of the palate or of the pharynx; it forms the palatopharyngeal arch. It **arises** from the soft palate and passes laterally and inferiorly, posterior to the palatine tonsil, to join the stylopharyngeus to be **inserted** on the posterior part of the thyroid cartilage.

These muscles are used mainly in the process of swallowing. The salpingopharyngeus, palatopharyngeus, and stylopharyngeus muscles aid in elevating the thyroid cartilage, and the three constrictor muscles (except for the cricopharyngeal portion of the inferior constrictor) aid in pushing the bolus of food inferiorly into the esophagus. The palatopharyngeus and the salpingopharyngeus muscles, in addition, are considered to be important in closing off the nasal cavity during the process of swallowing.

NERVES. The **glossopharyngeal** nerve innervates the stylopharyngeus muscle (derived from the third branchial arch), while the remaining muscles are innervated by the pharyngeal branch of the **vagus** (or the cranial division of the eleventh nerve) since they are derived from the fourth branchial arch. (The superior laryngeal branch of the vagus aids in innervating the inferior constrictor muscle.)

The mucous membrane receives most of its sensory nerve supply from the **glossopharyngeal** nerve. However, it is aided at the superior end of the pharynx by pharyngeal branches of the maxillary division of the **trigeminal** nerve and over the palatine tonsillar region by descending palatine branches of the same nerve (Fig. 7-126). The mucous membrane of the pharynx is of interest when one is classifying the components of the cranial nerves. Viscera usually do not respond to stimuli of touch as does skin or muscle. Violent contraction or distension induces pain, but viscera such as the intestines can be cut without the patient's feeling a painful sensation. If the pharynx is a viscus, why does the mucous membrane respond so rapidly to touch—the gag reflex? The same is true of the larynx—the cough reflex. Nevertheless, most anatomists consider these fibers as general visceral afferents.

The mucous membrane is innervated also by the **sympathetic** nervous system via the superior cervical ganglion. In fact, the ninth and tenth nerves and branches from the superior cervical ganglion form a plexus on the middle constrictor muscle of the pharynx—the pharyngeal plexus.

ARTERIES AND VEINS. The arteries of the pharynx are the ascending pharyngeal branch of the external carotid, the ascending palatine and tonsillar branches of the facial, and, to be studied later, the descending palatine and pharyngeal branches of the maxillary artery. The veins of the pharynx form an external plexus and a plexus between the constrictor muscles and the pharyngobasilar fascia, that fascia just deep to the mucous

membrane. These plexuses drain to the pterygoid plexus and into the internal jugular vein as well.

LYMPH DRAINAGE. **Lymphatic** vessels of the pharynx drain into the retropharyngeal and upper deep cervical nodes.

LARYNX

The **larynx**[*] is a hollow organ, located in the anterior portion of the neck. It is a direct connection between the pharynx superiorly and the trachea inferiorly. This complex structure forms part of the visceral tube we have mentioned previously. It has the following **relations: Anterior:** skin; superficial fascia containing the platysma muscle; enveloping deep fascia; infrahyoid muscles and their fasciae; pretracheal fascia; and often an extra lobe of the thyroid gland—pyramidal lobe. **Lateral:** the lobes of the thyroid gland deep to the infrahyoid muscles; contents of carotid sheath. **Posterior:** pharynx; prevertebral muscles; third to sixth cervical vertebrae except during swallowing or phonation.

The larynx is made up of cartilages, ligaments and membranes to connect the cartilages, and muscles to cause movement of various parts of the larynx. It is lined with a mucous membrane of columnar ciliated cells and contains many mucous glands. The nerve supply is very important.

CARTILAGES. There are three unpaired cartilages: (1) thyroid, (2) cricoid, and (3) epiglottic; and three paired cartilages: (1) arytenoid, (2) corniculate, and (3) cuneiform. These are pictured in Figure 7-116, except for the cuneiform cartilages.

The **thyroid cartilage** has a peculiar shape, with two large plates anteriorly and laterally placed and fused in the midline anteriorly but open posteriorly. From the posterior edges, projections extending superiorly and inferiorly are called **superior** and **inferior horns.** The thyroid cartilage, therefore, is horseshoe-shaped, with the opening posteriorly placed. The superior edge of the laminae presents a notch in the midline—the **thyroid notch.** A structure called the **oblique line** is found on the lateral surface of the thyroid cartilage; this line starts at the

---

[*]V. E. Negus, 1949, *The Comparative Anatomy and Physiology of the Larynx,* Heinemann, London.

POSTERIOR VIEW   LATERAL VIEW

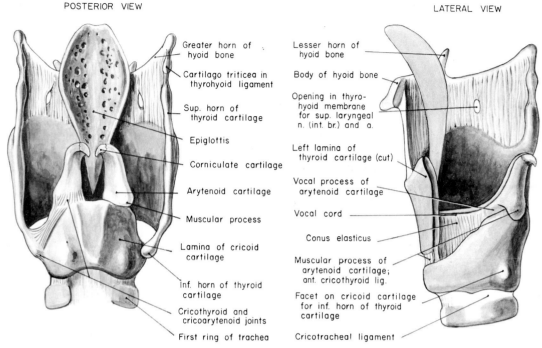

Greater horn of hyoid bone

Cartilago triticea in thyrohyoid ligament

Sup. horn of thyroid cartilage

Epiglottis

Corniculate cartilage

Arytenoid cartilage

Muscular process

Lamina of cricoid cartilage

Inf. horn of thyroid cartilage

Cricothyroid and cricoarytenoid joints

First ring of trachea

Lesser horn of hyoid bone

Body of hyoid bone

Opening in thyro-hyoid membrane for sup. laryngeal n. (int. br.) and a.

Left lamina of thyroid cartilage (cut)

Vocal process of arytenoid cartilage

Vocal cord

Conus elasticus

Muscular process of arytenoid cartilage; ant. cricothyroid lig.

Facet on cricoid cartilage for inf. horn of thyroid cartilage

Cricotracheal ligament

Figure 7-116.   *Posterior and lateral views of the cartilages and membranes of the larynx. The left side of the thyroid cartilage has been removed in the lateral view. In the posterior view the cricoarytenoid and cricothyroid joints are shown, with capsules on the left and without capsules on the right.*

base of the superior horn and extends anteriorly and inferiorly toward the inferior border of the thyroid cartilage. This line is important in the understanding of muscle attachments (Fig. 7-102).

The **cricoid cartilage** is the only complete ring found in the respiratory system. It is larger posteriorly than anteriorly, the posterior portion being formed by two laminae, while the anterior portion presents an arch. Near the base of each lamina is found the facet for articulation with the inferior horn of the thyroid cartilage. On the superior surface of the laminae are facets for articulation of the arytenoid cartilages. The posterior surfaces of the laminae show two hollowed areas on either side of the midline for the posterior cricoarytenoideus muscles.

The **epiglottic cartilage** is a leaf-shaped structure extending superiorly from the thyroid cartilage in a position just posterior to the hyoid bone and the base of the tongue. The base or stem of the epiglottic cartilage is

attached to the inside of the thyroid cartilage at the midline by the thyroepiglottic ligament.

The relatively small **arytenoid cartilages** rest upon the posterior portion of the cricoid cartilage. They possess an **apex,** a **base,** and **two processes.** The **base** is the point of attachment of the arytenoid cartilage to the cricoid cartilage, while the **apex** extends superiorly and is the point of attachment of the corniculate cartilages and the aryepiglottic fold. The **vocal process** is that which projects anteriorly and to which the vocal cord or vocal ligament is attached. The **muscular process** projects laterally and is the point of insertion of muscles that move the arytenoid cartilages.

The **corniculate cartilages** are small nodules on the apex of each arytenoid cartilage. They are, therefore, in the aryepiglottic fold near its point of attachment to the apex of the arytenoid cartilage.

The **cuneiform cartilages** are two small pieces of

cartilage also located in the aryepiglottic folds. They are of little importance.

JOINTS (Fig. 7-116).   There are two joints between the cartilages just described: (1) the cricothyroid joint between the inferior horn of the thyroid cartilage and the cricoid cartilage; and (2) cricoarytenoid joint between the arytenoid cartilages and the cricoid cartilage. These are freely movable joints, and both allow a rotary motion in addition to gliding movements.

LIGAMENTS AND MEMBRANES (Fig. 7-116).   The ligaments and membranes connecting the cartilages to each other and to the hyoid bone are important because they fill in the gaps between the cartilages, thereby completing the walls of the hollow organ. These ligaments and membranes are:

1. Thyrohyoid membrane
2. Hyoepiglottic ligament
3. Cricothyroid ligament
4. Vocal ligament
5. Conus elasticus
6. Vestibular ligament
7. Cricotracheal ligament

Note that, with the exception of the vocal and vestibular ligaments, and the conus elasticus, the names reveal the attachments of these structures.

The **thyrohyoid membrane** is attached inferiorly to the superior horns and superior edges of the laminae of the thyroid cartilage and superiorly to body and horns of the hyoid bone. It is thickened near the midline anteriorly, forming the median thyrohyoid ligament, and also thickened posteriorly, where the membrane attaches to the superior horn, to form the **lateral thyrohyoid ligament.** This ligament often contains a small cartilage—the **cartilago triticea.** It is obvious that this membrane is horseshoe-shaped, just as is the thyroid cartilage. The thyrohyoid membrane is pierced by the internal branch of the superior laryngeal branch of the vagus nerve and by the superior laryngeal blood vessels. The **hyoepiglottic ligament** (Fig. 7-118) is a strong band extending between the hyoid bone and the anterior surface of the epiglottis. We will see that this is an important ligament in attempting to understand the movement of the epiglottis during swallowing. The triangular **cricothyroid ligament**

is located near the midline anteriorly and extends from the arch of the cricoid cartilage superiorly and laterally, in a fan-shaped manner, to attach to the inferior border of the thyroid cartilage. The **vocal ligament** is attached anteriorly to the inside of the thyroid cartilage near the midline, and posteriorly to the vocal process of the arytenoid cartilage. This is the so-called vocal cord; it is covered with mucous membrane, which is raised up as the vocal fold. The **conus elasticus** extends superiorly from the superior border of the cricoid cartilage to attach to the vocal ligament. It is continuous with the cricothyroid ligament anteriorly. The weak **vestibular ligament** is situated just superior to the vocal ligament and extends from the thyroid cartilage anteriorly to the lateral surface of the arytenoid cartilage posteriorly. These are called the false vocal cords or vestibular cords. The mucous membrane forms a ridge over this ligament as well as over the vocal ligament and is termed the vestibular fold. The **cricotracheal ligament** is simply a ligament extending from the inferior border of the cricoid cartilage to the first tracheal ring.

INTRINSIC MUSCLES (Fig. 7-117).   Now that we have formed the skeleton of the larynx in its entirety, with the cartilages and the various ligaments and membranes between these cartilages, we can comprehend the attachments of the various intrinsic muscles of the larynx. These muscles are:

1. Cricothyroid
2. Posterior cricoarytenoid
3. Lateral cricoarytenoid
4. Transverse arytenoid
5. Oblique arytenoid
6. Thyroarytenoid
7. Thyroepiglotticus
8. Aryepiglotticus

Note that all these names reveal the muscle attachments. They are all derived from the fourth and sixth branchial arches and, therefore, are innervated by the vagus nerve (see Fig. 7-136).

   1. Cricothyroid.   This muscle is on the external surface of the larynx, arises from the lateral surface of the cricoid cartilage, and courses posteriorly and superiorly to insert onto the inferior horn and inferior border of the lamina of the thyroid cartilage. When it contracts, it pulls the thyroid cartilage anteriorly and inferiorly in a rocking

POSTERIOR VIEW

LATERAL VIEW

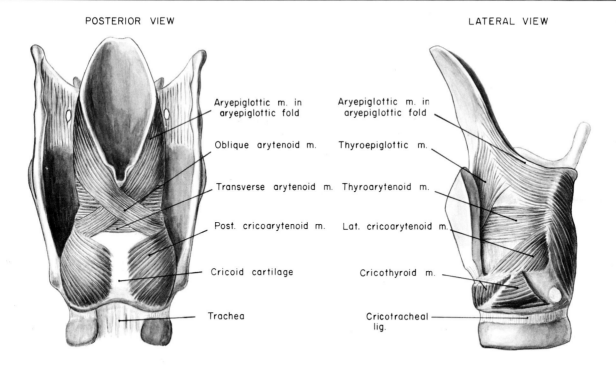

Aryepiglottic m. in aryepiglottic fold

Oblique arytenoid m.

Transverse arytenoid m.

Post. cricoarytenoid m.

Cricoid cartilage

Trachea

Aryepiglottic m. in aryepiglottic fold

Thyroepiglottic m.

Thyroarytenoid m.

Lat. cricoarytenoid m.

Cricothyroid m.

Cricotracheal lig.

MUSCLE ACTION

SUPERIOR VIEW

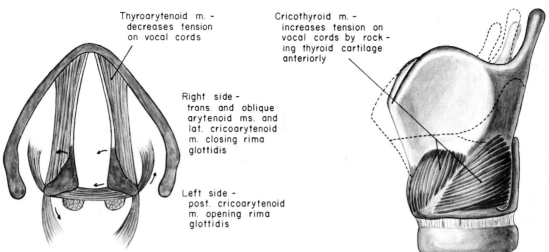

Thyroarytenoid m. - decreases tension on vocal cords

Right side - trans. and oblique arytenoid ms. and lat. cricoarytenoid m. closing rima glottidis

Left side - post. cricoarytenoid m. opening rima glottidis

Cricothyroid m. - increases tension on vocal cords by rocking thyroid cartilage anteriorly

Figure 7-117. *The intrinsic muscles of the larynx: upper left, posterior view; upper right, lateral view after removal of a portion of the thyroid cartilage; lower part displays the action of these muscles in opening and closing the rima glottidis and in increasing or decreasing tension on the vocal cords. (Partially after Jamieson.)*

action, thereby increasing the tension on the vocal cords. Figure 7-117 illustrates this action.

2. Posterior cricoarytenoid. This muscle arises from the posterior surface of the lamina of the cricoid cartilage and is directed superiorly and laterally to attach to the muscular process of the arytenoid cartilage. Its pull is naturally in a posterior medial direction, and when the muscular process is pulled in this direction, the vocal process moves in the opposite direction—laterally—due to a rotation of the arytenoid cartilage (Fig. 7-117). Therefore, this muscle widens the space between the two vocal cords—the rima glottidis.

3. Lateral cricoarytenoid. This muscle arises from the superior border of the lateral part of the arch of the cricoid cartilage. It is directed superiorly and posteriorly to be inserted onto the muscular process of the arytenoid cartilage. When it contracts, it rotates or pulls the muscular process anteriorly and laterally, and so the vocal process moves medially (Fig. 7-117). This muscle is used in decreasing the width between the two vocal cords.

4. Transverse arytenoid muscle. This muscle extends from the posterior surface of one arytenoid cartilage to the posterior surface of the opposite arytenoid cartilage. The muscle, when it contracts, tends to pull the arytenoid cartilages closer together by a gliding motion and, therefore, is used in closing the rima glottidis.

5. Oblique arytenoid muscles. These muscles lie superficial to the transverse arytenoid muscle and extend from the muscular process of one arytenoid cartilage to the apex of the arytenoid cartilage on the opposite side. Some muscular fibers continue into the aryepiglottic folds and finally reach the epiglottic cartilage. These muscles are used in bringing the arytenoid cartilages closer together and in tightening the aryepiglottic fold. They are used in the process of swallowing.

6. Thyroarytenoid. This muscle arises from the inside of the anterior part of the thyroid cartilage and extends posteriorly to insert on the arytenoid cartilage on its anterior and lateral surfaces. When these muscles contract, they bring the arytenoid cartilage closer to the thyroid cartilage and, therefore, decrease the tension on the vocal cords. A small, deep slip of this muscle is located on the lateral side of the vocal ligament and is called the **vocalis muscle.** It is thought that this muscle can be used in contracting a portion of a vocal cord in the production of very high tones.

7. Thyroepiglottic muscle. In addition to the muscle fibers just mentioned—the thyroarytenoid muscle—some of the superiorly located fibers of this muscle curve superiorly into the aryepiglottic folds and finally reach the lateral lip of the epiglottic cartilage. This muscle aids the aryepiglottic muscle in tightening the aryepiglottic fold and bringing the epiglottic cartilage into closer contact with the arytenoid cartilages during the process of closing the opening into the larynx during swallowing.

These muscle actions will be summarized after the mucous membrane has been described.

MUCOUS MEMBRANE.   Now that the larynx has been formed with its cartilages, ligaments, and muscles, the mucous membrane can be followed and the appearance of the inside of the larynx described (Fig. 7-118).

If the mucous membrane on the posterior surface of the tongue is followed in the midsaggital plane, it will be seen to be reflected onto the anterior surface of the tip of of the epiglottis as the **glossoepiglottic fold.** The mucous membrane reflects over the tip of the epiglottis to its posterior surface, on which it continues inferiorly. In this position, it aids in forming the anterior boundary of the inlet to the larynx. This mucous membrane can be followed further inferiorly onto the thyroid cartilage, the crico-thyroid ligament, the arch of the cricoid cartilage, the cricotracheal ligament, and on into the trachea.

If one starts again at the base of the tongue in a more lateral position, one can see that the mucous membrane is reflected onto the side walls of the pharynx, forming two folds called the **lateral glossoepiglottic folds.** The spaces between the median and lateral glossoepiglottic folds are the **epiglottic valleculae.** Just inferior to the lateral glossoepiglottic folds, a depression occurs between the larynx and the lateral wall of the pharynx—the **piriform recess** (see Fig. 7-113). The mucous membrane can be followed from this piriform recess onto the lateral surface of the aryepiglottic fold. From this lateral surface, the mucous membrane reflects over the aryepiglottic and thyroepiglottic muscles and enters the inside of the larynx (Fig. 7-113). The mucous membrane can be followed inferiorly until the vestibular fold is reached (Fig. 7-118). At this point, it dips into a depression between the vestibular fold and the vocal cord; this depression is the **laryngeal ventricle.** From this point, it reflects over the vocal cords themselves and then inferiorly on the conus

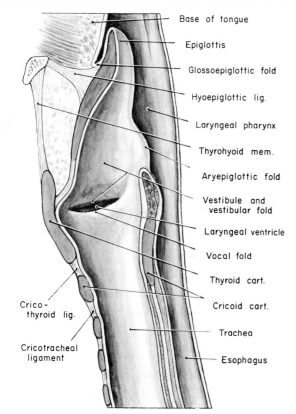

Base of tongue

Epiglottis

Glossoepiglottic fold

Hyoepiglottic lig.

Laryngeal pharynx

Thyrohyoid mem.

Aryepiglottic fold

Vestibule and vestibular fold

Laryngeal ventricle

Vocal fold

Thyroid cart.

Cricoid cart.

Trachea

Esophagus

Crico-thyroid lig.

Cricotracheal ligament

Figure 7-118. Sagittal section of pharynx and esophagus, and larynx and trachea. Mucous membrane is intact.

elasticus, the cricoid cartilage itself, the cricotracheal ligament, and thence into the trachea.

The **inlet of the larynx** is formed, therefore, by the epiglottis anteriorly, the aryepiglottic folds of mucous membrane on the aryepiglottic and thyroepiglottic muscles laterally, and the fold of mucous membrane between the arytenoid cartilages posteriorly. The **vestibule** of the larynx is the portion just inferior to the inlet between the inlet superiorly and the vestibular folds inferiorly. The **ventricle of the larynx** has already been described as the space between the vestibular fold and the true vocal fold, and the **rima glottidis** is the space between the two vocal folds.

SUMMARY OF MUSCLE ACTIONS. There are three distinct acts that must be accomplished by muscle action as far as the larynx is concerned: (1) opening and closing the rima glottidis during the process of breathing, swal-

lowing, or at times when an increased intra-abdominal pressure is needed; (2) regulating the tension on the vocal cords; and (3) closing of the larynx during the process of deglutition.

Opening and closing the rima glottidis (Fig. 7-117). The arytenoid cartilages rotate* on the cricoid cartilage to which they are attached; there is a gliding movement allowed as well. From its position it is obvious that the lateral cricoarytenoid muscle, by pulling laterally on the muscular process, will move the vocal process medially. This, because the vocal cords are attached to the vocal process, will close the space between the two vocal cords. The transverse and oblique arytenoid muscles aid the lateral cricoarytenoid muscles in this action by simply bringing the arytenoid cartilages closer to each other by gliding motion. The space between the two vocal cords is widened by the action of the posterior cricoarytenoid muscles. Because these muscles pull the muscular process posteriorly and medially, the vocal process by rotation moves laterally; this opens the rima glottidis.

Altering tension on vocal cords (Fig. 7-117). Tension on the vocal cords is decreased by the action of the thyroarytenoid muscles. They pull the arytenoid cartilage slightly closer to the thyroid cartilage, an act that can do nothing but decrease the tension on the vocal cords. Tension is increased on the vocal cords by the action of the cricothyroid muscles. These externally placed muscles arise from the arch of the cricoid cartilage and course posteriorly and superiorly to insert on the inferior horn and the inferior border of the lamina of the thyroid cartilage. When contracted, they pull the thyroid cartilage anteriorly and inferiorly in a "rocking" manner. This rocking uses the cricothyroid joint as a fulcrum. Because the thyroid cartilage is moved anteriorly while the arytenoid cartilages remain behind, the distance between the arytenoid cartilages and the thyroid cartilage increases, thus stretching the vocal cords; this naturally increases the tension on these cords. The vocalis muscle should be mentioned in this regard, as it is thought, because of its multiple attachments, to allow tensing of a portion of the vocal cord rather than the entire cord as is done by the cricothyroid muscle. Pressman and Kele-

---

*H. VonLeden, 1961, The mechanics of the cricoarytenoid joint, *Arch. Otolaryngol.* 73: 541. This report questions the "rotation" of the arytenoid cartilages.

men* (1955) have increased our knowledge of the physiology of the larynx.

Closing the larynx during swallowing (Fig. 7-117). The mechanism whereby the larynx is closed during the process of swallowing has been a point of argument for many years. The attachment of the rather heavy hyoepiglottic ligament to the hyoid bone anteriorly and to the epiglottic cartilage posteriorly shows that the epiglottis does not fall down over the opening of the larynx as a trap door. Furthermore, if one simply palpates the thyroid cartilage during swallowing, an upward motion of this cartilage can be felt. This upward thrust is brought about by the thyrohyoid muscles and three sets of muscles of the pharynx that insert onto the thyroid cartilage—the salpingopharyngeus, palatopharyngeus, and stylopharyngeus muscles. Therefore, contrary to the idea that the epiglottis simply drops down to cover the laryngeal inlet, the thyroid and cricoid cartilages are pulled superiorly against the epiglottis in the process of swallowing. Because the epiglottis is pushed up against the base of the tongue in this maneuver, the arytenoid cartilages are brought much closer to the epiglottis. According to Johnstone,† the entrance to the larynx is protected further by action of the thyroepiglottic and aryepiglottic muscles in pulling the superior tip of the epiglottis inferiorly. These combined actions effectively close the laryngeal inlet. In addition to this, because the breath is held during swallowing, the rima glottidis is closed by action of the transverse and oblique arytenoid muscles and the lateral cricoarytenoid muscles. In support of the above, it is well known that large portions of the epiglottis can be removed without interfering unduly with the process of swallowing.

NERVES. The nerve supply to the larynx can be divided into a sensory innervation to the mucous membrane, which is extremely important in inducing the cough reflex when food enters the larynx, and the motor nerve supply to the intrinsic muscles. The sensory nerve supply to the mucous membrane of the larynx as far inferiorly as the vocal cords is by the **internal branches of** the superior laryngeal branches of the vagus nerves. The mucous membrane from the vocal cords to the trachea derives its sensory nerve supply from the **recurrent laryngeal nerves.** The cricothyroid muscle is innervated by the **external branch of the superior laryngeal branch of the vagus nerve,** and it has been thought for years that all the remaining muscles of the larynx were innervated by the **recurrent laryngeal branches of the vagus nerve** (see Lemere‡). Recent evidence, however, indicates that in man part of the transverse arytenoid muscle is innervated by the internal branch of the superior laryngeal branch of the vagus.§ This helps explain the fact that in man the rima glottidis tends to close if both recurrent laryngeal nerves are accidentally cut during surgery. Since the only muscles remaining in such a situation are the cricothyroid muscles, whose action is to tense the vocal cords, and the transverse arytenoid muscle, which will bring the vocal cords closer together, it is easy to see that the cords should be tensed and approximated. In actuality, under these conditions the cords are flaccid. This is probably due to the inability of the cricothyroid muscle to induce enough movement to tense the cords, since the arytenoid cartilages can move forward because of a complete lack of any pull in the opposite direction by the paralyzed muscles. Nevertheless, the cords are approximated and the patient is unable to move them from this position. This might result in suffocation if a tracheotomy is not done immediately. The nerve supply to the larynx is extremely important from both a sensory and a motor point of view, and it must be kept in mind constantly during thyroid surgery or other surgery in the neck region.

ARTERIES AND VEINS. The blood supply to the larynx is via the superior and inferior laryngeal branches of the superior thyroid and inferior thyroid arteries, re-

---

*J. J. Pressman and G. Kelemen, 1955, Physiology of the larynx, *Physiol. Rev.* 35: 506. A long and thorough review of the structure and function of the larynx.

†A. S. Johnstone, 1942, A radiological study of deglutition, *J. Anat.* 77: 97. Radiological evidence used to show that the tip, at least, of the epiglottis turns down during swallowing. This is contrary to the theory that the epiglottis remains erect.

---

‡F. Lemere, 1932, Innervation of the larynx: I. Innervation of laryngeal muscles, *Am. J. Anat.* 51: 417. Evidence in support of the theory that the recurrent laryngeal nerve innervates all laryngeal muscles except the cricothyroid, which is innervated by the external branch of the superior laryngeal branch of the vagus.

F. Lemere, 1933, Innervation of the larynx: III. Experimental paralysis of the laryngeal nerve, *Arch. Otolaryngol.* 18: 413. Further evidence for claiming that the recurrent laryngeal nerve innervates all laryngeal muscles except the cricothyroid muscles.

§P. H. Vogel, 1952, The innervation of the larynx of man and the dog, *Am. J. Anat.* 90: 427. Convincing evidence that the internal laryngeal branch of the vagus does actually innervate part of the transverse arytenoid muscle in man. This is in contrast to the dog, in which the internal laryngeal nerve is purely sensory.

spectively. These arteries are accompanied by veins of the same name.

LYMPHATIC DRAINAGE.   The lymphatics of the larynx drain superiorly and inferiorly in a manner similar to the arterial supply. From the vocal cords superiorly the lymphatics pierce the thyrohyoid membrane and empty into the infrahyoid nodes and the upper deep cervical nodes. The lymphatic vessels draining inferiorly may pierce the cricothyroid membrane to reach the nodes anterior and lateral to the larynx and trachea—the prelaryngeal, pretracheal, and paratracheal nodes.

The larynx, in the male, undergoes secondary sexual changes after puberty. The thyroid cartilage enlarges and protrudes (Adam's apple), while the vocal cords become thicker and heavier. Varied growth at puberty is probably responsible for the production of tenor, baritone, and bass voices; females show a similar pattern of voice range.

DEGLUTITION

Throughout the description of the pharynx and larynx we have made frequent reference to deglutition or swallowing. This complex act is initiated at will, but many of the actions involved are accomplished at reflex level. After food has been chewed into small bits and moistened with saliva, it is ready to be swallowed. The act of swallowing consists of closing the mouth, holding the breath, and forcing the bolus of food into the pharynx by elevating the tongue against the roof of the mouth. The superior, middle, and inferior constrictor muscles of the pharynx then contract in sequence* (except for the cricopharyngeus muscle which must relax) and force the food into the esophagus.

Although swallowing can be initiated with the mouth open, it is more difficult than when the mouth is closed. If food is involved, it is almost mandatory that the mouth be closed. This is accomplished by the temporalis, masseter,

and medial pterygoid muscles. The tongue is raised against the hard palate by the action of the mylohyoid and styloglossus muscles, and the intrinsic muscles of the tongue.

In order to propel food into the esophagus, three openings must be closed: to the nasal cavity, to the mouth, and to the larynx. The opening to the nose is closed by elevating the soft palate and bringing the walls of the pharynx medially and anteriorly to meet the raised soft palate. The palate is elevated by the combined action of the levator veli palatini, tensor veli palatini, and musculus uvulae muscles, and the pharynx is moved medially and anteriorly by the superior constrictor muscle and the stylopharyngeus, salpingopharyngeus, and palatopharyngeus muscles.

The opening into the oral cavity is closed successfully by the elevation of the tongue against the hard palate as already mentioned; the palatoglossus muscle probably aids in this part of swallowing.

The mechanism of closing the entrance into the larynx during the process of swallowing has been a source of argument for many years.† There is complete agreement that the epiglottis does not fall down over the entrance to the larynx in the fashion of a trap door. In fact the hyoepiglottic ligament would make such a movement difficult. Actually the larynx is raised during swallowing, a fact easily checked by palpating the thyroid cartilage during this act. This forces the larynx (actually the edges of the aryepiglottic folds) against the epiglottis, which, in turn, is forced against the base of the tongue. The tip of the epiglottis is then turned down as a further protection.‡ In addition, holding the breath as already mentioned demands a closing of the rima glottidis.

Raising the larynx is accomplished by any muscles that insert on the hyoid bone as well as those attached to

---

*J. V. Basmajian and C. R. Dutta, 1961, Electromyography of the pharyngeal constrictors and soft palate in rabbits, *Anat. Record* 139: 443. In this animal the duration of the contraction in the constrictor muscles during swallowing decreases from superior to inferior; these muscles act in an "all or none" fashion, there being no tone between swallows.

---

†W. Lerche, 1950, *The esophagus and pharynx in action: A study of structure in relation to function,* Charles C Thomas, Springfield, Ill.

J. J. Pressman and G. Kelemen, 1955, Physiology of the larynx, *Physiol. Rev.* 35: 506. A long and thorough review of the structure and function of the larynx.

R. W. Doty and J. F. Bosma, 1956, An electromyographic analysis of reflex deglutition, *J. Neurophysiol.* 19: 44. Electromyographic pattern of activity was studied in twenty-two muscles that are involved in swallowing. Cat, dog, and monkey studied.

J. F. Bosma, 1957, Deglutition: Pharyngeal stage. *Physiol. Rev.* 37: 275. A complete review of data on swallowing up to year indicated.

‡A. S. Johnstone, 1942, A radiological study of deglutition, *J. Anat.* 77: 97. Radiological evidence used to show that the tip, at least, of the epiglottis turns down during swallowing. This is contrary to the theory that the epiglottis remains erect.

the larynx, for the hyoid bone is attached to the larynx by the firm thyrohyoid membrane. Therefore the larynx is raised by the digastric, mylohyoid, geniohyoid, stylohyoid and hyoglossus muscles elevating the hyoid bone, and the thyrohyoid, stylopharyngeus, palatopharyngeus, and salpingopharyngeus muscles elevating the larynx itself. The tip of the epiglottis is pulled inferiorly by the aryepiglotticus and thyroepiglotticus muscles, and the rima glottidis is closed by the transverse arytenoid and lateral cricoarytenoid muscles.

The formidable array of muscles enumerated above should not be memorized as lists; the various acts involved in swallowing should be remembered and then the muscles involved in each act determined by recalling muscles having attachments to the structure being moved. The first part of the act of swallowing is done at cortical levels—the mouth is closed, the breath is held, and the tongue forced against the hard palate. The remaining acts involved are done at subcortical levels—by reflex. The pharyngeal constrictor muscles propel the bolus inferiorly into the esophagus (the cricopharyngeus muscle relaxing to accomplish this), this destination being reached because other likely pathways into the nose, mouth, and larynx are successfully blocked.

NOSE AND NASAL CAVITY

GENERAL DESCRIPTION

The **external nose** is pyramidal in shape and has a **root** just inferior to the forehead and an **apex**—its free end. The inferiorly placed openings are the **nostrils** or **nares,** and these are separated from each other by a **septum.** There are **hairs** protecting the openings of the nares and preventing foreign matter from entering the nasal cavity. The nose is considered to have **lateral surfaces** and a **dorsum**—the most anterior crest of the nose. The superior part of the dorsum is made up of the two nasal bones and is called the **bridge** of the nose. The lateral surfaces end inferiorly in a rounded part—the **ala nasi.**

The **actual framework** of the nose consists of bones superiorly and laterally, and cartilage inferiorly. The **bony part** is made up of the nasal bones on the bridge of the

nose and the frontal process of the maxillary bones. The inferior half, or movable portion of the nose, is made up of several cartilages and fatty tissue. The **septal cartilage** makes up a good half of the septum of the nose, and the **lateral cartilages** are found on either side of the septal cartilage. The distal end of the nose contains the **greater alar cartilage** and several **lesser alar cartilages.** These are pictured in Figure 7-119. The skin, although thin near the base of the nose, is thick and firmly bound down by fibrous tissues to these cartilages in the distal end of the nose.

The muscles of the nose have already been described on page 480. Also note Figure 7-16.

The **nerves** to the external nose consist of branches

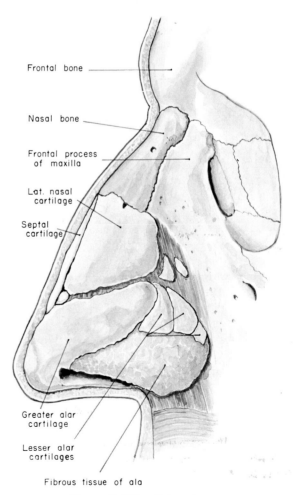

Frontal bone

Nasal bone

Frontal process of maxilla

Lat. nasal cartilage

Septal cartilage

Greater alar cartilage

Lesser alar cartilages

Fibrous tissue of ala

Figure 7-119.   External nose.

of the facial to the muscles, and sensory nerves to the skin. The latter include the infratrochlear and external nasal branches of the nasociliary branch of the ophthalmic division of the trigeminal nerve, and the nasal branches of the infraorbital branch of the maxillary division of the same nerve (Fig. 7-12).

The **nasal cavity** extends from the nares anteriorly to the posterior apertures of the nose—the choanae—which lead into the nasal pharynx. It is divided into two parts by the **nasal septum;** this septum is usually on one or the other side of the midline dividing the nasal cavity into two unequal portions. The lateral wall of the nose exhibits three shelves—the **conchae** (Fig. 7-120)—which are bony projections covered with mucous membrane.

The **superior concha** is small and rounded, the **middle concha** considerably larger, while the **inferior concha** is the largest of the three. The inferior and middle conchae are approximately horizontal in position and end posteriorly as small projections. Inferior and lateral to each of these bony shelves is a space—the **superior, middle, and inferior meatus.** The area superior to the superior concha is termed the **sphenoethmoid** recess. These conchae serve to rotate the air in such a manner that cold air from the outside does not enter the respiratory system before becoming warmed by the blood in the mucous membrane of the nasal cavity.

The **vestibule** of the nasal cavity is the dilated part superior to the nostril. The nasal cavity itself can be divided into **olfactory and respiratory regions,** the former being the superior concha and the superior edge of the septum, while the remainder of the cavity is the respiratory region.

If the bony conchae are removed, the openings of various structures into the nasal cavity can be seen (Fig. 7-120B). In the inferior meatus there is an opening, located about ½ inch from the anterior end of the concha, which is the inferior end of the **nasolacrimal canal.** There are no other openings in the inferior meatus. The middle meatus presents a rounded elevation superiorly—the **bulla ethmoidalis**—and a groove anterior and inferior to the **bulla**—the **hiatus semilunaris.** (This area can be understood by looking at Figure 7-121.) The **middle ethmoid sinuses** drain into the nasal cavity via an opening in the middle of the bulla ethmoidalis. The **anterior ethmoid sinuses** and **frontal sinuses** drain into the anterior superior end of the hiatus semilunaris, and the **maxillary sinus** drains into the most

posterior part of the hiatus. Therefore, all the sinuses, except the posterior ethmoid and the sphenoid, drain into the middle meatus. The **posterior ethmoid** sinuses drain into the superior meatus, while the **sphenoid sinus** drains at a point superior to the superior concha—into the sphenoethmoid recess.

PARANASAL SINUSES

The paranasal sinuses are cavities in the maxillary, ethmoid, sphenoid and frontal bones, and they take corresponding names. They are air-filled and communicate with the nasal cavity as just described. They are lined with a mucous membrane, which is continuous with that lining the nasal cavity. These structures are of great nuisance value because they become infected easily. Supposedly they serve as a means of decreasing the weight of the skull and giving resonance to the voice. Maresh* presents a series of examinations on the same individual that show the great increase in size of the paranasal sinuses during adolescence.

Ethmoidal sinuses (Fig. 7-122). These structures begin to form in the fetus during the second half of pregnancy. They are numerous, small, thin-walled sacs, which communicate with each other. They are contained in the entire ethmoid labyrinth. Their **relations** are as follows: **Superior**—sphenoidal and frontal sinuses, anterior cranial fossa, ethmoidal vessels and nerves. **Inferior**—superior and middle meatuses of nose. **Medial**—nasal cavity. **Lateral**—orbit. **Anterior**—frontal process of maxillary bone. **Posterior**—sphenoidal sinus. These air cells are divided into three sets: the posterior, opening into the superior meatus; the middle, into the middle meatus on the bulla ethmoidalis; and the anterior, into the middle meatus in the floor of the hiatus semilunaris.

Frontal sinus (Fig. 7-122). This sinus does not usually appear until approximately the seventh year of life, but may appear earlier. It is found in the frontal bone superior to the orbital margin in the root of the nose. It is separated from the sinus on the opposite side. It varies in size, but is usually about 1 inch in height and approximately the same in width. Occasionally it extends poste-

---

*M. D. Maresh, 1940, Paranasal sinuses from birth to late adolescence. I. Size of the paranasal sinuses as observed in routine posteroanterior roentgenograms, *Am. J. Diseases Children* 60: 55. A series of examinations on the same child allows a good study of sinus growth.

A

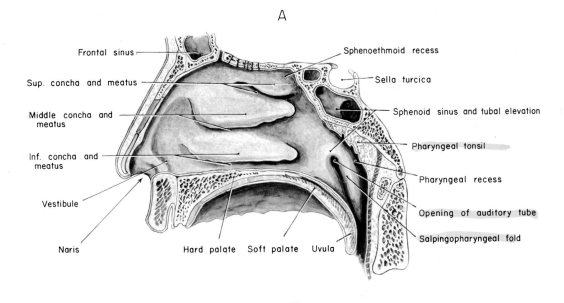

Frontal sinus

Sphenoethmoid recess

Sup. concha and meatus

Sella turcica

Middle concha and meatus

Sphenoid sinus and tubal elevation

Pharyngeal tonsil

Inf. concha and meatus

Pharyngeal recess

Vestibule

Opening of auditory tube

Naris

Salpingopharyngeal fold

Hard palate   Soft palate   Uvula

B

Arrow in opening of frontal sinus into semilunar hiatus of middle meatus

Opening of ant. ethmoid sinuses into semilunar hiatus of middle meatus

Arrow in opening of sphenoid sinus into sphenoethmoid recess

Opening of middle ethmoid sinuses onto bulla ethmoidalis in middle meatus

Openings of post. ethmoid sinuses into sup. meatus

Opening of naso-lacrimal duct into inf. meatus

Opening of maxillary sinus into semilunar hiatus of middle meatus

Auditory tube and salpingopharyngeus m.

Cut edge of mucous mem.

Tensor veli palatini m. a asc. palatine a.

Cut edge of mucous mem.

Levator veli palatini m.

Figure 7-120. *(A) Lateral wall of nasal cavity. The mucous membrane has been left intact. (B) Lateral wall of nasal cavity after removal of conchae and the mucous membrane around opening of the auditory tube. This reveals the openings of the paranasal sinuses, and the tensor and levator veli palatini, and salpingopharyngeus muscles.*

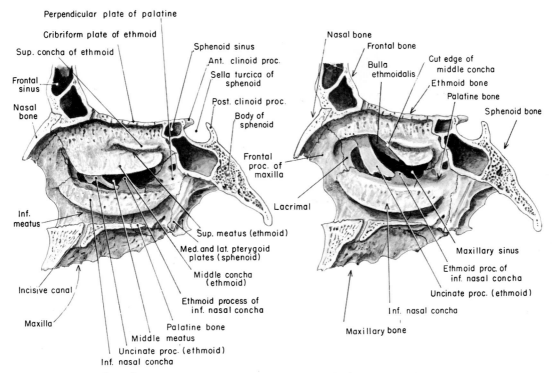

Figure 7-121. *Bones forming lateral wall of nasal cavity. In the picture to the right a portion of the middle concha has been removed.*

riorly between two bony tables in the roof of the orbit. Its **relations** are as follows: **Inferior**—orbit, nose, and anterior ethmoidal sinuses. **Posterior** and **superior**—cranial cavity and brain. This sinus opens into the superior end of the hiatus semilunaris in the middle meatus of the nose.

Maxillary sinus (Fig. 7-122). This sinus first appears in the maxillary bone of the fetus about the fourth month of pregnancy. It is extremely large in adults and has the following important relations. **Superior**—the orbit and its contents. **Inferior**—molar and premolar teeth. **Medial**—the lower half of the nasal cavity. **Posterior**—pterygopalatine and infratemporal fossae. The point of drainage of this sinus is at its superior end, which is an unfortunate location. Under normal conditions the cilia are powerful enough to push the mucous secretion out through the opening into the nasal cavity. However, when severe infection sets in, the cilia are powerless to accomplish this, and drainage is accomplished with diffi-

culty. The point of drainage is in the hiatus semilunaris of the middle meatus.

Sphenoidal sinuses (Figs. 7-114 and 7-122). These are found in the sphenoid bone and are separated by a bony septum, which is usually displaced from the midline. They occur approximately in the seventh year of life. Their **relations** are as follows. **Anterior**—the nasal cavity and ethmoid sinuses. **Posterior**—a thick layer of bone. **Inferior**—nasal pharynx, nasal cavity. **Superior**—the brain, optic chiasma, the intercavernous venous sinus, and the hypophysis. **Lateral**—the optic nerve, cavernous sinus and its contents. These sinuses open into the sphenoethmoid recess of the nasal cavity.

NERVE AND BLOOD SUPPLY

Before the nerve and blood supply to the nasal cavity can be considered, a complete description of the maxillary

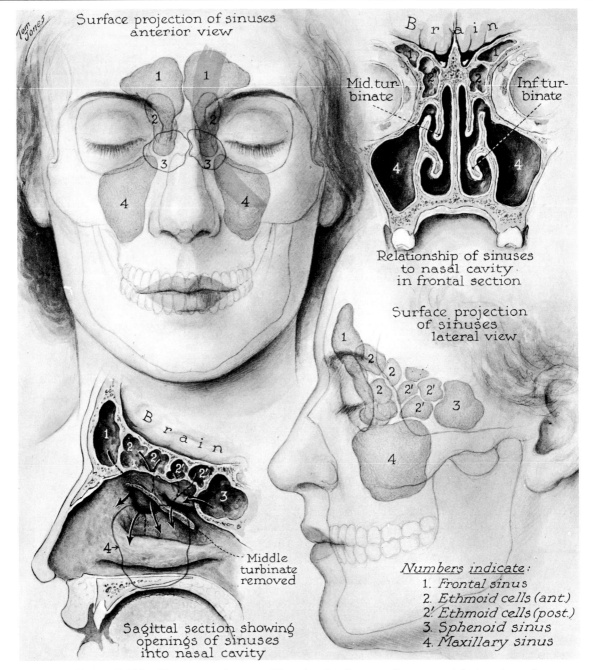

Figure 7-122. *The paranasal sinuses. (Note that this illustration has grouped the anterior and middle ethmoid sinuses into one set—the anterior ethmoids.) (Courtesy of Camp International, Inc.)*

division of the trigeminal nerve and the third part of the maxillary artery must be given. A quick review of the pterygopalatine fossa and the numerous openings to and from this fossa would be wise at this time (Figs. 7-123 and 7-2B).

MAXILLARY DIVISION OF TRIGEMINAL NERVE.

The maxillary nerve arises from the trigeminal ganglion and courses anteriorly along the inferior border of the cavernous sinus. At this point it is separated from the sphenoidal sinus medially by a very thin layer of bone. It continues anteriorly and leaves the skull via the foramen rotundum. After traversing the foramen rotundum, it appears in the superior part of the pterygopalatine fossa. It then leaves this fossa by curving anteriorly, passes through the deepest part of the infratemporal fossa to enter the inferior aspect of the orbit through the inferior orbital fissure, and becomes the infraorbital nerve, which courses through the infraorbital groove, the infraorbital canal, and the infraorbital foramen to terminate on the face as superior labial, external nasal, and inferior palpebral branches (note Fig. 7-124).

The **branches** of the maxillary nerve can be divided into those in the cranial cavity, those in the pterygopalatine fossa, those in the infratemporal fossa, those in the infraorbital groove and canal, and the terminal branches on the face.

In the cranial cavity
1. Meningeal
In the pterygopalatine fossa
2. Pterygopalatine (via pterygopalatine ganglion)
   a. Orbital
   b. Greater palatine
   c. Posterior superior nasal
   d. Pharyngeal
In infratemporal fossa
3. Zygomatic
   a. Communication with lacrimal of ophthalmic division
   b. Zygomaticofacial
   c. Zygomaticotemporal
4. Posterior superior alveolar
In infraorbital canal
5. Middle superior alveolar
6. Anterior superior alveolar
On the face
7. Inferior palpebral

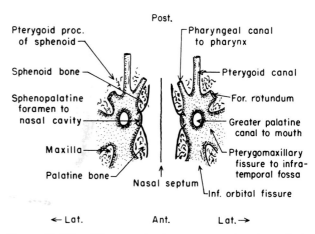

Figure 7-123. *Diagram of a transverse section through the pterygopalatine fossae (superior view), and the various entrances and exits to and from these fossae. Refer to Figure 7-2B.*

8. External nasal
9. Superior labial

1. Meningeal. This small nerve is the only branch of the maxillary division of the trigeminal nerve in the cranial cavity; it arises close to the ganglion and supplies the dura mater in that vicinity.

2. Pterygopalatine (Figs. 7-124 and 7-125). The **pterygopalatine** branches descend to the **pterygopalatine ganglion.** This ganglion is located in the pterygopalatine fossa surrounded by fat and by branches of the third part of the maxillary artery and their accompanying veins. The branches of the pterygopalatine ganglion are:

   a. Orbital
   b. Greater palatine
   c. Posterior superior nasal
   d. Pharyngeal

a. The **orbital branches** consist of two or three delicate filaments that enter the orbit via the inferior orbital fissure. They supply the periosteum of the orbit and possibly the mucous membrane of the posterior ethmoidal and sphenoidal sinuses.

b. The **greater palatine nerve** (Fig. 7-126) descends inferiorly through the greater palatine canal to reach the hard palate of the roof of the mouth. It sends branches to the soft and hard palates and also the gums. In the canal it gives rise to **nasal branches** (posterior

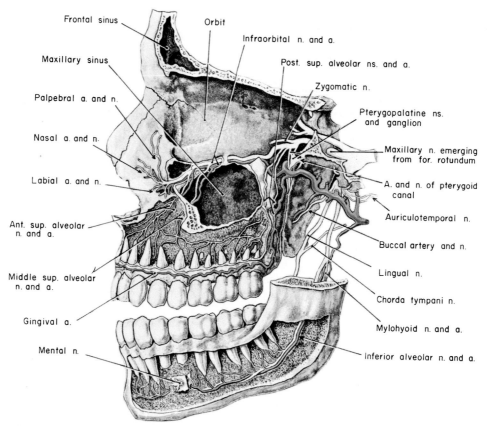

Figure 7-124. *Nerve and blood supply to upper and lower jaws. (From* Surgical Anatomy, *by Deaver © 1926 by McGraw-Hill, Inc. Used with the permission of the McGraw-Hill Book Company.)*

inferior nasal), which innervate the inferior and posterior part of the nose, and to the **lesser palatine** nerves, which descend through the lesser palatine canals to supply the soft palate, uvula, and the palatine tonsil.

c. The **posterior superior nasal** branches (Figs. 7-126 and 7-127) enter the posterior part of the nasal cavity via the sphenopalatine foramen and supply the mucous membrane covering the superior and middle conchae and the posterior part of the nasal septum. One branch, which is larger and longer than the others—the **nasopalatine nerve**—passes across the roof of the nasal cavity and descends anteriorly along the nasal septum to reach the incisive canal. It passes through this canal and gives nerve supply to the roof the mouth in its anterior part.

d. The **pharyngeal nerve** (Fig. 7-126) passes through the pharyngeal canal and is distributed to the mucous membrane of the nasal part of the pharynx posterior to the auditory tube.

These nerves arising from the pterygopalatine ganglion contain afferent fibers from the areas innervated; parasympathetic and sympathetic fibers to the lacrimal gland, mucous membranes of the nasal and oral cavities, and the superior end of the pharynx; and possibly some taste fibers. The afferent fibers simply pass through the ganglion, and there is no synapse involved. The taste fibers are from taste buds along the roof of the mouth and follow the nerve of the pterygoid canal and greater petrosal nerve back to the main facial nerve, and thence reach the brain stem. The ganglion cells are actually post-ganglionic parasympathetic cells to the areas mentioned above. In way of review, it will be recalled that the pre-

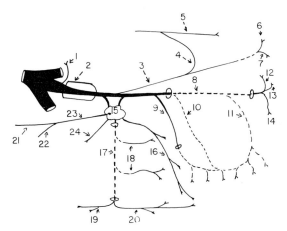

Figure 7-125. Diagram of maxillary division of trigeminal nerve: (1) meningeal branch, (2) foramen rotundum, (3) zygomatic, (4) communication of zygomatic with (5) lacrimal of ophthalmic division (secretory fibers for lacrimal gland), (6) zygomaticotemporal, (7) zygomaticofacial, (8) infraorbital, (9) posterior superior alveolar, (10) middle superior alveolar, (11) anterior superior alveolar, (12) palpebral, (13) nasal, (14) labial, (15) pterygopalatine ganglion, (16) nasopalatine, (17) greater palatine, (18) nasal, (19) lesser palatine, (20) greater palatine, (21) greater petrosal (parasympathetic), (22) deep petrosal (sympathetic), (23) nerve of the pterygoid canal, (24) pharyngeal.

ganglionic parasympathetic fibers to the lacrimal gland course with the facial nerve via its greater petrosal branch to reach the nerve of the pterygoid canal, and thence the pterygopalatine ganglion. Postganglionic fibers follow the short roots to the maxillary nerve, then along its zygomatic branch, which has an anastomosis with the lacrimal nerve, and then along this nerve to the lacrimal gland. You should recall also that sympathetic fibers are picked up via the deep petrosal, which joins with the greater petrosal nerve to form the nerve of the pterygoid canal. These are postganglionic fibers whose cell bodies are located in the superior cervical ganglion. They simply traverse the pterygopalatine ganglion and are distributed by its many branches to the blood vessels in the mucous membrane of the nasal and oral cavities and to the pharynx. Thus, the important concept of the pterygopalatine ganglion is to realize that it contains parasympathetic cell bodies and that the other nerves just described branching from it are simply attached to it with no functional significance.

3. Zygomatic (Fig. 7-125). This nerve branches from the maxillary in the infratemporal fossa. The zygomatic nerve passes through the inferior orbital fissure and along the lateral wall of the orbit, actually between the bone and the periosteum. It sends a filament to the lacrimal nerve (see Fig. 7-37), which has already been mentioned as the pathway for secretory fibers to the lacrimal gland, and then enters the zygomatic foramen. It divides in the substance of the zygomatic bone into its terminal branches—the **zygomaticofacial** and **zygomaticotemporal**—which emerge through separate foramina and give sensory nerve supply to the region over the cheekbones and the scalp on the anterior part of the temporal region (Fig. 7-12).

4. Posterior superior alveolar (Fig. 7-124). There are actually two posterior superior alveolar branches arising immediately before the maxillary nerve enters the orbit. These descend on the posterior part of the maxillary bone and supply the gums and perforate the bone to reach the molar teeth.

5 and 6. Middle and anterior superior alveolar (Fig. 7-124). These nerves arise in the infraorbital groove and canal, and descend in the wall of the maxillary sinus. They give twigs to the mucous membrane of this sinus, and then continue to the premolar teeth via the middle superior alveolar branch and the canine and incisor teeth via the anterior branch.

7, 8, and 9. Inferior palpebral, external nasal, and superior labial (Figs. 7-12 and 7-124). These branches on the face give cutaneous innervation to the lower eyelid, the lateral side of the nose, and the upper lip respectively.

THIRD PART OF MAXILLARY ARTERY. The branches of the maxillary nerve just described are accompanied by branches of the third part* of the maxillary artery. These branches are:

1. Posterior superior alveolar
2. Infraorbital
3. Descending palatine
4. Artery of the pterygoid canal
5. Pharyngeal
6. Sphenopalatine

1. The **posterior superior alveolar artery** (Fig.

---

*See page 551 for Parts I and II.

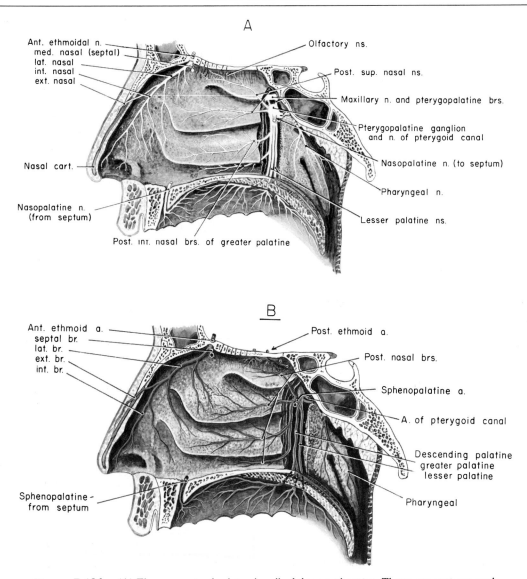

Figure 7-126. *(A) The nerves to the lateral wall of the nasal cavity. These are sensory and motor to the blood vessels and mucous membrane. (B) The arteries to the lateral wall of the nasal cavity. Note that those nerves and arteries shown as cut supply the nasal septum. These vessels are shown in Figure 7-127.*

7-124) descends on the posterior surface of the maxillary bone in company with the posterior superior alveolar nerves. These branches supply the gums, molar, and premolar teeth, and maxillary sinus.

2. The **infraorbital** (Fig. 7-124) enters the orbit through the inferior orbital fissure. It courses through the infraorbital groove and canal and exits via the infraorbital foramen to the face. It is accompanied in its extent by the infraorbital nerve. While in the floor of the orbit, it gives branches to the orbital structures, to the maxillary sinus, and its **anterior superior alveolar branch** supplies the canine and incisor teeth.

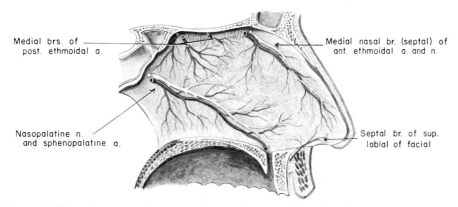

Medial brs. of
post. ethmoidal a.

Medial nasal br. (septal) of
ant. ethmoidal a. and n.

Nasopalatine n.
and sphenopalatine a.

Septal br. of sup.
labial of facial

Figure 7-127. *Arteries and nerves to the nasal septum. Note how these structures are continuous with those shown cut in Figure 7-126.*

3. The **descending palatine** artery (Fig. 7-126) divides into greater and lesser palatine arteries. The **greater palatine** descends through the greater palatine canal and foramen, runs anteriorly on the hard palate, and ends by passing superiorly through the incisive canal to enter the nasal cavity. The **lesser palatine** arteries pass through the lesser palatine canal and supply the soft palate as well as the posterior part of the hard palate.

4. The small **artery of the pterygoid canal** courses posteriorly through the pterygoid canal.

5. The **pharyngeal branch** (Fig. 7-126) runs posteriorly through the pharyngeal canal to the roof of the pharynx posterior to the opening of the auditory tube.

6. The **sphenopalatine artery** (Figs. 7-126 and 7-127) enters the nasal cavity through the sphenopalatine foramen and divides immediately into branches that supply the lateral wall and septum (**lateral and septal posterior nasal branches**). Its longest branch crosses the roof of the nasal cavity and descends anteriorly on the septum to anastomose with the greater palatine (via the incisive canal) and septal branch of the superior labial branch of the facial artery, which gives blood supply to the inferior end of the septum.

VEINS. All the above arteries are accompanied by **veins** of the same name. They drain into the pterygoid plexus of veins in the infratemporal fossa.

NERVES OF NASAL CAVITY. Now that the maxillary division of the trigeminal nerve has been described, the

nerve supply to the nasal cavity can be summarized. Since the frontonasal process as well as the maxillary process in the embryo forms the nose and nasal cavity, it should not be too surprising to find both the ophthalmic and maxillary divisions of the trigeminal nerve involved.

As can be seen in Figures 7-126 and 7-127, the **anterior ethmoidal branch** of the nasociliary branch of the ophthalmic division of the trigeminal nerve enters the nasal cavity through the anterior ethmoidal foramen and immediately gives off a **medial branch,** which ramifies on the anterior part of the nasal septum, and a **lateral branch,** which is distributed on the anterior part of the lateral nasal wall, and then continues between the nasal bones and cartilages to give cutaneous innervation to the tip of the nose (**external nasal branch**). In surgery of this area, it is important to anesthetize these nerves as well as those about to be described, which innervate the greater part of the nasal cavity.

The branches of the maxillary division of the trigeminal nerve that supply the nasal cavity (Figs. 7-126 and 7-127) emerge from the pterygopalatine ganglion. Those to the upper part of the lateral wall and those to the septum traverse the sphenopalatine foramen. Some of these **posterior superior nasal branches** ramify on the lateral wall, while others cross the roof of the cavity to join the septum. One of the latter branches is larger than the others—the **nasopalatine**—and finally terminates by traversing the incisive canal to reach the roof of the mouth. Returning to the lateral wall, the **posterior inferior nasal branches** are derived from the greater palatine nerve as it descends in the greater palatine canal. These branches destined to reach the nasal wall do so via small foramina in the palatine bone.

Anesthesia of these nerves can be obtained by treating the pterygopalatine ganglion, which can be reached via the sphenopalatine foramen in the nose, via the greater palatine canal in the mouth, or from a lateral approach via the pterygomaxillary fissure. This will eliminate the postganglionic parasympathetic and sympathetic fibers to the mucous membrane of the nasal and oral cavities as well as afferent fibers for general sensation of the trigeminal nerve, and will also interfere with secretion in the lacrimal gland.

The **olfactory nerves** should not be forgotten in a summary of the nerves of the nasal cavity. They originate in the superior part of the lateral wall and septum and traverse the cribriform plate of the ethmoid bone to reach the olfactory bulb.

BLOOD SUPPLY OF NASAL CAVITY. Since the nerve supply to the nasal cavity is derived from two sources, more than one blood supply might be expected. Actually the blood arrives in the nasal cavity from three sources: branches of the ophthalmic, of the maxillary, and of the facial. These are pictured in Figures 7-126 and 7-127. Any of these branches may be involved in hemorrhage into the nasal cavity.

The **posterior ethmoidal branch** of the ophthalmic artery, after supplying the posterior ethmoidal paranasal sinuses, enters the nasal cavity via the posterior ethmoidal foramen and reaches the posterior superior parts of the lateral wall and septum. The **anterior ethmoidal branch** of the ophthalmic, after giving off an anterior meningeal artery and supplying the anterior and middle ethmoid paranasal sinuses, enters the nasal cavity via the anterior ethmoidal foramen and gives branches to both the septum and the lateral wall. It then continues, in company with the external nasal nerve, to the tip of the nose as the **external nasal artery.**

The branch of the maxillary artery involved with the nasal cavity is the **sphenopalatine.** This enters via the sphenopalatine foramen and divides immediately into **lateral and septal posterior nasal branches,** which ramify on the lateral wall and septum respectively. The longest artery on the septum anastomoses with the septal branch of the superior labial branch of the facial, about to be described.

The third source of blood to the nasal cavity is the artery just mentioned—the **septal branch** of the superior labial of the facial. This small branch is the source of nosebleeds, particularly in children. It ramifies on a small part of the inferior end of the septum (Fig. 7-127).

The **veins** of the nasal cavity, which accompany the arteries just described, drain for the most part into the pterygoid plexus of veins; some may drain posteriorly into the cavernous sinus, especially those accompanying the ethmoidal arteries. Those on the surface of the middle and inferior conchae are very large, resembling erectile tissue; these aid in warming inspired air.

LYMPH DRAINAGE

The most inferior aspect of the nasal cavity drains into submandibular and superficial cervical nodes following the facial artery. The major part of the cavity drains posteriorly to reach the upper deep cervical and retropharyngeal nodes.

ORAL CAVITY

GENERAL DESCRIPTION

The oral cavity or mouth is actually divided into two parts—the **vestibule** and the **oral cavity proper.**

The **vestibule** is the narrow space between the lips and cheeks on the outside and the teeth on the inside. If the teeth are in contact, the vestibule communicates with the cavity proper just posterior to the last molar teeth. The **parotid ducts** open into the vestibule opposite the upper second molar teeth and there are folds of mucous membrane in the midline from the upper and lower lips to the gums—the **frenula** of the lips.

If the mouth is opened (Fig. 7-128), the **oral cavity proper** can be seen. Superiorly, the **upper teeth,** consisting of two incisors, one canine, two premolars, and three molars on either side, can be seen, and just medial to these teeth and their respective gums is the mucous membrane covering the **hard palate.** The **soft palate or velum** (*L.,* veil) **palatinum** extends posteriorly from the hard palate; the **uvula** is the centrally located extension of

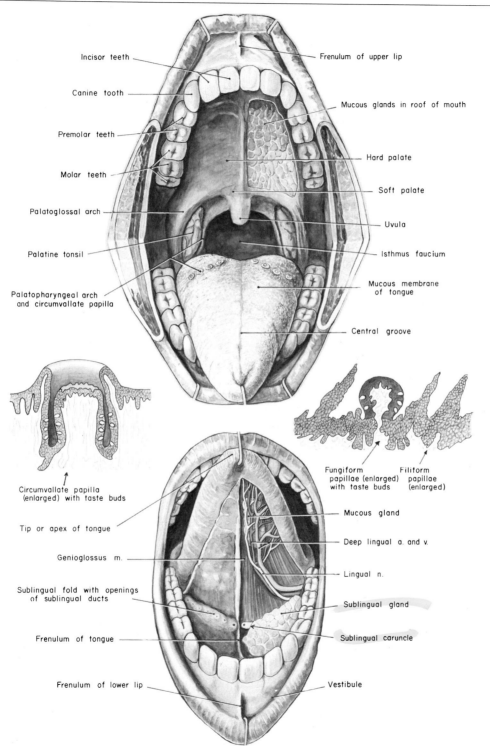

Incisor teeth

Canine tooth

Premolar teeth

Molar teeth

Palatoglossal arch

Palatine tonsil

Palatopharyngeal arch
and circumvallate papilla

Frenulum of upper lip

Mucous glands in roof of mouth

Hard palate

Soft palate

Uvula

Isthmus faucium

Mucous membrane
of tongue

Central groove

Circumvallate papilla
(enlarged) with taste buds

Fungiform
papillae (enlarged)
with taste buds

Filiform
papillae
(enlarged)

Tip or apex of tongue

Genioglossus m.

Sublingual fold with openings
of sublingual ducts

Frenulum of tongue

Frenulum of lower lip

Mucous gland

Deep lingual a. and v.

Lingual n.

Sublingual gland

Sublingual caruncle

Vestibule

Figure 7-128. Oral cavity.

the soft palate, which projects into the pharynx. Two folds can be seen extending laterally from the uvula to the lateral wall of the oral cavity; the more anterior one is the **palatoglossal arch,** which marks the posterior boundary of the mouth, and the more posterior one is the **palatopharyngeal arch.**

The **palatine tonsils,** if present, are located between these two arches. The entrance to the oral pharynx—the **isthmus faucium**—can be seen to be bounded by the palatoglossal arches, the uvula, and the base of the tongue.

The **tongue** is a muscular organ covered with mucous membrane; looking into the mouth, one can see the **central groove** of the tongue and posteriorly a line of **circumvallate papillae** in the form of a wide open V. The **foramen caecum** is at the base of this V. If the tongue is elevated, the fold of mucous membrane extending from the tongue to the floor of the mouth in the midline—the **frenulum** of the tongue—can be seen easily. On either side of the frenulum is the **sublingual caruncle** with its opening of the submandibular duct. The fold of tissue extending posteriorly and laterally from the caruncle—the **sublingual fold**—contains many openings of ducts from the sublingual gland. Just lateral and inferior to these structures is the row of lower teeth, also containing two incisors, one canine, two premolars, and three molars on either side.

The mouth is a very important part of the human body, for food is placed in it, ground with the teeth, moistened with saliva, and finally manipulated by the tongue into a position to be forced into the pharynx. The mouth also serves for oral expression, the tongue being extremely important in the formation of words.

LIPS

The **lips** are two mobile structures which, with the cheeks, form the lateral boundaries of the vestibule. These structures also are important in oral expression.

They consist of an outer layer of skin, and deep to this a layer of connective tissue containing the many muscles of facial expression; deep to these muscles is the submucous tissue containing mucous glands, and then the mucous membrane itself. The vertical depression in the middle of the upper lip is called the **philtrum.** The cheeks are very similar to the lips, and only two additional features need be mentioned. First, the **buccinator muscle** is covered with the **buccopharyngeal fascia,** and this fascia and the muscle are pierced by the **parotid duct.** Second, the **buccal pad of fat** is located just anterior to the masseter muscle. Decrease in volume of this fat pad gives people a hollow-cheeked look.

The lips are quite movable. The corners of the mouth are pulled posteriorly by the risorius and buccinator muscles, elevated by the levator anguli oris and the zygomaticus major and minor muscles, and lowered by the depressor anguli oris muscle. The upper lip is raised by the combined action of the levator anguli oris, zygomaticus major, zygomaticus minor, levator labii superioris, and levator labii superioris alaeque nasi muscles. The lower lip is lowered by the combined action of the depressor anguli oris, depressor labii inferioris, and platysma muscles. The lower lip is protruded by the mentalis muscles.

The **motor nerve** to the lips is the facial and the **sensory nerves** are the labial branch of the infraorbital, buccal, and mental nerves. These sensory nerves are all branches of the trigeminal, and they supply the mucous membrane on the inside as well as the skin on the outside.

The **blood supply** is via the superior and inferior labial branches of the facial artery, and similarly named **veins** drain into the facial vein.

The **lymphatics** of both lips end in the submental and submandibular nodes. It is important to note that the lymphatics from the lower lip can drain from one side to the opposite side, thereby crossing the midline.

TEETH AND GUMS

Any single tooth (Fig. 7-129) is composed of a **crown,** which is the visible portion, the **neck,** which is surrounded by the gums, and the **roots,** which are embedded in the bones. Each tooth possesses a cavity con-

---

Figure 7-129. *Teeth: (A) a single tooth, (B) deciduous and permanent teeth in a child. (Part A reproduced from Schour, in Noyes,* Oral Histology and Embryology; *part B from* Gray's Anatomy, *Charles M. Goss [ed.]; both courtesy of Lea and Febiger.)*

A

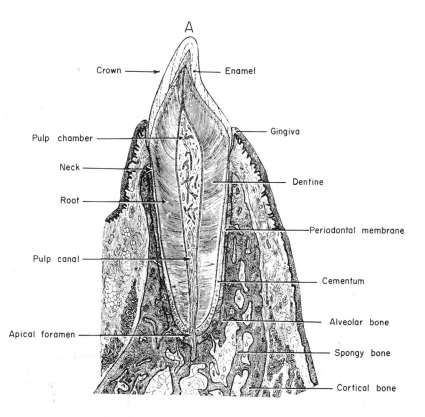

Crown — Enamel

Pulp chamber — Gingiva

Neck

Root

Dentine

Periodontal membrane

Pulp canal

Cementum

Alveolar bone

Spongy bone

Apical foramen

Cortical bone

B

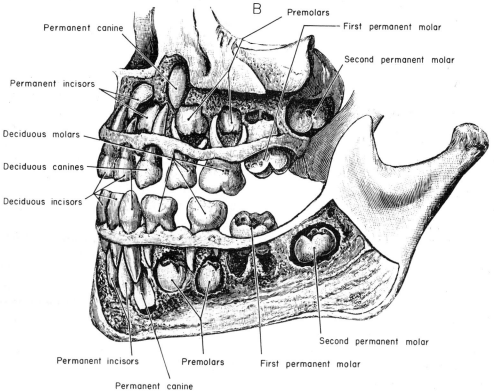

Permanent canine

Premolars

First permanent molar

Second permanent molar

Permanent incisors

Deciduous molars

Deciduous canines

Deciduous incisors

Second permanent molar

Permanent incisors

Premolars

First permanent molar

Permanent canine

taining the **pulp,** which is composed of connective tissue, blood vessels, and nerves. These vessels and nerves enter each root through the foramen in the apex of same. The basic structure of a tooth is the **dentine,** which contains the pulp canal. In the crown of the tooth the dentine is covered by **enamel;** in the root it is covered by **cementum,** which is surrounded by the **periodontal membrane.**

The crown takes different shapes in the different teeth. It is flattened, with a hollow area posteriorly, in the incisor teeth; it is conical in the canine; the premolars have two **tubercles** in most cases, while the molars have three or four tubercles.

The **deciduous teeth** (Fig. 7-129) are two incisors, one canine, no premolars, and two molars in each half of each jaw. The child, therefore, has a total of twenty teeth before the permanent teeth are erupted. The deciduous teeth erupt between the sixth and twenty-fourth months of life. Although the order is extremely variable, the lower central incisors usually erupt between the sixth and ninth months, the upper incisors between the eighth to tenth months, the lower lateral incisors and first molars between the fifteenth and twenty-first months; the canines between the sixteenth and twentieth months; and the second molars between the twentieth and twenty-fourth months.

The **dental formula** for the **permanent teeth** is two incisors, one canine, two premolars, and three molars on each half of each jaw, a total of thirty-two teeth. The permanent teeth erupt in the following order: the first molar about the sixth year, the medial incisors the seventh year, the lateral incisors the eighth, the first premolars the ninth, the second premolar the tenth, the canine teeth the eleventh, the second molar the twelfth, and the third molar the seventeenth to the twenty-fifth year or even later.

The **nerve supply** to the upper teeth is from the anterior, middle, and posterior superior alveolar branches of the maxillary nerve (Fig. 7-124); that to the lower teeth is by branches of the inferior alveolar nerve (Fig. 7-124), a branch of the mandibular division of the trigeminal. The **blood supply** to the upper teeth is by the anterior and posterior superior alveolar branches of the third part of the maxillary artery, the anterior being a branch of the infraorbital branch of the maxillary. The lower teeth obtain their blood supply from the inferior alveolar branch of the first part of the maxillary artery.

The **gums** are fibrous structures surrounding the necks of the teeth and are firmly attached to the alveolar margins of the jaws. They are composed of fairly dense connective tissue covered with a mucous membrane.

The **lymph drainage** of the lower teeth and gums is into the submental and submandibular nodes, while the upper teeth and gums drain more posteriorly into the submandibular, superficial cervical, and parotid nodes.

TONGUE

The **tongue** is a muscular organ divided into right and left halves by a **median fibrous septum.** It is composed of **intrinsic muscles** (Fig. 7-131), which intertwine with each other but can be divided into superior and inferior longitudinal, vertical, and transverse muscle fibers. The **extrinsic muscles**—the styloglossus, hyoglossus, and genioglossus—intertwine their insertions with the intrinsic muscles. These extrinsic muscles arise from the styloid process, hyoid bone, and genial tubercles of the mandible respectively; the base of the tongue, therefore, is attached to these structures.

The **dorsum of the tongue** consists of a **pharyngeal part,** which looks posteriorly into the oral pharynx, and an **oral part,** which is in relation with the hard palate. The dorsum is covered by a mucous membrane (Fig. 7-130), which continues around the projecting tip of the tongue to its inferior surface, where it is continuous with the floor of the mouth. The midline fold of mucous membrane is called the **frenulum** of the tongue. The mucous membrane on the pharyngeal part of the dorsum of the tongue has a pitted appearance produced by the underlying lymphoid tissue—the **lingual tonsils.** This portion of the mucous membrane is also connected with the epiglottis by a median ridge of mucous membrane—the **median glossoepiglottic fold.** Laterally, there are two similar folds—the **lateral glossoepiglottic folds.** The mucous membrane covering the oral part of the dorsum of the tongue is separated from the pharyngeal part by a V-shaped groove called the **sulcus terminalis,** the apex of which points posteriorly; the **foramen caecum** is located at this point. Just anterior to this sulcus terminalis is a row of **circumvallate papillae** containing many taste buds. Although not visible, the other types of papillae—the **fungiform and filiform**—are also located in great numbers on the dorsum of the tongue. These papillae are pictured in Figure 7-128. The **deep veins** can be seen

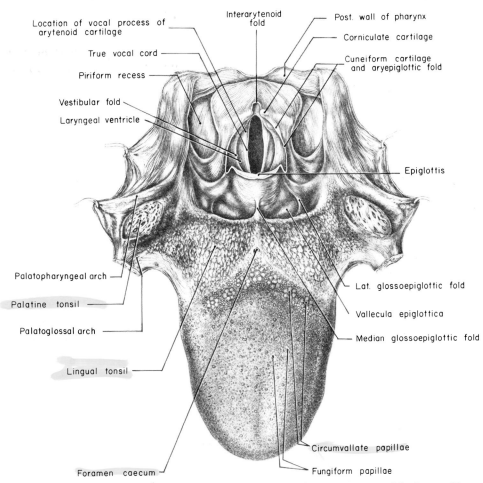

Location of vocal process of arytenoid cartilage

True vocal cord

Piriform recess

Vestibular fold

Laryngeal ventricle

Interarytenoid fold

Post. wall of pharynx

Corniculate cartilage

Cuneiform cartilage and aryepiglottic fold

Epiglottis

Palatopharyngeal arch

Palatine tonsil

Palatoglossal arch

Lingual tonsil

Foramen caecum

Lat. glossoepiglottic fold

Vallecula epiglottica

Median glossoepiglottic fold

Circumvallate papillae

Fungiform papillae

Figure 7-130.   *The mucous membrane of the tongue and a superior view of the larynx. (From Surgical Anatomy, by Deaver © 1926 by McGraw-Hill, Inc. Used with the permission of the McGraw-Hill Book Company.)*

under the mucous membrane on the inferior surface of the tongue, and more laterally the **deep arteries** are marked by fimbriated folds located superficial to them.

The **nerve supply** to the tongue consists of nerves of general sensation, nerves involved with taste, and motor nerves. In addition, the tongue is divided into an anterior two-thirds and a posterior one-third. **General sensation** for the anterior two-thirds is accomplished by the lingual branch of the trigeminal nerve, the posterior one-third by the glossopharyngeal nerve. **Taste** for the anterior two-thirds is carried in the chorda tympani branch of the facial nerve, for the posterior one-third in the glossopharyngeal.

The **motor nerve** to all muscles of the tongue, intrinsic as well as extrinsic, is the hypoglossal.* Figure 7-132 shows how the development of the mucous membrane on the tongue is closely associated with the branchial arches, and correlates the above nerve supply with this development.

The **blood supply** to the tongue is from the lingual

*M. N. Bates, 1948, The early development of the hypoglossal musculature in the cat, *Am. J. Anat.* 83: 329. The tongue musculature, according to this work, is definitely of somitic origin in the cat. A good account of the relation of the hypoglossal to C1 is given also.

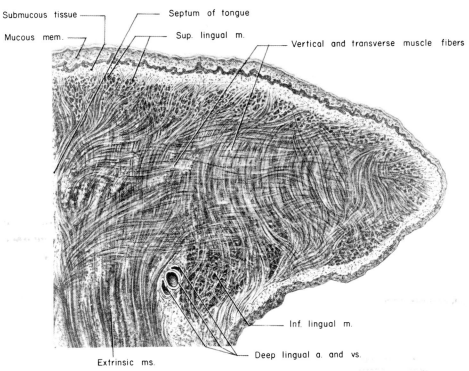

Submucous tissue

Mucous mem.

Septum of tongue

Sup. lingual m.

Vertical and transverse muscle fibers

Inf. lingual m.

Deep lingual a. and vs.

Extrinsic ms.

Figure 7-131. *A transverse section through the tongue to show its intrinsic musculature.* (*From* Surgical Anatomy, *by Deaver* © *1926 by McGraw-Hill, Inc. Used with the permission of the McGraw-Hill Book Company.*)

artery. This arises from the external carotid, courses on the deep surface of the hyoglossus muscle, and finally terminates as the **deep lingual artery,** which courses to the tip of the tongue. It gives off branches to the dorsum of the tongue—the **dorsal lingual**—and to the sublingual gland—the **sublingual artery.** The **veins** accompanying the lingual artery form venae comitantes, and there are veins of a similar nature accompanying the hypoglossal nerve. These join to empty either into the facial vein or into the internal jugular vein.

The **lymphatics** of the tongue are different for the anterior two-thirds and the posterior one-third, the anterior two-thirds draining into the submandibular and superficial cervical nodes, while the lymphatics of the posterior one-third drain into the upper deep cervical and retropharyngeal nodes.

The tongue is a very mobile structure,* the larger

movements being produced by the extrinsic and intrinsic muscles combined, and the more delicate movements involved in speech by the intrinsic muscles alone. The tongue can be protruded by the action of the genioglossus muscles, which arise from the genial tubercles of the mandible; Bennett and Hutchinson† feel that the intrinsic muscles are also necessary for real protrusion. It can be pulled posteriorly by the styloglossus muscles arising from the styloid processes, and pulled inferiorly and posteriorly by the hyoglossus muscles arising from the greater horns of the hyoid bone. It is an interesting clinical point that when a patient has an injured hypoglossal nerve and the tongue is protruded, the tongue turns toward the side of the injury. Accordingly, if a patient has an injured right hypoglossal nerve, the tongue when protruded turns toward the right. This is contrary to the effects of most nerve

---

*S. Abd-el-Malek, 1955, The part played by the tongue in mastication and deglutition, *J. Anat.* 89: 250. The remarkable contortions of the tongue during mastication are described.

†H. A. Bennett and R. C. Hutchinson, 1946, Experimental studies on the movements of the mammalian tongue II. The protrusion mechanism of the mammalian tongue. *Anat. Record* 94: 57. This work shows the importance of the intrinsic muscles in the act of protrusion.

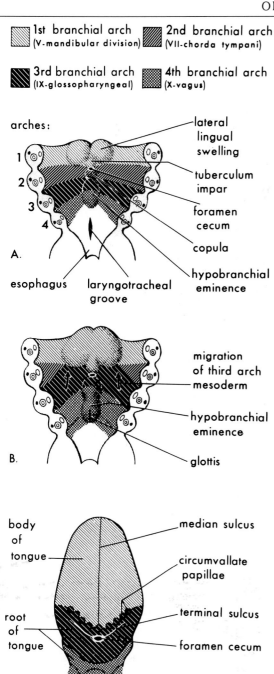

1st branchial arch
(V-mandibular division)

2nd branchial arch
(VII-chorda tympani)

3rd branchial arch
(IX-glossopharyngeal)

4th branchial arch
(X-vagus)

arches:

1
2
3
4

A.

lateral
lingual
swelling

tuberculum
impar

foramen
cecum

copula

hypobranchial
eminence

esophagus   laryngotracheal
groove

B.

migration
of third arch
mesoderm

hypobranchial
eminence

glottis

body
of
tongue

root
of
tongue

C.

median sulcus

circumvallate
papillae

terminal sulcus

foramen cecum

epiglottis

Figure 7-132. *(A and B) Schematic horizontal section through the pharynx showing successive stages in the development of the mucous membrane of the tongue (not the muscles) during the fourth and fifth weeks. Note how this explains the nerve supply of the adult tongue (C). (From Moore's* The Developing Human, *courtesy of W. B. Saunders Co.)*

injuries since the usual result is for the structure to move away from the point of injury. However, in this case the paralyzed muscles on the injured side tend to drag behind as the tongue is being protruded and therefore make the tongue turn toward that injured side. This is an important clinical test.

HARD AND SOFT PALATES

The dome-shaped roof of the mouth is known as the palate and consists of an anterior bony portion—the **hard palate**—and a posterior fleshy portion—the **soft palate.** The hard palate is made up of the palatine processes of the maxillary bones anteriorly and the horizontal part of the palatine bones posteriorly. It is covered with a mucous membrane, deep to which are found many **mucous palatine glands** (Fig. 7-133).

The **soft palate** is a posterior projection of the hard palate. It is constructed of skeletal muscles covered with a mucous membrane and contains many mucous glands. The **uvula** is a median projection from the posterior free edge of the soft palate. The soft palate extends posteriorly almost as far as the posterior pharyngeal wall.

The **muscles** associated with the soft palate are the palatoglossus, the palatopharyngeus, the tensor veli palatini, the levator veli palatini, and the musculus uvulae.

The **palatoglossus muscle** (Figs. 7-115 and 7-133) lies in the palatoglossal arch and forms the lowest muscular layer of the soft palate. It attaches to the inferior surface of the soft palate and, passing inferiorly and slightly anteriorly (anterior to the palatine tonsil), is attached to the side of the tongue. The **palatopharyngeus muscle** (Figs. 7-113 and 7-133) is contained in the palatopharyngeal arch, which is located posterior to the tonsil. This muscle attaches to the soft palate and passes inferiorly deep to the mucous membrane to attach to the posterior border of the thyroid cartilage. The palatoglossus muscle can be thought of as acting in both directions. In one manner it pulls the tongue superiorly and in another it tenses the palate; either way it closes off the opening into the oral pharynx. The palatopharyngeus muscle can also be thought of as playing a double role of either lifting the larynx in the process of swallowing or tensing the soft palate in this process. Both of these muscles are innervated by the vagus nerve.

Levator veli palatini muscle (Figs. 7-120 and 7-133). This muscle has its origin from the lower surface

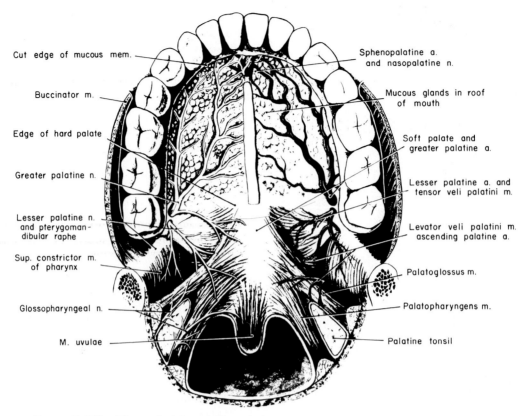

Figure 7-133. *The roof of the mouth, including muscles, and nerve and blood supply. The nerves shown are innervating the mucous membrane, not the muscles.*

of the petrous portion of the temporal bone and from the medial surface of the cartilage of the auditory tube. Its insertion is into the aponeurosis of the palate, and its main action is to lift the soft palate superiorly. Its nerve supply is via the vagus nerve.

The **tensor veli palatini muscle** (Figs. 7-120 and 7-133) arises from the sphenoid bone and from the lateral surface of the auditory tube. Its insertion is into the aponeurosis of the soft palate and onto the posterior edge of the bony hard palate; since it passes around the pterygoid hamulus, its action is to pull the soft palate laterally when it contracts. Recent evidence indicates that the portion of this muscle inserting on the hard palate arises from the lateral portion of the auditory tube; if the origin is considered to be the hard palate and the insertion the more movable auditory tube, it can be seen that it may serve to open the tube. Its nerve supply, since this

muscle happens to be derived from the first branchial arch, is the motor division of the trigeminal nerve.

The **musculus uvulae** consists of two small slips of muscle which are attached to the posterior nasal spine and unite into one muscle to enter the uvula itself. The nerve supply to this muscle is the vagus, and its main action is to close off the nasal cavity during the process of swallowing.

The **arterial supply** to the hard palate (Fig. 7-133) is by the greater palatine, which courses along the sides of the hard palate to anastomose with the arteries entering this area through the incisive foramina. The soft palate is given its blood supply via the lesser palatine vessels and the ascending palatine artery. These vessels are branches of the third part of the maxillary artery and the facial respectively. The **nerve supply** is similar, the greater palatine nerve providing the sensory supply to the hard palate, while the lesser palatine nerves innervate the soft palate. The very anterior part of the hard palate derives its sensory nerve supply from branches of the nasopalatine which course along the septum of the nose to get to this destination, and the area around the palatine tonsils obtains twigs from the glossopharyngeal nerve. You will re-call that there are taste fibers from this area, and that the mucous membrane in the hard and soft palates and the many glands associated with these membranes have a sympathetic and a parasympathetic nerve supply.

Figure 7-134 shows the development of the face, nasal cavity, and palate.

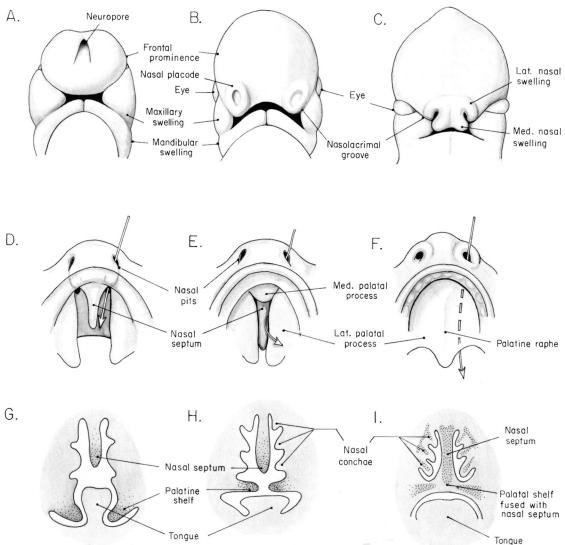

Figure 7-134. *Development of the face and nasal and oral cavities in a 6-week-old embryo. (A, B, and C) Changes in the face in development of the nose and mouth. Note how the swellings correspond to the three divisions of the trigeminal nerve. Failure of this process leads to harelips. (D, E, and F) Formation of the nasal septum and hard palate in the development of the two nasal cavities and the separation of the nasal from the oral cavity. Failure of complete development leads to cleft palate. (G, H, and I) Notice how the tongue can interfere with proper development of the palatine shelf. (Redrawn from Tuchmann-Duplessis and Haegel.)*

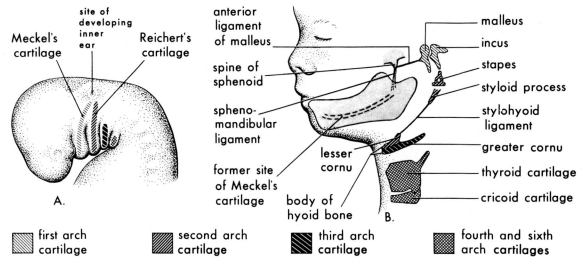

first arch cartilage

second arch cartilage

third arch cartilage

fourth and sixth arch cartilages

Figure 7-135. *(A) Schematic lateral view of the head and neck region of a 4-week-old embryo illustrating the location of the branchial arch cartilages. (B) The adult derivatives of these cartilages. (From Moore's* The Developing Human, *courtesy of W. B. Saunders Co.)*

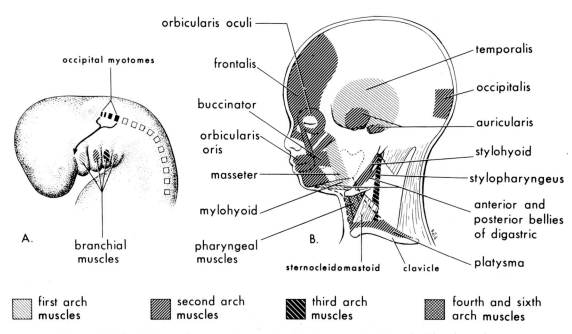

first arch muscles

second arch muscles

third arch muscles

fourth and sixth arch muscles

Figure 7-136. *(A) Lateral view of the head and neck region of a 4-week-old embryo showing primordia of the branchial muscles. (The arrow shows the pathway taken by myoblasts from the occipital myotomes to form the musculature of the tongue.) (B) The muscles derived from the branchial arches. The sternocleidomastoid and platysma muscles have been cut to show deeper structures. (From Moore's* The Developing Human, *courtesy of W. B. Saunders Co.)*

SYSTEMIC SUMMARY

Figures 7-135, 7-136, and 7-137 summarize the derivatives of the cartilages and muscle masses of the branchial arches, and depict the nerves involved with the arches. Study of these illustrations will aid in comprehending much of this systemic summary.

BONES

There is very little advantage in a systemic summary of the skeletal system since it was studied in this same manner on pages 455–469. You should reread these pages now that you have become thoroughly acquainted with the head and neck. Reading a description of the skull after having been exposed to the anatomy of the head and neck is a much more rewarding experience than reading it before having dissected this region. Now each structure mentioned will have a functional meaning as well as a structural one.

As an aid, the table on pages 638–642 presents a partial list of structures found in the skull, with the function of each. The approach is regional. This list should not be memorized as a list, but might serve as a quick reference.

JOINTS

The joints involved in the head and neck are the temporomandibular; sternoclavicular and acromioclavicular (although these are actually part of the upper limb); articulations between the cervical vertebrae and between the atlas and skull; and the sutures between the skull bones themselves. The temporomandibular joint is described on page 552, the sternoclavicular on page 576, the acromioclavicular on page 578, and the atlanto-occipital and atlantoaxial joints on page 598. The sutures are not complete at birth, the spaces thus formed—fontanelles—being points of danger in the infant. One occurs just posterior to the foramen magnum—the occipital fontanelle; another at the pterion—the anterolateral fontanelle; another at the asterion—posterolateral fontanelle; and still others in the midline at bregma—anterior

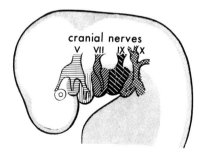

Figure 7-137. Lateral view of the head and neck region of a 4-week-old embryo showing the nerves supplying muscles derived from the branchial arches. (From Moore's The Developing Human, courtesy of W. B. Saunders Co.)

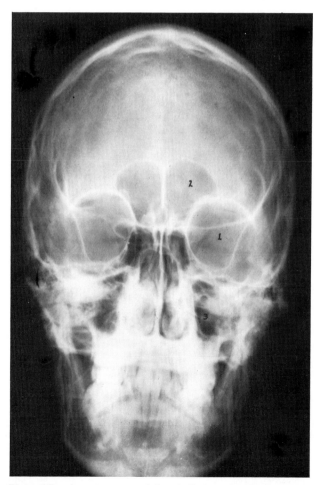

Plate 22. Anterior view of the skull: (1) orbit, (2) frontal sinus, (3) maxillary sinus.

Functions of Structures in the Skull—Regional Approach

Structure	Function
A. Anterior aspect (Fig. 7-1)	
1. Bones visible: frontal, nasal, zygomatic, maxilla, mandible, and bones forming orbit	
2. Frontal region	
a. Frontal eminence	Most prominent part of forehead
b. Temporal line	Outline of origin of temporalis muscle
c. Zygomatic process of frontal	Articulation with zygomatic bone
d. Supraorbital margin	Superior margin of orbit
e. Nasal margin of frontal	Articulation with nasal bones
f. Supraorbital notch or foramen	Transmits supraorbital n. and a.
g. Nasion	Center of frontonasal suture
h. Glabella	Smooth area just superior to nasion and between superciliary arches
i. Superciliary arches	Ridges just superior to supraorbital margins
3. Orbits	
a. Bones involved: frontal, zygomatic, maxilla, lacrimal, ethmoid, sphenoid, and palatine	
b. Lacrimal fossa (frontal)	For lacrimal gland
c. Trochlear fossa (frontal)	For pulley of superior oblique m.
d. Optic foramen (sphenoid)	Transmits optic n., ophthalmic a., sympathetic n.
e. Anterior and posterior ethmoidal foramina	Transmit anterior and posterior ethmoidal ns. and as.
f. Superior orbital fissure	Transmits oculomotor, trochlear, frontal, lacrimal, nasociliary, abducens, and sympathetic ns.; ophthalmic vs.; meningeal branch of ophthalmic a.
g. Inferior orbital fissure	Transmits maxillary and zygomatic ns.; infraorbital vessels; v. to pterygoid plexus
h. Nasolacrimal canal	Transmits nasolacrimal duct
i. Zygomatico-orbital foramen	Transmits zygomatic n. and a. and branch of lacrimal a.
j. Infraorbital groove, canal, and foramen	Transmits infraorbital n. and vessels
4. Facial region	
a. Nose	Made up of nasal bones
b. Zygomaticofacial foramen	For zygomaticofacial n. and a.
5. Mandible	
a. Mental protuberance and tubercle	Tip of chin
b. Oblique line	Origin for depressor labii inferioris and depressor anguli oris ms.
c. Incisor fossa	Origin for mentalis and oribicularis oris ms.
d. Mental foramen	Transmits mental n. and a.
B. Lateral aspect (Figs. 7-2A and 7-2B)	
1. Bones visible: nasal, frontal, parietal, occipital, temporal, sphenoid, zygomatic, maxilla, and mandible	
2. Pterion	Union of parietal, greater wing of sphenoid, and squamous portion of temporal
3. Asterion	Union of parietal, occipital, and mastoid process of temporal
4. Bregma	Union of sagittal and coronal sutures

Functions of Structures in the Skull—Regional Approach (Continued)

Structure	Function
5. Lambda	Union of sagittal and lambdoid sutures
6. Temporal line	Outline of temporalis m. attachment
7. External auditory meatus	Canal from auricle to tympanic membrane
8. Zygomatic arch	Origin masseter m.
9. Tympanic plate	Anterior to external auditory meatus
10. Temporal fossa	For temporal m.
11. Infratemporal fossa	Inferior to greater wing of sphenoid bone, deep to mandible
12. Zygomaticotemporal foramen	Transmits zygomaticotemporal a. and n.
13. Posterior superior alveolar foramen	Transmits n. and a. of same name to molar teeth
14. Lateral pterygoid lamina	Origin of lateral pterygoid m.
15. Pterygomaxillary fissure	Entrance into pterygopalatine fossa from infratemporal fossa; transmits maxillary n. and a.
16. Pterygopalatine fossa	Medial to pterygomaxillary fissure; contains pterygopalatine ganglion, maxillary n., and branches of maxillary a. and v.
17. Sphenopalatine foramen	From pterygopalatine fossa to nasal cavity; transmits posterior superior nasal n. and sphenopalatine a. and v.
18. Pharyngeal canals	From pterygopalatine fossa to superior end of pharynx; transmits pharyngeal n., a., and v.
19. Pterygoid canal	From area of foramen lacerum to pterygopalatine fossa; transmits pterygoid n. and vessels
20. Foramen rotundum	From pterygopalatine fossa to middle cranial fossa; transmits maxillary n.
21. Greater palatine canal	From pterygopalatine fossa to roof of mouth; transmits greater palatine vessels and n.
22. Lesser palatine canals	From greater palatine canal to roof of mouth; transmits lesser palatine n. and vessels
23. Inferior orbital fissure	From pterygopalatine fossa to orbit; transmits intraorbital n. and vessels; zygomatic n.; v. to pterygoid plexus
24. Ramus of mandible	Insertion of masseter m.
25. Angle of mandible	Joining of body and ramus
26. Condyloid process—head and neck	Articulation with temporal fossa
27. Coronoid process	Insertion of temporalis m.
28. Mandibular notch	Between coronoid and condyloid processes
C. Posterior aspect (Fig. 7-4)	
1. Bones visible: parietals, occipital, and mastoid processes of temporals	
2. Lambda	Union of sagittal and lambdoid sutures
3. External occipital protuberance	Origin of trapezius; attachment for ligamentum nuchae
4. Superior nuchal line	Insertion of sternocleidomastoid, splenius, and fronto-occipitalis ms.
5. Inferior nuchal line	No muscle attachments
6. Parietal foramen	Transmits parietal emissary v.
7. Mastoid foramen	Transmits mastoid emissary v. and meningeal branch of occipital a.
D. Base of skull (Fig. 7-4)	
1. Anteromedian area	
a. Hard palate	Maxillary and palatine bones
(i) Upper teeth	2-1-2-3 on each side
(ii) Posterior free margin	Attachment of soft palate

Functions of Structures in the Skull—Regional Approach (Continued)

Structure	Function
(iii) Posterior nasal spine	Attachment of uvula
(iv) Incisive foramen	Transmits sphenopalatine a. and nasopalatine n. from nasal cavity to hard palate
(v) Greater palatine foramen	Transmits greater palatine n. and vessels
(vi) Lesser palatine foramen	Transmits lesser palatine n. and vessels
b. Choanae	Passageway from nasal cavity to nasal pharynx
c. Vomer	Forms part of nasal septum
d. Pterygoid process	Extends inferiorly from sphenoid bone
(i) Medial lamina	Origin of superior constrictor m. of pharynx
(ii) Pterygoid hamulus	On inferior end of medial lamina; pterygomandibular raphe attached to it; tensor veli palatini m. bends around it
(iii) Pterygoid fossa	Between the pterygoid laminae; contains tensor veli palatini m.
(iv) Scaphoid fossa	Origin of tensor veli palatini m.
(v) Lateral lamina	Origin of medial pterygoid and inferior head of lateral pterygoid m.
(vi) Emissary sphenoidal foramen	Transmits emissary vein between cavernous sinus and pterygoid plexus
2. Anterolateral area	
a. Foramen ovale	Transmits mandibular division of trigeminal nerve including motor root; accessory meningeal a.; small veins; lymphatics from meninges
b. Foramen spinosum	Middle meningeal vessels; meningeal branch of mandibular n.; lymphatics from meninges
c. Sphenoid bone	Greater wing of this bone.
d. Mandibular fossa and tubercle	For articulation with mandible
3. Posteromedian area	
a. Basiocciput	Part of occipital bone anterior to foramen magnum
b. Pharyngeal tubercle	Attachment of superior constrictor m. of pharynx
c. Foramen magnum	Transmits spinal cord and its vessels and coverings; spinal accessory n.; vertebral as.; sympathetic fibers; lymphatics; membrana tectoria
d. Occipital condyle	Articulation with atlas
e. Posterior condylar canal	Transmits emissary v. between sigmoid sinus and suboccipital plexus
f. Hypoglossal canal	Transmits twelfth cranial n.; meningeal branch of ascending pharyngeal a.; small veins; lymphatics
g. Occipital crest	Attachment of ligamentum nuchae
4. Posterolateral area	
a. Foramen lacerum	Filled with cartilage, but does transmit meningeal as. from ascending pharyngeal; small veins; lymphatics; and greater petrosal n.
b. Posterior end of pterygoid canal	Transmits pterygoid n. and a.
c. Groove for auditory tube	Tube from middle ear to pharynx
d. Tympanic plate	Forms walls of external auditory meatus
e. Styloid process	Origin of stylopharyngeus, styloglossus, and stylohyoid muscles, and stylohyoid ligament
f. Stylomastoid foramen	Exit of facial nerve; contains a. of same name
g. Jugular foramen	Transmits internal jugular vein; ninth, tenth, eleventh cranial nerves; meningeal branches of ascending pharyngeal and occipital arteries; lymphatics
h. Mastoid canaliculus	Transmits auricular branch of vagus
i. Carotid canal	Transmits internal carotid a.; sympathetic ns.; lymphatics

Functions of Structures in the Skull—Regional Approach (Continued)

Structure	Function
E. Cranial fossae (Fig. 7-7)	
1. Anterior cranial fossa	
a. Bones involved: frontal, sphenoid, and ethmoid	
b. Crista galli	Attachment of falx cerebri
c. Frontal crest	Site of superior sagittal sinus
d. Foramen caecum	Transmits emissary v. between superior sagittal sinus and veins of nose
e. Olfactory foramina in cribriform plate	Transmit olfactory nerves
f. Ethmoidal foramina	Transmit anterior and posterior ethmoidal as. and ns.
g. Grooves	For anterior meningeal arteries
h. Orbital plate	Roof of orbit
2. Middle cranial fossa	
a. Bones involved: sphenoid, temporal, and small inferior portion of parietal	
b. Optic foramina	Transmit optic ns., ophthalmic as., sympathetic ns.
c. Sella turcica	Divided into tuberculum sellae anteriorly, hypophyseal fossa centrally, and dorsum sellae posteriorly; houses hypophysis
d. Anterior, posterior, and middle clinoid processes	Attachment of dura
e. Foramen lacerum	Transmits meningeal as.; small vs.; lymphatics; greater petrosal n.; otherwise filled with cartilage
f. Carotid canal	Transmits internal carotid a.; sympathetic ns.; lymphatics
g. Pterygoid canal	Transmits pterygoid n. and a.
h. Superior orbital fissure	Transmits third, fourth, ophthalmic division of fifth, and sixth cranial ns.; ophthalmic vs.; sympathetic ns.; meningeal branch of ophthalmic a.
i. Foramen rotundum	Transmits maxillary n.
j. Foramen ovale	Transmits mandibular n.; accessory meningeal a.; small vs.; lymphatics from meninges
k. Foramen spinosum	Transmits middle meningeal vessels; meningeal branch of mandibular n.; lymphatics
l. Groove	For middle meningeal a.
m. Arcuate eminence	Produced by superior semicircular canal; tentorium cerebelli attaches to it
n. Tegmen tympani	Anterior roof of middle ear
o. Hiatus and groove	For greater petrosal n.
3. Posterior cranial fossa	
a. Bones involved: occipital, temporal, and sphenoid	
b. Groove	For inferior petrosal sinus
c. Hypoglossal canal	Transmits twelfth n.; meningeal a.; small vs.; lymphatics
d. Posterior condylar canal	Transmits emissary v.
e. Foramen magnum	Transmits spinal cord and its vessels and coverings; spinal accessory nerves; vertebral as.; sympathetic fibers; lymphatics; membrana tectoria
f. Internal occipital protuberance	Site of confluens of sinuses; attachment of tentorium
g. Internal occipital crest	Attachment of falx cerebelli
h. Cerebellar fossae	For lobes of cerebellum

Functions of Structures in the Skull—Regional Approach **(Continued)**

Structure	Function
i. Grooves	For transverse and sigmoid sinuses
j. Mastoid foramen	Transmits emissary v. and meningeal a.
k. Internal auditory meatus	Transmits seventh and eighth cranial ns. and internal auditory a.
l. Jugular foramen	Transmits internal jugular v.; ninth, tenth, and eleventh cranial ns.; meningeal branches of ascending pharyngeal and occipital as.; lymphatics

F. Sagittal section of skull (Figs. 7-5 and 7-6)
  1. Nasal cavity and vicinity

Structure	Function
a. Bones involved: nasal, maxilla, ethmoid, inferior nasal concha, frontal, palatine, and sphenoid	
b. Superior and middle conchae	Part of ethmoid bone
c. Inferior nasal concha	A separate bone
d. Sphenoid sinus	Paranasal sinus
e. Frontal sinus	Paranasal sinus
f. Sphenoethmoid recess	Area above superior concha into which sphenoid sinuses drain
g. Superior meatus	Area under superior concha into which posterior ethmoid sinuses drain
h. Middle meatus	Area under middle concha into which frontal, maxillary, and anterior and middle ethmoid sinuses drain
i. Bulla ethmoidalis	Rounded elevation in middle meatus containing opening of middle ethmoid air cells
j. Hiatus semilunaris	Moonshaped slit into which frontal, anterior ethmoid, and maxillary sinuses drain
k. Uncinate process of ethmoid	Bony spicule that actually forms boundary of hiatus semilunaris
l. Inferior meatus	Area under inferior nasal concha into which nasolacrimal duct drains
m. Perpendicular plate of ethmoid	Forms upper part of nasal septum
n. Vomer	Forms lower part of nasal septum

  2. Around mandible

Structure	Function
a. Genial tubercle	Origin of genioglossus and geniohyoid muscles
b. Sublingual fossa	For sublingual gland
c. Mylohyoid line	Origin of mylohyoid muscle; attachment of fascia of submandibular gland, and, posteriorly, superior constrictor and pterygomandibular ligament
d. Submandibular fossa	For submandibular gland
e. Angle	Point of junction of body and ramus
f. Mandibular foramen	Transmits inferior alveolar n. and vessels
g. Lingula	Spicule of bone for attachment of sphenomandibular ligament
h. Mylohyoid groove	For mylohyoid n. and vessels

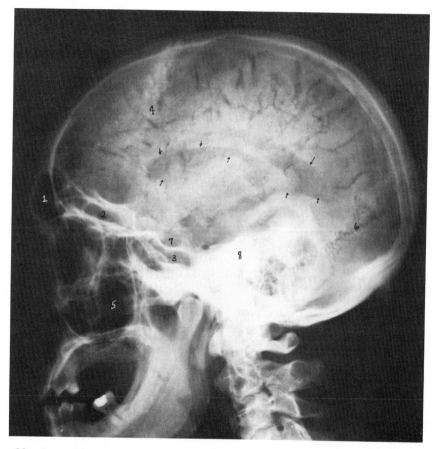

Plate 23. Lateral view of skull: (1) frontal sinus, (2) ethmoid sinuses, (3) sphenoid sinus, (4) coronoid suture, (5) maxillary sinus, (6) lambdoid suture, (7) sella turcica, (8) petrous portion of temporal bone. Arrows outline lateral ventricles of brain (a pneumoencephalogram).

fontanelle—and at lambda—posterior fontanelle. One can take advantage of these midline fontanelles as a source of blood, in babies, from the superior sagittal sinus. In addition, the sutures tend to disappear as age increases, the frontal suture rarely being found.

MUSCLES

The muscles of the head and neck can be divided into groups according to their development or in respect to their functions.

DEVELOPMENTAL GROUPING

I. From head somites
   A. From ocular myotomes (the ocular muscles arise from collections of mesenchyme found over each optic cup; it is thought that these represent derivatives of ancestral myotomes):

Levator palpebrae superioris
Superior rectus
Medial rectus ⎱ Oculomotor nerve
Inferior rectus
Inferior oblique

Superior oblique—Trochlear nerve
Lateral rectus   —Abducens nerve

B. From occipital myotomes (although a direct origin of the tongue muscles from distinct myotomes is difficult to determine, it is felt that they are actually derivatives of occipital somites):

Styloglossus
Hyoglossus
Genioglossus   } Hypoglossal nerve
Intrinsic muscula-
ture of tongue

II. From branchial arches

A. 1st (mandibular) arch, trigeminal nerve
Temporalis
Masseter
Medial pterygoid
Lateral pterygoid
Mylohyoid
Anterior belly of digastric
Tensor veli palatini
Tensor tympani

B. 2nd (hyoid) arch, facial nerve
Muscles of facial expression
Stapedius
Posterior belly of digastric
Stylohyoid

C. 3rd arch, glossopharyngeal nerve
Stylopharyngeus

D. 4th arch, pharyngeal and superior laryngeal branches of vagus
Constrictor muscles of pharynx
Musculus uvulae
Levator veli palatini
Palatopharyngeus
Palatoglossus
Salpingopharyngeus
Cricothyroid
Transverse arytenoid (in part)

E. 5th arch
No muscles

F. 6th arch—inferior (recurrent) laryngeal branch of vagus
All muscles of larynx except cricothyroid and part of transverse arytenoid

III. From postbranchial region, spinal accessory nerve and ventral rami C3 and C4
Trapezius
Sternocleidomastoid

IV. From cervical myotomes, cervical nerves

A. From epimeres, dorsal rami cervical nerves
Splenius

Semispinalis
Multifidus
Erector spinae (superior parts)
Rectus capitis posterior major
Rectus capitis posterior minor
Obliquus capitis superior
Obliquus capitis inferior

B. From hypomeres, ventral rami C1 to C4 (cervical plexus)
Prevertebral musculature
Scalene muscles
Infrahyoid muscles
Geniohyoid
Thyrohyoid
Trapezius (in part ?)
Sternocleidomastoid (in part ?)

FUNCTIONAL GROUPING.   Visualize the muscles and see if you agree with these actions.

I. The vertebral column

A. Flexion
Primary
Sternocleidomastoid
Scaleni
Longus cervicis
Longus capitis

B. Extension
Primary
Splenius cervicis
Semispinalis cervicis
Longissimus cervicis
Iliocostalis cervicis
Interspinalis
Multifidus
Rotators
Secondary
Splenius capitis
Semispinalis capitis

C. Rotation
To the right
Primary
Right splenius cervicis
Right longissimus cervicis
Right iliocostalis cervicis
Right longus colli
Left semispinalis cervicis
Left rotators
Left longus colli
Left scaleni

Secondary
Right splenius capitis
Left sternocleidomastoid
Left trapezius
To the left
Same muscles on opposite side of body

II. The head
A. Flexion
Primary
Longus capitis
Rectus capitis anterior
Sternocleidomastoid
Secondary
Suprahyoid muscles
Infrahyoid muscles
B. Extension
Primary
Splenius capitis
Semispinalis capitis
Obliquus capitis superior
Rectus capitis posterior major
Rectus capitis posterior minor
Secondary
Trapezius
Longissimus capitis
C. Rotation
To the right
Primary
Left sternocleidomastoid
Right splenius capitis
Right longissimus capitis
Right obliquus capitis inferior
Right rectus capitis posterior major
Secondary
Left trapezius
Left longus capitis
To the left
Same muscles on opposite side of body

III. Jaw, temporomandibular joint
A. Elevation
Temporalis
Masseter
Medial pterygoid
B. Depression
Lateral pterygoids
Geniohyoid
Mylohyoid
Digastric
Infrahyoid muscles
C. Protraction
Lateral pterygoids

D. Retraction
Temporalis (posterior part)
Digastric
E. Forceful lowering of jaw
Supra- and infrahyoid muscles
Note: In "grinding" the teeth, the protractors on either side contract alternately. In chewing, one side predominates over the other. For example, in chewing on the right side, muscles on this side will contract more forcefully than those on the left side.

IV. Movement of scalp
Fronto-occipitalis

V. Eyelids
Opening of eye
Levator palpebrae superioris
Smooth muscles in levator
Closing of eye
Primary
Orbicularis oculi
Secondary
Procerus
Corrugator supercilii

VI. Eye
Elevation
Superior rectus
Inferior oblique
Depression
Inferior rectus
Superior oblique
Abduction
Lateral rectus
Inferior oblique
Superior oblique
Adduction
Medial rectus
Superior rectus
Inferior rectus
Medial rotation
Superior rectus
Superior oblique
Lateral rotation
Inferior rectus
Inferior oblique

VII. Pupil of eye
Increase in diameter
Dilator pupillae
Decrease in diameter
Constrictor pupillae

VIII. Lens of eye
    Thickening
        Ciliary muscle, circular and meridional portions
    Flattening
        Muscles at rest
IX. Hearing
    Decrease in sound
        Tensor tympani
        Stapedius
    Increase in sound
        Above muscles at rest
X. Nose
    Dilation of nostrils
        Levator labii superioris alaeque nasi
        Dilator naris
    Contraction of nostrils
        Nasalis (pars transversus)
        Nasalis (pars alaris)
        Depressor septi
XI. Lips
    Elevation
        Zygomaticus major
        Zygomaticus minor
        Levator anguli oris
        Levator labii superioris
        Levator labii superioris alaeque nasi
    Depression
        Depressor anguli oris
        Depressor labii inferioris
        Platysma
    Retraction
        Risorius
        Buccinator
        Platysma
        Zygomaticus major
    Contraction
        Orbicularis oris
    Protraction
        Orbicularis oris
        Mentalis
XII. Tongue
    Elevation
        Styloglossus
        Palatoglossus
        Elevators of hyoid bone
    Depression
        Genioglossus
        Hyoglossus
        Depressors of hyoid bone
    Retraction
        Styloglossus

    Protrusion
        Genioglossus
  Note: The intrinsic musculature is involved in movements of the tongue involved with speech and with most gross movements of the tongue as well.
XIII. Swallowing
    Closing opening between pharynx and nasal cavity
        Levator veli palatini
        Tensor veli palatini
        Musculus uvulae
    Lifting pharynx
        Stylopharyngeus
        Palatopharyngeus
    Constricting pharynx
        Superior constrictor
        Middle constrictor
        Inferior constrictor
    Closing opening between pharynx and oral cavity
        Palatopharyngeus
        Palatoglossus
        Raising of tongue
    Raising hyoid bone
        Mylohyoid
        Digastric
        Geniohyoid
        Hyoglossus
        Stylohyoid
    Lowering hyoid bone
        Omohyoid
        Sternohyoid
        Thyrohyoid
    Elevating larynx
        Elevation of hyoid bone
        Thyrohyoid
        Salpingopharyngeus
        Stylopharyngeus
        Palatopharyngeus
    Closing entrance into larynx
        Transverse arytenoid
        Oblique arytenoid
        Aryepiglotticus
        Thyroepiglotticus
        Lateral cricoarytenoid
XIV. Larynx
    Closing rima glottidis
        Transverse arytenoid
        Oblique arytenoid
        Lateral cricoarytenoid
    Opening rima glottidis
        Posterior cricoarytenoid

Tensing vocal cords
   Cricothyroid
   Vocalis
Decreasing tension on vocal cords
   Thyroarytenoid

NERVES

The nerves of the head and neck consist of the twelve cranial nerves, the sympathetic and parasympathetic portions of the autonomic nervous system, and the first four cervical nerves.

CRANIAL NERVES

Olfactory nerve. This nerve consists of many branches on the superior concha of the lateral nasal wall and the superior part of the nasal septum. These nerve fibers traverse the cribriform plate of the ethmoid bone to enter the olfactory bulb. The fibers then follow the olfactory tract to the brain.

This nerve contains special afferent fibers for the sense of smell.

Optic nerve (Fig. 7-41). The optic nerve is made up of the axons of neurons contained in the retina. The optic nerve passes posteromedially through the center of the orbit, traverses the optic foramen, and joins the optic chiasma. From the optic chiasma two optic tracts course posteriorly to join the brain stem. Light from the left enters the eyes and is picked up by the nasal half of the retina in the left eye and the lateral half in the right eye. These nerves from the left eye cross in the optic chiasma to join those from the right in the right optic tract. Conversely, light from the right will enter the eyes and be picked up by the retina on the nasal half of the right eye and lateral half of the left; the nerve fibers so stimulated will be located in the left optic tract.

The fibers in this nerve are special afferent fibers for the special sense of sight.*

Pathology in an optic nerve leads to ipsilateral blindness; tract pathology leads to loss of vision in the contralateral visual field. The close relation of these nerves and tracts to the hypophysis should be noted; tumors of this gland often induce visual symptoms.

---

*A few efferent fibers have been found in the optic nerves; their function is problematical.

Oculomotor nerve (Fig. 7-42). This nerve arises from the brain stem just anterior to the pons. It proceeds anteriorly between the posterior cerebral and superior cerebellar arteries, pierces the dura, and enters the lateral wall of the cavernous sinus. It continues anteriorly and then enters the orbit via the superior orbital fissure. It immediately divides into superior and inferior divisions, the **superior division** bending superiorly and innervating the superior rectus and levator palpebrae superioris muscles. The **inferior division** courses anteriorly near the floor of the orbit to end in the inferior oblique muscle. It gives off a branch to the medial rectus and to the inferior rectus muscle. In addition, it gives off a branch to the ciliary ganglion, which is called the **motor root** of the ganglion; these are preganglionic parasympathetic fibers. Postganglionic fibers emerge from the ciliary ganglion as the short ciliary nerves that innervate the blood vessels of the eyeball, the ciliary muscles, and the constrictor pupillae muscle.

This nerve contains general somatic efferent fibers to all the orbital muscles except the lateral rectus and superior oblique, general somatic afferent fibers for proprioception from these muscles, and general visceral efferent fibers to the ciliary muscles and the constrictor pupillae muscle as well as blood vessels, and probably contains general visceral afferent fibers from these structures.

Pathology that involves the entire nerve results in (1) the eye moving laterally and inferiorly, leading to double vision (lateral rectus and superior oblique muscles are the only ones innervated); (2) a severe ptosis (loss of levator palpebrae superioris muscle); (3) a dilated pupil and (4) an inability to accommodate for near vision (loss of parasympathetic fibers to sphincter pupillae and ciliary muscles); and (5) a slight prominence of the eye (loss of tone in most orbital muscles). A patient may exhibit the above symptoms in varying degrees depending upon how much of the nerve is traumatized. In any pathology of the oculomotor nerve one should visualize into the body and thus realize that vascular pathology in the area of the cavernous sinus or the posterior cerebral or superior cerebellar arteries may be involved.

Trochlear nerve (Fig. 7-43). This nerve emerges from the posterior side of the brain stem, while all the others emerge from the anterior or lateral sides. It winds around the brain stem and finally emerges anteriorly, pierces the dura, and enters the lateral wall of the cavernous sinus. It proceeds into the orbit by way of the superior orbital fissure. It is quite superficially located and crosses medially at a plane superior to the levator palpebrae superioris muscle; it enters the superior border of the superior oblique muscle, which it innervates.

This nerve contains general somatic efferent fibers to the superior oblique muscle of the orbit, and general somatic afferent fibers for proprioception from that muscle.

If the trochlear nerve is interrupted the patient cannot move the eye laterally and inferiorly. The eye moves in the opposite direction when movement in the above direction is attempted, resulting in double vision. The proximity of this nerve to the posterior cerebral and superior cerebellar arteries and to the cavernous sinus should be considered as a possible source of the pathology. The patient may hold his or her head in such a position as to prevent diplopia.

Trigeminal nerve. This nerve arises from the middle of the pons and proceeds anteriorly to enter the trigeminal cave in the dura mater. Here the nerve spreads out to form the semilunar ganglion, a sensory ganglion. The three divisions of this nerve—the ophthalmic, maxillary, and mandibular—arise from the anterior and inferior surfaces of this ganglion. The mandibular division is accompanied on its medial side by the much smaller motor division of the trigeminal nerve, which makes this nerve a more difficult problem when the mandibular division has to be cut to relieve pain; this motor division must remain intact.

The **ophthalmic division** (Fig. 7-44) continues anteriorly in the lateral wall of the cavernous sinus and, after giving a branch to the meninges (mainly the tentorium cerebelli and falx cerebri), divides into lacrimal, frontal, and nasociliary branches, which enter the orbit through the superior orbital fissure.

The **lacrimal** nerve continues anteriorly and laterally to reach the lacrimal gland; it also gives sensory nerve supply to the lateral part of the eyelid and conjunctiva.

The **frontal** nerve is the most superior structure in the orbit. It divides into the **supraorbital** and **supratrochlear** nerves, the supraorbital passing through the supraorbital notch or foramen to reach the scalp, while the supratrochlear runs forward superior to the pulley of the superior oblique muscle, leaves the orbit, and proceeds superiorly to the scalp. Both these nerves give branches to the eyelids and the conjunctiva in passing through the orbital septum.

The **nasociliary** nerve turns medially between the optic nerve inferiorly and the superior rectus superiorly, and finally terminates, after coursing anteriorly for a certain distance, as the **anterior ethmoidal** nerve. This leaves the orbit via the anterior ethmoid foramen and enters the cranial cavity at the side of the cribriform plate of the ethmoid bone. After giving branches to the meninges, it enters the nasal cavity through a small slit—the nasal slit—and, after giving branches to the anterior part of the septum of the nose and the anterior part of the lateral wall, it leaves the nasal cavity as the external nasal branch, which supplies the skin of the lower half of the nose. The nasociliary nerve gives off a branch to the ciliary ganglion—its **sensory root**—and several **long ciliary** nerves that carry afferent fibers from the eyeball; some claim that these nerves also contain sympathetic fibers to supply the dilator pupillae muscle. The nasociliary nerve also gives off the **posterior ethmoidal** nerve, which leaves the orbit through the posterior ethmoidal foramen and gives sensory supply to the dura in this region, and, finally, it gives off the **infratrochlear** nerve, which passes anteriorly at a level inferior to the pulley of the superior oblique muscle and leaves the orbit just above the medial angle of the eye; it supplies the skin, upper half of the nose, eyelids, conjunctiva, lacrimal caruncle, and lacrimal sac.

The **maxillary division** (Fig. 7-125) of the trigeminal nerve emerges from the semilunar ganglion, enters the lateral wall of the cavernous sinus, and proceeds anteriorly to enter the foramen rotundum. After leaving the foramen rotundum, the nerve appears in the pterygopalatine fossa, is located for a short distance in the infratemporal fossa, enters the floor of the orbit via the inferior orbital fissure, proceeds anteriorly in the infraorbital canal, finally emerges on the face through the infraorbital foramen, and divides into its three **terminal branches**—the lateral nasal, superior labial, and palpebral—which give sensory innervation to the lateral side of the nose, upper lip, and lower eyelid respectively. Going back to the cranial cavity, the maxillary nerve gives off its **meningeal branch** to supply the meninges in that vicinity. In the pterygopalatine fossa the nerve gives off two small nerves—the **pterygopalatines**—which suspend the pterygopalatine ganglion. Several nerves branch from the ganglion including the nerve of the pterygoid canal, which brings sympathetic and parasympathetic fibers to the ganglion; the **pharyngeal** nerve to the superior lateral aspect of the pharynx; the **posterior superior nasal** nerves, which enter the nasal cavity through the sphenopalatine foramen and give sensory nerve supply to the lateral wall of the nose and the nasal septum (one large branch on the nasal septum—the **nasopalatine**— continues anteriorly and inferiorly to traverse the incisive foramen and reach the hard palate just posterior to the incisor and canine teeth); **greater palatine** nerves descending through the palatine canals and emerging through the greater and lesser palatine foramina to reach the hard and soft palates of the mouth (branches from these nerves course also to the posterior part of the lateral wall of the nasal cavity); and **orbital** branches to the periosteum surrounding the orbital contents. These nerves are sensory and also contain parasympathetic and sympathetic fibers to the mucous membranes of the nasal and oral cavities.

Two branches arise in the infratemporal fossa: zygomatic and posterior superior alveolar. The **zygomatic** branch continues anteriorly in the lateral part of the orbit and, after communicating with the lacrimal branch of the ophthalmic division (secretory fibers to the lacrimal gland), continues to emerge on the face through two separate foramina as the zygomaticotemporal and zygomaticofacial nerves. Another branch in the infratemporal fossa is the **posterior superior alveolar** nerve, which continues inferiorly and slightly anteriorly on the outside of the maxillary bone, and enters the bone to give nerve supply to the upper teeth.

Continuing into the infraorbital canal, the maxillary nerve, now called the **infraorbital** nerve, gives off the **anterior superior alveolar** and **middle superior alveolar** branches to the incisor, canine, and premolar teeth. These nerves course in the thin bony layer forming the lateral wall of the maxillary sinus.

The **mandibular division** (Fig. 7-74) of the trigeminal nerve arises from the semilunar ganglion and courses inferiorly through the foramen ovale accompanied by the motor division of the trigeminal nerve. Small nerves are given off from its undivided root while in the foramen ovale. These are the **meningeal** branch, the nerve to the tensor tympani muscle, and branches to the medial pterygoid and tensor veli palatini muscles. The meningeal nerve reenters the cranial cavity through the foramen spinosum and gives sensory nerve supply to a large area of the meninges. The **nerves to the tensor tympani,** the **medial pterygoid,** and the **tensor veli palatini** muscles course through the otic ganglion.

The mandibular division then divides into two divisions, anterior and posterior. The anterior contains the numerous **branches to the muscles of mastication**— two to the temporalis muscle, one to the masseter muscle, and one to the lateral pterygoid. In addition, the **buccal** branch of the mandibular division of the trigeminal nerve courses inferiorly and then anteriorly to give sensory nerve supply to the face near the corner of the mouth and the cheek on the inside of the mouth.

The posterior division gives off the lingual, inferior alveolar, and auriculotemporal nerves. The **auriculotemporal** nerve continues posteriorly, coursing around the middle meningeal artery and through the parotid gland to be distributed to the external ear and on the scalp over the temporal region. The postganglionic fibers from the otic ganglion to the parotid gland are contained in this nerve. The **lingual** nerve continues inferiorly in a plane superficial to the medial pterygoid muscle and deep to the lateral pterygoid muscle and to the mandible itself. It continues anteriorly to gain the superficial surface of the hyoglossus muscle and then is distributed to the tongue. This nerve carries general sensory fibers from the anterior two-thirds of the tongue and also contains the sensory fibers of taste for the anterior two-thirds of the tongue, contained in the chorda tympani nerve, a branch of the facial. This nerve also contains pre- and postganglionic parasympathetic fibers from the facial nerve to the submandibular and sublingual glands. The **inferior alveolar** nerve enters the mandibular foramen after giving off the mylohyoid branch and continues in the bony canal to terminate finally on the chin as the **mental** nerve, which gives sensory supply to the skin of the chin. The inferior alveolar nerve is the sensory nerve to all the lower teeth. Its **mylohyoid** branch continues medial to the mandible and reaches the chin region, coursing superficial to the mylohyoid muscle and deep to the anterior belly of the digastric muscle, innervating both.

The trigeminal nerve, therefore, contains general somatic afferent fibers to the skin of the head and neck, the mucous membranes of the nasal cavity and oral cavity including the teeth and gums, the meninges, and all areas associated with the eye and orbit. In addition, it is involved with general sensation to the anterior two-thirds of the tongue (general visceral afferent fibers). Since the muscles innervated by the trigeminal nerve—the temporalis, masseter, lateral pterygoid, medial pterygoid, mylohyoid, anterior belly of the digastric, tensor tympani, and tensor veli palatini—are derived from the first branchial arch, this nerve also carries branchial efferent fibers. Since those muscles must have proprioception, the nerve also contains branchial afferent fibers.

The trigeminal nerve is very important clinically; trigeminal neuralgia is a common occurrence. This may involve the entire nerve or divisions only. Pain is referred from one branch of a division to another of the same division, i.e., infection in teeth may be referred to the ear and temporomandibular joint, or pressure on the tentorium cerebelli may be referred to the forehead. In cases of intractible pain the trigeminal nerve may be cut, sections of the individual divisions or the ganglion being avoided due to their proximity to the cavernous sinus. However, when cutting the main nerve, the motor division must be preserved, and it is desirable to preserve the fibers of the ophthalmic division which are located in the upper medial side of the nerve; otherwise, the conjunctiva is denervated, which almost invariably results in ulceration. Individual divisions may be injected with alcohol, and surgery may be performed on the tract in the brain stem, which represents the continuation of trigeminal fibers, this procedure being performed when the ophthalmic division is the one involved; the corneal reflex is preserved in this procedure. Many areas of anesthesia can develop, the results depending on which nerve or nerves are involved and exactly where on the nerve the lesion occurs. If the motor division is cut, the patient's jaw moves toward the injured side when the mouth is opened.

Abducens nerve (Fig. 7-45). The sixth cranial nerve arises from the brain stem just inferior to the pons. It is quite close to the anterior inferior cerebellar artery. It pierces the dura mater, bends sharply forward over the superior edge of the petrous portion of the temporal bone, and enters the cavernous sinus. It continues anteriorly to enter the orbit through the superior orbital fissure, courses laterally, and immediately enters the medial surface of the lateral rectus muscle.

This nerve contains general somatic efferent fibers to this muscle and general somatic afferent fibers representing proprioception from this muscle.

The relation of this nerve at its origin to the anterior inferior cerebellar artery, and to the internal carotid in the cavernous sinus, makes it liable to vascular pathology. Furthermore, its closeness to the superior edge of the petrous portion of the temporal bone may involve it in fractures of the base of the skull. When this nerve is traumatized, the patient's eye will be drawn medially resulting in diplopia.

Facial nerve (Fig. 7-69). This nerve emerges from the medulla oblongata close to the inferior cerebellar peduncle. It courses anteriorly and immediately enters the internal auditory meatus. It continues anterolaterally until the geniculate ganglion is reached; it then curves posteriorly and inferiorly in the facial canal, having its characteristic relations to the middle ear. The nerve emerges from the stylomastoid foramen and, coursing through the parotid gland, finally divides into its terminal branches—the **temporal, zygomatic, buccal, mandibular,** and **cervical.** These terminal branches give nerve supply to the muscles of facial expression.

As we return to the geniculate ganglion, the first branch coming off the facial nerve is the **greater petrosal** (made up of preganglionic parasympathetic fibers destined for the pterygopalatine ganglion), which continues anteriorly and enters the middle cranial fossa through the facial hiatus. It picks up the **deep petrosal** nerve (sympathetic) and then becomes the **nerve of the pterygoid canal;** after traversing this canal, it enters the pterygopalatine fossa to end in the pterygopalatine ganglion. The postganglionic fibers carry on from this ganglion to the maxillary nerve, to its zygomatic branch and thence, by an anastomosis, proceeds to the lacrimal nerve and to the lacrimal gland. In addition, postganglionic fibers are carried in other branches of the pterygopalatine ganglion to the mucous membrane of the oral and nasal cavities. There are probably afferent fibers contained in this pathway for the sensation of taste around the region of the palate. Therefore, the greater petrosal nerve contains general visceral efferent fibers to the lacrimal gland and the mucous membranes mentioned, and special afferent fibers for taste. The next branch of the facial nerve is a contribution to the **lesser petrosal** nerve, which will be described with the glossopharyngeal. The next branch is the **branch to the**

stapedius muscle. Following that is the **chorda tympani** nerve, which enters the middle ear, exits from its anterior side, enters the infratemporal fossa, and joins the lingual nerve. These fibers continue along the lingual nerve, and end in the submandibular ganglion located on the superficial surface of the hyoglossus muscle. Postganglionic fibers join the lingual nerve again to enter the sublingual and submandibular glands or may enter the glands directly. The chorda tympani nerve is made up of preganglionic parasympathetic fibers to the salivary glands, but also contains fibers for taste from the anterior two-thirds of the tongue. The remaining branches emerge from the facial nerve after it exits from the stylomastoid foramen. The **posterior auricular** nerve gives motor supply to the superior and posterior auricular muscles, and also the occipital muscle in the scalp. In addition, a branch from this nerve tends to innervate the external ear; it joins with the vagus nerve and these fibers are general somatic afferent fibers to this region. The remaining branch is a double **branch to the stylohyoid muscle and the posterior belly of the digastric muscle,** both of which are innervated by the facial nerve.

The seventh nerve contains general somatic afferent fibers from the external auditory meatus; general visceral efferent fibers to the lacrimal, submandibular, and sublingual glands, and mucous membranes of the nasal and oral cavities; possibly general visceral afferent fibers from these structures; branchial efferent fibers to the muscles of facial expression, the stapedius, stylohyoid, and posterior belly of the digastric, all being derived from the second branchial arch; branchial afferent fibers for proprioception from these muscles; and special afferents for taste for the anterior two-thirds of the tongue.

The symptoms exhibited by patients with lesions of the facial nerve depend upon whether the pathology is located in a supranuclear position (in fibers leading to the facial nucleus), in the facial nucleus itself, or an infranuclear position (in facial nerve itself). Even if we restrict this description to the last category, these will vary with the location of the lesion along the pathway of the nerve. If the entire nerve is involved, the patient will have a lack of lacrimal fluid; a dryness of the mouth; loss of action of the stapedius muscle; loss of taste on anterior two-thirds of the tongue; paralysis of the stylohyoid muscle, posterior belly of the digastric, and the auricular muscles; and paralysis of all muscles of facial expression, leading to a lowered eyebrow, inability to blink (conjunctiva usually becomes ulcerous), inability to dilate nostrils, a one-sided smile, inability to contain food or saliva in the mouth, and a distinct droop of the face. All symptoms are ipsilateral. Actually most lesions either occur to the nerve in the petrous portion of the temporal bone (fractures) or in the stylomastoid foramen (Bell's palsy) where edema can put pressure on the nerve. Naturally, the symptoms will vary with the location of the lesion. This emphasizes the importance of knowledge of the location of the emergence of branches of the cranial nerves. Most cases of Bell's palsy are temporary in nature.

Vestibulocochlear nerve. The eighth nerve is a sensory nerve from the cochlea and semicircular canals. The cochlear portion originates in the cochlea from the organ of Corti; its sensory ganglion—the spiral ganglion—is located in the cochlea itself. The vestibular portion, from the semicircular canals, also has a ganglion—the vestibular ganglion—located close to the vestibule of the internal ear. These two portions join, traverse the internal auditory meatus, and enter the brain stem.

This nerve contains special afferent fibers for hearing and general somatic afferent fibers from the semicircular canals. *

Lesions of the vestibulocochlear nerve are usually the result of fractures of the petrous portion of the temporal bone (the facial nerve is also involved), in which case permanent deafness can occur on the affected side; or they can be inflammatory in nature, in which case the deafness may be temporary. Vertigo and nystagmus will usually occur in these cases due to the involvement of the vestibular portion of this nerve.

Glossopharyngeal nerve (Fig. 7-103). The ninth cranial nerve emerges from the medulla between the inferior cerebellar peduncle and the inferior olive and leaves the skull via the jugular foramen. It enters the carotid sheath, courses between the internal carotid artery and internal jugular vein, and then curves medially, crossing anterior or superficial to the internal carotid artery; it winds around the stylopharyngeus muscle and enters the wall of the pharynx and the base of the tongue.

---

*Efferent fibers in the cochlear nerve have been described. They seem to be involved with the hair cells and may influence hearing in a reflex manner.

The first branch of the glossopharyngeal nerve is the **tympanic** branch, which enters the tympanic cavity and, with the sympathetic fibers from the superior cervical ganglion, forms the tympanic plexus. A nerve emerges from this tympanic plexus and joins with a contribution from the facial to form the **lesser petrosal** nerve. This nerve ends in the otic ganglion, located in the infratemporal fossa; these are preganglionic parasympathetic fibers. Postganglionic fibers from this ganglion then join the auriculotemporal nerve to reach the parotid gland. The next branch is a sensory **nerve from the carotid sinus and body.** The next branch is **motor to the stylopharyngeus muscle,** the only muscle innervated by this cranial nerve. The terminal branches are those to the mucous membrane of the pharynx (afferent limb of gag reflex) and to the mucous membrane on the base of the tongue. These fibers are responsible for general sensation from the pharynx, palatine tonsils, and posterior one-third of the tongue, and for taste from the posterior one-third of the tongue.

The glossopharyngeal nerve, therefore, contains general visceral efferent fibers to the parotid gland; general visceral afferent fibers from the carotid sinus and body (this branch may join the vagus nerve instead), and the mucosa of the pharynx, auditory tube, middle ear, and the posterior one-third of the tongue; branchial efferent and afferent fibers to and from the stylopharyngeus muscle, which is derived from the third branchial arch; and special afferent fibers for taste from the posterior one-third of the tongue.

Lesions of the glossopharyngeal nerve do not induce obvious symptoms in patients (assuming the vagus is intact for contraction of the pharyngeal constrictors). However, the intactness of the nerve can be tested by the afferent limb of the gag reflex or by testing for taste on the posterior one-third of the tongue.

Vagus nerve (Fig. 7-104).   This nerve arises from the medulla between the inferior cerebellar peduncle and the inferior olive and leaves the skull via the jugular foramen. It enters the carotid sheath and occupies a position posterior and between the internal jugular vein and the internal carotid artery. It courses inferiorly in the carotid sheath throughout the entire neck, and leaves the area of the head and neck by entering the thoracic cavity.

The branches of the vagus nerve are numerous. The first one is the small auricular **branch to the external auditory meatus.** This is important because irritations here are frequently manifested as coughs or a nauseous feeling. The next branch is the **pharyngeal** nerve, which courses posterior to the internal carotid artery, thus leaving the carotid sheath, and enters the pharynx directly. It is responsible for nerve supply to the three constrictor muscles of the pharynx (efferent end of gag reflex), and all the muscles of the soft palate except the tensor veli palatini. The next branch from the vagus nerve, actually branching from the inferior ganglion, is the **superior laryngeal** nerve. This continues inferiorly and divides into internal and external branches. The internal enters through the thyrohyoid membrane and innervates the mucous membrane as far as the vocal cords (origin of cough reflex); it is thought also to be the nerve supply to part of the transverse arytenoid muscle. The external branch of the superior laryngeal branch of the vagus innervates the cricothyroid muscle. Branches from this nerve also innervate the inferior end of the inferior constrictor muscle of the pharynx. The next branches of the vagus are the **cardiac** nerves, which continue inferiorly in the carotid sheath and, after going to the cardiac plexuses, give nerve supply to the heart itself. The **recurrent laryngeal** branches of the vagus recur around the subclavian artery on the right and the aorta and ligamentum arteriosum on the left, and course superiorly to reach the larynx. These nerves are responsible for the sensory nerve supply to the mucous membrane inferior to the vocal cords and motor supply to all the muscles of the larynx except for part of the transverse arytenoid muscle and the cricothyroid. Branches from these nerves also innervate the trachea and esophagus.

In summary, the vagus nerve contains general somatic afferent fibers from the external auditory meatus; general visceral efferent fibers to the thoracic and abdominal viscera; general visceral afferent fibers from these same organs and the mucosa of the larynx; branchial efferent fibers to the muscles of the pharynx and larynx (fourth and sixth branchial arches); branchial afferent fibers for proprioception from these muscles; and special afferent fibers for taste around the epiglottis.

Lesions of the vagus nerve are not common but do occur. The symptoms are usually palpitation accompanied by an increased pulse rate, constant nausea, decreased rate of respiration, a sensation of suffocation, and a hoarse, low voice. The patient may complain of difficulty in swallowing. Tests used to confirm the diagnosis are the efferent end of the gag reflex (contraction of pharyngeal constrictors upon touching pharyngeal wall), faulty movement of the uvula, and a paralyzed vocal cord on the affected side as seen with a laryngoscope. Naturally the symptoms will vary depending on where the nerve is traumatized. This, once again, emphasizes the importance of knowledge of the location of the emergence of branches from the cranial nerves.

Spinal accessory nerve (Fig. 7-104). The eleventh cranial nerve arises from the spinal cord in the cervical region, courses superiorly through the foramen magnum, curves around, and exits from the cranial cavity again through the jugular foramen to enter the carotid sheath. It courses inferiorly between the internal jugular vein and the internal carotid artery, then crosses superficial to the internal jugular vein to enter the sternocleidomastoid muscle. It innervates this muscle, and then emerges from the posterior side of this same muscle, crosses the posterior triangle of the neck caught between the two layers of enveloping deep fascia, and reaches the trapezius muscle, which it innervates.

In summary, the spinal accessory nerve contains branchial efferent fibers to the trapezius and sternocleidomastoid muscles and branchial afferent fibers for proprioception from these muscles.*

If we assume that the spinal accessory nerve is the spinal portion only, it may be traumatized in skull fractures or by severe enlargement of the lymph nodes in the neck. In either case, torticollis (twisted neck) will result. If the nerve is merely irritated, the torticollis may result in the chin being turned to the opposite side (due to contraction of sternocleidomastoid on the affected side), but if the nerve is paralyzed, the head will turn toward the lesioned side and the patient cannot abduct the upper limb beyond the horizontal position. In surgical removal of infected nodes, the spinal accessory nerve should be located and preserved. In addition, its superficial location in the posterior triangle of the neck makes it liable to

---

*Previous descriptions of this nerve divided the spinal accessory nerve into cranial and spinal divisions. The spinal division is the one described here. The cranial division traverses the jugular foramen, but then joins the vagus nerve and is distributed with branches of this nerve. Actually, the vagal fibers distributed to the pharynx and larynx can be considered to be the cranial division of the spinal accessory nerve, and this is the reason why some books describe the innervation of the muscles of the pharynx and larynx via the spinal accessory nerve. It seems less confusing to forget the cranial division of the spinal accessory nerve (giving it to the vagus), and to limit the eleventh cranial nerve to the spinal portion, as the name implies.

injury; in this case, the patient cannot abduct the upper limb beyond the horizontal but no torticollis will be seen.

Hypoglossal nerve (Fig. 7-105). The twelfth cranial nerve emerges from the medulla oblongata between the inferior olive and the pyramid. It traverses the hypoglossal canal and then enters the carotid sheath. Coursing inferiorly in this sheath, it appears between the internal jugular vein and the internal carotid artery, and actually courses superficial to the external carotid artery as well. It curves anteriorly near the emergence of the occipital artery from the external carotid, passes superficial to the external carotid artery, and continues anteriorly on the superficial surface of the hyoglossus muscle. This nerve is distributed to the intrinsic and extrinsic muscles of the tongue and is the motor nerve to this organ.

It contains general somatic efferent fibers to the tongue musculature and general somatic afferent fibers of proprioception.

When the hypoglossal nerve is traumatized, a paralysis of the tongue on the affected side occurs. The tongue on that side loses bulk and the tongue when protruded turns toward the lesioned side due to a dragging effect of the paralyzed muscles. If the lesion is bilateral the patient has real problems with swallowing and with speech as well. All sensations for touch, temperature, and taste are intact.

PARASYMPATHETIC FIBERS. No summary of the nervous system in the head and neck would be complete without thinking of the parasympathetic supply as a unit. This system is found in the third, seventh, ninth, and tenth cranial nerves.

Preganglionic fibers start in the visceral nucleus of the oculomotor nerve and follow this nerve into the orbital cavity. They then continue with the inferior division of this nerve and branch off to the ciliary ganglion, forming the motor (parasympathetic) root of this ganglion. Synapse occurs here with postganglionic fibers, which carry on to the eyeball via the short ciliary nerves and innervate the ciliary muscle and the sphincter pupillae muscle.

The preganglionic fibers in the facial nerve start in the nucleus salivatorius of the brain stem and follow the facial nerve to the greater petrosal and chorda tympani branches. The greater petrosal nerve becomes the nerve of the pterygoid canal after being joined by the deep petrosal (sympathetic); its fibers synapse with postganglionic parasympathetic fibers in the pterygopalatine ganglion. Postganglionic fibers from this ganglion reach the lacrimal gland by coursing to the maxillary nerve and following its zygomatic branch, which anastomoses with the lacrimal nerve of the ophthalmic division of the fifth nerve. These postganglionic fibers also innervate the mucous membranes of the oral and nasal cavities by following the numerous branches of the ganglion to these structures. The preganglionic fibers in the chorda tympani nerve follow this nerve through the middle ear, into the infratemporal fossa, along the lingual nerve, to the submandibular ganglion, where synapse occurs with postganglionic fibers that innervate the sublingual and submandibular glands.

The parasympathetic system in the glossopharyngeal nerve also starts in the nucleus salivatorius in the brain stem. These preganglionic fibers follow the glossopharyngeal nerve to its tympanic branch. They then join the tympanic plexus on the medial wall of the middle ear, exit to form the lesser petrosal nerve, and finally reach the otic ganglion in the infratemporal fossa. Here synapse occurs with postganglionic fibers, which follow the auriculotemporal nerve to the parotid gland.

The vagus nerve contains many preganglionic parasympathetic fibers that start in the dorsal motor nucleus of the vagus nerve. These fibers continue in the vagus nerve and emerge from its many branches to synapse with ganglia in the walls of the viscera located in the thoracic and abdominal cavities. Postganglionic fibers from these ganglion cells innervate the smooth muscles in these organs.

The parasympathetic nervous system innervates the smooth muscles in blood vessels as well as the eye, the salivary glands, and the mucous membranes mentioned above.

Pathology in the parasympathetic system in the head and neck is not particularly obvious to the patient or physician when the trauma concerns that in the facial or glossopharyngeal nerves. The patient may complain of a dry mouth but the patient usually has a normal condition on the contralateral side. A drying of the conjunctiva can lead to serious conditions. Pathology of the vagus will not induce obvious changes in the head and neck. Of course, palpitation and respiratory changes will be present, but the interruption of the parasympathetic supply to the eye (oculomotor) is immediately obvious. The pupil is enlarged and will not respond to light, and the patient cannot accommodate for near vision.

SYMPATHETIC FIBERS. The sympathetic fibers for the head and neck are derived from the upper thoracic segments of the spinal cord. These preganglionic fibers leave the intermediolateral nucleus of these segments, exit from the spinal cord via the ventral roots to reach a segmental nerve, and then join the sympathetic chains via the white rami communicantes. These fibers do not synapse in the first ganglion reached, but continue superiorly in the sympathetic chain and synapse in the superior cervical ganglion. Postganglionic sympathetic fibers from the superior cervical ganglion then join the various blood vessels and follow these vessels to the organs to be innervated. Frequently these nerves leave the arteries and join other nerves to reach their destination. Examples of this are the deep petrosal joining with the greater petrosal to form the nerve of the pterygoid canal; the sympathetic contribution to the tympanic plexus, which ultimately ends in the lesser petrosal nerve; the sympathetic fibers contained in the long ciliary branches of the nasociliary nerve; and fibers traveling with the ophthalmic division of the trigeminal nerve, which enter directly into the ciliary ganglion (sympathetic root) and traverse this ganglion and short ciliary nerves to reach the blood vessels on the posterior part of the eyeball without synapsing. Some structures in the neck undoubtedly receive postganglionic sympathetic fibers from the middle and inferior cervical ganglia. It should be noted that gray rami communicantes branch from all of these ganglia to join the cervical spinal nerves.

Interruption of the sympathetic system anywhere along its pathway to the head and neck induces what is called Horner's syndrome. This consists of a decrease in size of the pupil on the affected side, a ptosis due to loss of the smooth muscle component of the levator palpebrae superioris muscle, and a flushing of the skin on the face. In addition a lack of sweating (anhidrosis) and a nasal congestion will be noticeable. Because cardiac branches arise from the cervical sympathetic ganglia, the heart rate may be decreased.

CERVICAL NERVES. In addition to the cranial nerves and the autonomic nervous system, the nerve supply to the head and neck must include the first four cervical nerves. Each nerve divides into dorsal and ventral rami, the ventral rami combining to form the cervical plexus.

You will recall that dorsal rami divide into medial and lateral branches, the former becoming cutaneous nerves in the upper part of the back, but remaining in the muscles in the lower back. The lateral branches naturally have the reverse arrangement. Therefore, the cervical dorsal rami should exhibit long medial branches which, in addition to innervating muscles, become cutaneous, and short lateral branches innervating muscles only.

The **dorsal ramus of C1** does not follow the above rule. It courses posteriorly between the posterior arch of the atlas and the vertebral artery, and ends in the muscles bounding the suboccipital triangle—obliquus capitis superior, obliquus capitis inferior, and rectus capitis major; the rectus capitis minor; and the semispinalis capitis. It usually does not have a sensory branch, although occasionally one can be found.

The **dorsal ramus of C2** does follow the rule. It divides into a small lateral branch, which innervates neighboring muscles, and a large medial branch. This medial branch, after sending twigs to muscles located nearby, becomes a long sensory nerve—the greater occipital—which appears just inferior to the obliquus capitis inferior muscle, turns superiorly in a plane superficial to the muscles forming the suboccipital triangle and deep to the semispinalis capitis muscle, pierces the latter muscle and usually the trapezius, and ramifies on the posterior surface of the scalp, reaching as far as the vertex.

The **dorsal ramus of C3** is typical, except that the medial branch does reach the scalp near the midline. Dorsal rami of C4, C5, and C6 are typical.

The ventral rami form the **cervical plexus,** which is pictured in Fig. 7-85. Besides muscular branches to the prevertebral and scalene muscles, it gives off sensory nerves—the great auricular, lesser occipital, transverse colli, and supraclaviculars—to the skin on the sides and anterior surface of the neck. In addition, a branch of C1 joins the hypoglossal nerve and ends in the thyrohyoid and geniohyoid muscles, and forms a superior limb of a loop—the **ansa cervicalis**—which, with the aid of the inferior limb from C2 and C3, innervates the infrahyoid muscles. And branches of C1 also innervate the dura mater around the foramen magnum, these fibers traveling via the vagus and hypoglossal nerves.

One other branch of the cervical plexus remains to be described—the **phrenic.** This important nerve courses inferiorly on the anterior surface of the anterior scalene muscle, enters the thoracic cavity, and follows the lateral surface of the mediastinal structures (just medial to the pleura on each side) to reach the diaphragm. An accessory phrenic nerve occurs frequently; it is a branch of C5 or of the nerve to the subclavius muscle.

The phrenic nerve has sensory fibers to the pleura and pericardium as well as both sensory and motor fibers to the diaphragm.

This origin of the phrenic nerve accounts for the fact that pathology around the diaphragm, particularly on its inferior surface, may be referred to the shoulder (supraclavicular nerves).

The accompanying table summarizes the nerve components of the cranial nerves and one segmental nerve. If you will glance at the table vertically, you will see that the nerves innervating the somatic muscles of the head and neck are the third, fourth, sixth, twelfth, and the cervical nerves, although much of the evidence for head somites has been derived from data on lower animals. The parasympathetic nervous system to the glands and smooth muscles located in the head are found in the third, seventh, ninth, and tenth cranial nerves. The postganglionic fibers of the sympathetic portion follow blood vessels for the most part, but do utilize several branches of the cranial nerves to reach their destination. Furthermore, these sympathetic fibers are found in the cervical nerves, which they follow to reach blood vessels, sweat glands, and arrector pili muscles. The nerves innervating muscles derived from branchial arches are the fifth for the first arch, seventh for the second, ninth for the third, tenth for the fourth and sixth arches, and eleventh for the sternocleidomastoid and trapezius muscles. (You will remember that these two muscles are innervated also by cervical nerves C3 and C4.)

The sensory components are not as easily classified. It is obvious that olfaction involves the first cranial nerve, and that taste is accomplished by fibers in the seventh, ninth, and tenth nerves. Sight is associated entirely with the second cranial nerve, and hearing with the eighth. The nerves concerned with touch, pain, and temperature,

Division of the Cranial Nerves and a

Nerve	General Somatic Afferent: *Touch, Pain, and Temperature; Proprioception, and Deep Pain from Trunk and Limbs and Nonvisceral Structures in Head*	General Somatic Efferent: *Motor to Nonbranchial Skeletal Muscles*	General Visceral Afferent: *Touch, Pain, and Temperature from Viscera*
1. Olfactory			
2. Optic			
3. Oculomotor	Orbital ms. except lateral rectus and superior oblique	Orbital ms. except lateral rectus and superior oblique	?
4. Trochlear	Superior oblique m.	Superior oblique m.	
5. Trigeminal	Face, scalp, nasal and oral cavities, meninges, conjunctiva, etc.		Anterior two-thirds of tongue
6. Abducens	Lateral rectus m.	Lateral rectus m.	
7. Facial	External auditory meatus		?
8. Vestibulocochlear	Semicircular canals		
9. Glossopharyngeal			Posterior one-third tongue, pharynx, auditory tube, middle ear, carotid body and sinus
10. Vagus	External auditory meatus		Larynx, pharynx, trachea, thoracic and abdominal viscera, aortic bodies, and possibly the carotid body
11. Spinal accessory			
12. Hypoglossal	Tongue musculature	Tongue musculature	
Segmental	Skin and ms. for that nerve	Skeletal ms.	Blood vessels

Segmental Nerve into Their Components

General Visceral Efferent: *Autonomic Nervous System to Glands, Smooth M., and Cardiac M.*	Branchial Afferent: *Proprioception and Deep Pain from Branchial Ms.*	Branchial Efferent: *Motor to Branchial Muscles*	Special Afferent: *From Organs of Special Sense*
			Olfaction: nasal mucosa
			Sight: retina
Ciliary ms., constrictor pupillae m.			
	Ms. of mastication, mylo-hyoid, anterior digastric, tensor tympani, tensor veli palatini	Ms. of mastication, mylo-hyoid, anterior digastric, tensor tympani, tensor veli palatini	
Lacrimal, submandibular sublingual glands; nasal and oral mucosae, etc.	Ms. of facial expression, posterior digastric, stylo-hyoid, stapedius	Ms. of facial expression, posterior digastric, stylo-hyoid, stapedius	Taste: anterior two-thirds of tongue and soft palate area
			Hearing: spiral organ (Corti)
Parotid gland	Stylopharyngeus m.	Stylopharyngeus m.	Taste: posterior one-third of tongue
Trachea, thoracic and abdominal viscera	Ms. of palate (except tensor veli palatini), pharynx, and larynx	Ms. of palate (except tensor veli palatini), pharynx, and larynx	Taste: around epiglottis
	Trapezius and sterno-cleidomastoid ms.	Trapezius and sterno-cleidomastoid ms.	
Blood vessels, sweat glands, arrector pili ms.			

proprioception, and deep pain for other parts of the body are the third, fourth, and sixth for proprioception from the orbital muscles (although these fibers probably join ultimately with the fifth nerve before entering the brain stem); the fifth nerve for vast areas of the head and neck such as the scalp, face, conjunctiva, nasal and oral cavities, parts of the external ear, and the meninges; the seventh and tenth for the external auditory meatus; the eighth for afferent fibers from the semicircular canals; the twelfth for proprioception from the tongue musculature; and the cervical nerves for proprioception from the muscles innervated by each nerve and sensation from the area of the skin innervated by each specific nerve.

Visceral sensations are found in the fifth nerve for the anterior two-thirds of the tongue; in the ninth nerve for the posterior one-third of the tongue, the pharynx, auditory tube, parts of the middle ear and the carotid sinus and body; in the tenth nerve for the trachea, aortic bodies (and possibly the carotid body), and all the thoracic viscera and many of the abdominal organs; and in the cervical nerves from blood vessels.

All that remains is proprioception from the branchial musculature, and these fibers are found in the nerves that innervate these muscles, i.e., the fifth, seventh, ninth, tenth, and eleventh.

ARTERIES AND VEINS

ARTERIES. You should avoid memorizing mere lists of arteries. It is more important to think of the course of these vessels and the areas or structures that receive their blood supply from them. Furthermore, if this is done, the names of branches are often recalled. Although occasional vessels may be omitted in using this system, the anatomy can be remembered for a longer time than when one resorts to mere memorization of lists.

The head and neck obtains its arterial supply from the common carotid and its internal and external carotid branches, from the vertebral, and from other branches of the subclavian. In general terms, the internal carotid supplies the contents of the cranial cavity and orbit; the vertebral supplies the spinal cord, and the brain; the remaining branches of the subclavian supply some of the deeper structures in the neck and the visceral tube inferior to the middle of the thyroid; and the external carotid supplies everything else.

The **common carotids** differ in their origins, the left

usually arising from the arch of the aorta as an independent vessel, while the right branches from the brachiocephalic. Both arteries course superiorly in the carotid sheath with the internal jugular vein (lateral) and the vagus nerve (posterior). The sheath and its contents are located just lateral to the centrally located visceral tube and anterior to the vertebral column and prevertebral musculature. There are no branches. The common carotids divide into internal and external carotid arteries.

The **internal carotid** continues superiorly in the carotid sheath and maintains the same relations to the visceral tube and the vertebral column as just mentioned for the common carotids. The internal jugular vein continues to be lateral in position to the artery and the vagus nerve posterior to both vessels. If you will become acquainted with this picture—the carotid sheath and its contents lying anterior to the vertebral column and lateral to the centrally located visceral tube—you will be able to visualize other structures in relation to the internal carotid artery and internal jugular vein. The styloid process and muscles attached to it, especially to its base, cross these vessels superficially, as does the digastric muscle. Several important nerves should be mentioned. The vagus remains posterior to both vessels; the spinal accessory starts initially in a posterior position, but soon appears between the vessels and crosses the internal jugular vein to reach the sternocleidomastoid muscle; from superior to inferior, the pharyngeal branch of the vagus, the glossopharyngeal, and the hypoglossal nerves all cross the internal carotid to reach their destinations; the superior laryngeal branch of the vagus remains deep to the artery; and the sympathetic chain has a posterior location to the contents of the carotid sheath.

The internal carotid artery continues through the base of the skull in an S-shaped manner in the carotid canal, passing very close to the floor of the middle ear in its course. It gives off branches to the middle ear (caroticotympanic), to the nerves in the cavernous sinus, the trigeminal ganglion, the hypophysis, and the choroid plexus, and then gives off the ophthalmic artery before dividing into the middle and anterior cerebral arteries. The posterior communicating artery may arise from either the carotid or the middle cerebral.

The **ophthalmic artery** has many branches. It courses through the optic foramen with the optic nerve—an important relation—and enters the orbit. Naturally it will nourish everything found in the orbit—the eyeball, the orbital muscles, lacrimal gland, orbital fat, optic nerve—but it also reaches into the anterior cranial fossa, into the nasal cavity, and onto the face and scalp. This artery accomplishes this as follows. Its first branch is the central retinal artery, which follows the optic nerve to give nourishment to the retina. The lacrimal artery nourishes the lacrimal gland and contiguous parts of the eyelid and eyeball. The ophthalmic artery then crosses the optic nerve, usually in a plane superior to it. It gives off its posterior and anterior ethmoidal branches, which enter the anterior cranial fossa through the anterior and posterior ethmoidal foramina, send branches to the dura mater and skull in this area, and then continue into the nasal cavity; and its supraorbital branch, which leaves the orbit by piercing the orbital septum to ramify on the scalp as far as the vertex. The ophthalmic artery continues anteriorly and divides into supratrochlear and dorsal nasal branches. The former ramifies on the forehead after giving off branches to the eyelids and eyeball, and the latter ends on the dorsum of the nose. Many ciliary branches are given off from the ophthalmic directly and from its several branches.

The **middle cerebral artery** courses laterally in the lateral fissure between the temporal and frontal lobes of the brain and ramifies on the lateral surface of the cerebrum. The **anterior cerebral** continues anteriorly, gives off a communicating artery to its mate of the opposite side, curls around the corpus callosum, remaining in the great cerebral fissure separating the two hemispheres. It gives off branches that ramify on the medial surfaces of the hemispheres and reach the superficial surfaces as well. Many small branches arise from these two cerebral vessels near their origins and penetrate the brain tissue to nourish very vital areas. These arteries are terminal— have no collateral connections—and can become occluded easily.

The **vertebral artery** is also involved with blood supply to the brain. This artery is a branch of the subclavian. It courses superiorly just anterior to the vertebral column and enters the transverse process of the sixth cervical vertebra. It continues superiorly through the foramina in the transverse processes of all cervical vertebrae, including the atlas. It then turns medially and slightly posteriorly on the posterior arch of the atlas, pierces the dura, traverses the foramen magnum, courses along the medulla for a short distance, and joins with the same vessel of the opposite side to form the basilar artery. This

artery courses on the pons and terminates by dividing into the two posterior cerebrals.

From its course, it is probable that the vertebral artery nourishes contiguous musculature, the vertebral column, spinal cord, meninges, medulla oblongata, pons, cerebellum, and the cerebrum. Small muscular branches are given off throughout its course in the neck. Spinal branches enter the intervertebral foramina, and divide into branches for the vertebrae and those that reinforce the blood supply to the spinal cord. As the vertebral artery nears the head it gives off branches, which anastomose with the descending branch of the occipital artery and with the deep cervical branch of the costocervical trunk; and as it penetrates the dura mater and traverses the foramen magnum, it gives off meningeal branches to aid in supplying the dura and skull in the posterior cranial fossa. It gives important anterior and posterior spinal arteries, which descend along the spinal cord, forming a single arterial channel anteriorly and double channels posteriorly. The last branch of the vertebral is the posterior inferior cerebellar, which arises near its termination and winds around the medulla to reach the inferior surface of the cerebellum.

The **basilar artery** nourishes the cerebellum, internal ear, and pons in addition to the cerebrum. It does this by giving off the anterior inferior cerebellar branch, which courses on the inferior surface of the cerebellum, the internal auditory branch, which accompanies the vestibulocochlear nerve into the internal auditory meatus, several pontine branches to that structure, a superior cerebellar artery (note the relation to the third nerve) to the superior surface of that organ, and then the posterior cerebrals.

The **posterior cerebrals** course laterally around the oculomotor nerves—an important relation—and ramify on the inferior surface of the temporal and occipital lobes. These arteries give off many terminal branches to vital centers in the brain stem, and also provide posterior communicating branches, which connect the posterior cerebrals with the middle cerebrals. This completes a circle—the cerebral arterial circle—made up of the anterior cerebrals and the anterior communicating, the middle cerebrals, the posterior communicating, and the posterior cerebrals. This circle is an important source of an alternative blood supply in cases involving occlusion of important channels.

The relations of all these arteries on the base of the

brain to the cranial nerves are particularly important. Aneurysms of these vessels often put pressure on the nerves. Realization of the presence and importance of these vessels is more vital than a knowledge of their nomenclature.

Other **branches of the subclavian artery** besides the vertebrals contribute to nourishment for head and neck structures. The deep cervical branch of the costocervical trunk ascends in the musculature of the neck and anastomoses with the descending branches of the vertebral and occipital. The thyrocervical trunk also contributes the inferior thyroid artery. This artery ultimately supplies the thyroid and parathyroid glands, but nourishes the prevertebral muscles, trachea, esophagus, and larynx in addition; this is accomplished by the ascending cervical, tracheal, esophageal, and inferior laryngeal branches.

The remaining structures in the head and neck are supplied by the **external carotid** and its branches. This artery, one of the terminal branches of the common carotid, courses superiorly alongside the internal carotid but at a more superficial plane. It pierces the carotid sheath and courses between the muscles arising from the styloid process and superficial to the styloid process itself. It is best to think of these branches as individuals unto themselves.

The first branch is the **superior thyroid.** As might be expected this artery supplies surrounding muscles, the infrahyoid region, and the larynx in addition to the thyroid gland. It has named branches corresponding to this role. The infrahyoid courses on the thyrohyoid ligament, the superior laryngeal enters the larynx along with the internal laryngeal nerve, and the muscular consist of a sternocleidomastoid branch, a cricothyroid branch, and several unnamed muscular twigs.

The **ascending pharyngeal** arises near the origin of the external carotid on its posterior surface. This vessel is not large. From the name one would expect this vessel to ascend on the pharynx, nourishing it and prevertebral muscles. In addition, it has branches that enter the skull through the jugular foramen or hypoglossal canal to supply meninges and skull in the posterior fossa, a tympanic branch (inferior tympanic), and palatine branches to the soft palate.

The **lingual artery** must certainly be involved with the tongue, and it gives branches to the suprahyoid region, sublingual gland, and palatine tonsils as well. It arises from the external carotid at the level of the hyoid bone, courses anteriorly on the middle constrictor muscle of the pharynx, runs along the superior surface of the hyoid bone just deep to the hyoglossus muscle and superficial to the genioglossus. Before reaching the hyoid bone it gives off a branch that courses on the superficial surface of the hyoglossus muscle (suprahyoid), another that reaches the dorsum of the tongue and palatine tonsil (dorsal lingual), and a third to the sublingual gland (sublingual), and then continues on the under surface of the tongue (deep lingual) to its tip.

The **facial artery** supplies the face, as would be expected; since it courses between the mandible and the pharynx, it supplies structures in the pharynx as well. It arises from the external carotid just above the lingual or in union with it, ascends deep to the digastric muscle on the posterior surface of the submandibular gland, and gains a position between the medial pterygoid muscle laterally and the middle constrictor muscle of the pharynx medially. It crosses the inferior edge of the mandible just anterior to the masseter muscle, coils superiorly and slightly medially in the face just lateral to the corner of the mouth and the nose, to end near the medial angle of the eye. Its ascending palatine branch ascends between the medial pterygoid and superior constrictor muscles and between the styloglossus and stylopharyngeus muscles to reach the base of the skull. It then follows the levator veli palatini muscle to the soft palate and palatine tonsil as well as the medial end of the auditory tube. The tonsillar branch is the main supply to the palatine tonsil and reaches it by merely piercing the superior constrictor muscle. As the facial artery curves around the mandible, it gives branches to the submandibular gland and submental region. It supplies branches to the lips (inferior and superior labial), to the nose (lateral nasal), to the many facial muscles (muscular), and the angular branch to the angle of the eye.

The **occipital,** from its name, must be involved with structures close to the occipital bone. It courses deep to the mastoid process and muscles attached to it, continues just deep to the splenius capitis muscle, and pierces the trapezius to reach the scalp. It gives muscular branches to the sternocleidomastoid muscle; meningeal branches, which traverse the mastoid and jugular foramina; and a descending branch, which anastomoses with the deep cervical, ascending cervical, and the superficial branch of the transverse cervical artery; and then terminates in branches that ramify on the scalp as far as the vertex.

The **posterior auricular artery** ascends deep to the styloid process and its muscles, and deep to the mastoid process, to reach the scalp just posterior to the ear. It supplies surrounding muscles (muscular), the tympanic membrane (stylomastoid branch, which follows the facial nerve through the stylomastoid foramen and facial canal to reach this organ), the external ear (auricular), and the scalp posterior to the ear (occipital).

The two terminal branches of the external carotid are the superficial temporal and the maxillary. The **superficial temporal artery** arises just medial to the neck of the condyloid process of the mandible. It ascends just anterior to the ear in company with its vein and the auriculotemporal nerve. In its course it is close to the parotid gland, external ear, face, zygomatic region, and the temporalis muscle. It has branches to all these structures, i.e., parotid, anterior auricular, transverse facial, zygomatic, and middle temporal. It terminates by dividing into frontal and parietal branches, which ramify on the sides of the head.

The **maxillary artery** courses medially deep to the mandible and through the infratemporal fossa. It may travel superfical or deep to the lateral pterygoid muscle. Once again, it is better to visualize the area and the structures present to be nourished, and the names of the branches will be recalled. The ear is close, and this should recall the deep auricular and anterior tympanic branches to the external auditory meatus and tympanic membrane respectively. The lower jaw is very near, and this makes one think of the inferior alveolar to the lower jaw and teeth and its mylohyoid branch to the muscle of the same name and the digastric muscle. The muscles of mastication are located in the infratemporal fossa also, and each gets its own artery, the temporalis receiving two. The buccinator muscle also has a branch, the buccal. The middle meningeal artery supplies the greater part of the meninges and bones of the skull by passing superiorly through the foramen spinosum. The accessory meningeal nourishes the dura around the area of the cavernous sinus, reaching its destination by way of the foramen ovale. The third part of the artery enters the pterygopalatine fossa through the pterygomaxillary fissure, and branches enter each of the foramina leaving this fossa. This should recall to mind the superior alveolar to the molar teeth of the upper jaw, infraorbital to the teeth more anteriorly located and to the face, the pharyngeal to the superior end of the pharynx, the pterygoid, the descending palatine to the hard and soft palates, and the sphenopalatine to the nasal cavity.

Figure 7-138 shows the branches of the subclavian that anastomose with branches of the carotid arteries. A good mental exercise is to determine the collateral circulation around the tie when arteries are tied off.

VEINS. The veins of the head and neck differ from the arteries in several respects. The presence of the dural venous sinuses and the resultant drainage of the brain, the presence of diploic veins and the important emissary veins, and the extreme complexity of the venous plexuses make the veins considerably different from the arteries.

Probably the best way to visualize the veins of the head and neck is to think of the basic organization first. There are three main drainage routes involved: the internal jugular, external jugular, and vertebral veins. In general terms, the areas drained by these three channels correspond to the areas supplied by the internal carotid, external carotid, and vertebral arteries respectively; but differences do occur since the external jugular vein does not receive as many tributaries as the external carotid has branches. Nevertheless, the **internal jugular** starts as a dilated bulb in the jugular fossa in the base of the skull. It occupies a lateral position in the carotid sheath, the internal and common carotid arteries being placed on its medial side. Thus it is located just anterior to the vertebral column and lateral to the visceral tube. It joins the subclavian veins on either side, to form the right and left brachiocephalic veins, the latter vessels joining to form the superior vena cava.

The **external jugular vein** is not as long as the internal. The superficial temporal vein joins with the maxillary to form the retromandibular; this vein, in turn, joins with the posterior auricular to form the external jugular vein approximately at a level with the angle of the jaw. The external jugular courses inferiorly in a plane superficial to the sternocleidomastoid muscle, pierces the deep fascia in so doing, continues in the superficial fascia, reenters the deep fascia, and empties into the subclavian vein.

The **vertebral veins** are not formed as one might expect; they start at the suboccipital venous plexuses, course in the foramina in the transverse processes of all the cervical vertebrae, and finally empty into the brachiocephalic veins.

With this basic pattern in mind, further details will be comprehensible.

The supraorbital and supratrochlear veins combine to form the angular vein; this vein becomes the facial and has tributaries that correspond to the facial artery. It courses inferiorly and laterally just posterior to the facial artery, and crosses the mandible just anterior to the masseter muscle. It does not curve around the mandible as does the artery (and therefore does not take a snakelike course through the face), but passes across the submandibular region to empty into the internal jugular vein. It usually does not drain the tonsillar region. This vein has important connections with the superior ophthalmic vein in the orbit, the pterygoid plexus in the infratemporal fossa, the retromandibular vein via the superior communicating vein, and the anterior jugular vein by means of the inferior communicating vein. One other connection should be mentioned: the anterior diploic vein drains into the supraorbital, so becoming a part of this system.

Returning to the scalp, the anterior and posterior tributaries of the superficial temporal vein form a single vein, which, after draining the temporalis muscle (middle temporal vein), and the face (transverse facial vein), joins the maxillary vein. The latter vein is short; it drains the pterygoid plexus of veins located in the infratemporal fossa. The pterygoid plexus is important because of its drainage of a rather large territory, but also because of its several connections with veins in other areas. Its tributaries correspond to the branches of the maxillary artery. It therefore drains the ear, meninges, muscles of mastication, the upper jaw, the nasal cavity (in part), mouth (in part), and the lower jaw. Its important connections are with the inferior ophthalmic vein in the orbit, with the facial vein, with the cavernous sinus via an emissary vein, and with the pharyngeal plexus of veins. As mentioned previously, the maxillary vein emerges from this plexus and joins with the superficial temporal to form the retromandibular vein.

Once again returning to the scalp—the part posterior to the ear—we find the posterior auricular vein draining the same territory as supplied by the artery of the same name. This vein joins with the abovementioned retromandibular to form the external jugular. Therefore, the external jugular drains blood from the lateral sides of the

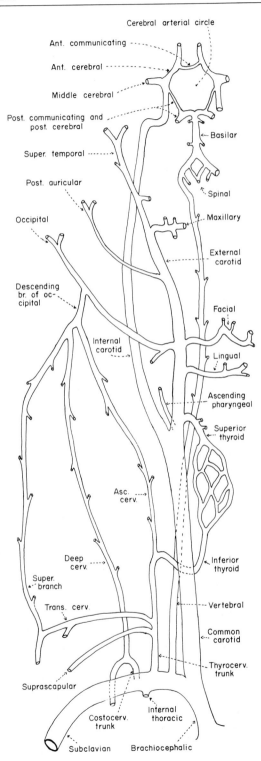

Figure 7-138. Important anastomotic channels between branches of the carotid and subclavian arteries.

scalp, the infratemporal fossa (the pterygoid plexus and all its tributaries), the external ear, and the posterior aspects of the face. The external jugular usually is the drainage site of the anterior jugular and a posterior jugular when present.

The last area of the scalp is that covered by the occipital artery. The occipital vein does not follow the course of the artery, for it drains into the suboccipital plexus in the suboccipital triangle. This plexus is drained by the vertebral vein already described and by the deep cervical vein. The occipital vein connects with the superior sagittal and sigmoid sinuses via the parietal and mastoid emissary veins, respectively. It also frequently drains the occipital diploic vein.

So far, we have been involved primarily with superficial veins. The deep veins are of equal importance.

The dura mater covering the brain splits at various locations to form blood-containing sinuses. These were described on pages 490–495 and do not need repeating. The superior sagittal, inferior sagittal, straight, intercavernous, and basilar are unpaired, while the sphenoparietal, cavernous, superior and inferior petrosals, transverse, sigmoid, and occipital are paired. The veins of the brain and of the orbit and two diploic veins drain into these sinuses.

All surface veins of the brain ultimately drain into these dural sinuses or into the lacunae on either side of the superior sagittal sinus. The veins draining the internal structures of the brain also drain into the sinuses. The internal cerebral veins join to form the great cerebral vein, while drains into the anterior end of the straight sinus. The ophthalmic veins drain into the cavernous sinus directly. Since the ophthalmic veins have tributaries that correspond to the branches of the ophthalmic artery, and the supraorbital and supratrochlear veins have direct connections with the facial vein and its branches, there is a direct connection between the cavernous sinus and the nasal cavity, the forehead, the nose itself, and the upper lip. Venous drainage can course internally, and infections on the face extend into the cavernous sinus with a resulting generalized septicemia. As many pimples occur in this region of the face, it is considered to be a dangerous area.

The **diploic veins** are four in number: frontal, anterior temporal, posterior temporal, and occipital. The frontal drains into the supraorbital veins, the anterior temporal usually into the sphenoparietal sinuses, the pos-

terior temporal into the sigmoid sinuses or into the mastoid emissary veins, and the occipital into the occipital veins or into the transverse sinuses.

All the dural sinuses ultimately drain into the internal jugular vein at its point of origin in the jugular fossa.

Although the **emissary veins** have been mentioned throughout this description, their importance in forming direct connections between veins outside the skull and the dural sinuses and, therefore, in being a source of trouble from the point of view of infection, warrants listing them again. There are eight recognized connections: (1) parietal, (2) mastoid, (3) condylar, (4) occipital, (5) sphenoidal, (6) veins in hypoglossal canal, (7) veins in the foramen ovale, and (8) veins following the internal carotid artery. Although the ophthalmic veins are not technically emissary veins, they serve the same dangerous function of connecting outside surfaces with dural sinuses, and they drain areas much more likely to become infected.

One area remains—the visceral tube. Pharyngeal veins form a plexus on the pharynx and ultimately drain into the internal jugular vein. Also, the lingual veins draining the tongue, the superior and middle thyroid veins, a vein from the sternocleidomastoid muscle, and the laryngeal veins all drain into the internal jugular vein. The inferior part of the thyroid gland, the trachea, and the esophagus all possess veins that drain into the brachiocephalic veins.

LYMPHATICS

The lymphatic drainage of the head and neck can be organized in such a fashion that the location of nodes draining any particular part can be determined by an intelligent guess.

There are superficial and deep nodes. The **superficial nodes** form a ring around the entire head. Starting posteriorly, there is a collection of nodes at the base of the occipital artery (occipital nodes) that drain all areas supplied by that artery. The posterior auricular nodes are located posterior and slightly inferior to the external ear and drain the external auditory meatus, the external ear, and the area of the scalp posterior to the ear—the areas supplied by the posterior auricular artery. The preauricular or parotid nodes are located at the base of the superficial temporal artery and drain all areas supplied by this artery. The superficial cervical, submandibular, and submental nodes are located just inferior to the mandible, the superficial cervical being located close to the angle, the submandibular close to the gland of the same name, and the submental in the submental triangle. These nodes drain all areas of the face supplied by the facial artery, the lymphatic channels taking a posterior and inferior course imitating the vascular system. There are buccal nodes in the cheek close to the buccal pad of fat that interrupt some channels. Lymphatics from midline structures such as the tongue and lower lip drain into nodes on the opposite side as well as into those on the same side.

The **deep nodes** are located along the internal jugular vein from the clavicle to the base of the skull. These nodes are divided into upper deep cervicals and lower deep cervicals by the omohyoid muscle. Extensions of these nodes occur in the retropharyngeal space, in front of (pretracheal) and alongside (paratracheal) the trachea, and anterior to the larynx (prelaryngeal).

With this basic knowledge and common sense, the drainage of any particular organ can be determined with remarkable accuracy. For example, the conjunctiva and eyelids drain to the preauricular and superficial cervical nodes; the external nose, upper lip, and upper jaw to the superficial cervical and submandibular nodes; and the lower lip, gums, and lower teeth to the submandibular and submental nodes, frequently with crossing to the other side. The ear can be divided through the middle ear, the lateral part draining into the pre- and postauricular nodes, the medial part draining into the upper deep cervicals. The nasal cavity can be divided into anterior and posterior parts, the anterior draining to the submandibular nodes, while the posterior part drains into the upper deep cervicals and retropharyngeals. Of course, the pharynx drains into the upper deep cervicals and retropharyngeals, the soft palate into the same area, and the palatine tonsils into the upper deep cervicals. The tongue, just as with its nerve supply, can be divided into anterior two-thirds and posterior one-third, the anterior part draining into the submandibular and sublingual (crossing the midline occurs), and the posterior portion

draining into the upper deep cervicals and retropharyngeals. As might be expected, the larynx drains into the upper deep cervicals and the prelaryngeal nodes. The trachea leads to the pre- and paratracheal nodes and into the lower deep cervicals.

It should be noted that lymph leaves the superficial nodes and reaches the deep nodes. From these deep nodes, channels empty into the venous system at the point where the subclavian and internal jugular veins become the brachiocephalics. Those on the right side empty directly, while those on the left are likely to drain into the thoracic duct.

Of great importance is the finding that cancer spreads via these lymphatic channels and prospers in nodes far removed from those expected to drain the particular area involved. Accordingly, surgeons are performing radical surgery, and all nodes in the area are removed.

SPECIAL SENSES

There is very little to be added to the accounts of vision given on pages 520–529, audition on pages 529–540, smell on page 626, and taste on page 631. The chart on pages 656 and 657 also is useful in summarizing the nerves involved in the special senses.

OTHER SYSTEMS

The remaining systems—respiratory, digestive, and endocrine—profit little from being summarized. The respiratory system was described as the nasal cavity, paranasal sinuses, nasal pharynx, and larynx. These are found on pages 617, 604, and 608, respectively. The digestive system was treated similarly, the oral cavity being found on page 626 and the oral pharynx on page 605.

The endocrine system in the head and neck consists of the hypophysis, thyroid, and parathyroid glands. Since all were described adequately on pages 505 and 573, there is no need for further description.

8

UPPER LIMB

Those students who have followed this text in the order written will find the following description repeats the appendicular muscles in the back and the pectoral region. This review will do no harm and will serve to bring the upper limb together as a whole.

Those who are starting with this section of the text will find the upper limb described in its entirety. Except for a description of the breast, there will be no need to look elsewhere in the text for this material.

The upper limb consists of the more superficial muscles in the back, the pectoral region, shoulder and axilla, arm (brachium), elbow, forearm (antebrachium), wrist, and hand. Care should be taken to refrain from using the term "arm" to include the entire upper limb.

It is helpful to keep the development of the upper limb in mind while studying this region. Its origin as a limb bud with dorsal and ventral areas and its subsequent lateral rotation are exemplified in many of the definitive arrangements. The development of the bipedal posture has created some interesting features. In contrast to the four-footed animal, where all limbs are used for locomotion, the human being has become upright and uses only lower limbs in locomotion (except for a natural swinging of the arms), leaving the upper limb for other acts. The

human being has taken advantage of this phenomenon to develop the many creative arts, such as music, painting, writing, and sculpture, and to create the many things we use every day of our lives. Clinically speaking, the hand is the most important part of the upper limb. Injuries here are very common, and repair of such damage is of vital importance to the patient. Circulatory problems are not as prevalent in this limb as in the lower, nor do joints seem subject to the weaknesses of the weight-bearing lower limb. Of course, one should not discount diseases that attack joints generally, for these conditions may be extremely painful.

We will attempt to correlate the structure of this limb with the foregoing comments on function and clinical significance.

BONY STRUCTURES

Before you can read intelligently about structures in the upper limb, you must become familiar with the more

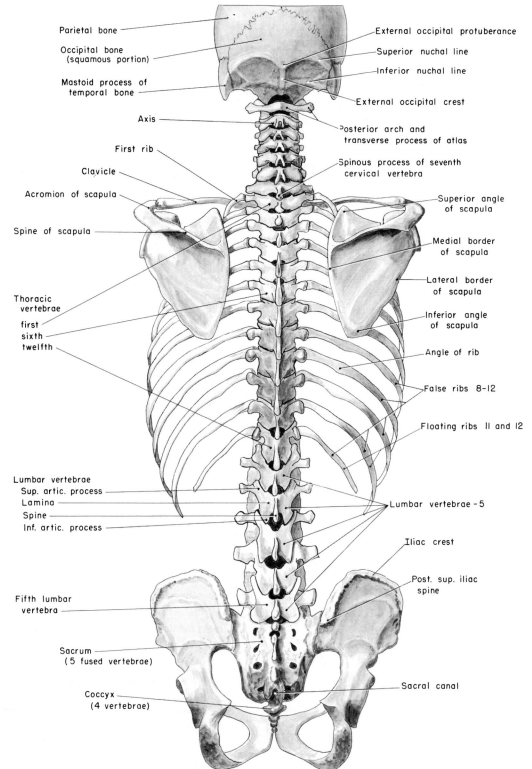

Figure 8-1. *Bones involved in a study of the back.*

important bony landmarks. Such knowledge is prerequisite to associating the various structures with those bones, a process that greatly enhances learning.

The bones of the upper limb are the clavicle, scapula, humerus, ulna, radius, carpals, metacarpals, and phalanges. However, the appendicular muscles located in the back are attached to parts of the axial skeleton and the coxal bone, so we should include parts of the skull, vertebral column, ribs, and coxal bone in our list. The parts of the axial skeleton that bear on our subject are shown in Figure 8-1.

The only features of the skull concerned are the **external occipital protuberance** and the **superior nuchal lines.** This protuberance is an elevation on the external surface of the occipital bone, and the ridges spreading laterally from this protuberance are the superior nuchal lines.

The appendicular muscles are also involved with the **spinous and transverse processes** of the vertebrae. The vertebral column consists of **33 vertebrae.** There are **7 cervical vertebrae** in the neck, **12 thoracic** in the upper back, **5 lumbar** in the lower part of the back, **5 sacral** fused into one bone—the sacrum, and **4 coccygeal** vertebrae which may or may not be fused.

Figure 8-2 shows the appearance of a **typical thoracic vertebra.** Such a vertebra consists of a large rounded **body** located anteriorly, from which a **vertebral arch** extends posteriorly. The arch is made up of two **pedicles** extending posteriorly from the body, and two **laminae** completing the arch. The foramen so formed is the **vertebral foramen,** through which pass the spinal cord and its coverings. At the point where the pedicles and laminae join are two projections, one on each side, called **transverse processes;** at the point where the two laminae come together a single **spinous process** extends more or less posteriorly. Each vertebra articulates with the one above and below; accordingly, there are two **articular processes,** containing two articular facets superiorly placed and two similar processes and facets inferiorly placed. The cervical, thoracic, and lumbar vertebrae have features that distinguish one from the other, but each set tends to resemble the one lower down as it approaches it. In other words, the lowest cervical vertebra resembles the upper thoracic; the lowest thoracic, the lumbar. Of course, there is no problem in distinguishing the sacrum or coccyx.

As one looks at the **posterior surface** of the verte-

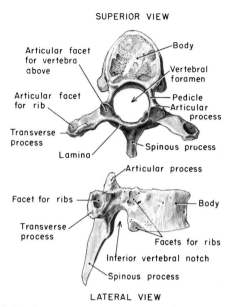

Figure 8-2. Superior and lateral views of a typical thoracic vertebra.

bral column (Fig. 8-3), one sees that the flattened cervical **transverse processes** change rapidly to the rather large thoracic type. These in turn gradually decrease in size until no processes can be seen at all on the last two thoracic vertebrae. The transverse processes of the lumbar vertebrae are quite long and amazingly slender. Differences in the appearance of the **spinous processes** in the various regions of the column are striking, as is the difference in the size of the space between the laminae. The bifid cervical spinous processes change rapidly into the long, thin, sloping variety found in the thoracic region; these in turn change to the short, blunt type found in the lumbar region. The sloping nature of the thoracic spines just mentioned makes the **interlaminal** space in the thoracic region very small, while this space is quite large in the lumbar region. This is fortunate for it provides a pathway for lumbar punctures and prevents the uninformed from entering the thoracic region, which would injure the spinal cord.

A portion of the **ilium** (Fig. 8-1) is involved in a study of the upper limb. Those parts concerned are the **posterior superior and posterior inferior iliac spines,** and the **iliac crest,** which starts posteriorly at this superior spine and flares superiorly, laterally, and anteriorly.

ANTERIOR   POSTERIOR   LATERAL

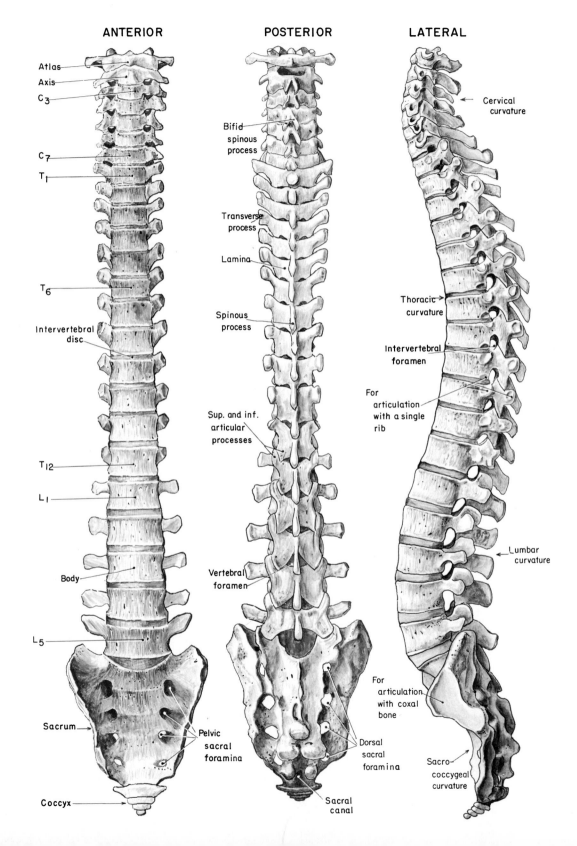

Atlas
Axis
C₃
C₇
T₁
T₆
Intervertebral disc
T₁₂
L₁
Body
L₅
Sacrum
Pelvic sacral foramina
Coccyx

Bifid spinous process
Transverse process
Lamina
Spinous process
Sup. and inf. articular processes
Vertebral foramen
Dorsal sacral foramina
Sacral canal

Cervical curvature
Thoracic curvature
Intervertebral foramen
For articulation with a single rib
Lumbar curvature
For articulation with coxal bone
Sacro-coccygeal curvature

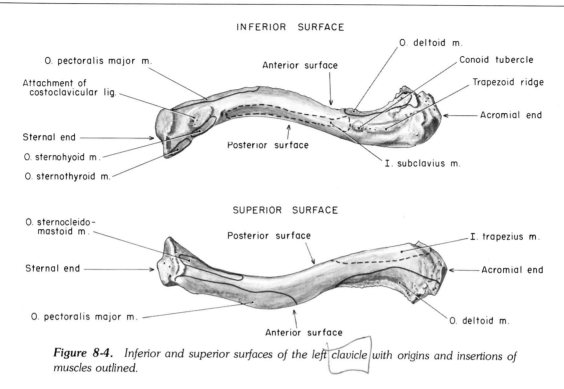

INFERIOR SURFACE

O. pectoralis major m.

Attachment of costoclavicular lig.

Sternal end

O. sternohyoid m.

O. sternothyroid m.

Anterior surface

Posterior surface

O. deltoid m.

Conoid tubercle

Trapezoid ridge

Acromial end

I. subclavius m.

SUPERIOR SURFACE

O. sternocleido-mastoid m.

Sternal end

O. pectoralis major m.

Posterior surface

Anterior surface

I. trapezius m.

Acromial end

O. deltoid m.

Figure 8-4. *Inferior and superior surfaces of the left clavicle with origins and insertions of muscles outlined.*

CLAVICLE
(Fig. 8-4)

This is an important bone, which can be felt along its entire length. Of all the bones in the body, this is fractured most frequently. It articulates with the sternum medially, at the only bony attachment of the upper limb to the trunk; it articulates with the acromion of the scapula laterally. It is responsible for keeping the scapula and shoulder joint in a lateral position, away from the important axillary structures. It has a curved form and is convex anteriorly at its medial two-thirds and concave anteriorly at its lateral one-third. The clavicle possesses a **shaft,** and **sternal** and **acromial ends.** The sternal end is rounded and smooth where it articulates with the sternum; the acromial end is rough superiorly but smooth where it articulates with the acromion process of the scapula. The shaft of the clavicle possesses **four surfaces:** superior, inferior, anterior, and posterior. The **superior surface** is smooth throughout the extent of the shaft. The **inferior**

surface exhibits a roughened area at the sternal end for attachment of the costoclavicular ligament, a smooth area in the medial third where the subclavius muscle inserts, and a rough area on the lateral third which is the conoid tubercle and trapezoid ridge for attachment of ligaments. The **anterior** and **posterior surfaces** exhibit no particular markings. The lateral third of this bone, being flattened, is said to have superior and inferior surfaces, but **anterior** and **posterior borders** rather than surfaces. See Figure 8-4 for the attachment of muscles. Some of these areas are roughened, the degree varying from clavicle to clavicle.

When the clavicle is fractured, pressure may be put upon structures in the axilla.

SCAPULA
(Fig. 8-5)

The scapula is a triangular bone that is attached to the clavicle laterally at the acromion process but otherwise is

---

Figure 8-3. *Anterior, posterior, and lateral views of the vertebral column.*

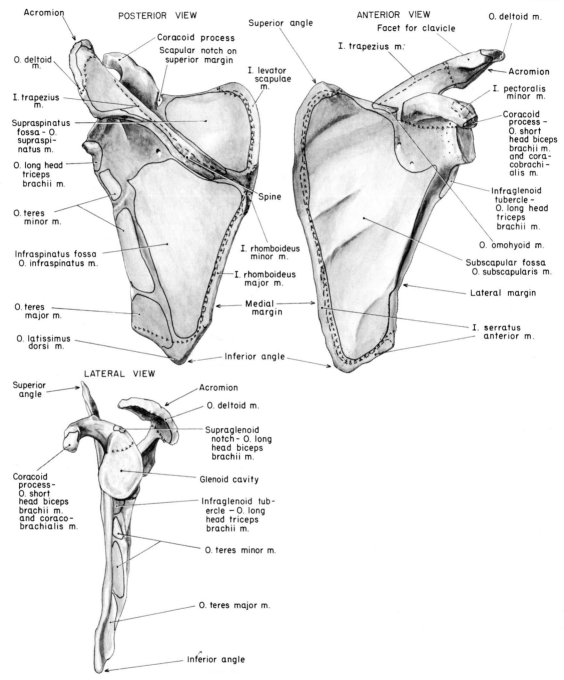

Figure 8-5. *Posterior, anterior, and lateral views of the left scapula, with origins and insertions of muscles outlined. (Crosses indicate location of epiphyseal lines.)*

suspended by muscles. It overlies ribs 2 to 7. The scapula has **medial, lateral, and superior margins** and **superior, inferior, and lateral angles.** The **spine** of the scapula separates the **supraspinatus fossa** from an **infraspinatus fossa** and extends from the medial margin to the enlarged **acromion process.** This ridge or spine gradually increases in height as it proceeds laterally. The superior margin possesses a **notch,** just at the base of a beaklike projection called the **coracoid process.** The **glenoid cavity,** for articulation with the head of the humerus, is located at the lateral angle. Anteriorly, the scapula presents a rather concave **fossa** for the subscapularis muscle.

HUMERUS

The **humerus** is the long bone found in the arm. It consists of a head, neck, shaft, and an inferior end (Figs. 8-6 and 8-7).

The **head** is a rounded structure that fits into the glenoid cavity of the scapula, a fossa made deeper by a rim of cartilage—the glenoid labrum. The head of the humerus differs from that of the femur in that it has no fossa for the attachment of any ligaments such as the ligamentum teres. The **anatomical neck** of the humerus immediately surrounds the base of the head, while the **surgical neck** is further down on the superior end of the shaft of the humerus. Just distal to the anatomical neck two tuberosities are found, one of which projects anteriorly—the **lesser tubercle**—and the other laterally—the **greater tubercle.** The groove between these tubercles is the **intertubercular groove;** the ridges of bone extending inferiorly on each side of this groove are the **major and minor crests** of the groove. The **shaft** of the bone has anteromedial, anterolateral, and posterior surfaces. On the anterolateral surface, about midway down the shaft of the bone, the **deltoid tuberosity** (for insertion of deltoid muscle) can be found, although this tuberosity often is not easily seen. The **groove for the radial nerve** occurs on the posterior surface of the shaft. Distally, **medial and lateral supracondylar ridges** are found extending inferiorly to medial and lateral projections, the **medial and lateral epicondyles.** Between these condyles is the pulley-like **trochlea,** placed medially, and the rounded **capitulum** (*L.,* little head) on the lateral side. The trochlea (*L.,* pulley) articulates with the ulna, while the capitulum articulates with the radius. Two fossae occur on the anterior surface; one is just proximal

to the trochlea—the **coronoid fossa**—and the other just proximal to the capitulum—the **radial fossa.** The coronoid process of the ulna and the head of the radius fit into these fossae when the elbow is flexed (see Fig. 8-12). On the posterior surface a larger fossa, the **olecranon fossa,** can be seen just superior to the trochlea. The olecranon process of the ulna fits into this fossa when the elbow joint is extended.

ULNA
(Fig. 8-8)

The **ulna** is a long bone, the proximal end of which exhibits a large **trochlear notch** that articulates with the trochlea of the humerus, and a smaller **radial notch** that articulates with the head of the other bone in the forearm, the radius. The **olecranon process** is superior to the trochlear notch; this articulates with the olecranon fossa of the humerus in extension (see Fig. 8-12). The **coronoid process** is inferior and anterior to the trochlear notch and this process enters the coronoid fossa during flexion at the elbow joint. Since the ulna is connected to the radius by the interosseus membrane, the ulna is said to present **anterior, posterior, and interosseus margins.** In between these margins are found the **anterior, posterior, and medial surfaces** of the bone. A small elevation just distal to the coronoid process is the **tuberosity of the ulna,** a point of attachment for a portion of the brachialis muscle. The **supinator fossa** is the large triangular depression distal to the radial notch and the **supinator crest** is the rather prominent posterior boundary of this fossa. The inferior end of the bone exhibits a **pronator ridge** on the distal end of the shaft of the bone, which starts anteriorly and winds around to the posterior side. This serves for the attachment of the pronator quadratus muscle and may be rather indistinct. The distal end of the bone also exhibits a medially placed **styloid process** and a laterally placed **head** of the ulna.

RADIUS
(Fig. 8-8)

The **radius** is the bone located in the lateral portion of the forearm consisting of a head, neck, shaft, and distal end. The **head** is flattened and articulates superiorly with the capitulum of the humerus and medially with the radial notch of the ulna. During the process of supination and

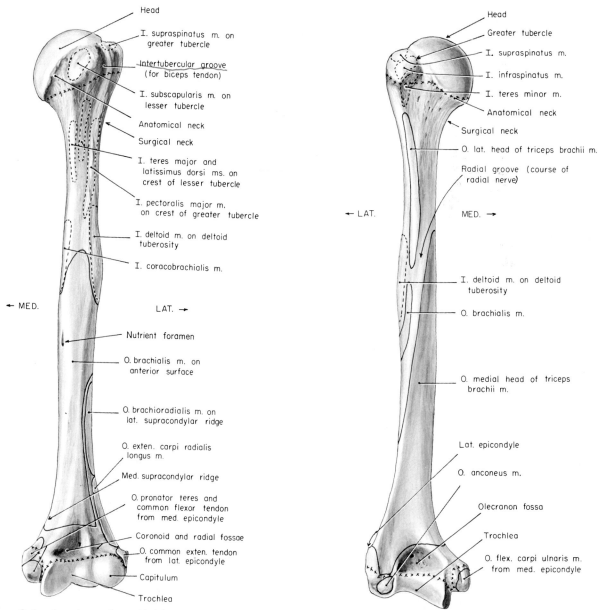

Head
I. supraspinatus m. on greater tubercle
Intertubercular groove (for biceps tendon)
I. subscapularis m. on lesser tubercle
Anatomical neck
Surgical neck
I. teres major and latissimus dorsi ms. on crest of lesser tubercle
I. pectoralis major m. on crest of greater tubercle
I. deltoid m. on deltoid tuberosity
I. coracobrachialis m.
← MED.
LAT. →
Nutrient foramen
O. brachialis m. on anterior surface
O. brachioradialis m. on lat. supracondylar ridge
O. exten. carpi radialis longus m.
Med. supracondylar ridge
O. pronator teres and common flexor tendon from med. epicondyle
Coronoid and radial fossae
O. common exten. tendon from lat. epicondyle
Capitulum
Trochlea

Head
Greater tubercle
I. supraspinatus m.
I. infraspinatus m.
I. teres minor m.
Anatomical neck
Surgical neck
O. lat. head of triceps brachii m.
Radial groove (course of radial nerve)
← LAT.
MED. →
I. deltoid m. on deltoid tuberosity
O. brachialis m.
O. medial head of triceps brachii m.
Lat. epicondyle
O. anconeus m.
Olecranon fossa
Trochlea
O. flex. carpi ulnaris m. from med. epicondyle

Figure 8-6. Anterior surface of left humerus, with origins and insertions of muscles outlined. (Crosses indicate location of epiphyseal lines.)

Figure 8-7. Posterior surface of left humerus, with origins and insertions of muscles outlined. (Crosses indicate location of epiphyseal lines.)

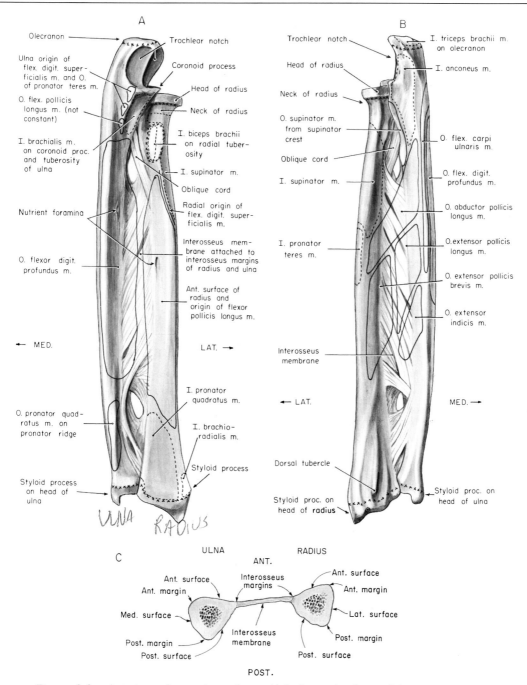

A

Olecranon

Ulna origin of flex. digit. super-ficialis m. and O. of pronator teres m.

O. flex. pollicis longus m. (not constant)

I. brachialis m. on coronoid proc. and tuberosity of ulna

Nutrient foramina

O. flexor digit. profundus m.

← MED.

O. pronator quad-ratus m. on pronator ridge

Styloid process on head of ulna

Trochlear notch

Coronoid process

Head of radius

Neck of radius

I. biceps brachii on radial tuber-osity

I. supinator m.

Oblique cord

Radial origin of flex. digit. super-ficialis m.

Interosseus mem-brane attached to interosseus margins of radius and ulna

Ant. surface of radius and origin of flexor pollicis longus m.

LAT. →

I. pronator quadratus m.

I. brachio-radialis m.

Styloid process

ULNA   RADIUS

B

Trochlear notch

Head of radius

Neck of radius

O. supinator m. from supinator crest

Oblique cord

I. supinator m.

I. pronator teres m.

Interosseus membrane

← LAT.

Dorsal tubercle

Styloid proc. on head of radius

I. triceps brachii m. on olecranon

I. anconeus m.

O. flex. carpi ulnaris m.

O. flex. digit. profundus m.

O. abductor pollicis longus m.

O. extensor pollicis longus m.

O. extensor pollicis brevis m.

O. extensor indicis m.

MED. →

Styloid proc. on head of ulna

C

ULNA   RADIUS

ANT.

Ant. surface

Ant. margin

Interosseus margins

Ant. surface

Ant. margin

Med. surface

Lat. surface

Post. margin

Interosseus membrane

Post. margin

Post. surface

Post. surface

POST.

Figure 8-8. *Anterior and posterior surfaces of left ulna and radius and the interosseus membrane. Origins and insertions of muscles have been outlined. (Crosses indicate location of epiphyseal lines.) A cross section is also shown.*

pronation (*see* Fig. 8-11), the head of the radius turns in the radial notch. The **neck** of the radius is just distal to the head, and just distal to the neck is a large elevation, the **radial tuberosity,** for the attachment of the biceps brachii muscle. The **shaft** of the radius possesses **anterior, posterior, and interosseus margins** and **anterior, posterior, and lateral surfaces.** The distal end exhibits **anterior, posterior, medial, lateral, and carpal surfaces.** The carpal surface articulates with the carpal bones. The radius also possesses a **styloid process** on its lateral side and, on the posterior side, a **dorsal tubercle,** which is the most prominent of the ridges in this location. The **ulnar**

notch of the radius articulates with the ulna and during pronation and supination glides on the surface of the distal end of the ulna.

CARPAL BONES
(Figs. 8-9, 8-10)

The **carpus** or wrist is made up of eight bones, which can be divided into a proximal row and a distal row. The proximal row, from thumb side to the little finger side, is made up of the **scaphoid** (*Gr.,* boat-shaped), **lunate,** (*L.,* moon-shaped), **triquetrum** (*L.,* three-cornered), and

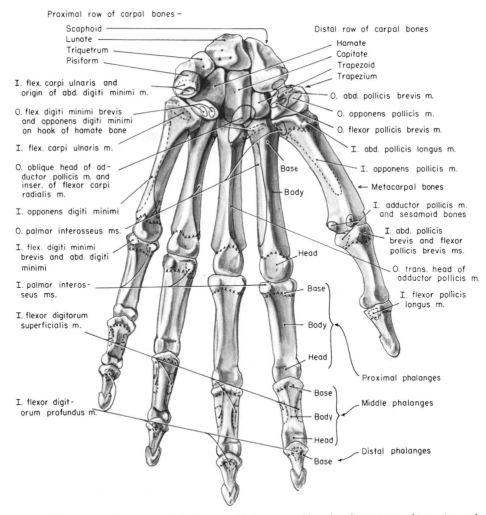

Figure 8-9. Anterior surface of the bones of left wrist and hand with origins and insertions of muscles outlined. (Crosses indicate location of epiphyseal lines.)

pisiform (*L.,* pea-shaped) bones. The distal row, from thumb side to little finger side, is made up of the **trapezium** (*Gr.,* small table), **trapezoid** (*Gr.,* table-shaped), **capitate** (*L.,* head-shaped), and **hamate** (*L.,* hooked) bones. The scaphoid and lunate bones articulate with the radius. The trapezium articulates with the metacarpal of the thumb, the trapezoid with the second metacarpal, the capitate with the third metacarpal, and the hamate with the fourth and fifth metacarpal bones.

The carpal bones should be learned as a group rather than as individuals. The carpus as a whole presents **proximal, distal, anterior, posterior, radial, and ulnar surfaces.** The scaphoid possesses a **tubercle** that should be noted, the trapezium a **ridge,** and the **hook** of the hamate is well worth investigation. The pisiform bone is in actuality a sesamoid bone.

The "anatomical time clock" is an interesting and useful concept. All bones of the upper limb are derived from cartilage; ossification occurs at various times after birth, that of the carpal bones occurring in such a sequence that the maturity of a child may be estimated. One should not try to be too exact with this concept, but a

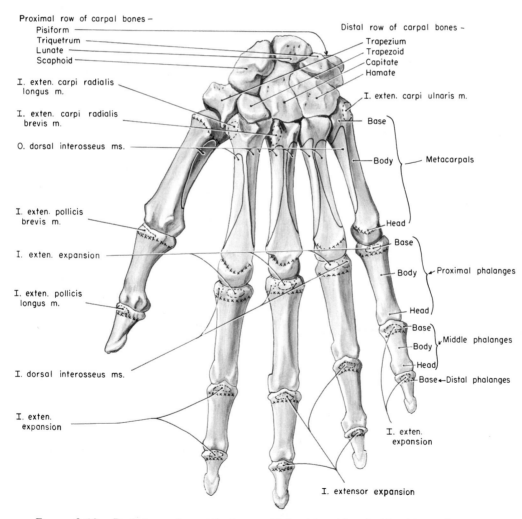

Figure 8-10. *Posterior surface of the bones of left wrist and hand with origins and insertions of muscles outlined. (Crosses indicate location of epiphyseal lines.)*

10-year-old child should not have ossification typical of a 3- to 4-year-old child.

Beginning ossification, according to Greulich and Pyle* is as follows:

	Males	*Females*
Capitate	Birth to 3 mo.	Birth to 3 mo.
Hamate	Birth to 3 mo.	Birth to 3 mo.
Triquetrum	2 yrs, 4 mo.	1 yr, 6 mo.
Lunate	3 yrs.	2 yrs, 6 mo.
Trapezium	3 yrs, 6 mo.	3 yrs.
Trapezoid	5 yrs.	3 yrs, 10 mo.
Scaphoid	5 yrs, 6 mo.	3 yrs, 10 mo.
Pisiform	10 yrs.	8 yrs, 4 mo.

METACARPALS
(Figs. 8-9, 8-10)

The **metacarpals,** five in number, possess a **base,** a **shaft,** and a **head.** The base is proximal and heads are distally located.

PHALANGES
(Figs. 8-9, 8-10)

The **phalanges** are two in number in the thumb,† and three in number in the other digits. The first row consists of the **proximal** phalanges; the next, the **middle;** and the third, the **distal** phalanges. Each phalange consists of a **base** proximally, a **shaft,** and a **head** distally. Figures 8-9 and 8-10 show the differences in shape among the three rows of bones, the distal ones being quite different in structure.

MOVEMENTS OF UPPER LIMB

Although the joints of the upper limb will be described when they are usually dissected—in the last phase of

---

Radiographic Atlas of Skeletal Development of the Hand and Wrist. Stanford University Press, 1959.

†Based on development, we would say that the thumb has no metacarpal bone, rather than only two phalanges. Note that the epiphyseal line in the metacarpal, as pictured in Figures 8-9 and 8-10, resembles those of the phalanges.

dissection—familiarity with the movements allowed in the joints is necessary at this time in order to understand muscle action. A review of the section on joints found in Chapter 1, page 14, would be helpful at this time.

The **shoulder joint** is a ball-and-socket joint, and all movements are allowed; these are flexion (moving arm anteriorly), extension (moving arm posteriorly), abduction (moving arm away from body), adduction (moving arm toward body), circumduction (a combination of the preceding), and medial (toward midline of body) and lateral (away from midline of body) rotation.

These movements occur around three axes: transverse, vertical, and anteroposterior (Fig. 8-12). The transverse axis of flexion and extension, as the name implies, passes from medial to lateral through the head of the humerus. The vertical axis of medial and lateral rotation continues inferiorly from the head of the humerus to the capitulum at its distal end. The anteroposterior axis for abduction and adduction passes through the head parallel to the glenoid cavity. Note that all three axes pass through the head of the humerus.

The **elbow joint** is strictly a hinge joint, allowing only two movements, flexion (moving forearm anteriorly) and extension (moving forearm posteriorly). These movements occur around a transverse axis which passes through the capitulum and trochlea.

The **superior radioulnar joint** and **inferior radioulnar joint** are gliding joints. The actions of pronation (palm of hand facing posteriorly) and supination (palm of hand facing anteriorly) occur at these two joints. (The superior radioulnar joint and the elbow joint are considered to be separate joints; the two should be clearly distinguished.) These movements occur around a vertical axis which is a continuation of the vertical axis of the shoulder joint. It passes vertically through the capitulum and head of the radius, and continues to the pit at the base of the styloid process of the ulna. In supination and pronation the radius rotates on this axis proximally and around it distally.

The **wrist joint** or radiocarpal joint is an ellipsoid joint and is the articulation between the distal end of the radius and the scaphoid and lunate bones. The gliding at this joint allows for flexion (moving hand anteriorly), extension (moving hand posteriorly), abduction (moving hand away from body), adduction (moving hand toward body), and circumduction (a combination of the above movements). There is no rotation at the wrist joint. These

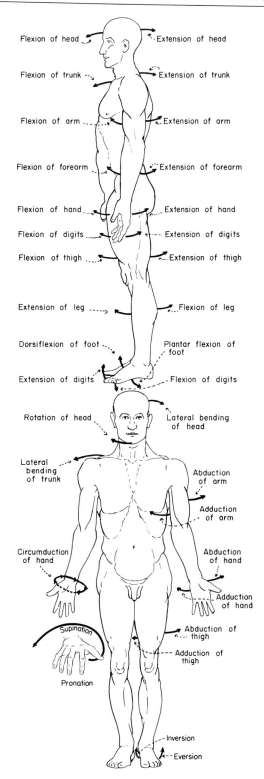

movements occur around two axes, transverse and palmardorsal (anteroposterior). The transverse axis of flexion and extension is complicated but can be simplified to a line passing transversely through the carpal bones (compromise transverse axis). The anteroposterior or palmardorsal axis for abduction and adduction passes through the base of the capitate bone.

Carpal joints. The movement allowed between the carpal bones is of a gliding nature, and the movement of the wrist is actually a combined movement of the wrist joint and the carpal joints.

Carpometacarpal joints. These are also gliding joints, and a great amount of motion is not necessary for free action. There is greater freedom of movement in the carpometacarpal joint of the thumb than in the other joints. The carpometacarpal joint of the thumb exhibits flexion, extension, abduction, adduction, and naturally, circumduction. In addition, a movement called **opposition** is possible. This is a combination of flexion, medial rotation, and adduction. This results in the metacarpal and the thenar eminence of the thumb being carried forward and medially to overhang the hollow of the palm. Slight flexion of the thumb allows it to touch the tips of the slightly flexed fingers in succession. Opposition is very important to us, for it is this movement that allows us to do so many things with our hands. In addition, the carpometacarpal joint of the little finger is slightly freer than the other digits, and opposition can be produced here as well.

Metacarpophalangeal joints. These are condyloid joints with the heads of the metacarpals articulating with the base of each proximal phalanx. Flexion (digits forward), extension (digits straight), abduction and adduction (away from and toward central axis of hand which is third digit), and circumduction are allowed at these joints—but no rotation occurs. These movements occur around a transverse axis for flexion and extension and palmardorsal or anteroposterior axes for abduction and adduction.

Interphalangeal joints. These are joints between the heads of one row of phalanges and the bases of the next row of phalanges. These are hinge joints, and only flexion (digits curled) and extension (digits straight) are allowed. These movements occur around transverse axes only.

Figure 8-11. *Diagram of the major movements of the body.*

In subsequent sections, the main action of muscles in the upper limb will be stated. These should be related to the axes mentioned above. For example, if a muscle is obviously anterior to the elbow joint, it will likely be anterior to the transverse axis also, thereby inducing a flexion at this joint.

In the anatomical position the hand is held away from the body. This is the result of the arrangement of the bones making up the elbow joint. The angle thus formed is known as the **carrying angle;** it varies from individual to individual.

BACK

SURFACE ANATOMY
(Fig. 8-13)

No matter how obese a person may be, a **median furrow** can be seen in the back. In muscular and lean persons the **scapula** can be outlined, the **acromion** and **spine** easily palpated. The outline of the **trapezius** muscle can be detected; the bulge on either side of the midline in the lower part of the back is made by the massive **erector spinae** muscles. Palpation easily reveals the **iliac crest, ribs,** and the **spinous processes** of all vertebrae from the seventh cervical to the sacrum.

It is helpful to realize that the base of the spine of the scapula is at the level of the third thoracic vertebra, while the inferior angle of the scapula is even with the seventh thoracic vertebra; the crest of the ilium is at the level of the spinous process of the fourth lumbar vertebra, while the posterior superior iliac spine is at the level of the second sacral vertebra.

SUPERFICIAL AND DEEP FASCIA

Removal of the skin of the back reveals the **superficial fascia.** This fascia varies in thickness from individual to

---

Figure 8-12. *The axes about which motion occurs in the upper limb; if the direction of pull of muscles is related to these axes, their actions become understandable.*

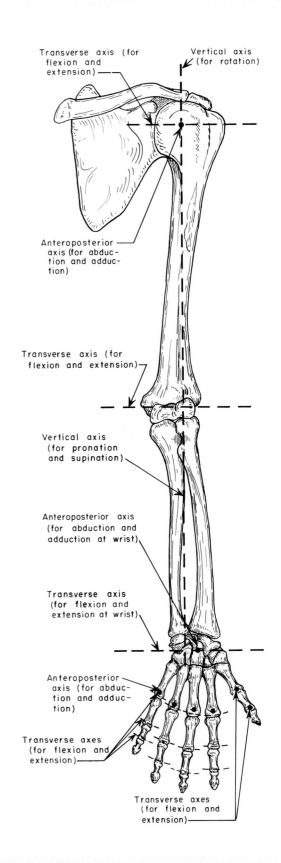

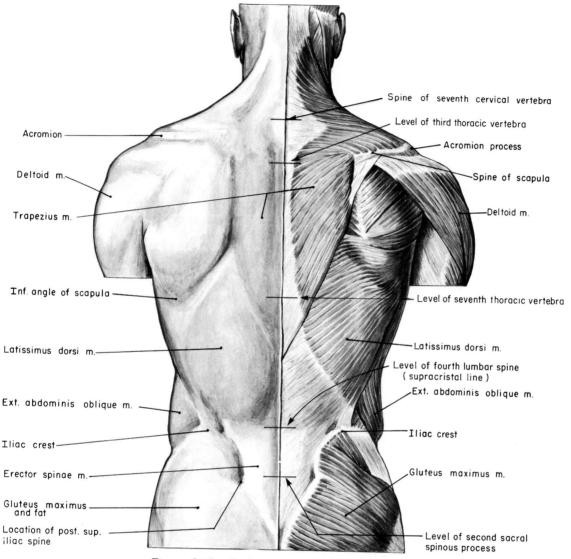

Acromion

Deltoid m.

Trapezius m.

Inf. angle of scapula

Latissimus dorsi m.

Ext. abdominis oblique m.

Iliac crest

Erector spinae m.

Gluteus maximus and fat

Location of post. sup. iliac spine

Spine of seventh cervical vertebra

Level of third thoracic vertebra

Acromion process

Spine of scapula

Deltoid m.

Level of seventh thoracic vertebra

Latissimus dorsi m.

Level of fourth lumbar spine (supracristal line)

Ext. abdominis oblique m.

Iliac crest

Gluteus maximus m.

Level of second sacral spinous process

Figure 8-13. *Surface anatomy of the back—male.*

individual, depending upon the obesity of the person. This layer tends to be thicker in females than in males, and in both sexes will be thicker in the lumbar part of the back than in other regions. Careful removal of this fascial layer reveals the thin but dense **deep fascia,** which rests directly upon the muscles and actually surrounds each muscle.

CUTANEOUS NERVES AND VESSELS

Coursing in this superficial fascia are the cutaneous nerves (Fig. 8-15) and vessels. They pierce the muscles and deep fascia covering the muscles, and then course in a lateral and slightly inferior direction to reach the skin.

Cutaneous nerves in the upper part of the back are found piercing the muscles and entering the superficial fascia quite close to the spines of the vertebrae. In the lower part of the back, the cutaneous nerves are some 3 to 4 inches away from the midline. This cannot be accounted for until we obtain a knowledge of a typical segmental spinal nerve.

Figure 8-14 shows such a **segmental nerve.** Each nerve is attached to the spinal cord by two **roots,** one **dorsal** and one **ventral.** The **dorsal root,** containing a ganglion known as the **dorsal root ganglion,** is the **sensory root;** the **ventral** is the **motor root.** The two roots converge as they emerge from an intervertebral foramen and form a **segmental nerve.** Almost immediately, this nerve divides into two branches, one coursing posteriorly into the back as the **dorsal ramus** of the spinal nerve, and the other continuing anteriorly as the **ventral ramus** of the spinal nerve. The latter portion of the spinal nerve continues around the body, giving off a **lateral cutaneous branch** and finally an **anterior cutaneous branch.**

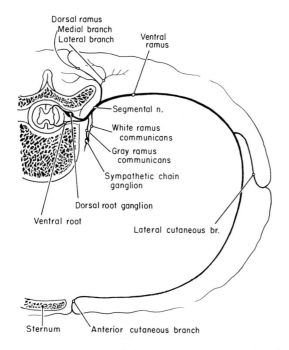

Figure 8-14. A typical segmental (spinal) nerve and its branches.

The **dorsal ramus** continues posteriorly and divides into **medial and lateral branches.** In the upper half of the body the lateral branch remains in the muscles, while the medial branch supplies muscles and then continues to the skin; in the lower part of the body, the medial branch remains in the muscles while the lateral branch innervates muscles and then continues to become a cutaneous branch. Therefore, the cutaneous nerves found in the upper part of the back are the medial branches, while those in the lower part of the back are the lateral branches.

Figure 8-15 shows the cutaneous nerves in the back. It is interesting to note that some of the cutaneous nerves in the cervical region have a distinctive pathway and that not all dorsal rami are represented in the skin area. The **first cervical nerve** has no sensory component except upon rare occasions. The medial branch of the dorsal ramus of the **second cervical nerve** is a very large cutaneous nerve and is called the **greater occipital nerve;** it ramifies over the scalp on the posterior part of the head. The medial branch of the dorsal ramus of the **third cervical** also reaches the more inferior aspect of the scalp. Cutaneous branches of the dorsal rami of the **fourth to sixth cervical** nerves do not exhibit any unusual characteristics. The **seventh and eighth cervical** nerves make no contribution to the back region.* Cutaneous branches of the **thoracic nerves** are quite usual in their distribution, those in the upper half of the body being derived from the medial branches of the dorsal rami, and those of the lower part of the body from the lateral branches. Cutaneous branches of the dorsal rami of the **first three lumbar** nerves are called the **superior cluneal** (L., buttock) nerves, and enter the superficial fascia just lateral to the erector spinae muscles to be distributed inferiorly over the gluteal region. The dorsal rami of the **fourth and fifth lumbar** nerves do not have cutaneous distribution.† The dorsal rami of the **first three sacral** nerves have cutaneous branches, which are called the **middle cluneal** nerves and are distributed over the gluteal region as well as on the posterior surface of the sacral region. The dorsal rami

---

*It seems that the skin that would have been innervated by these branches has simply disappeared in human beings.

†Here too, it would seem, the skin that would have been innervated by these branches has simply disappeared in human beings.

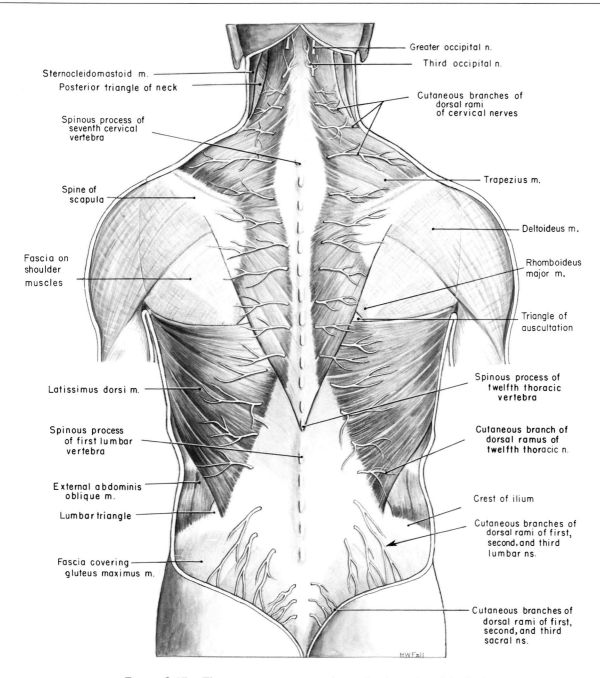

Sternocleidomastoid m.

Posterior triangle of neck

Spinous process of
seventh cervical
vertebra

Spine of
scapula

Fascia on
shoulder
muscles

Latissimus dorsi m.

Spinous process
of first lumbar
vertebra

External abdominis
oblique m.

Lumbar triangle

Fascia covering
gluteus maximus m.

Greater occipital n.

Third occipital n.

Cutaneous branches of
dorsal rami
of cervical nerves

Trapezius m.

Deltoideus m.

Rhomboideus
major m.

Triangle of
auscultation

Spinous process of
twelfth thoracic
vertebra

Cutaneous branch of
dorsal ramus of
twelfth thoracic n.

Crest of ilium

Cutaneous branches of
dorsal rami of first,
second, and third
lumbar ns.

Cutaneous branches of
dorsal rami of first,
second, and third
sacral ns.

H.W.Fall

Figure 8-15.   *The cutaneous nerves and superficial muscles of the back.*

of the **fourth and fifth sacral** nerves, and the **single coccygeal** nerve, do not divide into medial and lateral branches; a cutaneous nerve is formed from this complex, however, that is distributed in the region of the coccyx. The **inferior cluneal** nerves are described in Chapter 6, on the lower limb.

It should be understood that these cutaneous nerves are not purely sensory nerves; they all contain fibers of the autonomic nervous system to sweat glands, smooth muscles in the blood vessels, and arrector pili muscles.

Each of these nerves is accompanied by similar branches of the **cutaneous arteries and veins;** in fact, the vascular system has a distribution in the back similar to that of a regular segmental nerve (Fig. 8-16). Dorsal branches to the back are obtained from the vertebral artery in the neck, intercostal branches of the aorta in the thoracic region, lumbar branches of the aorta in the lower part of the back, and cutaneous branches of the lateral sacral arteries in the sacrococcygeal region.

MUSCLES INVOLVED WITH UPPER LIMB*

All striated or skeletal muscles are derived from myotomes of the mesodermal somites or from branchial arches. The former make up the somatic musculature while the latter group is branchiomeric.

Each myotome has its own spinal nerve, which remains with this particular myotome no matter where migration may take it. The myotomes divide longitudinally to form the back or dorsal musculature (epaxial trunk muscles; see Fig. 8-17) and the intercostal and ventrolateral trunk muscles (hypaxial trunk muscles), and the nerves divide in a similar manner. These myotomes may undergo other changes besides this longitudinal splitting. There may be a tangential splitting, a fusion of portions of successive myotomes, a migration of certain portions, and degeneration of parts or entire myotomes.

Using the bones as an axis, the limb buds grow in such a way that there are muscles located dorsally (postaxial) and ventrally (preaxial).

---

*As stated in Chapter 2, muscles in this text are described according to their location, origin, insertion, main action, and nerve supply. Enough information is presented on origins and insertions to allow an understanding of the main action of each muscle. In this connection, the student should realize that the insertion is usually more informative than the origin, unless the muscle happens to bridge over two joints.

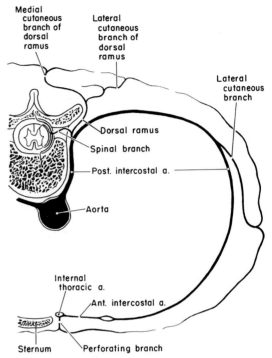

Figure 8-16. *A typical segmental artery and its branches (an intercostal artery is presented).*

In summary, the nonbranchial skeletal muscles are all derived from the myotomes, and the ultimate arrangement in the human depends upon the many changes that occur in the primary arrangement. It is important to realize that the innervations to the myotomes occur early, and that these nerves follow the myotomes or parts thereof no matter what happens to them. Therefore, the nerve supply is a good clue to the embryology of a particular muscle.

In spite of the above, at this time it seems simpler to divide the muscles of the back into three functional groups—those concerned with the upper limb, those involved with respiration, and those with main action on the back and head.

In this chapter we are involved with the first group only—those involved with the limb—and they are located most superficially. They are:

1. Trapezius
2. Latissimus dorsi
3. Levator scapulae

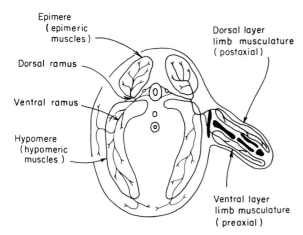

Figure 8-17. *Cross section of an embryo showing the division of trunk musculature into epimeres and hypomeres, and limb musculature into dorsal muscles (postaxial or posterior to axis of limb) and ventral muscles (preaxial or ventral to axis of limb). Note how the ventral rami of the nerves divide into preaxial and postaxial branches to innervate these preaxial and postaxial muscles.*

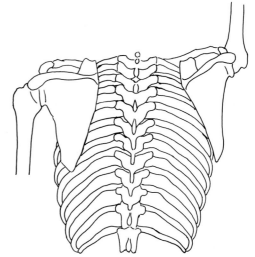

Figure 8-18. *Diagram showing rotation of the scapula in complete abduction at the shoulder joint.*

4. Rhomboideus major and minor
5. Serratus anterior

1 and 2. Trapezius and latissimus dorsi muscles. Removal of the superficial and deep layers of fascia in the back reveals two very large muscles, one located in the superior part and the other in the lower part of the back (Fig. 8-15). The superiorly placed muscle, the **trapezius,** has muscle bundles converging upon the scapula from the head, neck, and thoracic region; the lower muscle, the **latissimus dorsi,** exhibits fibers coursing in a lateral and superior direction toward the upper end of the humerus. These two muscles cover nearly the entire back.

The **trapezius** muscle **originates** from the medial end of the superior nuchal line, the external occipital protuberance, the ligamentum nuchae (Figure 2-25 on page 76 depicts this ligament clearly), and the spines of the seventh cervical and all thoracic vertebrae. The inferior portion of this muscle **inserts** on the base of the spine of the scapula,* the middle portion on the superior surface of the spine of the scapula and the acromion process, and the superior portion on the lateral third of the clavicle (Figs. 8-4 and 8-5). This muscle has several important functions. One of its most important **actions** is to hold the upper limb onto the trunk; when the muscle acts as a whole, it tends to adduct the scapula or bring it toward the midline. If the superior fibers alone contract, the scapula is raised, and if the inferior fibers contract alone the scapula is lowered. Another action of the trapezius muscle is to aid in rotation of the scapula in such a manner that the inferior angle is moved laterally. It is essential in complete abduction at the shoulder that the glenoid cavity of the scapula be tipped superiorly; the deltoid muscle (which forms the rounded portion of the shoulder—see Fig. 8-13) can abduct the humerus to a horizontal level, but further abduction requires a rotation of the scapula as shown in Figure 8-18. Patients with nerve injuries to the trapezius muscle have great difficulty in abducting the humerus above the horizontal plane, due partly to the fact that the scapula is not held firmly against the deeper structures.†

The **latissimus dorsi** muscle is a very broad, flat, triangular muscle in the lower part of the back; it is overlapped by the lower part of the trapezius muscle. It takes **origin** from spinous processes of the lower six thoracic

---

*A bursa intervenes between the tendon and the base of the spine.

†Note action of the serratus anterior muscle on page 686.

vertebrae, from spinous processes of all the lumbar and upper sacral vertebrae, and from the medial part of the iliac crest. There are additional slips of origin from the lower four ribs and a small slip of attachment to the inferior angle of the scapula. The **insertion** (Figs. 8-6 and 8-37) of this muscle is on the humerus in such a way (floor of intertubercular groove) that contraction of this muscle pulls the arm posteriorly, or extends it, adducts and rotates it medially; this large muscle is used in actions such as chopping wood or swimming with the crawl stroke.

Two small **triangles** should be noted at this time (Fig. 8-15). (1) The **lumbar triangle** is bounded medially by the latissimus dorsi muscle, laterally by a muscle just lateral to the latissimus dorsi—the external abdominis oblique—and inferiorly by the iliac crest. It is a fairly important triangle, for hernias and infections are occasionally presented here. (2) The **triangle of auscultation** is bounded by the trapezius muscle, the latissimus dorsi muscle, and the medial border of the scapula. The floor is formed by the rhomboideus major muscle but when the scapula is moved anteriorly, the sixth and seventh ribs can be seen; this is a good location at which to listen to internal structures.

If the trapezius and latissimus dorsi muscles are dissected and reflected laterally, the **nerve and blood supply** to these muscles is seen (Fig. 8-19). The trapezius muscle has a double **nerve supply** from the spinal accessory nerve (eleventh cranial) and from ventral rami of the third and fourth cervical nerves. The actual role of these nerves has not been determined completely, but it is thought that the cervical nerves may be sensory and the spinal accessory nerve may be the motor supply to the muscle.* The upper border of the trapezius muscle forms the posterior border of the posterior triangle of the neck. The spinal

accessory nerve is quite exposed in that it crosses the posterior triangle of the neck in the enveloping layer of the deep fascia and is quite superficially located. It then reaches the superior border of the trapezius muscle and ramifies on the deep surface throughout the entire length of the muscle; the nerve may enter the muscle independently, or it may join with the third and fourth cervical nerves and all three nerves ramify on the muscle as a single nerve.

The **blood supply** in this region is confusing because of frequent variation. An artery called the **transverse cervical** (Fig. 8-20) arises from the thyrocervical trunk, which is a branch of the subclavian artery. This artery finally reaches this region of the back and usually divides into two branches, one of which ramifies on the deep surface of the trapezius muscle, and the other on the deep surface of the next layer of muscles, the rhomboid muscles. The branch coursing on the deep surface of the trapezius is the **superficial** branch; the other is the **deep branch** of the transverse cervical artery. In approximately 50 percent of the cases, the superficial branch is the only one that arises from the thyrocervical trunk in the manner just described; when this occurs the name of **superficial cervical** artery is given to this branch, and the deep branch then arises independently from the third part of the subclavian artery and is known as the **descending scapular artery.**

The **nerve and artery to the latissimus dorsi** muscles are called **thoracodorsal nerve and artery;** this nerve is a branch of the brachial plexus (ventral rami forming a plexus of nerves that innervate the upper limb) and will be described when this plexus is studied. The artery is a branch of the subscapular from the axillary artery, which also will be described later.

You have now observed **three rules of the body.** (1) *Nerves and arteries are usually found on the deep side of the muscle to be innervated.* (2) *Nerves, once they have joined a muscle, remain with this muscle no matter where it migrates.* (3) *Arteries exhibit many variations from the so-called normal; they are quite variable in their arrangement and one side of the body need not resemble the other; textbook descriptions are based on the appearance of the majority of cases.*

3 and 4. Levator scapulae and rhomboideus major and minor muscles. Reflection of the trapezius

---

*K. N. Corbin and F. Harrison, 1939, The sensory innervation of the spinal accessory and tongue musculature in the rhesus monkey, *Brain* 62: 191. Evidence is presented indicating that proprioception from the trapezius, sternocleidomastoid, and the tongue musculature is carried via the cervical nerves, leaving the spinal accessory and hypoglossal nerves to these muscles as the motor supply, at least in the monkey.

W. L. Straus and A. B. Howell, 1936, The spinal accessory nerve and its musculature, *Quart. Rev. Biol.* 11: 387. A survey of the puzzling double nerve supply to the trapezius and sternocleidomastoid muscles. The authors feel that the cervical nerves are merely branchial fibers that have shifted to follow a spinal pathway.

McKenzie, J., 1962, The development of the sternomastoid and trapezius muscles, *Contrib. Embryol.* 37: 121. Based on a study of the human embryos of the Carnegie Institution, this author feels that the sternocleidomastoid and trapezius muscles are partially derived from myotomes. This would imply that the nerve supply from the third and fourth cervical nerves is motor as well as sensory.

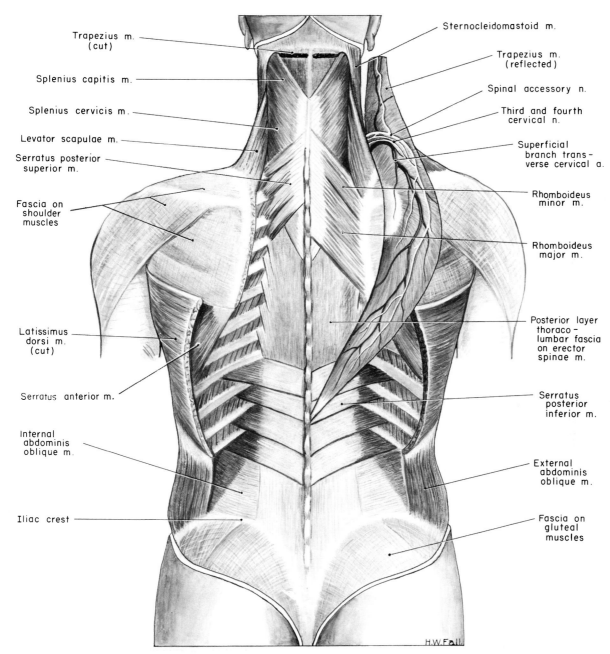

Figure 8-19. *Back musculature after cutting the latissimus dorsi muscle, and after reflecting the trapezius muscle (right side) or removing it (left side).*

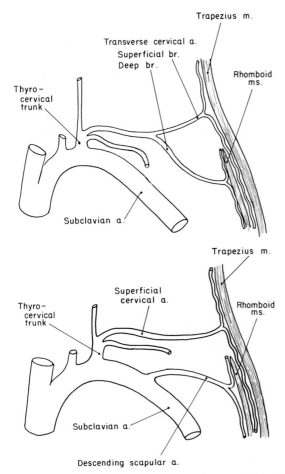

Figure 8-20. *Diagram of the two most frequently found arrangements of the blood supply to the trapezius and rhomboid muscles. Note difference in names in the two arrangements.*

muscles reveals the levator scapulae and rhomboideus major and minor muscles (Fig. 8-18). The more superiorly placed of the three, the levator scapulae, is seen high in the neck, while the rhomboids are in the form of a rhomboid and located between the vertebral column and the scapula.

The **levator scapulae muscle arises** from the transverse processes of the first four cervical vertebrae, and **inserts** on the superior angle of the scapula (Fig. 8-5). Its **main action** is obviously to raise the scapula. Its **nerve supply** is from ventral rami of the third and fourth cervical nerves.

The **rhomboideus major and minor muscles** are actually one sheet of muscle but are described separately; the **minor arises** from the inferior end of the ligamentum nuchae and the spine of the seventh cervical vertebra and **inserts** on the vertebral margin of the scapula just inferior to the point of attachment of the levator scapulae muscle. See Figure 8-5 for both of these insertions. The **rhomboideus major arises** from the first four thoracic spines and **inserts** on the medial margin of the scapula inferior to the base of the spine. In **action,** these muscles, in addition to holding the scapula onto the trunk, adduct the scapula and rotate it in such a manner that the inferior angle is moved medially. The **nerve supply** to the rhomboid muscles is by a specially named nerve—the dorsal scapular nerve—from the ventral ramus of the fifth cervical nerve. This nerve crosses the root of the neck and enters the deep surface of the rhomboid muscles. The deep branch of the transverse cervical artery occupies a similar position.

5. Serratus anterior. Although only a portion of this muscle is usually seen in a dissection of the back, its complementary action to the trapezius in rotating the scapula laterally in abduction of the arm warrants its inclusion at this point. It is illustrated in Figure 8-39 on page 705. This muscle is located on the lateral wall of the chest. Its **origin** is anteriorly placed and consists of eight digitations from the upper eight ribs. It hugs the lateral chest wall and **inserts** on the superior angle, the medial margin, and the inferior angle of the scapula on its ventral or anterior surface. The attachment to the inferior angle is particularly massive and is the part seen in this dissection (Fig. 8-19). Its **main action** is to pull the scapula forward in actions such as pushing, and, as just mentioned, to rotate the inferior angle laterally. Its **nerve supply** is by the long thoracic nerve, a branch of the brachial plexus. If this nerve is injured, the scapula is not held snugly to the chest wall, which results in a condition called winged scapula. Under this condition, total abduction of the arm cannot be accomplished. It is obvious that total abduction of the arm (above the horizontal) cannot occur if either the trapezius or serratus anterior muscles are denervated.

PECTORAL REGION

We come now to a study of that portion of the anterior chest wall that is concerned with movement of the upper limb. Figure 8-21 shows the bones involved in this region.

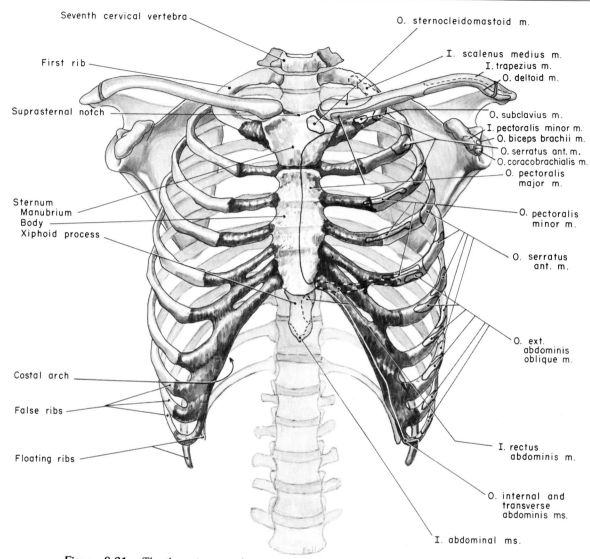

Seventh cervical vertebra

First rib

Suprasternal notch

Sternum
Manubrium
Body
Xiphoid process

Costal arch

False ribs

Floating ribs

O. sternocleidomastoid m.

I. scalenus medius m.
I. trapezius m.
O. deltoid m.

O. subclavius m.
I. pectoralis minor m.
O. biceps brachii m.
O. serratus ant. m.
O. coracobrachialis m.
O. pectoralis
major m.

O. pectoralis
minor m.

O. serratus
ant. m.

O. ext.
abdominis
oblique m.

I. rectus
abdominis m.

O. internal and
transverse
abdominis ms.

I. abdominal ms.

Figure 8-21. *The thoracic cage, showing manner of attachments of ribs to sternum, and the clavicle and scapula. The darker shaded areas are the cartilaginous portions of the ribs. Origins and insertions of muscles are indicated.*

SURFACE ANATOMY
(Fig. 8-22)

Many of the structures close to the surface can be seen by looking at the skin of a lean male body. Starting superiorly, in the midline one sees the **suprasternal notch.** Continuing in the midline, a groove marking the **sternum** is obvious; this ends as the **xiphoid process.**

The **clavicles,** extending laterally from the suprasternal notch, can be followed to their articulation with the **acromion processes** of the scapulae. The **bulges** on either side of the sternum are made by the large **pectoralis major muscles.** The inferior extent of this muscle is usually discernible and it can be followed to the axilla, where it forms the **anterior wall** or fold of that region. The **nipple** in the male is usually located superficial to the

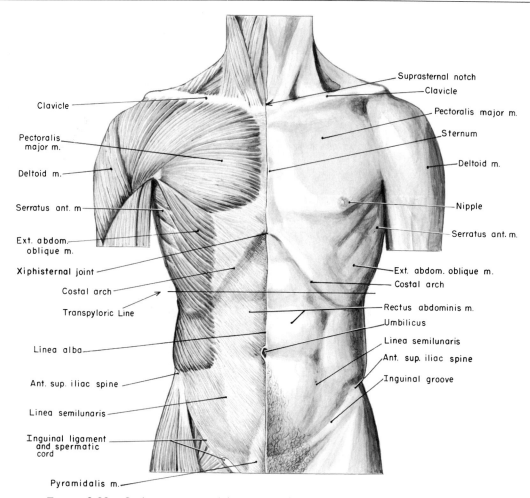

Figure 8-22. Surface anatomy of the anterior thoracic and abdominal walls in the male.

fourth intercostal space and about 4 inches from the mid-line. Sweeping inferiorly and laterally from the xiphister-nal joint is the **costal arch,** the inferior extent of the ribs. The **interdigitations** between serratus anterior and ex-ternal abdominal muscles can be seen in muscular indi-viduals.

Palpation reveals other structures. Although the first rib is overshadowed by the clavicles, the **second ribs** are easily palpated. They articulate with the sternum at the joint between the manubrium and body of the sternum, a point called the **sternal angle.** The seventh **costal carti-lage** is the lowest that attaches directly to the sternum and can also be palpated, as can the tips of the **eighth, ninth, and tenth ribs.** The **eleventh and twelfth ribs** can be

palpated behind the lowest part of the costal margin. In counting ribs it is wise to start with the second; the first is sometimes difficult to identify (except by palpation), as is the twelfth.

GENERAL DESCRIPTION

Upon removal of the skin, the **platysma muscle** can be seen to extend to a point inferior to the clavicle, as indi-cated in Figure 8-23. The **supraclavicular nerves,** branches of the cervical plexus, also extend onto the chest. Several anterior **cutaneous branches of inter-costal nerves** and **perforating branches of the internal thoracic artery** pierce the muscle and fascia just lateral to

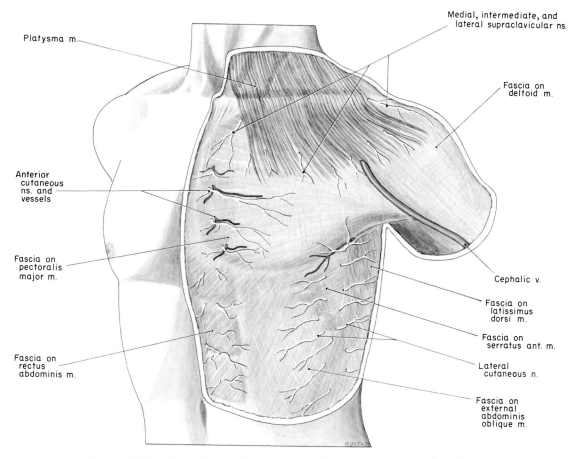

Platysma m.

Medial, intermediate, and
lateral supraclavicular ns.

Fascia on
deltoid m.

Anterior
cutaneous
ns. and
vessels

Fascia on
pectoralis
major m.

Cephalic v.

Fascia on
latissimus
dorsi m.

Fascia on
serratus ant. m.

Lateral
cutaneous n.

Fascia on
rectus
abdominis m.

Fascia on
external
abdominis
oblique m.

Figure 8-23. *Pectoral region I—after removal of the skin and superficial fascia.*

the sternum. Each intercostal nerve gives off a **lateral branch** which also becomes cutaneous; these branches divide further before emerging from the muscles and therefore appear as two branches, one coursing anteriorly and the other posteriorly. They are accompanied by **lateral cutaneous branches of the intercostal arteries.** A glance at Figures 8-14 and 8-16 will clarify the above. The arterial supply in this lateral region is supplemented by the **lateral thoracic artery,** a branch of the axillary.

Removal of the superficial fascia reveals the large **pectoralis major muscle** (Fig. 8-24), the muscle forming the large bulge on the anterior chest wall. The division between its clavicular and costosternal parts is readily seen; that between the costosternal and abdominal portions is not seen as easily. The **cephalic vein** marks the

dividing line between this muscle and its neighbor, the deltoideus.

If the pectoralis major is reflected laterally as in Figure 8-25, the **pectoralis minor muscle** can be seen through a layer of deep fascia. This fascia extends from the clavicle, where it surrounds the subclavius muscle, to the pectoralis minor muscle, surrounds it in turn, and then continues into the axilla, as diagramed in Figure 8-26. That portion between the clavicle and pectoralis minor muscles is the **clavipectoral fascia;** medially it is attached to the first costal cartilage and fascia on the first intercostal space, and laterally to the coracoid process. The clavipectoral fascia is **pierced** (1) by the **cephalic vein** on its way to drain into the axillary, (2) by the **lateral pectoral nerve** innervating the clavicular part of the pec-

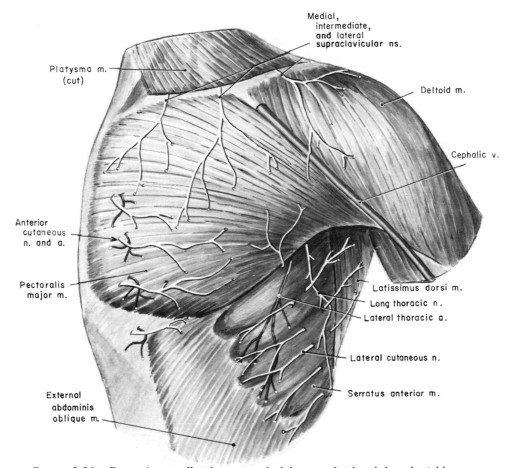

Medial, intermediate, and lateral supraclavicular ns.

Platysma m. (cut)

Deltoid m.

Cephalic v.

Anterior cutaneous n. and a.

Pectoralis major m.

Latissimus dorsi m.

Long thoracic n.

Lateral thoracic a.

Lateral cutaneous n.

Serratus anterior m.

External abdominis oblique m.

Figure 8-24. *Pectoral region II—after removal of the superficial and deep fascial layers.*

toralis major muscle, and (3) by the **thoracoacromial artery,** a branch of the axillary. The **medial pectoral nerve** pierces the pectoralis minor, which it innervates, before entering the substance of the costosternal and abdominal portions of the pectoralis major.

Removal of the clavipectoral fascia (Fig. 8-27) reveals the small subclavius muscle hugging the inferior surface of the clavicle, the axillary artery and vein, and portions of the brachial plexus.

MUSCLES

The muscles in the pectoral region are:

1. Pectoralis major

2. Pectoralis minor
3. Subclavius

1. Pectoralis major (Fig. 8-24). The clavicular portion of the pectoralis major muscle **arises** from the medial half of the clavicle (Fig. 8-21), the costosternal portion from the anterior surface of the sternum and costal cartilages of the upper six ribs (Fig. 8-21), and the abdominal portion from the aponeurotic tendon of the external abdominis oblique muscle. The **insertion** is on the crest of the greater tubercle of the humerus (Fig. 8-6), the clavicular portion overriding the costosternal part. Since this muscle passes anterior to the transverse and vertical axes and inferior to the anteroposterior axis of the shoulder joint, its **main action** is to flex, adduct, and medially

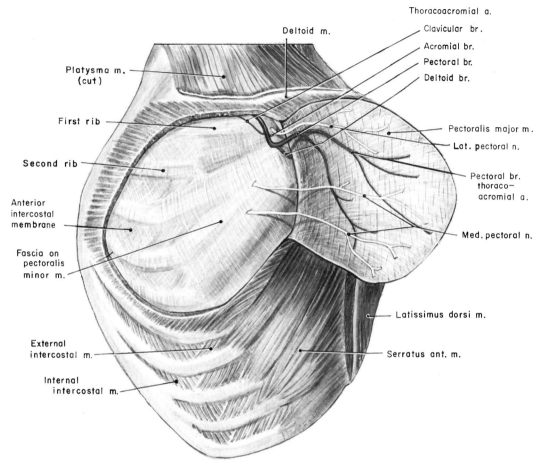

Figure 8-25. *Pectoral region III—after reflection of the pectoralis major muscle. Note fascia covering the pectoralis minor muscle.*

rotate the arm, an act which brings it across the chest. **Nerve supply** is by medial and lateral pectoral nerves from the medial and lateral cords of the brachial plexus.

2. The **pectoralis minor** (Fig. 8-27) muscle lies deep to the pectoralis major and **arises** from the external surface of three ribs, usually the second, third, and fourth (Fig. 8-21). It extends superiorly and laterally to **insert** on the coracoid process of the scapula (Fig. 8-5). Its **main action** is to lower the shoulder. **Nerve supply** is the medial pectoral nerve.

3. **Subclavius** (Fig. 8-27). The subclavius muscle is a small muscle **arising** from the first rib at the point of its junction with its cartilage (Fig. 8-21), coursing laterally, and **inserting** on the inferior surface of the clavicle (Fig.

8-4). Its **main action** is to lower the clavicle. **Nerve supply** is by the special nerve from the brachial plexus called the nerve to the subclavius.

ARTERIES

The arteries in the pectoral region are:

1. Perforating branches of internal thoracic
2. Lateral cutaneous of intercostal
3. Lateral thoracic of axillary
4. Thoracoacromial of axillary

1. **Perforating** (Figs. 8-23 and 8-24). These are branches of the internal thoracic artery (see Fig. 3-23 on page 121). They appear close to the sternum and usually divide into short medial and longer lateral branches. The latter branches of the third, fourth, and fifth perforating arteries are enlarged in the female to supply the breast.

2. Lateral cutaneous (Fig. 8-24). These are found more laterally placed and are branches of the intercostal arteries. The main stem is not visible but their anterior and posterior terminal branches can be seen emerging between the heads of the serratus anterior muscle. Several of these lateral branches are enlarged in the female to supply the breast.

3. Lateral thoracic (Figs. 8-24 and 8-27). Although its origin cannot be seen at this time, this artery is a branch of the axillary. It courses inferiorly and slightly anteriorly on the surface of the serratus anterior muscle. This artery also has mammary branches in the female.

4. The **thoracoacromial artery** (Figs. 8-25, 8-26, and 8-27), after piercing the clavipectoral fascia, immediately divides into four branches: (a) **pectoral,** to the pectoralis major and minor; (b) **acromial,** coursing laterally in a plane superficial to the coracoid process to reach the acromion process; (c) **clavicular,** turning medially to supply blood to the subclavius muscle and the sternoclavicular joint; and (d) **deltoid,** often arising from one of the other branches, accompanying the cephalic vein and supplying both pectoralis major and deltoid muscles.

VEINS

All of the foregoing arteries are accompanied by veins. One vein in this region, the **cepahlic,** has no accompanying artery. This vein drains the superficial parts of the arm and courses in the groove between the deltoid and pectoralis major muscles; in this position it actually crosses the pectoralis minor muscle. It empties into the axillary vein. It is often found wanting.

NERVES

The nerves in the pectoral region are:

1. Anterior and lateral cutaneous
2. Long thoracic
3. Lateral pectoral
4. Medial pectoral

1. Cutaneous (Figs. 8-23 and 8-24). The **an-**

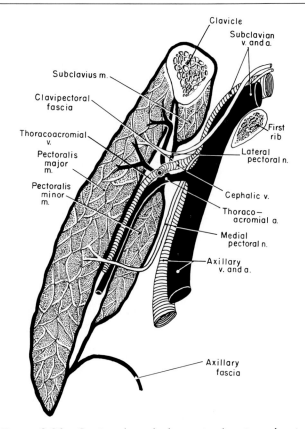

Figure 8-26. *Section through the pectoral region, showing structures piercing the clavipectoral fascia and their relation to the pectoral muscles. (Modified after Cunningham.)*

terior cutaneous nerves are distributed in a fashion similar to the arteries. They differ in that the nerves are the terminal branches of the intercostals, while the arteries arise from the internal thoracic. The **lateral cutaneous** branches are branches of the intercostals and divide into anterior and posterior branches just as the arteries do.

2. The **long thoracic** nerve accompanies the lateral thoracic artery (Figs. 8-24 and 8-27). Its origin from the rami of the brachial plexus is not visible; it courses inferiorly and slightly anteriorly on the surface of the serratus anterior muscle which it innervates. The relation of this nerve to its muscle is contrary to the usual arrangement.

3. The **lateral pectoral** nerve is named for the fact that it arises from the lateral cord of the brachial plexus, and courses to the pectoral region. It appears in this re-

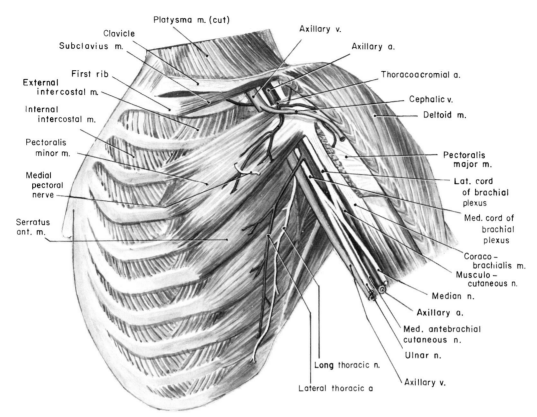

Figure 8-27. *Pectoral region IV—after removal of deep fascia. Note how the pectoralis minor muscle divides the vital structures in the axilla into three parts: proximal, deep, and distal to this muscle. The anterior intercostal membrane has been removed.*

gion by piercing the clavipectoral fascia (Figs. 8-25 and 8-26). It immediately enters the clavicular portion of the pectoralis major muscle.

4. Medial pectoral. This nerve is obviously named in a fashion similar to that described for the lateral pectoral nerve; it arises from the medial cord of the brachial plexus and enters the pectoral region but remains deep to the clavipectoral fascia and the pectoralis minor muscle; it then pierces this muscle, innervating it on the way; and enters the deep surface of the sternocostal and abdominal portions of the pectoralis major. There may be more than one branch of this nerve (Fig. 8-27).

The nerve supply to the subclavius muscle is not seen at this time.

It should be noted that all the muscles in this region

are concerned with the upper limb, the clavicle and scapula being parts of the appendicular skeleton. They are therefore innervated by branches from the brachial plexus, since that plexus innervates all muscles developed from the superior limb bud no matter where they may migrate.

POSTERIOR TRIANGLE OF NECK

GENERAL DESCRIPTION

Since many of the structures destined to reach the upper limb arise in the inferior end of the posterior triangle of the neck, this area will be described in a general way without going into details. This should aid in visualizing the origin of such structures as the brachial plexus, the transverse cervical and suprascapular arteries, and the subclavian artery and vein. This will be followed by a description of the sternoclavicular and acromioclavicular joints in case the clavicle is to be removed. If the head and neck have been studied, this section can be omitted.

In Figure 8-28 an area in the neck is seen bounded anteriorly by the sternocleidomastoid muscle, posteriorly by the trapezius muscle, and inferiorly by the clavicle; this area is the **posterior triangle of the neck.** In Figure 8-29 the platysma muscle has been removed. The **external jugular vein** is one of the most superficial structures in this triangle, coursing from superior to inferior across it. The triangle is filled in with a double-layered deep fascia, which is a portion of the fascia that envelops the neck. This fascia starts posteriorly from the ligamentum nuchae, surrounds the trapezius muscle, fuses over the triangle, and separates again when the posterior border of the sternocleidomastoid muscle is reached. These fascial layers then surround this muscle, come together anteriorly and meet the fascia from the opposite side in the midline. This is known as the enveloping layer of deep fascia.

Branches of the cervical plexus destined to become cutaneous nerves penetrate this fascia at the locations shown in Figure 8-29 in order to reach their destination. All these branches are portions of the cervical plexus, being made up of ventral rami, and appear in the triangle just posterior to the sternocleidomastoid muscle. The **great auricular nerve** turns superiorly and crosses the sternocleidomastoid muscle to ramify around the parotid gland and external ear. The **lesser occipital** follows the posterior border of the sternocleidomastoid muscle to ramify on the scalp posterior to the ear sending some branches to the auricle itself. The **transverse colli nerve** courses anteriorly around the sternocleidomastoid muscle to ramify on the anterior surface of the neck as far as the midline. The **medial, intermediate, and lateral supraclavicular nerves** are cutaneous to the skin on the lower part of the neck, on the shoulder, and to an area extending a short distance inferior to the clavicle.

The **spinal accessory nerve** innervates the sternocleidomastoid muscle and the trapezius with aid from branches of C3 and C4 of the cervical plexus. This important nerve pierces the sternocleidomastoid muscle and crosses the posterior triangle of the neck in an inferior and posterior direction to reach the deep surface of the

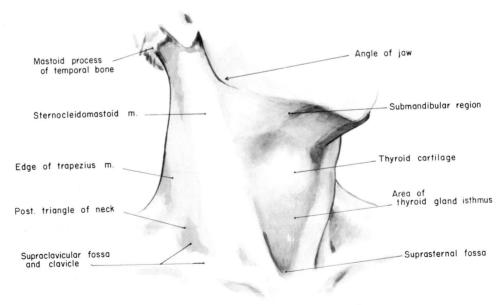

Mastoid process
of temporal bone

Sternocleidomastoid m.

Edge of trapezius m.

Post. triangle of neck

Supraclavicular fossa
and clavicle

Angle of jaw

Submandibular region

Thyroid cartilage

Area of
thyroid gland isthmus

Suprasternal fossa

Figure 8-28. *Surface anatomy of the neck.*

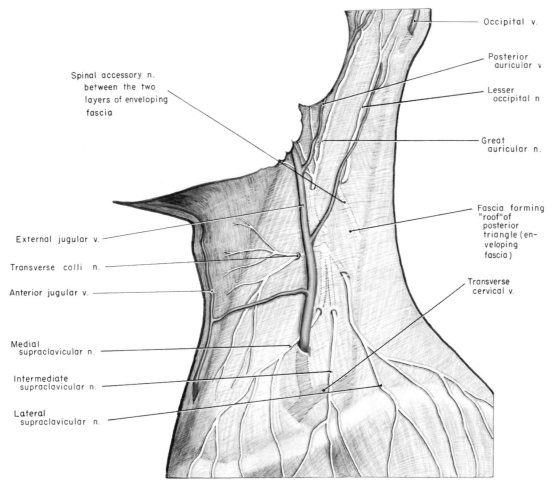

Spinal accessory n.
between the two
layers of enveloping
fascia

Occipital v.

Posterior
auricular v

Lesser
occipital n

Great
auricular n.

External jugular v.

Transverse colli n.

Anterior jugular v.

Fascia forming
"roof"of
posterior
triangle (en-
veloping
fascia)

Transverse
cervical v.

Medial
supraclavicular n.

Intermediate
supraclavicular n.

Lateral
supraclavicular n.

Figure 8-29. *Posterior triangle of the neck I. The platysma muscle and superficial fascia have been removed.*

trapezius. It is between the two layers of fascia just described and is therefore in a very superficial and vulnerable position in this triangle.*

If the enveloping layer of fascia in this posterior triangle is removed, the cutaneous nerves just described are revealed (Fig. 8-30). However, the muscles forming the so-called floor of this triangle are not revealed, because of a covering of deep fascia that spreads trans-

versely across the neck. This fascia attaches to the enveloping layer of deep fascia laterally and spreads toward the midline in the position mentioned. If this fascia is removed (Fig. 8-31), the muscles seen forming the floor of the posterior triangle from superior to inferior are the **splenius capitis, levator scapulae, scalenus posterior** (small portion only), **scalenus medius,** and **scalenus anterior.** See Figure 8-19 for another view of the splenius capitis and levator scapulae muscles.

The posterior belly of the **omohyoid muscle** courses from lateral to medial across the inferior end of the triangle, disappearing deep to the sternocleidomastoid muscle. The **transverse cervical** artery crosses the

*This position is understood easily if one realizes that the sternocleidomastoid and trapezius muscles are a single entity in some of the lower animals. Their separation in the human being has left the nerve caught between the two layers of fascia.

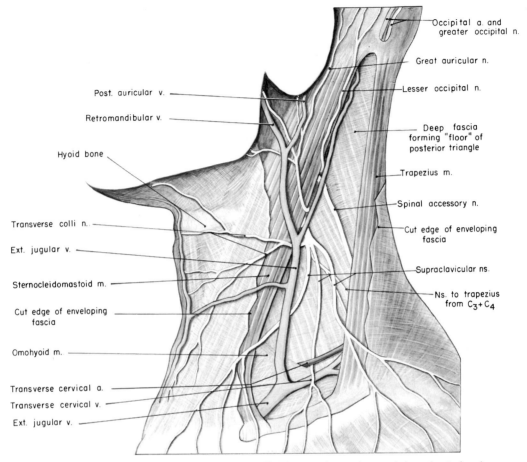

Occipital a. and greater occipital n.

Great auricular n.

Lesser occipital n.

Deep fascia forming "floor" of posterior triangle

Trapezius m.

Spinal accessory n.

Cut edge of enveloping fascia

Supraclavicular ns.

Ns. to trapezius from $C_3 + C_4$

Post. auricular v.

Retromandibular v.

Hyoid bone

Transverse colli n.

Ext. jugular v.

Sternocleidomastoid m.

Cut edge of enveloping fascia

Omohyoid m.

Transverse cervical a.

Transverse cervical v.

Ext. jugular v.

Figure 8-30. *Posterior triangle of the neck II. The enveloping layer of deep fascia has been removed without removing the spinal accessory nerve.*

triangle transversely at a level just superior to the omohyoid muscle, and another, the **suprascapular,** crosses the triangle at a level close to the clavicle.

The **phrenic nerve** (nerve supply to the diaphragm) is found on the anterior surface of the scalenus anterior muscle. A portion of the **brachial plexus** is seen between the scalenus anterior and scalenus medius muscles and the **dorsal scapular nerve** to the rhomboids can be seen emerging from the middle of the muscle belly of the scalenus medius muscle. These structures are depicted in Figure 8-32.

Therefore, from superficial to deep, the layers covering the posterior triangle of the neck are: (1) skin, (2) the superficial fascia with its contained platysma muscle and external jugular vein, (3) the enveloping deep fascia penetrated by the cutaneous branches of the cervical

plexus and containing the spinal accessory nerve, (4) deep fascia, and (5) the muscles forming the floor of the triangle. The student should be able to visualize the phrenic and dorsal scapular nerves, brachial plexus, transverse scapular and suprascapular vessels, and omohyoid muscle in their proper positions in this triangle.

STERNOCLAVICULAR JOINT

This joint (Fig. 8-33) is a double arthrodial joint consisting of the articulation between the sternal end of the clavicle

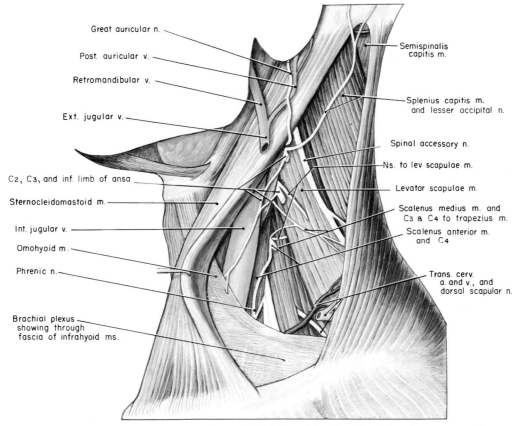

Great auricular n.

Post. auricular v.

Retromandibular v.

Ext. jugular v.

C₂, C₃, and inf. limb of ansa

Sternocleidomastoid m.

Int. jugular v.

Omohyoid m.

Phrenic n.

Brachial plexus
showing through
fascia of infrahyoid ms.

Semispinalis
capitis m.

Splenius capitis m.
and lesser occipital n.

Spinal accessory n.

Ns. to lev scapulae m.

Levator scapulae m.

Scalenus medius m. and
C₃ & C₄ to trapezius m.

Scalenus anterior m.
and C₄

Trans. cerv.
a. and v., and
dorsal scapular n.

Figure 8-31. *Posterior triangle of the neck III. Most of the deep fascia has been removed.*

and the articular disc, and between this same disc and the superior and lateral part of the manubrium and cartilage of the first rib.

The ligaments involved in this articulation are:

1. Articular capsule
2. Anterior sternoclavicular ligament
3. Posterior sternoclavicular ligament
4. Interclavicular ligament
5. Costoclavicular ligament
6. Articular disc

The **articular capsule,** as in other joints, is thick in some regions and thin in others; the thicker regions are actually the locations of the ligaments. The thick parts of the capsule are located anteriorly and posteriorly while the thin parts are located superiorly and inferiorly.

The **anterior sternoclavicular ligament** spreads from the clavicle medially and slightly inferiorly to attach to the manubrium of the sternum. This is accompanied

posteriorly by a similar ligament, the **posterior sterno-clavicular ligament.** These ligaments are quite firm and thick structures.

The **interclavicular ligament** extends from the medial end of one clavicle to the medial end of the other. It bridges across the notch of the manubrium and therefore is continuous from one clavicle to the other, just superior to the manubrium. This ligament has connections with both articular discs, and the **articular disc** stretches from these ligaments inferiorly to be attached to the cartilage of the first rib.

The **costoclavicular ligament** is a firm, short, and flat ligament that extends from the costal cartilage of the first rib superiorly and slightly laterally to the inferior surface of the medial end of the clavicle.

It can be seen from glancing at Figure 8-33 that the ligaments and articular disc are arranged in such a manner as to prevent the sternal end of the clavicle from escaping from the rather shallow fossa on the manu-

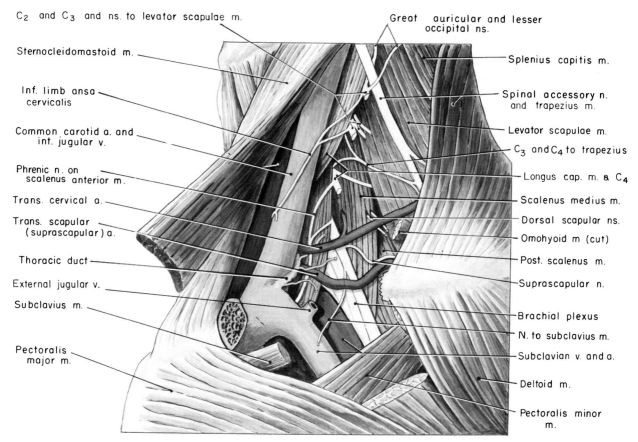

C₂ and C₃ and ns. to levator scapulae m.

Sternocleidomastoid m.

Inf. limb ansa cervicalis

Common carotid a. and int. jugular v.

Phrenic n. on scalenus anterior m.

Trans. cervical a.

Trans. scapular (suprascapular) a.

Thoracic duct

External jugular v.

Subclavius m.

Pectoralis major m.

Great auricular and lesser occipital ns.

Splenius capitis m.

Spinal accessory n. and trapezius m.

Levator scapulae m.

C₃ and C₄ to trapezius

Longus cap. m. & C₄

Scalenus medius m.

Dorsal scapular ns.

Omohyoid m (cut)

Post. scalenus m.

Suprascapular n.

Brachial plexus

N. to subclavius m.

Subclavian v. and a.

Deltoid m.

Pectoralis minor m.

Figure 8-32. *Posterior triangle of the neck IV. Removal of the clavicle has revealed the subclavian vessels and the brachial plexus entering the axilla. These are the vital structures protected by this bone.*

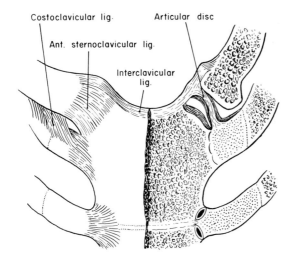

Costoclavicular lig.

Ant. sternoclavicular lig.

Articular disc

Interclavicular lig.

brium. The very fact that the articular disc is attached to the cartilage of the first rib prevents the sternal end of the clavicle from leaving the fossa and moving into the suprasternal notch. In addition, the anterior and posterior sternoclavicular ligaments prevent medial displacement of the clavicle. The angle of the costoclavicular ligament does not aid in this, although it would be very difficult, since this ligament is quite short, for the sternal end of the clavicle to travel any great distance.

The sternoclavicular joint represents the only bony attachment of the superior limb to the axial skeleton. Wide latitude of movement is allowed. Since the shoulder

Figure 8-33. *The sternoclavicular joint: right side intact; left side with a frontal cut.*

can be elevated, depressed, or moved anteriorly or posteriorly, this joint allows circumduction. It is thought that a certain amount of rotation also takes place in this joint which would make it almost a ball-and-socket type of articulation.

The **nerve supply** is from the medial supraclavicular nerve.

ACROMIOCLAVICULAR JOINT

This is a freely movable joint and is the articulation between the lateral end of the clavicle and the acromion process of the scapula (Fig. 8-34).

The ligaments associated with this joint are:

1. Articular capsule
2. Superior acromioclavicular
3. Inferior acromioclavicular
4. Articular disc
5. Coracoclavicular (accessory)
   a. Conoid
   b. Trapezoid

The **articular capsule** completely surrounds the joint and is strengthened superiorly and inferiorly by the **superior and inferior acromioclavicular ligaments.** The **articular disc** in this joint is attached to the superior part of the capsular ligament and projects into the cavity. This disc is usually incomplete and frequently is found wanting.

Although the **coracoclavicular ligaments** are not actually part of the acromioclavicular articulation, they are usually described with this joint since they serve to maintain the clavicle and the acromion process in juxtaposition, thereby supporting the scapula. Moreover, they are so situated as to prevent the scapula from moving medially, which restraint is the main function of the clavicle. The very vital structures in the axilla must be protected from pressure, and the clavicle does maintain

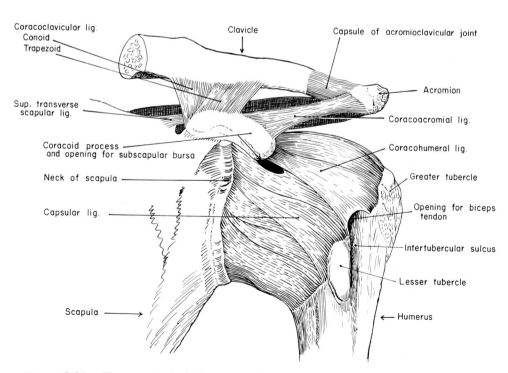

Figure 8-34. *The acromioclavicular joint and associated ligaments, and the capsule of the shoulder joint.*

the scapula in a normal lateral position. These ligaments are attached to the base of the coracoid process and to the under side or inferior surface of the clavicle. The lateral one is **trapezoid** in shape while the medial one is **conoid** in appearance, and these ligaments are often called by these names.

The movements allowed in this joint are of the gliding variety. As the scapula is moved anteriorly and posteriorly, a certain amount of rotation also occurs.

The **nerve supply** is by branches of the pectoral, suprascapular, and axillary nerves.

AXILLA AND POSTERIOR SHOULDER

GENERAL DESCRIPTION

The **axilla** or armpit is a dome-shaped area (covered with hair after puberty) between the arm and the lateral chest wall. If the arm is abducted as shown in Figure 8-35, the folds or walls of the axilla are visible. Its **anterior wall** is made up of the pectoralis major and minor muscles plus the clavipectoral fascia; the **posterior wall,** by the subscapularis, teres major, and latissimus dorsi muscles; its **medial wall,** by the upper four or five ribs and the intercostal spaces covered by the serratus anterior muscle; the **lateral wall,** by the coracobrachialis muscle and the humerus. These walls are depicted in cross section in Figure 8-36. The **base** of the axilla is made up of the axillary fascia, while the **apex** extends superiorly into the posterior triangle of the neck. The area through which the apex reaches the neck is called the **cervicoaxillary canal;** it is bounded anteriorly by the clavicle, posteriorly by the scapula, and medially by the first rib. A great majority of the structures coursing between neck and axilla or the reverse do so via this canal, the clavicle holding the shoulder in a lateral position to allow this. Infections can follow this same pathway.

When the **pectoralis major** muscle is removed and the fascia and fat surrounding the axillary structures is removed as well, the **pectoralis minor** muscle can be seen to cross at right angles a large vein, a large artery, and parts of the brachial plexus (Fig. 8-27). The vein is the axillary vein and is most medially placed; the artery is the axillary artery and is found completely surrounded by the cords of the brachial plexus. These three vital structures are close to the anterior wall proximally but, as the structures course distally, they become associated with the posterior wall of the axilla. For descriptive purposes, the pectoralis minor muscle may be said to divide the **axillary artery** into three parts. The **first part** extends from the lateral aspect of the first rib, where the artery is continuous with the subclavian, to the superior border of the pectoralis minor muscle; the **second part** of the artery is posterior to the pectoralis minor muscle; and the **third part** extends from the inferior border of the pectoralis minor muscle to the inferior border of the teres major tendon. At this point the artery becomes the brachial.

The **branches of the axillary artery** (Fig. 8-37) are visible at this time: the supreme thoracic arising from the first part of the axillary artery, the thoracoacromial and lateral thoracic arteries from the second part, and the subscapular and anterior and posterior humeral circumflex branches from the third part. The **supreme thoracic** courses to the first and second intercostal spaces, and the **thoracoacromial** divides into deltoid, acromial, clavicular, and pectoral branches. The **lateral thoracic** courses on the lateral thoracic wall, on the superficial surface of the serratus anterior muscle. The **subscapular** courses on the subscapularis muscle and divides into the **circumflex scapular artery,** which winds around the subscapularis muscle and disappears from view, and the **thoracodorsal artery,** which continues onto the latissimus dorsi muscle. Only the origins of the **anterior and posterior humeral circumflex** arteries are visible at this time.

The more medially placed **axillary vein** has tributaries that correspond to the arteries just mentioned. They are usually in the form of venae comitantes. In addition, the **cephalic vein,** which comes from the arm, courses between the pectoralis major and deltoid muscles and joins the axillary vein after piercing the clavipectoral fascia.

The axillary artery is completely surrounded by vari-

Figure 8-35. Surface anatomy of the axilla and shoulder. (From Appleton, Hamilton, and Tchaperoff, Surface and Radiological Anatomy, courtesy of The Macmillan Press Ltd.)

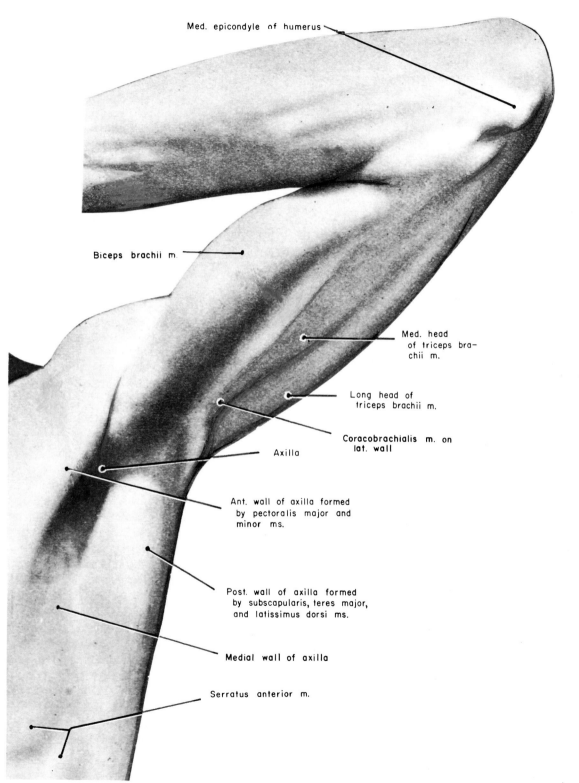

Med. epicondyle of humerus

Biceps brachii m.

Med. head
of triceps bra-
chii m.

Long head of
triceps brachii m.

Coracobrachialis m. on
lat. wall

Axilla

Ant. wall of axilla formed
by pectoralis major and
minor ms.

Post. wall of axilla formed
by subscapularis, teres major,
and latissimus dorsi ms.

Medial wall of axilla

Serratus anterior m.

ous parts of the brachial plexus (Fig. 8-37). In fact, the **cords** of the brachial plexus—the medial, lateral, and posterior—receive their names because of their relation to the artery. The most prominent nerves are the medial and lateral roots of the **median nerve,** which are on either side of the axillary artery; they combine to form the median nerve, which is anterior to the artery. **The medial antebrachial cutaneous nerve** can be seen between the axillary vein and artery and the **medial brachial cutaneous nerve** courses on the medial surface of the axillary vein. The **musculocutaneous** and **ulnar nerves** are not noticeable unless the artery is pulled to one side. The same can be said for the entire posterior cord and its terminal branches, the **radial** and **axillary nerves.** Other nerves, however, can be seen. The **long thoracic nerve** to the serratus anterior courses on the superficial surface of the serratus anterior muscle. In addition, the **thoracodorsal nerve** to the latissimus dorsi is evident, and the two nerves to the subscapularis muscle—the upper and lower subscapulars—are also visible on the surface of that muscle. Both the **medial and lateral pectoral nerves** are evident going to the pectoralis minor and major, and one of the smaller branches of the plexus—the **branch to the subclavius muscle**—can also be seen coursing to this muscle.

If the trapezius and deltoid muscles are removed, the important muscles located on the back of the shoulder and attached to the posterior surface of the scapula can be seen (Fig. 8-38). The **supraspinatus fossa** contains the **supraspinatus muscle,** while the **infraspinatus muscle** occupies a large part of the **infraspinatus fossa.** Attached to the inferior border of the scapula and often inseparable from the infraspinatus muscle is the **teres minor muscle** and, inferior to that, the **teres major muscle.** The **long head of the triceps brachii** muscle separates the teres minor and teres major from each other, and with these muscles forms two spaces. The **quadrilateral** space is located between the teres minor muscle superiorly, the humerus laterally, the teres major muscle inferiorly, and the long head of the triceps medially. The axillary nerve and the posterior humeral circumflex artery course through this quadrilateral space. The other space formed by the long head of the triceps is the **triangular space,** which is bounded superiorly by the teres minor muscle, laterally by the long head of the triceps, and inferiorly by the teres major muscle. Although a branch of the circumflex scapular artery often traverses this trian-

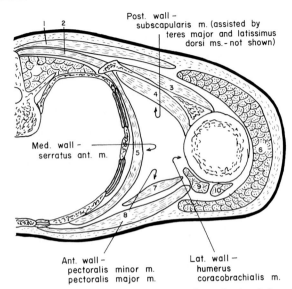

Figure 8-36. *A transverse section through the left axilla, showing the makeup of its walls: (1) trapezius muscle; (2) rhomboid muscles; (3) infraspinatus muscle; (4) subscapularis muscle; (5) serratus anterior muscle; (6) deltoid muscle; (7) pectoralis minor muscle; (8) pectoralis major muscle; (9) coracobrachialis muscle; (10) biceps brachii muscle. (Modified after Buchanan.)*

gular space, the circumflex scapular artery itself does not; it can be seen through the space, however.

If the supraspinatus and infraspinatus muscles are cut (Fig. 8-38), the **suprascapular nerve** and **artery** can be followed into the supraspinatus fossa, around the spine of the scapula, and into the infraspinatus fossa. The suprascapular artery anastomoses with the circumflex scapular branch of the subscapular artery, this anastomosis occurring in the infraspinatus fossa. This is an important collateral pathway around an occlusion of the axillary artery. The suprascapular nerve innervates both the supraspinatus and infraspinatus muscles.

The suprascapular, circumflex scapular, and posterior humeral circumflex arteries are accompanied by corresponding veins.

MUSCLES

The muscles involved in the axilla and shoulder are:

1. Pectoralis major
2. Pectoralis minor

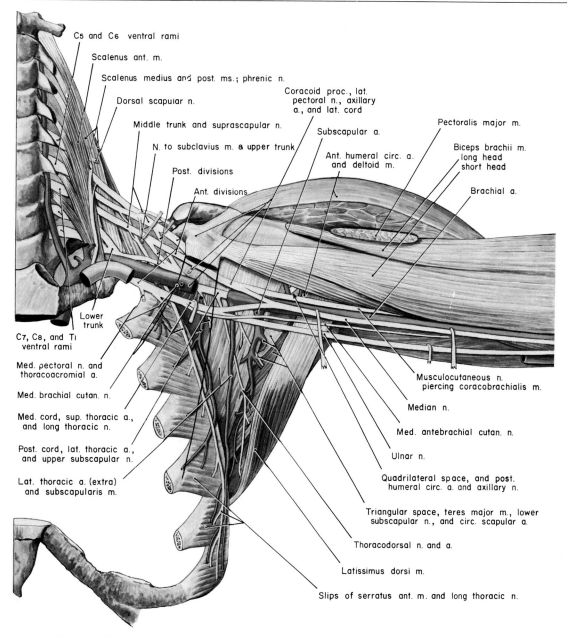

C5 and C6 ventral rami

Scalenus ant. m.

Scalenus medius and post. ms.; phrenic n.

Dorsal scapular n.

Middle trunk and suprascapular n.

N. to subclavius m. & upper trunk

Post. divisions

Ant. divisions

Coracoid proc., lat. pectoral n., axillary a., and lat. cord

Subscapular a.

Ant. humeral circ. a. and deltoid m.

Pectoralis major m.

Biceps brachii m.
long head
short head

Brachial a.

Lower trunk

C7, C8, and T1 ventral rami

Med. pectoral n. and thoracoacromial a.

Med. brachial cutan. n.

Med. cord, sup. thoracic a., and long thoracic n.

Post. cord, lat. thoracic a., and upper subscapular n.

Lat. thoracic a. (extra) and subscapularis m.

Musculocutaneous n. piercing coracobrachialis m.

Median n.

Med. antebrachial cutan. n.

Ulnar n.

Quadrilateral space, and post. humeral circ. a. and axillary n.

Triangular space, teres major m., lower subscapular n., and circ. scapular a.

Thoracodorsal n. and a.

Latissimus dorsi m.

Slips of serratus ant. m. and long thoracic n.

Figure 8-37. *Axilla after removal of the pectoralis minor muscle (outlined). The upper parts of the brachial plexus have been pulled superolaterally and the median and ulnar nerves pulled medially for the sake of clarity. The axillary vein lies just medial to the artery, with the medial cord of the brachial plexus and the ulnar nerve lying between the two vessels.*

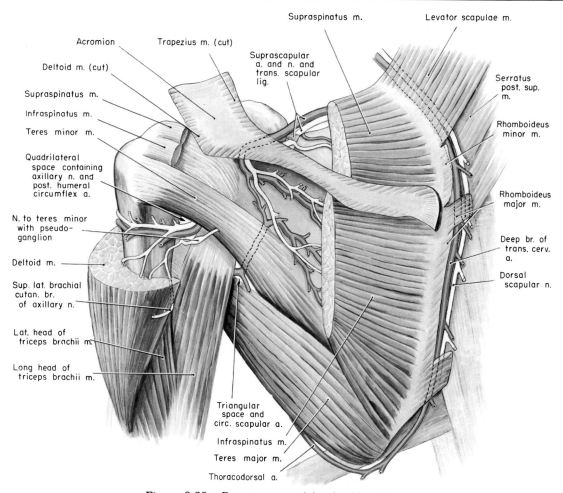

Figure 8-38. *Posterior view of the shoulder—left side.*

3. Deltoideus
4. Serratus anterior
5. Subscapularis
6. Supraspinatus
7. Infraspinatus
8. Teres minor
9. Teres major
10. Latissimus dorsi

Since most of these muscles have an action on the shoulder joint, it might be recalled that this joint is a ball-and-socket joint allowing all types of movement. It has three axes around which these actions take place: transverse, vertical, and anteroposterior (Fig. 8-12).

1 and 2. The pectoralis major and minor have just been described (pages 690 and 691).

3. Deltoid muscle (Figs. 8-13 and 8-38). This muscle forms the rounded prominence of the shoulder. It **arises** from the lateral third of the clavicle, the acromion, and the spine of the scapula (Figs. 8-4 and 8-5). It should be noted that these origins are directly in line with the insertions of the trapezius muscle. The deltoid **inserts** on the deltoid tuberosity of the humerus, about a third of the way down on the lateral side of the shaft (Figs. 8-6 and 8-7). Since its direction of pull is superior to the anteroposterior axis, its **main action** is to abduct the arm to the horizontal plane. Further abduction involves rotation of the scapula, which is accomplished by the serratus anterior and trapezius muscles. All parts of the deltoid muscle need not work at the same time. If only the anterior fibers are used, the humerus is flexed and medially rotated (pull is anterior to transverse and vertical axes); if the posterior fibers are used only, the humerus is extended and laterally rotated (pull is posterior to transverse and vertical axes). **Nerve supply** is the axillary.

4.   **Serratus anterior** (Fig. 8-39).   This important muscle forms the medial wall of the axilla and is located on the lateral chest wall. Its **origin** is anteriorly placed and consists of eight digitations from the upper eight ribs (Fig. 8-21) and fascia covering the intercostal muscles. It hugs the lateral chest wall and **inserts** onto the superior angle, the medial margin, and the inferior angle of the scapula on its ventral or anterior surface (Fig. 8-5). Its **main action** is to pull the scapula away from the midline in such actions as pushing, and it rotates the scapula so that the inferior angle is moved laterally. This is particularly important when the arm is abducted because this movement beyond the horizontal cannot occur without a rotation of the scapula.* **Nerve supply** is by the long thoracic.

5.   The **subscapularis** muscle (Fig. 8-36) lies in the subscapular fossa on the deep or anterior side of the scapula. Its **origin** is from the subscapular fossa (Fig. 8-5) and its **insertion** is onto the lesser tubercle of the humerus (Fig. 8-6). Its direction of pull lies inferior to the anterior-posterior axis and anterior to the vertical axis; thus, its **main action** is to adduct the arm and rotate it medially. Its **nerve supply** is by the upper and lower subscapular nerves from the posterior cord of the brachial plexus.

6.   **Supraspinatus** (Fig. 8-38).   This muscle lies in the supraspinatus fossa and **arises** from its walls (Fig. 8-5). It **inserts** into the superior facet of the greater tubercle of the humerus (Fig. 8-6). Its direction of pull is superior to the anteroposterior axis, so its **main action** is to abduct the humerus, many claiming it is responsible for initiating this action. It also plays an important role in holding the head of the humerus in the glenoid cavity of the scapula. **Nerve supply** is by the suprascapular.

---

*Recall that the trapezius muscle aids in this action.

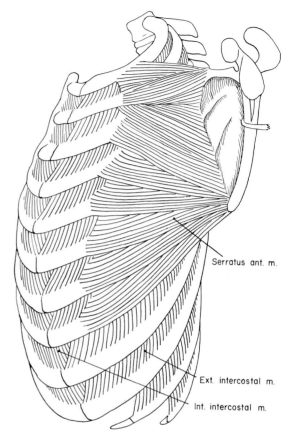

Figure 8-39.   *The serratus anterior muscle. The scapula has been pulled laterally to show the insertion of this muscle.*

7.   **Infraspinatus** (Fig. 8-38).   Occupying the infraspinatus fossa and **arising** from its walls (Fig. 8-5), this muscle **inserts** onto the middle facet of the greater tubercle of the humerus (Fig. 8-6 and 8-7). Its **main action** is to rotate the arm laterally, since its direction of pull passes posterior to the vertical axis. **Nerve supply** is the suprascapular.

8.   The **teres minor muscle** (Fig. 8-38) lies along the inferior border of the scapula intimately associated with the infraspinatus muscle. It **arises** from the inferior or lateral margin of the scapula (Fig. 8-5) and **inserts** onto the most inferior facet on the greater tubercle of the humerus (Fig. 8-7). Its **main action** is to rotate the arm laterally, since it passes posterior to the vertical axis. **Nerve supply** is the axillary.

9.   **Teres major** (Fig. 8-38).   This muscle lies just

inferior to the scapula and forms part of the posterior wall of the axilla. It **arises** from the dorsal surface of the scapula near the inferior angle (Fig. 8-5) and **inserts** onto the medial lip of the intertubercular groove of the humerus (Fig. 8-6). Its **main action** is to extend and adduct the arm and rotate it medially. To accomplish these actions its direction of pull must pass posterior to the transverse and medial to the anteroposterior and vertical axes. The **nerve supply** is the lower subscapular.

10.   Latissimus dorsi (Figs. 8-15 and 8-37).   This broad sheetlike muscle is the most superficial muscle in the lower part of the back. It was described on page 683.

Musculotendinous or rotator cuff.   The tendons of the subscapularis, supraspinatus, infraspinatus, and teres minor muscles strengthen the capsule of the shoulder joint and hold the head of the humerus into the glenoid cavity. They prevent displacement anteriorly, superiorly, and posteriorly. Since at least three of these muscles are involved in rotation of the humerus, this musculotendinous cuff around the head of the humerus is often called a **rotator cuff.**

Bursae.   There are several bursae associated with the muscles around the shoulder joint. They become inflamed frequently and in this state induce considerable pain. A subscapular bursa occurs deep to this muscle. This bursa communicates with the shoulder joint. Another occurs between the infraspinatus muscle and the joint capsule. The subacromial bursa lies deep to the deltoid muscle and the acromion. Another occurs superficial to the acromion. Others are associated with the tendons of the coracobrachialis, teres major, long head of the triceps, and with the latissimus dorsi muscles. Some of these bursae are pictured in Figure 8-75 found on page 755.

NERVES

The nerves involved in the axilla and shoulder consist of the brachial plexus and its branches. The **brachial plexus** is made up of ventral rami from C5, C6, C7, and C8 and of the first thoracic nerves. It is located partially in the neck—in the posterior triangle between the scalenus anterior and scalenus medius muscles—and continues deep

to the clavicle and superficial to the first rib to enter the axilla. Because of this relation to the clavicle, the plexus is often divided into **supraclavicular** and **infraclavicular** portions. It completely surrounds the axillary artery and is lateral to the axillary vein except for the medial brachial cutaneous nerve, which gains the medial surface of this vein.

Although many variations occur, the basic structure of the brachial plexus is as follows (Fig. 8-40): the ventral rami of C5 and C6 join to form the upper trunk; the ventral ramus of C7 continues alone to form the middle trunk; and the ventral rami of C8 and T1 join to form the lower trunk. Each of these trunks divides into anterior and posterior divisions. All three posterior divisions join to form the posterior cord; the anterior divisions from the upper and middle trunks join to form the lateral cord; and the anterior division of the lower trunk forms the medial cord. These cords then divide into their terminal branches, the terminal branches of the lateral cord being the lateral root to the median nerve and the musculocutaneous, those of the medial cord the medial root to the median nerve and the ulnar, and those of the posterior cord the radial and axillary nerves. The sequence, therefore, is rami, trunks, divisions, cords, and terminal branches.

BRANCHES FROM RAMI AND FROM TRUNKS. These nerves are:

1. Root of the phrenic from the fifth cervical
2. Nerve to the subclavius muscle
3. Dorsal scapular nerve
4. Long thoracic nerve
5. Suprascapular nerve
6. Nerves to prevertebral muscles

1.   The small **root to the phrenic** nerve is simply a contribution from the fifth cervical to the phrenic which arises mainly from the fourth cervical segment of the spinal cord. This nerve continues inferiorly on the anterior surface of the scalenus anterior muscle, enters the thoracic cavity, traverses this cavity, and finally innervates the diaphragm.

2.   The **nerve to the subclavius** muscle (Figs. 8-37 and 8-40) is one of the two nerves that are actually anterior or superficial to the plexus itself. This nerve arises from the fifth and sixth ventral rami or from the upper trunk and courses anterior to the plexus and the subcla-

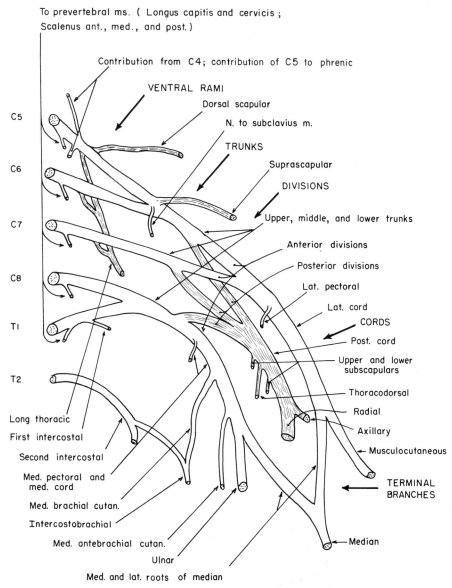

To prevertebral ms. (Longus capitis and cervicis ;
Scalenus ant., med., and post.)

Contribution from C4; contribution of C5 to phrenic

VENTRAL RAMI

Dorsal scapular

N. to subclavius m.

TRUNKS

Suprascapular

DIVISIONS

Upper, middle, and lower trunks

Anterior divisions

Posterior divisions

Lat. pectoral

Lat. cord

CORDS

Post. cord

Upper and lower
subscapulars

Thoracodorsal

Radial

Axillary

Musculocutaneous

TERMINAL
BRANCHES

Median

C5

C6

C7

C8

T1

T2

Long thoracic

First intercostal

Second intercostal

Med. pectoral and
med. cord

Med. brachial cutan.

Intercostobrachial

Med. antebrachial cutan.

Ulnar

Med. and lat. roots of median

Figure 8-40. *Diagram of the brachial plexus—left side. The clear nerves are preaxial; shaded, postaxial.*

vian vessels to end in the subclavius muscle. Since it arises from the upper trunk, it naturally contains fibers from the fifth and sixth cervical segments.

3. Dorsal scapular (Figs. 8-37 and 8-40). This nerve arises from the ventral ramus of the fifth cervical nerve, actually courses through the scalenus medius

muscle, and descends on the deep surface of the scapula and the rhomboid muscles, giving nerve supply to the levator scapulae muscle as well as to the rhomboideus major and minor.

4. The **long thoracic nerve** (Figs. 8-37 and 8-40) emerges from the ventral rami of the fifth, sixth, and seventh cervical nerves. The branches from the fifth and sixth rami pierce the scalenus medius muscle and join the branch from the seventh to form a single nerve. This nerve then descends into the axilla and courses on the superficial surface of the serratus anterior muscle which it innervates.

5. Suprascapular (Figs. 8-37 and 8-38). This nerve arises from the upper trunk (C5 and C6) and courses laterally and posteriorly across the scalenus medius muscle. It joins the suprascapular artery and enters the supraspinatus fossa, coursing below the transverse scapular ligament in doing so. It innervates the supraspinatus muscle, curls around the spinoglenoid notch to enter the infraspinatus fossa. It innervates the infraspinatus muscle and gives a sensory branch to the shoulder joint.

The remaining branches of the brachial plexus are from the cords.

BRANCHES FROM LATERAL CORD. These nerves are:

1. Lateral pectoral
2. Musculocutaneous
3. Lateral root of the median nerve

1. Lateral pectoral (Figs. 8-37 and 8-40). This nerve courses anterior to the brachial plexus and reaches the clavipectoral fascia, which it pierces to enter the pectoralis major muscle. It contributes to the medial pectoral nerve before piercing the fascia. Since the lateral cord is derived from C5, C6, and C7, it is not surprising to find this nerve containing fibers from these same segments.

2. Musculocutaneous (Figs. 8-37 and 8-40). This nerve (C5, C6, C7) is one of the terminal branches of the lateral cord and it courses between the axillary artery and the coracobrachialis muscle. It then pierces the coracobrachialis muscle and obtains a position between that muscle and the biceps brachii. This nerve, after innervating the coracobrachialis, biceps brachii, and a part of the brachialis muscle, continues into the forearm to become a cutaneous nerve.

3. Lateral root of the median. This is the other terminal branch of the lateral cord. This root joins with the median root from the medial cord to form the median nerve (Fig. 8-40) which continues into the upper limb.

BRANCHES FROM MEDIAL CORD. These nerves are:

1. Medial pectoral
2. Medial brachial cutaneous
3. Medial antebrachial cutaneous
4. Ulnar
5. Medial root of the median

1. Medial pectoral (Figs. 8-37 and 8-40). This nerve springs from the medial cord and, after picking up a contribution from the lateral pectoral nerve, courses between the axillary artery and vein, pierces the pectoralis minor muscle, and ends in the pectoralis major supplying both muscles. Since the medial cord is derived from the eighth cervical and first thoracic segments, it is not surprising that this nerve contains contributions from those two segments.

2. The **medial brachial cutaneous nerve** (Figs. 8-37 and 8-40), from the first thoracic segment, courses posterior to the axillary vein and then on its medial side. After piercing the deep fascia, it supplies skin on the medial side of the arm. This nerve frequently joins the intercostobrachial nerves, which will be described at the end of this section.

3. Medial antebrachial cutaneous (Figs. 8-37 and 8-40). This nerve from the eighth cervical and first thoracic segments is given off from the medial cord and then courses inferiorly on the anterior surface of the axillary artery between it and the axillary vein. It then courses on the medial side of the brachial artery and finally becomes a cutaneous nerve for the forearm.

4. Ulnar nerve (Figs. 8-37 and 8-40). This is one of the terminal branches of the medial cord and it contains fibers from the eighth cervical and first thoracic segments. It courses between the axillary artery and vein and descends along the medial side of the brachial artery. This will be described later.

5. The **medial root of the median** simply joins the lateral root of the median to form the median nerve (Fig. 8-40).

BRANCHES FROM POSTERIOR CORD. These nerves are:

1. Upper subscapular
2. Thoracodorsal
3. Lower subscapular
4. Axillary
5. Radial

1. Upper subscapular (Figs. 8-37 and 8-40). This nerve arises from the fifth and sixth cervical segments, descends on the anterior surface of the sub-scapularis muscle, and sinks into its substance to give it innervation. It also sends some filaments to the shoulder joint.

2. The **thoracodorsal nerve** (Figs. 8-37 and 8-40) arises from cervical segments 6, 7, and 8 and runs inferiorly and posteriorly to sink into the latissimus dorsi muscle, which it innervates.

3. Lower subscapular (Figs. 8-37 and 8-40). This arises from the fifth and sixth cervical segments, follows the anterior surface of the subscapularis muscle, sinking into its substance, and then ends in the teres major muscle, which it also innervates.

4. The **axillary nerve** (Figs. 8-38 and 8-40) is one of the terminal branches of the posterior cord. This nerve courses inferiorly between the axillary artery and the subscapularis muscle and disappears into the quadrilateral space. After giving off a branch to the shoulder joint it splits into two branches; one of these, the **posterior branch,** innervates the teres minor muscle and the deltoid muscle and then winds around the posterior border of the deltoid muscle to become the **upper lateral cutaneous nerve of the arm.** The branch to the teres minor muscle possesses a small swelling that has the appearance of a ganglion.* The **anterior branch** follows the posterior humeral circumflex artery and winds around the surgical neck of the humerus deep to the deltoid muscle. This branch is mainly a motor branch to the deltoid muscle, although some branches do pierce the muscle to become cutaneous.

5. Radial nerve. This nerve, containing fibers from spinal cord segments C5, C6, C7, C8 and sometimes the first thoracic, is the other terminal nerve of the posterior cord and will be described subsequently.

It is interesting to note that, except for the teres major and teres minor muscles, all muscles in this region have direct branches coursing to them. The teres major and the teres minor muscles borrow branches from other nerves, the teres minor a branch from the axillary and the teres major from the lower subscapular. Actually the subscapularis muscle, both teres muscles, and the deltoid muscle are all one muscle mass in some lower animal forms, which explains this arrangement in the human.

*G. Gitlin, 1957, Concerning the gangliform enlargement (pseudo-ganglion) on the nerve to the teres minor muscle, *J. Anat.* 91: 466. This structure is simply a condensation of connective tissue, according to this author.

INTERCOSTOBRACHIAL NERVES (Fig. 8-40). The description of the nerves in the axilla would not be complete without mentioning that the lateral cutaneous branches of the second and often the third intercostal nerves, instead of remaining on the thoracic wall, extend into the superior limb. These nerves are called the **intercostobrachial nerves** and give sensory nerve supply to the medial surface of the arm. They course through the fat in the axilla to reach their destination and are frequently joined by the medial brachial cutaneous nerve.

ARTERIES

The arteries involved in the axilla and shoulder are the **axillary** and **suprascapular** and their branches.

AXILLARY. The **axillary** artery is a continuation of the subclavian artery. It begins at the outer border of the first rib and continues through the axilla until the inferior border of the teres major tendon is reached. At this point it becomes the brachial artery. The artery is divided into three parts, the first part being between the subclavian artery and the superior border of the pectoralis minor muscle, the second part being deep to the pectoralis minor, and the third part between the inferior border of the pectoralis minor and the brachial artery.

Relations. If you know the walls of the axilla and the relations of the axillary artery (omitting nerves), you will have a good conception of this area. Then the nerves can be added. Anterior: from proximal to distal, the cephalic vein, clavipectoral fascia, pectoralis major muscle, pectoralis minor (second part only), and then pectoralis major muscle again. Distal to the inferior edge of the latter muscle, the artery is covered with skin and fascia only. Posterior: serratus anterior muscle and intercostal space; subscapularis, latissimus dorsi, and teres major muscles. Lateral: mostly the coracobrachialis muscle, the long head of the biceps, and the humerus. Medial: the axillary vein, fascia, and skin.

Relating the nerves to the above, the cords are medial, lateral, and posterior to the artery and thence the names. The terminal branches of the posterior cord—radial and axillary—remain posterior, the radial nerve gaining a more lateral position distally. The ulnar, the medial cutaneous nerves of the forearm and arm, and the medial head of the median are medial to the artery, the medial cutaneous nerve of the arm gaining an intimate contact with the axillary vein. The musculocutaneous nerve is lateral, and the lateral head of the median nerve courses from lateral to anterior to join the medial head to form the median nerve, which courses on the anterior surface of the third part of the artery.

The **branches** of the axillary artery can be organized according to their origins, as follows:

From the first part:
1. Supreme thoracic
From the second part:
2. Thoracoacromial
3. Lateral thoracic
From the third part:
4. Subscapular
5. Anterior humeral circumflex
6. Posterior humeral circumflex

1. Supreme thoracic (Fig. 8-37). The supreme thoracic artery is small and ramifies over the first intercostal muscle. It is frequently found wanting.

2. The **thoracoacromial** artery (Figs. 8-25, 8-27, and 8-37) courses anteriorly and pierces the clavipectoral fascia. It then divides into its four branches. (a) The **pectoral branch** descends between the pectoralis major and minor muscles. It is distributed to these muscles and to the mammary gland. (b) The **acromial branch** courses laterally deep to the deltoid muscle and finally reaches the acromion. It gives blood supply to the deltoid muscle before breaking up into many branches on the acromion process. (c) The **clavicular branch** courses superiorly and medially to the sternoclavicular joint, supplying this joint and the subclavius muscle. (d) The **deltoid branch,** often arising in common with the acromial branch, crosses the pectoralis minor and accompanies the cephalic vein in the groove between the pectoralis major and deltoid muscles. It gives blood supply to both these muscles.

3. Lateral thoracic (Fig. 8-37). This artery courses inferiorly and medially along the inferior border of the pectoralis minor muscle to the wall of the thorax. It supplies the serratus anterior muscle and the pectoralis minor muscle, and also gives blood supply to the mammary gland. There is often more than one such artery.

4. Subscapular (Fig. 8-37). This large vessel arises from the axillary and courses inferiorly on the subscapularis muscle. It divides into the **circumflex scapular** artery and the **thoracodorsal** artery. The thoracodorsal

artery continues on the latissimus dorsi while the circumflex scapular artery winds around the lateral margin of the scapula to enter the infraspinatus fossa deep to the infraspinatus muscle (Fig. 8-38). Here the branches anastomose with the suprascapular artery and the deep branch of the transverse cervical artery. It can readily be seen that this anastomosis among these three vessels in the infraspinatus fossa will provide a means of getting around a tie in the second or third parts of the subclavian artery or the first and second parts of the axillary artery (Fig. 8-41).

5. The **anterior humeral circumflex artery** (Fig. 8-41) is quite small and courses laterally in a plane anterior to the surgical neck of the humerus. It sends branches superiorly and inferiorly in the intertubercular groove. This artery often arises in common with the posterior humeral circumflex. Although small, this artery is considered to be the main source of blood supply to the superior end of the humerus.

6. Posterior humeral circumflex (Fig. 8-38). Larger than the anterior humeral circumflex, this artery may arise from the subscapular artery or may be found wanting, being replaced by the deltoid branch of the profunda brachii (see page 720). It passes posteriorly with the axillary nerve through the quadrilateral space. It winds around the surgical neck of the humerus deep to the deltoid muscle. It sends branches to the adjoining muscles and an important anastomotic branch with the profunda branch of the brachial artery.

SUPRASCAPULAR. The **suprascapular artery** is very conspicuous in the shoulder region but is not a branch of the axillary. Arising from the thyrocervical trunk, this artery courses laterally and posteriorly across the root of the neck (Fig. 8-31). In doing so, it is deep to the sternocleidomastoid muscle, clavicle, and a portion of the trapezius muscle, and anterior to the scalenus anterior muscle, phrenic nerve on the anterior surface of that muscle, third part of the subclavian artery, and the brachial plexus. It reaches the supraspinatus fossa by crossing the transverse scapular ligament (Fig. 8-38). After giving branches to the supraspinatus muscle, it passes into the infraspinatus fossa through the spinoglenoid notch. In the infraspinatus fossa, it gives blood supply to the infraspinatus muscle and has the rich anastomosis already mentioned with the circumflex scapular branch of the subscapular artery and the deep branch of the transverse cervical artery.

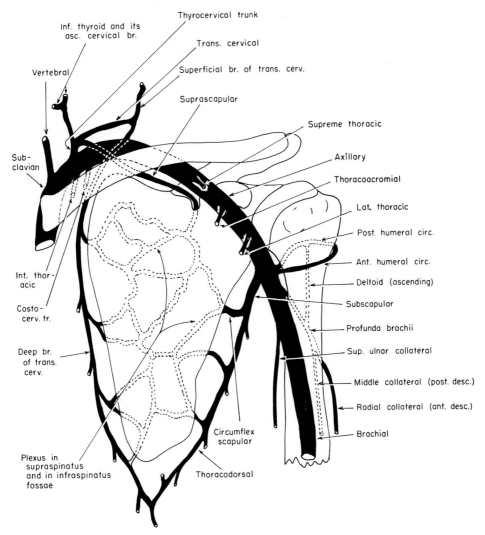

Inf. thyroid and its
asc. cervical br.

Thyrocervical trunk

Trans. cervical

Vertebral

Superficial br. of trans. cerv.

Suprascapular

Supreme thoracic

Sub-
clavian

Axillary

Thoracoacromial

Lat. thoracic

Post. humeral circ.

Ant. humeral circ.

Int. thor-
acic

Deltoid (ascending)

Subscapular

Costo-
cerv. tr.

Profunda brachii

Sup. ulnar collateral

Deep br.
of trans.
cerv.

Middle collateral (post. desc.)

Radial collateral (ant. desc.)

Brachial

Circumflex
scapular

Plexus in
supraspinatus
and in infraspinatus
fossae

Thoracodorsal

Figure 8-41. *Diagram of the anastomosis of arteries around shoulder joint.*

VEINS

The veins correspond to the arteries with the exception that the cephalic vein has no counterpart among the arteries. The veins are usually in the form of venae comitantes.

LYMPH NODES

The axillary nodes, some twenty to thirty in number, are extremely important, since they drain the upper limb and thoracic walls including the breast. The nodes are divided into groups, and it is important to note that every group is connected with every other group. For example, although a statement may be made to the effect that the breast drains into certain groups, it should be pointed out that, in radical surgery for breast cancer, all nodes are removed from the axilla.

The groups of nodes are:

1. Lateral group. These nodes are named for their relation to the axillary vein and are located along the lateral wall of the axilla. They drain the entire upper limb.

2. Pectoral group. These nodes lie along the inferior border of the pectoralis minor muscle and in rela-

[handwritten margin note: ←mmn: closest place for i/cb nerves to go]

tion to the lateral thoracic artery. They drain the anterior and lateral thoracic walls and also the mammary gland.

3. Subscapular group. This group of nodes is located on the subscapularis muscle and in relation to the subscapular artery. They drain the lower part of the back of the neck and the posterior aspect of the thoracic wall.

4. Central group. This group consists of four or five large nodes located in the fat near the base of the axilla. All the other nodes seem to drain into this set.

5. Apical group. These nodes are situated posterior to the pectoralis minor and superior to the border of this muscle; in other words, they are located in the apex of the axilla. They drain the mammary gland but in addition receive connections from all of the other axillary lymph nodes. Efferent vessels from this group unite to form the subclavian trunk, which opens either directly into the junction of the internal jugular and subclavian veins or into the jugular lymphatic trunk. On the left side, it may end in the thoracic duct directly.

CUTANEOUS NERVES AND VEINS

As the skin of the entire upper limb is usually removed at one time, it is perhaps best to describe the cutaneous nerves and veins of the entire limb rather than concentrating on its various regions.

NERVES

Although there is considerable variation in the areas covered by these cutaneous nerves, Figure 8-42 shows their usual distribution on the anterior and posterior sides of the upper limb; the areas innervated are outlined by dotted lines.

From these illustrations it can be seen that the anterior aspect of the shoulder derives its cutaneous nerve supply from the **supraclavicular branches** of the cervical plexus. The tip of the shoulder also receives its cutaneous supply from these nerves. The posterior aspect of the shoulder has contributions from the supraclavicular nerves but they are aided by branches of the **dorsal rami of the upper thoracic** nerves.

The medial side of the arm, from the axilla inferiorly, receives its nerve supply from the **intercostobrachial nerves,** the anterior surface by branches of the **medial brachial cutaneous nerve.** This nerve aids in innervating the medial surface of the arm as well. On the posterior side of the arm, the area just inferior to the tip of the shoulder derives its cutaneous nerve supply from the **superior lateral brachial cutaneous,** a branch of the axillary nerve. The remaining part of the posterior aspect of the arm derives its nerve supply from the **inferior lateral brachial cutaneous** and **posterior brachial cutaneous** nerves, which are branches of the radial.

Before leaving the arm, the student should note that cutaneous branches supplying the forearm pierce the deep fascia while still in the arm. On the anterior side, following the basilic vein, is the **medial antebrachial cutaneous** nerve, derived from the medial cord of the brachial plexus. Just superior to the elbow, this nerve divides into anterior and posterior branches, which course on the anterior and posterior surfaces of the forearm, respectively. The lateral surface of the forearm is innervated by the cutaneous portion of the **musculocutaneous** nerve. This nerve pierces the deep fascia just anterior to the elbow joint and also has anterior and posterior branches. The posterior surface of the forearm is innervated additionally by the **posterior antebrachial cutaneous** nerve, a branch of the radial that is found piercing the deep fascia about halfway down the arm (Fig. 8-42).

The cutaneous nerve supply of the hand is by branches of the median, ulnar, and radial nerves. The **median** nerve provides cutaneous nerve supply to the anterior surfaces of the thumb (with help from the radial nerve), the palm of the hand (palmar cutaneous branch), and the anterior surfaces of the second, third, and half of the fourth digits (digital branches). These digital branches wind around to the dorsal surface of the middle and terminal phalanges of these same digits. The **ulnar** nerve supplies cutaneous nerves to the hypothenar side of the palm of the hand (**palmar cutaneous branch),** to the anterior surface of the ulnar side of the fourth digit, and to the entire fifth digit (**digital branches).** These digital branches supply the dorsal surface of the middle and terminal phalanges of these same digits. The ulnar nerve also supplies the dorsal side of the hand in this same area (**dorsal branch)** and the same digits on the posterior surface of the proximal phalanges. The posterior surface

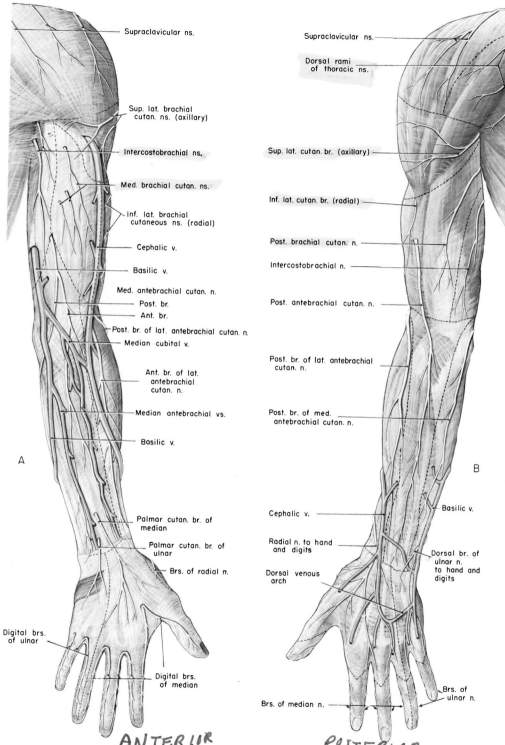

Supraclavicular ns.

Sup. lat. brachial cutan. ns. (axillary)

Intercostobrachial ns.

Med. brachial cutan. ns.

Inf. lat. brachial cutaneous ns. (radial)

Cephalic v.

Basilic v.

Med. antebrachial cutan. n.

Post. br.

Ant. br.

Post. br. of lat. antebrachial cutan. n.

Median cubital v.

Ant. br. of lat. antebrachial cutan. n.

Median antebrachial vs.

Basilic v.

Palmar cutan. br. of median

Palmar cutan. br. of ulnar

Brs. of radial n.

Digital brs. of ulnar

Digital brs. of median

A

ANTERIOR

Supraclavicular ns.

Dorsal rami of thoracic ns.

Sup. lat. cutan. br. (axillary)

Inf. lat. cutan. br. (radial)

Post. brachial cutan. n.

Intercostobrachial n.

Post. antebrachial cutan. n.

Post. br. of lat. antebrachial cutan. n.

Post. br. of med. antebrachial cutan. n.

Cephalic v.

Radial n. to hand and digits

Dorsal venous arch

Basilic v.

Dorsal br. of ulnar n. to hand and digits

Brs. of median n.

Brs. of ulnar n.

B

POSTERIOR

Figure 8-42. Cutaneous nerves and veins on the A. anterior surface of the upper limb, and B. on the posterior side. Areas innervated are outlined by dotted lines.

of the radial side of the hand, the thumb, and proximal ends of the second, third, and radial side of the fourth digit are all supplied by cutaneous branches of the **radial** nerve.

Figure 8-43 shows the **dermatome** arrangement of the upper limb. A dermatome represents the area of the skin supplied by a single segment of the spinal cord. The cutaneous nerve supply differs from this in that these cutaneous nerves may have contributions from more than one segment of the cord.

By way of summarizing the cutaneous nerve supply of the upper limb, it can readily be seen that the shoulder is supplied by the supraclavicular nerves, plus some contribution from the dorsal rami of the upper thoracic nerves. The arm is innervated by a special nerve, the medial brachial cutaneous, by the intercostobrachial nerves, and by branches of the axillary and radial nerves. The forearm is supplied by the terminal branches of the musculocutaneous nerve, the radial nerve, and the me-dial antebrachial cutaneous nerve on the lateral, posterior, and medial surfaces respectively. The hand is innervated by the median, ulnar, and radial nerves. Note that the dermatome arrangement in general terms starts at the shoulder with C4 and continues down the radial side of the arm and forearm as C5; the hand includes C6, C7, and C8; and C8 continues up the ulnar side to meet the first thoracic (T1) dermatome.

VEINS

The cutaneous veins vary from individual to individual. A typical pattern is presented in Figure 8-42. It should suffice for the student to realize that there is a **dorsal venous arch** on the hand as well as a **ventral arch;** that these start as the metacarpal veins and form two main channels, the **cephalic vein** on the lateral side of the forearm and the **basilic vein** on the medial side—there may be one or more **median antebrachial** veins on the anterior

mmy: head; basl

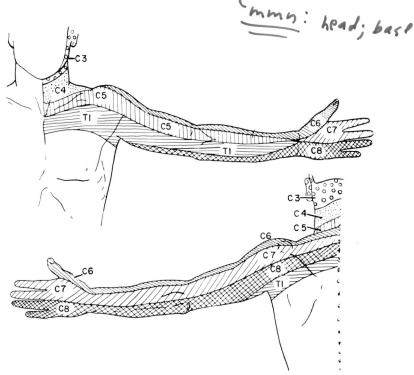

Figure 8-43. *Dermatome chart of the upper limb, based on hyposensitivity to painful stimuli after loss of a single nerve root. (From J. J. Keegan and F. D. Garrett, Anat. Record 102: 409, 1948, courtesy of the authors and the Wistar Institute.) Other workers, using different techniques, have found different patterns.*

surface of the forearm; that the cephalic vein continues superiorly through the cubital fossa to the lateral side of the arm, courses in the groove between the deltoid and pectoralis major muscles, and finally empties into the axillary vein; that the basilic vein courses superiorly, gains the anterior surface of the cubital fossa, and finally pierces the deep fascia about two-thirds of the way down the arm and joins the brachial vein to form the axillary vein; and finally that there is a **median cubital** vein, which connects the cephalic with the basilic in the cubital fossa, a vein which is frequently used to obtain blood from patients.

As mentioned previously, the superficial venous pattern of the upper limb is extremely variable; the veins mentioned, however, are usually present. You will soon see that the limb has a deep set of veins as well.

ARM

GENERAL DESCRIPTION

The surface anatomy of the arm (Fig. 8-44) in a muscular individual reveals many of the underlying structures. The rounded prominence of the shoulder is produced by the deltoid muscle, and the bulge anteriorly is made by the biceps brachii. Just posterior to the bulge of the biceps, on the lateral surface of the arm, a swelling can be seen produced by the brachialis muscle. The triceps brachii muscle produces a large swelling on the posterior surface of the arm. All of these muscles can be seen in Figures 8-45, 8-46, and 8-47.

When the skin is removed, the fascia can be seen to form not only an enveloping layer of fascia around all surfaces of the arm and around each muscle, but **two intermuscular septa**—one medial and one lateral. These attach to the sides of the humerus and divide the arm into anterior and posterior compartments (Fig. 8-48). The most anteriorly placed muscle is the **biceps brachii;** there are two heads of origin. Along the same plane but more superiorly located is the **coracobrachialis** muscle. If the biceps is pulled to one side, one can see the musculocutaneous nerve coursing inferiorly between the biceps and the muscle just deep to the biceps, the

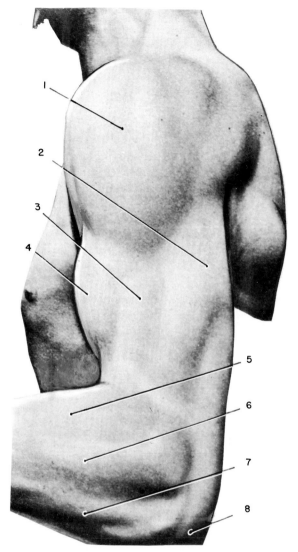

Figure 8-44. *Surface anatomy of the arm: (1) deltoid muscle, (2) triceps brachii muscle, (3) brachialis muscle, (4) biceps brachii muscle, (5) brachioradialis muscle, (6) extensor carpi radialis longus muscle, (7) extensor muscles, (8) olecranon. (From Appleton, Hamilton, and Tchaperoff,* Surface and Radiological Anatomy, *courtesy of The Macmillan Press Ltd.)*

716

brachialis. The **brachial artery** also follows these muscles, coursing first on the medial side of the biceps and coracobrachialis and then gradually gaining a position between the biceps and the brachialis muscle. This artery is followed by the **median nerve,** which is at first lateral to the artery (between it and the coracobrachialis muscle), then anterior to the artery, and finally medial to it near the elbow. The **ulnar nerve** can be seen just medial to the artery; it disappears from the scene by piercing the medial intermuscular septum to gain the posterior aspect of the arm.

The posterior side of the arm is made up almost entirely of the **triceps brachii** muscle. This three-headed muscle has a long head arising from the infraglenoid tubercle of the scapula and a lateral head arising from the upper end of the humerus, both of which are quite obvious; the medial head is difficult to see unless the lateral head is bisected and reflected. The **radial nerve** winds around the posterior side of the humerus between the medial and lateral heads of the triceps. It is accompanied by the profunda **brachial artery** throughout its extent in this region. After giving off branches to all heads of the triceps muscle, the radial nerve pierces the lateral intermuscular septum to gain the anterior side of the elbow joint.

A review of Figures 8-6, 8-7, 8-8, and 8-9 at this time would be helpful.

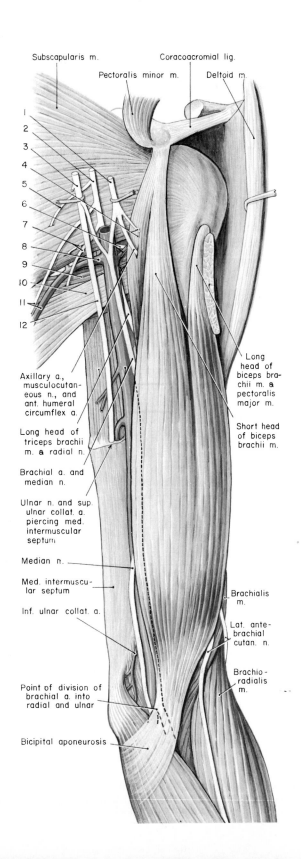

Figure 8-45. *Anterior surface of left arm I. All deep fascia has been removed except for a portion of the medial intermuscular septum. Note that the ulnar nerve and superior ulnar collateral artery pierce this septum to gain the posterior compartment: (1) lateral cord of brachial plexus, (2) posterior cord of brachial plexus, (3) medial cord of brachial plexus, (4) coracobrachialis muscle and its nerve, (5) upper subscapular nerve, (6) axillary nerve and posterior humeral circumflex artery, (7) lower subscapular nerve, (8) subscapular artery, (9) teres major muscle, (10) latissimus dorsi muscle, (11) thoracodorsal nerve and artery, (12) ulnar nerve.*

Labels on figure:
Subscapularis m.
Coracoacromial lig.
Pectoralis minor m.
Deltoid m.
Axillary a., musculocutaneous n., and ant. humeral circumflex a.
Long head of triceps brachii m. a radial n.
Brachial a. and median n.
Ulnar n. and sup. ulnar collat. a. piercing med. intermuscular septum
Median n.
Med. intermuscular septum
Inf. ulnar collat. a.
Point of division of brachial a. into radial and ulnar
Bicipital aponeurosis
Long head of biceps brachii m. a pectoralis major m.
Short head of biceps brachii m.
Brachialis m.
Lat. antebrachial cutan. n.
Brachioradialis m.

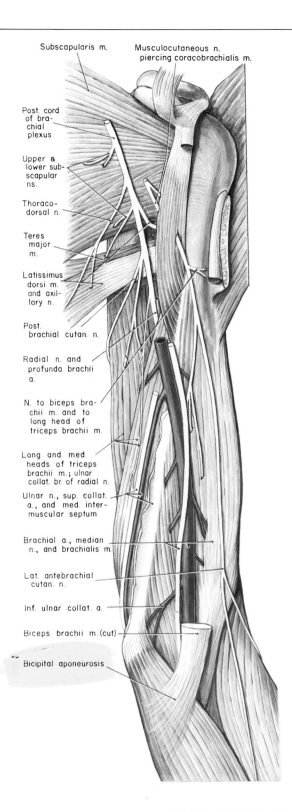

Subscapularis m.

Musculocutaneous n. piercing coracobrachialis m.

Post. cord of brachial plexus

Upper & lower subscapular ns.

Thoracodorsal n.

Teres major m.

Latissimus dorsi m. and axillary n.

Post. brachial cutan. n.

Radial n. and profunda brachii a.

N. to biceps brachii m. and to long head of triceps brachii m.

Long and med. heads of triceps brachii m.; ulnar collat. br. of radial n.

Ulnar n., sup. collat. a., and med. intermuscular septum

Brachial a., median n., and brachialis m.

Lat. antebrachial cutan. n.

Inf. ulnar collat. a.

Biceps brachii m. (cut)

Bicipital aponeurosis

MUSCLES

The muscles located in the arm are:

1. Biceps brachii
2. Coracobrachialis
3. Brachialis
4. Triceps brachii
5. Anconeus

1. Biceps brachii (Fig. 8-45). This is the most superficial muscle on the anterior part of the arm. It **arises** by two heads*—a short head from the coracoid process of the scapula and a long head from the supraglenoid tubercle of the scapula. The long head is a rounded tendon located, near its origin, in the superior aspect of the capsule of the shoulder joint, and then in the intertubercular groove of the humerus. The biceps muscle extends throughout the entire length of the arm and **inserts** on the radial tuberosity (Fig. 8-8).† In addition to this bony attachment, the muscle is attached to a fascial expansion—the **bicipital aponeurosis**—which courses medially and covers the muscles arising from the medial epicondyle; it blends with the fascia on these muscles (Fig. 8-45). Since the direction of pull is anterior to the axis of the elbow joint, the **main action** is to flex the forearm. It also has some action in flexing the arm due to the long head. The long head also serves in keeping the head of the humerus in the glenoid cavity. Since the radial tuberosity occupies a more posterior position when the forearm is pronated, the biceps has a powerful supinating action, although Travill and Basmajian‡ claim that this occurs under resistance only. Nevertheless, this accounts for the hardware industry creating screws and nuts and bolts that tighten clockwise. The nerve supply is the musculocutaneous.

---

*It may exhibit more than two heads.

†It actually inserts on half of the radial tuberosity, the remaining half being occupied by a bursa for the tendon.

‡A. Travill and J. V. Basmajian, 1961, Electromyography of the supinators of the forearm, *Anat. Record* 139: 557. This work reveals that the supinator muscle is the primary mover in the act of supination, and that the biceps acts in an auxiliary fashion when resistance is met.

Figure 8-46. Anterior surface of left arm II. The biceps brachii muscle has been removed. Note that the posterior cord of the brachial plexus and its branches are shown. All fascia, including the medial intermuscular septum, has been removed.

2. Coracobrachialis (Fig. 8-46). This small muscle aids in formation of the lateral wall of the axilla and is located on the medial side of the arm. Its **origin** is from the coracoid process of the scapula; its **insertion** is into an impression on the middle of the medial margin of the humerus (Fig. 8-6).* Since its direction of pull is anterior to the transverse axis of the shoulder joint and medial to the anteroposterior axis, its **main action** is to flex the arm and adduct it. **Nerve supply** is the musculocutaneous.

3. The **brachialis muscle** (Fig. 8-46) lies on the anterior surface of the humerus deep to the biceps. Its **origin** is from a large area on the anterior surface of the distal two-thirds of the humerus (Fig. 8-6) and its **insertion** is onto the coronoid process and tuberosity of the ulna (Fig. 8-8A). Its **main action** is to flex the forearm. Its **nerve supply** is by both the musculocutaneous and radial nerves; the function of the radial nerve is questionable.

4. Triceps brachii (Fig. 8-47). This large muscle is in the posterior part of the arm and its **origin** is in three parts: the long head by a tendon from the infraglenoid tubercle of the scapula, the lateral head from the lateral border of the humerus superior to the radial groove, and the medial head from a large triangular area on the posterior surface of the humerus extending from the olecranon fossa to the point of insertion of the teres major muscle (Figs. 8-5 and 8-7). The radial nerve courses between the medial and lateral heads. The muscle **inserts** on the olecranon of the ulna (Fig. 8-8B), and its **main action** is to extend the forearm because its direction of pull is posterior to the axis of the elbow joint. Since the long head bridges more than one joint, it has a slight action at the shoulder of extension and adduction. Its **nerve supply** is the radial.

5. The **anconeus** muscle (Fig. 8-47) is a small, triangular muscle located on the posterior surface of the elbow.† Its **origin** is from the lateral epicondyle (Fig. 8-7),

---

*According to W. E. Adams (*Notes on Anatomy,* 1959, N. M. Peryer, Ltd., New Zealand), this muscle is the remnant of the adductor group of muscles which is largely suppressed in the human forelimb, although remaining as a powerful sheet in the thigh. The muscle in its full development consists of three parts: (1) proximal, attached to the coracoid process and to the humerus just distal to the lesser tubercle; (2) intermediate, to the middle of the shaft of the humerus; and (3) distal, which reaches as far as the medial epicondyle. The intermediate portion and some of the distal portion persists in man, the musculocutaneous nerve separating the two parts.

†According to W. E. Adams (*Notes on Anatomy,* N. M. Peryer, Ltd., New Zealand), this muscle is actually part of the medial head of the triceps brachii muscle.

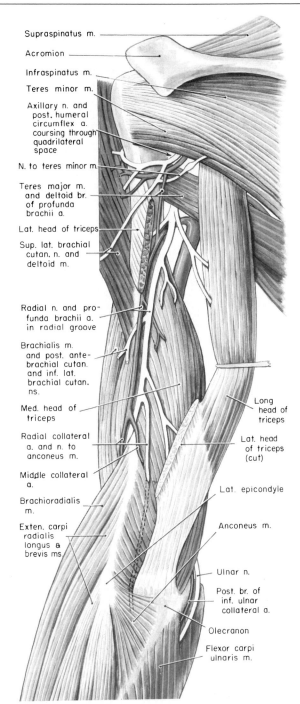

Figure 8-47. *Posterior surface of left arm. The lateral head of the triceps brachii muscle has been cut. All fascia has been removed.*

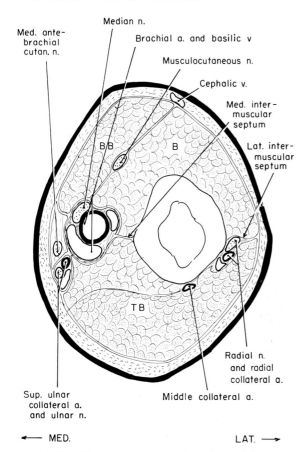

Med. ante-
brachial
cutan. n.

Median n.

Brachial a. and basilic v

Musculocutaneous n.

Cephalic v.

Med. inter-
muscular
septum

Lat. inter-
muscular
septum

BB   B

TB

Radial n.
and radial
collateral a.

Sup. ulnar
collateral a.
and ulnar n.

Middle collateral a.

← MED.   LAT. →

Figure 8-48. *Transverse section through the left arm a little distal to the middle. You are looking superiorly. Note that the medial and lateral intermuscular septa divide the arm into anterior and posterior compartments: (BB) biceps brachii muscle, (B) brachialis muscle, (TB) triceps brachii muscle. (Modified after Gray.)*

and its **insertion** is into a triangular area on the proximal end of the posterior surface of the ulna (Fig. 8-8B). Its **main action** is to extend the forearm. Its **nerve supply** is likewise radial.

NERVES

The nerves in the arm are the (1) median, (2) musculocutaneous, (3) ulnar, and (4) radial, but these nerves are by no means limited to the arm alone.

MEDIAN (Figs. 8-45 and 8-46). The median nerve starts by a combination of the lateral and medial roots,

which branch from the lateral and medial cords of the brachial plexus, respectively. The nerve is on the lateral side of the axillary artery and the brachial artery as far as the middle of the arm. At this point, the nerve crosses anterior to the artery and continues inferiorly on its medial side, where it remains until the forearm is reached. In its course it is just medial and slightly posterior to the coracobrachialis and biceps brachii muscles. There are no branches of the median nerve in the arm.

ie only innervates forearm

MUSCULOCUTANEOUS (Figs. 8-45 and 8-46). This nerve arises from the lateral cord of the brachial plexus and immediately pierces the coracobrachialis muscle. It then continues inferiorly in a position deep to the biceps muscle and superficial to the brachialis. It continues in this plane to reach the region of the elbow, where it becomes the lateral antebrachial cutaneous nerve (Fig. 8-42).

The branches of the musculocutaneous nerve are:

1. Muscular
2. Terminal—lateral antebrachial cutaneous

1. The **muscular** branch to the coracobrachialis muscle arises just as the musculocutaneous emerges from the brachial plexus. Those to the biceps brachii and brachialis muscles arise as the nerve courses between these muscles.

2. The musculocutaneous nerve terminates as the **lateral antebrachial cutaneous** nerve, which was described on page 712. (Note that this nerve also sends a small branch to the elbow joint and another to the humerus, entering the nutrient foramen.)

In summary, the musculocutaneous nerve in the arm innervates the coracobrachialis, biceps brachii, and most of the brachialis muscle (muscles derived from one muscle mass), sends twigs to the humerus and to the elbow joint, and terminates as the lateral antebrachial cutaneous nerve.

ULNAR (Figs. 8-45 and 8-46). This important nerve arises from the medial cord of the brachial plexus and is located between the axillary artery and axillary vein. It descends along the medial side of the brachial artery to the middle of the arm, where it pierces the medial intermuscular septum. (It is accompanied in this route by the superior ulnar collateral artery.) It continues toward the elbow posterior to the medial intermuscular septum and is actually on the medial head of the triceps muscle. It crosses the posterior surface of the medial epicondyle—

where it is likely to be hit (crazy bone)—and the medial ligament of the elbow to enter the forearm. The ulnar nerve has no branches in the arm.

RADIAL (Fig. 8-47). The radial nerve is one of the terminal branches of the posterior cord of the brachial plexus. It is located between the brachial artery and the long head of the triceps. It then courses inferiorly between the long head and medial head of the triceps to enter the radial groove on the back of the humerus. At this point it is liable to be injured in fractures of the humerus. After coursing through the groove, it pierces the lateral intermuscular septum accompanied by the radial collateral branch of the deep brachial artery. It continues inferiorly between the brachialis and brachioradialis muscles and courses into the forearm by crossing anterior to the capsule of the elbow joint.

The branches of the radial nerve in the area are:

1. Posterior brachial cutaneous
2. Inferior lateral brachial cutaneous
3. Posterior antebrachial cutaneous
4. Muscular

1. Posterior brachial cutaneous (Fig. 8-42). This nerve arises from the radial nerve close to the axilla. It pierces the deep fascia inferior to the posterior fold of the axilla and continues inferiorly. It innervates a large area on the lateral and posterior parts of the arm.

2. The **inferior lateral brachial cutaneous** nerve (Fig. 8-42) gives cutaneous innervation to the lateral surface of the arm aiding the superior lateral brachial cutaneous branch of the axillary in covering this territory.

3. The **posterior antebrachial** cutaneous (Figs. 8-42 and 8-47) arises from the radial nerve as it lies in the radial groove. It pierces the deep fascia, courses posterior to the lateral epicondyle, and continues on the posterior surface of the forearm as far as the wrist. This nerve innervates a small area in the arm as well.

4. Muscular. The radial nerve gives off branches to the long, medial, and lateral heads of the triceps in its course between these muscles, and gives off a small branch to the anconeus muscle, which arises about the middle of the arm and descends in the substance of the medial head of the triceps muscle to reach its destination. The radial also gives a branch to a small part of the brachialis muscle, and the branches to the brachioradialis and extensor carpi radialis longus—muscles of the forearm—also arise while the nerve is still in the arm.

Near the elbow, the radial nerve gives off an **articular branch** to the capsule of the joint.

———

In summary, the flexor muscles on the anterior surface of the arm are innervated by the musculocutaneous nerve, the extensors on the posterior surface by the radial. The only exception to this rule is the dual supply to the brachialis muscle.

ARTERIES

The arteries in the arm are the brachial artery and its branches.

BRACHIAL. The brachial artery (Figs. 8-45 and 8-46) starts as a continuation of the axillary artery at the lower border of the teres major tendon. It continues inferiorly and slightly laterally in a plane anterior to the long head of the triceps, the medial head of the triceps, and then the brachialis muscle; it is at first medial to the coracobrachialis and biceps brachii, but courses posterior to the biceps to reach the cubital fossa, where it divides into radial and ulnar arteries. The ulnar nerve is medial to the artery, the median nerve is at first lateral and then crosses to its medial side, and the musculocutaneous nerve, although not in direct relation to it, is lateral to the artery. The radial nerve is posterior to the artery but soon is separated from it by the medial head of the triceps muscle.

The branches of the brachial artery are:

1. Profunda brachii
   a. Nutrient
   b. Deltoid (ascending)
   c. Middle collateral (posterior descending)
   d. Radial collateral (anterior descending)
2. Superior ulnar collateral
3. Nutrient
4. Muscular branches
5. Inferior ulnar collateral
   a. Anterior branch
   b. Posterior branch
6. Two terminal branches

1. Profunda brachii (Fig. 8-47). This artery arises very close to the origin of the brachial and follows the path taken by the radial nerve. It gives off a deltoid (ascending) branch which anastomoses with the posterior humeral circumflex artery, and gives muscular branches to the triceps and a nutrient branch to the humerus. It then divides into radial collateral (anterior) and middle collateral (posterior) descending branches, the radial collateral accompanying the radial nerve through the lateral intermuscular septum to the anterior side of the elbow. The middle collateral branch courses on the back of the lateral epicondyle. Both of these vessels take part in the anastomosis around the elbow joint. These anastomotic channels are shown in Figure 8-49.

2. **Superior ulnar collateral** (Fig. 8-46). This branch arises rather high up in the arm and immediately pierces the medial intermuscular septum. In doing this, it accompanies the ulnar nerve in its pathway thus passing posterior to the elbow joint on the medial side. This vessel also takes part in the anastomosis around the elbow (Fig. 8-49).

3. The **nutrient** artery arises close to the ulnar collateral and enters the humerus directly.

4. **Muscular.** These are numerous small branches supplying the muscles of the arm.

5. **Inferior ulnar collateral** (Figs. 8-45 and 8-46). Arising just superior to the elbow joint, this branch runs medially on the brachialis muscle posterior to the median nerve. It divides into anterior and posterior branches which course on the anterior and posterior surfaces of the medial side of the elbow joint and take part in the anastomosis of vessels around the elbow joint.

6. The brachial artery terminates by dividing into **radial** and **ulnar** arteries at the elbow joint. This division frequently occurs higher up in the arm.

VEINS

Most of the above arteries are accompanied by veins, multiple channels often being found. Furthermore, the cutaneous veins play an important role in this area (Fig. 8-42). The **cephalic** vein courses on the lateral side of the arm to gain that area between the pectoralis major and deltoid muscles, and the **basilic** vein(s) join the **brachial** vein to form the **axillary**. The **median cubital** vein is found at the distal end of the arm in the cubital fossa.

Many structures in the arm are shown in cross section in Figure 8-48. Study of this figure at this time should be helpful.

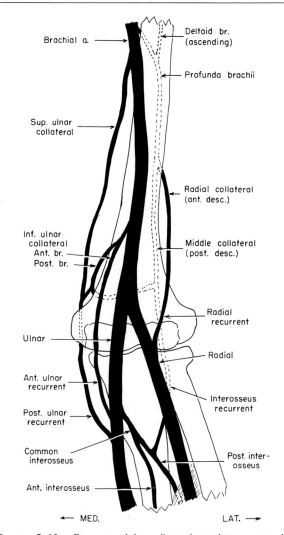

Figure 8-49. *Diagram of the collateral circulation around the elbow joint—left side.*

ANTERIOR SURFACE OF FOREARM

GENERAL DESCRIPTION

The anterior surface of the proximal end of the forearm (Figs. 8-50 and 8-51) shows a swelling laterally made by the brachioradialis muscle which arises from the lateral

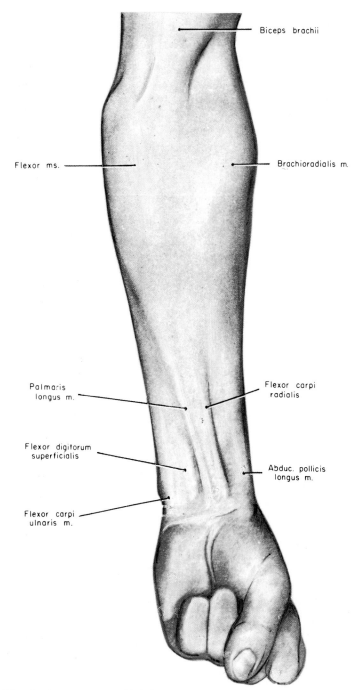

Figure 8-50. *Surface anatomy of left forearm—anterior aspect. (From Appleton, Hamilton, and Tchaperoff,* Surface and Radiological Anatomy, *courtesy of The Macmillan Press Ltd.)*

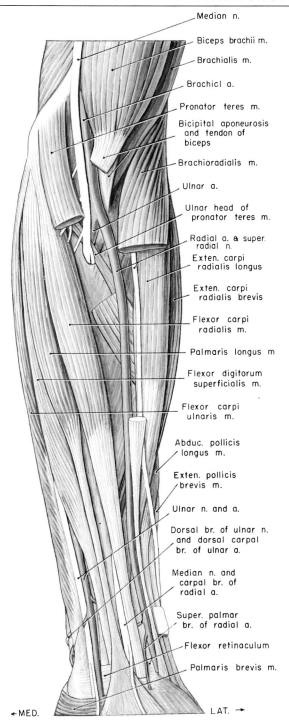

Median n.

Biceps brachii m.

Brachialis m.

Brachial a.

Pronator teres m.

Bicipital aponeurosis
and tendon of
biceps

Brachioradialis m.

Ulnar a.

Ulnar head of
pronator teres m.

Radial a. & super.
radial n.

Exten. carpi
radialis longus

Exten. carpi
radialis brevis

Flexor carpi
radialis m.

Palmaris longus m

Flexor digitorum
superficialis m.

Flexor carpi
ulnaris m.

Abduc. pollicis
longus m.

Exten. pollicis
brevis m.

Ulnar n. and a.

Dorsal br. of ulnar n.
and dorsal carpal
br. of ulnar a.

Median n. and
carpal br. of
radial a.

Super. palmar
br. of radial a.

Flexor retinaculum

Palmaris brevis m.

← MED.   LAT. →

Figure 8-51. *Anterior surface of left forearm I. The humeral head of the pronator teres muscle and the brachioradialis muscle have been cut, and all deep fascia removed.*

supracondylar ridge, and another medially produced by the pronator teres, flexor carpi radialis longus, and palmaris longus muscles arising from the medial epicondyle. Distally, the tendons are prominent, particularly if the fist is clenched; the one producing the most marked elevation is that of the palmaris longus, if present. The flexor carpi radialis also produces a ridge, and the flexor carpi ulnaris and a portion of the flexor digitorum superficialis can also be seen.

You should review at this time the cutaneous nerves and superficial veins described on page 712.

The deep fascia of the forearm completely invests it and is continuous with the fascia in the arm and in the hand. There are many septa that extend between the muscles. There are two thickenings in the deep fascia at the wrist. The one on the anterior side is the **flexor retinaculum,** that on the posterior side the **extensor retinaculum.** These serve to hold the tendons in place and are important structures. The flexor retinaculum is stronger and thicker than the extensor and bridges across the groove in the carpal bones, converting it into a tunnel. This retinaculum is attached to the pisiform and hamate bones medially and to the scaphoid and trapezium laterally. These retinacula are illustrated in Figures 8-51 and 8-57.

When the fascia is removed from the forearm, a large number of muscles and tendons appear (Fig. 8-51). The most proximal portion of the anterior surface of the forearm is known as the **cubital fossa.** It is a triangular area with a line drawn between the two epicondyles of the humerus as its base, the brachioradialis muscle as its lateral boundary, and the pronator teres muscle as its medial boundary; its apex is the point at which the brachioradialis and pronator teres muscles meet.

On the lateral side of the forearm, the **brachioradialis** muscle is the most prominent structure. The **radial artery** can be seen to course inferiorly just on the medial side of this muscle; it may be partially hidden by it. The radial artery at the wrist is not covered by muscles, and this is the reason why the pulse is customarily taken at this location. If the brachioradialis muscle is pulled to one side or cut, the **radial nerve** is also seen to be coursing just deep to this muscle and, therefore, superficial to the muscle just posterior to the brachioradialis, the **extensor carpi radialis longus.** It is lateral to the radial artery, as might be expected from their positions in the region of the elbow. The radial nerve continues inferiorly between the extensor carpi radialis longus posteriorly and the brachioradialis muscle anteriorly and finally winds around the lateral surface of the wrist to gain a posterior position, as indicated in Figure 8-51. Just posterior to the extensor carpi radialis longus muscle is the **extensor carpi radialis brevis.**

Looking at the medial side of the forearm, several muscles can be seen to have a common origin from the medial epicondyle of the humerus (Fig. 8-51). These are, from lateral to medial, the **pronator teres, flexor carpi radialis, palmaris longus** and **flexor digitorum superficialis,** and **flexor carpi ulnaris.** The **pronator teres** consists of two portions that are divided from each other by the median nerve; this nerve disappears from view between the two heads of this muscle. The **ulnar artery** also disappears deep to the muscles just mentioned and does not appear again until near the wrist, where it is seen to course between the tendons of the flexor digitorum superficialis and the flexor carpi ulnaris.

If one reflects the pronator teres, flexor carpi radialis, and palmaris longus, the entire extent of the rather large flexor digitorum superficialis can be seen (Fig. 8-52). If the flexor digitorum superficialis muscle is removed (Fig. 8-53), the ulnar artery and ulnar nerve can be seen to join each other and course on the superficial surface of the flexor digitorum profundus muscle, between it and the flexor digitorum superficialis.

As we return to the lateral side of the forearm, if the brachioradialis and extensor carpi radialis longus and brevis are removed, the muscle most proximally located—the **supinator**—becomes visible, and distal to that the **flexor pollicis longus** muscle (Fig. 8-53). Although the supinator muscle, the flexor pollicis longus, and the flexor digitorum profundus muscles are the deepest muscles on the anterior surface of the forearm and form a complete layer, there is one muscle deep to the flexor pollicis longus and the flexor digitorum profundus. This is located near the wrist and is a muscle—the **pronator quadratus**—whose fibers course from side to side (Fig. 8-54).

Removal of the flexor digitorum superficialis muscle also reveals the **interosseus membrane** and the **anterior interosseus nerve and artery.** The anterior interosseus artery arises from the **common interosseus** which is a branch of the ulnar. The common interosseus divides into anterior and posterior interosseus branches, and the posterior can be seen disappearing behind the interos-

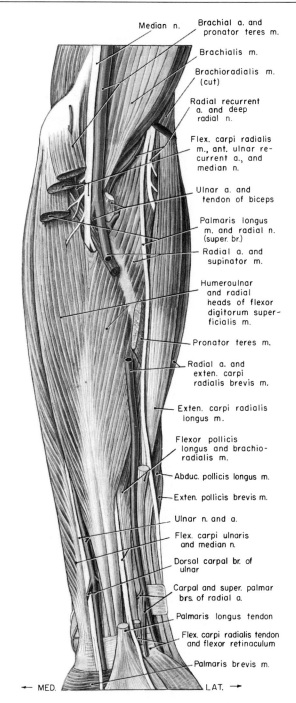

Median n.
Brachial a. and pronator teres m.
Brachialis m.
Brachioradialis m. (cut)
Radial recurrent a. and deep radial n.
Flex. carpi radialis m., ant. ulnar recurrent a., and median n.
Ulnar a. and tendon of biceps
Palmaris longus m. and radial n. (super. br.)
Radial a. and supinator m.
Humeroulnar and radial heads of flexor digitorum superficialis m.
Pronator teres m.
Radial a. and exten. carpi radialis brevis m.
Exten. carpi radialis longus m.
Flexor pollicis longus and brachioradialis m.
Abduc. pollicis longus m.
Exten. pollicis brevis m.
Ulnar n. and a.
Flex. carpi ulnaris and median n.
Dorsal carpal br. of ulnar
Carpal and super. palmar brs. of radial a.
Palmaris longus tendon
Flex. carpi radialis tendon and flexor retinaculum
Palmaris brevis m.

← MED.   LAT. →

Figure 8-52. *Anterior surface of left forearm II. The pronator teres, flexor carpi radialis, palmaris longus, and brachioradialis muscles have been removed to reveal the large flexor digitorum superficialis muscle.*

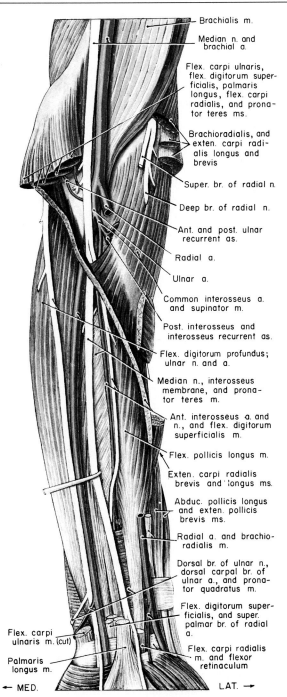

Figure 8-53. Anterior surface of left forearm III. Shown after removal of the superficial muscle, including the flexor digitorum superficialis muscle.

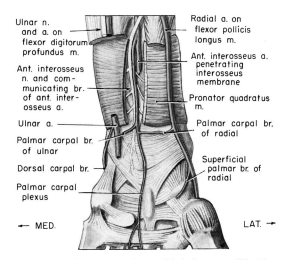

Figure 8-54. Anterior surface of left forearm IV. The pronator quadratus muscle and structures deep to it are shown.

seus membrane to gain the posterior aspect of the forearm. The anterior interosseus nerve is a branch of the median; it innervates the flexor pollicis longus and the radial half of the flexor digitorum profundus, and then continues inferiorly and gives nerve supply to the pronator quadratus muscle, the ligaments around the carpal bones, and the carpal joints.

MUSCLES

The muscles located on the anterior surface of the forearm are:

1. Pronator teres
2. Flexor carpi radialis
3. Palmaris longus
4. Palmaris brevis (not actually in forearm)
5. Flexor carpi ulnaris
6. Flexor digitorum superficialis
7. Flexor digitorum profundus
8. Flexor pollicis longus
9. Pronator quadratus

You will soon note that many of these muscles flex the forearm, since they arise from the medial epicondyle of the humerus and pass anterior to the transverse axis of the elbow joint.

1. Pronator teres (Fig. 8-51). This muscle lies across the superior part of the anterior surface of the forearm and actually forms the medial boundary of the **cubital fossa.** Its **origin** is by two heads, a humeral head from the medial epicondyle of the humerus and an ulnar head from the coronoid process of the ulna (Figs. 8-6 and 8-8A). The median nerve passes between these two heads. The **insertion** is onto the lateral surface of the radius approximately halfway down the shaft (Fig. 8-8B). Its **main action** is to pronate the hand (only under resistance, according to Basmajian and Travill*) and flex the forearm. **Nerve supply** is the median.

2. **Flexor carpi radialis** (Fig. 8-51). Arising from the medial epicondyle, this muscle extends throughout the length of the forearm to **insert** onto the base of the second metacarpal bone (Fig. 8-10). Its tendon pierces the flexor retinaculum. Its **main action** is to flex and abduct the hand, since its direction of pull is anterior to the transverse axis of the wrist and on the radial side of the anteroposterior axis (Fig. 8-12); in addition, it has a slight action in flexing the forearm and in pronation. **Nerve supply** is by the median.

3. The **palmaris longus** muscle (Fig. 8-51), although frequently absent, **arises** from the medial epicondyle of the humerus and **inserts** into the palmar aponeurosis, a heavy layer of fascia (tendon) in the palm of the hand. Its tendon is superficial to the flexor retinaculum. Its **main action** is to flex the wrist (pull is anterior to transverse axis); it has a slight action in flexing the forearm. **Nerve supply** is the median.

4. Although the **palmaris brevis** muscle is actually part of the hand (Fig. 8-51 and 8-62), it must be kept in mind at this time or it will be missed. This small muscle **arises** from the anterior surface of the flexor retinaculum and from the ulnar margin of the palmar aponeurosis. It passes medially at a level superficial to the ulnar vessels and nerve to be **inserted** into the skin on the ulnar border of the hand. This muscle is said to aid the hand in taking a firm grasp of an object by raising the skin over the hypothenar eminence. **Nerve supply** is the ulnar.

5. The **flexor carpi ulnaris muscle** (Fig. 8-51) lies along the ulnar side of the forearm and has its **origin** by two heads, one being attached to the medial epicondyle

and the other to the medial side of the olecranon process; the latter attachment, in common with the flexor digitorum profundus, extends down the upper two-thirds of the dorsal border of the ulna (Fig. 8-9). This muscle extends through the entire length of the forearm and **inserts** (by a tendon attached to the pisiform bone) to the hook of the hamate bone and the base of the fifth metacarpal (Fig. 8-10). Its **main action** is to flex and adduct the hand (pull is anterior to transverse axis and on the ulnar side of the anteroposterior axis); it plays a slight role in flexion of the forearm. The **nerve supply** is by the ulnar. This muscle and half of the flexor digitorum profundus muscle are the only muscles in the forearm receiving their nerve supply from the ulnar.

6. **Flexor digitorum superficialis** (Fig. 8-52). This muscle is a rather extensive muscle, with two heads of **origin,** one (the humeroulnar) from the medial epicondyle of the humerus and the coronoid process of the ulna, the other (the radial head) from a rather long line on the radius between the supinator muscle and the flexor pollicis longus muscle (see Figs. 8-8A and 8-53). The tendons of this muscle course deep to the flexor retinaculum and then enter the fibrous flexor sheaths of each finger. Each tendon splits to transmit the tendon of the flexor digitorum profundus muscle and then **inserts** onto the margins of the middle phalanx. The **main action** of this muscle is to flex the middle and proximal phalanges and to flex the hand; it has a slight action in flexing the forearm. **Nerve supply** is by the median.

7. **Flexor digitorum profundus** (Fig. 8-53). Located on the ulna, deep to the flexor digitorum superficialis and the flexor carpi ulnaris, this muscle has an **origin** from the anterior and medial surfaces of the ulna (see Fig. 8-8) which is quite extensive. Although not important, it also arises from the medial surface of the coronoid process and the olecranon, and from a small portion of the interosseus membrane. It forms four tendons which enter the fibrous sheaths of the digits and, piercing the tendons of the flexor digitorum superficialis, finally **insert** onto the base of the distal phalanges. Because of its anterior relation to the transverse axes of the wrist and digits, its **main action** is to flex all phalanges and the hand. Its **nerve supply** is from two sources, the anterior interosseus branch of the median to the lateral half, and the ulnar to the medial half.

8. **Flexor pollicis longus** (Fig. 8-53). This muscle also has an extensive origin **arising** from the anterior

---

*J. V. Basmajian and A. Travill, 1961, Electromyography of the pronator muscles in the forearm, *Anat. Record* 139:45. The pronator quadratus muscle seems to play a much more important role in pronation than does the pronator teres, the latter muscle acting as an auxiliary that reinforces the action of the pronator quadratus.

surface of the radius and part of the interosseus membrane (Fig. 8-8A).* It **inserts** onto the base of the distal phalanx in the thumb after traversing the carpal tunnel deep to the flexor retinaculum and entering the fibrous sheath of the thumb (Fig. 8-9). Its **main action** is to flex the phalanges of the thumb, to flex the metacarpal bone of the thumb, and to flex the hand. Its **nerve supply** is by the anterior interosseus branch of the median.

9. Pronator quadratus (Fig. 8-54). The pronator quadratus muscle lies at the distal end of the forearm deep to the flexor tendons and other structures crossing the anterior surface of the wrist. Its **origin** is from the anterior surface of the ulna and its **insertion** is onto the anterior surface of the radius (Fig. 8-8A). Its **main action** is to pull the radius across the ulna, thus inducing pronation. It is the primary pronator, according to Basmajian and Travill (see footnote on page 726). **Nerve supply** is by the anterior interosseus branch of the median.

Note that the great majority of these muscles arise from a common origin—the medial epicondyle—and that all the muscles listed are either flexors or pronators.†

Although the most superficial muscles on the lateral side of the forearm, such as the brachioradialis, are actually on the anterior surface, these muscles arise from the region of the lateral epicondyle and will be described with the other muscles in this group located on the posterior surface of the forearm.

NERVES

The nerves on the anterior surface of the forearm are the median and ulnar. Although the radial nerve appears on the anterior surface and innervates the muscles arising from the lateral epicondyle, it will be described after the muscles on the posterior side of the forearm are studied.

MEDIAN (Figs. 8-51 and 8-53). The median nerve ap-

pears in the cubital fossa just medial to the brachial artery. After coursing between the two heads of the pronator teres muscle, it disappears from view by coursing between the flexor digitorum superficialis and flexor digitorum profundus muscles. It finally appears again near the wrist, between the tendon of the flexor pollicis longus and tendons of the flexor digitorum superficialis muscle, where it is easily mistaken for a tendon.

The branches of the median nerve in the forearm are:

1. Muscular
2. Anterior interosseus
   a. Muscular
   b. Articular
3. Palmar cutaneous

1. The **muscular branches** are to the pronator teres, flexor carpi radialis, palmaris longus, and flexor digitorum superficialis muscles. These branches arise in the cubital fossa from the medial side of the nerve (a fact worth noting and one that might be expected) and sink immediately into their respective muscles. Some of these branches are said to give sensory nerve supply to the elbow joint.

2. The **anterior interosseus nerve** (Fig. 8-53) accompanies the anterior interosseus artery on the anterior surface of the interosseus membrane. It gives off its (a) **muscular branches** to the flexor pollicis longus and radial half of the flexor digitorum profundus proximally and to the pronator quadratus distally as the nerve courses deep to this muscle (Fig. 8-54). The (b) **articular branch** is the termination of this nerve; it ramifies on the ligaments on the anterior surface of the wrist and gives sensory innervation to the inferior radioulnar joint as well as the wrist joint.

3. Palmar cutaneous (Fig. 8-42). This small branch of the median nerve arises above the wrist, pierces the antebrachial fascia, and enters the hand. It divides into a medial branch to the center of the palm and a lateral branch that innervates the area of the thenar eminence.

In summary, the median nerve in the forearm innervates all flexors and pronators except the flexor carpi ulnaris and the ulnar half of the flexor digitorum profundus muscle. It has sensory branches to the elbow, inferior radioulnar, and wrist joints, and cutaneous branches to the skin in the palm and over the thenar eminence.

---

*An attachment to the medial epicondyle is a frequent occurrence.
†J. P. McMurrich, 1903, The phylogeny of the forearm flexors, *Am. J. Anat.* 2: 177. The professional student would benefit greatly if this paper were read, not to remember the exact details presented, but to obtain the important concepts it contains.
  W. L. Straus, Jr., 1942, The homologies of the forearm flexors: Urodeles, lizards, mammals, *Am. J. Anat.* 70: 281. The phylogeny of the forearm flexor muscles is presented. This gives a clear picture of how some muscles have disappeared while others have appeared.

ULNAR (Figs. 8-53). The ulnar nerve appears in the forearm after coursing posterior to the medial epicondyle of the humerus. It is on the superficial surface of the flexor digitorum profundus muscle and deep to the flexor digitorum superficialis muscle (note that the median nerve is also located between these two muscles), and also lateral to the flexor carpi ulnaris muscle. It continues distally and finally enters the hand in company with the ulnar artery between the tendons of the flexor digitorum superficialis muscle and the flexor carpi ulnaris muscle, in a plane that is superficial to the flexor retinaculum.

The branches of the ulnar nerve in the forearm are:

1. Articular
2. Muscular
3. Palmar cutaneous
4. Dorsal branch

1. The **articular branches** are several to the elbow joint, given off as the nerve winds around the medial epicondyle.

2. The **muscular branches** of the ulnar are to the flexor carpi ulnaris and the ulnar half of the flexor digitorum profundus muscles. These arise at the proximal end of the forearm and sink immediately into the respective muscles.

3. The **palmar cutaneous nerve** (Fig. 8-42) arises approximately in the middle of the forearm; this slender nerve accompanies the ulnar artery into the hand. It terminates in the palm and aids the median nerve in innervating this particular area of the skin.

4. **Dorsal branch** (Figs. 8-42 and 8-53). This branch is superficially located and is occasionally forgotten by the surgeon. It arises at various points in the distal half of the forearm, accompanies the main ulnar nerve in its position deep to the tendon of the flexor carpi ulnaris muscle to just proximal to the pisiform, and then courses posteriorly and distally across the ulnar side of the wrist joint at a plane just deep to the skin. It divides into two dorsal digital nerves, one of which sends branches to the dorsum of the hand and the dorsal side of the fifth digit, while the other innervates the dorsum of the hand and contiguous sides of the dorsal areas of the little and ring fingers. These branches innervate the skin over the proximal phalanges only.

In summary, the ulnar nerve in the forearm innervates the flexor carpi ulnaris and half of the flexor digitorum profundus muscle, gives sensory branches to the elbow joint, and cutaneous branches to the palm, to the dorsal side of the hand, and to the dorsal side of the ulnar half of the ring finger and all of the dorsal surface of the little finger but only as far as the base of the middle phalanges.

ARTERIES

The arteries in the forearm are the ulnar and the radial and their branches.

ULNAR (Figs. 8-52 and 8-53). The ulnar artery begins as a terminal branch of the brachial in the cubital fossa. It immediately disappears deep to the muscles arising from the medial epicondyle and courses inferiorly between the flexor digitorum superficialis and the flexor digitorum profundus muscles. The artery finally escapes from the flexor digitorum superficialis muscle and comes to lie between the tendon of this muscle and that of the flexor carpi ulnaris. The ulnar nerve accompanies the artery throughout its extent in the forearm.

The branches of the ulnar artery are:

1. Anterior ulnar recurrent
2. Posterior ulnar recurrent
3. Common interosseus
   a. Anterior interosseus
      i. Median
      ii. Nutrient
      iii. Muscular
      iv. Communicating branch to palmar carpal arch
   b. Posterior interosseus
4. Palmar and dorsal carpals
5. Muscular

1. The **anterior ulnar recurrent** (Figs. 8-49, and 8-52) ascends on the anterior side of the medial epicondyle and anastomoses with the anterior branch of the inferior ulnar collateral artery.

2. The **posterior ulnar recurrent** (Fig. 8-49) courses superiorly between the flexor digitorum superficialis and profundus posterior to the medial epicondyle and anastomoses with the superior ulnar collateral and the posterior branch of the inferior ulnar collateral.

3. The **common interosseus artery** (Figs. 8-49 and 8-53) is quite short and arises from the ulnar about an inch from its origin. This artery almost immediately divides into **anterior and posterior interosseus arteries.**

The posterior gives off its interosseus recurrent branch (Fig. 8-53) and then disappears from view (see page 737), but the anterior interosseus continues inferiorly on the anterior surface of the interosseus membrane between the flexor pollicis longus and the flexor digitorum profundus muscles. It is accompanied throughout its extent by the anterior interosseus branch of the median nerve. Just superior to the pronator quadratus muscle it dives through the interosseus membrane and communicates with the dorsal carpal arch. Before it pierces the membrane, however, it gives rise to a **communicating branch** which descends, anterior to the interosseus membrane but deep to the pronator quadratus muscle, to join the vessels forming the palmar carpal arch (Fig. 8-54). Other branches of this artery are the **median,** which courses into the median nerve (this branch can be enlarged), the **nutrients** to the ulna and radius, and the usual **muscular** branches to adjoining muscles.

4. The **palmar carpal** (Fig. 8-54) arises near the pisiform bone, passes laterally just anterior to the bone, and anastomoses with a similar branch from the radial to form the anterior or palmar carpal arch, while the **dorsal carpal** turns posteriorly and inferiorly to anastomose with similar branches from the radial on the posterior surface of the carpal bones.

5. **Muscular branches** nourish adjoining muscles.

RADIAL (Fig. 8-51). The radial artery also arises in the cubital fossa as one of the terminal branches of the brachial. It courses inferiorly and slightly laterally throughout the length of the forearm to reach the styloid process of the radius. It is deep to the brachioradialis muscle but is not covered by this muscle to any great degree. It is superficial, from proximal to distal, to the tendon of the biceps, the supinator muscle, the pronator teres muscle, the radial head of the flexor digitorum superficialis muscle, the flexor pollicis longus muscle, the pronator quadratus muscle, and then the styloid process of the radius. This can be clearly understood by glancing at Figure 8-51.

The branches of the radial artery in the forearm are:

1. Radial recurrent
2. Superficial palmar
3. Palmar carpal
4. Muscular branches

1. **Radial recurrent** (Figs. 8-49 and 8-52). This vessel arises close to the origin of the radial and courses superiorly on the supinator muscle to anastomose with the radial collateral branch of the profunda brachial artery.

2. The **superficial palmar branch** (Fig. 8-51) arises near the wrist joint and continues into the palm of the hand by piercing the muscles of the thenar eminence (base of thumb). It may or may not continue to complete the superficial palmar arch (see Fig. 8-63).

3. The **palmar carpal branch** (Figs. 8-51 and 8-54) courses medially on the anterior surface of the distal end of the radius. This anastomoses with a similar branch from the outer side of the wrist (from the ulnar) and also with the communicating branch from the anterior interosseus artery.

4. **Muscular branches.** These are small arteries to adjoining muscles.

VEINS

All these arteries are accompanied by one or more veins of the same names.

Many of the structures in the forearm are shown in cross section in Figure 8-55.

POSTERIOR SURFACE OF FOREARM AND HAND

GENERAL DESCRIPTION

Except for the tendons on the posterior surface of the hand, the surface anatomy of the posterior surface of the forearm and hand (Fig. 8-56) is not particularly revealing.

The posterior surface of the forearm is considered to be the extensor side, and the muscles arising from the lateral epicondyle and supracondylar ridge to be the extensor muscles.

The superficial nerves and veins found in the deep fascia have already been described (see Fig. 8-42). The only matter concerned with the deep fascia on the posterior surface of the forearm that requires mention is the presence of an **extensor retinaculum** (Fig. 8-57). This fascia is longer but not as dense as the flexor retinaculum and occurs at a higher level. Its direction is oblique from superior to inferior, from the radial to the ulnar side. It is attached to the radius on the lateral side and to the triquetrum and pisiform bones on the ulnar side. There are five septa that divide the underlying space into six fascial compartments. The extensor retinaculum aids in holding the tendons in place.

The posterior surface of the forearm is made up mainly of muscles and their tendons (Fig. 8-57). Starting on the lateral side, the **muscles** in order are the **brachioradialis, extensor carpi radialis longus, extensor carpi radialis brevis, extensor digitorum, extensor digiti minimi, extensor carpi ulnaris,** and, close to the elbow joint, the **anconeus** muscle. These muscles are considered to be the superficial group of extensor muscles. The deeper muscles (Fig. 8-58) can be seen coursing between the extensor carpi radialis brevis and the extensor digitorum muscles. These **muscles** are the **abductor pollicis longus, extensor pollicis brevis, extensor pollicis longus,** and **extensor indicis** muscles. The **supinator** muscle is also a deep muscle and is located deep to the other muscles arising from the lateral epicondyle.

In addition, the posterior interosseus nerve and artery course between the superficial and deep sets of muscles. The **posterior interosseus nerve** is a continuation of the deep radial nerve after it emerges from the substance of the supinator muscle and courses on the abductor pollicis longus muscle; it is called posterior interosseus when it reaches the interosseus membrane. The **posterior interosseus artery,** a branch of the common interosseus from the ulna, courses on the surface of the abductor pollicis longus and extensor pollicis longus muscles and is not really in contact with the intersosseus membrane. The **anterior interosseus artery** pierces the interosseus membrane in the distal part of the forearm, where it joins the posterior interosseus nerve.

Since all the muscles on the posterior surface of the forearm, except the brachioradialis and anconeus, have an action on the hand or the digits, the posterior surface of the hand will be described at this time.

The skin on the dorsal surface of the hand is quite thin and quite movable. The **dorsal metacarpal veins** and the dorsal metacarpal arch or network have already

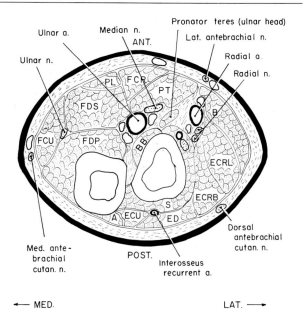

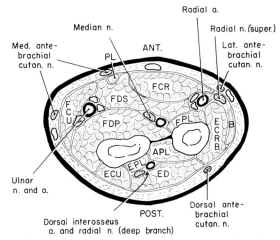

Figure 8-55. *Transverse sections through the left forearm between the proximal and middle thirds and between the middle and distal thirds. You are looking superiorly or proximally: (A) anconeus, (APL) abductor pollicis longus, (B) brachioradialis, (BB) biceps brachii, (ECRB) extensor carpi radialis brevis, (ECRL) extensor carpi radialis longus, (ECU) extensor carpi ulnaris, (ED) extensor digitorum, (EPL) extensor pollicis longus, (FCR) flexor carpi radialis, (FCU) flexor carpi ulnaris, (FDP) flexor digitorum profundus, (FDS) flexor digitorum superficialis, (FPL) flexor pollicis longus, (PL) palmaris longus, (PT) pronator teres, (S) supinator. (Modified after Gray.)*

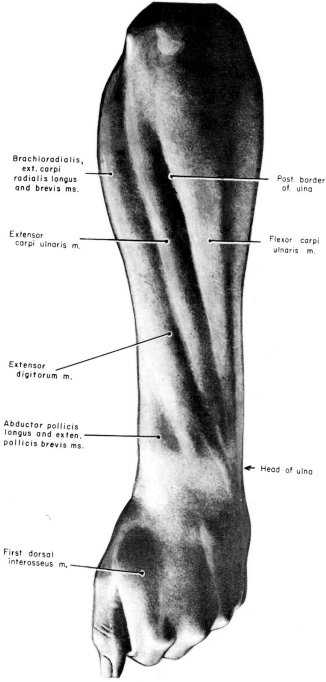

Brachioradialis,
ext. carpi
radialis longus
and brevis ms.

Post. border
of. ulna

Extensor
carpi ulnaris m.

Flexor carpi
ulnaris m.

Extensor
digitorum m.

Abductor pollicis
longus and exten.
pollicis brevis ms.

← Head of ulna

First dorsal
interosseus m.

Figure 8-56. *Surface anatomy of the posterior surface of left forearm. (From Appleton, Hamilton, and Tchaperoff,* Surface and Radiological Anatomy, *courtesy of The Macmillan Press Ltd.)*

been studied (Fig. 8-42). In addition, the **dorsal branch** of the ulnar nerve and the **dorsal cutaneous branches** of the radial nerve have been discussed (page 712).

Of the nine muscles that extend from the posterior surface of the forearm into the hand, three are involved with the thumb, three with the fingers, and three with the hand itself. In crossing the wrist joint, these tendons are protected by six **synovial membranes** (Fig. 8-57), which correspond to the six connective tissue compartments made by the extensor retinaculum. Starting laterally: (1) the abductor pollicis longus and extensor pollicis brevis are contained within one synovial sheath; (2) the extensor carpi radialis longus and brevis in the second; (3) the extensor pollicis longus in a third,* which takes an oblique course from the midline of the forearm to the thumb crossing the sheath for the extensor carpi radialis brevis and longus just mentioned; (4) the extensor digitorum and extensor indicis; (5) the extensor digiti minimi; and (6) the extensor carpi ulnaris. The tendons of the extensor carpi radialis longus and brevis and the extensor carpi ulnaris, those muscles having their point of action at the wrist joint, extend only to the base of the metacarpal bones. The remaining tendons extend to the phalanges of the fingers and thumb. The extensor tendons are attached to the phalanges in a rather complex manner. Each tendon broadens to form the **dorsal extensor expansion.** On the proximal phalanx the extensor tendon divides into three slips, the middle slip continuing to insert on the base of the middle phalanx. The other two take a more lateral position and join on the dorsal surface of the middle phalanx and insert on the base of the distal phalanx. These lateral portions are joined by the tendons of the lumbricals (Fig. 8-64) and the interossei (Fig. 8-58). This expansion and the attachments of the abovementioned muscles are shown in Figure 8-59. The thumb does not have any such arrangement as described for the fingers.

The **dorsal metacarpal arteries,** branching from the dorsal carpal arch, course in the spaces between the metacarpal bones just deep to the abovementioned tendons. These arteries are superficial to the **dorsal interossei muscles.**

---

*The depression between the combined abductor pollicis longus and extensor pollicis brevis tendons anterolaterally and the extensor pollicis longus tendon posteromedially is the "anatomical snuffbox." Snuff users utilize this depression.

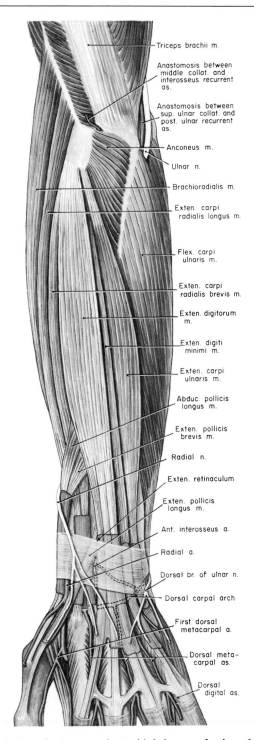

Figure 8-57. Posterior surface of left forearm I—deep fascia removed; muscles intact.

Triceps brachii m.

Anastomosis between middle collat. and interosseus recurrent as.

Anastomosis between sup. ulnar collat. and post. ulnar recurrent as.

Anconeus m.

Ulnar n.

Brachioradialis m.

Exten. carpi radialis longus m.

Flex. carpi ulnaris m.

Exten. carpi radialis brevis m.

Exten. digitorum m.

Exten. digiti minimi m.

Exten. carpi ulnaris m.

Abduc. pollicis longus m.

Exten. pollicis brevis m.

Radial n.

Exten. retinaculum

Exten. pollicis longus m.

Ant. interosseus a.

Radial a.

Dorsal br. of ulnar n.

Dorsal carpal arch

First dorsal metacarpal a.

Dorsal metacarpal as.

Dorsal digital as.

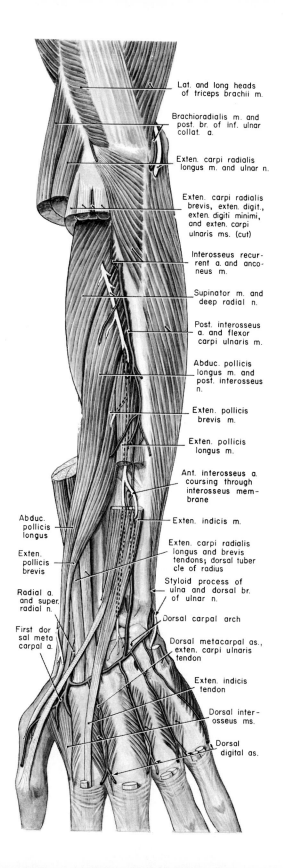

Lat. and long heads of triceps brachii m.

Brachioradialis m. and post. br. of inf. ulnar collat. a.

Exten. carpi radialis longus m. and ulnar n.

Exten. carpi radialis brevis, exten. digit., exten. digiti minimi, and exten. carpi ulnaris ms. (cut)

Interosseus recurrent a. and anconeus m.

Supinator m. and deep radial n.

Post. interosseus a. and flexor carpi ulnaris m.

Abduc. pollicis longus m. and post. interosseus n.

Exten. pollicis brevis m.

Exten. pollicis longus m.

Ant. interosseus a. coursing through interosseus membrane

Exten. indicis m.

Exten. carpi radialis longus and brevis tendons; dorsal tubercle of radius

Styloid process of ulna and dorsal br. of ulnar n.

Dorsal carpal arch

Dorsal metacarpal as., exten. carpi ulnaris tendon

Exten. indicis tendon

Dorsal interosseus ms.

Dorsal digital as.

Abduc. pollicis longus

Exten. pollicis brevis

Radial a. and super. radial n.

First dorsal meta carpal a.

MUSCLES*

The muscles on the extensor surface of the forearm and hand are:

1. Brachioradialis
2. Extensor carpi radialis longus
3. Extensor carpi radialis brevis
4. Extensor digitorum
5. Extensor digiti minimi
6. Extensor carpi ulnaris
7. Anconeus
8. Abductor pollicis longus
9. Extensor pollicis brevis
10. Extensor pollicis longus
11. Extensor indicis
12. Supinator
13. Dorsal interossei

1. Brachioradialis (Fig. 8-57). This interesting muscle lies on the lateral side of the forearm and is really an anteriorly placed muscle; it forms the lateral boundary of the cubital fossa. It **arises** from the lateral supracondylar ridge of the humerus (Fig. 8-6) and **inserts** onto the lateral surface of the distal end of the radius (Fig. 8-8A). Its **main action** is said to be to initiate supination and pronation (denied by de Sousa *et al.*†) and to flex the forearm (direction of pull is anterior to the transverse axis of elbow joint). Although this muscle is located in the forearm, it has no action at the wrist joint. **Nerve supply** is the radial.

2. Extensor carpi radialis longus (Fig. 8-57). This muscle, located just posterior to the brachioradialis, also **arises** from the lateral supracondylar ridge (Fig. 8-6). It is called "longus" because it extends further up into the arm than its companion muscle, the extensor carpi radialis brevis. Its tendon is enclosed in a sheath in common with the extensor carpi radialis brevis

---

*W. L. Straus, Jr., 1941, The phylogeny of the human forearm extensors, *Human Biol.* 13: 23 and 203. A quick reading, however, will give the student understanding of why so many variations and anomalies in musculature occur in the human.

†O. M. de Sousa, J. L. de Moraes, and F. L. de Moraes Vieira, 1961, Electromyographic study of the brachioradialis muscle, *Anat. Record* 139: 125. According to these authors, the brachioradialis muscle acts in flexion of the forearm, but plays no role in pronation or supination.

Figure 8-58. Posterior surface of left forearm II—after removal of the more superficial muscles.

and then is **inserted** onto the base of the second metacarpal bone (Fig. 8-10). Its **main action** is to extend and abduct the hand, this action being explained by a pull which is posterior to the transverse axis of the wrist and on radial side of the anteroposterior axis. Its **nerve supply** is by the deep branch of the radial.

3. **Extensor carpi radialis brevis** (Fig. 8-57). Located just posterior to the extensor carpi radialis longus, this muscle **arises** from the lateral epicondyle. After joining the extensor carpi radialis longus in a common tendon sheath, it **inserts** onto the base of the third metacarpal bone (Fig. 8-10). Its **main action** is to extend and abduct the hand for the same reasons as described for the longus. Its **nerve supply** is the deep branch of the radial.

4. **Extensor digitorum** (Fig. 8-57). This muscle **arises** from the lateral epicondyle of the humerus. After crossing the wrist enclosed in a synovial sheath, the various tendons break up to be **distributed** to the fingers. The tendon to the fifth digit is often wanting or may arise at a more inferior point than do the remaining tendons. The tendons also have connections that prevent completely independent use of the fingers. This is particularly true of the ring finger, where there are almost always connections of this tendon to that of the third and fifth digits. This prevents the fourth digit being raised very far if the third and fifth digits are not allowed to move. The tendons of the extensor digitorum join the extensor expansion and **insert** primarily onto the base of the proximal and middle phalanges. These tendons on the index and little fingers are joined by other muscles, that on the ring finger splits, and that on the middle finger remains single. The **main action** of the extensor digitorum is to extend the fingers and secondarily to extend the hand (pull is posterior to all axes). **Nerve supply** is by the deep branch of the radial.

5. The **extensor digiti minimi** (Fig. 8-57) also **arises** from the lateral epicondyle, and after passing through the extensor retinaculum in its own compartment and its own tendon sheath, is **inserted** into the common extensor expansion with the tendon of the extensor digitorum. Due to its position its **action** is to extend the fifth digit and secondarily to extend the hand. **Nerve supply** is the deep branch of the radial.

6. **Extensor carpi ulnaris** (Fig. 8-57). This muscle **arises** from the lateral epicondyle of the humerus, is enclosed in its own compartment in the extensor ret-

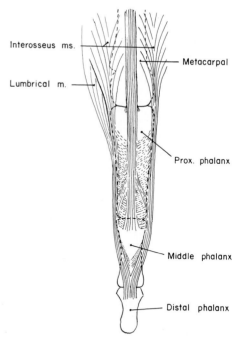

Figure 8-59. *Diagram of the extensor expansion.*

inaculum, and has its own tendon sheath. It is **inserted** onto the base of the fifth metacarpal bone (Fig. 8-10). Its **main action** is to extend and adduct the hand (pull is posterior to transverse axis of wrist and on ulnar side of anteroposterior axis). **Nerve supply** is by the deep branch of radial.

7. **Anconeus** (Fig. 8-58). This small triangular muscle is located on the posterior side of the elbow joint.* Its **origin** is the lateral epicondyle and its **insertion** is onto a triangular area on the posterior surface of the ulna and the lateral surface of the olecranon process. Its **main action** is to extend the forearm. Its **nerve supply** is the radial.

The actions of the muscles on the thumb will be understandable if one realizes that this digit has twisted in such a manner that flexion and extension are at right angles to these same movements in the fingers. This necessitates a different movement for abduction and

*See footnote on page 718.

adduction as well, abduction raising the metacarpal anterolaterally and adduction bringing the metacarpal posteromedially (in contact with the lateral edge of the palm).

---

8. Abductor pollicis longus (Fig. 8-58). This muscle lies deep in the forearm and has its **origin** from approximately the middle third of the posterior surface of the ulna and from the radius and intervening interosseus membrane (Fig. 8-8B). After crossing the extensor carpi radialis brevis and longus, it enters the extensor retinaculum in common with the extensor pollicis brevis muscle and in a synovial sheath with this muscle. It is **inserted** onto the base of the first metacarpal (Fig. 8-9). Its **main action** is to draw the metacarpal bone of the thumb anterolaterally away from the palm. **Nerve supply** is by the deep branch of the radial.

9. Extensor pollicis brevis (Fig. 8-58). This deep muscle **arises** from the middle third of the posterior surface of the radius and from the interosseus membrane (Fig. 8-8B). It courses through the extensor retinaculum in common with the abductor pollicis longus and in the same synovial sheath. It then **inserts** (Fig. 8-10) onto the base of the proximal phalanx of the thumb (earning the name ''brevis'' because of this insertion on the proximal phalanx). Its **main action** is to extend the thumb and secondarily to extend and abduct the hand. Its **nerve supply** is by the posterior interosseus branch of the radial.

10. Extensor pollicis longus (Fig. 8-58). This muscle too is located deep on the posterior surface of the forearm and **arises** from the posterior surface of the middle third of the ulna and the adjoining interosseus membrane (Fig. 8-8B). The tendon of this muscle takes an oblique course crossing the tendons of the extensor carpi radialis longus and brevis muscles to reach the thumb. It **inserts** onto the base of the distal phalanx, thus being longer than the brevis (Fig. 8-10). Its **main action** is to extend the thumb and secondarily to extend and abduct the hand. The **nerve supply** is the posterior interosseus branch of the radial.

11. Extensor indicis (Fig. 8-58). This muscle **arises** from the posterior surface of the ulna and the interosseus membrane adjoining it (Fig. 8-8B). Its tendon goes through a special opening in the extensor retinaculum, and has its own synovial sheath that joins with the sheath for the tendon of the extensor digitorum. Its

main action is to extend the index finger and secondarily to extend the hand. **Nerve supply** is the posterior interosseus branch of the radial.

12. Supinator (Fig. 8-58). This muscle completely surrounds the upper third of the radius. It **arises** from the lateral epicondyle of the humerus, and from the supinator crest and fossa of the ulna (Fig. 8-8B); after curving around posterior to the radius, it is **inserted** onto the back, lateral side, and front of the upper third of this bone (Fig. 8-8B). Its **main action** is to supinate the hand and forearm. **Nerve supply** is the deep branch of the radial, which pierces the muscle in its course.

13. Dorsal interossei (Figs. 8-58 and 8-60). The

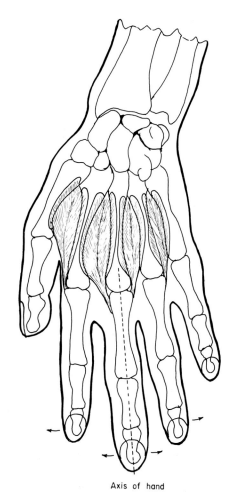

Axis of hand

Figure 8-60. *Diagram illustrating the action of the dorsal interosseus muscles in abducting the fingers.*

dorsal interossei are four muscles located between the metacarpal bones on the dorsum of the hand. They **arise** by two heads from adjacent sides of the metacarpals between which they lie. They **insert** on the proximal phalanges and on the extensor expansions. The first is inserted onto the radial side of the index finger, the second onto the radial side of the middle, the third onto the ulnar side of the middle finger, and the fourth onto the ulnar side of the ring finger. These insertions need not be memorized, for the following reasons. Since the axis of the hand is considered to be the third digit and the dorsal interossei are known as abductors of the fingers, it is mandatory that they insert in the locations just described. The second digit is moved away from the third in abduction, the middle digit is moved either way from the axis, and the ring finger is moved away from the third digit. The little finger has its own abductor and therefore does not have an interosseus muscle. It does have a ventral interosseus muscle to move it toward the midline of the hand. In addition to the action of the dorsal interossei in abducting the fingers, the insertion of these muscles into the extensor expansion as well as onto the base of the first phalanx allows an additional action of extending the middle and distal phalanges. Therefore, the **main action** of the dorsal interossei is to abduct at the metacarpophalangeal joints and to extend at the middle and distal interphalangeal joints.* The **nerve supply** is the deep branch of the ulnar. Abduction of the fingers is a good test for intactness of the ulnar nerve.

Note that all of the above muscles (except interossei) were innervated by the radial nerve and its branches.

NERVES

The nerves involved with the posterior surface of the forearm and hand are the superficial and deep branches of the radial, and the dorsal cutaneous branch of the ulnar.

---

*S. Sunderland, 1945, The actions of the extensor digitorum (communis), interosseus and lumbrical muscles, *Am. J. Anat.* 77: 189. A fine example of synergistic action of muscles; this author makes a plea for thinking of actions of muscles in relation to others rather than as isolated actions. Although the lumbrical and interosseus muscles will flex the proximal phalanges and extend the others, they really serve to aid the extensor digitorum in producing well-coordinated extension of the digits, according to this worker.

RADIAL (Figs. 8-57 and 8-58).   After giving branches to the anconeus, brachialis, brachioradialis, and extensor carpi radialis longus muscles, the radial nerve divides into its terminal superficial and deep branches and enters the forearm by crossing the anterior surface of the elbow joint on the radial side between the brachialis and the brachioradialis muscles.

In the forearm, its **superficial branch** proceeds inferiorly at a position slightly lateral to the radial artery, deep to the brachioradialis muscle and anterior to the extensor carpi radialis longus muscle (Fig. 8-51). When the distal part of the forearm is reached, the nerve courses posteriorly across the tendons of the extensor carpi radialis longus, abductor pollicis longus, and extensor pollicis brevis muscles. It then pierces the deep fascia to gain a more superficial position, crosses the extensor retinaculum at the lateral side of the radius and breaks up at this level into its terminal cutaneous branches. They cross the wrist joint at the depression between tendons on the radial side of the wrist, the so-called anatomical snuffbox. They are distributed to the skin over the radial side of the back of the hand, the radial side of the thenar eminence, and the backs of the proximal phalanges of the thumb, index, and middle fingers, and half of the ring finger. The skin on the posterior surfaces of the middle and distal phalanges is innervated by branches of the median nerve, which course around to this posterior surface from the anterior side.

The **deep branch** of the radial arises just proximal to the elbow joint, crosses the anterior surface of the joint, gives nerve supply to the extensor carpi radialis brevis and supinator muscles, winds around the lateral side of the radius while piercing the supinator muscle, and enters the deep part of the posterior surface of the forearm posterior to the interosseus membrane and abductor pollicis longus muscle. After innervating the extensor digitorum, extensor digiti minimi, extensor carpi ulnaris, and abductor pollicis longus muscles, it continues inferiorly, as the **posterior interosseus** nerve, to innervate the remaining deep muscles of the forearm—the extensor pollicis brevis, extensor pollicis longus, and extensor indicis—and to give sensory branches to the wrist and intercarpal joints.

ULNAR.   The branch of the ulnar nerve involved with the posterior surface of the forearm and hand is the **dorsal branch** of the ulnar nerve, which arises near the distal end of the forearm, and winds around to the posterior surface in a plane deep to the tendon of the flexor carpi ulnaris muscle. It then continues into the hand superficial to the extensor retinaculum and in company with the dorsal carpal branch of the ulnar artery. It is the cutaneous nerve for the posterior part of the hand on the ulnar side, and the posterior surface of the proximal phalanx of the entire little finger and the medial side of the ring finger. The exact location of these cutaneous nerves is of utmost clinical importance (Fig. 8-42).

In summary, cutaneous nerves are provided for the dorsal surface of the forearm and radial side of the hand by the superficial branch of the radial nerve and for the dorsal surface of the ulnar side of the hand by the dorsal branch of the ulnar.

The deep branch of the radial nerve and its continuation, the posterior interosseus, is entirely muscular and articular. It innervates the extensor carpi radialis brevis, supinator, extensor digitorum, extensor digiti minimi, extensor carpi ulnaris, abductor pollicis longus, extensor pollicis brevis, extensor pollicis longus, and extensor indicis muscles; and provides sensory innervation for the wrist and intercarpal joints.

ARTERIES

The arteries found on the posterior surface of the forearm and hand are (1) the posterior interosseus, (2) the termination of the anterior interosseus, (3) the dorsal carpal arch and its branches, and (4) the first dorsal metacarpal.

1. Posterior interosseus artery (Fig. 8-58). This artery is a branch of the common interosseus. It passes into the posterior aspect of the arm between the oblique cord and the superior edge of the interosseus membrane. It appears just inferior to the supinator muscle and immediately gives off an **interosseus recurrent branch** which anastomoses with the middle collateral branch of the deep brachial (Fig. 8-49). The interosseus artery continues on the posterior surface of the deep muscles on the posterior surface of the forearm and finally terminates by anastomosing with the anterior interosseus artery, after the latter has pierced the interosseus membrane, and then with the dorsal carpal arch.

2. Anterior interosseus artery (Fig. 8-58). The anterior interosseus artery appears on the posterior surface of the distal end of the forearm and wrist after piercing the interosseus membrane. It continues inferiorly in a plane deep to the deep muscles on the posterior surface of the forearm, and joins the **dorsal carpal arch.**

3. Dorsal carpal arch (Fig. 8-58). The dorsal carpal arch is primarily a branch of the radial artery after it winds around to the dorsal surface of the radial side of the hand. At this point it reaches the dorsal aspect of the first intermetacarpal space. The **dorsal carpal arch** courses medially on the carpal bones to join a smaller but similar branch from the ulnar artery. From this artery three dorsal metacarpal arteries descend to the webs of the fingers, in the second, third, and fourth intermetacarpal spaces. They divide and are distributed to the sides of the fingers. They anastomose with arteries in the palm by means of perforating arteries, of which there may be a single set or two sets—one proximal and one distal. In addition, a dorsal digital artery for the fifth finger may arise from this arch.

4. The **first dorsal metacarpal** is a branch of the radial that divides into a vessel for the index finger and one for the thumb.

VEINS

Each of the arteries just described is accompanied by a vein or veins. Combined with the superficial veins (Fig. 8-42), they provide adequate venous drainage for this area.

Many of the structures found on the posterior surface of the forearm are depicted in cross section in Figure 8-55.

HAND*

GENERAL FEATURES

Clinically the hand is of the utmost importance. Hands are constantly being injured and restoration is extremely

*F. W. Jones, 1942, *The Principles of Anatomy as Seen in the Hand,* Williams & Wilkins, Baltimore.
 E. B. Kaplan, 1953, *Functional and Surgical Anatomy of the Hand,* Lippincott, Philadelphia.
 S. Bunnell, 1956, *Surgery of the Hand,* Lippincott, Philadelphia.

important to the patient. Treatment of injuries to the hand is almost a specialty unto itself.

The surface anatomy of the hand (Fig. 8-61) reveals much of its contents. The **thenar eminence** at the base of the thumb is the location of the short muscles involved in moving the thumb, while the eminence on the ulnar side—the **hypothenar eminence**—consists of muscles involved in moving the little finger. The creases in the hand vary from individual to individual but usually consist of prominent transverse creases, which mark the position of the metacarpophalangeal joints, and of longitudinal creases, which are located between the thenar and hypothenar eminences. The **creases** in the fingers mark the location of the distal and middle interphalangeal joints, but the point where the palm and fingers join on the anterior side is not the location of the metacarpophalangeal joints. Turning the hand over reveals that this point corresponds to approximately the middle of the first phalanx. It should be noted that the thumb has only two phalanges* while the fingers have three and that the metacarpal bone of the thumb seems to be more movable than the metacarpals of the fingers.

If the skin is removed, the cutaneous nerves in the superficial fascia are seen, as shown in Figures 8-62 and 8-42. It will be recalled that branches of the median nerve are distributed to the palm on the radial side, and to the anterior surface of the thumb, index finger, middle finger, and half of the ring finger (Fig. 8-42); and further, that branches of these digital nerves wind around to the posterior surface of these particular digits to innervate the dorsal surface over the last two phalanges. The rest of the palm of the hand on the ulnar side, the entire little finger, and the ulnar side of the ring finger are innervated by the anterior cutaneous branch of the ulnar nerve. These nerves also reach the dorsal surface of the middle and distal phalanges.

Palmar aponeurosis (Fig. 8-62).   This heavy and dense layer is triangular, with apex proximally and base distally located, is continuous with the tendon of the palmaris longus muscle, and consists of heavy, longitudinally running fibers plus similar fibers coursing transversely. The lateral edges of this aponeurosis meet the deep fascia surrounding the thenar and hypothenar muscles. The aponeurosis ends, approximately opposite the metacar-

pophalangeal joints, in connective tissue strands that course longitudinally toward the base of each digit. It is continuous with deep fascial coverings of the tendons on the anterior side of the digits. Furthermore, it winds around the base of each proximal phalanx and blends with the dense fascia on the posterior surface of the fingers.

The palmar aponeurosis covers the long tendons of the digital muscles in the forearm as they stream through the hand, and also covers the arteries and nerves destined to reach the fingers. The vessels and nerves escape from the aponeurosis to proceed to the web between each two digits before dividing to reach contiguous sides of the two digits. The tendons, however, do not appear, for they are held down by the dense connective tissue bands which, as just mentioned, are directly continuous with the palmar aponeurosis.

If the deep fascia is removed from the **thenar eminence,** the elevation can be seen to be formed by three short muscles, the two most superficial muscles being the **abductor pollicis brevis** and the **flexor pollicis brevis** (Fig. 8-62). These muscles are close together and in a single plane, the abductor being placed more laterally than the flexor. The third muscle, the small **opponens pollicis,** lies deep to the other two.

Deep to the palmaris brevis muscle, the **hypothenar** eminence is made up of three muscles which function comparably to those just given for the thenar eminence. The **abductor digiti minimi** and **flexor digiti minimi** are on the same superficial plane (Fig. 8-62), and deep to these is the **opponens digiti minimi** muscle.

When the palmar aponeurosis is removed, the most superficial structure to be found in the palm of the hand is the **superficial palmar arterial arch** (Fig. 8-63). This arterial arch has its main contribution from a direct continuation of the **ulnar artery.** The smaller contribution, frequently wanting, is a branch of the **radial artery,** which arises just before this artery enters the posterior side of the thenar musculature. This superficial palmar arch usually gives rise to **four palmar digital branches.** The first goes to the ulnar side of the little finger and the remaining three go to the webs of the hand between the remaining digits. At this point they divide into two branches for contiguous sides of the fingers. The radial side of the index finger has its own branch from the radial, as we will see later.

Just deep to the palmar arterial arch and in the same

---

*From the developmental point of view, the thumb is lacking a metacarpal rather than a phalanx.

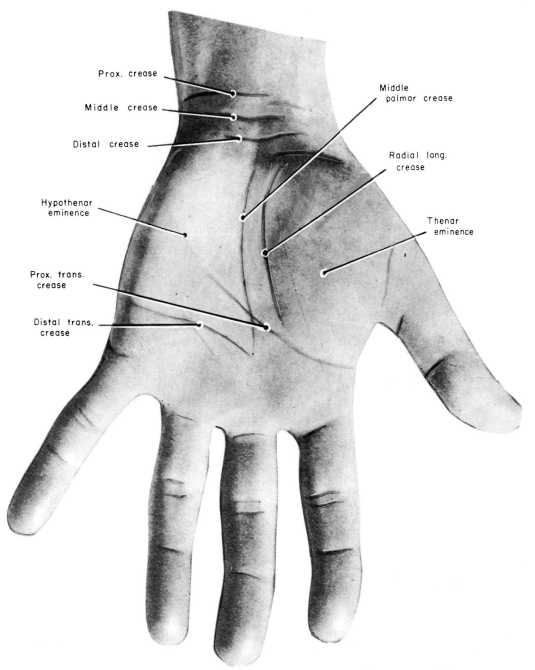

Figure 8-61. *Surface anatomy of left hand. (From Appleton, Hamilton, and Tchaperoff,* Surface and Radiological Anatomy, *courtesy of The Macmillan Press Ltd.)*

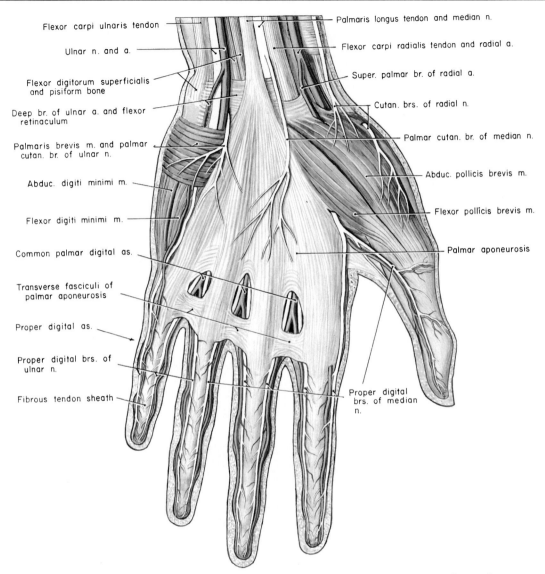

Flexor carpi ulnaris tendon

Ulnar n. and a.

Flexor digitorum superficialis and pisiform bone

Deep br. of ulnar a. and flexor retinaculum

Palmaris brevis m. and palmar cutan. br. of ulnar n.

Abduc. digiti minimi m.

Flexor digiti minimi m.

Common palmar digital as.

Transverse fasciculi of palmar aponeurosis

Proper digital as.

Proper digital brs. of ulnar n.

Fibrous tendon sheath

Palmaris longus tendon and median n.

Flexor carpi radialis tendon and radial a.

Super. palmar br. of radial a.

Cutan. brs. of radial n.

Palmar cutan. br. of median n.

Abduc. pollicis brevis m.

Flexor pollicis brevis m.

Palmar aponeurosis

Proper digital brs. of median n.

Figure 8-62. *Palm of left hand I—the palmar aponeurosis and superficial muscles in the thenar and hypothenar eminences.*

plane with the digital branches of this arch are the branches of the median and ulnar nerves to the hand (Fig. 8-63). As the **median nerve** enters the hand, it gives off a recurrent branch which supplies the muscles of the thenar eminence, and then divides into its three remaining branches. The first, the branch to the thumb and to the radial side of the index finger, also gives a branch to the first lumbrical muscle. The second branch, which goes

to the web between the index and middle fingers and then divides to go to contiguous sides of those fingers, gives off a branch to the second lumbrical. The median nerve, therefore, supplies the three muscles of the thenar eminence plus the first and second lumbricals. We will see later that all the remaining muscles of the hand are innervated by the ulnar nerve. The third branch of the median nerve joins with a branch from the ulnar to course to the

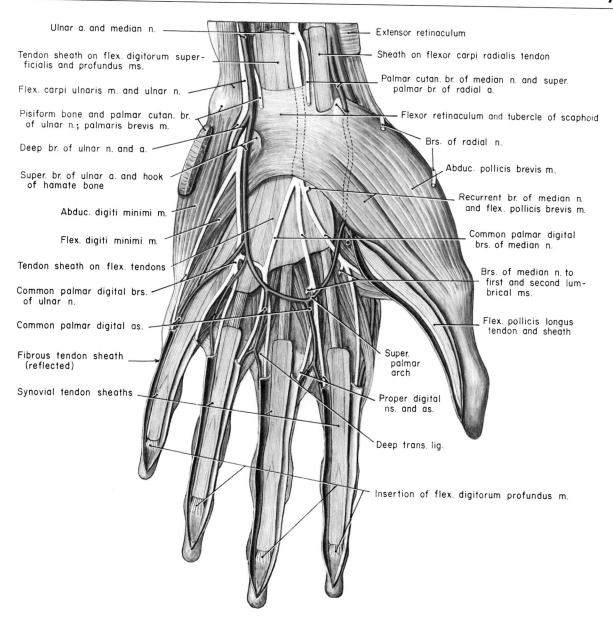

Ulnar a. and median n.

Tendon sheath on flex. digitorum superficialis and profundus ms.

Flex. carpi ulnaris m. and ulnar n.

Pisiform bone and palmar cutan. br.
of ulnar n.; palmaris brevis m.

Deep br. of ulnar n. and a.

Super. br. of ulnar a. and hook
of hamate bone

Abduc. digiti minimi m.

Flex. digiti minimi m.

Tendon sheath on flex. tendons

Common palmar digital brs.
of ulnar n.

Common palmar digital as.

Fibrous tendon sheath
(reflected)

Synovial tendon sheaths

Extensor retinaculum

Sheath on flexor carpi radialis tendon

Palmar cutan. br. of median n. and super.
palmar br. of radial a.

Flexor retinaculum and tubercle of scaphoid

Brs. of radial n.

Abduc. pollicis brevis m.

Recurrent br. of median n.
and flex. pollicis brevis m.

Common palmar digital
brs. of median n.

Brs. of median n. to
first and second lum-
brical ms.

Flex. pollicis longus
tendon and sheath

Super.
palmar
arch

Proper digital
ns. and as.

Deep trans. lig.

Insertion of flex. digitorum profundus m.

Figure 8-63. *Palm of left hand II—note the superficial arterial arch, the branches of the median and ulnar nerves, and the synovial sheaths.*

web between the middle and ring fingers. Here the nerve divides into two digital nerves, one for each side of these fingers. Hence, the median nerve supplies the anterior surface of the hand on the thumb side, including the thumb, index and middle fingers, and the radial side of the ring finger. The ulnar nerve has branches that take care of the ulnar side of the ring finger and the entire little finger. You will recall that these cutaneous nerves also innervate the skin on the dorsal side of the middle and terminal phalanges as well.

Just deep to the nerves are the tendons of the **flexor digitorum superficialis** and **flexor digitorum profundus muscles.** Each of these tendons is surrounded by a **synovial membrane;** these membranes take the form presented in Fig. 8-63. If these sheaths are removed, the tendons can be seen streaming distally across the hand to reach the digits (Fig. 8-64). Since the flexor digitorum profundus inserts on the base of the distal phalanx whereas the flexor digitorum superficialis inserts on the base of the middle phalanx, the deeper tendons have to

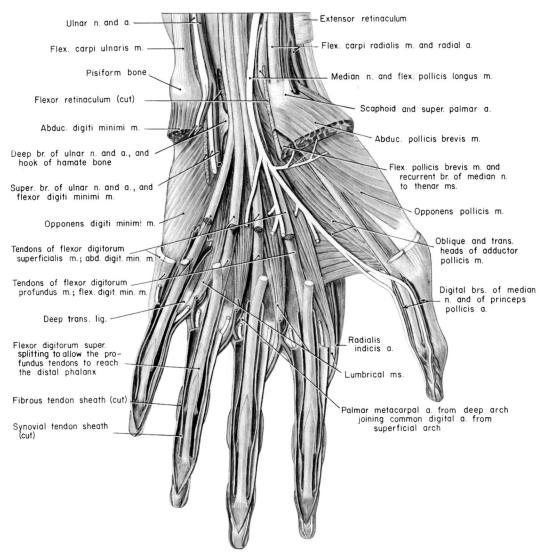

Ulnar n. and a.
Flex. carpi ulnaris m.
Pisiform bone
Flexor retinaculum (cut)
Abduc. digiti minimi m.
Deep br. of ulnar n. and a., and hook of hamate bone
Super. br. of ulnar n. and a., and flexor digiti minimi m.
Opponens digiti minimi m.
Tendons of flexor digitorum superficialis m.; abd. digit. min. m.
Tendons of flexor digitorum profundus m.; flex. digit. min. m.
Deep trans. lig.
Flexor digitorum super. splitting to allow the profundus tendons to reach the distal phalanx
Fibrous tendon sheath (cut)
Synovial tendon sheath (cut)

Extensor retinaculum
Flex. carpi radialis m. and radial a.
Median n. and flex. pollicis longus m.
Scaphoid and super. palmar a.
Abduc. pollicis brevis m.
Flex. pollicis brevis m. and recurrent br. of median n. to thenar ms.
Opponens pollicis m.
Oblique and trans. heads of adductor pollicis m.
Digital brs. of median n. and of princeps pollicis a.
Radialis indicis a.
Lumbrical ms.
Palmar metacarpal a. from deep arch joining common digital a. from superficial arch

Figure 8-64. *Palm of left hand III—after removal of the tendon sheaths, and the superficial muscles of the thenar and hypothenar eminences.*

go further distally on the digits. There must, therefore, be some means for the deeper-placed tendons to escape from their location deep to the tendons of the superficialis muscle; this is accomplished by the splitting of the tendons of the superficialis muscle to form a tunnel that is used by the profundus tendons to get to the more distal location. This ingenious arrangement is pictured in Figure 8-64. On the same level with these tendons in the palm, the four small, wormlike **lumbrical muscles** can be seen with their origins from the tendons in the palm of the

hand and their insertions onto the proximal phalanx and the extensor expansions. The lumbrical muscles course on the radial side of each metacarpophalangeal joint.

If the superficial arch is cut and the tendons of the flexor digitorum superficialis and profundus reflected, the lumbricals will be reflected as well. This reveals the deeper portions of the palm of the hand and the **adductor muscle of the thumb—the adductor pollicis** (Fig. 8-65). This muscle consists of an oblique and a transverse head, between which the deep arterial arch enters the palm of

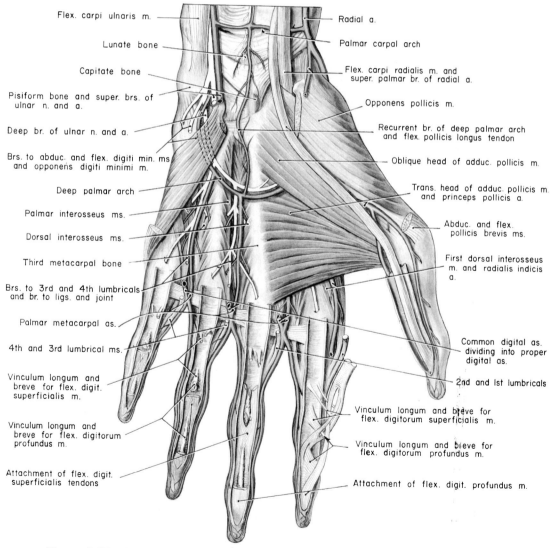

Flex. carpi ulnaris m.
Lunate bone
Capitate bone
Pisiform bone and super. brs. of ulnar n. and a.
Deep br. of ulnar n. and a.
Brs. to abduc. and flex. digiti min. ms and opponens digiti minimi m.
Deep palmar arch
Palmar interosseus ms.
Dorsal interosseus ms.
Third metacarpal bone
Brs. to 3rd and 4th lumbricals and br. to ligs. and joint
Palmar metacarpal as.
4th and 3rd lumbrical ms.
Vinculum longum and breve for flex. digit. superficialis m.
Vinculum longum and breve for flex. digitorum profundus m.
Attachment of flex. digit. superficialis tendons

Radial a.
Palmar carpal arch
Flex. carpi radialis m. and super. palmar br. of radial a.
Opponens pollicis m.
Recurrent br. of deep palmar arch and flex. pollicis longus tendon
Oblique head of adduc. pollicis m.
Trans. head of adduc. pollicis m. and princeps pollicis a.
Abduc. and flex. pollicis brevis ms.
First dorsal interosseus m. and radialis indicis a.
Common digital as. dividing into proper digital as.
2nd and 1st lumbricals
Vinculum longum and breve for flex. digitorum superficialis m.
Vinculum longum and breve for flex. digitorum profundus m.
Attachment of flex. digit. profundus m.

Figure 8-65. *Palm of left hand IV—the long flexor tendons have been removed after cutting the superficial arterial arch; this reveals the adductor muscle of the thumb.*

the hand. It then arches superiorly to join the branch from the ulnar artery that aids in the formation of this deep arterial arch. The arterial arch is also accompanied by the **deep branch of the ulnar nerve,** the nerve supply to the adductor muscles of the thumb and all of the interossei muscles, soon to be mentioned. In the palm of the hand the deep arterial arch gives rise to (1) the **princeps pollicis artery,** which courses distally and divides into two branches to be distributed on either side of the thumb; (2) the **radialis indicis artery,** which goes to the radial side

of the index finger; and (3) **palmar metacarpal branches,** which go to the webs between the fingers and join the common palmar digital branches of the superficial arch (Fig. 8-72). The princeps pollicis, the radialis indicis, and the first palmar metacarpal artery are deep to the adductor pollicis muscle, and the transverse head of this muscle has to be removed before these can be seen.

The spaces between the metacarpal bones are occupied by the **interosseus muscles** (Fig. 8-66). These muscles are divided into two sets, dorsal interossei and

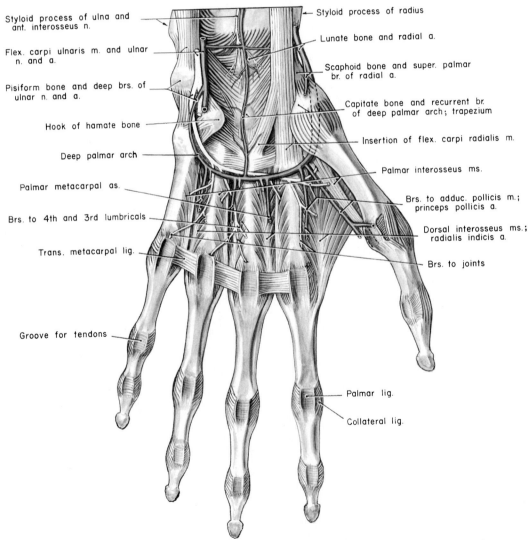

Styloid process of ulna and ant. interosseus n.

Flex. carpi ulnaris m. and ulnar n. and a.

Pisiform bone and deep brs. of ulnar n. and a.

Hook of hamate bone

Deep palmar arch

Palmar metacarpal as.

Brs. to 4th and 3rd lumbricals

Trans. metacarpal lig.

Groove for tendons

Styloid process of radius

Lunate bone and radial a.

Scaphoid bone and super. palmar br. of radial a.

Capitate bone and recurrent br. of deep palmar arch; trapezium

Insertion of flex. carpi radialis m.

Palmar interosseus ms.

Brs. to adduc. pollicis m.; princeps pollicis a.

Dorsal interosseus ms.; radialis indicis a.

Brs. to joints

Palmar lig.

Collateral lig.

Figure 8-66. *Palm of left hand V—the adductor pollicis muscle has been removed.*

palmar interossei. The bellies of these muscles are in the spaces between the metacarpal bones; the tendons can be seen on either side of the metacarpophalangeal joints.

MUSCLES

The muscles of the hand can be divided into those forming the thenar eminence, those forming the hypothenar eminence, the adductor muscle of the thumb, the long flexor tendons and the attached lumbrical muscles, and the interossei.

THENAR MUSCLES. The muscles forming the thenar eminence are the (1) abductor pollicis brevis, (2) flexor pollicis brevis, and (3) the opponens pollicis. The insertions of these muscles are more important than their origins. And, as stated previously, the action of the muscles of the thumb are understood more clearly if it is noted that the thumb flexes and extends at right angles to the directions taken by the fingers when they flex and extend. Therefore, abduction is moving the thumb anteriorly and slightly laterally if it is in the anatomical position, while adduction is moving the thumb toward the second metacarpal bone (posteromedially). Flexion is bending the thumb across the palm of the hand, while extension is movement in the opposite direction, away from the palm.

1. Abductor pollicis brevis (Fig. 8-62). This muscle is the superficial muscle on the radial side of the thenar eminence. It **arises** from the scaphoid and trapezium (Fig. 8-9) and is **inserted** onto the radial side of the base of the proximal phalanx of the thumb (Fig. 8-9). Its **main action** is to abduct the thumb at the carpometacarpal joint and to slightly flex the thumb. This type of abduction may be called "short abduction" and is a little different from the abduction induced by the abductor pollicis longus muscle—"long abduction." True abduction is a combination of the "short" and "long" abduction. **Nerve supply** is the median.

2. Flexor pollicis brevis (Fig. 8-62). This muscle is also superficially placed on the thenar eminence but on the ulnar side. It **arises** from the flexor retinaculum and also the trapezium (Fig. 8-9). It is **inserted** onto the proximal phalanx in a common tendon with the abductor pollicis brevis.* Its **main action** is to flex the thumb, the

---

*Some describe a medial or deep portion of this muscle, but its attachments and nerve supply resemble the palmar interossei. If present, it may be better to consider it as a first palmar interosseus muscle.

action taking place at both the carpometacarpal and metacarpophalangeal joints. A sesamoid bone is found in the tendon of this muscle. **Nerve supply** is the median.

3. Opponens pollicis (Fig. 8-64). This muscle is deep to the preceding two muscles; it **arises** from the flexor retinaculum and the trapezium and is **inserted** onto practically the entire length of the radial border of the first metacarpal bone (Fig. 8-9). Its **main action** is to oppose the thumb or to bring the first metacarpal bone across the surface of the palm. Such an action allows the thumb to touch the ends of each of the fingers. The action of opposition is extremely important to us. **Nerve supply** is by the median.

HYPOTHENAR MUSCLES. The muscles of the hypothenar eminence are the (1) abductor digiti minimi, (2) flexor digiti minimi, and (3) opponens digiti minimi.

1. Abductor digiti minimi (Fig. 8-63). This muscle is quite superficially placed on the ulnar side of the hypothenar eminence, **arises** from the pisiform bone, and **inserts** onto the base of the proximal phalanx of the little finger (Fig. 8-9). Its **main action** is to abduct and slightly flex the little finger at the metacarpophalangeal joint. **Nerve supply** is the ulnar.

2. Flexor digiti minimi (Fig. 8-63). This muscle lies next to the abductor digiti minimi and is on the radial side of the hypothenar eminence. It **arises** from the hamate bone and is inserted onto the base of the proximal phalanx of the fifth digit in a common tendon with the abductor digiti minimi muscle (Fig. 8-9). Its **main action** is to flex the fifth digit at the metacarpophalangeal joint. **Nerve supply** is the ulnar.

3. Opponens digiti minimi (Fig. 8-64). This muscle lies deep to the other two muscles of the hypothenar eminence, **arises** from the hamate bone, and is **inserted** onto the ulnar border of the fifth metacarpal bone (Fig. 8-9). It **pulls** the metacarpal bone toward the middle of the palm. **Nerve supply** is the ulnar.

Note the similarity in muscle arrangement in the thenar and hypothenar eminences.

ADDUCTOR MUSCLE OF THE THUMB. The **adductor pollicis** muscle (Fig. 8-65) lies deep in the radial side of the palm and consists of two heads, an oblique and a transverse. The oblique head **arises** from the carpal bones and from the bases of the metacarpal bones, and the transverse head **arises** from the shaft of the third metacarpal bone. The two heads join together and are **inserted** onto the base of the proximal phalanx of the thumb (Fig. 8-9). The **main action** is to adduct the thumb or draw it toward the second metacarpal bone. A sesamoid bone is often found in the tendon of the adductor muscle. **Nerve supply** is the ulnar.

MIDPALMAR MUSCLES. The midpalmar muscles consist of (1) the tendons of the flexor digitorum superficialis, (2) those of the flexor digitorum profundus, and (3) the lumbrical muscles.

1. The **tendons of the flexor digitorum superficialis** muscle course between the flexor retinaculum anteriorly and the flexor profundus tendons posteriorly (Fig. 8-64). The tendons for the middle and ring fingers are anterior to those for the index and little fingers. Each tendon courses through the palm towards its proper digit anterior to the tendon of the flexor digitorum profundus muscle. These tendons are posterior to the palmar aponeurosis, the superficial arterial arch, and the branches of the median nerve. Both sets of tendons enter the fibrous flexor sheath opposite the head of each metacarpal bone. Anterior to the proximal phalanx, each of the flexor digitorum superficialis tendons splits to allow the profundus tendon to continue distally through it. Opposite the base of the middle phalanx, the two halves of the superficial tendons reunite and form a gutter in which

the profundus tendon lies. The two halves of each tendon of the superficialis muscle insert onto the sides of the middle phalanx (Fig. 8-9).

2. The **flexor digitorum profundus tendons,** as they course through the wrist, are more or less all on the same plane, that to the index finger being separated from the other three. These tendons course posterior to those of the flexor digitorum superficialis muscle and are anterior to the deep palmar arch, the adductor muscle, and the interossei muscles and metacarpal bones. Each enters the fibrous flexor sheath with a flexor digitorum superficialis tendon, passes through the tunnel formed by the superficialis tendon, and finally inserts on the base of the distal phalanx (Fig. 8-9).

Figure 8-67 shows the tendons in a finger. Within the fibrous sheath the two flexor tendons are connected to each other and to the phalanges by slender tendinous bands called **long and short vinculae.** Blood vessels course in these bands.

3. The **lumbrical muscles** are four in number and arise from the tendons of the flexor digitorum profundus muscle (Fig. 8-64). Each passes distally in the same plane with the tendons, crosses anterior to the deep transverse ligament (between the heads of the metacarpal bones), and ends in a tendon that passes on the radial side of the metacarpophalangeal joint to be inserted onto the radial side of the extensor tendon of the finger. These muscles

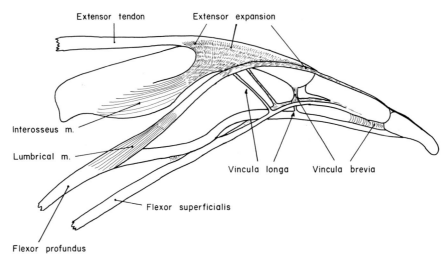

Figure 8-67. *Diagram showing the arrangement of the flexor and extensor tendons. The flexor tendons have been pulled away from the bone.*

flex the metacarpophalangeal joints while simultaneously extending the interphalangeal joints.* The first action is accomplished by the fascial attachment onto the first phalanx and the second by the insertion onto the extensor expansion. The **nerve supply** to the first two lumbricals is by the median nerve, the last two by the ulnar nerve.

INTEROSSEI MUSCLES

INTEROSSEI MUSCLES (Figs. 8-66 and 8-68). The muscles located between the metacarpal bones are very appropriately called the interossei muscles. They are divided into dorsal and palmar interossei although they are practically at the same plane. If we bear in mind that the middle digit is the axis of the hand, and that dorsal interossei abduct while the palmar interossei adduct the fingers, it is easy to place the interossei in their proper locations. The dorsal interossei (Fig. 8-60) have been described on page 735.

Since the palmar interossei adduct the fingers, one will be needed between the second and third metacarpals to insert on the ulnar side of the index finger to move that toward the middle finger, one between the third and fourth metacarpal bones to move the ring finger toward the middle finger, and one between the fourth and fifth metacarpal bones to move the little finger toward the ring finger (Fig. 8-68). There is frequently a small remnant of an additional palmar interosseus muscle† associated with the thumb.

The interossei muscles are located between the metacarpal bones and **arise** from the sides of these bones. They cross the metacarpophalangeal joints and the tendons pass posterior to the deep transverse ligament. They **insert** onto the bases of the proximal phalanges and also onto the extensor expansions. Their **main action,** therefore, is to abduct (dorsal interossei) and adduct (palmar interossei) the fingers and also to extend the interphalangeal joints because of their attachment to the extensor expansions. The **nerve supply** to all the interossei muscles is the deep branch of the ulnar nerve; abduction and adduction of the fingers is a good test for intactness of this nerve.

---

*See S. Sunderland (footnote on page 736).

†This muscle (called the medial or deep part of the flexor pollicis brevis muscle by some) arises from the ulnar side of the base of the first metacarpal bone and inserts with the oblique head of the adductor pollicis muscle on the first phalanx of the thumb.

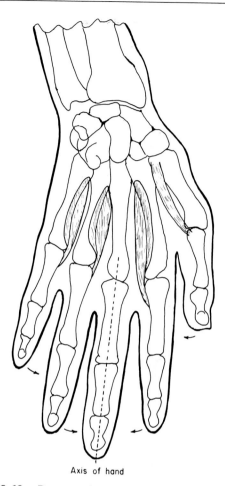

Axis of hand

Figure 8-68. *Diagram showing how the palmar interossei adduct the fingers.*

TENDON SHEATHS

TENDON SHEATHS. Each of the long tendons in the hand is surrounded (1) by a tendon sheath, which is a double-layered, elongated sac composed of a synovial membrane and containing synovial fluid (Fig. 8-63), and (2) a fibrous sheath outside the synovial sheath, which serves to hold the tendon onto the phalanges and their joints during flexion (Fig. 8-62).

The synovial sheath on the flexor pollicis longus tendon starts at the base of the terminal phalanx of the thumb and extends proximally to an inch above the crease in the wrist. This is known clinically as the **radial bursa.**

The synovial sheaths on the remaining digits commence at the base of the terminal phalanges and extend proximally to about the middle of the palm. At this point the sheaths of the index, middle, and ring fingers usually terminate. However, that of the little finger joins a large dilation that is made up of the combined tendon sheaths of all the flexor tendons, both superficial and deep. This continues into the wrist deep to the flexor retinaculum, and is known as the **ulnar bursa.** A study of Figure 8-69 will show that the tendons have indented the ulnar side of this sheath and thereby obtain a layer intimately associated with them, which is separated from the outside layer by the synovial fluid; the tendons are not actually inside the bursa.

It is easily seen that infections in these synovial sheaths can spread rapidly throughout their entirety. Infections in the thumb, if within this sheath, will spread to the wrist. Similar results are obtained with infection in the little finger. Those in the first, middle, or ring fingers, however, will spread only to the palm, owing to the interruption in these synovial sheaths.

COMPARTMENTS. Frequent reference to Figure 8-69, showing a cross section of the hand, will greatly aid your comprehension of the following description. Those who have studied the lower limb will recall that the muscles to the great toe were contained in a medial compartment; those to the fifth toe, in a lateral compartment; the flexor digitorum longus, flexor digitorum brevis, quadratus plantae, and lumbricals, in a central compartment; and the adductor hallucis and interosseus muscles, in an adductor-interosseus compartment. It should be no surprise to find the hand has a similar arrangement.

The **deep fascia of the hand** is directly continuous with that of the wrist and with the flexor and extensor retinacula. This fascia has definite attachments to the deep muscles and the metacarpal bones, and these attachments occur in such a manner as to divide the hand into **fascial compartments.** That bounding the muscles in the thenar eminence forms the **thenar compartment,** that bounding the muscles in the hypothenar, the **hypothenar compartment.** There is a **midpalmar compartment,** which is formed by the palmar aponeurosis anteriorly and deep fascia on the interosseus muscles and the adductor muscles of the thumb posteriorly. This midpalmar compartment is divided into two areas, an anterior one containing nerves and vessels, and a posterior occupied by the tendons of the long muscles of the

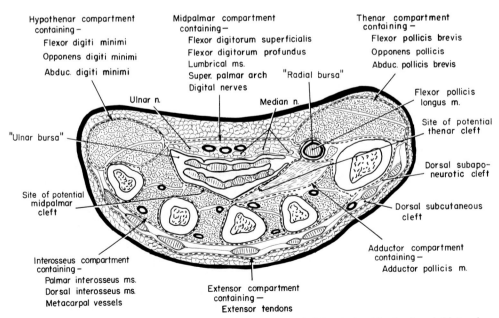

Figure 8-69. *Diagram of the compartments (surrounded by dashes) in the hand. Note also the potential clefts or spaces.*

forearm plus their tendon sheaths. The deep fascia makes still another compartment in the palm of the hand containing the adductor muscle of the thumb—**the adductor compartment.** Thus there are four compartments in the palm: (1) thenar, (2) hypothenar, (3) midpalmar, and (4) adductor. Furthermore, another compartment, the **interosseus,** contains the interosseus muscles and metacarpal vessels, and still another on the extensor surface contains the extensor tendons and is termed the **extensor compartment.** These are important from the point of view of containment or spread of infections.

FASCIAL CLEFTS OR SPACES. In addition to tendon sheaths and fascial compartments, there are clefts in the hand (sometimes called spaces but in actuality only potential spaces) which may be located between fascial compartments or within them. The following is based on the description by Kanavel; Grodinsky and Holyoke disagree on several points.*

If one lifts the combined tendons of the flexor digitorum superficialis and profundus muscles, a membrane is found to be attached to the fascia around these tendons and to the third metacarpal bone. This membrane separates the cleft made by this reflection into two parts, one part on the ulnar side just anterior to the interosseus muscles—the **middle palmar or midpalmar cleft,** and another just anterior to the adductor pollicis muscle—the **thenar cleft** (Fig. 8-70). These clefts are continuous with the fascial tubes surrounding the lumbrical muscles. Another cleft occurs between the adductor muscle of the thumb and the interosseus muscles. There are two additional clefts on the dorsum of the hand. The **dorsal subcutaneous cleft** is located between the superficial and deep fascial layers, in other words, superficial to the extensor tendons. The **dorsal subaponeurotic fascial cleft** lies between these long tendons and the fascia on the dorsal interossei muscles. These clefts are shown in Figure 8-69.

These clefts are important in understanding location and potential spread of infection in the hand. Infections can be found in these clefts, in muscle compartments, or in tendon sheaths and such infections will be contained or

---

*A. B. Kanavel, 1939, *Infections of the Hand,* Lea and Febiger, Philadelphia.

M. Grodinsky and E. A. Holyoke, 1941, The fascia and fascial spaces of the palm, *Anat. Record* 79: 435.

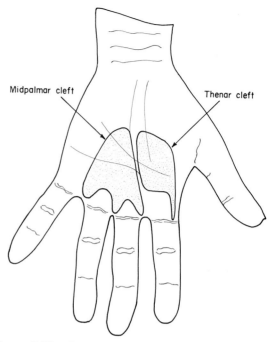

Figure 8-70. *Diagram of the midpalmar and thenar clefts.*

spread accordingly. Of course, further spread can be expected if the infection invades lymphatics or the vascular system.

NERVES

The nerves in the hand are branches of the median, ulnar, and radial.

MEDIAN (Fig. 8-63). The median nerve descends into the hand deep to the flexor retinaculum. It divides immediately into lateral and medial divisions which present themselves in the palm of the hand superficial to the tendons and deep to the superficial palmar arch.

The **lateral division** immediately gives a **recurrent branch** to supply the three muscles of the thenar eminence and then divides into three **digital** nerves. Two of these nerves course on the anterior surface of the adductor pollicis muscle and are distributed to the thumb in company with branches of the princeps pollicis artery on either side of the tendon of the flexor pollicis longus muscle. The remaining digital branch passes distally on the anterior surface of the first lumbrical and is distributed on the radial side of the index finger. This digital nerve has a branch that supplies the first lumbrical muscle.

The **medial division** divides into two **digital** nerves that pass distally on the superficial surfaces of the middle two lumbrical muscles. They supply the second lumbrical. Each of the digital nerves splits into two, which supply contiguous sides of the index and middle fingers, and of the middle and ring fingers.

All these digital nerves have branches that wind around the fingers to reach the dorsal side of the middle and distal phalanges. Furthermore, there is a communication between the more medial digital branch and a digital branch of the ulnar nerve that accounts for the considerable amount of variation in cutaneous nerve distribution.

The median nerve, therefore, is the motor nerve supply to the muscles of the thenar eminence as well as the first and second lumbricals. It is the sensory nerve to the anterior surface of the palm, the thumb, index, middle, and radial side of the ring fingers. The cutaneous nerves of the median extend onto the dorsal surface of the middle and terminal phalanges. Those who have studied the lower limb will note that this nerve corresponds to the medial plantar nerve.

ULNAR (Fig. 8-63). The ulnar nerve enters the hand on the anterior surface of the flexor retinaculum, in contrast to the median nerve, which passes deep to it. The ulnar nerve courses deep to the palmaris brevis muscle and then divides into its superficial and deep branches.

The **superficial branch** innervates the palmaris brevis muscle and then passes distally and divides into two digital branches. One of these digital branches innervates the ulnar side of the little finger; the other follows the fourth lumbrical muscle to reach the web between the ring and little fingers, where it divides into the branches for contiguous sides of those fingers. These nerves supply the distal ends of the phalanges on the dorsal surface, as described for the median nerve.

The **deep branch** follows the course taken by the deep branch of the ulnar artery: between the abductor digiti minimi and flexor digiti minimi, and deep to the opponens digiti minimi. It supplies these muscles and then crosses the palm following the deep palmar arch. It gives branches to all the interosseus muscles, the third and fourth lumbrical muscles, and to joints and bones of the hand, and terminates in branches to the oblique and

transverse heads of the adductor pollicis muscle. The ulnar supplies all the muscles in the hand not supplied by the median nerve.

RADIAL. The posterior side of the hand receives the radial nerve on the radial side and the ulnar nerve on the ulnar side of the hand. These dorsal branches have already been described (page 712). Figure 8-71 summarizes the blood and nerve supply to a digit.

ARTERIES

The arteries in the hand are the distal ends of the radial and ulnar arteries.

RADIAL. The radial artery at the wrist crosses the distal end of the radius and then winds around the trapezium and first metacarpal bone to gain its dorsal surface (Fig. 8-72). It then pierces the first dorsal interosseus muscle, thereby gaining the space between the first and second metacarpal bones. After penetrating the first dorsal interosseus muscle, the artery enters the palm by passing between the oblique and transverse heads of the adductor pollicis muscle (Fig. 8-65). The deep arch is formed by joining a smaller branch from the ulnar artery. This arch is just anterior to the interossei muscles and metacarpal bones, and just posterior to the combined tendons of the flexor digitorum profundus and superficialis muscles and their attached lumbricals. This arch is more proximally located than the superficial arch and is accompanied by the deep branch of the ulnar nerve.

The branches of the radial artery at the wrist (already described for the forearm, but repeated for the sake of completeness) and in the hand are the following:

1. Palmar radial carpal
2. Superficial palmar
3. Dorsal carpal
    a. Dorsal metacarpal and dorsal digital
4. First dorsal metacarpal
5. Princeps pollicis
6. Radialis indicis
7. Deep palmar arch
    a. Recurrent
    b. Perforating
    c. Palmar metacarpals

1. The **palmar radial carpal artery** (Figs. 8-65

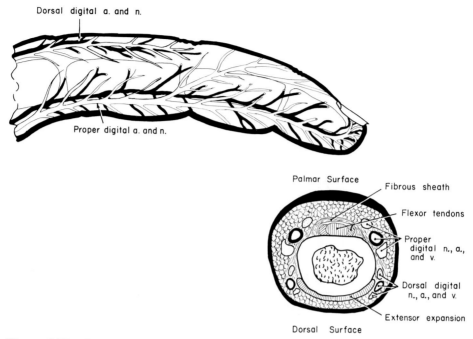

Dorsal digital a. and n.

Proper digital a. and n.

Palmar Surface

Fibrous sheath

Flexor tendons

Proper digital n., a., and v.

Dorsal digital n., a., and v.

Extensor expansion

Dorsal Surface

Figure 8-71. *Diagram of blood vessels and nerves of a finger. Transverse section is through distal end of proximal phalanx.*

and 8-72) runs medially on the distal end of the radius and deep to the flexor tendons, and forms the anterior carpal arch by joining a similar branch from the ulnar artery.

2. The **superficial palmar artery** (Figs. 8-63 and 8-72) descends across the wrist joint and either passes through or superficial to the thenar muscles near their origins, to terminate by completing the superficial palmar arch.

3. The **dorsal carpal branch** (Figs. 8-58 and 8-72). This small vessel crosses the wrist and forms the dorsal carpal arch and network with a similar branch from the ulnar artery, and with contributions from the anterior interosseus artery. Three rather slender **dorsal metacarpal** arteries arise from this network and course distally on the second, third, and fourth dorsal interosseus muscles and then bifurcate into the dorsal digital branches to supply adjacent sides of the middle, ring, and little fingers. They communicate with the palmar digital branches of the superficial palmar arch.

4. First dorsal metacarpal (Figs. 8-57 and 8-72). This branch arises just before the radial artery en-

ters the first dorsal interosseus muscle; it divides into two branches, which supply adjacent sides of the thumb and index finger.

5. The **princeps pollicis artery** (Figs. 8-65 and 8-66) arises from the radial after it penetrates the first dorsal interosseus muscle and reaches a position deep to the oblique and transverse heads of the adductor pollicis muscle. The princeps immediately turns distally and divides into two branches that continue to the thumb on either side of the flexor pollicis longus tendon.

6. The **radialis indicis artery** (Figs. 8-64 and 8-65) arises from the deep arch very close to the origin of the princeps pollicis; in fact, it may arise as a common stem with this artery. The radialis indicis courses distally between the first dorsal interosseus muscle posteriorly and the transverse head of the adductor pollicis anteriorly. It reaches the metacarpophalangeal joint and courses distally on the radial side of the index finger.

7. The continuation of the radial artery in the hand is considered to be the **deep palmar arch** (Figs. 8-66 and 8-72). This arch has three sets of branches: (a) recurrent, (b) perforating, and (c) palmar metacarpal. (a) The **recur-**

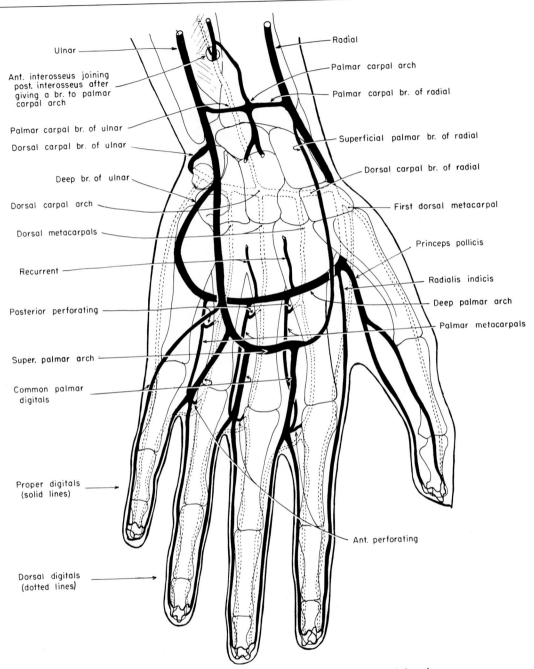

Figure 8-72. *Diagram summarizing the arteries of left hand.*

rent **branches** course superiorly over the front of the wrist joint and anastomose with the anterior carpal arch. (b) The **perforating branches** pass posteriorly through the upper part of the spaces between the metacarpal bones by passing through the two heads of the dorsal interossei muscles. They join the dorsal metacarpal arteries. (c) The **palmar metacarpal** branches are three in number; they descend on the anterior surface of the interosseus muscles of the medial three spaces and finally join the common digital branches of the superficial arch just before the latter branches bifurcate.

ULNAR. The ulnar artery passes anterior to the flexor retinaculum and terminates at this point by dividing into deep and superficial branches (Figs. 8-63 and 8-72).

The branches of the ulnar artery at the wrist and in the hand are:

1. Palmar carpal
2. Dorsal ulnar carpal
3. Deep branch
4. Superficial branch—the superficial palmar arch
   a. Common digital and proper digitals

1. The **palmar carpal branch** (Fig. 8-72) joins with the palmar carpal branch of the radial to form the palmar carpal arch.

2. The **dorsal ulnar carpal branch** (Fig. 8-72) courses deep to the flexor carpi ulnaris muscle, passes posteriorly by crossing the ulnar side of the wrist joint, and continues laterally in a plane deep to the extensor carpi ulnaris to join a similar branch of the radial artery to form the dorsal carpal arch.

3. The **deep branch** (Figs. 8-63 and 8-72) of the ulnar courses between the abductor digiti minimi and flexor digiti minimi muscles near their origins and then passes deep to the opponens digiti minimi to join the deep palmar arch.

4. The **superficial palmar arch** (Figs. 8-63 and 8-72) is the direct continuation of the ulnar artery. This arch is superficial to the tendons in the hand and also superficial to the lumbrical muscles. The nerves—branches of the median and ulnar—are also posterior to the superficial palmar arch. The arch is more distally located than is the deep palmar arch. The branches of the superficial palmar arch are the **four common palmar digital arteries,** the first being distributed to the ulnar side of the little finger and the other three lying in the

intervals between the long flexor tendons. They reach the webs of the fingers and then divide into two branches, the **proper digitals,** one for each of the contiguous sides of the fingers. Each artery, just before it divides into its two terminal branches, is joined by the metacarpal branches of the deep palmar arch; the **anterior perforating arteries** also emerge at this point, joining the common palmar digital arteries with the dorsal metacarpals.

Figure 8-71 shows the exact relation of the digital arteries to the digital nerves and shows the termination of the arteries forming a plexus in the bed of the nail and in the pulp of the fingers.

Figure 8-72 summarizes the vascular supply to the hand.

VEINS

All the above-described arteries are accompanied by venae comitantes.

JOINTS OF UPPER LIMB

Joints are extremely important, for it is only at these locations that motion of the limb occurs. Furthermore, they are important clinically. Dislocations, ligamentous tears, and inflammations from various causes are but a few of the disease conditions that can occur. A review of the structure of freely movable joints, found on page 14, would be helpful at this time.

The joints of the upper limb are:

1. Sternoclavicular
2. Acromioclavicular
3. Shoulder
4. Elbow
5. Superior and inferior radioulnar
6. Wrist
7. Intercarpal
8. Carpometacarpal
9. Intermetacarpal
10. Metacarpophalangeal
11. Interphalangeal

The sternoclavicular joint was described on page 696, the acromioclavicular on page 699.

SHOULDER JOINT

The shoulder joint is a synovial articulation of the ball-and-socket variety. The head of the humerus articulates with the glenoid cavity of the scapula. This cavity is deepened by a ring of cartilage, the **glenoid labrum,** which is attached to the edges of the glenoid cavity and is triangular in cross section. All movements are allowed: flexion and extension on a transverse axis, abduction and adduction on an anteroposterior axis, circumduction, and medial and lateral rotation on a vertical axis.

The **ligaments** of the shoulder joint are:

1. Capsular
2. Coracohumeral
3. Glenohumeral
4. Transverse
5. Coracoacromial

1. Capsular ligament (Fig. 8-73). This thin but strong capsule completely surrounds the shoulder joint. It is attached to the edges of the glenoid labrum medially and to the anatomical neck of the humerus laterally. It is also attached to the **transverse ligament** (see item 4) as it bridges the intertubercular groove. The **long tendon of the biceps** is contained within the capsular ligament and attached to the superior aspect of the glenoid labrum. The capsular ligament usually has at least one opening in it; we will see later that the synovial membrane protrudes through this opening, located near the base of the coracoid process, to form the subscapular bursa. Other openings occasionally occur which allow formation of the infraspinatus bursa and a communication between the subacromial bursa and the shoulder. Another opening occurs for the tendon of the biceps muscle.

2. Coracohumeral ligament (Fig. 8-74). This is a strong, thick band that starts on the base of the coracoid process and stretches laterally to attach to the capsular ligament near the greater tubercle. It is a thickening in the capsule and is superiorly located between the supraspinatus and subscapularis muscles.

3. The **glenohumeral ligaments** (Fig. 8-74) are anteriorly located and are best seen on the inside of the capsular ligament. There are three of these ligaments, all of which arise from the glenoid labrum. The superior glenohumeral ligament attaches laterally to an area just superior to the lesser tubercle, the middle to the lesser tubercle itself, and the inferior to the medial side of the

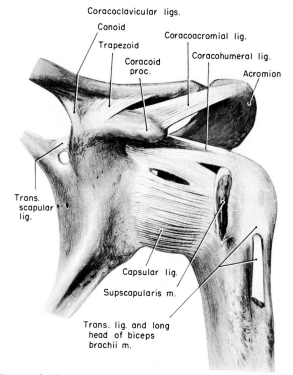

Figure 8-73. *Anterior view of left shoulder. Apertures in capsule are for connections between the joint cavity and surrounding bursae. (From Quain,* Elements of Anatomy, *courtesy of Longman Group Ltd.)*

neck of the humerus. These ligaments are often difficult to identify.

4. Transverse ligament (Fig. 8-73). This bridges the bicipital groove and is directly continuous with the capsular ligament. It has the important function of holding the biceps tendon in place.

5. The **coracoacromial ligament** (Figs. 8-73 and 8-74), although not exactly part of the joint, is an accessory ligament of the shoulder. It aids in forming an arch for protection of the shoulder joint. It is attached to the lateral border of the coracoid process and to the tip of the acromion, and is in contact with the deltoid muscle superiorly and the infraspinatus inferiorly. The subacromial bursa lies between the infraspinatus muscle and this ligament.

Synovial membrane and bursae. The inside of the capsule is lined with synovial membrane that folds back on itself to cover the glenoid labrum medially and

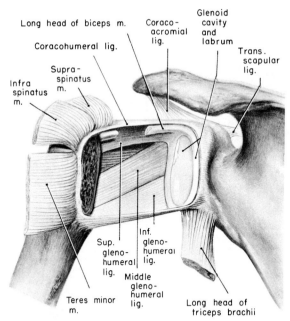

Figure 8-74 labels:
Long head of biceps m. — Coraco-acromial lig. — Glenoid cavity and labrum — Trans. scapular lig. — Coracohumeral lig. — Supra-spinatus m. — Infra spinatus m. — Sup. gleno-humeral lig. — Inf. gleno-humeral lig. — Middle gleno-humeral lig. — Teres minor m. — Long head of triceps brachii

Figure 8-74. *A posterior view of the anterior capsule of the left shoulder joint, made possible by removal of the head of the humerus. (From Quain, Elements of Anatomy, courtesy of Longman Group Ltd.)*

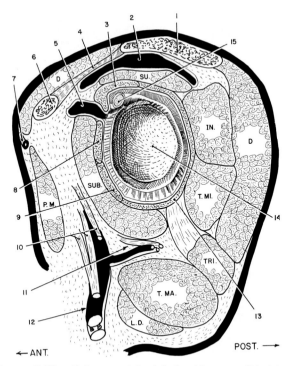

Figure 8-75. *Relations of the left shoulder joint: (D) deltoid, (SU.) supraspinatus, (IN.) infraspinatus, (P.M.) pectoralis major, (SUB.) subscapularis, (T.MI.) teres minor, (T.MA.) teres major, (TRI.) long head of triceps brachii, (L.D.) latissimus dorsi, (1) acromion, (2) subacromial bursa, (3) superior glenohumeral ligament, (4) coracoacromial ligament, (5) subscapular bursa, (6) coracoid process of scapula, (7) cephalic vein, (8) middle glenohumeral ligament, (9) inferior glenohumeral ligament, (10) parts of brachial plexus and its branches, (11) axillary nerve and posterior humeral circumflex vessels traversing quadrilateral space, (12) axillary vessels, (13) capsular ligament, (14) glenoid cavity, (15) biceps tendon and synovial sheaths. (Modified after Jamieson.)*

the neck of the humerus distally; it sends a tube along the biceps tendon which follows it into the bicipital groove, and extends out through the apertures in the capsular ligament to form bursae deep to the subscapularis muscle and deep to the infraspinatus muscle, the subscapular and infraspinatus bursae, respectively. These two bursae are, therefore, connected directly with the shoulder joint cavity; the subacromial bursa may be connected but usually is not.

In addition to the bursae mentioned above, a bursa occurs superficial to the acromion, and there are bursae associated with the coracobrachialis, teres major, long head of the triceps, and latissimus dorsi muscles.

Bursae become inflamed and are very painful.

Nerves and arteries. The suprascapular, upper subscapular, and axillary nerves all supply the shoulder joint, while the arteries are branches of the suprascapular and posterior humeral circumflex arteries.

Relations. The structures in relation to the shoulder joint are shown in Figure 8-75. Those in close relation to the capsular ligament are the subscapularis muscle and

its bursa, anteriorly; the supraspinatus muscle, subacromial bursa, and coracoacromial ligament and acromion, superiorly; the infraspinatus muscle and its bursa and the teres minor muscle, posteriorly; and the subscapularis, teres major muscles, the long head of the triceps brachii, and the contents of the quadrilateral space, inferiorly. These muscles serve to support the freely movable shoulder joint by keeping the head of the humerus in close contact with the glenoid cavity. The location of least muscle support is inferior to the joint; dislocation, a fairly common event, usually occurs in this direction. Disloca-

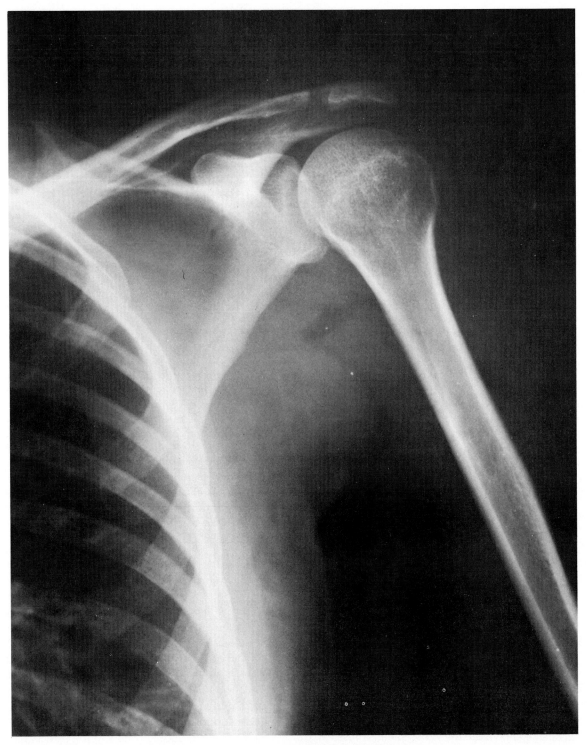

Plate 24. *Anterior view of normal shoulder joint.*

tion usually occurs when the arm is abducted; this position should be avoided after reduction has been accomplished. In ankylosis of the joint, the movement is taken over by the scapula.

ROTATOR CUFF

As mentioned on page 706, the tendons of the subscapularis, supraspinatus, infraspinatus, and teres minor muscles strengthen the capsule of the shoulder joint and hold the head of the humerus in the glenoid cavity. They prevent displacement anteriorily, superiorly, and posteriorly. Since at least three of these muscles are involved in rotation at the shoulder joint, this musculotendinous cuff is called a rotator cuff. Many problems around the shoulder are related to this cuff.

ELBOW JOINT

The elbow joint consists of an articulation between the trochlea of the humerus and the trochlear notch of the ulna, and the capitulum of the humerus and the head of the radius. It is a synovial joint of the hinge variety. Only two movements are allowed—flexion and extension—both on a transverse axis.

The joint is surrounded by a capsule which is thickened medially and laterally to form (1) ulnar collateral and (2) radial collateral ligaments respectively. The areas between these two ligaments form the (3) anterior and (4) posterior ligaments.

1. The **ulnar collateral ligament** is triangular (Fig. 8-76). It is attached to the medial epicondyle of the humerus proximally and the coronoid process and olecranon of the ulna distally; another band of tissue stretches between the two latter points, completing the triangle.

2. Radial collateral ligament (Fig. 8-77). This ligament extends from the lateral epicondyle proximally to the annular ligament (surrounding the head of the radius) distally.

3. Anterior ligament (Fig. 8-76). This thin part of the capsule attaches to the humerus just superior to the radial and coronoid fossae and to the medial epicondyle, and inferiorly to the coronoid process and the annular ligament.

4. The **posterior ligament** (Fig. 8-77) is also quite thin and attaches to the margins of the olecranon fossa and the lateral epicondyle superiorly, and the olecranon process inferiorly.

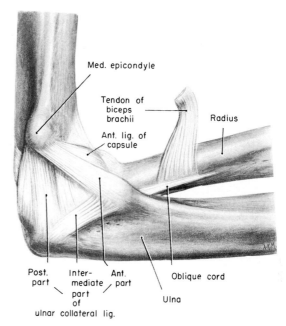

Figure 8-76. *Elbow joint—medial view. (From Quain,* Elements of Anatomy, *courtesy of Longman Group Ltd.)*

Synovial membrane. The synovial membrane lines the inside of the capsule and reflects onto the bones until the articulatory surfaces are reached. The synovial cavity of the elbow joint proper is directly continuous with the cavity of the superior radioulnar joint. These two joints should not be regarded as a single joint; the elbow flexes and extends only, around a single transverse axis.

Nerves and arteries. The elbow joint receives branches from the median, musculocutaneous, ulnar, and radial nerves. Its arterial supply comes from branches of those vessels forming an anastomosis around this joint, i.e., the superior and inferior ulnar collateral, and radial recurrent and interosseus recurrent arteries.

Relations. The structures in relation to the elbow joint are pictured in Figure 8-78. Those muscles most intimately concerned are the brachialis, anteriorly; the common extensor tendon and extensor carpi radialis longus, laterally; the common flexor muscles, medially; and the triceps (there is a bursa between this muscle and the olecranon), anconeus, and flexor carpi ulnaris muscles, posteriorly. All the vessels taking part in the anastomosis around the elbow are in intimate contact with the capsule, as is the ulnar nerve.

The elbow joint is a comparatively firm articulation

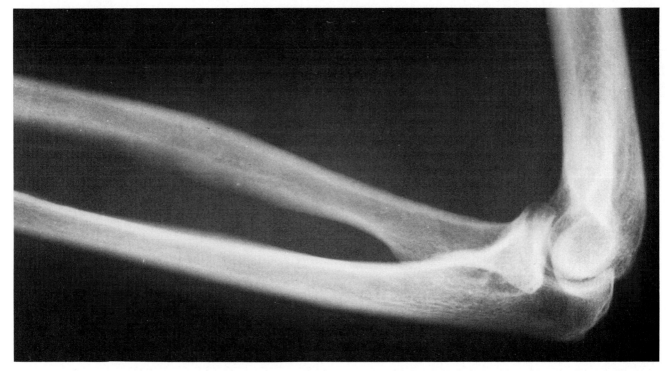

Plate 25. *Left elbow joint: in flexion (lateral view, above) and in anatomical position (anterior view, right).*

and dislocations are not common. But when they do occur, the dislocation is posteriorly. The coronoid process is frequently fractured in these cases. Abduction dislocations can occur; in these cases, owing to the strength of the ligaments, the medial epicondyle is torn away. Inflammations of the elbow joint occur and are quite painful. "Tennis elbow" is actually an inflammation of the periosteum at the site of origin of the extensor muscles located in the forearm.

RADIOULNAR JOINTS

These joints are articulations between the radius and the ulna, one being located superiorly and one inferiorly. Pronation and supination take place at these joints around a vertical axis that extends from the capitulum of the humerus to the pit at the base of the styloid process of the ulna.

SUPERIOR. This consists of an articulation between the head of the radius and the radial notch of the ulna (Fig.

8-79). The head of the radius glides in this notch as the acts of pronation and supination take place.

This articulation is held together by two ligaments, (1) annular and (2) quadrate.

The **annular ligament** is a circular arrangement of connective tissue fibers that surround the head of the radius; it attaches to the ulna at points just anterior and posterior to the notch. The inferior lip of this ligament curls in toward the neck of the radius and therefore aids in preventing the radius from leaving the confines of the ligament in an inferior direction. This ligament is strengthened further by the attachment to it of the anterior and lateral ligaments of the elbow joint.

The **quadrate ligament** is a small collection of fibers coursing between the two points of attachment of the annular ligament; it is located just inferior to the radial notch of the ulna.

The **synovial membrane** of the superior radioulnar joint is directly continuous with that of the elbow joint proper. It lines the inside of the annular ligament and then reflects superiorly as far as the articular cartilage.

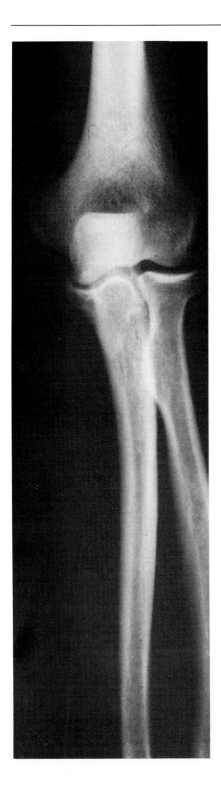

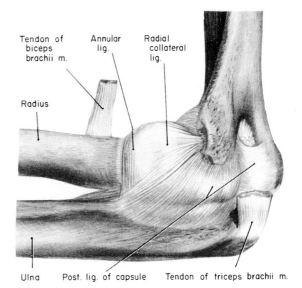

Figure 8-77. *Elbow joint—lateral view. (From Quain,* Elements of Anatomy, *courtesy of Longman Group Ltd.)*

INFERIOR.   The inferior radioulnar joint consists of the articulation between the head of the ulna and the ulnar notch of the radius. At this end, the radius moves around the fixed ulna.

The ligaments of this articulation consist of (1) a capsular ligament and (2) an articular disc.

The **capsular ligament** cannot play too great a role in maintaining the integrity of the joint, for it is quite lax. It is attached to anterior and posterior surfaces of the articular disc and also to the same surfaces of the bones themselves. Some fibers do extend superiorly to cover the superior prolongation of the synovial sac, the **recessus sacciformis** (Fig. 8-84).

The **articular disc** is located at the distal end of the ulna and separates this bone from the carpal bones. It is attached to the distal end of both the ulna and radius and forms the superior surface of the wrist joint along with the radius (see Fig. 8-84).

The **synovial membrane** of the inferior radioulnar joint lines the inside surface of the disc, and the capsular ligament, and extends superiorly between the radius and ulna.

INTEROSSEUS MEMBRANE.   The two radioulnar joints are maintained by the interosseus membrane as

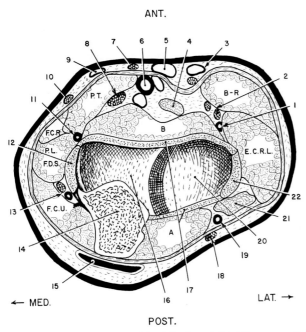

ANT.

POST.

← MED.   LAT. →

Figure 8-78. *Relations of the left elbow joint (diagrammatic). You are looking superiorly: (A) anconeus, (B) brachialis, (B-R) brachioradialis, (E.C.R.L.) extensor carpi radialis longus, (F.C.R.) flexor carpi radialis, (F.C.U.) flexor carpi ulnaris, (P.M. and F.D.S.) palmaris longus and flexor digitorum superficialis, (P.T.) pronator teres, (1) anastomosis between radial collateral and radial recurrent arteries, (2) superficial and deep branches of radial nerve, (3) cephalic vein and lateral antebrachial cutaneous nerve, (4) biceps brachii tendon, (5) median cubital vein, (6) brachial artery and vena comitans, (7) anterior branch of medial antebrachial cutaneous nerve, (8) median nerve, (9) basilic vein, (10) posterior branch of medial antebrachial cutaneous nerve, (11) anastomosis between inferior ulnar collateral and anterior ulnar recurrent arteries, (12) ulnar collateral ligament, (13) ulnar nerve and anastomosis between superior ulnar collateral and posterior ulnar recurrent arteries, (14) olecranon, (15) olecranon bursa, (16) trochlea, (17) anterior and posterior ligaments, (18) posterior antebrachial cutaneous nerve, (19) anastomosis between middle collateral and interosseus recurrent arteries, (20) capitulum, (21) extensor tendon, (22) radial collateral ligament. (Modified after Jamieson.)*

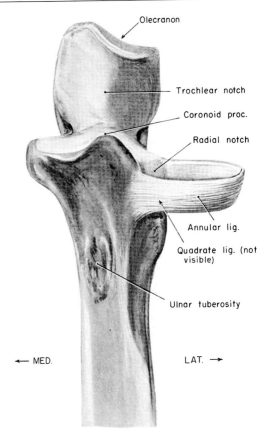

← MED.   LAT. →

Figure 8-79. *The annular ligament. (From Quain,* Elements of Anatomy, *courtesy of Longman Group Ltd.)*

well as by their proper ligaments. This membrane stretches between the interosseus borders of the radius and ulna, and most fibers course inferiorly and medially from the radius to the ulna; a few fibers—the **oblique cord**—extend in the opposite direction. The superior border of the interosseus membrane is approximately one inch inferior to the radial tuberosity, and it extends inferiorly as far as the inferior radioulnar joint (see Figure 8-8).

The action of the interosseus membrane is interesting (Fig. 8-80). The radiocarpal joint is a strong articulation, while the ulnocarpal joint is quite weak. Contrariwise, the radiohumeral joint is weak, while the ulnohumeral joint is quite strong. Therefore, any force placed against the hand has to be transmitted from the radius to the ulna and thence to the humerus; the radius is being forced superiorly, the ulna inferiorly. The interos-

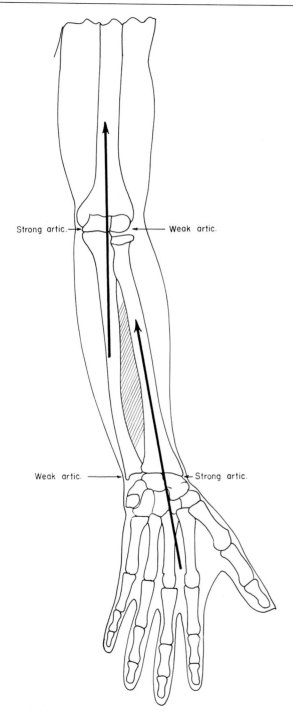

Figure 8-80. *Diagram demonstrating the action of the interosseus membrane in absorbing force placed against the hand.*

seus membrane prevents the displacement that these forces try to create. In other words, the membrane is constructed to aid in walking as a four-footed animal. This is the reason why carrying heavy loads, such as a loaded suitcase, is quite painful; the weight is pulling the forearm apart and the interosseus membrane—except for the small oblique cord—is so constructed that it cannot aid. Heavy loads can be carried more easily on the palm of the hand with the elbow flexed (as waiters carry heavy trays of dishes), for this uses the interosseus membrane.

This membrane also serves to increase the area for attachment of muscles.

Dislocation of the head of the radius does occur. This is accompanied by a rupture of the annular ligament.

WRIST JOINT

The wrist joint is also known as the radiocarpal joint, for the radius is the bone that actually takes part in articulation with the carpal bones. The ulna is separated from these carpal bones by a disc. Therefore, the wrist joint is an articulation between the radius and disc superiorly and the scaphoid, lunate, and triquetral carpal bones inferiorly. It is obvious that the radiocarpal joint is of considerable strength, while the ulnocarpal articulation is very weak.

The wrist joint is a synovial articulation of the ellipsoid type. Movements allowed are flexion and extension around the transverse axis, abduction (radial abduction) and adduction (ulnar abduction) around the anteroposterior axis, and circumduction. Movement at the wrist joint is augmented by movement between the carpal bones.

The ligaments supporting the wrist joint are thickenings in the **capsule** which completely surrounds the joint (Figs. 8-81 and 8-82). These thickenings are the (1) **palmar** and (2) **dorsal radiocarpal ligaments,** and (3) **ulnar collateral** and (4) **radial collateral ligaments.** The capsule attaches to the radius and ulna superiorly and to the proximal end of the carpals (except the pisiform) inferiorly.

The **synovial membrane** of the wrist joint lines the entire capsular ligament and extends onto the interosseus ligaments between the carpal bones. It also covers the disc and is directly continuous with the synovial membrane lining the inferior radioulnar joint.

The **blood supply** to the wrist joint is from adjoining

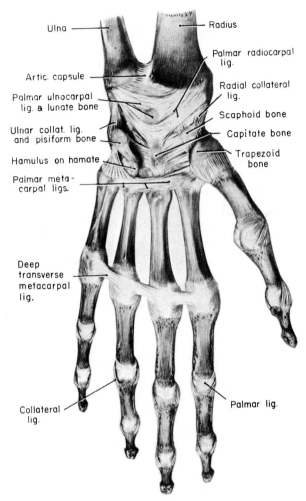

Figure 8-81. Joints of the wrist of left hand—anterior view. (From Quain, Elements of Anatomy, *courtesy of Longman Group Ltd.)*

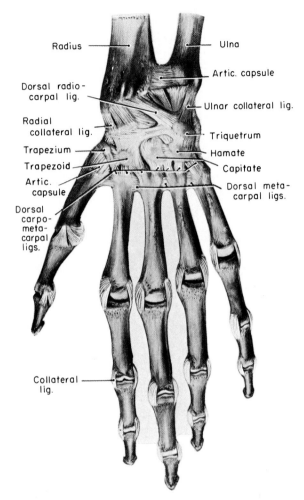

Figure 8-82. Joints of the wrist of left hand—posterior view. (From Quain, Elements of Anatomy, *courtesy of Longman Group Ltd.)*

blood vessels, i.e., carpal branches of the radial and ulnar, branches of the interosseus vessels, and branches from the deep palmar arch. The **nerve supply** is by the interosseus nerves and the dorsal and deep branches of the ulnar.

Relations. The many structures related to the wrist joint are shown in Figure 8-83.

INTERCARPAL JOINTS

The articulations between the carpal bones are synovial joints of the plane variety. A gliding motion occurs, increasing latitude of movement at the wrist joint.

Ligaments. Members of each row of bones are connected to each other by (1) dorsal, (2) palmar, and (3) interosseus ligaments; in addition, the proximal row is joined to the distal by (1) dorsal, (2) palmar, (3) medial, and (4) lateral ligaments. These ligaments are small but numerous. Because of the special arrangement of the pisiform bone (a sesamoid bone), two ligaments should perhaps be mentioned by name. The tendon of the flexor carpi ulnaris muscle inserts on the pisiform bone which is, in turn, attached to the back of the hamate and the base of the fifth metacarpal by **pisohamate** and **pisometacarpal** ligaments. In other words, thinking of the pisiform bone as a sesamoid bone, the actual insertion of the

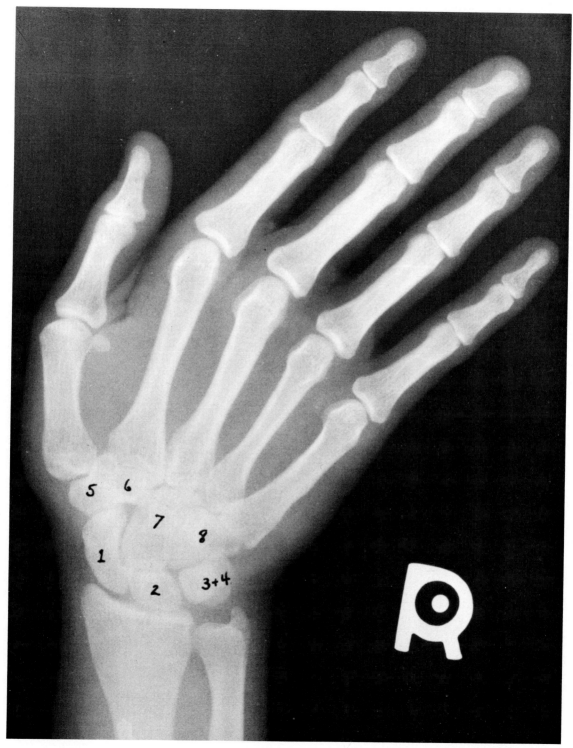

Plate 26. *Left hand in adduction: (1) scaphoid, (2) lunate, (3) triquetrum, (4) pisiform, (5) trapezium, (6) trapezoid, (7) capitate, (8) hamate. Note how the ulna fails to participate in the formation of the wrist joint.*

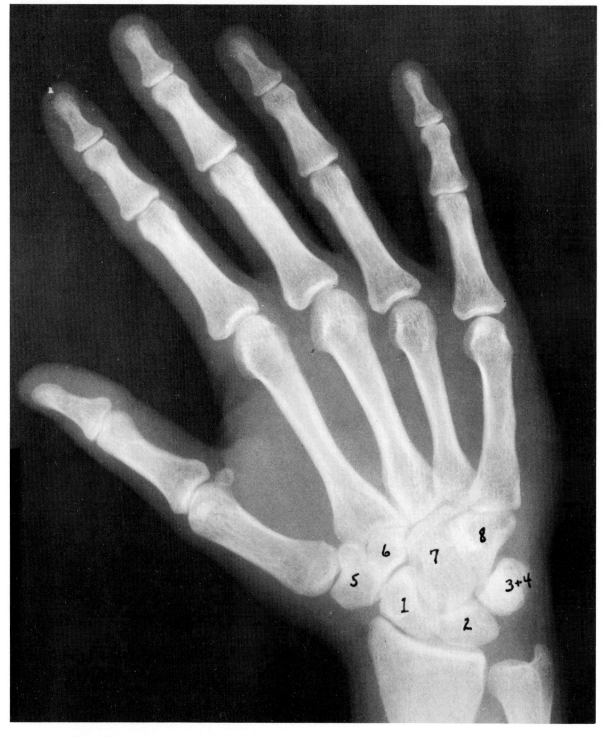

Plate 27. *Left hand in abduction: (1) scaphoid, (2) lunate, (3) triquetrum, (4) pisiform, (5) trapezium, (6) trapezoid, (7) capitate, (8) hamate. Note how the ulna fails to participate in the formation of the wrist joint.*

flexor carpi ulnaris muscle is to the hamate and fifth metacarpal bones. The joints formed by the carpal articulations are shown clearly in Figure 8-84.

Nerve supply to these joints is by the interosseus nerves as well as the deep and dorsal branches of the ulnar.

CARPOMETACARPAL AND INTERMETACARPAL JOINTS

These are all synovial joints of the plane variety, which allow a gliding movement.

Both carpometacarpal and intermetacarpal joints are maintained by **dorsal, palmar,** and **interosseus** ligaments.

The **joint cavity** between the trapezium and first metacarpal is separate from the others; all those remaining form one joint cavity and are lined with a synovial membrane continuous with that lining the intercarpal joints (Fig. 8-84).

The **nerve supply** is similar to that of the intercarpal joints: the interosseus nerves and the deep and dorsal branches of the ulnar.

Movement at the carpometacarpal joints varies, that of the thumb and little finger being greater in degree than the others. In fact, at the carpometacarpal joint of the thumb, flexion, extension, abduction, adduction, circumduction, and opposition (a combination of flexion, adduction, and medial rotation) can occur to a marked degree; and the same movements can occur in the carpometacarpal joint of the fifth digit but to a lesser degree.

METACARPOPHALANGEAL JOINTS

These articulations are synovial and of the condyloid variety. Flexion and extension (transverse axis), abduction and adduction (palmardorsal axes), and circumduction occur at these joints, but no rotation. The head of each metacarpal articulates with the base of a proximal phalanx (Figs. 8-81 and 8-82).

Ligaments. The ligaments involved in this joint form a capsule that is thin dorsally because of the presence of the extensor tendon, but is thickened anteriorly to form the **palmar ligaments** and laterally to form the **collateral ligaments.** The latter ligaments start on the dorsum of the metacarpal and course distally and an-

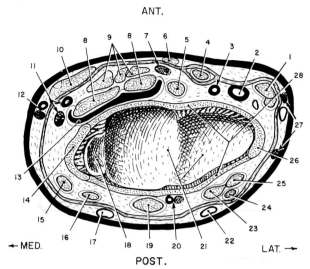

Figure 8-83. *Relations of the left wrist joint (diagrammatic). You are looking superiorly: (1) abductor pollicis longus and extensor pollicis brevis in a common tendon sheath, (2) radial artery, (3) superficial palmar branch of radial artery, (4) flexor carpi radialis tendon and sheath, (5) flexor pollicis longus tendon and sheath, (6) palmaris longus tendon, (7) median nerve, (8) flexor digitorum profundus tendons, (9) flexor digitorum superficialis tendons, (10) flexor carpi ulnaris, (11) ulnar artery and nerve, (12) dorsal branch of ulnar nerve and dorsal carpal branch of ulnar artery, (13) common flexor tendon sheath, (14) ulnar collateral ligament, (15) extensor carpi ulnaris tendon and sheath, (16) extensor digiti minimi tendon and sheath, (17) basilic vein, (18) articular disc, (19) extensor digitorum, (20) posterior interosseus nerve and artery, (21) distal end of radius, (22) cephalic vein, (23) extensor carpi radialis brevis tendon and sheath, (24) extensor pollicis longus tendon and sheath, (25) extensor carpi radialis longus tendon and sheath, (26) radial collateral ligament, (27) branches of radial nerve, (28) anterior and posterior ligaments. (Modified after Jamieson.)*

teriorly to attach to the phalanx. The palmar ligaments are very strong and quite wide; the edges of the fibrous tendon sheath attach to these ligaments. In the metacarpophalangeal joints of the thumb, the tendons of the adductor brevis and flexor pollicis brevis are fused with the palmar ligaments, and at this point small sesamoid bones are likely to develop. Although such bones may be found around the other metacarpophalangeal joints, they are not common.

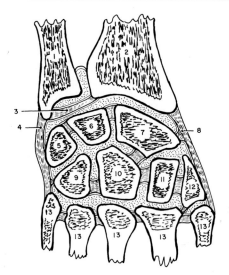

Figure 8-84. *A frontal section through the left wrist: (1) ulna, (2) radius, (3) articular disc, (4) ulnar collateral ligament, (5) triquetrum, (6) lunate, (7) scaphoid, (8) radial collateral ligament, (9) hamate, (10) capitate, (11) trapezoid, (12) trapezium, (13) metacarpal bones. (Modified after Quain.)*

Additional ligaments course transversely between the heads of the metacarpal bones. These ligaments—**the deep transverse metacarpal**—prevent spreading of the metacarpal bones from one another. The lumbrical muscles and digital nerves and vessels course anterior to these ligaments, while the interosseus muscles are found posterior to them. (See Fig. 8-65 on page 743).

The nerve supply to each of these joints is the corresponding digital nerve.

INTERPHALANGEAL JOINTS

These articulations (synovial joints of the ginglymus variety) join the head of one phalanx and the base of the more distal one. Hence, only flexion and extension are allowed around transverse axes.

The **ligaments** are exactly the same as found around the metacarpophalangeal joints, namely, a capsule thickened by palmar and collateral ligaments (Figs. 8-81 and 8-82).

The **nerve supply** to each of the interphalangeal joints is the proper digital nerve.

SYSTEMIC SUMMARY

The limbs first make their appearance in the embryo measuring about 4 mm in length. The limb buds increase in length parallel to the long axis of the embryo. Indentations develop that represent the future joint areas, and at the same time the limbs take a more anterior position, their axes being at right angles to that of the trunk. In this situation both the thumb and radial side of the forearm, and the large toe and tibial side of the leg are facing toward the cranium, just as is shown in an animal such as the adult salamander; the palmar surface of the hand and the plantar surface of the foot face medially.

Something has happened to the limbs in the human being, for in the upper limb just studied the thumb and radius are laterally placed, while the great toe and tibia are medially placed. This difference results from a 90-degree lateral rotation in the forelimb and a simultaneous medial rotation in the hindlimb of equal degree.

BONES AND JOINTS

There is no particular advantage derived from describing the bones of the upper limb at this point for they were described systemically at the beginning of the section on the upper limb. Now that you know more about structures attached to and associated with these bones, a review of pages 665–676 would be advantageous at this point.

The articulations of the upper limb have been described systemically on pages 753–766 and do not demand repetition. Those students who have studied the lower limb should note the similarities and differences between corresponding parts of the two limbs. There is no doubt that the coxal bone is attached to the axial skeleton in a firmer manner than found in the upper limb where the only attachment, the clavicle with the sternum, is quite movable. Although the hip and shoulder joints are both freely movable, ball-and-socket joints, simple observation gives convincing evidence that the hip joint is a more stable articulation than the shoulder joint. The acetabulum is considerably deeper than the glenoid cavity, and the head of the femur is held in the acetabulum by the ligamentum teres, while the head of the humerus has no such attachment, although many feel this ligament plays a minor role. The femur itself is a more massive structure than the humerus. The elbow and knee joints are both hinge joints; but of the two, it would seem that the elbow is a more stable articulation than the knee. The rotation allowed in the knee, in spite of its being a locking mechanism, would seem to weaken this joint as far as weight-bearing is concerned. In contrast is the fact that the superior and inferior radioulnar joints in the upper limb allow considerable movement not found in the lower limb. The tibia is a more massive bone than either bone in the forearm, and this bone bears the entire weight of the body, the fibula being a bone for muscle attachment only. Any weight put against the foot, as in standing, is transferred to the tibia and then directly to the femur. In contrast, weight put against the hand is transferred to the radius, from the radius to the ulna, and from this bone to the humerus. The interosseus membrane is very important in this sequence, for it prevents the upward thrust of the radius and the simultaneous downward thrust of the ulna. Thus the upper limb uses two bones and an interosseus membrane, while the lower limb achieves the same result with a single strong bone. The ankle joint is a hinge joint, while the wrist is a condyloid articulation allowing considerably more movement than in the ankle. The tarsal bones are large compared with the small carpal bones, and these bones in the foot are connected with one another by much stronger ligaments than occur in the wrist. Finally, the long, slender fingers and the opposable thumb are much more usable in performing fine movements than the shorter, less movable toes.

In summary, the lower limb, with its strong articulation with the axial skeleton, its strong hip joint, massive femur, stable and massive tibia, relatively stable ankle joint, massive foot bones, short toes, and strong ligaments, is well suited for locomotion and weight-bearing; while the upper limb, with its more movable attachment to the axial skeleton, freely movable shoulder joint, less massive bone structure, more movable forearm, more movable wrist joint, and long and opposable digits, is well suited to perform the many fine acts that have been partially responsible for the human being enjoying a status well above that enjoyed by lower animals. Thus our limbs have evolved from being almost identical in the lower animal forms to the specialized extremities just described.

There is still room for improvement, however. For example, the knee joint might be more efficient if constructed in a manner that would give it more stability, and the interosseus membrane in the upper limb would probably be more useful to us if reversed in its direction.

MUSCLES

As the limb buds grow,* two muscle masses develop, one dorsal and one ventral. Each of these muscle masses receives nerves from the ventral rami of the segmental nerves associated with the particular somites giving rise to limb buds. The dorsal muscle mass develops into postaxial muscles, which are the extensors and supinators, and the corresponding branch of the ventral ramus innervating these muscles is called a postaxial nerve; conversely, the ventral muscle mass develops into the preaxial muscles, the flexors and pronators, and these are innervated by preaxial nerves.†

While the limbs are in the quadripedal position, the preaxial muscles face in a medial direction in both extremities whereas the postaxial muscles are on the lateral surface. As indicated previously, the upper limb undergoes a rotation laterally that causes the postaxial muscles to face posteriorly and laterally, the preaxial muscles anteriorly and medially. (This arrangement is the opposite to that found in the lower limb since that limb rotates medially.)

Therefore, the extensor muscles on the posterior surface of the arm are postaxial, as is the deltoid muscle on the lateral surface. The muscles on the posterior and lateral surfaces of the forearm and hand are also postaxial, and their extensor action on the digits and supination of the hand are characteristic.

The preaxial muscles in the arm are those on the anterior surface. The preaxial muscles continue into the forearm, being located on the anteromedial surfaces. Their action on digits (as flexors), and pronation of the

---

*W. H. Lewis, 1910, The development of the muscular system, Chap. 12, pp. 454-522, in Keibel and Mall, *Human Embryology*, J. B. Lippincott Co., Philadelphia. This chapter is a thorough elaboration of the development of muscles. It is interesting to compare the development with the phylogeny of these muscles.

†In this text the terms preaxial and postaxial are related to an axis of the limb when in the anatomical position; dorsal musculature becomes postaxial; ventral, preaxial.

hand are proper actions for such muscles. Furthermore, all muscles located on the palmar surface of the hand are preaxial in origin.

The muscles in the back associated with the upper limb (latissimus dorsi, the rhomboids, and levator scapulae) are postaxial muscles, as are the serratus anterior, teres major and minor, and subscapularis muscles. The trapezius does not fit into this scheme and the origin of the supraspinatus and infraspinatus muscles is debatable; they are probably postaxial muscles.

It should be noted that muscles arising from the medial end of the clavicle, sternum, coracoid process, and medial epicondyle of the humerus are preaxial, while those arising from the vertebral column, main body of the scapula, lateral end of clavicle, and lateral epicondyle are postaxial in origin.

It is obvious from this description that the dorsal scapular, suprascapular, and long thoracic nerves, and all branches of the posterior cord of the brachial plexus (upper and lower subscapulars, thoracodorsal, axillary, and radial) are postaxial nerves, whereas all remaining branches of the brachial plexus are preaxial. This concept will be reexamined in the summary on the nerves to the upper limb.

MOVEMENT OF SCAPULA AND SHOULDER. The muscles that move the scapula and the shoulder girdle also serve to attach the upper limb to the trunk. The scapula is **adducted** by the trapezius and rhomboid muscles, **abducted** by the serratus anterior muscle, **elevated** by the superior fibers of the trapezius and the levator scapulae muscle, and **depressed** by the inferior fibers of the trapezius and pectoralis minor muscle. The scapula is **rotated medially** by the rhomboids and **rotated laterally** by the trapezius and serratus anterior muscles. The entire shoulder is **raised** by the trapezius muscle and **depressed** by gravity and the subclavius and pectoralis minor muscles, this action taking place at the sternoclavicular joint.

MOVEMENT AT SHOULDER JOINT. Since the shoulder joint is a ball-and-socket joint, all movements are allowed.

Flexion (direction of pull must pass anterior to transverse axis). This movement is accomplished mainly by the strong pectoralis major muscle and the anterior fibers of the deltoid muscle. They are assisted by the coracobrachialis muscle and the long head of the biceps brachii.

Extension (direction of pull must pass posterior to transverse axis). The main extensors are the latissimus dorsi, teres major, and the posterior part of the deltoid muscle. These are assisted by the long head of the triceps.

Abduction (direction of pull must pass superior to anteroposterior axis). This movement is complicated by the fact that abduction at the shoulder joint can be carried to the horizontal position without movement of the scapula, but beyond this point a rotation of the scapula must occur in such a way as to tilt the glenoid cavity superiorly; this is a lateral rotation, or, in other words, a swinging of the inferior angle laterally. Abduction to the horizontal is done by the supraspinatus and deltoid muscles, but further abduction requires the combined action of the serratus anterior and trapezius muscles. When the nerves to these latter muscles are cut, the horizontal position is the limit of abduction, one reason being that the scapula will not rotate properly unless held firmly onto the body.

Adduction (direction of pull must pass inferior to anteroposterior axis). The powerful adductors are the pectoralis major and latissimus dorsi muscles. They are aided by teres major, subscapularis, and coracobrachialis muscles.

Medial rotation (direction of pull must pass anterior and/or medial to vertical axis). This movement is done mainly by the subscapularis, pectoralis major, teres major, and latissimus dorsi muscles, which are aided by the anterior fibers of the deltoideus.

Lateral rotation (direction of pull must pass posterior to vertical axis). This is accomplished by the infraspinatus, teres minor, and posterior fibers of the deltoideus.

MOVEMENT AT ELBOW JOINT. The elbow joint is a hinge joint allowing flexion and extension only.

Flexion (direction of pull must pass anterior to transverse axis). As indicated by the work of Basmajian and Latif,* the most important flexor of the elbow joint is the brachialis muscle. The biceps brachii, although playing an important role in flexion when the arm is supinated and

---

*J. V. Basmajian and A. Latif, 1957, Integrated actions and functions of the chief flexors of the elbow. A detailed electromyographic analysis, *J. Bone Joint Surg.* 39A: 1106. All students should read this paper. Individual variation in use of muscles is one of its major contributions.

under resistance, does not function when arm and hand are in pronation. In fact, there seems to be a reflex inhibition of the biceps brachii under these conditions that accounts for the ability to flex the elbow without previous supination. Since the biceps does function to its maximum when arm and hand are supinated, this position is by far the most efficient when powerful flexion of the elbow is required. This action is also aided secondarily by all the muscles arising from the medial epicondyle and also by the brachioradialis muscle, this muscle being used primarily when resistance is met. Therefore, the brachialis muscle is the primary flexor of the elbow joint, the biceps and brachioradialis muscles playing an equally important role when the arm and hand are supinated and resistance is met.

Extension (direction of pull must pass posterior to transverse axis). This act is accomplished by the triceps brachii assisted by the anconeus muscle.

MOVEMENT AT RADIOULNAR JOINTS. These movements are supination and pronation (vertical axis involved), both of which are said to be initiated by the brachioradialis muscle. (This is denied by de Sousa *et al.;* see footnote page 733.) Supination is done mainly by the supinator muscle, the biceps brachii muscle coming into play when resistance is met, while pronation is accomplished primarily by the pronator quadratus, the pronator teres coming into action only when resistance appears. Supination is considerably stronger than pronation, a point taken into consideration in manufacturing screws, door locks and knobs, etc.

MOVEMENT AT WRIST JOINT. The wrist joint, or radiocarpal joint, is an ellipsoid articulation that allows flexion, extension, abduction, adduction, and circumduction.

Flexion (direction of pull must pass anterior to transverse axis). This motion is accomplished mainly by the flexor carpi radialis and the flexor carpi ulnaris. These two muscles are assisted by the flexors of the digits: the flexor digitorum superficialis and profundus, and the flexor pollicis longus. When a palmaris longus is present, its main action is also to flex this joint.

Extension (direction of pull must pass posterior to transverse axis). Extension of the wrist joint is accomplished primarily by the combined action of the extensor carpi radialis brevis and longus, and the extensor carpi ulnaris. The extensors of the digits play a secondary role in this same action.

Abduction (direction of pull must pass lateral to palmardorsal axis). Movement away from the midline of the body is accomplished by the combined action of the flexor carpi radialis and the extensor carpi radialis brevis and longus. The long extensor and abductor muscles of the thumb will play a secondary role in this action.

Adduction (direction of pull must pass medial to palmardorsal axis). Movement toward the body is made by the flexor carpi ulnaris and the extensor carpi ulnaris in a combined act.

Circumduction is simply a combination of the four actions just described.

MOVEMENT OF THUMB. The **carpometacarpal** joint of the thumb, called a saddle joint, allows more movement than any other articulation in the thumb; the actions of flexion, extension, abduction, adduction, circumduction, and opposition occur here. However, in contrast to the **metacarpophalangeal** joints in the fingers, where abduction and adduction can occur in addition to flexion and extension, only flexion and extension occur at this joint in the thumb.* The **interphalangeal** joint of the thumb allows the same movement as occurs at these joints in the fingers.

If we think of the thumb as a whole, the following actions occur:

Flexion. This is movement of the thumb at right angles to the fingers and is accomplished by the flexor pollicis longus and brevis.

Extension. This is movement in the opposite direction and is accomplished by the extensor pollicis longus and brevis muscles.

Abduction. This is a movement anteriorly and laterally, brought about by the abductor pollicis brevis and longus muscles. This movement can be divided into "short" and "long" abduction, the brevis pulling the thumb anteriorly and the longus moving it laterally. True abduction is a neutral position between these two.

Adduction. This action is movement of the thumb toward the index finger and is accomplished by a special muscle, the adductor pollicis.

---

*Possibly this should not be too surprising for, based on ossification (Figs. 8-9 and 8-10), the bone missing in the first digit is not a phalanx but the metacarpal bone.

Opposition. This is a movement of the thumb—actually the metacarpal bone of the thumb—across the palm in such a manner that the tip of the thumb meets the tips of the fingers. This action is very important. It is one of the main differences between man and the apes. This act is accomplished by the opponens pollicis muscle.

MOVEMENT OF FINGERS. Flexion, extension, abduction, and adduction (and therefore circumduction) occur at the metacarpophalangeal joints, while flexion and extension only occur at the interphalangeal joints.

Flexion (direction of pull must pass anterior to transverse axis). This is accomplished by the flexor digitorum superficialis and profundus muscles, aided by the lumbrical and interosseus muscles. Actually, all the muscles named can be involved in flexion at the metacarpophalangeal joints, the lumbrical and interosseus muscles inducing an extension of the interphalangeal joints at the same time. The flexor digitorum superficialis flexes the first interphalangeal joint, while the flexor digitorum profundus flexes the distal joint. Flexion of the little finger is also done by the flexor digiti minimi muscle.

Extension (direction of pull must be posterior to transverse axis). Extension of the fingers is accomplished by the extensor digitorum muscle, aided by the interossei and lumbricals; the index finger has an additional extensor, the extensor indicis, as does the little finger—the extensor digiti minimi.

Abduction. This movement occurs at the metacarpophalangeal joints (palmardorsal axes involved). Since the dorsal interossei are abductors and the axis of the hand is considered to be the third digit (it is the second in the foot), a dorsal interosseus muscle must insert on the radial side of the index finger, both sides of the middle finger, and the ulnar side of the ring finger. The little finger is abducted by a special muscle, the abductor digiti minimi.

Adduction. This is movement (palmardorsal axes involved) toward the middle finger, made by the palmar interossei. These are three in number, and they insert on the ulnar side of the index finger and the radial sides of the ring and little fingers.

Opposition. The only finger that can be opposed is the little finger. This action brings the fifth metacarpal bone toward the palm of the hand, aiding in grasping things like poles. This movement is done by the opponens digiti minimi muscle.

MUSCLE COMPARTMENTS AND CLEFTS

In addition to the deep fascia that surrounds the arm, there is a direct continuation of the prevertebral fascia of the neck into the axilla and arm, following the brachial plexus and axillary and brachial vessels. Furthermore two intermuscular septa, a medial and a lateral, divide the flexors on the anterior surface from the extensors on the posterior surface. The brachioradialis muscle, although a postaxial muscle innervated by the radial nerve, actually is located anterior to the lateral septum; this muscle is closely associated with the brachialis muscle and was probably a part of this muscle in lower forms.

Along with the bones and interosseus membrane, deep fascia in the forearm separates the flexor muscles arising from the medial epicondyle and anterior surface of the forearm from those arising from the lateral epicondyle and posterior surface of this area. Thickenings in this fascia also produce the flexor and extensor retinacula.

The compartments in the hand were thoroughly described on page 748; as noted, there are midpalmar, thenar, hypothenar, and adductor compartments. Clefts occur between these compartments, one being found between the midpalmar compartment and the interosseus muscles—the midpalmar cleft; another between the midpalmar compartment and the adductor muscle—the thenar cleft; a third between the adductor muscle of the thumb and the interosseus muscles; and two clefts, one superficial and the other deep to the extensor tendons on the dorsum of the hand. The tendon sheaths should not be forgotten when thinking of spread of infection (Fig. 8-63).

NERVES

Except for the trapezius muscle and an area of the skin in the arm near the axilla, the nerve supply to the entire upper limb is derived from the brachial plexus and its branches (Fig. 8-40).

There is considerable variation and disagreement on the cutaneous nerve supply, but Figure 8-85 presents a generally accepted distribution of the cutaneous nerves and the arrangement of the dermatomes—those regions of the skin that are innervated only by a single spinal cord segment. This knowledge is important clinically, for if an area of anesthesia is found to fit one of those presented

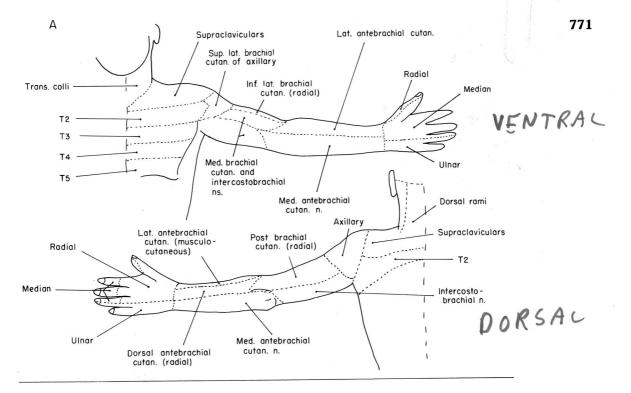

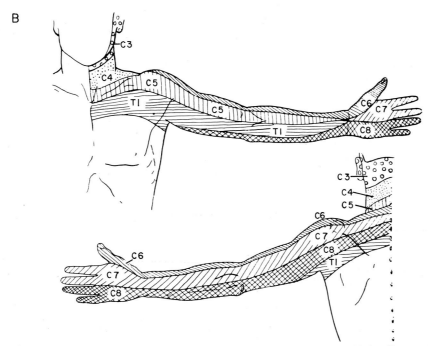

Figure 8-85. *Diagram of cutaneous innervation of the upper limb: (A) distribution of cutaneous nerves. (B) dermatome chart based on hyposensitivity from loss of function of a single nerve root. (Part B from J. J. Keegan and F. D. Garrett,* Anat. Record *102: 409, courtesy of the authors and the Wistar Institute.)*

in Figure 8-85A, it is undoubtedly due to an injury to cutaneous nerves; if, however, the pattern fits Figure 8-85B, it is indicative of dorsal root or spinal cord injury. The general arrangement of dermatomes is easily remembered, for they start at the shoulder, go down the lateral side, across the hand, and up the medial side to the axilla.

Under the systemic summary of muscles it was noted that two muscle masses developed, pre- and postaxial. The nerves to these muscle masses are established early and are carried along to wherever these muscles migrate. In Figure 8-40 the preaxial nerves are clear, while the postaxial are shaded.

In contrast to the lumbosacral plexus, the brachial plexus is subject to frequent damage and its architecture should be learned. This plexus exhibits considerable variation. It originates from the ventral **rami** of C5, C6, C7, C8, and T1. The rami of C5 and C6 join to form the **upper trunk,** C7 forms the **middle trunk,** and C8 and T1 join to form the **lower trunk.** Each trunk divides into anterior and posterior **divisions.** The anterior divisions from the upper and middle trunks join to form the **lateral cord,** the anterior division from the lower trunk forms the **medial cord,** and all three posterior divisions combine to form the **posterior cord.** These cords are named for their relation to the axillary artery.

The nerves arising from the rami and trunks are the **long thoracic** (C5, C6, C7) to the serratus anterior muscle, the **dorsal scapular** (obviously C5) to the rhomboids and levator scapulae muscle, the nerve to the subclavius muscle (obviously C5 and C6), the **suprascapular** (obviously C5 and C6) to the supraspinatus and infraspinatus muscles, and several twigs to the prevertebral muscles and scaleni. Since these nerves arise from the plexus at points superior to the clavicle, they are spoken of as **supraclavicular branches.**

The branches of the medial cord are the **medial brachial cutaneous** (T1), which is distributed, very often in company with the intercostobrachial nerves, to the medial side of the arm; the **medial antebrachial cutaneous** (C8, T1), which is distributed to the medial side of the forearm; the **medial pectoral** (C8, T1) to the pectoralis major and minor muscles; and the terminal branches, the ulnar and the medial root of the median nerve.

The branches of the lateral cord are the **lateral pectoral** (C5, C6, C7) to the pectoralis major muscle; and the terminal branches—the musculocutaneous and lateral root of the median.

The posterior cord has an **upper subscapular** (C5, C6) nerve to the subscapularis muscle; a **lower subscapular** (C5, C6) to this same muscle and the teres major muscle; a thoracodorsal (C6, C7, C8) to the latissimus dorsi; and the terminal branches—the axillary and radial nerves.

If one knows the course taken by the terminal branches of the brachial plexus, it is often quite easy to remember the muscles innervated by each nerve. This is a much better method than merely memorizing lists of muscles.

The **musculocutaneous** nerve (C5, C6) (Fig. 8-86) courses in the midst of the muscles found in the anterior compartment of the arm—the coracobrachialis which it pierces in its route, the biceps brachii, and most of the brachialis muscle—and naturally innervates this group. It then becomes **cutaneous** for the **lateral side** of the forearm.

All remaining preaxial muscles are located in the forearm and hand and are innervated by the median and ulnar nerves. They consist of those arising from the medial epicondyle on the anterior surface of the radius or ulna and the interosseus membrane, and the muscles of the hand. Of this group the ulnar innervates only one and one-half muscles in the forearm—the flexor carpi ulnaris and half of the flexor digitorum profundus—while all the remaining are innervated by the median. In the hand, there is an opposite trend, with the ulnar innervating more mucles. Simply stated, the median nerve innervates the muscles of the thenar eminence and the first two lumbricals, while all remaining muscles are innervated by the ulnar. Let us see how this can be visualized.

The **median nerve** (obviously C5, C6, C7, C8, T1; see Fig. 8-87), as it courses through the arm, is at first lateral to the brachial artery, but then passes anterior to it to gain its medial side. It innervates no muscles in the arm. After sending a branch to the elbow joint, it continues into the forearm anterior to the elbow, gives off **muscular branches** on its medial side to the muscles arising from the medial epicondyle—the pronator teres, flexor carpi radialis, palmaris longus, and flexor digitorum superficialis, and a branch on its lateral side—the **anterior interosseus.** This branch courses anterior to the interosseus membrane, innervating the muscles on this

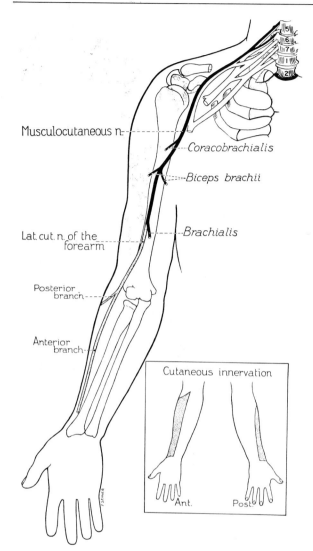

Musculocutaneous n.

Coracobrachialis

Biceps brachii

Brachialis

Lat. cut. n. of the forearm

Posterior branch

Anterior branch

Cutaneous innervation

Ant.   Post.

Figure 8-86. *Diagram of the origin and distribution of the musculocutaneous nerve (lat. cut. n. of the forearm = lateral antebrachial cutaneous nerve). (From Haymaker and Woodhall,* Peripheral Nerve Injuries, *courtesy of W. B. Saunders Co.)*

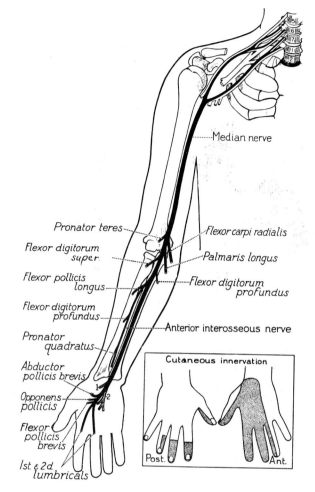

Median nerve

Pronator teres

Flexor digitorum super.

Flexor pollicis longus

Flexor digitorum profundus

Pronator quadratus

Abductor pollicis brevis

Opponens pollicis

Flexor pollicis brevis

1st & 2d lumbricals

Flexor carpi radialis

Palmaris longus

Flexor digitorum profundus

Anterior interosseous nerve

Cutaneous innervation

Post.   Ant.

Figure 8-87. *Diagram of the origin and distribution of the median nerve: (1) palmar cutaneous branch; (2) palmar digital branches. The branch to the thenar muscles is the recurrent branch; the branches to the lumbricals actually arise from the digital nerves. (From Haymaker and Woodhall,* Peripheral Nerve Injuries, *courtesy of W. B. Saunders Co.)*

membrane—the lateral half of the flexor digitorum profundus, flexor pollicis longus, and pronator quadratus—and then continues to the wrist and intercarpal joints. Returning to the main median nerve, it courses between the two heads of the pronator teres muscle and then continues distally in a plane between the flexor digitorum

superficialis and profundus muscles. Near the wrist it gives off a small **palmar cutaneous branch,** then courses deep to the flexor retinaculum (on radial side of palmaris longus tendon) to enter the palm. It immediately breaks up into three branches one of which gives off a **recurrent branch** to the three muscles of the thenar eminence and then continues as **digital** branches to give cutaneous nerve supply to the thumb and radial side of the index finger, and to innervate the first lumbrical muscle. The middle branch innervates the second lumbrical and then continues as **digital** branches to the contiguous sides of the index and middle fingers. The third branch has a **communication** with similar branches of the ulnar nerve and then continues as **digital** branches to contiguous sides of the middle and ring fingers. It should be noted that these digital branches of the median wind around the digits to innervate the dorsal sides over the middle and terminal phalanges; this, of course, includes the nail beds.

Those who have studied the lower limb will note that the median nerve in the hand corresponds to the medial plantar nerve in the foot,* the main difference being the number of lumbrical muscles innervated (only one in the foot).

The **ulnar** nerve (Fig. 8-88) (C8, T1) arises from the medial cord of the brachial plexus and immediately comes to lie between the axillary artery (lateral) and axillary vein (medial). It continues distally and pierces the medial intermuscular septum, in company with the superior ulnar collateral artery, about halfway down the arm. It continues inferiorly, to pass posterior to the medial epicondyle where it gives off an **articular branch** to the elbow joint. It then courses into the muscles arising from the medial epicondyle in a plane between the flexor carpi ulnaris anteriorly and the flexor digitorum profundus posteriorly, naturally innervating these two muscles (only half of profundus), and joins the ulnar artery, both lying anterior to the flexor digitorum profundus and between the flexor carpi ulnaris and flexor digitorum superficialis. The **palmar cutaneous branch** is given off in the forearm. The **dorsal branch** of the ulnar is given off near the wrist, winds around to the posterior side of the hand in a plane just deep to the flexor carpi ulnaris tendon, and gives branches to the wrist, hand, and **digitals** to the little finger and ulnar side of the ring finger. (The amount of coverage varies considerably from body to body.) After giving off its dorsal branch, the ulnar soon divides into its superficial and deep branches. The **superficial branch** innervates the palmaris brevis muscle, gives off a branch that terminates as a **digital** branch to the ulnar side of the little finger, **communicates** with branches of the median nerve, and then divides into **digital** branches which in-

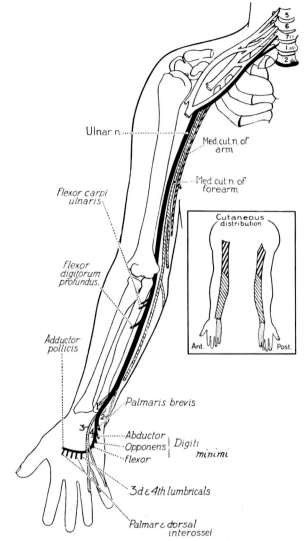

Figure 8-88. *Diagram of the origin and distribution of the ulnar nerve. The medial brachial cutaneous nerve (med. cut. n. of arm) and the medial antebrachial cutaneous nerve (med. cut. n. of forearm) are also included. (1) palmar cutaneous branch, (2) dorsal branch, (3) superficial terminal branch, (4) deep terminal branch. (From Haymaker and Woodhall, Peripheral Nerve Injuries, courtesy of W. B. Saunders Co.)*

---

*This point may be fixed in mind by noting the initial *m*'s of "median" and "medial."

nervate contiguous sides of the little and ring fingers. These branches, as we noted with the median nerve, wind around to the dorsal surface of the middle and terminal phalanges. The **deep branch** of the ulnar accompanies the deep arterial arch between the abductor digiti minimi medially and the flexor digiti minimi and opponens digiti minimi laterally, naturally innervating all three muscles, courses across the palm just anterior to the interosseus muscles, innervating all of them and the third and fourth lumbricals, and then terminates in the adductor pollicis muscle.

Since the median nerve innervates the muscles of the thenar eminence and the first two lumbricals, from the description above it is obvious that the ulnar innervates all the remaining muscles in the hand. It corresponds to the lateral plantar nerve of the foot.

The remaining nerves are the postaxial axillary and radial nerves, both terminal branches of the posterior cord. The **axillary** (Fig. 8-89) (C5, C6) courses inferiorly between axillary artery and subscapularis muscle and then traverses the quadrilateral space. It gives off a branch to the shoulder and then divides into anterior and posterior divisions. The **anterior division** courses with the posterior humeral circumflex artery, winding around the humerus deep to the deltoid muscle which it innervates. Some twigs of this nerve reach the skin. The **posterior division** gives off branches to the deltoid and the teres minor muscles and then winds around the posterior border of the deltoid muscle to become the **superior lateral cutaneous nerve of the arm.**

The **radial nerve** (Fig. 8-90) (C5, C6, C7, C8, and T1) must innervate the remaining postaxial muscles; these are the extensors and supinator muscle, which arise from the lateral epicondyle, supracondylar ridge, or from the posterior surface of the ulna, radius, and the interosseus membrane. The radial emerges from the posterior cord in a position posterior to the axillary artery. Then, as might be expected, it gives off branches to all three heads of the triceps brachii muscle, and winds around the humerus in the radial groove where it is vulnerable when fractures of the humerus occur. In this position it courses between the medial and lateral heads of the triceps brachii muscle, and is accompanied by the deep brachial artery. While in this position posterior to the lateral intermuscular septum, the radial nerve gives off a long slender branch which courses inferiorly on or in the substance of the medial head of the triceps to reach the anconeus

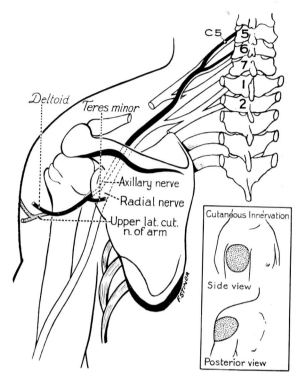

Figure 8-89. *Diagram of the origin and distribution of the axillary nerve. (From Haymaker and Woodhall, Peripheral Nerve Injuries, courtesy of W. B. Saunders Co.)*

muscle, and then gives off two branches, the **inferior lateral brachial cutaneous nerve** and the **posterior antebrachial cutaneous nerve,** which are distributed to the posterior surface of the distal end of the arm and the entire forearm respectively. The radial nerve pierces the lateral intermuscular septum, accompanied by the anterior descending branch of the profunda artery, and comes to lie between the brachialis and brachioradialis muscles sending branches to both and to the capsule of the elbow joint. The radial nerve then divides into its **deep and superficial** branches. The deep branch innervates the extensor carpi radialis brevis and supinator muscles, pierces the latter, gives branches to the extensor digitorum, extensor digiti minimi, extensor carpi ulnaris, and abductor pollicis longus muscles, and courses on the posterior surface of the abductor pollicis longus muscle to reach the posterior surface of the interosseus membrane where it is called the **posterior interosseus** nerve. This nerve innervates the remaining three muscles which arise from the membrane and contiguous bones—extensor pollicis longus and brevis, and the extensor indicis—and then continues to the wrist and intercarpal joints. The superficial branch of the radial nerve courses distally between the brachioradialis muscle anteriorly and the extensor carpi radialis longus muscle posteriorly. In the distal third of the forearm, it escapes from these muscles and winds around the wrist (superficial to the tendons of the abductor pollicis longus and extensor pollicis brevis muscles) to reach the posterior surface of the hand and digits. It does this by dividing into two main **digital** branches, one of which innervates the thumb and radial side of the index finger over the proximal phalanx, and the other the ulnar side of the index finger, the whole middle finger, and the radial side of the ring finger. (This distribution is quite variable.)

NERVE INJURIES. Determining the results of nerve injuries is not only good mental exercise but reveals whether one really can use the information already learned. You should not memorize these symptoms but figure them out when needed. It must be remembered that the appearance of patients with nerve injuries depends upon how recently the injury was obtained. If a recent injury, the appearance will depend on the injury alone and the position taken will be dependent on antagonistic action of muscles. If the injury is of long standing, the atrophy of involved muscles will not only be evident from a loss of volume (such as a thenar atrophy) but the atrophy may exert a pull on the part involved, thereby confusing the picture. Furthermore, patients tend to vary in how muscles are used. Nevertheless, a general pattern can be described.

Musculocutaneous. Naturally, if this nerve is cut there will be an anesthesia on the lateral side of the forearm. Furthermore, there will be a great loss of flexion at the elbow since biceps and brachialis muscles are involved, and there will be a decreased power of supination from loss of the supinating power of the biceps muscle. Flexion at the elbow will not be completely eliminated, for some flexion can be accomplished with the brachioradialis muscle and by others arising from the two epicondyles. In addition, the brachialis muscle has some innervation from the radial nerve, although this may be sensory only. Flexion at the shoulder will be slightly involved, from loss of coracobrachialis and the long head of the biceps.

Median. Cutting this nerve as it emerges from the

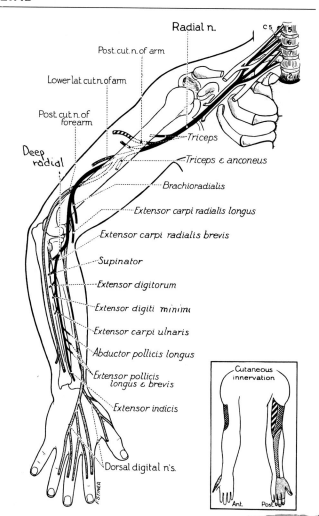

Figure 8-90. *Origin and distribution of the* radial nerve. *(Post. cut. n. of arm and forearm = post. brachial and antebrachial cutaneous nerves respectively.) (From Haymaker and Woodhall,* Peripheral Nerve Injuries, *courtesy of W. B. Saunders Co.)*

brachial plexus will have practically no effect on the shoulder or elbow joints; the effect on the wrist and hand is profound. To begin with, there will be anesthesia of the radial side of the palm and of the palmar side of the thumb, index, and middle fingers, and half of the ring finger, and this anesthesia will extend to the dorsal surface of these digits over the middle and terminal phalanges. Since both pronators have been denervated, the hand will be supinated. The only flexors of the wrist remaining are the flexor carpi ulnaris and the ulnar half of the flexor digitorum profundus. Therefore the wrist will be slightly extended. The index and middle fingers will be extended since there are no flexors remaining to these digits except a slight amount allowed by the interossei muscles. The ring and little fingers will be in a natural position—slightly flexed—for the profundus tendons to these fingers are still intact. The thumb has lost all flexors and one abductor (abductor pollicis brevis), and so will be extended and adducted. The patient will be unable to oppose the thumb and will gradually develop an atrophy of the thenar eminence.

Ulnar. Loss of this nerve will result in anesthesia on both palmar and dorsal surfaces of the ulnar half of the hand, including the ulnar half of the ring finger and the entire little finger. Since this nerve innervates no muscles in the arm, the shoulder and elbow will remain essentially unaffected. Loss of the ulnar half of the flexor digitorum profundus and the flexor carpi ulnaris will leave the wrist very slightly extended and abducted. Loss of the last two lumbricals, all interosseus muscles, and those in the hypothenar eminence causes an abnormal extension in the metacarpophalangeal joints and an abnormal flexion in the interphalangeal joints, owing to loss of the normal extension provided by the interossei muscles. This is less pronounced on the radial side since the first two lumbricals are intact. The acts of abduction and adduction of the fingers are abolished as is adduction of the thumb from loss of the adductor pollicis muscle. The patient gradually develops an atrophy of the hypothenar eminence.

Radial. Paralysis of this nerve will result in anesthesia of the dorsolateral portions of the distal end of the arm, posterior surface of the forearm, and the dorsum of the hand and proximal phalanges on the radial side. The elbow will be flexed through loss of the triceps brachii; the wrist will be flexed, for all extensors of the wrist are denervated; and, since loss of the supinator muscle will result in pronation, the flexion of the wrist will result in a "wrist drop." The thumb will be flexed and adducted, and there will be a tendency, due to loss of the extensor digitorum, for the fingers to be flexed; this is not severe, however, for the interossei are still functional, and they extend the interphalangeal joints. Radial nerve paralysis is very definite and there is no difficulty in recognizing it.

As we return to the brachial plexus, paralysis of its other branches can occur but may be difficult to diagnose. Impairment of medial rotation of the scapula will result from paralysis of the dorsal scapular nerve; a winged scapula (the scapula is not held snugly against chest wall and therefore sticks out too far posteriorly) results from paralysis of the long thoracic, making abduction at the shoulder joint beyond the horizontal difficult; paralysis of the suprascapular nerve hinders the initiation of abduction at the shoulder (but this has been questioned) and lateral rotation will be weakened. Paralysis of the pectoral nerves will result in severe loss of flexion at the shoulder joint but loss of the nerve to the subclavius muscle might go undetected. Loss of the cutaneous nerves would show anesthesia in the proper places. This leaves the branches of the posterior cord. Loss of the subscapular nerves might impair shoulder joint function, because of a loss of the holding power of the subscapularis muscle; the head of the humerus would not be held to the glenoid cavity as snugly as under normal conditions. Loss of the teres major or the latissimus dorsi muscles, or of both, would greatly impair extension at the shoulder.

The student can continue this mental exercise by thinking about injury to various parts of the brachial plexus (a common occurrence), to various branches of the long nerves of the upper limb, such as the interosseus branches, and injury to these long nerves at various locations along their pathways. A knowledge of the location where branches emerge is prerequisite to such drill.

AUTONOMIC NERVOUS SYSTEM. The autonomic nervous system reaches the smooth muscles in the vessels of the upper limb, the sweat glands, and the arrector pili muscles. These structures are innervated entirely by the sympathetic portion of the autonomic nervous system; no parasympathetic fibers have, as yet, been described going to either limb or the trunk. Preganglionic sympathetic fibers start in the intermediolateral cell column of the upper thoracic spinal cord segments, leave the cord via ventral roots, reach the spinal nerves, and then course to the sympathetic chain via white rami communicantes. All these fibers do not synapse at these ganglia but ascend in the chain to ganglia in the cervical region (superior, middle, and inferior cervical ganglia), where a synapse with postganglionic fibers occurs. These postganglionic fibers reach the nerves of the brachial plexus via gray rami communicantes and are distributed with these nerves to the upper limb. Some of these postganglionic fibers course on the arteries as well.

Stimulation of this system causes a vasodilation in the muscles, a vasoconstriction in the skin, a contraction of the arrector pili muscles, and an increase in activity of the sweat glands. Visceral afferent fibers from the blood vessels are present and follow the nerves and dorsal roots back to the cord.

ARTERIES

Arteries grow into limb buds as several channels,* but some disappear; thus only one main vessel remains, usually on the flexor side of the joints (arteries on the extensor side would be stretched during flexion). Because of this type of growth and the fact that structures can obstruct normal development so that more than one channel can remain, there are many variations in arterial pattern, and it should be remembered that the arterial pattern of one limb does not necessarily resemble that of the opposite limb. The following is a description of the arrangement found in the majority of cases (note also Figure 8-91).

The subclavian artery is the main channel leading to the upper limb, and it enters the limb by passing posterior to the scalenus anterior muscle and superior to the first rib. The branches of the subclavian involved in the upper limb are the transverse cervical and suprascapular arteries. The **transverse cervical** artery arises from the thyrocervical trunk and courses across the neck to reach the edge of the levator scapulae muscle. Here it divides into **superficial** and **deep branches,** the former ramifying on the deep surface of the trapezius muscle and the latter on the deep surface of the rhomboids. These branches, especially the deep, anastomose with the circumflex scapular and thoracodorsal branches of the subscapular branch of the axillary. Approximately 50 per cent of limbs exhibit the superficial and deep branches as independent arteries, in which case the former is called a superficial cervical branch (arising from the thyrocervical trunk) and the deep arises from the third part of the subclavian and is called the descending scapular artery. The **suprascapular** artery also arises from the thyrocervical trunk and immediately courses to the suprascapular notch and enters the supraspinatus fossa deep to the supraspinatus muscle. It descends through the spinoglenoid notch to terminate in the infraspinatus fossa where it anastomoses with the circumflex scapular branch of the subscapular artery (a branch of the axillary).

As the subclavian artery crosses the lateral border of the first rib, it becomes the **axillary.** This artery enters into the complexity of the brachial plexus, where it is surrounded by the medial, lateral, and posterior cords. It continues through the axilla until it reaches the inferior border of the teres major muscle.

The first branch of the axillary artery is the rather inconstant and relatively unimportant **supreme thoracic,** which ramifies over the first intercostal muscle. The **thoracoacromial** artery courses anteriorly, pierces the clavipectoral fascia, and immediately divides into four branches. Its **pectoral** branch descends between the pectoralis major and minor muscles, the **acromial** branch courses deep to the deltoid muscle to reach the acromion, the **clavicular** branch runs superiorly and medially to the sternoclavicular joint, and the **deltoid** branch (often arising in common with the acromial branch) courses in the groove between the deltoid and pectoralis major muscles in company with the cephalic vein.

There may be more than one **lateral thoracic** branch which ramifies on the superficial surface of the serratus anterior muscle. The **subscapular** artery is a large branch of the axillary, which almost immediately divides into the **circumflex scapular** artery, entering the infraspinatus fossa to anastomose with the suprascapular artery, and the **thoracodorsal** artery, ramifying on the deep surface of the latissimus dorsi muscle. Both of these vessels anastomose with branches of the transverse cervical artery.

The **anterior and posterior humeral circumflex** arteries may arise from the axillary or from the subscapular. The **anterior** is usually quite small and winds around the anterior side of the surgical neck of the humerus. The **posterior** passes through the quadrilateral space and winds around the posterior side of the humerus in a position deep to the deltoid muscle. It anastomoses with the deltoid branch of the profunda brachial.

As the axillary artery passes the inferior border of the teres major muscle, it becomes the **brachial.** This artery continues distally on the anterior surface of the arm at first on the triceps muscle and then on the brachialis muscle. It is partially covered by the biceps brachii. It continues across the anterior side of the elbow joint and then divides into its terminal branches.

*H. L. Woolard, 1922, The development of the principal arterial stems in the forelimb of the pig, *Carnegie Contrib. Embryol.* 14: 139.

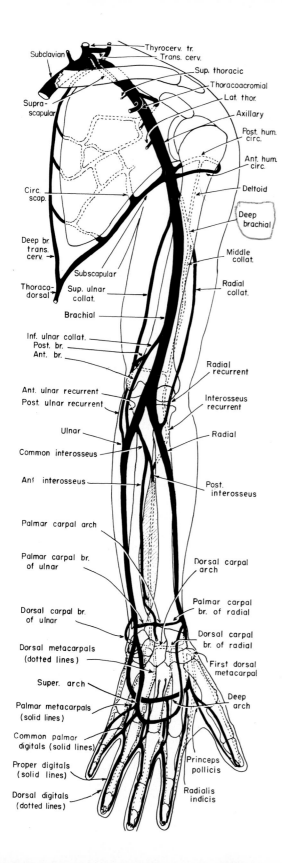

The first branch of the brachial artery is the **profunda** or **deep brachial.** This artery arises quite proximally, follows the radial nerve around the posterior side of the humerus, and divides into **radial collateral** (anterior) and **middle collateral** (posterior) descending branches which participate in the anastomosis around the elbow joint. The profunda artery also sends a **deltoid** branch superiorly, which anastomoses with the posterior humeral circumflex artery. This anastomotic channel is the only arterial pathway around an occlusion of the brachial artery proximal to the emergence of the profunda brachii. Another branch of the brachial is the **superior ulnar collateral,** which arises fairly high up in the arm and follows the same pathway as taken by the ulnar nerve. It therefore penetrates the medial intermuscular septum, where it participates in the anastomosis around the elbow. In addition to a **nutrient** artery and several **muscular** branches, the brachial gives off the **inferior ulnar collateral** artery near the elbow, which also participates in the anastomosis around the elbow joint.

After crossing the elbow joint the brachial artery divides into the radial and ulnar arteries. This division can occur as high in the arm as the brachial plexus, frequently resulting when a part of the plexus seems to interfere with the distal growth of the artery.

The **radial** artery courses distally along side and partially deep to the brachioradialis muscle. It crosses several muscles, including the supinator, pronator teres, flexor digitorum superficialis, flexor pollicis longus, and pronator quadratus. It is quite superficial in position at the wrist and can be felt easily just lateral to the tendon of the flexor carpi radialis muscle. Its branches are a **radial recurrent,** which courses superiorly in a position anterior to the elbow joint to participate in the anastomosis around the joint, several **muscular** branches, a **palmar carpal** branch that courses on the anterior surface of the wrist bones to anastomose with a similar branch of the ulnar, and a **superficial palmar** branch that courses through the thenar muscles to participate in formation of the superficial palmar arch.

The **radial** artery continues by winding around the trapezium and the first metacarpal bone to gain its dorsal surface. It penetrates the first interosseus muscle between

---

Figure 8-91. Diagrammatic summary of the arteries to the upper limb.

the thumb and second metacarpal, turns medially, and enters the palm between the two heads of the adductor pollicis muscle to form the deep palmar arch. In this course it gives off a **dorsal carpal branch** which, with its partner from the ulnar, forms the **dorsal carpal arch** which is joined by the interosseus vessels. This arch gives off three dorsal **metacarpal** arteries which proceed to the webs of the fingers and divide into **digital** arteries for contiguous sides of the fingers. These metacarpal arteries communicate with the palmar arteries. Returning to the radial, before penetrating the interosseus muscle, it gives off the **first dorsal metacarpal** which proceeds distally to be distributed to contiguous sides of the thumb and index finger. The **princeps pollicis** artery arises from the radial after the latter has penetrated the first interosseus muscle. This artery proceeds distally to split and course to each side of the thumb. The **radialis indicis** artery courses to the radial side of the index finger. The remaining portion of the main artery is called the **deep arch.** It gives off **recurrent** branches which return to join the palmar carpal arch, **communicating** branches which join the dorsal metacarpal branches, and **anterior metacarpal** branches (three in number) which descend to the webs of the fingers and join similar branches of the superficial arch.

The **ulnar** artery arises just distal to the elbow and continues distally between the flexor digitorum superficialis and profundus muscles where it accompanies the ulnar nerve. Its first branches are the **anterior** and **posterior ulnar recurrents,** which course superiorly in a position anterior or posterior to the medial epicondyle to anastomose with the ulnar collateral branches of the brachial. The **common interosseus** artery arises next and almost immediately divides into **anterior and posterior interosseus** vessels which course on either side of the membrane as the names indicate. The **anterior interosseus** runs distally with the nerve of the same name, and after giving off a branch to the palmar carpal arch, penetrates the interosseus membrane to join the posterior interosseus artery. The latter artery, after branching from the common interosseus, gives off an **interosseus recurrent** artery which joins in the anastomosis around the elbow, continues distally in company with the posterior interosseus nerve to provide blood for the dorsal musculature, and then joins the anterior interosseus artery.

The ulnar artery at the wrist gives off a **dorsal carpal branch** which winds around the wrist deep to the ten-

dons of the flexor carpi ulnaris to join a similar branch of the radial artery. It then divides into a **deep** and **superficial** branch; the latter branch, the direct continuation of the ulnar, forms the **superficial palmar arch** which courses across the tendons and cutaneous nerves of the hand and joins with the palmar cutaneous branch of the radial. This arch reaches further distally than the deep arch. Its branches are four **common digitals,** three of which course to the webs between the fingers, join the metacarpal branches of the deep arch, and divide to course along the sides of the fingers as the **proper digitals.** The fourth branch courses to the ulnar side of the little finger.

One of the best ways of reviewing the arterial supply to a region of the body such as the upper limb is to ligate the main artery anywhere along its pathway and determine how blood could possibly get around this tie. Since arteries have no valves, blood can course distally in a branch, flow through an anastomosis, continue proximally in the second vessel, and return to the main channel at a point distal to the ligature. For example, a tie of the axillary artery approximately in the middle of the axilla will not be too serious, for there are side tracks or collateral pathways for blood flow. In this instance blood can flow from the subclavian, to the thyrocervical trunk, to the suprascapular, to the circumflex scapular, to the subscapular, and to the axillary distal to the tie. A similar pathway would use the transverse cervical branch of the thyrocervical trunk and the thoracodorsal branch of the subscapular. In addition, the deltoid branch of the thoracoacromial trunk and the circumflex vessels of the axillary may form a connecting link. Some areas may have only one possible bypass, such as a tie placed at the point at which the axillary becomes the brachial. The only way around such a tie is that via the posterior humeral circumflex, to the deltoid (ascending) branch of the profunda brachii, to the profunda artery itself, and to the brachial distal to the tie.

In an area such as the elbow joint, the anastomotic channels are numerous but often small. These small vessels will enlarge considerably if the occlusion of the main channel is gradual in development; if sudden, they sometimes do not suffice. Unlike vital organs such as the brain and heart, muscle will survive under temporary conditions of oxygen want. You should have no difficulty finding pathways around a tie of the radial or ulnar arteries.

VEINS

The veins of the limbs are characterized by being divided into two sets, superficial and deep. The **deep veins** follow the same pathways as just described for the arteries and often form a concomitans of vessels. The **superficial veins** course in the superficial fascia, and here the pattern is quite variable. The vessels constant enough to be named are the **digitals, metacarpals,** and **arches** on the dorsal surface of the hand; the **cephalic,** on the radial side of the forearm and arm, which ultimately drains into the axillary vein; the **basilic,** which is located on the ulnar side and drains into the brachial about midway up the arm; a **median antebrachial** vein that courses proximally in the midline on the anterior surface of the forearm; and the **median cubital,** which joins the basilic with the cephalic, coursing across the anterior surface of the elbow (cubital fossa) to do so. This last vein is used very frequently for intravenous injection and for obtaining blood.

LYMPHATICS

The lymphatic drainage of each area or organ in the body is very important; the nodes become involved in the spread of cancer in addition to serving as barriers to the spread of infection.

The lymphatics of the upper limb are also divided into superficial and deep groups, the **superficial nodes** being located only along the basilic vein in the distal third of the arm. Naturally, these nodes drain the areas covered by the basilic vein and its tributaries. The remaining nodes are **deep** and lie in the cubital fossa and in the axilla. The axillary nodes are named for their position in the axilla, namely, lateral, pectoral, subscapular, central, and apical; they are quite important. Although not palpable when normal, they are readily felt when swollen. They serve as a temporary barrier to spread of infection, but only a temporary one. Infection or cancer can travel from these nodes to secondary collections in the neck.

Furthermore, the axillary nodes may become involved in mammary cancer; they are removed entirely in a radical mastectomy.

CONCEPTS: GENERAL RULES ON STRUCTURE OF THE HUMAN BODY*

Those students who started with a study of the upper limb have been exposed to several concepts or general rules about the structure of the human body that should be very helpful in learning the remaining parts.

From the arrangement of the vertebrae, the nerves, and the blood vessels, it is quite obvious that the human being has been derived from and is a segmentally arranged organism. We have seen that the skin does indeed serve as an efficient covering of the body; that it has a segmentally arranged sensory nerve supply, a dermatome being an area innervated by one segment of the spinal cord; that the skin contains arrector pili muscles, smooth muscles in blood vessels and sweat glands which are innervated by the sympathetic portion of the autonomic nervous system, and therefore the so-called "sensory nerves" to the skin are in reality combined sensory and motor nerves and might better be called "cutaneous nerves"; that the skin is bound down to underlying structures by a layer of superficial fascia containing a varying amount of fat in which these cutaneous nerves course.

We have found that many muscles, in their development, migrate to other parts of the body but carry their previously established nerve supply along with them; that this nerve supply, and the vascular supply as well, is likely to be found on the deep side of the muscle; that the nerve supply is dependent upon the embryological origin of the muscle; that muscles attach to the osse-

---

*This section is similar to those presented in other chapters of this textbook, but contains special features that relate to the upper limb.

ous elements and by their pull not only cause movement of these osseous structures, and thereby of the body, but also cause elevations to develop on these bones; that the main action of a muscle can usually be determined by a simple mechanical sense if the attachments of the muscle are known, the more movable attachment (the insertion) usually being more revealing; that each muscle is surrounded by a layer of deep fascia and that groups of muscles are often surrounded by a thicker enveloping fascial layer.

We have also found that arteries, in addition to being accompanied by a vein and a nerve, exhibit frequent variations in pattern on one side of the body or on both sides; that these variations are particularly prevalent in the limbs, since multiple channels develop in the limb bud, most of which disappear with further growth of the limb; that the definitive main arterial channel is located on the flexor side of joints; that anastomoses of arteries, particularly around joints, provide important sidetracks around an occluded main channel, a fact made possible by the absence of valves in arteries; that veins are even more variable, often form intricate plexuses, are divided into a superficial set and a deep set of venous channels in the limbs, and frequently exhibit multiple (a concomitans) rather than single channels.

Differentiation between the lymphatic drainage of various parts of the upper limb has revealed the important fact that lymphatic drainage seems to follow vascular pathways.

Study of the upper limb has provided us an opportunity to see that human beings take advantage of their structure to place themselves well above the apes; that the upper limb (evolved from a structure designed primarily for locomotion) with its light bone structure, comparatively weak muscles, great freedom of movement, long tapering fingers, and opposable thumb, correlates well with its functional role in providing art, music, and writing as well as material things.

Lastly, as stated several times in this book, I hope you are convinced that the parts of the limb you remember best are those you can visualize easily and those you have thought about enough to understand.

INDEX

Care has been taken to avoid giving too many page references to any individual structure; reference is not given to material presented in the General Descriptions unless this material is not covered elsewhere. If more than one page reference is given and there is a difference in their importance, the more valuable is printed in boldface. In using this index, find the noun followed by the adjective, e.g., "aorta, abdominal" not "abdominal aorta."